AN INTRODUCTION TO ANALYSIS
DIFFERENTIAL CALCULUS

Part I

AN INTRODUCTION TO ANALYSIS
DIFFERENTIAL CALCULUS
Part I

Ram Krishna Ghosh
Formerly Professor of Mathematics, St Xavier's College, Kolkata
Sometime Senior Lecturer in Mathematics, University of Cape Coast, Ghana
Sometime Part-time Lecturer in Mathematics, University of Kalyani, West Bengal

Kantish Chandra Maity
Formerly Professor of Mathematics, St Xavier's College, Kolkata
Formerly Principal, Raja Peary Mohan College, Uttarpara, West Bengal
Formerly Head of the Dept of Mathematics, Asutosh College, Kolkata

Edited
by

Dr Kartick Chandra Pal MSc (Gold Medallist) PhD
Head and Chairman of the Board of Studies of Mathematics (Postgraduate Department)
Ramakrishna Mission Vidyamandira (an Autonomous A+ College)
Belurmath, Howrah
Visiting Professor of Calcutta University

New Central Book Agency (P) Ltd
LONDON
HYDERABAD ERNAKULAM BHUBANESWAR
DELHI **KOLKATA** PUNE GUWAHATI

NCBA

REGD OFFICE
8/1 Chintamoni Das Lane, Kolkata 700 009, India
ncbapvtltd@eth.net

OVERSEAS
NCBA (UK) Ltd, 149 Park Avenue North, Northampton, NN3 2HY, UK
ncbauk@yahoo.com

EXPORT
NCBA Exports Pvt. Ltd, 212 Shahpur Jat, New Delhi 110 049
ncbaexports@ncbaindia.com

BRANCHES

212, Shahpur Jat
New Delhi 110 049
ncbadel@ncbapvtltd.com

House No. 3-1-315, 1st Floor
Nimboliadda, Kachiguda, **Hyderabad** 500 027
ncbahydb@ncbapvtltd.com

Shop Nos. 3 & 4, Vinayak Towers
681/B Budhwar Peth, Appa Balwant Chowk
Pune 411 002
ncbapune@ncbapvtltd.com

GSS Shopping Complex, 1st Floor
Opposite College Post Office, Convent Road
Ernakulam 682 035
ncbaernk@ncbapvtltd.com

Shop No. 15, Bharati Towers, Ground Floor
Forest Park, **Bhubaneswar** 751 001
ncbabub@ncbapvtltd.com

Radhamayee Complex (Second Floor)
Opp. SBI, Panbazar, Jaswanta Road
Guwahati 781 001
ncbaguwahati@ncbapvtltd.com

AN INTRODUCTION TO ANALYSIS: DIFFERENTIAL CALCULUS [Part I]: Ghosh, Maity, Pal

© Copyright reserved by the Authors

Publication, Distribution, and Promotion Rights reserved by the Publisher

All rights reserved. No part of the text in general, and the figures, diagrams, page layout, and cover design in particular, may be reproduced or transmitted in any form or by any means—electronic, mechanical, photocopying, recording, or by any information storage and retrieval system—without the prior written permission of the Publisher

First Published: 1960, Twelfth Edition: 2001, Revised Reprint: 2002
Reprinted: 2003, 2004, 2006, 2007
Revised and Updated Thirteenth Edition: 2011
Reprinted: 2012, 2013

PUBLISHER AND TYPESETTER
New Central Book Agency (P) Ltd
8/1 Chintamoni Das Lane, Kolkata 700 009

PRINTER
New Central Book Agency (P) Ltd
Web-Offset Division, Dhulagarh, Sankrail, Howrah

COVER DESIGNER
Soumen Paul

TECHNICAL EDITOR
Dipan Roy

PROJECT SUPERVISOR
Abhijit Sen

ISBN: 978-81-7381-437-2

Price: ₹735.00 MAXIMUM RETAIL PRICE INCLUSIVE OF ALL TAXES

Contents

Preface to the Thirteenth Edition — xix
Preface to the First Edition — xx

Section I : Chapters 0–13 (Elements of Analysis)

Chapter 0: ELEMENTS OF MATHEMATICAL LOGIC — 3–13
- 0.1. Introduction — 3
- 0.2. Fundamental Logical terms and Notions — 4–10
  [Statement—Compound—Disjunction—Negation—Conditional—Implications—Biconditional (iff)—Universal and Existential Statements. Proofs in Mathematics—Direct: Detachment and Syllogism—Indirect]
- Exercise on Chapter 0 — 10-11
- Main Points: A Recapitulation — 11
- Symbols — 12-13

Chapter 1: BASIC CONCEPTS: SETS, RELATIONS AND MAPPINGS — 17–38
- 1.1. Introduction — 17
- Basic Notions of Set Theory — 17-19
  [Universal set—Null set—Finite and Infinite set—Subsets—Equality of two sets—Families of sets—Power set]
- 1.2. Operations on Sets (Set Algebra) — 19-21
  [Union—Intersection—Complement—Difference—Symmetric difference—Laws of Algebra of sets]
- Ordered Pairs: Cartesian Products: Relations on a set — 21-22
  [Equivalence relation—Partial order—Total order]
- 1.3. Solved Examples — 23-25
- 1.4. Relation from A to B — 25-28
- Exercises: I(A) — 28-31
- 1.5. Mapping or Function — 31-35
- Exercises: I(B) — 35-38

Chapter 2: THE REAL NUMBER SYSTEM — 41–115
- 2.1. Introduction — 41
- 2.2. The Natural Numbers — 42-45
  [Peano's Axioms—Symbols for natural numbers—Algebraic operations on $\mathbb{N}$—Order relation on $\mathbb{N}$—Subtraction and Division on $\mathbb{N}$—Well-Ordering Principle]
- 2.3. Solved Examples — 45-49
- Exercise for self practice — 49-51
- 2.4. Natural Numbers ($\mathbb{N}$) and Integers ($\mathbb{Z}$) — 51-55

v

2.5. Rational Numbers and their Main Properties   56–65

[Definition—$\mathbb{Q}$: Ordered Field—Unbounded—Dense and Archimedean—Decimal representation—Countability—Not order complete—Geometrical Representation of Rational numbers]

Illustrative Examples   65–67

Exercises: II(A)   67–70

2.6. Existence of numbers other than rational numbers   70–73

[Linear Continuum—Arithmetic Continuum—Cantor-Dedekind Axiom—Gaps on a directed line]

Section II: The Real Number System $\mathbb{R}$ 73–93

2.7. Real Number System from Axiomatic approach   73–80

[Algebraic structure of $\mathbb{R}$—Order structure—LUB axiom of $\mathbb{R}$—Set of rational numbers NOT order complete]

2.8. Some Important Properties of the Real Field   81–86

[Archimedean Property—Density Property—Decimal Representation—Geometrical Representation—Countability of Real Numbers]

2.9. The Modulus of a Real Number   86–88

[Intervals: Closed and Open—Extended Real Number System]

2.10. Illustrative Examples   88–93

Exercises: II(B)   93–100

APPENDIX: Dedekind's Concept of Real Numbers   101–114

Exercises: II(C)   114–115

Chapter 3: REAL-VALUED FUNCTIONS OF A SINGLE REAL VARIABLE   **119–172**

3.1. Introduction   119

3.2. Real Variables: Continuous and Discrete   120–121

[Constants: Arbitrary and Fixed]

3.3. Real-Valued Functions of a Single Real Variable   121–124

[Functions using Analytic Expressions, using Tables and using Graphs]

3.4. Illustrative Examples   124–132

[Signum Function: Dirichlet's Function]

3.5. Different classes of Functions: Their graphical representations   132–163

Exercises: III   163–171

Main points: Chapter 3   172

Chapter 4: ELEMENTS OF POINT SET THEORY   **173–238**

4.1. Introduction   173

4.2. Neighbourhood, Interior Points and Open sets   173–179

4.3. Theorems on Open Sets	180–185
4.4. Important Definitions	186–195

[Accumulation Point or Limit Point of a set—Isolated point—Derived set—Closed set—Dense-in-itself—Perfect set—Adherent Point—Closure of a set]

4.5. Characteristic Properties of Closed Sets	195–200
4.6. Bolzano Weierstrass Theorem for Infinite Point Sets	200–208
Exercises: IV(I)	208–212
4.7. The Cantor Set	212–217
4.8. Borel Sets	217–219
4.9. Compact Sets in $\mathbb{R}$	219–227
4.10. The Cantor Intersection Theorem	227–233
4.11. Other Characterizations of Compact Sets in $\mathbb{R}$	233–235
Exercises: IV(II)	235–238
Chapter 5: SEQUENCE OF REAL NUMBERS	**241–333**
5.1. Introduction	241
5.2. Examples of Sequences in $\mathbb{R}$	242–243

[Recursive definition of a sequence—Geometrical representations of sequence points on a line or on a plane]

5.3. Bounded Sequences [lub–glb]	244–246
5.4. Monotone Sequences	246–247

[M. increasing—M. decreasing—neither increasing nor decreasing sequence]

5.5. Limit of a Sequence	247–257

[Geometrical representations—Varied examples based on definition of limit of a sequence—Null sequence]

5.6. Theorems on limits of Sequences	257–268

[Convergent Sequence bdd—Converse not true—Sum, Difference, Product and Quotient of two convergent sequences, Sandwich Theorem]

5.7. Non-Convergent Sequences	268–271

[Properly divergent, Improperly divergent (or Oscillating) sequences]

5.8. A very useful Criterion of Covergence	271–282

[A monotone sequence can never oscillate, examples, Theorem on nested intervals]

5.9. Cauchy Sequences	282–287

[Cauchy's Convergence Criterion or Cauchy's General Principle of Convergence]

5.10. Cauchy's Limit Theorems — 287–294

[Cauchy's First and Second Theorems on Limits]

5.11. Miscellaneous Illustrative Examples — 295–303

Exercises: VA — 303–308

Main Points: A Recapitulation — 309–312

5.12. Sub-sequences — 313–320

[Existence of monotone subsequence. Every bounded sequence has a convergent subsequence. Theorem of Nested Intervals, Bolzano-Weierstrass Theorem]

5.13. Cluster Points of a Sequence — 320–322

5.14. Upper and Lower Limits of Sequences of Real Numbers — 322–333

[Theorems concerning Limit Superior and Limit Inferior]

Chapter 6: THEORY OF CONVERGENCE: INFINITE SERIES OF REAL CONSTANTS — 337–425

6.1. Introduction — 337–339

6.2. Two Important Series — 339–342

[Geometric Series and p-Series]

6.3. Cauchy's Convergence criterion applied to infinite series — 342–345

[Abel—Pringsheim's Theorem, Abel's series]

6.4. Applications of Cauchy's Principle — 345–347

6.5. A Few General Theorems on Convergence or Divergence — 347–350

[Associativity or Grouping in parentheses]

6.6. Series of positive terms: Tests of Convergence or Divergence — 350–394

[A positive series can never oscillate
- I. Comparison Tests: for convergence and for divergence. Another form: Limit form of Comparison Test. Further Applications of Comparison Tests.
- II. Cauchy's Root Test: Applications.
- III. Ratio Tests: D'Alembert's, Raabe's and De Morgan and Bertrand: Their Proof and Applications.
- IV. Kummer's Test.
- V. Gauss' Test (Order Notations)—Proof and Applications: Hypergeometric Series, Binomial Series.
- VI. Logarithmic Test.
- VII. Another Logarithmic Test: An alternative to Bertrand's Test.
- VIII. Cauchy's Condensation Test: $\sum_{n=2}^{\infty} \dfrac{1}{n(\log n)^p}$.

Summary: Tests of Positive Series]

6.7. Series of Positive and Negative Terms	394–399

[Absolutely convergent series: Conditionally convergent series, and Alternating series: Leibnitz's Test—Properties of Mixed Series]

6.7.1. Power Series	399–400
6.8. Illustrative Examples	400–403
6.9. Rearrangement of Series	403–407

[Rearrangement of non-absolutely convergent series. Riemann's Theorem]

6.10. Multiplication of Two Infinite Series	408–411

[Merten's Theorem, Abel's Theorem]

Miscellaneous Exercises	411–425

Chapter 7: THE LIMIT OF A FUNCTION OF A CONTINUOUS REAL VARIABLE   429–461

7.1. Introduction	429
7.2. Meanings of the symbols: $x \to a$, $x \to \pm\infty$	429–430
7.3. Limit of a Function	430–436

[$\lim_{x \to a} f(x) = l$; Negation of the statement $\lim_{x \to a} f(x) = l$; Geometric interpretation; Right-hand and Left-hand limits; Infinite limits]

7.4. Theorems on Limits	436–444

[Sum, Product, Quotient: Limit of a function of a function; Neighbourhood property; Sandwich Theorem]

7.5. Illustrative Examples using Theorems on Limits	444–446
7.6. Cauchy's criterion for the existence of limit of a Function	446–450

[including illustrative examples]

7.7. Some Important Limits	450–455

$$\left[\lim_{x \to 0} \frac{\sin x}{x} = 1;\ \lim_{x \to \infty} \left(1 + \frac{1}{x}\right)^x = e;\ \lim_{x \to 0}(1+x)^{1/x} = e; \right.$$
$$\lim_{x \to 0} \frac{a^x - 1}{x} = \log a;\ \lim_{x \to 0} \frac{e^x - 1}{x} = 1;\ \lim_{x \to 0} \frac{\log(1+x)}{x} = 1$$
$$\left. \lim_{x \to a} \frac{x^n - a^n}{x - a} = na^{n-1}\ (n\ \text{any real number}) \right]$$

7.8. Limit of a function defined in the language of a sequence	455–457
Main Points: A Recapitulation	457–458
Exercises: VII	458–461

Chapter 8: CONTINUITY OF FUNCTIONS OF A SINGLE REAL VARIABLE   463–513

8.1. Introduction	463
8.2. Continuity at a point and on an interval I [example, note]	463–471

ix

8.3. Piecewise Continuous Functions: Discontinuous Functions   471–476

[Discontinuity of First kind and Second kind. Removable and Non removable Discontinuity. Infinite Discontinuity. Oscillatory Discontinuity. Jump Discontinuity. Illustrative Examples]

8.4. Elementary Properties of Continuous Functions   476–480

[Neighbourhood Property]

8.5. Deeper Properties of Continuous Functions   480–500

[Intermediate Value Property (Bolzano's Theorem); Boundedness Property (Attainment of Bounds); Uniform Continuity Property; Continuity of Inverse functions; Continuity and Monotonicity]

Main Points: A Recapitulation   501–502

Exercises: VIII   502–513

Chapter 9: DIFFERENTIATION: DERIVATIVES OF FIRST ORDER   517–578

9.0. Definition   518–520

[Illustrative examples]

9.1. Relation between Continuity and Derivability   521–523

9.2. Standard Formulae and General Rules of Differentiation   523–529

9.3. Implicit Functions   529

9.4. Parametric Functions   529–533

[Miscellaneous problems]

Exercises: IX (I)   533–542

9.5. Concepts of Differentiability and Differentials   542–544

9.6. Geometrical Interpretation of the derivative and differential   545–550

[Use of differentials in small errors, Relative and Percentage error]

9.7. Meaning of the sign of the derivative at a point: Geometrical Considerations   550–551

9.8. Important Thorems on Derivatives   552–559

[Lipschitz condition, Proof of Chain Rule, Derivative of an inverse function, Darboux's Theorem, Intermediate Value Property for Derivatives]

9.9. Derivative as a Rate-Measurer   559–565

[Related Rates]

Main Points: A Recapitulation   566–567

Exercises: IX (II)   567–578

Chapter 10: REPEATED DIFFERENTIATION: SECOND AND HIGHER ORDER DERIVATIVES   581–619

10.1. Introduction   581

10.2. A Standard Result: nth Derivative of x^k	582–585
10.3. Other Standard Results of nth order Derivatives	585–587
10.4. Second-Order Derivatives: Typical Problems	587–592
[Implicit Functions, Parametric Functions; Function of a Function]	
10.5. Use of Partial Fractions in Finding nth Derivatives	592–595
10.6. Leibnitz's Theorem on Successive Derivatives	595–602
10.7. A Few Important Problems of Harder Type	602–606
Main Points: A Recapitulation	606–607
Exercises: X	608–619
Chapter 11: MEAN VALUE THEOREMS	**623–687**
11.1. Introduction	623
11.2. Rolle's Theorem	623–628
[Geometric Interpretation]	
11.3. Mean-Value Theorem: Lagrange's Form	628–634
[Geometric Interpretation: Important Deductions; Derived function $f'(x)$ cannot have a discontinuity of the first kind]	
11.4. Mean-Value Theorem: Cauchy's Form	634–636
[Geometric Interpretation]	
11.5. Illustrative Examples	637–645
Exercises: XI (I)	645–651
Hints & Answers of Exercises XI (I)	651–656
11.6. Generalised Mean-Value Theorem: Taylor's Theorem	656–662
[Taylor's Theorem with remainder due to Lagrange, Cauchy and Schlömilch-Roche. Taylor's Theorem deduced from Cauchy's MVT. Maclaurin's Theorem. Young's form of Taylor's Theorem]	
11.7. Illustrative Examples	662–666
Exercises: XI (II)	666–668
11.8. Taylor's Infinite Series	668–681
[Series Expansion of a function. Cauchy Function. Power Series Expansions of some standard functions—Exponential, Sine and Cosines, Logarithmic and Binomial. Method of Formation of a differential equation and then expansion of functions]	
Main Points: A Recapitulation (Chap 11A and 11B)	681–684
Exercises: XI (III)	684–687
Chapter 12: INDETERMINATE FORMS AND MAXIMA AND MINIMA	**691–735**
12.1. Introduction	691–692

12.2. Indeterminate Forms: L'Hospital's Rule	692–702

[Forms: $0/0$; ∞/∞; $0.\infty$; $\infty - \infty$; 0^0; ∞^0; $1^{\pm\infty}$]

Exercises: XII (I)	702–707
12.3. Maxima and Minima: Extreme Values	708–711

[Sufficient conditions for the existence of extreme values: Use of first derivatives only]

12.4. Illustrative Examples: Use of first derivative only	711–714
12.5. Higher Derivative Test for the existence of Extreme Values	714–717
12.6. Illustrative Examples: Use of Higher Order Derivatives	717–724

[including Applied Problems]

Main Points: A Recapitulation	724–725
Exercises: XII (II)	725–735
Chapter 13: FUNCTIONS OF SEVERAL VARIABLES	**739–894**
13.1. Introduction: Point sets in Higher Dimensions	739–747

[Points in $\mathbb{R}^n$; Norm of $\vec{x} \in \mathbb{R}^n$; open balls in $\mathbb{R}^n$; open sets in $\mathbb{R}^n$; Adherent Point and Accumulation point of a set; Derived set; closed set; closure of a set; Bolzano-Weierstrass Theorem]

13.2. Functions in Higher Dimensional Space	747–749

[Linear, Rational and Algebraic functions in $\mathbb{R}^2$]

13.3. Limits of Functions	749–758

[Functions of two variables: Simultaneous limit or Double limit; Non-existence of limit; Repeated or Iterated limits; Another approach of limit based on the notion of limits of sequences; Algebra of limits]

13.4. Continuity of a function of several variables	758–765

[Continuity in $\mathbb{R}^2$; Examples on Continuity Functions continuous on compact sets]

13.5. Partial Derivatives	765–773

[Directional derivatives in $\mathbb{R}^2$; Geometric Interpretations of Partial derivatives; A sufficient condition for continuity (in $\mathbb{R}^2$); Mean Value Theorem for a function of two variables]

13.6. Differentiability: Total differential	773–783

[Notion of differentiability for functions of two variables; Total differential; Sufficient condition for differentiability of a function of two variables; Notion of differentiability for functions of three variables—Total differential; Use of total differential in Approximations and Small Errors]

13.7. Composite Functions: Chain Rules	783–788

[Chain Rules for functions of two and three variables]

13.8. Homogeneous Functions: Euler's Theorem   788–793

[Euler's Theorem for homogeneous functions of two and three variables; Converse of Euler's Theorem]

13.9. Partial Derivatives of Higher Order   794–802

[Second Order Mixed derivative for functions of two variables are not, in general, commutative. Examples: Schwarz Theorem. Young's Theorem]

13.10. Change of Variables: Calculation of second order partial derivatives using Chain Rules   803–816

[Various Illustrated Examples]

13.11. Jacobians: Their Important Properties   816–828

[Chain Rules for Jacobians, Jacobian of Implicit Functions; Illustrated Examples]

13.12. Functional Dependence   828–833

[Necessary and Sufficient condition for functional dependence]

13.13. Implicit Functions   834–845

[Existence Theorem—Proof. On Functional Equation $F(x,y,z) = 0$: To solve z in terms of x and y. Two functional equations $F(x,y,u,v) = 0$ and $G(x,y,u,v) = 0$. To solve for u and v in terms of x and y. Three functional Equations $F(x,y,z,u,v,w) = 0$, $G(x,y,z,u,v,w) = 0$ and $H(x,y,z,u,v,w) = 0$: To solve for u,v,w in terms of x,y,z]

13.14. Proofs of the Theorems on Functional Dependence   845–847

13.15. Illustrated Examples (Miscellaneous Types)   848–869

Exercises: XIII   869–894

Section II: Chapters 14–21 (Applications of Derivatives)

Chapter 14: **SOME IMPORTANT CURVES: THEIR EQUATIONS AND SHAPES**   899–925

14.1. Introduction   899

14.2. Rectangular Cartesian Equations of Plane Curves   899–901

[Parametric, Implicit and Explicit Representations]

14.3. The Equations of some well-known Curves in Parametric Form   901–910

[Circle, Ellipse, Parabola, Hyperbola, Cycloid, Astroid, Epicycloid and Hypocycloid]

14.4. Implicit Cartesian Equations of Curves   910–913

[Semicubical Parabola, Folium of Descartes, Cissoid of Diocles, Strophoid]

14.5. Curves in Polar Co-ordinates	913–920
[Circle, Line, Cardiode, Lemniscate of Bernoulli, Spirals: Equiangular, Archimedes, Reciprocal, Rose Petals]	
14.6. Catenary: $y = c\cosh x/c$ (Cartesian); $s = c\tan\psi$ (Intrinsic)	920–923
Main Points: A Recapitulation	923–925

Chapter 15: TANGENTS AND NORMALS: ASSOCIATED CURVES: DIFFERENTIAL OF ARC LENGTH 929–987

15.1. Introduction	929
Section I: Tangents and Normals	930–967
15.2. Equations of Tangents and Normals: Various Forms	930–935
15.3. Sub tangent and Sub normal: Lengths of Tangent and Normal	935–938
15.4. Angle of Intersection of two Curves	938–939
15.5. Tangents at the Origin	939–940
15.6. Illustrative Examples (Miscellaneous Type)	940–947
Exercises: XV (A)	947–952
15.7. Equations of the Tangent and Normal to a Polar Curve	953–956
15.8. Angle between the Tangent and Radius Vector at any point of a Polar Curve	956–957
[Angle of intersections of two Polar Curves]	
15.9. Subtangent and Subnormal, Perpendicular from Pole to the Tangent: Lengths of Tangent and Normal	958
15.10. Pedal Equation	958–959
[From Cartesian to Pedal, From Polar to Pedal]	
15.11. Illustrative Examples	959–964
Exercises: XV (B)	964–967
Section II: Associated Curves	967–979
15.12. Pedals, Inverses and Reciprocal Polars	967–977
Exercises: XV (C)	977–979
Section III: Arc-Length and its Differentials	980–983
15.13. The Differential of Arc Length	980–983
[Cartesian, Polar]	
Main Points: A Recapitulation	983–986
Exercises: XV (D)	986–987

Chapter 16: CURVATURE 991–1017

16.1. Introduction	991
16.2. Measure of Bending: Definitions	991–993

16.3. Formulae for the Radius of Curvature — 993–1000

[Curves in Intrinsic form, Curves given in Cartesian (Explicit, Implicit, Parametric), Pedal, Polar, Tangential Polar forms]

Exercises: XVI (A) — 1000–1003

16.4. Theorem on Centre of Curvature — 1004

16.5. Concept of Curvature: Newton's approach — 1004–1008

16.6. Co-ordinates of Centre of Curvature: Equation of Circle of Curvature — 1008–1010

16.7. Evolute and Involute — 1010–1012

16.8. Chord of Curvature — 1012

Main Points: A Recapitulation — 1013–1014

Exercises: XVI (B) — 1014–1017

Chapter 17: RECTILINEAR ASYMPTOTES — 1021–1056

17.1. Preliminary Assumptions — 1021–1022

17.2. Asymptotes: Definition — 1022–1023

17.3. Asymptotes Parallel to the Axes — 1023–1026

17.4. Oblique Asymptotes — 1026–1031

Exercises: XVII (A) — 1032–1033

17.5. Determinations of Asymptotes Non-Parallel to Y-axis of the General Rational Algebraic Curve $F(x,y) = 0$ — 1033–1038

17.6. An Alternative Method for Finding Asymptotes of Algebraic Curves — 1038–1043

Exercises: XVII (B) — 1043–1046

17.7. Intersection of a Curve with its Asymptotes — 1046–1048

Exercises: XVII (C) — 1048–1050

17.8. Asymptotes in Polar Co-ordinates — 1051–1052

Main Points: A Recapitulation — 1052–1054

Exercises: XVII (D) — 1054–1055

17.9. Position of a Curve with regard to its Asymptotes — 1055–1056

Chapter 18: ENVELOPES: EVOLUTES — 1059–1085

18.1. Introduction — 1059

18.2. Family of Curves — 1059–1060

18.3. Envelope: Definition — 1060–1064

18.4. Relation between Envelope and Curves Enveloped — 1064–1066

18.5. Remarks on Definitions — 1066–1069

Exercises: XVIII (A) — 1069–1077

18.6. Evolutes — 1077–1080

Main Points: A Recapitulation　　　　　　　　　　　　1080–1081

Exercises: XVIII (B)　　　　　　　　　　　　　　　　1081–1085

Chapter 19: CONCAVITY, CONVEXITY: POINTS OF INFLEXION　　　　1089–1105

19.1. Introduction　　　　　　　　　　　　　　　　　　1089

19.2. Sense of Concavity and Convexity　　　　　　　　1089–1090

[With respect to a line]

19.3. Concave Upwards/Downwards: Convex Upwards/Downwards　　　1090–1094

19.4. Illustrative Examples　　　　　　　　　　　　　1095–1098

19.5. Concavity and Convexity of a Polar Curve　　　1098–1102

Main Points: A Recapitulation　　　　　　　　　　　　1102–1103

Exercises: XIX　　　　　　　　　　　　　　　　　　　1103–1105

Chapter 20: SINGULAR POINTS: MULTIPLE/DOUBLE POINTS　　　1109–1126

20.1. Introduction　　　　　　　　　　　　　　　　　　1109

20.2. Definitions　　　　　　　　　　　　　　　　　　 1110

20.3. Singular points at the Origin of a Rational Algebraic Curve　1110–1113

20.4. Singular points at points other than the Origin　1113–1114

20.5. Condition for the existence of Double Points　 1114–1116

20.6. Discrimination of Species of a Cusp　　　　　　1116–1118

20.7. Systematic Procedure　　　　　　　　　　　　　　1118–1120

20.8. Radii of Curvature at Multiple Points　　　　　1121–1122

Main Points: A Recapitulation　　　　　　　　　　　　1122–1124

Exercises: XX　　　　　　　　　　　　　　　　　　　　1124–1126

Chapter 21: CURVE TRACING: SYSTEMATIC METHOD　　1129–1149

21.1. Introduction　　　　　　　　　　　　　　　　　　1129

21.2. Important Curves　　　　　　　　　　　　　　　　1129–1130

21.3. Systematic Procedure of Curve Tracing　　　　　1130–1131

[Curve $f(x, y) = 0$]

21.4. Illustrative Examples　　　　　　　　　　　　　1131–1135

21.5. Polar Curves: Procedure; Curve given by $f(r, \theta) = 0$　　1135

21.6. Well-known Polar Curves　　　　　　　　　　　　1136–1139

[Class of Curves: $r = a \sin n\theta$; $r^n = a^n \cos n\theta$. Spirals, Limacon and Cardiode]

xvi

21.7. A Few more Special Curves   1140–1145

    [Folium of Descartes, Parametric equations of different types of Cycloids, Astroid, Catenary, Further Properties of Cycloid]

Main Points: A Recapitulation   1145–1146

Exercises: XXI   1146–1149

Applications of Derivatives   1153–1157

[Recapitulation: For Self Practice]

Appendix: Chapterwise Important Problems   1161–1225

[with Hints/Solutions]

Miscellaneous Revision Exercises   1229–1239

Hints/Solutions of Miscellaneous Revision Exercises   1243–1257

Index   1259–1265

Preface to the Thirteenth Edition

In the present edition first few chapters on Analysis part have been modified thoroughly and enlarged wherever necessary according to the new concepts and notations. The application part is rich enough and so almost no modification has been done.

A large number of miscellaneous problems with hints have been introduced. Additional problems based on 'Model Question Booklet' published by the Calcutta University and recent Honours Question papers have been considered in appendix. Many of them have been solved in detail and a few of them left out with hints.

I hope the book in its present form is self-sufficient and is of international standard. I further hope that the book is student-friendly and most of the students will be able to follow it independently. I spent a lot of time to make the book error-free. Any constructive suggestion for further development of the book will be thankfully accepted from all quarters.

I thank Mr Amitabha Sen, Director of New Central Book Agency (P) Ltd, who gave me the opportunity to carry out such a prestigious and responsible task.

Kolkata
19 August 2010

Kartick Chandra Pal

Preface to the First Edition

This book treats the whole of the subject-matter covered by the syllabi of Differential Calculus for undergraduate studies of different Universities. The motive behind the preparation of this book is a simple one, namely, to set forth the processes of thinking by which the concepts and methods of Calculus are developed. We have neither evaded the vital issues nor burdened them with formal demonstrations that might deflect attention away from the main current of ideas.

To obtain the greatest possible clarity, the student has been kept in mind all through. We realise that a first course of Calculus like this stands on the border-line between Elementary Mathematics (such as Algebra and Plane Geometry) and the more advanced abstract Analysis. The notion of a function approaching a limit is studied in Elementary Mathematics through occasional illustrations of problems only. In contrast with this, the concept of limit has appeared in this book as a fundamental principle which, apart from any other feature, demands that the student should enlarge his mathematical concepts and operations and become familiar with the new mode of approach to a problem.

Rigour is not our primary aim, though every attempt has been made to maintain the highest standards of rigour appropriate to textbooks in this particular field. Our aim is to instil insight into the minds of the students, to make them see that the book can be read, and for the most part understood, without the help of classroom explanations.

Theorems and their limitations have been stated carefully and wherever possible, complete proofs have been given. The student has always been taken into authors confidence, and has been told why a particular method has or has not been used and what practical as well as theoretical pitfalls have had to be avoided. In order that the students may have a thorough grasp of the subject-matter, a large variety of examples have been worked out and set as exercises in the body of the book. To assist the students to organise the materials, each chapter begins with an introduction outlining the work to be undertaken and indicating its relation to what has gone before. For a further aid we have added an appendix at the end of the book and a list of formulae and results concerning Algebra, Trigonometry and Analytical Geometry together with a glossary containing a few mathematical and physical terms.

Starting with an introductory analysis of the historical background of the subject of the book and its purpose, we have discussed number system, function of an integral variable, function of a continuous variable, limit, continuity and derivative. This has been followed by the application of the derivative in Mechanics, Theory of Equations and Geometry. We have next entered into detailed discussions of Taylor's Theorem with the estimation of the remainder after n terms, which in turn has been naturally supplemented by the representation of a function by series and the problem on maxima and minima. In view of its growing importance in applied branches of mathematics, we have dealt in detail with the subject of functions of two variables, hyperbolic functions, the theory of curves and its associated concepts like curvature, asymptotes, singular points etc. and finally we have given systematic procedure of curve tracing where the sketches of well-known curves have been shown.

Acknowledgement is here made of the author's indebtedness to the many colleagues, students, and former teachers, and in particular to Mr. P. K. Bagchi, Professor of Mathematics, Vidyasagar College, Calcutta, for the assistance and inspiration, direct and indirect, that they received for the preparation of this book. Any suggestions to ensure further improvement of the book will be gratefully appreciated by the authors.

Calcutta   **Ghosh**
the 25th July, 1960   **Maity**

Contents of
INTRODUCTION TO ANALYSIS
(Differential Calculus)
(Part II)

0. Recapitulations
1. Metric Spaces
2. Compactness: Heine-Borel Theorem: Continuity on a Compact Set
3. Uniform Convergence of Sequence and Series of Functions
4. Power Series
5. Functions of Several Variables: Selected Topics only including Taylor's Theorem in R^2 Extremes of Functions of two and three variables.
6. Double Sequence and Double Series

Contents of the Companion Volume
INTRODUCTION TO ANALYSIS
(Integral Calculus)

1. Basic Methods of Integration
2. Integration by Partial Fractions
3. Integration of Trigonometric and Hyperbolic Functions
4. Trigonometric and Hyperbolic Substitutions
5. Reduction Formulae
6. Integration by Special Devices
7. Definite Integral as the Limit of a sum
8. The Riemann Integral
9. The Logarithmic and Exponential Functions
10. Sequence and Series of Function: Power Series
11. Improper Integrals
12. Fourier Series
13. Integral containing an Arbitrary Parameter
14. Gamma and Beta Functions
15. Multiple Integrals
16. Evaluation of Area
17. Lengths of Plane Curves
18. Volumes and Surfaces of Revolution
19. Centroids and Moments of Inertia
20. Line and Surface Integrals

AN INTRODUCTION TO
ANALYSIS

(Differential Calculus—Part I)

Chapter 0

Elements of Mathematical Logic

Chapter 0

Elements of Mathematical Logic

0.1 Introduction

What is Mathematics?

Not an easy question to answer. A rather partial, certainly not adequate, answer might be the following:

Mathematics is the language, techniques and results of logical deductions based on axioms and definitions used in studying

(i) *Natural phenomena* in Physics, Engineering, Astronomy, Chemistry, Biology, Genetics, Medicine, Psychology, etc.

(ii) *Human endeavours* such as Economics, Actuary, Accountancy, Architecture, Town planning, etc.

(iii) *Logical structures, Special investigations* such as Abstract concepts derived from intuitive notions of numbers and Space, Computer programming, Linguistics, etc.

There is a similarity in the basic ideas underlying the mathematics used for all the aforesaid disciplines and that a common education (Mathematical Logic) provides a basis for any mathematical activity. Some of the words and concepts, language and punctuation that have to be mastered are: Sets, Relations, Mappings or Functions, operations on a set and algebraic structures (Groups, Rings, Fields, Vector spaces), properties of functions defined on a set of real numbers (i.e., Calculus).

This initial chapter is concerned with certain notions of mathematical logic. The symbols and concepts introduced here will be constantly used in later chapters. They will be better understood when the reader will become familiar with their frequent uses.

It should, however, be remembered that mathematics is concerned with new ideas, new concepts rather than with the new symbols. The symbols are the tools for transference of ideas—they shorten the language and often bring in more clarity in exposition. But mathematics becomes more exciting and more appreciated only when the underlying ideas and concepts become clear to the mind.

0.2 Fundamental logical terms and notions

It is not our intention to introduce at this level the concepts of mathematical logic with utmost rigour. We shall rather very briefly and informally describe the meanings and significances of some frequently used logical terms and symbols with suitable illustrations.

1. **Statement**.

    By *a statement* we mean a collection of symbols (or sounds) which is meaningful in the sense that is either TRUE (T) or FALSE (F) but not both (not ambiguous also). A statement must be assertive—not interrogative, not imperative, not exclamatory.

    e.g., $2 + 2 = 4$ is a mathematical statement. It is assertive.
    It has the truth value T (true).
    $2 + 2 = 5$ is also a mathematical statement. It is also assertive but it has a truth value F (not true or, false).
    Where are you going? Stand up! Oh! how said it is! $\longrightarrow$ They are not assertions. One cannot assert whether they are TRUE or FALSE.

    Open sentence.

    Consider the relation $x + 3 = 8$. This equation, by itself, is not a statement since it is not possible to determine whether it is true or it is false unless the symbol x has been replaced by some specific number. This means that we must have in the background a collection of possible replacements for x, e.g., the set of natural numbers or the set of real numbers, such that, for each possible replacement a for x, either $a + 3 = 8$ is true or $a + 3 = 8$ is false.
    For example, $5 + 3 = 8$ is true but $1 + 3 = 8$ is false. Such a relation $x + 3 = 8$ is called an *open sentence*.
    An equation such as $x + 3x = 4x$, which is true for every replacement of x, is considered to be an open sentence because no specific value of x is indicated. We may, in many cases like this call an open sentence a *statement* so that each statement is either true or false in each case in which it makes sense.

2. **Compound Statement**.

    A statement formed by joining two or more statements is called a compound statement. Different compound statements may be formed by joining two or

more (simple) statements by different rules. Such as:
(a) *Conjunction of two statements p and q:*
If p and q are two statements, we form a single statement by joining them with the word AND, and denote this Compound Statement by the symbol $p \wedge q$ (read: p and q).
Remember: $\wedge$ is the symbol for AND, e.g., Let p be the statement $2 > 1$ and q be the statement $2 > 3$.
Then $p \wedge q$ will stand for $2 > 1$ and $2 > 3$.
Note: p is true but q is false and clearly $p \wedge q$ is FALSE.
Remember: $p \wedge q$ is true only when both p and q are true.
We put these ideas in a Tabular form (we call it a truth table):

p	q	$p \wedge q$
T	T	T
T	F	F
F	T	F
F	F	F

(b) *Disjunction of two statements p and q:*
If p and q are two statements we form a single statement by joining them with the word OR, and denote this compound statement by the symbol $p \vee q$ (read: p or q).
Remember: $\vee$ is the symbol for OR
e.g., $p: 2 > 1$; $q: 2 > 3$. Then $p \vee q$ will stand for '$2 > 1$ or $2 > 3$'.
Note: p is true, q is false and clearly $p \vee q$ is true since at least p is true.
Remember: $p \vee q$ is true when at least one of p or q is true. The truth table in this case is given below:

p	q	$p \vee q$
T	T	T
T	F	T
F	T	T
F	F	F

(c) *Negation of a statement p:*
The statement obtained by prefixing a NOT with the statement p is called the *negation* of p, denoted by $\sim p$ (read: not p).
Remember: $\sim$ is the symbol for *not*
e.g., if p be the statement:
In the $\triangle ABC$, $\angle A + \angle B + \angle C = 180°$;
then $\sim p$ will stand for:
In the $\triangle ABC$, $\angle A + \angle B + \angle C \neq 180°$.

The truth table for $\sim p$:

p	$\sim p$
T	F
F	T

We show below truth tables of different compound statements:

Example 0.2.1. *Truth tables for* $\sim (p \wedge q)$ *and also for* $(\sim p) \vee (\sim q)$.

p	q	$p \wedge q$	$\sim (p \wedge q)$	$\sim p$	$\sim q$	$(\sim p) \vee (\sim q)$
T	T	T	F	F	F	F
T	F	F	T	F	T	T
F	T	F	T	T	F	T
F	F	F	T	T	T	T

Observe that the truth values for $\sim (p \wedge q)$ and $(\sim p) \vee (\sim q)$ are identical; we say in such a case that the two propositions are logically equivalent ($\equiv$). We write $\sim (p \wedge q) \equiv (\sim p) \vee (\sim q)$.

Students may now try to prove: $\sim (p \vee q) \equiv (\sim p) \wedge (\sim q)$.

These two results are known as **De Morgan's Laws**.

Example 0.2.2. *Truth table for* $(p \wedge q) \vee \sim r$ (p, q, r *are three statements*).

p	q	r	$\sim r$	$p \wedge q$	$(p \wedge q) \vee \sim r$
T	T	T	F	T	T
T	T	F	T	T	T
T	F	T	F	F	F
F	T	T	F	F	F
T	F	F	T	F	T
F	T	F	T	F	T
F	F	T	F	F	F
F	F	F	T	F	T

3. Conditional Statement.

Implications between two statements p and q:

The statement 'If p, then q', is called the *implication* of q by p, denoted by $p \Rightarrow q$ (read: p implies q). Here p is the hypothesis and q is the conclusion.

Remember: $p \Rightarrow q$ stands for any of the following expressions: (i) p implies q; (ii) p, only if q; (iii) p is sufficient for q; (iv) q is necessary for p.

Note: $p \Rightarrow q$ is always true except in the case when p is true, and q is false (in this latter case $p \Rightarrow q$ is false).

Truth table for $p \Rightarrow q$:

p	q	$p \Rightarrow q$
T	T	T
T	F	F
F	T	T
F	F	T

4. **Associated Implications.**

   Associated with $p \Rightarrow q$ we have three other implications:
   (a) *Converse* of $p \Rightarrow q$ is $q \Rightarrow p$.
   (b) *Inverse* of $p \Rightarrow q$ is $\sim p \Rightarrow \sim q$.
   (c) *Contrapositive* of $p \Rightarrow q$ is $\sim q \Rightarrow \sim p$.

   Remember: $p \Rightarrow q$ and $\sim q \Rightarrow \sim p$ are logically equivalent.
   [This can be verified by drawing up truth values of $p \Rightarrow q$ and $\sim q \Rightarrow \sim p$ in the same truth table.]

5. **Biconditional.**

   In a situation where $p \Rightarrow q$ and the converse $q \Rightarrow p$ are both true, we write $p \Longleftrightarrow q$. $p \Longleftrightarrow q$ is read as: p implies and implied by q. $p \Longleftrightarrow q$ stands for: p if and only if q.
   [We shall write 'iff' in place of 'if and only if'.]
   In this statement the part 'if' corresponds to $q \Rightarrow p$ but the part 'only if' corresponds to $p \Rightarrow q$.
   Alternatively $p \Longleftrightarrow q$ stands for the following statement.
   A necessary and sufficient condition for p to be true is that q is true.
   $p \Rightarrow q$ corresponds to "q is true is a necessary condition for p to be true".
   $q \Rightarrow p$ corresponds to "q is true is a sufficient condition for p to be true".

   Example 0.2.3.
   p: The $\triangle ABC$ is right-angled at A.
   q: The $\triangle ABC$ has the area $\frac{1}{2} AB \cdot AC$.
   Here $p \Rightarrow q$ stands for
   If the triangle ABC is right-angled at A, then it has the area $\frac{1}{2} AB \cdot AC$.
   If p is true, then we know that q is also true and $p \Rightarrow q$ has the truth value T.

   Example 0.2.4.
   p: The integer n is odd.
   q: The integer n^2 is even.
   $p \Rightarrow q$ stands for "If the integer n is odd, then n^2 is even." See that if p is true then q is false and $p \Rightarrow q$ is false.

Example 0.2.5.
p: A is an equilateral triangle.
q: A is an isosceles triangle.
Here $p \Rightarrow q$ is true but observe that the converse $q \Rightarrow p$ is false.
Now see that the contrapositive statement $\sim q \Rightarrow \sim p$ is true.
[If A is not an isosceles triangle then A is not an equilateral triangle.]

Example 0.2.6. *If x^2 is odd, then prove that x must be odd.*
We prove the contrapositive statement: "If x is not odd (i.e., if x is even), then x^2 is not odd (i.e., x^2 is even)".
Proof. Let $x = 2k$ (k is an integer). Then $x^2 = 4k^2 = 2(2k^2)$. That is, x^2 is even.

Now since the contrapositive statement is true the given statement must be true ($\because$ they are logically equivalent).

Example 0.2.7.
p: A quadrilateral ABCD is cyclic.
q: A quadrilateral ABCD has opposite angles supplementary.
We know $p \Rightarrow q$ is true. Therefore the contrapositive statement $\sim q \Rightarrow \sim p$ must also be true, that is,

If the opposite angles of a quadrilateral $ABCD$ are not supplementary, then the quadrilateral is not cyclic.

See that in this case the converse $q \Rightarrow p$ is also true. Hence, here $p \iff q$.

Example 0.2.8.
p: A positive integer n is divisible by 9.
q: The sum of the digits of n is divisible by 9.
Here $p \iff q$ so that a positive integer n is divisible by 9 iff the sum of its digits is divisible by 9.

6. **Universal and Existential Statements.**

   (a) We know that for all real values of x, $\cos^2 x + \sin^2 x = 1$.
   This is an example of an *universal statement* in the sense that it involves all the objects of a given collection of objects (in the present case, collection of all real numbers).
   Remember: The symbol $\forall x$ stands for "for all x".

   (b) Consider the statement "there exists a real number x whose square is 2".
   This is an example of an *existential statement*, since it involves the possible existence of an object with a particular property in a given collection of objects.
   Remember: The symbol $\exists x$ stands for "there exists x".

The two symbols $\forall$ and $\exists$ are known as *quantifiers* (Note $\sim \forall \Rightarrow \exists$).

Example 0.2.9.

$p : \forall$ *real numbers* x, $\cos x + \sin x = 1$.

$\sim p : \exists$ *a real number* x *such that* $\cos x + \sin x \neq 1$.

Prove that $\sim p$ is true.

Solution. We shall prove that p is false by proving that $\sim p$ is true.

To prove: $\sim p$ is true, it is sufficient to find a real number x such that $\cos x + \sin x \neq 1$.

Take $x = \pi/4$ (say). Then $\cos(\pi/4) + \sin(\pi/4) = \sqrt{2} \neq 1$.

Thus $\sim p$ is true. Consequently p is false.

Observation: We have established the falsity of p by constructing what we call a counter example. The students should observe that the negation of an universal statement is an existential statement and conversely. Replace $\forall$ by $\exists$ and p by $\sim p$. See No. 5 (Exercises on Chapter 0).

7. Methods of Proofs in Mathematics.

We usually prove mathematical propositions either by direct proof, or by indirect proof.

Direct Method:

(a) *Rule of detachment*:

From
$$\left. \begin{array}{c} p \text{ true} \\ \text{and } p \Rightarrow q \text{ true} \end{array} \right\} \text{ we infer that } q \text{ is true.}$$

Example 0.2.10. $p : x$ *and* y *are even integers, and* $q : x + y$ *is an even integer.*

Suppose p is true. Then $p \Rightarrow q$ is also true. Therefore, by rule of detachment q is true, i.e., $x + y$ is an even integer.

(b) *Law of syllogism* (*Transitive property of implications*):

From $p \Rightarrow q$ and $q \Rightarrow r$, we conclude $p \Rightarrow r$, i.e., $\{p \Rightarrow q \wedge q \Rightarrow r\} \Rightarrow (p \Rightarrow r)$.

Example 0.2.11.

$p : x$ *is an even integer.*

$q : x^2$ *is an even integer.*

$r : x^2 - 1$ *is an odd integer.*

Here $p \Rightarrow q$ and $q \Rightarrow r$. Hence the conclusion is $p \Rightarrow r$ (i.e., if x is an even integer, then $x^2 - 1$ is odd).

Indirect Method:

If p is a statement, then either p is true or $\sim p$ is true. Both cannot be simultaneously true. If by logical arguments one can prove that both p and $\sim p$ are true, then one of the assumptions made must be wrong (**method of contradiction**).

Example 0.2.12. *Let*

$p : xy$ *is an odd integer.*

$q : x$ *and* y *are both odd.*

Suppose that p is true. To prove that q is true.

Assume that q is false (i.e., Suppose that $\sim q$ is true).

It then follows that at least one of the integers x or y must be even; say x is even $= 2k$. Then $xy = 2(ky) =$ even, i.e., p is false (or $\sim p$ is true).

Thus we see that we arrive at the result : both p and $\sim p$ are true. This cannot happen.

∴ Our assumption that q is false is not tenable (leads to a contradiction).

∴ When p is true, q must be true.

EXERCISES ON CHAPTER 0

1. Draw up truth tables and then prove that
    (a) $\sim (p \wedge q) \equiv \sim p \vee \sim q$;
    (b) $\sim (\sim p) \equiv p$;
    (c) $p \Rightarrow q \equiv (\sim p) \vee q$;
    (d) $\sim (p \Rightarrow q) \equiv p \wedge (\sim q)$;
    (e) $p \Rightarrow (q \wedge r) \equiv (p \Rightarrow q) \wedge (p \Rightarrow r)$;
    (f) $(p \vee q) \Rightarrow r \equiv (p \Rightarrow r) \wedge (q \Rightarrow r)$.
2. Which of the following conditions are:
    (a) necessary;
    (b) sufficient;
    (c) both necessary and sufficient
        for the positive integer n to be divisible by 6?
        i. n is divisible by 3;
        ii. n is divisible by 9;
        iii. n is divisible by 12;
        iv. n^2 is divisible by 12;
        v. n is even and divisible by 3;
        vi. $n = m^3 - m$ for some positive integer m.
    Ans. (a) (i), (iv), (v); (b) (iii), (iv), (v), (vi); (c) (iv), (v).

ELEMENTS OF MATHEMATICAL LOGIC

3. Use the *method of contradiction* to prove the following:
   (a) If n^3 is odd, then n is odd, n being a positive integer.
   (b) If $\sin \theta \neq 0$, then $\theta \neq k\pi$, for any integer k.
   (c) If x and y are integers such that xy^2 is even, then at least one of x, y is even.

4. Write down the contrapositive statement for each of the following:
   (a) If he has the courage, he will win.
   (b) It is necessary to be strong in order to be a soldier.
   (c) It is sufficient for it to be a square in order to be a rectangle.
   (d) If x is less than zero, then x is not positive.

   Ans. No answers are given. Try yourself.

5. [See the definition of limit of a function in a later Chapter]
   $\lim_{x \to a} f(x) = l$ is defined as
   $\forall \varepsilon > 0; \exists \delta > 0$ s.t. $\forall x \in 0 < |x - a| < \delta \Rightarrow |f(x) - l| < \varepsilon$.
   Let us call $p : 0 < |x - a| < \delta \Rightarrow |f(x) - l| < \varepsilon$.
   Then the definition of limit is as: $\forall \varepsilon > 0; \exists \delta > 0$ s.t. p is true.

   What is the negation of this statement?

   Ans. $f(x)$ does not tend to l as $x \to a$ if $\exists \varepsilon > 0$ s.t. $\forall \delta > 0 \exists x$ s.t. $\sim p$ is true (replacing $\exists$ by $\forall$ and $\forall$ by $\exists$, p by $\sim p$).

MAIN POINTS: CHAPTER 0

1. Statement—its truth value is either T or F.
2. Meanings of $\wedge, \vee, \sim$ (and, or, negation).
3. If p then q or $p \Rightarrow q$ or p only if q or q is necessary for p or p is sufficient for q—Equivalent.
4. $p \Longleftrightarrow q$, p iff q; interpret.
5. Meanings of $\forall, \exists$ (for all, there exists).
6. Proofs in mathematics—Direct and Indirect (method of contradiction).

FREQUENTLY USED SYMBOLS

$\in$	belongs to
$\notin$	does not belong to
$\subseteq$	subset
$\subset$	proper subset
$\cup$	union
$\cap$	intersection
$\forall$	for all
$\exists$	there exists
$\Rightarrow$	implies
$\Leftrightarrow$	implies and is implied by (if and only if or iff)
$\{\ \}$	set
ϕ	empty set
$A - B$	difference of two sets A and B
A^c or A'	complement of A relative to the universal set $\cup$
int. A or A^i	interior of A
A'	set of all limit points of A (or derived set of A)
$\bar{A}$	closure of A (set of all adherent points of A)
$\mathbb{N}$	set of natural numbers $\{1, 2, 3, \cdots\}$
$\mathbb{Z}$	set of all integers $\{0, \pm 1, \pm 2, \cdots\}$
$\mathbb{Q}$	set of all rational numbers
$\mathbb{R}$	set of all real numbers $(-\infty, \infty)$
$\mathbb{C}$	set of all complex numbers
$[a, b]$	closed interval
(a, b) or $]a, b[$	open interval
$f : A \to B$	f maps A to B (or function from A to B)
$g \circ f$	composite function
$f \setminus E$	restriction of f to the set E
$d(x, y)$	distance between x and y
$d(x, A)$	distance of x from the set A
$d(A, B)$	distance between sets A and B
$d(A)$	diameter of A
$nbdA$ or $nbd \cdot A$	neighbourhood of A
$\inf A$	infimum (or glb) of A
$\sup A$	supremum (or lub) of A
$\mathbb{R}^n$	Euclidean n-space
$B[a, b]$	space of bounded functions on $[a, b]$
$C[0, 1]$	space of continuous functions on $[0, 1]$
$\mathcal{P}[a, b]$	set of polynomials on $[a, b]$
$\mathcal{R}[a, b]$	space of integrable functions in the sense of Riemann
$\triangle$	triangle

THE GREEK ALPHABET

A	α	Alpha	I	ι	Iota	Σ	σ	Sigma
B	β	Beta	K	κ	Kappa	T	τ	Tau
Γ	γ	Gamma	Λ	λ	Lambda			
Δ	δ	Delta	M	μ	Mu	Y	υ	Upsilon
E	ϵ	Epsilon	N	ν	Nu	Φ	ϕ	Phi
Z	ζ	zeta	Ξ	ξ	Xi	X	χ	Chi
H	η	Eta	Π	π	Pi	Ψ	ψ	Psi
Θ	θ	Theta	R	ρ	Rho	Ω	ω	Omega

AN INTRODUCTION TO ANALYSIS

(Differential Calculus—Part I)

Chapter 1

Basic Concepts: Sets, Relations and Mappings

Chapter 1

Basic Concepts: Sets, Relations and Mappings

1.1 Introduction

In this initial chapter we recall the concepts of *sets* and *functions* which are fundamental in the study of real analysis.

In every language there are certain terms which are basic and remain undefined but whose meanings are universally accepted. In Mathematics the word *set* is such a term: a set is understood to be a well-defined collection of distinct objects called *elements*. The term well-defined is that property of the set by which one is able to determine whether a given element belongs to the set, or not.

Some authors prefer to take the word *set* as a primary and an undefined concept and then develop it axiomatically (as in the book *Axiomatic Set Theory* by P. Suppes).

Our understanding of *well-defined collection of distinct objects* is intuitive and naive and adequate for our purpose.

About Sets. We shall identify a set by stating its *members* (or *elements*). We denote sets by capital letters A, B, C, etc. and use lower case letters a, b, c, etc. to represent their elements.

If an element x is in the set A, we write $x \in A$ and say that x is a *member* of A or x belongs to A. If x is not in A, we write $x \notin A$ (x does not belong to A).

We write $\{x\}$ to denote a *singleton set* whose only member is x.

We write $\{x_1, x_2, \cdots, x_n\}$ to denote a finite set of n elements $x_1, x_2, \cdots, x_n$.

We may write an infinite set like $\mathbb{N} = \{1, 2, 3, \cdots\}$, the set of all natural numbers, where we use a curly bracket to enclose some elements and three dots to imply the existence of other elements.

When it is possible to list all the elements of a finite set, we call it *Roaster Method of representation* of sets. e.g., $\{a, e, i, o, u\}$, the set of five vowels in the English alphabet. But most often a set is represented by some specific property $P(x)$ common to all elements of the set. We write

$$X = \{x : x \text{ obeys } P(x)\} \text{ or } simply\ X = \{x : P(x)\}.$$

We shall, throughout this text, use the following notations for some specific sets:

$\mathbb{N}$ = set of all natural numbers or positive integers = $\{1, 2, 3, \cdots\}$.

$\mathbb{Z}$ = set of all integers = $\{0, \pm 1, \pm 2, \cdots\}$.

$\mathbb{Q}$ = set of all rational numbers = $\{p/q : p \in \mathbb{Z} \text{ and } q \in \mathbb{N}\}$

$\mathbb{R}$ = set of all real numbers.

We write $\mathbb{Z}^+, \mathbb{Q}^+, \mathbb{R}^+$ to denote the positive elements in the respective sets.

Subsets. If each element in the set A is also a member of the set B, then we say that A is a *subset* of B and we write $A \subseteq B$ (A is included in B) or equivalently, $B \supseteq A$ (B includes A).

We say that a set A is a *proper subset* of B if $A \subseteq B$ and $\exists$ at least one element of B that is not in A.

We then write: $A \subset B$ (read: A is a proper subset of B).

Equality of two sets. Two sets A and B are said to be equal if the sets consist of precisely the same elements. We then write $A = B$.

It is easy to see that $A = B$, provided $A \subseteq B$ and $B \subseteq A$ and conversely,

$$A \subseteq B \text{ and } B \subseteq A \Longrightarrow A = B.$$

Universal set U and Empty set ϕ. In any discussion involving sets we consider a fixed set U which is the set of all elements under discussion. Thus every set in that discussion is a subset of U. We call U the *universal set* or *the universe*.

In real analysis the set $\mathbb{R}$ of all real numbers is taken as the universe. We deal, therefore, with subsets of real numbers.

Note: We also use S for an universal set.

The symbol ϕ denotes what we call the *empty set* or *null set* which contains no element. For e.g., the set whose elements are common elements of $\{2, 3, 4\}$ and $\{5, 8, 7\}$ is the null set ϕ. The set of all students of Class V of a school who secured more than 90 out of 100 is ϕ, if the highest mark is 90. We do not know in advance whether any student secured more than 90 or not. That is why, such a conceptual set is necessary in theory of sets. For logical consistency ϕ is a subset of any set A. It is called an improper subset of any set A.

Remember: For every set A we have $\phi \subseteq A \subseteq U$ and $A \subseteq A$. [A is called trivial subset of A]

An useful result. *Set inclusion is transitive*, i.e., if $A \subseteq B$ and $B \subseteq C$, then $A \subseteq C$.

[For, $A \subseteq B \Longrightarrow x \in A \Longrightarrow x \in B$; $B \subseteq C \Longrightarrow x \in B \Longrightarrow x \in C$.
∴ two together $\Longrightarrow x \in A \Longrightarrow x \in C$, i.e., $A \subseteq C$]

Family or Collection of Sets. Let I be any set. Suppose for each member $i \in I$ we can associate a set A_i. Then the collection $\{A_i : i \in I\}$ form a *family of sets indexed by I* (I is known as an *index set*).

Power set. Given a set A. We collect the family of all subsets of A (this family includes the set A itself and ϕ). This family of sets is called the *power set* of A, denoted by $P(A)$.

As for example, let $A = \{1, 2, 3\}$. Then
$$P(A) = \{\{1\}, \{2\}, \{3\}, \{1,2\}, \{1,3\}, \{2,3\}, \{1,2,3\}, \phi\}.$$
In fact, if A is a finite set of n elements, then $P(A)$ has 2^n elements.

[Hints: $^nC_0 + {^nC_1} + {^nC_2} + \cdots + {^nC_n} = 2^n$]

1.2 Operations on Sets (Set Algebra)

We now define methods of obtaining new sets from given ones—these methods are called *operations on sets*. Some of those operations—*Union, Intersection, Complementation* and *Difference of two sets*, are described below:

I. Union: The *union* of two sets A and B is the set $A \cup B = \{x : x \in A \text{ or } x \in B\}$.

(The word *or* is to be used in the *inclusive sense* allowing the possibility that x may belong to both the sets).

For the collection of sets A_i indexed by $i \in I$ we define the union of this collection by $\bigcup_{i \in I} A_i = \{x : x \in A_i \text{ for some } i \in I\}$, where I is an index set.

[In case $I = \mathbb{N}$ = set of all positive integers n, the union is denoted by $\bigcup_{n=1}^{\infty} A_n = \{x : x \in A_n \text{ for some } n \in \mathbb{N}\}$; it has a special name—*countable union*]

II. Intersection: The intersection of two sets A and B is the set
$$A \cap B = \{x : x \in A \text{ and } x \in B\}.$$

For the collection of sets A_i indexed by $i \in I$, the intersection
$$\bigcap_{i \in I} A_i = \{x : x \in A_i \text{ for all } i \in I\}$$
and the countable intersection (i.e., when $I = \mathbb{N}$)
$$\bigcap_{i=1}^{\infty} A_i = \{x : x \in A_i \text{ for all } i \in \mathbb{N}\}.$$

Examples (i) Let $A_n = \{n\}$. Then $\bigcup_{n=1}^{\infty} A_n = \mathbb{N}$ and $\bigcup_{n=-\infty}^{\infty} A_n = \mathbb{Z}$, $\bigcap_{n=1}^{\infty} A_n = \phi$.

(ii) Let $A_n = (-1/n, 1/n)$, $n \in \mathbb{N}$. Then $\bigcup_{n=1}^{\infty} A_n = (-1, 1)$ and $\bigcap_{n=1}^{\infty} A_n = \{0\}$.

(iii) Let $A_n = \{1, 2, 3, \cdots, n\}$. Then $\bigcup_{n \in \mathbb{N}} A_n = \mathbb{N}$ but $\bigcap_{n \in \mathbb{N}} A_n = \{1\}$.

Disjoint sets. Two sets A and B are said to be disjoint, if they have no elements in common, i.e., if $A \cap B = \phi$.

The family of sets is called *pairwise disjoint*, if each distinct pair of elements of the collection are disjoint. Thus an *indexed collection* $\{A_i\}_{i \in I}$ is pairwise disjoint, if $A_i \cap A_j = \phi$ for all $i, j \in I$ and $i \neq j$.

III. *Complementation.* Let U be the universal set. Suppose that A and B are two subsets of U. Then we define *the complement (or difference) of B relative to A*, denoted by $A - B$ or $A \backslash B$ (A slash B), to be the set
$$A - B = \{x : x \in U, x \in A \text{ and } x \notin B\}.$$

By A' or A^c (complement of A) we mean $U - A$, i.e.,
$$A^c = \{x : x \in U, x \notin A\} = \{x : x \notin A\} \ (\because x \text{ always belongs to } U, \text{ no need to mention}).$$

In *Real Analysis*, the universe is $\mathbb{R}$, the set of all real numbers. If $A \subseteq \mathbb{R}$, then
$$A' \text{ or } A^c = \mathbb{R} - A = \mathbb{R} \backslash A = \{x : x \in \mathbb{R} \text{ and } x \notin A\} = \{x : x \notin A\}.$$

Example (i) Let $A = \{2, 4, 6\}$ and $B = \{2, 6, 10, 14\}$. Then complement of B relative to A is the set
$$A \backslash B = A - B = \{x : x \in A \text{ and } x \notin B\} = \{4\} \text{ and}$$
$$B \backslash A = B - A = \{x : x \in B \text{ and } x \notin A\} = \{10, 14\}.$$

Note that $A - B$ and $B - A$ are *two disjoint sets*.

(ii) Let $\mathbb{Q} \subset \mathbb{R}$ ($\mathbb{Q}$ is the set of all rational numbers). Then $\mathbb{Q}' = \mathbb{R} - \mathbb{Q} = \{\text{set of all irrational numbers}\}$.

CHAPTER 1: BASIC CONCEPTS: SETS, RELATIONS AND MAPPINGS

IMPORTANT CONSEQUENCES

Let U be the universal set and $A, B, C \subseteq U$. Then

1. $A \cup B = B \cup A$ and $A \cap B = B \cap A$. Union and Intersection are commutative.
2. $A \cup (B \cup C) = (A \cup B) \cup C$; $A \cap (B \cap C) = (A \cap B) \cap C$.
   Union and intersection are associative.
3. $A \cup (B \cap C) = (A \cup B) \cap (A \cup C)$; $A \cap (B \cup C) = (A \cap B) \cup (A \cap C)$.
   Union is distributive over intersection and intersection is distributive over union, i.e., each is distributive over the other.
4. De Morgan's laws: $(A \cup B)' = A' \cap B'$ and $(A \cap B)' = A' \cup B'$.
5. $A - B = A \cap B'$.

Besides Union, Intersection and Complementation we introduce *two important operations*:

IV. *Symmetric difference*: Symmetric difference of two sets A and B is denoted by $A \triangle B$ and is defined by
$$A \triangle B = (A - B) \cup (B - A).$$

V. *Cartesian product*: If A and B are two non-empty sets then the *Cartesian product* $A \times B$ of A and B is the set of all ordered pairs (a, b) with $a \in A$ and $b \in B$, that is, $A \times B = \{(a, b) : a \in A, b \in B\}$.

For e.g., let $A = \{1, 2, 3\}$ and $B = \{1, 4\}$, then the Cartesian product $A \times B$ is the set whose members are $(1,1)$, $(1,4)$, $(2,1)$, $(2,4)$, $(3,1)$, $(3,4)$.

We may visualize that the set $A \times B$ corresponds to six points on the plane with coordinates that we have listed above.

We may draw a diagram (Fig. 1.2.1) to exhibit the elements of $A \times B$.

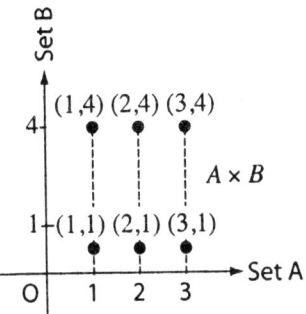

Fig 1.2.1

It is interesting to draw the diagram of $A \times B$, if

$A = \{x : x \in \mathbb{R} \text{ and } 1 \leq x \leq 2\}$
and $B = \{y : y \in \mathbb{R} \text{ and } 0 \leq y \leq 1 \text{ or } 2 \leq y \leq 3\}$.

The diagram of $A \times B$ is the adjoined.

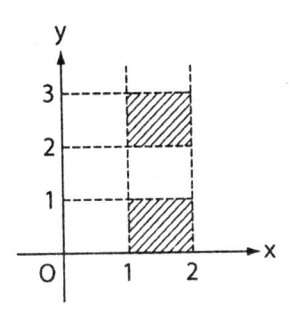

Fig 1.2.2

Properties: Symmetric difference and Cartesian products

1. Symmetric difference is both commutative and associative:
   $A \triangle B = B \triangle A$;
   $A \triangle (B \triangle C) = (A \triangle B) \triangle C$.
2. We have defined $A \triangle B = (A - B) \cup (B - A)$. From this we can show that $A \triangle B = (A \cup B) - (A \cap B)$ (see Ex. 1.3.5). Draw a diagram of $A \triangle B$.
3. $A \cap (B \triangle C) = (A \cap B) \triangle (A \cap C)$.
4. $A \triangle \phi = A$ and $A \triangle A = \phi$.
5. *Cartesian products*:
   (a) $A \times B \neq B \times A$; $A \times B = B \times A \iff A = B$;
   (b) $(A \cup B) \times C = (A \times C) \cup (B \times C)$; $A \times (B \cup C) = (A \times B) \cup (A \times C)$;
   (c) $(A \cap B) \times C = (A \times C) \cap (B \times C)$; $A \times (B \cap C) = (A \times B) \cap (A \times C)$;
   (d) $(A - B) \times C = (A \times C) - (B \times C)$; $A \times (B - C) = (A \times B) - (A \times C)$;
   (e) $A \neq \phi$, $A \times B = A \times C \implies B = C$.
6. *Extension*: Let $A_1, A_2, \cdots, A_n$ be a finite collection of n sets. Then
   $A_1 \times A_2 \times A_3 \times \cdots \times A_n = \{(a_1, a_2, \cdots, a_n) : a_1 \in A_1, a_2 \in A_2 \cdots a_n \in A_n\}$.
   In case $A_1 = A_2 = \cdots = A_n = A$ (say), the Cartesian product
   $$A^n = \{(a_1, a_2, \cdots, a_n) : a_i \in A\}.$$
   Let $A = \mathbb{R}$. Then $\mathbb{R}^n = \{(a_1, a_2, \cdots, a_n); a_i \in \mathbb{R}\}$.
   We call $(a_1, a_2, \cdots, a_n)$, where each $a_i \in \mathbb{R}$, an *n-tuple* of real numbers.

Venn diagram. For the purpose of illustrations we may often use *Venn diagrams*. Given below Venn diagram representations of the different operations on sets:

(a) $A \cup B$ (b) $A \cap B$ (c) $A - B$

(d) $U - A = A'$ (e) $A \triangle B = (A - B) \cup (B - A)$

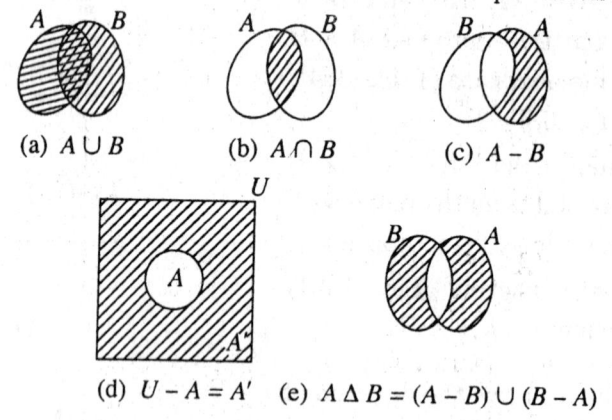

Fig 1.2.3 Diagrammatic representation of operations of sets.

CHAPTER 1: BASIC CONCEPTS: SETS, RELATIONS AND MAPPINGS

1.3 Solved Examples

Example 1.3.1. Let $A = \{1, 1/2, 1/3, 1/4, 1/5\}$ and $B\{1/2, 1/5, 1/8, 1/100\}$. Determine the elements of the set $A \cup (A \cap B)$.

Solution: $A \cap B = \{1/2, 1/5\}$ and hence $A \cup (A \cap B)$,

i.e., $\{1, 1/2, 1/3, 1/4, 1/5\} \cup \{1/2, 1/5\} = \{1, 1/2, 1/3, 1/4, 1/5\}$.

Example 1.3.2. If the universe $S = \{1, 2, 3, 4, 5, 6\}$ and if A, B, C are three subsets of S, where $A = \{1, 3, 4, 6\}$ and $B \cap C = \{1, 2, 6\}$, then determine the set:
(i) $(A \cup B) \cap (A \cup C)$; (ii) $B' \cup C'$.

Solution: (i) $(A \cup B) \cap (A \cup C) = A \cup (B \cap C)$ (using distributive law)
$= \{1, 3, 4, 6\} \cup \{1, 2, 6\}$
$= \{1, 2, 3, 4, 6\}$.

(ii) $B' \cup C' = (B \cap C)' = S - (B \cap C) = \{1, 2, 3, 4, 5, 6\} - \{1, 2, 6\} = \{3, 4, 5\}$.

Note: In solving problems on set equalities, e.g., to prove $X = Y$, we show $X \subseteq Y$ and $Y \subseteq X$. Another useful relation is $A - B = A \cap B'$.

Example 1.3.3. If A, B, C are three sets, then (i) $A - (B \cup C) = (A - B) \cap (A - C)$; (ii) $A - (B \cap C) = (A - B) \cup (A - C)$.

Solution: To prove (i), we shall prove that every element of LHS $A - (B \cup C)$ is contained in both $(A - B)$ and $(A - C)$ and conversely.

Let $x \in A - (B \cup C)$, then $x \in A$ and $x \notin B \cup C$. Hence, $x \in A$ and x is neither in B nor in C. Therefore, $x \in A$ but $x \notin B$ and $x \in A$ but $x \notin C$.

Then $x \in A - B$ as well as $x \in A - C$. $\therefore x \in (A - B) \cap (A - C)$.

$$\therefore A - (B \cup C) \subseteq (A - B) \cap (A - C). \tag{1}$$

Conversely, if $x \in (A - B) \cap (A - C)$, then $x \in A - B$ and $x \in A - C$. Hence, $x \in A$ and $x \notin B$ and $x \notin C$. Therefore, $x \in A$ and $x \notin (B \cup C)$, i.e., $x \in A - (B \cup C)$.

$$\therefore (A - B) \cap (A - C) \subseteq A - (B \cup C). \tag{2}$$

Relations (1) and (2)

$$A - (B \cap C) = (A - B) \cap (A - C)$$

To prove (ii), proceed exactly in a similar manner.

Example 1.3.4. Let S be the universal set. A, B, C are any three subsets of S. Then

(i) $A \cap (B - C) = (A \cap B) - (A \cap C)$; [CH]
(ii) $(A' \cup B') \cup (A \cap B \cap C') = A' \cup B' \cup C'$;
(iii) $(A' \cap B' \cap C) \cup (B \cap C) \cup (A \cap C) = C$;

where A', B', C' are respectively the complements of A, B, C relative to S.

Solution: (i) RHS $= (A \cap B) - (A \cap C) = (A \cap B) \cap (A \cap C)'$
$= (A \cap B) \cap (A' \cup C')$ (De Morgan's law)
$= \{(A \cap B) \cap A'\} \cup \{(A \cap B) \cap C'\}$ (Distribution law)
$= \phi \cup \{(A \cap B) \cap C'\} = (A \cap B) \cap C'$

LHS $= A \cap (B - C) = A \cap (B \cap C') = (A \cap B) \cap C'$ (Associative law)

$\therefore A \cap (B - C) = (A \cap B) - (A \cap C)$, proved.

Note: $(A \cap B) \cap A' = A' \cap (A \cap B)$ (Commutative property)
$= (A' \cap A) \cap B$ (Associative law)
$= \phi \cap B = \phi$.

(ii) LHS $= (A' \cup B') \cup (A \cap B \cap C')$
$= (A \cap B)' \cup \{(A \cap B) \cap C'\}$ (De Morgan's law)
$= \{(A \cap B)' \cup (A \cap B)\} \cap \{(A \cap B)' \cup C'\}$ (Distributive law)
$= S \cap \{(A \cap B)' \cup C'\} = (A \cap B)' \cup C'$
$= (A' \cup B') \cup C'$ (De Morgan's law)
$= A' \cup B' \cup C'$ (Associative property)
$=$ RHS (proved).

(iii) First, we observe that
$(B \cap C) \cup (A \cap C) = (C \cap B) \cup (C \cap A)$ (Commutative law)
$= C \cap (B \cup A)$ (Distributive law)
$= (B \cup A) \cap C$ (Commutative law)
$= (A \cup B) \cap C$. (Commutative law)

We also have $A' \cap B' \cap C = (A' \cap B') \cap C$ (Associative law)
$= (A \cup B)' \cap C$. (De Morgan's law)

Now, the given LHS $= (A' \cap B' \cap C) \cup (B \cap C) \cup (A \cap C)$
$= \{(A \cap B)' \cap C\} \cup \{(A \cup B) \cap C\}$
$= \{(A \cup B)' \cup (A \cup B)\} \cap C$ (Distributive law)
$= S \cap C = C =$ RHS (proved).

Example 1.3.5. *Prove the following for any three subsets A, B, C of the universal set S:*

(i) $(A - C) \cap (B - C) = (A \cap B) - C$;
(ii) $(A - B) \cup B = A$ iff $B \subseteq A$.
(iii) $(A - B) \cup (B - A) = (A \cup B) - (A \cap B)$, i.e., $A \triangle B = (A \cup B) - (A \cap B)$.

(iv) If $A \triangle B = A \triangle C$, prove that $B = C$.

Solution: (i) $A - C = A \cap C'$ and $B - C = B \cap C'$.
$$\begin{aligned}
\text{LHS} &= (A - C) \cap (B - C) \\
&= (A \cap C') \cap (B \cap C') \\
&= (A \cap C') \cap (C' \cap B) && \text{(Commutative law for } \cap\text{)} \\
&= \{(A \cap C') \cap C'\} \cap B && \text{(Associative law)} \\
&= \{A \cap (C' \cap C')\} \cap B && \text{(Associative law)} \\
&= (A \cap C') \cap B && (\because C' \cap C' = C') \\
&= B \cap (A \cap C') && \text{(Commutative law for } \cap\text{)} \\
&= (B \cap A) \cap C' && \text{(Associative law)} \\
&= (A \cap B) \cap C' && \text{(Commutative law for } \cap\text{)} \\
&= (A \cap B) - C \\
&= \text{RHS (proved)}.
\end{aligned}$$

(ii) $\begin{aligned}[t](A - B) \cup B &= (A \cap B') \cup B \\
&= (A \cup B) \cap (B' \cup B) && \text{(Distributive law)} \\
&= (A \cup B) \cap S && (S = \text{universe} = B' \cup B) \\
&= A \cup B && [\because A \cup B \subseteq S].
\end{aligned}$

Now if $B \subseteq A$, $A \cup B = A$ and conversely $A \cup B = A \implies B \subseteq A$.
$\therefore (A - B) \cup B = A$ iff $B \subseteq A$.

(iii) $\begin{aligned}[t](A - B) \cup (B - A) &= (A \cap B') \cup (B \cap A') \\
&= \{(A \cap B') \cup B\} \cap \{(A \cap B') \cup A'\} && \text{(Distributive law)} \\
&= \{(A \cup B) \cap (B' \cup B)\} \cap \{(A \cup A') \cap (B' \cup A')\} \\
& && \text{(Distributive law)} \\
&= \{(A \cup B) \cap S\} \cap \{S \cap (B' \cup A')\} && (S = \text{universe}) \\
&= (A \cup B) \cap (B' \cup A') \\
&= (A \cup B) \cap (A \cap B)' && \text{(De Morgan's law)} \\
&= (A \cup B) - (A \cap B).
\end{aligned}$

Thus $A \triangle B = (A \cup B) - (A \cap B)$.

(iv) $B = \phi \triangle B = (A \triangle A) \triangle B = A \triangle (A \triangle B) = A \triangle (A \triangle C) = \phi \triangle C = C$.

1.4 Relation from A to B: Relation on a set A

Let A and B be two non-empty sets. A relation $\Re$ from A to B is any subset of the Cartesian product $A \times B$.

We often speak of a relation $\Re$ from A to A: we call it *relation on the set A*.

Definition. Let A be a non-empty set. A subset $\Re$ of the Cartesian product $A \times A$ is called a *relation on A*. If $(x, y) \in \Re$, we say that x is related to y by the relation $\Re$ and we write $x\Re y$.

e.g., $\Re = \{(1, 2), (1, 3), (2, 3)\}$ is a relation on the set $A = \{1, 2, 3\}$. Here $1\Re 2$, $1\Re 3$, $2\Re 3$. Actually it is the usual relation $<$ (less than) because $1 < 2$, $1 < 3$, $2 < 3$. On the same set $A = \{1, 2, 3\}$, relation $\leq$ is described by the set $\{(1, 1), (2, 2), (3, 3), (1, 2), (1, 3), (2, 3)\}$.

We note that $<, =, >, \leq, \geq$ are relations on the sets of numbers $\mathbb{R}, \mathbb{N}, \mathbb{Z}, \mathbb{Q}$. *Is the mother of*, *Is the brother of*, *Is married to* are relation on the set of all human beings.

Definition. Let $\Re$ be a relation on a set A. For any three elements $x, y z \in A$

(i) If $x\Re x$ (i.e., $(x, x) \in \Re$) then $\Re$ is called a *reflexive relation*;

(ii) If $x\Re y \implies y\Re x$ (i.e., whenever $(x, y) \in \Re$, (y, x) also belongs to $\Re$), then $\Re$ is said to be a *symmetric relation*;

(iii) If $x\Re y$ and $y\Re z \implies x\Re z$ (i.e., if $(x, y) \in \Re$ and $(y, z) \in \Re$, then $(x, z) \in \Re$), then $\Re$ is said to be a *transitive relation*.

A relation $\Re$ on a set A is called an *equivalence relation on A* if $\Re$ is *reflexive, symmetric* and *transitive*.

Note: Sometimes such relations are called *binary relations*.

Moreover, $\Re$ is called *antisymmetric*, if $a\Re b$ and $b\Re a$ together imply $a = b$.

We consider a few examples on relations:

Example 1.4.1. *Let $A = \mathbb{Z}$, the set of all integers. Consider the subset $\Re$ of $\mathbb{Z} \times \mathbb{Z}$ defined by $\Re = \{(x, y) : x - y \text{ is divisible by } 3\}$.*

Solution: Here $x\Re y$, if $(x - y)$ is divisible by 3.

(a) $\Re$ is *reflexive* ($\because x - x$ is divisible by 3 for every $x \in \mathbb{Z}$);

(b) $\Re$ is *symmetric* (because $x - y$ is divisible by $3 \implies y - x$ is divisible by 3, i.e., $x\Re y \implies y\Re x$).

(c) If $x - y$ is divisible by 3 and $y - z$ is divisible by 3, then it is certainly true that $x - z = (x - y) + (y - z)$ is also divisible by 3.

Thus $x\Re y$ and $y\Re z \implies x\Re z$. Hence $\Re$ is *transitive*.

Thus here $\Re$ is an equivalence relation.

Example 1.4.2. *Let $A = \mathbb{R}$ (the set of all real numbers). It is easy to prove that the relation '=' is an equivalence relation (exactly as in the previous problem).*

Example 1.4.3. *Let $A = \mathbb{Z} - \{1\}$. We define the relation $\Re$ on this set by the rule $\{x\Re y$, iff x and y have common factor other than $1\}$. Verify that this relation is reflexive* and symmetric but not transitive (R and S but not T). For e.g., $x = 12, y = 15, z = 25$, $x\Re y, y\Re x, x\bar{\Re} z$.

Example 1.4.4. *Let $A = \mathbb{Z} =$ set of all integers. On $\mathbb{Z}$, we define the relation $\Re$ to mean $x > y$. This relation $\Re$ is neither reflexive nor symmetric but it is transitive. (T but not R or S)*

Example 1.4.5. *On $\mathbb{Z}$ we define $x\Re y$ to mean $x \leq y$. This relation is reflexive and transitive but not symmetric. (R and T but not S). For e.g., $3 \leq 7$ but $y \leq 3$ is not true.*

Example 1.4.6. *On $\mathbb{Z}$ we define $x\Re y$ to mean $x \leq y + 1$. This relation is reflexive, but neither symmetric nor transitive. (R but not S and T)*

Example 1.4.7. *On $\mathbb{Z}$ we define $x\Re y$ to mean $x = -y$. This relation is neither reflexive, nor transitive but it is symmetric. (S but not R and T)*

Example 1.4.8. *On the set A of all fractions of the form a/b, where a, b are integers with $a, b \neq 0$, we define $a/b \,\Re\, c/d$, iff $b = c$. This relation is neither reflexive, nor symmetric, nor transitive. (Not R, S, T)*

Example 1.4.9. *Let $A = \{1, 2, 4, 6, \cdots\}$. We define the relation $\Re$ by $x\Re y$, iff x and y have a common factor other than 1. This relation is symmetric and transitive, but it is not reflexive because $1\Re 1$ is not true. (S and T but not R)*

Example 1.4.10. *Let A be the set of all complex numbers. We define the relation $\Re$ on this set by $z\Re w$ (where z and w are two complex numbers) to mean $\operatorname{Re}(z) \leq \operatorname{Re}(w)$ and $\operatorname{Im}(z) \leq \operatorname{Im}(w)$. Then this relation is reflexive and transitive but not symmetric. (R and T but not S)*

Example 1.4.11. *On $\mathbb{Z}$ define $a\Re b$, if $a - b$ is even. This relation $\Re$ is an equivalence relation (i.e., it is reflexive symmetric and transitive), but it is not anti-symmetric.*

Example 1.4.12. *On $\mathbb{N}$ define $a\Re b$, if and only if a is a divisor of b. Then the relation $\Re$ is not symmetric but it is reflexive, anti-symmetric and transitive.*

Partial order relation. A relation $\Re$ on a set A is said to be a *partial order relation*, if $\Re$ is reflexive, anti-symmetric and transitive.

e.g., *set inclusion:* $A \subseteq A$, $A \subseteq B$ and $B \subseteq A \Longrightarrow A = B$ and $A \subseteq B$, $B \subseteq C \Longrightarrow A \subseteq C$ defines a *partial order relation* on the set of all subsets of a given set $\mathcal{A}$.

A set A with a partial order relation $\mathcal{R}$ is called a *partially ordered set* (or a POSET).

The symbol $\leq$ is used to indicate partial order relation; thus a set $\mathcal{A}$ with a partial order relation is written as $(\mathcal{A}, \leq)$.

If the partial order on a set $\mathcal{A}$ be such that for any two elements $a, b \in \mathcal{A}$, either $a \leq b$ or $b \leq a$, then the partial order is said to be a *total order* or a *linear order*.

ORDERED SETS

Q. *What is an Ordered set?*

Let S be a set. An *order* on S is a relation, denoted by $<$, with the following two properties:

(i) If $x \in S$ and $y \in S$, then one and only one of the statements: $x < y$, $x = y$, $y < x$ is true (*Law of Trichotomy*);

(ii) If $x, y, z \in S$, if $x < y$ and $y < z$, then $x < z$ (*Law of Transitivity*).

An ordered set is a set S in which an order is defined.

e.g., $\mathbb{Q}$ is an *ordered set*, if $r < s$ is defined to mean that $s - r$ is a positive rational number.

Bounds of an ordered set: *For an ordered set S, we introduce the concepts of upper bound and lower bound.*

Let $E \subseteq S$. If $\exists \, \beta \in S$ such that $x \leq \beta$ for $\forall x \in E$, then we say that E is *bounded above* and we call β, an *upper bound* of E.

Lower bounds are defined in the same way (with $\geq$ in place of $\leq$).

Definition. Suppose S is an ordered set and $E \subseteq S$, and suppose that E is bounded above. Let $\alpha \in S$ with the following two properties:

(i) α is an upper bound of E;

(ii) If $\gamma < \alpha$, then γ is *not an upper bound* of E.

Then α is called the *least upper bound* (*lub*) of E or the supremum of E and we write $\alpha = \sup E$.

The greatest lower *bound* or infimum of E which is bounded below is defined in a similar manner, i.e., α is a lower bound of E and that no β with $\beta > \alpha$ is a lower bound of E, then $\alpha = $ glb of E or inf E.

These concepts will be used in real analysis.

EXERCISES ON CHAPTER 1–I(A)

(On Basic Concepts)

1. Are the following statements true? Give reasons.

(a) $2 \in \{2, 3\}$;
(b) $3 \in \{4, 1, 5\}$;
(c) $5 \in \{x : x \text{ is a positive integer}\}$;
(d) $\{3, 4\}$ is a subset of $\{3, 4\}$;
(e) $\{a, e, i, o, u\} = \{u, o, i, e, a\}$;
(f) $\{2, 3\} = \{3, 4\}$;
(g) If $a = 3$, and $A = \{x : 3x = 9\}$, then $a = A$.

[Ans. (a), (c), (d), (e) are true, others are not true]

2. Let $A = \{1, 2, 3\}$. Check whether the following statements are true or not.

   (a) $2 \in A$;   (b) $2 \subset A$;   (c) $\{2\} \in A$;   (d) $\{2\} \subset A$.

   [Ans. (a) and (d) are true; others are not]

3. Given A is any arbitrary set and ϕ is the null set. State whether the following are true or false.

   (a) $\{0\} = \phi$;   (c) $\{\phi\} = \{0\}$;   (e) $\phi \subseteq A$;   (g) $A \in \{A\}$.
   (b) $\{\phi\} = \phi$;   (d) $\phi \in A$;   (f) $A \subseteq A$;

   [Ans. True $\to$ (e), (f) and (g); others are false]

4. $A = \{1, 2\}$, $B = \{2, 4, 6\}$, $C = \{1, 3, 4, 6\}$, $D = \{1, 2, 3, 4, 6\}$. Write down a set whose subsets are A, B, C, D.

5. Let $A = \{1, 2, 3\}$, $B = \{1, 2, 4\}$. Obtain the members of the sets: $A \cup B$, $A \cap B$, $A \times B$, $B \times A$.

6. Let S = universal set = $\{1, 2, 3, 4, 5, 6, 7, 8\}$ and A, B, C are its three subsets given by $A = \{1, 5, 6\}$, $B = \{2, 3, 5, 7, 8\}$ and $C = \{1, 3, 6, 8\}$. Obtain the following sets: $A \cap B' \cap C$ and $(A \cup C) \cap (B \cup C')$.

7. Let $A = \{1, 2, 3, \{4, 5, 6\}\}$ and $B = \{1, 2, \{4, 6\}\}$. Find $A \cap B$ and $A \cup B$.

8. Let $A = \{1, 2, 3\}$ and $B = \{1, 5\}$. Obtain the Cartesian products $A \times B$, $B \times A$, $A \times A$, $B \times B$, $(A \times B) \times A$ and $A \times (B \times A)$.

9. Define equality of two sets. For n sets let $A_1, A_2, \cdots, A_n$. Let $A_1 \subseteq A_2 \subseteq A_3 \subseteq \cdots \subseteq A_n$ and also let $A_1 \supseteq A_n$. Prove that the n sets are all equal.

10. Let $A = \{1, 2, 3, 4, 5\}$, $B = \{2, 4, 6, 8, 10\}$ and $C = \{3, 4, 5, 6\}$. Obtain the sets: $(A - B)$; $(B - C)$; $(C - A)$; $(B - A)$; $(B - B)$.

11. Let $A = \{1, 1/2, 1/3, 1/4\}$, $B = \{1/2, 1/4, 1/6, 1/8\}$, $C = \{1/3, 1/4, 1/5, 1/6\}$ and suppose that the universe is $S = \{1, 1/2, 1/3, 1/4, \cdots, 1/9\}$. Obtain: $(A \cap C)'$, $(A \cup B)'$, $(A')'$, $(B - C)'$, $B - A$, $B' - A'$, $A' \cap B$, $A \cup B'$, $A' \cap B'$.

(On applications of the Laws of Algebra of Sets)

1. A, B, C are any three sets. Verify the following properties:
   (a) $A - B = A - (A \cap B) = (A \cup B) - B$;

(b) $(A - B) - C = A - (B \cup C)$;
(c) $A - (B - C) = (A - B) \cup (A \cap C)$;
(d) $(A \cup B) - C = (A - C) \cup (B - C)$;
(e) $A - (B \cup C) = (A - B) \cap (A - C)$;
(f) $A \cap (B - C) = (A \cap B) - (A \cap C)$;
(g) $(A - C) \cap (B - C) = (A \cap B) - C$.

Remember: $A - B = A \cap B'$; use this fact whenever necessary.

2. The *symmetric difference* of two sets A and B (denoted by $A \Delta B$) is given by $A \Delta B = (A - B) \cup (B - A)$. Verify the following properties:
   (a) $A \Delta B = (A \cup B) - (A \cap B)$;
   (b) $A \Delta (B \Delta C) = (A \Delta B) \Delta C$;   (Associativity)
   (c) $A \Delta \Phi = A$; $A \Delta A = \Phi$;
   (d) $A \Delta B = B \Delta A$;   (Commutativity)
   (e) $A \cap (B \Delta C) = (A \cap B) \Delta (A \cap C)$.   (Distributive Law)

3. Prove by using laws of Algebra of sets:
   (a) $(A \cap B) \cup (A \cap B') = A$;
   (b) $(A \cap B') \cup B = A \cup B$;
   (c) $A \cap \Phi = \Phi$; $(A \cap B') \cap A' = \Phi$;
   (d) $(A' \cup C) \cap (B' \cup D') = (B' \cup A') \cup (B' \cap C) \cup (D' \cap A') \cup (D' \cap C)$;
   (e) $A \cup (A \cap B) = A$; $A \cup (A' \cap B) = A \cup B$;
   (f) $A \cap (A' \cup B) = A \cap B$; $A \cap (A \cup B) = A$;
   (g) $(A \cup B) \cap (B \cup C) \cap (C \cup A) = (A \cap B) \cup (A \cap C) \cup (B \cap C)$.

(On Relations on Sets)

1. Define a relation $\Re$ on the set P of all people by taking $x \Re y$ to mean x and y have the same age.
   Is it an equivalence relation? Justify your assertion.

2. Show that the relation $>$ (greater than) on the set of real numbers is *Transitive* but neither *Reflexive* nor *Symmetric*.

3. Give an example of a relation on a set:
   (a) which is symmetric and transitive but not reflexive.
   (b) which is reflexive and symmetric but not transitive.
   (c) which is symmetric but neither reflexive nor transitive.

[Ans. (a) $x \Re y$, only when $x - y \neq 0$ $(x, y \in \mathbb{R})$ | $\mathbb{R}$ is the set of all real numbers.
Another example: Ex. 1.4.9 | $\mathbb{Z}$ is the set of all integers]
(b) Ex. 1.4.3.
(c) Ex. 1.4.7.]

4. Let $\mathbb{Z}$ be the set of all integers. We define a relation "≡" (called *Congruence relation*): If $a, b \in \mathbb{Z}$, then $a \equiv b \pmod 5$, iff $a - b$ is divisible by 5. Prove that '≡' is an equivalence relation on $\mathbb{Z}$.

5. Find which of the following relations are reflexive, symmetric, transitive or equivalence relation: [$\mathbb{R}, \mathbb{S}, \mathbb{T}$ or $\mathbb{EQ}$]

Set	Relation
I. Set of all triangles in a plane	(a) is congruent to
	(b) is similar to
II. Set of all lines in a plane	(a) is perpendicular to
	(b) is parallel to
III. Set of all integers	(a) $a\Re b$, iff $|a - b| \leq b$
	(b) $a\Re b$, iff $3a + 4b$ is divisible by 7
	(c) $a\Re b$, iff $a - b$ divisible by 5

[*Ans.* I. (a) $\mathbb{EQ}$; (b) $\mathbb{EQ}$. II. (a) $\mathbb{S}$ (not $\mathbb{R}, \mathbb{T}$); (b) $\mathbb{EQ}$. III. (a) Not $\mathbb{R}, \mathbb{S}, \mathbb{T}$; (b) $\mathbb{EQ}$; (c) $\mathbb{EQ}$.]

1.5 Mapping or Function

We now discuss the most fundamental notion of analysis, namely *function* (or *mapping*).

A *function* f from a set A into a set B is a *rule of correspondence* that assigns to each element $x \in A$, a uniquely determined element $f(x)$ in B. We also call it a *mapping* from A into B and write $f : A \to B$. (*read:* f maps A into B).

In the definition of function given above we have used a phrase *rule of correspondence* which needs further clarification. So the following definition of function is more widely accepted:

Definition. Let A and B be two non-empty sets. Then *a function* (or *a mapping*) from A to B is a set f of ordered pairs in $A \times B$ such that for each $x \in A$ there exists a unique $y \in B$ with $(x, y) \in f$. (This means that if $(x, y) \in f$ and $(x, y') \in f$, then $y = y'$). A function f from A to B is a relation from A to B such that no two elements of f have the same first component.

The set A of the first elements of a function is called the *domain* of f and the set of all the second elements of f is called the *range* of f. Note that dom $f = A$ but range of $f \subseteq B$. Range of f is also denoted by $f(A)$.

Example 1.5.1. *Let $\mathbb{R}$ be the set of all real numbers. Suppose f maps $\mathbb{R}$ into $\mathbb{R}$ (i.e., $f : \mathbb{R} \to \mathbb{R}$) defined by $f(x) = x^2$, $x \in \mathbb{R}$. What are the values of the function at $x = 0, -1, 2, -3$?*

Clearly, $f(0) = 0$, $f(-1) = (-1)^2 = 1$, $f(2) = 4$, $f(-3) = 9$. The domain of f is $\mathbb{R}$ and the co-domain is also $\mathbb{R}$. Notice that $x^2 \geq 0$ $\forall x \in \mathbb{R}$. So the range of f is $\mathbb{R}^+$, the set of all positive real numbers and the singleton set $\{0\}$, i.e., $f(\mathbb{R}) = \mathbb{R}^+ \cup \{0\}$.

1.6 Different types of Mappings

I. ONTO Mapping (or Surjective Mapping). The map $f : A \to B$ is called *surjective* (or ONTO) *iff* $f(A) = B$, i.e., range of $f =$ co-domain of f, i.e., $\forall y \in B$, $\exists$ $x \in A$ such that $f(x) = y$.

II. One-One mapping (or Injective mapping). The map $f : A \to B$ is called *one-one mapping* (or *injective mapping*), if and only if distinct members of A have distinct images in B, i.e., for $x_1, x_2 \in A$, $f(x_1) = f(x_2) \implies x_1 = x_2$ or the contra positive statement: $x_1 \neq x_2 \implies f(x_1) \neq f(x_2)$.

III. Bijective mapping. The map $f : A \to B$ is said to be a *bijective mapping* or a *one-one* onto mapping or *one-one correspondence* if it is both *injective* and *surjective*, i.e., if every $x \in A$ has a unique image $y \in B$ and every $y \in B$ has a unique pre-image $x \in A$. A bijective mapping is also called a *bijection*.

Example 1.6.1. *Let $f : \mathbb{R} \to \mathbb{R}$, defined by $f(x) = x^2$, $\forall x \in \mathbb{R}$.*

- This mapping is *not injective* (see that $f(1) = 1$ and $f(-1) = 1$; again $f(2) = 4$ and $f(-2) = 4$, i.e., it is not true that distinct members of the domain have distinct images).
- This mapping is not *surjective* (see that $\exists$ $-1 \in$ co-domain $\mathbb{R}$ which has no pre-image x in the domain $\mathbb{R}$ because every image is non-positive).

Thus the mapping cannot be a *bijective mapping*.

Example 1.6.2. *Let $f : \mathbb{R}^+ \cup \{0\} \to \mathbb{R}$, defined by $f(x) = x^2$, $x \in \mathbb{R}^+ \cup \{0\}$. Verify that this mapping is injective but not surjective.*

Example 1.6.3. *Let $f : \mathbb{R} \to \mathbb{R}^+ \cup \{0\}$, defined by $f(x) = x^2$, $\forall x \in \mathbb{R}$. Check: This mapping is not injective but it is surjective.*

Example 1.6.4. *Let $f : \mathbb{R}^+ \cup \{0\} \to \mathbb{R}^+ \cup \{0\}$ defined by $f(x) = x^2$, $\forall x \in \mathbb{R}^+ \cup \{0\}$. Verify that this mapping is both injective and surjective, i.e., it is a bijective mapping.*

Remember: (i) In order to prove that the mapping $f : A \to B$ is an injective mapping, we must establish that $\forall x_1, x_2 \in A$, if $f(x_1) = f(x_2)$, then $x_1 = x_2$. To do this, start with $f(x_1) = f(x_2)$ and show that $x_1 = x_2$. f is injective if each element of B has at most one pre-image.

CHAPTER 1: BASIC CONCEPTS: SETS, RELATIONS AND MAPPINGS 33

(ii) To prove $f : A \to B$ is a surjective mapping, we must show that for any $y \in B$, $\exists$ at least one $x \in A$ such that $f(x) = y$. f is surjective if each element of B has at least one pre-image.

(iii) To prove $f : A \to B$ is a bijection, we are to show that each element $y \in B$ has exactly one pre-image $x \in A$.

Example 1.6.5. Let $A = \{x : x \in \mathbb{R} \text{ but } x \neq 1\}$. Define $f(x) = 2x/(x-1)$, $x \in A$.

Solution: To prove that the function f is injective, we start with $f(x_1) = f(x_2)$, where $x_1, x_2 \in A$, i.e., $2x_1/(x_1-1) = 2x_2/(x_2-1) \implies x_1(x_2 - 1) = x_2(x_1 - 1) \implies x_1 = x_2$.

$\therefore f$ is injective.

To determine the range of f we solve for x the equation $\frac{2x}{(x-1)} = y$. We obtain $x = \frac{y}{(y-2)}$ which is defined for $y \neq 2$.

Thus range of $f = \{y : y \in \mathbb{R} \text{ and } y \neq 2\} = B$ (say). Then $f : A \to B$ is a bijection.

Example 1.6.6. Let $f : \mathbb{Z} \to \mathbb{Z}$ defined by $f(x) = x + 2$, $x \in \mathbb{Z}$. See that this mapping is both injective and surjective, i.e., f is bijective.

IV. Inverse mappings (or Inverse functions)

Definition. If $f : A \to B$ is a bijection of A ONTO B, then the set

$$g = \{(b, a) \in B \times A : (a, b) \in A \times B\}$$

is a function on B into A. This function is called the *inverse function* (or *inverse mapping*) of f, and is denoted by f^{-1}. The function f^{-1} is also called the *inverse* of f.

Note: In order to define inverse of $f : A \to B$, f must be bijective and domain $(f) =$ range of f^{-1} and range of $f =$ domain of f^{-1}. Also $y = f(x)$, if and only if $x = f^{-1}(y)$.

Example 1.6.7. We have observed in Example 1.6.4, that the function

$$f : R^+ \cup \{0\} \to R^+ \cup \{0\}$$

defined by $f(x) = x^2$, $x \in R^+ \cup \{0\}$ is both injective and surjective (i.e., here f is a bijection on $R^+ \cup \{0\}$) and hence f^{-1} exists.

The function inverse to f is given by (solving $y = x^2$ give $x = \sqrt{y}$)

$f^{-1}(y) = \sqrt{y}$ for $y \in R^+ \cup \{0\}$ (range of f).

We may write $f^{-1}(x) = \sqrt{x}$ for $x \in R^+ \cup \{0\}$ (replacing y by x).

Example 1.6.8. See Example 1.6.5. $f : A = \{x \in \mathbb{R} : x \neq 1\} \to B = \{y \in \mathbb{R} : y \neq 2\}$ defined by $f(x) = 2x/(x-1)$, $x \in A$.

This function f is a bijection of A ONTO B. Hence f^{-1} exists and f^{-1} is given by $f^{-1}(y) = y/(y-2)$, $y \in B$ [Solving $y = 2x/(x-1)$ gives $x = y/(y-2)$].

We may also write $f^{-1}(x) = x/(x-2)$, $x \in B$ (replacing y by x).

V. Identity mapping (or Identity function). The function $f : A \to A$ defined by $f(x) = x$, $x \in A$ is called the *identity function* on A, denoted by I_A. Then I_A keeps every element of A fixed.

VI. Equality of two mappings. Let $f : A \to B$ and $g : A \to B$ be two functions, both defined on A. Then f and g are called *equal* (written as $f = g$), if $f(x) = g(x)$ $\forall x \in A$.

Note: For equality, f and g must have the same domain and for each $x \in A$, $f(x) = g(x)$.

Example 1.6.9. Let $f : \mathbb{R} \to \mathbb{R}$ defined by $f(x) = |x|$, $\forall x \in \mathbb{R}$ and let $g : \mathbb{R} \to \mathbb{R}$ be defined by

$$g(x) = \begin{cases} x, & \text{when } x \geq 0 \\ -x, & \text{when } x < 0. \end{cases}$$

Both f and g have the same domain and moreover, $f(x) = g(x)$ $\forall x \in \mathbb{R}$.

$\therefore$ in this case, we write $f = g$.

VII. Composition of function (or Composite mappings). To compose two functions f and g, we first find $f(x)$ and then apply g-rule to $f(x)$ and obtain $g\{f(x)\}$. Obviously this is not possible unless $f(x)$ is an element in the domain of g. In order to be able to do this *for all* $f(x)$ we are to assume that the range of f is contained in the domain of g, i.e., range of $f \subseteq$ domain of g.

Definition. Let $f : A \to B$ and $g : C \to D$ be two mappings. If the range $f(A)$ of f is a subset of C then the mapping $g \circ f : A \to D$ defined by $(g \circ f)(x) = g(f(x))$, $x \in A$ is called a *composite mapping*.

In particular if $f : A \to B$ and $g : B \to C$ then the composite mapping $g \circ f$ is always possible as $f(A) \subseteq B$. The order of the compositions should be maintained strictly, because $g \circ f$ and $f \circ g$ are different functions, in general, when both are defined.

Example 1.6.10. Let f and g be two functions defined on $\mathbb{R}$, given by $f(x) = 3x$ and $g(x) = 2x^2 - 1$.

Since dom $g = \mathbb{R}$ and range of $f \subseteq$ dom $g = \mathbb{R}$, the domain of $g \circ f$ is also equal to $\mathbb{R}$ and the composite function $g \circ f$ is defined by

$$(g \circ f)(x) = g\{f(x)\} = g(3x) = 2(3x)^2 - 1 = 18x^2 - 1.$$

On the other hand, $(f \circ g)x = f\{g(x)\} = f(2x^2 - 1) = 3(2x^2 - 1) = 6x^2 - 3$.

Thus we observe $g \circ f \neq f \circ g$.

Most important point to construct $g \circ f$ is to see that the range of f is contained in the domain of g.

e.g., let $f(x) = 1 - x^2$ and $g(x) = \sqrt{x}$, domain of $g = \{x : x \in \mathbb{R}, x \geq 0\}$.

The composite function $g \circ f$ is given by $(g \circ f)x = g\{f(x)\} = g(1-x^2) = \sqrt{1-x^2}$. Only for $x \in$ domain of f that satisfies $f(x) \geq 0$, i.e., for those x which satisfies $-1 \leq x \leq 1$.

VIII. Restrictions of functions. Suppose f maps A into B and let $A_1 \subset A$. We define $f_1 : A_1 \to B$ by the rule $f_1(x) = f(x)\ \forall x \in A_1$. This function f_1 is called the *restriction of f to A_1*, denoted by f_{A_1}. We thus *cut down in size* the domain of a function. Of course, there are good reasons for restricting the domain in this manner.

A very common example: Let $f : \mathbb{R} \to \mathbb{R}$ defined by $f(x) = x^2$, for $x \in \mathbb{R}$. This function is certainly not injective $\{\because f(-1) = 1, f(1) = 1\}$ and hence not bijection and so it cannot have an inverse.

However, if we restrict f to set $A_1 = \{x : x \in \mathbb{R}, x \geq 0\}$ or $A_1 = \mathbb{R}^+ \cup \{0\}$, then the restriction function f_{A_1} is both injective and surjective so that the restriction function f_{A_1} has an *inverse function (positive square root function)*.

The trigonometric functions $\sin x$ and $\cos x$ are not injective for all $x \in \mathbb{R}$. However, by making suitable restrictions of these functions, one can obtain the *inverse sine* and *inverse cosine* functions. Sine-function can be restricted to $-\frac{\pi}{2} \leq x \leq \frac{\pi}{2}$ and cosine function can be restricted to $0 \leq x \leq \pi$. In these restrictive domains $\sin^{-1} x$ and $\cos^{-1} x$ are defined.

Note: The function $f : A \to B$ is an extension of its restriction
$$f_1 : A_1 \to B \text{ then } A_1 \subset A$$

EXERCISES ON CHAPTER 1: I(B)

(On Mappings)

(Hints are given at the end of this Exercise for *-marked problems)

1. Define mapping of a set X into a set Y. Do the following correspondences conform to your definition? If so, mention the Range or Image set.
    (a) $f : \mathbb{Z}^+ \to E$, defined by $f(x) = 2x, \forall x \in \mathbb{Z}^+$. ($\mathbb{Z}^+$ is the set of all positive integers and E is the set of even positive integers.)
    (b) $g : \mathbb{R} \to \mathbb{R}$, defined by $g(x) = e^x, \forall x \in \mathbb{R}$.
    (c) $f : X \to Y$ (X = set of all students of your college and Y = set of ages of the students in years).
    (d) $f : \mathbb{R} \to \mathbb{R}$, defined by $f(x) = \log x, x \in \mathbb{R}$.

(e) $f : \mathbb{R}^+ \to \mathbb{R}$, defined by $f(x) = \log x$, $x \in \mathbb{R}^+$.
(f) $g : \mathbb{R}^+ \to \mathbb{R}$, defined by $f(x) = \pm\sqrt{x}$, $x \in \mathbb{R}^+$.

2. Satisfy yourself: $f : \mathbb{R} \to \mathbb{R}$, defined by $f(x) = \sin x$, $\forall x \in \mathbb{R}$ is neither injective nor surjective.

3. X and Y are two finite sets each having n elements. Prove that $f : X \to Y$ is injective iff f is an ONTO-mapping.

*4. Establish that the set $\mathbb{N}$ of natural numbers and the set of all integers have one-to-one correspondence and ONTO-mapping (i.e., $\exists$ a bijection between the two sets).

[We shall see in Chapter 2, that if any set A is in one-to-one correspondence and onto mapping with $\mathbb{N}$ we call it a COUNTABLE set; thus here the set of all integers is a countable set.]

*5. Let f map A into B. Prove that
  (a) $A \subset B \Longrightarrow f(A) \subset f(B)$;
  (b) $f(A \cup B) = f(A) \cup f(B)$ but $f(A \cap B) \neq f(A) \cap f(B)$.
      In fact, $f(A \cap B) \subset f(A) \cap f(B)$.

*6. Let $f : A \to B$. Prove that
  (a) $A \subseteq B \Longrightarrow f^{-1}(A) \subseteq f^{-1}(B)$;
  (b) $f^{-1}(A \cup B) = f^{-1}(A) \cup f^{-1}(B)$;
  (c) $f^{-1}(A \cap B) = f^{-1}(A) \cap f^{-1}(B)$.

Note: Union and Intersection are preserved under Inverse mapping.

7. Let $A = \{1, 2, 3\}$ and $B = \{0, 1\}$. List all the mappings $f : A \to B$. How many of these mappings are surjective? How many injective? Is there any bijective mapping? Give a restriction of f which is bijective.

8. Let $f : \mathbb{R} \to \mathbb{R}$ be a mapping s.t. $f(x) = 1/2(e^x + e^{-x})$, $x \in \mathbb{R}$. Give a restriction of f which is bijective.

9. (a) Let $f : \mathbb{R} \to \mathbb{R}$, defined by $f(x) = e^x$, $x \in \mathbb{R}$ and let $g : \mathbb{R} \to \mathbb{R}$, defined by $g(x) = \sin x$, $x \in \mathbb{R}$. Verify: $g \circ f \neq f \circ g$.
  (b) Let $f : X \to Y$, $g : Y \to Z$, $h : Z \to W$ be three maps. Prove that $h \circ (g \circ f) = (h \circ g) \circ f$.

*10. (a) Let $f : A \to B$ be a bijection. Prove that $f^{-1} \circ f = I_A$ and $f \circ f^{-1} = I_B$.
  (b) Let $f : A \to B$ and $g : B \to C$ be two bijective mappings. Prove that $g \circ f$ is a bijection and that $(g \circ f)^{-1} = f^{-1} \circ g^{-1}$.
  (c) Let $\mathbb{Z}$ be the set of all integers. Define
      $f : \mathbb{Z} \to \mathbb{Z} \times \mathbb{Z}$; by $f(m) = (m-1, 1)$, $m \in \mathbb{Z}$
      and $g : \mathbb{Z} \times \mathbb{Z} \to \mathbb{Z}$ by $g(m, n) = m + n$, $(m, n) \in \mathbb{Z} \times \mathbb{Z}$.
      Prove that $g \circ f = I_z$. What can you say about $f \circ g$?

11. (a) Let $f : A \to B$ and $g : B \to A$ be two bijective mappings s.t. $g \circ f = I_A$ and

CHAPTER 1: BASIC CONCEPTS: SETS, RELATIONS AND MAPPINGS

$f \circ g = I_B$. Prove that $g = f^{-1}$ and $f = g^{-1}$.

(b) Let $f : A \to B$ and $g : B \to C$. Prove that

$g \circ f$ injective $\Longrightarrow f$ is injective, and $g \circ f$ surjective $\Longrightarrow g$ is surjective.

HINTS AND ANSWERS ON EXERCISES I(B)

1. (a) $f(\mathbb{Z}^+) = E$;   (c) Yes, a mapping;   (e) Yes, a mapping;
   (b) $g(\mathbb{R}) = \mathbb{R}^+$;   (d) Not a mapping;   (f) No, not a mapping.

4. $\mathbb{N}$ 1   2   3   4   5   6   ⋯   ⋯
        ↕   ↕   ↕   ↕   ↕   ↕
   $\mathbb{Z}$ 0   1   −1   2   −2   3   ⋯   ⋯

5. (a) If $A \subset B$, then $x \in A \Longrightarrow x \in B$.
   Now $y \in f(A) \Longrightarrow \exists\, x \in A$ s.t. $f(x) = y$
   $\Longrightarrow \exists\, x \in B$ s.t. $f(x) = y$
   $\Longrightarrow y = f(x) \in f(B)$ [$\because x \in B \Longrightarrow f(x) \in f(B)$],
   i.e., $y \in f(A) \Longrightarrow y \in f(B)$ or $f(A) \subseteq f(B)$.

   (b) $y \in f(A \cup B) \Longleftrightarrow y$ is the image of an element in $A \cup B$
   $\Longleftrightarrow y$ is the image of some element in A or some element in B
   $\Longleftrightarrow y \in f(A)$ or $y \in f(B)$
   $\Longleftrightarrow y \in f(A) \cup f(B)$; hence etc.

   Note that we can prove $f(A \cap B) \subset f(A) \cap f(B)$, but equality may not hold. For example, suppose A is the set of all positive real numbers and B is the set of all non-positive real numbers and let $f : A \to B$ defined by $f(x) = x^2$. Here $A \cap B = \Phi$ and $f(A \cap B) = \Phi$ but $f(A) \cap f(B) \neq \Phi$.

6. (a) $x \in f^{-1}(A) \Longrightarrow f(x) \in A$, i.e., $f(x) \in B$ ($\because A \subset B$)
   $\Longrightarrow x \in f^{-1}(B)$, i.e., $f^{-1}(A) \subset f^{-1}(B)$.

   (b) $x \in f^{-1}(A \cup B) \Longleftrightarrow f(x) \in A \cup B$
   $\Longleftrightarrow f(x) \in A$ or $f(x) \in B$
   $\Longleftrightarrow x \in f^{-1}(A)$ or $x \in f^{-1}(B)$
   $\Longleftrightarrow x \in f^{-1}(A) \cup f^{-1}(B)$; hence etc.

   (c) $x \in f^{-1}(A \cap B) \Longleftrightarrow f(x) \in A \cap B$
   $\Longleftrightarrow f(x) \in A \wedge f(x) \in B$
   $\Longleftrightarrow x \in f^{-1}(A) \wedge x \in f^{-1}(B)$
   $\Longleftrightarrow x \in f^{-1}(A) \cap f^{-1}(B)$; hence etc.

10. (a) Let $f(x) = y$, $x \in A$, $y \in B$. Then $x = f^{-1}(y)$. Now $f^{-1} \circ f : A \to A$, given by $(f^{-1} \circ f)(x) = f^{-1}(f(x)) = f^{-1}(y) = x = I_A(x)$.
    $\therefore (f^{-1} \circ f)(x) = I_A(x) = x$, $\forall x \in A$, hence $f^{-1} \circ f = I_A$.
    The other part can be proved in a similar manner.

(b) If $x_1, x_2 \in A$, then
$$(g \circ f)(x_1) = (g \circ f)(x_2) \Longrightarrow g\{f(x_1)\} = g\{f(x_2)\}$$
$$\Longrightarrow f(x_1) = f(x_2) \; (\because g \text{ is injective})$$
$$\Longrightarrow x_1 = x_2 \; (\because f \text{ is injective}).$$
Hence $g \circ f$ is injective. Now to prove that $g \circ f$ is surjective: $g \circ f : A \to C$ and
$$(g \circ f)(A) = g(f(A)) = g(B) \; (\because f(A) = B, f \text{ being surjective})$$
$$= C \; (\because g(B) = C, g \text{ being surjective}).$$
$\therefore g \circ f$ surjective. Hence we obtain that $g \circ f$ is bijective.
Now $(g \circ f)^{-1}$ and $f^{-1} \circ g^{-1}$ map $C \to A$. Let $(g \circ f)^{-1}(z) = x$, where $z \in C$, $x \in A$.
Also $(f^{-1} \circ g^{-1})(z) = f^{-1}(g^{-1}(z))$
$$= f^{-1}(y) \; (\because g(y) = z \Longrightarrow g^{-1}(z) = y)$$
$$= x \; (\because f(x) = y \Longrightarrow f^{-1}(y) = x).$$
$\therefore (g \circ f)^{-1}(z) = (f^{-1} \circ g^{-1})(z), \forall z \in C.$
Consequently, $(g \circ f)^{-1} = f^{-1} \circ g^{-1}$.

MAIN POINTS: CHAPTER 1

1. Basic ideas on Sets. Universal Set. Null Set. Set Inclusion.
2. Operations on Sets: $\cup, \cap$ and Complements. Symmetric Difference.
3. Cartesian Product. Relation. Equivalence Rel. Partial Order. Ordered Set.
4. Function or Mapping. Different Kinds of Mapping. Injective. Surjective. Bijective. Inverse Mappings. Composite Mappings. Restrictions of a Mapping.

AXIOM OF CHOICE

There are many results in Set Theory, (in particular Measure Theory) which make use of the *Axiom of Choice*, the statement of which is given below:

Axiom of Choice: *If a non-empty set is divided into a class of non-empty pairwise disjoint subsets, then $\exists$ at least one set which contains exactly one element of each one of these sets.* Such a set is called **cross-section** of the class.

AN INTRODUCTION TO ANALYSIS

(Differential Calculus—Part I)

Chapter 2

The Real Number System

Natural Numbers: Peano's Axioms—Well ordering Principle: Principle of Mathematical Induction

Rational Numbers: Main Properties including Countability
Real Numbers—Axiomatic Approach
Appendix: Dedekind's Theorem of Real Numbers

Chapter 2

The Real Number System

2.1 Introduction

The study of the main concepts of real analysis, namely, convergence, continuity, differentiability, integrability, etc., has its basis on an accurately defined number-concept—more specifically, on the concept of the real number system. Our approach, in the present text, is not to give a formal method of construction of real numbers. Instead, we shall exhibit a list of fundamental properties which will characterise the system of real numbers and use these properties in learning the tools of real analysis.

We assume that our student-readers have the initial acquaintance of the primitive systems like the set $\mathbb{N}$ of natural numbers, the set $\mathbb{Z}$ of all integers and the set $\mathbb{Q}$ of all rational numbers (i.e., numbers of the form $\frac{p}{q}$, where p and q are integers, $q \neq 0$). Addition and multiplication of the elements of these sets will be supposed to be known.

Our approach in introducing real number system is, what we call, *axiomatic*: we assume that:

> There exists an ordered field $\mathbb{R}$ which is complete (i.e., which has the least-upper-bound property). Further, $\mathbb{R}$ contains $\mathbb{Q}$ as subfield. The members of $\mathbb{R}$ are called *real numbers*.

We shall discuss the existence-statement in considerable details in the following sections.

Richard Dedekind (1831–1916), however, made a completely different approach: He introduced the concept of cut (*Dedekind cut*) of **rational numbers** and thereby generated not only *real rational numbers* but also new type of numbers called *real irrational numbers*. Finally he showed that the section of real numbers does not lead to any further generalisation (**Dedekind's theorem**). (See Appendix at the end of this chapter 2).

> Elementary Approx: Natural Numbers—Integers and Rational Numbers

First Approach: We begin with Peano's Axioms: Well ordering Prime. Principle of Mathematical Induction. Introduce $\mathbb{Z}$, the set of all integers and then finally define a rational number as the ratio of two integers where the denominator in different from zero. The totality of all rational numbers form the system $\mathbb{Q}$ of rational numbers.

Another Approach: Define a Field—an ordered field. Define Archimedian Property, Density property, Countability Property and *then $\mathbb{Q}$ is defined as an ordered field obeying Density property, Archimedian property and Countability property.* Every rational number is then made to correspond to a unique point on a directed line (*Geometrical Representation of Rational Numbers*). What is most important to remember in these two approaches of introducing Rational numbers is that $\mathbb{Q}$ *is not order complete* this notion will be explained in due course.

Section I: Natural Numbers, Integers and Rational Numbers

2.2 The Set $\mathbb{N}$ of Natural Numbers: Peano's Axioms

We assume familiarity with the set of natural numbers $\mathbb{N} = \{1, 2, 3, 4, 5, 6, \cdots\}$, along with the usual arithmetic operations of *addition* and *multiplication* of two natural numbers and with the meaning of one natural number being less than another (*order relation*).

In the following discussions we shall include three important notions:

I. **Peano's axioms** (All the known properties of natural numbers can be shown to be consequences of these axioms).

II. **Well ordering property** of the system $\mathbb{N}$ of natural numbers.

III. **Principle of mathematical induction** which is a part of Peano's axioms.

I. **Peano's axioms**: Let $\mathbb{N}$ be a set whose members we shall call natural numbers. We take the statements P_1 to P_5 as our axioms (called *Peano's axioms* or *Peano's postulates*):

P1. $1 \in \mathbb{N}$; that is, $\mathbb{N}$ is a non-empty set and contains an element which we designate as 1.

P2. For each element $n \in \mathbb{N}$, there exists a unique element $n' \in \mathbb{N}$, called the *successor of n*.

P3. For each element $n \in \mathbb{N}$, $n' \neq 1$; that is, 1 is not the successor of any element in $\mathbb{N}$.

CHAPTER 2: THE REAL NUMBER SYSTEM

P4. For each pair $n, m \in \mathbb{N}$, with $n \neq m$, $n' \neq m'$; that is, distinct elements in $\mathbb{N}$ have distinct successors.

P5. If (a) $M \subseteq \mathbb{N}$; (b) $1 \in M$ and (c) $n \in M \Longrightarrow n' \in M$, then $M = \mathbb{N}$.

The last axiom **P5** is known as the **Principle of mathematical induction**. This principle is an important tool in many mathematical proofs. It often appears in the following alternative form:

Alternative form of the principle of mathematical induction:

If $P(n)$ is a *statement*[1] about $n \in \mathbb{N}$, then $P(n)$ may be true for some values of n and not true for some other values of n, e.g., let $P(n)$ be the statement '$n^2 = n$'. Then for $n = 1$, $P(1)$ is true, while $P(n)$ is not true for any $n > 1$, $n \in \mathbb{N}$.

With above background in mind, we can formulate the principle of mathematical induction in the following language:

For each $n \in \mathbb{N}$, let $P(n)$ be some statement about n. Suppose that

1. *$P(1)$ is true.*

2. *For every $k \in \mathbb{N}$, if $P(k)$ is true, then $P(k+1)$ is also true.*

Then $P(n)$ is true for all $n \in \mathbb{N}$.

Take $M = \{n : n \in \mathbb{N}$ and $P(n)$ is true$\}$. Then $M \subseteq \mathbb{N}$. Then the conditions (b) and (c) of **P5** correspond exactly to the conditions (1) and (2) stated above. The conclusion $M = \mathbb{N}$ in **P5** corresponds to the conclusion '$P(n)$ is true for all $n \in \mathbb{N}$.'

Observation: In (2) the assumption "if $P(k)$ is true" is called the *induction hypothesis*. In establishing (2) *we assume $P(k)$ to be true* and then establish $P(k+1)$ is true. In fact, $P(k)$ may not be true. For example, let $P(k) : k = k + 3$, then (2) is logically correct because we can simply add 1 to both sides of $P(k)$ to obtain $P(k+1)$. However, $P(1)$ is not true (since $1 = 4$ is false). Therefore, we cannot use the principle of mathematical induction to conclude that $n = n + 3$, for all $n \in \mathbb{N}$. [See Page 51. Q.4]

A second version of the principle of mathematical induction:

It may happen that the statement $P(n)$ are false for certain natural numbers, but they are true for all $n \geq$ some particular natural number m. The principle of mathematical induction can be suitably modified in such a case. m is the basis in this case.

Statement: *Second version of the principle of mathematical induction*:

Let m be a fixed natural number.

Let $P(n)$ be a statement for each natural number $n \geq m$.

Suppose that

1. *The statement $P(m)$ is true.*

---
[1] A statement means an expression which has a truth value, i.e., either it is true or false.

2. For all $k \geq m$, the truth of $P(k)$ implies the truth of $P(k+1)$.

Then $P(n)$ is true for all $n \geq m$.

We have also another useful form of the principle of mathematical induction known as the *Second Principle of Induction* [see Example 2.3.7, Page 48].

II. Well ordering principle on $\mathbb{N}$: A fundamental property of the system $\mathbb{N}$ of natural numbers is what we call *well-ordering property of* $\mathbb{N}$. It states:

Every non-empty subset of $\mathbb{N}$ has a least element.

This means: If S is a subset of $\mathbb{N}$ and if $S \neq \phi$, then there exists an element $m \in S$ such that $m \leq k$, for all $k \in S$; m is then called the least element of S.

Explanation: 1 is the least element of $\mathbb{N} \subseteq \mathbb{N}$. 2 is the least of $\mathbb{N} - \{1\} \subset \mathbb{N}$ and $1 < 2$, 3 is the least element of $\mathbb{N} - \{1,2\}$ and $1 < 2 < 3$ etc.

Important Deductions

A. *If we assume the truth of well-ordering principle, the principle of mathematical induction follows:*

The Principle of Mathematical Induction states:

Let M be a subset of the system $\mathbb{N}$ of natural numbers with two conditions:

1. $1 \in M$ and

2. For any $n \in \mathbb{N}$, if $n \in M$, then $(n+1) \in M$.

Then the principle states: $M = \mathbb{N}$.

Proof. Subject to the conditions (1) and (2) we shall prove that $M = \mathbb{N}$. Suppose to the contrary that $M \neq \mathbb{N}$. Then the set $F = \mathbb{N} - M$ is not empty. So, by the well-ordering principle F has a least element m (say) (note that $m \in F$ and so $m \notin M$). Now, by hypothesis (1), $1 \in M$, so $m > 1$ (no integer in $\mathbb{N}$ can be < 1). But this implies that $(m-1)$ is a natural number and $m - 1 < m$. Since m is the least natural number such that $m \notin M$, we conclude $(m-1) \in M$.

We now apply condition (2) to the element $m - 1 \in M$. We thus obtain

$$(m-1) + 1 \in M, \quad \text{i.e.,} \quad m \in M.$$

But this statement contradicts the fact that $m \notin M$. This m was obtained on the assumption that F is not empty and by well-ordering principle $m \in F$. The contradiction proves that we cannot accept the assumption. Therefore, we must have F is empty so that $M = \mathbb{N}$

B. We shall now deduce well-ordering principle using the principle of mathematical induction.

CHAPTER 2: THE REAL NUMBER SYSTEM

Well-ordering Principle States: Every non-empty subset $S \subseteq \mathbb{N}$ has a least element.

Proof. Assume that S is a non-empty subset of $\mathbb{N}$ and suppose that S has no least element. We shall prove that this supposition leads to a contradiction and we can then infer that S has a least element proving the well-ordering principle.

We construct $M \subseteq \mathbb{N}$ such that
$$M = \{x \in \mathbb{N} : x < a \text{ for each } a \in S\}.$$

By the laws of trichotomy, $M \cap S = \phi$. Now, $1 \notin S$; otherwise 1 would be the least element of S. Hence for each $a \in S$, $a > 1$ and so $1 \in M$.

Now assume $p \in M$; then $p < a$ for each $a \in S$. If $p+1 \in S$, then $p+1$, which is the first natural number larger than p would be the least element of S, in contradiction to our assumption that S has no least element.

Thus $p+1 \notin S$ and so $p+1 < a$ for each $a \in S$.

Hence, $p+1 \in M$. Thus we get two conditions: $1 \in M$ and $p \in M$ implies $p+1 \in M$. Hence, by the principle of mathematical induction $M = \mathbb{N}$. But $M \cap S = \phi$ and so $S = \phi$, which is a contradiction. Therefore, S must have a least element.

Conclusion from the two deductions A and B:

Well-ordering principle for the system $\mathbb{N}$ is logically equivalent to the principle of mathematical induction in the sense that any one of them can be deduced, if we assume the other.

2.3 Solved Examples:

(Problems on Principle of Mathematical Induction)

Example 2.3.1. *Prove the formula: For each $n \in \mathbb{N}$, the sum of the first n natural numbers is given by $1 + 2 + 3 + \cdots + n = \frac{n(n+1)}{2}$.*

Solution: Let S be the set of all $n \in \mathbb{N}$ for which the formula is true. If $n = 1$, then we have $1 = \frac{1 \cdot (1+1)}{2}$ so that $1 \in S$.

Next, we assume that $k \in S$, i.e., we assume that
$$1 + 2 + 3 + \cdots + k = \frac{k(k+1)}{2}.$$

Add $(k+1)$ to both sides; we then obtain
$$(1 + 2 + 3 + \cdots + k) + (k+1) = \frac{k(k+1)}{2} + (k+1)$$
$$= \frac{(k+1)(k+2)}{2} \quad \text{(This is the formula for } n = k+1\text{).}$$

Thus, if $k \in S$, we have verified $(k+1) \in S$.

Thus $S \subseteq \mathbb{N}$ having two properties: (1) $1 \in S$ and (ii) $k \in S \implies (k+1) \in S$.

Consequently, by the principle of mathematical induction we infer that $S = \mathbb{N}$ and hence the formula holds for all $n \in \mathbb{N}$.

Example 2.3.2. *For each $n \in \mathbb{N}$, prove the inequality*

$$1 + \tfrac{1}{4} + \tfrac{1}{9} + \cdots + \tfrac{1}{n^2} \leq 2 - \tfrac{1}{n}.$$ [CH 1985]

Solution: Let S be the set of all $n \in \mathbb{N}$ for which the inequality

$$1 + \frac{1}{4} + \frac{1}{9} + \cdots + \frac{1}{n^2} \leq 2 - \frac{1}{n} \text{ holds.}$$

If $n = 1$, then $\frac{1}{1^2} \leq 2 - \frac{1}{1}$ is true so that we write $1 \in S$.

Next we assume that $k \in S$ and we wish to infer from this assumption that $k+1 \in S$.

Thus $1 + \tfrac{1}{4} + \tfrac{1}{9} + \cdots + \tfrac{1}{k^2} \leq 2 - \tfrac{1}{k}$ (assumed)

Adding $\frac{1}{(k+1)^2}$, we obtain

$$1 + \frac{1}{4} + \frac{1}{9} + \cdots + \frac{1}{k^2} + \frac{1}{(k+1)^2} \leq 2 - \frac{1}{k} + \frac{1}{(k+1)^2}.$$

Now,

$$2 - \frac{1}{k} + \frac{1}{(k+1)^2} = 2 - \frac{1}{k+1} + \frac{1}{k+1} - \frac{1}{k} + \frac{1}{(k+1)^2}$$

$$= 2 - \frac{1}{k+1} + \frac{k^2 + k - k^2 - 2k - 1 + k}{k(k+1)^2}$$

$$= 2 - \frac{1}{k+1} - \frac{1}{k(k+1)^2}$$

$$< 2 - \frac{1}{k+1}.$$

$$\therefore \quad 1 + \frac{1}{4} + \frac{1}{9} + \cdots + \frac{1}{k^2} + \frac{1}{(k+1)^2} < 2 - \frac{1}{k+1}.$$

Thus, if $k \in S$, then $(k+1) \in S$. We also have $1 \in S$.

$\therefore$ by the principle of mathematical induction $S = \mathbb{N}$, i.e., the inequality holds for all $n \in \mathbb{N}$.

Example 2.3.3. *For each $n \in \mathbb{N}$, the sum of the squares of the first n natural numbers is given by the formula*

$$1^2 + 2^2 + 3^2 + \cdots + n^2 = \frac{n(n+1)(2n+1)}{6}.$$

CHAPTER 2: THE REAL NUMBER SYSTEM

Solution: To establish this formula, we see that it is true for $n = 1$, since

$$1^2 = \frac{1 \cdot (1+1)(2.1+1)}{6}.$$

If we assume that it is true for $n = k$, then

$$1^2 + 2^2 + 3^2 + \cdots + k^2 = \frac{k(k+1)(2k+1)}{6}.$$

Then adding $(k+1)^2$ to both sides, we obtain

$$\begin{aligned}
1^2 + 2^2 + 3^2 + \cdots + k^2 + (k+1)^2 &= \frac{k(k+1)(2k+1)}{6} + (k+1)^2 \\
&= \frac{1}{6}(k+1)\left[k(2k+1) + 6(k+1)\right] \\
&= \frac{1}{6}(k+1)\left(2k^2 + 7k + 6\right) \\
&= \frac{1}{6}(k+1)(k+2)(2k+3),
\end{aligned}$$

i.e., the formula is valid for $n = k+1$, if we assume it to be true for $n = k$ and we have already proved that the formula is valid for $n = 1$.

Consequently, the formula is valid for all $n \in \mathbb{N}$ (by the principle of mathematical induction).

Example 2.3.4. *Prove, by induction for each $n \geq 2$, $(n+1)! > 2^n$.*

Solution: The inequality holds for $n = 2$, since $(2+1)! > 2^2$.

We assume that the inequality holds for some natural number $k \geq 2$. Then

$$(k+1)! > 2^k. \tag{1}$$

Now,

$$\begin{aligned}
(k+2)! = (k+2)\left\{(k+1)!\right\} &> (k+2)2^k, \quad \text{by (1)} \\
&> 2 \cdot 2^k \quad (\because k+2 > 2),
\end{aligned}$$

i.e., $(k+2)! > 2^{k+1}$, i.e., $(\overline{k+1}+1)! > 2^{k+1}$

$\therefore$ the inequality holds for $n = k+1$, if we assume it to be true for $n = k$ $(k \geq 2)$.

$\therefore$ by the principle of induction, the inequality holds for all natural numbers $n \geq 2$.

Note: The inequality is not true for $n = 1$. So we start with the basis $n = 2$.

Example 2.3.5. *The inequality $2^n > 2n + 1$ is not true for $n = 1, 2$, but it is true for $n = 3$. We take the basis $n = 3$. We can easily prove that $2^n > 2n + 1$, for all $n \in \mathbb{N}$, where $n \geq 3$.*

Example 2.3.6. *Given two positive real numbers x and y : prove by induction that $x - y$ is a factor of $x^n - y^n$ for all natural numbers n.*

Solution: The statement is true for $n = 1$. If we now assume that $x - y$ is a factor of $x^k - y^k (k \geq 1)$, then

$$x^{k+1} - y^{k+1} = x^{k+1} - xy^k + xy^k - y^{k+1}$$
$$= x\left(x^k - y^k\right) + y^k(x - y).$$

By our assumption, $x - y$ is a factor of $\left(x^k - y^k\right)$ and clearly $x - y$ is a factor of $y^k(x - y)$.

∴ $x - y$ is a factor of $x^{k+1} - y^{k+1}$ (whenever $x - y$ is a factor of $x^k - y^k$).

∴ it follows from the principle of mathematical induction that $x - y$ is a factor of

$$x^n - y^n \text{ for all } n \in \mathbb{N}.$$

As a particular case, see that $11^n - 4^n$ is divisible by $11 - 4 = 7$ for all $n \in \mathbb{N}$.

Example 2.3.7. *Principle of strong induction (also known as second principle of mathematical induction).*

Let S be a subset of $\mathbb{N}$ such that
(a) $1 \in S$ and
(b) If for every $k \in \mathbb{N}$,
$\{1, 2, 3, \cdots, k\} \subset S$, then $k + 1 \in S$.
Then $S = \mathbb{N}$.

Proof. Let $F = \mathbb{N} - S$. To prove $F = \phi$.

If $F \neq \phi$, then by well-ordering principle, F will have a least element m (say). Since $1 \in S$, $1 \notin F$.

As m is the least element of F and $1 \notin F$, $m > 1$.

All natural numbers $1, 2, \cdots, m - 1$ (less than m) belong to S.

Then, by hypothesis (b), $m \in S$ which implies $m \notin F$. This is a contradiction. Hence we infer $F = \phi$, i.e., $S = \mathbb{N}$. This proves the second principle of mathematical induction.

Use this principle to prove that for all $n \in \mathbb{N}$, $\left(3 + \sqrt{5}\right)^n + \left(3 - \sqrt{5}\right)^n$ is an even integer.

Proof. The statement is clearly true for $n = 1$.

$$\left(\because \left(3 + \sqrt{5}\right)^1 + \left(3 - \sqrt{5}\right)^1 = 6 = \text{an even integer}\right)$$

CHAPTER 2: THE REAL NUMBER SYSTEM

Let us assume that the statement is true for $n = 1, 2, 3, \cdots, k$.
Then

$$\begin{aligned}(3+\sqrt{5})^{k+1} + (3-\sqrt{5})^{k+1} &= a^{k+1} + b^{k+1} \quad \left(\text{where } a = 3+\sqrt{5} \text{ and } b = 3-\sqrt{5}\right) \\ &= \left(a^k + b^k\right)(a+b) - a^k b - b^k a \\ &= \left(a^k + b^k\right)(a+b) - ab\left(a^{k-1} + b^{k-1}\right) \\ &= 6\left(a^k + b^k\right) - 4\left(a^{k-1} + b^{k-1}\right) \quad (\because a+b = 6, ab = 4).\end{aligned}$$

This is an even integer because $a^k + b^k$ and $a^{k-1} + b^{k-1}$ are even integers (by assumption). Hence the statement for $n = k+1$ is true whenever it is true for $n = 1, 2, 3, \cdots, k$.

∴ by the second principle of induction, the statement is true for all $n \in \mathbb{N}$.

On Principle of Mathematical Induction
Exercises for Self-Practice

Q1. Use the principle of mathematical induction to prove that for every natural number n,

(i) $2^n < n!$ for all $n \in \mathbb{N}$, $n \geq 4$.

(ii) $10^{n+1} + 10^n + 1$ is divisible by 3. [CH 1995]

(iii) $2^{n+1} < 1 + (n+1)2^n$ for all natural numbers $n \geq 1$.

(iv) $n^2 < n!$ for all natural numbers $n \geq 4$.

(v) $1^2 - 2^2 + 3^2 - \cdots + (-1)^{n+1}n^2 = (-1)^{n+1}\dfrac{n(n+1)}{2}$ for all $n \in \mathbb{N}$.

(vi) $\dfrac{1}{\sqrt{1}} + \dfrac{1}{\sqrt{2}} + \dfrac{1}{\sqrt{3}} + \cdots + \dfrac{1}{\sqrt{n}} > \sqrt{n}$ for all $n \in \mathbb{N}$.

$\left[\text{Hints}: \sqrt{k} + \dfrac{1}{\sqrt{k+1}} = \dfrac{\sqrt{k}\sqrt{k+1} + 1}{\sqrt{k+1}} > \dfrac{k+1}{\sqrt{k+1}}, \text{ i.e., } > \sqrt{k+1}.\right]$

(vii) $n^3 + 5n$ is divisible by 6 for all $n \in \mathbb{N}$.

[Hints: $(k+1)^3 + 5(k+1) = (k^3 + 5k) + 3k(k+1) + 6$ and $k(k+1)$ is always even.]

(viii) $\dfrac{1}{2} + \dfrac{1}{2^2} + \cdots + \dfrac{1}{2^n} = 1 - \dfrac{1}{2^n}$, for all $n \in \mathbb{N}$.

(ix) Suggest a formula for the sum of first n odd natural numbers

$$1 + 3 + 5 + \cdots + (2n-1)$$

and establish your conjecture by using mathematical induction.

(x) $\dfrac{1}{1.2} + \dfrac{1}{2.3} + \cdots + \dfrac{1}{n(n+1)} = \dfrac{n}{n+1}$, for all $n \in \mathbb{N}$.

Q2. Applications of the principle of mathematical induction:

Prove, by using the principle of mathematical induction, that the following inequality holds for all $n \in \mathbb{N}$:
$$(1+a)^n \geq 1 + na + \dfrac{n(n-1)}{2}a^2 \quad (a > 0).$$
[CH 1980]

[For $n = 1$, LHS $= 1 + a$ and RHS $= 1 + a$, i.e., equality holds and the statement is true for $n = 1$.

Assume that for $n =$ a certain fixed positive integer $k \geq 1$, the inequality is true, i.e., let
$$(1+a)^k \geq 1 + ka + \dfrac{k(k-1)}{2}a^2 (a > 0) \text{ be true}. \tag{1}$$

Now multiplying both sides by $(1+a)$ which is positive, we get

$$(1+a)^{k+1} = (1+a)^k \cdot (1+a)$$
$$\geq \left(1 + ka + \dfrac{k(k-1)}{2}a^2\right)(1+a) \quad \text{by (1)}$$
$$= 1 + (k+1)a + \left\{\dfrac{k(k-1)}{2} + k\right\}a^2 + \dfrac{k(k-1)}{2}a^3$$
$$\geq 1 + (k+1)a + \dfrac{(k+1)k}{2}a^2, \text{ as } \dfrac{k(k-1)}{2}a^3 \geq 0 \; \because k \geq 1 \text{ and } a > 0$$

Thus the statement is true for $n = 1$ and when it is assumed to be true for $n = k$, it is found to be true for $n = k+1$. Hence by the principle of mathematical induction the statement is true for all $n \in \mathbb{N}$.]

Q3. Inductive or Recursive Definition (using induction principle)

(a) Definition of positive integral exponents (using the principle of induction).

If x is any real number, then we define $x^1 = x$ and $x^{n+1} = x^n \cdot x$ for all $n \geq 1$ $(n \in \mathbb{N})$.

Fit in the principle of mathematical induction to see that the above definition can define x^n for every $n \in \mathbb{N}$.

[Let $M =$ the set of all natural numbers n for which x^n is defined. Then $1 \in M$ (given condition $\longrightarrow$ first part).

For every $n \in M$, $(n+1) \in M$ (by the second part of the given condition). Hence, by induction $M = \mathbb{N}$. Thus x^n is defined for every $n \in \mathbb{N}$.]

CHAPTER 2: THE REAL NUMBER SYSTEM

(b) If x is a real number ($x \neq 0, 1$) and m is any arbitrary natural number, then prove that $x^{m+n} = x^m \cdot x^n$ for all $n \in \mathbb{N}$.

[When $n = 1$, the result becomes $x^{m+1} = x^m \cdot x^1 = x^m \cdot x$ ($\because x^1 = x$) which gives the second part of the definition of exponents (see (a) above) and hence true, i.e., $1 \in$ set of natural numbers for which the result is true.

Now assume that k is a value of n for which the result is true, i.e., the induction hypothesis is $x^{m+k} = x^m \cdot x^k$,

whence $\quad x^{m+k} \cdot x = \left(x^m \cdot x^k\right) \cdot x = x^m \cdot \left(x^k \cdot x\right) \quad$ (Assoc. Law)

i.e., $\quad x^{m+k+1} = x^m \cdot x^{k+1}$,

i.e., the result is true for $n = k+1$, if we assume it to be true for $n = k$, i.e., $(k+1)$ is in the set of natural numbers for which the result holds.

∴ by the principle of mathematical induction the result holds for every natural number n.]

Q4. Suppose we wish to prove, by induction, $2^n = 0$ for all $n \in \mathbb{N}$.

We first prove the *induction hypothesis*: Suppose k is a positive integer for which the statement $2^n = 0$ is true, i.e., suppose $2^k = 0$. Then, $2^{k+1} = 2^k \cdot 2 = 0 \cdot 2 = 0$, i.e., if the statement is true for $n = k$, then it is also true for $n = k+1$. Can we now assert: by the principle of induction, the statement $2^n = 0$ for all $n \in \mathbb{N}$?

Certainly not. See that we have not verified the result for $n = 1$ (in fact, for $n = 1$ the result is **not true**). Thus, for the conclusion of principle of induction, both the conditions ($n = 1$ and induction hypothesis) are to be verified.

2.4 The Set of Natural Numbers ($\mathbb{N}$) and the Set of all Integers ($\mathbb{Z}$)

A. The set $\mathbb{N}$ of natural numbers: Algebraic structure and Order structure.

We have the set $\mathbb{N}$ of natural numbers: $\mathbb{N} = \{1, 2, 3, \cdots\}$.

Algebraic structure on $\mathbb{N}$: To each pair $a, b \in \mathbb{N}$ we associate

(i) a natural number $a + b$, called the **sum** of a and b and

(ii) a natural number $a \cdot b$ (or ab), called the **product** of a and b.

The first operation is called **addition composition** and the second one is the composition of **multiplication** on the set of natural numbers. These compositions along with the basic properties given below are referred to as the **set $\mathbb{N}$ possessing an algebraic structure.**

Basic properties of the two compositions in $\mathbb{N}$:

1. Addition and multiplication are both commutative and associative:
   $a + b = b + a; \qquad a \cdot b = b \cdot a$ **(commutativity)**
   $a + (b + c) = (a + b) + c; \quad a \cdot (b \cdot c) = (a \cdot b) \cdot c$ **(associativity)**
   (a, b, c are any three natural numbers.)

2. Cancellation laws hold: $a + c = b + c \Longrightarrow a = b$
   $a \cdot c = b \cdot c \Longrightarrow a = b$ [as $c \in \mathbb{N} \therefore c \neq 0$]

3. Distributive law of multiplication over addition:
   $a \cdot (b + c) = a \cdot b + a \cdot c$ for all $a, b, c \in \mathbb{N}$.

4. Identity property of 1:
   $a \cdot 1 = a$ for all $a \in \mathbb{N}$ (1 is called the **multiplicative identity.**)

   Order structure on $\mathbb{N}$:

   We define a relation in $\mathbb{N}$, called **an order relation,** symbolized by '$<$' and read as 'less than'.

   For any $a, b \in \mathbb{N}$ we shall write $a < b$ (read: a is less than b), if $\exists$ a natural number c such that $a + c = b$.

   We shall write: $a > b$ (a is greater than b) to mean $b < a$.

   The presence of such a relation in $\mathbb{N}$ is referred to as $\mathbb{N}$ having *an order relation.*

Properties of the order relation:

1. **Trichotomy:** Given two natural numbers a, b either $a < b$ or $a = b$ or $b < a$ (exclusive or).

2. **Transitivity:** Given any three natural numbers a, b, c for which $a < b$ and $b < c$, then $a < c$.

3. **Compatibility:** $a, b, c \in \mathbb{N}$.
   $a < b \Longrightarrow a + c < b + c$ **(compatibility w.r.t. addition composition)**
   $a < b \Longrightarrow a \cdot c < b \cdot c$ **(compatibility w.r.t. multiplication composition)**

Inverse operations: Subtraction and Division in $\mathbb{N}$

Subtraction: Given two natural numbers a, b, there need not always exist some natural number x such that $a + x = b$. (e.g., if $a = 5$, $b = 2$, then for no $x \in \mathbb{N}$, we obtain $5 + x = 2$, i.e., we do not get the desired x, if $a > b$).

Remember: Given $a, b \in \mathbb{N}$ such that $a < b$, $\exists$ a unique $x \in \mathbb{N}$ such that when x is added to a we obtain b. We denote x by $b - a$. For two natural numbers a, b, the symbol $b - a$ *is meaningful,* only if $a < b$.

CHAPTER 2: THE REAL NUMBER SYSTEM

Division: Given two natural numbers a, b, there need not always exist some natural number c such that $a \cdot c = b$ (e.g., if $a = 3$, $b = 13$, $\exists$ no natural number c such that $3 \cdot c = 13$). But when such a number c exists, it is unique, follows as a consequence of cancellation law: $a \cdot c = a \cdot d \implies c = d$.

Note that c exists if a is a divisor of b.

> **Remember:** Given two natural numbers a and b so that a is a divisor of b, then $\exists$ a unique natural number x such that $a \cdot x = b$. We denote x by $b \div a$ (or by $\frac{b}{a}$). x is the unique natural number which when multiplied by a gives b. The symbol $b \div a$ is meaningful, if and only if a is a divisor of b, (when we are concerned with natural numbers only).

Thus subtraction and division are not always possible with any pair of natural numbers. In order to give meanings to these operations in all cases, we are to extend our number system—with the introduction of negative integer, zero and the rational number system $\mathbb{Q}$.

B. The Set $\mathbb{Z}$ of all integers: Algebraic structure and Order structure

We introduce the symbols $\{-1, -2, -3, \cdots\}$ and call them **negative integers**.

We next define the symbol 0 (zero) with two properties:

$$0 + a = a + 0 = a$$
$$\text{and} \quad 0 \cdot a = a \cdot 0 = 0,$$

where a is an integer.

The next step is to introduce the set $\mathbb{Z}$ of all integers

$$\mathbb{Z} = \{0, 1, -1, 2, -2, 3, -3, \cdots\}.$$

Algebraic structure of the set $\mathbb{Z}$ of all integers:

Addition composition in $\mathbb{Z}$: To each pair $a, b \in \mathbb{Z}$ there corresponds a sum $a + b \in \mathbb{Z}$. This addition composition has the following basic properties:

(i) $a + b = b + a$, for all $a, b \in \mathbb{Z}$: Commutativity.

(ii) $a + (b + c) = (a + b) + c$, for all $a, b, c \in \mathbb{Z}$: Associativity.

(iii) The symbol $0 \in \mathbb{Z}$ (0 is an integer) such that $a + 0 = 0 + a = a$, for all $a \in \mathbb{Z}$ (0 is known as the **additive identity**).

(iv) To each $a \in \mathbb{Z}$ there corresponds an integer $-a \in \mathbb{Z}$ such that

$$a + (-a) = (-a) + a = 0.$$

The integer $-a$ is called the *negative* of the integer a (or the **additive inverse** of a).

(v) **Inverse of addition:** The equation $a + x = b$, $a \in \mathbb{Z}$, $b \in \mathbb{Z}$, admits of a unique solution x, namely $x = b - a \in \mathbb{Z}$. $b - a$ is the **subtraction** of a from b.

Subtraction is always possible in $\mathbb{Z}$: Given two integers $a, b \in \mathbb{Z}$, $\exists$ another integer $b - a$ which, when added to a, gives b.

Remember: Subtraction $b - a$ is possible in $\mathbb{N}$, only when $a < b$.

(vi) **Cancellation law holds,** i.e., $a + c = b + c \implies a = b$ for all $a, b, c \in \mathbb{Z}$.

Multiplicative composition in $\mathbb{Z}$: To each pair $a, b \in \mathbb{Z}$ we associate an integer denoted by $a \cdot b$ (or ab), called the *product* of a and b.

This multiplicative composition has the following basic properties:

(i) $a \cdot b = b \cdot a$, for all $a, b \in \mathbb{Z}$: Commutativity.

(ii) $a \cdot (b \cdot c) = (a \cdot b) \cdot c$, for all $a, b, c \in \mathbb{Z}$: Associativity.

(iii) The integer $1 \in \mathbb{Z}$ is such that $a \cdot 1 = 1 \cdot a = a$, for all $a \in \mathbb{Z}$.

The integer '1' is, therefore, known as the *multiplicative identity*.

(iv) Cancellation law for multiplication of integers holds *with an important constraint;* thus $a \cdot b = a \cdot c$ (where $a \neq 0$) $\implies b = c$.

The two compositions—addition and multiplication—are connected by the *distributive law:* $a \cdot (b + c) = a \cdot b + a \cdot c$, for all $a, b, c \in \mathbb{Z}$.

Deductions: From these basic properties of addition and multiplication on the set $\mathbb{Z}$ of integers it can be deduced:

(i) $a \cdot b = 0 \implies a = 0$ or $b = 0$, for all $a, b \in \mathbb{Z}$ [inclusive or].

(ii) $a \cdot (-b) = -(a \cdot b)$; $(-a) \cdot b = -(a \cdot b)$; $(-a) \cdot (-b) = a \cdot b$ for all $a, b \in \mathbb{Z}$.

Division in $\mathbb{Z}$: The operation division is defined thus: If a and b are two members of $\mathbb{Z}$, with $b \neq 0$, then dividing a by b (denoted by $a \div b$ or $\frac{a}{b}$) is to obtain a **unique** integer c such that $a = b \cdot c$.

Clearly, $\frac{a}{b}$ is not defined for any two integers with $b \neq 0$.

e.g., for $\frac{3}{4}$, no integer exists such that $4 \cdot c = 3$. But there are cases like $\frac{27}{9}$, where a unique integer 3 exists such that $9 \cdot 3 = 27$.

Observe that no integer exists if $b = 0$ (whatever a may be).

e.g., for $\frac{5}{0}$, no integer c exists such that $c \cdot 0 = 5$

and for $\frac{0}{0}$, no unique integer c exists such that $c \cdot 0 = 0$ ($1 \cdot 0 = 0, 2 \cdot 0 = 0$, etc.).

Once for all, we agree to assume that:

DIVISION BY ZERO IS AN UNDEFINED OPERATION.

(e.g., $\frac{2}{0}, -\frac{7}{0}, \frac{0}{0}$, etc. are all undefined operations.)

CHAPTER 2: THE REAL NUMBER SYSTEM

Note:

1. In simplifying an expression like $\frac{(x^2-4)}{(x-2)}$ we shall proceed thus:

   When $x \neq 2$, $\frac{x^2-4}{x-2} = \frac{(x+2)(x-2)}{x-2} = x+2$.

   But when $x = 2$, $\frac{x^2-4}{x-2}$ is **undefined**.

2. In case a and b are two **non-zero members** of $\mathbb{Z}$ and if $\exists\, c \in \mathbb{Z}$ such that $a = b \cdot c$ (i.e., $c = \frac{a}{b}$ or $c = a \div b$), *then we say that b is a factor of 'a' or we say that b is a divisor of 'a'*.

Order structure of $\mathbb{Z}$: Let S be a set. An order on S is a **relation**, denoted by $<$ (read as *less than*) with the following two properties:

1. **Trichotomy:** For $a, b \in S$, either $a < b$ or $b < a$ unless $a = b$.
2. **Transitivity:** If $a, b, c \in S$, $a < b$ and $b < c \implies a < c$.

The statement $a < b$ may be read as *a is less than b* or *a precedes b*.

It is often convenient to write $b > a$ (*b is greater than a*) in place of $a < b$.

An **ordered set** is a set in which an **order** is defined. The set $\mathbb{Z}$ is an ordered set.

If $a, b \in \mathbb{Z}$, then we define the relation '$<$' in $\mathbb{Z}$ thus:

$a < b$, if $b - a$ is a positive integer

$b < a$, if $b - a$ is a negative integer.

[e.g., $-3 < -2$, since $(-3) - (-2) = -1$ (a negative integer)]

$a = b$, if $b - a$ is zero.

Note: An integer a is a positive integer, if $a > 0$ and a is a negative integer, if $a < 0$. Zero (0) is neither a positive integer nor a negative integer but we include 0 in the system $\mathbb{Z}$ of all integers.

Our definition of $<$ in $\mathbb{Z}$ obeys the two properties:

1. Given $a, b \in \mathbb{Z}$, either $a < b$ or $b < a$ unless $a = b$ (**Law of Trichotomy**)
2. Given $a, b, c \in \mathbb{Z}$, $a < b$ and $b < c \implies a < c$ (**Law of Transitivity**)
3. **Compatibility:** $a < b \implies a + c < b + c$ (**Addition composition**)

   $a < b$ and $c > 0$ (**Multiplication composition**)
   $\implies ac < bc$

Conclusion: The system $\mathbb{Z}$ of all integers has an order structure compatible with its algebraic structure.

2.5 Rational Numbers and their Main Properties

A number of the form $\frac{p}{q}$, when p, q are integers and $q \neq 0$ is called a rational number. The totality of all rational numbers forms a system called the system $\mathbb{Q}$ of rational numbers. In general, we shall take q to be a positive integer, i.e., $q \in \mathbb{N}$. With this understanding, $\frac{p}{q}$ is a positive rational number, if p is a positive integer, and $\frac{p}{q}$ is a negative rational number, if p is a negative integer. However, if $p = 0$, then the rational number $\frac{0}{q} = 0$ (zero of the rational number system). Taking $q = 1$, it can be easily seen that the set $\mathbb{Z}$ of all integers is a proper subset of $\mathbb{Q}$, i.e., $\mathbb{Z} \subset \mathbb{Q}$.

The system $\mathbb{Q}$: based on its Fundamental properties

We list below the main properties of the system $\mathbb{Q}$ of rational numbers:

I. The system $\mathbb{Q}$ forms an **ordered field**. (i.e., $\mathbb{Q}$ forms a field in which an order relation is defined).

II. The system $\mathbb{Q}$ is *dense* as well as *Archimedean*.

III. Any member of $\mathbb{Q}$ can be expressed as a *decimal* which is either terminating (e.g., $\frac{11}{4} = 2.75$) or recurring (e.g., $\frac{1}{3} = 0.\dot{3} = 0.333\cdots$).

IV. The system $\mathbb{Q}$ is *countably infinite*, i.e., $\mathbb{Q}$ can be put in one-one correspondence with the infinite set $\mathbb{N}$ of natural numbers. A set which is either finite or countably infinite is called a countable set. Therefore, Q is a *countable set*.

V. Every rational number r can be made to correspond to a point on a *directed line* but the converse is not true, i.e., every point on a directed line may not correspond to a rational number. This indicates that there are gaps between rational numbers (these gaps give rise to the existence of irrational numbers).

VI. $\mathbb{Q}$ is *unbounded*, both above and below.

VII. Lastly, the system $\mathbb{Q}$ is *not order-complete* (In other words, $\mathbb{Q}$ does no obey LUB-property)

We have, in the aforesaid list of properties of rational numbers, used certain terms which require clarification.

We begin with the first property:

Property I of Rational Numbers: The system $\mathbb{Q}$ of all rational numbers forms an ordered field.

1. What is a field?

A *field* is a non-empty set F in which two operators, called *addition* ($+$) and *multiplication* ($\cdot$), are defined and they satisfy the following axioms, known as *field axioms*.

CHAPTER 2: THE REAL NUMBER SYSTEM

A. Field axioms for addition:

A1. Closure property: If $x \in F$, $y \in F$, then their sum $x + y \in F$.

A2. Associative property: $(x+y) + z = x + (y+z)$, for all $x, y, z \in F$.

A3. Commutative property: $x + y = y + x$, for all $x, y \in F$.

A4. Existence of additive identity: $\exists$ a unique element 0 (zero), called *additive identity*, such that $0 + x = x + 0 = x$, for every $x \in F$.

A5. Existence of additive inverse: To every $x \in F$ corresponds an element $-x \in F$ (called *additive inverse* of x or *negative of* x) such that: $x + (-x) = (-x) + x = 0$.

In short, **under addition F is an Abelian group.**

M. Field axioms for multiplication:

M1. Closure property: If $x \in F$ and $y \in F$, then their product $x \cdot y \in F$ (in place of $x \cdot y$ we may write xy).

M2. Associative property: $(x \cdot y) \cdot z = x \cdot (y \cdot z)$, or all $x, y, z \in F$.

M3. Commutative property: $x \cdot y = y \cdot x$, for all $x, y \in F$.

M4. Existence of multiplicative identity: $\exists$ a unique element $1 \neq 0$ such that $1 \cdot x = x \cdot 1 = x$, for all $x \in F$.

M5. Existence of multiplicative inverse for a non-zero element of F: For every $x \in F$, $x \neq 0$, $\exists$ an element x^{-1} (or $\frac{1}{x}$) $\in F$ such that $x \cdot x^{-1} = x^{-1} \cdot x = 1$.

(Note that multiplicative inverse exists for **any non-zero element** of F but additive inverse exists **for every element** of F.)

D. The distributive property: If x, y, z be three elements of F, then

$x \cdot (y + z) = x \cdot y + x \cdot z$ (**Left Distributive Law**)

$(y + z) \cdot x = y \cdot x + z \cdot x$ (**Right Distributive Law**).

(Because of commutative property, any one can be taken as an axiom—the other clearly follows).

2. What is an ordered field?

In order to understand the meaning of 'ordered field' we first define an **ordered set**.

Order: Let S be a set. An **order** on S is a **relation** (denoted by '<', read as 'less than') with the following two properties:

(i) **Trichotomy:** For any two elements $x, y \in S$, either $x < y$ or $x = y$ or $y < x$.

(ii) **Transitivity:** For any three elements $x, y, z \in S$, $x < y$ and $y < z \implies x < z$.

[$x < y$ may be read as 'x is less than y' or 'x precedes y'. We shall write $y > x$ to mean $x < y$ (read: $y > x$ as 'y is greater than x'). The notation $x \leq y \implies x < y$ or $x = y$]

Ordered Set: An ordered set is a set S in which an order is defined.

Ordered Field: A field F is called an *ordered field*, if:

(i) F obeys all the field axioms: $A_1 - A_5$, $M_1 - M_5$ and D. (We often say that F has **an algebraic structure.**)

(ii) F is an ordered set (an order relation '<' is defined on F obeying **Trichotomy** and **Transitivity** and the two compatibility conditions namely

(iii) If $x, y, z \in F$ and $y < z$, then $x + y < x + z$ \qquad (Addition composition)

(iv) If $x, y \in F$ with $x < y$ and $z > 0$, then $xz < yz$ \qquad (Multiplication composition).

Property I of Rational Numbers:

The set $\mathbb{Q}$ of all rational numbers forms an ordered field.

We define **addition** and **multiplication** of two rational numbers $\frac{p}{q}$ and $\frac{r}{s}$ in the following way:

Addition: $\dfrac{p}{q} + \dfrac{r}{s} = \dfrac{ps + qr}{qs}$ \qquad ($qs \neq 0$ as $q \neq 0$ and $s \neq 0$)

Multiplication: $\dfrac{p}{q} \cdot \dfrac{r}{s} = \dfrac{p \cdot r}{q \cdot s}$ $\left(\text{or we write } \dfrac{pr}{qs}\right)$.

With these definitions we can verify that all the field axioms for addition and multiplication along with the distributive properties are satisfied. Hence $\mathbb{Q}$ is a field.

[As a sample, we prove the **associative property for addition:** We take three rational numbers $\frac{p}{q}, \frac{r}{s}$ and $\frac{t}{u}$.

$$\frac{p}{q} + \left(\frac{r}{s} + \frac{t}{u}\right) = \frac{p}{q} + \left(\frac{ru + st}{su}\right) = \frac{psu + qru + qst}{qsu}$$

$$(qsu \neq 0 \text{ as } q \neq 0, s \neq 0 \text{ and } u \neq 0)$$

Again,

$$\left(\frac{p}{q} + \frac{r}{s}\right) + \frac{t}{u} = \frac{ps + qr}{qs} + \frac{t}{u} = \frac{psu + qru + qst}{qsu}$$

$$\therefore \quad \frac{p}{q} + \left(\frac{r}{s} + \frac{t}{u}\right) = \left(\frac{p}{q} + \frac{r}{s}\right) + \frac{t}{u}.]$$

Now **we define an order on** $\mathbb{Q}$: If $a, b \in \mathbb{Q}$, then $a < b$, if $b - a$ is a positive rational number. Now we observe that Law of Trichotomy and Law of Transitivity both hold.

[We verify law of transitivity: $a < b \Longrightarrow b - a$ is a positive rational number

$$b < c \Longrightarrow c - b \text{ is a positive rational number.}$$

CHAPTER 2: THE REAL NUMBER SYSTEM

∴ the sum $(b-a)+(c-b)$ must be a positive rational number, i.e., $c-a$ is a positive rational number, i.e., $a < c$.]

Remember: If a is a positive rational number, then we write $a > 0$ and if a is a negative rational number, then we write $a < 0$.

Now check the two compatibility conditions:

$a < b \Longrightarrow a+c < b+c,$ for all $a, b, c \in \mathbb{Q}$

$a < b$ and $c > 0 \Longrightarrow ac < bc,$ for all $a, b \in \mathbb{Q}$ and $c > 0$.

Thus we have established that:

The system of rational numbers $\mathbb{Q}$ is a field and this field is an ordered field (order being defined as $r < s$, if $s - r$ is a positive rational number).

Property II of Rational numbers:

The system $\mathbb{Q}$ of rational number is dense as well as Archimedean.

A. $\mathbb{Q}$ is a dense set: By this we mean that between any two unequal rational numbers a and b $(a \neq b)$, there exist infinitely many rational numbers.

Since $a \neq b$, by the law of trichotomy **either** $a < b$ or $b < a$.

Let us take the case $a < b$. Then $\exists$ a **rational number** $c = \frac{a+b}{2}$.

We prove: $a < c < b$.

For, $a < b \Longrightarrow a + b < b + b$ (by compatibility condition)
$\Longrightarrow \frac{a+b}{2} < b$, i.e., $c < b$.

Again, $a < b \Longrightarrow a + a < b + a$ (by compatibility condition)
$\Longrightarrow a < \frac{b+a}{2}$ or $a < c$.

∴ it follows $a < c < b$.

Now we see that *existence of one rational number c between a and b*
$\Longrightarrow$ *existence of infinite number of rational numbers between them.*

[$c = \frac{a+b}{2}$ lies between a and b; also $d = \frac{a+c}{2}$ lies between a and c and $e = \frac{c+b}{2}$ lies between c and b and so on. Thus we may obtain infinitely many rational numbers between a and b]

Thus we have established that $\mathbb{Q}$ is a dense set.

B. $\mathbb{Q}$ is Archimedean: *Statement:* If $a, b \in \mathbb{Q}$ and $a > 0$, then there exists a natural number n such that $na > b$. [CH 1985]

Proof. When $(b < 0)$ or $(b > 0$ and $a > b)$, the results holds for $n = 1$. Now, we consider $b > 0$ and $a < b$. Let $a = \frac{p}{q}$ and $b = \frac{r}{s}$, where $p, q, r, s \in \mathbb{N}$ (a and b are given to be two positive rational numbers).

We assert that the natural number $n = qr + 1$ will serve our purpose; because

$$na = (qr+1)\frac{p}{q} = pr + \frac{p}{q} > pr \geq r \geq \frac{r}{s} = b.$$

So we find some n, namely $n = qr + 1$ such that $na > b$.

The existence of n is assured by these arguments.

An alternative proof (A proof by contradiction): Suppose, on the contrary, $\mathbb{Q}$ is non-Archimedean, i.e., if a and b are positive rational numbers and $a < b$, then **there exists no natural number n such that $na > b$.**

This implies that for all $n \in \mathbb{N}$, $na \leq b$.

Now, $na \leq b \implies na \cdot b^{-1} \leq b \cdot b^{-1}$ (b^{-1} is the multiplicative inverse of b)
$\implies (na) \cdot b^{-1} \leq 1 < m$, where m is any natural number other than 1.

Thus the assumption $na \leq b$ implies $(na) \cdot b^{-1} < m \implies \frac{n}{m} < \frac{b}{a}$.

But $\frac{b}{a}$ is a fixed rational number and $\frac{n}{m}$ is any arbitrary positive rational number. Thus we have arrived at the following conclusion:

Any arbitrary positive rational number < a fixed rational number,
which is clearly **not** true. Thus we arrive at a contradiction. Hence we cannot assume that $\mathbb{Q}$ is non-Archimedean.

In other words, **$\mathbb{Q}$ must possess the Archimedean property.**

Property III of Rational numbers: *Decimal representation of a rational number.*

Any member of $\mathbb{Q}$ is either a *Terminating Decimal* or a *Recurring Decimal* (also called a *Periodic Decimal*).

The rational fraction $\frac{a}{b}$ can be expressed as a decimal by *long division*. If the denominator b contains no prime factors other than 2 or 5, the decimal for $\frac{a}{b}$ will *terminate*. Otherwise, the decimal will be *recurring* or *periodic*, i.e., eventually a group of digits will repeat without end.

This is clear from the process of long division of a by b; for after the digits in a have been exhausted and zeros are carried down, only the $b - 1$ remainders of b can appear. After at most $b - 1$ divisions, a remainder r will appear for a second time and thereafter all remainders will repeat infinitely in the same order.

If there are $n(< b)$ different remainders $r_1, r_2, \cdots, r_n$ and

$$10 r_i = b a_i + r_i + 1 \quad (i = 1, 2, 3, \cdots, n), \quad r_{n+1} = r_1,$$

then the period of $\frac{a}{b}$ will consist of the digits $a_1, a_2, \cdots, a_n$.

For example, $\frac{1}{3} = 0.3333\cdots = 0 \cdot \dot{3}$, $\frac{1}{5} = 0 \cdot 2\dot{0}$, $\frac{1}{7} = 0 \cdot \dot{1}4285\dot{7}\cdots$ the dots above marking the period (the digit 0 is the period of terminating decimal).

CHAPTER 2: THE REAL NUMBER SYSTEM

Conversely, a periodic decimal is an infinite geometric series whose sum to infinity is a rational fraction of the form $\frac{a}{b}$, *e.g.*,

$$0 \cdot 1\dot{8} = \frac{18}{100} + \frac{18}{10000} + \cdots = \frac{0.18}{1 - 0.01} = \frac{0.18}{0.99} = \frac{2}{11}.$$

Property IV. The system $\mathbb{Q}$ of all rational numbers is countably infinite.

Definition 1. Countable and Uncountable sets

A set S is called *countably infinite* (or *denumerable* or *enumerable*) set if there exists a one-to-one function f which maps $\mathbb{N}$ **onto** S. We then write

$$S \sim \{1, 2, 3, \ldots, n, \ldots\}$$

(S is *equinumerous* or *equivalent* to the set $\mathbb{N}$ of natural numbers)

In this case $\exists$ a function f which establishes a one-to-one correspondence between natural number and the elements of the set S. Hence S can be displayed thus:

$$S = \{f(1), f(2), f(3), \ldots\}$$
$$\text{or} \quad S = \{a_1, a_2, a_3, \ldots\}$$

where $f(k)$ is denoted by a_k.

A countably infinite set is said to have a **Cardinal number** $\aleph_0$ (read: *aleph nought*)

Definition 2. A set S is said to be countable (or *at most countable*), if it is either a finite set or it is a countably infinite set.

A set which is not countable is called *uncountable*. The terms *denumerable* (or *enumerable*) and *nondenumerable* (or *non-enumerable*) are used in place of *countable* and *uncountable respectively*.

Summary

1. A set S is said to be *denumerable* or *countably infinite* if there exists a one-to-one function f which maps $\mathbb{N}$ onto S, i.e., if $S \sim \mathbb{N}$.

2. A set S is said to be *countable* (or *at most countable*) if it is either *finite* or *countably infinite*.

   Thus S is countable if there exists a one-to-one function f from $\mathbb{N}$ onto S. The elements of S are then the images of $\{1, 2, 3, \cdots\}$ which we can write as

   $$S = \{f(1), f(2), f(3), \cdots\} \quad \text{or,} \quad \{a_1, a_2, a_3, \cdots\}.$$

3. S is said to be *uncountable if it is not countable*.

Examples on Countability

Example 2.5.1. *Let E be the set of all possible even positive integers. Then E is countably infinite and hence countable. The function $f : \mathbb{N} \longrightarrow E$ defined by $f(n) = 2n$ for each $n \in \mathbb{N}$ gives the one-to-one correspondence. Hence, $E \sim \mathbb{N}$ and therefore, E is countably infinite. See the pictorial representation given below:*

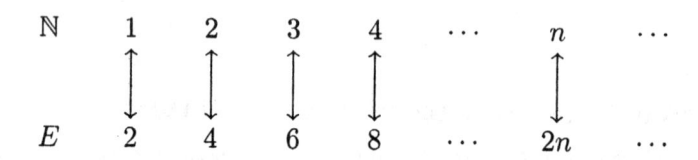

$$
\begin{array}{ccccccc}
\mathbb{N} & 1 & 2 & 3 & 4 & \cdots & n & \cdots \\
& \updownarrow & \updownarrow & \updownarrow & \updownarrow & & \updownarrow & \\
E & 2 & 4 & 6 & 8 & \cdots & 2n & \cdots
\end{array}
$$

Example 2.5.2. *The set $\mathbb{Z}$ of all integers is countable. The required one-to-one correspondence $f : \mathbb{N} \longrightarrow \mathbb{Z}$ is*

$$f(n) = \frac{n-1}{2} \quad (n = 1, 3, 5, \cdots)$$

$$= -\frac{n}{2} \quad (n = 2, 4, 6, \cdots).$$

Pictorial representation is the following:

$$
\begin{array}{ccccccccc}
\mathbb{N} & 1 & 2 & 3 & 4 & \cdots & 2n & , & 2n+1 & \cdots \\
& \updownarrow & \updownarrow & \updownarrow & \updownarrow & & \updownarrow & & \updownarrow & \\
\mathbb{Z} & 0 & -1 & 1 & -2 & \cdots & -n & , & n & \cdots
\end{array}
$$

Important results on countable sets because of its special importance have been given separately in Chapter 3A. Three such results are given below:

    I. Every infinite set has a denumerable subset.

    II. Let $A_1, A_2, A_3, \ldots$ be a sequence of countable sets then their union $\bigcup_{n=1}^{\infty} A_n$ is countable.

    With these results in mind one can prove **Property IV of rational numbers** mentioned earlier:

Property IV of Rational Numbers: The set $\mathbb{Q}$ of all rational numbers is countably infinite and hence $\mathbb{Q}$ is a countable set.   [C.H. 2006]

CHAPTER 2: THE REAL NUMBER SYSTEM

Proof. Let E_n be the set of all rational numbers which can be written with denominator n. Then the set $\mathbb{Q}$ of all rational numbers is $\bigcap_{n=1}^{\infty} E_n$. Now consider

$$E_n = \left\{\frac{0}{n}, -\frac{1}{n}, +\frac{1}{n}, -\frac{2}{n}, +\frac{2}{n}, -\frac{3}{n}, +\frac{3}{n}, \cdots\right\}$$

$$= \left\{\frac{0}{n} = 0\right\} \cup \left\{\frac{1}{n}, \frac{2}{n}, \frac{3}{n}, \cdots\right\} \cup \left\{-\frac{1}{n}, -\frac{2}{n}, -\frac{3}{n}, \cdots\right\}$$

i.e., E_n is the union of three countable sets and hence their union is countable.

E_n being a countable set for each n, by the same theorem, $\bigcup_{n=1}^{\infty} E_n$ is countable, i.e., $\mathbb{Q}$ itself is countable.

III. Suppose A and B are two infinite sets such that $B \subseteq A$ (or $A \supseteq B$).

(a) **An infinite subset B of a countable set A is countable;**

(b) **If B is an uncountable set, then A is also an uncountable set.**

Proof. Let $A = \{a_1, a_2, a_3, \cdots\}$ be a countable set and let B be an infinite subset of A. We have to prove that B is countable. From the hypothesis, each element of B is some a_j.

Let n_1 be smallest subscript for which $a_{n_1} \in B$.

Let n_2 be the least positive integer such that $n_2 > n_1$ and $a_{n_2} \in B$, and so on.

Then $B = \{a_{n_1}, a_{n_2}, \cdots\}$.

Since the set $\{n_1, n_2, n_3, \cdots\}$ is countable, $\{a_{n_1}, a_{n_2}, \cdots\}$ is countable, i.e., B is countable.

As a corollary see that **the set of all rational numbers in $[0, 1]$ is countable** (because the set of rational numbers in $[0, 1]$ is an **infinite subset** of the countable set of all rational numbers).

Property IV of Rational Numbers, namely set $\mathbb{Q}$ of all rational numbers is countable, can also be proved if we proceed in the following way:

First step. We prove that the Cartesian product $\mathbb{N} \times \mathbb{N}$ is countable.

Proof. We have $\mathbb{N} \times \mathbb{N} = \{(a, b)/a, b \in \mathbb{N}\}$.

First consider all the ordered pairs (a, b) such that $a + b = 2$. There is only one such pair, namely $(1, 1)$.

Next consider all the ordered pairs with $a + b = 3$.

In this case we have $(1, 2)$ and $(2, 1)$.

All the ordered pairs (a, b) with sum $a + b = 4$ are $(3, 1), (2, 2), (1, 3)$.

Proceeding in this manner, all the elements of $\mathbb{N} \times \mathbb{N}$ are written as $(1, 1), (1, 2), (2, 1), (3, 1), (2, 2), (1, 3), \cdots$

This set contains every ordered pair belonging to $\mathbb{N} \times \mathbb{N}$.

Starting from $(1,1)$, we can enumerate them as $1, 2, 3, 4, \cdots$. Hence, $\mathbb{N} \times \mathbb{N}$ is countable.

Second step. Let $\mathbb{Q}^+$ be the set of all positive rational numbers and $\mathbb{Q}^-$ be the set of all negative rational numbers.

Then, $\mathbb{Q} = \mathbb{Q}^+ \cup \{0\} \cup \mathbb{Q}^-$ is the set of all rational numbers.

Let $\frac{p}{q} \in \mathbb{Q}^+$. Define $f : \mathbb{Q}^+ \longrightarrow \mathbb{N} \times \mathbb{N}$ by the rule $f(\frac{p}{q}) = (p, q)$.

It is easy to see that f is one-to-one and $\mathbb{Q}^+$ is equivalent to a subset of $\mathbb{N} \times \mathbb{N}$.

Since $\mathbb{N} \times \mathbb{N}$ is countable, $\mathbb{Q}^+$ being an infinite subset of $\mathbb{N} \times \mathbb{N}$ is countable. In a similar manner $\mathbb{Q}^-$ is countable. Hence $\mathbb{Q} = \mathbb{Q}^+ \cup \{0\} \cup \mathbb{Q}^-$ is countable.

Note: One may prove that the set $\mathbb{Z}$ of all integers is countable, thus:

If $\mathbb{N}$ is the set of all natural numbers, then let $(-\mathbb{N}) = \{-1, -2, -3, -4, \cdots\}$. Hence we have

$$\mathbb{Z} = (-\mathbb{N}) \cup \{0\} \cup \mathbb{N}.$$

Since $-\mathbb{N}$ is countable, $\mathbb{Z}$ is the union of countable sets. Hence, $\mathbb{Z}$ is countable.

Property V of Rational Numbers (Geometric representation of rational numbers)

This property asserts that *to every rational number there corresponds a unique point on a directed line*. Is the converse true? Does every point on the line represents a rational number? The answer is no. We shall show that there are points on the line which do not represent rational numbers. But before that we explain how a given rational number corresponds to a unique point on a directed line.

We take a directed line—a line in which a direction (positive or negative) is indicated. The positive sense is indicated by an arrow (Fig. 2.5.1). Points 0 and 1 are chosen arbitrarily on this line.

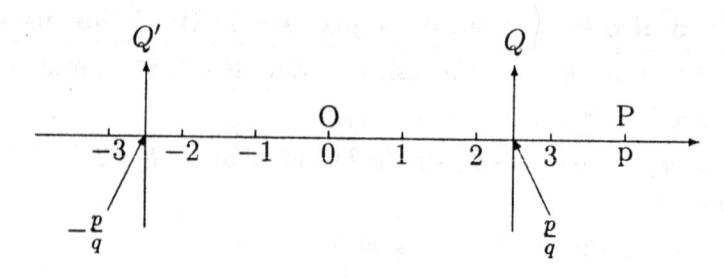

Fig 2.5.1

Then integers $\{\cdots, -3, -2, -1, 0, 1, 2, 3, \cdots\}$ are represented by an endless set of equidistant points.

CHAPTER 2: THE REAL NUMBER SYSTEM

To represent a positive rational number $\frac{p}{q}$ (p and q are both positive integers), we first represent the positive integer p by the point P on the line on the right side of O (length of $OP = p$ times the length from 0 to 1).

Next divide the length OP into q equal parts. The portion OQ representing the first of these q equal parts (or the point Q) represents the rational number $\frac{p}{q}$. A symmetric point Q' on the left of O represents the negative rational number $-\frac{p}{q}$ (p is also a positive integer and q is also a positive integer). Thus for each rational number $\frac{p}{q}$ we obtain a unique point on the line.

Examples showing the existence of numbers other than rational numbers

Example 2.5.3. *We define $\sqrt{2}$ as that number x whose square $x^2 = 2$. Prove that $\sqrt{2}$ is not a rational number.*

Proof. (by contradiction) Assume that there exists a rational number x such that $x^2 = 2$. Then x is of the form $\frac{p}{q}$, where p, q are integers, $q \neq 0$ and let p, q are in their lowest terms, i.e., there is no common factor other than 1 between p and q. This assumption can be made without any loss of generality.

Then $x^2 = 2 \implies (\frac{p}{q})^2 = 2$ or, $p^2 = 2q^2$ which implies that p^2 is an even integer and hence p is an even integer (since the square of an odd integer is always odd).

Let $p = 2m$. Then $p^2 = 2q^2$ gives $(2m)^2 = 2q^2$ or $q^2 = 2m^2$.

$\therefore q^2$ is an even integer and hence q is even.

Thus the assumption that x is rational of the form $\frac{p}{q}$ leads to the conclusion that p and q have a common factor 2, which is contrary to our hypothesis that p and q have no common factor other than 1. The contradiction proves that x must not be a rational number, i.e., there exists no rational number which satisfies $x^2 = 2$.

A more general problem is the following:

Example 2.5.4. *Show that no positive integer m other than a square number (like 4, 9, 16, 25, 49, etc.) has a square root within the system $\mathbb{Q}$ of rational numbers.*

[CH 1984, 1989]

Solution: To prove that $\exists$ no rational number x such that $x^2 = m$ where m is a non-square positive integer. Suppose to the contrary $\exists$ a rational number x satisfying $x^2 = m$. Then such an x can be written as $\frac{p}{q}$, where p and q are both integers, $q \neq 0$ and let p, q are in their lowest terms.

Then $x^2 = m \implies (\frac{p}{q})^2 = m$, i.e., $p^2 = mq^2$.

Corresponding to the positive integer m we can find a positive integer n such that

$$n^2 < m < (n+1)^2$$

or, $\quad n^2 < \left(\dfrac{p}{q}\right)^2 < (n+1)^2$

or, $\quad n^2 q^2 < p^2 < \{(n+1)q\}^2$

or, $\quad nq < p < (n+1)q$

or, $\quad 0 < p - nq < q.$

We next consider

$$\begin{aligned}(mq-np)^2 &= m^2 q^2 - 2mnpq + n^2 p^2 \\ &= m(mq^2) - 2mnpq + n^2(mq^2) \\ &= mp^2 - 2mnpq + m(n^2 q^2) \\ &= m(p^2 - 2npq + n^2 q^2) \\ &= m(p-nq)^2\end{aligned}$$

i.e., $\quad \left(\dfrac{mq-np}{p-nq}\right)^2 = m.$

This shows that m is the square of a fraction $\dfrac{mq-np}{p-nq}$, whose denominator $p-nq$ is less than q, i.e., $\dfrac{p}{q}$ is not in its lowest terms, which is a contradiction to our hypothesis. The contradiction proves that $\sqrt{m}$ cannot be a rational number.

Alt. Solution: Let m be a positive integer and is not a perfect square, so we can find positive integer n such that $n^2 < m < (n+1)^2$, i.e., $n < \sqrt{m} < n+1$

$\therefore \sqrt{m}$ cannot be an integer.

If possible, let $\sqrt{m} = \dfrac{p}{q}, p \in \mathbb{Z}, q \in \mathbb{N} - \{1\}$ and gcd $(p,q) = 1$.

$\therefore m = \dfrac{p^2}{q^2}$, i.e., $mq = \dfrac{p^2}{q}$, which is not possible as LHS is an integer whereas RHS is not an integer as gcd $(p^2, q) = 1$ and $q > 1$. $\therefore \sqrt{m} \neq \dfrac{p}{q} \in \mathbb{Q}$.

Example 2.5.5. *Show that $\sqrt{23}$ is not a rational number.*

Solution: We have $4^2 < 23 < 5^2$ $\therefore$ $4 < \sqrt{23} < 5$ $\therefore$ $\sqrt{23}$ is not an integer.

Let $\sqrt{23} = \dfrac{p}{q}, p \in \mathbb{Z}, q \in \mathbb{N} - \{1\}$ and gcd $(p,q) = 1$.

$\therefore 23q = \dfrac{p^2}{q}$, which is not possible as LHS is an integer but RHS is not an integer, for gcd $(p^2, q) = 1$ and $q > 1$. $\therefore \sqrt{23} \neq \dfrac{p}{q}$, i.e., $\sqrt{23}$ is not rational.

Note: One can try this method to prove that $\sqrt{2}, \sqrt{3}, \sqrt{5}, \sqrt{7}, \sqrt{11}, \sqrt{13}, \sqrt{19}$, etc. are not rational numbers.

[CH 1989]

Example 2.5.6. (*Gauss' theorem on the nature of the roots of a polynomial equation*). Any rational root of the equation $x^n + a_1 x^{n-1} + a_2 x^{n-2} + \cdots + a_{n-1} x + a_n = 0$ whose coefficients $a_1, a_2, \cdots, a_n$ are all integers (n is a positive integer), must be an integer which divides a_n exactly.

Solution: Let $x = \frac{p}{q}$ be a root of the equation where p and q are integers, $q \neq 0$, and p, q are in their lowest terms (no common factor except 1). Then putting $x = \frac{p}{q}$ in the equation and multiplying through by q^{n-1}, we obtain

$$-\frac{p^n}{q} = a_1 p^{n-1} + a_2 p^{n-2} q + \cdots + a_{n-1} p q^{n-2} + a_n q^{n-1}.$$

Here LHS $-\frac{p^n}{q}$ is a fraction in its lowest term and RHS is a sum of several integers and, therefore, RHS itself is an integer. This is not possible unless $q = 1$ and $x = p$ simultaneously.

$\therefore x = p$, an integer, is a root. So $p^n + a_1 p^{n-1} + a_2 p^{n-2} + \cdots + a_{n-1} p + a_n = 0$, i.e., $-p\left(p^{n-1} + a_1 p^{n-2} + a_2 p^{n-3} + \cdots + a_{n-1}\right) = a_n$,

p is a divisor of a_n. This proves the theorem.

Note: If m is an integer which is not a perfect square, then $x^2 - m = 0$ can have no rational root. That is, $\sqrt{m}$ is not rational. (e.g., $\sqrt{2}, \sqrt{3}, \sqrt{5}$, etc. are not rational). [See **Example 2.5.4** given above]

> The notions of boundedness and the completeness properties of real numbers will be discussed under the heading: **The Real Number System** $\mathbb{R}$ (Art. 2.7): Axiomatic Approach.

EXERCISES ON CHAPTER 2: IIA
(On Rational Numbers)

1. Give the value, if any, of the following expressions:

   (a) $\dfrac{x^2 + 8x}{x}$, when $x = 0$;

   (b) $\sin \dfrac{1}{x}$, when $x = 0$.

   [*Ans.* (a) and (b) Undefined]

2. For what values of x are the following expressions undefined?

   (a) $\dfrac{x^2 + a^2}{x^2 - a^2}$;

   (b) $\tan x$;

   (c) $\dfrac{3x + 5}{(x-5)(x+3)}$;

   (d) $\dfrac{x^2}{(x+1)(x+2)(x+3)}$.

[Ans. (a) $x = \pm a$; (b) $x = (2n+1)\frac{\pi}{2}$, $n \in \mathbb{Z}$; (c) $x = 5$, $x = -3$; (d) $x = -1$, $x = -2$, $x = -3$.]

3. State Peano's axioms on natural numbers. [CH 1988, 1991]

4. State the axiom of induction and well-ordering principle of the set $\mathbb{N}$ of natural numbers. [CH 1994]

5. Define a rational number. Prove that no positive integer other than a square number has a square root which is rational. Prove that $\sqrt{7}$ is not a rational number. [CH 1989]

6. Prove that there is no rational number whose square is 12.

7. If α and β are two rational numbers, verify that $\alpha+\beta$, $\alpha-\beta$, $\alpha\cdot\beta$ and $\frac{\alpha}{\beta}(\beta \neq 0)$ are all rational numbers. (You can assume that integers are closed under addition, subtraction and multiplication.)

8. Prove that between any two distinct rational numbers there exist infinitely many rational numbers.

9. State and prove the Archimedean property of rational numbers. [CH 1988]

10. When is a set S said to be *denumerable* or *countably infinite*? What is a *countable set* (or *at most countable set*)? When is a set S said to be *uncountable*? (A *denumerable set* is also called an *enumerable set*.)

11. When do you call two sets A and B are *equipotent* or *similar* or *equinumerous* ($A \sim B$)? Prove that equipotency is an equivalence relation.

12. Justify:

    (a) The set $E = \{2n : n \in \mathbb{N}\}$ of even positive integers is *denumerable*.

    (b) The set $O = \{2n - 1 : n \in \mathbb{N}\}$ of odd natural numbers is *denumerable*.

    (c) The set $\mathbb{Z}$ of all integers is *denumerable*.

    (d) The union of two disjoint denumerable sets is *denumerable*.

    Remember: A set S is said to be *denumerable* (or *countably infinite*), if $\exists$ a bijection of $\mathbb{N}$ onto S. S is said to be *countable* if it is either finite or denumerable. S is *uncountable* if it is not countable.

13. Define countability of a set. Show that the set $\mathbb{Q}$ of rational numbers is countable. [CH 2006]

CHAPTER 2: THE REAL NUMBER SYSTEM

14. Define denumerability of a set. Prove that any subset of a denumerable set is either finite or denumerable. [CH 2005]

15. If p, q, r, s are rational numbers such that $(ps - qr)^2 + 4(p-r)(q-s) = 0$, prove that either

    (a) $p = r$ and $q = s$, or

    (b) $(1 - pq)$ and $(1 - rs)$ are squares of rational numbers.

    [Hints. Put $X = p - r$ and $Y = q - s$. Then the given relation becomes

    $$(qX - pY)^2 + 4XY = 0.$$

    This relation is satisfied, if both $X = 0$ and $Y = 0$. But if both X and Y are not zero, then solving for $\frac{X}{Y}$, show that $(1 - pq)$ must be a perfect square using the fact that $\frac{X}{Y}$ must be rational.

    The given equation is unaltered, if we interchange r and p, s and q.

    Hence, $1 - rs$ must also be a perfect square. Hence, etc.]

16. If all the values of x and y obtained from

    $$ax^2 + 2hxy + by^2 = 1$$
    $$\text{and} \quad a'x^2 + 2h'xy + b'y^2 = 1$$

    are rational numbers, then prove that

    $$(h - h')^2 - (a - a')(b - b')$$
    $$\text{and} \quad (ab' - a'b)^2 + 4(ah' - a'h)(bh' - b'h)$$

    are both squares of rational numbers.

    (Given: a, a', b, b', h, h' are rational numbers.)

    [Hints. Let $(\alpha_1, \beta_1), (\alpha_2, \beta_2), (-\alpha_1, -\beta_1), (-\alpha_2, -\beta_2)$ be the solutions. Prove that $\frac{\alpha_1}{\beta_1}, \frac{\alpha_2}{\beta_2}$ are the roots of $(a - a')t^2 + 2(h - h')t + b - b' = 0$.

    Since t is rational, the first result follows at once.

    Also show $\alpha, \beta, \alpha_2 \beta_2$ are roots of

    $$z^2 \left\{(ab') - 4(ah')(hb')\right\} + 2z\left\{(a - a')(hb') - (b - b')(ah')\right\}$$
    $$+ (a - a')(b - b') = 0,$$

    where $(ab') = \begin{vmatrix} a & a' \\ h & h' \end{vmatrix} = ab' - a'b.$

Hence show that
$$\alpha_1^2 \alpha_2^2 = \frac{(b-b')^2}{(ab')^2 - 4(ah')(hb')},$$
from which the second result follows.]

17. If $x^6 + x^5 - 2x^4 - x^3 + x^2 + 1 = 0$ and $y = x^4 - x^2 + x - 1$, then show that y satisfies a quadratic equation with rational coefficients.

[Required quadratic equation is $y^2 + y + 1 = 0$.]

2.6 Existence of numbers other than Rational Numbers

1. Though the system $\mathbb{Q}$ of rational numbers is **dense**, there exists gaps between the rational numbers, however close they may be. We shall fill in these gaps by introducing what we call **irrational numbers**.

   The totality of rational and irrational numbers forms the system $\mathbb{R}$ of **real numbers**.

2. We have observed that every rational number can be expressed as a decimal which is either *terminating* ($\frac{3}{4} = 0.75$) or *recurring* ($\frac{1}{3} = 0.\dot{3}$). But there exist decimals like
$$0.10100100010000100000 1 \cdots,$$
which are neither terminating nor recurring. They are called *irrational numbers*. Rationals and irrationals together form the system of **real numbers** $\mathbb{R}$.

   Thus every real number is a decimal.

   > A real rational number can be expressed as *a terminating* or *a recurring decimal*.
   >
   > A real irrational number can be expressed as *a non-terminating* and *non-recurring decimal*.

3. We have seen that a rational number can be represented by a unique point on the directed line. But the converse is *not true*. There are points on the line which do not correspond to rational numbers. Such points are called *irrational points*—they represent irrational numbers.

4. The totality of all points on a directed line is called *linear continuum* or *geometric continuum*. The totality of all rational and irrational number is called *arithmetic continuum*.

CHAPTER 2: THE REAL NUMBER SYSTEM

We have an important axiom known as

Cantor-Dedekind Axiom: The set $\mathbb{R}$ of all real numbers can be put in one-to-one correspondence with the points of a directed line (called the *real line*), i.e., **arithmetic continuum has one-to-one correspondence with geometric continuum.**

The following examples should be carefully studied:

Example 2.6.1. *Prove that for every positive rational number satisfying the inequality $r^2 < 2$, there exists a larger rational number $r + h$ ($h > 0$) for which $(r + h)^2 < 2$.*

Solution: We can assume $h < 1$. Then $h^2 < h$ and

$$(r + h)^2 = r^2 + 2rh + h^2 < r^2 + 2rh + h$$
$$= r^2 + (2r + 1)h.$$

It is now sufficient to find h such that

$$r^2 + (2r + 1)h = 2, \quad \text{i.e.,} \quad h = \frac{2 - r^2}{2r + 1}.$$

Observation: What we find here is that any rational approximation of $\sqrt{2}$ (such approximation being less than $\sqrt{2}$) will always give rise to a better approximation to $\sqrt{2}$ (remaining still less than $\sqrt{2}$).

Example 2.6.2. *Prove that for every positive rational number s satisfying $s^2 > 2$, there exists a smaller rational number $s - k$ ($k > 0$) for which $(s - k)^2 > 2$.*

Solution: $(s - k)^2 = s^2 - 2sk + k^2 > s^2 - 2sk$ ($\because k^2 > 0$).

It is now sufficient to find k such that

$$s^2 - 2sk = 2, \quad \text{i.e.,} \quad k = \frac{s^2 - 2}{2s}.$$

We can now put these ideas together in the next example.

Example 2.6.3. *Let A and B be two sets of positive rational numbers defined by*

$$A = \{p : p \in \mathbb{Q}^+ \text{ and } p^2 < 2\}$$
$$B = \{p : p \in \mathbb{Q}^+ \text{ and } p^2 > 2\}.$$

Prove that: (i) *For every $p \in A$, $\exists$ an element $q \in A$ such that $q > p$ and*

(ii) *For every $p \in B$, $\exists$ an element $q \in B$ such that $q < p$.*

If we can justify (i) and (ii) it will mean that A has no largest element and B has no least element.

Justifications: (i) Given an element $p \in \mathbb{Q}^+$, let us construct $q \in \mathbb{Q}^+$ such that

$$p - q = \frac{p^2 - 2}{p + 2}, \quad \text{i.e.,} \quad q = p - \frac{p^2 - 2}{p + 2} = \frac{2p + 2}{p + 2}.$$

Since $p \in A$, therefore, $p^2 < 2$ and so $p - q < 0$, i.e., **$q > p$**.

Again,

$$q^2 - 2 = \frac{(2p+2)^2}{(p+2)^2} - 2 = \frac{2(p^2 - 2)}{(p+2)^2}.$$

Since $p \in A$, therefore, $p^2 < 2$ and so $q^2 - 2 < 0$, i.e., **$q^2 < 2$**. Thus $q \in A$ and $q > p$, i.e., *we cannot fix a largest element in* **A**.

Next, let $p \in B$, then $p^2 > 2$. In that case $p - q > 0$, i.e., $q < p$.

Moreover, $q^2 - 2 > 0$, i.e., **$q^2 > 2$** $\left(\because q^2 - 2 = \frac{2(p^2-2)}{(p+2)^2} \text{ and } p^2 > 2\right)$, i.e., $q \in B$.

So we have an element $q \in B$ and $q < p$, i.e., *we cannot fix a smallest element in* **B**.

Observation: We find that the irrational number $\sqrt{2}$ is such that we can always find a rational number as close to $\sqrt{2}$ as we please (rational approximation to $\sqrt{2}$).

This shows that between two rational numbers which may be as close we desire, there exists an irrational number. See that $\sqrt{2}$ fills the gap between A and B.

We have chosen $q = \dfrac{2p + 2}{p + 2}$. This choice is *not unique*. We could take

$$q = \frac{p(p^2 + 6)}{3p^2 + 2} \quad \text{or,} \quad q = \frac{4 + 3p}{3 + 2p}.$$

The conclusion will be similar.

Example 2.6.4. (*Geometrical representation of $\sqrt{2}$*) *Take a directed line OX (positive sense indicated by the arrow) with O as origin and U the unit point. Measure off on the line a length $OA = OP =$ the hypotenuse of an isosceles right-angled triangle (each of the two equal sides OU and $UP = 1$). Then the point A represents $\sqrt{2}$.* ($\because OP^2 = 1^2 + 1^2$, i.e., $OP = \sqrt{2}$) [Vide Fig 2.6.1].

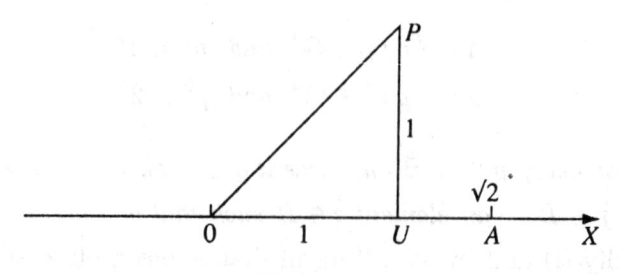

Fig 2.6.1

CHAPTER 2: THE REAL NUMBER SYSTEM

Observation: We have already proved that $\sqrt{2}$ is not a rational number. Thus the point A represents not a rational number. Hence we conclude: *All points of the directed line OX do not correspond to rational numbers.* Those points of the line which do not represent rational numbers are called *irrational points*—they represent *irrational numbers*.

[2.7–2.10] The Real Number System $\mathbb{R}$-Axiomatic Approach

---

2.7 The Real Number System $\mathbb{R}$: Axiomatic Approach

We assume the following:

Existence theorem: There exists a non-empty set $\mathbb{R}$ of objects (called *real numbers*) which satisfy the following three classes of axioms:

I. The Field Axioms;

II. The Order Axioms;

III. The Completeness Axioms.

I. We begin with **field axioms** or **algebraic properties** of $\mathbb{R}$.

We introduce *two operations* (or *two compositions*) on $\mathbb{R}$.

1. **Addition:** To every ordered pair (x, y) of real numbers there corresponds a unique real number $x + y$, called their *sum*.

2. **Multiplication:** To every ordered pair (x, y) of real numbers there corresponds a unique real number $x \cdot y$ (often written as xy), called their *product*.

The processes of obtaining *the sum* and *product* of two real numbers are called *addition* and *multiplication* respectively.

These operations obey the properties listed below—they are called *field axioms* or *algebraic properties of* $\mathbb{R}$:

With respect to addition:

(A1): $x + y = y + x$, for all $x, y \in \mathbb{R}$. (Commutative property of addition)

(A2): $x + (y + z) = (x + y) + z$, for all $x, y, z \in \mathbb{R}$.
(Associative property of addition)

(A3): $\exists$ a unique real number 0 (zero) such that
$0 + x = x + 0 = x$, for all $x \in \mathbb{R}$.   (Existence of additive identity)

(A4): For each element $x \in \mathbb{R}$, $\exists$ a unique real number $-x \in \mathbb{R}$ such that
$x + (-x) \equiv (-x) + x = 0$.   (Existence of additive inverse)

Note: We shall write $x - y$ for $x + (-y)$. This process is known as *subtraction of y from x*.

With respect to multiplication:

(M1): $x \cdot y = y \cdot x$, for all $x, y \in \mathbb{R}$. (**Commutative property of multiplication**)

(M2): $x \cdot (y \cdot z) = (x \cdot y) \cdot z$, for all $x, y, z \in \mathbb{R}$.

(**Associative property of multiplication**)

(M3): $\exists$ a unique real number 1 (one) (other than the element zero) such that for any real number x, we have
$$x \cdot 1 = 1 \cdot x = x.$$
(**Existence of multiplicative identity**)

(M4): For any real number x (other than zero, i.e., $x \neq 0$), there exists a unique real number x^{-1} (or $\frac{1}{x}$) such that
$$x^{-1} \cdot x = x \cdot x^{-1} = 1.$$
(**Existence of multiplicative inverse**)

(We call x^{-1} or $\frac{1}{x}$ as the *reciprocal* of x or *multiplicative inverse* of x).

Note: We shall write $x \div y$ or $\frac{x}{y}$ or x/y to mean $x \cdot y^{-1}$. This process is called *division* of x by y.

The two operations—addition and multiplication—are connected by a rule, called **distributive law:**

(D1): $x \cdot (y + z) = x \cdot y + x \cdot z$.

(**Distributive property of multiplication over addition**)

These nine properties **A1–A4, M1–M4** and **D1** characterise the system $\mathbb{R}$ of real numbers—they are the *field properties* or *algebraic properties of* $\mathbb{R}$. All familiar results of algebra can be derived from these nine properties. We list below some very frequently used results, the proofs of which are given as worked examples of **Art. 2.10**.

Let $x, y, z \in \mathbb{R}$. Then

1. (a) $x + y = x + z \Longrightarrow y = z$. (**Cancellation law for addition**)
   (b) $x + y = x \Longrightarrow y = 0$.
   (c) $x + y = 0 \Longrightarrow y = -x$.
   (d) $-(-x) = x$. [See Example 2.10.1 under Art. 2.10]

2. (a) $x \neq 0$ and $x \cdot y = x \cdot z \Longrightarrow y = z$. (**Cancellation law for multiplication**)
   (b) $x \neq 0$ and $x \cdot y = x \Longrightarrow y = 1$.
   (c) $x \neq 0$ and $x \cdot y = 1 \Longrightarrow y = \frac{1}{x}$.
   (d) $x \neq 0 \Longrightarrow \frac{1}{(\frac{1}{x})} = x$.

3. (a) $x > 0 \Longleftrightarrow -x < 0$.
   (b) $x > 0$ and $y < z \Longrightarrow x \cdot y < x \cdot z$.
   (c) $x < 0$ and $y < z \Longrightarrow x \cdot y > x \cdot z$.
   (d) $x \neq 0 \Longrightarrow x^2 > 0$ (and hence $1 > 0$).
   (e) $0 < a < b \Longrightarrow \frac{1}{b} < \frac{1}{a}$. (See Example 2.10.3 under Art. 2.10)

4. (a) $0 \cdot x = 0$.
   (b) $x \neq 0, y \neq 0 \implies x \cdot y \neq 0$.
   (c) $(-x) \cdot y = x \cdot (-y) = -(x \cdot y)$.
   (d) $(-x) \cdot (-y) = x \cdot y$.
   (e) $(-1) \cdot x = -x$.   (See Example 2.10.2 under Art. 2.10)
5. $x \cdot y = 0 \implies$ either $x = 0$ or $y = 0$.

[Suppose $x \neq 0$. Then $\frac{1}{x} \in \mathbb{R}$ and $x \cdot \frac{1}{x} = 1$.

Now assuming $x \neq 0$, let $xy = 0$.

i.e., $\frac{1}{x} \cdot (xy) = \frac{1}{x} \cdot 0 = 0$,

or $\left(\frac{1}{x} \cdot x\right) \cdot y = 0$ (Associativity of multiplication)

$\implies 1.y = 0$, i.e., $y = 0$.

$\therefore x \neq 0, xy = 0 \implies y = 0$, which also gives $y \neq 0$ and $xy = 0 \implies x = 0$ (contra-positivity).

Therefore, when $xy = 0$, either $x = 0$ or $y = 0$.]

II. The order Axioms of $\mathbb{R}$:

We assume that there exists a relation $<$ (*read*: less than) in $\mathbb{R}$. This relation establishes an ordering among the real numbers and satisfies the following axioms:

O1. For every ordered pair of real numbers (x, y) exactly one of the relations: $x = y, x < y, x > y$ holds. (*Law of Trichotomy*)

Note:

(i) $x > y$ is the same as $y < x$. $x < y$ is read as 'x is less than y'. $x > y$ is read as 'x is greater than y'.

(ii) A real number x is called *positive*, if $x > 0$; *negative*, if $x < 0$. The set of all positive real numbers is denoted by $\mathbb{R}^+$ and the set of all negative real numbers is denoted by $\mathbb{R}^-$.

O2. If x, y, z are three real numbers and if $x < y$ and $y < z$, then $x < z$ (This is known as *Law of transitivity*).

O3. Compatibility conditions:

(i) We have a property pertaining jointly to the addition composition and the 'less than' relation by saying the addition composition is *compatible with less than relation*. If $x < y$, then for every z we have $x + z < y + z$.

(ii) We have also multiplication composition compatible with *less than* relation:

If $x > 0, y > 0$, then their product $x \cdot y > 0$.

If $x < y$ and $z > 0$, then $xz < yz$.

If $x < y$ and $z < 0$, then $xz > yz$.

Note: The twelve properties **A1–A4, M1–M4, D1** and **01, 02, 03** make the system $\mathbb{R}$ of real numbers **an ordered field**.

Note: We shall often use the following notations:
1. $a \geq b$ means $a > b$ or $a = b$.
2. $a \leq b$ means $a < b$ or $a = b$.

It is easy to deduce: $a \geq b$ and $b \geq a \iff a = b$.

Note: The following result is often used in proofs in analysis:

Given two real numbers x and y satisfying $x \leq y + \epsilon$, for every $\epsilon > 0$. Then $x \leq y$.

Justifications: If $x > y$, then $x \leq y + \epsilon$ is violated for $\epsilon = \frac{x-y}{2}$, because

$$y + \epsilon = y + \frac{x-y}{2} = \frac{x+y}{2} < \frac{x+x}{2} = x.$$

Hence, by the law of trichotomy, we must have $x \leq y$.

III. On completeness Axiom of $\mathbb{R}$:

We have discussed in considerable details the two axioms of real number system—namely, *field axioms* and *order axioms*. These axioms also hold for the system $\mathbb{Q}$ of rational numbers. But there is an additional property to characterise the real number system—this additional property is known as the *completeness property* (or *Supremum property*). It is an essential property of $\mathbb{R}$ (not true for the system $\mathbb{Q}$) and thus we shall arrive at the statement:

$\mathbb{R}$ is a complete ordered field.

There are several different ways to describe the completeness property: The following results are equivalent to each other and any one can be used as a *completeness property*:

1. **Dedekind property:** Let $\mathbb{R}$ be decomposed into non-empty disjoint sets A and B such that $a \in A$ and $b \in B \implies a < b$.
   Then either A has the last element or greatest element or B has the first element or least element.
2. **Least upper bound property:** Every non-empty subset of $\mathbb{R}$ which is bounded above has the **least upper bound** or **supremum** in $\mathbb{R}$.
3. **Greatest lower bound property:** Every non-empty subset of $\mathbb{R}$ which is bounded below has the **greatest lower bound** or **infimum** in $\mathbb{R}$.
4. **Cauchy's criterion:** Every Cauchy sequence is convergent.
5. **Principle of monotone convergence:** Every bounded monotone sequence is convergent.

CHAPTER 2: THE REAL NUMBER SYSTEM

> 6. **Nested spheres property:** A nest of non-empty closed bounded spheres has a non-empty intersection.
> 7. **Bolzano-Weierstrass property:** Every infinite bounded set has a limit point.
> 8. **Heine-Borel property:** Every open covering of a closed and bounded set has a finite subcovering.
>
> * *Ref: Introduction to Real Variable Theory: Saxena and Shah.*

Note: Properties **4** and **7** are always implied by any one of the remaining properties. Assuming the Archimedean property, all the properties mentioned above are equivalent to each other. The readers will find the discussions of these properties at various places of the present text. A *curious reader* after reading the whole text will find it interesting to establish the equivalence of these statements.

Remember: In the present text we shall describe the completeness property of $\mathbb{R}$ by assuming that each non-empty bounded above subset of $\mathbb{R}$ has a supremum in $\mathbb{R}$. It is also known as the LUB-property of $\mathbb{R}$ (No. 2 in the list given above).

We first introduce the notions of **upper bound** and **lower bound** of a set of real numbers.

Definition 2.7.1. *Let S be a non-empty subset of $\mathbb{R}$.*

(a) The set S is said to be bounded above if $\exists$ a number $b \in \mathbb{R}$ such that for all $x \in S$, $x \leq b$. Each such number b is called an upper bound of S.

(b) The set S is said to be bounded below if $\exists$ a number $c \in \mathbb{R}$ such that for all $x \in S$, $x \geq c$. Each such number c is called a lower bound of S.

(c) A set S is said to be bounded if it is both bounded above and bounded below.

(d) A set S is said to be unbounded if it is not bounded.

An upper bound b of a set S may or may not belong to S. If it does belong to S, then b is the **largest element** of S. But a set S may or may not have a *largest element*, even when S is bounded above.

Example 2.7.1. *Let $S = \{x : x \in \mathbb{R} \text{ and } x < 5\}$. Then S is bounded above and 5 is an upper bound of S. The set has no lower bound and hence it is not bounded below. S is unbounded below (even though it is bounded above).*

If a set S has one upper bound b, then it has infinitely many upper bounds—any number greater than b is an upper bound. Similar observations can be made for lower bounds. In the set of upper bounds of S and the set of lower bounds of S we look to the *least element* and *greatest element* respectively. They define *supremum* and *infimum* respectively of S. We call them *least upper bound* (lub) M and *greatest lower bound* (glb) m of the set S. We give these definitions more precisely:

Definition 2.7.2. *Let S be a non-empty subset of $\mathbb{R}$.*

(a) *If $\mathbb{S}$ is bounded above, then a number M is called a supremum [or a least upper bound (lub)] of S if it satisfies the following two conditions:*

1. *M is an upper bound of S, i.e., for all $x \in S$, $x \leq M$.*

2. *No number $M' < M$ is an upper bound of S, i.e., for each $\epsilon > 0$, $\exists$ a member $y \in S$ such that $y > M - \epsilon$.*

[*It is easy to show that there can be only one supremum of a given subset S of $\mathbb{R}$. So when a supremum exists we refer it to **the** supremum instead of **a** supremum.*]

When a supremum M of a set S exists, we write $M = \sup S$.

(b) *If S is bounded below, then a number m is called an infimum [or a greatest lower bound (glb)] of S if it satisfies the following two conditions:*

1. *m is a lower bound of S, i.e., for all $x \in S$, $x \geq m$.*

2. *No number $m' > m$ is a lower bound of S, i.e., for every $\epsilon > 0$, $\exists$ a member $y \in S$ such that $y < m + \epsilon$.*

[*Here also an infimum, when exists, is unique and we write $m = \inf S$.*]

Note: If M' is an arbitrary upper bound of a non-empty set S, then $\sup S \leq M'$, i.e., $\sup S$ is the least of all the upper bounds of S.

If m' is an arbitrary lower bound of a non-empty set S, then $\inf S \geq m'$, i.e., $\inf S$ is the greatest of all the lower bounds of S.

It is important to remember that the supremum of a set S may or may not belong to S. If it does belong to S, then it is the greatest element of S. Similar observations are noted for the infimum of S.

III. The completeness property of $\mathbb{R}$ (LUB-axiom of $\mathbb{R}$) [CH 2006]

STATEMENT

1. Every non-empty subset S of real numbers, which is bounded above, has a supremum (or a least upper bound) in $\mathbb{R}$.
   This property is called the **completeness axiom** or **LUB-property** or **supremum property** of $\mathbb{R}$. An analogous property can be deduced in the language of infimum.

2. Every non-empty subset S of real numbers, which is bounded below, has an infimum (or a greatest lower bound) in $\mathbb{R}$.
   This property is called the GLB-property or infimum property.

Theorem 2.7.1. *GLB-property follows if LUB-property is assumed.*

CHAPTER 2: THE REAL NUMBER SYSTEM

Given: A set S of real numbers, which is bounded below.

To prove: S has the infimum (or GLB) in $\mathbb{R}$.

Proof. We construct a set S' of those real numbers x such that $-x \in S$, i.e.,

$$S' = \{x : x \in \mathbb{R} \text{ and } -x \in S\}.$$

Since S is bounded below, S has a lower bound k, (say $k \in \mathbb{R}$).

Now, if x be any member of S', then $-x \in S$ and $-x \geq k$.

$\therefore \ x \leq -k$ for all $x \in S'$.

This proves that $-k$ is an upper bound of S', i.e., S' is bounded above.

Hence we can use LUB-property on S', i.e., $\exists$ a supremum M of S' in $\mathbb{R}$.

By definition of supremum, we obtain for every $x \in S'$, $x \leq M$ and hence $-x \geq -M$,

$$\text{i.e., each member of } S \geq -M. \tag{2.7.1}$$

Also $\exists$ at least one member $y \in S'$ such that $y > M - \epsilon$ (ϵ is any positive number, no matter how small).

This means that $\exists -y \in S$ such that

$$-y < -M + \epsilon. \tag{2.7.2}$$

(1) and (2) together establish that $-M$ is the infimum or GLB of S.

Note: The axioms **I, II** and **III** lead to the assertion: **$\mathbb{R}$ is a complete ordered field.**

$\mathbb{Q}$ is an important subset of $\mathbb{R}$. We have already shown that $\mathbb{Q}$ is an **ordered field**. That $\mathbb{Q}$ is not **complete** can be shown by the theorem given below:

Theorem 2.7.2. *The set $\mathbb{Q}$ of rational numbers is not order complete.*   [CH 2006]

The statement of the theorem is true if we can give an example of a non-empty set S which is a subset of $\mathbb{Q}$ and which is bounded above (i.e., $\exists$ an upper bound) but does not have the supremum in $\mathbb{Q}$, i.e., no member of $\mathbb{Q}$ is the supremum of S.

Such an example is the set S where $S = \{x : x \in \mathbb{Q}, x > 0 \text{ and } x^2 < 2\}$, i.e., S contains those positive rational numbers whose square is less than 2.

Clearly, $S \subseteq \mathbb{Q}$, $S \neq \phi$ ($\because 1 \in S$) and S has an upper bound 2. Thus S is a non-empty subset of $\mathbb{Q}$ which is bounded above.

We now assert that no rational number can become supremum of S. If possible, let k be a **rational number** which is the supremum of S. Then $k > 0$ and $k \in \mathbb{Q}$. By the law of trichotomy, exactly one of the following holds: either $k^2 < 2$ or $k^2 = 2$ or $k^2 > 2$.

(i) If $k^2 < 2$, then let us take a rational number
$$y = \frac{4+3k}{3+2k}.$$

Then $y > 0$;
$$k - y = k - \frac{4+3k}{3+2k} = \frac{2(k^2-2)}{3+2k} < 0 \quad (\because k^2 < 2).$$

Thus we get
$$0 < k < y. \tag{2.7.3}$$

Also
$$2 - y^2 = 2 - \left(\frac{4+3k}{3+2k}\right)^2 = \frac{2-k^2}{(3+2k)^2} > 0.$$

Therefore, we have
$$\mathbf{y > 0} \quad \text{and} \quad \mathbf{y^2 < 2}. \tag{2.7.4}$$

(2.7.4) Show that $y \in S$ and (2.7.3) implies $k <$ one element y of S. Therefore, k is not the supremum of S—a contradiction to our assumption. Therefore, $k^2 \not< 2$.

(ii) If $k^2 = 2$, then k is not rational ($\because \sqrt{2}$ is not rational). This contradicts our assumption that k is a rational number. Therefore, $\mathbf{k^2 \neq 2}$.

(iii) If $k^2 > 2$, let us again take a rational number y, where
$$y = \frac{4+3k}{3+2k}.$$

Then $y > 0$;
$$k - y = \frac{2(k^2-2)}{3+2k} > 0 \quad (\because k^2 > 2)$$
$$\text{i.e.,} \quad 0 < y < k. \tag{2.7.5}$$

Also
$$2 - y^2 = \frac{2-k^2}{(3+2k)^2} < 0, \quad \text{i.e.,} \quad y^2 > 2. \tag{2.7.6}$$

(2.7.6) shows that y is an upper bound of S and (2.7.5) shows that k is not the supremum of S.

This is a contradiction to our assumption that $k = \sup S$. Therefore, $k^2 \not> 2$.

None of three possibilities: $k^2 > 2$, $k^2 = 2$, $k^2 < 2$ can hold. Hence our assumption that $\sup S$ is a rational number is not correct. Therefore, no rational number can become supremum of S.

We conclude: $\mathbb{Q}$ is not order complete.

2.8 Some Important Properties of the Real Field

I. Archimedean Property: If $x, y \in \mathbb{R}$ and $x > 0$, then there exists a positive integer n such that $nx > y$. [CH 2000]

Trivial Cases: If $y < 0$ or if $0 \leq y < x$, there is nothing to prove: because, then obviously $x > y$, i.e., Archimedean property holds for $n = 1$.

We, therefore, prove Archimedean property in the form: If $x, y \in \mathbb{R}$, $x, y > 0$ and $x < y$, then there exists a positive integer n such that $nx > y$.

Proof. Let $A = \{nx : n = 1, 2, 3, \cdots\}$.

If the Archimedean Property were not TRUE, i.e., if for all $n \in \mathbb{N}$, $nx \leq y$, then y would become an upper bound of the set A, i.e., A is a non-empty set (because $y \in A$), bounded above. Therefore, by lub-axiom, sup A exists $= M$ (say) and $M \in \mathbb{R}$.

Since $x > 0$, $M - x < M$ and $M - x$ is not an upper bound of A ($\because M$ is the lub of A), i.e., $M - x < mx$, for some positive integer m, i.e., $M < (m+1)x$; but $(m+1)x \in A$.

This means that M is not the sup A, a contradiction. Hence our assumption that the Archimedean property is NOT TRUE is not correct. So $\exists$ some n for which $nx > y$, i.e., the Archimedean property is TRUE in $\mathbb{R}$.

Alternative proof: Archimedean Property can also be deduced in the following manner:

Step 1. The set $\mathbb{N}$ of all natural numbers is unbounded above.

Otherwise, if $\mathbb{N}$ were bounded above, then by lub-axiom there would exist a supremum of $\mathbb{N} = a$ (say). Choose any positive number ϵ such that $0 < \epsilon < 1$. Then by the property of supremum there would exist a natural number n such that $n > a - \epsilon$, i.e., $n + 1 > a + (1 - \epsilon)$ and so the natural number $n + 1 > a$ ($\because 1 - \epsilon$ is positive). Now $n + 1 \in \mathbb{N}$, i.e., $\exists$ a member $n + 1$ of $\mathbb{N}$ which is greater than the supremum of $\mathbb{N}$. This would then contradict the fact that $a = \sup \mathbb{N}$.

$\therefore$ $\mathbb{N}$ cannot be bounded above. In other words, $\mathbb{N}$ is unbounded above.

Step 2. Given a positive real number x, $\exists$ a positive integer n such that $n > x$. If this were not true, then some x would be an upper bound of $\mathbb{N}$, contradicting the result '$\mathbb{N}$ is unbounded above' derived in step 1.

Step 3. If x, y are two positive real numbers such that $x < y$, then $\exists$ a positive integer n such that $nx > y$ (Archimedean property).

Use step 2 for the real number $\frac{y}{x}$. Then $\exists$ a positive integer n such that $n > \frac{y}{x}$ or $nx > y$.

As an immediate consequence of Archimedean property we prove the following *interesting* result:

Theorem 2.8.1. *For any positive real number x there exists a unique positive integer n such that $n - 1 \leq x < n$.*
[CH 1990, 1999]

Proof. Given $x > 0$, by the Archimedean property there exists a positive integer k such that $k > x$. (Consider two positive real numbers 1 and x, and by Archimedean property $\exists$ a natural number k such that $k \cdot 1 > x$, i.e., $x < k$.)

We construct a set S of all those natural numbers p for which $p > x$, i.e.,
$$S = \{p : p \in \mathbb{N} \text{ and } p > x\}.$$
Then $S \neq \phi$ ($\because k \in S$).

$\therefore$ by well-ordering property of the set $\mathbb{N}$ of natural number, S has a least member n (i.e., $n \in S$ and $n \leq$ any member of S).

Since $n \in S$, $n > x$ (by construction).

Since n is the least member of S, $(n-1) \notin S$.

So $n - 1 \leq x$ (negation of $n - 1 > x$).

$\therefore$ we finally obtain that for a positive real number x, $\exists$ a unique positive integer n such that $n - 1 \leq x < n$ (unique because n is the least member of S).

Note:
1. The result of this theorem can be stated as: If x is any positive real number, then $\exists$ a non-negative integer denoted by $[x]$ such that $[x] \leq x < [x] + 1$ (here $[x] = n - 1$).
2. If $x \in \mathbb{R}$ and $x > 0$, then $\exists$ a natural number n such that $0 < \frac{1}{n} < x$.
[In the result of Archimedean property (namely $nx > y$), put $y = 1$. So, $\exists$ a natural number, n such that $nx > 1$ or $\frac{1}{n} < x$. Since n is a natural number, $n > 0$ and hence $\frac{1}{n} > 0$. $\therefore 0 < \frac{1}{n} < x$.]

Note: When ϵ is any arbitrary positive number $0 < \frac{1}{n} < \epsilon \Longrightarrow \lim_{n \to \infty} \frac{1}{n} = 0$.

II. Density property of real numbers

A. If x and y are two real numbers with $x < y$, then there exists a rational number r such that $x < r < y$.
[CH 1990]

Observation: Existence of one rational number between x and y implies the existence of infinitely many rational numbers between x and y. This proves that $\mathbb{R}$ is dense with rational numbers.

Proof of A. Suppose $x > 0$ and the given condition states that $x < y$, i.e., $y - x > 0$. Therefore (see Note 2 above), $\exists$ a natural number n such that
$$0 < \tfrac{1}{n} < y - x.$$

Therefore, we have
$$nx + 1 < ny.$$

Now, see that $nx > 0$ and hence by Theorem 2.8.1, we obtain $m \in \mathbb{N}$ with
$$m - 1 \leq nx < m.$$

Therefore, $m \leq nx + 1 < ny$,
whence, $nx < m < nx + 1 < ny$
or $nx < m < ny$
or $x < \frac{m}{n} < y$.

Thus the rational number $r = \frac{m}{n}$ lies between x and y.

i.e., $\exists$ a rational number in the open interval (x, y), where $0 < x < y$.

Observation: We have taken $x > 0$ and $x < y$. Even when $x < 0$, we can find a rational number between x and y:

1. Suppose $x \leq 0 < y$, then by the Archimedean property, $\exists$ a positive integer n with $\frac{1}{n} < y$. Clearly, $\frac{1}{n}$, a rational number, lies in the open interval (x, y).
2. Suppose $x < y \leq 0$. Then we can write $0 \leq -y < -x$ (i.e., $-y$ and $-x$ are positive). So by the previous cases, $\exists$ a rational number $r \in (-y, -x)$ and so the rational number $-r \in (x, y)$.

What we observe is that in proving **A** there is no loss of generality to assume that $x > 0$ and $x < y$.

Remember: *Every open interval (x, y) contains a rational number.*

B. If x, y are real numbers with $x < y$, then there exists an irrational number z such that $x < z < y$. Here also we can add the following: [CH 1990]

Observation: Existence of one irrational number between x and y implies the existence of infinitely many irrational numbers between them. Hence this result proves that $\mathbb{R}$ is dense with irrational numbers.

Note that the results **A** and **B** together with the observations stated lead to the conclusion:

> Real numbers are dense with rational and irrational numbers, i.e., between two real numbers x and y, $\exists$ infinitely many real numbers. This is known as the density property of the system $\mathbb{R}$.

Proof of B. If we apply the density property **A** to real numbers $\frac{x}{\sqrt{2}}$ and $\frac{y}{\sqrt{2}}$, then we obtain a rational number $r(\neq 0)$ such that $\frac{x}{\sqrt{2}} < r < \frac{y}{\sqrt{2}}$.

Then $z = r\sqrt{2}$ is irrational and satisfies $x < z < y$.

See that $r\sqrt{2}$ is irrational because if $r\sqrt{2}$ is assumed to be rational, then the product of $\frac{1}{r} \cdot r\sqrt{2} = \sqrt{2}$ should be rational. But it is known that $\sqrt{2}$ is not rational. So we must have to accept that $r\sqrt{2}$ is irrational.

III. Decimal representations of real numbers

Let $x \in \mathbb{R}$ and let $x > 0$. Let n' be the largest integer s.t. $n' \leq x$. [Such an n' clearly exists by the Archimedean property of $\mathbb{R}$.]

Let there be largest integers $n_1, n_2, n_3, \cdots, n_k$ s.t.

$$n' + \frac{n_1}{10} \leq x; \quad n' + \frac{n_1}{10} + \frac{n_2}{10^2} \leq x; \quad \cdots; \quad n' + \frac{n_1}{10} + \frac{n_2}{10^2} + \cdots + \frac{n_k}{10^k} \leq x; \quad \cdots.$$

Let A be the set of numbers

$$\left\{ n', \; n' + \frac{n_1}{10}, \; n' + \frac{n_1}{10} + \frac{n_2}{10^2}, \; n' + \frac{n_1}{10} + \frac{n_2}{10^2} + \frac{n_3}{10^3}, \cdots \right\}.$$

This set A is bounded above. In fact, $\sup A = x$. We then say that the decimal expansion of x is the infinite series

$$n' + \frac{n_1}{10} + \frac{n_2}{10^2} + \frac{n_3}{10^3} + \cdots$$

written in a shorter form, $x = n' \cdot n_1 n_2 n_3 \cdots$.

Conversely, for any infinite decimal $n' \cdot n_1 n_2 n_3 \cdots$ the set A of numbers given before is bounded above and we shall say that $n' \cdot n_1 n_2 n_3 \cdots$ = decimal expansion of $\sup A$.

Particular cases

1. If at any stage, $n_i = 0$ for $i \geq k$, the decimal expansion gives a finite series. We call it a **terminating decimal.**
2. If a group of integers $n_1, n_2, n_3, \cdots$ repeats without end, we call it a **recurring decimal** or a **periodic decimal.** Suppose, for example, $n' = 0$, a_i are all equal to a, then the series becomes

$$0 + \frac{a}{10} + \frac{a}{10^2} + \frac{a}{10^3} + \cdots = a \left(\frac{\frac{1}{10}}{1 - \frac{1}{10}} \right) = \frac{a}{9}.$$

(summing up an infinite G.P. with common ratio $\frac{1}{10}$.)

If $a = 9$, this sum is 1, i.e., the number 1 has two decimal representations $1 = 1 \cdot 0000 \cdots$ or $1 = 0 \cdot 9999 \cdots$. More generally $\frac{1}{8} = 0.125000 \cdots = 0.124999 \cdots$. Except for such situations the decimal representations of real numbers are UNIQUE.

IV. Geometrical representation of real numbers

The real numbers can be represented by POINTS on a directed line (called the *real axis*). Our choice of positive sense of the line is arbitrary. We take a point 0 on the

CHAPTER 2: THE REAL NUMBER SYSTEM

line to represent the real number *zero* and another point in the positive sense of the line is taken to represent the real number *one*. These two points determine the scale for representing *any real rational number*. Other points are *real irrational numbers*. Then each point on the real axis corresponds to one and only one real number and *conversely*, each real number is represented by an unique point on the line. (This is an axiom, called **Cantor-Dedekind Axiom**). The real numbers represented by points on a line will be also called *Points in one dimension*.

The set of all real numbers is also called the *Arithmetical Continuum* and the set of all points on a straight line is called the *Linear* or *Geometric Continuum*. Hence,

Cantor-Dedekind Axiom states that $\exists$ a one-to-one correspondence between the two **continua**—Arithmetic Continuum and Geometric Continuum.

In view of such a one-to-one correspondence it will be convenient to use the word POINT for a REAL NUMBER and A LINEAR POINT SET for a subset of Real Numbers. Or A POINT SET OF ONE DIMENSION.

V. The Uncountability of the system $\mathbb{R}$ of real numbers

The concept of a countable set was discussed in Art. 2.5 under the heading "*On Countability*" and the countability of the set $\mathbb{Q}$ of rational numbers was established there. We shall now prove that the system $\mathbb{R}$ is not a countable set. The proof given here is based on the the decimal representation of real numbers. We shall first prove:

Theorem 2.8.2. *The set of all real numbers between 0 and 1 is not a countable set.*

[CH 1990]

Once we prove this theorem, we can conclude that $\mathbb{R}$ *itself is not a countable set, for if* $\mathbb{R}$ *were countable, the subset* $I = (0, 1)$ *would also be countable (see Theorem 2.5.3 under Art. 2.5).*

Proof. If the real numbers between 0 and 1 were countable, then they could be written as a succession:
$$x_1, x_2, x_3, x_4, \cdots, x_n, \cdots \tag{2.8.1}$$

Let us express each x_n as a decimal. If we agree not to use recurring 9s, this can be done in only one way. Let us agree to this and write the decimals, thus:

$$x_1 = 0.a_1 \, a_2 \, a_3 \, a_4 \cdots$$
$$x_2 = 0.b_1 \, b_2 \, b_3 \, b_4 \cdots$$
$$x_3 = 0.c_1 \, c_2 \, c_3 \, c_4 \cdots$$
$$\cdots \cdots$$
$$\cdots \cdots$$

Now let us take the diagonal $a_1 b_2 c_3 \cdots$ and form a decimal $0.\alpha\beta\gamma\delta\cdots$ by defining

$$\alpha = \begin{cases} 1, & \text{if } a_1 \neq 1 \\ 2, & \text{if } a_1 = 1 \end{cases} ; \quad \beta = \begin{cases} 1, & \text{if } b_2 \neq 1 \\ 2, & \text{if } b_2 = 1 \end{cases} ; \quad \gamma = \begin{cases} 1, & \text{if } c_3 \neq 1 \\ 2, & \text{if } c_3 = 1 \end{cases} \quad \text{and so on.}$$

Then $y = 0.\alpha\beta\gamma\delta\cdots$ (which contains no $\dot{9}$) denotes a real number between 0 and 1 and so must itself appear somewhere in the succession (2.8.1), if this succession is to contain ALL REAL NUMBERS between 0 and 1.

But y is not x_1 for it differs from x_1 in the first place after the decimal point; it is not x_2 for it differs from x_2 in the second place after the decimal point; it is not x_3 for it differs from x_3 in the third place after the decimal point; and so on. Thus $y \notin$ the succession (2.8.1) at all but y is a real number between 0 and 1. Hence, **no succession** $x_1, x_2, x_3, \cdots$ can include **all real numbers between 0 and 1**, i.e., the set of real numbers between 0 and 1 is **not countable**.

Observations. We have thus proved that the open interval $(0, 1)$ [The definition of an open interval is given in the next article, **Art. 2.9**] is not enumerable. If we write

$$w = \frac{x - a}{b - a} \quad (b > a),$$

then as x goes from a to b, w goes steadily from 0 to 1. Thus enumerability of all real numbers x between $x = a$ and $x = b$ would lead to an enumeration of the numbers between $x = 0$ and $x = 1$. Hence by what we have proved before:

The set of all real numbers between any two given real numbers cannot be enumerated. In fact, the system $\mathbb{R}$ is not a countable set.

2.9 The Modulus of a Real Number

If x is any real number, then we define **absolute value** or **modulus** of x, denoted by $|x|$, in the following manner:

$$|x| = \begin{cases} x, & \text{if } x \geq 0 \\ -x, & \text{if } x < 0. \end{cases}$$

Since $|x| = x \geq -x$, if $x \geq 0$ and $|x| = -x > x$, if $x < 0$, therefore in either case $|x|$ is greater of the two numbers $x, -x$, i.e., $|x| = \max\{x, -x\}$.

Remember: For any real number, $-|x| \leq x \leq |x|$.

Check the following almost-evident results:

$|x| = |-x|$; $|x \cdot y| = |x| \cdot |y|$; $\left|\frac{x}{y}\right| = \frac{|x|}{|y|}$ $(y \neq 0)$; $|x^2| = |x|^2$.

[CHAPTER 2: THE REAL NUMBER SYSTEM]

A fundamental inequality: If $k > 0$, then the inequality $|x| < k$ holds,
$$\text{iff } -k < x < k.$$

Proof. $|x| = \max\{x, -x\} < k \implies x < k \wedge -x < k$ ($\wedge$ means 'and')
$\implies x < k \wedge -k < x \implies -k < x < k$.

Conversely, let $-k \leq x \leq k$.
Then if $x \geq 0$, we have $|x| = x < k$, whereas if $x < 0$, then $|x| = -x < k$.
In either case, $|x| < k$ holds.

Theorem 2.9.1. *For any two real numbers x and y: $|x \pm y| \leq |x| + |y|$.*

Proof. $-|x| \leq x \leq |x|$ and $-|y| \leq y \leq |y|$.
Adding, $-(|x| + |y|) \leq x + y \leq (|x| + |y|)$.
Hence, by the previous fundamental inequality, $|x+y| \leq |x| + |y|$.
Since $|-y| = |y|$, we also have $|x - y| \leq |x| + |-y|$. i.e., $\leq |x| + |y|$.

> **Remember:** $|x + y| < |x| + |y|$, if x, y are of opposite signs and $|x+y| = |x| + |y|$, if x, y are of same signs or if one of x, y is zero. Verify the truth of this statement.

Corollary. $|x - y| \geq |x| - |y|$ (assuming $|x| > |y|$).
For, $|x| = |x - y + y| \leq |x - y| + |y|$, i.e., $|x| - |y| \leq |x - y|$; hence, etc.
Again, $|x - y| \geq |y| - |x|$ (assuming $|y| > |x|$)
For, $|y| = |x + y - x| \leq |x| + |y - x| \implies |y| - |x| \leq |x - y|$.

One can remember a more compact result: $||x| - |y|| \leq |x \pm y| \leq |x| + |y|$, for all $x, y \in \mathbb{R}$.

Generalisation: $|x_1 + x_2 + x_3 + \cdots + x_n| \leq |x_1| + |x_2| + |x_3| + \cdots + |x_n|$ and also
$$|x_1 + x_2 + x_3 + \cdots + x_n| \geq |x_1| - |x_2| - \cdots - |x_n|.$$

Use the method of induction:
An immediate consequence of the fundamental inequality is the following:
$$|x - y| < k \implies y - k < x < y + k.$$
[**Try to prove from the fact:** $|x| = \max(x, -x)$]

Note:
1. $0 < |x - y| < k \implies y - k < x < y + k$, but $x \neq y$.
2. $a \leq x \leq b \implies |x - \frac{1}{2}(b + a)| \leq \frac{1}{2}(b - a)$ (e.g., $-4 \leq x \leq 10 \implies |x - 3| \leq 7$).

Intervals: closed and open

Definition. Let a and b be two real numbers and let $a < b$. We now define **intervals** as follows:

Closed Interval $\quad [a, b] = \{x : x \text{ is real and } a \leq x \leq b\}$
Open Interval $\quad (a, b) = \{x : x \text{ is real and } a < x < b\}$
Semi-open Intervals $[a, b) = \{x : x \text{ is real and } a \leq x < b\}$ [open on the right]
$\quad\quad\quad\quad\quad\quad\quad (a, b] = \{x : x \text{ is real and } a < x \leq b\}$ [open on the left].

The points a and b are called the end-points of the intervals, a being the left-end point and b being the right-end point.

Each interval is a bounded set. a, the left-end point, is the **glb** and b, the right-end point, is the **lub**, of the interval (set).

It is easy to see that a **set of real numbers is bounded, iff it can be enclosed in a certain interval.**

Extended Real Number System

The extended real number system $\mathbb{R}_e$ consists of the Real field $\mathbb{R}$ and **two symbols** $-\infty$ and $+\infty$, i.e., $\mathbb{R}_e = \mathbb{R} \cup \{-\infty, +\infty\}$.

For every $x \in \mathbb{R}$, we shall assume the order $-\infty < x < +\infty$.

We shall use symbols $(-\infty, a)$, (a, ∞), $(-\infty, a]$, $[a, +\infty)$ and $(-\infty, \infty)$ as intervals of the extended real number system $\mathbb{R}_e$.

Note: $+\infty$ is an upper bound of every subset of $\mathbb{R}_e$ and then every non-empty subset has a lub.

If A is a non-empty subset of real numbers which is not bounded above in $\mathbb{R}$, then we write $\sup A = +\infty$. Similar remarks apply to lower bounds.

2.10 Illustrative Examples

We shall now give a number of illustrative examples involving all the concepts introduced in this chapter.

Example 2.10.1. *On the assumption that the Real number system $\mathbb{R}$ is a field, prove the following:* $(a, b, c \in \mathbb{R})$:

(i) *If $a + b = a + c$, then $b = c$ [Cancellation Law for addition];*
(ii) *If $a + b = a$, then $b = 0$.*
(iii) *If $a + b = 0$, then $b = -a$;*
(iv) $-(-a) = a.$

CHAPTER 2: THE REAL NUMBER SYSTEM

Proof. (i) If $a + b = a + c$, then using the field axioms we may write

$$b = 0 + b = (-a + a) + b = (-a) + (a + b)$$
$$= -a + (a + c) = (-a + a) + c$$
$$= 0 + c = c \quad \text{(Cancellation Law proved)}.$$

Take $c = 0$ in (i) and obtain (ii); take $c = -a$ in (i) to obtain (iii). Since $-a + a = 0$, (iv) can be deduced from (iii).

Example 2.10.2. *Using the field axioms of $\mathbb{R}$, prove the following $(a, b \in \mathbb{R})$:*
(i) $0 \cdot a = 0$;
(ii) *If $a \neq 0$ and $b \neq 0$, then $a \cdot b \neq 0$;*
(iii) $(-a) \cdot b = -(a \cdot b) = a \cdot (-b)$;
(iv) $(-a) \cdot (-b) = a \cdot b$;
(v) $(-1) \cdot a = -a$.

Proof. (i) We may write $0 \cdot a + 0 \cdot a = (0 + 0) \cdot a = 0 \cdot a$.
∴ by Example 2.10.1 (ii), it follows, $0 \cdot a = 0$.
(ii) Suppose, if possible, $a \neq 0$, $b \neq 0$, but $a \cdot b = 0$, where $a, b \in \mathbb{R}$. Then a^{-1}, b^{-1} both lie in $\mathbb{R}$ and we have

$$\left(b^{-1} \cdot a^{-1}\right) \cdot (a \cdot b) = b^{-1} \cdot \left(a^{-1} \cdot a\right) \cdot b = \left(b^{-1} \cdot 1\right) \cdot b = b^{-1} \cdot b = 1.$$

Thus $(b^{-1} \cdot a^{-1}) \cdot 0 = 1$, i.e., $0 = 1$, a contradiction.
Hence (ii) is TRUE.
(iii) $(-a) \cdot b + a \cdot b = (-a + a) \cdot b = 0 \cdot b = 0$.
∴ $(-a) \cdot b = -(a \cdot b)$ [cf. Example 2.10.1.(iii)].
Similarly prove the other part.
(iv) $(-a) \cdot (-b) = -[a \cdot (-b)] = -[-(a \cdot b)] = a \cdot b$; hence, etc.
(v) We have

$$1 + (-1) = 0 \Longrightarrow [1 + (-1)] \cdot a = 0$$
$$\Longrightarrow a + (-1) \cdot a = 0 \quad \text{(Distributive Law)}$$
$$\Longrightarrow (-a) + [a + (-1)a] = (-a) + 0$$
$$\Longrightarrow [-a + a] + (-1)a = -a$$
$$\Longrightarrow 0 + (-1) \cdot a = -a, \text{ i.e., } (-1)a = -a.$$

Example 2.10.3. *Assuming that $\mathbb{R}$ is an ordered field, prove the following:* $(a, b, c \in \mathbb{R})$:

(i) *If $a > 0$, then $-a < 0$ and vice versa;*
(ii) *If $a > 0$, and $b < c$, then $a \cdot b < a \cdot c$;*
(iii) *If $a < 0$ and $b < c$, then $a \cdot b > a \cdot c$;*
(iv) *If $a \neq 0$, then $a^2 > 0$ in particular $1 > 0$;*
(v) *If $0 < a < b$, then $0 < \frac{1}{b} < \frac{1}{a}$.*

Proof. (i) If $a > 0$, then $(-a) + (a) > (-a) + 0$ (compatibility condition)
i.e., $0 > -a$, or $-a < 0$.
If $a < 0$, then $(-a) + (a) < (-a) + 0$, i.e., $0 < -a$, or $-a > 0$.
(ii) $b < c \Longrightarrow c > b$; hence $c + (-b) > b + (-b)$, i.e., $c - b > 0$.
Since $a > 0$, $c - b > 0$, therefore, $a \cdot (c - b) > 0$.   (compatibility condition)
Then $a \cdot c = a \cdot c - a \cdot b + a \cdot b = a \cdot (c - b) + a \cdot b > 0 + a \cdot b = a \cdot b$.
(iii) $-[a \cdot (c - b)] = (-a) \cdot (c - b) > 0$ so that $a \cdot (c - b) < 0$, hence $a \cdot c < a \cdot b$.
(iv) If $a > 0$, then $a \cdot a > 0$, i.e., $a^2 > 0$.   (compatibility condition).
If $a < 0$, then $-a > 0$ and $(-a)^2 > 0$.
But $(-a)^2 = a^2$, because $(-a)^2 = (-a) \cdot (-a) = (a \cdot a)$ [Example 2.10.2. (iv)]; hence, etc.

In particular, take $a = 1$ and $1 = 1^2$, hence $1 > 0$.
(v) If $b > 0$ and $d \leq 0$, then $b \cdot d \leq 0$. But $b \cdot b^{-1} = 1 > 0$, hence $b^{-1} > 0$.
(If $b^{-1} \leq 0$, then $b \cdot b^{-1}$ would be ≤ 0.)
Similarly, $a^{-1} > 0$. Now from $a < b$, we obtain $a \left(a^{-1} \cdot b^{-1} \right) < b \left(a^{-1} \cdot b^{-1} \right)$, i.e., $b^{-1} < a^{-1}$.

Example 2.10.4. *Prove that $\sqrt{2}$ is not a rational number and hence deduce that $\sqrt{2} + \sqrt{3}$ cannot be rational.*   [CH 1981]

Solution: That $\sqrt{2}$ is not a rational number has been proved in Example 2.5.3.
Suppose, if possible, $\sqrt{2} + \sqrt{3} = $ a rational number.
Then the number $\sqrt{3} - \sqrt{2} = \frac{1}{\sqrt{3}+\sqrt{2}}$ is also a rational number ($\because$ it is the quotient of two rational numbers: 1 and $\sqrt{3} + \sqrt{2}$.)
Hence, $\sqrt{2} = \frac{1}{2} \left[(\sqrt{3} + \sqrt{2}) - (\sqrt{3} - \sqrt{2}) \right]$ is rational, which contradicts the irrational nature of $\sqrt{2}$ (already proved).
$\therefore$ the supposition is not tenable, i.e., $\sqrt{3} + \sqrt{2}$ is not rational.

Example 2.10.5. *If k is an approximation to $\sqrt{2}$, then prove that $\frac{4+3k}{3+2k}$ is a nearer one.*

Solution: Suppose $k < \sqrt{2}$, i.e., $k^2 \not< 2$ or $k^2 - 2 < 0$.

Then $k - \dfrac{4+3k}{3+2k} = \dfrac{2(k^2-2)}{3+2k} < 0$ and $2 - \left(\dfrac{4+3k}{3+2k}\right)^2 = \dfrac{2-k^2}{(3+2k)^2} > 0$,

i.e., $\dfrac{4+3k}{3+2k} > k$ but $\left(\dfrac{4+3k}{3+2k}\right)^2 < 2$, i.e., $k < \dfrac{4+3k}{3+2k} < \sqrt{2}$.

This shows that $\dfrac{4+3k}{3+2k}$ is a better approximation to $\sqrt{2}$ than k.

Similarly, see that when $k^2 > 2$, i.e., when $k > \sqrt{2}$, $\sqrt{2} < \dfrac{4+3k}{3+2k} < k$, i.e., in this case also $\dfrac{4+3k}{3+2k}$ is a better approximation to $\sqrt{2}$ than k.

Example 2.10.6. *(A very important result)* Let S be a set of real numbers. $T \subset S$. If S is bounded, prove that T is also bounded and show that
(i) $m \leq m' \leq M' \leq M$, where $m = \inf S$, $m' = \inf T$, $M = \sup S$ and $M' = \sup T$;
(ii) $0 \leq M' - m' \leq M - m$.

> **Remember:** $\sup S - \inf S$ is called the Oscillation of a bounded set S.

Solution: To prove that T is bounded, if S is bounded.

Given that $T \subset S$; and also S is a bounded set.

Hence, $\exists\ K, l$ s.t. $x \leq K$, for all $x \in S$ and $x \geq l$, for all $x \in S$.

But every member of T is a member of S ($\because T \subset S$).

$\therefore$ every member of $T \leq K$ and is also $\geq l$.

$\therefore$ T is bounded both above and below, i.e., T is a bounded set.

Since S and T are both bounded sets of real numbers, $\exists$ infimum and supremum of these two sets.

Let $m' = \inf T$, $M' = \sup T$, $m = \inf S$ and $M = \sup S$.

Next, to prove, (i) $m \leq m' \leq M' \leq M$.

Since $M' = \text{lub } T$ and M is one of its upper bounds, we may write $M' \leq M$.

Similarly, $m \leq m'$.

Clearly, for any bounded set infimum $\leq$ supremum.

Hence, $m' \leq M'$.

Thus, we get $m \leq m' \leq M' \leq M$.

Finally to prove: (ii) $0 \leq M' - m' \leq M - m$, i.e., oscillation of $T \leq$ oscillation of S.

Since, from (i) $m \leq m'$, it follows, $-m \geq -m'$ or $-m' \leq -m$ and $M' \leq M$.

$\therefore M' - m' \leq M - m$.

Since $M' \geq m'$, we have $0 \leq M' - m' \leq M - m$.

This proves (ii).

Example 2.10.7. *If $x \in$ a bounded set S of real numbers, then prove that the set T formed by elements $-x$, i.e., $T = \{-x : x \in S\}$ is also bounded and $\sup T = -\inf S$ and $\inf T = -\sup S$.*

Solution: Since S is a bdd. set of real numbers, $\exists\ M$ and m s.t.; $M = \sup S$ and $m = \inf S$.

Now, $x \in S \Longrightarrow m \leq x \leq M \Longrightarrow -M \leq -x \leq -m \Longrightarrow T$ is a bounded set.

Let $\sup T = M_1$, $\inf T = m_1$ (their existence is assured, since T is also a bounded set of real numbers).

By definition, (i) $x \leq M$, for all $x \in S$;

(ii) if $M_1 < M$, $\exists\ x' \in S$ s.t. $x' > M_1$,

whence it follows

(i)$'$ $-x \geq -M$, for all $-x \in T$;

(ii)$'$ $\exists\ -x' \in T$ s.t. when $-M_1 > -M$, $-x' < -M_1$.

In other words, $-M$ is the exact l.b. of T, i.e., $\inf T = -\sup S$.

Similarly, $\sup T = -\inf S$.

Example 2.10.8. *Show that if x and y are any two members of bounded sets of real numbers S_1 and S_2 respectively, then prove that the set S whose elements are of the form $x + y$ is also bounded and $\sup S_1 + \sup S_2 = \sup S$, $\inf S_1 + \inf S_2 = \inf S$.*

Solution: Since S_1 and S_2 are two bounded sets of real numbers, $\exists\ M_1, m_1$ and M_2, m_2 s.t.

$$\inf S_1 = m_1, \quad \inf S_2 = m_2,$$
$$\sup S_1 = M_1, \quad \sup S_2 = M_2.$$

By definition, (i) $\forall\ x \in S_1$, $x \leq M_1$

(ii) $\exists\ x' \in S_1$, s.t. $x' > M_1'$, where $M_1' < M_1$

(iii) $\forall\ y \in S_2$, $y \leq M_2$

(iv) $\exists\ y' \in S_2$, s.t. $y' > M_2'$, where $M_2' < M_2$.

Whence it follows: $x + y \leq M_1 + M_2$, $x \in S_1$, $y \in S_2$ and hence $x + y \in S$ and $x' + y' > M_1' + M_2'$, where $M_1' + M_2' < (M_1 + M_2)$ and $x' + y' \in S$, i.e., the set S is bdd. above and $M_1 + M_2$ is its supremum. Similarly, we can prove that the set S is bounded below and its infimum $= m_1 + m_2$.

Example 2.10.9. *x and y are any two members of a bounded set S of real numbers, find the exact upper bound of the set S_1 whose members are of the type $|x - y|$.*

[CH 1999]

CHAPTER 2: THE REAL NUMBER SYSTEM   93

Solution: We shall actually prove that $\sup S_1 = $ oscillation of $S = \sup S - \inf S$.

For the bounded set S of real numbers, let $\sup S = M$, $\inf S = m$.

$\therefore$ by definition,
 (i) For any two elements x and y of S (suppose $x > y$), $x \leq M$, $y \geq m$.
    Then $x - y \leq M - m$ and $x - y \in S_1$.
 (ii) $\exists\ x' \in S$ s.t. $x' > M_1$, when $M_1 < M$ and also $\exists\ y' \in S$ s.t. $y' < m_1$, when $m_1 > m$, i.e., $x' - y' > M_1 - m_1$, when $M_1 - m_1 < M - m$ and $x' - y' \in S_1$.

This shows that the set S_1 of numbers of the type $x - y$ is bounded above and its upper bound is $M - m = $ oscillation of S_1. We could get the same result by taking members of S_1 in the form $y - x$, where $y > x$; hence, etc.

Example 2.10.10. *The set S whose elements are of the form $\frac{1}{p} + \frac{1}{q}$, when p and q are positive integers is a bounded set for which $\inf S = 0$; $\sup S = 2$.*

Solution: The members of the set S are

$(p = 1)$ $\qquad 1 + \frac{1}{1}, 1 + \frac{1}{2}, 1 + \frac{1}{3}, 1 + \frac{1}{4}, 1 + \frac{1}{5}, 1 + \frac{1}{6}, 1 + \frac{1}{7}, \cdots$;

$(p = 2)$ $\qquad \frac{1}{2} + \frac{1}{2}, \frac{1}{2} + \frac{1}{3}, \frac{1}{2} + \frac{1}{4}, \frac{1}{2} + \frac{1}{5}, \frac{1}{2} + \frac{1}{6}, \cdots$;

$(p = 3)$ $\qquad \frac{1}{3} + \frac{1}{3}, \frac{1}{3} + \frac{1}{4}, \frac{1}{3} + \frac{1}{5}, \cdots$;

$\qquad \cdots\cdots\cdots\cdots\cdots\cdots\cdots\cdots\cdots\cdots\cdots$

$(p = n)$ $\qquad \frac{1}{n} + \frac{1}{n}, \frac{1}{n} + \frac{1}{n+1}, \frac{1}{n} + \frac{1}{n+2}, \cdots$

Clearly, no member of S can exceed $1 + \frac{1}{1} = 2$ and $\exists$ members $> k$, if $k < 2$. i.e., $\sup S = 2$. Also observe that 2 is the greatest element of S, $\therefore \sup S = 2$. Similarly, see that $\inf S = 0$.

EXERCISES ON CHAPTER 2: IIB
(On Real Numbers)

1. Let $\mathbb{Q}^+$ be the set of all positive rational numbers. Subdivide $\mathbb{Q}^+$ into two disjoint sets A and B such that A has no greatest element and B has no least member.

   (See Example 2.6.3)

   List all the *field axioms* and *order axioms* of the system $\mathbb{R}$ of real numbers. With the help of these axioms prove the following statements:

   (Examples 2–5; given below)

   Assume $x, y, z \in \mathbb{R}$:

2. (a) If $x + y = x + z$, then $y = z$.   (Cancellation Law)
   (b) If $x + y = x$, then $y = 0$.

(c) If $x + y = 0$, then $y = -x$.

(d) $-(-x) = x$.

3. (a) If $x \neq 0$ and $xy = xz$, then $y = z$. (Cancellation Law)

   (b) If $x \neq 0$ and $x \cdot y = x$, then $y = 1$.

   (c) If $x \neq 0$ and $xy = 1$, then $y = \frac{1}{x}$.

   (d) If $x \neq 0$, then $\frac{1}{\left(\frac{1}{x}\right)} = x$.

4. (a) $0 \cdot x = 0$.

   (b) If $x \neq 0$, $y \neq 0$, then $x \cdot y \neq 0$.

   (c) $(-x) \cdot y = -(x \cdot y) = x \cdot (-y)$.

   (d) $(-x) \cdot (-y) = x \cdot y$.

5. (a) If $x > 0$, then $-x < 0$ and vice versa.

   (b) If $x > 0$, and $y < z$, then $x \cdot y < x \cdot z$.

   (c) If $x < 0$, and $y < z$, then $x \cdot y > x \cdot z$.

   (d) If $x \neq 0$, then $x^2 > 0$. In particular, $1 > 0$.

   (e) If $0 < x < y$, then $0 < \frac{1}{y} < \frac{1}{x}$.

6. State the least upper bound axiom for the set $\mathbb{R}$ of real numbers. Prove that the set of all rational numbers is not order complete. [CH 2003]

   Another way of asking the same question: State completeness axiom of the set of real numbers. Can it be justified for the set of rational numbers? Give reasons for your answer. [CH 2006]

7. If $a, b \in \mathbb{R}$, $0 < a < b$, then prove that there exists a positive integer n such that $na > b$. [CH 1996, 2004]

   Otherwise: State and prove the Archimedean property of real numbers.

   [CH 2002]

8. (a) State the least upper bound axiom in $\mathbb{R}$. Prove that every non-empty subset of $\mathbb{R}$, which is bounded below, has the greatest lower bound. [CH 2001]

   (b) Assuming the lub-axiom for the set $\mathbb{R}$ of real numbers, deduce the Archimedean property of $\mathbb{R}$. [CH 2005]

9. (a) Using Archimedean property of real numbers prove that if x is a real number, then there exists a unique integer n such $n \leq x < n + 1$. [CH 1990]

CHAPTER 2: THE REAL NUMBER SYSTEM 95

(b) Prove that for every real number x there exists a positive integer n such that $n > x$. [CH 1999]

10. If A is a non-empty set of real numbers, bounded below and if the set $(-A)$ is defined by $-A = \{-x : x \in A\}$, then show that $\inf A = -\sup(-A)$. [CH 1995]

11. (a) Prove that between two real numbers there exists infinite number of rational numbers.

    (b) Prove that every open interval (a, b) contains infinitely many rational numbers. [CH 2003]

12. Using lub-axiom for $\mathbb{R}$, prove that the set $\mathbb{N}$ of natural numbers is unbounded above. [CH 1997]

13. Let a and b be two real numbers. Assuming that there exists a rational number r such that $a < r < b$, prove that there exists an irrational number α such that $a < \alpha < b$. [CH 2003]

14. Find the rational number whose decimal expansion is

    (a) $0.334444\cdots$; $\left[Ans.\ \frac{301}{900}\right]$
    (b) $0.217\dot{1}\dot{3}$. $\left[Ans.\ \frac{2687}{12375}\right]$

15. Prove that $\sqrt{2} + \sqrt{3}$ cannot be a rational number.

    (assuming that $\sqrt{2}$ is an irrational number.)

16. If $\frac{p}{q} < \frac{r}{s}$ with $q > 0$, $s > 0$, prove that $\frac{p+r}{q+s}$ lies between $\frac{p}{q}$ and $\frac{r}{s}$.

    (p, q, r, s are integers.)

17. Let m and n be positive integers. Show that $\sqrt{2}$ always lies between the two fractions of the form $\frac{m}{n}$ and $\frac{m+2n}{m+n}$. Which fraction is closer to $\sqrt{2}$? [CH 1991]

18. Show that if m is an approximation to $\sqrt{2}$, then $\frac{(4+3m)}{(3+2m)}$ is a nearer one.

19. Show that the supremum and the infimum of a set S are uniquely determined whenever they exist and that $\inf S \leq \sup S$. When the equality holds?

20. Define the least upper bound of a bounded set and obtain it for the set

    $$A = \left\{\frac{1}{2}, \frac{2}{3}, \frac{3}{4}, \cdots, \frac{n}{n+1}, \cdots\right\}.$$ [Ans. 1]

21. Define the least upper bound of a set of real numbers and state the least upper bound axiom for real numbers. Deduce that a set of real numbers which is bounded below has the greatest lower bound. [CH 1980, 1991, 1993, 2001]

22. Determine the exact bounds (lub and glb) of $\{10, 3, 3\frac{1}{3}, 5\frac{2}{3}, 3\frac{1}{9}, 5\frac{8}{9}, 3\frac{1}{27}, 5\frac{26}{27}, \cdots\}$ justifying that they satisfy the definitions of exact bounds. [Ans. 10, 3]

23. Find the sup and inf of the set of real numbers of the form $2^{-p} + 3^{-q} + 5^{-r}$, where p, q, r take on all positive integral values.

24. Let A and B be two sets of real numbers bounded above. Let $a = \sup A$ and $b = \sup B$. Let C be the set of real numbers formed by considering all products of the form $x \cdot y$, where $x \in A$ and $y \in B$. Show that in general $a \cdot b \neq \sup C$. [$a \cdot b = \sup C$, if the members of A and B are all positive.]

25. If x and y are members of bounded sets A and B of real numbers, prove that bounds of the set C of numbers $\frac{y}{x}$ are $\frac{\sup B}{\inf A}$ and $\frac{\inf B}{\sup A}$, provided $\inf A \neq 0$ and $\sup A \neq 0$ and the members of A and B are all positive.

26. If S_1 and S_2 are two bounded sets of real numbers prove that the bounds of the set $S_1 \cup S_2$ are $\max\{\sup S_1, \sup S_2\}$ and $\min\{\inf S_1, \inf S_2\}$.

27. Determine which of the following sets of real numbers are bounded above, bounded below or bounded; also determine the glb and lub, wherever they exist:

    (a) The set of all even integers starting from 2; [Ans. Unbdd. above $(m = 2)$]
    (b) The set of all negative integers; [Ans. Unbdd. below $(M = -1)$]
    (c) $\{0, 1, 2, 4, 19, -5\}$; [Ans. $m = -5$, $M = 19$]
    (d) $\{\frac{3}{2}, -\frac{4}{3}, \frac{5}{4}, -\frac{6}{5}, \frac{7}{6}, -\frac{8}{7}, \cdots\}$; [Ans. Bdd.]
    (e) $\{2, -2, \frac{1}{2}, -\frac{1}{2}, \frac{2}{3}, -\frac{2}{3}, \cdots, \frac{n}{n+1}, -\frac{n}{n+1}, \cdots\}$; [Ans. Bdd.]
    (f) $\{1, 2, \frac{1}{2}, 3, \frac{1}{3}, 4, \frac{1}{4}, 5, \frac{1}{5}, \cdots\}$; [Ans. Unbdd. above $(m = 0)$]
    (g) $\{\frac{3}{2}, \frac{5}{2}, \frac{5}{3}, \frac{7}{3}, \frac{7}{4}, \frac{9}{4}, \frac{9}{5}, \frac{11}{5}, \cdots\}$; [Ans. $m = \frac{3}{2}$, $M = \frac{5}{2}$]
    (h) $\{-2, \frac{3}{2}, -\frac{4}{3}, \frac{5}{4}, -\frac{6}{5}, \frac{7}{6}, -\frac{8}{7}, \frac{9}{8}, \cdots\}$. [Ans. $m = -2$, $M = \frac{3}{2}$]

28. Check the equivalence of the following statements:

    (a) $-4 \leq x \leq 10 \equiv |x - 3| \leq 7$;
    (b) $-7 \leq x \leq -3 \equiv |x + 5| \leq 2$;
    (c) $|x + 3| \leq 2 \equiv -5 \leq x \leq -1$;
    (d) $0 < |x - a| < \delta \equiv a - \delta < x < a + \delta (x \neq a)$.

29. Solve the following inequalities (when the solution exists):

CHAPTER 2: THE REAL NUMBER SYSTEM

(a) $|2x - 3| < 1$;   [Ans. Open interval $(1, 2)$]

(b) $|x^2 - 7x + 12| > x^2 - 7x + 12$;   [Ans. Open interval $(3, 4)$]

(c) $|x| = x + 5; |x| = x - 5$;   [Ans. $-\frac{5}{2}$; No solution]

(d) $|(x^2 + 4x + 9) + (2x - 3)| = |x^2 + 4x + 9| + |2x - 3|$;   [Ans. $x \geq \frac{3}{2}$]

(e) $|(x^4 - 4) - (x^2 + 2)| = |x^4 - 4| |x^2 + 2|$;   [Ans. $|x| \geq \sqrt{3}$]

(f) $|\sin x| = \sin x + 1$.   [Ans. $x = n\pi - (-1)^n \frac{\pi}{6}$. $n = 0, \pm 1, \pm 2$, etc.]

(g) $x^2 - 2|x| - 3 = 0$.   [Ans. $-3, 3$]

[**Suggestions:** (b) Valid if $x^2 - 7x + 12 < 0$; (d) $|a + b| = |a| + |b|$, if a, b are of the same signs; (e) $|a - b| = |a| - |b|$, iff a, b are of the same signs and $|a| \geq |b|$; (f) Will be true only for those x for which $\sin x < 0$, i.e., when $-\sin x = \sin x + 1$, $\sin x = -\frac{1}{2}$, etc.]

30. If $x, y \in \mathbb{R}$ such that $x < y$, then obtain some irrational number α such that $x < \alpha < y$.

    Density property: It gives a rational number r s.t. $x < r < y$. See that $y - r > 0$; hence by Archimedean property, $\exists$ a positive integer n s.t. $n(y - r) > \sqrt{2}$, where $\sqrt{2}$ is some positive real irrational number and $y - r$ is a positive real number. Then $r < r + \frac{\sqrt{2}}{n} < y$, or $x < \alpha < y$, where $\alpha = r + \frac{\sqrt{2}}{n}$ and α is irrational.

31. Prove that the set of all positive integers is unbounded above.

32. If H is a bounded non-empty set of real numbers, then show that the non-negative real number $d(H) = \sup\{|x - y| : x \in H, y \in H\}$ exists. (See **Example 2.10.9**)   [CH 1990]

33. Prove that the the set of real numbers in $[0, 1]$ is non-denumerble.   [CH 2004]

    A few important inequalities:

34. **Cauchy-Schwarz inequality:** If $a_1, a_2, \cdots, a_n$ and $b_1, b_2, \cdots, b_n$ are arbitrary real numbers, then

$$\left(\sum_{i=1}^n a_i b_i\right)^2 \leq \left(\sum_{i=1}^n a_i^2\right)\left(\sum_{i=1}^n b_i^2\right).$$

Proof. The square of any real number ≥ 0. Hence, if x and y are arbitrary real numbers,
$$(x - y)^2 = x^2 - 2xy + y^2 \geq 0 \Longrightarrow x^2 + y^2 \geq 2xy.$$

Now, write

$$x = \frac{|a_i|}{\sqrt{a_1^2 + a_2^2 + \cdots + a_n^2}} \quad \text{and} \quad y = \frac{|b_i|}{\sqrt{b_1^2 + b_2^2 + \cdots + b_n^2}},$$

for $i = 1, i = 2, \cdots, i = n$ successively and add the resulting inequalities. On the left we obtain the sum 2, for

$$\left(\frac{a_1}{\sqrt{S_1}}\right)^2 + \left(\frac{a_2}{\sqrt{S_1}}\right)^2 + \cdots + \left(\frac{a_n}{\sqrt{S_1}}\right)^2 = 1 \quad (S_1 = a_1^2 + a_2^2 + \cdots + a_n^2)$$

and $\left(\frac{b_1}{\sqrt{S_2}}\right)^2 + \left(\frac{b_2}{\sqrt{S_2}}\right)^2 + \cdots + \left(\frac{b_n}{\sqrt{S_2}}\right)^2 = 1 \quad (S_2 = b_1^2 + b_2^2 + \cdots + b_n^2).$

We thus get (on dividing both sides of the inequality by 2)

$$\frac{|a_1 b_1| + |a_2 b_2| + \cdots + |a_n b_n|}{\sqrt{S_1} \cdot \sqrt{S_2}} \leq 1$$

or, $\quad |a_1 b_1| + |a_2 b_2| + \cdots + |a_n b_n| \leq \sqrt{S_1 \cdot S_2}.$

Since the expressions on both sides are positive we may square and then delete the modulus signs (which are not then necessary):

$$(a_1 b_1 + a_2 b_2 + \cdots + a_n b_n)^2 \leq \left(a_1^2 + a_2^2 + \cdots + a_n^2\right)\left(b_1^2 + b_2^2 + \cdots + b_n^2\right).$$

Another Method: The sum of squares of n real quantities is always greater than or equal to zero. Hence, for every real value of λ,

$$\sum_{i=1}^{n}(a_i \lambda + b_i)^2 \geq 0$$

$$\Longrightarrow A\lambda^2 + 2B\lambda + C \geq 0 \quad \left(\text{where } A = \sum_{i=1}^{n} a_i^2,\ C = \sum_{i=1}^{n} b_i^2,\ B = \sum_{i=1}^{n} a_i b_i\right).$$

Case 1. If $A > 0$, put $\lambda = -\frac{B}{A}$. Obtain $B^2 - AC \leq 0$. **(desired inequality)**

Case 2. If $A = 0$, then each of $a_i = 0$.

Then the equality

$$\sum a_i b_i = \sum a_i^2 \sum b_i^2 \quad \text{holds, for each side} = 0.$$

$\therefore$ in general, $B^2 \leq AC$.

CHAPTER 2: THE REAL NUMBER SYSTEM

Third method: Prove Lagrange's Identity for real numbers

$$\left(\sum_{i=1}^{n} a_i b_i\right)^2 = \sum_{i=1}^{n} a_i^2 \cdot \sum_{i=1}^{n} b_i^2 - \sum_{1 \leq k < j \leq n} (a_k b_j - a_j b_k)^2$$

and hence Cauchy-Schwarz inequality follows at once.

Another method: Let $\vec{a} = (a_1, a_2, \ldots, a_n)$ and $\vec{b} = (b_1, b_2, \ldots, b_n)$ be two elements of the real vector space $\mathbb{R}^n$. Consider the dot product

$$\vec{a} \cdot \vec{b} = |\vec{a}||\vec{b}| \cos \theta \quad \text{i.e.} \quad \left(\sum_{i=1}^{n} a_i b_i\right)^2 = \sum_{i=1}^{n} a_i^2 \sum_{i=1}^{n} b_i^2 \cos^2 \theta$$

$$\therefore \left(\sum_{i=1}^{n} a_i b_i\right)^2 = \sum_{i=1}^{n} a_i^2 \sum_{i=1}^{n} b_i^2$$

35. If $a_1 \geq a_2 \geq \cdots \geq a_n$ and $b_1 \geq b_2 \geq \cdots \geq b_n$, prove that

$$\frac{1}{n}\left(\sum_{i=1}^{n} a_i\right)\left(\sum_{i=1}^{n} b_i\right) \leq \sum_{i=1}^{n} a_i b_i.$$

36. Prove the inequality

$$\sqrt{(a_1 - b_1)^2 + (a_2 - b_2)^2 + \cdots + (a_n - b_n)^2}$$
$$\leq \sqrt{a_1^2 + a_2^2 + \cdots + a_n^2} \cdot \sqrt{b_1^2 + b_2^2 + \cdots + b_n^2}$$

and state its geometrical interpretation.

[Square both sides and then use Cauchy-Schwarz Inequality. The sum of the lengths of two sides of a triangle is not less than the third side.]

37. Show that if $a > 0$, $ax^2 + 2bx + c \geq 0$ for all values of x, when $b^2 - ac \leq 0$.

38. Prove the following inequalities:

    (a) $x + \frac{1}{x} \geq 2$, if $x > 0$;
    (b) $x + \frac{1}{x} \leq -2$, if $x < 0$;
    (c) $\left|x + \frac{1}{x}\right| \geq 2$, if $x \neq 0$.

39. **Weierstrass' Inequalities:** If $a_1, a_2, \cdots, a_n$ are n positive real numbers, each less than 1 and $s_n = a_1 + a_2 + \cdots + a_n$, then prove that

(a) $1 - s_n < (1 - a_1)(1 - a_2) \cdots (1 - a_n) < \frac{1}{(1+s_n)}$;

(b) $1 + s_n < (1 + a_1)(1 + a_2) \cdots (1 + a_n) < \frac{1}{(1-s_n)}$ (where $s_n < 1$).

40. If a and b are positive and unequal and n is any rational number, then prove that $na^{n-1}(a - b) > a^n - b^n > nb^{n-1}(a - b)$, unless $0 < n < 1$, in which case the inequalities are reversed.

41. Prove that the arithmetic mean of n positive real numbers, which are not all equal to one another, is greater than their geometric mean. Hence prove:

(a) $n^n > 1 \cdot 3 \cdot 5 \cdots (2n - 1)$;
(b) $(n+1)^n > 2^n n!$.

Proof. Suppose $S = \{a_1, a_2, a_3, \cdots, a_n\}$, where $a_i > 0$ and $a_i \neq a_j$.

Let a_r and a_s be the greatest and the least of the a's and let G be their GM and A be their AM. We now replace the elements of S by the elements $b_1, b_2, \cdots, b_n$, where every b is the same as the corresponding a except that a_r is replaced by G and a_s is replaced by $\frac{(a_r a_s)}{G}$. Thereby we do not alter the value of GM (Why?), but certainly we do not increase the AM.

$\therefore G + \frac{a_r a_s}{G} - a_r - a_s = \frac{(a_r - G)(a_s - G)}{G} \leq 0$, i.e., AM of b-set elements $\leq$ AM of a-set elements.

We repeat the arguments until we have replaced each of $a_1, a_2, \cdots, a_n$ by G in at most n steps.

Since the final value of the AM is clearly G,

$\therefore G \leq$ AM of the initial set of elements S, i.e., $G < A$.

Appendix to Chapter 2: Dedekind's Concept of Real Numbers

Introduction: Richard Dedekind (1831–1916), a famous mathematician of nineteenth century in his memorable work entitled *Was Sund Und was Sollen die Zahlen* (*What are and what should be numbers*) introduced what we call Dedekind's cut of rational numbers. With the help of such cuts he developed the real number system $\mathbb{R}$. We present here the essential points of this concept. Interested readers may consult

A course in Pure Mathematics—**G H Hardy** for further details.

I. Definition of a Dedekind cut of rational numbers: The set $\mathbb{Q}$ of all rational numbers can be subdivided into two *non-empty disjoint sets* L and U satisfying the following conditions: (i) $x \in L$ and $y \in U \implies x < y$; (ii) L-class does not contain a *greatest member*.

Instead of (ii) we could take: U-class does not contain a *least member*.

Such a subdivision of $\mathbb{Q}$ is known as *Dedekind's cut of rational numbers*. We denote the cut by (L, U) or by L/U – L is called the *lower class* and U, the *upper class*.

More explicitly, for a cut L/U, we must have the following *salient points*:

1. $L \neq \phi$, $U \neq \phi$. This means that *each class is non-empty*, i.e., all rational numbers do not belong to the same class so that other class remains void.
2. Each rational number has a class so that no rational number escapes classification ($L \cup U = \mathbb{Q}$).
3. Every member of L-class is less than every member of U-class.
4. L-class has no greatest member (instead we could take U-class has no least member).
5. If a rational number $r \in L$, then $r' < r$ also belongs to L.
   If $s \notin L$, then any $s' > s$ cannot belong to L.

With all these points in mind we obtain two types of cuts:

A. This type of cuts corresponds to a rational number α. We say, The cut L/U generates a **rational number** α.

We give an example: Let Q be divided into two classes L and U, where

$$L = \{x : x \in \mathbb{Q} \text{ and } x < 3\}$$
$$U = \{x : x \in \mathbb{Q} \text{ and } x \geq 3\}.$$

It is easy to prove that L/U satisfies all the **cut-properties** and U-class has a least member 3. We say, the cut L/U corresponds to the least member of U-class yielding **rational number 3**.

B. A cut L/U whose U-class has no least member: This type of cut corresponds to an irrational number. We say, The cut L/U generates a **real irrational number**.

We give an example of such a cut. Let the set $\mathbb{Q}$ be subdivided into two classes L and U such that

$$L = \{x : x \in \mathbb{Q}, \ x \leq 0 \text{ and } x > 0 \text{ with } x^2 < 2\}$$
$$U = \{x : x \in \mathbb{Q}, \ x > 0 \text{ with } x^2 > 2\}.$$

We first prove: L/U satisfies all the conditions of a Dedekind cut.
1. **Check:** No class is empty $\left(\frac{5}{3} \in L, 3 \in U\right)$.
2. No rational numbers escapes classification. The only number which we have not classified is that number x whose square $= 2$. We have shown earlier that no $x \in \mathbb{Q}$ satisfies $x^2 = 2$. So, we have classified all numbers $\in \mathbb{Q}$ into two classes L and U–none is left out.
3. Since any positive rational number x which belongs to L obeys $x^2 < 2$ and any $y \in U$ obeys $y > 0$ and $y^2 > 2$, hence $x < y$. Negative rational numbers and zero are clearly less than any member of y ($\because y > 0$).
   $\therefore$ each $x \in L$ is less than every $y \in U$.
4. We now justify: L has no greatest member. If possible let r be the greatest member of L. Then r should be positive and $r^2 < 2$. Now consider another positive rational number $\frac{4+3r}{3+2r}$.
   We have

$$2 - \left(\frac{4+3r}{3+2r}\right)^2 = \frac{2-r^2}{(3+2r)^2} > 0 \quad (\because r^2 < 2).$$

Therefore,

$$\left(\frac{4+3r}{3+2r}\right)^2 < 2, \quad \text{i.e.,} \quad \frac{4+3r}{3+2r} \in L.$$

Also

$$\frac{4+3r}{3+2r} - r = \frac{2(2-r^2)}{2+2r} > 0 \quad (\because r^2 < 2)$$

Therefore,

$$\frac{4+3r}{3+2r} > r.$$

Thus we have a positive rational number

$$\frac{4+3r}{3+2r} \in L \text{ but } \frac{4+3r}{3+2r} > r.$$

This contradicts our assumption that r is the greatest member of L. Hence we conclude L has no greatest member.

CHAPTER 2: THE REAL NUMBER SYSTEM

We have so far established that the cut L/U obeys all the conditions of a Dedekind cut.

Now we shall prove that for this cut L/U, *the upper class U has no least member*: If possible, let k be the least member of U. Then $k > 0$ and $k^2 > 2$.

Take a positive rational number $\dfrac{4+3k}{3+2k}$.

We see that

$$2 - \left(\frac{4+3k}{3+2k}\right)^2 = \frac{2-k^2}{(3+2k)^2} < 0 \quad (\because k^2 > 2).$$

Therefore,

$$\left(\frac{4+3k}{3+2k}\right)^2 > 2, \quad \text{i.e.,} \quad \frac{4+3k}{3+2k} \in U.$$

Also

$$\frac{4+3k}{3+2k} - k = \frac{2(2-k^2)}{3+2k} < 0 \quad (\because k^2 > 2)$$

and so $\dfrac{4+3k}{3+2k} < k$, i.e., for a given $k \in U$, $\exists$ a rational number $\dfrac{4+3k}{3+2k} \in U$, which is less than k, i.e., k cannot because the least member of k.

Since k is arbitrary, we conclude **U-class has no least member**.

This cut generates an **irrational number** $\sqrt{2}$, which was a gap between L and U.

Remember: If the set $\mathbb{Q}$ of rational numbers be divided into two classes L and U such that each class exists, no rational number escapes classification, every $x \in L <$ any $y \in U$, the lower class L has no greatest member. Such a division of $\mathbb{Q}$ is called a *Dedekind cut of rational numbers*. Each such cut gives a *real number*—

1. Rational number, if the upper class U has a least member.
2. Irrational number, if the upper class U has no least member.

The totality of all Dedekind cuts of $\mathbb{Q}$ forms the real number system $\mathbb{R}$.

Based on the definition of a Dedekind cut of rational numbers we know that every real number α is a cut (L, U). In the following two sections we introduce **order structure** and **algebraic structure** (or **field structure**) of the system $\mathbb{R}$. Then $\mathbb{R}$ will form an **ordered field**.

II. ℝ is an ordered field

A. Order structure of $\mathbb{R}$: Let $\alpha = (L_1/U_1)$, $\beta = (L_2/U_2)$ and $\gamma = (L_3/U_3)$ be three real numbers defined by Dedekind cuts of rational numbers.

(i) α is said to be equal to β ($\alpha = \beta$) if the sets L_1 and L_2 contain same elements (which implies U_1 and U_2 have also same elements).

(ii) α is said to be less than β ($\alpha < \beta$), if $L_1 \subset L_2$, i.e., $p \in L_1 \Longrightarrow p \in L_2$, but $\exists$ at least one $q \in L_2$ s.t. $q \notin L_1$.

$\alpha < \beta$ (α is less than β) and $\beta > \alpha$ (β is greater than α) are equivalent statements, i.e., $\alpha < \beta \Longleftrightarrow \beta > \alpha$. If $\alpha \leq \beta$, then $\sigma < \beta$ or $\alpha = \beta$ and $\alpha \geq \beta$, then $\alpha > \beta$ or $\alpha = \beta$.

(iii) Real number 0 (zero) is defined by the cut (M/P), where $M = \{x : x \in \mathbb{Q} \text{ and } x < 0\}$ and $P = \{x : x \in \mathbb{Q} \text{ and } x \geq 0\}$.

(iv) A real number α is said to be positive or negative according as $\alpha > 0$ or $\alpha < 0$.

(v) (*Law of Trichotomy*): For any two real numbers $\alpha = (L_1, U_1)$ and $\beta = (L_2, U_2)$, we have either $\alpha = \beta$ or $\alpha < \beta$ or $\alpha > \beta$.

(vi) (*Law of Transitivity*): If $\alpha = (L_1, U_1)$, $\beta = (L_2, U_2)$, $\gamma = (L_3, U_3)$ be three real numbers defined by Dedekind cuts and $\alpha < \beta$ and $\beta < \gamma$, then $\alpha < \gamma$.

Proof. Since

$$\alpha < \beta, \quad L_1 \subset L_2 \tag{2.10.1}$$

and since

$$\beta < \gamma, \quad L_2 \subset L_3. \tag{2.10.2}$$

By (1), if $x \in L_1$, then $x \in L_2$ and hence by (2) $x \in L_3$, i.e., *every element of L_1 is an element of L_3*.

Again, by (2), $\exists\, y \in L_3$ s.t. $y \notin L_2$; then this y cannot also belong to L_1, for if $y \in L_1$, it would have belonged to L_2 by (1), but $y \notin L_2$, i.e., $\exists$ an element $y \in L_3$ which does not belong to L_1.

We have thus proved L_1 is a proper subset of L_3, i.e., $\alpha < \gamma$.

From these results (i)–(vi) we conclude that the system $\mathbb{R}$ of real numbers defined by Dedekind cuts with the definition of the relation $<$ given in (ii) is **an ordered set**.

An Important Result: Let (L, U) be a Dedekind cut of rational numbers and let δ be any positive rational number. Then there exist rational numbers x and y, $x \in L$ and $y \in U$ such that $y - x = \delta$.

Proof. Two cases may arise: δ may belong to L or δ may belong to U.

Case 1. Let $\delta(>0) \in L$. If z is any rational number in U, then $\delta < z$. By Archimedean property, $\exists$ a positive integer n s.t. $n\delta > z$ and hence $n\delta \in U$. Now consider the following set of rational numbers

$$\{n\delta, (n-1)\delta, (n-2)\delta, (n-3)\delta, \cdots, 3\delta, 2\delta, \delta\}.$$

Observe that $\delta \in L$ but $n\delta \in U$ and among other members of this set some are in L and some in U. Therefore, $\exists$ two consecutive members of the set, say $(n-r)\delta$ and $(n-r-1)\delta$, such that

$$x = (n-r-1)\delta \in L$$
$$\text{but} \quad y = (n-r) \quad \delta \in U$$

with their difference $y - x = \delta$. The result follows.

Case 2. Let $\delta(> 0)$ be in U. If $0 \in L$, then let $x = 0$, and $y = \delta$; again it follows $x \in L$, $y \in U$ and $y - x = \delta$.

But if 0 is in U, then as in Case 1, we can find a negative integer $-n$ s.t. $-n\delta \in L$ and $(-n+1)\delta \in U$. Letting $x = -n\delta$ and $y = (-n+1)\delta$, the result follows.

The proof of the theorem is now complete.

B. Algebraic Structure of the Real Number System $\mathbb{R}$: Algebraic Operations:

(i) Definition of sum of two real numbers, both defined by Dedekind cut:
Let

$$\alpha = (L_1, U_1)$$
$$\text{and} \quad \beta = (L_2, U_2)$$

be two real numbers defined by Dedekind cut of rational numbers. Then their sum $\alpha + \beta$ is defined by a cut (L, U) where

$$L = \{x : x = a + b, a \in L_1, b \in L_2\}$$
$$U = \mathbb{Q} - L.$$

This definition can be accepted only if we can justify that our construction of L and U satisfies all the conditions of a Dedekind cut of rational numbers.

Justifications:

1. By construction $L \cup U = \mathbb{Q}$, i.e., every rational number is classified and obviously $L \cap U = \Phi$.
2. Since L_1 and L_2 are not empty, $\exists\ x \in L_1$ and $y \in L_2$ so that $\exists$ an $x + y \in L$, i.e., L is *non-void*.

   If $p \in U_1$ and $q \in U_2$, then $p + q > a + b$ for all $a \in L_1$ and $b \in L_2$. This implies $p + q \notin L$ and hence $p + q \in U$, i.e., U is also non-void.

3. If $p \in L$, $q \in U$, then $p < q$, for, if possible, let $q \leq p$.
   Then $q \leq p \Longrightarrow q \leq a + b$, $a \in L_1$, $b \in L_2$, where $a + b = p$.
   This will mean that $q - b \leq a \Longrightarrow q - b \in L_1$, i.e., $c \in L_1$, where $q - b = c$.
   Thus $q = b + c$. Since $b \in L_2$, $c \in L_1$, we must conclude $q \in L$, which contradicts our assumption that $q \in U$.
   $\therefore$ we cannot assume $q \leq p$, i.e., $q > p$ or $p < q$.
4. Finally, L does not contain a largest member.
   For, any $z \in L \Longrightarrow z = w + k$, $w \in L_1$ and $k \in L_2$.
   But $\exists$ $w_1 \in L_1$ and $k_1 \in L_2$ s.t. $w < w_1$ and $k < k_1$, since neither L_1 nor L_2 contains a largest number.
   Let $z_1 = w_1 + k_1$. We thus obtain a rational number $z_1 \in L$ such that

$$z < z_1 \quad (\because w + k < w_1 + k_1).$$

Since $z \in L$ was arbitrary, it follows that the L-class does not contain a largest rational number.

Conclusion: (L, U) defined above is a Dedekind cut of rational numbers. We may then write $(L, U) = (L_1, U_1) + (L_2, U_2)$.

(ii) **Negative of a Real Number:** If $\alpha = (L_1, U_1)$ be any real number defined according to Dedekind. Then the *negative of* α (written as $-\alpha$) is the cut (L, U) where
$L = \{-x : x \in U_1$ and $x \neq$ the least element of U_1, if it exists$\}$ and $U = \mathbb{Q} - L$.

We may justify that (L, U) satisfies all the conditions of a Dedekind cut of rational numbers.

With the definition of the real number 0 (zero), by the cut (M, P) and the relation ($<$ or $>$) as defined above we can verify the following results:

If $\alpha > 0$, then $-\alpha < 0$. If $\alpha < 0$, then $-\alpha > 0$.

(iii) **Difference between two Real Numbers:** The difference $\alpha - \beta$ of two real numbers $\alpha = (L_1, U_1)$ and $\beta = (L_2, U_2)$ is defined as the sum of α and the negative of β.

Thus, $\alpha - \beta = \alpha + (-\beta) = (L_1, U_1) + [-(L_2, U_2)]$.

(iv) **Product of two Real Numbers:**

P-A If $\alpha = (L_1, U_1)$ and $\beta = (L_2, U_2)$ be **two positive real numbers** defined by Dedekind cuts of rational numbers then their product $\alpha\beta$ or $\alpha \cdot \beta$ is the cut (L, U) where $L = M \cup \{xy : x \geq 0, y \geq 0, x \in L_1$ and $y \in L_2\}$; $U = \mathbb{Q} - L$ (M is the set of all negative rational numbers.)

P-B If α and β are **both negative real numbers,** then their product is defined by $(-\alpha) \cdot (-\beta)$.

CHAPTER 2: THE REAL NUMBER SYSTEM

P-C If α be a **positive real number** and β be a **negative** real number, then their product is defined by $-[\alpha \times (-\beta)]$.

Similarly, we may define the product, if α is *negative* but β is positive.

P-D If either α or β or both be **Zero**, i.e., the cut (M, P), then their product is always the real number $0 = (M, P)$.

It is clear that all the definitions **P-B, P-C, P-D** will be acceptable provided we can prove that the cut (L, U) given in Definition **P-A** is a Dedekind cut. We give below the justification:

1. Our construction of L and U shows that every rational number is either in L or in U (i.e., $L \cup U = \mathbb{Q}$), but not in both.

L is not empty because at least all negative rational numbers are in L-class.

2. *To prove that U is not also empty.*

Since (L_1, U_1) and (L_2, U_2) are Dedekind cuts, by the important result proved above, $\exists$ rational numbers $x \in L_1$ and $y \in L_2$ s.t. $x + 1 \in U_1$, $y + 1 \in U_2$.

We can now prove $(x+1)(y+1) \in U$.

If possible, let $(x+1)(y+1) \in L$.

Then $(x+1)(y+1) = ab$, where $a \in L_1$ and $b \in L_2$ and $a, b > 0$.

But since $a \in L_1, a < x+1$ and since $b \in L_2, b < y+1$. Hence $a \cdot b < (x+1)(y+1)$, a contradiction.

$\therefore (x+1)(y+1)$ cannot belong to L. In other words, the rational number $(x+1)(y+1) \in U$ so that U-class is not certainly empty.

3. *To prove L contains no largest rational number.*

Let z be a positive rational number which belongs to L.

Then $z = xy$, where x and y are positive rational numbers s.t. $x \in L_1$ and $y \in L_2$. But $(L_1, U_1), (L_2, U_2)$ being Dedekind cuts, there exists no largest element in L_1 or in L_2. Hence, $\exists$ some positive rational number $w \in L_1$ s.t. $w > x$.

Then $wy > xy$, i.e., $wy > z$ and $wy \in L$.

Since z is arbitrary, it follows that L-class does not contain a largest element.

4. *To prove that if $p \in L, q \in U$, then $p < q$.*

If possible, let $p > q$. Then we may assume $p = qk, k > 1$.

Now since $p \in L$, $p = a \cdot b$ where $a \in L_1, b \in L_2$.

Clearly, then $\frac{a}{k} \in L_1$ and since

$q = \frac{p}{k} = \left(\frac{a}{k}\right) \cdot b$ $\left(\frac{a}{k} \in L_1, b \in L_2\right)$, $q \in L$, which is a contradiction.

Hence the conclusion is that p cannot be $> q$, i.e., $p < q$.

Conclusion: The cut (L, U) defined in **P-A** is a Dedekind cut.

(v) **Reciprocal of a non-zero Real Number:**

R-A If $\alpha = [L_1, U_1]$ be a positive real number, then the cut (L, U) where
$L = M \cup \{0\} \cup \{\frac{1}{x} : x \in U_1$ and $x \neq$ the least element of U_1, if exists$\}$; $U = \mathbb{Q} - L$
is a Dedekind cut and this cut defines the reciprocal of the real number α, denoted by α^{-1} or $\frac{1}{\alpha}$.

Suggestions for Justification:

[L has no largest element: For let $z \in L(z > 0)$, then $z = \frac{1}{x}$, $x \in U$, and $x > 0$. Now $\exists\ y < x$ and $y \in U_1$ and hence $\frac{1}{y} \in L$ and $\frac{1}{y} > z$; hence, etc.

If $z \in L$, $w \in U(z, w > 0)$, then $z < w$. For, if possible, let $z > w$. Then $z = wk(k > 1)$.

But $z \in L$ and hence $z = \frac{1}{x}$, $x \in U_1$. Clearly, xk also belongs to U_1.

Therefore, $w = \frac{z}{k} = \frac{1}{x} \cdot \frac{1}{k} \in L$, a contradiction; hence, etc.]

R-B. If α be a negative real number, then its reciprocal is defined by

$$\alpha^{-1} = -\left(\frac{1}{-\alpha}\right).$$

R-C. The reciprocal of the real number $0 = (M, P)$ is not defined.

(vi) **Division of any real number by a non-zero Real Number:** If $\alpha = (L_1, U_1)$ be any real number and $\beta = (L_2, U_2)$ be a non-zero real number, then the quotient on dividing α by β, denoted by $\alpha \div \beta$ or α/β is defined by the product of α and the reciprocal of β, i.e., $\alpha/\beta = \alpha \cdot (1/\beta)$.

Fundamental Laws connecting the aforesaid operations: Let $\alpha = (L_1, U_1)$, $\beta = (L_2, U_2)$ and $\gamma = (L_3, U_3)$ be three real numbers.

Let the real number $0 = (M, P)$ and the real number $1 = (L_4, U_4)$, where $L_4 = M \cup \{0\} \cup \{x : x \in \mathbb{Q}, x > 0$ but $x < 1\}$ and $U_4 = \mathbb{Q} - L_4$.

Then we have the following laws:

1. Addition and Multiplication obey Commutative and Associative Laws:

$$\left.\begin{array}{l}\alpha + \beta = \beta + \alpha \\ \alpha + (\beta + \gamma) = (\alpha + \beta) + \gamma\end{array}\right\} \quad \begin{array}{l}\alpha \cdot \beta = \beta \cdot \alpha \\ \alpha \cdot (\beta \cdot \gamma) = (\alpha \cdot \beta) \cdot \gamma\end{array}$$

2. For every real number α, we have $\alpha + \bar{0} = \alpha$, where $\bar{0} = (M, P)$. $\bar{0}$ is now called the *additive identity*.

3. For every real number α, we have $\alpha \cdot 1 = \alpha$, where $1 = (L_4, U_4)$. 1 is then called the *Multiplicative identity*.

4. For every real number α, we have $\alpha + (-\alpha) = \bar{0}$. $-\alpha$ is then called the *additive inverse* of α.

CHAPTER 2: THE REAL NUMBER SYSTEM

5. For any non-zero real number α, we have $\alpha \times \alpha^{-1} = 1$. α^{-1} is then called *the multiplicative inverse* of α.

6. Addition and Multiplication are connected by the *Distributive Law*:

$$\alpha \cdot (\beta + \gamma) = \alpha \cdot \beta + \alpha \cdot \gamma.$$

7. **On Inequalities:** (i) If $\alpha > \beta$, then $\alpha + \gamma > \beta + \gamma$; (ii) If $\alpha > \beta$, then $\alpha \cdot \gamma > \beta \cdot \gamma$, if $\gamma > \bar{0}$.

We prove **2, 4, 6** and **7**(i) and leave others as exercises.

Proof of 2. Let $\alpha = (L_1, U_1)$, $\bar{0} = (M, P)$ and $\alpha + \bar{0} = (L, U)$.

To prove $(L, U) = (L_1, U_1)$.

Let $x \in L$. Then $x = x_1 + x_2$, $x_1 \in L_1$, $x_2 \in M$. (Note $x_2 < 0$).

Therefore, $x_1 + x_2 < x_1$ i.e., $x < x_1$.

Now $x_1 \in L_1 \Longrightarrow x \in L_1$, i.e., *every member of L belongs to L_1*.

Again, let $y \in L_1$. Since L_1 has no largest element (by Dedekind's method of construction), $\exists$ a $w > y$ s.t. $w \in L_1$. Take $w = y + k$, where $k > 0$. Then $y = w - k = w + (-k)$, where $w \in L_1$ and $-k \in M$. Hence, $y \in L$, i.e., *every member of L_1 is a member of L*.

This proves that $L = L_1$ and hence, $U = Q - L = Q - L_1 = U_1$.

Thus $(L, U) = (L_1, U_1)$.

Proof of 4. (Existence of Additive Inverse). Let $\alpha = (L_1, U_1)$, $-\alpha = (L_2, U_2)$ and $\alpha + (-\alpha) = (L, U)$. We shall prove $(L, U) = \bar{0} = (M, P)$.

Let $z \in L$. Then $z = x + y$, where $x \in L_1$, $y \in L_2$.

By definition of $-\alpha$, we have $y = -w$, where $w \in U_1$. Since $x \in L_1$, $w \in U_1$, we conclude, $x < w$, i.e., $x < -y$ or $z = x + y < 0$, i.e., every element of L is negative, i.e., $L \subseteq M$.

Next take any negative rational number z.

Then $\exists\ x, y$ where $x \in L_1$, $y \in U_1$ s.t. $y - x = -z$ (See III, Important Theorem).

Hence, $z = x + (-y)$ $(x \in L_1, -y \in L_2) \in L$, i.e., every negative rational number belongs to $M \subseteq L$.

Hence, $M = L$, so that $(L, U) = (M, P) = \bar{0}$.

Proof of 6. Distributive Law: $\alpha \cdot (\beta + \gamma) = \alpha \cdot \beta + \alpha \cdot \gamma$.

First, suppose that α, β, γ are all positive real numbers.

Let $\alpha = (L_1, U_1)$, $\beta = (L_2, U_2)$, $\gamma = (L_3, U_3)$ and let $\alpha \cdot \beta + \alpha \cdot \gamma = (L', U')$, $\alpha \cdot (\beta + \gamma) = (L, U)$.

To prove $(L, U) = (L', U')$.

All negative rational numbers and the rational number zero are necessarily members of L as well as of L'.

The positive rational numbers of L are of the form $x_1(x_2 + x_3)$ and the positive rational numbers of L' are of the type $x_1x_2 + y_1x_3$, where x_1 and y_1 are any positive members of L_1 and x_2, x_3 are any positive members of L_2, L_3 respectively.

Since $x_1(x_2 + x_3) = x_1x_2 + y_1x_3$, if we take $y_1 = x_1$, we see that *every positive member of L is also a member of L'*.

Again any member $x_1x_2 + y_1x_3$ of L' is clearly a member of L, if $y_1 = x_1$. In general, we can however accept $x_1 > y_1$ so that $y_1/x_1 < 1$.

Now $y_1x_3 = x_1[(y_1/x_1) \cdot x_3] = x_1x_3'$, say. See that $x_3' = (y_1/x_1) \cdot x_3 < 1 \cdot x_3 = x_3$, and hence $x_3' \in L_3$.

Since $x_1x_2 + y_1x_3 = x_1x_2 + x_1x_3' = x_1(x_2 + x_3')$, we observe that *every positive member of L' is also a member of L*.

Hence, $(L, U) = (L', U')$, i.e., $\alpha \cdot (\beta + \gamma) = \alpha \cdot \beta + \alpha \cdot \gamma$, where α, β, γ are all positive.

When α, β, γ are not all positive, the proof will depend on the definition of the product of two real numbers when they are not both positive.

e.g., let α and β be negative and γ be positive and suppose $\beta + \gamma$ is positive. Then

$$\alpha \cdot (\beta + \gamma) = -[(-\alpha) \cdot (\beta + \gamma)]$$
$$= -[(-\alpha) \cdot \{-(-\beta) + \gamma\}]$$
$$= -[-\alpha \cdot \{\gamma - (-\beta)\}] \quad \text{(Commutative Law)}$$
$$= -[-\alpha \cdot \gamma - (-\alpha) \cdot (-\beta)] \quad \text{(Distributive Law proved before)}$$
$$= -[-\alpha\gamma - \alpha\beta]$$
$$= \alpha\gamma + \alpha\beta$$
$$= \alpha\beta + \alpha\gamma.$$

Proof of 7(i). Let $\alpha = (L_1, U_1)$, $\beta = (L_2, U_2)$, $\gamma = (L_3, U_3)$ and suppose that $\alpha + \gamma = (L, U)$, $\beta + \gamma = (L', U')$.

$\alpha > \beta \Longrightarrow$ every member of L_2 is a member of L_1; and hence *every member of L' is a member of L*.

Also, $\alpha > \beta \Longrightarrow \exists$ at least one element $x \in L_1$ s.t. $x \notin L_2$ and hence, $x \in U_2$. Since L_1 has no largest member, $\exists\ y > x$ and $y \in L_1$. We can find $z \in L_3$, $w \in U_3$ s.t. $w - z = \delta$, where $\delta = y - x > 0$.

(By III. Important Theorem). This gives $w + x = z + y$.

Now $y \in L_1$, $z \in L_3 \Longrightarrow y + z \in L$ and $x \in U_2$, $w \in U_3 \Longrightarrow x + w \in U'$. Thus, $\exists$ a member common to L and U', i.e., $\exists$ a rational number which belongs to L but not to L', i.e., L' is a proper subset of L, i.e., $\alpha + \gamma > \beta + \gamma$.

CHAPTER 2: THE REAL NUMBER SYSTEM

Observation: We defined Real numbers by Dedekind Cuts of Rational numbers. We then introduced two compositions or operations–addition and multiplication–on the system $\mathbb{R}$ of real numbers.

We have then verified that

(i) both addition and multiplication obey Commutative and Associative Laws;

(ii) ∃ two distinct identity elements–0 (**Additive Identity**) and 1 (**Multiplicative Identity**);

(iii) for every real number, ∃ an unique additive inverse;

(iv) for every real number except for additive identity (0), ∃ an unique multiplicative inverse;

(v) the two operations are connected by the Distributive Laws.

In the language of Abstract Algebra these facts are stated as under:

The system of real numbers R forms a field under addition and multiplication.

We have also defined an order relation ($<$) on R which obeys the Law of Trichotomy and Law of Transitivity. We have also the following two results:

(i) $\alpha < \beta \implies \alpha + \gamma < \beta + \gamma$; (ii) $\alpha < \beta, \gamma > \bar{0} \implies \alpha\gamma < \beta\gamma$.

These facts lead us to state that the set $\mathbb{R}$ of real numbers is *ordered*.

Conclusion: The set of real numbers defined by Dedekind is an *ordered field*. In fact, the set $\mathbb{Q}$ of rational numbers is also an *ordered field* and $\mathbb{Q} \subseteq \mathbb{R}$. The two sets, however, differ in respect of what we call **order-completeness**. The set $\mathbb{R}$ is an **ordered-complete field** while that set $\mathbb{Q}$ is an **ordered field** but not an ordered-complete field.

V. Order-Completeness of the Real Number System $\mathbb{R}$: We have the following theorem, known as **Dedekind's Theorem on Real Numbers** which can be stated as under:

STATEMENT: *Let R be the set of all real numbers (defined by cuts of rational numbers) and let R be decomposed into two non-empty disjoint sets A and $B (A \cup B = R)$ such that if $a \in A$ and $b \in B$, then $a < b$.*

*Then ∃ **one and only one** real number c such that $a \leq c$ for all $a \in A$, and $c \leq b$ for all $b \in B$. Furthermore, c can be taken as either the greatest element of A or the least element of B.*

Before we prove Dedekind's Theorem on real numbers we establish the following two lemmas:

Lemma 1. If $(L_1, U_1) < (L_2, U_2)$, then ∃ a rational number $b = (L, U)$ such that $(L_1, U_1) < (L, U) < (L_2, U_2)$, i.e., between any two distinct Dedekind cuts there lies at least one rational cut.

Proof. By hypothesis, L_1 is a proper subset of L_2, i.e., $\exists$ a rational number $a \in L_2$ s.t. $a \notin L_1$.

Since a cannot be the largest rational number of L_2, $\exists$ a rational number $b \in L_2$ s.t. $b > a$.

The cut $(L, U) = b$ is defined by $L = \{x : x < b \text{ and } x \in \mathbb{Q}\}$ and $U = \mathbb{Q} - L$.

Now $b \notin L$ and $b \in L_2 \implies (L, U) < (L_2, U_2)$.

Also $a \in L \cdot (a < b)$ and $a \notin L_1 \implies (L_1, U_1) < (L, U)$; hence, etc.

Lemma 2. For any cut (L_1, U_1), a rational number $r \in L_1$, iff $(L, U) < (L_1, U_1)$, where $(L, U) =$ rational cut determining r.

The proof of this lemma is simple and is left as an exercise.

Proof of Dedekind's Theorem: Either A has the largest element *or* A does not have the largest element.

Case 1. If A has a greatest member, say c, then $a \leq c$, for all $a \in A$ and since $a \in A$, it follows from the given hypothesis that $c < b$, for all $b \in B$. This proves the truth of the theorem.

Case 2. If A does not have a greatest member, then we construct a cut (L, U) of rational numbers by using the following rule:

Let $L = \{x : x \in \mathbb{Q} \text{ and also } x \in A\}$ and $U = \mathbb{Q} - L$.

To prove that (L, U) is a Dedekind cut of rational numbers.

By hypothesis, $A \neq \Phi$; hence, $\exists$ a real number $a \in A$. Suppose $a = (L_1, U_1)$. $\exists$ one rational number $r \in L_1$. Then by Lemma 2, $r < a \implies r \in L \implies L \neq \Phi$.

Since $B \neq \Phi$, $\exists$ at least one real number $b \in B$.

Let $b = (L_2, U_2)$. $\exists$ at one *rational number* $p \in U_2$.

Using Lemma **2** again, $p \geq b \implies p \in B \implies p$ is a rational number not in A.

This implies that $p \in U$ and, therefore, $U \neq \Phi$.

Moreover, U consists of all rational numbers in B because L consists of all rational numbers in A.

It follows immediately that $q \in L$ and $p \in U \implies q < p$.

Next L does not have a rational number; for if r is a rational number in L, then $r \in A$ and since A contains no largest real number, $\exists$ a real number $a \in A$ s.t. $r < a$. By Lemma 1, $\exists$ a rational number q s.t. $r < q < a \implies q \in L$, and, therefore, r cannot be the largest rational number in L. This shows that (L, U) is a cut.

Let us write $c = (L, U)$. Let a be any real number of A.

Since A does not have a largest real number (assumed), $\exists$ a real number $a' > a$ s.t. $a' \in A$. By Lemma 1, $\exists$ a rational number d s.t. $a < d < a'$; clearly, $d \in A \implies d \in L$. By Lemma 2, we then have $d < c \implies a < c$.

CHAPTER 2: THE REAL NUMBER SYSTEM

Now if $b \in B$ and r is any rational number s.t. $r < c$, then by Lemma 2, $r \in A \implies r \notin B \implies r < b$.

We have thus shown that for any $b \in B$, $r < c \implies r < b$. Thus $c \leq b$.

Moreover, $c \in B$ because c could not be in A which proves that c is the smallest element of B. We now prove the uniqueness of c.

If there were two such numbers c and c' s.t. $c \neq c'$, then either $c < c'$ or $c' < c$. In either case, $\exists$ a rational number r lying between c and c'. Suppose $c < r < c'$.

$$r > c \implies r \notin A$$
$$\text{and} \quad r < c' \implies r \notin B.$$

Therefore, we have a contradiction. Hence, $c = c'$. This completes the proof.

Observation: This theorem illustrates the most significant difference between $\mathbb{Q}$ (the set of all rationals) and $\mathbb{R}$ (the set of all real numbers). If, for example, we partition the set of all rational numbers into the two sets as follows:

$$A = \{x : x^3 < 2 \text{ and } x \in \mathbb{Q}\}$$
$$\text{and} \quad B = \mathbb{Q} - A,$$

then A and B satisfy the hypothesis of Dedekind's Theorem but fail to satisfy the conclusion. For we cannot find a rational number c, $c \geq x$ for all $x \in A$ and $c \leq y$ for all $y \in B$. However, there is a real number $\sqrt[3]{2}$, which would fill the gap in this example.

The sections of real numbers do not lead to any further generalisation of our number-system $\mathbb{R}$. The real-number system has, therefore, a kind of **completeness** which the rational number system $\mathbb{Q}$ lacks.

The system $\mathbb{R}$ of real numbers is order-complete but the system $\mathbb{Q}$ of rational numbers is not order-complete.

VI. Density Property of $\mathbb{R}$:

STATEMENT: Between any two distinct real numbers there exists an infinite number of real numbers.

In Lemma 1 above, we have already shown that $\exists$ a rational number (and hence, an infinite number) between two distinct real numbers. We now prove that *between two distinct real numbers, $\exists$ an infinite number of irrational numbers*.

Note: We shall use a bar to denote a real-rational number obtained by Dedekind cuts. No bar will be put over an ordinary rational number.

Proof. Let α and β be two distinct real numbers. Suppose $\alpha < \beta$. Then by Lemma 1 we can find two real rational numbers, say $\bar{p}$ and $\bar{q}$, such that $\alpha < \bar{p} < \bar{q} < \beta$. We shall now prove that $\exists$ an irrational real number between $\bar{p}$ and $\bar{q}$ and then the desired result will be true.

Let $\gamma = (L, U)$ be any arbitrary irrational number. Then either γ lies between $\bar{p}$ and $\bar{q}$ or γ does not lie between them.

In the latter case we prove that by adding to γ a suitable real-rational number it is possible to make this new number lie between $\bar{p}$ and $\bar{q}$.

Let $\delta = q - p > 0$, where p and q are ordinary rational numbers corresponding to $\bar{p}$ and $\bar{q}$ respectively. Then $\exists\, m \in L,\, n \in U$ s.t. $n - m < \delta$ [See **Example 10**, under **Exercises** given below.]

Now since $m \in L$, $\bar{m} < \gamma$ and since $n \in U$, $\bar{n} > \gamma$, that is $\bar{m} < \gamma < \bar{n}$. Then it follows $\bar{0} < \Gamma - \bar{m}$, i.e., $\bar{0} < \Gamma - \bar{m} < \bar{q} - \bar{p}$ ($\because \delta = q - p$ and $n - m < \delta$).

This gives $\bar{0} + \bar{p} < \bar{p} < \Gamma - \bar{m} < \bar{q}$ or $\bar{p} < \Gamma + (\bar{p} - \bar{m}) < \bar{q}$.

Note that $\gamma + (\bar{p} - \bar{m})$ is an irrational number [$\because$ sum of an irrational number and a rational number cannot be a rational number (Why?)] and it lies between $\bar{p}$ and $\bar{q}$.

EXERCISES ON CHAPTER 2: IIC
(Theoretical Questions)

1. How did Dedekind define sections of rational numbers?
   Explain the concepts of a real-rational number $\bar{3}$ and ordinary rational number 3.

2. Indicate the sections corresponding to $\sqrt{3}$, and show that the sections satisfy all the requirements of a Dedekind Section of rational numbers.

3. Define a real number according to Dedekind and state when the number so defined is irrational. If α and β be two such real numbers, when will you say that α is greater than β? And then deduce that $\sqrt{3}$ is greater than $\sqrt{2}$.

4. If α, β, γ be three real numbers defined according to Dedekind, and if $\alpha < \beta$ and $\beta < \gamma$, then prove that $\alpha < \gamma$. Deduce that $1 < \sqrt{2} < \sqrt{3}$.

5. When are two real numbers (defined according to Dedekind) said to be equal? Define (a) the real number $\bar{0}$; (b) the sum of two real numbers and hence show that if α be any real number then $\alpha + \bar{0} = \alpha$.

6. Define the negative of a real number and also the reciprocal of a non-zero real number. Next define the product and quotient of two real numbers.

7. Prove that between any two distinct real numbers there lie both rational and irrational numbers.

8. Comment on the following statement: "There are gaps between rational numbers while the arithmetic continuum is free from gaps."

9. State and prove Dedekind's Theorem on Sections of real numbers and indicate how the sections of real numbers differ from those of rational numbers so far as generalisation of number concept is concerned.

10. Let (L, U) be a Dedekind Section of $\mathbb{Q}$, and d be an ordinary positive rational number in $\mathbb{Q}$. Show that $\exists$ an $r \in L$ and $s \in U$ s.t. $s - r < d$.
    [Take a p in L and a q in U. Choose $n \in \mathbb{N}$ s.t. $(q - p)/n = t < d$. Now consider the numbers $p, p + t, p + 2t, \cdots, p + nt = q$. There would be two consecutive numbers in the above succession, with the first one in L and the second one in U; hence, etc.]

MAIN POINTS: CHAPTER 2
(On Real Number System)

1. **Peano's axioms:** Set $\mathbb{N}$ of a natural numbers. Two useful properties of $\mathbb{N}$—Principle of mathematical induction and Well-ordering principle. Set $\mathbb{Z}$ of all integers: algebraic structure and order relation defined on $\mathbb{N}$ as well as on $\mathbb{Z}$.

2. **Rational numbers:** Set $\mathbb{Q}$. Field axioms and order axioms defined $\longrightarrow \mathbb{Q}$ is an ordered field. $\mathbb{Q}$ is dense as well as Archimedean. Each member of $\mathbb{Q}$ is a decimal—either terminating or recurring. All rationals can be represented by points on a directed line but all points on the line are not rational numbers.

3. **Countable sets:** Finite or denumerable. Union of countable sets is countable. Subset of a countable set is countable $\longrightarrow \mathbb{Q}$ is a countable set.

4. **Concept of bounds of an ordered set:** Definition of infimum and supremum (glb and lub).
   $A \subset$ an ordered set S, bounded above, has its supremum in S. This is known lub-property or completeness property.

5. **Axiomatic approach** of real number system $\longrightarrow \exists$ an ordered field $\mathbb{R}$ which obeys lub-property, i.e., $\mathbb{R}$ is a complete ordered field ($\mathbb{Q}$ is not complete thought it is an ordered field). $\mathbb{R}$ is dense, Archimedean; an element of $\mathbb{R}$ can be expressed in decimals. $\mathbb{R}$ has one-to-one correspondence with points of directed line.

6. **Dedekind's approach** for the development of real number system (given in the Appendix). **Dedekind's theorem.**

AN INTRODUCTION TO
ANALYSIS

(Differential Calculus—Part I)

Chapter 3

Real Valued Functions of a Real Variable

Chapter 3

Real-Valued Functions of a Real Variable

3.1 Introduction

What is a function? What is a Real-Valued Function of a real variable?

The word *'function'* was introduced into Mathematics by the great German Mathematician (inventor of Calculus) Gottfried Leibnitz (1646–1716) who used the term primarily to refer to certain kinds of mathematical formulas, e.g.,

The volume V of a cube whose each edge has the length x units obeys the formula: $V = x^3$.

Here both x and V can assume different values. We call them variables. If we know the value of x, then we can determine V, i.e., V depends on x.

We say: *V is a funciton of x.*

We call x, the *independent variable* and V, the *dependent variable.*

Later it was realised that Leibnitz's idea of function was limited in scope and the meaning of the term *function* can be understood from more generalised concepts.

With the introduction of Set Theory mathematicians accept the following definition of a function.

Let A and B be two non-empty sets.

A function f from A to B is a rule of correspondence that assign to each element $x \in A$, a unique element $y \in B$.

We write this as $f : A \longrightarrow B$ (**read**: f maps A into B).

The set A is called **the domain** and B, **the co-domain** of f.

Functions are generally denoted by $f, g, h, F, G, H, \phi, \psi, \chi$, etc.

Note: If $f(x)$ is not defined when $x \notin A$, A is called natural domain. A domain which is not natural is called *restricted*. Normally domain means natural domain.

If $y \in B$ is associated with $x \in A$, we write $y = f(x)$ (*read: f of x*), *the value of the function* at $x \in A$. In general, every $y \in B$ need not be an $f(x)$ for some $x \in A$. The subset of B containing those $y \in B$ for which $\exists x \in A$ such that

$$f(x) = y$$

will be called the **Range** of f or **Image set of** f, denoted by $f(A)$, i.e., $f(A) = \{f(x) : x \in A\}$ and $f(A) \subseteq B$.

In the present text we shall consider the special case where both the **domain** and the **range** are members of the real number system $\mathbb{R}$.

The function where the **Range** is a set of real numbers will be called a **Real-Valued Function** and when the domain and the range are both sets of real numbers, we call it a **real-valued function** of a **real variable**. Such functions are briefly called **Real functions**.

We shall, in this chapter, discuss about various ways of specifying real functions, their classifications and graphical representations.

3.2 Real Variables: Continuous and Discrete

Let S denote a set of real numbers (i.e., $S \subseteq \mathbb{R}$). Individually each member of S is a real number with a fixed value. We say that each member of S is a **fixed constant** (or, simply, **a constant**). Let x be any *unspecified member* of S and suppose that x represents every member of S *in turn*. Then x is called a *real variable* over the *domain* S. A particular member of S which x assumes at some stage is called *the value of x*. The symbols $x, y, z, u, v, w, s, t, \cdots$ are used for real variables. Note that *with any real variable x there is associated a domain S over which it varies (i.e., assumes values)*.

Usually the domains of real variables are $(-\infty, \infty)$, $[a, b]$, (a, b), $[a, b)$, $(a, b]$ or any other subset of the real number system. If a variable x takes on all real numbers from a to b (a, b are two fixed real numbers and $a < b$), we say that x varies *continuously* from a to b : x is then called a *continuous real variable*.

A real variable need not always be continuous. Its domain may contain only discrete real numbers (e.g., the set of positive integers or the set of all rational numbers, etc.). We call such a variable a *discrete real variable*. When the domain is the set of positive integers we shall always use the symbol n (NOT x) and call n, the *positive integral variable*.

CHAPTER 3: REAL-VALUED FUNCTIONS OF A REAL VARIABLE

Geometrically, a real variable x varying continuously over the closed interval $[a, b]$, i.e., $a \leq x \leq b$ represents a segment AB on the *real axis;* the point A representing the number a and the point B, the number b. If the point P corresponds to any particular value of x, we often say that the variable point P generates the segment AB when x varies over $[a, b]$. In case x is a discrete real variable, we obtain a set of discrete points on the real axis for the domain of x.

Constants: Arbitrary and Fixed

The specified numbers like $2, -3, 5/2, \pi$, etc. are called **Fixed constants**. They remain unchanged under any set of mathematical operations/investigations. Sometimes we use symbols like $a, b, c, d, \cdots, m, n, \cdots$ to represent some numerical magnitudes which are supposed to remain fixed throughout a particular *mathematical investigation*, but for a different mathematical investigation the same symbols may represent different numerical magnitudes. We call them **arbitrary constants**, e.g., $y = mx$ represent, a **particular** straight line passing through the origin in two-dimensional Coordinate Geometry, when m is a *fixed constant* but $y = mx$ represents a family of straight lines through the origin, if m is an *arbitrary constant*. (for different numerical values of m we get different lines through the origin).

3.3 Discussions on Real-valued Functions of a Single Real Variable

We have defined a function in the following way:

A function f from a set A into a set B is a rule of correspondence that assigns to each element x in A, a uniquely determined element $f(x)$ in B.

This idea of function may be illustrated in more than one way. For example, in **Figure 3.3.1** the sets A and B are thought of as sets of points, and an arrow is used to suggest a *pairing* of a typical point x in A with its image point $f(x)$ in B.

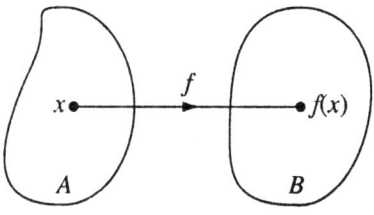

Fig 3.3.1

Another scheme is shown in **Figure 3.3.2**. Here the function f is imagined to be a transformer into which objects of the set A (inputs) are fed and objects of the set B

(outputs) are produced. When an object $x \in A$ is fed into the transformer, the output is the object $f(x) \in B$.

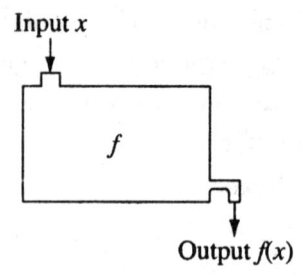

Fig 3.3.2

Real-Valued Functions of a Real Variable

We shall mainly discuss *real-valued functions of a real variable* (or more briefly, *real functions*) where the *domain* A is a set of real numbers and *range* (a subset of B) is also a set of real numbers.

Such real functions may be illustrated geometrically by **a graph** in the xy-plane. We plot the domain A on the x-axis and for each point $x \in A$ we plot the point (x, y), where $y = f(x)$. The totality of such points (x, y) is the **graph of the function**.

N.B. Often we write $D(f) =$ domain of f $(D(f) = A)$
and $R(f) =$ Range of $f (R(f) \subseteq B)$

A Formal definition of function based on Set Concept

Observation: The phrase 'rule of correspondence' which we have used in the definition of function requires clarification. We, therefore, reformulate the whole idea in a different way, based on set concept only.

DEFINITION OF FUNCTION: *Let A and B be two non-empty sets. Then a **function** from A to B is the set f of ordered pairs in $A \times B$ such that for each $a \in A$, there exists a **unique** $b \in B$ with $(a, b) \in f$. In other words, no two ordered pairs can have the same first member, i.e., $(a, b) \in f$ and $(a, b') \in f$ is not possible for $b \neq b'$.*

Note that if f is a function, the set of all elements a that occur as first members of (a, b) in f is **the domain of** f. The set of second members b is **the range of** f or **the set of values of** f.

Methods of Specifying Functions

A. Using Analytic Expressions

Most of the functions which we meet are given by some type of mathematical formulae or analytic expressions connecting the dependent variable y and the independent variable x.

CHAPTER 3: REAL-VALUED FUNCTIONS OF A REAL VARIABLE

When the function f is given by a formula like

$$y = f(x) = x^2 + 5x + 6.$$

The domain is usually the whole system of real numbers $(-\infty, \infty)$. But if f is given by a formula like

$$y = f(x) = \frac{x}{x^2 - 4},$$

then the domain of f is the the system of real numbers except $x = -2$ and $x = +2$ (because $\exists$ no real number y corresponding to these two values of x), i.e., the domain of f is

$$(-\infty, \infty) - \{-2, 2\}$$
$$\text{or,} \quad \{-\infty < x < -2\} \cup \{-2 < x < +2\} \cup \{2 < x < \infty\}.$$

We say that the function is defined for all real values of x, except for two points $x = 2$ and $x = -2$.

Note. *The definition of function does not presuppose an analytic expression connecting dependent variable (y) and independent variable (x).*

e.g., y is a function of n, where the domain is the set of all positive integers n, defined thus:

$$y = 0, \text{ if } n \text{ is prime}$$
$$= 1, \text{ if } n \text{ is composite.}$$

This function is well-defined over its domain ($\mathbb{N}$), but no analytic expression for it is known.

B. Using Tables

Often we give a tabular specification of a function by writing down the series of values of the independent variable and the corresponding values of the function (or dependent variable).

Example 3.3.1. *Suppose while studying the dependence of electrical resistance (r) of a copper rod on its temperature (t) we obtain the following Table:*

t	19.10	25.00	30.10	36.00	40.00	45.10	50.00
r	76.30	77.80	79.75	80.80	82.35	83.90	85.10

We have two variables here, namely, t, the independent variable and r, the dependent, one. The resistance r is a function of the temperature t. No analytic expression between r and t is given. The set of values of t given above is the domain of definition of the function and the set of values of r gives the range. Notice that here t is a *discrete variable*.

C. Using Graphs

A function f has been defined as a set of ordered pairs $(x, f(x))$, where $x \in \text{dom}(f)$. On the Euclidean plane each such ordered pair corresponds to a point having co-ordinates $(x, f(x))$ with reference to a suitably chosen system of rectangular axes. The collection of all such points of the Euclidean plane is called the **graph of the function** f. The two terms "the function f" and "the graph of f" practically mean the same thing—the first term has *analytical connotation* and the second a *geometrical one*. Plotting the points $(x, f(x))$ on the Euclidean plane we obtain the geometrical representation of f.

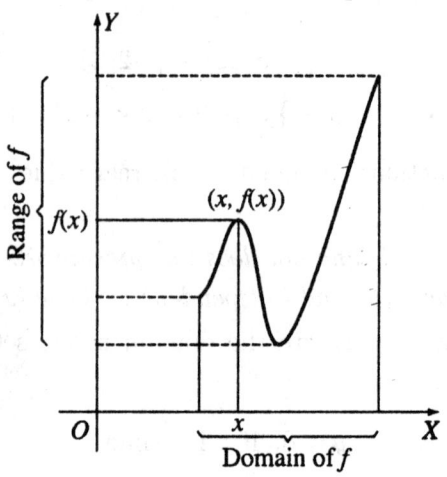

Fig. 3.3.3 Graph of f (as a subset of the Euclidean plane)

See Fig. 3.3.3, where the graph of f as a subset of the Euclidean plane has been shown (domain and range both are taken continuous here).

Observation. It should be observed that a function f assigns a unique value y to each independent variable x (also called the **argument** x) means that the graph of f meets each line parallel to the Y-axis in at most one point.

Again a function f is one-to-one iff the graph meets each line parallel to the X-axis in at most one point.

3.4 Illustrative Examples

Example 3.4.1. *Given a function f defined by $f(x) = (x+1)/(x-1)$. Find the expressions for $f(3x), 3f(x), f(x^3), [f(x)]^3$.*

CHAPTER 3: REAL-VALUED FUNCTIONS OF A REAL VARIABLE

Solution:

$$f(3x) = \frac{3x+1}{3x-1}, \quad 3f(x) = 3\frac{x+1}{x-1} = \frac{3x+3}{x-1}$$

$$f(x^3) = \frac{x^3+1}{x^3-1}, \quad \{f(x)\}^3 = \left(\frac{x-1}{x+1}\right)^3.$$

Example 3.4.2. *A function ϕ is defined over $(-\infty, \infty)$ given by*

$$\phi(x) = \begin{cases} 2x^3 + 1, & \text{if } x \leq 2 \\ \frac{1}{x-2}, & \text{if } 2 < x \leq 3 \\ 2x - 5, & \text{if } x > 3. \end{cases}$$

Find the values of $\phi(\sqrt{2})$, $\phi(2\sqrt{2})$, $\phi\left(\sqrt{\log_2 1024}\right)$.

Solution:

$$\phi(\sqrt{2}) = 2(\sqrt{2})^3 + 1 \quad (\because x = \sqrt{2} < 2)$$
$$= 4\sqrt{2} + 1.$$

$$\phi(2\sqrt{2}) = \frac{1}{2\sqrt{2} - 2} \quad (\because 2\sqrt{2} > 2, \text{ but } 2\sqrt{2} < 3)$$

$$\frac{1}{2(\sqrt{2}-1)} = \frac{\sqrt{2}+1}{2(\sqrt{2}-1)(\sqrt{2}+1)} = \frac{1}{2}(\sqrt{2}+1).$$

$$\log_2 1024 = \log_2 2^{10} = 10 \log_2 2 = 10; \quad \sqrt{\log_2 1024} = \sqrt{10} > 3.$$

$$\therefore \quad \phi(\sqrt{\log_2 1024}) = \phi(\sqrt{10}) = 2 \cdot \sqrt{10} - 5.$$

Example 3.4.3. *If $f(x) = x + 1, \phi(x) = x - 2$, then solve the equation:*

$$|f(x) + \phi(x)| = |f(x)| + |\phi(x)|.$$

Solution: The relation $|(x+1) + (x-2)| = |x+1| + |x-2|$ is valid, iff both $x + 1$ and $x - 2$ have the same signs.

Both ≥ 0, if $x > 2$ and both ≤ 0, if $x < -1$.

Hence, the equation is satisfied by those real values of x for which $x \leq -1$ or for which $x \geq 2$.

Example 3.4.4. *Find a function of the form*

$$f(x) = ax^2 + bx + c,$$

if it is known that $f(0) = 5, f(-1) = 10, f(1) = 6$.

Solution:

$$f(0) = a \cdot 0 + b \cdot 0 + c = 5 \text{ (given)} \quad \text{i.e.,} \quad c = 5.$$
$$f(-1) = a \cdot (-1)^2 + b \cdot (-1) + c = a - b + 5 = 10 \text{ (given)}, \quad \text{i.e.,} \quad a - b = 5.$$
$$f(1) = a \cdot (1)^2 + b \cdot (1) + c = a + b + 5 = 6 \text{ (given)}, \quad \text{i.e.,} \quad a + b = 1.$$

Solving $a - b = 5$, $a + b = 1$, we get $a = 3, b = -2$.

$$\therefore \quad f(x) = 3x^2 - 2x + 5.$$

Example 3.4.5. *Find the natural domain of definition of the real-valued function defined by*
(i) $f(x) = \sqrt{x-2} + \sqrt{4-x}$;
(ii) $f(x) = \frac{1}{\sqrt{|x|-x}}$.

Solution: (i) $f(x)$ takes on real values, if

$$x - 2 \geq 0 \text{ as well as } 4 - x \geq 0.$$

This will happen if $x \geq 2$ and $x \leq 4$.
$\therefore$ the natural domain of definition of f will be $[2, 4]$.
(ii) $f(x)$ has real value if $|x| > x$, i.e., when $x < 0$.
$\therefore$ the natural domain of definition of f will be $(-\infty, 0)$.

Example 3.4.6. *A constant function*

A function whose range consists of a single number c is called a constant function: $f(x) = c \ \forall x \in (-\infty, \infty)$.

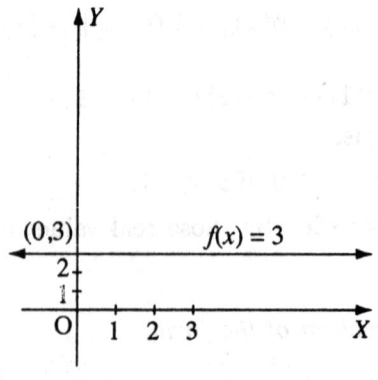

Fig 3.4.1 The constant function $f(x) = 3$

An example is shown in Figure 3.4.1, where $f(x) = 3$ for every real x. The domain is $(-\infty, \infty)$ and range is a singleton set $\{3\}$. The graph is a horizontal line (i.e., parallel to X-axis) cutting the Y-axis at the point $(0, 3)$. When $c = 0$, i.e., when $f(x) = 0$, the graph is the X-axis itself.

Example 3.4.7. *The Identity Function*

Suppose that $f(x) = x$, $x \in (-\infty, \infty)$. This function is called the *identity function*. Its *domain* is the Real line (X-axis), i.e., the set of all real numbers $(-\infty, \infty)$.

Here $x = y$ for each point (x, y) on the graph of f.

The graph is a straight line through the origin making equal angles with the co-ordinate axis (**Figure 3.4.2**). *The range* of f is also the set of all real numbers $(-\infty, \infty)$.

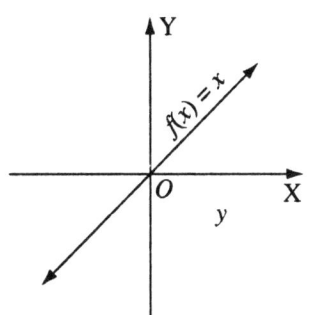

Fig 3.4.2 Identity Function $f(x) = x$

Example 3.4.8. *The Absolute-value function* or *The Modulus function*

$y = f(x) = |x|$ (**or** $y = x$, if $x > 0, y = -x$, if $x < 0, y = 0$, if $x = 0$).

This function assigns to each real number x, the non-negative number $|x|$. The graph of this function consists of two half-lines $y = x$ and $y = -x$ bisecting the axes and meeting at the origin (**Figure 3.4.3**). The points of the graph lie on the first and the second quadrants only; for when $x > 0, y = x(> 0)$ and when $x < 0, y = -x(> 0)$.

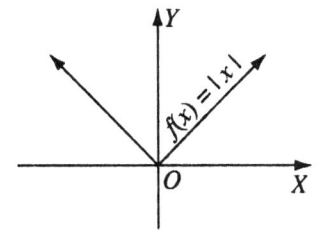

Fig 3.4.3 Absolute-Value Function: $f(x) = |x|$

We give below some properties of absolute-value functon $f(x) = |x|$:

(i) $f(-x) = f(x)$ (Even function).
(ii) $f(x^2) = x^2$.
(iii) $f(x) = +\sqrt{x^2}$.
(iv) $f(x+y) \leq f(x) + f(y)$ (Triangle inequality).

The Prime-number function (an example of step-function)

For any $x > 0$, let $\phi(x)$ be the *number of primes* less than or equal to x.

The domain of ϕ is the set of all positive real numbers.

Its range is the set of all non-negative integers 0, 1, 2, 3, ...

Check the following results:

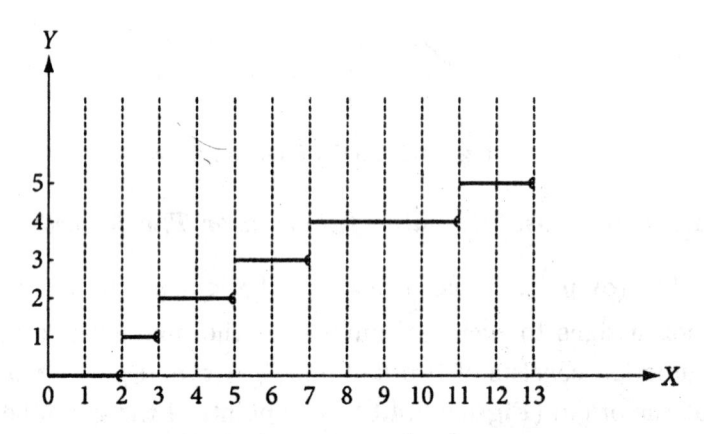

$$0 < x < 2, \quad \phi(x) = 0$$
$$2 \leq x < 3, \quad \phi(x) = 1$$
$$3 \leq x < 5, \quad \phi(x) = 2$$
$$5 \leq x < 7, \quad \phi(x) = 3$$
$$7 \leq x < 11, \quad \phi(x) = 4$$
$$11 \leq x < 13, \quad \phi(x) = 5 \text{ etc.}$$

Fig 3.4.4 Prime-number funtion

The graph of ϕ, as shown in **Figure 3.4.4**, consists of horizontal line segments. As x increases, the values of $\phi(x)$ remains constant until x reaches a prime, at which point the functional value increases by 1.

This is an example of a class of function called **step-functions**.

A familiar example of a step function is the "*Postage Function*".

Postal charges for a letter on ordinary mail when the weight does not exceed 20 gm is Rs. 5.00 and for every additional 20 gm or fractions thereof the extra charges are Rs. 5.00.

We define $P(x) =$ total postal charges for the weight of x gms.

Then $0 < x \leq 20$, $P(x) = $ Rs. 5
$20 < x \leq 40$, $P(x) = $ Rs. 10
$40 < x \leq 60$, $P(x) = $ Rs. 15 and so on.

The domain of $P(x)$ is the set of all positive real numbers.

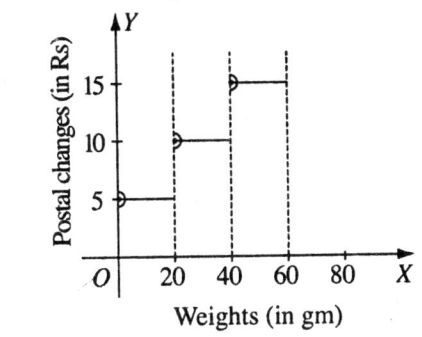

Fig 3.4.5 The Postage Function
(A Step function)

The graph of $y = P(x)$ (see **Fig 3.4.5**) consists of horizontal line segments (left end point of each segment excluded).

Note: Fare function for Fare versus distance may be constructed in the same way as step functions.

Example 3.4.9. *The Signum Function*

The *signum function* f is defined by

$$f(x) = \operatorname{sgn} x = \begin{cases} 1, & \text{if } x > 0 \\ 0, & \text{if } x = 0 \\ -1, & \text{if } x < 0. \end{cases}$$

Another way of writing this function is

$$\operatorname{sgn} x = \begin{cases} \frac{|x|}{x}, & \text{if } x \neq 0 \\ 0, & \text{if } x = 0. \end{cases}$$

An analytic form of this function is given by

$$\operatorname{sgn} x = \frac{2}{\pi} \lim_{n \to \infty} \tan^{-1}(nx).$$

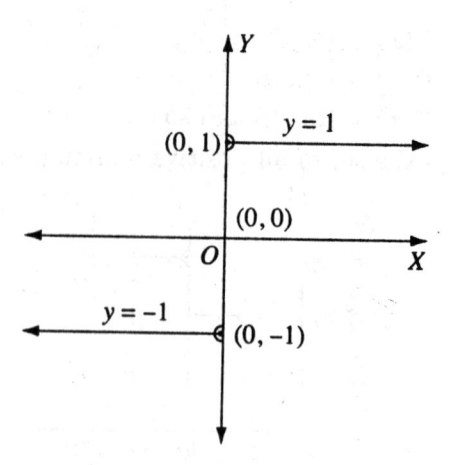

Fig 3.4.6 The Signum Function

The graph of the signum function has been shown in **Fig 3.4.6**.

Note that the domain is the whole real line $(-\infty, \infty)$ and the range consists of three discrete points $-1, 0, 1$.

See that the graph consists of three parts:

(i) the portion of the line $y = 1$ that lies in the first quadrant excluding the point $(0, 1)$;

(ii) the protion of the line $y = -1$ that lies in the third quadrant excluding the point $(0, -1)$

and (iii) the point $(0, 0)$.

Example 3.4.10. *The Greatest integer function:* $y = f(x) = [x]$

We consider a function f defined for all real x such that $f(x) = [x]$ = greatest integer $\leq x$. This is one of the frequently used example of *step function*.

For every real x, $\exists$ an integer n such that $n - 1 \leq x < n$.

$$\therefore [x] = n - 1 \quad \text{i.e.,} \quad \cdots$$
i.e., $\cdots \quad \cdots \quad \cdots$
$\cdots \quad \cdots \quad \cdots$
$-3 \leq x < -2, \quad f(x) = -3$
$-2 \leq x < -1, \quad f(x) = -2$
$-1 \leq x < -0, \quad f(x) = -1$
$0 \leq x < -1, \quad f(x) = 0$
$1 \leq x < -2, \quad f(x) = 1$
$2 \leq x < -3, \quad f(x) = 2$
$\cdots\cdots\cdots\cdots \quad \cdots\cdots\cdots$ etc.

The graph of this function is shown in **Fig. 3.4.7**. We obtain segments of parallel lines (all parallel to X-axis). Each line-segment including its **left-end** point (excluding the **right-end** point) belongs to the graph. All these line-segments lie on the first or on the third quadrant. See that these line-segments look like steps (hence the name *step-function*).

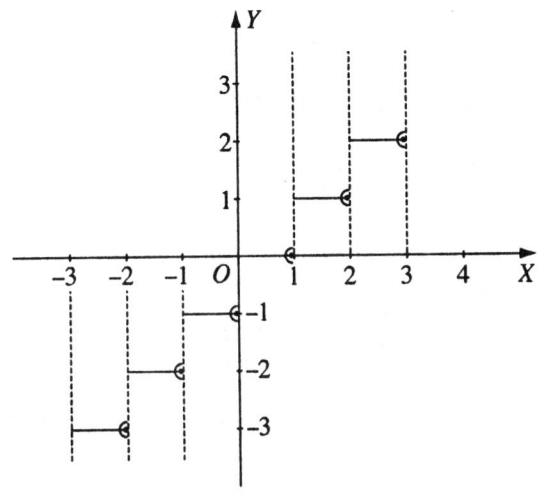

Fig 3.4.7 Greatest Integer Function

Example 3.4.11. *Dirichlet's Function:*

$$y = f(x) = \begin{cases} 1, & \text{when } x \text{ is rational} \\ 0, & \text{when } x \text{ is irrational.} \end{cases}$$

The function f defined above has its domain $(-\infty, \infty)$. The values of f *jump infinitely often* from 1 to 0 and back, in any interval of x, however small. This function has an analytic form, namely,

$$y = f(x) = 1 - \lim_{n \to \infty} sgn\,(\sin^2 n!\pi x),$$

but cannot be represented by a graph on the Euclidean plane.

A Function of Elementary Geometry

Example 3.4.12. *A rectangle with altitude x is inscribed in a triangle ABC with the base b and altitude h. Express the perimeter P and area S of the rectangle as functions of x.*

Solution: See **Fig. 3.4.8**: $LMNP$ is a rectangle of height x and base LM inscribed in a triangle of base $b(=BC)$ and height $h(=AD)$.

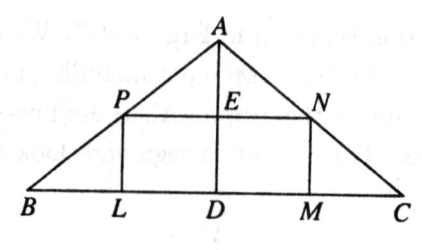

From the similar triangles APE and ABD, we get

$$\frac{AE}{PE} = \frac{AD}{BD}, \text{ i.e., } PE = \frac{(AE)(BD)}{AD} = \frac{(h-x)BD}{h}.$$

Again, from similar triangles AEN and ADC, we get

$$\frac{AE}{EN} = \frac{AD}{DC}, \text{ i.e., } EN = \frac{(h-x)}{h}DC.$$

$\therefore\ PE + EN = PN = \dfrac{h-x}{h}(BD + DC) = \dfrac{h-x}{h}b\ (\because BC = b).$

$\therefore\ $ The perimeter P of $PLMN = 2(x + PN) = 2\left(x + \dfrac{h-x}{h}b\right)$

i.e., $\ P(x) = 2b + 2x - \dfrac{2bx}{h}$

and Area A of $PLMN = x.PN = x\left(\dfrac{h-x}{h}\right)b = b\left(1 - \dfrac{x}{h}\right)x,$

i.e., $A(x) = b\left(1 - \dfrac{x}{h}\right)x.$

3.5 Different Classes of Functions: Their Graphical Representations

I. Power Functions: $y = f(x) = x^n$, for any natural number n.

[*Remember the Inductive Definition:* $x^0 = 1,\ x^1 = x^0 \cdot x,\ x^2 = x^1 \cdot x = x \cdot x, \cdots\cdots,$ $x^{n+1} = x^n \cdot x,\ \cdots$]

The domain of definition of any such power function is $(-\infty, \infty)$.

Since $x^0 = 1\ \forall$ real $x \neq 0$, the function $y = f(x) = x^0$ is a *constant function* and its graph is parallel to the X-axis with a discontinuity at the origin.

Example 3.5.1. *The graphs of odd powers* $y = x,\ y = x^3,\ y = x^5$ *are shown in* Fig. 3.5.1.

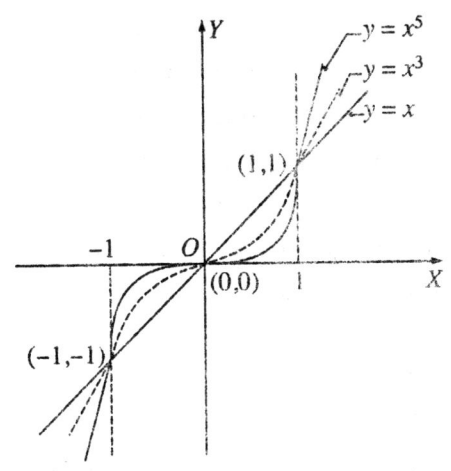

CHAPTER 3: REAL-VALUED FUNCTIONS OF A REAL VARIABLE

Fig 3.5.1 Graphs of odd powers: $y = x, y = x^3, y = x^5$.

Solution: (i) **In** each case the graph consists of the points $(0,0), (1,1)$ and $(-1,-1)$.

(ii) No point of the graph lies in the second or in the fourth quadrant.

($\because$ y is positive, when x is positive and y is negative, when x is negative.)

(iii) In $-\infty < x < 0$, as x increases in numerical value, the values of y also increase in numerical value (y always remaining negative).

In $0 < x < \infty$, as x increases, y also increases (y remaining positive).

(iv) As n increases the graph of $y = x^{2n+1}$ **approximates more and more closely** to the **step-shape** indicated by dotted lines in Fig. 3.5.1. The step is given by two vertical half-lines $x = 1, y \geq 0$ and $x = -1, y \leq 0$ and the segment of the X-axis $-1 < x < 1$.

Example 3.5.2. *The graphs of even powers* $y = x^2$, $y = x^4$, $y = x^6$ *are shown in* **Fig. 3.5.2.**

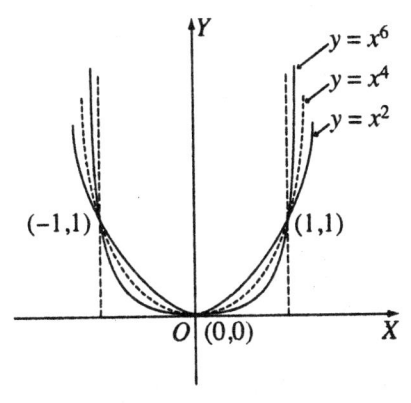

Fig 3.5.2 Graphs of even powers: $y = x^2, y = x^4, y = x^6$.

Solution: (i) In each case, the graph contains the points $(0,0)$, $(1,1)$ and $(-1,1)$.

(ii) No point of any of these graphs lies in the 3rd or in the 4th quadrant. ($\because$ y is positive when both $x > 0$ and $x < 0$).

(iii) In $-\infty < x < 0$, y decreases as x increases.
In $0 < x < \infty$, y increases as x increases.

(iv) As n increases, the graph of $y = x^{2n}$ approximates more and more closely to the U-shaped dotted lines consisting of two vertical half-lines $x = \pm 1, y \geq 0$ together with the segment $-1 < x < 1$ of the X-axis.

II. Reciprocals of Power Functions

$y = f(x) = x^{-n} = \frac{1}{x^n}$; for all natural numbers $n \geq 1$.

The domain of definition of any such function is

$$\{-\infty < x < 0\} \cup \{0 < x < \infty\}$$
i.e., $\{-\infty < x < \infty\} - \{0\}$.

Example 3.5.3. *The graphs of $y = x^{-n}$ for odd natural numbers n.*

Solution: The graphs of $y = x^{-1}$ and $y = x^{-3}$ are shown in **Fig. 3.5.3**.

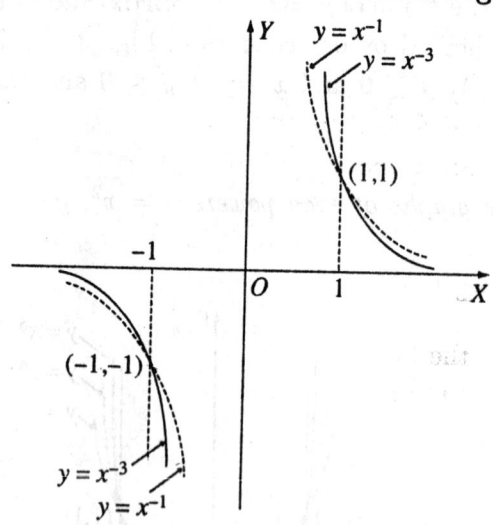

Fig 3.5.3 Graphs of $y = 1/x, y = 1/x^3$.

(i) y is not defined at $x = 0$.

(ii) The graph contains the points $(1,1)$ and $(-1,-1)$.

(iii) All points lie either in the first or in the third quadrant.

(iv) In $-\infty < x < 0$, y decreases as x increases. In $0 < x < \infty$, y again decreases as x increases.

(v) As n increases, the graphs approximate to the shape indicated by dotted lines of **Fig. 3.5.3**.

Example 3.5.4. *The graphs of $y = x^{-n}$ for even natural numbers n.*

Solution: The graphs of $y = x^{-2}$ and $y = x^{-4}$ are shown in **Fig. 3.5.4**.

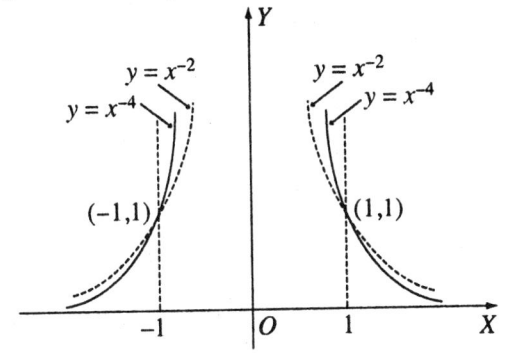

Fig 3.5.4 Graphs of $y = 1/x^2$, $y = 1/x^4$.

(i) y is undefined at $x = 0$.
(ii) The graph contains $(1, 1)$ and $(-1, 1)$.
(iii) All points of the graph lie either in the first quadrant or in the second quadrant.
(iv) As x increases in $-\infty < x < 0$, y increases.
As x increases in $0 < x < \infty$, y decreases.

(v) As n increases, the graphs approximate to the shape indicated by the dotted lines of **Fig. 3.5.4**.

III. Linear Functions

For any real number a, the function f with domain $\mathbb{R}$ given by $f(x) = ax$, $x \in \mathbb{R}$ is called a **linear function** or a **linear mapping** from $\mathbb{R}$ to $\mathbb{R}$.

> **Remember:** Any linear function from $\mathbb{R}$ into $\mathbb{R}$ satisfies the linearity relations $f(x + y) = f(x) + f(y)$; $f(\lambda x) = \lambda f(x)$ for all real x, y and λ and conversely, any function f from $\mathbb{R}$ into $\mathbb{R}$ which satisfies at least the **second** of these relations is linear, since it is necessarily of the form $f(x) = ax$ with $a = f(1)$. Hence the linear function from $\mathbb{R}$ into $\mathbb{R}$ are characterised by the linearity relations (and indeed by the second linearity relation alone). The two relations can be put in the single for $f(cx + dy) = cf(x) + df(y) \; \forall \; x, y \in \mathbb{R}$ and for all real scalars c, d.

IV. Polynomial Functions

For any natural number n and any real numbers $a_0, a_1, a_2, \cdots, a_n$ with $a_0 \neq 0$ the function f given by

$$y = f(x) = a_0 x^n + a_1 x^{n-1} + a_2 x^{n-2} + \cdots\cdots + a_{n-1} x + a_n \quad (a_0 \neq 0)$$

is called a *polynomial (function)* of nth degree.

Special Case: $f(x) = 0 \; \forall \, x \in \mathbb{R}$, is a polynomial but we do not assign it a degree. $f(x) = c \, (\neq 0), \; \forall \, x \in \mathbb{R}$ is a polynomial of degree zero.

The domain of any polynomial function is $-\infty < x < \infty$ (Set of all real numbers).

(i) The graph of $y = a_0 x + a_1 (a_0 \neq 0, a_1 \neq 0)$ is a line not parallel to either axis and we do call it a **linear function** though this function does not satisfy the linearity relations stated above. Here we use the term linear as its degree is one. The name linear function is usually reserved for functions which obey linearity relations.

(ii) **Quadratic function:** $f(x) = a_0 x^2 + a_1 x + a_2 \, (a_0 \neq 0)$. (*Parabolic curve*)

Its graph is a **parabola** with the axis parallel to Y-axis and with *vertex lowest point* and concavity upwards, if $a_0 > 0$ and *with vertex highest point* and concavity downwards, if $a_0 < 0$.

Example 3.5.5. $y = -x^2 + 3x + 2 = -\left[x^2 - 2 \cdot \frac{3}{2} \cdot x + \frac{9}{4}\right] + 2 + \frac{9}{4} = -\left(x - \frac{3}{2}\right)^2 + \frac{17}{4}$, i.e., $y - 17/4 = -(x - 3/2)^2$ or, $X^2 = -Y$, where $X = x - 3/2, Y = y - 17/4$.

Transfer the origin to $(3/2, 17/4)$, which is the vertex and the concavity is downwards, the axis is parallel to Y-axis. See **Fig. 3.5.5**.

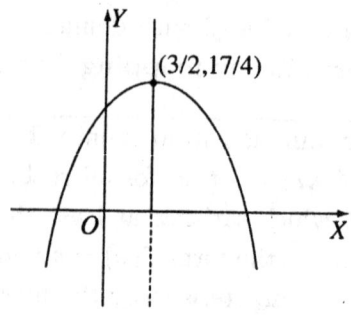

Fig 3.5.5 Graphs of the parabola $y = -x^2 + 3x + 2$.

V. Rational Functions

The quotient of two polynomials is called a *Rational Function*, i.e., f is a rational function if it is of the form $f(x) = g(x)/h(x)$, where g and h are polynomials, not necessarily of the same degree.

Since $g(x)/h(x)$ is uniquely determined whenever $h(x) \neq 0$ and is undefined when $h(x) = 0$, the domain of a rational function g/h is *the set of all real numbers* excepting those x for which $h(x) = 0$.

Note that *the class of rational functions contains the class of polynomials*, since in our definition of f we can take h to be a constant function, and then f becomes a polynomial.

We have already considered special rational functions $1/x, 1/x^2, 1/x^3$, etc.

VI. Explicit and Implicit Functions

The division of functions into explicit and implicit usually refers to functions specified analytically:

We may express y directly in terms of x (*the argument*) by an analytical expression in x. We then call y an *explicit function of* x. (e.g., $y = x^2, y = x + \sqrt{x}, y = x - 3$, etc.)

However, a relation between two variables (say, x and y) which is solved for either of them, can define more than one functions. For example, from the relation $\frac{x^2}{a^2} + \frac{y^2}{b^2} = 1$, y can be regarded as two functions of x.

In fact, for each x in $(-a, a)$ there are two values of y. We say that the relation $\frac{x^2}{a^2} + \frac{y^2}{b^2} = 1$ is an *implicit relation* and it defines two functions (solving for y, say)

$$y = +\frac{b}{a}\sqrt{a^2 - x^2} \quad \text{and} \quad y = -\frac{b}{a}\sqrt{a^2 - x^2}.$$

Note: In fact, if we are given an equation of the form $F(x, y) = 0$, this does not necessarily represent a function (for example, $x^2 + y^2 - 7 = 0$). The equation $F(x, y) = 0$ always represents a relation, viz., the set of all pairs (x, y) which satisfy the equation $F(x, y) = 0$. Under some suitable condition $F(x, y) = 0$ represents a function.

Examples of implicit functions:

(i) $x^3 + y^3 = 3axy$; (ii) $xy - 2^x + 2^y = 0$; (iii) $x^2 + y^2 + 2gx + 2fy + c = 0$.

In each case they are relations and they may determine more than one function of x.

VII. Parametric Functions: $x = f(t), y = \phi(t)$ (t is the parameter)

If both dependent variable (y) and the independent variable (x) are expressed as a function of a third variable t (or θ), we say that the function has been represented parametrically.

This third variable (t or θ) is called a **parameter.**

e.g., $x = at^2, y = 2at$ represent parametrically $y^2 = 4ax$ (a parabola).

VIII. Odd and Even Functions

(i) A function $f : A \to B$ is said to be an odd function

$$\text{if } f(-x) = -f(x), \forall\ x \in A.$$

(ii) A function $f : A \to B$ is said to be an even function

$$\text{if } f(-x) = f(x), \forall\ x \in A.$$

Examples of odd functions are the odd powers of $x (y = x^{2n+1})$ and their reciprocals ($y = 1/x^{2n+1}$), and the Trigonometric functions $y = \sin x, y = \tan x$.

Examples of even functions are the even powers of $x (y = x^{2n})$ and their reciprocals ($y = 1/x^{2n}$), and the Trigonometric functions $y = \cos x, y = \sec x$; also the absolute value function $f(x) = |x|$. Domain of an odd or even function is $(-a, a), a \in \mathbb{R}$, i.e., the domain is symmetric about origin.

Geometric Significance: *Symmetry of the graph about the Y-axis* and *Symmetry of the graph about the origin.*

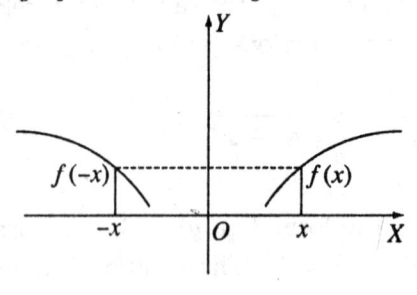

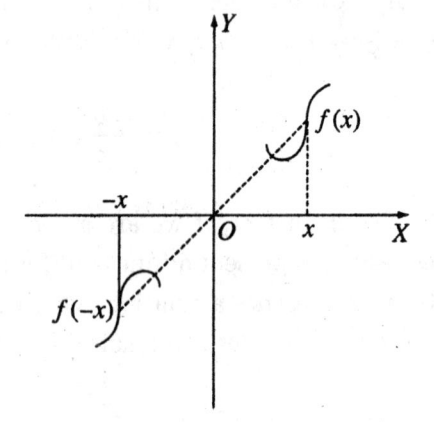

Fig 3.5.6a Symmetry about Y-axis: Even function

Fig 3.5.6b Symmetry about the origin: Odd function

Symmetry of the graph of f about the Y-axis (**Fig. 3.5.6a**) means that

$$(x, y) \in f, \text{ iff } (-x, y) \in f \quad (Even\ function\ f(x) = f(-x) = y).$$

Symmetry of the graph of f about the origin (**Fig. 3.5.6b**) means that
$(x, y) \in f$, iff $(-x, -y) \in f$ (*Odd function* $f(x) = -f(-x)$).

Remember:

A function f is even, if whenever $x \in$ dom. (f), then $-x \in$ dom. (f) and $f(x) = f(-x)$.

A function f is odd, if whenever $x \in$ dom. f, then $-x \in$ dom. f and $f(-x) = -f(-x)$.

[dom. (f) stands for the domain of f]

Examples of odd and even functions

Example 3.5.6

(i) $f(x) = \log(x + \sqrt{1 + x^2})$ is an odd function; for

$$\begin{aligned} f(x) + f(-x) &= \log(x + \sqrt{1 + x^2}) + \log(-x + \sqrt{1 + x^2}) \\ &= \log(1 + x^2 - x^2) = \log 1 = 0, \end{aligned}$$

hence $f(-x) = -f(-x)$, $\forall x \in$ dom f and the function is odd.

(ii) $\phi(x) = \log_e \frac{(1-x)}{(1+x)}$ is also an odd function, because

$$\phi(-x) = \log_e \frac{1+x}{1-x} = -\log_e \frac{1-x}{1+x} = -\phi(x), \ \forall x \in dom \ \phi$$

(iii) $g(x) = 4 - 2x^4 + \sin^2 x$ is clearly an even function,
since $g(-x) = 4 - 2(-x)^4 + \{\sin(-x)\}^2 = 4 - 2x^4 + \sin^2 x = g(x)$.

(iv) $\theta(x) = \sin x + \cos x$ is neither odd nor even.

(v) $h(x) =$ constant c (a *constant function*) is always even ($\because h(-x) = h(x) = c$)

IX. Partition and Step function

Suppose $[a, b]$ is a given closed interval. We decompose $[a, b]$ into n non-overlapping sub-intervals by inserting $n-1$ points of subdivision, say $x_1, x_2, x_3, \cdots \cdots, x_{n-1}$ subject to the restriction

$$a < x_1 < x_2 < \cdots \cdots < x_{n-1} < b. \tag{1}$$

It is found convenient to denote the point a itself by x_0 and the point b by x_n.

A collection of points $a = x_0, x_1, x_2, \cdots, x_{n-1}, x_n = b$ satisfying (1) is called a partition P of $[a, b]$ and we write

$$P = \{x_0, x_1, x_2, \cdots \cdots, x_{n-1}, x_n\}$$

to denote this partition. The partition P determines n closed non-overlapping sub-intervals: $[x_0, x_1], [x_1, x_2], [x_2, x_3], \cdots, [x_{n-1}, x_n]$.

A typical closed interval $[x_{k-1}, x_k]$ is called the kth closed sub-interval of P. The corresponding open interval (x_{k-1}, x_k) is called the kth open sub-interval of P.

With these ideas in mind we can now define what we call a **Step function** (also called *piecewise constant function*).

A function f, whose domain is a closed interval $[a, b]$, is called a *step function*, if there is a partition
$$P = \{x_0, x_1, \cdots, x_n\}$$
of $[a, b]$ such that f is constant on each half open sub-interval of P, i.e., for each $k = 1, 2, 3, \cdots, n$ there exists a real number a_k, such that

$$f(x) = a_k, \quad \text{if} \quad x_{k-1} \leq x < x_k.$$

We have already given some examples of step functions:

See **Ex. 3.4.9**, **Ex. 3.4.10**.

X. Real Functions: Injective, Surjective and Bijective

In **chapter 1** (**art. 1.6**) we defined **Injective**, **Surjective** and **Bijective** Functions in a general set up. In the present paragraph, we define such funcitons taking domain and range both subsets of $\mathbb{R}$ (Real valued functin of a real variable).

Definitions: (a) Let $D \subseteq \mathbb{R}$.

Suppose that f maps D to $\mathbb{R}$ (i.e., $f : D \longrightarrow \mathbb{R}$). D is called domain and $\mathbb{R}$ is the co-domain of the function.

Then f is said to be **injective** (i.e., **one-to-one**) if distinct members of the domain have distinct images in the codomain, i.e., whenever $x_1, x_2 \in D$ and $x_1 \neq x_2$, then $f(x_1) \neq f(x_2)$.

If f is an injective function, we also say that f is an **injection**.

[*] In order to prove that a function f is injective, we must establish that
$\forall x_1, x_2 \in D$, if $f(x_1) = f(x_2)$, then $x_1 = x_2$, i.e., assume that $f(x_1) = f(x_2)$ and show that $x_1 = x_2$.

If we consider the **graph of** f, then every horizontal line $y = b$ will intersect the graph f in **at the most one point**.

Definition: (b) Let $D \subseteq \mathbb{R}$ and $f : D \to \mathbb{R}$. The function f is called surjective if $f(D) = \mathbb{R}$, where $f(D) = \{f(x) : x \in D\}$, the range of f. That is a function is surjective if the range $f(D)$ becomes identical with the co-domain $\mathbb{R}$.

CHAPTER 3: REAL-VALUED FUNCTIONS OF A REAL VARIABLE

> [*] To prove that a function f is surjective, we must show that for any member $b \in$ codomain of f there exists at least one $x \in A$ such that $f(x) = b$.

If we consider the graph of f, then every horizontal line $y = b$ with $b \in$ codomain of f intersects the graph of f in at **least one point**. We give an example:

Example 3.5.7. Let $D = \{x : x \in \mathbb{R} \text{ and } x \neq 2\}$ and Range $(f) \subseteq \mathbb{R}$.
Suppose f maps D to Range (f) by the rule $f(x) = \frac{x}{x-2}, \forall x \in D$.
We shall establish that f is a bijection of D onto Range (f).

To show that f is injective, we take x_1 and x_2 in D and assume that $f(x_1) = f(x_2)$.
Thus we have $\frac{x_1}{x_1-2} = \frac{x_2}{x_2-2}$, which implies

$$x_1(x_2 - 2) = x_2(x_1 - 2), \quad \text{i.e.,} \quad x_1 = x_2.$$

Therefore, f is injective.

To show that f is surjective, we first determine the Range (f).

For that purpose we solve the equation $y = \frac{x}{x-2}$ for x in terms of y. We obtain $x = \frac{2y}{y-1}$ which gives real values of x, if $y \neq 1$. Thus the range of f is the set of all real values of y except $y = 1$. Thus each value of the range $\exists$ a well-defined x, i.e., f is surjective.

Hence, finally, we conclude that **f is a bijection of D ONTO Range (f)**.

XI. Equal Functions: Restriction function: Composite function

A. Equal Functions:

Two functions f and g are equal, if and only if
(a) f and g have the same domain
and (b) $f(x) = g(x)$ for each $x \in$ domain of f (or g).

Example 3.5.8. $f(x) = |x|, x \in \mathbb{R}$ and $g(x) = x$, when $x > 0$
$\qquad\qquad\qquad\qquad\qquad\qquad\quad = -x$, when $x < 0$
$\qquad\qquad\qquad\qquad\qquad\qquad\quad = 0$, when $x = 0$.

Here f and g have same domain $\mathbb{R}$ and $f(x) = g(x) \forall x \in \mathbb{R}$.
Therefore, $f = g$.

But $f(x) = \sqrt{x(x-2)}$ and $g(x) = \sqrt{x} \cdot \sqrt{x-2}$ have not the same domain (for all $x \leq 0$ $f(x)$ is defined, but for all $x \leq 0$ $g(x)$ is not defined) and, therefore, $f \neq g$.

B. Restriction Function

Let $f : D \longrightarrow \mathbb{R}$, where $D \subset \mathbb{R}$. Suppose that D_1 is a non-empty subset of D. Then we can define a function $f_1 : D_1 \longrightarrow \mathbb{R}$ by the rule $f_1(x) = f(x) \forall x \in D_1$.

This function f_1 is called **the restriction of f to D_1**.

Sometimes it is denoted by $f_1 = f/D_1$.

Example 3.5.9. (a) *Let $f : \mathbb{R} \longrightarrow \mathbb{R}$ be defined by $f(x) = \operatorname{sgn} x$, $x \in \mathbb{R}$. (See* **Ex. 3.4.9**).

Let $D_1 = \{x : x \in \mathbb{R} \text{ and } x > 0\}$. Then the restriction function f/D_1 is defined by $f/D_1(x) = 1$, $x > 0$.

Let $D_2 = \{x : x \in \mathbb{R} \text{ and } x < 0\}$. Then the restriction function f/D_2 is defined by $f/D_2(x) = -1$, $x < 0$.

(b) Let $f : \mathbb{R} \longrightarrow \mathbb{R}$ by the rule $f(x) = x^2, x \in \mathbb{R}$.

Let $D_1 = \{x : x \in \mathbb{R} \text{ and } x \geq 0\}$. Then the restriction function f/D_1 is defined by $f/D_1(x) = x^2, x \geq 0$.

Note. For many purposes, we are to restrict a function to a subset of its domain, e.g., in this example (b);

f is not injective, so it cannot have an **inverse function.**

However, if we restrict f to D_1, then the restriction function f/D_1 is a bijection of D_1 onto D_1. Therefore, this restriction function f/D_1 has an inverse function which is the positive **square root function,** D_1 being set of non-negative real numbers.

C. Composition of Functions

Sometimes we wish to *compose* two functions f, g by first finding $f(x)$ and then applying g to get $g(f(x))$; clearly this is possible only when $f(x)$ is a member of the domain of g. In order to be able to do this for all $f(x)$, we must have to assume that the range of f is contained in the domain of g.

Definition. If $f : D \longrightarrow \mathbb{R}$ (where $D \subseteq \mathbb{R}$) and if

$g : E \longrightarrow \mathbb{R}$ (where Range of $f = f(D) \subset E \subseteq \mathbb{R}$),

then the **composite function** $g \circ f$ is the function from D into $\mathbb{R}$ defined by

$$(gof)(x) = g(f(x)), \quad \forall x \in D.$$

Example 3.5.10. *(a) In the composition of function, the order of the composition must be carefully observed. For, let f and g be the functinos whose values at $x \in \mathbb{R}$ are given by*

$$f(x) = 3x \quad \text{and} \quad g(x) = 4x^2 - 1.$$

Some Dom.$(g) = \mathbb{R}$ and Range $(f) \subseteq \mathbb{R} = D(g)$, then the domain $D(gof)$ is also equal to $\mathbb{R}$ and the composite function gof is given by

$$(gof)(x) = g\{f(x)\} = g\{3x\} = 4(3x)^2 - 1 = 36x^2 - 1.$$

On the other hand, the domain of the composite function $f \circ g$ is also $\mathbb{R}$, but

$$(f \circ g)(x) = f\{g(x)\} = f(4x^2 - 1) = 3(4x^2 - 1) = 12x^2 - 3.$$

See that here $g \circ f \neq f \circ g$. So, composition of two functions is not commutative. In fact, $f \circ g$ and $g \circ f$ are different functions. They are equal only in some particular cases. Moreover, the existence of one does not imply the existence of the other.

(b) In considering $g \circ f$, it is necessary to make sure that the range of f is contained in the domain of g. For example, if

$$f(x) = 1 - x^2 \quad \text{and} \quad g(x) = \sqrt{x},$$

then since $\text{dom}(g) = \{x : x \in \mathbb{R} \text{ and } x > 0\}$, the composite function $g \circ f$ is given by

$$(g \circ f)(x) = g\{f(x)\} = g(1 - x^2) = \sqrt{1 - x^2},$$

only for $x \in D(f)$ that satisfy $f(x) \geq 0$, i.e., for x satisfying

$$-1 \leq x \leq 1.$$

But the composition $f \circ g$ is given by

$$(f \circ g)(x) = f\{g(x)\} = f(\sqrt{x}) = 1 - (\sqrt{x})^2 = 1 - x,$$

only for x is in the domain $(g) = \{x : x \geq 0\}$.

XII. Monotonic Functions

Let us consider two points x_1, x_2 in the domain of definition D of the function f and let $x_1 < x_2$. Then the function f

is *increasing* in D, if	$f(x_1) \leq f(x_2)$;
is *decreasing* in D, if	$f(x_1) \geq f(x_2)$;
is *strictly increasing* in D, if	$f(x_1) < f(x_2)$;
is *strictly decreasing* in D, if	$f(x_1) > f(x_2)$.

We say that f is *monotone*, if it is either an increasing or a decreasing function and that f is *strictly monotone*, if it is either a *strictly increasing function* or a *strictly decreasing function*.

For any subset E of D we say that f is *increasing on E*, if the restriction of f to E is increasing.

Note: In fact, it is better to say f is non-decreasing in D
if $f(x_1) \leq f(x_2) \Longleftrightarrow x_1 < x_2$ and non-increasing in D
if $f(x_1) \geq f(x_2) \Longleftrightarrow x_1 < x_2$
and f is increasing if $f(x_1) < f(x_2) \Longleftrightarrow x_1 < x_2$
and f is decreasing if $f(x_1) > f(x_2) \Longleftrightarrow x_1 < x_2$

Example 3.5.11. (i) $f(x) = c$ (*constant function*) *is both increasing and decreasing (though not strictly). Conversely, any function which is both increasing and decreasing is a constant function.*

(ii) *For any positive integer n, the power function $y = f(x) = x^n$ is strictly increasing on $[0, \infty)$. See that*
$$x_2^n - x_1^n = (x_2 - x_1)(x_2^{n-1} + x_2^{n-2} \cdot x_1 + x_2^{n-3} \cdot x_1^2 + \cdots + x_2 \cdot x_1^{n-2} + x_1^{n-1}).$$
Now, if $0 \leq x_1 < x_2$, both factors on the right are positive and hence
$$x_2^n > x_1^n, \text{ i.e., } f(x_2) > f(x_1) \quad [\textit{i.e., } f \textit{ is strictly increasing}].$$

(iii) *For any non-negative integer n, the odd power function $y = x^{2n+1}$ (with domain $\mathbb{R}$) is strictly increasing (Fig. 3.5.1). To justify this assertion consider separately the cases where*
$$0 \leq x_1 < x_2, \quad x_1 < 0 < x_2, \quad x_1 < x_2 \leq 0.$$

The first case follows from (ii) and the third case reduces to the first by the substitution $x = -t$. The second case is immediate from the consideration of the sign of x^{2n+1}.

(iv) *For any positive integer n, the even power function $y = x^{2n}$ (with domain $\mathbb{R}$) is not monotone (Fig. 3.5.2). It is, however, strictly increasing on $[0, \infty)$ and strictly decreasing on $(-\infty, 0]$.*

(v) *The **greatest integer function** $f(x) = [x]$, is increasing, but not strictly (Fig. 3.4.7).*

XIII. Inverse Function

The Concept of inverse function:

The concept was introduced in chapter **1**.

Let f be a function whose domain is A and whose range ($\equiv$ CODOMAIN) is B. Then, by definition, for each $x \in A, \exists$ an unique $y = f(x) \in B$; x is called the *argument* and y or $f(x)$ is the *value of the function at x*.

If now, **the function f is one-to-one**, then we shall get for each $y \in B$, a unique x in A. This correspondence is called the *inverse mapping* or *inverse function*, denoted by f^{-1}. We may consider that f^{-1} maps each $y \in B$ to a unique $x = f^{-1}(y) \in A$. Thus, when f^{-1} exists, B is its *domain* and A is its *range*. If we call *an argument of f^{-1} by x* (NOT by y), then
$$f : x \to f(x); \quad f^{-1} : x \to f^{-1}(x),$$
where in each case $x \in$ (domain of the function).

An Important Theorem: *If f is strictly increasing, then f is one-to-one (and hence f^{-1} exists) and that the inverse function f^{-1} is strictly increasing. Similarly, if f is strictly decreasing, then f is one-to-one (f^{-1} exists) and that f^{-1} is strictly decreasing.*

We first note that if f is strictly decreasing, then clearly $-f$ is strictly increasing and conversely. So the "strictly decreasing" case follows from the "strictly increasing" case. We, therefore, prove only the strictly increasing case:

Proof: If f is strictly increasing and if x_1, x_2 belong to the domain of f s.t. $x_1 \neq x_2$, then either $x_1 < x_2$ or $x_1 > x_2$, whence either
$$f(x_1) < f(x_2) \quad \text{or} \quad f(x_1) > f(x_2) \quad \text{so that} \quad f(x_1) \neq f(x_2).$$
Hence f is one-to-one and f^{-1} exists.

Also let y_1, y_2 belong to the range of f and $y_1 < y_2$.

If $x_1 = f^{-1}(y_1)$ and $x_2 = f^{-1}(y_2)$, then $f(x_1) = y_1, f(x_2) = y_2$.

Now we have assumed that $y_1 < y_2$, i.e., we have assumed that $f(x_1) < f(x_2)$. This implies that $x_1 < x_2$, because if $x_1 \geq x_2$, then $f(x_1) \geq f(x_2)$.

$\therefore$ we see that $y_1 < y_2 \Rightarrow x_1 = f^{-1}(y_1) < f^{-1}(y_2) = x_2$.

i.e., f^{-1} is a strictly increasing function.

> **Remember:** f is one-to-one, if $f(x) = f(x')$ implies $x = x'$ or equivalently, $x \neq x' \Rightarrow f(x) \neq f(x')$.

Example 3.5.12. *(i) For any non-empty set X, the identity function $f : x \to x$, with domain X is one-to-one.*

(ii) *The square-root function $f : x \to +\sqrt{x}$, whose domain is the set of all non-negative real numbers is one-to-one.*
$$[\because \sqrt{x} = \sqrt{x'} \Rightarrow x = (\sqrt{x})^2 = (\sqrt{x'})^2 = x']$$

(iii) *The power function $f : x \to x^2$, whose domain is the set of all real numbers is not one-to-one.*
$$[\because x^2 = x'^2 \text{ may imply } x = -x'.]$$

On the other hand, *the restriction of the function to the set of all non-negative real numbers is one-to-one.*

[∵ if $x \geq 0, x' \geq 0$, then we cannot have $x = -x'$ unless $x = x' = 0$.]

Example 3.5.13. *For which of the functions in* **Ex. 3.5.12** *above you think that f^{-1} exists?*

[Use the important theorem stated above.]

Geometrical Representation

If we regard y as a function of x, we may write $y = f(x)$.

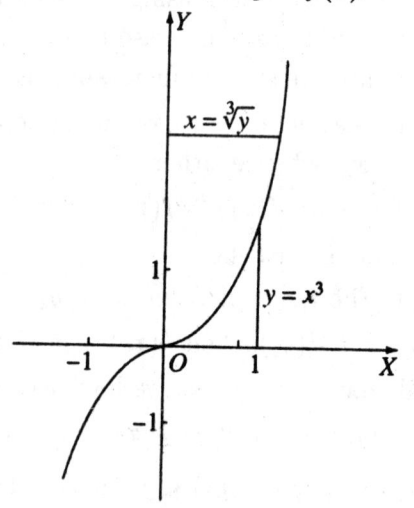

Fig. 3.5.7 The curve $y = x^3, x = \sqrt[3]{y}$ on the same system of axes

In case, inverse function exists we may write $x = f^{-1}(y)$ or $\phi(y)$, say. Both $y = f(x)$ and $x = \phi(y)$ express the **same curve** in the same system of axes OXY, though the axis of the argument for the first function is OX while for the second it is OY.

The dependence of x on y, expressed by ϕ, will be, in general, different from the dependence of y on x, expressed by f.

For instance, if $y = x^3$, then $x = \sqrt[3]{y}$. The graph of these relationships is the *cubical parabola* (See **Fig. 3.5.7**).

Whereas the first function is found from the argument by raising it to the power three, the second is found by taking the cube root of the argument. The nature of the dependence of the function on the argument is thus found to be different.

If we always stick to the notation that the argument is to be denoted by x, then the inverse function of $y = x^3$ should be written as $y = \sqrt[3]{x}$.

There is a simple means of drawing the graph of the inverse function based on the following property:

The graph of the inverse function $y = \phi(x)$ (taking the argument as x) is symmetrical to the graph of the direct function $y = f(x)$ with respect to the bisector of the first and the third quadrants.

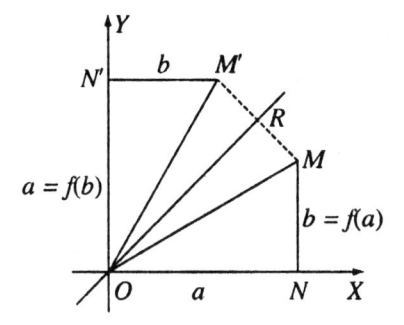

Fig 3.5.8

In fact, for $x = a$ (See **Fig. 3.5.8**) let $y = f(a) = b$. The point $M(a, b)$ belongs to the graph of the direct function. The point $M'(b, a)$ now belongs to the graph of the inverse function, since $a = \phi(b)$. It is easy to see that [consider the congruences of the triangles ONM and $ON'M'$] the bisector OR of the $\angle MOM'$ is the axis of symmetry of $\triangle MOM'$.

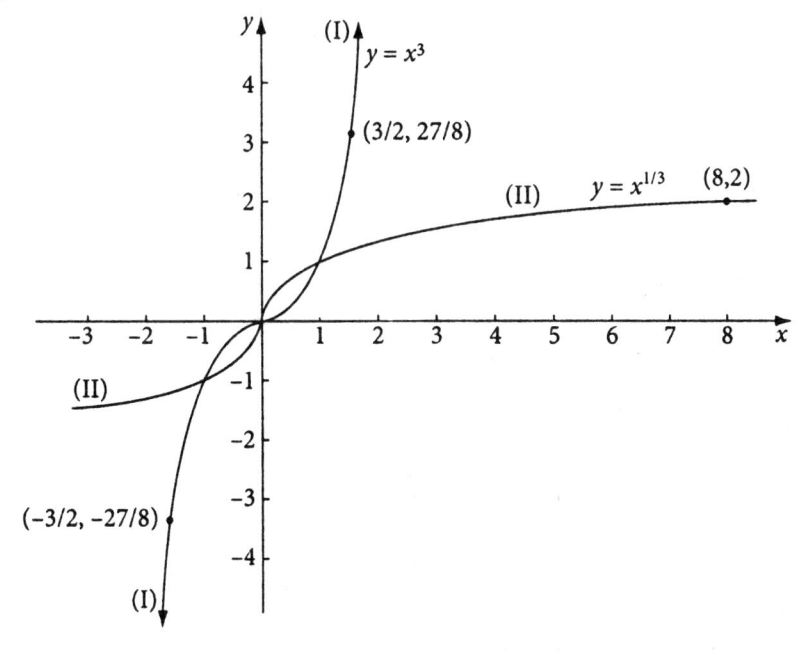

Fig 3.5.9 Graph of $y = x^3$ (I) and $y = x^{1/3}$ (II)

Thus to each point of the graph of the direct function there corresponds a point on the graph of the inverse, situated symmetrically to the first with respect to the bisector of the first and the third quadrants.

Knowing the graph of a function, we can find a graph of its inverse simply by rotating the figure about the bisector of the first and the third quadrants. **Fig. 3.5.9** illustrates the graphs of $y = x^3$ (See No. I of the figure) and $y = \sqrt[3]{x}$ (See No. II of the figure).

Example 3.5.14. (i) $y = x^2$ $(0 \leq x < \infty)$ and $y = \sqrt{x}$ $(0 \leq x < \infty)$ are inverse functions.

(ii) $y = x^4$ $(0 \leq x < \infty)$ and $y = \sqrt[4]{x}$ $(0 \leq x < \infty)$ are inverse functions.

The graphs of (i) and (ii) are shown in Fig. 3.5.10. See that the inverses are symmetric to the line $y = x$.

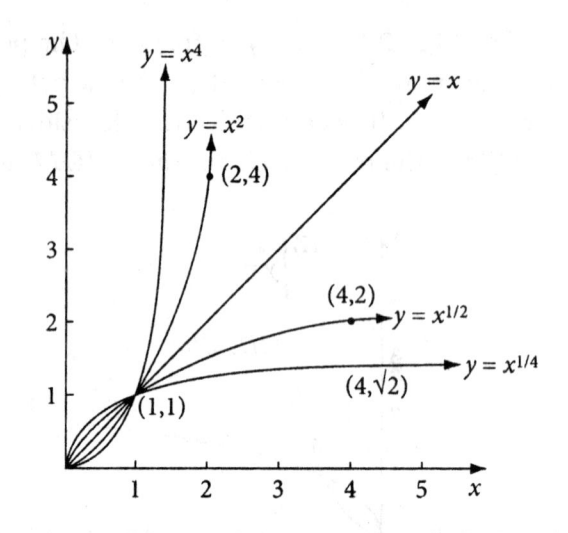

Fig 3.5.10 Graphs of $y = x^2$, $y = x^4$, $x \geq 0$ and their inverses

XIV. Periodic Functions

Definition. For any function f, a real number $\lambda > 0$ is called **a period**, if whenever $x \in$ dom. f, then $x + \lambda$ also belongs to dom. f and $f(x) = f(x + \lambda)$ and λ is least possible.

Example 3.5.15. (i) The functions, $\sin x$ and $\cos x$ are periodic and both functions have period 2π.

The function $\tan x$ is periodic and its period is π.

(ii) *Greatest integer function* $[x]$ *has been defined in* **Ex. 3.4.10**.
Consider the function f defined by $y = f(x) = x - [x]$.
Here, if $0 \le x < 1, y = x$; if $1 \le x < 2, y = x - 1$, etc.
but, if $-1 \le x < 0, y = x + 1$; if $-2 \le x < -1, y = x + 2$, etc.
The graph of this function is shown on **Fig. 3.5.11**.
The function $x - [x]$ is **periodic** and the period is 1.
See that $f(0) = f(1) = f(-1) = f(2) = f(-2)$ etc. $= 0$.

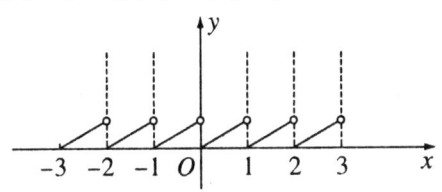

Fig 3.5.11 Graph of $x - [x]$.

(iii) *Let us define a function f over the set of all integers n $\{0, \pm 1, \pm 2, \pm 3,$ etc.$\}$ given by $f(n) = (-1)^n$.*

(iv) *A constant function is periodic and every real number is its period.*

(v) *The function f defined by*

$$f(x) = \begin{cases} 1, & \text{if } x \text{ is rational} \\ 0, & \text{if } x \text{ is irrational} \end{cases}$$

is periodic and every rational number is a period, for if λ is rational, then $x + \lambda$ is rational or irrational according as x is rational or irrational.

XV. Algebraic and Transcendental Functions

(i) Functions which are obtained by carrying out a finite number of algebraic operations on the independent variable x are called *Explicit Algebraic Functions* of x (e.g., $y = (\sqrt{x} - 1)/(x^2 + 1)$).

In a more general way, an *algebraic function* is a root of an algebraic equation whose coefficients are polynomials in the independent variable x:

$$P_0(x)y^n + P_1(x)y^{n-1} + \cdots + P_{n-1}(x)y + P_n(x) = 0,$$

where $P_i(x)$'s are polynomials in x.
See that $y = (\sqrt{x} - 1)/(x^2 + 1)$ is a root of $(x^4 + 2x^2 + 1)y^2 + 2(x^2 + 1)y + (1 - x) = 0$.

Example 3.5.16. *Polynomials, Rational functions, Irrational functions like $\sqrt{x}$ or $\sqrt{x} + 3\sqrt{x^3} + \sqrt{x}, x \ge 0$ are all examples of algebraic functions.*

(ii) Transcendental Functions[1] (non-algebraic functions):

We shall now introduce some functions other than the aforesaid algebraic functions. They will be called *Transcendental functions*. They are:

[A] Exponential Functions (a^x), Logarithmic Functions ($\log_a x$),

[B] Circular (or Trigonometric) Functions ($\sin x, \cos x$, etc.),

[C] Inverse circular (or Inverse Trigonometric Functions) ($\sin^{-1} x, \cos^{-1} x$, etc.),

[D] Hyperbolic and Inverse Hyperbolic Functions ($\sinh x, \cosh x, \sinh^{-1} x$, etc.).

[A] Exponential Functions:

An exponential function has the form $y = a^x$. When $a > 0, a \neq 1, y$ is defined for all x in $(-\infty, \infty)$. If $a < 0$, we are to restrict the domain (x should be of the form p/q with q odd). So we once for all assume that we shall always take exponential function a^x with $a > 0$ and $a \neq 1$. The following results are assumed (without proof):

$a^0 = 1, a^{p/q} = \sqrt[q]{a^p}, a^{-r} = 1/a^r, a^m \cdot a^n = a^{m+n}, (a^m)^n = a^{mn}, \forall m, n \in \mathbb{R}$.

a^x increases from 0 to ∞ as x increases from $-\infty$ to $+\infty$ (if $a > 1$).

a^x decreases from ∞ to 0 as x increases from $-\infty$ to $+\infty$ (if $0 < a < 1$).

a^x is everywhere continuous (no break in the curve).

When $a = e = 1 + \frac{1}{1!} + \frac{1}{2!} + \frac{1}{3!} + \cdots$, the exponential function takes the form $y = e^x$.

Logarithmic Functions:

Since $a^x (a > 1)$ strictly increases in $-\infty < x < \infty$, hence inverse of $y = a^x$ exists. We call it Logarithmic function

$$x = \log_a y \quad (0 < y < \infty, -\infty < x < \infty)$$
$$\text{or,} \quad y = \log_a x \quad (0 < x < \infty, -\infty < y < \infty).$$

The inverse function is also strictly increasing in $0 < x < \infty$.

When $a = e$, the inverse function is called the natural logarithm of x, written as

$$y = \log_e x = \ln x \quad (0 < x < \infty, -\infty < y < \infty).$$

Since a^x and $\log_a x$ are mutually inverses, we have

$$\log_a a^x = x, \quad a^{\log_a x} = x.$$

See that a power function

$$y = x^n (x > 0)$$

---

[1] Axiomatic approach is desirable. We shall, in a later chapter, introduce postulation development of these functions. For the present we give the traditional approach.

can be written as
$$y = x^n = a^{n \log_a x}.$$

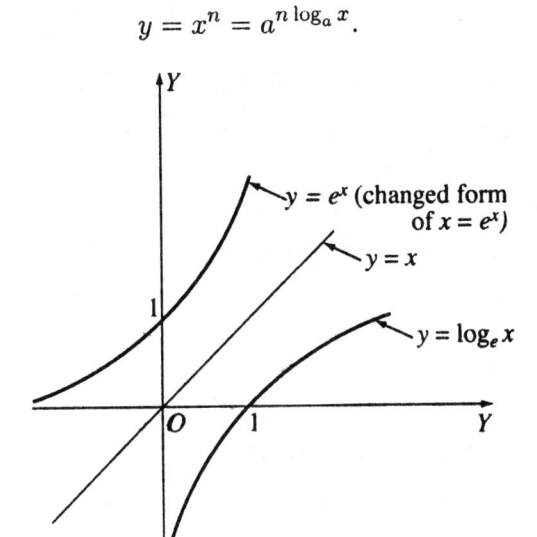

Fig 3.5.12 Graphs of $y = \log_e x$ and $y = e^x$.

In Fig. 3.5.12 we have shown the graph of $y = \log_e x$ and its inverse e^x. The following table may help the reader to sketch the graph:

x	0.05	0.08	0.14	0.22	0.37	0.61	1	1.65	2.23	e	3.32	3.67	4.06	4.48
$y = \log_e x$	-3	-2.5	-2	-1.5	-1	-0.5	0	0.5	0.8	1	1.2	1.3	1.4	1.5

Example 3.5.17. *Draw the curves $y = 2^x (a = 2)$ and $y = (1/2)^x (a = 1/2)$ and by reflection on the line $y = x$, draw the inverse curves.*

Check. The negative semi-axis Ox' is an *asymptote* of $y = a^x (a > 1)$ and the positive semi-axis Ox is an *asymptote* of $y = a^x (0 < a < 1)$.

A Practical Example. Suppose A rupees is deposited in a bank which gives $r\%$ compound interest per annum. That is, after x years the amount y will be
$$y = A\left(1 + \frac{r}{100}\right)^x,$$
which is an exponential function of the argument x.

[B] Trigonometric (or Circular) Functions:

The trigonometric (or circular) functions are
$$y = \sin x, \quad y = \cos x, \quad y = \tan x,$$
and their reciprocals $y = \operatorname{cosec} x, \quad y = \sec x, \quad y = \cot x.$

We shall always take the radian measure of the angle as the argument x, e.g., the value of $y = \sin x$ at $x = x_0$ is equal to the sine of the angle of x_0 radians.

[One radian is approximately $57°17'44 \cdot 8$; it is sometimes convenient to write x_0 radians as part of π radians: $x_0 = \frac{x_0}{\pi} \cdot \pi \simeq \frac{x_0}{3 \cdot 14}\pi$.]

We shall assume all the familiar formulae connected with $\sin x, \cos x, \tan x$, etc.

Observation. Many authors attempted to free the trigonometric functions from the concept of angle-measure by defining them as functions of the set of real numbers representing arc-length along the unit circle measured from a fixed point on the circle. Such a procedure merely shifts the emphasis from angle-measure in radians to the measure in arc-length along the unit circle. What is desirable is that trigonometric functions should be defined as functions of a real variable, free from any prescribed measure. This is done by employing a postulation development of these functions. We shall come back to these discussions in a later chapter.

The trigonometric functions are *periodic*. Functions $\sin x$ and $\cos x$ have a period 2π, functions $\tan x$ and $\cot x$ have period π.

(a) The graph of $y = \sin x$ is supposed to be familiar to our readers. The following table may help them to draw the curve: [**Fig. 3.5.13(a)**]

x	0	$\pi/6$	$\pi/4$	$\pi/3$	$\pi/2$	π	$3\pi/2$	$-\pi/2$	$-\pi$
$y = \sin x$	0	$1/2$	$1/\sqrt{2}$	$\sqrt{3}/2$	1	0	-1	-1	0

Observations.

(i) $y = \sin x$ *increases* from 0 to 1 as x increases from 0 to $\pi/2$.

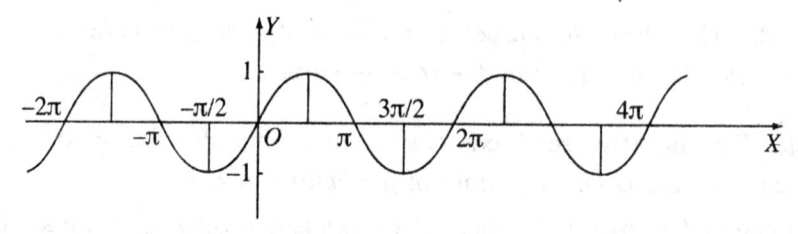

Fig 3.5.13(a) The graph of $y = \sin x$.

(ii) $y = \sin x$, then *decreases* from 1 to -1 as x further increases from $\pi/2$ to $3\pi/2$; at $x = \pi, y = 0$.

(iii) $y = \sin x$ finally *increases* from -1 to 0 in $[3\pi/2, 2\pi]$.

(iv) $y = \sin x$ has period 2π. The entire graph (*sine wave*) is obtained by shifting the interval $[0, 2\pi]$ together with corresponding part of the graph to the left and to the right by $2\pi, 4\pi, 6\pi, \cdots$.

(v) Since $\sin(-x) = -\sin x, y$ is an odd function—the graph is symmetrical about the *origin*.

CHAPTER 3: REAL-VALUED FUNCTIONS OF A REAL VARIABLE

(b) The graph of $y = \cos x$ [**Fig. 3.5.13(b)**] is not essentially new, since

$$y = \cos x = \sin(x + \pi/2).$$

Hence, if the co-ordinate system is transformed by $x' = x + \pi/2, y' = y$, the equation takes the form $y' = \sin x'$ in the system $O'XY'$.

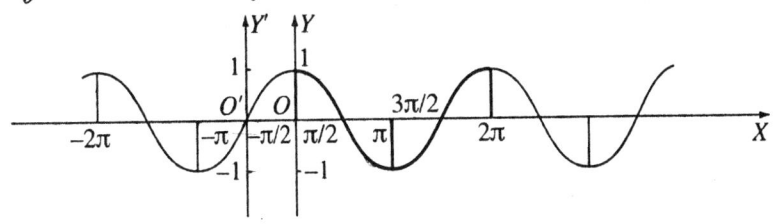

Fig 3.5.13(b) The graph of $y = \cos x$.

The curve $y = \cos x$ is thus the sine wave shifted to the left along OX by $\pi/2$.

Observations.

(i) $y = \cos x$ *decreases* from 1 to -1 as x increases from 0 to π, at $x = \pi/2, y = 0$.

(ii) $y = \cos x$, then increases from -1 to $+1$ as x further increases from π to 2π; at $x = 3\pi/2, y = 0$.

(iii) Both to the left of $x = 0$ and to the right of $x = 2\pi$, in the course of each interval of length 2π, $y = \cos x$ takes the same values in the same sequence as in the basic interval $[0, 2\pi]$.

(iv) $\cos x = \cos(-x)$; hence $y = \cos x$ is an even function and so its graph is symmetrical about *the axis of ordinates*.

(c) $y = \tan x$ is defined throughout $[0, \pi]$ except for $x = \pi/2$ (where it is undefined). When x approaches $\pi/2$ from the left (See **Fig. 3.5.14**) y is positive but its values grow indefinitely large. But when x approaches $\pi/2$ from the right the values of y are negative but the absoluté values grow infinitely large.

$y = \tan x$ has the period π and hence the same picture is seen in the neighbourhood of any point $x = (2k+1)\pi/2$ ($k = \pm 1, \pm 2, \pm 3$, etc.).

The vertical lines $x = \pi/2$, $x = 3\pi/2$, etc., $x = -\pi/2$, $x = -3\pi/2, \cdots$ are all *asymptotes* of the curve for different branches.

The complete graph of $y = \tan x$ is obtained from the graph of the part $[0, \pi]$ by simple repetition using the property of periodicity.

Observations:

(i) $y = \tan x$ increases from $-\infty$ to 0 as x increases from $-\pi/2$ to 0.

(ii) Then $y = \tan x$ further increases from 0 to ∞ in $[0, \pi/2)$.

(iii) Again, $y = \tan x$ increases from $-\infty$ to 0 as x increases from $\pi/2$ to π.

(iv) $y = \tan x = \tan(-x)$; hence an odd function. The graph is symmetric about the origin.

(v) We take the basic interval $(-\pi/2, \pi/2)$ where $y = \tan x$ increases from $-\infty$ to $+\infty$.

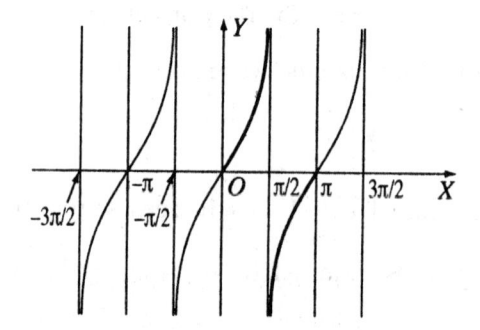

Fig 3.5.14 The graph of $y = \tan x$.

(d) $y = \cot x$ is defined for all x except for $x = n\pi$ ($n = 0, \pm 1, \pm 2$, etc.).
(i) The function decreases from $+\infty$ to $-\infty$ as x increases from 0 to π.
(ii) Again it decreases from $+\infty$ to $-\infty$ as x increases from π to 2π and so on.
(iii) The lines $x = n\pi$ are asymptotes.
(iv) $y = \cot x$ has a period π and is an odd function. Now draw the graph of $y = \cot x$.

(e) $y = \sec x$ is defined for all x except for $x = (2n+1)\frac{\pi}{2}$ ($n = 0, \pm 1, \pm 2$, etc.).
(i) The absolute values of $\sec x$ are ≥ 1, whatever x may have.
(ii) $y = \sec x$ increases from 1 to ∞ as x increases from 0 to $\pi/2$ and then it increases from $-\infty$ to -1 as x increases from $\pi/2$ to π and so on.
(iii) $y = \sec x$ has asymptotes $x = \pi/2, x = 3\pi/2$, etc.
(iv) $y = \sec x$ is periodic (period 2π).
(v) $y = \sec x$ is an even function.

Now see the graph of $y = \sec x$ (Fig. 3.5.15). and draw other branches.

Draw the graph of $y = \operatorname{cosec} x$ from the following information:

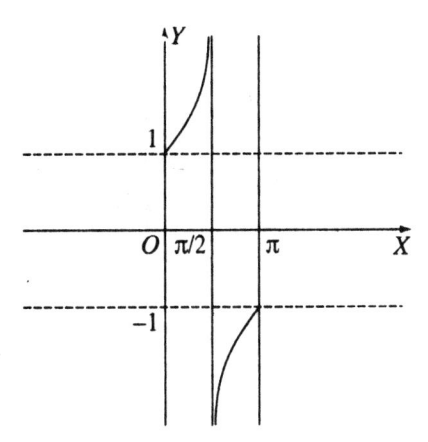

CHAPTER 3: REAL-VALUED FUNCTIONS OF A REAL VARIABLE 155

Fig 3.5.15 Graph of $y = \sec x$.

(f) $y = \operatorname{cosec} x$ is defined for all x except for $x = n\pi$.

The absolute values of $\operatorname{cosec} x \geq 1$, whatever x may be.

$y = \operatorname{cosec} x$ decreases from -1 to $-\infty$ in $[-\pi/2, 0)$, it decreases from ∞ to $+1$ in $(0, \pi/2]$. The lines $x = n\pi$ are asymptotes.

[C] Inverse Circular (or Trigonometric) Functions:

The functions $y = \operatorname{Sin}^{-1} x$ (or Arc sin x), $y = \operatorname{Cos}^{-1} x$ (or Arc cos x), $y = \operatorname{Tan}^{-1} x$ (or Arc tan x), etc., inverse to trigonometric functions $\sin x, \cos x, \tan x$, etc. are called inverse trigonometric (or circular) functions: $y = \operatorname{Sin}^{-1} x \Rightarrow$ for a given x, to find the angle y (or the angles y) s.t. $\sin y = x$. The values of these functions express radian measures of the angles or the lengths of the arcs of a unit circle.

Since the trigonometric functions are periodic, $\exists$ an infinite set of angles (or arcs) for which the given function has the same value. It follows that the corresponding inverse trigonometric functions are infinitely many-valued. We have used capital letters S, C, T, A to indicate the many-valued characters of these functions. From the graph (say **Fig. 3.5.13**) it is also apparent that a line parallel to OX cuts the graph of $\operatorname{Sin}^{-1} x$ at many points, i.e., to a given y $\exists$ many x.

Now the graph of an inverse trigonometric function can be obtained by drawing a curve symmetrical about $y = x$ with the graph of the corresponding trigonometric function.

The graph of $y = $ Arc sin x is a sin wave (**Fig. 3.5.16**). This function is defined only in $-1 \leq x \leq 1$ (No angle exists whose sine has an absolute value > 1).

In this interval the function is many-valued.

Any line parallel to Y-axis and passing inside $-1 \leq x \leq 1$ cuts the curve $y = $ Arc sin x in an infinity of points.

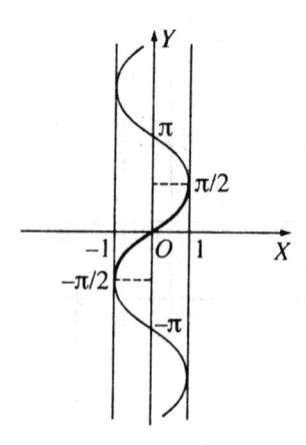

Fig. 3.5.16 Graph of $y = \text{Sin}^{-1} x$.
Heavy line: $y = \sin^{-1} x$.

We shall distinguish the single-valued branch of this many-valued function by choosing the longest part of the graph which cuts any line parallel to OY in not more than one point. We take the part joining the points $(-1, -\pi/2)$ and $(1, \pi/2)$ (this part has been shown with a heavy line on the graph). Thus, from all the possible values of Arc $\sin x$, corresponding to a given value of x, we distinguish the one which has an absolute value not exceeding $\pi/2$; it is written as $\sin^{-1} x$ (or arc $\sin x$) [*small letter s, a*] and is called the **principal value** of Arc $\sin x$ (for the given x); thus

$$-\frac{\pi}{2} \leq \text{arc sin } x \leq \frac{\pi}{2}.$$

Now $y = \text{arc sin } x$ is a single-valued and increasing function, defined in $[-1, 1]$.

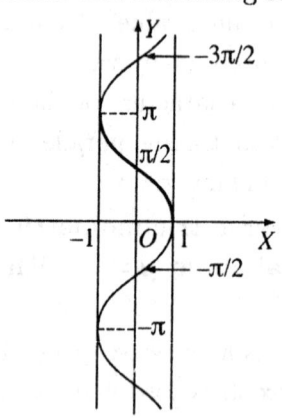

Fig. 3.5.17 Graph of $y = \text{Arc Cos } x$.
Heavy line: $y = \text{arc cos } x$.

We obtain the graph of $y = \text{Arc cos } x$, if the graph of $y = \text{Arc sin } x$ is displaced $\pi/2$ downwards (Fig. 3.5.17). The function $y = \text{Arc cos } x$ is defined in $-1 \leq x \leq 1$

and is infinitely many-valued. We distinguish from all the values of $y = \text{Arc}\cos x$ corresponding to a given x, the one which lies between O and π; it is known as the principal value of arc cos x and is written as $\cos^{-1} x$ or arc cos x:

$$0 \leq \text{arc}\cos x \leq \pi.$$

This function $y = \text{Arc}\cos x$ is single-valued, positive and decreasing; it is defined in $[-1, 1]$. Its graph is the heavy line of **Fig. 3.5.17**.

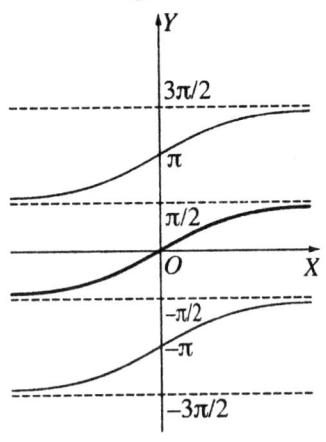

Fig. 3.5.18 Graph of $y = \text{Arc}\tan x$.
Heavy line: $y = \text{arc}\tan x$.

The function $y = \text{Arc}\tan x$ is defined throughout the X-axis and is infinitely many-valued. Its graph consists of an infinite set of parallel branches lying in the strips (see **Fig. 3.5.18**),

$$\ldots\ldots, \frac{-3}{2}\pi \leq y \leq \frac{-\pi}{2}, \ \frac{-\pi}{2} \leq y \leq \frac{\pi}{2}, \ \frac{\pi}{2} \leq y \leq \frac{3\pi}{2}, \ldots\ldots.$$

Every straight line $y = (2k+1)\frac{\pi}{2}$ (where k is any integer) is an asymptote of the corresponding branch of the function $y = \text{Arc}\tan x$. From all the value of this function, corresponding to a given x, we choose the one which lies between $-\pi/2$ and $\pi/2$; it is called the *principal value* and is written as arc tan x or $\tan^{-1} x$:

$$-\frac{\pi}{2} < \text{arc}\tan x < \frac{\pi}{2}.$$

The function $y = \text{arc}\tan x$ is single-valued, increasing and defined for all x, in $-\infty < x < \infty$. Its graph is distinguished by a heavy line in **Fig. 3.5.18**.

Example 3.5.18. *Draw the graphs of $y = \text{Arc}\cot x$ and arc cot x, $y = \text{Arc}\sec x$ and arc sec x, $y = \text{Arc}\cosec x$ and arc cosec x.*

> **Remember:** $0 < \text{arc cot } x < \pi,$
> $0 \leq \text{arc sec } x \leq \pi,$
> $-\pi/2 \leq \text{arc cosec } x \leq \pi/2.$
> These give principal values of Arc cot x, Arc sec x and Arc cosec x respectively.

[D] Hyperbolic Functions:

We define: hyperbolic sine of x (written as $\sinh x$) $= \frac{1}{2}(e^x - e^{-x})$

hyperbolic cosine of x (written as $\cosh x$) $= \frac{1}{2}(e^x + e^{-x})$

$$\tanh x = \frac{\sinh x}{\cosh x} = \frac{e^x - e^{-x}}{e^x + e^{-x}}; \quad \coth x = \frac{1}{\tanh x} = \frac{e^x + e^{-x}}{e^x - e^{-x}};$$

$$\text{sech } x = \frac{1}{\cosh x} = \frac{2}{e^x + e^{-x}}; \quad \text{cosech } x = \frac{1}{\sinh x} = \frac{2}{e^x - e^{-x}}.$$

The following two results follow from our definitions:

$$\cosh x + \sinh x = e^x \quad \text{and} \quad \cosh x - \sinh x = e^{-x}$$

and hence

$$\boxed{\cosh^2 x - \sinh^2 x = 1.} \tag{1}$$

It also follows,

$$\text{sech}^2 x = 1 - \tanh^2 x \quad \text{[divide (3.2) throughout by } \cosh^2 x\text{]}$$
$$\text{cosech}^2 x = \coth^2 x - 1 \quad \text{[divide (3.2) throughout by } \sinh^2 x\text{]}$$

Addition Theorems:

$$\cosh(x \pm y) = \cosh x \cosh y \pm \sinh x \sinh y \tag{2}$$
$$\sinh(x \pm y) = \sinh x \cosh y \pm \cosh x \sinh y \tag{3}$$
$$\tanh(x \pm y) = \frac{\tanh x \pm \tanh y}{1 \pm \tanh x \tanh y}. \tag{4}$$

[See that $\cosh(x+y) = \frac{1}{2}[e^{x+y} + e^{-(x+y)}] = \frac{1}{2}(e^x \cdot e^y + e^{-x} \cdot e^{-y})$

$= \frac{1}{2}\{(\cosh x + \sinh x)(\cosh y + \sinh y) + (\cosh x - \sinh x)(\cosh y - \sinh y)\}$

$= \cosh x \cosh y + \sinh x \sinh y.$

Hence (3.3) follows. Similarly others can be verified.]

From Addition Theorems, putting $y = x$, we may obtain

$$\cosh 2x = \cosh^2 x + \sinh^2 x = 2\cosh^2 x - 1 = 1 + 2\sinh^2 x.$$

CHAPTER 3: REAL-VALUED FUNCTIONS OF A REAL VARIABLE

$$\sinh 2x = 2\sinh x \cosh x \quad \text{and} \quad \tanh 2x = (2\tanh x)/(1 + \tanh^2 x).$$

In fact, one may remember the **Osborn's Rule** which states:

In any formula connecting trigonometric functions, replace each trigonometric function by the corresponding hyperbolic functions **and change the sign of every product (or, implied product) of two sines.**

For example, in the trigonometric formula

$$\cos(A + B) = \cos A \cos B - \sin A \sin B,$$

replace cos by cosh, sin by sinh and change the sign before the product $\sin A \sin B$; thus obtain:

$$\cosh(x + y) = \cosh x \cosh y + \sinh x \sinh y.$$

Properties of hyperbolic functions:

Hyperbolic Sine:

1. $\sinh(-x) = -\sinh x; \quad \sinh 0 = 0.$
2. $y = \sinh x$ is defined for every real value of x in $-\infty < x < \infty$ and the function is continuous (no break in the curve).
3. $y = \sinh x$ steadily increases as x increases from $-\infty$ to $+\infty$.

The following table may help us to draw the graph of $y = \sinh x$.

x	0	0.05	0.10	0.20	0.40	0.60	0.80	1.00	1.25
$y = \sinh x$	0.00	0.0500	0.1002	0.2013	0.4108	0.6367	0.8881	1.1752	1.6019

x	1.50	2.00	2.50	3.00	4.00	6.00	8.00	9.00	10.00
$y = \sinh x$	2.1293	3.6269	6.0502	10.018	27.290	201.71	1490.5	4051.5	11013.00

[*To construct the table for negative values of x, remember:* $\sinh x$ *is an odd function.*]

The graph of $y = \sinh x$ has been shown in Fig. 3.5.19.

Hyperbolic cosine:

1. $\cosh(-x) = \cosh x; \quad \cosh 0 = 1.$
2. $y = \cosh x$ is defined for every value of x in $-\infty < x < \infty$ and is continuous throughout; it is always positive being half of the sum of two positive quantities e^x and $1/e^x$.
3. $y = \cosh x$ decreases as x increases from $-\infty$ to 0 and then increases as x increases from 0 to ∞.
4. $\cosh x > \sinh x$ for all values of x ($\because \cosh x - \sinh x = 1/e^x > 0$).

5. The least value (= 1) of $\cosh x$ occurs when $x = 0$.

$$\left[\because \cosh x = \frac{1}{2}(e^{x/2} - e^{-x/2})^2 + 1\right].$$

6. Now see the graph of $y = \cosh x$ (Fig. 3.5.19). This curve is called a **Catenary** (it is the curve in which a uniform flexible chain with fixed ends hangs).

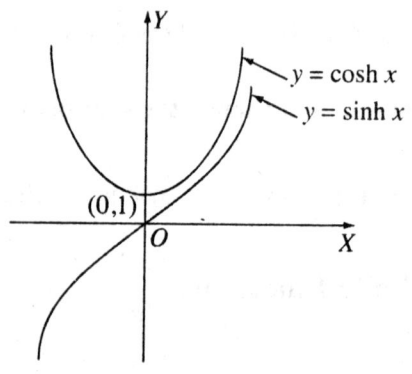

Fig. 3.5.19 Graphs of $y = \sinh x$ and $y = \cosh x$.

Table of values of $y = \cosh x$:

x	0	0.05	0.10	0.20	0.40	0.60	0.80	1.00	1.25
$y = \cosh x$	1.000	1.0013	1.0050	1.0201	1.0811	1.1855	1.3374	1.5431	1.8884

x	1.50	2.00	2.50	3.00	4.00	6.00	8.00	9.00	10.00
$y = \cosh x$	2.3524	3.7622	6.1323	10.068	27.308	201.72	1490.5	4051.5	11013

[*To construct the table for negative values, see that* $\cosh(-x) = \cosh(x)$.]

Hyperbolic Tangent and Cotangent:

1. $\tanh x$ and $\coth x$ are both odd functions of x.

2. $\tanh 0 = 0$.

3. $y = \tanh x$ is defined for every real x but $\coth x$ is defined for every real x except when $x = 0$.

4. As x increases from 0 to ∞, $\tanh x$ increases from 0 to 1. As x increases from $-\infty$ to 0, $\tanh x$ increases from -1 to 0. $y = \pm 1$ are the asymptotes. Remember: $|\tanh x| < 1$, for all x.

It then follows, $|\coth x| > 1$ ($\because \coth x = 1/\tanh x$).

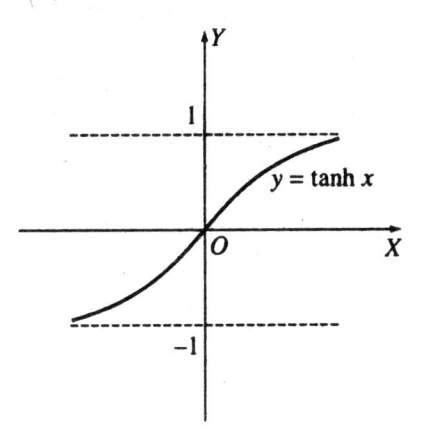

CHAPTER 3: REAL-VALUED FUNCTIONS OF A REAL VARIABLE

Fig. 3.5.20 Graphs of $y = \tanh x$.

In **Fig. 3.5.20**, we have shown the graph of $y = \tanh x$. The students should now draw the graph of $y = \coth x$.

Hyperbolic secant *and* cosecant:

1. $0 < \text{sech } x \leq 1$. sech x has its maximum value 1 at $x = 0$. sech $(-x) = $ sech x (even function).

2. cosech x has the same sign as x, since
   $$\text{cosech } (-x) = -\text{cosech } x.$$

3. When x grows large in absolute values, sech x approaches zero.

When x grows large in absolute values, cosech x also approaches zero.

Exercise: Draw the graphs of coth x, sech x and cosech x.

Inverse Hyperbolic Functions

A. $y = \sinh x$ is strictly increasing in $(-\infty, \infty)$ and its inverse $x = \sinh^{-1} y$ or taking x as independent variable $y = \sinh^{-1} x$ exists for all x.

The inverse function is also increasing. We shall express the inverse hyperbolic in terms of logarithmic functions.

Let $y = \sinh^{-1} x$. That is, $x = \sinh y = \frac{1}{2}(e^y - e^{-y})$.

We have, then $e^{2y} - 2x.e^y - 1 = 0$ and solving for e^y we get $e^y = x \pm \sqrt{x^2 + 1}$.

Since e^y is always positive and $\sqrt{x^2 + 1} > x$, so we shall take

$$e^y = x + \sqrt{x^2 + 1} \text{ (positive value)}$$

or $y = \sinh^{-1} x = \log(x + \sqrt{x^2 + 1})$ **(defined for all real x).**

B. Let $y = \cosh^{-1} x$. That is, $x = \cosh y = \frac{1}{2}(e^y + e^{-y})$.

We have, then $e^{2y} - 2x.e^y + 1 = 0$, whence $e^y = x \pm \sqrt{x^2 - 1}$.

Either sign is admissible ($\because$ both $x + \sqrt{x^2 - 1}$ and $x - \sqrt{x^2 - 1}$ are positive and real when $x \geq 1$). The quantities $x + \sqrt{x^2 - 1}$ and $x - \sqrt{x^2 - 1}$ are reciprocals to one another and hence their logarithms differ in sign.

We observe that x can have no value less than 1 and that for any $x \geq 1$, there are two values of y, equal in magnitude but opposite in sign.

So, we shall assume that $y = \cosh^{-1} x$ is defined only when $x \geq 1$, and to make it single-valued let us agree to write
$$\cosh^{-1} x = \log\left(x + \sqrt{x^2 - 1}\right) \ (x \geq 1).$$

Some authors prefer to write
$$\cosh^{-1} x = \pm \log\left(x + \sqrt{x^2 - 1}\right) \ (x \geq 1)$$

and call them as two branches of $\cosh^{-1} x$.

C. We leave to the students to proceed in a similar manner and obtain the following logarithmic expressions for inverse hyperbolic functions. Check in each case, the domain of definition:

$\tanh^{-1} x = \frac{1}{2} \log \frac{1+x}{1-x} (-1 < x < 1$ or $|x| < 1)$. $\coth^{-1} x = \frac{1}{2} \log \frac{x+1}{x-1} (|x| > 1)$

$\operatorname{sech}^{-1} x = \log \frac{1+\sqrt{1-x^2}}{x} (0 < x < 1)$. $\operatorname{cosech}^{-1} x = \log \frac{1 \pm \sqrt{1+x^2}}{x}$ $\begin{bmatrix} +ve \text{ sign, if } x > 0 \\ -ve \text{ sign, if } x < 0 \end{bmatrix}$

The graphs of $y = \sinh^{-1} x, y = \cosh^{-1} x$ and $y = \tanh^{-1} x$ can be drawn by rotation of the graphs of $y = \sinh x, y = \cosh x, y = \tanh x$ respectively about the line $y = x$.

Observation. The equation $\cosh^2 \theta - \sinh^2 \theta = 1$ shows that $x = a \cosh \theta, y = b \sinh \theta$ satisfy the equation of the hyperbola $\frac{x^2}{a^2} - \frac{y^2}{b^2} = 1$. The expressions $x = a \cosh \theta, y = b \sinh \theta$, therefore, give the parametric representation of the hyperbola $\frac{x^2}{a^2} - \frac{y^2}{b^2} = 1$, θ being the parameter.

Interpretation of θ. If O be the centre and A the vertex $(a, 0)$ of the hyperbola, then the area of the sector AOP is $\frac{1}{2} ab\theta$, P being any point with parameter θ.

[See Authors' Integral Calculus—An Introduction to Analysis.]

XVI. Function of a Function

Let $u = f(x)$ and $y = \phi(u)$ be two functions s.t. f is defined over a set S of real numbers and ϕ is defined over a set T of real numbers. Suppose every $f(x)$ for $x \in S$ is a member of T. Then clearly the two relations $u = f(x)$ and $y = \phi(u)$ determine y as a function of x defined over S. We call y as a *function of a function* (or composite function) *of* x.

e.g., (i) $u = x^3$, $y = \sin u$ determine $y = \sin x^3$,
(ii) $u = \sin x$, $y = \log u$ determine $y = \log(\sin x)$.

CHAPTER 3: REAL-VALUED FUNCTIONS OF A REAL VARIABLE

[Note that in (ii) u is defined $\forall\, x, y = \log u$, then associates a value of y to that value of u which is positive; but since u is positive, iff x lies in $(-4\pi, -3\pi)$, $(-2\pi, -\pi)$, $(0, \pi)$, $(2\pi, 3\pi)$, etc. Then the composite function is defined over these set of intervals only.]

To Remember: Hyperbolic Functions—their Properties:

$$\cosh x = \frac{e^x + e^{-x}}{2}, \quad \sinh x = \frac{e^x - e^{-x}}{2}.$$

They are defined for all real x and connected by

$$\cosh^2 x - \sinh^2 x = 1, \quad \cosh^2 x + \sinh^2 x = \cosh 2x, \text{ etc.}$$

$$\tanh x = \frac{\sinh x}{\cosh x}, \quad \operatorname{sech} x = \frac{1}{\cosh x}, \quad \operatorname{cosech} x = \frac{1}{\sinh x},$$

$$\coth x = \frac{1}{\tanh x}, \text{ etc.}$$

Using *Osborn's Rule,* the familiar identities of Trigonometry will give formulae connecting hyperbolic functions.

Inverse hyperbolic functions:

$$\sinh^{-1} x = \log(x + \sqrt{x^2 + 1})\ \forall\, x \in \mathbb{R},$$

$$\cosh^{-1} x = \log(x + \sqrt{x^2 - 1})\ (x \geq 1),$$

$$\tanh^{-1} x = \frac{1}{2} \log \frac{1+x}{1-x} \text{ for } |x| < 1,$$

$$\coth^{-1} x = \frac{1}{2} \log \frac{x+1}{x-1}, |x| > 1,$$

$$\operatorname{sech}^{-1} x = \log \frac{1 + \sqrt{1 - x^2}}{2}\, (0 < x < 1),$$

$$\operatorname{cosech}^{-1} x = \log \frac{1 \pm \sqrt{1 + x^2}}{x}\, (+\text{ve sign if } x > 0, -\text{ve sign if } x < 0).$$

EXERCISES ON CHAPTER 3

(On real-valued functions of a single real variable)

[A]

[*Functional Values*]

1. If $f(x) = \dfrac{1-x}{1+x}$, find $f(2), f(x^2), f(g), f(\cos x) \cdot [f(x)]^2$.

2. Let $f(x) = \log_e \sin x,\ \phi(x) = \log_e \cos x$, verify that

a) $f(x) + \phi(x) = f(2x) - \log_e 2$; b) $e^{2\phi(x)} = \frac{1}{2}(1 + \cos 2x)$;
c) $e^{2f(x)} + e^{2\phi(x)} = 1$; d) $e^{2\phi(x)} - e^{2f(x)} = \cos 2x$.

3. If $y = f(x) = \dfrac{lx + m}{nx - l}$, prove that $x = f(y)$.

4. If $f(x) = ax^2 + bx + c$, $f(1) = 3$, $f(2) = 7$, $f(3) = 13$. Find the values of a, b and $f(0)$. (each = 1).

5. $f(x) = \dfrac{x-1}{x+1}$, verify that $\dfrac{f(x) - f(y)}{1 + f(x) \cdot f(y)} = \dfrac{x - y}{1 + xy}$.

6. If $\phi(x) = m\dfrac{x-l}{m-l} + l\dfrac{x-m}{l-m}$, prove that $\phi(l + m) = \phi(l) + \phi(m)$.

7. If a, b, c are three distinct real numbers, and
$$f(x) = \frac{(x-a)(x-b)}{(c-a)(c-b)} + \frac{(x-b)(x-c)}{(a-b)(a-c)} + \frac{(x-c)(x-a)}{(b-c)(b-a)},$$
prove that $f(a) = f(b) = f(c) = f(0)$, each being equal to 1.

8. If $f(x) = [x]$, where $[x]$ denotes the greatest integer in x but not greater than x, find
$$f(0), f(1), f(2), f\left(\frac{1}{2}\right), f(3/2), f(7/5), f(3.01), f(-3), f(-2.5).$$

9. If $f(x) = \sin x$, $\phi(x) = \cos x$, verify that
a) $f(x) = \phi(\frac{\pi}{2} - x)$; b) $f(x) = -f(-x)$; c) $\phi(x) = \phi(-x)$;
d) $f^2(x) + \phi^2(x) = 1$, where $f^2(x) = [f(x)]^2$ and $\phi^2(x) = [\phi(x)]^2$;
e) $\phi(2x) = 1 - 2f^2(x)$; f) $f(x + y) = f(x)\phi(y) + f(y)\phi(x)$.

10. If $f(x) = \dfrac{x}{1-x}$, verify that
$$\frac{f(x+h) - f(x)}{h} = \frac{1}{(1-x)(1-x-h)} \quad (h \neq 0).$$

11. If $f(x) = \begin{bmatrix} \cos x & \sin x \\ -\sin x & \cos x \end{bmatrix}$, then verify that $f(x)f(y) = f(x + y) = f(y)f(x)$.

12. If $f(x) = x^3$, prove that $\displaystyle\int_a^b f(x)dx = \dfrac{b-a}{6}\left\{f(a) + 4f\left(\dfrac{a+b}{2}\right) + f(b)\right\}$.

CHAPTER 3: REAL-VALUED FUNCTIONS OF A REAL VARIABLE

[B]

[Hints are given for** marked problems]

[Different types of Functions]

1. Indicate the **domains** of the following functions:

   (a) $f(x) = \dfrac{1}{x-2}$; (b) $\phi(x) = \sqrt{x}$; (c) $\psi(x) = \sqrt{x-2}$;

   (d) $\theta(x) = \sqrt{(x-2)(x-3)}$; (e) $g(x) = \dfrac{x^2 - 4}{x-2}$; (f) $h(x) = \tan x$;

   (g) $F(x) = \sqrt{1-x^2}$ (h) $G(x) = \sin \dfrac{1}{x}$; (i) $H(x) = \dfrac{1}{\sin \pi x}$.

   **(j) $f(x) = \operatorname{arc cos} 3/(4 + 2\sin x)$; (k) $f(x) = \sqrt{\operatorname{arc sin} \log_2 x}$.

   > Remember: **arc** stands for **inverse**. e.g., $\operatorname{arc sin} x = \sin^{-1} x$.

2. Find the **range** of f, where

   a) $f(x) = \begin{cases} 0, & \text{when } x \text{ is rational} \\ 1, & \text{when } x \text{ is irrational.} \end{cases}$ dom. $f = \mathbb{R}$.

   b) $f(x) = [x]$ (dom. $f = \mathbb{R}$), where $[x]$ denotes the greatest integer in x but not greater than x.

   **c) $y = f(x) = \dfrac{1}{2 - \cos 3x}$;   **d) $y = f(x) = \dfrac{x}{1+x^2}$.

**3. If f is a function defined over the entire set $\mathbb{Q}$ of rational numbers, satisfying $f(x+y) = f(x) + f(y)$, show that $f(x) = ax$, where $a = f(1)$.

4. Define: An odd function; An even function; A periodic function; A monotonic function; A strictly monotonic function; An Inverse function; A parametric function; Explicit and Implicit algebraic functions; Transcendental functions.

   [Give one example in each case]

**5. Prove that no periodic function can be expressed as $P(x)/Q(x)$, where P and Q are polynomials unless it is a constant.

6. Prove that any function of x, defined over $\mathbb{R}$, is the sum of an even and an odd function of x. Is the constant function even or odd?

   [Hints. $f(x) = \tfrac{1}{2}\{f(x) + f(-x)\} + \tfrac{1}{2}\{f(x) - f(-x)\}$]

7. Show that if x is a rational function of y and y is a rational function of x, then
$$Axy + Bx + Cy + D = 0.$$

8. Construct a polynomial in x of degree two such that

   a) when $x = a, b, c$, then the respective values of the polynomial $= \alpha, \beta, \gamma$.

   b) when $x = 0, 1, 2$, then the respective values of the polynomial become $1/c$, $1/(c+1)$ and $1/(c+2)$. Verify that the value is $1/(c+1)$ when $x = c + 2$.

9. Which of the following functions are even and which are odd?

   a) $f(x) = \sqrt{1+x+x^2} - \sqrt{1-x+x^2}$; b) $f(x) = \sin x + \cos x$;
   c) $f(x) = 4 - 2x^4 + \sin^2 x$; d) $f(x) = $ constant $\forall\, x$.

10. Find the period of

    a) $f(x) = \tan 2x$; b) $f(x) = \cot(x/2)$; **c) $f(x) = |\cos x|$.

[C]
[Graphs; hyperbolic identities]

1. Draw the graphs of

   a) $f(x) = \dfrac{|x|}{x}$;

   b) $f(x) = \dfrac{x^2 - 1}{x - 1}$;

   c) $f(x) = 1$;

   d) $f(x) = \begin{cases} 2, & \text{when } x \geq 0 \\ 0, & \text{when } x < 0 \end{cases}$;

   e) $f(x) = \begin{cases} x, & \text{when } 0 \leq x \leq \frac{1}{2} \\ 1-x, & \text{when } \frac{1}{2} < x \leq 1. \end{cases}$

2. Represent graphically the following functions:

   (a) $y = [x]$, where $[x]$ has the usual significance;
   (b) $y = x - [x]$; c) $y = \sqrt{x - [x]}$.

3. On the same graph draw the diagrams of

   (a) $y = x^2$, $y = \dfrac{1}{x^2}$, $y = \sqrt{x}$, $y = \dfrac{1}{\sqrt{x}}$.

   (b) $y = x^3$, $y = \dfrac{1}{x^3}$, $y = \sqrt[3]{x}$, $y = \dfrac{1}{\sqrt[3]{x}}$.

CHAPTER 3: REAL-VALUED FUNCTIONS OF A REAL VARIABLE

4. Draw the graphs of the following functions within the indicated intervals:
   a) $\sin x$ $(-\pi/2 \le x \le \pi/2)$; b) $\cos x$ $(0 \le x \le \pi)$;
   c) $\tan x$ $(-\pi/2 < x < \pi/2)$; d) $\sin^{-1} x$ $(-1 \le x \le 1)$;
   e) $\cos^{-1} x$ $(-1 \le x \le 1)$; f) $\tan^{-1} x$ $(-\infty < x < \infty)$.

5. What are the principal values of

$$\sin^{-1}(1/2); \quad \tan^{-1} 1; \quad \cot^{-1}(-1); \quad \sec^{-1} 2;$$
$$\cos^{-1} 0; \quad \text{cosec}^{-1}\sqrt{2}; \quad \cos^{-1}(-1/2); \quad \sin^{-1} 1.$$

6. Verify the following relations (for hyperbolic functions):

   (a) $1 + \cosh x = 2\cosh^2(x/2)$;

   (b) $1 - \cosh x = -2\sinh^2(x/2)$;

   (c) $\sinh x = \dfrac{2\tanh(x/2)}{1 - \tanh^2(x/2)}$;

   (d) $\cosh x = \dfrac{1 + \tanh^2(x/2)}{1 - \tanh^2(x/2)}$;

   (e) $\tanh x = \dfrac{2\tanh(x/2)}{1 + \tanh^2(x/2)}$;

   (f) $\sinh x - \sinh y = 2\cosh\dfrac{x+y}{2}\sinh\dfrac{x-y}{2}$;

   (g) $\cosh x - \cosh y = 2\sinh\dfrac{x+y}{2}\sinh\dfrac{x-y}{2}$;

   (h) $\coth\dfrac{x}{2} - \coth x = \text{cosech}\, x$;

   (i) $\sin^2\theta \cosh^2\phi + \cos^2\theta \sinh^2\phi = \dfrac{1}{2}(\cosh 2\phi - \cos 2\theta)$.

7. (a) If $\sinh x = \tan\theta$, express $\cosh x$ and $\tanh x$ in terms of θ and show that

$$x = \log_e\left\{\tan\left(\frac{\pi}{4} + \frac{\theta}{2}\right)\right\}.$$

   (b) Prove that $\tanh^{-1}\dfrac{x^2 - 1}{x^2 + 1} = \log x$.

8. Draw the graphs of

   (a) $\sinh x$ $(-\infty < x < \infty)$; $\sinh^{-1} x$ $(-\infty < x < \infty)$.

   (b) $\cosh x$ $(-\infty < x < \infty)$; $\cosh^{-1} x$ $(1 \le x < \infty)$.

(c) $\tanh x$ $(-\infty < x < \infty)$; $\tanh^{-1} x$ $(-1 < x < 1)$.

9. (a) $y = \sec^{-1} x$ increases from 0 to $\pi/2$ as x increases from 1 to ∞ and $y = \sec^{-1} x$ increases from $\pi/2$ to π as x increases from $-\infty$ to -1; also $\sec^{-1} 1 = 0$ and $\sec^{-1}(-1) = \pi$. From these considerations draw the graph of $y = \sec^{-1} x$.

(b) $y = \operatorname{cosec}^{-1} x$ decreases from 0 to $-\pi/2$ in $-\infty < x \leq -1$ and decreases from $\pi/2$ to 0 in $1 \leq x < \infty$.

Now draw the graph of $y = \operatorname{cosec}^{-1} x$.

(c) $y = \cot^{-1} x$ decreases from π to 0 in $-\infty < x < \infty$ and $\cot^{-1} 0 = \pi/2$. Draw the graph of $\cot^{-1} x$.

10. Verify the following:

a) $\sinh^{-1} x = \log_e(x + \sqrt{x^2 + 1})$; $(-\infty < x < \infty)$

b) $\cosh^{-1} x = \log_e(x + \sqrt{x^2 - 1})$; $(1 \leq x < \infty)$

c) $\tanh^{-1} x = \dfrac{1}{2} \log_e \dfrac{1+x}{1-x}$; $(-1 < x < 1)$

d) $\coth^{-1} x = \dfrac{1}{2} \log \dfrac{x+1}{x-1}$; $(|x| > 1)$

e) $\operatorname{sech}^{-1} x = \log \dfrac{1 + \sqrt{1-x^2}}{x}$; $(0 < x < 1)$

f) $\operatorname{cosech}^{-1} x = \log \dfrac{1 \pm \sqrt{1+x^2}}{x}$. (+ve sign, if $x > 0$; −ve sign, if $x < 0$)

[D]

[Miscellaneous]

(Harder)

Hints are given for **-marked Problems

1. Draw sketch-graphs of the following functions:
   a) $f(x) = (-1)^{[x]}(x - [x])$; b) $f(x) = |x| - [|x|]$;
   c) $f(x) = (x - [x])^2$; d) $f(x) = \sqrt{[x]}$.

   $[[x] =$ greatest integer in x but not greater than $x]$

2. The function f is defined by

$$f(x) = \begin{cases} x, & \text{when } -1 \leq x \leq 1 \\ 1/x, & \text{when } x > 1 \text{ and when } x < -1. \end{cases}$$

Prove that the composite function fof (or simply $f \cdot f$) $= f$.

3. Decide whether the following statements are True or False, giving a general proof for a **True statement** and a counter example for a **false statement**:

   a) If the function f is odd, then $|f|$ is odd.
   b) For any function f, $\frac{1}{2}[f(x) + f(-x)]$ is an even function.
   c) If the function f is monotone, then it is not even.
   d) If the function f is periodic, then it is not monotone.
   e) If the function f is non-empty and one-to-one, then it is not even.
   f) If the function f is non-empty and one-to-one, then it is not periodic.

**4. The function f is odd and g is a function whose domain is same as the range of f and the composite function $g o f$ is odd. Prove that g is odd.

5. The functions f and g are odd and even respectively. State whether the following functions are odd, even, not necessarily either. Justify your assertion:

   a) $f + g$; b) fg; c) f/g; d) $f o g$.

6. Prove that if f is periodic and has a period λ, then $|f|$ is periodic and has a period λ.

7. a) Give an example of a one-to-one function which is not monotone.
   b) In what ranges the following functions are strictly increasing?—strictly decreasing?
   $$f(x) = x(x-1); \quad g(x) = x(x^2-1); \quad h(x) = |x|.$$

**8. If $f(x)$ is a periodic function with period T, then prove that $f(ax+b)$ when $a > 0$, is periodic with period T/a.

**9. Check that the function $y = f(x) = x^2 - x + 1$ is an increasing function of x in $1/2 \leq x < \infty$ and then $3/4 \leq y < \infty$. Hence find the inverse function $x = \phi(y)$; state its domain and range. Write down this inverse function with y as dependent and x as independent variable. Can you then suggest a method to solve the equation
$$x^2 - x + 1 = \frac{1}{2} + \sqrt{x - \frac{3}{4}}?$$

10. Sketch the graphs of

   a) $f(x) = \begin{cases} \sin x, & -\pi \leq x \leq 0 \\ 2, & 0 < x \leq 1 \\ \frac{1}{x-1}, & 1 < x \leq 4; \end{cases}$

b) $f(x) = \cos x + |\cos x|$;

c) $f(x) = 2|x - 2| - |x + 1| + x$;

d) $y = f(x) = x^2 \operatorname{sgn} x$ where $\operatorname{sgn} x = $ signum function of x (**Ex. 3.4.9**).

Hints and Answers

[A]

1. $-\dfrac{1}{3}, \quad \dfrac{1-x^2}{1+x^2}, \quad \dfrac{1-g^2}{1+g^2}, \quad \tan^2 \dfrac{x}{2}, \quad \left(\dfrac{1-x}{1+x}\right)^2$

8. $0, 1, 2, 0, 1, 1, 3, -3, -3$.

[B]

$[\mathbb{R} = \{-\infty < x < \infty\}]$

1. a) $\mathbb{R} - \{2\}$;   b) $\mathbb{R}^+ \cup \{0\}$; c) $\mathbb{R} - \{x : x < 2\}$;

d) $\mathbb{R} - \{x : 2 < x < 3\}$; e) $\mathbb{R} - \{2\}$; f) $\mathbb{R} - \{0, \pm 1, \pm 2, \cdots\}$;

g) $-1 < x < 1$;   h) $\mathbb{R} - \{0\}$; i) $\mathbb{R} - \{0, \pm 1, \pm 2, \cdots\}$;

j) The function is defined for the values of x for which $-1 \leq \dfrac{3}{4+2\sin x} \leq 1$. Since $4 + 2\sin x > 0$ at any x, the problem is reduced to solving the inequality $3/(4 + 2\sin x) \leq 1$, whence $\sin x \geq -\dfrac{1}{2}$, i.e., $-\dfrac{\pi}{6} + 2n\pi \leq x \leq \dfrac{7\pi}{6} + 2n\pi (n = 0, \pm 1, \pm 2,$ etc.).

k) $1 \leq x \leq 2$.

2. a) $\{0, 1\}$; b) $\{0, \pm 1, \pm 2, \cdots$ etc.$\}$;

c) We have $\cos 3x = \dfrac{2y-1}{y}$. Since $-1 \leq \cos 3x \leq 1$, we have $-1 \leq \dfrac{2y-1}{y} \leq 1$, whence, taking into account that $y > 0$, we obtain $-y \leq 2y - 1 \leq y$ or, $1/3 \leq y \leq 1$;

d) $x = (1 \pm \sqrt{1 - 4y^2})/2y$, whence we should have $1 - 4y^2 \geq 0$ or, $-1/2 \leq y \leq 1/2$.

3. Put $x = y = 0$ in $f(x + y) = f(x) + f(y)$; then $f(0) = 0$.

Put $x = -y$; then $f(0) = f(x) + f(-x)$ or $f(x) = -f(-x)$ (odd function).

Put $x = 1, y = 1$; then $f(2) = 2f(1) = 2a$, where $a = f(1)$.

Then $f(3) = f(2) + f(1) = 3f(1) = 3a$; by induction, we can verify that $\forall n \in \mathbb{N}$, $f(n) = an$. If n be a negative integer, then $n = -m$, where $m \in \mathbb{N}$.

$$\therefore f(n) = f(-m) = -f(m) = -mf(1) = an.$$

Thus $f(x) = ax$, where x is any integer or zero.

Let $x = p/q$, where $q \in \mathbb{N}$.

Then $f(p/q) + f(p/q) + \cdots$ to q terms

$= f(p/q + p/q + \cdots$ to q terms$) = f(p/q \cdot q) = f(p) = ap$.

CHAPTER 3: REAL-VALUED FUNCTIONS OF A REAL VARIABLE 171

Thus $qf(p/q) = ap$, i.e., $f(p/q) = ap/q$.
Hence, $f(x) = ax$ for all $x \in \mathbb{Q}$.

5. Assume $f(x) = P(x)/Q(x), P, Q$ are polynomials in x and let $f(x) = f(x+a)$, for all values of x for which $f(x)$ is defined. Let $f(0) = k$. Then $P(x) - kQ(x) = 0$ is satisfied by $x = 0, a, 2a$, etc. (i.e., for an infinite number of values of x). Thus it becomes an identity. Hence $f(x) = k$ for all x.

8. a) $\alpha \frac{(x-b)(x-c)}{(a-b)(a-c)} + \beta \frac{(x-c)(x-a)}{(b-c)(b-a)} + \gamma \frac{(x-a)(x-b)}{(c-a)(c-b)}$. [Take $f(x) = \lambda x^2 + \mu x + \nu$].

9. a) odd; b) neither even nor odd; c) even; d) even.

10. a) $\pi/2$; b) 2π; c) $f(x) = |\cos x| = \sqrt{\cos^2 x} = \sqrt{\frac{1}{2}(1+\cos 2x)}$; since $\cos 2x$ has a period π, hence the given function has the same period.

[C]

5. $\frac{\pi}{6}$; $\frac{\pi}{4}$; $\frac{3\pi}{4}$; $\frac{\pi}{3}$; $\frac{\pi}{2}$; $\frac{\pi}{4}$; $\frac{2\pi}{3}$; $\frac{\pi}{2}$.

[D]

4. Let $x \in$ dom. (g). Then $\exists\, y \in$ dom. (f) s.t. $f(y) = x$.
Hence $-y \in$ dom. (f) and $f(-y) = -f(y)$, so that
$-x = -f(y) = f(-y) \in$ Range $(f) =$ dom. (g).
Further, $g(-x) = g(f(-y)) = -g(f(y)) = -g(x)$.

7. b) $(\frac{1}{2}, \infty)$ Inc. and $(-\infty, \frac{1}{2})$ Dec.; $(1/\sqrt{3}, \infty)$ and $(-\infty, 1/\sqrt{3})$ Inc.; $(-1/\sqrt{3}, 1/\sqrt{3})$ Inc.; $(0, \infty)$ Inc. and $(-\infty, 0)$ Dec.

8. $f\{a(x+T/a)+b\} = f\{(ax+b)+T\} = f(ax+b)$, since T is a period of $f(x)$.
Next, let T_1 be a positive number s.t. $f\{a(x+T_1)+b\} = f(ax+b)$.
Let us take an arbitrary point x' from the domain of definition of the function $f(x)$ and put $x' = (x-b)/a$. Then
$f(ax'+b) = f\left\{a\frac{x-b}{a} + b\right\} = f(x) = f\{a(x'+T_1)+b\} = f(ax'+b+aT_1) = f(x+aT_1)$,
whence it follows that the period $T \le aT_1$, i.e., $T_1 \ge T/a$ and T/a is the period of the function $f(ax+b)$.

9. $y = x^2 - x + 1 = (x - \frac{1}{2})^2 + \frac{3}{4}$ increases in $1/2 \le x < \infty$ and there we have $3/4 \le y < \infty$. Hence, defined in $3/4 \le y < \infty$ is the inverse function $x = g(y)(x \ge 1/2)$ which is obtained from $x^2 - x + (1-y) = 0$, i.e., $x = g(y) = \frac{1}{2} + \sqrt{y - \frac{3}{4}} = \phi(y)$.

Let us now solve the eqn. $x^2 - x + 1 = \frac{1}{2} + \sqrt{x - \frac{3}{4}}$.
Since the graphs of the original and the inverse function can intersect only on the line $y = x$, solving the eqn. $x^2 - x + 1 = x$ we get $x = 1$.

MAIN POINTS: CHAPTER 3

1. • What is a Function?
   • What is a Real-valued Function of a Real Variable? (**art. 3.1**)
   • Single-Valued & Multiple-Valued Functions (**art. 3.3**)
2. Continuous and Discrete Real Variables. (**art. 3.2**)
3. Methods of Specifying Functions
   [Using Analytic expressions, Using Tables and Using Graphs] (**art. 3.3**)
4. **Graph**: Constant Function; Identity Function; Absolute-Value Function; **Step-Functions** (Prime number function, Greatest integer functions); Signum Function: Dirichlet's Function (**art. 3.4**)
5. **Different classes of functions—their graphs:** (**art. 3.5**)
   (i) Power Functions: x^n (n odd, n even) and their Reciprocals
   (ii) Polynomial Functions: $f(x) = a_0 x^n + a_1 x^{n-1} + \cdots + a_{n-1} x + a_n (a_0 \neq 0)$
   (iii) Rational Functions: $\frac{f(x)}{g(x)}$. Not defined for $g(x) = 0$
   (iv) Explicit and Implicit Functions
   (v) Parametric functions
   (vi) Odd and Even functions
   (vii) Step Functions
   (viii) Restriction Functions
   (ix) Composition of Functions: Function of a Function
   (x) Monotonic Functons
   (xi) Inverse Functions
   (xii) Periodic Functions
   (xiii) Algebraic Functions
   (xiv) Transcendental Functions
   Exponential; Logarithmic; Circular; Inverse Circular Hyperbolic; Inverse Hyperbolic Functions.

AN INTRODUCTION TO
ANALYSIS

(Differential Calculus—Part I)

Chapter 4

Elements of Point Set Topology

Chapter 4

Elements of Point Set Topology

4.1 Introduction

In the present Chapter we shall study some very important sets of real numbers like open sets, closed sets, compact sets etc. Since real numbers can be represented by points on a line, we may call these sets **linear points sets** and their study is known as the study of **point set topology**. The concepts—neighbourhoods, interior points, exterior points, boundary points, limit points, adherent points etc.—will be introduced in a systematic manner.

4.2 Intervals: Neighbourhoods: Interior points: Interior of a set

We shall be concerned with **Real numbers only**.

1. Intervals: open, closed, half-open

The most common type of sets of real numbers are

(i) **Open intervals** like (a, b) or $]a, b[$ or $a < x < b$. The set $\{x : x \in \mathbb{R}$ and $a < x < b\}$ is called an *open interval*.

(ii) **Closed intervals** like $[a, b]$ or $a \leq x \leq b$.

The set $\{x : x \leq \mathbb{R}$ and $a \leq x \leq b\}$ is called *a closed interval*.

(iii) **Half-open intervals** like

$$[a, b) = \{x : x \in \mathbb{R} \text{ and } a \leq x < b\} \text{ (open on the right)}$$
$$\text{and } (a, b] = \{x : x \in \mathbb{R} \text{ and } a < x \leq b\} \text{ (open on the left)}$$

We introduce two symbols: $+\infty$ (plus infinity) and $-\infty$ (minus infinity).

They are symbols (**not real numbers**) having meaning explained below:
(iv) **Infinite Intervals**:

$$(a, +\infty) = \{x : x \in \mathbb{R} \text{ and } x > a\} \text{ (open on the left)}$$
$$[a, +\infty) = \{x : x \in \mathbb{R} \text{ and } x \geq a\} \text{ (closed on the left)}$$
$$(-\infty, b) = \{x : x \in \mathbb{R} \text{ and } x < b\} \text{ (open on the right)}$$
$$(-\infty, b] = \{x : x \in \mathbb{R} \text{ and } x \leq b\} \text{ (closed on the right)}$$

(v) **Real number system**—all real numbers, +ve or −ve or zero $= (-\infty, \infty)$

(vi) **Extended Real number System**

$$\mathbb{R}_e = (-\infty, \infty) \cup \{\text{Two symbols } -\infty \text{ and } +\infty\}$$

2. Neighbourhoods

A. Definition of ϵ-negibourhood of a point $a \in \mathbb{R}(N_\epsilon(a))$

Let $a \in \mathbb{R}$ and $\epsilon > 0$.

Then the ϵ-**neighbourhood** of a is the set

$$N_\epsilon(a) = \{x : x \in \mathbb{R} \text{ and } |x - a| < \epsilon\}.$$

For $a \in \mathbb{R}$, the expression "x belongs to $N_\epsilon(a)$" is equivalent to

$$-\epsilon < x - a < \epsilon \text{ or } a - \epsilon < x < a + \epsilon$$

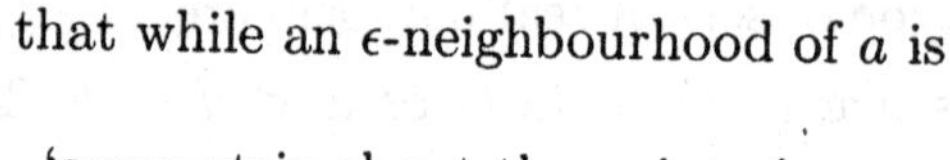

Fig. 4.3.1 ε-nbd. of a on the real line

B. A **neighbourhood** of a point $a \in \mathbb{R}$ is any set N that contains ϵ-**neighbourhood** $[N_\epsilon(a) = (a - \epsilon, a + \epsilon)]$ of the point a for some $\epsilon > 0$. We denote neighbourhood of a point a by $N(a)$. [observe that while an ϵ-neighbourhood of a is required to be

'symmetric about the point a',

the general neighbourhood does not require this special feature of symmetry about a]

C. Deleted neighbourhood of a point $a \in \mathbb{R}$.

We have

ϵ-**neighbourhood** of $a = N_\epsilon(a) = \{x : x \in \mathbb{R} \text{ and } a - \epsilon < x < a + \epsilon \text{ or } |x - a| < \epsilon\}$

If from the open interval $(a - \epsilon, a + \epsilon)$ we exclude or delete the point a we get what we call a **deleted ϵ-neighbourhood** of a and is denoted by $N'_\epsilon(a)$,

i.e., $N'_\epsilon(a) = \{x : x \in \mathbb{R} \text{ and } a - \epsilon < x < a + \epsilon (x \neq a) \text{ or } 0 < |x - a| < \epsilon\}.$

CHAPTER 4: ELEMENTS OF POINT SET TOPOLOGY 175

We may write $N'_\epsilon(a)$ as $(a - \epsilon, a) \cup (a, a + \epsilon)$

Note. Hence forward we shall write *nbd.* or *nbd* for the word **neighbourhood**.

Examples.

Example 4.2.1. $.01 - nbd$ *of the point* $3 \equiv (3 - .01, 3 + .01)$

The set x of real number satisfying $3 - .01 < x < 3 + .01$

When the point 3 is excluded, we get deleted $.01 - nbd.$ of 3.

Example 4.2.2. *Let* $a \in \mathbb{R}$. *If* x *belongs to the neighbourhood* $N_\epsilon(a)$ *for every* $\epsilon > 0$, *then* $x = a$.

For, if x satisfies $|x - a| \leq \epsilon$, for every $\epsilon > 0$, then $|x - a| = 0$ and hence $x = a$.

3. Interior point of a set $S \subset \mathbb{R}$: **Interior of** S

Let S be a subset of $\mathbb{R}$. Suppose x is any point of $S (x \in S)$.

Then x is called **an interior point of** S if $\exists$ some neighbourhood $N(x)$ of x such that $N(x) \subset S$.

i.e. $\exists$ **some** $nbd. N(x)$, **all of whose points belong to** S.

Formal Definition: $x \in S$ (where $S \subset \mathbb{R}$) **is an interior point** of S if $\exists$ an open interval (a, b) containing x and contained in S i.e.
$$x \in (a, b) \subset S$$

Example 4.2.3. *Let* $S = $ *open interval* $(1, 2)$.

Every point of S *is* **an interior point** *of* S *because for each* $x \in (1, 2)$ *we can find some nbd which lies entirely within* $(1, 2)$.

Definition of Interior of S

The set of all interior points of $S \subset \mathbb{R}$ is called the **Interior** of S, denoted by **int. S** or S^i or **int S**.

Example 4.2.4. *The interior of the closed interval* $[a, b]$ *is the open interval* (a, b) *(Note* a, b *are* **not interior points**)

Example 4.2.5. *The interior of a finite set of real numbers* $\{x_1, x_2, \cdots, x_n\}$ *is the null set* ϕ. *No point of the set is an interior point.*

Illustrative Examples: (On Neighbourhoods)

Example 4.2.6. *The set* $\mathbb{R}$ *of all real numbers is the neighbourhood of each of its points.*

Example 4.2.7. *The set $\mathbb{Q}$ of all rational numbers is not the neighbourhood of any of its points.*

Example 4.2.8. *The open interval $S = (a, b)$ is a neighbourhood of each of its points.*

Example 4.2.9. *The closed interval $T = [a, b]$ is the neighbourhood of each point of (a, b) but not a nbd of the end points a and b.*

Example 4.2.10. *The null set ϕ is a neighbourhood of each of its points in the sense that there is no point in ϕ of which it is not a neighbourhood.*

Example 4.2.11. *A non-empty finite set is not a nbd. of any point*

[A set S can be a *nbd* of a point if it contains an open interval containing the point. Since an interval necessarily contains an infinite number of points, therefore in order that a set may become a *nbd.* of a point it must necessarily contain an infinity of points. Thus a finite set cannot become a *nbd.* of any point.]

Example 4.2.12. *A set containing a nbd. of a point x is also a nbd. of x. i.e. if N is nbd. of a point x and $M \supset N$ then M is also nbd. of x.*

[Since N is a *nbd* of x, therefore, $\exists$ an open interval $(x - \epsilon, x + \epsilon), \epsilon > 0$ such that

$$x \in (x - \epsilon, x + \epsilon) \subset N$$
$$\implies x \in (x - \epsilon, x + \epsilon) \subset N \subset M \ (\because M \supset N)$$
$$\implies x \in (x - \epsilon, x + \epsilon) \subset M$$

Hence M is a neighbourhood of x]

Example 4.2.13. *Let $x \in \mathbb{R}$. The union of two neighbourhoods of x is a neighbourhood of x.*

[Let $N_1 \subset \mathbb{R}$ and $N_2 \subset \mathbb{R}$ be two *nbd.s* of x.
Then $\exists$ open intervals $(a_1.b_1)$ and (a_2, b_2) such that

$$x \in (a_1, b_1) \subset S_1$$
$$\text{and } x \in (a_2, b_2) \subset S_2$$

Then $a_1 < b_1, a_2 < b_1, a_1 < b_2, a_2 < b_2$.
Let $a_3 = \text{minimum } \{a_1, a_2\}$ and
$b_3 = \text{maximum } \{b_1, b_2\}$.

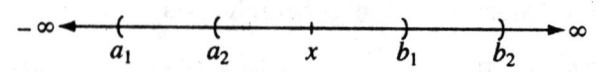

Fig. 4.3.2

Then $(a_1, b_1) \subset S_1 \cup S_2$ and $(a_2, b_2) \subset S_1 \cup S_2$

$$\Rightarrow (a_3, b_3) = (a_1, b_1) \cup (a_2, b_2) \subset S_1 \cup S_2.$$

$\therefore x \in (a_3, b_3) \subset S_1 \cup S_2$

Therefore $S_1 \cup S_2$ is a nbd. of x.]

Note. *The union of a finite number of neighbourhoods of a point $x \in \mathbb{R}$ is a nbd. of x.*

In fact, we can prove that the union of an arbitrary number of neighbourhoods of a point $x \in \mathbb{R}$ is a nbd. of x.

Example 4.2.14. *If M and N are two neighbourhoods of a point x, then $M \cap N$ is also neighbourhood of x.*

[Since M and N are neighbourhoods of x, therefore, there exist $\epsilon_1 > 0$ and $\epsilon_2 > 0$ such that

$$x \in (x - \epsilon_1, x + \epsilon_1) \subset M$$
$$\text{and } x \in (x - \epsilon_2, x + \epsilon_2) \subset N$$

Let $\epsilon = \min(\epsilon_1, \epsilon_2)$ so that $\epsilon \leq \epsilon_1$ and $\epsilon \leq \epsilon_2$. Thus

$$\left. \begin{array}{l} x \in (x - \epsilon, x + \epsilon) \subset (x - \epsilon_1, x + \epsilon_1) \subset M \\ x \in (x - \epsilon, x + \epsilon) \subset (x - \epsilon_2, x + \epsilon_2) \subset N \end{array} \right\}. \tag{4.1}$$

$\therefore x \in (x - \epsilon, x + \epsilon) \subset M \cap N$. Hence $M \cap N$ is nbd. of x.]

Note. (a) The intersection of a finite number of neighbourhoods of a point $x \in \mathbb{R}$ is a neighbourhood of x.

Let $M_1, M_2, \cdots, M_n$ be a finite number (say n) nbds of $x \in \mathbb{R}$.
Then $\exists \delta_1, \delta_2, \cdots, \delta_n$ (all > 0) such that $x \in (x - \delta_i, x + \delta_i) \subset M_i$ ($i = 1, 2, 3, \cdots, n$)
Let $\delta = \min(\delta_1, \delta_2, \cdots, \delta_n)$. Then

$$x \in (x - \delta, x + \delta) \subset M_1 \cap M_2 \cap \cdots M_n$$

$\therefore M_1 \cap M_2 \cap M_3 \cap \cdots \cap M_n$ is a nbd of x.

(b) But the intersection of an arbitrary collection of neighbourhoods of a point may not always be a neighbourhood of that point. Consider the following example.
Let

$$G_n = \left(-\frac{1}{n}, \frac{1}{n} \right) \quad \forall n \in \mathbb{N}.$$

Then G_n is a neighbourhood of the point '0' for each n. But clearly

$$\bigcap_{n=1}^{\infty} G_n = \{0\},$$

where $\{0\}$ is not a neighbourhood of 0. **Hence the statement (b) follows.**

Example 4.2.15. *Define a neighbourhood of a point. Show that the intersection of all the neighbourhoods of a point $a \in \mathbb{R}$ is the singleton set $\{a\}$.*

Definition. A subset $S \subset \mathbb{R}$ is called a neighbourhood of a point $x \in \mathbb{R}$ if $\exists$ some $\epsilon > 0$ such that $(x - \epsilon, x + \epsilon) \subset S$.

Second part.
Let $\{N(a)\}$ be a family of all neighbourhoods of a point $a \in \mathbb{R}$. We have to show that

$$\cap N(a) = \{a\}$$

Consider any point $b \neq a$. Let $\epsilon = |a - b|$.
Then $(a - \epsilon, a + \epsilon)$ is a **nbd.** of a

and $b \notin (a - \epsilon, a + \epsilon)$ since
$$b \in (a - \epsilon, a + \epsilon)$$
$$\implies a - \epsilon < b < a + \epsilon$$
$$\implies |a - b| < \epsilon, \text{ a contradiction.}$$

Since $b \notin (a - \epsilon, a + \epsilon)$. So $b \notin N(a)$

Hence $\cap N(a) = \{a\}$.

Example 4.2.16. *Can a set N be a neighbourhood of any point of its complement $\mathbb{R} - N$?*

Ans. We shall prove that N cannot be a neighbourhood of any point of $R - N$.
Let $a \in \mathbb{R} - N$. Then $a \notin N$.
Suppose N is a neighbourhood of a. Then $\exists$ some $\epsilon > 0$ such that

$$a \in (a - \epsilon, a + \epsilon) \subset N \implies a \in N,$$

which is a contradiction.

Example 4.2.17. *Can a non-empty finite set be a neighbourhood of any of its points?*

Ans. Let $a \in N$ where N is a non-empty finite set. If N is a neighbourhood of a, then

$$\text{for some } \epsilon > 0, a \in (a - \epsilon, a + \epsilon) \subset N$$

This is impossible, since N is finite and $(a - \epsilon, a + \epsilon)$ contains infinite number of points.

On Interior points:

Example 4.2.18. Let $S = \{1, \frac{1}{2}, \frac{1}{3}, \cdots, \frac{1}{n}, \cdots\}$

Let $x \in S$. Then x is a **not an interior point of** S become every nbd. of x contains points not belonging to S.

So x cannot be interior point of S.

$$\therefore \text{int } S \quad i.e. \quad S^i = \phi.$$

CHAPTER 4: ELEMENTS OF POINT SET TOPOLOGY

Using similar arguments we can prove that

$$\text{int } \mathbb{N} = \phi \quad (\mathbb{N} \text{ is the set of all natural numbers})$$
$$\text{int } \mathbb{Z} = \phi \quad (\mathbb{Z} \text{ is the set of all integers})$$
$$\text{int } \mathbb{Q} = \phi \quad (\mathbb{Q} \text{ is the set of all rational numbers})$$

But see that int $\mathbb{R} = \mathbb{R}$ ($\mathbb{R}$ is the set of all real numbers) ($\because$ every **nbd.** of any $x \in \mathbb{R}$ contains points of $\mathbb{R}$ only and so every point of $\mathbb{R}$ is an interior point)

Example 4.2.19. *Let $S =$ open interval (a, b).*

Every point of this set S is an interior point of S.

Therefore ***int*** $S = (a, b) = S$

*But if $S = $ closed interval $[a, b]$, then the points of (a, b) are all interior points but the end points a, b are **not interior points**.*

Therefore interior of $[a, b]$ is (a, b).

Example 4.2.20. *The interior of a finite set is ϕ (why?)*

A Few Exercises on Neighbourhoods

Try Yourself

Q1. Define a neighbourhood of a point $x \in \mathbb{R}$. Write down any two neighbourhoods of each of the following points: $0, -1, +1, -3/5, \sqrt{2}$.

Q2. Which of the following statements are true?

(i) $[3, 4]$ is **nbd.** of 2.

(ii) $[3, 4]$ is a **nbd.** of 4.

(iii) $(2, 3)$ is a **nbd.** of 3.

(iv) $(1, 2) \cup (3, 4)$ is a **nbd.** of 1.

Q3. Give an example to justify that arbitrary intersection of the **nbds** of a point may not be a **nbd.** of the point.

Q4. Give an example of each of the following:

(i) A set which is a **nbd.** of each of its points.

(ii) A set which is not a **nbd.** of any of its points.

(iii) A set which is a **nbd.** of each of its points with the exception of one point.

(iv) A set which is not an interval but is a neighbourhood of each of its points.

(Hint. Consider the union of two open intervals which is not an interval but is an open set and so it is a *nbd.* of each of its pts.)

4.3 Open Sets in $\mathbb{R}$: Theorems on Open Sets

Definition. A set $S \subseteq \mathbb{R}$ is said to be an **open set** if all its points are interior points.

In other words, a set $S \subseteq \mathbb{R}$ is an open set for each $x \in S$ there exists an open interval I_x such that
$$x \in I_x \subset S,$$
i.e., S is a *nbd.* of each of its points.

Evidently S is an open set if and only if
$$\text{int.}S (\text{i.e., } S^i) = S$$

Note. To show that $S \subseteq \mathbb{R}$ is an open set, it suffices to prove that each point $x \in S$ has an $\epsilon - nbd$ contained in S, i.e., for each $x \in S \exists \epsilon > 0$ such that
$$(x - \epsilon, x + \epsilon) \subseteq S.$$

Of course, S **is not an open set** if it is not a **nbd.** of one of its points or that *there is at least one point of the set which is not an interior point.*

Illustrative Examples

Example 4.3.1. *The set $\mathbb{R} = (-\infty, \infty)$ of real numbers is an open set because $\mathbb{R}$ is nbd. of each of its points (or each point $\mathbb{R}$ is an interior point).*

Example 4.3.2. *The null set ϕ is open; for there is no point in ϕ of which ϕ is not a neighbourhood.*

Example 4.3.3. *An open interval (a, b) is an open set or Every open interval (a, b) is a nbd. of each of its points.*

[*Proof.* Let x be any point of (a, b) so that
$$a < x < b$$

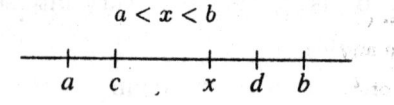

$$\begin{array}{c|c|c|c|c} \hline a & c & x & d & b \end{array}$$

Fig. 4.3.3

Let c, d be two real numbers such that $a < c < x$ and $x < d < b$ so that we have $a < c < x < d < b$
$$\implies x \in (c, d) \subset (a, b)$$

Thus the given interval (a, b) contains an open interval containing the point x, and is therefore a **nbd** of x.

Hence the open interval (a, b) is a **nbd** of each of its points and is therefore an open set.]

Example 4.3.4. *The closed interval $[a, b]$ is not an open set, because the two end points a and $b \in [a, b]$ but they are not interior points.*

CHAPTER 4: ELEMENTS OF POINT SET TOPOLOGY

Example 4.3.5. *A non-empty finite set $S \subset \mathbb{R}$ is not an open set because S is not a neighbourhood of any of its points.*

Example 4.3.6. *No point of the set $\mathbb{Q}$ of all rational numbers is an interior point; hence $\mathbb{Q}$ is not an open set.*

Example 4.3.7. *The discrete set $S = \{1, \frac{1}{2}, \frac{1}{3}, \frac{1}{4}, \cdots, \frac{1}{n}, \cdots\}$ is not an open set, because no point of S is an interior point (By definition, in order that S is open, every point of S should be an interior point).*

Example 4.3.8. *Every open set is an union of open intervals*

[Let S be an open set.
Then for each point $a \in S$, $\exists$ some $\epsilon_a > 0$, such that

$$a \in (a - \epsilon_a, +\epsilon_a) \subset S$$

It follows that

$$S \subset \bigcup_{a \in S}(a - \epsilon_a, a + \epsilon_a) \subset S$$

Hence

$$S = \bigcup_{a \in S}(a - \epsilon_a, a + \epsilon_a),$$

which proves the desired result]

Example 4.3.9. *The interior of a set $S \subseteq \mathbb{R}$ is an open set.*

[**Recall:** The set of all interior points of a given set $S \subseteq \mathbb{R}$ is called the **Interior of S (int S)**]
[Let S be a given set and S^i its interior.
  To prove: S^i is an open set.
  If $S^i = \phi$, then S^i is open ($\because \phi$ is an open set)
  Suppose $S^i \neq \phi$. Let x be any arbitrary point of S^i.
  As x is an interior point of S, $\exists$ an open Interval I_x such that $x \in I_x \subset S$
  But I_x being an open interval, is a **nbd.** of each of its points,
  $\implies$ every point of I_x is an interior point of I_x and $I_x \subset S \implies$ every point of I_x is an interior point of S.
  $\therefore I_x \subset S^i$
  $\implies x \in I_n \subset S^i \implies$ any point x of S^i is an interior point of S^i
  $\implies S^i$ is an open set.]

Theorems on open sets.

Theorem 4.3.1. *S is open iff $S = \textbf{int}S$, where $S \subseteq \mathbb{R}$*

Proof. (i) Let $S \subseteq \mathbb{R}$. **To prove:**

If S is open, then int. $S = S$.

Case 1. For $S = \phi$, the statement is clearly true because int $\phi = \phi$ and ϕ is an open sent.

Case 2. Let $S \neq \phi$ be an open set.

Suppose $x \in S$. Since S is open, x must be an interior point of S, i.e. $x \in int.S$

$$\therefore S \subset int.S \tag{4.1}$$

Let $y \in intS$. Since y is an interior point of S,

$$\therefore y \in S$$

Thus $y \in intS \implies y \in S$

$$\therefore int.S \subseteq S$$

From (1) and (2) we conclude $S = intS$

(ii) *Conversely* we prove that for a non-empty set S,

If int $S = S$, then S is open.

Let $x \in S$. Then $x \in intS (\because S = intS)$

Thus every point of S is an interior point of S

$\therefore S$ is an open set.

Hence we have proved the following result:

$$S \text{ open} \implies S = intS$$
$$\text{and} \quad S = intS \implies S \text{ is an open set.}$$

Theorem 4.3.2. *The union of two open sets A and B in $\mathbb{R}$ is an open set.*

Proof. Let A and B be two open sets in $\mathbb{R}$.

Let $S = A \cup B$. To prove that S is an open set.

Let $x \in S$. Then $x \in A$ or $x \in B$.

Suppose, $x \in A$. Since A is an open set, x must be an interior point of A, i.e., $\exists$ some $\delta - $ **nbd** of x such the

$$N_\delta(x) \subset A \subset S \quad (\because A \subset A \cup B)$$

$\therefore$ for any point $x \in S$, $N_\delta(x) \subseteq S$.

CHAPTER 4: ELEMENTS OF POINT SET TOPOLOGY

$\therefore$ S is a *nbd.* of each of its points.

$\therefore$ $S = A \cup B$ is an open set.

If, however, $x \in B$, then we can, with similar arguments, arrive at the conclusion that $S = A \cup B$ is an open set.

This completes the proof of the theorem.

Theorem 4.3.3. *The union of an arbitrary collection of open sets is an open set.*

Proof. Let $\{S_\alpha : \alpha \in A\}$ be an arbitrary family of open sets S_α (A is the index set). Let

$$S = \bigcup_{\alpha \in A} S_\alpha.$$

To prove that S is an open set

Suppose $x \in S$. Then x must belong at least one of sets of the collection $\{S_\alpha : \alpha \in A\}$. Suppose $x \in S_\alpha$ for some fixed $\alpha \in A$.

Since S_α is an open set, $\exists$ some **nbd** $N(x)$ such that

$$N(x) \subseteq S_\alpha$$

But S_α is a subset of S

Hence $N(x) \subseteq S$ i.e. x is an interior point of S.

Thus every point $x \in S$ is also an interior point of S.

Hence S is an open set.

Theorem 4.3.4. *The intersection of two open sets in $\mathbb{R}$ is an open set.*

Proof. Let S_1 and S_2 be two open sets in $\mathbb{R}$.

Case 1. Suppose that $S_1 \cap S_2 = \phi$.

Since ϕ is an open set, **$S_1 \cap S_2$ is an open set when $S_1 \cap S_2 = \phi$.**

Case 2. Suppose that $S_1 \cap S_2 \neq \phi$.

Let $x \in S_1 \cap S_2$. Then $x \in S_1$ and also $x \in S_2$.

Since S_1 is an open set and $x \in S_1$, x is an interior point of S_1. Hence $\exists$ a positive ϵ_1 such that $\epsilon_1 - $ **nbd.** of x i.e. $N_{\epsilon_1}(x) \subset S_1$.

Again S_2 is an open set and $x \in S_2$, x is an interior point of S_2. Hence $\exists$ a positive ϵ_2 such that

$$\epsilon_2 - \textbf{nbd} \text{ of } x_2 \text{ i.e. } N_{\epsilon_2}(x) \subset S_2.$$

Let $\epsilon = \min\{\epsilon_1, \epsilon_2\}$. Then $\epsilon > 0$ and

$$N_\epsilon(x) \subset N_{\epsilon_1}(x) \subset S_1$$
$$\text{and } N_\epsilon(x) \subset N_{\epsilon_2}(x) \subset S_2$$

Hence $N_\epsilon(x) \subset S_1 \cap S_2$.

$\Rightarrow x$ is an interior point of $S_1 \cap S_2$.

Since x is an arbitrary point of $S_1 \cap S_2$, it follows that

$$S_1 \cap S_2 \text{ is an open set, when } S_1 \cap S_2 \neq \phi$$

This completes the proof of the theorem.

Theorem 4.3.5. *The intersection of a finite number of open sets in $\mathbb{R}$ is an open set.*

Proof. Let $S_1, S_2, S_3, \cdots, S_n$ be m open sets in $\mathbb{R}$.

Let
$$S = \bigcap_{k=1}^{m} S_k,$$

where each S_k is an open set.

Case 1. If $S = \phi$, then S is open since ϕ is an open set.

Case 2. If $S \neq \phi$, we take any $x \in S$. Then $x \in S_k$ for each $k = 1, 2, 3, \cdots, m$. Since each S_k is an open set, $\exists\ \epsilon_1, \epsilon_2, \cdots, \epsilon_k$ (all > 0) such that

$$N_{\epsilon_1}(x) \subset S_1, N_{\epsilon_2}(x) \subset S_2, \cdots, N_{\epsilon_k}(x) \subset S_k.$$

Let $\epsilon = \min\{\epsilon_1, \epsilon_2, \epsilon_3, \cdots, \epsilon_k\}$ and consider $N_\epsilon(x)$.

Then clearly $N_\epsilon(x) \subseteq S_k$ for each $k = 1, 2, 3, \cdots, m$.

$\therefore N_\epsilon(x) \subseteq S$ i.e. x is an interior point of S.

Since x is an arbitrary element of S it follows that every member of S is an interior point of S.

Hence S is an open set.

This proves the theorem.

An Important observation:

We have observed that from given open sets we can construct new open sets by taking their *arbitrary union* (**Theorem 4.3.3**) or by taking their *finite intersection* (**Theorem 4.3.5**).

However, *by taking arbitrary intersection of open sets* we may or may not get an open set, e.g.

$$\bigcap_{n=1}^{\infty} \left(-\frac{1}{n}, \frac{1}{n}\right) = \{0\} = \textbf{Not an open set} \quad (\text{See Ex. 5 given above}),$$

though each open interval $\left(-\frac{1}{n}, \frac{1}{n}\right)$ is an open set but

$$\bigcap_{n=1}^{\infty}\left(-\frac{n}{n+1}, \frac{n}{n+1}\right) = \left(-\frac{1}{2}, \frac{1}{2}\right) \cap \left(-\frac{2}{3}, \frac{2}{3}\right) \cap \left(-\frac{3}{4}, \frac{3}{4}\right) \cap \cdots$$
$$= \left(-\frac{1}{2}, \frac{1}{2}\right) = \text{An open set}$$

($\because$ every open interval is an open set)

Here also the open intervals $\left(-\frac{n}{n+1}, \frac{n}{n+1}\right)$ are open sets.

Q. What can you say about the infinite intersection of open sets $\bigcap_{n=1}^{\infty}\left(0, 1+\frac{1}{n}\right)$?

Ans. Not an open set. Here intersection $= (0, 1]$

Note: The intersection and union of two open sets are open are sufficient conditions but not necessary. For example:

$[0, 2) \cap (1, 3] = (1, 2)$. Here $(1, 2)$ is open whereas $[0, 2)$ and $(1, 3]$ are not open and $(0, 2] \cup [1, 3) = (0, 3)$. Here $(0, 3)$ is open whereas $[1, 3)$ is not open.

Theorem 4.3.6. *The interior of a set $S \subseteq \mathbb{R}$ is the largest open subset of S*

OR

The interior of a set $S \subseteq \mathbb{R}$ contains every open subset of S.

Proof. We have already proved that the interior S^i of a set S is an open subset of S.

We shall now prove that any open subset S_1 of S is contained in S^i.

Let x be any point of S_1.

Since an open set is a **nbd.** of each of its points, therefore S_1 is a **nbd.** of x. But S is a superset of S_1.

$\therefore S$ is also **nbd.** of x

$\implies x$ is an interior point of S

$\implies x \in S^i$

Thus $x \in S_1 \implies x \in S^i$ i.e., $S_1 \subset S^i$.

Hence every open subset of S is contained in its interior $S^i \implies S^i$, the interior of S, is the largest open subset of S.

Corollary. Interior of a set S is the union of all open subsets of S.

4.4 Important Definitions

[Point of Accumulation—Isolated Point—Boundary point/Exterior point—Derived Set S^i—Adherent Point—Closure of a set S—Closed Set]

I. Definition of an Accumulation Point or Limit Point of a set $S \subset \mathbb{R}$.

Let S be a subset of real number system $\mathbb{R}(S \subseteq \mathbb{R})$. A real number ξ, which may or may not belong to S is called an **Accumulation Point of S** (also called a **Limit Point** of S) if every **nbd.** of ξ contains at least one point of S other than the point ξ.

This definition implies: ξ is a limit point of $S \subseteq \mathbb{R}$ if every deleted **nbd** of ξ contains a point of S i.e. if

$$N'_\epsilon(\xi) \cap S = [N_\epsilon(\xi) - \{\xi\}] \cap S \neq \phi \text{ for each } \epsilon > 0$$

We shall use the terms: *Limit Point* or *Limiting Point* or *Accumulation point* or *Point of Accumulation* or *Cluster point*. They are all equivalent.

Note. (i) A limit point of a set may or many not belong to S.

(ii) A set may have no limit point, a unique limit point, a finite number of limit points or an infinite number of limit points. See the examples given below.

Example 4.4.1. *The set $S = \{\frac{1}{n} : n \in \mathbb{N}\} = \{1, \frac{1}{2}, \frac{1}{3}, \frac{1}{4}, \cdots\}$ has a **a unique limit point** 0 (zero). This limit point $\notin S$.*

Example 4.4.2. *Every point of the set S = closed interval $[a, b] \subset \mathbb{R}$ is a limit point of S and no point not belonging to the interval is a limit point, i.e.*

the set of limit of a closed interval coincides with the interval.

This set has **infinitely many limit points,** each such limit point belong to the set.

In the set S = open interval (a, b), every point of S is a limit point and the two end-points a, b are **also limit points of S**.

Thus an open interval has two limit points (viz, a and b) which do not belong to the interval and other limit points belong to the interval.

Example 4.4.3. *The set $S = \mathbb{Q}$ of all rational numbers has every real number a limit point of $\mathbb{Q}$—the rational limit points belong to the set and irrational limit points do not belong to the set.*

Our definition of a limit point of a set S requires at least one point of S other than the point itself.

The theorem given below shows that every neighbourhood of a limit point of S contains infinitely many points of S.

Theorem 4.4.1. *If ξ is a point of accumulation of a set $S \subseteq \mathbb{R}$, then every **nbd** of ξ contains infinitely many points of S.*

Proof. Assume the contrary i.e. suppose a **nbd.** of ξ contains only a finite number of points of S, say $x_1, x_2, \cdots, x_m$ (all $\in S$).

If r is the minimum of the quantities

$$|x_1 - \xi|, |x_2 - \xi|, \cdots, |x_n - \xi|$$

then $r/2$-neighbourhood of ξ i.e. $N_{r/2}(\xi)$ will contains no point of S.

This is contradiction to our definition of limit point of a set S.

Hence we assert that the theorem must be true.

Observations. This theorem implies that a set S cannot have a limit point unless it contains infinitely many points to begin with.

The converse, however, is not true.

See that the set S of all positive integers is an infinite set but it has no limit point.

We shall presently show that an infinite set of real numbers, **if bounded,** must have at least one point of accumulation.

This is an extremely important result in Real Analysis: it is known as

Bolzano-Weierstrass Theorem.

II. Definition of an Isolated point

We have observed that if $S \subseteq \mathbb{R}$ and $x \in \mathbb{R}$, then x is called an accumulation point of S if every **nbd** of x contains at least one point of S distinct from x.

If $x \in S$, **but x is not an accumulation point** of S then x is called an **isolated point** of S.

In other words, any point of S is either a point of *accumulation* of S or it is *an isolated point* of S.

Examples

Example 4.4.4. *Every point of a finite set is an isolated point because no such point is an accumulation point.*

Example 4.4.5. *Every member of the infinite set $\{1, 2, 3, \cdots\}$ is an isolated point.*

Example 4.4.6. *Since every point of a closed interval $[a, b]$ is a point of accumulation of the interval. We see that there exists no isolated point of the closed interval $[a, b]$.*

Boundary point of S

A point $x_0 \in \mathbb{R}$ is called **a boundary point** of $S \subseteq \mathbb{R}$ (or **a frontier point** of S) if *every neighbourhood* of x_0 i.e. $N_\delta(x_0)$, $\delta > 0$ contains at least one points of S and at least one points of the complement S^c of S. The point x_0 may or may not belong to S.

e.g., In the closed interval $[a,b]$, a,b are boundary points of $[a,b]$. See that in the open interval (a,b), a and b are also boundary points of (a,b). In the first case the boundary points belongs to the set and in the second case the boundary points do not belong to the set.

Boundary of S

The set containing all the boundary points of S is known as **the boundary** of S. In the above example the set $\{a,b\}$ is the boundary of $[a,b]$ or (a,b).

Exterior point of S

A point $\alpha \in \mathbb{R}$ is called an exterior point of $S \subseteq \mathbb{R}$ if there exist some **nbd** of α (i.e. $N_\delta(\alpha)$ for some $\delta > 0$) which lies entirely outside S i.e. which lies entirely in the complement S^c of S.

A point to Remember

Let S be a set of real numbers ($S \subseteq \mathbb{R}$) and suppose $x_0 \in \mathbb{R}$. Then there are three mutually exclusive possibilities:

Either x_0 is an *interior point* of S

or x_0 is a *boundary point* of S

or x_0 is an *exterior point* of S.

[**Recall:** x_0 is an interior point of S if $\exists$ a **nbd** of x_0 which is a subset of S]

Note: Every interior point of a set is a limit point of the set but the converse is not true. For e.g., in the set $[a,b]$, points a and b are limit points of $[a,b]$ but not interior points of $[a,b]$.

III. Derived Set: Definition

The set of *all points of accumulation* (or *limiting points*) of a set S of real number ($S \subseteq \mathbb{R}$) is called the derived set of S, denoted by S'.

Remember: When we write $x \in S'$, we mean

$$x \text{ is a limit point of } S.$$

Illustrative Examples on Derived Sets

In order that a real number x_0 may be a limit point of a set $S \subset \mathbb{R}$, each and every **nbd**. of x_0 must contain at least one member of S, other than the point x_0 or alternatively every **nbd** of x_0 must contain infinitely many points of S (**Theorem 4.4.1**). We shall use these facts in the examples given below:

Example 4.4.7. *The set $\mathbb{N}$ of natural numbers has no limit point so that $\mathbb{N}'$ (derived set of $\mathbb{N}$) $= \phi$.*

[(i) Let n be any natural number. Then the open interval $(n - \frac{1}{2}, n + \frac{1}{2})$ is a neighbourhood of n which contains any member of $\mathbb{N}$ except n. So n is not a limit point of $\mathbb{N}$

(ii) Let $n \in \mathbb{R} - \mathbb{N}$, i.e. n is a real number but not a natural number. Then $\exists$ some integer k such that $k < n < k + 1$.

Consequently $(k, k+1)$ is a **nbd** of n which does not contain any point of $\mathbb{N}$. Hence $n \in \mathbb{R} - \mathbb{N}$ is not a limit point of $\mathbb{N}$.

In any case, $\mathbb{N}$ has no limit point, i.e. $\mathbb{N}' = \phi$.]

Example 4.4.8. *The set $\mathbb{Z}$ of all integers has no limit point, for a **nbd** $(m - \frac{1}{2}, m + \frac{1}{2})$ of $m \in \mathbb{Z}$ contains no point of $\mathbb{Z}$ other than m. Also, for any $p \in \mathbb{R}$ other than $m \in \mathbb{Z}$ $\exists$ a **nbd** which contains no point of $\mathbb{Z}$. Thus $\mathbb{Z}$ has no limit point i.e. $\mathbb{Z}' = \phi$.*

Example 4.4.9. *Every point of $\mathbb{R} = (-\infty, \infty)$ is a limit point, for every **nbd**. of any of its point contains infinitely many members of $\mathbb{R}$. Thus $\mathbb{R}' = \mathbb{R}$.*

Example 4.4.10. *Every point of the set $\mathbb{Q}$ of all rational numbers is a limit point, for, between any two rationals $\exists$ an infinity of rationals. Further every irrational number is also limit point of $\mathbb{Q}$ for between any two irrationals there are infinitely many rationals. Thus every real number is a limit point of $\mathbb{Q}$, so that $\mathbb{Q}' = \mathbb{R}$.*

Example 4.4.11. *The set $\{\frac{1}{n} : n \in \mathbb{N}\}$ has only one limit point, namely, 'the point 0' and 0 is not a member of the set.*

The set $\{0, 1, \frac{1}{2}, \frac{1}{3}, \cdots, \frac{1}{n}, \cdots\}$ has also one limit point, namely 'the point zero' but here 0 is a member of the set. Thus in both cases the derived set $= \{0\}$.

Example 4.4.12. *Every point of the closed interval $[a, b]$ is its limit point and a point not belonging to the interval is **not a** limit point. Therefore*

$$[a, b]' = [a, b].$$

Example 4.4.13. *Every point of the open interval (a, b) is its limit point. Also the boundary points a and b which are not members of the set (a, b) are also its limit parts. Thus*

$$(a, b)' = [a, b].$$

Example 4.4.14. *A finite set S clearly has no limit point. Because in order that x_0 may be a limit point, any **nbd** of x_0 must contain infinitely many points of the set S. Therefore $S' = \phi$.*

Example 4.4.15. *Let $S = \{\frac{1}{m} + \frac{1}{n} : m, n \in \mathbb{N}\}$.*

Show that

(i) *0 is a limit point of S*

(ii) *If $k \in \mathbb{N}$, then $\frac{1}{k}$ is a limit point of S*

(iii) *$S' = \{0\} \cup \{\frac{1}{n} : n \in \mathbb{N}\}$*

Proof. (i) Let ϵ be any positive number.

But Archimedean property of $\mathbb{R} \ni$ natural numbers p, q such that $0 < \frac{1}{p} < \frac{\epsilon}{2}$ and $0 < \frac{1}{q} < \frac{\epsilon}{2}$; hence $0 < \frac{1}{p} + \frac{1}{q} < \epsilon$. This shows that the ϵ-**nbd** of 0 [i.e. open intrval $(-\epsilon, \epsilon)$] contains a point $\frac{1}{p} + \frac{1}{q}$ of S, **other than zero.**

So 0 is a limit point of S.

(ii) We take any point $\frac{1}{k}$ where $k \in \mathbb{N}$.

Suppose ϵ is a positive number.

By Archimedean property of $\mathbb{R}$, $\exists$ a natural number s such that $0 < \frac{1}{s} < \epsilon$.

Then we can write $-\epsilon < \frac{1}{s} < \epsilon$

$$\text{or } \frac{1}{k} - \epsilon < \frac{1}{k} + \frac{1}{s} < \frac{1}{k} + \epsilon$$

i.e. ϵ-**nbd** of $\frac{1}{k}$ contains a point $\frac{1}{k} + \frac{1}{s}$ of the set S, other than $\frac{1}{k}$.

So $\frac{1}{k}$ is a limit point of S.

Since $\frac{1}{k}, k \in \mathbb{N}$ is a limit point, **the set $\{\frac{1}{n} : n \in \mathbb{N}\}$ has all its points limit points of S.**

(iii) Let $T = \{0\} \cup \{\frac{1}{n} : n \in \mathbb{N}\}$.

We have observed: $T \subset S'$ ($\because$ members of T are all limit points of S)

From the definition of S we have $S \subset (0, 2]$ and hence $S' \subset (0, 2]' = [0, 2]$.

We can now prove $T = S'$ i.e. no point of $(0, 2] - T$ can become a limit point of S.

$\therefore S'$ (derived set of S) $= T = \{0\} \cup \{\frac{1}{n} : n \in \mathbb{N}\}$.

Example 4.4.16. (a) *Prove that the set of all limit points of*

$$S = \left\{\frac{1}{2m} + \frac{1}{3n} : m, n \in \mathbb{N}\right\}$$

is given by

$$S' = \left\{\frac{1}{2m} : m \in \mathbb{N}\right\} \cup \left\{\frac{1}{3n} : n \in \mathbb{N}\right\} \cup \{0\}$$

CHAPTER 4: ELEMENTS OF POINT SET TOPOLOGY

(b) *Let* $S = \left\{(-1)^m + \dfrac{1}{n}\right\}, m, n \in \mathbb{N}$

Then

$$S = \left\{(-1)^{2k} + \frac{1}{n}, k, n \in \mathbb{N}\right\} \cup \left\{(-1)^{2k-1} + \frac{1}{n}, k, n \in \mathbb{N}\right\}$$

$$= \left\{1 + \frac{1}{n} : n \in \mathbb{N}\right\} \cup \left\{-1 + \frac{1}{n}, n \in \mathbb{N}\right\}$$

1 is a limit point of $\left\{1 + \dfrac{1}{n} : n \in \mathbb{N}\right\}$

−1 is a limit point of $\left\{-1 + \dfrac{1}{n} : n \in \mathbb{N}\right\}$

No other point can be a limit point of S.

$\therefore S' = \{-1, 1\}.$

Some Properties of Derived Sets

If S and T are two sets of real numbers, then the following two results will follow:

(i) $S \subset T \implies S' \subset T'$

(ii) $(S \cup T)' = S' \cup T'$

(iii) $(S \cap T)' \subset S' \cap T'$

[**Proofs:** (i) If $S' = \phi$, then evidently $S' \subset T'$.
When $S' \neq \phi$, let $\xi \in S'$ and $N(\xi)$ be any **nbd** of ξ.
$\implies N(\xi)$ contains an infinite number of members of S.
But $S \subset T$. Therefore $N(\xi)$ contains infinitely many members of T
$\implies \xi$ is a limit point of T i.e. $\xi \in T'$
Thus $\xi \in S' \implies \xi T'$. Hence $S' \subset T'$.
(ii) Now $S \subset S \cap T \implies S' \subset (S \cup T)'$
and $T \subset S \cap T \implies T' \subset (S \cup T)'$
Consequently $S' \cup T' \subset (S \cup T)'$ (1)
Now we prove: $(S \cup T)' \subset S' \cup T'$.]
If $(S \cup T)' = \phi$, then evidently $(S \cup T)' \subset S' \cup T'$.
When $(S \cup T)' \neq \phi$, let $\xi \in (S \cup T)'$

Now since ξ is a limit point of $(S \cup T)$, therefore every **nbd.** of ξ contains an infinite number of points of $S \cup T$

$\implies$ every **nbd** of ξ contains infinitely many points of S or T or of both
$\implies \xi$ is a limit point of S or ξ is a limit point of T
$\implies \xi \in S'$ or $\xi \in T' \implies \xi \in S' \cup T'$
Thus $\xi \in (S \cup T)' \implies \xi \in S' \cup T'$
Consequently $(S \cup T)' \subset S' \cup T'$ (2)
From (1) and (2) it follows that

$$(S \cup T)' = S' \cup T'$$

i.e. derived set of the union = the union of derived sets.

(iii) $S \cap T \subset S \Longrightarrow (S \cap T)' \subset S'$

and $S \cap T \subset T \Longrightarrow (S \cap T)' \subset T'$

Consequently $(S \cap T)' \subset S' \cap T'$]

Note. In general it is not true that $(S \cap T)' = S' \cap T'$.

See the following two examples:

Example 4.4.17. Let $S = \{0, 1, \frac{1}{2}, \frac{1}{3}, \cdots\}$
and $T = \{0, -1, -1\frac{1}{2}, -\frac{1}{3}, \cdots\}$

Then $S \cap T = \{0\}$ and hence $(S \cap T)' = \phi$.
But $S' = \{0\}, T' = \{0\}$ and therefore $S' \cap T' = \{0\}$.
$\therefore (S \cap T)' \neq S' \cap T' (\because \phi \neq \{0\})$

Example 4.4.18. Let $S = (1, 2)$ and $T = (2, 3)$ so that $S \cap T = \phi$ whence

$$(S \cap T)' = \phi' = \phi$$

Also $S' = [1, 2], T' = [2, 3]$ and $S' \cap T' = \{2\}$.
Thus $(S \cap T)' \neq S' \cap T'$.

IV. Definition of an Adherent Point

Let S be a subset of $\mathbb{R}$ and let x be a real number (x is not necessarily a member of S).

Then x is said to **adhere to** S (or x is **adherent** to S) if every **nbd** of x contains at least one point of S.

Thus 'x adheres to S' $\Longrightarrow$ For every $\epsilon > 0$, $N\epsilon(x) \cap S \neq \phi$.

Example 4.4.19. If $x \in S$, then 'x adheres to S' because every **nbd.** of x contains at least one point of S, namely, the point x itself.

Example 4.4.20. If S is a subset of $\mathbb{R}$ which is bounded above, then $\sup S$ exists by lub property of $\mathbb{R}$ and $\sup S$ is adherent to S.

V. Definition of closure of a set S

The set of all adherent points of a set $S \subseteq \mathbb{R}$ is called the **closure** of S and is denoted by $\bar{S}$.

Theorem 4.4.2. To establish: $\bar{S} = S \cup S'$ for any set S.

($\bar{S}$ is the closure of S and S' is the derived set of S)

CHAPTER 4: ELEMENTS OF POINT SET TOPOLOGY

Proof. Let $x \in S$. Then every **nbd.** of x contains x (which mean that every **nbd** of x contain a point of S). Therefore x is an adherent point of S.

Thus $x \in S \implies x \in \bar{S}$ i.e., $S \subseteq \bar{S}$ \hfill (1)

Now let $x \in S'$. Then x is a point of accumulation of S and hence every **nbd.** of x, $N(x)$ contains a point of S distinct from x. i.e. $N(x) \cap S \neq \phi$. Therefore x is an adherent point of S i.e., $x \in \bar{S}$.

Thus $x \in S' \implies x \in \bar{S}$ i.e. $S' \subset \bar{S}$ \hfill (2)

From (1) and (2) we write $S \cup S' \subset \bar{S}$ \hfill (3)

Now suppose $y \notin S \cap S'$.

Then $y \notin S$ as well as $y \notin S'$.

Since $y \notin S'$. $\exists$ some **nbd** $N(y)$ of y such that

$$N'(y) \cap S = \phi.$$

Also since $y \notin S$, $N'(y) \cap S = \phi \implies N(y) \cap S = \phi$.

This proves that y is not an adherent point of S.

Therefore $y \notin S \cup S' \implies y \notin \bar{S}$.

Therefore $y \in \bar{S} \implies y \in S \cup S'$ (Taking contra positively)

Hence $\bar{S} \subset S \cup S'$ \hfill (4)

$\therefore$ from (3) and (4), we conclude

$$\bar{S} = S \cup S'.$$

Hence the theorem.

Note. Often we require the application of **Negativity statement:**

x is not a limiting point of S

if there exists even one nbd of x not containing a point of S distinct from x.

Observations. What we have observed is the following:

Every point x of a set $S \subset \mathbb{R}$ is an adherent point.

Some points adhere to S because every **nbd** of such points contains points of S distinct from x.

They are called *limit points* or *points of accumulation*. In fact, in every *nbd.* of a limit point of S there exists infinitely many points of S.

$\therefore$ An adherent point is either an **isolated member of S** or *a point of accumulation of S*.

Therefore: $\bar{S} = S \cup S'$. (**Theorem 4.4.2** is a formal proof of this statement).

VI. Closed Sets: Definitions

We give two equivalent statements as the definitions of a closed set.

Definition (i) A set $S \subseteq \mathbb{R}$ is said to be **closed** if it contains all *its points of accumulation (or limit points)*.

i.e. every limit point of a closed set S is a member of S.

or a set $S \subseteq \mathbb{R}$ is closed if the derived set S' is a subset of S (i.e. if $S' \subset S$)

Definition (ii) A set $S \subseteq \mathbb{R}$ is called a **closed** set if its complement S^c i.e. $\mathbb{R} - S$ is an open set.

We can use any one of the above two statements as the definition of a closed set after proving that one follows from the other.

(See Theorems 4.4.3 and 4.4.4)

Theorem 4.4.3. *Let S be a subset of $\mathbb{R}$ ($S \subseteq \mathbb{R}$).*

Suppose S contains all its points of accumulation (i.e. assume that $S' \subset S$).

Then S^c is an open set.

Alt. If S is a closed subset of $\mathbb{R}$ then its complement S^c is open in $\mathbb{R}$.

Proof. Case 1. When $S = \mathbb{R}, S^c = \phi$, which is an open set.

Case 2. When $S \subset R$, i.e., S is a proper subset of $\mathbb{R}$.

Let $x_0 \in S^c$.

Then x_0 is not an accumulation point of S because it is given that S contains all its points of accumulation.

But $x_0 \neq$ an accumulation point of S

$\implies \exists$ a **nbd.** of x_0 say $N(x_0)$ which does not contain any point of S

$\implies N(x_0)$ consists of points of S^c only

$\implies N(x_0) \subset S^c$

$\implies x_0$ is an interior point of S^c

Now x_0 is an arbitrary point of S^c, therefore all points of S^c are interior points of S^c i.e. S^c is an open set.

Note. What we have proved is that

If S is a closed set in the sense that S contains all its points of accumulation, then it follows that S is closed also in the sense that its complement S^c is open. (i.e. Definition (i) of a closed set $\implies$ Definition (ii) of the closed set)

Theorem 4.4.4. *Let S be a subset of $\mathbb{R}$ ($S \subseteq \mathbb{R}$).*

Suppose the complement S^c of S is an open set (i.e. $\mathbb{R} - S$ is open).

Then S contains all its points of accumulation.

Alt. If S^c, the complement of S in $\mathbb{R}$, be open then S is closed.

Proof. Case 1. Let $S^c = \phi$. $\therefore$ its complement is $\mathbb{R}$ which is closed.

Case 2. Let $S^c \neq \phi$.

Let $x_0 \in S^c$ where it is given that S^c is an open set.

We prove that x_0 is not an accumulation point of S.

Given: S^c is an open set

$\implies x_0$ is an interior point of S^c

$\implies \exists$ some δ-**nbd** of x_0 such that $N_0(x_0) \subset S^c$

$\implies x_0$ cannot be a point of accumulation of S

Now x_0 is an arbitrary point of S^c, therefore no point of S^c can become a point of accumulation of S. So all points of accumulation of S are in S i.e. S contains all its points of accumulation (i.e., S is closed).

4.5 Characteristic Properties of Closed Sets:

Theorem 4.5.1. *A set $S \subseteq \mathbb{R}$ is closed if and only if it contains all its adherent points.*

Proof. (i) *Assume that S is closed* (i.e. S^c is open).

To prove: S contains all its adherent points.

Let x be adherent to S. We shall prove that $x \in S$.

If possible, let $x \notin S$. Then $x \in S^c$.

Since we have assumed that S^c is open $\exists$ some **nbd** $N(x)$ of x lies entirely in S^c. Thus $N(x)$ contains no points of S, contradicting the fact that x adheres to S. Then $x \notin S$ is not tenable. Therefore $x \in S$.

x was taken to be an arbitrary adherent point of S, it follows that **S contains all its adherent points.**

(ii) Now, in order to prove the converse, we assume that S contains all its adherent points and show that S is closed (i.e. S^c is open)

Let $x \in S^c$. Then $x \notin S$ and therefore does not adhere to S. Hence $\exists$ some **nbd.** $N(x)$ of x which does not contain any point of S and so $N(x) \subset S^c$.

$\therefore x$ is an interior point of S^c. As x is arbitrary, so every point of S^c is an interior point of S^c.

Therefore S^c is open or **S is a closed set.**

Theorem 4.5.2. *A set $S \subset \mathbb{R}$ is closed if and only if $S = \bar{S}$.*

Let S be closed. Then its derived set $S' \subseteq S$. $\therefore S \cup S' = S$ or $\bar{S} = S$.

Conversely, $S \subset \mathbb{R}$ be such that $\bar{S} = S$. $\therefore S \cup S' = S \Longrightarrow S' \subseteq S$ and so S is a closed set.

Theorem 4.5.3. *The union of two closed sets in $\mathbb{R}$ is closed.*

Proof. Let A, B be two closed sets in $\mathbb{R}$. So, $A' \subset A, B' \subset B, A' \cup B' \subset A \cup B$ or $(A \cup B)' \subset A \cup B \Rightarrow A \cup B$ is closed.

Theorem 4.5.4. *The union of any finite number of closed set is a closed set*

Proof. We first prove this theorem using the definition of a closed set as the complement of an open set.

Let $A_1, A_2, \cdots, A_n$ be a finite collection of closed sets in $\mathbb{R}$. Let $S = A_1 \cup A_2 \cup A_3 \cup \cdots \cup A_n$.

By De Morgan's Laws the complement of S is given by

$$S^c = A_1^c \cap A_2^c \cap A_3^c \cap \cdots \cap A_n^c$$

Since each $A_1, A_2, A_3, \cdots, A_n$ is a closed set, each of $A_1^c, A_2^c, \cdots, A_n^c$ is an open set and their intersection

$$S^c = \bigcap_{i=1}^{n} A_i^c$$

is an *open set* (**Theorem 4.3.5**) and therefore *complement of S^c* i.e. **S is a closed set.**

Alternative Proof:

Using the fact that S is a closed set if S contains all its adherent points we can give an alternate proof of **Theorem 4.5.3**.

Let $A_1, A_2, \cdots, A_n$ be a finite number of closed sets.
Let $S = A_1 \cup A_2 \cup A_3 \cup \cdots \cup A_n$.
Let x be an adherent point of S.
Then for a given $\epsilon > 0$, $N_\epsilon(x) \cap S \neq \phi$.
Let $y \in N_\epsilon(x) \cap S$ i.e. $y \in N_\epsilon(x)$ as well as $y \in S$.
Now $y \in S \Longrightarrow y$ belongs to at least one A_i, say A_k.
$\therefore y \in N_\epsilon(x) \cap A_k \Longrightarrow x$ is an adherent point of A_k.
Since A_k is a closed set $x \in A_k$.
$\therefore x \in A_1 \cup A_2 \cup \cdots \cup A_n = S$

Thus $x \in \bar{S} \Longrightarrow x \in S$ i.e. $\bar{S} \subset S$. Therefore, S is a closed set. This completes the alternative proof of **Theorem 4.5.4**.

CHAPTER 4: ELEMENTS OF POINT SET TOPOLOGY

Theorem 4.5.5. *The intersection of an arbitrary collection of closed sets in $\mathbb{R}$ is a closed set.*

Proof. Let $T = \bigcap_{\alpha \in F} A_\alpha$, where F is any arbitrary collection of closed sets A_α. Then A_α^c is open.

We have already proved that for any arbitrary collection of open sets, the union is an open set.

By De Morgan's Law we have

$$T^c = \left(\bigcap_{\alpha \in F} A_\alpha\right)^c = \bigcup_{\alpha \in F} A_\alpha^c$$

$$\therefore T^c = \left(\bigcap_{\alpha \in F} A_\alpha\right)^c \text{ is an open set}$$

$\Longrightarrow T$ is a closed set i.e. $\bigcap_{\alpha \in F} A_\alpha$ is a closed set.

An alternative Proof.

We can define a closed set in terms of its limit points: thus a set S in $\mathbb{R}$ is closed if it contains all its limit points (i.e. if $S' \subset S$).

Using this definition we can prove Theorem 4.5.15.

Let $\{A_\alpha : \alpha \in I\}$, I being the index set, be a collection of closed sets in $\mathbb{R}$. Then $(A_\alpha)' \subset A_\alpha$ for each $\alpha \in I$ (each A_a being a closed set).

Let $T = \bigcap_{\alpha \in I} A_\alpha$.

In case, $T' = \phi$. Then clearly $T' \subset T$ (since $\phi \subset$ any set T). Hence T is closed.

So we discuss the case: $T' \neq \phi$.

Let $p \in T'$. Then p is a limit point of T.

$\Longrightarrow$ For a given $\epsilon > 0, \exists$ a deleted ϵ-nbd. of p when contains a point q of T i.e. $q \in N'_\epsilon(p) \cap T$

which $\Longrightarrow q \in N'_\epsilon(p) \cap A_\alpha$, for each $\alpha \in I$.

In other words, p is a limit point of A_α for each $\alpha \in I$.

Since each A_α is a closed set $p \in A_\alpha$ for each $\alpha \in I$.

Hence $p \in \bigcap_{\alpha \in T} A_\alpha$ i.e. $p \in T$.

Thus $p \in T' \Longrightarrow p \in T$ and therefore $T' \subset T$.

This proves that T is a closet set and hence **Theorem 4.5.5** follows.

Corollary 4.5.5. *The intersection of a finite number of closed sets is a closed set.*

In particular, let A and B be two closed sets. Let $T = A \cap B$.

(i) If T has no limit point, then $T' = \phi$ and so $T' \subseteq T$. $\therefore$ T is a closed set.

(ii) Let T have a limit point α. Then every *nbd* of α contains an infinite number of points of T and hence an infinite number of points of each of A and B i.e. α is a limit point of A we well as of B. Since A and B are closed sets, $\alpha \in A, \alpha \in B$. Therefore $\alpha \in A \cap B$. Hence T is a closed set. Exactly similar arguments could be given for proving **Theorem 4.5.5** (*Try yourself*)

> REMEMBER: The union of an arbitrary family of closed sets may or may not be a closed set.

(i) Let $A_n = \left\{ x : \frac{-n}{n+1} \leq x \leq \frac{n}{n+1} \right\}, n = 1, 2, 3, \cdots$

Then $\bigcup_{n=1}^{\infty} = \{x : -1 \leq x \leq 1\}$

Here each A_n is a closed set and their union $[-1, 1]$ is also a closed set.

(ii) Let $A_n = \left[\frac{1}{n}, 2\right], n = 1, 2, 3, \cdots$

$\bigcup_{n=1}^{\infty} A_n = (0, 2]$ is not a closed set.

(iii) Let $A_n = \left[\frac{n}{n+1}, \frac{n+1}{n+2}\right], n = 1, 2, 3, \cdots$

$\bigcup_{n=1}^{\infty} A_n = [\frac{1}{2}, 1)$ is not a closed set.

REMEMBER:	
Open set Properties	**Closed set Properties**
The union of an arbitrary collection of open sets in $\mathbb{R}$ is **open**.	The intersection of open arbitrary collection of closed sets in $\mathbb{R}$ is **closed**.
The intersection of any finite collection of open sets in $\mathbb{R}$ is **open**.	The union of any finite collection of closed sets in $\mathbb{R}$ is closed.
The intersection of infinitely many open sets in $\mathbb{R}$ need not be open. $G_n = \left(0, 1 + \frac{1}{n}\right) : n \in \mathbb{N}$, each is open but $\bigcap_{n=1}^{\infty} G_n = (0, 1]$ (Not an open set)	The union of infinitely many closed sets in $\mathbb{R}$ need not be closed. $F_n = \left[\frac{1}{n}, 1\right], n \in \mathbb{N}$, each is closed but $\bigcup_{n=1}^{\infty} F_n = (0, 1]$ (Not a closed set)

Closed Set: Dense Set: Perfect Set

Let S be a subset of $\mathbb{R}$ and let S' be the derived set of S i.e. S' contains all limit points of S.

CHAPTER 4: ELEMENTS OF POINT SET TOPOLOGY

S is a **closet set** if $S' \subset S$.

S is said to be **Dense-in-iteself** if $S \subset S'$.

S is called a **perfect set** if S is **closed** as well as **Dense-in-itself** i.e. if $S = S'$.

S is **dense** or **every where dense** if every point of $\mathbb{R}$ (i.e. every real number) is a limit point of S. or equivalently $S' = \mathbb{R}$.

Illustrative Examples.

Example 4.5.1. *Let $S = \{x \in \mathbb{R} : 2 < x < 3\}$. Then $S' = [2,3]$. Then $S \subset S'$ and therefore, S is **dense-in-itself**.*

Example 4.5.2. *The set $\mathbb{Q}$ of all rational numbers is **dense-in-itself** because $\mathbb{Q} \subset \mathbb{Q}'$. (**Note:** $\mathbb{Q}' = \mathbb{R}$ and so $\mathbb{Q}$ is everywhere dense)*

Example 4.5.3. *The set $\mathbb{R}$ is clearly dense-in-itself and also $\mathbb{R}$ is a closed set. Hence the set $\mathbb{R}$ is a **perfect set**.*

Example 4.5.4. *The set $S = \{x \in \mathbb{R} : 2 \leq x \leq 3\}$ is a **perfect set** because it is both dense-in-itself and closed. The set ϕ is a **perfect set**.*

Example 4.5.5. *Let $S \subseteq \mathbb{R}$ be a dense-in-itself set. Then S' is a perfect set.*

Proof. $S \subset S'$ ($\because S$ is dense-in-itself)

Now $S \subset S' \implies S' \subset (S')'$ and $S' \subset (S')' \implies S'$ is dense-in-itself.

Also S' is a closed set. Therefore, S' is both dense-in-itself and closed and hence S' is a perfect set.

Example 4.5.6. *A more general definition of a Dense set:*

Let $S \subseteq \mathbb{R}$ and $T \subseteq S$. Then T in Dense in S if $S \subseteq T'$.

> e.g., Let $S = \{x : x \in \mathbb{R} \text{ and } 2 \leq x \leq 3\}$
> and $T = \{x : x \in \mathbb{R} \text{ and } 2 < x < 3\}$. Then $T' = [2,3]$.
> Here $S \subseteq T'$ and therefore T is dense in S.

Example 4.5.7. *(i) The set $\mathbb{Q}$ of all rational number is dense-in-itself but not closed.*

*(ii) A finite set S, however, **is closed** ($S' = \phi$) but not dense-in-itself. None of these two sets is a perfect set.*

An important Example:

Example 4.5.8. *(a) If C is an open set and B is a closed set and $C \supset B$ then $C - B$ is an open set.*

(b) If C is an open set and B is a closed set and $B \supset C$ then $B - C$ is a closed set.

Soln. We can easily check: $C - B = C \cap B^c$.

When C is open and B is closed (i.e. B^c is open), then the intersection $C \cap B^c$ is open i.e. $C - B$ is open.

Again $B - C = B \cap C^c$ = closed set ($\because$ B and C^c are both closed sets, therefore their intersection is closed)

Example 4.5.9. *If S is a closed set, then the complement of S (relative to any open set A containing S) is open.*

*If S is an open set, then the complement of S (relative to any **closed set containing** S) is closed.*

Soln. Let us suppose that the sets involved are linear sets, the universe being the system $\mathbb{R}$ of all real numbers.

Assume $S \subset A$. Then $A - S = \mathbb{R} - (S \cup A^c)$

[see that $\mathbb{R} - (S \cup A^c) = \mathbb{R} \cap (S \cup A^c)^c = \mathbb{R} \cap (S^c \cap A = S^c \cap A = A \cap S^c = A - S$]

We have assumed that S is a closed set and A is an open set, then A^c is closed and $S \cup A^c$ is a closed set (union of two closed sets) and hence $\mathbb{R} - (S \cup A^c)$ is an open set i.e. $A - S$ is an open set.

The section part can be similarly proved.

4.6 Bolzano Weierstrass Theorem for Infinite Point Sets

A finite set S has no limit point. If $p \in \mathbb{R}$ is a limit point of the finite set then every **nbd** of p must contain infinitely many points of S which is not possible because S has only a finite number of elements.

An infinite set may or may not have a limit point, e.g., The infinite set $\mathbb{Z}$ of all integers has no limit point but the infinite set $\mathbb{Q}$ of all rational number has every real number as one of its limit points.

We shall prove in this section a sufficient condition for the existence of at least one limit point of an infinite set.

Theorem 4.6.1. *Bolzano-Weierstrass Theorem.*

Every bounded infinite set S of real numbers (i.e. $S \subset \mathbb{R}$) has at least one limit point in $\mathbb{R}$.

First Proof. (*Using lub-axiom of $\mathbb{R}$*)

Let S be an infinite bounded subset of real numbers.

Let m, M be its infimum and supremum respectively.

CHAPTER 4: ELEMENTS OF POINT SET TOPOLOGY

We construct a set H of real numbers whose members obey the following property: $x \in H$ iff it exceeds at the most a finite number of elements of S.

Clearly H is non-empty, for at least $m \in H$.

Also M is an upper bound of H, for no number $\geq M$ can belong to H.

Thus the set H of real numbers is non-empty and bounded above.

∴ By the lub-axiom, H has a supremum ξ in $\mathbb{R}$.

We shall now prove that ξ is a limit point of S.

Consider any ε-nbd $(\xi - \varepsilon, \xi + \varepsilon)$ of ξ (where $\varepsilon > 0$).

Since $\xi = \sup H$, $\exists$ at least one member $\eta \in H$ s.t. $\eta > \xi - \varepsilon$. Now since $\eta \in H$, therefore it exceeds at the most a finite number of members of S, and consequently $\xi - \varepsilon$ (which is $< \eta$) can exceed at the most a finite number of members of S.

Again since $\xi = $ Supremum of H, $\xi + \varepsilon$ cannot belong to H and consequently $\xi + \varepsilon$ must exceed an infinite number of members of S.

We, therefore, conclude that $(\xi - \varepsilon, \xi + \varepsilon)$ contains an infinite number of members of S. In other words, ξ is a limit point of S.

Second Proof of Bolzano Weierstrans Theorem (*Using Nested Interval Theorem*)

In the next chapter i.e. **in Chapter 5** we shall introduce the concept of **Nested intervals** and **prove the Theorem of Nested intervals**:

A sequence $\{I_1, I_2, I_3, \cdots, I_n, \cdots\}$ of closed intervals is called a *nest of intervals* (or a set of nested intervals) if

(i) $I_1 \supset I_2 \supset I_3 \supset \cdots \supset I_n \supset \cdots$.

If, more over, we find that

(ii) the sequence $\{d_n\}$ be such that $d_n \to 0$ as $n \to \infty$, where $d_n = l(I_n) = $ length of the interval I_n, then $\exists$ **exactly one point** which is common to all the intervals. This is known as **Nested Interval Theorem.**

Based on this theorem of Nested intervals we have an alternative proof of **Bolzano-Weierstrass Theorem:**

Proof. Let S be an infinite bounded subset of real numbers.

Then $\exists$ two finite real numbers a and b such that $S \subset [a, b]$.

Divide $[a, b]$ into two closed intervals $[a, c]$ and $[c, b]$, where $c = \frac{1}{2}(a + b)$.

Since S is an infinite set, at least one of the sub-intervals $[a, c], [c, b]$ must contain an infinite number of members of S.

We select the one which contains infinitely many points of S.

(If both $[a, c]$ and $[c, b]$ contain infinitely many points of S, then we select any one of them).

We rename this chosen interval as $[a_1, b_1] = I_1$ (say). Thus

I_1 contains infinitely many points of S;

$$[a, b] \supset I_1 \quad \text{and} \quad b_1 - a_1 = \tfrac{1}{2}(b - a) = \tfrac{1}{2} \{\text{length of } [a, b]\}.$$

We now repeat the process on $[a_1, b_1]$: i.e., let $c_1 = \tfrac{1}{2}(b_1 + a_1)$; then $[a_1, c_1]$ or $[c_1, b_1]$ or both must contain infinitely many points of S. We select that subinterval which contains infinitely many members of S and rename it as $[a_2, b_2] = I_2$ (say). Then

I_2 contains infinitely many points of S;

$$I_1 \supset I_2 \quad \text{and} \quad b_2 - a_2 = \tfrac{1}{2}(b_1 - c_1) = \tfrac{1}{2^2}(b - a).$$

The process is again repeated on $[a_2, b_2]$ and so on *ad infinitum*. We thus obtain a sequence of closed intervals:

$$[a, b], [a_1, b_1] = I_1, [a_2, b_2] = I_2, \cdots, [a_n, b_n] = I_n, \cdots$$

such that

$$[a, b] \supset I_1 \supset I_2 \supset \cdots \supset I_n \supset I_{n+1} \cdots \supset \cdots\cdots;$$

each subinterval contains infinite number of points of S;

and $b_n - a_n = \tfrac{1}{2^n}(b - a)$ and hence tends to zero as $n \to \infty$.

Thus $\{I_n\}$ form a nest of intervals and hence, by the theorem of Nested intervals their intersection is a single point ξ (say).

We now verify that ξ is a limit point of S :

Let ε be any given positive number. Since $\tfrac{1}{2^n}(b - a) \to 0$ as $n \to \infty, \exists$ sufficiently large values of n (say, $\forall\, n > N$),

$$\tfrac{1}{2^n}(b - a) < \varepsilon \quad \text{i.e.,} \quad b_n - a_n < \varepsilon.$$

For such values of n, $I_n = [a_n, b_n] \subset (\xi - \varepsilon, \xi + \varepsilon)$ and consequently the ε-nbd of ξ contains infinitely many points of S (namely, those points which lie in I_n) and hence ξ is a limit point of S.

This proves the theorem.

Third Proof of Bolzano-Weierstrass Theorem [*using Dedikind's theorem of Real numbers:* See Appendix to Chapter 2 (Page 105)]

Let S be a bounded infinite set of real numbers.

We divide the set $\mathbb{R}$ of all real numbers into two sets A and B in the following manner:

Let $x \in B$ if there are infinitely many points of S less than x, otherwise $x \in A$. Now B is non-empty, since S is infinite and bounded, and as such every upper bound

CHAPTER 4: ELEMENTS OF POINT SET TOPOLOGY

of S belongs to B; moreover, every lower bound of S is in A, guaranteeing that A is non-empty.

Consider any two points $a \in A$ and $b \in B$. There are at most finite number of points of S less than a, yet there must be infinitely many points of S greater than a but less than b. Consequently $a < b$, i.e. every point of A is less than every point of B. By the Dedekind theorem we may now conclude that $\exists$ a point c such $a \leq c$ and $c \leq b$ for all $b \in B$.

We now prove that the point c is a limit point of S. Assuming the contrary, there would then exist a nbd. (α, β) of c such that $(\alpha, \beta) - \{c\}$ does not contain any pointing S. Choose a_0 such that $\alpha < a_0 < c$. This implies $a_0 \in (\alpha, \beta) \cap A$. Since $c < \beta$, it follows $\beta \in B$, so that there must exist infinitely many points of S less than β. None of these points, however, can be in $(\alpha, \beta) - \{c\}$, a contradiction. Therefore c is a limit point on c which completes the proof.

Observations:

1. The boundedness of an infinite set of real numbers is a **sufficient condition but not a necessary condition for the infinite set possessing a limit point.**

See that the set $\mathbb{Q}$ of rational numbers in certainly not a bounded set but it has many limit points.

2. Bolzano-Weierstrass Theorem does not hold in the system of rational numbers i.e. it is not true that infinite bounded set of rational numbers has a limit point in the system $\mathbb{Q}$ of all rational numbers e.g.

The infinite set of rational numbers

$$\left\{1+1, \left(1+1+\frac{1}{2!}\right), \left(1+1+\frac{1}{2!}+\frac{1}{3}\right), \cdots \left(1+1+\frac{1}{2!}+\frac{1}{3!}+\cdots+\frac{1}{n!}\right), \cdots\right\}$$

is bounded but **it has no limit point in** $\mathbb{Q}$.

The set, however, has a limit point e in the system $\mathbb{R}$ of **real numbers.**

3. An infinite bounded set of $\mathbb{R}$ may have more than one limit point in $\mathbb{R}$.

Theorem 4.6.2. *An infinite bounded set of real numbers has greatest and least limit points.*

Let S be an infinite bounded set of real numbers so that $\exists$ a finite interval $[a, b]$ such that $S \subset [a, b]$.

Clearly, **no limit point** of S can be greater than b or smaller than a i.e., every limit point of $S \in [a, b]$ as $S' \subset [a, b]' = [a, b]$.

The derived set i.e. is a bounded set S'.

Let $G = \sup S'$ and $g = \inf S'$.

We show that G and g are members of S' i.e., they are limit points of S.

Since G is the supremum of S', $\exists$ at least one member of S' which lies inside any left-nbd of G and consequently in that nbd $\exists$ infinitely many points of S, i.e., G is a limit point of S (or, $G \in S'$).

We may similarly prove that $g \in S'$.

Observations.

1. The greatest and least limit points of an infinite bounded set S are often called the **upper limit** μ and the **lower limit** λ.

2. μ and λ have the following characteristic properties:

   (i) For any $\varepsilon > 0$, the set S has infinitely many elements $> \mu - \varepsilon$ and has only a finite number of elements $> \mu + \varepsilon$. [For if $\exists$ infinitely elements of S, then they form an infinite subset A, say, of S. As S is bounded and so A is bounded. $\therefore$ by Bolzano-Weierstrass theorem A has at least one limit point. Any such limit point is also a limit point of S and is greater than μ, which contradicts the fact that μ is the upper limit of S].

   (ii) For any $\varepsilon > 0$, the set S has infinitely many elements $< \lambda + \varepsilon$ and has only a finite number of elements $< \lambda - \varepsilon$. [Similar argument as in (i)]

3. We may write $\overline{\lim} S$ for greatest limit μ and $\underline{\lim} S$ for least limit λ.

4. If $m = \text{glb} S$ and $M = \text{lub} S$, then all the elements of S lie in $[m, M]$, not necessarily in $[\lambda, \mu]$. In fact,

$$\boxed{m \leq \lambda \leq \mu \leq M.}$$

Examples:

Example 4.6.1. Let $S = \{1, 1/2, 1/3, \cdots, 1/n, \cdots\}$.

S is an infinite bounded set of real numbers $m = 0, M = 1; \lambda = \mu = 0$. The derived set $S' = \{0\}$. S' is not a subset of S and hence S is not a closed set. Here $m = \lambda = \mu < M$.

Example 4.6.2. Let $S = \{1/2, 2/3, 3/4, 4/5, \cdots, n/(n+1), \cdots\cdots\}$.

This set is also an infinite bounded set of real numbers. $m = 1/2, M = 1; \lambda = \mu = 1$. Here $m < \lambda = \mu = M$. $S' = \{1\}$. $\therefore$ S' is not a subset of S and hence S is not a closed set.

Example 4.6.3. $S = \{1 + \frac{1}{n} : n \in \mathbb{N}\} \cup \{-1 - \frac{1}{n} : n \in \mathbb{N}\}$

i.e., $S = \{2, -2, 3/2, -3/2, 4/3, -4/3, 5/4, -5/4, \cdots\}$.

This set is bounded. $m = -2, M = 2, \lambda = -1, \mu = 1$.

Here $m < \lambda < \mu < M$. $S' = \{-1, 1\}$ and S is not a closed set ($\because$ S' is not a subset of S).

Example 4.6.4. $S = \{-1 - \frac{1}{n} : n \in \mathbb{N}\} \cup \{-\frac{1}{n} : n \in \mathbb{N}\}$.

i.e., $S = \{-2, -1, -3/2, -1/2, -4/3, -1/3, \cdots\cdots\}$.

In this bounded set, $m = -2, M = 0; \lambda = -1, \mu = 0$.

Here $m < \lambda < \mu = M$. $S' = \{-1, 0\}$ and S is not a closed set.

Example 4.6.5. $S = \{\frac{1}{n} : n \in \mathbb{N}\} \cup \{1 + \frac{1}{n} : n \in \mathbb{N}\}$.

i.e., $S = \{1, 2, \frac{1}{2}, \frac{3}{2}, \frac{1}{3}, \frac{4}{3}, \cdots\cdots\}$.

See that $m = 0, M = 2; \lambda = 0, \mu = 1$. Here $m = \lambda < \mu < M$.

$S' = \{0, 1\}$ and S is not a closed set.

Example 4.6.6. $S = \{-2, 2\} \cup [1/n : n \in \mathbb{N}\} \cup \{-1/n : n \in \mathbb{N}\}$.

i.e., $S = \{-2, 2, -1, 1, -\frac{1}{2}, \frac{1}{2}, -\frac{1}{3}, \frac{1}{3}, \cdots\cdots\}$.

Here $m = -2, M = 2; \lambda = \mu = 0$. Thus $m < \lambda = \mu < M$.

$S' = \{0\}$ and S is not a closed set.

Example 4.6.7. $S = \{-1 + \frac{1}{n} : n \in \mathbb{N}\} \cup \{1 - \frac{1}{n} : n \in \mathbb{N}\}$

i.e., $S = \{0, -\frac{1}{2}, \frac{1}{2}, -\frac{2}{3}, \frac{2}{3}, -\frac{3}{4}, \frac{3}{4}, \cdots\cdots\}$

$m = -1, M = 1, \lambda = -1, \mu = 1$. Thus $m = \lambda < \mu = M$.

$S' = \{-1, 1\}$ and S is not a closed set.

Example 4.6.8. *Construct an example of a set no element of which lies between its upper and lower limits.*

Soln. The set $= \{-1/n : n \in \mathbb{N}\} \cup \{1 + \frac{1}{n} : n \in \mathbb{N}\} = \{-1, 2, -\frac{1}{2}, \frac{3}{2}, -\frac{1}{3}, \frac{4}{3}, \cdots\cdots\}$.

Here $\lambda = 0, \mu = 1$. The elements crowd near 0 and 1 but no element lies between 0 and 1.

Example 4.6.9. *Let S be a bounded infinite set of real numbers. Let M be the least upper bound of the set S. Prove that M is either the greatest element of S or else M is a limit point of S (i.e., $M = \mu$).* [C.H. 1996, 99]

Soln. The lub of a set may or may not belong to S. In case $M \in S$, then every member of S, is either less than or equal to M, M being a member of S is the greatest element of S.

But suppose $M \notin S$. By definition
for any $x \in S, x \leq M$ and
$\exists\, y \in S$ s.t. $y \in (M - \varepsilon, M)(\varepsilon > 0$ is any arbitrary positive number).

Since ε is arbitrary, it proves that in any nbd of $M\ \exists$ at least one element of S, other than M i.e., M is a limit point of S. In fact $\mu = M$.

A similar result can be stated for the glb m of the set S.

Example 4.6.10. *The set* $S = \left\{\dfrac{1}{p} + \dfrac{1}{q} : p \in \mathbb{N}, q \in \mathbb{N}\right\}$.

Soln. This set is bounded and its lub $M = 2$ and glb $m = 0$. See that $M \in S$ but $m \notin S$.

The derived set $S' = \{0,$ all fractions like $1/p$ where $p \in \mathbb{N}\}$.

[Take p, a fixed positive integer and then take $q = 1, 2, 3, 4, \cdots$. See that in any nbd of $1/p\ \exists$ infinite number of elements of S.]

Excepting 0, all other limit points belong to S. In any case it is not a closed set. It may be verified that the set is not open. Thus this is an example of a set which is *neither open nor closed.*

Example 4.6.11. *Find all the accumulation points of the set*

$$S = \left\{\frac{(-1)^m}{m} + \frac{1}{n} : m, n = 1, 2, 3, 4, \cdots\right\}.$$
[C.H. 1981]

Soln.

Case 1. $m = a$ fixed even positive integer.
The points of the set $S\left\{\dfrac{1}{m} + 1, \dfrac{1}{m} + \dfrac{1}{2}, \dfrac{1}{m} + \dfrac{1}{3}, \cdots\right\}$
will determine $1/m$ as its point of accumulation.
[Since in the right ε-nbd of $\frac{1}{m}\ \exists$ infinitely many points.]

Case 2. $m = a$ fixed odd positive integer.
The points of the set $S\left\{\dfrac{-1}{m} + 1, \dfrac{-1}{m} + \dfrac{1}{2}, \dfrac{-1}{m} + \dfrac{1}{3}, \cdots\right\}$
will determine $-1/m$ as its point of accumulation.

Case 3. $n = a$ fixed positive integer (odd or even).
The points of the set $\left\{\dfrac{1}{n} - 1, \dfrac{1}{n} + \dfrac{1}{2}, \dfrac{1}{n} - \dfrac{1}{3}, \dfrac{1}{n} + \dfrac{1}{4}, \cdots\right\}$
will determine $\frac{1}{n}$ as its point of accumulation.

CHAPTER 4: ELEMENTS OF POINT SET TOPOLOGY

(Since $\exists$ infinitely many points of the set on the right as well as left nbd of $\frac{1}{n}$).

Case 4. Finally
$$\left|\frac{(-1)^m}{m} + \frac{1}{n}\right| \leq \frac{1}{m} + \frac{1}{n} < \frac{2}{n} \text{ if } m > n$$
$$< \varepsilon \ \forall \ n > \frac{2}{\varepsilon} \text{ and } m > n.$$

$\therefore$ Zero is also a limiting point of S.

Conclusion. {All fractions of the form $\frac{1}{p}, p = 1, 2, 3, \cdots$}
{All fractions of the form $\frac{-1}{p}, p = 1, 3, 5, 7, \cdots$}
and {Zero}

are all points of accumulation of the set S.
i.e., the derived set is $\{1, \frac{1}{2}, \frac{1}{3}, \frac{1}{4}, \frac{1}{5}, \cdots\} \cup \{\frac{-1}{1}, \frac{-1}{3}, \frac{-1}{5}, \frac{-1}{7}, \cdots\} \cup \{0\}$.

Example 4.6.12. *For any set S of real numbers, prove that the derived set S' and the closure set $\bar{S}$ are both closed sets.* [C.H. 1996, 99]

Soln. (a) Let α be an accumulation point of S'.

If $N_\delta(\alpha)$ be any nbd of α, then $N_\delta(\alpha)$ contains a point $\beta \in S'$ (where $\beta \neq \alpha$) so that $|\alpha - \beta| < \delta$. Since $\beta \in S'$, β is an accumulation point of S. Let $\delta' = \delta - |\alpha - \beta| > 0$. Then $N_{\delta'}(\beta) \subseteq N_\delta(\alpha)$ and in $N_{\delta'}(\beta)$ $\exists$ infinite number of points of S and so $N_\delta(\alpha)$ contains infinite number of points of S i.e., α is a point of accumulation of S i.e., $\alpha \in S'$. Thus any point of accumulation of S' belongs to S' i.e., S' is a closed set.

```
                               δ'
─────────┼──────────┼────┼──────────[──────────
       α − δ        α    β        α + δ
```

(b) $(\bar{S})' = (S \cup S')' = S' \cup (S')' \subseteq S' \cup S' = S' \subseteq \bar{S}$
i.e., $\bar{S}$ is a closed set.

Example 4.6.13. *To prove (i) If $A \subseteq B$ then $A' \subseteq B'$ and (ii) $(A \cup B)' = A' \cup B'$.*

Soln. (i) If $A' = \phi$ then evidently $A' \subseteq B'$.

But if $A' \neq \phi$, let $\alpha \in A'$ and $N(\alpha)$ be some nbd of α.

Then $N(\alpha)$ contains infinite number of members of A. But $A \subseteq B$.

Hence $N(\alpha)$ contains infinite number of members of B, which means α is an accumulation point of B i.e., $\alpha \in B'$.

Thus $\alpha \in A' \Rightarrow \alpha \in B'$. Hence $A' \subseteq B'$.

(ii) To prove $(A \cup B)' \subseteq A' \cup B'$ i.e., to prove $\alpha \notin A' \cup B' \Rightarrow \alpha \notin (A \cup B)'$.

Now $\alpha \notin A' \cup B' \Rightarrow \alpha \notin A'$ and $\alpha \notin B' \Rightarrow \alpha$ is not an accumulation point of A or B.

$\therefore$ $\exists$ some *nbds* N_1, N_2 of α s.t., N_1 contains no point of A other than α and N_2 contain no point of B other than A.

Again since $N_1 \cap N_2 \subseteq N_1$, $N_1 \cap N_2 \subseteq N_2$, therefore $\exists$ a *nbd* $N_1 \cap N_2$ of α which contains no point other than α of A of B and hence of $A \cup B$.

i.e., α is not a point of accumulation of $A \cup B$ i.e., $\alpha \notin (A \cup B)'$ i.e., $\alpha \notin A' \cup B' \Rightarrow \alpha \notin (A \cup B)'$. Hence etc.

Exercises IV(I)

[A]
(Theory)

**Sets mentioned here are all linear point sets.
A point **is just a** *real number.*

1. (a) Define the following terms (with examples):

    Neighbourhood of a Point;   Deleted Neighbourhood of a Point;
    Interior Point of a Set;   Interior of a Set;
    Open Set;   Boundary Point of a Set;
    Boundary of a Set;   Exterior Point of a Set;
    Point of Accumulation or Limit Point or Limiting Point;
    Derived Set of a given Set;   Closed Set;
    Dense-in-itself Set;   Perfect Set;
    Adherent Point of a Set;   Closure of a Set.

    (b) From the definition of limiting point of a set, show that a finite set cannot have a limiting point. Give an example of an infinite set which does not have any limiting point. [C.H. 1992]

    (c) Define Accumulation Point of a set S of real numbers. Show that the set of all accumulation points of S is a closed set. [C.H. 1996]

2. (a) If A and B are open sets prove that $A \cup B$ and $A \cap B$ are also open sets.

    (b) Prove that $\bigcup_{k=1}^{n} G_k$ is an open set assuming that each G_k is an open set.

    [C.H. 1980]

    (c) Prove that the union of an arbitrary collection of open sets is an open set.

    [C.H. 1996]

(d) Prove that the intersection of a finite number of open sets is an open set.
[C.H. 1992, 95]

(e) With the help of suitable examples show that intersection of an arbitrary collection of open sets may or may not be open. [C.H. 1991, 95]

3. (a) Prove that the derived set S' of a given set is closed.

   (b) Prove that $S \subset T \Rightarrow S' \subset T'$ and $(S \cup T)' = S' \cap T'$.

   (c) Prove that the closure $\bar{S}$ (i.e., $S \cup S'$) is a closed set. [C.H. 1995]

4. (a) Prove that if A and B are two closed sets then $A \cup B$ and $A \cap B$ are both closed sets. [C.H. 1983, 94]

   (b) Prove that the intersection of an arbitrary collection of closed sets is a closed set.
   [C.H. 1991]

   (c) Justify (by examples) that the union of an arbitrary collection of closed sets may or may not be a closed set.

5. State and Prove Bolzano-Weierstrass Theorem for an infinite bounded set of real numbers.
   Verify it for the sets $\left\{\dfrac{n}{n+1}, n = 1, 2, 3, \cdots\right\}$ and $\left\{1 + \dfrac{1}{n} : n \in \mathbb{N}\right\}$.
   [C.H. 1995, 99]

6. (a) Let E be a linear point set, bounded above.
   If the lub M of the set E is not a member of E, then prove that M must be a limit point of E. [C.H. 1992]

   (b) If S be a bounded infinite set of real numbers, prove that the lub of S is the greatest element of the derived set S' when it belongs to S'. [C.H. 1991]

   (c) If a set E of real numbers is closed and bounded above, then prove that lub M of E belongs to E. [C.H. 1983]

   (d) Prove that a closed and bounded set has a greatest as well as a least member.

   (e) Give an example of a set which is neither open nor closed.
   [Ans. $\{(-1)^n(1 + \frac{1}{n}), n = 1, 2, 3, \cdots\}$]

[B]

1. Find m(glb), M(lub), λ (least limit point), μ (greatest limit point) for the following sets:

(a) $\left\{1, 2, \dfrac{1}{2}, 3, \dfrac{1}{3}, 4, \dfrac{1}{4}, 5, \dfrac{1}{5}, \cdots\cdots\right\}$;

(b) $\left\{\dfrac{3}{2}, \dfrac{5}{2}, \dfrac{5}{3}, \dfrac{7}{3}, \dfrac{7}{4}, \dfrac{9}{4}, \dfrac{9}{5}, \dfrac{11}{5}, \dfrac{11}{6}, \cdots\cdots\right\}$;

(c) $\left\{10, 3, 5, 3\dfrac{1}{3}, 5\dfrac{2}{3}, 3\dfrac{1}{9}, 5\dfrac{8}{9}, 3\dfrac{1}{27}, 5\dfrac{26}{27}, \cdots\cdots\right\}$;

(d) $\left\{\dfrac{3}{2}, -\dfrac{4}{3}, \dfrac{5}{4}, -\dfrac{6}{5}, \dfrac{7}{6}, -\dfrac{8}{7}, \dfrac{9}{8}, \cdots\cdots\right\}$;

(e) $\left\{-2, \dfrac{3}{2}, -\dfrac{4}{3}, \dfrac{5}{4}, -\dfrac{6}{5}, \dfrac{7}{6}, -\dfrac{8}{7}, \dfrac{9}{8}, \cdots\cdots\right\}$.

2. Find the derived set of

(a) $\left\{0, 1 - \dfrac{1}{2}, 1 - \dfrac{1}{3}, 1 - \dfrac{1}{4}, \cdots\cdots\right\} \cup \left\{[1, 2] - (1, \dfrac{3}{2})\right\}$;

(b) $\left\{\dfrac{2}{p} + \dfrac{3}{q} \,\middle/\, p, q = 1, 2, 3, \cdots\right\}$;

(c) $\left\{\dfrac{(-1)^m}{m} + \dfrac{1}{n} \,\middle/\, m, n = 1, 2, 3, \cdots\cdots\right\}$; [C.H. 1989]

(d) The set containing all rational numbers in $[1, 2]$;

(e) $\{x / x \in \mathbb{R} \text{ and } x < 0 \text{ or } x > 1\}$; [C.H. 1992]

(f) $\left\{\dfrac{1}{2^m} + \dfrac{1}{3^n}, m, n = 1, 2, 3, \cdots\right\}$;

(g) $S = \{x / 3x^2 - 10x + 3 > 0\}$; [C.H. 1984]

(h) $S = \left\{\dfrac{1}{m} + \dfrac{1}{n} \,\middle/\, m, n = 1, 2, 3, 4, \cdots\cdots\right\}$; [C.H. 1993]

(i) $\left\{(-1)^n + \dfrac{1}{m} \,\middle/\, m, n = 1, 2, 3, \cdots\cdots\right\}$;

(j) $\left\{-\dfrac{1}{2}, 1\dfrac{1}{2}, -\dfrac{2}{3}, 1\dfrac{2}{3}, 1\dfrac{2}{3}, -\dfrac{3}{4}, 1\dfrac{3}{4}, \cdots\right\}$. [C.H. 1994]

3. Sets A, B, C are defined by

$A = \{1, \tfrac{1}{2}, \tfrac{1}{3}, \cdots\}, \quad B = \{x / \tfrac{1}{2} < x < 1\}, \quad C = \{1, 2, 3, \cdots\}$.

Let X denote the set of all accumulation points of the set $A \cup B \cup C$ and Y the complement of X in $A \cup B \cup C$. Find the lub and glb of X and Y. [C.H. 1982]

4. (a) Show that the set $\{(0, 1) - \{\tfrac{1}{n}, n = 2, 3, \cdots\}\}$ is open.

CHAPTER 4: ELEMENTS OF POINT SET TOPOLOGY   211

(b) Given $S = \left\{\dfrac{n-1}{n+1} \big/ n = 1, 2, 3, \cdots \right\} \cup (2,3)$; find the set of all accumulation points and the set of all isolated points of S. How are the results modified if instead of $(2,3)$ we consider $\{2,3\}$? [C.H. 1986, 99]

(c) Show that the set $\left\{\dfrac{1+(-1)^n}{n} \big/ n = 1, 2, 3, \cdots \right\}$ is a closed set.

(d) Show that $S = \{x \in R / 0 < x \leq 1\}$ is not a closed set. [C.H. 1996]

5. (a) Justify $\{x/x = (-1)^n(1+\frac{1}{n}), n = 1, 2, 3, \cdots\}$
and $\{x/x = 3^{-n} + 5^{-n}, n = 1, 2, 3, \cdots\}$
are examples of *neither open* nor *closed sets*. [C.H. 1988]

(b) Prove that $(0,1) \cup (2,3)$ is an open set, using the definition of open set.

(c) $A = \{1, \frac{1}{2}, \frac{1}{3}, \cdots\}$, $B_n = (n-1, n], n = 1, 2, 3, \cdots$.

Examine if A is a closed set and $\bigcup\limits_{n=1}^{\infty} B_n$ is an open set.

(d) Prove that $(0,1) - \{\frac{1}{2^n}, n = 1, 2, 3, \cdots\}$ is an open set.

6. Find the upper and lower limits of (Take $n = 1, 2, 3, \cdots$)

(a) $\{[2 - (-1)^n]n\}$;   (b) $\{(-1)^n + \cos\frac{n\pi}{4}\}$;

(c) $\{(-1)^n(2+\frac{5}{n})\}$;   (d) $\{(-1)^n(2+\frac{3^n}{n!}+\frac{4}{n^2})\}$;

(e) $\{(-1)^n + \sin\frac{n\pi}{4}\}$;   (f) $\{n^{(-1)^n}\}$;

(g) $\{(-1)^n n^{(-1)^n}\}$.

Answers
[B]

1. a) $\overset{m}{0}, \overset{M}{\infty}, \overset{\lambda}{0}, \overset{\mu}{\infty}$;   b) $\overset{m}{3/2}, \overset{M}{3/2}, \overset{\lambda}{2}, \overset{\mu}{2}$;   c) $\overset{m}{3}, \overset{M}{10}, \overset{\lambda}{3}, \overset{\mu}{5}$;

d) $-\frac{4}{3}, \frac{3}{2}, -1, 1$;   e) $-2, \frac{3}{2}, -1, 1$.

2. (a) $\{1\} \cup [3/2, 2]$;

(b) $\{2/p, p = 1, 2, 3, \cdots\} \cup \{\frac{3}{p}, p = 1, 2, 3, \cdots\} \cup \{0\}$;

(c) $\{1, \frac{1}{2}, \frac{1}{3}, \frac{1}{4}, \cdots\} \cup \{-1, -\frac{1}{3}, -\frac{1}{5}, \cdots\} \cup \{0\}$;

(d) $[1, 2]$;

(e) $\{x : x \in \mathbb{R}/x \leq 0\} \cup \{x \in \mathbb{R}/x \geq 1\}$;

(f) $\{\frac{1}{2^m}, m = 1, 2, 3, \cdots\} \cup \{\frac{1}{3^n}, n = 1, 2, 3, \cdots\} \cup \{0\}$;

(g) $\{x : x \in \mathbb{R}, x \geq 3\} \cup \{x : x \in \mathbb{R}, x \leq \frac{1}{3}\}$;

(h) $\{\frac{1}{m}, m = 1, 2, 3, \cdots\} \cup \{0\}$;

(i) $\{-1, 1\}$;

(j) $\{-1, 1\}$.

3. $\mathrm{lub}\, X = 1, \mathrm{glb}\, X = 0; \mathrm{lub}\, Y = \infty, \mathrm{glb}\, Y = 0$.

4. b) $\{1\} \cup [2, 3]$; Isolated points are all points of the set $\{\frac{n-1}{n+1}, n = 1, 2, 3, \cdots\}$.
   Modified Answers: $\{1\}$. Isolated points $\{\frac{n-1}{n+1} : n \in \mathbb{N}\}$ and $\{2, 3\}$.

6. (a) $\lambda = \mu = \infty$; (b) $\lambda = -1 - \frac{1}{\sqrt{2}}, \mu = 2$; (c) $\lambda = -2, \mu = +2$; (d) $\lambda = -2, \mu = 2$; (e) $\lambda = -1 - \frac{1}{\sqrt{2}}, \mu = 2$; (f) $\lambda = 0, \mu = \infty$; (g) $\lambda = 0, \mu = \infty$.

> In the following sections we shall introduce three very interesting sets:
> **Cantor Set: Borel Set: Compact Sets**
> They are very useful in the study of Real Analysis, *Measure Theory*, in particular.

4.7 The Cantor Set

We construct *the Cantor Set* in the following manner: Let I be the unit closed interval $[0, 1]$. From this interval remove the open interval $(\frac{1}{3}, \frac{2}{3})$ which is called the '*open middle third*' of I. Then $[0, \frac{1}{3}]$ and $[\frac{2}{3}, 1]$ and left.

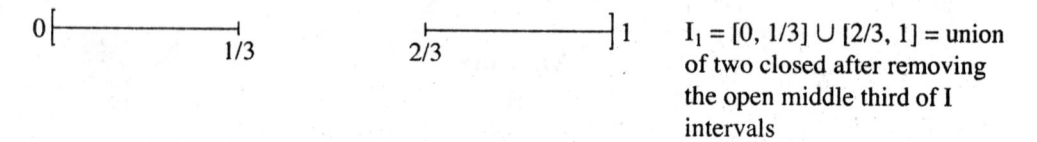

I = unit closed interval $[0, 1]$

$I_1 = [0, 1/3] \cup [2/3, 1]$ = union of two closed after removing the open middle third of I intervals

$I_2 = [0, 1/9] \cup [2/9, 1/3] \cup [8/9, 1]$ = union of two closed intervals after removing the open middle thirds of $[0, 1/3]$ and $[2/3, 1]$

Next the open intervals $(\frac{1}{9}, \frac{2}{9})$ and $(\frac{7}{9}, \frac{8}{9})$, *the open middle thirds* of $[0, \frac{1}{3}]$ and $[\frac{2}{3}, 1]$ are removed, leaving

$$I_2 = \left[0, \frac{1}{3}\right] \cup \left[\frac{2}{9}, \frac{1}{3}\right] \cup \left[\frac{2}{3}, \frac{7}{9}\right] \cup \left[\frac{8}{9}, 1\right]$$

At each succeeding state remove the *open middle thirds* of each remaining closed intervals.

CHAPTER 4: ELEMENTS OF POINT SET TOPOLOGY

If this process is carried out denumerably many times, the result is the **Cantor set** (also called **Cantor Ternary Set** or **Cantor middle third set**) $C = \bigcap_{n=1}^{\infty} I_n$ ($n \in \mathbb{N}$) where I_n = union of remaining 2^n closed intervals after removing open middle thirds starting with $[0, 1]$.

Since C is the intersection of arbitrary number of closed sets, C is itself a closed set. Hence we have

Theorem 4.7.1. *The Cantor set C is a closed set.*

The reader can also state that C is

"The complement of the union of removed open intervals and the intervals $(-\infty, 0)$ and $(1, \infty)$."

and hence C **is a closed set.**

Another interesting result which follows from the process of construction of the Cantor set C is provided by the following theorem.

Theorem 4.7.2. *The total length L of the removed intervals in the construction of the Cantor Set is unity.*

The first middle third has length $\frac{1}{3}$, the next two middle thirds have lengths whose sum $= \frac{1}{9} + \frac{1}{9} = \frac{2}{9} = \frac{2}{3^2}$

the next 2^2 middle thirds have lengths whose sum $= \frac{1}{27} + \frac{1}{27} + \frac{1}{27} + \frac{1}{27} = \frac{2^2}{3^3}$ etc.

$\therefore$ Total lengths of the removed intervals

$$L = \frac{1}{3} + \frac{2}{3^2} + \frac{2^2}{3^3} + \frac{2^3}{3^4} + \cdots$$

$$= \frac{1}{3} \frac{1}{1-\frac{2}{3}} \text{ (sum of an infinite GP with common ratio} = \tfrac{2}{3})$$

$$= 1.$$

Apparently, it may appear that the only points left in the process of construction of the Cantor Set are the end points

$$\frac{1}{3}, \frac{2}{3}, \frac{1}{9}, \frac{2}{9}, \frac{7}{9}, \frac{8}{9}, \text{ etc. etc.}$$

and these are denumerable in number, but this apparent conclusion is not true. In fact, we can prove the next theorem.

Theorem 4.7.3. *The Cantor set C is non-denumerable.*

Before going to a formal proof this theorem we shall introduce the notion of
Ternany expansion of real numbers in $[0, 1]$

Preliminaries on Ternary expansions:

In the *ternary expansion of a real number x*, we use the digits $0, 1$ and 2. Thus in the ternary scale
$$x = 0 \cdot a_1 a_2 a_3 \cdots$$
means that
$$x = \frac{a_1}{3} + \frac{a_2}{3^2} + \frac{a_3}{3^3} + \cdots$$
where $a_1, a_2, a_3, \cdots$ will take any one of the values $0, 1, 2$. Except for numbers like $\frac{1}{3}, \frac{2}{3}$, etc. the *ternary expansion of a real number is unique*.

But $\frac{1}{3}$ has two ternary expansions:
$$\frac{1}{3} = 0.1000 \cdots \quad \text{or,} \quad \frac{1}{3} = 0.02222 \cdots$$

In case of $\frac{2}{3}$ we have the following two expansions:
$$\frac{2}{3} = 0.1222 \cdots \quad \text{or,} \quad \frac{2}{3} = 0.2000 \cdots$$

The numbers $\frac{1}{2}$ and $\frac{1}{4}$ have the following expansions:
$$\frac{1}{2} = 0.11111 \cdots \quad \text{and} \quad \frac{1}{4} = 0.20202 \cdots$$

A Formal Proof of Theorem 4.7.3

Let us express all real numbers in $[0, 1]$ in their ternary expansions **e.g.** $\frac{2}{9}$ becomes $(0.02)_3$
or $\frac{5}{27}$ becomes $(0.012)_3$ etc.

1st stage:

When we remove the interval $(\frac{1}{3}, \frac{2}{3})$, we actually remove all those real numbers between 0.1 and 0.2 (in *ternary form*) *i.e.* all numbers having the digit 1 in the first place of their ternary expansion.

2nd stage:

Removing the intervals $(\frac{1}{9}, \frac{2}{9})$ and $(\frac{7}{9}, \frac{8}{9})$ means discarding all those numbers having ternary expansion between $.01$ and $.02$ **or** between $.11$ and $.12$. Each of these numbers has the digit 1 in second place (in ternary form).

3rd stage:

The third removal discards all those numbers with 1 in the third place.

CHAPTER 4: ELEMENTS OF POINT SET TOPOLOGY

4th stage:

The fourth removal discards those numbers with 1 in the fourth place etc.

We see that the points in $[0, 1]$ which do not belong to the Cantor set have the digit 1 in at least one place in their ternary expansion. The Cantor set must then consist of those numbers in the interval $[0, 1]$ which contain only the digits 0 and 2 in their ternary expansions.

The end point of the removed intervals in the construction of Cantor set have two different ternary representations: one containing the digit 1 at a certain place and the other having the digits 0 to 2 recurring, e.g.

$$\frac{1}{3} = (.1000\cdots)_3 = (.0222\cdots)_3$$
$$\text{and } \frac{2}{3} = (.1222\cdots)_3 = (.2000\cdots)_3$$

If we agree to use that representation which does not contain the digit 1, then all the points in the Cantor set can be expressed in ternary expansion by the digits 0 and 2.

There is an obvious one-to-one correspondence between those numbers and the set of all numbers in $[0, 1]$ expressed in binary form.

This $1-1$ correspondence merely interchanges the digits 1 and 2.

e.g., $(0.02)_3 \longleftrightarrow (0.01)_2$

and $(0.0202)_3 \longleftrightarrow (0.0101)_2$

In other words, the set of all real numbers in $[0, 1]$ can be put in $1-1$ correspondence with a subset of the Cantor set, and by the Schroeder-Bernstein theorem it follows that the Cantor set has power c i.e. it is *non-denumerable*. This proves the theorem.

Statement of Schroeder-Bernstein Theorem:

Let A and B be two sets.

If there exists an injective mapping of A into B and also an injective mapping of B into A, then

$A \sim B$ (read: A is equivalent to B)

Note. $A \sim B \implies \exists$ a bijective mapping of A ONTO B. In particular, $A \sim \mathbb{N} \implies A$ is a **denumerable set** or **an enumerable set.**

All sets which are equivalent to each other have a common characteristic—called **Cardinal number** or **the power** of each one of these sets.

For an empty set, the cardinal number is taken as *Zero*. We assume 1 as the Cardinal number of all singleton sets. Again, all sets equivalent to $\{a_1, a_2\}$ $(a_1 \neq a_2)$ have the **power** 2 or the **cardinal number** 2.

We shall write the Cardinal number of the class of all denmerable sets by the symbol $\aleph_0$ (read as aleph nought). $\aleph_0$ is the first letter of the Hebrew alphabet.

Theorem 4.7.4. *The Cantor set is dense-in-itself. (every point of the Cantor set is a limit point)*

Proof. Let x_0 be any point of the Cantor set.

Let us express x_0 in ternary expansion: say

$$x_0 = \cdot t_1 t_2 t_3 \cdots t_n \cdots \text{ where } t_i = 0 \text{ or, } 2.$$

We then construct a sequence of points in the following manner:

$$x_1 = \cdot t'_1 t_2 t_3 t_4 \cdots t_n \cdots$$
$$x_2 = \cdot t_1 t'_2 t_3 t_4 \cdots t_n \cdots$$
$$x_3 = \cdot t_1 t_2 t'_3 t_4 \cdots t_n \cdots$$
$$\cdots\cdots\cdots\cdots\cdots$$
$$x_n = \cdot t_1 t_2 t_3 t_4 \cdots t_{n-1} t'_n t_{n+1} \cdots$$
$$\cdots\cdots\cdots\cdots\cdots$$

where $t'_n = 0$ if $t_n = 2$ and $t'_n = 2$ if $t_n = 0$.

In this way we obtain a sequence of a distinct points, $\{x_n\}$, all belonging to the Contor set, such that x_n differs from x_0 in the nth place in the ternary expansion.

We construct this sequence so that $\lim_{n \to \infty} x_n = x_0$ and then we may conclude that x_0 is a limit point of the Cantor set. Thus every point of the Cantor set is a limit point and the **set is therefore dense-in-itself.**

Now we have observed that the Cantor set is closed. Hence it follows that

Theorem 4.7.5. *The Contor set is perfect. (closed and dense-in-itself)*

Examples.

Example 4.7.1. *The Cantor set C contains no non-empty open interval as a subset.*

In fact if C contains a non-empty open interval $I = (a, b)$ then since $I \subset I_n$ for all $n \in \mathbb{N}$ we must have $0 < b - a < \left(\frac{2}{3}\right)^n$ for all $n \in \mathbb{N}$. Therefore $b - a = 0$, when I is empty, a contradiction.

Example 4.7.2. *The Cantor set C has infinitely many points.*

The Cantor set C contains all the *end-pounds* of the removed open intervals, and these are points of from $\frac{2^k}{3^n}$, where $k = 0, 1, \cdots, n$ for each $n \in \mathbb{N}$. There are infinitely many points of this form.

Example 4.7.3. *Construct a generalized Cantor set as follows:*

Let $0 < a < 1$

(i) *Remove from $[0, 1]$ an open interval of length $\frac{1}{2}a$ with mid point $\frac{1}{2}$.*

(ii) From the remaining closed intervals remove two open intervals each of length $\frac{1}{8}a$ with mid points $\frac{1}{4} - \frac{1}{4}a$ and $\frac{3}{4} + \frac{1}{4}a$ respectively. Repeat this process indefinitely as the case of the Cantor set.

Show that the sum of the length of the intervals removed from $[0,1]$ is 'a' and the remaining set is a perfect nowhere dense set.

[Such a set is called a Cantor set of positive measure since the 'length' of what is left is $1 - a > a$]

4.8 Borel Sets

An interesting class of sets, known as **Borel Sets** will be discussed in this section.

We recall that the union of any number of open sets is an open set and the intersection of an arbitrary number of closed sets is a closed set.

We have given examples to show that the union of any number of closed sets may or may not be closed; also the intersection of an arbitrary number of open sets may or may not be open.

Borel sets are constructed basically by taking the union of closed sets and intersection of open sets. We define two types of sets—one is the type F_σ and the other is of type G^δ.

Definition 4.8.1. *A set is said to be of type F_σ if it can be obtained as the union of a denumerable number of closed sets.*

Definition 4.8.2. *A set is said to be of type G^δ if it can be obtained as the intersection of a denumerable number of open sets.*

Illustrative Examples

Example 4.8.1. *The set $\mathbb{Q}$ of all rational numbers is an F_σ set.*

Since the set of all rational numbers $\mathbb{Q}$ is denumerable, $\mathbb{Q}$ can be described as the set $\{x_1, x_2, x_3, \cdots\}$

$\therefore \mathbb{Q} = \{x_1\} \cup \{x_2\} \cup \{x_3\} \cup \cdots$ = union of a denumerable number of closed sets, each singleton set being a closed set.

Hence Q is of type F_σ.

Example 4.8.2. *A closed and bounded interval $[a,b]$ is a G^δ set.*

Since $[a,b]$ can be expressed as $\bigcap\limits_{n=1}^{\infty} \left(a - \frac{1}{n}, b + \frac{1}{n} \right)$.

For each $n \in \mathbb{N}$, $\left(a - \frac{1}{n}, b + \frac{1}{n}\right)$ is an open set.

i.e. $[a, b]$ = intersection of denumerable number of open sets.

Hence $[a, b]$ is of type G^δ.

Example 4.8.3. *An open bounded interval (a, b) is a set of type F_σ.*

Sol. (a, b) can be expressed as the arbitrary union

$$\bigcup_{n=1}^{\infty} \left[a + \frac{1}{n}, b - \frac{1}{n}\right]$$

and for each $n \in \mathbb{N}$, $\left[a + \frac{1}{n}, b - \frac{1}{n}\right]$ is a closed set.

Thus (a, b) is the union of a denumerable number of closed sets. **So (a, b) is an F_σ set.**

Example 4.8.4. *Every semi-open bounded interval $(a, b]$ is of type G^δ.*

Sol. We can write

$$(a, b] = \bigcap_{n=1}^{\infty} \left(a, b + \frac{1}{n}\right), \quad n \in \mathbb{N}$$

= intersection of a denumerable number of open sets.

Hence $(a, b]$ is a G^δ set.

Important Result I. *Every interval is of type F_σ or G^δ*

Open interval (a, b) is a set of type F_σ

Closed interval $[a, b]$ is a set of type G^δ

half-open interval $(a, b]$ is a set of type G^δ

Important Result II. *The complement of a set of type F_σ is a set of type G^δ and the complement of a set of type G^δ is a set of type F_σ.*

[Use De Morgan's laws and the fact the complement of an open set is closed and the complement of a closed set is open.]

Important Result III. (i) *The set of irrationals is of type G^δ*, **not of type F_σ.**

(ii) **The set of rationals is of type F_σ, not of type G^δ.**

Important Result IV. *Every open set is type F_σ and*

Every closed set is of type G^δ.

By allowing the denumerable intersections and unions of sets of type F_σ or G^δ we construct other Borel sets:

CHAPTER 4: ELEMENTS OF POINT SET TOPOLOGY

Definition 4.8.3. *A set S is of type F_σ^δ if it is the intersection of a denumerable class of sets of type F_σ; S is of type $F_{\sigma\sigma}^\delta$ (or $F_{\sigma 2}^\delta$) if it is the union of a denumerable class of sets of type F_σ^δ and so on.*

Definition 4.8.4. *A set S is of type G_σ^δ if it is the union of a denumerable class of sets of type G^δ; S is of type $G_\sigma^{\delta 2}$ (or $\gamma_\sigma^{\delta 2}$). if it is the intersection of a denumerable class of sets of type G_σ^δ; and so on.*

Definition 4.8.5. *A class of sets (A_α) is called a σ-field (or σ-algebra or Borel Field) if*

(i) the entire set $\mathbb{R}$ and the empty set ϕ are members of the class,

(ii) the union of a countable number of sets of the class is the class

and (iii) the complement of every set of the class is in the class.

From Def. 4.8.5 it can be easily proved that the intersection of a countable number of sets in a Borel field is a member of that Borel field.

An important result which can be proved is that the **class of Borel sets is the smaller Borel field containing open sets (open intervals)**. This means that any other Borel field containing open intervals will contain the class of Borel sets.

4.9 Compact Sets in $\mathbb{R}$

In the present article we shall first introduce the concept of a **covering** of a linear point set S (i.e. $S \subseteq \mathbb{R}$) and then introduce the notion of a compact set in $\mathbb{R}$.

Definitions: Covering, Open Covering, Sub covering

Let S be a subset of $\mathbb{R}(S \subseteq \mathbb{R})$.

Definition 4.9.1. *A collection F of sets is called a **covering** of S if*

$$S \subseteq \bigcap_{A \in F} A.$$

(This means that for each $x \in S, x \in$ some A of the collection F. We then say: F covers S).

Definition 4.9.2. *If F is a collection of **open sets**, then F is called an **open covering** of S.*

Definition 4.9.3. *If F' is a subcollection of sets from F such that the union of sets in F' also contains S (i.e. F' also covers S) then F' is called a **sub-cover** of F.*

Definition 4.9.4. *If F' contains only a finite number of sets, then F' is called a **finite sub covering** of F.*

Illustrative Examples of Coverings

Example 4.9.1. *We recall that in $\mathbb{R}$ every open interval is an open set. So the collection of all intervals of the form $\frac{1}{n} < x < \frac{2}{n}$ ($n = 2, 3, 4, \cdots$) is an **open covering** of the set*

$$S = \{x : 0 < x < 1\}$$

*This is an example of a **countable covering**.*

Example 4.9.2. *The real line $\mathbb{R}$ (i.e. set of x where $-\infty < x < \infty$) is covered by the collection of all open intervals (a, b). This covering is also an **open covering** but it is **not a countable covering**. But it contains a countable covering of $\mathbb{R}$, namely, all intervals of the form $(n, n+2)$ where n runs through all integers, positive or negative or zero.*

Example 4.9.3. *There can be many different open covers for a given set S where $S \subseteq \mathbb{R}$. Suppose $S = [1, \infty)$. See that the following collections of open intervals are all open covers of S:*

(a) $F_0 = \{(0, \infty)\}$
(b) $F_1 = \{I_r = (r-1, r+1) \text{ where } r > 0 \text{ and } r \in \mathbb{Q}\}$
(c) $F_2 = \{I_n = (n-1, n+1), n \in \mathbb{N}\}$
(d) $F_3 = \{I_n = (0, n), n \in \mathbb{N}\}$
(e) $F_4 = \{I_n = (0, n), n \in \mathbb{N}, n \geq 20\}$

Clearly F_2 is a sub cover of F_1;

F_4 is a sub cover of F_3.

Many other open covers of S can be described.

Example 4.9.4. *Given $S \subseteq \mathbb{R}$, we can always cover S by an infinite collections of open intervals.*

But in every case S may not be covered by a finite sub cover.

Def. 4.8.5: Definition of a compact set in $\mathbb{R}$

*A subset S of $\mathbb{R}$ is said to be **compact** if an only if every open cover of S has a finite sub covering.*

If a set $S \subseteq \mathbb{R}$ is such that every open covering of S has a finite subcovering then we say:

$$\textbf{S obeys Heine-Borel property.}$$

CHAPTER 4: ELEMENTS OF POINT SET TOPOLOGY

Hence our definition of compact set can be stated in the following language:

A set $S \subseteq \mathbb{R}$ is called a compact set iff it obeys the Heine-Borel property.

Theorem 4.9.1. *MAIN THEOREM OF THE PRESENT CHAPTER:*

Heine-Borel Theorem: Statement

Every open covering of a closed and bounded set $S \subseteq \mathbb{R}$ has a finite sub-covering

or equivalently

A closed and bounded set $S \subseteq \mathbb{R}$ is compact.

☐ **Converse of Heine-Borel Theorem: Statement**

A compact set in $\mathbb{R}$ is both bounded and closed.

Many authors state **Heine-Borel Theorem** in the following form:

A subset S of $\mathbb{R}$ is compact if only if it is closed and bounded.

This statement involves two steps:

Step 1. If S is compact, then S is both closed and bounded.

(**Converse of Heine-Borel Theorem**)

Step 2. If S is closed and bounded, then S is compact.

(**Heine-Borel Theorem**)

Remember: 'Compact' $\implies$ Every open covering has a finite sub-covering

Before we give formal proof of the Heine-Borel Theorem and its converse we shall discuss a few solved problems on Compact sets:

Remember:

A. **How to prove that a given set $S \subseteq \mathbb{R}$ is compact?**

Examine any open cover of S. See that it as a finite subcover: Conclusion: S is compact

B. **How to prove that a given set $S \subseteq \mathbb{R}$ is not a compact set?**

Exhibit one specific open cover of S which has no finite subcover: Conclusion: S is not compact

(See the Examples given below)

Solved problems

Example 4.9.5. *Prove that a finite subset of real numbers*

$$S = \{x_1, x_2, x_3, \cdots, x_n\}$$

is a compact set.

Soln. Let the set of open intervals $\{I_\alpha\}$ be an open cover of S.

Each member x_i is contained in some $I_{\alpha_i} \in \{I_\alpha\}$.

Then the union of the sets in the collection

$$\{I_{\alpha_1}, I_{\alpha_2}, \cdots, I_{\alpha_n}\}$$

Contains S, so that it is a finite subcover of $\{I_\alpha\}$.

Thus an arbitrary open cover $\{I_\alpha\}$ has a finite subcover. Hence it follows that **the finite set S is compact.**

[Otherwise: See that $S = \{x_1, x_2, \cdots, x_n$ is clearly bounded; also S is closed because the derived set $S' = \phi \subset S$. Hence S is closed and bounded. Therefore S is compact]

Example 4.9.6. *Prove that the set $T = \{x : x \in \mathbb{R} \text{ and } 0 < x < \infty\}$ is not compact.*

Soln. Consider a collection F of open intervals given by

$$F = \{(-1, n), n \in \mathbb{N}\}$$

Clearly,

$$T \subset \bigcup_{n=1}^{\infty} (-1, n)$$

i.e. F is some open cover of T.

However, let $\{(-1, n_1), (-1, n_2), \cdots, (-1, n_k)\}$ be any finite sub-collection of T, n_1, n_2, being positive integers. Suppose $m = \max\{n_1, n_2, \cdots, n_k\}$. Then

$$\bigcup_{n=1}^{k} (-1, n_k) = (-1, m)$$

which certainly does not cover all the points of $T = [0, \infty[$. Thus no finite subcollection of a given open cover F can cover T.

Hence **T is not a compact set in $\mathbb{R}$.**

[See that $T = [0, \infty[$ is unbounded but it a **closed set** since all limit points of T are in T]

Example 4.9.7. *Prove that the set S of real numbers in the open interval $(0, 1)$ is not compact.*

Soln. Let $I_n = \left(\frac{1}{n}, 1\right), n \in \mathbb{N}$.

CHAPTER 4: ELEMENTS OF POINT SET TOPOLOGY

Then
$$W = \bigcup_{n=1}^{\infty} I_n = (0,1).$$

The collection $F = \{I_n, n \in \mathbb{N}\}$ is an open cover of W. Let $\{I_{n_1}, I_{n_2}, \cdots, I_{n_k}\}$ be any finite collection of F $n_1, n_2, \cdots, n_k$ being k positive integers and suppose
$$r = \max\{n_1, n_2, n_3, \cdots, n_k\}$$

Then
$$\bigcup_{i=1}^{k} I_{n_i} = I_r = \left(\frac{1}{r}, 1\right)$$

It is clear that an element $\frac{1}{r} \in W$ but $\frac{1}{r} \notin I_r$ ($\because I_r$ is open)

$\therefore$ The union of any finite collection of F does not contain W.

$\therefore W$ is not compact.

[Again see that $W = (0,1)$ is bounded but not closed (because the limit point 0 and 1 do not belong to W and hence W is not a compact set by **Heine-Borel Theorem**)].

Note: In order to apply Heine-Borel Theorem **both the conditions of Heine-Borel Theorem—Boundedness and Closed Set property—must be satisfied**.

Example 4.9.8. *Prove that the set $\mathbb{R}$ of all real numbers is not compact.*

Soln. Let $I_n = (-n, n), n \in \mathbb{N}$.

Then $\sigma = \{I_n : n \in \mathbb{N}\}$ is an open cover of the set $\mathbb{R} = (-\infty, \infty)$

Let $\{I_{n_1}, I_{n_2}, \cdots, I_{n_k}\}$ be any finite subcollection of σ. Let $n_0 = \max\{n_1, n_2, \cdots, n_k\}$. The n_0 itself cannot belong to any I_{n_i} ($i = 1, 2, 3, \cdots, k$).

Thus no finite subclass of σ can cover $\mathbb{R}$. **Hence $\mathbb{R}$ is not compact**

(Here $\mathbb{R}$ is closed but not bounded. Hence $\mathbb{R}$ cannot be compact by Heine-Borel Theorem also)

Example 4.9.9. *Prove that an open interval (a, b) in $\mathbb{R}$ is not compact*

Soln. Let $I_n = (a + \frac{1}{n}, b), n \in \mathbb{N}$.

Then $\sigma = \{I_n : n \in \mathbb{N}\}$ is an open cover of (a, b), since $\bigcup_{n=1}^{\infty} I_n = (a, b)$.

But it is not possible to find a subcollection of σ which covers (a, b). For, if $\sigma' = \{I_{n_1}, I_{n_2}, \cdots, I_{n_k}\}$ be any finite sub-collection of σ, and let $n_0 = \max\{n_1, n_2, \cdots, n_k\}$, then it is evident that $(a, a + \frac{1}{n_0})$ of (a, b) is not covered by σ'. Thus there exists an open cover of (a, b) which does not admit of a finite sub-cover. Hence the **open interval (a, b) is not compact in $\mathbb{R}$**.

[Here (a,b) is evidently bounded but **not closed** ($\because a, b$ are limit points of (a,b) but they do not belong to (a,b)) and hence the set (a,b) is not compact].

Example 4.9.10. *Exhibit an open cover of the semi-open interval $(1,2]$ that has no finite sub-cover.*

Soln. Let $I_n = (1 + \frac{1}{n}, 3)$ for $n \in \mathbb{N}$.

Then $\sigma = \{I_n : n \in \mathbb{N}\}$ is an open cover of $(1,2]$, since $(1,2] \supseteq \bigcup_{n=1}^{\infty} I_n$.

Consider a finite subcollection

$$\left\{ \left(1 + \frac{1}{n_1}, 3\right), \left(1 + \frac{1}{n_2}, 3\right), \cdots, \left(1 + \frac{1}{n_k}, 3\right) \right\}$$

Then it does not cover all the points of $(1,2]$.

e.g., if $n_0 = \max\{n_1, n_2, \cdots, n_k\}$, then $(1, 1 + \frac{1}{n_0}) \in (1,2)$ but not covered by the subcollection.

Note. The half-open interval $(1,2]$ is neither open nor closed though it is bounded and so it cannot be a compact set. The point 2 is not an interior point and 1 is a limit point but does not belong to the set.

Example 4.9.11. *If S is a closed subset of a compact set K in $\mathbb{R}$, then prove that S is compact.*

Soln. Given: $S \subseteq K \subseteq \mathbb{R}$.

S is a closed set

and K is a compact set in $\mathbb{R}$.

To prove: S is a compact.

Let σ be an open cover of S.

Since S is closed, its complement $C(S)$ is an open set.

Therefore $\sigma \cup C(S)$ is an open cover of K.

But K is compact. Therefore $\exists$ a finite subcover of this open cover of K, which is also a finite subcover of its subset S. Then S is also compact.

Example 4.9.12. *Find an infinite collection $\{I_n : n \in \mathbb{N}\}$ of compact sets in $\mathbb{R}$ such that the union $\bigcup_{n=1}^{\infty} I_n$ is not compact.*

Soln. Let $I_n = [0, n], n \in \mathbb{N}$. Each one of $[0, n]$ is a compact set, but their union, namely $[0, \infty)$ is not compact.

CHAPTER 4: ELEMENTS OF POINT SET TOPOLOGY

Example 4.9.13. *Prove that the intersection of an arbitrary collection of compact sets in $\mathbb{R}$ is compact.*

Soln. Let σ be a collection of an arbitrary family of compact sets in $\mathbb{R}$.

Each member of this collection being a compact set, is closed and bounded (*converse of Heine-Borel Theorem*).

We know that the intersection of an arbitrary family of closed sets is a *closed set*.

Also if each member of this collection is bounded then their intersection is also bounded.

Hence, by Heine-Borel Theorem, the intersection of an arbitrary collections of compact sets, is a compact set.

Example 4.9.14. *Let A and B be two subsets of $\mathbb{R}$. Given that A is a closed set and B is a compact set. Prove that $A \cap B$ is a compact set.* [C.H. 1997, 2002]

Soln. Given: B is a compact set in $\mathbb{R}$

Hence B is **closed** and **bounded**.

Again, **Given:** A is a closed set in $\mathbb{R}$.

The intersection of two closed sets is always a closed set.

Hence $A \cap B$ contains common elements of A and B and B is bounded. Therefore $A \cap B$ must be **bounded**.

$\therefore$ $A \cap B$ **being closed and bounded, is compact set in $\mathbb{R}$ (Heiene-Borel Theorem)**

Example 4.9.15. *$S \subseteq \mathbb{R}$ is a compact set if every open cover of S has a finite subcover. Using this definition of a compact set prove that the union of two compact sets in $\mathbb{R}$ is a compact set. With the help of an example show that the union of an infinite number of compact sets in $\mathbb{R}$ is not necessarily a compact set.*

Soln. Let A and B be two compact subsets in $\mathbb{R}$.

To prove: $A \cup B$ is a compact set in $\mathbb{R}$.

Let F_1 be an open cover of A and let F_2 be an open cover of B.

Then clearly $F_1 \cup F_2$ is an open cover of $A \cup B$.

Now A being a compact set $\exists$ a finite collection of sets of F_1 which covers A. Similarly $\exists$ a finite collection of sets of F_2 which covers B. Then union of these two finite collections is a finite collection i.e. $F_1 \cup F_2$ has a finite cover of $A \cup B$ i.e. $A \cup B$ **is a compact set.**

But see that $\exists$ an infinite collection of compact sets

$$\{I_n = [0, n]; n \in \mathbb{N}\}$$

such that the union $\bigcup_{n=1}^{\infty} I_n = [0, \infty)$ has no finite subcover and hence the union is not compact.

Therefore, the union of an infinite collection of compact sets **is not always a compact set**.

Example 4.9.16. *Correct or Justify: The set $\mathbb{N}$ of all natural numbers is not a compact set in $\mathbb{R}$.* [C.H. 2004]

Soln. The collection $F = \{(n - \frac{1}{2}, n + \frac{1}{2}) : n \in \mathbb{N}\}$ is certainly an open cover of $\mathbb{N}$, because each nature number n is contained in $(n - \frac{1}{2}, n + \frac{1}{2})$. But a finite number of open intervals $(n - \frac{1}{2}, n + \frac{1}{2})$ cannot cover $\mathbb{N}$ because many of the integers will be outside the union of this finite number of open intervals like $(n - \frac{1}{2}, n + \frac{1}{2})$.

∴ the conclusion is that the set $\mathbb{N}$ of natural numbers **is not compact**.

(See that if $S = \mathbb{N}$ then the derived set $S' = \phi \subset S$; so S is closed but certainly S is not bounded)

Example 4.9.17. *Let S be a set of real numbers.*

What is meant by saying that 'S possesses Heine-Borel property'? Does the set $S = \{\frac{1}{n} : n \in \mathbb{N}\}$ possess this property? Justify your assertion. [C.H. 1997]

Soln. If a subset S of $\mathbb{R}$ is such that every open cover of S has a finite subcover then we say

'S obeys Heine-Borel property'

Now consider $S = \{1, \frac{1}{2}, \frac{1}{3}, \cdots, \frac{1}{n}, \cdots\}$.

The point $\frac{1}{n}$ is covered by $\left(\frac{1}{n+\frac{1}{2}}, \frac{1}{n-\frac{1}{2}}\right)$

$$\therefore \left\{\left(\frac{1}{n+\frac{1}{2}}, \frac{1}{n-\frac{1}{2}}\right) ; n \in \mathbb{N}\right\}$$

is an open cover of S.

But it is not possible to choose a finite number of these intervals which will cover S

(Choose a finite number of intervals, many points of the form $\frac{1}{n}$ will be left out i.e. not covered)

Example 4.9.18. *Given an example of a set S of real numbers which is closed but not bounded. It cannot therefore be a compact set.*

Now exhibit a countable open covering F of S such that no finite subset of F covers S so that S is not compact. [C.H. 1991]

Thereoms related to Compact Sets

We begin with **Nested Interval Theorem:**

A sequence $\{I_1, I_2, \cdots, I_n, \cdots\}$ of closed intervals is called a nest of intervals if

(i) $I_1 \supset I_2 \supset I_3 \supset \cdots \supset I_n \supset I_{n+1} \supset \cdots$

and (ii) the sequence $\{d_n\}$ be such that $d_n \longrightarrow 0$ as $n \longrightarrow \infty$, where

$$d_n = l(I_n) = \text{length of the interval } I_n$$

then $\bigcap_{n=1}^{\infty} I_n$ has exactly one point i.e. $\exists$ exactly one point common to all the intervals)

This is known as the **Nested Interval Theorem.**

4.10 The Cantor Intersection Theorem

This theorem is an application of Bolzano-Weierstrass Theorem.

Statement: Let $\{C_1, C_2, C_3, \cdots\}$ be a countable collection of non-empty sets in $\mathbb{R}$ such that

(i) $C_{K+1} \subseteq C_K (K = 1, 2, 3, \cdots)$, i.e., $C_1 \supseteq C_2 \supseteq C_3 \supseteq \cdots$

(ii) Each set C_K is closed and C_1 is a bounded set.

Then the intersection $\bigcap_{K=1}^{\infty} C_K$ is closed and non-empty.

Proof. Let $S = \bigcap_{K=1}^{\infty} C_K$. Since each C_K is closed, S is closed.

($\because$ it is known that the intersection of an arbitrary collection of closed sets is closed.)

To prove that S is non-empty, we exhibit a point x in S. Assume that each C_K contains infinitely many points (otherwise the proof is trivial).

Now form a collection of distinct points

$$A = \{x_1, x_2, x_3, \cdots\}, \text{ where } x_K \in C_K.$$

Clearly A is an infinite set contained in the bounded set C_1.

$\therefore$ A has an accumulation point, say x.

We shall show that $x \in S$, by verifying that for each K, $x \in C_K$.

It will be enough to show that x is an accumulation point of each C_K, since they are all closed sets. But every neighbourhood of x contains infinitely many points of A and since all except perhaps a finite number of points of A belong to C_K, this neighbourhood also contains infinitely many points of C_K. Therefore, x is an accumulation point of C_K and the theorem is proved.

Theorem 4.10.1. *The Lindelöf Covering Theorem* (*in* $\mathbb{R}$)

Statement: Every open covering of a set of real numbers has a countable subcovering (or an at most countable subcovering).

Proof. Let S be a set of real numbers. Let $\{I_\alpha\}$ be an open covering (we cover by open intervals) of the set S and let $x \in S$. Then there exists some I_α such that $x \in I_\alpha$. Suppose $I_\alpha = (a_\alpha, b_\alpha)$. Now there are rational numbers p_α and q_α such that

$$a_\alpha \leq p_\alpha \leq x \leq q_\alpha \leq b_\alpha$$

i.e., $\quad x \in (p_\alpha, q_\alpha) \subset I_\alpha.$

Now we know that the class of open intervals with rational end points is **denumerable**. Consequently the class of open intervals of the form (p_α, q_α) is **denumerable**.

Since the set S can be covered by intervals of this form, it follows that S can be covered by a countable subclass of $\{I_\alpha\}$, which completes the proof.

The set of all open intervals having rational end points is denumerable.
Proof. Let the set of all rational numbers be written as $\{x_1, x_2, \cdots, x_n, \cdots\}$. The set in question is the set of open intervals (x_m, x_n) for all m, n ($m < n$) such that $x_m < x_n$. For a fixed n, let $A_n = \{(x_m, x_n) : x_m < x_n\}$. This set is equivalent to the set $\{m : x_m < x_n\} \subset \mathbb{N}$. Hence A_n is at most countable. But A_n is infinite so that A_n is denumerable and hence

$$\bigcup_{n=1}^{\infty} A_n \text{ is denumerable.}$$

We have already introduced.

Heine-Borel Theorem (Theorem 4.8.1)

Statement: Every open covering of a closed and bounded set $S \subseteq \mathbb{R}$ has a finite subcover. (*see Art. 1.2*)

In other words, *any closed and bounded set $S \subseteq \mathbb{R}$ is* **compact**.

[If a set S is such that every open covering of S has a finite subcovering, then S is said to have the Heine-Borel Property. A set S is called a **compact set** if it has the Heine-Borel property.]

Proof of Heine-Borel Theorem in $\mathbb{R}$

Let $S \subseteq \mathbb{R}$ be a closed and bounded set.

Let the class of open intervals $\{I_\alpha : \alpha \in \lambda, \lambda$ being an index set$\}$ be any open covering of S.

To prove: There exists a finite subcollection of $\{I_\alpha : \alpha \in \lambda\}$ which also covers S.

If the collection $\{I_\alpha : \alpha \in \lambda\}$ contains a finite number of open intervals, then the conclusion of the theorem becomes obvious. We assume, therefore, that the class

CHAPTER 4: ELEMENTS OF POINT SET TOPOLOGY

$\{I_\alpha : \alpha \in \lambda\}$ is infinite. By the Lindelöf covering theorem there exists a **countable** subclass of $\{I_\alpha : \alpha \in \lambda\}$ which is a subcovering of S. Again if this countable class is finite, then the result is obvious.

We, therefore, assume that we have a countably infinite subcovering $\{I_1, I_2, I_3, \cdots, I_n, \cdots\}$ of S. We shall show that it is possible to cover S by a finite number of these open intervals.

Suppose by way of contradictions that it is impossible to cover S by a finite number of these open intervals; then $S - \left(\bigcup_{K=1}^{n} I_K\right)$ is always non-empty. Each I_K is open, so the union of a finite number of I_K is open, i.e., $\bigcup_{K=1}^{n} I_K$ is open and S is bounded and closed, it follows that $S - \left(\bigcup_{K=1}^{n} I_K\right)$ is closed and bounded.

Note that if A is a closed set and B is an open set, then $A - B = A \cap B^c$ is a closed set. Again if A is bounded, then $A - B$ is a bounded set.

Moreover, if we take

$$S_1 = S - I_1,$$
$$S_2 = S - (I_1 \cup I_2),$$
$$\cdots \cdots \cdots \cdots \cdots \cdots$$
$$S_n = S - \left(\bigcup_{K=1}^{n} I_K\right),$$
$$\cdots \cdots \cdots \cdots \cdots \cdots$$

then $S_1 \supset S_2 \supset S_3 \supset \cdots \supset S_n \supset \cdots$.

Now using Cantor's Intersection Theorem $\bigcap_{n=1}^{\infty} S_n$ is not empty. Let $x \in \bigcap_{n=1}^{\infty} S_n$. Since x is in every S_n, x cannot belong to any I_n; yet x belongs to S. This is a contradiction, since $\{I_n : n \in \mathbb{N}\}$ is a covering of S. Hence the theorem is established.

Another proof of Heine-Borel Theorem on $\mathbb{R}$: Using the Theorem on Nested Intervals

Let S be a **closed** and **bounded** set of real numbers and let $\sigma = \{I_\alpha : \alpha \in \lambda\}$ be a collection of open intervals which covers S.

To prove: $\exists$ a finite subcollection of σ which covers S, i.e., we wish to prove that *S is contained in the union of a finite subcollection of σ.* The proof will be again by contradiction.

Assume that

$$\left.\begin{array}{r}S \text{ is not contained in the union of any}\\ \text{finite number of open intervals of } \sigma.\end{array}\right\} \quad (4.1)$$

Since S is bounded, there exists a positive real number M such that

$$S \subseteq [-M, M].$$

We now consider two closed intervals $[-M, 0]$ and $[0, M]$. At least one of the two collections $S \cap [-M, 0]$ and $S \cap [0, M]$ must be non-empty and has the property that

it is not contained in the union of a finite number of open intervals of σ.

[For, if both of $S \cap [-M, 0]$ and $S \cap [0, M]$ are contained in the union of some finite number of open intervals of σ, then

$$S = \{S \cap [-M, 0]\} \cup \{S \cap [0, M]\}$$

is contained in the union of some finite number of open intervals in σ, which goes against our assumption (1).]

If $S \cap [-M, 0]$ is not contained in the union of some finite number of open intervals of σ, then we shall call $I_1 = [-M, 0]$; otherwise $S \cap [0, M]$ has this property and then we shall call $I_1 = [0, M]$. We note the length of $I_1 = M$.

Next we shall bisect I_1 into two *closed intervals* I_1' and I_1''. If $S \cap I_1'$ is non-empty and is not contained in the union of some finite number of open intervals of σ, then we shall call $I_2 = I_1'$; otherwise $S \cap I_1''$ will have this property and we shall call $I_2 = I_1''$. The length of $I_2 = \frac{M}{2^1}$.

We can continue this process of subdivision and obtain thereby a sequence of closed intervals

$$\{I_1, I_2, I_3, \cdots, I_n, \cdots\}$$

with the length of $I_n = \frac{M}{2^{n-1}}$ which tends to zero as $n \to \infty$ and $S \cap I_n$ is not contained in the union of some finite number of open intervals of σ. Our construction shows that

$$I_n \supseteq I_{n+1} \quad (n = 1, 2, 3, \cdots).$$

Thus the closed intervals are nested and length of $I_n \to 0$ as $n \to \infty$.

$\therefore$ by Nested Interval property, there exists exactly one point z which belongs to all the I_n's ($n \in \mathbb{N}$).

Since each interval I_n contains points of S, the point z is a point of accumulation of S. Moreover, since S is assumed to be closed, $z \in S$.

CHAPTER 4: ELEMENTS OF POINT SET TOPOLOGY

∴ there exists an open interval I_λ of $\sigma = \{I_\alpha : \alpha \in \lambda\}$ with $z \in I_\lambda$. Since I_λ is an open set, there exists $\varepsilon > 0$ such that

$$(z - \varepsilon, z + \varepsilon) \subseteq I_\lambda.$$

Since the intervals I_n are obtained by repeated bisections of $[-M, M]$, with the length of $I_n = \frac{M}{2^{n-1}}$, if follows that if n is so large that $\frac{M}{2^{n-1}} < \varepsilon$, then $I_n \subseteq (z-\varepsilon, z+\varepsilon) \subseteq I_\lambda$.

But this means that if n is such that $\frac{M}{2^{n-1}} < \varepsilon$, then $S \cap I_n$ is contained in the *single set* I_λ of σ, contrary to our construction of I_n. This contradiction shows that the assumption (1) that the *closed bounded set S requires an infinite number of open intervals in σ to cover it, is untenable*, i.e., a finite subset of σ is capable of covering the closed bounded set S. We, therefore, conclude that S is compact.

Converse of Heine-Borel Theorem (Theorem 4.9.1)

A compact set in $\mathbb{R}$ is both closed and bounded.

(i) We shall first prove that a compact set S in $\mathbb{R}$ is bounded.

Given: $S \subseteq \mathbb{R}$ is compact, i.e., every open cover of S has a finite subcover.

To prove: S is bounded.

Proof. For each positive integer m, let $I_m = (-m, m)$ so that each I_m is open and since

$$S \subseteq \bigcup_{m=1}^{\infty} I_m = \mathbb{R},$$

we see that the collection $F = \{I_m : m \in \mathbb{N}\}$ is an open cover of S.

Since S is compact, F has a finite subcover, i.e., there exists some positive integer M such that

$$S \subseteq \bigcup_{m=1}^{M} I_m = I_M = (-M, M)$$

i.e., S is contained in a bounded interval $(-M, M)$ and as such S is a *bounded set*.

(ii) We next prove that a compact set S is closed (i.e., all limit points of S will lie on S).

Given: $S \subseteq \mathbb{R}$ is compact, i.e., every open cover of S has a finite subcover.

To prove: S is a closed set, i.e., we shall prove that every limit point of S is a member of S.

Proof. (By contradiction). If possible, suppose S is not a closed set. Then there must exist at least one limit point of S (say, α) where $\alpha \notin S$.

We construct some open cover of S by the following rule:

For each x, let $r_x = \frac{1}{2}|x - \alpha|$. Since no $x \in S$ coincides with α, therefore, $r_x > 0$. Now, corresponding to $x \in S$, let

$$I_{r_x} = (x - r_x, x + r_x).$$

The collection $F = \{I_{r_x} : x \in S\}$ certainly is an open cover of S. Thus S is a compact set and F is an open cover of S.

$\therefore$ there exists a finite subcollection of F (say F') such that

$$F' = \left\{I_{r_{x_1}}, I_{r_{x_2}}, \cdots, I_{r_{x_k}}\right\}$$

also covers S, i.e.,

$$S \subseteq \bigcup_{p=1}^{k} I_{r_{x_p}}.$$

Let $r = \min\{r_{x_1}, r_{x_2}, \cdots, r_{x_k}\}$. Then the open interval $(\alpha - r, \alpha + r)$ has no points in common with any of the open intervals of F' (proof given below).

Therefore, $(\alpha - r, \alpha + r) \cap S = \phi$, contradicting the fact that α is a limit point of S. This contradiction leads to the conclusion that S cannot but be closed. In other words, S must be a closed set.

[**Justification** of the assertion that $(\alpha - r, \alpha + r)$ has no point in common with any of the open intervals of F':

Let $x \in (\alpha - r, \alpha + r)$.

Then $|x - \alpha| < r \leq r_{x_p} \quad (p = 1, 2, 3, \cdots, k)$

$$(\because r = \min\{r_{x_1}, r_{x_2}, \cdots, r_{x_k}\}.)$$

Hence, $\begin{aligned}[t]|x - x_p| &= |(x - \alpha) + \alpha - x_p| \\ &= |(\alpha - x_p) - (\alpha - x)| \\ &\geq |\alpha - x_p| - |x - \alpha| \\ &> 2r_{x_p} - |x - \alpha| \quad (\because r_{x_p} = \tfrac{1}{2}|x_p - \alpha|) \\ \text{i.e.,} \quad &> 2r_{x_p} - r_{x_p} \quad (\because |x - \alpha| < r_{x_p} \text{ and so } -|x - \alpha| > -r_{x_p}) \\ \text{i.e.,} \quad &> r_{x_p}.\end{aligned}$

Thus for any $p = 1, 2, 3, \cdots, k$ we get $|x - x_p| > r_{x_p}$ and hence $x \notin I_{r_{x_p}}$.]

Observations. We can prove that if S is a compact set, then S is closed by showing that its complement S^c is open. To do so, let $u \in S^c$ be arbitrary and for each $n \in \mathbb{N}$,

we let
$$I_n = \left\{ y \in R : |y - u| > \frac{1}{n} \right\}.$$

Then each I_n is open and $R - \{u\} = \bigcup_{n=1}^{\infty} I_n$, since $u \notin S$ we have $S \subseteq \bigcup_{n=1}^{\infty} I_n$. Since S is compact, then exist $m \in \mathbb{N}$ such that
$$S \subseteq \bigcup_{n=1}^{m} I_n = I_m.$$

Now it follows from this that $S \cap \left(u - \frac{1}{m}, u + \frac{1}{m}\right) = \phi$, so that $\left(u - \frac{1}{m}, u + \frac{1}{m}\right) \subseteq S^c$. Since u was arbitrary point in S^c, we conclude that S^c is open.

4.11 Other Characterizations of Compact Sets in $\mathbb{R}$

I. Sequential Characterization or Cluster Point Characterization

Theorem 4.11.1. *Let S be an infinite set of real numbers. S is compact if and only if every sequence in S has a subsequence that converges to a point $\alpha \in S$.*

Proof. (i) Suppose that S is compact. Let $\{x_n\}_n$ be a sequence of real numbers with every $x_n \in S$.

By the converse of the Heine-Borel theorem (see Art. 1.7(i)] the set S is bounded, so that the sequence $\{x_n\}_n$ must also be bounded ($\because$ every $x_n \in S$).

By Bolzano-Weierstrass theorem on sequences there exists a sub-sequence
$$\{x_{n_K}\}_{K=1}^{\infty} \text{ that converges.}$$
Again, by Art. 1.7(ii), S is a closed set.

$\therefore$ the cluster point $\lim_{K \to \infty} x_{n_K} = x$ (say) belongs to S. Thus every sequence of a compact set S has a subsequence that converges to a point of S.

(ii) To establish the converse, we will show that if S is either **not closed** or **not bounded**, then there must exist a sequence in S that has no subsequence converging to a point of S.

First, if S is not closed, there is a cluster point α of S that does not belong to S. Since α is a cluster point of S, there is a sequence $\{x_n\}_n$ with $x_n \in S$ and $x_n \neq \alpha \, \forall \, n \in \mathbb{N}$ such that $\lim_{n \to \infty} x_n = \alpha$. Then every subsequence of $\{x_n\}_n$ also converges to α and since $\alpha \notin S$, there is no subsequence that converges to a point of S.

Secondly, if S is not bounded, then there exists a sequence $\{x_n\}_n$ in S such that $|x_n| > n$ for all $n > N$. Then every subsequence of $\{x_n\}_n$ is unbounded, so that no subsequence of it can converge to a point in S.

II. Limit Point Characterization

Theorem 4.11.2. *Let S be an infinite set of real numbers.*

(a) If S is closed and bounded, i.e., if S is compact, then every infinite subset of S has a limit point in S.

(b) If every infinite subset of S has a limit point in S, then S is both closed and bounded (and hence compact).

Proof. (a) Let T be any infinite subset of S ($T \subseteq S$), where S is a given bounded and closed set. $\because$ S is bounded and $T \subseteq S$, T is also bounded. By Bolzano-Weierstrass theorem for sets there exists a limit point (or a point of accumulation) α of T. This point α must also be a limit point of S ($\because S \supseteq T$). Again S is a closed set, therefore, $\alpha \in S$. We assumed T as any arbitrary subset of S, so we infer that every infinite subset of a closed and bounded set S has a limit point in S.

(b) *Given:* S is an infinite set of real numbers with the property that every infinite subset of S has a limit point in S.

To prove: (i) S is bounded and (ii) S is closed and hence S is compact.

Proof of (i), i.e., to prove that S is bounded. Suppose, if possible, S is not bounded. Then for every positive integer N, there exists a number $x_n \in S$ such that $|x_n| > N$. The collection $T = \{x_1, x_2, \cdots, x_n, \cdots\}$ of such points of S is certainly an infinite subset of S and by our given hypothesis T has a limit point, say $\beta \in S$.

Now for $N > 1 + |\beta|$, we have

$$|x_n - \beta| \geq |x_n| - |\beta| > N - |\beta| > 1,$$

contradicting the fact that β is a limit point of T, i.e., our assumption that S is not bounded, is not tenable.

In other words, S must be bounded.

Proof of (ii), i.e, to prove that S is closed, i.e., to prove every limit point of S is a member of S. Let α be a limit point of S.

Since every neighbourhood of α contains infinitely many points of S

$\exists\, x_1 \in S$ in the open interval $(\alpha - 1, \alpha + 1)$,

$\exists\, x_2 \in S$ in the open interval $\left(\alpha - \frac{1}{2}, \alpha + \frac{1}{2}\right)$,

$\cdots \quad \cdots \quad \cdots \quad \cdots \quad \cdots \quad \cdots \quad \cdots \quad \cdots \quad \cdots$

$\exists\, x_k \in S$ in the open interval $\left(\alpha - \frac{1}{K}, \alpha + \frac{1}{K}\right)$, etc., etc.

Then the collection $T = \{x_1, x_2, x_3, \cdots, x_K, \cdots\}$ is an infinite subset of S and α is a limit point of T. Since T is an infinite subset of S, by our assumption, T has a limit

CHAPTER 4: ELEMENTS OF POINT SET TOPOLOGY

point in S. If we can prove that α is the **only** limit point of T, then this α is the limit point which would have to be belonged to S. Then it is proved that any limit point of S belongs to S. Hence S is closed and our required result will be proved.

To prove that α is the only limit point of T, suppose, if possible, $\beta(\neq \alpha)$ is another limit point of T. Then for $x_K \in T$, we have

$$\begin{aligned} |x_k - \beta| &= |\alpha - \beta + x_k - \alpha| \\ &= |(\alpha - \beta) - (\alpha - x_k)| \geq |\alpha - \beta| - |x_k - \alpha| \end{aligned}$$

$[\because x_k \in T$, by our construction $x_k \in \left(\alpha - \frac{1}{K}, \alpha + \frac{1}{K}\right)$,

therefore, $|x_k - \alpha| < \frac{1}{K}$ or, $-|x_k - \alpha| > -\frac{1}{K}]$
$\therefore |x_k - \beta| > |\alpha - \beta| - \frac{1}{K}.$

If K_0 be taken so large that $\frac{1}{K} < \frac{1}{2}|\alpha - \beta|$, whenever $K \geq K_0$, then the above inequality leads to

$$|x_k - \beta| > \tfrac{1}{2}|\alpha - \beta|, \text{ whenever } K \geq K_0.$$

This shows that $x_k \notin (\beta - r, \beta + r)$ when $K \geq K_0$ and $r = \frac{1}{2}|\alpha - \beta|$. Thus β is not a limit point of T, a contradiction. So, as stated earlier, α is the only limit point of T.

This completes the proof of the **Theorem 4.9.5**.

Observations. We are now in a position to state the following result: *Let S be a set of real numbers: Then the following three statements are equivalent:*

1. *S is compact.*

2. *S is closed and bounded.*

3. *Every infinite subset of S has an accumulation point in S.*

Heine-Borel theorem gives $(2) \Longrightarrow (1)$. Converse of Heine-Borel theorem gives $(1) \Longrightarrow (2)$. **Theorem 4.9.5** gives $(2) \Longleftrightarrow (3)$. This establishes the equivalence of all three statements.

Exercises on Chapter IV (II)

1. Exhibit on open cover of the semi-closed interval $(2, 3]$ which has no finite sub-cover.

   [Let $\sigma_n = (2 + \frac{1}{n}, 4)$ for all $n \in \mathbb{N}$.]

2. Exhibit on open cover of the set S whose members are of the form $\frac{1}{n}$, $n \in \mathbb{N}$ that has no finite subcover. [Take $\sigma_n = \left(\frac{1}{2n}, 2\right)$ for $n \in \mathbb{N}$.]

3. Use the definition: "A subset S of $\mathbb{R}$ is said to be a **compact set** if every open cover of S has a finite subcover" to prove that if A and B are two compact sets in $\mathbb{R}$, then their union $A \cup B$ is compact.

4. Find an infinite collection

$$\{A_n : n \in \mathbb{N}\}$$

of compact sets in $\mathbb{R}$ such that their union $\bigcup_{n=1}^{\infty} A_n$ is not compact.

(Take $A_n = [0, n]$ for $n \in \mathbb{N}$.)

5. For a non-empty compact set S in $\mathbb{R}$, prove that $\inf S$ and $\sup S$ both exist and both belong to S.

[Since S is non-empty and bounded, $\inf S$ exists in $\mathbb{R}$. If $\sigma_n = \{a \in S \text{ and } a \leq \inf S + \frac{1}{n}\}$, $n \in \mathbb{N}$. Then σ_n is closed and bounded and hence compact. Then $\bigcap \sigma_n$ is not empty (see the exercise **No. 16**, given below) but if $x_0 \in \bigcap \sigma_n$, then $x_0 \in S$ and $x_0 = \inf S$.

Otherwise use Theorem 1.8.1].

6. The collection F of open intervals of the form $\left(\frac{1}{n}, \frac{2}{n}\right)$ where $n = 2, 3, \cdots$ is an open covering of the open interval $(0, 1)$. Prove that no finite subcollection of F covers $(0, 1)$.

7. Give an example of a set S which is closed but not bounded and exhibit a countable open covering F such that no finite subset of F covers S. [C.H. 1991]

8. Define covering of a set of real numbers. What is an open covering? Countable covering? Give an example of a countable covering and another example of non-countable covering.

9. If $S \subseteq \mathbb{R}$ is covered by a collection σ of open intervals, show that **an at most enumerable** (or **countable**) subcollection of σ shall cover S. (**Art. 1.5**)

10. What is known as Heine-Borel property of a set S of real numbers? Does the set $S = \{\frac{1}{n} : n \in \mathbb{N}\}$ possess Heine-Borel property? Justify your answer. [C.H. 1997]

11. Prove that if S is a closed and bounded subset of real numbers, then every infinite subset of S has an accumulation point in S.

Prove the converse of this result, if true. [C.H. 1998]

CHAPTER 4: ELEMENTS OF POINT SET TOPOLOGY 237

12. Give an example of a set S in $\mathbb{R}$ which is bounded but not closed. Show that S is not compact in the sense that no open cover of S has a finite subcover.

    $[S = (0, 1]$. Take $I_n = \left(\frac{1}{n}, 2\right) : n \in \mathbb{N}$ and consider the collection $\{I_n\}_n$.]

    (*compare Ex. 2 given above*)

13. Prove that the union of a finite number of compact sets is compact. Give an example of a non-compact set which is the union of an infinite sequence of compact sets,

    $$\left[\bigcup [0, n)\right].$$

14. Let A and B be two subsets of $\mathbb{R}$. Suppose A is a closed set but B is a compact set. Prove that $A \cap B$ is compact. [C.H. 1994, '97]

15. Let S be a compact subset of $\mathbb{R}$ and F is a subset of S. If F is a closed set in $\mathbb{R}$, then prove that F is compact in $\mathbb{R}$.

16. Let $\{A_n : n \in \mathbb{N}\}$ be a sequence of non-empty sets in $\mathbb{R}$ such that

    $$A_1 \supseteq A_2 \supseteq A_3 \cdots \supseteq A_n \supseteq \cdots.$$

    Prove that there exists at least one point $x \in \mathbb{R}$ such that $x \in A_n$ for all $n \in \mathbb{N}$; that is, the intersection of $\bigcap_{n=1}^{\infty} A_n$ is not empty.

17. Let S be a non-empty set compact set in $\mathbb{R}$. Show that S has a least element. [C.H. 2002]

    [**Soln.** If possible, let S have no least element. For each $a \in S$ let $I_a = \{x : x \in \mathbb{R} \text{ and } x > a\}$. Then I_a becomes an open interval.

    The collection σ of all such open intervals will **be an open cover of S**, because for **any** $b \in S$ there is an a in S such that $a < b$ and therefore, $b \in I_a$.

    Since S is compact, there is a finite subcollection of σ, say $\{I_{a_1}, I_{a_2}, \cdots, I_{a_m}\}$ such that $S \subseteq \bigcup_{i=1}^{m} I_{a_i}$. Let $\alpha = \min\{a_1, a_2, \cdots, a_n\}$. Then $a_0 \in S$ but $a_0 \notin \bigcup_{i=1}^{m} I_{a_i}$.

    i.e., the collection σ is not an open cover of S. We arrive at a contradiction and hence we must take that S has a least element.]

18. If $A = \left[\frac{1}{2}, \frac{7}{2}\right]$ and $B = \left(1, \frac{9}{2}\right)$, show whether $A \cup B$ is compact or not. [C.H. 2002]

    [*Ans.* $A \cup B$ is not a compact set—both A and B are bounded and hence $A \cup B$ is bounded. In $A \cup B$, $\frac{9}{2}$ is a limit point but does not belong to it and therefore, $A \cup B$ is not a closed set.]

19. Let $H = [0, \infty]$ and $G = \bigcup_{n \in \mathbb{N}} G_n$, where $G_n = (-1, n)$. Prove that G is an open cover of H but no finite subcollection of G can cover H. Explain why H fails to satisfy Heine-Borel Property. [C.H. 2002]

20. (a) Define 'Open Covering' of a point-set of real numbers and hence define a compact set in $\mathbb{R}$. Using this definition of compact set, show that if C be a compact set of real numbers, then C is a bounded set. [C.H. 2003]

   (b) Correct or justify: The $\mathbb{N}$ of natural numbers is not a compact set. [C.H. 2004]

AN INTRODUCTION TO
ANALYSIS

(Differential Calculus—Part I)

Chapter 5

Sequence of Real Numbers

5(A)   Elementary Notions on Sequence of Points in One Dimensions: Main Points

5(B)   Subsequences: Upper and Lower Limits

Chapter 5

Sequence of Real Numbers

5(A): Sequence of Points in One Dimension (Elementary Topics)

5.1 Introduction

A sequence in a set S is a function whose **domain** is the set $\mathbb{N}$ of natural numbers, and whose range is contained in the set S. In this chapter we will be concerned with sequence in $\mathbb{R}$ and their convergence.

Definition 1. A sequence in $\mathbb{R}$ (or a **Sequence of points in One Dimension**) or **a sequence of real numbers** is a function defined on the set $\mathbb{N} = \{1, 2, 3, \ldots\}$ of natural numbers whose range is contained in the set $\mathbb{R}$ of real numbers.

For a sequence in $\mathbb{R}$ we assign, by some determinable manner a well-defined real number to each natural number $n = 1, 2, 3, \ldots$

If the function $x : \mathbb{N} \to \mathbb{R}$ is a sequence, we shall usually denote the value of x at n by the symbol x_n instead of the usual symbol $x(n)$. The values x_n are known as the **terms** or **elements** of the sequence. We shall denote this sequence by any one of the following notations:

$$\{x_n\}_{n=1}^{\infty} \quad \text{or,} \quad \{x_n\}_n \quad (n \in \mathbb{N}) \quad \text{or,} \quad \{x_n\}$$

or simply $\{x_1, x_2, x_3, \ldots, x_n, \ldots\}$.

5.2 Examples of Sequences in $\mathbb{R}$

Sequences are often defined by giving a formula for the nth term x_n. But it is frequently convenient to list the terms of a sequence in order e.g., $\{x_n\}_n = \{\frac{1}{2}, \frac{1}{4}, \frac{1}{6}, \frac{1}{8}, \ldots\}$; or $\{x_n\}_n = \{\frac{1}{2n}, n \in \mathbb{N}\}$.

Another way of defining a sequence is to specify the value of x_1 and then give a formula for x_{n+1} ($n \geq 1$) in terms of x_n. (This process is known as Recursive Definition of a sequence: See **Fibonascci sequence** below).

Example 5.2.1. $\{x_n\}_n$ defined by $x_n = 2n, n \in \mathbb{N}$ is the sequence $\{2, 4, 6, 8, \ldots, 2n, \ldots\}$.

Example 5.2.2. $\{1, 2, 3, \ldots, n, \ldots\}$ is a sequence of positive integers. We may write this sequence as $\{x_n\}_n$, where $x_n = n$ ($n \in \mathbb{N}$).

Example 5.2.3. $\{x_n\}_n$, where $x_n = 2n - 1$ ($n \in \mathbb{N}$) is the sequence of odd positive integers: $\{1, 3, 5, 7, \ldots\}$.

Example 5.2.4. The sequence $\{a_n\}$, where $a_n = 1/n, n \in \mathbb{N}$ is the well-known **harmonic sequence** $\{1, \frac{1}{2}, \frac{1}{3}, \frac{1}{4}, \ldots, \frac{1}{n}, \ldots\}$.

Example 5.2.5. The sequence $\{p_n\}_n$, where $p_n = 1 + \frac{1}{2} + \frac{1}{3} + \cdots + \frac{1}{n}$ has the first term $p_1 = 1$, the second term $p_2 = 1 + \frac{1}{2} = \frac{3}{2}$, the third term $p_3 = 1 + \frac{1}{2} + \frac{1}{3} = \frac{11}{6}$, etc.

Example 5.2.6. The sequence $\{x_n\}_n$, where $x_n = 4, \forall n \in \mathbb{N}$ is the **constant sequence** $\{4, 4, 4, 4, \ldots\}$ whose every term $= 4$.

We must be careful to distinguish between a sequence $\{x_n\}$ and its associated range (also called Trace). By Trace we shall mean the set of *distinct points* that appear as terms of the sequence.

Example 5.2.7. $\{x_n\}_n$, where $x_n = (-1)^{n-1}$, $n \in \mathbb{N}$. The sequence is $\{1, -1, 1, -1, 1, -1, \ldots\}$ but the trace is the set $\{1, -1\}$ containing only two elements 1 and -1.

Example 5.2.8. $\{x_n\}_n$, where $x_n = \frac{1}{2}\{1 + (-1)^{n-1}\}$, $n \in \mathbb{N}$. The sequence is $\{1, 0, 1, 0, 1, 0, \ldots\}$ and its trace is the set $\{1, 0\}$.

We may not always obtain an explicit analytical formula to describe x_n; but the terms of the sequence may be well-determined.

Example 5.2.9. $\{x_n\}_n$, where $x_n = n$th prime number (No known analytical formula exists here) gives the sequence $\{2, 3, 5, 7, 11, 13, \ldots\}$.

** **Example 5.2.10. Recursive Definition of a sequence:** Let $x_1 = 1$, $x_2 = 1$, $x_n = x_{n-1} + x_{n-2}$ $(n \in \mathbb{N}, n \geq 3)$. The sequence may be written as

$$\{1, 1, 2, 3, 5, 8, 13, 21, \ldots\}.$$

This particular sequence is called **Fibonacci's sequence** used in *Phyllotaxy*, the study of arrangement of leaves in the plant system.

Example 5.2.11. The sequence $\{x_n\}_n$ defined by $x_n = 3^{n-1}, n \in \mathbb{N}$ can also be described by the recurrence relation $x_1 = 1$, $x_{n+1} = 3x_n (n \geq 1, n \in \mathbb{N})$.

Example 5.2.12. Let a_1 and a_2 be any two positive real numbers and let

$$a_{n+1} = \tfrac{1}{2}(a_n + a_{n-1}) (n \geq 2, n \in \mathbb{N}).$$

The sequence $\{a_n\} = \{a_1, a_2, \tfrac{1}{2}(a_2 + a_1), \tfrac{1}{2}\{a_2 + \tfrac{1}{2}(a_2 + a_1)\}, \ldots\}$.

Example 5.2.13. The sequence $\{x_n\}_n$ defined by $x_1 = \sqrt{2}$ and $x_{n+1} = \sqrt{2x_n}, n \geq 1$ is $\{\sqrt{2}, \sqrt{2\sqrt{2}}, \sqrt{2\sqrt{2\sqrt{2}}}, \ldots, \}$.

Example 5.2.14. $\{x_n\}$, where $x_n = \sec(\tfrac{1}{2}\pi\sqrt{n})$ does not define a sequence since x_n is not defined for infinite number of values of n (whenever n is the square of an odd positive integer).

** **Example 5.2.15.** The rational numbers between 0 and 1 can be arranged in the form of a sequence: $\{\tfrac{1}{2}, \tfrac{1}{3}, \tfrac{2}{3}, \tfrac{1}{4}, \tfrac{3}{4}, \tfrac{1}{5}, \tfrac{2}{5}, \tfrac{3}{5}, \tfrac{4}{5}, \ldots\}$.

[First arrange in increasing magnitudes of the denominators and then in increasing magnitude of numerators.]

$$x_1 \to 1/2; \quad x_2 \to 1/3; \quad x_3 \to 2/3; \quad x_4 \to 1/4 \text{ and so on.}$$

There is a one-to-one correspondence between the fractions and the set of natural numbers $\{1, 2, 3, \ldots\}$. The set of rational numbers between 0 and 1 thus forms a countable set.

Geometrical Representations

Each term of a sequence $\{x_n\}_n$ of real numbers corresponds to a point on the real axis. We then call the collection of such points a *sequence of points on the real line*.

On the plane determined by two mutually perpendicular lines OX and OY, we may plot the points (n, x_n) and thus obtain *a sequence of points on the plane*. OX is the axis representing n and OY is the axis representing x_n.

5.3 Bounded Sequences

We have already discussed (**Def. 2.7.1**) about the bounds of a set of real numbers. For a sequence of real numbers we apply those definitions on the *range set* (or *trace*). Thus $\{x_n\}_n$ *is bounded above or below according as the range set of $\{x_n\}_n$ is bounded above or bounded below. Also $\{x_n\}_n$ is a bounded sequence if the range set of $\{x_n\}_n$ is bounded both above and below.*

The exact upper bound (or Supremum) and the exact lower bound (or Infimum) of a bounded sequence are the corresponding numbers of the range set of the sequence. For ready references we put the following definitions in a systematic manner:

Definitions 5.3.1

1. (a) A sequence $\{x_n\}_n$ is said to be *bounded above* if $\exists$ a real number U s.t. $x_n \leq U$, $\forall\, n \in \mathbb{N}$. U is called an *upper bound* of $\{x_n\}_n$.

   (b) A sequence $\{x_n\}_n$ is said to be *unbounded above* (i.e., not bounded above) if, for any given $G > 0$, $\exists$ at least one member of the sequence, say, x_p, s.t. $x_p > G$.

2. (a) A sequence $\{x_n\}_n$ is said to be *bounded below* if $\exists$ a real number L s.t. $x_n \geq L$, $\forall\, n \in \mathbb{N}$. L is then called *a lower bound* of $\{x_n\}_n$.

   (b) A sequence $\{x_n\}_n$ is said to be *unbounded below*, if for any given real number $G > 0$, $\exists$ at least one member x_k (say) s.t. $x_k < -G$.

3. A sequence $\{x_n\}_n$ is said to be *bounded* if it is bounded both above and below. Thus $\{x_n\}_n$ is bounded if $\exists$ a real positive number K s.t. $\forall\, n \in \mathbb{N}$,

$$|x_n| \leq K \quad \text{i.e.} \quad -K \leq x_n \leq K.$$

4. If $\{x_n\}_n$ is *bounded above*, then by the lub-property of the real number system $\exists$ a real number M (called, *Supremum* or *exact upper bound* or *least upper bound*) with following two properties:

   (a) $\forall\, n \in \mathbb{N}, x_n \leq M$

   (b) Given any $\varepsilon > 0$ (no matter, how small) $\exists$ at least one member of $\{x_n\}_n$, say, x_p s.t. $x_p > M - \varepsilon$.

5. If $\{x_n\}_n$ is *bounded below*, then by the glb-property of the real number system $\exists$ a real number m (called, *Infimum* or *exact lower bound* or *greatest lower bound*) with following two properties:

(a) $\forall n \in \mathbb{N}, x_n \geq m$

(b) Given any $\varepsilon > 0$ (no matter how small) $\exists$ at least one member of $\{x_n\}_n$, say, x_q s.t. $x_q < m + \varepsilon$.

In most of our ordinary problems we wish to know whether a sequence is bounded or not (i.e., we may not require to find lub M or glb m for a bounded sequence). The definition stated under No. **3** above may be used in such cases (see the examples given below).

Examples:

Example 5.3.1. *To show that* $\left\{\dfrac{3n+1}{n+2}\right\}$ *is a bounded sequence.*

Solution: Writing $\dfrac{3n+1}{n+2} = \dfrac{3 + 1/n}{1 + 2/n}$, we may **guess** that $\forall n, \dfrac{3n+1}{n+2}$ cannot exceed 3.

To justify that the guess is correct we require to show that $\forall n$, $\left|\dfrac{3n+1}{n+2}\right| \leq 3$.

In order that this inequality may be true we should have $3n + 1 \leq 3n + 6$, $\forall n$ and $3n + 1 \leq 3n + 6 \Rightarrow 1 \leq 6$ which is true, whatever n we may take. Hence the given sequence is bounded (both above and below).

Example 5.3.2. $\{1, \frac{1}{2}, \frac{1}{3}, \frac{1}{4}, \ldots, \frac{1}{n}, \ldots\}$ *is a bounded sequence, because*

$$|x_n| = |1/n| \leq 1, \quad \forall n \in \mathbb{N}.$$

Example 5.3.3. $\{(-1)^{n-1}\}$ *is bounded, because*

$$|(-1)^{n-1}| \leq 1, \ \forall n \in \mathbb{N}. \ [\text{In fact}, |(-1)^{n-1}| = 1, \ \forall n].$$

Example 5.3.4. $\{n^2\}$ *is bounded below* $(\because n^2 \geq 1, \forall n)$ *but unbounded above* $(\because n^2 > G$, *whenever* $n > \sqrt{G}$; *so at least one* n *exists for which* $n^2 > G$).

Example 5.3.5. $\{-n^2\}$ *is bounded above but not bounded below* (**Check**).

Example 5.3.6. $\{(-1)^n n\}$ *is neither bounded above nor bounded below.* (*Why?*).

[For any given $G > 0, \exists n$ s.t. $|(-1)^n n| = n > G$].

Example 5.3.7. $\left\{\dfrac{n-1}{2n}\right\}$ *is bounded above, because* $\dfrac{n-1}{2n} < \dfrac{1}{2}$ $\forall n \in \mathbb{N}$. *Thus*, $\frac{1}{2}$ *is an upper bound of the given sequence. What is its exact upper bound?*

We shall prove that $\frac{1}{2}$ is also the exact upper bound. For any given $\varepsilon > 0$, we see that $\frac{n-1}{2n} > \frac{1}{2} - \varepsilon$, if $n > \frac{1}{2\varepsilon}$.
So, certainly $\exists$ at least one member of the sequence $> \frac{1}{2} - \varepsilon$.
Thus $\frac{1}{2}$ is the exact upper bound, i.e., $\sup x_n = 1/2$.
One can easily verify that the sequence is also bounded below, its glb being 0.

5.4 Monotone Sequences

Definitions 5.4.1

1. (a) A sequence $\{x_n\}_n$ is said to be *monotone increasing* (sometimes called *non-decreasing*), iff $\forall n$, $x_{n+1} \geq x_n$.

   e.g., $\left\{\dfrac{1}{2}, \dfrac{2}{3}, \dfrac{3}{4}, \ldots, \dfrac{n}{n+1}, \ldots\right\}$ is a monotone increasing (m.i.) sequence.

   $$\left[\because \frac{n+1}{n+2} - \frac{n}{n+1} = \frac{1}{(n+2)(n+1)} > 0.\right]$$

   (b) If for every n, $x_{n+1} > x_n$ (no two members are equal), then the sequence $\{x_n\}$ is often called *strictly monotone increasing*. The sequence $\left\{\dfrac{n}{n+1}\right\}$ is strictly m.i.

2. (a) A sequence $\{x_n\}$ is said to be *monotone decreasing* (also called *non-increasing*) if, for every n, $x_{n+1} \leq x_n$.

   e.g., $\left\{1, \dfrac{1}{2}, \dfrac{1}{3}, \ldots, \dfrac{1}{n}, \ldots\right\}$ is monotone decreasing.

   $$\left[\text{since, } \frac{1}{n+1} - \frac{1}{n} = \frac{-1}{n(n+1)} = \text{negative, } \forall\, n.\right]$$

   (b) A sequence $\{x_n\}$ is called *strictly monotone decreasing* if, for every n, $x_{n+1} < x_n$ (no two members are equal). The sequence $\{1/n\}$ is a strictly monotone decreasing sequence.

3. A sequence $\{x_n\}$ is said to be *simply* **monotone**, if it is either monotone increasing or monotone decreasing.

4. If a sequence $\{x_n\}$ is **m.i.**, then the sequence $\{-x_n\}$ is **m.d.**

CHAPTER 5: SEQUENCE OF REAL NUMBERS

Examples:

Example 5.4.1. $\{2, 4, 8, 16, \ldots, 2^n, \ldots\}$ *is a strictly m.i. sequence.*

Example 5.4.2. $\{1, 1, 2, 2, 2, 3, 3, 4, 4, 5, 5, 5, 6, \ldots\}$ *is a m.i. sequence but not a strictly m.i. sequence.*

Example 5.4.3. *Let* $x_n = \dfrac{3n+1}{n+2}$. *Is* $\{x_n\}$ *a m.i. sequence?*

Solution: We obtain,

$$x_{n+1} - x_n = \frac{3n+4}{n+3} - \frac{3n+1}{n+2} = \frac{(3n+4)(n+2) - (3n+1)(n+3)}{(n+3)(n+2)}$$

$$= \frac{5}{(n+3)(n+2)} = \text{positive, for all values of } n.$$

i.e., $x_{n+1} > x_n$; $\forall\, n$.

∴ the sequence is *m.i.* (*in fact, strictly m.i.*).

Example 5.4.4. *Let* $S_n = \dfrac{n+1}{2n+1}$. *Verify whether the sequence* $\{S_n\}$ *is a m.d. sequence?*

Solution:

$$S_{n+1} - S_n = \frac{n+2}{2n+3} - \frac{n+1}{2n+1} = \frac{(n+2)(2n+1) - (2n+3)(n+1)}{(2n+3)(2n+1)} = \frac{-1}{(2n+3)(2n+1)},$$

i.e., $S_{n+1} - S_n < 0$, $\forall\, n$ and hence $\{S_n\}$ is a monotone decreasing (in fact, strictly m.d.) sequence.

** Example 5.4.5. A sequence can be neither increasing nor decreasing. *i.e., a sequence need not always be monotone, e.g.,*

$$\{1, 0, 1, 0, 1, 0, \ldots\} \text{ is neither } m.i.\ \text{nor}\ m.d.$$

5.5 Limit of a Sequence

Definition 5.5.1. *A sequence* $\{x_n\}_n$ *of real numbers is said to tend to a finite real number* l, *iff for any given positive number* ε (*no matter, how small*) *there exists a positive integer* n_0 (n_0 *will usually depend on* ε) *such that* $\forall\, n \geq n_0$, *the inequality* $|x_n - l| < \varepsilon$ *holds.*

In this case, we write: $\lim\limits_{n \to \infty} x_n = l$ **or**, $x_n \to l$ as $n \to \infty$.

Symbolically, $\{x_n\}_n$ has a limit $l \Rightarrow$ Given $\varepsilon > 0$; $\exists\, n_0 \in \mathbb{N}$ s.t. $|x_n - l| < \varepsilon$, $\forall\, n \geq n_0$.

[For "$\varepsilon > 0$"; read "given any arbitrary positive number ε" no matter how small].

Note: In the definition we may also consider $\forall n > n_0$ for some suitable n_0.

Note 5.5.1. $\lim\limits_{n \to \infty} x_n = l \Rightarrow$ (i) $\lim\limits_{n \to \infty} x_n$ *exists and* (ii) *the limit is* l.

Definition 5.5.2 *A sequence $\{x_n\}$ of real numbers is said to be **convergent** if it has a finite real number l as its limit. We then say that the sequence $\{x_n\}_n$ converges to l. Thus $\{x_n\}_n$ converges to $l \Rightarrow$ Given $\varepsilon > 0$; $\exists\, n_0 \in \mathbb{N}$ s.t. $|x_n - l| < \varepsilon$, $\forall\, n \geq n_0$.*

Observation: In order to establish the limit l of a sequence $\{x_n\}_n$, we are to start with an arbitrary positive number ε and then find **some positive integer** n_0 (need not always be the least possible positive integer) such that the numerical magnitude of the difference $x_n - l$ will remain less than ε, for every x_n, where $n \geq n_0$.

The statement $\forall\, n \geq n_0$ is extremely important—n_0 should be so determined that for any one $n > n_0$, $|x_n - l|$ is never $> \varepsilon$. If such an n_0 cannot be obtained, then the sequence does not tend to l.

Geometrical representations

Let OX be the real axis (Fig. 5.1). We mark points x_n for each n and let P be the point corresponding to the real number l. $x_n \to l$ as $n \to \infty$ means that after some stage (say, after n has attained n_0) all members of $\{x_n\}_n \in$ the open interval $(l - \varepsilon, l + \varepsilon)$.

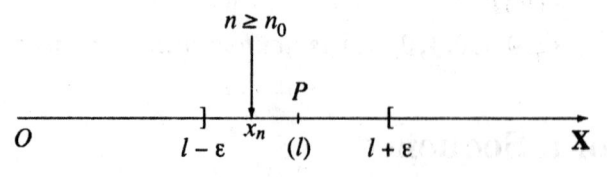

Fig 5.1

In Fig. 5.2 each term x_n corresponds to a dot given by (n, x_n).

If $x_n \to l$ as $n \to \infty$, then what we should see is that all the dots to the right of **a certain point** (n_0, x_{n_0}) lie within the strip of breadth 2ε. Such a point (n_0, x_{n_0}) should be always available, however narrow the strip may be, if $\{x_n\}$ has a limit l.

CHAPTER 5: SEQUENCE OF REAL NUMBERS

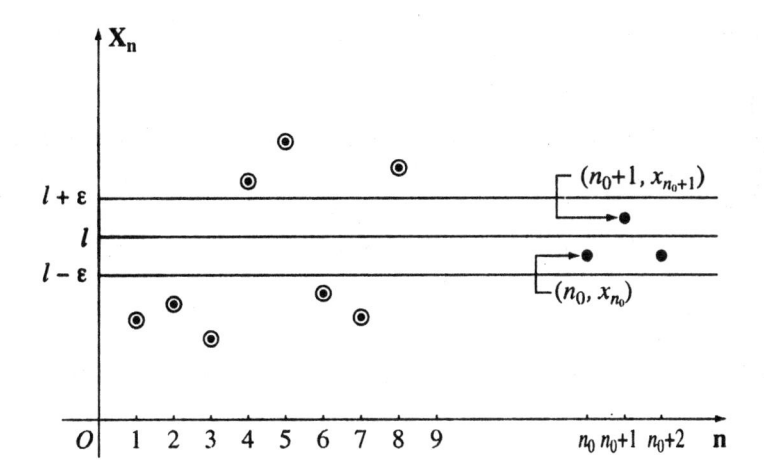

Before we proceed further, the following **uniqueness theorem** should be established:

Theorem 5.5.1. *A sequence $\{x_n\}_n$ of real numbers can have at most one limit, if at all $\{x_n\}$ has a limit. [An equivalent statement: A convergent sequence determines its limit uniquely.]*

Proof. What we have to prove is the following:

$$\left.\begin{array}{l} \text{If} \quad x_n \to l \quad \text{as} \quad n \to \infty \\ \text{and also} \quad x_n \to l' \quad \text{as} \quad n \to \infty \end{array}\right\} \text{ then } l = l'.$$

Let $\varepsilon > 0$ be any arbitrary positive number.

Since $x_n \to l$ as $n \to \infty$, corresponding to $\varepsilon/2$, $\exists\, n_0' \in \mathbb{N}$ s.t. $|x_n - l| < \varepsilon/2$, $\forall\, n \geq n_0'$.

Again, since $x_n \to l'$ as $n \to \infty$, corresponding to $\varepsilon/2$,

$$\exists\, n_0'' \in \mathbb{N} \text{ s.t. } |x_n - l'| < \varepsilon/2,\ \forall\, n \geq n_0''.$$

Let $n_0 = \max.\{n_0', n_0''\}$ = greater of the two integers n_0' and n_0''.
Then $\forall\, n \geq n_0$, $|x_n - l| < \varepsilon/2$ and also $|x_n - l'| < \dfrac{\varepsilon}{2}$.

$$\therefore\ |l - l'| = |l - x_n + x_n - l'| \leq |l - x_n| + |x_n - l'|$$
$$< \varepsilon/2 + \varepsilon/2,\ \forall\, n \geq n_0.$$
$$\text{i.e., } |l - l'| \leq \epsilon,\ \forall n \geq n_0.$$

But l and l' are fixed real numbers and ε is any arbitrary positive number, no matter how small. Therefore, the above inequality says that $|l - l'|$ is less than every positive real number. But by definition, $|l - l'| \geq 0$. The only way out is for $|l - l'|$ to equal to zero, i.e., $l = l'$.

Examples:

Example 5.5.1.

1. $\lim\limits_{n\to\infty} \dfrac{2n+1}{n+1} = 2.$

2. $\lim\limits_{n\to\infty} x_n = c$ if $\forall\, n, x_n = c$ (constant sequence).

Solution:

1. Let ε be any arbitrary positive number, no matter how small. Then
$$\left|\frac{2n+1}{n+1} - 2\right| < \varepsilon \Rightarrow \frac{1}{n+1} < \varepsilon.$$

   This last inequality is true whenever $n+1 > \frac{1}{\varepsilon}$, i.e., whenever $n > \frac{1}{\varepsilon} - 1$.

   We take $n_0 = \left[\frac{1}{\varepsilon} - 1\right] + 1$.

   Thus whenever $n \geq n_0$, $\left|\frac{2n+1}{n+1} - 2\right| < \varepsilon$, i.e., the limit is established.

2. $|x_n - c| = |c - c| = 0 < \varepsilon,\ \forall\, n \geq 1.$

   Thus, by definition $\lim\limits_{n\to\infty} x_n = c$.

Example 5.5.2. *Verify that the harmonic sequence $\left\{\frac{1}{n}\right\}_n$ converges to 0.*

Solution: The statement is true, if, for any given positive number ε, no matter how small, we can find a positive integer n_0 s.t.
$$\forall\, n \geq n_0,\ \left|\frac{1}{n} - 0\right| < \varepsilon,\ \text{which is true, if } \frac{1}{n} < \varepsilon.$$

Clearly, $\frac{1}{n} < \varepsilon$, whenever $n > \frac{1}{\varepsilon}$. Taking $n_0 = \left[\frac{1}{\varepsilon}\right] + 1$, our definition requirement is met and we establish $\lim\limits_{n\to\infty} \frac{1}{n} = 0$, i.e., the sequence $\left\{\frac{1}{n}\right\}_n$ converges to 0.

Alternatively; given $\varepsilon > 0, \left|\frac{1}{n} - 0\right| < \varepsilon$ is true if $n > \frac{1}{\varepsilon}$
i.e., if $n > n_0$ whenever $n_0 = \left[\frac{1}{\varepsilon}\right]$, the integral part of $\frac{1}{\varepsilon}$.

$\therefore$ By defn. $\lim\limits_{n\to\infty} \dfrac{1}{n} = 0.$

[For e.g., take $\varepsilon = 0.11$. $\therefore \left|\frac{1}{n} - 0\right| < 0.11$ if $n > \frac{1}{0.11}$, i.e., $n > 9\frac{1}{11}$, i.e., $n > \left[9\frac{1}{11}\right] = 9$.

$\therefore \left|\frac{1}{n} - 0\right| < 0.11$ whenever $n > 9$ or $n \geq \left[9\frac{1}{11}\right] + 1$, i.e., $n \geq 10$.

Note that as $n \in \mathbb{N}$. $\therefore n > 9$ and $n \geq 10$ are the same.

Even box function is not necessary, i.e., $\left|\frac{1}{n} - 0\right| < 0.11$ whenever $n > 9\frac{1}{11}$, i.e., whenever $n \geq 10$.]

CHAPTER 5: SEQUENCE OF REAL NUMBERS

Example 5.5.3. *To discuss the convergence of $\left\{\dfrac{n-1}{2n}\right\}$.*

Solution: See that
$$\frac{n-1}{2n} = \frac{1}{2} - \frac{1}{2n}.$$
Intuition shows that when n increases indefinitely $\frac{n-1}{2n}$ should approach to $\frac{1}{2}$. We now verify
$$\lim_{n\to\infty} \frac{n-1}{2n} = \frac{1}{2}.$$
Now,
$$\left|\frac{n-1}{2n} - \frac{1}{2}\right| = \left|-\frac{1}{2n}\right| = \frac{1}{2n}.$$
Hence
$$\left|\frac{n-1}{2n} - \frac{1}{2}\right| < \text{any given positive number } \varepsilon,$$
whenever $\frac{1}{2n} < \varepsilon$, i.e., whenever $n > \frac{1}{2\varepsilon}$.

Choosing $n_0 = [\frac{1}{2\varepsilon}]+1$, our definition-requirement is met and the limit is established.

Example 5.5.4. $x_n = \dfrac{n^2+1}{2n^2+3}$. *Verify*: $x_n \to \frac{1}{2}$ as $n \to \infty$.

Solution:
$$\left|\frac{n^2+1}{2n^2+3} - \frac{1}{2}\right| = \left|\frac{-1}{4n^2+6}\right| = \frac{1}{4n^2+6} < \text{any given positive number } \varepsilon,$$
whenever $4n^2+6 > \frac{1}{\varepsilon}$, i.e., whenever $n > \sqrt{\frac{1}{4\varepsilon} - \frac{3}{2}}$.

Choose
$$n_0 = \left[\sqrt{\frac{1}{4\varepsilon} - \frac{3}{2}}\right] + 1.$$
Then for $n \geq n_0$,
$$\left|\frac{n^2+1}{2n^2+3} - \frac{1}{2}\right| < \varepsilon.$$
$\therefore$ by definition, $x_n \to \frac{1}{2}$ as $n \to \infty$.

Example 5.5.5. $x_n = 2 + \left(-\frac{1}{2}\right)^n$. *Verify*: $x_n \to 2$ as $n \to \infty$.

Solution:
$$|x_n - 2| = \left|2 + \left(-\frac{1}{2}\right)^n - 2\right| = \frac{1}{2^n} < \text{any given positive number } \varepsilon;$$

whenever $n > \{\log 1/\varepsilon\}/\log 2$. Choose $n_0 = \left[\frac{\log 1/\varepsilon}{\log 2}\right]$ and then $\forall\, n > n_0, |x_n - 2| < \varepsilon$. Thus by def. $x_n \to 2$ as $n \to \infty$.

Observe that no member of the sequence is 2. The successive terms are alternately greater or less than 2.

Example 5.5.6. $\lim\limits_{n\to\infty} \dfrac{n^2 - 1}{n^2 + n + 1}$. *Guess the limit and then verify its truth by definition test.*

Solution:
$$\frac{n^2 - 1}{n^2 + n + 1} = \frac{1 - \frac{1}{n^2}}{1 + \frac{1}{n} + \frac{1}{n^2}}.$$

Since for large values of n, $1/n^2, 1/n$ approach zero, we guess that the limit may be **1**.

We now verify
$$\lim_{n\to\infty} \frac{n^2 - 1}{n^2 + n + 1} = 1$$

$$\left|\frac{n^2 - 1}{n^2 + n + 1} - 1\right| = \left|\frac{-n - 2}{n^2 + n + 1}\right| = \frac{n + 2}{n^2 + n + 1}.$$

Estimation: For, $n > 2$, $n + 2 < n + n$

and $n^2 + n + 1 > n^2$, i.e., $1/(n^2 + n + 1) < 1/n^2$.

$$\therefore \quad \frac{n+2}{n^2+n+1} < \frac{2n}{n^2} = \frac{2}{n},$$

which is less than any given positive number ε, whenever $n > 2/\varepsilon$. Choosing $n_0 = \max.\{2, [2/\varepsilon]\}$, we see that

$$\forall\, n > n_0, \left|\frac{n^2 - 1}{n^2 + n + 1} - 1\right| < \varepsilon$$

and the definition requirement is met and the limit is established.

Try in a similar manner, to establish
$$\lim_{n\to\infty} \frac{2n^2 + 1}{n^2 + 3n} = 2.$$

Example 5.5.7. *To establish:* $\lim\limits_{n\to\infty} \dfrac{n^2 + n + 1}{n^2 - 5n - 3} = 1.$

Solution:
$$\left|\frac{n^2 + n + 1}{n^2 - 5n - 3} - 1\right| = \frac{6n + 4}{|n^2 - 5n - 3|}.$$

CHAPTER 5: SEQUENCE OF REAL NUMBERS

Estimation: For

$$n \geq 1, 6n + 4 \leq 6n + 4n = 10n,$$
$$\text{but } |n^2 - (5n+3)| \geq n^2 - |5n+3|$$
$$\geq n^2 - \frac{1}{2}n^2, \text{ whenever } 5n + 3 < \frac{1}{2}n^2.$$
$$= \frac{1}{2}n^2, \quad \text{whenever } 5n + 3 < \frac{1}{2}n^2.$$

Now, $5n + 3 < \frac{1}{2}n^2$, whenever $5n + 3n < \frac{1}{2}n^2$, i.e., whenever $n > 16$.
∴ for $n > 16$,

$$\frac{6n+4}{|n^2 - 5n - 3|} < \frac{10n}{\frac{1}{2}n^2} = \frac{20}{n} < \varepsilon \ (\varepsilon > 0)$$
$$\text{i.e., } \left| \frac{n^2 + n + 1}{n^2 - 5n - 3} - 1 \right| < \varepsilon,$$

whenever $n > \frac{20}{\varepsilon}$.

Take $n_0 = \max.\{17, [20/\varepsilon]\}$. Then

$$\forall \, n > n_0, \left| \frac{n^2 + n + 1}{n^2 - 5n - 3} - 1 \right| < \varepsilon$$

and by definition, the limit is established.

Example 5.5.8. $\lim\limits_{n \to \infty} x_n = 3$, where $x_n = \dfrac{3n}{n + 5\sqrt{n}}$.

Solution:

$$\left| \frac{3n}{n + 5\sqrt{n}} - 3 \right| = \frac{15\sqrt{n}}{n + 5\sqrt{n}} < \frac{15\sqrt{n}}{n} \ (\because n + 5\sqrt{n} > n).$$

∴ $\dfrac{15\sqrt{n}}{n} <$ any given positive number ε, if $n > \dfrac{225}{\varepsilon^2}$.

Choose $n_0 = \left[\dfrac{225}{\varepsilon^2} \right] + 1$. Then
$\forall \, n \geq n_0, |x_n - 3| < \varepsilon$, which establishes the limit.

REMEMBER. A sequence $\{x_n\}_n$ is called a **Null sequence**, iff $\lim\limits_{n \to \infty} x_n = 0$.
i.e., iff, given $\varepsilon > 0, \exists \, n_0 \in \mathbb{N}$ s.t. $\forall \, n \geq n_0, |x_n| < \varepsilon$.

Example 5.5.9. *Prove that the sequence $\left\{ \frac{1}{n^p} \right\}$, where $p > 0$, is a null sequence.*

Solution: Let $\varepsilon > 0$ be given. Then

$$\left|\frac{1}{n^p} - 0\right| = \frac{1}{n^p} < \varepsilon, \text{ whenever } n^p > 1/\varepsilon \text{ [or } n > (1/\varepsilon)^{1/p}].$$

Choose

$$n_0 = \left[\left(\frac{1}{\varepsilon}\right)^{1/p}\right] + 1 = [e^{1/p \log 1/\varepsilon}] + 1.$$

Then

$$\forall\, n \geq n_0, \left|\frac{1}{n^p} - 0\right| < \varepsilon.$$

This proves that $\lim_{n \to \infty} \frac{1}{n^p} = 0$. i.e., $\{\frac{1}{n^p}\}_n, p > 0$ is a null sequence.

We give below discussions of some very important null sequences: (marked **):

** Example 5.5.10. *The sequence $\{x^n\}_n$ is a null sequence, if $-1 < x < 1$ (i.e., if $|x| < 1$), i.e., to prove: $\lim_{n \to \infty} x^n = 0$ (where $|x| < 1$).*

Solution: Let ε be a given positive number.

Since $|x| < 1$, we may write

$$|x| = \frac{1}{1+h} \quad (h > 0).$$

Then

$$|x^n - 0| = |x|^n = \frac{1}{(1+h)^n} < \frac{1}{1+nh}.$$

[$\because (1+h)^n = 1 + nh + n\frac{(n-1)}{2!}h^2 + \cdots + h^n > 1 + nh$, h being positive.]

$\therefore |x^n - 0| < \varepsilon$, whenever $\frac{1}{1+nh} < \varepsilon$, i.e., whenever $n > \left(\frac{1}{\varepsilon} - 1\right)/h$.

In fact, we may take $n > \frac{1}{\varepsilon h}$. Choosing $n_0 = \left[\frac{1}{\varepsilon h}\right]$, we may then state $\forall\, n > n_0, |x^n - 0| < \varepsilon$; hence, by definition, $x^n \to 0$ as $n \to \infty$. (where $|x| < 1$).

Note 5.5.2. *In particular, remember that the sequences*

$$\left\{\frac{1}{2^n}\right\}, \left\{\frac{1}{3^n}\right\}, \left\{\frac{1}{10^n}\right\}, \left\{\left(\frac{4}{5}\right)^n\right\}, \text{ etc.}$$

are all null sequences.

** Example 5.5.11. *The sequence $\{nx^n\}$, where $|x| < 1$ is a null sequence. i.e., to prove:*

$$\lim_{n \to \infty} nx^n = 0 \text{ (if } |x| < 1).$$

CHAPTER 5: SEQUENCE OF REAL NUMBERS

Solution: Let $\varepsilon > 0$ be a given positive number.

Since $|x| < 1$, we shall write
$$|x| = \frac{1}{1+h} \ (h > 0).$$

Now
$$|nx^n - 0| = n|x|^n = n\frac{1}{(1+h)^n}$$

$$= \frac{n}{1 + nh + \frac{n(n-1)}{2}h^2 + \cdots + h^n} \ \text{(using Binomial theorem)}.$$

If $n > 1$, we may retain only the third term in the binomial expansion of $(1+h)^n$ and thus obtain
$$|nx^n - 0| < \frac{n}{\frac{n(n-1)}{2}h^2} < \frac{2}{(n-1)h^2}.$$

$\therefore |nx^n - 0| < \varepsilon$, whenever $\frac{2}{(n-1)h^2} < \varepsilon$, i.e., whenever $n > 1 + \frac{2}{\varepsilon h^2}$.

Choosing $n_0 = \left[1 + \frac{2}{\varepsilon h^2}\right]$, we see that

$$|nx^n - 0| < \varepsilon, \ \forall\, n > n_0$$

i.e., $\lim\limits_{n \to \infty} nx^n = 0$ (if $|x| < 1$).

Particular Case: If $0 < a < 1$, prove that $na^n \to 0$ as $n \to \infty$. [C.H. 1980]

** Example 5.5.12. *The sequence* $\left\{\dfrac{\log n}{n}\right\}$ *is a null sequence, i.e.,* $\lim\limits_{n \to \infty} \dfrac{\log n}{n} = 0$, (*the logarithm to the base* e).

Solution: For any given positive integer $n(\geq 3)$, we can obtain a positive integer m (called, *the characteristic of the logarithm*) such that
$$m \leq \log_e n < m+1, \quad \text{i.e., } e^m \leq n < e^{m+1}$$

$$\therefore \quad \frac{\log n}{n} \leq \frac{m+1}{e^m} = (m+1)x^m, \text{ where } x = 1/e < 1.$$

As $n \to \infty, m$ and hence $(m+1) \to \infty$.

By the previous example $(m+1)x^m(|x| < 1) \to 0$ as $(m+1) \to \infty$ (i.e., when $n \to \infty$) and hence $(m+1)x^m < \varepsilon$ (any given positive number) for sufficiently large m (and hence for sufficiently large n, i.e., $\forall\, n > n_0$).

A.I.T.D.C.[P-I]—17

Thus
$$\frac{\log n}{n} < \varepsilon, \forall n > n_0$$
i.e., $\dfrac{\log n}{n} \to 0$ as $n \to \infty$.

More generally, one can similarly prove,
$$\lim_{n \to \infty} \frac{\log_b n}{n^h} = 0,$$
if the base b of the logarithm > 1 and $h > 0$.

** Example 5.5.13. *To prove* $\lim\limits_{n \to \infty} \sqrt[n]{n} = 1$. [C.H. 1983, '94]

Solution: It is sufficient to prove here that the sequence, $\{\sqrt[n]{n} - 1\}$ is a null sequence, i.e., to prove $\lim\limits_{n \to \infty} (\sqrt[n]{n} - 1) = 0$, i.e., to prove $\lim\limits_{n \to \infty} h_n = 0$, where $h_n = \sqrt[n]{n} - 1$.

Now, for $n > 1$, $\sqrt[n]{n} > 1$ and $h_n = \sqrt[n]{n} - 1 > 0$.

But
$$n = (1 + h_n)^n = 1 + nh_n + \frac{n(n-1)}{2}h_n^2 + \cdots + h_n^n > \frac{n(n-1)}{2}h_n^2 \; (h_n \text{ being positive}),$$

so that $h_n^2 < \frac{2}{n-1}$ or $h_n < \sqrt{\frac{2}{n-1}} < $ a given positive number ε, whenever $\frac{2}{n-1} < \varepsilon^2$, i.e., whenever $n > \frac{2}{\varepsilon^2} + 1$.

Choosing $n_0 = \left[\frac{2}{\varepsilon^2} + 1\right]$, we see that $|h_n - 0 < \varepsilon, \forall n > n_0$.

Consequently, $\lim\limits_{n \to \infty} h_n = 0$, i.e., $\lim\limits_{n \to \infty} \sqrt[n]{n} = 1$.

** Example 5.5.14. *To prove* $\lim\limits_{n \to \infty} x^{1/n} = 1$, *whenever* $x > 0$.

Solution: We take three cases separately:

Case 1. Let **x = 1**. Then every member of the sequence $\{x^{1/n}\}$ is 1.

$\therefore |x^{1/n} - 1| = |1 - 1| = 0 <$ any given positive number $\varepsilon \; \forall \, n \geq 1$.

$\therefore$ by definition, $\lim\limits_{n \to \infty} x^{1/n} = 1$, when $x = 1$.

Case 2. Let **0 < x < 1**. Then $x^{1/n} < 1, \forall \, n$. We can, therefore, take
$$x^{1/n} = \frac{1}{1 + h_n}, \text{ where } h_n > 0.$$

We shall conclude, $\lim\limits_{n \to \infty} x^{1/n} = 1$, if we can prove $\lim\limits_{n \to \infty} h_n = 0$.

But see that
$$x = \frac{1}{(1 + h_n)^n} = \frac{1}{1 + nh_n + \frac{n(n-1)}{2}h_n^2 + \cdots + h_n^n} < \frac{1}{1 + nh_n}.$$

$\therefore h_n < \frac{\frac{1}{x}-1}{n} <$ any given number ε, if $\frac{\frac{1}{x}-1}{n} < \varepsilon$, i.e., if $n > \frac{\frac{1}{x}-1}{\varepsilon}$.

Hence, choosing $n_0 = \left[\frac{\frac{1}{x}-1}{\varepsilon}\right] + 1$, we observe that
$\forall n \geq n_0, h_n < \varepsilon$, which implies $h_n \to 0$ as $n \to \infty$.
Consequently, $x^{1/n} \to 1$ as $n \to \infty$ $(0 < x < 1)$.

Case 3. Let **x > 1**. Then we write,

$$x^{1/n} = 1 + p_n \quad (p_n > 0).$$

To prove $x^{1/n} \to 1$ as $n \to \infty$ is the same as to prove $p_n \to 0$ as $n \to \infty$.
But $x = (1 + p_n)^n > 1 + np_n$ i.e.,

$$p_n < \frac{x-1}{n} < \varepsilon, \quad \text{if } n > \frac{x-1}{\varepsilon}.$$

Choosing $n_0 = \left[\frac{x-1}{\varepsilon}\right] + 1$, we see that $p_n < \varepsilon \ \forall \ n \geq n_0$ and hence, $p_n \to 0$ as $n \to \infty$. Consequently, $x^{1/n} \to 1$ as $n \to \infty$ $(x > 1)$.

**Example 5.5.15 *To prove:* $\lim_{n \to \infty} (\sqrt{n+1} - \sqrt{n}) = 0.$

Solution:

$$\left|\sqrt{n+1} - \sqrt{n} - 0\right| = \left|\frac{(\sqrt{n+1} - \sqrt{n})(\sqrt{n+1} + \sqrt{n})}{\sqrt{n+1} + \sqrt{n}}\right|$$

$$= \frac{1}{\sqrt{n+1} + \sqrt{n}} < \frac{1}{2\sqrt{n}} \quad (\because \sqrt{n+1} > \sqrt{n}).$$

Let ε be any given positive number.
Then $\left|\sqrt{n+1} - \sqrt{n}\right| < \varepsilon$, whenever $\frac{1}{2\sqrt{n}} < \varepsilon$, i.e., whenever $n > \frac{1}{4\varepsilon^2}$.
Choosing $n_0 = \left[\frac{1}{4\varepsilon^2}\right]$, we see that $\forall n > n_0$,

$$\left|\sqrt{n+1} - \sqrt{n} - 0\right| < \varepsilon, \quad \text{i.e., } \lim_{n \to \infty} \{\sqrt{n+1} - \sqrt{n}\} = 0.$$

5.6 Theorems on limits of Sequences

Theorem 5.6.1. *Every convergent sequence is bounded.* [C.H. 1995]
Proof. Let $\{x_n\}$ be a convergent sequence with limit l. Then, given any $\varepsilon > 0$ *(for sake of definiteness, take $\varepsilon = 1$)*, $\exists$ a positive integer n_0 such that

$$\forall n > n_0, \quad |x_n - l| < 1.$$

It follows that $|x_n| = |x_n - l + l| \leq |x_n - l| + |l| < 1 + |l|, \ \forall n > n_0$.
Let $k = \max.\{|x_1|, |x_2|, \ldots, |x_{n_0}|, |l| + 1\}$.
Then clearly, $|x_n| \leq k, \ \forall n$. i.e., $\{x_n\}$ is a bounded sequence.

> REMEMBER: "$\{x_n\}$ **converges** $\Rightarrow$ $\{x_n\}$ **bounded**" can be stated also as
>
> "$\{x_n\}$ **not bounded** $\Rightarrow$ $\{x_n\}$ **not convergent**."

Observation: The converse of this theorem is not always true:

EVERY BOUNDED SEQUENCE IS NOT CONVERGENT.

For e.g., $\{x_n\}$, where $x_n = 1 + (-1)^n$ defines the sequence $\{0, 2, 0, 2, 0, 2, \ldots\}$.

The sequence is bounded (*no member can exceed 2, and also no member can be below 0*), but it does not converge to any finite limit. We shall prove later that the sequence oscillates between 0 and 2 without converging to a unique limit.

Theorem 5.6.2. *If a sequence $\{x_n\}$ converges to a finite number l, then the sequence $\{|x_n|\}$ must converge to $|l|$, i.e., $x_n \to l$ as $n \to \infty \Rightarrow |x_n| \to |l|$ as $n \to \infty$.*

Proof. Let ε be any given positive number.

Since $x_n \to l$ as $n \to \infty$, $\exists\, n_0$ s.t. $|x_n - l| < \varepsilon$, $\forall\, n \geq n_0$.

Now $||x_n| - |l|| \leq |x_n - l|$ and hence $< \varepsilon$, $\forall\, n \geq n_0$.

This proves: $\lim_{n \to \infty} |x_n| = |l|$.

Remark 5.6.1. *The converse of this theorem is not always true. Take, for example, the sequence $\{y_n\}$, where $y_n = (-1)^n$. We have $\{|y_n|\} = \{1, 1, 1, \ldots\}$ which converges to 1, but $\{y_n\}$ does not converge (in fact, it oscillates between -1 and $+1$, to be proved later).*

Sum, Difference, Product and Quotient of two Convergent Sequences.

Theorem 5.6.3. *Let $\{x_n\}$ and $\{y_n\}$ be two convergent sequences.*

Let $x_n \to l$ as $n \to \infty$ and $y_n \to m$ as $n \to \infty$. Then we shall prove:

1. $\lim_{n \to \infty} (x_n \pm y_n) = l \pm m = \lim_{n \to \infty} x_n \pm \lim_{n \to \infty} y_n$.

2. $\lim_{n \to \infty} (x_n y_n) = l \cdot m = \lim_{n \to \infty} x_n \cdot \lim_{n \to \infty} y_n$.

3. $\lim_{n \to \infty} \dfrac{x_n}{y_n} = \dfrac{l}{m} = \dfrac{\lim_{n \to \infty} x_n}{\lim_{n \to \infty} y_n}$, provided, $\lim_{n \to \infty} y_n = m$, i.e., $m \neq 0$.

CHAPTER 5: SEQUENCE OF REAL NUMBERS

Proof. **1. Sum.**

Let $\varepsilon > 0$ be given. Since $x_n \to l$ as $n \to \infty$ and $y_n \to m$ as $n \to \infty$, by definition of limit, $\exists$ two positive integers n'_0 and n''_0 s.t.

$$|x_n - l| < \varepsilon/2, \ \forall \, n \geq n'_0$$
$$\text{and} \quad |y_n - m| < \varepsilon/2, \ \forall \, n \geq n''_0.$$

Let $n_0 = \max.\{n'_0, n''_0\}$. Then $\forall \, n \geq n_0$ both the inequalities $|x_n - l| < \varepsilon/2$ and $|y_n - m| < \varepsilon/2$ hold.

Then

$$|(x_n + y_n) - (l + m)| = |(x_n - l) + (y_n - m)|$$
$$\leq |x_n - l| + |y_n - m|$$
$$\text{i.e.,} \quad < \varepsilon/2 + \varepsilon/2 = \varepsilon, \ \forall \, n \geq n_0.$$

$\therefore$ by definition,

$$\lim_{n \to \infty} (x_n + y_n) = l + m = \lim_{n \to \infty} x_n + \lim_{n \to \infty} y_n.$$

Exactly in a similar way we prove:

$$\lim_{n \to \infty} (x_n - y_n) = l - m = \lim_{n \to \infty} x_n - \lim_{n \to \infty} y_n.$$

Proof. **2. Product.**

Since $\{x_n\}_n$ converges, $\{x_n\}_n$ is bounded, i.e., $\exists$ a positive number K such that

$$|x_n| \leq K, \ \forall \, n. \tag{5.6.1}$$

Let $\varepsilon > 0$ be any given arbitrary positive number. Since $\lim_{n \to \infty} x_n = l$. Corresponding to the arbitrary positive number $\frac{\varepsilon/2}{1+|m|}$, $\exists$ a positive integer n'_0 s.t.

$$|x_n - l| < \frac{\varepsilon/2}{1 + |m|}, \ \forall \, n \geq n'_0. \tag{5.6.2}$$

Since $\lim_{n \to \infty} y_n = m$, corresponding to the arbitrary positive number $\frac{\varepsilon/2}{K}$, $\exists$ a positive integer n''_0 s.t.

$$|y_n - m| < \frac{\varepsilon/2}{K}, \ \forall \, n \geq n''_0. \tag{5.6.3}$$

Choose $n_0 = \max.\{n_0', n_0''\}$. Then both the inequalities (5.6.2) and (5.6.3) hold $\forall\, n > n_0$. For such values of n,

$$|x_n y_n - lm| = |x_n(y_n - m) + m(x_n - l)|$$
$$\leq |x_n|\,|y_n - m| + |m|\,|x_n - l|$$
$$< K \cdot \frac{\varepsilon/2}{K} + |m|\frac{\varepsilon/2}{1 + |m|} \quad \text{[using (5.6.1),(5.6.2),(5.6.3)]}$$

i.e., $\quad < \dfrac{\varepsilon}{2} + \dfrac{\varepsilon}{2}\left(\dfrac{|m|}{1 + |m|}\right)$

i.e., $\quad < \dfrac{\varepsilon}{2} + \dfrac{\varepsilon}{2} \quad \left(\because \dfrac{|m|}{1 + |m|} < 1\right)$

$$= \varepsilon,\ \forall\, n \geq n_0.$$

This proves that

$$\lim_{n \to \infty} x_n y_n = l \cdot m = \lim_{n \to \infty} x_n \cdot \lim_{n \to \infty} y_n.$$

Note that in (5.6.2) the arbitrary positive number has been chosen as $\frac{\varepsilon/2}{1+|m|}$ instead of $\frac{\varepsilon/2}{|m|}$ because m may be zero.

Alternative Proof. Let us take two positive numbers ε and k of which ε is any arbitrary positive number and k is a certain definite positive number which we shall specify later. Then ε/k is an arbitrary positive number.

Since $x_n \to l$ as $n \to \infty$ and $y_n \to m$ as $n \to \infty$, it follows that corresponding to ε/k, $\exists$ two positive integers n_0' and n_0'' s.t.

$$|x_n - l| < \varepsilon/k,\ \forall\, n \geq n_0' \quad \text{and} \quad |y_n - m| < \varepsilon/k,\ \forall\, n \geq n_0''.$$

Both the inequalities will certainly hold for all $n \geq n_0$, where $n_0 = \max.\{n_0', n_0''\}$. Then,

$$|x_n y_n - lm| = |x_n(y_n - m) + m(x_n - l)|$$
$$\leq |x_n||y_n - m| + |m||x_n - l|$$
$$< \frac{\varepsilon}{k}\{|x_n| + |m|\},\ \forall\, n \geq n_0.$$

But the sequence $\{x_n\}$ is convergent and, therefore, bounded, i.e., $\exists$ a positive number K s.t. $|x_n| \leq K,\ \forall\, n$.

$$\therefore\ \forall\, n \geq n_0,\ |x_n y_n - lm| < \frac{\varepsilon}{k}\{K + |m|\}.$$

CHAPTER 5: SEQUENCE OF REAL NUMBERS

Now, as stated earlier, we specify the value of k; we take k greater than $K + |m|$ i.e., $\dfrac{K + |m|}{k} < 1$ and then we have at once

$$|x_n y_n - lm| < \varepsilon, \quad \forall\, n \geq n_0.$$

∴ by definition,
$$\lim_{n \to \infty} x_n y_n = l.m. = \lim_{n \to \infty} x_n \cdot \lim_{n \to \infty} y_n.$$

Corollary 5.6.1. $\{a_n\}_n \to 0$ as $n \to \infty$ and $\{b_n\}$ is bounded $\Rightarrow \{a_n b_n\} \to 0$ as $n \to \infty$.

[C.H. 2006]

Corollary 5.6.2. $x_n \to l$ as $n \to \infty \wedge y_n \to m$ as $n \to \infty \wedge z_n \to t$ as $n \to \infty$

$$\Rightarrow x_n \cdot y_n \cdot z_n \to l.m.t. \text{ as } n \to \infty.$$

The law is true for the product of a finite number of sequences.

Proof. **3. Quotient.**

Since $\lim_{n \to \infty} y_n = m (\neq 0)$, $\exists$ a positive integer m_1 such that when $n \geq m_1$ (corresponding to $\varepsilon = \tfrac{1}{2}|m|$)

$$|y_n - m| < \frac{1}{2}|m| \; (\forall\, n \geq m_1).$$

Hence, if $n \geq m_1$, then

$$\frac{1}{2}|m| > |y_n - m| = |m - y_n| \geq |m| - |y_n|$$

and so $\quad |y_n| > |m| - \dfrac{1}{2}|m| = \dfrac{1}{2}|m|, \; \forall\, n \geq m_1$

$$\text{i.e., } \frac{1}{|y_n|} < \frac{2}{|m|}, \; \forall\, n \geq m_1 \tag{5.6.4}$$

We have, for every value of n

$$\left| \frac{x_n}{y_n} - \frac{l}{m} \right| = \left| \frac{m x_n - l y_n}{m y_n} \right| = \left| \frac{m(x_n - l) - l(y_n - m)}{m y_n} \right|$$

$$\leq \frac{|m||x_n - l| + |l||y_n - m|}{|m||y_n|}. \tag{5.6.5}$$

From (5.6.4) and (5.6.5) we find that $\forall\, n \geq m_1$

$$\left| \frac{x_n}{y_n} - \frac{l}{m} \right| \leq \frac{|m||x_n - l| + |l||y_n - m|}{|m|} \cdot \frac{2}{|m|}$$

$$= \frac{2}{|m|} |x_n - l| + \frac{2|l|}{|m|^2} |y_n - m|. \tag{5.6.6}$$

Let ε be any positive number, no matter how small.

Since $x_n \to l$ as $n \to \infty$, corresponding to $\frac{1}{4}|m|\varepsilon$ $\exists$ a positive integer m_2 such that

$$|x_n - l| < \frac{|m|}{4}\varepsilon, \quad \forall\, n \geq m_2.$$

Again, since $y_n \to m$ $(m \neq 0)$, corresponding to the positive number $\frac{|m|^2\varepsilon}{4(|l|+1)}$, $\exists$ a positive integer m_3 such that

$$|y_n - m| < \frac{|m|^2}{4(|l|+1)}\varepsilon, \quad \forall\, n \geq m_3.$$

Let $n_0 = \max.\ \{m_1, m_2, m_3\}$.

Then $\forall\, n \geq n_0$ the relation (5.6.6) becomes

$$\left|\frac{x_n}{y_n} - \frac{l}{m}\right| < \frac{2}{|m|} \cdot \frac{|m|}{4}\varepsilon + \frac{2|l|}{|m|^2} \cdot \frac{|m|^2\varepsilon}{4(|l|+1)}$$

i.e., $\qquad < \dfrac{\varepsilon}{2} + \dfrac{\varepsilon}{2}\dfrac{|l|}{|l|+1}$

i.e., $\qquad < \dfrac{\varepsilon}{2} + \dfrac{\varepsilon}{2} = \varepsilon.\quad \left(\because\ \dfrac{|l|}{|l|+1} < 1\right).$

This proves

$$\lim_{n \to \infty} \frac{x_n}{y_n} = \frac{l}{m}\,(m \neq 0).$$

Alternative Proof. We set down two positive numbers ε and k of which ε is arbitrary and k is some definite positive number to be specified later. Thus ε/k is an arbitrary positive number.

Since $x_n \to l$ as $n \to \infty$ and $y_n \to m$ as $n \to \infty$, there exists a suitable positive integer n_0' s.t. $\forall\, n \geq n_0'$,

$$|x_n - l| < \frac{\varepsilon}{k} \quad \text{and} \quad |y_n - m| < \frac{\varepsilon}{k}.$$

Then

$$\left|\frac{x_n}{y_n} - \frac{l}{m}\right| = \left|\frac{mx_n - ly_n}{my_n}\right|$$

$$= \left|\frac{m(x_n - l) + l(m - y_n)}{my_n}\right|$$

$$\leq \frac{|m||x_n - l| + |l||y_n - m|}{|m||y_n|}.$$

Hence,
$$\left|\frac{x_n}{y_n} - \frac{l}{m}\right| < \frac{\varepsilon}{k} \frac{\{|l| + |m|\}}{|m||y_n|}, \quad \forall\, n \geq n_0'.$$

Since $y_n \to m (m \neq 0$, given), it follows that $|y_n| \to |m|$ (Theorem **5.6.2**). Let $0 < H < |m|$. Then $|y_n| > H$ when $n \geq$ a certain positive integer n_0''.

Suppose, now $n_0 = \max.\{n_0', n_0''\}$. Then $\forall\, n \geq n_0$,
$$\left|\frac{x_n}{y_n} - \frac{l}{m}\right| < \frac{\varepsilon}{k} \frac{\{|l| + |m|\}}{|m|H}.$$

Now, as stated earlier, we give a definite value of k; namely, k is such that
$$\frac{|l| + |m|}{k|m|H} < 1, \quad \text{i.e.,} \quad k > \frac{|l| + |m|}{|m|H}.$$

Then
$$\left|\frac{x_n}{y_n} - \frac{l}{m}\right| < \varepsilon, \quad \forall\, n \geq n_0,$$

and consequently,
$$\lim_{n \to \infty} \frac{x_n}{y_n} = \frac{l}{m} \quad (m \neq 0).$$

Observation: What we have proved is that, if $\{x_n\}_n$ and $\{y_n\}_n$ are two convergent sequences, then the sequences $\{x_n \pm y_n\}, \{x_n \cdot y_n\}$ are also convergent and the sequence $\{x_n/y_n\}$ is convergent, if $\lim y_n \neq 0$. The **converse**, however, may not be always true. e.g.,

1. $x_n = (-1)^n$ and $y_n = (-1)^{n+1}$ are not convergent sequences, but see that $x_n + y_n = 0$, $\forall\, n$ and hence $\{x_n + y_n\}$ converges to zero.

2. $x_n = (-1)^n$ and $y_n = (-1)^n$ are two non-convergent sequences but $\{x_n - y_n\}$, $\{x_n \cdot y_n\}, \{x_n/y_n\}$ converge to $0, 1, 1$ respectively (*can be easily verified*).

Theorem 5.6.4. *If* $\lim_{n \to \infty} x_n = l$ *and* $x_n \geq 0$ $\forall\, n$, *then* $l \geq 0$.

Proof. We prove by contradiction.

If possible, let $l < 0$.

Since $x_n \to l$ as $n \to \infty$, for a given positive number $\varepsilon = -\frac{1}{2}l$ (l negative and hence $\varepsilon > 0$), $\exists$ a positive integer n_0 s.t. $\forall\, n \geq n_0$.
$$l - \varepsilon < x_n < l + \varepsilon$$
$$\text{i.e.,} \quad \frac{3}{2}l < x_n < \frac{1}{2}l.$$

i.e., $\forall\ n \geq n_0$, x_n lies between two negative numbers and hence itself is negative, contradicting the given condition $x_n \geq 0\ \forall\ n$. Hence, l cannot be negative, i.e., $l \geq 0$.

Theorem 5.6.5. *If $x_n \to l$ as $n \to \infty$ and $y_n \to m$ as $n \to \infty$ and if $\forall\ n, x_n < y_n$; then $l \leq m$.*

[What is significant is that l is not just less than m always; l may be sometimes equal to m.]

Proof. Since $\forall\ n, x_n - y_n$ is positive, by the previous theorem, the result follows.

Otherwise proceed as given below:

Given any $\varepsilon > 0$, $\exists$ a suitable positive integer n_0 s.t. $\forall\ n \geq n_0$,

$$l - \varepsilon < x_n < l + \varepsilon$$
$$\text{and}\quad m - \varepsilon < y_n < m + \varepsilon.$$

Take any n for which both the inequalities hold. Then

$$l - \varepsilon < x_n < y_n < m + \varepsilon$$

i.e., $l - \varepsilon < m + \varepsilon$ or $l - m < 2\varepsilon$ which $\Rightarrow m - l > -2\varepsilon$.

It then follows that $m - l \geq 0$ for ε is arbitrary and so the last inequality shows that $m - l$ exceeds **every negative number**. Thus $l \leq m$.

Example 5.6.1. Let $x_n = \dfrac{1}{n+1}$, $y_n = \dfrac{1}{n}$. Then $x_n < y_n$, $\forall\ n$, but

$$y_n - x_n = \frac{1}{n(n+1)} \to 0;\ \text{hence}\ \lim_{n \to \infty} x_n = \lim_{n \to \infty} y_n.$$

Theorem 5.6.6. *(The Sandwich Theorem).* *Suppose that there are three sequences $\{x_n\}, \{y_n\}$ and $\{z_n\}$ s.t. $\forall\ n, x_n \leq y_n \leq z_n$. Moreover, assume that $x_n \to l$ as $n \to \infty$ and $z_n \to l$ as $n \to \infty$. Then the Sandwich Theorem asserts that*

$$\lim_{n \to \infty} y_n = l.\qquad\text{[C.H. 1982, '95]}$$

Proof. Let ε be any given positive number.

Since $x_n \to l$ and $z_n \to l$, as $n \to \infty$, there exists a *suitable* positive integer n_0 s.t. $\forall\ n \geq n_0$,

$$l - \varepsilon < x_n < l + \varepsilon$$
$$\text{and}\quad l - \varepsilon < z_n < l + \varepsilon.$$

And it is given that: $x_n \leq y_n \leq z_n$, $\forall\ n$.

Hence $\forall\, n \geq n_0$, we may write,
$$l - \varepsilon < x_n \leq y_n \leq z_n < l + \varepsilon$$
i.e., $\forall\, n \geq n_0$, $\quad l - \varepsilon < y_n < l + \varepsilon \quad$ or, $\quad |y_n - l| < \varepsilon$
$$\therefore \lim_{n \to \infty} y_n = l.$$

The following result is extremely useful in applications. We put the result in the form of a theorem:

Theorem 5.6.7. *If $\{x_n\}$ be a sequence such that*
$$\lim_{n \to \infty} \left| \frac{x_{n+1}}{x_n} \right| = l, \quad \text{where } 0 \leq l < 1, \quad \text{then } \lim_{n \to \infty} x_n = 0.$$

Proof. Since $0 \leq l < 1$, we can choose a positive number ε, so small that
$$0 < l + \varepsilon < 1.$$

Corresponding to such ε, $\exists$ a positive integer n_0 s.t. $\forall\, n \geq n_0$.
$$\left| \left| \frac{x_{n+1}}{x_n} \right| - l \right| < \varepsilon, \quad \text{i.e., } \quad l - \varepsilon < \left| \frac{x_{n+1}}{x_n} \right| < l + \varepsilon.$$

Thus, we have, $\forall\, n \geq n_0$,
$$\left| \frac{x_{n+1}}{x_n} \right| < l + \varepsilon = k \ \ (\text{say}). \tag{5.6.7}$$

In particular, consider a positive integer n much larger than n_0. Then
$$\left| \frac{x_{n_0+1}}{x_{n_0}} \right| < k, \quad \left| \frac{x_{n_0+2}}{x_{n_0+1}} \right| < k, \quad \cdots, \quad \left| \frac{x_n}{x_{n-1}} \right| < k.$$

[In (5.6.7) we have replaced n by $n_0, n_0 + 1, \ldots, (n-1)$.]

Multiplying these $(n - n_0)$ inequalities we get
$$\left| \frac{x_{n_0+1}}{x_{n_0}} \cdot \frac{x_{n_0+2}}{x_{n_0+1}} \cdots \frac{x_n}{x_{n-1}} \right| < k^{n - n_0} \quad (\text{where } k < 1)$$
$$\Rightarrow \left| \frac{x_n}{x_{n_0}} \right| < \frac{k^n}{k^{n_0}} \quad \text{i.e., } |x_n| < \frac{|x_{n_0}|}{k^{n_0}} \cdot k^n.$$

Since $0 < k < 1$, $k^n \to 0$ as $n \to \infty$ and hence, $\frac{|x_{n_0}|}{k^{n_0}} \cdot k^n \to 0$ as $n \to \infty$ and then $|x_n|$ can be made arbitrarily small for sufficiently large n.
$$\therefore \lim_{n \to \infty} x_n = 0.$$

Applications of Theorem 5.6.7: Important Results:

Example 5.6.2. $\lim_{n\to\infty} \dfrac{x^n}{n!} = 0$, *for any real x. For,*

$$\left| \frac{x^{n+1}}{(n+1)!} \div \frac{x^n}{n!} \right| = \frac{1}{n+1} |x| \to 0 \quad \text{as } n \to \infty,$$

for any real x.

Example 5.6.3. $\lim_{n\to\infty} \dfrac{n^p}{x^n} = 0$, *provided $|x| > 1$ and $p > 0$.*

For,

$$\left| \frac{(n+1)^p}{x^{n+1}} \div \frac{n^p}{x^n} \right| = \left(1 + \frac{1}{n}\right)^p \frac{1}{|x|} \to \frac{1}{|x|} < 1, \text{ if } |x| > 1.$$

$$\therefore \lim_{n\to\infty} \frac{n^p}{x^n} = 0, \quad \text{for } |x| > 1, p > 0.$$

** Example 5.6.4. $\lim_{n\to\infty} \dfrac{m(m-1)(m-2)\cdots(m-n+1)}{\lfloor n} x^n = 0$, *if $|x| < 1$ and m, any real number.*

Solution: Let

$$x_n = \frac{m(m-1)(m-2)\cdots(m-n+1)}{\lfloor n} x^n.$$

Then

$$x_{n+1} = \frac{m(m-1)(m-2)\cdots(m-n+1)(m-n)}{\lfloor n+1} x^{n+1}.$$

$$\therefore \left| \frac{x_{n+1}}{x_n} \right| = \left| \frac{m-n}{n+1} x \right| = \left| \frac{\frac{m}{n} - 1}{1 + \frac{1}{n}} \right| |x| \to |x| \text{ as } n \to \infty.$$

Hence, if $|x| < 1$,

$$\lim_{n\to\infty} x_n = 0$$

Example 5.6.5. $\lim_{n\to\infty} \dfrac{x^n}{n} = 0$, *if $|x| \leq 1$.*

1. $\left| \dfrac{x^{n+1}}{n+1} \div \dfrac{x^n}{n} \right| = \dfrac{n}{n+1} |x| \to |x|$ as $n \to \infty$.

    $\therefore \lim_{n\to\infty} \dfrac{x^n}{n} = 0$, if $|x| < 1$.

2. For $x = 1$, we have the harmonic sequence $\{1/n\}$ which tends to 0 as $n \to \infty$.

3. For $x = -1$, we can use our definition and see that $\lim_{n\to\infty} \dfrac{(-1)^n}{n} = 0$.

CHAPTER 5: SEQUENCE OF REAL NUMBERS

Example 5.6.6. If $x_n = \dfrac{3.5.7\cdots(2n+1)}{2.5.8\cdots(3n-1)}$, then $x_n \to 0$ as $n \to \infty$.

For,
$$\frac{x_{n+1}}{x_n} = \frac{2n+3}{3n+2} = \frac{2+\frac{3}{n}}{3+\frac{2}{n}} \to \frac{2}{3}$$

as $n \to \infty$; hence, etc.

Applications of Theorems 5.6.3 and 5.6.6.

Example 5.6.7. $\lim\limits_{n\to\infty} \dfrac{n^2+3n}{2n^2+n-1} = \dfrac{1}{2}$. [Theorem 5.6.3]

Solution: We have

$$\lim_{n\to\infty} \frac{n^2+3n}{2n^2+n-1} = \lim_{n\to\infty} \frac{1+\frac{3}{n}}{2+\frac{1}{n}-\frac{1}{n^2}} = \frac{\lim\limits_{n\to\infty}\left(1+\frac{3}{n}\right)}{\lim\limits_{n\to\infty}\left(2+\frac{1}{n}-\frac{1}{n^2}\right)}$$

$$= \frac{1+\lim\limits_{n\to\infty}\frac{3}{n}}{2+\lim\limits_{n\to\infty}\frac{1}{n}-\lim\limits_{n\to\infty}\frac{1}{n^2}} = \frac{1+0}{2+0-0} = \frac{1}{2}.$$

** Example 5.6.8. $\lim\limits_{n\to\infty}\left(\dfrac{1}{\sqrt{n^2+1}} + \dfrac{1}{\sqrt{n^2+2}} + \cdots + \dfrac{1}{\sqrt{n^2+n}}\right) = 1.$

[C.H. 1973, '82]

Solution: For large values of n, we may write

$$\frac{1}{\sqrt{n^2+1}} > \frac{1}{\sqrt{n^2+n}}, \quad \frac{1}{\sqrt{n^2+2}} > \frac{1}{\sqrt{n^2+n}}, \quad \text{etc. etc.}$$

and also

$$\frac{1}{\sqrt{n^2+2}} < \frac{1}{\sqrt{n^2+1}}, \quad \frac{1}{\sqrt{n^2+3}} < \frac{1}{\sqrt{n^2+1}}, \quad \text{etc. etc.}$$

Therefore, if we write

$$x_n = \frac{1}{\sqrt{n^2+1}} + \frac{1}{\sqrt{n^2+2}} + \cdots + \frac{1}{\sqrt{n^2+n}}.$$

Then for all values of n

$$\frac{n}{\sqrt{n^2+n}} \leq x_n \leq \frac{n}{\sqrt{n^2+1}}.$$

Now,
$$\lim_{n\to\infty} \frac{n}{\sqrt{n^2+n}} = \lim_{n\to\infty} \cdot \frac{1}{\sqrt{1+\frac{1}{n}}} = 1$$
$$\text{and } \lim_{n\to\infty} \frac{n}{\sqrt{n^2+1}} = \lim_{n\to\infty} \frac{1}{\sqrt{1+\frac{1}{n^2}}} = 1.$$

$\therefore$ by Sandwich Theorem, $\lim_{n\to\infty} x_n = 1$.

5.7 Non-Convergent Sequences

Definition 5.7.1. *A sequence $\{x_n\}$ is said to diverge to $+\infty$, if for any positive number K, there corresponds a positive integer N such that*

$$x_n > K \text{ for all } n > N.$$

This situation is indicated by writing
$$\lim_{n\to\infty} x_n = \infty \ (\text{ or } \ x_n \to \infty \ \text{ as } \ n \to \infty).$$

Note 5.7.1. *In this case $\{x_n\}$ is called a **divergent sequence**. Just as we think of ε in the limit definition of Art. 5.5 as being small positive, we think of the K in this definition as being large positive. Thus when $x_n \to \infty$ as $n \to \infty$, then all terms of the sequence except the first few are large. The reader should be sure to verify that if $x_n \to \infty$ as $n \to \infty$, then x_n definitely does not approach a finite number. This justifies our use of the phrase 'diverges to infinity'. We never refer to infinity as a limit of a sequence. A limit of a sequence must be a finite real number.*

Example 5.7.1. $x_n = n$. Show that $\{x_n\}$ diverges to ∞.

For this we are to find a positive integer N for a given K such that $n > K$ for $n > N$.

Thus $n = $ integral part of K and the sequence $\{n\}$ diverges to infinity.

Example 5.7.2. $x_n = \sqrt{n} \to \infty$ as $n \to \infty$.

Here $\sqrt{n} > K$, when $n > K^2$. Hence, $N = $ int. part of K^2, etc.

See that the following sequences diverge to ∞ :

1. $\{2, 4, 6, \ldots, 2n, \ldots\}$;

2. $\{3, 3^2, 3^3, \ldots, 3^n, \ldots\}$;

CHAPTER 5: SEQUENCE OF REAL NUMBERS

3. $\{x, x^2, x^3, \ldots, x^n, \ldots\}$, where $x > 1$.

Definition 5.7.2. *A sequence $\{x_n\}$ is said to diverge to $-\infty$ if, for any negative number K, there always exists a positive integer N such that*

$$x_n < K \quad \text{for all } n > N.$$

In symbols, we write,
$$\lim_{n \to \infty} x_n = -\infty.$$

Here K is *generally* chosen s.t. its absolute value is large enough.

Example 5.7.3. *Show that* $\lim\limits_{n \to \infty} x_n = -\infty$, *if* $x_n = \log \frac{1}{n}$.

Solution: To find N for a given K such that

$$\log(1/n) < K \quad \text{for all } n > N,$$
i.e., $\quad \log n > -K \quad \text{for } n > N,$
or, $\quad n > e^{-K} \quad \text{for } n > N,$

and choose N = integral part of e^{-K} and the result follows.

The following sequences diverge to $-\infty$ (*Verify*):

1. $\{-2, -4, -6, \ldots, -2n, \ldots\}$

2. $\{-x, -x^2, -x^3, \ldots, -x^n, \ldots\}$, where $x > 1$.

Definition 5.7.3. *A sequence which neither converges nor diverges to $+\infty$ or $-\infty$ will be called oscillatory. If in an oscillatory sequence $\{x_n\}$, a constant C exists such that for all n, $|x_n| < C$, then the sequence is said to oscillate finitely, otherwise it is said to oscillate infinitely.*

An oscillatory sequence is also called an **indefinitely** (or **improperly**) divergent sequence and a sequence tending to $+\infty$ or $-\infty$ is sometimes called **definitely** (or **properly**) divergent.

Example 5.7.4. *Prove that $\{x_n\}$ is not convergent, where $x_n = (-1)^n$.*

Assume the contrary. Then $x_n \to l$ (a finite quantity) and, therefore, for a given $\varepsilon = \frac{1}{2}$, there exists a positive integer n_0 such that
$$|(-1)^n - l| < \tfrac{1}{2} \text{ for } n > n_0.$$
For n even and $> n_0$, $|1 - l| < \frac{1}{2}$, or, $\frac{1}{2} < l < \frac{3}{2}$.

and for n odd and $> n_0, |-1-l| < \frac{1}{2}$, i.e., $|1+l| < \frac{1}{2}$, or, $-\frac{3}{2} < l < -\frac{1}{2}$.

And, we come to a contradiction. Hence the sequence is non-convergent.

Observation: The sequence $\{x_n\} = \{(-1)^n\}$ is not convergent. Also it can be verified that it does not diverge to $+\infty$ or to $-\infty$. Hence it is oscillatory. Also see that $|x_n| = 1 < 2$, for every n. The sequence, therefore, *oscillates finitely*.

Example 5.7.5. $x_n = n/2$, *if n be even;*
$= 0$, *if n be odd,*

i.e., $\{x_n\} = \{0, 1, 0, 2, 0, 3, 0, 4, 0, 5, 0, \ldots\}$.

It can easily be verified that the sequence $\{x_n\}$ *oscillates infinitely*.

The following sequences are oscillating: (1) finitely; (2) and (3) infinitely.

1. $\{(-1)^n 3\} = \{-3, 3, -3, 3, -3, \ldots\}$.
2. $\{(-1)^n n\} = \{-1, 2, -3, 4, -5, \ldots\}$.
3. $\{1, \frac{1}{2}, 3, \frac{1}{4}, \ldots, 2n-1, 1/2n, \ldots\}$.

Example 5.7.6. *To show that the sequence* $\{x_n\} = \{1 + (-1)^n\}$ *oscillates finitely.*

The members of the sequence $= \{0, 2, 0, 2, 0, 2, \ldots\}$.

1. We first prove that the sequence cannot converge to a finite limit. If possible, let the sequence $\{x_n\}$ converge to **a finite limit** l. i.e., suppose $x_n \to l$ as $n \to \infty$. Then, by definition, corresponding to $\varepsilon = 1/2$ (say), $\exists$ a positive integer n_0 s.t.

$$\forall n > n_0, \quad |x_n - l| < \frac{1}{2}$$

i.e., $|1 + (-1)^n - l| < \frac{1}{2}$.

For n even and $> n_0$, we then get $\frac{3}{2} < l < \frac{5}{2}$.

For n odd and $> n_0$, we get $-\frac{1}{2} < l < \frac{1}{2}$.

The arguments lead to contradiction. In other words, $\{x_n\}$ does not converge to a finite limit.

2. The range set of $\{x_n\}$ contains only two elements $\{0, 2\}$.

$\therefore$ $\{x_n\}$ cannot diverge to $+\infty$ or $-\infty$, but $\{x_n\}$ is bounded.

Hence, $\{x_n\}$ is oscillating finitely between 0 and 2.

The following model can be easily remembered about the behaviour of a sequence w.r.t. convergence or divergence:

CHAPTER 5: SEQUENCE OF REAL NUMBERS

```
                          Sequences
                             ↓
              ┌──────────────┴──────────────┐
              ↓                             ↓
         Convergent                   Non-convergent
         (→ a finite limit)           (↛ a finite limit)
                                            ↓
                              ┌─────────────┴─────────────┐
                              ↓                           ↓
                          Properly                    Improperly
                          divergent                   divergent
                              ↓                           ↓
                      ┌───────┴───────┐           ┌───────┴───────┐
                      ↓               ↓           ↓               ↓
                  Diverges        Diverges    Oscillates      Oscillates
                  to + ∞          to − ∞      finitely        infinites
```

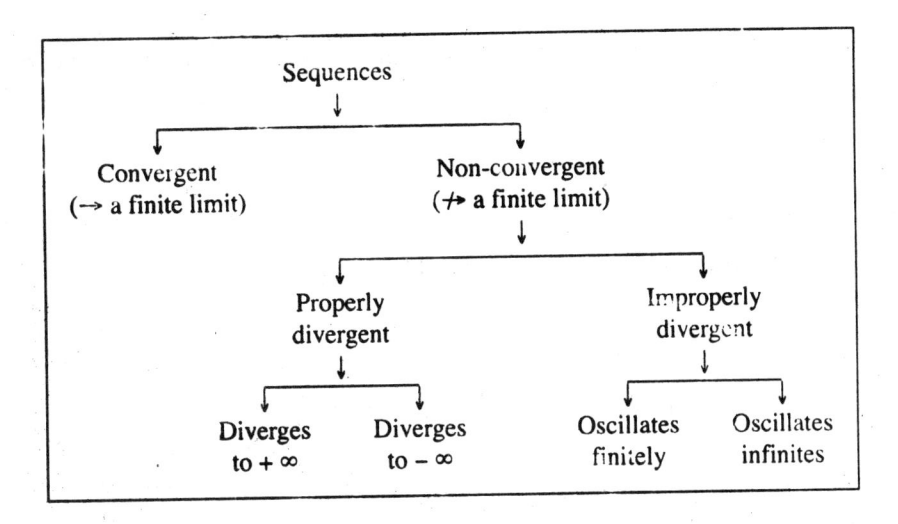

5.8 A very useful Criterion of Convergence

Given a sequence it is usual to consider two problems:

Problem 1. Does the sequence converge or diverge?

If it does not converge, does it diverge to $+\infty$ or $-\infty$ or does it oscillate? If it oscillates, does it oscillate finitely or infinitely?

Problem 2. If the sequence converges, what is the finite real number to which it converges?

The first problem is to investigate the *existence of limit* and the second is *to evaluate the limit, if it exists*. In the present article we shall confine our attention to the first problem only.

We shall prove in the theorems that follow:

A monotone sequence, when bounded, converges.

A monotone sequence, when unbounded, properly diverges.

i.e., **A monotone sequence can never oscillate.**

Theorem 5.8.1. *If a sequence $\{x_n\}$ of real numbers is monotone increasing (m.i.) and bounded above, then it converges to its exact upper bound (i.e., $x_n \to \sup x_n$).*

1. We first note that

    A m.i. sequence $\{x_n\}$ is always bounded below, since $\forall n \in \mathbb{N}, x_n \geq x_1$.

    ∴ when we state that the sequence $\{x_n\}$ is m.i. and *bounded above*, we may assume $\{x_n\}$ is m.i. and *bounded* (both above and below).

2. Also, we recall that the real number system $\mathbb{R}$ obeys the lub-property. So, if $\{x_n\}$ is bounded above, $\sup x_n$ exists in $\mathbb{R}$. Let $M = \sup x_n =$ exact u.b. of $\{x_n\}$.

Proof. What we are going to prove is that

A bounded m.i. sequence $\{x_n\}$ of real numbers converges to its exact upper bound M.

The two characteristic properties of M are

1. For all n, $x_n \leq M$.

2. Given any $\varepsilon > 0$, no matter how small, $\exists$ at least one member, say $x_p > M - \varepsilon$.

From **1**, it follows $x_n < M + \varepsilon$, $\forall\, n$.

Since $\{x_n\}$ is m.i., the members of the sequence beginning from x_{p+1} cannot be less than x_p and hence they are all $> M - \varepsilon$ ($\because$ by **2**, $x_p > M - \varepsilon$). Thus

$$x_n < M + \varepsilon, \; \forall\, n$$
$$\text{and} \quad x_n < M - \varepsilon, \; \forall\, n \geq p.$$

$\therefore\; M - \varepsilon < x_n < M + \varepsilon, \; \forall\, n \geq p.$

This proves: $\lim\limits_{n \to \infty} x_n = M$.

Theorem 5.8.2. *If sequence $\{x_n\}$ of real numbers is monotone decreasing (m.d.) and bounded below, then it converges to its exact lower bound (i.e., $x_n \to \inf x_n$).*

We leave the proof of this theorem to our students. The proof is exactly similar to Theorem **5.8.1** above (*only use the two-characteristic properties of m, the exact l.b. of $\{x_n\}$*).

Theorem 5.8.3. *If a sequence $\{x_n\}$ of real numbers is monotone increasing but not bounded above (i.e., unbounded above), then it diverges to $+\infty$.* [C.H. 1993]

Proof. Since $\{x_n\}$ is unbounded above, *corresponding to any positive number G, no matter how large, $\exists$ a positive integer m (say) such that $x_m > G$.*

Again, since $\{x_n\}$ is m.i., none of the members beginning from x_{m+1}, i.e., x_{m+1}, x_{m+2}, x_{m+3}, ... can be less than x_m and hence they are all $> G$ ($\because x_m > G$).

Thus, we obtain m s.t. $\forall\, n \geq m, x_n > G$.

i.e., $x_n \to \infty$ as $n \to \infty$.

Theorem 5.8.4. *If a sequence $\{x_n\}$ of real numbers is monotone decreasing and unbounded below then it diverges to $-\infty$.*

CHAPTER 5: SEQUENCE OF REAL NUMBERS

The proof is exactly similar to Theorem **5.8.3** above.

Conclusions. The four theorems together lead to the conclusion:

A monotone sequence has always a *definite behaviour*—either it tends to a finite limit or it tends to $+\infty$ or $-\infty$, i.e.,

$$\boxed{\text{A monotone sequence can never oscillate.}}$$

Example 5.8.1. *If* $x_n = \dfrac{3n-1}{n+2}$, *prove that* $\{x_n\}$ *is m.i. and bounded.*

What can you say about the behaviour of $\{x_n\}$ with respect to convergence?

1.
$$x_{n+1} - x_n = \frac{3(n+1)-1}{(n+1)+2} - \frac{3n-1}{n+2} = \frac{3n+2}{n+3} - \frac{3n-1}{n+2}$$

$$= \frac{(3n+2)(n+2) - (3n-1)(n+3)}{(n+3)(n+2)}$$

$$= \frac{(3n^2+8n+4) - (3n^2+8n-3)}{(n+3)(n+2)}$$

$$= \frac{7}{n^2+5n+6}, \text{ which is positive } \forall\, n \geq 1.$$

Conclusions. $\forall\, n,\, x_{n+1} > x_n$, i.e., $\{x_n\}$ is m.i. (*in fact, strictly m.i.*).

2. $x_n = \dfrac{3n-1}{n+2} = \dfrac{3(n+2)-7}{n+2} = 3 - \dfrac{7}{n+2} < 3,\ \forall\, n \geq 1.$

Thus the sequence $\{x_n\}$ is m.i. and bounded above; being m.i. $\{x_n\}$ is certainly bounded below ($x_n > x_1 = 2/3$). Hence *the sequence $\{x_n\}$ is m.i. and bounded and hence the sequence $\{x_n\}$ is convergent.*

Note that 3 is **an upper bound**; we have not proved that 3 is the least upper bound. So we can only conclude:

$$\lim_{n \to \infty} x_n \leq 3.$$

Note 5.8.1. *To evaluate* $\lim\limits_{n\to\infty} \dfrac{3n-1}{n+2}$, *we can now use the limit theorems.*

Thus,
$$\lim_{n\to\infty} \frac{3n-1}{n+2} = \lim_{n\to\infty} \frac{3 - \frac{1}{n}}{1 + \frac{2}{n}} = \frac{3-0}{1+0} = 3.$$

It is easy to justify, by definition, that the limit is 3.

$$\left[|x_n - 3| = \left|\frac{3n-1}{n+2} - 3\right| = \frac{7}{n+2} < \varepsilon,\ \text{if } n > -2 + \frac{7}{\varepsilon};\ \text{hence, etc.}\right]$$

Note 5.8.2. *This particular example shows that a sequence whose members are all rational numbers has a limit which is also a rational number. But it is not always true as will be evident from the following example.*

** **Example 5.8.2. A very important example.**

To discuss the behaviour of the sequence $\{x_n\}$, where

$$x_n = \left(1 + \frac{1}{n}\right)^n.$$

1. *The sequence is monotone increasing.*

For a fixed positive integer n, we see that

$$x_n = \left(1 + \frac{1}{n}\right)^n = 1 + n \cdot \frac{1}{n} + \frac{n(n-1)}{2!} \cdot \frac{1}{n^2} + \frac{n(n-1)(n-2)}{3!} \cdot \frac{1}{n^3} + \cdots$$

$$\cdots + \frac{n(n-1)\cdots 1}{n!} \cdot \frac{1}{n^n}$$

$$= 1 + 1 + \frac{1}{2!}\left(1 - \frac{1}{n}\right) + \frac{1}{3!}\left(1 - \frac{1}{n}\right)\left(1 - \frac{2}{n}\right) + \cdots$$

$$\cdots + \frac{1}{n!}\left(1 - \frac{1}{n}\right)\left(1 - \frac{2}{n}\right)\cdots\left(1 - \frac{n-1}{n}\right).$$

Then

$$x_{n+1} = 1 + 1 + \frac{1}{2!}\left(1 - \frac{1}{n+1}\right) + \frac{1}{3!}\left(1 - \frac{1}{n+1}\right)\left(1 - \frac{2}{n+1}\right) + \cdots$$

$$\cdots + \frac{1}{(n+1)!}\left(1 - \frac{1}{n+1}\right)\left(1 - \frac{2}{n+1}\right)\cdots\left(1 - \frac{n}{n+1}\right).$$

We observe that x_{n+1} contains one more term than x_n and further since

$$1 - \frac{1}{n} < 1 - \frac{1}{n+1}; \quad 1 - \frac{2}{n} < 1 - \frac{2}{n+1}, \text{ etc.,}$$

each term of x_{n+1} (after the first two terms) is greater than the corresponding term of x_n.

Hence we conclude: $x_{n+1} > x_n$, $\forall\, n$, i.e., $\{x_n\}$ is monotone increasing (*in fact, strictly m.i.*).

Alternative Proof: $\{x_n\} = \left\{\left(1 + \frac{1}{n}\right)^n\right\}$ is monotone increasing. Consider $(n + 1)$ positive numbers:

CHAPTER 5: SEQUENCE OF REAL NUMBERS

$1 + \frac{1}{n}, 1 + \frac{1}{n}, \ldots$ (n times) and 1 and apply the formula. A.M. > G.M.

$$\frac{n\left(1+\frac{1}{n}\right)+1}{n+1} > \left\{\left(1+\frac{1}{n}\right)^n\right\}^{\frac{1}{n+1}}$$

or $\left(1+\frac{1}{n}\right)^{n+1} > \left(1+\frac{1}{n}\right)^n$ i.e., $x_{n+1} > x_n$

i.e., $\{x_n\}$ is monotone increasing.

2. *The sequence $\{x_n\}$ is bounded.*

Clearly, $x_n > x_1 = 2$, for $n > 1$, i.e., 2 is a lower bound.

We shall prove that $\{x_n\}$ has an upper bound 3.

$$x_n = 1 + 1 + \frac{1}{2!}\left(1-\frac{1}{n}\right) + \frac{1}{3!}\left(1-\frac{1}{n}\right)\left(1-\frac{2}{n}\right) + \cdots$$
$$\cdots + \frac{1}{n!}\left(1-\frac{1}{n}\right)\left(1-\frac{2}{n}\right)\cdots\left(1-\frac{n-1}{n}\right)$$
$$\leq 1 + 1 + \frac{1}{2!} + \frac{1}{3!} + \cdots + \frac{1}{n!}$$
$$\leq 1 + 1 + \frac{1}{2} + \frac{1}{2^2} + \frac{1}{2^3} + \cdots + \frac{1}{2^{n-1}}$$
$$(\because 3! = 3\cdot 2\cdot 1 > 2\cdot 2, \text{ i.e., } 1/3! < 1/2^2, \text{ etc.})$$
$$= 1 + \frac{1-(1/2)^n}{1-1/2} \quad \text{(summing up the G.P.)}$$

i.e., $x_n \leq 3 - 2\cdot\left(\frac{1}{2}\right)^n$.

Hence, $x_n < 3$, $\forall\, n$. Thus $2 \leq x_n < 3$, $\forall\, n$. i.e., $\{x_n\}$ is bounded.

Conclusions. Since $\{x_n\}$ is m.i. and bounded, it is convergent and

$$2 \leq \lim x_n \leq 3.$$

> REMEMBER: Limit operations make strong inequalities weak.

We shall, in future, denote the limit of $(1+\frac{1}{n})^n$ as $n \to \infty$ by the letter e (*exponential e*).

Note 5.8.3. *It can be proved that this limit e is* NOT *a rational number. Up to eleven significant figures, the value of e is*

$$e = 2.7182818285.$$

That e is not a rational number will be proved in a later Chapter.

** Example 5.8.3. Another sequence whose limit is also e.

Let $\{y_n\}$ be a sequence defined by

$$y_n = 1 + \frac{1}{1!} + \frac{1}{2!} + \frac{1}{3!} + \cdots + \frac{1}{n!}.$$

To prove that $\{y_n\}$ is m.i. and bounded and hence convergent.

Denoting $\lim_{n \to \infty} y_n = e'$, to prove that $e = e'$.

Solution:

1. Clearly, $y_{n+1} > y_n, \forall n \left(\because y_{n+1} - y_n = \frac{1}{(n+1)!} > 0 \right)$.

   Also

   $$\begin{aligned} y_n &= 1 + \frac{1}{1!} + \frac{1}{2!} + \frac{1}{3!} + \cdots + \frac{1}{n!} \\ &\leq 1 + 1 + \frac{1}{2} + \frac{1}{2^2} + \cdots + \frac{1}{2^{n-1}} \\ &= 1 + \frac{1 - \frac{1}{2^n}}{1 - \frac{1}{2}}. \end{aligned}$$

   i.e., $y_n < 3, \forall n$. Further $y_n > y_1 = 2, \forall n > 1$.

   $$\therefore \ 2 \leq y_n < 3, \forall n.$$

   Hence, $\{y_n\}$ converges ($\because$ it is m.i. and bounded).

2. Suppose $\lim_{n \to \infty} y_n = e'$. To prove $e = e'$.

In Ex. 5.8.2 We have proved: $x_n \leq 1 + \frac{1}{1!} + \frac{1}{2!} + \frac{1}{3!} + \cdots + \frac{1}{n!} = y_n$.

$\therefore$ taking limits (both limits exist here) on both sides,

$$\lim_{n \to \infty} x_n \leq \lim_{n \to \infty} y_n, \text{ i.e., } \mathbf{e} \leq \mathbf{e'}.$$

CHAPTER 5: SEQUENCE OF REAL NUMBERS

Take a positive integer $m < n$. Then
$$x_n = 1 + \frac{1}{1!} + \frac{1}{2!}\left(1 - \frac{1}{n}\right) + \cdots + \frac{1}{m!}\left(1 - \frac{1}{n}\right)\left(1 - \frac{2}{n}\right)\cdots\left(1 - \frac{m-1}{n}\right)$$
$$+ \frac{1}{(m+1)!}\left(1 - \frac{1}{n}\right)\left(1 - \frac{2}{n}\right)\cdots\left(1 - \frac{m}{n}\right) + \cdots$$
$$+ \frac{1}{n!}\left(1 - \frac{1}{n}\right)\left(1 - \frac{2}{n}\right)\cdots\left(1 - \frac{n-1}{n}\right).$$

Clearly, $1 + \frac{1}{1!} + \frac{1}{2!}\left(1 - \frac{1}{n}\right) + \frac{1}{3!}\left(1 - \frac{1}{n}\right)\left(1 - \frac{2}{n}\right) + \cdots$
$$+ \frac{1}{m!}\left(1 - \frac{1}{n}\right)\left(1 - \frac{2}{n}\right)\cdots\left(1 - \frac{m-1}{n}\right) < x_n.$$

Suppose m remains fixed and we make $n \to \infty$ and take limits of both sides. Then we get
$$1 + \frac{1}{1!} + \frac{1}{2!} + \frac{1}{3!} + \cdots + \frac{1}{m!} \leq \lim_{n \to \infty} x_n = e,$$
for all value of m.

Hence, as $m \to \infty$, $\lim_{m \to \infty} y_m \leq e$, i.e., $e' \leq e$.

The two results $e \leq e'$ and $e' \leq e$ cannot hold simultaneously unless we take $e = e'$.

Note 5.8.4. $\{x_n\}$ and $\{y_n\}$ contain members which are all rational but their common limit e is not a rational number.

** Example 5.8.4. Prove that $\{x_n\}$ where
$$x_n = \frac{1}{n+1} + \frac{1}{n+2} + \frac{1}{n+3} + \cdots + \frac{1}{2n},$$
is a convergent sequence. What is your estimation of the value of the limit of this sequence?

$$x_{n+1} - x_n = \left(\frac{1}{n+2} + \frac{1}{n+3} + \cdots + \frac{1}{2n+2}\right) - \left(\frac{1}{n+1} + \frac{1}{n+2} + \cdots + \frac{1}{2n}\right)$$
$$= \frac{1}{2n+1} + \frac{1}{2n+2} - \frac{1}{n+1} = \frac{1}{2n+1} - \frac{1}{2(n+1)}$$
$$= \frac{(2n+2) - (2n+1)}{(2n+1)(2n+2)} = \frac{1}{(2n+1)(2n+2)}, \text{ which is positive } \forall\, n \geq 1.$$

$\therefore\ x_{n+1} > x_n\ \forall\, n \geq 1$, i.e., $\{x_n\}$ is monotone increasing.

Now see that
$$x_n = \frac{1}{n+1} + \frac{1}{n+2} + \frac{1}{n+3} + \cdots + \frac{1}{n+n}$$
$$\leq \frac{1}{n+1} + \frac{1}{n+1} + \frac{1}{n+1} + \cdots + \frac{1}{n+1} = \frac{n}{n+1}.$$

i.e., $x_n \leq 1 - \frac{1}{n+1}$ and hence $x_n < 1$, $\forall\, n$, i.e., $\{x_n\}$ is bounded above and one upper bound is 1.

$\therefore\ \lim_{n \to \infty} x_n$ exists and that limit is ≤ 1.

Example 5.8.5. *To show that the sequence $\{x_n\}$ defined by*

$$x_n = 1 + \frac{1}{2} + \frac{1}{3} + \cdots + \frac{1}{n}$$

is a divergent sequence.

1. Clearly, $x_{n+1} - x_n = 1/(n+1)$, which is positive $\forall\, n$.

   Hence, $x_{n+1} > x_n$, $\forall\, n$, i.e., $\{x_n\}$ is a strictly m.i. sequence.

2. We shall prove that $\{x_{n_j}\}$ is unbounded above.

   Choose a positive number G, no matter how large.
   Next take a positive integer $m > 2G$. Then, whenever $n > 2^m$, $x_n > x_{2^m}$,

   i.e., $x_n > \left(1 + \frac{1}{2}\right) + \left(\frac{1}{3} + \frac{1}{2^2}\right) + \left(\frac{1}{5} + \frac{1}{6} + \frac{1}{7} + \frac{1}{2^3}\right) + \cdots$

   $\cdots + \left(\frac{1}{2^{m-1}+1} + \frac{1}{2^{m-1}+2} + \cdots + \frac{1}{2^m}\right)$

   $> \frac{1}{2} + \left(\frac{1}{2^2} + \frac{1}{2^2}\right) + \left(\frac{1}{2^3} + \frac{1}{2^3} + \frac{1}{2^3} + \frac{1}{2^3}\right) + \cdots$

   $+ \cdots + \left(\frac{1}{2^m} + \frac{1}{2^m} + \cdots + \frac{1}{2^m}\right)$

   $= \frac{1}{2} + 2 \cdot \frac{1}{2^2} + 2^2 \cdot \frac{1}{2^3} + \cdots + 2^{m-1} \cdot \frac{1}{2^m}$

   $= m \cdot \frac{1}{2} > G.$

   Hence $x_n > G$.
   Thus, given any $G > 0$, $\exists$ a positive integer $n > 2^m$ $\forall\, m > 2G$ s.t. $x_n > G$.
   In other words, $\{x_n\}$ is unbounded above.
   Thus $\{x_n\}$ is m.i. and unbounded above; hence it diverges to $+\infty$.

Example 5.8.6. *Prove that the sequence $\{x_n\}$ where $x_{n+1} = \dfrac{6(1+x_n)}{7+x_n}$, $x_1 = c > 0$ is monotone increasing or decreasing according as $c < 2$ or $c > 2$ and that, in either case, the sequence converges to 2. Discuss the case $c = 2$.*

CHAPTER 5: SEQUENCE OF REAL NUMBERS

Solution:

$$x_{n+1} - x_n = \frac{6(1+x_n)}{7+x_n} - x_n = \frac{6 - x_n - x_n^2}{7+x_n} = \frac{(2-x_n)(3+x_n)}{7+x_n}$$

and $\quad 2 - x_{n+1} = 2 - \dfrac{6(1+x_n)}{7+x_n} = \dfrac{4(2-x_n)}{7+x_n}.$

Hence, if $x_n < 2$, then $x_{n+1} > x_n$ and $x_{n+1} < 2$, for every n, i.e., if $x_n < 2$, then $x_n < x_{n+1} < 2$, for all n.

1. So, if $x_1 = c < 2$, $x_1 < x_2 < x_3 < \cdots < 2$, i.e., the sequence is monotone increasing (when $c < 2$).

2. Similarly, if $x_1 = c > 2$, $x_1 > x_2 > x_3 > \cdots > 2$ and so the sequence is monotone decreasing.
   (See that in this case, if $x_n > 2$, then $x_{n+1} < x_n$ and $x_{n+1} > 2$).

3. If $c = 2$, then $\forall\, n, x_n = x_{n+1} = 2$, i.e., the sequence is

$$\{2, 2, 2, 2, \ldots\}$$

and it clearly converges to 2.

In case (1) we have proved that $\{x_n\}$ is m.i. and bounded above (one u.b. being 2) and hence it converges to a limit l (say), i.e., $\lim\limits_{n \to \infty} x_n = l\,(l \leq 2)$.

Hence taking limits

$$\lim_{n \to \infty} x_{n+1} = \frac{6 + 6\lim\limits_{n \to \infty} x_n}{7 + \lim\limits_{n \to \infty} x_n}$$

or, $\quad l = \dfrac{6 + 6l}{7 + l}\quad$ (since $\lim\limits_{n \to \infty} x_n = l \Rightarrow \lim\limits_{n \to \infty} x_{n+1} = l$)

or, $\quad 7l + l^2 = 6 + 6l$, i.e., $l^2 + l - 6 = 0$

or, $\quad (l+3)(l-2) = 0$ i.e., $l = -3$ or 2.

But the terms of the sequence are all positive and hence the limit cannot be negative. Hence, the required limit $= \mathbf{2}$.

Similarly, in case (2) the limit is also **2**.

Example 5.8.7. *To prove that the sequence $\{x_n\}$ defined by*

$$x_1 = \sqrt{2},\, x_{n+1} = \sqrt{2x_n}\ \text{for}\ n \geq 1,$$

converges to 2. [C.H. 1999]

Solution: We first observe that the sequence may be written as

$$\left\{\sqrt{2}, \sqrt{2\sqrt{2}}, \sqrt{2\sqrt{2\sqrt{2}}}, \ldots\right\}.$$

1. Here $x_{n+1}^2 = 2x_n$ and $x_n^2 = 2x_{n-1}$ (for $n \geq 2$).

$$\therefore \quad x_{n+1}^2 - x_n^2 = 2(x_n - x_{n-1}).$$

Hence, $x_{n+1} \gtreqless x_n$ according as $x_n \gtreqless x_{n-1}$.

Similarly,

$$x_n \gtreqless x_{n-1} \quad \text{according as} \quad x_{n-1} \gtreqless x_{n-2}$$
$$\ldots\ldots\ldots\ldots$$
$$\ldots\ldots\ldots\ldots$$
$$x_3 \gtreqless x_2 \quad \text{according as} \quad x_2 \gtreqless x_1.$$

But, $x_1 = \sqrt{2}$ and $x_2 = \sqrt{2\sqrt{2}}$; clearly, $x_2 > x_1$ and consequently,

$$x_3 > x_2, \; x_4 > x_3, \; \cdots, \; x_{n+1} > x_n, \ldots$$

i.e., $\quad x_1 < x_2 < x_3 < x_4 < \cdots < x_n < x_{n+1} < \cdots$

i.e., $\{x_n\}$ is monotone increasing.

2. Since $x_{n+1} > x_n$, we may write $\sqrt{2x_n} > x_n$, or, $2x_n > x_n^2$, i.e., $x_n(x_n - 2) < 0$.

Since every $x_n > 0$, we conclude, $x_n < 2 \; \forall \; n$.

Thus the sequence is bounded above (one u.b. being 2).

From (1) and (2), we now conclude that $\{x_n\}$ is convergent. Suppose, $x_n \to l$ as $n \to \infty$.

Then taking limits, $\lim_{n \to \infty} x_{n+1}^2 = 2 \lim_{n \to \infty} x_n$, i.e., $l^2 = 2l$.

$\therefore \quad l = 2 \, (l \neq 0$, since the first term $= \sqrt{2}$ and the sequence is *m.i.*).

Hence, the limit of the given sequence is 2.

Before we proceed further we shall introduce the notion of *Nested Interval Property* followed by *Theorem of Nested Intervals*:

Let $\{I_1 = [a_1, b_1], I_2 = [a_2, b_2], \ldots, I_n = [a_n, b_n], \ldots\}$ be a sequence of closed intervals. If $I_n \supset I_{n+1}, \; \forall \; n$, (i.e., each interval contains ALL its successors), then we say that these intervals are nested.

CHAPTER 5: SEQUENCE OF REAL NUMBERS

Example 5.8.2. (*Theorem on Nested Intervals*). *Suppose that $\{I_n\}$ is a sequence of closed intervals such that*

(a) $I_n \supset I_{n+1}$, $\forall\, n = 1, 2, 3, 4, \ldots$, *i.e., the sequence of closed intervals $\{I_n\}$ is nested;*

(b) *the length d_n of I_n tends to zero as $n \to \infty$,*

*then there exists **exactly one point** which is common to all the intervals.*

[C.H. 1983, '90]

Proof. Let $\{I_n = [a_n, b_n], n = 1, 2, 3, 4, \ldots\}$ be a sequence of closed intervals.
Given : $I_n \supset I_{n+1}$. Hence,

$$a_n \leq a_{n+1} < b_{n+1} \leq b_n, \ \forall\, n.$$

Again, all the intervals are contained in $[a_1, b_1]$.

$$\therefore \quad a_1 \leq a_n \leq a_{n+1} < b_{n+1} \leq b_n \leq b_1.$$

This shows that both the sequences $\{a_n\}$ and $\{b_n\}$ are convergent.
Moreover,

$\{a_n\}$ is monotone increasing, bounded above by b_1,

and $\{b_n\}$ is monotone decreasing, bounded below by a_1.

$\therefore$ both the sequences $\{a_n\}$ and $\{b_n\}$ must be convergent,

$$\lim_{n \to \infty} a_n = \sup a_n = l_1 \text{ (say)} \quad \text{and} \quad \lim_{n \to \infty} b_n = \inf b_n = l_2 \text{ (say)}.$$

But by the given condition (ii), $b_n - a_n \to 0$ as $n \to \infty$.

$$\therefore \quad \lim_{n \to \infty} b_n = \lim_{n \to \infty} a_n, \text{ i.e., } l_1 = l_2 = l \text{ (say)}.$$

Since $a_n \to \sup a_n = l$ and $b_n \to \inf b_n = l$, for all n

$$a_n \leq l \quad \text{and} \quad b_n \geq l, \quad \text{i.e., } a_n \leq l \leq b_n\ \forall\, n.$$

$\therefore$ l lies in every interval $I_n = [a_n, b_n]$.

Uniqueness: There cannot be more than one point which lies in every interval I_n. If possible, let there be two points l, m ($l \neq m$) such that

$$a_n \leq l \leq b_n \quad \text{and} \quad a_n \leq m \leq b_n \quad \forall\, n.$$

so that $d_n = b_n - a_n \geq |m - l|$, $\forall\, n$, i.e., d_n remains greater than a fixed positive number $\forall\, n$, contrary to the hypothesis that $d_n \to 0$ as $n \to \infty$.

Note 5.8.5. If $I_n = (0, 1/n)$ are open intervals for $n = 1, 2, 3, \ldots$, then we observe that $I_n \supset I_{n+1}$ and $d_n \to 0$ as $n \to \infty$. But we see that $\exists$ no point which is common to all the intervals I_n. This example shows that for Nested Interval property it is essential that the intervals should be **closed. The theorem may or may not be true if the intervals are not closed.** e.g., consider the nest of open intervals $\{(0, \frac{1}{n})\}$. These intervals have empty intersection. But, the intersection of nest of open intervals $\{(-\frac{1}{n}, \frac{1}{n})\}$ have one point in common, namely the point 0.

5.9 Cauchy Sequences

Definition 5.9.1. A sequence $\{x_n\}$ of real numbers is said to be a **Cauchy Sequence**, if for any $\varepsilon > 0, \exists$ a positive integer N such that

$$|x_m - x_n| < \varepsilon, \ \forall \, m, n > N.$$

Replacing m by $n + p$ we may write the inequality as

$$|x_{n+p} - x_n| < \varepsilon, \ \forall \, n > N \ \text{and} \ \forall \, p = 1, 2, 3, 4, \ldots.$$

We shall refer to the inequality of this definition as **Cauchy's Criterion**.

Note 5.9.1. Intuitively this definition implies that for sufficiently large values of n, the terms of $\{x_n\}$ get closer and closer to each other. If $\{x_n\}$ is a Cauchy sequence, then we may write $\lim(x_m - x_n) = 0$ as $m \to \infty, n \to \infty$.

** Example 5.9.1. The sequence $\left\{\dfrac{1}{n}\right\}$ is a Cauchy sequence: \hfill [C.H. 1988]

Justification: Let us take positive integers m, n such that $m > n$.
Then
$$\left|\frac{1}{m} - \frac{1}{n}\right| = \frac{m-n}{mn} = \frac{1}{n}\left(1 - \frac{n}{m}\right) < \frac{1}{n}.$$

Now, for any given $\varepsilon > 0, \left|\frac{1}{m} - \frac{1}{n}\right| < \varepsilon$, if $\frac{1}{n} < \varepsilon$ i.e., if $n > \frac{1}{\varepsilon}$ and $m > n$.

Thus the definition of Cauchy's criterion is satisfied and $\{1/n\}$ is a Cauchy sequence.

** Example 5.9.2. (i) The sequence $\{n^2\}$ is not a Cauchy sequence:

Justification: As in Ex. **5.9.1** take m, n positive integers such that $m > n$.
Now $|m^2 - n^2| = (m+n)(m-n) > 2n > 1$, whatever n may be.
Take $\varepsilon = 1$, we see that no positive integer N exists for which

$$|m^2 - n^2| < \varepsilon \ \ \forall \, m, n > N.$$

Hence, by definition, $\{n^2\}$ is not a Cauchy sequence.

(ii) *The sequence $\{1 + \frac{1}{2} + \frac{1}{3} + \cdots + \frac{1}{n}\}$ is not a Cauchy sequence.*

The sequence $\{x_n\}$, where $x_n = 1 + \frac{1}{2} + \frac{1}{3} + \cdots + \frac{1}{n}$ is not Cauchy, not convergent and not bounded. In fact, $\lim x_n = \infty$.

On the other hand the sequence

$$x_n = 1 + \frac{1}{2} + \frac{1}{3} + \cdots + \frac{1}{n} - \log n \text{ (m.d.)}$$
$$\text{and} \quad y_n = 1 + \frac{1}{2} + \frac{1}{3} + \cdots + \frac{1}{n-1} - \log n \text{ (m.i.)}$$

converge to a real number r known on Euler's constant γ where $0.3 < \gamma < 1$ (**see Ex. 5.11.4**).

Justification: $\{x_n\}$ is not Cauchy.

Let $m > n$ and $m = 2n$. Then

$$|x_m - x_n| = \frac{1}{n+1} + \frac{1}{n+2} + \cdots + \frac{1}{m} \geq (m-n)\frac{1}{m} = \frac{1}{2}.$$

Hence $\{x_n\}$ is not Cauchy and so it is not convergent. It is clear that

$$x_{n+1} \geq x_n \geq 0 \quad [\text{consider } m = n+1 \text{ in above inequality}].$$

Hence, $\{x_n\}$ is unbounded above; for if it were bounded, by monotonicity it would be convergent. **Hence, $\lim x_n = \infty$.**

We shall now prove:

1. Every convergent sequence of real numbers is a Cauchy sequence.

2. Every Cauchy sequence of real numbers converges.

These two results can be put in the following language which is known as **Cauchy's General Principle of Convergence (or Cauchy's Convergence Criterion):**

A necessary and sufficient condition for the convergence of a sequence $\{x_n\}$ of real numbers is that $\{x_n\}$ should be a Cauchy sequence. [C.H. 1983, 1992]

In a more compact language:

> A sequence $\{x_n\}$ of real numbers converges, iff it is a Cauchy sequence.

Proof.

Step 1. To prove that Cauchy's criterion is a necessary condition for the convergence of a sequence of real numbers. i.e., to prove that every convergent sequence is a Cauchy sequence.

Let $\{x_n\}$ be a convergent sequence of real numbers converging to the limit l. Then, by definition of limit, for every $\varepsilon > 0$, $\exists$ a positive integer N s.t.

$$|x_n - l| < \varepsilon/2, \ \forall \, n > N.$$

Therefore, if $m, n > N$, then we have

$$|x_m - x_n| = |(x_m - l) + (l - x_n)| \leq |x_m - l| + |x_n - l|, \ \text{i.e.,} \ < \varepsilon/2 + \varepsilon/2 = \varepsilon.$$

This is precisely the Cauchy's criterion. Thus $x_n \to l \Rightarrow \{x_n\}$ is a Cauchy sequence.

Step 2. To prove that Cauchy's criterion is a sufficient condition for the convergence of a sequence of real numbers, i.e., To prove that every Cauchy sequence of real numbers converges.

Let $\{x_n\}$ be a Cauchy sequence of real numbers.

1. We shall first prove that *every such Cauchy sequence is bounded*. Since $\{x_n\}$ is a Cauchy sequence, for $\varepsilon = p$ (p being a fixed positive number) $\exists$ a positive integer N such that

   $$|x_m - x_n| < p \ \forall \, m, n > N.$$

   Choose a positive integer $n_0 > N$. Now for $n > N$,

   $$|x_n| = |x_n - x_{n_0} + x_{n_0}| \leq |x_n - x_{n_0}| + |x_{n_0}| < p + |x_{n_0}|.$$

   Let $M = \max.\{|x_1|, |x_2|, \ldots, |x_N|, |x_{n_0}| + p\}$.

   Then, clearly for every positive integer n, $|x_n| \leq M$. i.e., $\{x_n\}$ is bounded.

2. Next we prove that *every Cauchy sequence of real numbers converges (i.e., it has a finite limit)*.

   Let $\{x_n\}$ be a Cauchy sequence of real numbers.

   By (1), $\{x_n\}$ is bounded and hence it has a convergent sub-sequence, by Bolzano-Weierstrass Theorem for sequences.

   Suppose that $\{x_n\}$ has a sub-sequence $\{x_{n_k}\}$, which converges. Let $\lim\limits_{k \to \infty} x_{n_k} = l$.

   We now prove that $\lim\limits_{n \to \infty} x_n = l$.

   Let ε be any given positive number, no matter how small. Since $\{x_n\}$ is a Cauchy sequence, $\exists$ a positive integer N such that $|x_m - x_n| < \varepsilon/2$, whenever $m, n > N$. Since

   $$\lim_{k \to \infty} x_{n_k} = l. \tag{5.9.1}$$

CHAPTER 5: SEQUENCE OF REAL NUMBERS

It is possible to choose a positive integer n_k so large that

$$|x_{n_k} - l| < \varepsilon/2 \quad \text{and} \quad n_k > N. \tag{5.9.2}$$

Now for $n > N$, we have

$$\begin{aligned}|x_n - l| &= |x_n - x_{n_k} + x_{n_k} - l| \\ &\leq |x_n - x_{n_k}| + |x_{n_k} - l| \\ &< \frac{\varepsilon}{2} + \frac{\varepsilon}{2} = \varepsilon. \quad \text{[using (5.9.1) and (5.9.2)]}\end{aligned}$$

Thus $\{x_n\}$ converges to l.

3. Finally we prove that *this unique cluster point l (say) of $\{x_n\}$ is the limit of the sequence.*

Take a δ-nbd. of l, namely $(l - \delta, l + \delta)$.

There can be only a finite number of terms of $\{x_n\}$ which can lie outside this *nbd.* For, if there were infinite number of members of $\{x_n\}$ outside this *nbd.*, then those terms (being bounded) would have a cluster point other than l. We, therefore, have for any $\delta > 0$, a positive integer n_0 such that

$$|x_n - l| < \delta, \ \forall \, n \geq n_0$$

i.e., $x_n \longrightarrow l$ as $n \longrightarrow \infty$.

We have thus completely established the sufficiency condition;
i.e., $\{x_n\}$ is a Cauchy sequence $\Rightarrow$ $\{x_n\}$ converges to a finite limit l.

Applications

**1. *Use Cauchy's Criterion to prove that $\{x_n\}$ does not converge, where*

$$x_n = 1 + \frac{1}{2} + \frac{1}{3} + \frac{1}{4} + \cdots + \frac{1}{n}.$$

Solution: Suppose, if possible, $\{x_n\}$ converges to a finite limit.
Then, by Cauchy's criterion, for $\varepsilon = 1/2, \exists$ a positive integer n_0 s.t.

$$|x_m - x_n| < \frac{1}{2}, \ \forall \, m, n \geq n_0.$$

We suppose $m > n \geq n_0$. Then we have

$$\left|\frac{1}{n+1} + \frac{1}{n+2} + \cdots + \frac{1}{m}\right| < \frac{1}{2}, \ \forall \, m > n \geq n_0.$$

Taking $m = 2n$ we now see that

$$\frac{1}{n+1} + \frac{1}{n+2} + \cdots + \frac{1}{2n} > n \cdot \frac{1}{2n}, \text{ i.e., } > \frac{1}{2}.$$

We arrive at a contradiction. *Hence the given sequence cannot converge.*

Note 5.9.2. *This sequence is clearly m.i.* ($\because x_{n+1} > x_n \; \forall \; r \in \mathbb{N}$). *Therefore, if it does not converge, it must diverge to* $+\infty$.

**2. *Use Cauchy's General principle of convergence to prove that* $\{y_n\}$ *is convergent, where*

$$y_n = 1 + \frac{1}{2!} + \frac{1}{3!} + \cdots + \frac{1}{n!}.$$

Solution: We observe that

$$\frac{1}{n!} = \frac{1}{2 \cdot 3 \cdot 4 \cdots n} < \frac{1}{2 \cdot 2 \cdots 2} = \frac{1}{2^{n-1}}.$$

Then

$$|y_m - y_n| = \left| \frac{1}{(n+1)!} + \frac{1}{(n+2)!} + \cdots + \frac{1}{m!} \right| \quad \text{(where } m > n\text{)}$$
$$< \frac{1}{2^n} + \frac{1}{2^{n+1}} + \cdots + \frac{1}{2^{m-1}}$$
$$= \frac{1}{2^n} \cdot \frac{1 - (\frac{1}{2})^{m-n}}{1 - \frac{1}{2}}, \text{ i.e., } < \frac{1}{2^n} \cdot \frac{1}{1 - \frac{1}{2}} = \frac{1}{2^{n-1}}.$$

Since $\frac{1}{2^{n-1}} \to 0$ as $n \to \infty$, we have for every $\varepsilon > 0$, a positive integer n_0 exists for which

$$\frac{1}{2^{n-1}} < \varepsilon, \; \forall \, n \geq n_0 \quad \text{where } n_0 = \max\{1, [-\log_2 \varepsilon + 1] + 1\}.$$

$\therefore \; |y_m - y_n| < \varepsilon \; \forall \, n \geq n_0$ and $m > n$.

i.e., Cauchy's criterion is satisfied and hence $\{y_n\}$ converges to a finite limit.

**3. *Use Cauchy's criterion to prove that* $\{x_n\}$, *where* $x_n = 1 - \frac{1}{2} + \frac{1}{3} - \frac{1}{4} + \cdots + (-1)^{n-1}\frac{1}{n}$, *is a convergent sequence.* [C.H. 1992]

Solution: Take $m > n$. Then

$$|x_m - x_n| = \left|(-1)^n \frac{1}{n+1} + (-1)^{n+1}\frac{1}{n+2} + \cdots + (-1)^{m-1}\frac{1}{m}\right|$$

$$= \frac{1}{n+1} - \frac{1}{n+2} + \frac{1}{n+3} - \frac{1}{n+4} + \cdots + (-1)^{m-n-1}\frac{1}{m}$$

$$= \frac{1}{n+1} - \left(\frac{1}{n+2} - \frac{1}{n+3}\right) - \left(\frac{1}{n+4} - \frac{1}{n+5}\right) - \cdots$$

i.e., $< \dfrac{1}{n+1}$, which is less than any given $\varepsilon > 0$,

if $n >$ integral part of $\left[\dfrac{1}{\varepsilon} - 1\right] = n_0$.

$\therefore \ |x_m - x_n| < \varepsilon \ \forall \ m > n > n_0$ and hence by Cauchy's criterion the sequence converges.

Observation: We have proved that if a sequence of real numbers is a Cauchy sequence, then it must have a limit (which is also a real number). This property is called the **property of completeness : Real number system is complete.** In fact, any system S is called **complete** if every Cauchy sequence in S has a limit l, where $l \in S$. But note that the system $\mathbb{Q}$ of rational numbers is not complete since,

the sequence $\left\{\left(1 + \dfrac{1}{n}\right)^n\right\}_n$

is a sequence of rational numbers and its limit is e which is *not a rational number*. We may further remark that the **lub property** and the **property of completeness** are logically equivalent.

5.10 Cauchy's Limit Theorems

Theorem 5.10.1. *(Cauchy's First Theorem). If the sequence $\{x_n\}$ has a limit l, then the sequence $\{y_n\}$, where $y_n = \dfrac{x_1 + x_2 + \cdots + x_n}{n}$ also has the same limit l.*

[C.H. 1991, 95, 99]

Proof.

First Step. Let $l = 0$; then, given any positive number ε, however small, we can find an integer m such that

$$|x_n| < \varepsilon/2, \text{ for all } n > m.$$

Now
$$y_n = \frac{x_1 + x_2 + x_3 + \cdots + x_m + x_{m+1} + x_{m+2} + \cdots + x_n}{n}$$

or, $|y_n| \leq \dfrac{|x_1 + x_2 + \cdots + x_m|}{n} + \dfrac{|x_{m+1}| + |x_{m+2}| + \cdots + |x_n|}{n}.$

As m is fixed, we can choose N_1 so that

(a) $\dfrac{|x_1 + x_2 + \cdots + x_m|}{n} < \dfrac{\varepsilon}{2}$ for $n > N_1$; and

(b) $\dfrac{|x_{m+1}| + |x_{m+2}| + \cdots + |x_n|}{n} < \dfrac{\varepsilon}{2} \cdot \dfrac{n-m}{n}$

[as each of $(n-m)$ terms of the numerator $< \varepsilon/2$].

Thus for $n > \max. (N_1, m)$, we have

$$|y_n| < \frac{\varepsilon}{2} + \frac{\varepsilon}{2} \cdot \frac{n-m}{n} < \varepsilon.$$

Therefore, $y_n \to 0$ as $n \to \infty$.

Second Step. Let $l \neq 0$, then $(x_n - l) \to 0$ and by the result proved in the first step:

$$\frac{(x_1 - l) + (x_2 - l) + \cdots + (x_n - l)}{n} = (y_n - l) \to 0$$

i.e., $y_n = \dfrac{x_1 + x_2 + \cdots + x_n}{n} \to l$ as $n \to \infty.$

Hence the theorem.

An Important Corollary

If the terms $x_1, x_2, x_3, \ldots, x_n, \ldots$ of the sequence $\{x_n\}$ are all positive and $x_n \to l(\neq 0)$, then

$$y_n = \sqrt[n]{x_1 x_2 \cdots x_n} \to l. \qquad \text{[C.H. 1983]}$$

Proof.
We know that $\log x (x > 0)$ is a continuous function.
So, since $x_n \to l$, $\log x_n \to \log l$ as $n \to \infty$.

$$\therefore \frac{\log x_1 + \log x_2 + \cdots + \log x_n}{n} \to \log l, \text{ as } n \to \infty$$

or, $\log(x_1 \cdot x_2 \cdot x_3 \cdots x_n)^{1/n} \to \log l$, as $n \to \infty$.

$\therefore (x_1 x_2 x_3 \cdots x_n)^{1/n} \to l$ as $n \to \infty$ (again, by using continuity).

CHAPTER 5: SEQUENCE OF REAL NUMBERS

Applications

1. Since $\frac{1}{n} \to 0$, as $n \to \infty$. $\therefore \frac{1+\frac{1}{2}+\frac{1}{3}+\cdots+\frac{1}{n}}{n} \to 0$.

2. Since $\frac{n}{n-1} \to 1$, as $n \to \infty$. $\therefore \sqrt[n]{\frac{2}{1} \cdot \frac{3}{2} \cdots \frac{n}{n-1}} \to 1$ i.e., $\sqrt[n]{n} \to 1$.

3. $\frac{1+\sqrt[2]{2}+\sqrt[3]{3}+\cdots+\sqrt[n]{n}}{n} \to 1$, (using 2).

4. Since $\left(1+\frac{1}{n}\right)^n \to e$, as $n \to \infty$.

$$\therefore \sqrt[n]{\frac{2}{1} \cdot \left(\frac{3}{2}\right)^2 \cdot \left(\frac{4}{3}\right)^3 \cdots \left(\frac{n+1}{n}\right)^n} = \frac{n+1}{\sqrt[n]{n!}} \to e \text{ and hence } \frac{\sqrt[n]{n!}}{n} \to \frac{1}{e}.$$

5. If $x_n \to \infty$ or $-\infty$, then also $y_n = \frac{x_1+x_2+\cdots x_n}{n} \to \infty$ or $-\infty$.

Note 5.10.1. *The converse of this theorem is not necessarily true.*

1. thus with $x_{2n-1} = 1, x_{2n} = 0, \{x_n\}$ oscillates, but $\left\{\frac{x_1+x_2+\cdots+x_n}{n}\right\} \to \frac{1}{2}$ as $n \to \infty$ either through odd or even integers.

2. with $x_{2n} = +1, x_{2n-1} = -1$ {i.e., $x_n = (-1)^n$}, $\{x_n\}$ oscillates but

$$\frac{x_1 + x_2 + \cdots + x_n}{n} \to 0 \text{ as } n \to \infty.$$

Theorem 5.10.2. *(Cauchy's Second Theorem).* If all the terms of a sequence $\{x_n\}$ are positive and if $\lim\limits_{n\to\infty} \left(\frac{x_{n+1}}{x_n}\right)$ exists, so also does $\lim\limits_{n\to\infty} x_n^{\frac{1}{n}}$, and the two limits are equal.

Proof.

1. Suppose $\lim\limits_{n\to\infty} \frac{x_{n+1}}{x_n} = l$ (finite and $\neq 0$).

   For any given positive number ε, we can find an integer m, s.t.

   $$\left|\frac{x_{n+1}}{x_n} - l\right| < \frac{1}{2}\varepsilon \text{ for } n \geq m.$$

   Thus, each of the ratios

   $$\frac{x_{m+1}}{x_m}, \frac{x_{m+2}}{x_{m+1}}, \ldots, \frac{x_n}{x_{n-1}}$$

   lies between $l - \varepsilon/2$ and $l + \varepsilon/2$. Then, on multiplication, we have

   $$\left(l - \frac{\varepsilon}{2}\right)^{n-m} < \frac{x_n}{x_m} < \left(l + \frac{\varepsilon}{2}\right)^{n-m} \qquad (5.10.1)$$

Now first part of the inequality (5.10.1) can be written as

$$\left(l-\frac{\varepsilon}{2}\right)^n < \frac{x_n}{x_m}\left(l-\frac{\varepsilon}{2}\right)^m, \text{ which is } < \frac{x_n}{x_m}l^m$$

and the second part of the inequality (5.10.1) gives

$$\left(l+\frac{\varepsilon}{2}\right)^n > \frac{x_n}{x_m}\left(l+\frac{\varepsilon}{2}\right)^m, \text{ which is } > \frac{x_n}{x_m}l^m.$$

$$\therefore \left(l-\frac{\varepsilon}{2}\right)^n < \frac{x_n}{x_m}l^m < \left(l+\frac{\varepsilon}{2}\right)^n$$

or, $\left(\frac{x_m}{l^m}\right)^{\frac{1}{n}}\left(l-\frac{\varepsilon}{2}\right) < (x_n)^{\frac{1}{n}} < \left(l+\frac{\varepsilon}{2}\right)\left(\frac{x_m}{l^m}\right)^{\frac{1}{n}}.$ \hfill (5.10.2)

Now we know that as $n \to \infty$, $\left(\frac{x_m}{l^m}\right)^{\frac{1}{n}} \to 1$ and hence

$$\left(\frac{x_m}{l^m}\right)^{\frac{1}{n}}\left(l-\frac{\varepsilon}{2}\right) \to \left(l-\frac{\varepsilon}{2}\right) \text{ as } n \to \infty$$

and $\left(\frac{x_m}{l^m}\right)^{\frac{1}{n}}\left(l+\frac{\varepsilon}{2}\right) \to \left(l+\frac{\varepsilon}{2}\right)$ as $n \to \infty$.

$\therefore \exists\, n_0 \geq m$ such that

$$\left(l-\frac{\varepsilon}{2}\right) - \frac{\varepsilon}{2} < \left(\frac{x_m}{l^m}\right)^{\frac{1}{n}}\left(l-\frac{\varepsilon}{2}\right) < \left(l-\frac{\varepsilon}{2}\right) + \frac{\varepsilon}{2}, \text{ (if } n > n_0);$$

and $\left(l+\frac{\varepsilon}{2}\right) - \frac{\varepsilon}{2} < \left(\frac{x_m}{l^m}\right)^{\frac{1}{n}}\left(l+\frac{\varepsilon}{2}\right) < \left(l+\frac{\varepsilon}{2}\right) + \frac{\varepsilon}{2}$, (if $n > n_0$).

Hence using (5.11.2) also

$$l - \varepsilon < \left(\frac{x_m}{l^m}\right)^{\frac{1}{n}}\left(l-\frac{\varepsilon}{2}\right) < (x_n)^{\frac{1}{n}} < \left(\frac{x_m}{l^m}\right)^{\frac{1}{n}}\left(l+\frac{\varepsilon}{2}\right) < l+\varepsilon \text{ (if } n > n_0)$$

or, $l - \varepsilon < (x_n)^{\frac{1}{n}} < l + \varepsilon$, $\forall\, n > n_0$

or, $\lim_{n\to\infty} x_n^{\frac{1}{n}} = l$.

2. Again suppose $\lim \frac{x_{n+1}}{x_n} = \infty$, we can then find m, so that $\frac{x_{n+1}}{x_n} > 2G$ (an arbitrary large positive number given in advance), if $n \geq m$.

Hence, as above, $\frac{x_n}{x_m} > (2G)^{n-m}$, or, $x_n^{\frac{1}{n}} > (2G)\left\{\frac{x_m}{(2G)^m}\right\}^{\frac{1}{n}}$.

But as $n \to \infty$, $\lim \left\{\frac{x_m}{(2G)^m}\right\}^{\frac{1}{n}} = 1$, so that we can find $n_0 \geq m$ such that

$$\left\{\frac{x_m}{(2G)^m}\right\}^{\frac{1}{n}} > \frac{1}{2}, \text{ if } n > n_0.$$

CHAPTER 5: SEQUENCE OF REAL NUMBERS

Then $x_n^{\frac{1}{n}} > G$, if $n > n_0$ or, $\lim_{n \to \infty} x_n^{\frac{1}{n}} = \infty$.

3. The case when $\lim \frac{x_{n+1}}{x_n} = 0$ can be reduced to case (2) by writing $x_n = 1/y_n$.

4. A more general result is given below:

$$\underline{\lim} \frac{x_{n+1}}{x_n} \leq \underline{\lim} \sqrt[n]{x_n} \leq \overline{\lim} \sqrt[n]{x_n} \leq \overline{\lim} \frac{x_{n+1}}{x_n}.$$

The inner inequality is obvious and hence it is usual to write more shortly:

$$\underline{\lim} \frac{x_{n+1}}{x_n} \leq \underline{\lim} \sqrt[n]{x_n} \leq \overline{\lim} \frac{x_{n+1}}{x_n}.$$

Proof. If $\underline{\lim} \frac{x_{n+1}}{x_n} = \lambda$ and $\overline{\lim} \frac{x_{n+1}}{x_n} = \mu$, we can find a suitable positive integer m s.t.

$$\lambda - \varepsilon < \frac{x_{n+1}}{x_n} < \mu + \varepsilon, \text{ if } n > m.$$

Repeating the previous arguments, we find that

$$x_n^{\frac{1}{n}} > \left(\frac{x_m}{l^m}\right)^{\frac{1}{n}} (\lambda - \varepsilon) > \lambda - 2\varepsilon, \text{ if } n > n_0.$$

and $x_n^{\frac{1}{n}} < \left(\frac{x_m}{l^m}\right)^{\frac{1}{n}} (\mu + \varepsilon) < \mu + 2\varepsilon$, if $n > n_0$.

$$\therefore \underline{\lim} x_n^{\frac{1}{n}} \geq \lambda; \quad \text{and} \quad \overline{\lim} x_n^{\frac{1}{n}} \leq \mu.$$

Applications

**1. $\lim_{n \to \infty} \frac{1}{n} (n!)^{\frac{1}{n}}$.

Write $x_n = \frac{n!}{n^n}$ so that $\frac{x_{n+1}}{x_n} = \frac{n^n}{(n+1)^n} = \left(1 + \frac{1}{n}\right)^{-n}$.

$$\therefore \lim_{n \to \infty} \frac{1}{n} \sqrt[n]{n!} = \lim_{n \to \infty} x_n^{\frac{1}{n}} = \lim_{n \to \infty} \frac{x_{n+1}}{x_n} = \lim_{n \to \infty} \frac{1}{\left(1 + \frac{1}{n}\right)^n} = \frac{1}{e}.$$

2. $\lim_{n \to \infty} \frac{1}{n} \{(m+1)(m+2) \cdots (m+n)\}^{\frac{1}{n}} = \frac{1}{e}$.

Write $x_n = \frac{(m+1)(m+2) \cdots (m+n)}{n^n}$ and proceed as before.

3. $\lim_{n \to \infty} \frac{1}{n} \{(n+1)(n+2) \cdots (2n)\}^{\frac{1}{n}} = \frac{4}{e}$

for, $\frac{x_{n+1}}{x_n} = \frac{2(2n+1)}{n+1} \left(1 + \frac{1}{n}\right)^{-n}$.

**4. $\lim_{n\to\infty} \left(\dfrac{2n!}{n!n!}\right)^{\frac{1}{n}} = 4$

for, $\dfrac{x_{n+1}}{x_n} = \dfrac{(2n+2)(2n+1)}{(n+1)(n+1)} = \dfrac{\left(2+\frac{2}{n}\right)\left(2+\frac{1}{n}\right)}{\left(1+\frac{1}{n}\right)\left(1+\frac{1}{n}\right)} \to 4$ as $n \to \infty$.

Note 5.10.2. *The converse of this theorem is not necessarily true. Thus* $\lim \sqrt[n]{x_n}$ *may exist even when* $\lim \dfrac{x_{n+1}}{x_n}$ *does not. The sequence*

$$1, a, ab, a^2b, a^2b^2, \ldots, a^n b^{n-1}, a^n b^n, \ldots \text{ where } a \neq b.$$

the ratio of any term to the preceding is alternately a and b but $\sqrt[n]{x_n} \to \sqrt{ab}$ *as* $n \to \infty$.

Theorem 5.10.3. *If* $\{x_n\}$ *and* $\{y_n\}$ *be two null sequences and if* $\{y_n\}$ *is a **strictly monotone decreasing** sequence so that* $y_n > y_{n+1}$, *for all n, then*

$$\lim_{n\to\infty} \frac{x_n}{y_n} = \lim_{n\to\infty} \frac{x_n - x_{n+1}}{y_n - y_{n+1}},$$

provided the right-hand limit exists (limit may be finite or infinite).

Proof:

Case 1. Let $\lim_{n\to\infty} \dfrac{x_n - x_{n+1}}{y_n - y_{n+1}} = l$ (finite).

$\therefore$ given any $\varepsilon > 0$, there exists a positive integer m s.t.

$$l - \varepsilon < \frac{x_n - x_{n+1}}{y_n - y_{n+1}} < l + \varepsilon \text{ for all } n \geq m.$$

or, $(l-\varepsilon)(y_n - y_{n+1}) < x_n - x_{n+1} < (l+\varepsilon)(y_n - y_{n+1})$

[note that $y_n - y_{n+1} > 0$].

For n, write $n+1, n+2, \ldots, n+p-1$ and then add these p inequalities. We then have

$$(l-\varepsilon)(y_n - y_{n+p}) < x_n - x_{n+p} < (l+\varepsilon)(y_n - y_{n+p}).$$

Keep n fixed and let $p \to \infty$, then $x_{n+p} \to 0$ as $p \to \infty$ and also $y_{n+p} \to 0$ as $p \to \infty$.

$\therefore$ $(l-\varepsilon)y_n \leq x_n \leq (l+\varepsilon)y_n$, for $n \geq m$

or, $l - \varepsilon \leq \dfrac{x_n}{y_n} \leq l + \varepsilon$, for $n \geq m$

or, $\lim_{n\to\infty} \dfrac{x_n}{y_n} = l$.

Case 2. Let $\lim_{n\to\infty} \dfrac{x_n - x_{n+1}}{y_n - y_{n+1}} = \infty$.

CHAPTER 5: SEQUENCE OF REAL NUMBERS

Let G be any arbitrary positive number, no matter how large. Then there exists a positive integer m such that

$$x_n - x_{n+1} > G(y_n - y_{n+1}), \quad \text{for } n \geq m.$$

As in Case 1, we obtain $x_n - x_{n+p} > G(y_n - y_{n+p})$ and then make $p \to \infty$ (keeping n fixed); thus

$$x_n \geq Gy_n, \quad \text{for } n \geq m$$
$$\text{or,} \quad \frac{x_n}{y_n} \geq G, \quad \text{for } n \geq m$$
$$\text{i.e.,} \quad \lim_{n \to \infty} \frac{x_n}{y_n} = \infty.$$

Theorem 5.10.4. *If $\{y_n\}$ is a **strictly monotone increasing** sequence diverging to ∞ and $\{x_n\}$ is any arbitrary sequence, then*

$$\lim_{n \to \infty} \frac{x_n}{y_n} = \lim_{n \to \infty} \frac{x_{n+1} - x_n}{y_{n+1} - y_n}$$

provided the limit on the right exists (finite or infinite).

Proof:

Case 1. Let $\displaystyle\lim_{n \to \infty} \frac{x_{n+1} - x_n}{y_{n+1} - y_n} = l$ (finite).

$\therefore$ given any $\varepsilon > 0$, there corresponds a positive integer m_1 s.t.

$$l - \frac{1}{2}\varepsilon < \frac{x_{n+1} - x_n}{y_{n+1} - y_n} < l + \frac{1}{2}\varepsilon, \quad \text{for } n \geq m_1.$$

$$\text{or,} \quad \left(l - \frac{1}{2}\varepsilon\right)(y_{n+1} - y_n) < x_{n+1} - x_n < \left(l + \frac{1}{2}\varepsilon\right)(y_{n+1} - y_n).$$

For n, write $n+1, n+2, \ldots, n+p-1$ and then add these p inequalities. We then get

$$\left(l - \frac{1}{2}\varepsilon\right)(y_{n+p} - y_n) < x_{n+p} - x_n < \left(l + \frac{1}{2}\varepsilon\right)(y_{n+p} - y_n)$$

$$\text{or,} \quad \left(l - \frac{1}{2}\varepsilon\right)\left(1 - \frac{y_n}{y_{n+p}}\right) < \frac{x_{n+p}}{y_{n+p}} - \frac{x_n}{y_{n+p}} < \left(l + \frac{1}{2}\varepsilon\right)\left(1 - \frac{y_n}{y_{n+p}}\right)$$

$$\text{or,} \quad \left(l - \frac{1}{2}\varepsilon\right)\left(1 - \frac{y_n}{y_{n+p}}\right) + \frac{x_n}{y_{n+p}} < \frac{x_{n+p}}{y_{n+p}} < \left(l + \frac{1}{2}\varepsilon\right)\left(1 - \frac{y_n}{y_{n+p}}\right) + \frac{x_n}{y_{n+p}}$$

for all $n \geq m_1$ and for $p > 0$.

Keep n fixed and let $p \to \infty$. Then, since

$$\left(l - \frac{1}{2}\varepsilon\right)\left(1 - \frac{y_n}{y_{n+p}}\right) + \frac{x_n}{y_{n+p}} \to l - \frac{1}{2}\varepsilon$$

$$\left(l + \frac{1}{2}\varepsilon\right)\left(1 - \frac{y_n}{y_{n+p}}\right) + \frac{x_n}{y_{n+p}} \to l + \frac{1}{2}\varepsilon.$$

We can find a positive integer m_2 s.t., for every $p \geq m_2$ we have

$$l - \frac{1}{2}\varepsilon - \frac{1}{2}\varepsilon < \left(l - \frac{1}{2}\varepsilon\right)\left(1 - \frac{y_n}{y_{n+p}}\right) + \frac{x_n}{y_{n+p}} < l - \frac{1}{2}\varepsilon + \frac{1}{2}\varepsilon$$

$$l + \frac{1}{2}\varepsilon - \frac{1}{2}\varepsilon < \left(l + \frac{1}{2}\varepsilon\right)\left(1 - \frac{y_n}{y_{n+p}}\right) + \frac{x_n}{y_{n+p}} < l + \frac{1}{2}\varepsilon + \frac{1}{2}\varepsilon$$

whence it follows

$$l - \varepsilon < \frac{x_{n+p}}{y_{n+p}} < l + \varepsilon, \quad \text{for every } n \geq m_1 \text{ and } p \geq m_2.$$

$$\therefore\ l - \varepsilon < \frac{x_n}{y_n} < l + \varepsilon, \quad \text{for every } n \geq (m_1 + m_2).$$

$$\therefore\ \lim_{n \to \infty} \frac{x_n}{y_n} = l.$$

Case 2. Let $\lim\limits_{n \to \infty} \dfrac{x_{n+1} - x_n}{y_{n+1} - y_n} = \infty.$

Proceed as in case 2 of Theorem **5.13.3**.

Applications

Apply the above results to prove

(i) $\lim\limits_{n \to \infty} \dfrac{\log n}{n} = 0;$ (ii) $\lim\limits_{n \to \infty} \dfrac{e^n}{n} = \infty.$

Observation: Cauchy's first theorem (i.e., Theorem **5.13.1**) is a particular case of the present theorem (i.e., Theorem **5.13.4**). Replace x_n by $(x_1 + x_2 + \cdots + x_n)$ and y_n by n in Theorem **5.13.4** and thereby obtain Theorem **5.13.1**.

For Cauchy's second theorem write,

$$x_n = \frac{x_1}{1} \cdot \frac{x_2}{x_1} \cdot \frac{x_3}{x_2} \cdots \frac{x_n}{x_{n-1}}.$$

$$\therefore\ \log \sqrt[n]{x_n} = \frac{\log(x_1/1) + \log(x_2/x_1) + \cdots + \log(x_n/x_{n-1})}{n}.$$

Now deduce, if $x_n/x_{n-1} \to l$, then $\sqrt[n]{x_n} \to l$.

5.11 Miscellaneous Illustrative Examples

Example 5.11.1. *Given two real numbers x_1, x_2 such that $0 < x_1 < x_2$. Construct a sequence $\{x_n\}$, where x_1, x_2 are given as above and for $n > 2$,*

$$x_n = \frac{x_{n-1} + x_{n-2}}{2}.$$

Prove that the sequence $\{x_n\}$ is convergent and that it converges to $\frac{1}{3}(x_1 + 2x_2)$.

Solution: Step 1. Given $0 < x_1 < x_2$ and $x_3 = \frac{1}{2}(x_1 + x_2) = $ A.M. of x_1, x_2.

$$\therefore \quad x_1 < x_3 < x_2.$$

Now $x_3 < x_2$ similarly implies $x_3 < x_4 < x_2$
 $x_3 < x_4$ implies $x_3 < x_5 < x_4$
 $x_5 < x_4$ implies $x_5 < x_6 < x_4$
 etc. etc. etc.

$\therefore$ we have $x_1 < x_3 < x_5 < x_7 < \cdots < x_6 < x_4 < x_2$. i.e., the sequence $\{x_1, x_3, x_5, x_7, \ldots\}$ is m.i. and the sequence $\{x_2, x_4, x_6, x_8, \ldots\}$ is m.d.

Both are bounded by x_1 and x_2. Hence both the sequence converge to finite limits.

Step 2. Now see that

$$x_2 - x_1 > 0$$

$$x_3 - x_2 = \frac{1}{2}(x_1 + x_2) - x_2 = -\frac{1}{2}(x_2 - x_1)$$

$$x_4 - x_3 = \frac{1}{2}(x_3 + x_2) - x_3 = -\frac{1}{2}(x_3 - x_2) = \left(-\frac{1}{2}\right)^2 (x_2 - x_1)$$

$$\cdots\cdots\cdots\cdots\cdots\cdots\cdots\cdots\cdots\cdots\cdots\cdots\cdots\cdots\cdots\cdots$$

$$x_n - x_{n-1} = \left(-\frac{1}{2}\right)^{n-2} (x_2 - x_1).$$

$$\therefore \quad x_n - x_2 = (x_2 - x_1)\left[-\frac{1}{2} + \left(-\frac{1}{2}\right)^2 + \left(-\frac{1}{2}\right)^3 + \cdots + \left(-\frac{1}{2}\right)^{n-2}\right]$$

or, $(x_n - x_2) + (x_2 - x_1) = (x_2 - x_1)\left[1 - \frac{1}{2} + \frac{1}{4} - \frac{1}{8} + \cdots + (-1)^{n-2}\frac{1}{2}\right]$

$$= (x_2 - x_1)\frac{1 - (-\frac{1}{2})^{n-1}}{1 - (-\frac{1}{2})} = \frac{2}{3}(x_2 - x_1)\left[1 + \left(\frac{1}{2}\right)^n\right]$$

i.e., $\quad x_n - x_1 = \frac{2}{3}(x_2 - x_1)\left[1 + \left(\frac{1}{2}\right)^n\right].$

Since $\lim\limits_{n\to\infty} x_n$ exists and $\lim\limits_{n\to\infty} \left(\dfrac{1}{2}\right)^n = 0$, we obtain

$$\lim_{n\to\infty} x_n = x_1 + \frac{2}{3}(x_2 - x_1) = \frac{x_1 + 2x_2}{3}.$$

Note 5.11.1. *Observe that we first prove that $\{x_n\}$ converges and then show that the limit to which it converges is $\frac{1}{3}(x_1 + 2x_2)$. But instead of using the method of Step 1 to prove that the sequence converges, we could prove that the given sequence is a Cauchy sequence and we know every Cauchy sequence converges. Hence the sequence will converge.*

To prove that the sequence $\{x_n\}$ defined in **Ex. 5.14.1** is a Cauchy sequence. As in Step 2, we obtain

$$x_2 - x_1 > 0$$

$$x_3 - x_2 = \left(-\frac{1}{2}\right)(x_2 - x_1)$$

$$x_4 - x_3 = \left(-\frac{1}{2}\right)^2 (x_2 - x_1)$$

$$\cdots\cdots\cdots\cdots\cdots\cdots$$

$$x_n - x_{n-1} = \left(-\frac{1}{2}\right)^{n-2} (x_2 - x_1).$$

Consider two natural numbers m, n and take $m > n$.
Then

$$|x_m - x_n| = |x_m - x_{m-1} + x_{m-1} - x_{m-2} + x_{m-2} - x_{m-3} + \cdots + x_{n+1} - x_n|$$

$$\leq |x_m - x_{m-1}| + |x_{m-1} - x_{m-2}| + \cdots + |x_{n+1} - x_n|$$

$$= \left\{\left(\frac{1}{2}\right)^{m-2} + \left(\frac{1}{2}\right)^{m-3} + \cdots + \left(\frac{1}{2}\right)^{n-1}\right\} |x_2 - x_1|$$

$$= \left(\frac{1}{2}\right)^{n-1} \left[1 + \frac{1}{2} + \frac{1}{2^2} + \cdots + \left(\frac{1}{2}\right)^{m-n-1}\right] |x_2 - x_1|$$

$$= \left(\frac{1}{2}\right)^{n-1} \frac{1 - \left(\frac{1}{2}\right)^{m-n}}{1 - \frac{1}{2}} |x_2 - x_1| = \frac{4}{2^n}\left[1 - \left(\frac{1}{2}\right)^{m-n}\right] |x_2 - x_1|$$

$$< \frac{4(x_2 - x_1)}{2^n}.$$

Given any $\varepsilon > 0$, we can find a natural number k such that

$$\frac{4}{2^n}(x_2 - x_1) < \varepsilon, \quad \forall n \geq k.$$

$$\therefore \quad |x_m - x_n| < \varepsilon, \quad \forall m, n \geq k.$$

This proves that the sequence $\{x_n\}$ is a Cauchy sequence.

We give below *two* similar problems and ask the students to solve them by method similar to **Ex. 5.11.1** above.

** **Example 5.11.2.** *Given $0 < x_1 < x_2$; establish the following results:*
(a) *if each $x_n =$ the geometric mean of $x_{n-1}, x_{n-2} = \sqrt{x_{n-1} x_{n-2}}$ $(n > 2)$, then $\{x_n\}$ converges and the limit is $\sqrt[3]{x_1 x_2^2}$.*
(b) *if each $x_n =$ the harmonic mean of $x_{n-1}, x_{n-2} = 2 \Big/ \left(\dfrac{1}{x_{n-1}} + \dfrac{1}{x_{n-2}} \right)$, then $\{x_n\}$ converges to $\dfrac{\frac{1}{3} x_1 x_2}{2x_1 + x_2}$.*

** **Example 5.11.3.** *If $x_{n+1} = \sqrt{k + x_n}$, where x_1 and k are positive, show that the sequence $\{x_n\}$ is increasing or decreasing according as x_1 is less or greater than the positive root of $x^2 - x - k = 0$ and has, in either case, this root as its limit.*

Solution: Here $x_{n+1}^2 = k + x_n$ and hence, $x_n^2 = k + x_{n-1}$.

$$\therefore \quad x_{n+1}^2 - x_n^2 = x_n - x_{n-1}.$$

This implies that $x_{n+1} \gtreqless x_n$ according as $x_n \gtreqless x_{n-1}$.

Similarly, $x_n \gtreqless x_{n-1}$ according as $x_{n-1} \gtreqless x_{n-2}$ and so on.

Thus, if $x_2 > x_1$, then the sequence $\{x_n\}$ is monotone increasing and, if $x_2 < x_1$, then it is monotone decreasing.

Now, suppose $x_2 > x_1$, i.e., $\sqrt{k + x_1} > x_1$; i.e., $x_1^2 - x_1 - k < 0$, which is true, if x_1 lies between the roots of $x^2 - x - k = 0$.

The constant term being negative, one of its roots will be positive and the other negative. Since x_1 is given to be positive, it is less than the positive root of the equation. Exactly, in a similar way we can show that if x_1 is greater than the positive root, then $x_2 < x_1$ and so $\{x_n\}$ is decreasing.

Now supposing that the sequence is increasing, we have

$$x_{n+1} > x_n, \quad \text{i.e.,} \quad \sqrt{x_n + k} > x_n,$$

i.e., x_n is less than the positive root of the equation, as before.

Thus, although $\{x_n\}$ is increasing, still its terms are all less than the positive root of the equation $x^2 - x - k = 0$. In other words, $\{x_n\}$ is bounded above.

$\therefore$ $\{x_n\}$ tends to a finite limit.

To find the actual limit, let $x_n \to l$; and then a *fortiori* $x_{n+1} \to l$.

Hence the relation $x_{n+1} = \sqrt{k + x_n}$ becomes in the limit $l = \sqrt{l + k}$, i.e., $l^2 = l + k$.

In other words, l is the positive root of the equation $x^2 - x - k = 0$.

The case for $\{x_n\}$ decreasing can be treated in exactly similar manner.

Hence our result.

Corollary 5.11.1. *Exactly, in a similar way we can prove that the sequence*

$$\sqrt{2}, \sqrt{2\sqrt{2}}, \sqrt{2\sqrt{2\sqrt{2}}}, \ldots \text{ tends to } 2.$$ [C.H. 1981, '99]

(Here $x_n = \sqrt{2x_{n-1}}$)

** Example 5.11.4. *If*

$$x_n = 1 + \frac{1}{2} + \frac{1}{3} + \cdots + \frac{1}{n} - \log n$$

$$y_n = 1 + \frac{1}{2} + \frac{1}{3} + \cdots + \frac{1}{n-1} - \log n,$$

then prove that $\{x_n\}$ is monotone decreasing and $\{y_n\}$ is monotone increasing and that each converges to the same limit γ (Euler's Constant), where $0.3 < \gamma < 1$.

Solution:

$$x_{n+1} - x_n = \frac{1}{n+1} - \log(n+1) + \log n = \frac{1}{n+1} - \log\left(1 + \frac{1}{n}\right)$$

and $\quad y_{n+1} - y_n = \frac{1}{n} - \log\left(1 + \frac{1}{n}\right).$

We shall use the inequality

$$\frac{1}{n+1} < \log\left(1 + \frac{1}{n}\right) < \frac{1}{n}.$$

[See that

$$e^{1/n} > 1 + \frac{1}{n}$$

i.e., $\dfrac{1}{n} > \log\left(1 + \dfrac{1}{n}\right)$

and $\log \dfrac{n+1}{n} = -\log \dfrac{n}{n+1} = -\log\left(1 - \dfrac{1}{n+1}\right)$

$$= \frac{1}{n+1} + \frac{1}{2}\frac{1}{(n+1)^2} + \cdots > \frac{1}{n+1}]$$

CHAPTER 5: SEQUENCE OF REAL NUMBERS

Then
$$x_{n+1} - x_n < 0 \quad \text{and} \quad y_{n+1} - y_n > 0.$$

$\therefore$ $\{x_n\}$ is monotone decreasing and $\{y_n\}$ is monotone increasing.

Also since
$$\frac{1}{n} > \log(n+1) - \log n$$
$$\sum_{r=1}^{n} \frac{1}{r} > \log(n+1) > \log n.$$

$\therefore$ x_n is positive and since $\{x_n\}$ is monotone decreasing it converges to a limit γ where $\gamma \geq 0$. But $x_1 = 1$ and so $\gamma < 1$.

Also $x_n - y_n = \frac{1}{n}$ and hence $\lim_{n \to \infty} (x_n - y_n) = 0$. Therefore, $\{y_n\}$ converges to the same limit γ. But $y_2 = 1 - \log 2 > 0.3$.

$$\therefore \gamma > 0.3.$$

The common limit of the two sequences is γ, called *Euler's Constant*. Its value is $0.5772157\cdots$. See that $1 + \frac{1}{2} + \frac{1}{3} + \cdots + \frac{1}{n} \longrightarrow \infty$, and $\log n \longrightarrow \infty$ but γ turns out to be finite and its value is approximately 0.577216.

Example 5.11.5. $x_n = \dfrac{2.4.6 \cdots 2n}{1.3.5 \cdots (2n-1)}$. To prove $x_n \to \infty$ as $n \to \infty$.

Solution: If $x > 0, y > 0$, then $\dfrac{x}{y} > \dfrac{x+1}{y+1}$, if $x > y$.

$$\therefore \frac{2}{1} > \frac{3}{2}, \quad \frac{4}{3} > \frac{5}{4}, \ldots, \frac{2n}{2n-1} > \frac{2n+1}{2n}.$$

$$\therefore x_n^2 = x_n \cdot x_n$$
$$> x_n \cdot \frac{3}{2} \cdot \frac{5}{4} \cdots \frac{2n+1}{2n}$$
$$> \frac{2.4.6 \cdots 2n}{1.3.5 \cdots (2n-1)} \cdot \frac{3}{2} \cdot \frac{5}{4} \cdots \frac{2n+1}{2n}$$
$$> (2n+1);$$
or, $x_n > \sqrt{2n+1}$,

which can be made greater than any large positive number G, if $2n + 1 > G^2$, i.e., if $n > \frac{1}{2}(G^2 - 1)$. That is $x_n \to \infty$ as $n \to \infty$.

Example 5.11.6. a) If $u_n = \dfrac{1.3.5\cdots(2n-1)}{2.4.6\cdots 2n}$ and $v_n = \dfrac{3.5.7\cdots(2n+1)}{2.4.6\cdots 2n}$, then prove that
$$u_n \to 0, v_n \to \infty \text{ and } \frac{1}{2} < u_n v_n < 1.$$

b) If $w_n = \dfrac{1.3.5\cdots(2n-1)}{2.4.6\cdots 2n}\sqrt{2n+1}$, then $w_n \to l$, where $\dfrac{1}{\sqrt{2}} < l < 1$.

Proceed as in **Ex. 5.11.5** above.

** Example 5.11.7. If $x_{n+1} = \dfrac{12}{1+x_n}$ $(n \geq 1)$ and $x_1 > 0$, prove that the two sequences $\{x_1, x_3, x_5, \ldots\}$ and $\{x_2, x_4, x_6, \ldots\}$ are monotonic-one increasing and the other decreasing. Also prove that $\{x_n\}$ converges to the positive root of the quadratic equation $x^2 + x - 12 = 0$.

Solution:

$$x_n - x_{n-2} = \frac{12}{1+x_{n-1}} - \frac{12}{1+x_{n-3}} = \frac{-12(x_{n-1}-x_{n-3})}{(1+x_{n-1})(1+x_{n-3})} \qquad (5.11.1)$$

$$= \frac{12^2(x_{n-2}-x_{n-4})}{(1+x_{n-1})(1+x_{n-2})(1+x_{n-3})(1+x_{n-4})}. \qquad (5.11.2)$$

The result (5.13.2) shows that even and odd terms form separate monotonic subsequences: Again (5.12.1) shows that if odd terms form a m.d. sub-sequence, even terms will form m.i. sub-sequence and vice versa. Since every term of the sequence is positive, $12 - x_n = \frac{12 x_{n-1}}{1+x_{n-1}} > 0 \Rightarrow 0 < x_n < 12$. Thus, m.i. sub-sequence is bounded below by 0.

$\therefore$ two sub-sequences are convergent (let even terms $\to l$ and odd terms $\to l'$).

For n even, taking limit $x_n = \dfrac{12}{1+x_{n-1}}$, we get $ll' + l = 12$. For n odd, $ll' + l' = 12$.

$$\therefore\ l = l'.$$

Thus $l = \dfrac{12}{1+l}$, i.e., l is a root of $x^2 + x - 12 = 0$ (l is positive).

Example 5.11.8. *(Cesaro's Theorem).* If $x_n \to x$ as $n \to \infty$ and $y_n \to y$ as $n \to \infty$, then prove that the sequence $\{z_n\}$, where

$$z_n = \frac{1}{n}(x_1 y_n + x_2 y_{n-1} + x_3 y_{n-2} + \cdots + x_n y_1),$$

converges to xy.

Solution: Write $x_n = x + a_n$ and let $|a_n| = A_n$. Then $x_n \to x \Rightarrow a_n \to 0$ and so A_n also tends to zero.

$$\frac{1}{n}(x_1 y_n + x_2 y_{n-1} + \cdots + x_n y_1)$$
$$= \frac{1}{n}\{(x + a_1)y_n + (x + a_2)y_{n-1} + \cdots + (x + a_n)y_1\}$$
$$= \frac{x}{n}(y_1 + y_2 + y_3 + \cdots + y_n) + \frac{1}{n}(a_1 y_n + a_2 y_{n-1} + \cdots + a_n y_1). \quad (5.11.3)$$

$$\left|\frac{1}{n}(a_1 y_n + a_2 y_{n-1} + \cdots + a_n y_1)\right| \le \frac{1}{n}\{|a_1||y_n| + \cdots + |a_n||y_1|\}$$
$$< \frac{k}{n}\{|a_1| + |a_2| + \cdots + |a_n|\}$$
$$(\because |y_n| < k, \; \forall\, n \text{ the seq. being convgt.})$$
$$< \frac{k}{n}(A_1 + A_2 + \cdots + A_n) \to 0 \text{ as } n \to \infty.$$

[$\because A_n \to 0$, therefore, $\frac{A_1 + A_2 + \cdots + A_n}{n} \to 0$ as $n \to \infty$ by Cauchy's First Theorem on limits.]

Now from (5.14) the result follows $\left(\because \frac{y_1 + y_2 + y_3 + \cdots + y_n}{n} \to y \text{ as } n \to \infty\right)$.

** Example 5.11.9. *Two sequences $\{x_n\}$ and $\{y_n\}$ are defined by*

$$x_n = \sqrt{x_{n-1} y_{n-1}} \quad \text{and} \quad \frac{2}{y_n} = \frac{1}{x_n} + \frac{1}{y_{n-1}} \quad (n \ge 2).$$

For $n = 1$ x_1 and y_1 are two fixed positive real numbers such that $0 < x_1 < y_1$. Prove that both the sequences $\{x_n\}$ and $\{y_n\}$ are convergent and tend to the same limit l where $x_1 < l < y_1$.

Solution: We know that for two positive real numbers a, b (say $a < b$)

$$\text{Geometric mean} = \sqrt{ab} = G \text{ (say)}$$
$$\text{and} \quad \text{Harmonic mean} = \frac{2}{\left(\frac{1}{a} + \frac{1}{b}\right)} = H \text{ (say)}.$$

They are connected by $a < H < G < b$.
In the present problem, $0 < x_1 < y_1$.

$$\therefore \quad x_1 < \sqrt{x_1 y_1} < y_1,$$
$$\text{i.e.,} \quad x_1 < x_2 < y_1. \;\; (\because x_2 = \sqrt{x_1 y_1})$$

Also

$$x_2 < \frac{2}{\frac{1}{x_2} + \frac{1}{y_1}} < y_1$$
$$\text{i.e.,} \quad x_2 < y_2 < y_1$$
$$\text{i.e.,} \quad x_1 < x_2 < y_2 < y_1.$$

Again
$$x_2 < y_2$$
$$\Rightarrow x_2 < x_3 < y_1 \quad \text{and} \quad x_3 < y_3 < y_1,$$
$$\text{i.e., } x_1 < x_2 < x_3 < y_3 < y_1.$$

It follows by induction that
$$x_{n-1} < x_n < y_n < y_{n-1} \quad (n = 2, 3, 4, \cdots).$$

Thus the sequence $\{x_n\}$ increases and bounded above by y_1 and the sequence $\{y_n\}$ decreases and bounded below by x_1. Hence, both sequences converge.

Suppose $x_n \to l$ as $n \to \infty$ and $y_n \to m$ as $n \to \infty$, then
$$l^2 = lm \quad \text{and} \quad \frac{2}{m} = \frac{1}{l} + \frac{1}{m}.$$

Both the sequences yield $l = m$. Clearly, $x_1 < l < y_1$.

** Example 5.11.10. *Arithmetico-Geometric Mean of two positive numbers. If u_1, v_1 are two given unequal positive numbers and*
$$u_n = \frac{1}{2}(u_{n-1} + v_{n-1}) \quad \text{and} \quad v_n = \sqrt{u_{n-1}v_{n-1}} \quad \text{for } n \geq 2,$$
then prove that the sequences $\{u_n\}$ and $\{v_n\}$ converge to a common limit l.

Proof. Here
$$u_n - v_n = \frac{1}{2}(u_{n-1} + v_{n-1}) - \sqrt{u_{n-1}v_{n-1}}$$
$$= \frac{1}{2}(\sqrt{u_{n-1}} - \sqrt{v_{n-1}})^2 > 0$$
i.e., $u_n > v_n$ for all n.

Again
$$u_n = \tfrac{1}{2}(u_{n-1} + v_{n-1}) < u_{n-1} \quad (\because v_{n-1} < u_{n-1})$$
$$\text{and } v_n = \sqrt{u_{n-1}v_{n-1}} > v_{n-1}. \quad (\because u_{n-1} > v_{n-1})$$

∴ it follows that
$$v_1 < v_2 < v_3 < \cdots < v_n < \cdots < u_n < u_{n-1} < \cdots < u_2 < u_1,$$

i.e., $\{u_n\}$ and $\{v_n\}$ are both monotone–the first decreasing and the second increasing and both are bounded. Hence both converge to limits, say, l and l' respectively.

Also
$$\lim_{n\to\infty} v_n = \lim_{n\to\infty} v_{n-1}$$
$$= \lim_{n\to\infty} (2u_n - u_{n-1})$$
$$= \lim_{n\to\infty} u_{n-1}, \quad \text{i.e.,} \quad l' = l.$$

Hence, etc.

Example 5.11.11. *If x_1 and y_1 are positive and $x_{n+1} = \frac{1}{2}(x_n + y_n)$, $y_{n+1} = \sqrt{x_n y_n}$ $(n \geq 1)$, show that the sequences $\{x_n\}$, $\{y_n\}$ converge to a common limit, where*
$$x > y > 0 \quad \text{and} \quad x_1 = \frac{1}{2}(x+y) \quad \text{and} \quad y_1 = \sqrt{xy}.$$

[*Hints.* Use A.M. $\geq$ G.M., the sequences $\{x_n\}$ and $\{y_n\}$ are m.d. and m.i. respectively.]

Example 5.11.12. *If x_1, y_1 are positive and if, for $n \geq 1$,*
$$x_{n+1} = \frac{x_n + y_n}{2}, \quad \frac{2}{y_{n+1}} = \frac{1}{x_n} + \frac{1}{y_n},$$
show that $\{x_n\}$ and $\{y_n\}$ are monotonic sequences which converge to the common limit $l = \sqrt{x_1 y_1}$.

[Proceed as in previous problems. Remember $A \geq G \geq H$.]

Exercises on Chapter 5A
[A]

1. Define the limit of a sequence. Use your definition to establish the following:

(a) $\lim_{n\to\infty} \dfrac{2n+5}{6n-11} = \dfrac{1}{3}$.

(b) $\lim_{n\to\infty} \dfrac{3n}{n+5\sqrt{n}} = 3$.

(c) $\lim_{n\to\infty} \dfrac{n^2+n+1}{3n^2+1} = \dfrac{1}{3}$.

2. *(a) Use Sandwich Theorem to prove
$$\lim_{n\to\infty} \left(\frac{1}{\sqrt{n^2+1}} + \frac{1}{\sqrt{n^2+2}} + \cdots + \frac{1}{\sqrt{n^2+n}} \right) = 1. \qquad \text{[C.H. 1982]}$$

*(b) Prove that $\lim_{n\to\infty}\left(\frac{1}{1}+\frac{1}{2}+\frac{1}{3}+\cdots+\frac{1}{n}\right) = \infty$, using the inequality

$$\frac{n}{2} < \frac{1}{2}+\frac{1}{3}+\frac{1}{4}+\cdots+\frac{1}{2^n} < n.$$

[Group the terms thus: $\frac{1}{2} + \left(\frac{1}{3}+\frac{1}{4}\right) + \left(\frac{1}{5}+\frac{1}{6}+\frac{1}{7}+\frac{1}{8}\right) + \cdots$]

3. (a) Define a monotone sequence. Prove that a monotone sequence is convergent, iff it is bounded. What happens when a monotone sequence is unbounded? Prove that the sequence $\{x_n\}$, where $x_n = (1+1/n)^n$, converges, and that

$$2 \leq \lim_{n\to\infty}\left(1+\frac{1}{n}\right)^n \leq 3.$$

(b) Prove that a m.i. sequence $\{x_n\} \to \sup x_n$ as $n \to \infty$. [C.H. 1980]

*4. **Important limits:**

(a) Evaluate $\lim_{n\to\infty} \sqrt[n]{n}$ and $\lim_{n\to\infty} \frac{\log n}{n^p}\,(p>0)$. [C.H. 1994]

[*Ans.* a) 1 and 0.]

(b) Prove that if $0 < a < 1$, then $na^n \to 0$ as $n \to \infty$.

(c) If $a > 0$, prove that $a^{1/n} \to 1$ as $n \to \infty$.

5. Using limit theorems prove that

(a) $\lim_{n\to\infty} \frac{6n^3+2n+1}{n^3+n^2} = 6$.

(b) $\lim_{n\to\infty} (\sqrt{n+1}-\sqrt{n})\sqrt{n+1/2} = \frac{1}{2}$.

(c) $\lim_{n\to\infty} (\sqrt[3]{n+1}-\sqrt[3]{n}) = 0$.

*(d) $\lim_{n\to\infty}\left(\frac{1}{n^2}+\frac{2}{n^2}+\cdots+\frac{n}{n^2}\right) = \frac{1}{2}$.

(e) $\lim_{n\to\infty}(\sqrt{n^2+n}-n) = \frac{1}{2}$.

(f) $\lim_{n\to\infty} \frac{3+2\sqrt{n+1}}{\sqrt{n+1}} = 2$.

6. Define a Cauchy sequence. Prove that the sequence $\{x_n\}$ converges, iff it is a Cauchy sequence. Use this criterion to prove that [C.H. 1992]

(a) $\{x_n\}$ does not converge, where $x_n = 1+\frac{1}{2}+\frac{1}{3}+\cdots+\frac{1}{n}$.

(b) $\left\{\dfrac{n-1}{n+1}\right\}$ converges. [C.H. 1983]

*(c) $\left\{1+\dfrac{1}{3}+\dfrac{1}{5}+\cdots+\dfrac{1}{2n-1}\right\}$ does not converge.

7. Point out with justifications which of the following sequences are Cauchy sequences and which are not:

(a) $\left\{\dfrac{n}{n+1}\right\}$.

*(b) $\left\{1+\dfrac{1}{1!}+\dfrac{1}{2!}+\cdots+\dfrac{1}{n!}\right\}$.

(c) $\{(-1)^n\}$.

(d) $\{(-1)^n \cdot n\}$. [Ans. a) and b) only are Cauchy sequences.]

8. Define upper limit μ and lower limit λ of a sequence. Determine these limits for the following sequences:

*(a) $\left\{(-1)^n + \dfrac{1}{n+1}\right\}$.

(b) $\left\{\dfrac{n}{n+1}\right\}$.

(c) $\left\{(-1)^n\left(1+\dfrac{1}{n}\right)\right\}$.

(d) $\{a + n(-1)^n\}$.

(e) $\left\{a + \dfrac{(-1)^n}{n}\right\}$.

(f) $\{(-1)^n n\}$.

*(g) $\left\{a - n^{(-1)^n}\right\}$.

(h) $\{x_n\}$, where $x_1 = 0$, $x_{2k} = \tfrac{1}{2}x_{2k-1}$, $x_{2k+1} = \tfrac{1}{2} + x_{2k}$.

[Ans. a) $\lambda = -1, \mu = +1$; b) $\lambda = \mu = 1$; c) $\lambda = -1, \mu = +1$; d) $\lambda = a, \mu = +\infty$; e) $\lambda = \mu = a$; f) $\lambda = -\infty, \mu = +\infty$; g) $\lambda = -\infty, \mu = a$; h) $\lambda = 0, \mu = 1/2$.]

9. Prove that a bounded sequence $\{x_n\}$ is convergent iff $\underline{\lim}\, x_n = \overline{\lim}\, x_n$.

10. If $x_n \to l$ as $n \to \infty$, then prove that $\dfrac{x_1 + x_2 + \cdots + x_n}{n} \to l$ as $n \to \infty$.

[C.H. 1991, '95]

If $\forall n, x_n > 0$, then $\sqrt[n]{x_1 \cdot x_2 \cdots x_n}$ also tends to l as $n \to \infty$.

Deduce that if $\dfrac{x_{n+1}}{x_n} \to l$ as $n \to \infty$, then $\sqrt[n]{x_n} \to l$ as $n \to \infty$. [C.H. 1983]

11. Prove that

(a) $\lim\limits_{n\to\infty} \left\{ \dfrac{2}{1}\left(\dfrac{3}{2}\right)^2 \left(\dfrac{4}{3}\right)^3 \cdots \left(\dfrac{n+1}{n}\right)^n \right\}^{1/n} = e.$

*(b) $\lim\limits_{n\to\infty} \left\{ \dfrac{3n!}{(n!)^3} \right\}^{1/n} = 27.$

(c) $\lim\limits_{n\to\infty} \dfrac{n}{(n!)^{1/n}} = e.$

(d) $\lim\limits_{n\to\infty} \left\{ \dfrac{(n+1)(n+2)\cdots(n+n)}{n} \right\}^{1/n} = \dfrac{4}{e}.$

*(e) $\lim\limits_{n\to\infty} \left(\dfrac{2n!}{n!n!}\right)^{1/n} = 4.$

12. Define sub-sequence of a given sequence $\{x_n\}$.

   (a) Show that if any sub-sequence of $\{x_n\}$ converges to a limit α, then any arbitrary neighbourhood of α contains x_n for an infinite number of values of n.

   *(b) Prove that if $\{x_n\}$ converges to l, then every sub-sequence of $\{x_n\}$ also converges to l.
   [C.H. 1994]

   (c) Prove that every sequence has a monotonic sub-sequence.

   *(d) Prove that every bounded sequence has a convergent sub-sequence.
   [C.H. 1990, '92, '96]

13. Define a nest of intervals. State and prove the theorem on Nested intervals. Does the theorem necessarily hold for a nest of open intervals?

   [Hints: $\bigcap\limits_{n=1}^{\infty} (0, 1/n) = \Phi$; $\bigcap\limits_{n=1}^{\infty} (-1/n, 1/n) = \{0\}$.]

*14. (a) If $x_n \to l$ ($l \neq 0$) as $n \to \infty$, then prove that $|x_n| \to |l|$ as $n \to \infty$. Show that, in general, the converse is not true.

   (b) If $x_n \to l$ as $n \to \infty$, and if $x_n \geq k$ (a fixed positive number), then show that $l \geq k$.

   (c) Give an example of a sequence $\{a_n\}$ which does not converge, but $\{|a_n|\}$ converges.

   (d) Construct a sequence with four distinct sub-sequential limits such that $\limsup x_n = -1$ and $\liminf x_n = -2$.

   (e) If $x_n \to l$ as $n \to \infty$, prove that $(1/x_n) \to 1/l$ as $n \to \infty$ ($l \neq 0$).

(f) If $(x_n \pm y_n)$ or $(x_n y_n)$ or (x_n/y_n) is convergent, does it always mean (x_n) and (y_n) are both convergent?

[Hints: For (f): (i) $x_n = n^2, y_n = -n^2, (x_n + y_n) \to 0$ and $(x_n/y_n) \to -1$, but neither (x_n) nor (y_n) converges. (ii) $x_n = y_n = (-1)^n$, $(x_n - y_n) \to 0, (x_n/y_n) \to 1$ and $(x_n y_n) \to 1$, but (x_n) and (y_n) do not converge.]

(g) Show that $\{a_n\}_n$ defined by $a_n = (1 - \frac{1}{n})\sin\frac{n\pi}{2}$ $(n = 1, 2, 3, \ldots)$ has a convergent subsequence but the sequence itself is not convergent.

[C.H. 1999]

[B]

1. Find upper and lower limits of the following sequences:

    (a) $\{[3 - (-1)^n]^n\}$.
    (b) $\{(-1)^n + \cos\frac{n\pi}{4}\}$.
    (c) $\{\cos\frac{n\pi}{4} + \cos n\pi\}$.
    (d) $\{(-1)^n (3 + \frac{5}{n})\}$.
    (e) $\{1, 2, \frac{1}{2}, 3, \frac{1}{3}, 4, \frac{1}{4}, 5, \frac{1}{5}, \ldots\}$.

   In each case state whether the sequence converges, diverges or oscillates.

2. If $x_{n+1} = \frac{k}{1+x_n}$, where x_1 and k are both positive, prove that $\{x_n\}$ converges to the positive root of $x^2 + x - k = 0$.

*3. A sequence $\{s_n\}$ is defined as follows:
$$s_1 = a > 0; \quad s_{n+1} = \sqrt{\frac{ab^2 + s_n^2}{a + 1}} \quad \forall\, n \in \mathbb{N}.$$

   Prove that

   (a) $\{s_n\}$ is m.d. and bounded, if $s_1 > b$.
   (b) $\{s_n\}$ is m.i. and bounded, if $s_1 < b$.
   (c) in either case the limit is b.

*4. The sequence $\{x_n\}$ is defined by
$$x_1 = \sqrt{7} \text{ and } \forall\, n \geq 2,\ x_n = \sqrt{7 + x_{n-1}}.$$

   Prove that the sequence $\{x_n\}$ converges to the positive root of the quadratic equation $t^2 - t - 7 = 0$.

[C.H. 1993]

5. Prove that

   *(a) $\lim_{n\to\infty} \left(\dfrac{1}{\sqrt{n}} + \dfrac{1}{\sqrt{n+1}} + \dfrac{1}{\sqrt{n+2}} + \cdots + \dfrac{1}{\sqrt{n+n}} \right) = \infty.$

   (b) $\lim_{n\to\infty} \dfrac{n!}{n^n} = 0.$

   (c) $\lim_{n\to\infty} \dfrac{n}{\alpha^n} = 0 \ (\alpha > 1).$

   (d) $\lim_{n\to\infty} \left\{ \dfrac{1}{n^2} + \dfrac{1}{(n+1)^2} + \cdots + \dfrac{1}{(2n)^2} \right\} = 0.$

*6. If $\{x_n\}$ and $\{y_n\}$ are bounded sequences, prove that

   (a) $\overline{\lim} \ x_n \cdot \overline{\lim} \ y_n \geq \overline{\lim} \ (x_n \cdot y_n).$

   (b) $\underline{\lim} \ x_n \cdot \underline{\lim} \ y_n \leq \underline{\lim} \ (x_n \cdot y_n).$

   (c) $\underline{\lim} \ x_n - \underline{\lim} \ y_n \leq \underline{\lim} \ (x_n - y_n) \leq \overline{\lim} \ (x_n - y_n) \leq \overline{\lim} \ x_n - \overline{\lim} \ y_n.$

7. Let $\{u_n\}$ and $\{v_n\}$ be two sequences ($u_n, v_n > 0, \forall \ n$), the first sequence being bounded above and for the second sequence $\lim_{n\to\infty} v_n = 1$. Prove that

$$\overline{\lim} \ (u_n \cdot v_n) = \overline{\lim} \ u_n.$$

*8. Given $0 < x_1 < x_2$, establish the following results:

   (a) if each $x_n = \dfrac{x_{n-1} + x_{n-2}}{2}$, then $\{x_n\} \to \dfrac{1}{3}(x_1 + 2x_2)$ as $n \to \infty$.

   (b) if each $x_n = \sqrt{x_{n-1} \cdot x_{n-2}}$, then $\{x_n\} \to \sqrt[3]{x_1 \cdot x_2^2}$ as $n \to \infty$.

   (c) if each $x_n = $ the harmonic mean of x_{n-1} and x_{n-2}, then

$$\{x_n\} \to \dfrac{(1/3)x_1 x_2}{2x_1 + x_2} \text{ as } n \to \infty.$$

*9. Prove that the two sequences $\{x_n\}$ and $\{y_n\}$ have a common limit, where

   (a) $x_{n+1} = \sqrt{x_n y_n}$ and $\dfrac{2}{y_{n+1}} = \dfrac{1}{x_{n+1}} + \dfrac{1}{y_n}$ $(n \geq 1, x_1 > 0, y_1 > 0)$.

   (b) $x_{n+1} = \dfrac{1}{2}(x_n + y_n)$ and $\dfrac{2}{y_{n+1}} = \dfrac{1}{x_n} + \dfrac{1}{y_n}$ $(n \geq 1, x_1 > 0, y_1 > 0)$.

   [C.H. 1990]

   In the last case prove that the limit is $\sqrt{x_1 y_1}$.

10. If $\limsup x_n = M$, then prove that the limit superior of no sub-sequence of $\{x_n\}$ can exceed M.

Main Points: Chapter 5A

Topic of Chapter 5A

1. **Definition of a sequence in $\mathbb{R}$:**

   It is a **map x** or a **function** $x : \mathbb{N} \to \mathbb{R}$.

   If we write $x(n)$, value of x at n, by the notation x_n, then the sequence is often written as $\{x_n\}_{n=1}^{\infty}$ or $\{x_n\}_n$ or $\{x_n\}$.

2. x_n are given by **formula** or by a **recursive** definition, e.g., Fibonacci Sequence $\{1, 1, 2, 3, 5, 8, \ldots\}$.

3. $\{x_n\}_n$ is a **Bounded Sequence** if $\exists K > 0$ such that $|x_n| < K, \forall n \in \mathbb{N}$.

4. **Limit of a sequence**: A sequence $\{x_n\}_n$ in $\mathbb{R}$ is said to converge to a *finite limit* $l \in \mathbb{R}$ or l is said to *the limit* of $\{x_n\}_n$ as $n \to \infty$ if for each $\epsilon > 0$, $\exists$ a natural number m such that $\forall n \geq m$, $|x_n - l| < \epsilon$. We then write $\lim\limits_{n \to \infty} x_n = l$. When such a finite limit l exists for a sequence $\{x_n\}_n$ then we say that the sequence is *convergent*. If it has no such limit then it is **non-convergent** (or *divergent*).

   A sequence in $\mathbb{R}$ can have at most one limit. When the limit $l = 0$, we call the sequence a null sequence.

5. A sequence $\{x_n\}$ may be *monotone increasing* (i.e., $x_{n+1} \geq x_n \; \forall n \in \mathbb{N}$) or *monotone decreasing* (i.e., $x_{n+1} \leq x_n \; \forall n \in \mathbb{N}$). However, a sequence $\{x_n\}_n$ need not always be monotone (e.g., $1, 0, 1, 0, \ldots$ is neither m.i. nor m.d.).

6. Important Examples of *Null Sequence* (Justify by definition in each case):

   $\{x^n, \text{ where } |x| < 1\}$; $\{nx^n, \text{ where } |x| < 1\}$; $\left\{\frac{\log n}{n}\right\}$; $\{\sqrt[n]{n} - 1\}$; $\{x^{1/n} - 1, x > 0\}$.

7. Every Convergent Sequence is bounded but not every bounded sequence is convergent, e.g., $\{(-1)^n\}$ bounded but not convergent.

8. **Non-convergent Sequences** are of two types:

   (a) *Properly Divergent* (**either** $x_n \to +\infty$ or $x_n \to -\infty$).

   (b) *Improperly Divergent* (bounded and oscillatory, e.g., $\{(-1)^n\}_n$) or unbounded and oscillatory, e.g., $\{(-1)^n 2^n\}$.

9. **Operations on Limits**: Sum, Difference, Product of two convergent sequences is also convergent but in case of **Quotient**:

remember: $\lim\limits_{n\to\infty} \dfrac{x_n}{y_n} = \dfrac{\lim\limits_{n\to\infty} x_n}{\lim\limits_{n\to\infty} y_n}$, provided $\lim\limits_{n\to\infty} y_n \neq 0$.

What about the converse? $x_n = (-1)^n$ and $y_n = (-1)^{n+1}$ are both not convergent sequences but $\{x_n + y_{n+1}\}$ converges to zero.

Inequalities: If $\lim\limits_{n\to\infty} x_n = l$ and every $x_n \geq 0$, then $l \geq 0$. If $\lim\limits_{n\to\infty} x_n = l$ and $\lim\limits_{n\to\infty} y_n = m$ and $\forall n$, $x_n < y_n$, then $\lim\limits_{n\to\infty} x_n \leq \lim\limits_{n\to\infty} y_n$ (i.e., $l \leq m$), i.e., limit operation makes the inequality weak (equality is possible, not just less than).

Sandwich Theorem: Three sequences $\{x_n\}, \{y_n\}_n, \{z_n\}_n$ where $\forall n$, $x_n \leq y_n \leq z_n$. If $x_n \to l$ and $z_n \to l$ as $n \to \infty$ then the sandwiched sequence $\{y_n\}$ converges to l as $n \to \infty$.

As an application try: $\lim\limits_{n\to\infty}\left(\dfrac{1}{\sqrt{n^2+1}} + \dfrac{1}{\sqrt{n^2+2}} + \cdots + \dfrac{1}{\sqrt{n^2+n}}\right) = 1$.

10. An useful result for evaluation of limits:

    If $\lim\limits_{n\to\infty}\left|\dfrac{x_n+1}{x_n}\right| = l$ when $0 < l < 1$, then $\lim\limits_{n\to\infty} x_n = 0$.

    Try: $\lim\limits_{n\to\infty} \dfrac{x^n}{\angle n} = 0$ or $\lim\limits_{n\to\infty} \dfrac{m(m-1)(m-2)\cdots(m-n+1)}{\angle n} x^n = 0$ if $|x| < 1$.

11. A very useful criterion:

    A monotone sequence can never oscillate—either it will converge to a infinite limit (when it is bounded) or it will properly diverge (when it is unbounded).

    Most Important Example: $\left\{\left(1+\frac{1}{n}\right)^n\right\}_n$ is m.i. and bounded and hence convergent. We write this limit by the letter e (exponential e).

    Another importance example: $\{x_n\}_n$ defined recursively $x_1 = \sqrt{2}, x_{n+1} = \sqrt{2x_n}$ $\forall n \geq 1$, then the sequence converges to 2.

12. **Nested Interval Property:** Theorem of Nested Intervals (The proof should be carefully read).

13. **Cauchy sequence:** $\{x_n\}_n$ is a Cauchy sequence if for any $\epsilon > 0$ $\exists$ a natural number $\mathbb{N}$ such that whenever two natural numbers $m, n > \mathbb{N}$, we have $|x_m - x_n| < \epsilon$.

14. The following results about Cauchy sequences should be known:

    (a) Every Cauchy sequence is bounded.

    (b) Every Convergent sequence is a Cauchy sequence.

    (c) Every Cauchy sequence is a convergent sequence.

(d) The results (b) and (c) combined together is known as **Cauchy's General Principle of Convergence:** *A necessary and sufficient condition for a sequence $\{x_n\}$ in $\mathbb{R}$ may be convergent is that it must be a Cauchy sequence.* Using this criterion to prove that $\{1 + \frac{1}{2} + \frac{1}{3} + \cdots + \frac{1}{n}\}_n$ does not converge, $\{1 + \frac{1}{2!} + \frac{1}{3!} + \cdots + \frac{1}{n!}\}_n$ does converge and $\{1 - \frac{1}{2} + \frac{1}{3} + \frac{1}{4} + \cdots + (-1)^{n-1}\frac{1}{n}\}_n$ is a convergent sequence.

15. **Construction of New Sequences:**

    Cauchys Limit Theorems: First Theorem: If $x_n \to l$ as $n \to \infty$ then $\frac{x_1 + x_2 + \cdots + x_n}{n}$ also tends to l and if all x_n are positive then $\sqrt[n]{x_1 x_2 \cdots x_n}$ also tends to l.

    Second Theorem: If every $x_n > 0$ and if $\frac{x_{n+1}}{x_n} \to l$ as $n \to \infty$, then $\sqrt[n]{x_n}$ also converges to the same limit l.

16. **A More General Result** (Upper and Lower limit concepts being known)

    $$\underline{\lim} \frac{x_{n+1}}{x_n} \leq \underline{\lim} \sqrt[n]{x_n} \leq \overline{\lim} \sqrt[n]{x_n} \leq \overline{\lim} \frac{x_{n+1}}{x_n}$$

    Remember: $\lim_{n \to \infty} \sqrt[n]{x_n}$ may exist even when $\lim_{n \to \infty} \frac{x_{n+1}}{x_n}$ does not exist.

    Applications: $\sqrt[n]{n!}/n \to \frac{1}{e}$; $\sqrt[n]{\frac{2n!}{n!n!}} \to 4$

17. **Some Miscellaneous Examples:** Given $0 < x_1 < x_2$ and for $n > 2$, $x_n = \frac{x_{n-1} + x_{n-2}}{2}$. To prove $x_n \to \frac{1}{3}(x_1 + 2x_2)$; Given $0 < x_1 < x_2$ and each $x_n = \sqrt{x_{n-1} x_{n-2}}$ $(n > 2)$ then prove that $x_n \to \sqrt[3]{x_1 \cdot x_2^2}$;

    Given $0 < x_1 < x_2$ and each $x_n = \dfrac{2}{\dfrac{1}{x_{n-1}} + \dfrac{1}{x_{n-2}}}$ then prove that $x_n \to \dfrac{\frac{1}{3}x_1 x_2}{2x_1 + x_2}$.

18. If $x_n = 1 + \frac{1}{2} + \frac{1}{3} + \cdots + \frac{1}{n} - \log n$ and $y_n = 1 + \frac{1}{2} + \frac{1}{3} + \cdots + \frac{1}{n-1} - \log n$ then $\{x_n\}$ is m.d. and $\{y_n\}$ is m.i. and both converge to the same limit γ (*Euler's constant*) $(0.3 < \gamma < 1)$.

19. **Cesaro's Theorem:** If $\{x_n\} \to x$ as $n \to \infty$ and $\{y_n\} \to y$ as $n \to \infty$ then

    $$z_n = \frac{x_1 y_n + x_2 y_{n-1} + x_3 y_{n-2} + \cdots + x_n y_1}{n} \to xy \text{ as } n \to \infty$$

20. **Arithmetico-Geometric Mean of two positive real numbers.** Given two unequal positive numbers u_1, v_1 and also it is given that

$$u_n = \tfrac{1}{2}(u_{n-1} + v_{n-1}) \quad \text{and} \quad v_n = \sqrt{u_{n-1}v_{n-1}} \quad (n \geq 2)$$

then both $\{u_n\}$ and $\{v_n\}$ converge to a common limit.

CHAPTER 5: SEQUENCE OF REAL NUMBERS   313

5(B): Subsequences: Upper and Lower Links

5.12 Sub-sequences

We shall write a sequence of real numbers either by $\{x_1, x_2, x_3, \ldots\}$ or by $\{x_n\}_{n=1}^{\infty}$ or by $\{x_n\}_n$ or simply by a function X $(X : \mathbb{N} \to \mathbb{R})$.

Definition 5.12.1. Let $X = \{x_n\}_n$ be a sequence of real numbers and let $n_1 < n_2 < n_3 < \cdots < n_k < n_{k+1} < \cdots$ be a strictly increasing sequence of natural numbers. Then the sequence $X' = \{x_{n_k}\}_{k=1}^{\infty} = \{x_{n_1}, x_{n_2}, x_{n_3}, \ldots, x_{n_k}\}$ is called a subsequence of X.

Actually, a subsequence of a sequence is a selection of terms from the given sequence such that the selected terms form a new sequence.

Note: Subsequence of a sequence $\{u_n\}$ is also a sequence, since it is the composite mapping $n_o r: \mathbb{N} \to \mathbb{R}$ defined by $(n_o r)(n) = u(r_n) = u_{r_n}$, where $r : \mathbb{N} \to \mathbb{N}$ is an increasing sequence, i.e., $r_1 < r_2 < \cdots$

Example 5.12.1. If $X = \{\frac{1}{1}, \frac{1}{2}, \frac{1}{3}, \ldots\}$, then the **selection** of even indexed terms gives the subsequence

$$X' = \left\{\frac{1}{2}, \frac{1}{4}, \frac{1}{6}, \ldots, \frac{1}{2k}, \ldots\right\},$$

where $n_1 = 2, n_2 = 4, \ldots, n_k = 2k, \ldots$

Other subsequences of $X = \{1/n\}_{n=1}^{\infty}$ are the following:

$$\left(\frac{1}{1}, \frac{1}{3}, \frac{1}{5}, \ldots, \frac{1}{2n-1}, \ldots\right)$$

or $\quad \left(\frac{1}{2!}, \frac{1}{4!}, \frac{1}{6!}, \ldots, \frac{1}{2k!}, \ldots\right).$

But $(\frac{1}{2}, \frac{1}{1}, \frac{1}{4}, \frac{1}{3}, \frac{1}{6}, \frac{1}{5}, \ldots)$ or $(\frac{1}{1}, 0, \frac{1}{3}, 0, \frac{1}{5}, 0, \ldots)$ are not subsequences of $X = \{\frac{1}{n}\}_{n=1}^{\infty}$.

Example 5.12.2. $\{1, 2, 3, 4, 5, \ldots\}$ has the two subsequences (a) $\{1, 3, 5, 7, \ldots\}$ (b) $\{2, 4, 6, 8, \ldots\}$. But see that $\{4, 2, 8, 6, 12, 10, \ldots\}$ is NOT a subsequence of $\{1, 2, 3, 4, 5, \ldots\}$.

Example 5.12.3. Let $\{x_n\}_n$ be a sequence defined by $x_n = \left(1 - \frac{1}{n}\right)\sin\frac{n\pi}{2}$.

We give below three subsequences of this sequence

(a) $\{x_{2n}\} = \{x_2, x_4, x_6, \ldots\}$

$$= \left\{\left(1-\frac{1}{2}\right)\sin\frac{2\pi}{2}, \left(1-\frac{1}{4}\right)\sin\frac{4\pi}{2}, \ldots\right\}$$

$$= \{0, 0, 0, \ldots\}$$

(Note that this subsequence converges to 0)

(b) $\left\{x_{4n+1}\right\}_{n=1}^{\infty} = \{x_5, x_9, x_{13}, x_{17}, \ldots\}$

$$= \left\{\left(1-\frac{1}{5}\right)\sin\frac{5\pi}{2}, \left(1-\frac{1}{9}\right)\sin\frac{9\pi}{2}, \ldots\right\}$$

$$= \left\{\frac{4}{5}, \frac{8}{9}, \frac{12}{13}, \ldots\right\}$$

(Note that this subsequence converges to one)

(c) $\{x_{4n-1}\} = \{x_3, x_7, x_{11}, x_{15}, x_{19}, \ldots\}$

$$= \left\{\left(1-\frac{1}{3}\right)\sin\frac{3\pi}{2}, \left(1-\frac{1}{7}\right)\sin\frac{7\pi}{2}, \ldots\right\}$$

$$= \left\{-\frac{2}{3}, -\frac{8}{7}, -\frac{10}{11}, \ldots\right\}$$

(Note that this subsequence converges to minus one)

Observations: $0, 1, -1$, to which (a), (b), (c) converge are known as **subsequential limits**. Here all subsequences do not converge to the same real number. In that case one can conclude that the original sequence $\left\{\left(1-\frac{1}{n}\right)\sin\frac{n\pi}{2}\right\}_{n=1}^{\infty}$ does not converge.

Example 5.12.4. Let $x_n = (-1)^n$, $n_1 = 2$, $n_2 = 4$, $\ldots$, $n_k = 2k \cdots$. Then

$$\{x_{n_1}, x_{n_2}, x_{n_3}, \ldots, x_{n_k} \cdots\}$$

is the subsequence $\{1, 1, 1, \ldots, 1, \ldots\}$ which converges to 1.

We shall now prove:

Theorem 5.12.1. *If a sequence $\{x_n\}$ converges to a limit l, then every sub-sequence of $\{x_n\}$ must converge to l and conversely.* [C.H. 1994]

Proof. If $\{x_n\}$ converges to l, then $\lim_{n\to\infty} x_n = l$ i.e., for any pre-assigned positive number ε, there exists a positive integer m such that

$$|x_n - l| < \varepsilon, \; \forall \, n \geq m.$$

CHAPTER 5: SEQUENCE OF REAL NUMBERS

Let $\{x_{n_k}\}$ be a certain sub-sequence of $\{x_n\}$. Now whenever $n_k \geq m$ we must have $|x_{n_k} - l| < \varepsilon$. This means that $\{x_{n_k}\}$ converges to l.

The converse follows almost trivially: we use the fact that $\{x_n\}$ is itself a subsequence of $\{x_n\}$. So when we say that every sub-sequence of $\{x_n\}$ converges to the same limit l, we of course include that the sequence $\{x_n\}$ converges to l.

Observation: If two subsequences of the sequence $\{x_n\}$ converge to two different real numbers, then the sequence $\{x_n\}$ cannot converge.

If $\{x_n\}$ has a divergent subsequence then the sequence $\{x_n\}$ itself is a divergent sequence.

Theorem 5.12.2. *If the two subsequences $\{x_{2n}\}$ and $\{x_{2n-1}\}$ of a sequence $\{x_n\}$ converge to the same limit l then the sequence $\{x_n\}$ is convergent and $\lim_{n \to \infty} x_n = l$.*

Proof: Choose a positive number ϵ. Since $\lim_{n \to \infty} x_n = l$. Corresponding to ϵ, $\exists$ a natural number m, such that $|x_{2n} - l| < \epsilon \; \forall n \geq m_1$. Since $\lim_{n \to \infty} x_{2n-1} = l$. Corresponding to the same ϵ, $\exists$ a natural number m_2 such that $|x_{2n-1} - l| < \epsilon, \; \forall n \geq m_2$.

Let $m = \max\{m_1, m_2\}$. Then $\forall n \geq m$

$$l - \epsilon < x_{2n} < l + \epsilon$$
$$\text{and} \quad l - \epsilon < x_{2n-1} < l + \epsilon$$

That is $l - \epsilon < x_n < l + \epsilon \; \forall n \geq 2m - 1$.

As $2m - 1$ is a natural number, it follows that $\lim_{n \to \infty} x_n = l$.

Note: If two of the subsequences of a given sequence converge to the same limit l, the sequence itself may not always be convergent, e.g., $\{x_n\} = \{\sin \frac{n\pi}{4}\}_n$ is a given sequence. The subsequence $\{x_{8n-7}\} = \{\sin \frac{\pi}{4}, \sin \frac{9\pi}{4}, \sin \frac{17\pi}{4}, \ldots\}$ which converges to $\frac{1}{\sqrt{2}}$ and subsequence $\{x_{8n-5}\} = \{\sin \frac{3\pi}{4}, \sin \frac{11\pi}{4}, \sin \frac{19\pi}{4}, \ldots\}$ which converges to $\frac{1}{\sqrt{2}}$. But this sequence $\{\sin \frac{n\pi}{4}\}_n$ is not convergent. This is justified by considering another subsequence $\{x_{4n}\} = \{\sin \pi, \sin 2\pi, \sin 3\pi, \ldots\} = \{0, 0, 0, \ldots\}$ which converges to 0. So, every subsequence does not converge to the same limit.

Monotone Subsequences

Theorem 5.12.3. (a) *Every subsequence of a monotone increasing sequence of real numbers is monotone increasing.* (b) *Every subsequence of monotone decreasing sequence of real number is monotone decreasing.* (c) *A monotone sequence of real numbers having convergent subsequence with sub-sequential limit l is convergent to the same limit l.* (d) *A monotone sequence of real numbers having a divergent subsequence is properly divergent.*

Proof: (a) Let $\{x_n\}_n$ be a monotone increasing sequence of real numbers. Then for any two natural numbers m_1 and m_2 with $m_2 > m_1, x_{m_2} \geq x_{m_1}$. Let $\{x_{n_k}\}$ be a subsequence of $\{x_n\}$. Then $\{n_k\}$ is a strictly increasing sequence of natural numbers. This implies $n_{k+1} > n_k$ for all $k \in \mathbb{N}$.

$\therefore n_{k+1} > n_k \Rightarrow x_{n_{k+1}} \geq x_{n_k}$.

This proves that $\{x_{n_k}\}$ is a monotone increasing subsequence.

(b) Proof is similar as that of (a).

(c) Let $\{x_n\}_n$ be a monotone sequence of real numbers and let $\{x_{n_k}\}_{k=1}^{\infty}$ be a subsequence of $\{x_n\}_n$ such that $\lim\limits_{k \to \infty} x_{n_k} = l$.

Since $\{x_n\}_n$ is a monotone increasing sequence, the subsequence $\{x_{nk}\}_{k=1}^{\infty}$ is also monotone increasing [proved in (b)] and since $\{x_{n_k}\}_{k=1}^{\infty}$ has a limit l, it must be bounded above.

[We assert that the sequence $\{x_n\}_n$ must be bounded above. If not, let $\{x_n\}_n$ must be unbounded above. Then being a monotone increasing sequence it must diverge to $+\infty$. Therefore, given any $G > 0$ (however large), there must exist a natural number p such that $x_n > G \,\forall n \geq p$. Since G is arbitrary, the sequence $\{x_{n_k}\}$ must diverge to ∞, which is a contradiction. So our assertion is established and the sequence $\{x_n\}$ being a monotone increasing sequence, bounded above, is convergent.]

Thus the sequence $\{x_n\}$ being a monotone increasing sequence, bounded above, is convergent.

Let $\lim\limits_{n \to \infty} x_n = m$. Then $\{x_{n_k}\}$ being a subsequence of $\{x_n\}$ converges to also m. Therefore $l = m$. This completes the proof.

(d) Let $\{x_n\}_n$ be a monotone increasing sequence of real numbers having a divergent subsequence $\{x_{nk}\}_{k=1}^{\infty}$. Since $\{x_n\}_n$ is monotone increasing, therefore it is a properly divergent subsequence.

Consequently $\{x_{n_k}\}_{k=1}^{\infty}$ is unbounded above.

Hence the sequence $\{x_n\}$ must be unbounded above and therefore it is properly divergent.

Similarly the case of monotone decreasing sequence can be dealt with.

Example 5.12.5. *Using the notions given above, prove that the sequence $\{x_n = 1 + \frac{1}{2} + \frac{1}{3} + \cdots + \frac{1}{n}\}_{n=1}^{\infty}$ is divergent.*

Solution: Here $x_{n+1} - x_n = \frac{1}{n+1} > 0$. Therefore $(x_n)_n$ is a monotone increasing sequence.

Let $n_k = 2^k$. Then $\{n_k\}$ is a strictly increasing sequence of natural numbers and so $\{x_{n_k}\}$ is a subsequence of $\{x_n\}$.

Now
$$x_{n_k} = x_{2^k} = 1 + \frac{1}{2} + \frac{1}{3} + \cdots + \frac{1}{2^k}$$
$$= 1 + \frac{1}{2} + \left(\frac{1}{3} + \frac{1}{4}\right) + \left(\frac{1}{5} + \cdots + \frac{1}{8}\right) + \cdots + \left(\frac{1}{2^{k-1}+1} + \cdots + \frac{1}{2^k}\right)$$
$$> 1 + \frac{1}{2} + \left(\frac{1}{4} + \frac{1}{4}\right) + \left(\frac{1}{8} + \cdots + \frac{1}{8}\right) + \cdots + \left(\frac{1}{2^k} + \cdots + \frac{1}{2^k}\right)$$
$$= 1 + \frac{1}{2} + 2 \cdot \left(\frac{1}{2}\right)^2 + 2^2 \cdot \left(\frac{1}{2}\right)^2 + \cdots + 2^{k-1} \cdot \frac{1}{2^k}$$
$$= 1 + \frac{k}{2}.$$

Let $v_k = 1 + \frac{k}{2}$. Then $x_{n_k} > v_k \; \forall k > 2$ and $\lim_{k \to \infty} v_k = \infty$. Therefore $\lim_{n \to \infty} x_{n_k} = \infty$.

Thus the sequence is a monotone increasing sequence having a properly divergent subsequence $\{x_{nk}\}$ and therefore the sequence $\{x_n\}$ is properly divergent.

Theorem 5.12.4. *Every sequence has a monotone sub-sequence.*

In order to prove this theorem let us define what we call *a greatest term* of a given sequence:

Definition 5.12.2. *A term x_k of a sequence $\{x_n\}$ is called a greatest term of $\{x_n\}$ if for every positive integer n, $x_k \geq x_n$.*

[A sequence may have no greatest term e.g., $\{2, 4, 6, 8, \ldots\}$ has no greatest term. Some sequences may have many terms, each of which is a greatest term, e.g., in $\{x_n\}$, where $x_n = (-1)^{n-1}$ each of the terms $x_1, x_3, x_5, \ldots$ is a greatest term.]

If $\{x_{n_k}\}$ be a sub-sequence of $\{x_n\}$ and if x_{n_p} and x_m be the greatest terms of $\{x_{n_k}\}$ and $\{x_n\}$ respectively, then
$$x_{n_p} \leq x_m.$$

Proof of Theorem 5.9.2. Let $\{x_n\}$ be a given sequence.

To prove: $\exists$ a monotonic sub-sequence of $\{x_n\}$.

We call:
$$s_0 = \{x_1, x_2, x_3, x_4, \cdots\cdots\cdots\}.$$
$$s_1 = \{x_2, x_3, x_4, \cdots\cdots\cdots\}.$$
$$s_2 = \{x_3, x_4, x_5, \cdots\cdots\cdots\}.$$
$$s_3 = \{x_4, x_5, x_6, \cdots\cdots\cdots\}.$$

and so on.

1. Suppose that each of the sequences $s_0, s_1, s_2, s_3, \ldots$ has a greatest term.

   Let x_{n_1} be a greatest term of s_0; $\quad x_{n_2}$ be a greatest term of s_1;

   $\quad x_{n_3}$ be a greatest term of $s_2, \ldots$ etc. etc.

   Then $n_1 \leq n_2 \leq n_3 \leq \cdots$ and $x_{n_1} \geq x_{n_2} \geq x_{n_3} \geq \cdots$ so that $\{x_{n_1}, x_{n_2}, x_{n_3}, \ldots\}$ is a monotonically decreasing sub-sequence of $\{x_n\}$.

2. Suppose that at least one of the sequences $s_0, s_1, s_2, \ldots$ has no greatest term. Let s_m have no greatest term. Then each term of s_m is ultimately followed by some term of s_m that exceeds it (*otherwise, s_m will have a greatest term*).

Clearly, the first term of s_m is x_{m+1}.

Let x_{n_2} be the first term of s_m that exceeds x_{m+1}.

Let x_{n_3} be the first term of s_m that follows x_{n_2} and exceeds x_{n_2}.

Let x_{n_4} be the first term of s_m that follows x_{n_3} and exceeds it; and so on. Then

$$\{x_{m+1}, x_{n_2}, x_{n_3}, x_{n_4}, \ldots\}$$

is a monotonically increasing sub-sequence of $\{x_n\}$.

The main result of the present chapter is

> Every bounded sequence has some Convergent sub-sequence.

This result is known as Bolzano-Weierstrass Theorem for sequences in $\mathbb{R}$.

Bolzano-Weierstrass Theorem for Sequences (Using Nested Interval Property)

Theorem 5.12.5. *Every bounded sequence of real numbers has a convergent subsequence.* [C.H. 1984, '90, '92, '96]

Proof. Let $\{x_n\}$ be a bounded sequence of real numbers.

To prove that it has at least one convergent sub-sequence.

Since $\{x_n\}$ is a bounded sequence there exists a positive number M such that $|x_n| \leq M$, for every natural number n. Hence, $x_n \in [-M, M]$ for all $n \in \mathbb{N}$, where $\mathbb{N}$ is the set of all natural numbers.

Consider two closed intervals $[-M, 0]$ and $[0, M]$.

At least one of these two intervals must **contain x_n for infinitely many natural numbers** n [observe that x_n's may not be necessarily all distinct]. Call such an interval I_0.

CHAPTER 5: SEQUENCE OF REAL NUMBERS

Next divide I_0 into two sub-intervals of equal length; at least one of these sub-intervals must contain x_n for infinitely many natural numbers n. Call such an interval I_1. We next divide I_1 into two sub-intervals of equal length; at least one of these two sub-intervals must contain x_n for infinitely many n. Call it I_2.

We continue in this manner and obtain a sequence of closed intervals
$$\{I_0, I_1, I_2, \cdots, I_n, I_{n+1}, \ldots\} \text{ such that}$$
$$I_0 \supset I_1 \supset I_2 \supset \cdots \supset I_n \supset I_{n+1} \supset \cdots$$
i.e., the intervals are nested.

Moreover, it is easy to see that the length of I_n is
$$d_n = \frac{M}{2^n} \text{ which tends to } 0 \text{ as } n \to \infty.$$

$\therefore$ by the Nested-Interval property there is exactly one point l which is common to all these intervals.

Choose points $x_{n_1}, x_{n_2}, x_{n_3}, \ldots$ such that
$$x_{n_1} \in I_1,$$
$$x_{n_2} \in I_2 \text{ with } n_2 > n_1,$$
$$x_{n_3} \in I_3 \text{ with } n_3 > n_2,$$
$$x_{n_4} \in I_4 \text{ with } n_4 > n_3.$$
$$\cdots\cdots\cdots\cdots\cdots\cdots$$

[Such selections are possible since each I_k contains x_n, for infinitely many natural numbers n.]

Thus we have collected a sub-sequence
$$\{x_{n_1}, x_{n_2}, x_{n_3}, \ldots, x_{n_k}, \ldots\}$$
of the sequence $\{x_n\}$, with the property that x_{n_k} and l are both contained in I_k. Thus
$$|x_{n_k} - l| < \frac{M}{2^k}.$$

Since $\frac{M}{2^k} \to 0$ as $k \to \infty$, we may make $\frac{M}{2^k} <$ any given positive number ε, no matter how small for all $k > $ a certain positive integer m. Thus
$$\lim_{k \to \infty} x_{n_k} = l.$$

We have thus shown that the bounded sequence $\{x_n\}$ has one convergent sub-sequence $\{x_{n_k}\}$.

Example 5.12.6. *Take, for example, the bounded sequence*

$$\left\{(-1)^n\left(1+\frac{1}{n}\right)\right\} = \left\{-(1+1), +\left(1+\frac{1}{2}\right), -\left(1+\frac{1}{3}\right), +\left(1+\frac{1}{4}\right), -\left(1+\frac{1}{5}\right), \ldots\right\}$$

has two convergent sub-sequences—one $\{x_1, x_3, x_5, \ldots\}$ converging to -1 and the other $\{x_2, x_4, x_6, \ldots\}$ converging to $+1$.

5.13 Cluster Points of a Sequence

For an infinite set S of real numbers we defined *limit point* thus: A point ξ (which may or may not belong to S) is called a *limit point* (also called a *point of accumulation*) of S if to each $\varepsilon > 0$, there exists $x \in S$ such that $x \neq \xi$ and $|x - \xi| < \varepsilon$.

This definition implies that in any neighbourhood N of ξ there exist infinite number of members of the set. So a finite set has no limit point. In order that a set of real numbers can have a limit point, the set must be **infinite**. But all infinite sets may not have a limit point (e.g., the set of natural numbers has no limit point, but it is an infinite set).

In case of a **sequence** of real numbers we define **cluster point** (an analog of *limit point* of an infinite set).

Definition 5.13.1. *Let $\{x_n\}$ be a sequence of real numbers. A real number ξ which may or may not belong to $\{x_n\}$ is called a **cluster point** of the sequence $\{x_n\}$, if $\exists$ some sub-sequence of $\{x_n\}$ which converges to ξ.*

Thus a sub-sequential limit of a sequence is a cluster point of the sequence.

By Bolzano-Weierstrass Theorem we then have:

> Every bounded sequence has at least one cluster point.

What about Unbounded Sequences?

Unbounded sequences may or may not have convergent sub-sequences, i.e., unbounded sequences may or may not have cluster points.

Thus the sequence $\{\sqrt{n}\}$ has no cluster point but the sequence

$$\{1, 1, 2, 1, 2, 3, 1, 2, 3, 4, 1, 2, 3, 4, 5, 1, \ldots\}$$

has infinitely many cluster points—namely, each natural number $1, 2, 3, 4, \ldots$. Both the sequences are, of course, unbounded sequences.

We now give another equivalent definition of cluster point of a sequence.

CHAPTER 5: SEQUENCE OF REAL NUMBERS

Definition 5.13.2. *A real number ξ is called a cluster point of the sequence $\{x_n\}$ if to each $\varepsilon > 0$ and to each positive integer m, $\exists$ a positive integer $k > m$ such that $|x_k - \xi| < \varepsilon$, i.e., $\xi - \varepsilon < x_k < \xi + \varepsilon$.*

This definition implies that in every *nbd.* of ξ, $\exists$ infinitely many natural numbers n such that the corresponding x_n's belong to that *nbd.*

Both the definitions give that the cluster points of a sequence include

(a) **all those points of the sequence which recur infinitely often; and**

(b) **the limit points of the range set.**

e.g., $\{(-1)^n\}$ has *two cluster points* -1 and 1 as they recur infinitely often.

Note that the range set contains only two points $\{1, -1\}$ and a finite set has no *limit point*.

$\therefore$ the only cluster points of $\{(-1)^n\}$ are -1 and $+1$.

Using the previous Definition of a cluster point of a sequence we can also prove:

Theorem 5.13.1. *Every bounded sequence of real numbers has at least one cluster point.*

Proof. Let $\{x_n\}$ be an bounded sequence of real numbers.

Let the range set $S = \{x_n : n \in \mathbb{N}\}$.

Since the sequence $\{x_n\}$ is bounded, S is either a finite set or an infinite bounded set.

1. If S is a finite set, there must exist infinitely many suffixes n, for which $x_n =$ some real number q. If m is any given positive integer, then in any *nbd.* of q $\exists$ another positive integer $k > m$ s.t. $x_k = q \in$ that *nbd.*, i.e., q is a cluster point.

2. If S is an infinite bounded set, then by Bolzano-Weierstrass Theorem for Sets (see Chap. 4), S has at least one limit point p. Therefore, any *nbd.* of p contains infinitely many points of S. So if m is a given positive integer and N be any *nbd.* of p, then for some positive integer $k > m$, $x_k \in N$ and consequently, by Definition **5.13.2**, p is a cluster point of $\{x_n\}$.

We shall now prove that Definitions **5.13.1** and **5.13.2** are equivalent.

Theorem 5.13.2. *A real number ξ is a cluster point of a sequence $\{x_n\}$ (in the sense of Definition* **5.13.2***) iff there exists a sub-sequence of $\{x_n\}$ converging to ξ.*

Proof.

1. Definition **5.13.2** $\Longrightarrow$ Definition **5.13.1** [*cluster point is a sub-sequential limit*]

   Given: ξ is a cluster point of a sequence $\{x_n\}$ of real numbers.

   To prove: $\exists$ a sub-sequence of $\{x_n\}$ which converges to ξ.

Choose $\varepsilon = 1$ and $m = 1$. Then $\exists$ a positive integer $n_1 > 1$ such that $\xi - 1 < x_{n_1} < \xi + 1$.

Next choose $\varepsilon = 1/2$ and $m = n_1$. Then $\exists$ a positive integer $n_2 > n_1$ such that $\xi - \frac{1}{2} < x_{n_2} < \xi + \frac{1}{2}$.

Proceeding in this manner we can construct a sub-sequence $\{x_{n_1}, x_{n_2}, x_{n_3}, \ldots, x_{n_k}, \ldots\}$ such that $\xi - \frac{1}{k} < x_{n_k} < \xi + \frac{1}{k}$.

We shall now show that this sub-sequence converges to ξ. Given any $\varepsilon > 0$, we can find a positive integer p such that $1/p < \varepsilon$. For such a choice of p

$$|x_{n_k} - \xi| < \frac{1}{p} < \varepsilon, \text{ for all } k \geq p$$

and hence, $x_{n_k} \to \xi$ as $k \to \infty$.

i.e., the sub-sequence $\{x_{n_k}\}$ converges to ξ.

2. Definition 5.13.1 $\implies$ Definition 5.13.2 [*A sub-sequential limit is a cluster point*].

Given: $\{x_{n_k}\}$ is a sub-sequence of $\{x_n\}$ converging to a real number ξ.

To prove: ξ is a cluster point of $\{x_n\}$.

Let $\varepsilon > 0$ be any given positive real number.

Let m be a given positive integer.

Since $\{x_{n_k}\}$ converges to ξ, $\exists$ a positive integers p such that $|x_{n_k} - \xi| < \varepsilon$, whenever $k \geq p$.

Let q be a positive integer $\geq p$ such that $n_q > m$. Then

$$|x_{n_q} - \xi| < \varepsilon.$$

i.e., we have found a positive integer $n_q > m$ such that

$$\xi - \varepsilon < x_{n_q} < \xi + \varepsilon.$$

This is exactly what we require for ξ to be a cluster point of $\{x_n\}$ by Definition 5.13.2.

5.14 Upper and Lower Limits of Sequences of real Numbers

We know that every convergent sequence is bounded but not every bounded sequence is convergent. With every bounded sequence, however, we associate two real numbers—one we call *lower limit* (λ) and the other *upper limit* (μ). A bounded sequence converges only when $\lambda = \mu$.

CHAPTER 5: SEQUENCE OF REAL NUMBERS

We shall give two equivalent definitions for λ and μ—one definition can be deduced if the other is assumed. For unbounded sequences we also define upper and lower limits but then we shall have to consider the extended Real number system (i.e., all real numbers and $\pm\infty$).

A. For a **Bounded sequence** $\{x_n\}$ of real numbers:

First definition of λ and μ	Second definition of λ and μ
1. We call a real number μ, the *upper limit* or *limit superior* of the bounded sequence $\{x_n\}$ if μ satisfies the following two conditions: (i) For every $\varepsilon > 0$, $\exists$ only a finite number of members of $\{x_n\}$ which are $> \mu + \varepsilon$, i.e., for every $\varepsilon > 0$, $\exists$ a positive integer N such that $$n > N \implies x_n \leq \mu + \varepsilon.$$ (ii) Given $\varepsilon > 0$ and given positive number m, $\exists$ a positive integer $n > m$ such that $x_n > \mu - \varepsilon$. [i.e., $x_n > \mu - \varepsilon$ for infinitely many n] We write, $\mu = \limsup x_n$ or $\overline{\lim} x_n$. 2. A real number λ is called the *lower limit* of $\{x_n\}$, if λ satisfies the following two conditions: (i) For every $\varepsilon > 0$, $\exists$ only a finite number of members of x_n less than $\lambda - \varepsilon$, i.e., for every $\varepsilon > 0$, $\exists$ a positive integer n_0 such that $$n > n_0 \implies x_n \geq \lambda - \varepsilon.$$ (ii) Given $\varepsilon > 0$ and given $m > 0$, $\exists$ a positive integer $n > m$ such that $x_n < \lambda + \varepsilon$, i.e., for infinitely many n, $$x_n < \lambda + \varepsilon.$$ We write, $\lambda = \liminf x_n$ or $\underline{\lim} x_n$.	We know that every bounded sequence of real numbers has a convergent sub-sequence. We call the limit of any such sub-sequence a *cluster point* of the sequence. Now given a bounded sequence $\{x_n\}$ of real numbers we collect all the sub-sequential limits of $\{x_n\}$. This collection is denoted by E, i.e., each member of E is the limit of a convergent sub-sequence. The set E is clearly non-empty and bounded; hence E has a glb and a lub. We now define: (i) The lub or supremum of E as the upper limit μ of the sequence $\{x_n\}$; we write $\mu = \limsup x_n$ or $\overline{\lim} x_n$. (ii) The glb or infimum of E is called the *lower limit* or *limit inferior* λ of the sequence $\{x_n\}$. We then write $$\lambda = \liminf x_n \quad \text{or} \quad \underline{\lim} x_n.$$

Note: The second definition is precise and impressive.

B. For Unbounded Sequences we introduce the following definitions:

(a) If $\{x_n\}$ is unbounded above, then $\limsup x_n = +\infty$.

(b) If $\{x_n\}$ is unbounded below, then $\liminf x_n = -\infty$.

(c) If $\{x_n\}$ is unbounded below, and there is no other sub-sequential limit then its limit superior is also $-\infty$, so that

$$\overline{\lim} \, x_n = \underline{\lim} \, x_n = -\infty.$$

(d) If $\{x_n\}$ is unbounded above, and there is no other sub-sequential limit, then its limit inferior is also $+\infty$ so that

$$\overline{\lim} \, x_n = \underline{\lim} \, x_n = +\infty.$$

Observations:

1. Every real sequence has a *limit superior* (μ) and a limit inferior (λ) in the *extended* real number system.

2. For any real sequence, $\lambda \leq \mu$.

3. A real sequence converges, if and only if μ and λ are both finite and $\lambda = \mu$. This common value is the limit of the sequence.

4. (a) A real sequence diverges to $+\infty$, if and only if

$$\liminf x_n = \limsup x_n = +\infty.$$

   (b) A real sequence diverges to $-\infty$, if and only if

$$\liminf x_n = \limsup x_n = -\infty.$$

5. A real sequence for which

$$\liminf x_n \neq \limsup x_n$$

is said to oscillate—oscillates finitely, if $\lambda \neq \mu$, but λ, μ are both finite and oscillates infinitely, if either λ or μ is infinite ($\lambda \neq \mu$).

6. For a bounded sequence of real numbers we have given two definitions of μ and λ. Assuming any one of these two definitions other can be deduced.

CHAPTER 5: SEQUENCE OF REAL NUMBERS

The reader is expected to be acquainted with the proofs of these observations. Before giving these proofs we discuss some illustrative examples:

Example 5.14.1. *The bounded sequence* $\{(-1)^n\} = \{-1, +1, -1, +1, \ldots\}$ *has lower limit* $\lambda = -1$ *and upper limit* $\mu = +1$. *There are only two sub-sequential limits* $\{-1, 1\}$ *for the sequence* $\{(-1)^n\}$. *Here* $\lambda \neq \mu$, *the sequence oscillates finitely between* -1 *and* $+1$.

Example 5.14.2. *Consider the sequence* $\{-\frac{1}{2}, 0, \frac{1}{2}, 0, -\frac{2}{3}, 0, \frac{2}{3}, 0, -\frac{3}{4}, 0, \frac{3}{4}, 0, \ldots\}$.

This sequence is also bounded. The sub-sequential limits are $-1, 1$ and 0.
Lower limit $\lambda = -1$, Upper limit $\mu = 1$. ($\lambda \neq \mu$; oscillates finitely)

Example 5.14.3. *The sequence* $\{-2, 2, -\frac{3}{2}, \frac{3}{2}, -\frac{4}{3}, \frac{4}{3}, \ldots\}$ *has* $\lambda = -1$ *and* $\mu = 1$.

Observe that none of the terms of the sequence lies between -1 and $+1$, i.e., -1 is not a lower bound and $+1$ is not an upper bound.

Observation: The lower limit of a sequence is not necessarily the glb of the sequence, nor is the upper limit the lub. In fact, the lower limit may not even be a lower bound, as can be seen from **Ex. 5.11.3.** above.

** Example 5.11.4. $x_n = (-1)^n(1 + \frac{1}{n})$ $(n = 1, 2, 3, \ldots)$.

The members of the sequence are

$$\left\{-2, -\frac{3}{2}, -\frac{4}{3}, \cdots\right\} \quad \text{for } n = 1, 3, 5, 7, \ldots$$

$$\text{and} \left\{+\frac{3}{2}, +\frac{5}{4}, +\frac{7}{6}, \cdots\right\} \quad \text{for } n = 2, 4, 6, 8, \ldots$$

The two sub-sequential limits are -1 and 1.

$$\underline{\lim} x_n = \text{Lower limit } \lambda = -1; \quad \overline{\lim} x_n = \text{Upper limit } \mu = +1.$$

Example 5.14.5. $\{x_n\}$ *where* $x_n = \sin \frac{n\pi}{3}$ $(n = 1, 2, 3, 4, \ldots)$.

The sequence $\{x_n\}$ is

$$\left\{\frac{\sqrt{3}}{2}, \frac{\sqrt{3}}{2}, 0, -\frac{\sqrt{3}}{2}, -\frac{\sqrt{3}}{2}, 0, \frac{\sqrt{3}}{2}, \frac{\sqrt{3}}{2}, 0, -\frac{\sqrt{3}}{2}, -\frac{\sqrt{3}}{2}, \ldots\right\}$$

The sub-sequential limits are $\frac{\sqrt{3}}{2}, 0$ and $-\frac{\sqrt{3}}{2}$ as those terms occur infinitely many times in the sequence.

$$\underline{\lim} x_n = -\frac{\sqrt{3}}{2} \quad \text{and} \quad \overline{\lim} x_n = +\frac{\sqrt{3}}{2}.$$

** Example 5.11.6. $\{x_n\}$ where $x_n = \sin\dfrac{n\pi}{2} + \dfrac{(-1)^n}{n+1}$ $(n = 0, 1, 2, 3, 4, 5, \ldots)$.

$$|x_n| \leq \left|\sin\dfrac{n\pi}{2}\right| + \left|\dfrac{(-1)^n}{n+1}\right| \leq 1 + \dfrac{1}{2}$$

∴ The sequence in bounded.

The convergent sub-sequences are

$$\{x_{4k}\}, \{x_{4k+1}\}, \{x_{4k+2}\} \quad \text{and} \quad \{x_{4k+3}\}, \quad k = 0, 1, 2, \ldots$$

where $x_{4k} = \dfrac{1}{4k+1}$, $x_{4k+1} = 1 - \dfrac{1}{4k+2}$, $x_{4k+2} = \dfrac{1}{4k+3}$ and $x_{4k+3} = -1 - \dfrac{1}{4k+4}$.
Hence the sub-sequential limits are $1, -1,$ and 0.

$$\therefore \quad \underline{\lim} \, x_n = -1 \quad \text{and} \quad \overline{\lim} \, x_n = 1.$$

Example 5.14.6. *Check the following assertions:*

	Sequence $\{x_n\}$ where $x_n =$ $(n = 1, 2, 3, \ldots)$	Lower limit $\lambda = \underline{\lim}\, x_n$	Upper limit $\mu = \overline{\lim}\, x_n$	Convergent, Divergent or Oscillating
(i)	$1 + (-1)^n$	0	2	$\lambda \neq \mu, \lambda, \mu$ finite, hence Oscillating
(ii)	$\dfrac{(-1)^n}{n^2}$	0	0	$\lambda = \mu, \lambda, \mu$ finite, hence Convergent
(iii)	$n(1 + (-1)^n)$	0	∞	$\lambda \neq \mu, \lambda$ finite and μ not finite, hence Oscillating infinitely.
(iv)	$(-1)^n n$	$-\infty$	∞	Oscillating infinitely
(v)	n^2	∞	∞	Diverges to $+\infty$.

** Example 5.11.8. $\{x_n\}$ where

$$x_n = 2, \text{ if } n = 2, 4, 6, 8, \ldots$$
$$= 1, \text{ if } n = 1$$
$$= \text{least prime factor of } n, \text{ if } n = \text{odd} > 1$$

The sequence terms are

$$\{1, 2, 3, 2, 5, 2, 7, 2, 3, 2, 11, 2, 13, 2, 3, \ldots\}.$$

CHAPTER 5: SEQUENCE OF REAL NUMBERS   327

The sequence is bounded below but not bounded above.
The sub-sequential limits are infinite in number.

$$E = \{2, 3, 5, 7, 11, \ldots\}.$$

The set E has glb $= 2$, but it is not bounded above.

$$\therefore \quad \underline{\lim} \, x_n = 2, \quad \overline{\lim} \, x_n = \infty.$$

The sequence oscillates infinitely.

Example 5.14.9. *Rational number system is countable, i.e., we can enumerate all rational numbers in the form of a sequence $\{r_n\}$, then every real number is a subsequential limit and $\mu = +\infty$ and $\lambda \neq -\infty$.*

The sequence oscillates infinitely.

Example 5.14.10. *If $\{x_n\}$ be an enumeration of all rational numbers between 0 and 1, then $\underline{\lim} \, x_n = 0$ and $\overline{\lim} \, x_n = 1$. $\lambda \neq \mu$ but λ, μ are finite. Hence the sequence oscillates finitely.*

Example 5.14.11. *Check the following results:*

(a) $\{-1, -2, -3, \ldots, -n, \ldots\} : \lambda = \mu = -\infty$ (diverges to $-\infty$).

(b) $\{1, 0, 2, 0, 3, 0, \ldots\}$ has $\mu = +\infty, \lambda = 0$ (oscillates infinitely).

(c) $\{1, 2, 3, \ldots, n, \ldots\}; \lambda = \mu = +\infty$ (diverges to $+\infty$).

This sequence is unbounded above and hence $\mu = +\infty$ and it has no finite limit inferior. So we take $\lambda = +\infty$. Hence, etc.

(d) $\{-1, 2, -3, 4, -5, \ldots\}$ has $\mu = \infty, \lambda = -\infty$ (oscillates infinitely).

** **Example 5.14.12.** *Find the upper and lower limits of the following sequences:*

1. $\{(-1)^n + \sin \frac{n\pi}{4}\}$,

2. $\{(-1)^n (2 + \frac{3^n}{n!} + \frac{4}{n^2})\}$.

Solution: **1.** We first find the terms of the sequence for different values of n.

(a) When n is of the form $4k$ $(k = 1, 2, 3, \ldots)$

$$x_n = (-1)^{4k} + \sin \frac{4k\pi}{4} = 1 \quad (n = 4, 8, 12, 16, \ldots).$$

(b) When n is of the form $4k+1$ $(k = 0, 1, 2, 3, \ldots)$

$$x_n = (-1)^{4k+1} + \sin(4k+1)\frac{\pi}{4} = -1 + (-1)^k \sin\frac{\pi}{4} = -1 + (-1)^k \frac{1}{\sqrt{2}}$$

$$x_n = -1 - \frac{1}{\sqrt{2}} \quad (n = 5, 13, 21, 29, \ldots)$$

$$= -1 + \frac{1}{\sqrt{2}} \quad (n = 1, 9, 17, 25, 33, \ldots).$$

(c) When n is of the form $4k+2$ $(k = 0, 1, 2, 3, \ldots)$

$$x_n = x_{4k+2} = (-1)^{4k+2} + \sin(4k+2)\frac{\pi}{4} = 1 + (-1)^k \sin\frac{\pi}{2} = 1 + (-1)^k,$$

i.e., $x_n = 2$ $(n = 2, 10, 18, 26, \ldots)$

$$= 0 \quad (n = 6, 14, 22, \ldots).$$

(d) When n is of the form $4k+3$ $(k = 0, 1, 2, 3, \ldots)$

$$x_n = x_{4k+3} = (-1)^{4k+3} + \sin(4k+3)\frac{\pi}{4} = -1 + (-1)^k \sin\frac{3\pi}{4}$$

$$= -1 + (-1)^k \sin\frac{\pi}{4} = -1 + (-1)^k \frac{1}{\sqrt{2}}$$

i.e., $x_n = -1 + \frac{1}{\sqrt{2}} \quad (n = 3, 11, 19, 27, \ldots)$

$$= -1 - \frac{1}{\sqrt{2}} \quad (n = 7, 15, 23, \ldots).$$

The terms of the sequence, when arranged in order, are

$\{x_1 = -1 + \frac{1}{\sqrt{2}},\ x_2 = 2,\ x_3 = -1 + \frac{1}{\sqrt{2}},\ x_4 = 1,\ x_5 = -1 - \frac{1}{\sqrt{2}},\ x_6 = 0,$
$x_7 = -1 - \frac{1}{\sqrt{2}},\ x_8 = 1,\ x_9 = -1 + \frac{1}{\sqrt{2}},\ x_{10} = 2,\ x_{11} = -1 + \frac{1}{\sqrt{2}},\ x_{12} = 1,$
$x_{13} = -1 - \frac{1}{\sqrt{2}},\ x_{14} = 0,\ x_{15} = -1 - \frac{1}{\sqrt{2}},\ x_{16} = 1,\ x_{17} = -1 + \frac{1}{\sqrt{2}},\ x_{18} = 2,$
$x_{19} = -1 + \frac{1}{\sqrt{2}},\ x_{20} = 1,\ x_{21} = -1 - \frac{1}{\sqrt{2}},\ x_{22} = 0,\ x_{23} = -1 - \frac{1}{\sqrt{2}},\ \text{etc., etc.}\}.$

We can easily form convergent sub-sequences whose limits are

$$0, 1, -1 + \frac{1}{\sqrt{2}}, -1 - \frac{1}{\sqrt{2}} \text{ and } 2,$$

which give the exhaustive list of sub-sequential limits.

$$\therefore \quad \mu = \limsup x_n = 2 \quad \text{and} \quad \lambda = \liminf x_n = -1 - \frac{1}{\sqrt{2}}.$$

CHAPTER 5: SEQUENCE OF REAL NUMBERS

2. Verify : $\limsup x_n = +2$ and $\liminf x_n = -2$.

Theorems concerning Limit Superior and Limit Inferior

I. *Let $\{x_n\}$ be a bounded sequence of real numbers. Suppose μ is the upper limit of $\{x_n\}$ in the sense that μ is the lub of the set E, each member of which is the limit of a certain sub-sequence of $\{x_n\}$. To establish Theorems 5.14.1 and 5.14.2, given below:*

Theorem 5.14.1. *Given any $\varepsilon > 0$, $\exists$ a natural number N such that*

$$x_n \leq \mu + \varepsilon, \text{ for all } n > N,$$

i.e., $\exists$ only a finite number of members of $\{x_n\} > \mu + \varepsilon$.

Proof. If the theorem were not true, then we would have

$$x_n > \mu + \varepsilon$$

for infinitely many natural numbers. From these infinite number of members of the bounded sequence $\{x_n\}$ we could construct a suitable sub-sequence which would converge to a value $p > \mu + \varepsilon$. This would contradict the fact that μ is the lub of the set E.

Hence, $x_n > \mu + \varepsilon$ for only finitely many natural numbers n. Consequently, $\exists$ a natural number N such that $x_n \leq \mu + \varepsilon$, whenever $n > N$. This proves Theorem **5.14.1**.

Theorem 5.14.2. *Given any $\varepsilon > 0$, $x_n > \mu - \varepsilon$, for infinitely many natural numbers n.*

[C.H. 1981]

Proof. If this statement were not true, then for the given $\varepsilon > 0$ there would exist only a finite number of members of $\{x_n\}$, greater than $\mu - \varepsilon$ and then we shall not obtain any member of E in $(\mu - \varepsilon, \mu)$ for suitably small ε. This will contradict the fact that μ being the lub of the set E, there must exist a member of E in any left ε-nbd. of μ. Hence, $\exists x_n > \mu - \varepsilon$ for *infinitely many* n. Consequently given any positive number ε, $\exists$ a natural number $n > m$ such that $x_n > \mu - \varepsilon \; \forall \; n > m$.

II. *Let $\{x_n\}$ be a bounded sequence of real numbers. Suppose λ is the lower limit of $\{x_n\}$ in the sense that λ is the glb of the set E of all sub-sequential limits of $\{x_n\}$. To establish Theorems 5.14.3 and 5.14.4 given below:*

Theorem 5.14.3. *Given any $\varepsilon > 0$, $\exists$ a natural number N such that*

$$x_n \geq \lambda - \varepsilon, \; \forall \, n > N.$$

i.e., $\exists$ only a finite number of members of $\{x_n\} < \lambda - \varepsilon$.

Theorem 5.14.4. *Given any $\varepsilon > 0$, $\exists$ infinite number of natural numbers n for which $x_n < \lambda + \varepsilon$.*

The proofs of Theorems **5.14.3** and **5.14.4** are exactly similar to those of Theorems **5.14.1** and **5.14.2** (*Proofs by contradiction*).

Observation: What we have proved is that

Second definitions of λ and $\mu \implies$ first definitions of λ and μ.

III. We shall now establish the converse proposition i.e.,

First definitions of λ and $\mu \implies$ their second definitions.

Theorem 5.14.5. *Let $\{x_n\}$ be a bounded sequence of real numbers. Suppose $\exists$ a real number μ having two properties* **1(i)** *and* **1(ii)** *of the first definition. Then $\mu =$ lub of the set E which consists of all sub-sequential limits of the sequence $\{x_n\}$.*

Proof. By the property **1(i)**, no point of E can be greater than μ. If possible, let there be a point $p > \mu$ such that $p \in E$. Then we can choose $\varepsilon > 0$ suitably so that there would be an infinite number of natural numbers n for which $x_n > \mu + \varepsilon$, contradicting property **1(i)**.

By property **1(ii)** $\exists$ a member $q \in E$ where $q > \mu - \varepsilon$. [Since for infinitely many $n, x_n > \mu - \varepsilon$, we can construct a sub-sequence of $\{x_n\}$ converging to q. It follows from the fact that $\{x_n\}$ is bounded and hence a subset of it is also bounded and a bounded sequence has a convergent sub-sequence.]

$\therefore$ it follows that $\mu =$ lub of the set E.

Theorem 5.14.6. *Let $\{x_n\}$ be a bounded sequence of real numbers. Suppose $\exists$ a real number λ having two properties* **2(i)** *and* **2(ii)** *of the first definition.*

Then $\lambda =$ glb of the set E of all sub-sequential limits of $\{x_n\}$.

Proof is similar to that of Theorem **5.14.5** and is left as an Exercise for the students.

Observation: See that our first definition of μ suggests that **ultimately all terms** of the sequence lie to the left of $\mu + \varepsilon$ and $\exists$ infinitely many terms to the right of $\mu - \varepsilon$, i.e., μ is the **greatest of all cluster points** of the sequence. Similarly, λ is the **least of all the cluster points** of $\{x_n\}$.

With the help of the first definitions of λ and μ, we prove the following theorem:

Theorem 5.14.7. *Let $\{x_n\}$ be a sequence of real numbers. Then we have*

(a) $\underline{\lim} \, x_n \leq \overline{\lim} \, x_n$.

(b) The sequence $\{x_n\}$ converges if and only if λ and μ are both finite and equal. In this case,

$$\lim_{n\to\infty} x_n = \underline{\lim} x_n = \overline{\lim} x_n. \qquad \text{[C.H. 1988]}$$

(c) The sequence $\{x_n\}$ diverges to $+\infty$, if and only if

$$\underline{\lim} x_n = \overline{\lim} x_n = +\infty \quad (\text{i.e., } \lambda = \mu = +\infty).$$

(d) The sequence $\{x_n\}$ diverges to $-\infty$, if and only if

$$\underline{\lim} x_n = \overline{\lim} x_n = -\infty.$$

[The results of this theorem imply that if $\lambda \neq \mu$, then $\{x_n\}$ oscillates.]

Proof. a) *To prove $\lambda \leq \mu$ for the sequence $\{x_n\}$ of real numbers.*

If this statement be not true, then λ should be greater than μ (i.e., $\lambda - \mu > 0$). Choose $\varepsilon = \frac{1}{2}(\lambda - \mu)$ (i.e., $\lambda - \varepsilon = \mu + \varepsilon$).

By definition of λ, $\exists$ only finite number of terms of $\{x_n\}$ less than $\lambda - \varepsilon$; but by definition of μ, there are infinite number of terms of $\{x_n\} < \mu + \varepsilon (= \lambda - \varepsilon)$. The two statements are clearly contradictory. Hence $\lambda \not> \mu$, i.e., $\lambda \leq \mu$.

Proof. b) Let $\{x_n\}$ converge to a finite limit l.

To prove $\underline{\lim} x_n = \overline{\lim} x_n = l$. (i.e., $\lambda = \mu = l$).

Since $\{x_n\}$ converges, it must be bounded and $\overline{\lim} x_n = \mu$ and $\underline{\lim} x_n = \lambda$ both exist, where $\lambda \leq \mu$. We shall prove that $\mu = \lambda = l$.

If possible, let $\mu > l$. Choose $\varepsilon = \frac{1}{2}(\mu - l)$.

(i) Then $\exists$ a positive integer N_1 such that $x_n < l + \varepsilon$, $\forall n > N_1$ (by definition of limit).

(ii) But by definition of μ, $\exists$ infinite number of terms of $\{x_n\} > \mu - \varepsilon$. See that we have chosen ε in such a way that $\mu - \varepsilon = l + \varepsilon$.

Then the two statement (i) and (ii) cannot be simultaneously true.

Hence, $\mu \not> l$, i.e., $\mu \leq l$.

In a similar manner, $\lambda \not< l$ i.e., $\lambda \geq l$. But $\mu \geq \lambda$.

$\therefore$ the only conclusion is $\lambda = \mu = l$.

Inversely, suppose $\mu = \lambda = l$ (say). Then for any $\varepsilon > 0$, $\exists$ positive integers N_1 and N_2 such that

$$x_n < \mu + \varepsilon, \quad \forall n > N_1 \quad (\text{definition of } \mu)$$
$$x_n > \lambda - \varepsilon, \quad \forall n > N_2 \quad (\text{definition of } \lambda).$$

Since we have assumed $\lambda = \mu = l$, it follows that $\forall\, n > N = \max\{N_1, N_2\}$,

$$l - \varepsilon < x_n < l + \varepsilon$$

or, $|x_n - l| < \varepsilon$, i.e., $\lim_{n \to \infty} x_n - l.$

In other words, $\{x_n\}$ converges to l.

Proof. c) If $\lambda = \mu = +\infty$, then for any $G > 0$ arbitrarily chosen, $\exists$ positive integer N s.t. $\forall\, n > N, x_n > G$, i.e., we then have $\lim_{n \to \infty} x_n = +\infty$.

Conversely, if $\lim_{n \to \infty} x_n = +\infty$, then given any $G > 0$, we have a positive integer N s.t. $\forall\, n > N, x_n > G$; therefore, there are at most a finite number of n's for which $x_n < G$ and infinite number of n's for which $x_n > G$.

The sequence $\{x_n\}$ is then bounded below but not bounded above and $\exists$ no finite upper limit; hence, $\liminf x_n = \lambda = +\infty$ and *ipso facto* also $\mu = +\infty$. Hence, $\mu = \lambda = +\infty$.

Proof. d) In a similar manner we may prove that if $\lambda = \mu = -\infty$, then the sequence diverges to $-\infty$ and conversely.

Important Inequalities

Theorem 5.14.8. *If $\{x_n\}$ and $\{y_n\}$ are bounded sequences of real numbers, then*

1. $\overline{\lim}\, (x_n + y_n) \leq \overline{\lim}\, x_n + \overline{\lim}\, y_n.$

2. $\underline{\lim}\, (x_n + y_n) \geq \underline{\lim}\, x_n + \underline{\lim}\, y_n.$

Proof.

a) Let $\overline{\lim}\, x_n = \mu_1$, $\overline{\lim}\, y_n = \mu_2$, $\overline{\lim}\, (x_n + y_n) = \mu$.

To prove:

$$\mu \leq \mu_1 + \mu_2.$$

Let ε be any arbitrary positive number; $\exists$ positive integers m_1 and m_2 such that

$$x_n < \mu_1 + \tfrac{1}{2}\varepsilon, \quad \forall\, n \geq m_1$$

$$y_n < \mu_2 + \tfrac{1}{2}\varepsilon, \quad \forall\, n \geq m_2.$$

Hence, $x_n + y_n < (\mu_1 + \mu_2) + \varepsilon$, $\forall\, n > m$, where $m = \max.(m_1, m_2)$.

This proves that $\overline{\lim}\, (x_n + y_n)$ cannot exceed $\mu_1 + \mu_2$ (ε being arbitrary)

or, $\overline{\lim}\, (x_n + y_n) \leq \mu_1 + \mu_2$ i.e., $\leq \overline{\lim}\, x_n + \overline{\lim}\, y_n.$

CHAPTER 5: SEQUENCE OF REAL NUMBERS

b) Proof is similar to (a).

Note 5.14.1. *It is **not** always true that*

$$\overline{\lim}(x_n + y_n) = \overline{\lim} x_n + \overline{\lim} y_n,$$

even if the sequences $\{x_n\}$ and $\{y_n\}$ are bounded.

e.g., if $x_n = (-1)^n$ and $y_n = (-1)^{n+1}$, then $x_n + y_n = 0$, $\forall\, n$.
Here $\overline{\lim} x_n = +1$ and $\overline{\lim} y_n = +1$ but $\overline{\lim}(x_n + y_n) = 0$.

c) If $x_n \leq y_n \;\forall\, n > N$ (where N is a fixed positive integer), then

$$\underline{\lim} x_n \leq \underline{\lim} y_n \quad \text{and} \quad \overline{\lim} x_n \leq \overline{\lim} y_n.$$

d) $\overline{\lim}(x_n y_n) \leq \overline{\lim} x_n \cdot \overline{\lim} y_n, \quad$ but $\underline{\lim}(x_n y_n) \geq \underline{\lim} x_n \cdot \underline{\lim} y_n.$

e) $\underline{\lim} x_n - \underline{\lim} y_n \leq \overline{\lim}(x_n - y_n) \leq \overline{\lim} x_n - \overline{\lim} y_n.$

Here the notation $\overline{\underline{\lim}}(x_n - y_n)$ means $\underline{\lim}(x_n - y_n) \leq \overline{\lim}(x_n - y_n)$ which is obviously true.

The proofs of the results (c)–(e) are left as challenging exercises for the students.

AN INTRODUCTION TO ANALYSIS

(Differential Calculus—Part I)

Chapter 6

Infinite Series of Real Constants

6(A) Convergence of Infinite Series, Divergence, Cauchy sequence etc

6(B) Tests for Series of Positive Terms

6(C) Test for series of Arbitrary Terms

Chapter 6

Infinite Series of Real Constants

6(A) Convergence of Infinite Series, Divergence, Cauchy sequence etc

6.1 Introduction

Let $\{a_n\}$ be a sequence of real numbers.

Then the symbolic expression

$$a_1 + a_2 + a_3 + \cdots + a_n + \cdots$$
$$\text{or, } \sum_{n=1}^{\infty} a_n \text{ or, simply } \sum a_n \qquad (6.1.1)$$

will be called *an infinite series* of real numbers (also called *infinite series of real constants*).

The numbers $a_1, a_2, \cdots, a_n, \cdots$ are called its *terms* (a_n is the nth term of the series) and the sequence $\{S_n\}$, where $S_n = \sum_{r=1}^{n} a_r = a_1 + a_2 + \cdots + a_n$, is called the *sequence of partial sums* of the series $\sum_{n=1}^{\infty} a_n$.

1. The infinite series (6.1.1) is said to be *convergent* if

$$\lim_{n \to \infty} S_n \text{ exists and } = \text{a finite real number } S.$$

We then say that the series $\sum a_n$ has the sum S and we write

$$S = \sum_{n=1}^{\infty} a_n.$$

e.g., To prove that

$$\sum_{r=1}^{\infty} \frac{1}{r(r+1)} = \frac{1}{1.2} + \frac{1}{2.3} + \frac{1}{3.4} + \cdots\cdots + \frac{1}{n(n+1)} + \cdots$$

is convergent and that its sum is 1.

We consider,

$$\begin{aligned}
S_n &= \frac{1}{1.2} + \frac{1}{2.3} + \frac{1}{3.4} + \cdots\cdots + \frac{1}{n(n+1)} \\
&= \left(\frac{1}{1} - \frac{1}{2}\right) + \left(\frac{1}{2} - \frac{1}{3}\right) + \left(\frac{1}{3} - \frac{1}{4}\right) + \cdots + \left(\frac{1}{n} - \frac{1}{n+1}\right) \\
&= \frac{1}{1} - \frac{1}{n+1}.
\end{aligned}$$

Hence, as $n \to \infty$, $S_n \to 1$. This proves that the series is convergent and that its sum is 1.

2. If $\lim_{n \to \infty} S_n = +\infty$ or $-\infty$, then the series $\sum a_n$ is said to diverge properly to $+\infty$ or $-\infty$ as the case may be.

e.g., $\sum_{r=1}^{\infty} r = 1 + 2 + 3 + \cdots + n + \cdots$ diverges to $+\infty$.

For,

$$\begin{aligned}
S_n &= 1 + 2 + 3 + \cdots + n \\
&= \frac{n(n+1)}{2} \longrightarrow \infty; \text{ as } n \to \infty.
\end{aligned}$$

3. Again, if $\{S_n\}$ is oscillating finitely then the series $\sum a_n$ is said to oscillate finitely.

e.g., $\sum_{r=1}^{\infty} (-1)^{r+1} = 1 - 1 + 1 - 1 + 1 - \cdots$.

Here

$$S_n = \begin{cases} 1, & \text{if } n \text{ is odd}; \\ 0, & \text{if } n \text{ is even}. \end{cases}$$

So $\{S_n\}$ oscillates between 0 and 1 and consequently $\sum_{r=1}^{\infty} (-1)^{r+1}$ also oscillates between 0 and 1.

4. Lastly if $\{S_n\}$ is oscillating infinitely then $\sum a_n$ is said to oscillate infinitely.

e.g., $\sum_{r=1}^{\infty}(-1)^{r+1}r = 1 - 2 + 3 - 4 + 5 - 6 + \cdots$ oscillates infinitely.

Here
$$S_n = \begin{cases} \frac{1}{2}(n+1), & \text{if } n \text{ is odd;} \\ -\frac{1}{2}n, & \text{if } n \text{ is even.} \end{cases}$$

Hence $\{S_n\}$ oscillates between $-\infty$ and ∞ and consequently $\sum a_n$ does so.

The present chapter will investigate the convergence or divergence of infinite series of real constants.

6.2 Two Important Series

I. Geometric Series: *The infinite Geometric Series*

$$a + ar + ar^2 + \cdots + ar^{n-1} + \cdots \quad (a > 0) \quad \text{is}$$

(a) *convergent if the common ratio r lies between -1 and $+1$ (i.e., $|r| < 1$); in this case the sum of the series is $a/(1-r)$.*

(b) *properly divergent (to $+\infty$) if $r \geq 1$.*

(c) $\left.\begin{array}{l}\text{oscillates finitely,} \quad \text{if } r = -1 \text{ and} \\ \text{oscillates infinitely,} \quad \text{if } r < -1\end{array}\right\}$ **Improperly divergent.**

Proof:
$$\begin{aligned} S_n &= a + ar + ar^2 + \cdots + ar^{n-1} \\ &= a\frac{r^n - 1}{r - 1} \quad (r \neq 1). \end{aligned}$$

(a) Now if $|r| < 1$ then $r^n \to 0$ as $n \to \infty$.
$$\therefore \quad S = \lim_{n \to \infty} S_n = \frac{a}{1-r}.$$

(b) (i) If $r > 1, r^n \to \infty$ as $n \to \infty$ and then $S_n \to \infty$ as $n \to \infty$; hence the geometric series diverges to $+\infty$, if $r > 1$.

(ii) If $r = 1$, $S_n = a + a + \cdots + a = na \to \infty$ as $n \to \infty$.
$\therefore$ the series properly diverges to $+\infty$.

(c) (i) If $r < -1$, $\{r^n\}$ oscillates infinitely.

∴ $\{S_n\}$ oscillates infinitely and the series also behaves in the same way.

(ii) If $r = -1$, we have

$$S_n = a - a + a - a + \cdots \text{ to } n \text{ terms.}$$

$$\therefore S_n = \begin{cases} a, & \text{if } n \text{ is odd;} \\ 0, & \text{if } n \text{ is even.} \end{cases}$$

∴ the series oscillates finitely.

Thus the Geometric series $\sum_{r=1}^{\infty} ar^{n-1}$ converges only when $|r| < 1$.

II. The p-series: *The infinite series*

$$\sum_{n=1}^{\infty} \frac{1}{n^p} = \frac{1}{1^p} + \frac{1}{2^p} + \frac{1}{3^p} + \cdots + \frac{1}{n^p} + \cdots$$

(a) *converges if $p > 1$.*

(b) *diverges if $p \leq 1$.*

Proof: a) Suppose $p > 1$. To prove that $\sum_{m=1}^{\infty} \frac{1}{m^p}$ converges.

We examine partial sums of order $2^n - 1$ where n is a certain positive integer.

$$S_{2^n-1} = 1 + \left(\frac{1}{2^p} + \frac{1}{3^p}\right) + \left(\frac{1}{4^p} + \frac{1}{5^p} + \cdots + \frac{1}{7^p}\right)$$

$$+ \left(\frac{1}{8^p} + \frac{1}{9^p} + \cdots + \frac{1}{15^p}\right) + \cdots$$

$$+ \left\{\frac{1}{(2^{n-1})^p} + \frac{1}{(2^{n-1}+1)^p} + \cdots + \frac{1}{(2^n-1)^p}\right\}$$

$$\leq 1 + \left(\frac{1}{2^p} + \frac{1}{2^p}\right) + \left(\frac{1}{4^p} + \frac{1}{4^p} + \cdots + \frac{1}{4^p}\right) + \left(\frac{1}{8^p} + \frac{1}{8^p} + \cdots + \frac{1}{8^p}\right)$$

$$+ \cdots + \left\{\frac{1}{(2^{n-1})^p} + \frac{1}{(2^{n-1})^p} + \cdots + \frac{1}{(2^{n-1})^p}\right\}$$

$$= 1 + 2 \cdot \frac{1}{2^p} + 4 \cdot \frac{1}{4^p} + 8 \cdot \frac{1}{8^p} + \cdots + 2^{n-1} \cdot \frac{1}{(2^{n-1})^p}$$

$$= 1 + \frac{1}{2^{p-1}} + \frac{1}{(2^{p-1})^2} + \frac{1}{(2^{p-1})^3} + \cdots + \frac{1}{(2^{p-1})^{n-1}}$$

$$= \frac{1 - \left(\frac{1}{2^{p-1}}\right)^n}{1 - \frac{1}{2^{p-1}}} \quad \text{i.e.,} \quad < \frac{1}{1 - \frac{1}{2^{p-1}}}, \text{ which is a constant } K \text{ (say) } \forall n.$$

CHAPTER 6: INFINITE SERIES OF REAL CONSTANTS

Now for any positive integer m, $\exists$ a positive integer n such that

$$2^n - 1 > m.$$

$\{S_m\}$ is clearly monotone increasing ($\because$ the terms are all positive).

$$\therefore \quad S_m < S_{2^n-1} < K, \ \forall \, m.$$

i.e., $\{S_m\}$ is m.i. and bounded above and hence convergent.

$$\therefore \text{ the series } \sum \frac{1}{m^p} \text{ converges if } p > 1.$$

b) Suppose $p \leq 1$. To prove $\sum_{m=1}^{\infty} \frac{1}{m^p}$ diverges.

First observe, when $p \leq 1$, $n^p \leq n$ (i.e., $1/n^p \geq 1/n$), n being a positive integer.
We now examine the sequence of partial sums of order 2^n.

$$\begin{aligned}
S_{2^n} &= 1 + \frac{1}{2^p} + \left(\frac{1}{3^p} + \frac{1}{4^p}\right) + \left(\frac{1}{5^p} + \frac{1}{6^p} + \frac{1}{7^p} + \frac{1}{8^p}\right) + \cdots \\
&\quad \cdots + \left\{\frac{1}{(2^{n-1}+1)^p} + \cdots + \frac{1}{(2^n)^p}\right\} \\
&\geq \left(1 + \frac{1}{2}\right) + \left(\frac{1}{3} + \frac{1}{4}\right) + \left(\frac{1}{5} + \frac{1}{6} + \frac{1}{7} + \frac{1}{8}\right) + \cdots \\
&\quad \cdots + \left(\frac{1}{2^{n-1}+1} + \frac{1}{2^{n-1}+2} + \cdots + \frac{1}{2^n}\right) \\
&\geq \left(1 + \frac{1}{2}\right) + \left(\frac{1}{4} + \frac{1}{4}\right) + \left(\frac{1}{8} + \frac{1}{8} + \frac{1}{8} + \frac{1}{8}\right) + \cdots \\
&\quad \cdots + \left(\frac{1}{2^n} + \frac{1}{2^n} + \cdots + \frac{1}{2^n}\right) \\
&= 1 + \frac{1}{2} + 2 \cdot \frac{1}{4} + 4 \cdot \frac{1}{8} + \cdots + 2^{n-1} \frac{1}{2^n}.
\end{aligned}$$

i.e., $S_{2^n} \geq 1 + \frac{n}{2}$.

$\therefore$ Given any $G > 0$, we have $S_{2^n} > G$ whenever $1 + \frac{n}{2} > G$ i.e., whenever $n > 2G - 2$.
The partial sums are obviously monotone increasing (m.i.).

$$\therefore \text{ if } m > 2^n, \text{ then } S_m > S_{2^n} > G \text{ for } m > 2^{2G-2}.$$

i.e., The sequence of partial sums $\{S_m\}$ is m.i. and unbounded above and hence tends to $+\infty$ as $n \to \infty$.

Consequently the series $\sum(1/m^p)$ diverges to $+\infty$, when $p \leq 1$.

In particular, **the Harmonic series.**

$$\sum_{n=1}^{\infty} \frac{1}{n} = 1 + \frac{1}{2} + \frac{1}{3} + \cdots + \frac{1}{n} + \cdots$$

diverges to $+\infty$. (**Write down an independent proof**).

Q. Which of the three series — $\sum \frac{1}{\sqrt{n}}, \sum \frac{1}{n}, \sum \frac{1}{n^2}$ is convergent?

6.3 Cauchy's Convergence Criterion applied to infinite series

Theorem 6.3.1. *The infinite series $\sum_{n=1}^{\infty} a_n$ converges if and only if given any $\varepsilon > 0$, there exists a positive integer $N(\varepsilon)$, such that whenever $n \geq N(\varepsilon)$*

$$|a_{n+1} + a_{n+2} + \cdots + a_{n+p}| < \varepsilon \quad (p = 1, 2, 3, \cdots).$$

Alternative Statement: A necessary and sufficient condition that an infinite series $\sum_{n=1}^{\infty} a_n$ converges is that, for each $\varepsilon > 0$, there exists a positive integer $N(\varepsilon)$ for which

$$|a_{n+1} + a_{n+2} + \cdots + a_m| < \varepsilon$$

whenever $m, n \geq N$. (one may take $m > n \geq N$).

[For m, if we take $n + p$, the previous form of the statement follows.]

Proof: **Theorem 1.** The series $\sum a_n$ converges if and only if the sequence of partial sums $\{S_n\}$ converges. By Cauchy's General Principle of Convergence of a sequence, the sequence $\{S_n\}$ converges if and only if, given any $\varepsilon > 0$, $\exists$ a positive integer $N(\varepsilon)$ such that whenever $n \geq N(\varepsilon)$ and $p = $ a positive integer, the inequality

$$|S_{n+p} - S_n| < \varepsilon \text{ holds.}$$

i.e., $\left| \sum_{k=1}^{n+p} a_k - \sum_{k=1}^{n} a_k \right| < \varepsilon$ holds

i.e., $|a_{n+1} + a_{n+2} + \cdots + a_{n+p}| < \varepsilon$ holds.

Taking $p = 1$ in Theorem 6.3.1 we obtain the following result:

Theorem 6.3.2. *If $\sum_{n=1}^{\infty} a_n$ converges then $\lim_{n \to \infty} a_n = 0$.*

CHAPTER 6: INFINITE SERIES OF REAL CONSTANTS

Proof: In fact, putting $p = 1$ in Cauchy's convergence criterion we get:
If $\sum a_n$ converges, then for any $\varepsilon > 0$, $\exists N(\varepsilon)$ such that

$$|a_{n+1}| < \varepsilon, \ \forall \, n \geq N(\varepsilon)$$

which implies, $\lim_{n \to \infty} a_{n+1} = 0$, or $\lim_{k \to \infty} a_k = 0$ or we can write $\lim_{n \to \infty} a_n = 0$.

Alternative Proof:

$$\lim_{n \to \infty} a_n = \lim_{n \to \infty} (S_n - S_{n-1})$$
$$= \lim_{n \to \infty} S_n - \lim_{n \to \infty} S_{n-1} = S - S = 0.$$

$$\left(\because \sum_{n=1}^{\infty} a_n \text{ converges} \Rightarrow \lim_{n \to \infty} S_n = S \right)$$

Observations:

This theorem is useful in its contrapositive form.

1. According to Theorem **6.3.2**,

   If $\sum_{n=1}^{\infty} a_n$ converges, then $a_n \to 0$ as $n \to \infty$.

   $\therefore$ the contra-positive statement must be true:

   > If a_n does not tend to zero as $n \to \infty$, then the series $\sum a_n$ does not converge.

Note: $a_n \to 0$ as $n \to \infty$ is the necessary condition of convergence of the series $\sum a_n$.

Example 6.3.1. $\sum_{n=1}^{\infty} \dfrac{n^n}{n!}$ *cannot converge, because*

$$a_n = \frac{n^n}{n!} = \frac{n \cdot n \cdot n \cdots n}{1 \cdot 2 \cdot 3 \cdots n} > 1 \text{ whenever } n > 1.$$

$\therefore \ a_n$ cannot tend to zero as $n \to \infty$ and consequently $\sum a_n$ cannot converge.

[C.H. 1992]

Example 6.3.2. $\sum_{n=1}^{\infty} \dfrac{n}{n+1}$ *does not converge because* $u_n = \dfrac{n}{n+1} \to 1$ as $n \to \infty$.

2. The condition that a_n must tend to zero when $\sum a_n$ converges is not a *sufficient* condition.

For e.g., the Harmonic Series $\sum_{n=1}^{\infty} \frac{1}{n}$ diverges, but its nth term $\frac{1}{n} \to 0$ as $n \to \infty$.

Consider the Harmonic series $\sum_{n=1}^{\infty} \frac{1}{n}$.

For any $n \in \mathbb{N}$.

$$\begin{aligned}
S_{2^n} &= \left(1 + \frac{1}{2} + \frac{1}{3} + \frac{1}{4} + \frac{1}{5} + \cdots + \frac{1}{2^n}\right) \\
&= \left(1 + \frac{1}{2}\right) + \left(\frac{1}{3} + \frac{1}{4}\right) + \left(\frac{1}{5} + \frac{1}{6} + \frac{1}{7} + \frac{1}{6}\right) \\
&\quad + \cdots + \left(\frac{1}{2^{n-1}+1} + \frac{1}{2^{n-1}+2} + \cdots + \frac{1}{2^n}\right) \\
&> 1 + \frac{1}{2} + \left(\frac{1}{4} + \frac{1}{4}\right) + \left(\frac{1}{8} + \frac{1}{8} + \frac{1}{8} + \frac{1}{8}\right) + \cdots + \left(\frac{1}{2^n} + \frac{1}{2^n} + \cdots + \frac{1}{2^n}\right) \\
&= 1 + \left(\frac{1}{2} + \frac{1}{2} + \cdots + \frac{1}{2}\right) = 1 + \frac{n}{2}.
\end{aligned}$$

It follows that $\{S_n\}$ is monotonic increasing and unbounded. For obvious details see art. 6.2 — the proof of p-series (b) and hence diverges. Therefore $\sum_{n=1}^{\infty} \frac{1}{n}$ diverges yet,

$$\lim_{n \to \infty} \frac{1}{n} = 0.$$

Q. What can you say about the series $\sum_{n=1}^{\infty} \frac{n}{2n+1}$?, $\sum_{n=1}^{\infty} \frac{n}{n+10^3}$?

(Ans. They do not converge since $\frac{n}{2n+1} \to \frac{1}{2}$ as $n \to \infty$ and $\frac{1}{1+10^3/n} \to 1 \neq 0$ as $n \to \infty$)

Theorem 6.3.3. Abel's Theorem: (*also known as* **Abel-Pringsheim's Theorem**).

If $\{a_n\}$ is m.d., the condition $\lim_{n \to \infty} na_n = 0$ is a necessary condition for the convergence of the positive series $\sum a_n$.   [C.H. 1983, 92]

For, if $\sum a_n$ converges, corresponding to any $\varepsilon > 0$, $\exists$ a positive integer m such that

$$a_{m+1} + a_{m+2} + \cdots + a_n < \varepsilon/2 \text{ if } n > m.$$

Now each term of the left side $\geq a_n$.

Therefore $(n-m)a_n \leq a_{m+1} + a_{m+2} + \cdots + a_n < \varepsilon/2$, if $n > m$.

We know that if $\sum a_n$ converges, then $a_n \to 0$ as $n \to \infty$.

$\therefore$ we can choose $N > m$ so that $m\, a_n < \varepsilon/2$ if $n > N$.

Thus $n\, a_n - m\, a_n < \varepsilon/2$ $(n > m)$ gives $n\, a_n < \varepsilon/2 + \varepsilon/2$ $\forall n > N$.

$\therefore \quad n\, a_n \to 0$ as $n \to \infty$.

Note 6.3.1. *That this condition is not sufficient, follows from Abel's Series* $\sum \dfrac{1}{n \log n}$ *which is a divergent series (See Example 6.6.27) although* $\lim\limits_{n \to \infty} n \cdot \dfrac{1}{n \log n} = 0$ *and* $\left\{\dfrac{1}{n \log n}\right\}$ *is m.d.*

Example 6.3.3. *The series* $\sum \dfrac{1}{an+b}$ $(a, b > 0)$ *diverges.*

Let $a_n = \dfrac{1}{an+b}$. Here $\{a_n\}$ is m.d.

$$\lim_{n\to\infty} n\, a_n = \lim_{n\to\infty} \frac{n}{an+b} = \lim_{n\to\infty} \frac{1}{a+b/n} = \frac{1}{a} \neq 0.$$

$\therefore \sum a_n$ **cannot converge,** for if it would, then $n\, a_n$ would tend to zero as $n \to \infty$.

Being a positive series $\sum a_n$ cannot oscillate, so the only conclusion is that $\sum a_n$ diverges.

This theorem gives a negative test:

> If $n\, a_n$ does not tend to zero as n tends to infinity, then $\sum a_n$ does not converge.

6.4 Applications of Cauchy's Principle

Example 6.4.1. *Prove by using Cauchy criterion, that the series:*

$$\sum_{n=0}^{\infty} \frac{1}{n!} = 1 + \frac{1}{\lfloor 1} + \frac{1}{\lfloor 2} + \frac{1}{\lfloor 3} + \frac{1}{\lfloor 4} + \cdots + \frac{1}{\lfloor n} + \cdots \quad \text{converges.}$$

Solution.

$$\left|\sum_{k=n+1}^{n+p} \frac{1}{k!}\right| = \frac{1}{(n+1)!} + \frac{1}{(n+2)!} + \frac{1}{(n+3)!} + \cdots + \frac{1}{(n+p)!}$$

$$< \frac{1}{2^n} + \frac{1}{2^{n+1}} + \cdots + \frac{1}{2^{n+p-1}}$$

$$= \frac{1}{2^n} \frac{1-(1/2)^p}{1-(1/2)}, \text{ which is } < \frac{1}{2^n} \frac{1}{1-\frac{1}{2}}, \forall p.$$

i.e., $\left|\sum_{k=n+1}^{n+p} \frac{1}{k!}\right| < 2\left(\frac{1}{2}\right)^n, \forall p.$

$\because \left(\frac{1}{2}\right)^n \to 0$ as $n \to \infty$, therefore for each $\varepsilon > 0 \ \exists \ N(\varepsilon)$ for which $2\left(\frac{1}{2}\right)^n < \varepsilon$ if $n > N$.

$\therefore \left|\sum_{k=n+1}^{n+p} \frac{1}{k!}\right| < \varepsilon$, whenever $n > N$ and $p \geq 1$ and the series converges by Cauchy's criterion.

Example 6.4.2. *Use Cauchy's criterion to prove that the series*

$$1 - \frac{1}{2} + \frac{1}{3} - \frac{1}{4} + \frac{1}{5} - \frac{1}{6} + \cdots + (-1)^{n-1}\frac{1}{n} + \cdots \text{ is convergent.}$$

Solution.

$\left|\sum_{k=n+1}^{n+p} (-1)^{k-1}\frac{1}{k}\right| = \left|(-1)^n\frac{1}{n+1} + (-1)^{n+1}\frac{1}{n+2} + \cdots + (-1)^{n+p-1}\frac{1}{n+p}\right|$

$= |(-1)^n|\left|\frac{1}{n+1} - \frac{1}{n+2} + \frac{1}{n+3} - \cdots + (-1)^{p-1}\frac{1}{n+p}\right|$

$= \frac{1}{n+1} - \left(\frac{1}{n+2} - \frac{1}{n+3}\right) - \cdots$, which is $< \frac{1}{n+1} \ \forall \ p.$

$\therefore \left|\sum_{k=n+1}^{n+p} (-1)^{k-1}\frac{1}{k}\right| < \varepsilon$ if $n + 1 > \frac{1}{\varepsilon}$ and $\forall \ p = 1, 2, \cdots$.

i.e., if $n > N$ and $p = 1, 2, 3, \cdots$ where N = integral part of $(1/\varepsilon - 1)$.

Example 6.4.3. *Using Cauchy's principle prove that the Harmonic series*

$$\sum_{n=1}^{\infty} \frac{1}{n} = 1 + \frac{1}{2} + \frac{1}{3} + \frac{1}{4} + \cdots + \frac{1}{n} + \cdots$$

cannot converge.

Solution.

$\left|\frac{1}{n+1} + \frac{1}{n+2} + \cdots + \frac{1}{n+p}\right| = \left|\frac{1}{n+1} + \frac{1}{n+2} + \cdots + \frac{1}{n+n}\right|$, when $p = n$

$= \frac{1}{n+1} + \frac{1}{n+2} + \cdots + \frac{1}{2n}$

$> \frac{1}{2n} + \frac{1}{2n} + \cdots + \frac{1}{2n}$ (n terms) $= n \cdot \frac{1}{2n}.$

i.e., $\left|\sum_{k=n+1}^{n+p} \frac{1}{k}\right| > \frac{1}{2} \ \forall \ n$ and $p = n.$

∴ if we choose $\varepsilon < \frac{1}{2}$ then $\exists\, n$ and p s.t. $|S_{n+p} - S_n| > \varepsilon$.

In other words, Cauchy's condition is not satisfied.

i.e., the series does not converge.

[Being a positive series it cannot oscillate also; the only conclusion is that the series $\sum(1/n)$ diverges to $+\infty$.]

6.5 A Few General Theorems on Convergence or Divergence

Theorem 6.5.1. *Let N be a fixed positive integer.*

(a) *The series* $\displaystyle\sum_{k=1}^{\infty} a_k$ *and* $\displaystyle\sum_{k=N+1}^{\infty} a_k$ *are both convergent or both divergent.*

(b) *If the series* $\displaystyle\sum_{k=1}^{\infty} a_k$ *converges to S, then*

$$\sum_{k=N+1}^{\infty} a_k \text{ converges to } S - (a_1 + a_2 + a_3 + \cdots + a_N).$$

Proof: (a) Let $n > N$ and $T_n = \displaystyle\sum_{k=N+1}^{n} a_k$ and

let
$$K = a_1 + a_2 + \cdots + a_N = \text{ a fixed number.}$$

Then,
$$S_n = a_1 + a_2 + \cdots + a_N + a_{N+1} + \cdots + a_n$$
$$= K + T_n.$$

Letting $n \to \infty$, the results of the theorem can be easily verified. In fact, taking limit as $n \to \infty$ we obtain $\displaystyle\sum_{k=N+1}^{\infty} a_k = S - (a_1 + a_2 + \cdots + a_N)$.

i.e., *Removal or Addition of a finite number of terms at the beginning of an infinite series does not affect the convergence or divergence of a series.*

Theorem 6.5.2. *Let λ be any positive constant. Then the infinite series $\sum_{k=1}^{\infty} a_k$ and $\sum_{k=1}^{\infty} (\lambda a_k)$ are both convergent or both divergent.*

Proof: Let $S_n = a_1 + a_2 + \cdots + a_n$ and $T_n = \lambda a_1 + \lambda a_2 + \cdots + \lambda a_n$. Then, $T_n = \lambda S_n$. Now make $n \to \infty$. The results will follow from definitions.

Corollary 6.5.1. *If λ be any non-zero constant (i.e., $\lambda > 0$ or $\lambda < 0$) and $\sum_{k=1}^{\infty} a_k$ diverges, then $\sum_{k=1}^{\infty} \lambda a_k$ also diverges.*

Thus, *multiplication of each term of a series by a non-zero constant does not affect the behaviour of the series.*

Theorem 6.5.3. (Associativity). *If the terms of a convergent series are grouped in parentheses in any manner to form new terms (the order of the terms remaining unaltered), then the resulting series will converge and converges to the same sum.*

i.e., if the series $a_1 + a_2 + a_3 + \cdots + a_n + \cdots$ converges to S, then the series

$$\underbrace{(a_1 + a_2 + \cdots + a_{n_1})}_{u_1} + \underbrace{(a_{n_1+1} + a_{n_1+2} + \cdots + a_{n_2})}_{u_2} + \cdots$$

also converges to S.

Proof: Let $S_n = a_1 + a_2 + \cdots + a_n$. Then, since $\sum a_n$ converges to S, we have $S_n \to S$ as $n \to \infty$. Hence $S_{n_p} \to S$ as $n_p \to \infty$.

The partial sum of the second series is

$$u_1 + u_2 + \cdots + u_p = S_{n_p}; \text{ hence etc.}$$

Caution. It is not true, however, that if a series converges after parenthesis are inserted, it will still converge when they are removed:

For example,

1. The series $(1-1) + (1-1) + (1-1) + (1-1) + \cdots$ converges to zero but the series

$$1 - 1 + 1 - 1 + 1 - 1 + 1 - 1 + \cdots$$

does not converge (it oscillates between 0 and 1).

CHAPTER 6: INFINITE SERIES OF REAL CONSTANTS

2. The series
$$\left(\frac{3}{2}-\frac{4}{3}\right)+\left(\frac{5}{4}-\frac{6}{5}\right)+\cdots+\left(\frac{2n+1}{2n}-\frac{2n+2}{2n+1}\right)+\cdots$$
is a convergent series; because
$$S_n = \sum_{k=1}^{n} a_k, \text{ where } a_k = \left(\frac{2k+1}{2k}-\frac{2k+2}{2k+1}\right) = \left(\frac{1}{2k}-\frac{1}{2k+1}\right)$$
$$= a_1 + a_2 + \cdots + a_n$$
$$= \left(\frac{3}{2}-\frac{4}{3}\right)+\left(\frac{5}{4}-\frac{6}{5}\right)+\cdots+\left(\frac{2n+1}{2n}-\frac{2n+2}{2n+1}\right)$$
$$= \left(\frac{1}{2}-\frac{1}{3}\right)+\left(\frac{1}{4}-\frac{1}{5}\right)+\cdots+\left(\frac{1}{2n}-\frac{1}{2n+1}\right)$$
$$= \frac{1}{2}-\frac{1}{3}+\frac{1}{4}-\frac{1}{5}+\cdots+\frac{1}{2n}-\frac{1}{2n+1}$$

(since the sum is finite we can remove parenthesis).

$$S_{n+1} - S_n = \frac{1}{2n+2} - \frac{1}{2n+3} > 0 \text{ i.e., the sequence } \{S_n\} \text{ is } m.i.$$
and $$S_n = \frac{1}{2} - \left(\frac{1}{3}-\frac{1}{4}\right) - \left(\frac{1}{5}-\frac{1}{6}\right) - \cdots - \left(\frac{1}{2n-1}-\frac{1}{2n}\right) - \frac{1}{2n+1}$$
$$< \frac{1}{2} \quad \forall\, n.$$

i.e., $\{S_n\}$ is bounded.

∴ $\{S_n\}$ converges and consequently the series with parenthesis converges.

But the series (removing the parenthesis) becomes:
$$\frac{3}{2}-\frac{4}{3}+\frac{5}{4}-\frac{6}{5}+\cdots+(-1)^{n-1}\frac{n+2}{n+1}+\cdots.$$

Its nth term does not tend to zero as $n \to \infty$ and the series, therefore, cannot converge. Rather it is finitely oscillating.

3. It is to be remembered, however, that if a series with parenthesis diverges, then the series without parenthesis cannot converge: for if it converged, it would still converge after grouping the terms in parenthesis.

What is interesting to note is that:

If a series whose separate terms are sums in parenthesis converges to S, then the series with parenthesis omitted will converge to the same sum S, if it converges at all.

[This follows from Theorem **6.5.3** above; if the series without parenthesis would converge to S', then inserting parenthesis will not change its sum by this Theorem; hence $S' = S$.]

The next result, is of great utility, as will be observed in some of the theorems that follow.

Theorem 6.5.4. *Let $\{x_n\}_n$ be a sequence of postive real numbers. Then the series $\sum_{n=1}^{\infty} x_n$ converges if and only if the sequence $\{S_n\}$ of partial sums is bounded. In fact*

$$\sum_{n=1}^{\infty} x_n = \lim_{n \to \infty} S_n = \sup\{S_n : n \in \mathbb{N}\}$$

[Since $x_n > 0$, the sequence S_n of partial sums is monotone increasing:

$$S_1 < S_2 < S_3 < \cdots < S_n < \cdots,$$

and hence by theorem of convergence of monotone increasing sequence, the sequence $\{S_n\}$ converges iff it is bounded. $\therefore \sum_{n=1}^{\infty} x_n$ converges iff $\{S_n\}$ is bounded.]

6(B) Series of Positive Terms

6.6 Series of Positive Terms: Tests of Convergence or Divergence

We shall, in this section, consider series of the form $\sum^{\infty} a_n$, where $a_n > 0$.

For such a series every a_n is a positive real number and hence the sequence $\{S_n\}$ of partial sums is always monotone increasing. So if we can prove that $\{S_n\}$ is bounded above, then S_n must tend to a finite limit S as $n \to \infty$ and the series becomes convergent. [In fact $S = $ lub of the set partial sums $\{S_n : n = 1, 2, 3, \ldots\}$]. On the other hand, if $\{S_n\}$ is unbounded above, then $S_n \to +\infty$ as $n \to \infty$ and the series $\sum_{n=1}^{\infty} a_n$ becomes divergent.

> REMEMBER: A positive series can never oscillate; either it will converge or if will diverge to $+\infty$.

Example 6.6.1. *Prove that the series $\sum_{n=1}^{\infty} \dfrac{1}{n^2}$ converges.*

Solution: Here, the series is a series of positive terms and

$$S_n = \frac{1}{1^2} + \frac{1}{2^2} + \cdots + \frac{1}{n^2} < \frac{1}{1} + \frac{1}{1.2} + \frac{1}{2.3} + \frac{1}{3.4} + \cdots + \frac{1}{(n-1)n}$$
$$(\because 2^2 = 2.2 > 1.2; 3^2 = 3.3 > 2.3; \cdots n^2 = nn > (n-1)n)$$

i.e. $\quad S_n < \frac{1}{1} + \left(\frac{1}{1} - \frac{1}{2}\right) + \left(\frac{1}{2} - \frac{1}{3}\right) + \left(\frac{1}{3} - \frac{1}{4}\right) + \cdots + \left(\frac{1}{n-1} - \frac{1}{n}\right)$

or, $\quad S_n < 1 + 1 - \frac{1}{n} \quad$ i.e., $\quad S_n < 2, \quad \forall n = 1, 2, 3, \ldots$

i.e., $\{S_n\}$ is bounded above (one upper bound is evidently 2) and hence the series $\sum_{n=1}^{\infty} \frac{1}{n^2}$ is convergent.

The sum of this is, of course, ≤ 2.

In fact, this sum has been found to be $\pi^2/6$

(See *Fourier Series* expansion of $f(x) = x + x^2$ in $-\pi \leq x \leq \pi$: See Authors' **Integral Calculus**)

Our next step is to introduce a number of **Tests for a series** of positive terms.

Most useful Test is the Comparison Test.

I. Comparison Test

We shall state this Test in various ways:

I. Comparison Tests: Various ways of Stating the Tests

Statement A: *Let* $\sum_{n=1}^{\infty} a_n$ *and* $\sum_{n=1}^{\infty} b_n$ *be two series of positive terms. Suppose* $\sum_{n=1}^{\infty} b_n$ *is known to be a convergent series.*

If $\exists$ a positive constant k and a positive integer N such that

$$\frac{a_n}{b_n} \leq k, \quad \forall n \geq N$$

then $\sum a_n$ also converges. [**Case for Convergence**]

Statement B: *Let* $\sum_{n=1}^{\infty} a_n$ *and* $\sum_{n=1}^{\infty} b_n$ *be two series of positive terms. Suppose* $\sum_{n=1}^{\infty} b_n$ *is known to be a divergent series.*

If $\exists$ is a positive constant l and a positive integer N such that

$$\frac{a_n}{b_n} \geq l, \quad \forall n \geq N$$

then $\sum a_n$ also diverges. [**Case for divergence**]

Statement C (for convergence): A more General Form of A and B (introducing upper and lower limits).

Let $\sum_{n=1}^{\infty} a_n$ and $\sum_{n=1}^{\infty} b_n$ be two series of positive terms. Suppose $\sum_{n=1}^{\infty} b_n$ is known to be convergent. If now $\overline{\lim} \dfrac{a_n}{b_n}$ is positive and not infinite (i.e., $0 < \overline{\lim} \dfrac{a_n}{b_n} < \infty$) then Σa_n is convergent.

Statement D (for Divergence): Let $\sum_{n=1}^{\infty} a_n$ and $\sum_{n=1}^{\infty} b_n$ be two series of positive terms.

Suppose $\sum_{n=1}^{\infty} b_n$ is known to be divergent. If now, $\underline{\lim} \dfrac{a_n}{b_n} > 0$ then $\sum_{n=1}^{\infty} a_n$ diverges.

Statement E: LIMIT FORM OF COMPARISON TEST (A Special Case):
$\lim \dfrac{a_n}{b_n} = l$ i.e., $\overline{\lim} \dfrac{a_n}{b_n} = \underline{\lim} \dfrac{a_n}{b_n} = l$:

If $\sum_{n=1}^{\infty} a_n$ and $\sum_{n=1}^{\infty} b_n$ be two series of positive terms such that $\lim_{n \to \infty} \dfrac{a_n}{b_n}$ exists and equals to l say, where $0 < l < \infty$, then either both $\sum_{n=1}^{\infty} a_n$ and $\sum_{n=1}^{\infty} b_n$ converge or both diverge.

[C.H. 1995]

Proofs of all these statements should be learnt carefully:

Proof of Statement A:

We take a positive integer $n \geq N$. Then from the given condition it follows.

$$a_{N+1} + a_{N+2} + \cdots + a_n$$
$$\leq k(b_{N+1} + b_{N+2} + \cdots + b_n)$$

Writing $s_n = a_1 + a_2 + \cdots + a_n$ and $\sigma_n = b_1 + b_2 + b_2 + \cdots + b_n$ it follows from the above inequality

$$s_n - s_N \leq k(\sigma_n - \sigma_N)$$
$$\text{or } \quad s_n \leq \underbrace{(s_N - k\sigma_N)}_{\text{a fixed number}} + k\sigma_N \, (n > N)$$

Now it is given that $\sum_{n=1}^{\infty} b_n$ converges (to σ, say) then every partial sum $\sigma_n \leq \sigma$.

Hence, from above, $s_n \leq$ a fixed number $+ k\sigma$ $\forall n$ i.e. $\{s_n\}$ is a bounded sequence and $\{s_n\}$ is clearly monotone increasing so that $\{s_n\}$ is convergent and hence $\sum_{n=1}^{\infty} a_n$ converges.

CHAPTER 6: INFINITE SERIES OF REAL CONSTANTS

Proof of Statement B:

A's before, taking $n \geq N$

$$a_{N+1} + a_{N+2} + \cdots + a_n \geq l \ (b_{N+1} + b_{N+2} + \cdots + b_n)$$

or $\quad s_n - s_N \geq l(\sigma_n - \sigma_N)$

or $\quad s_n \geq (s_N - l\sigma_N) + l\sigma_n$

Since $\sum_{n=1}^{\infty} b_n$ divergences, $\{\sigma_n\}$ is unbounded above; consequently $\{s_n\}$ is also unbounded and $\{s_n\}$ is clearly m.i. Therefore $\{s_n\}$ diverges and the series $\sum_{n=1}^{\infty} a_n$ diverges.

Proof of Statement C:

Let $\overline{\lim} \frac{a_n}{b_n} = \mu (0 < \mu < \infty)$.

Let $\epsilon > 0$ be any given positive number. Then by the property of upper limit of sequence, $\exists$ a positive integer N such that

$$\forall n \geq N, \quad \frac{a_n}{b_n} \leq \mu + \epsilon$$

$|\mu + \epsilon| = k$ say, k is then a positive constant.

$\therefore \ \forall n \geq N, \frac{a_n}{b_n} \leq k(> 0)$. So, by the statement A, Σa_n is convergent whenever $\sum b_n$ is convergent.

Proof Statement D

Let $\underline{\lim} \frac{a_n}{b_n} = \lambda (\lambda > 0)$.

Let $(\epsilon > 0)$ be an arbitrary positive number.

Then, by using the property of lower limit of a sequence, $\exists$ a positive integer N such that

$$\frac{a_n}{b_n} \geq \lambda - \epsilon = l \text{ (say)}, \quad \forall n \geq N$$

(ϵ is so chosen that l is a positive constant). So, $\frac{a_n}{b_n} \geq l > 0 \ \forall \ n \geq N$.

So, by the statement B it follows that whenever $\sum b_n$ is divergent $\sum a_n$ is also divergent.

Proof of Statement E

Here we are assuming $\overline{\lim} \frac{a_n}{b_n} = \underline{\lim} \frac{a_n}{b_n} = \lim \frac{a_n}{b_n}$.

Let $\lim_{n \to \infty} \frac{a_n}{b_n} = l \ (l > 0$ and finite; $0 < l < \infty)$.

Therefore corresponding to a positive number ϵ (so chosen that $l - \epsilon$ is > 0), $\exists$ a positive integer N such that

$$\forall n \geq N, \quad \left|\frac{a_n}{b_n} - l\right| < \epsilon \quad \text{or} \quad (l-\epsilon)b_n < a_n < (l+\epsilon)b_n.$$

Now if $\sum_{n=1}^{\infty} b_n$ converges, then take $a_n < (l+\epsilon)b_n, \forall n \geq N$ so that by **A**, $\sum_{n=1}^{\infty} a_n$ converges.

But if $\sum_{n=1}^{\infty} b_n$ diverges, then we take $a_n > (l-\epsilon)b_n, \forall n \geq N$ so that by **B**, $\sum_{n=1}^{\infty} a_n$ diverges.

Hence we see that in this case $\sum_{n=1}^{\infty} a_n$ and $\sum_{n=1}^{\infty} b_n$.

Converge or diverge together.

Statement F: Another type of Comparison Test

If $\sum_{n=1}^{\infty} a_n$ and $\sum_{n=1}^{\infty} b_n$ be two series of positive terms and if $\exists$ a positive integer m such that

$$\frac{a_n}{a_{n+1}} \geq \frac{b_n}{b_{n+1}}, \quad \forall n \geq m$$

then (i) $\sum_{n=1}^{\infty} a_n$ is convergent if $\sum_{n=1}^{\infty} b_n$ is convergent.

(ii) $\sum_{n=1}^{\infty} b_n$ is divergent if $\sum_{n=1}^{\infty} a_n$ is divergent.

Proof: As usual take

$$S_n = a_1 + a_2 + \cdots + a_n$$
$$\sigma_n = b_1 + b_2 + \cdots + b_n$$

Then for $n \geq m$ we have

$$\frac{a_m}{a_n} = \frac{a_m}{a_{m+1}} \cdot \frac{a_{m+1}}{a_{m+2}} \cdot \frac{a_{m+2}}{a_{m+3}} \cdots \frac{a_{n-1}}{a_n}$$
$$\geq \frac{b_m}{b_{m+1}} \cdot \frac{b_{m+1}}{b_{m+2}} \cdot \frac{b_{m+2}}{b_{m+3}} \cdots \frac{b_{n-1}}{b_n} = \frac{b_m}{b_n}$$
$$\Rightarrow a_n \leq \frac{a_m}{b_m} \cdot b_n = kb_n \text{ (say)}$$

where $k = \frac{a_m}{b_m} =$ a fixed positive constant, since m is a fixed positive integer.

Thus $a_n \leq kb_n, \forall n \geq m$.

CHAPTER 6: INFINITE SERIES OF REAL CONSTANTS

Now see the statements **A** and **B**: If $\sum_{n=1}^{\infty} b_n$ converges then $\sum_{n=1}^{\infty} a_n$ converges and if $\sum_{n=1}^{\infty} a_n$ diverges then $\sum_{n=1}^{\infty} b_n$ diverges.

> **Remember:** Limit Form of Comparison Test is frequently used (i.e., Statement **E** should be used, if found convenient.)

APPLICATIONS OF COMPARISON TESTS

Example 6.6.2. *Using Comparison Test prove that the p-series* $\sum_{n=1}^{\infty} \frac{1}{n^p}$ *converges if $p > 1$ and diverges if $p \leq 1$.*

Proof:

(i) $p > 1$, $a_n = \frac{1}{n^p}$. Let $\sum b_n$ be obtained by grouping the terms of $\sum a_n$ thus:

$$\sum a_n = 1 + \left(\frac{1}{2^p} + \frac{1}{3^p}\right) + \left(\frac{1}{4^p} + \frac{1}{5^p} + \frac{1}{6^p} + \frac{1}{7^p}\right) + \cdots.$$

Take, $b_1 = 1$, $b_2 = \frac{1}{2^p} + \frac{1}{3^p}$, $b_3 = \frac{1}{4^p} + \frac{1}{5^p} + \frac{1}{6^p} + \frac{1}{7^p}$,

$b_4 = \frac{1}{8^p} + \frac{1}{9^p} + \cdots + \frac{1}{15^p}$, etc.

See that $b_2 < \frac{1}{2^p} + \frac{1}{2^p} = \frac{2}{2^p} = \frac{1}{2^{p-1}}$;

$b_3 < \frac{1}{4^p} + \frac{1}{4^p} + \frac{1}{4^p} + \frac{1}{4^p} = \frac{1}{(2^{p-1})^2}$ etc. etc.

In general, $b_n < \left(\frac{1}{2^{p-1}}\right)^{n-1}$ $\forall\, n \geq 2$. Now when $p > 1$, $\frac{1}{2^{p-1}} < 1$.

$\therefore$ $\sum b_n$ is convergent by comparing it with the convergent geometric series $\sum \left(\frac{1}{2^{p-1}}\right)^{n-1}$ whose common ratio is < 1. Therefore $\sum a_n$ converges since $\sum b_n$ is obtained from $\sum a_n$ by introduction of brackets.

(ii) $p = 1$, we get the harmonic series which diverges (using Cauchy's principle)

(iii) $p < 1$. Then $\frac{1}{2^p} > \frac{1}{2}$, $\frac{1}{3^p} > \frac{1}{3}, \cdots \frac{1}{n^p} > \frac{1}{n}$. $\sum \frac{1}{n^p}$ diverges by comparing it with the divergent harmonic series.

Example 6.6.3. *Prove that the series* $\sum a_n = 1 + \dfrac{2}{1!} + \dfrac{2^2}{2!} + \dfrac{2^3}{3!} + \dfrac{2^4}{4!} + \cdots$ *is convergent.*

Solution. Compare the series with $\sum b_n = 1 + \dfrac{2}{1!} + \dfrac{2^2}{2!} + \dfrac{2^2}{2!}\left(\dfrac{2}{3}\right) + \dfrac{2^2}{2!}\left(\dfrac{2}{3}\right)^2 + \cdots$.

From the 5th term onwards, each term of the given series $\sum a_n$ is less than the corresponding term of the series $\sum b_n$.

That is, $a_n < b_n \ \forall n \geq 5$.

But if we omit the first two terms of $\sum b_n$, we have a G.P. with common ratio $2/3(<1)$, which is therefore convergent and hence $\sum b_n$ converges.

So (by Comparison Test) $\sum a_n$ (given series) converges.

Example 6.6.4. *Prove that the series*

$$\dfrac{1}{1.3} + \dfrac{2}{3.5} + \dfrac{3}{5.7} + \dfrac{4}{7.9} + \cdots \text{ is divergent.}$$

Solution. The given series is $\sum a_n$, where $a_n = \dfrac{n}{(2n-1)(2n+1)}$.

Here
$$a_n > \dfrac{n}{2n \cdot 3n} \quad (\because 2n-1 < 2n; \ 2n+1 < 2n+n), \ \forall n.$$

Now we compare the series $\sum a_n$ with the series $\sum b_n$ where $b_n = \dfrac{1}{6n} \cdot a_n > b_n \ \forall n$.

But $\sum b_n$ diverges (Harmonic Series) and so by Comparison Test $\sum a_n$ also diverges.

Use $\lim\limits_{n\to\infty} \dfrac{a_n}{b_n}$, where it exists, in the following examples:

Example 6.6.5. *Examine the series*

$$\dfrac{5}{1.2.4} + \dfrac{7}{2.3.5} + \dfrac{9}{3.4.6} + \dfrac{11}{4.5.7} + \cdots.$$

Solution. The given series is $\sum a_n$, where $a_n = \dfrac{2n+3}{n(n+1)(n+3)}$.

Consider a series $\sum b_n$ where $b_n = \dfrac{n}{n^3} = \dfrac{1}{n^2}$.

[In choosing b_n, we take the highest power of n in the Numerator and also in the Denominator. This choice is useful if both Numerator and Denominator of the General term of the given series are polynomials in n.]

Now

$$\dfrac{a_n}{b_n} = \dfrac{2n+3}{n(n+1)(n+3)} \bigg/ \dfrac{1}{n^2}$$

$$= \dfrac{n^2(2n+3)}{n(n+1)(n+3)} = \dfrac{n(2n+3)}{(n+1)(n+3)} = \dfrac{2+3/n}{(1+1/n)(1+3/n)}.$$

CHAPTER 6: INFINITE SERIES OF REAL CONSTANTS

$\therefore \dfrac{a_n}{b_n} \to 2$ (a finite positive quantity) as $n \to \infty$.

$\therefore \sum a_n$ and $\sum b_n$ will either both converge or will both diverge. But $\sum b_n = \sum \dfrac{1}{n^2}$ converges (p-series with $p > 1$). Therefore the given series $\sum a_n$ also converges.

Example 6.6.6. (i) $\sum \dfrac{n}{4n^3 - 2}$ converges; compare with $\sum b_n$ where $b_n = \dfrac{n}{n^3} = \dfrac{1}{n^2}$.

(ii) $\sum_{n=2}^{\infty} \dfrac{\log n}{\sqrt{n+1}}$ diverges; since we can compare it with the divergent series $\sum \dfrac{1}{\sqrt{n}}$ and then $\dfrac{a_n}{b_n} = \dfrac{\log n}{\sqrt{n+1}} \times n^{1/2}$ which tends to $+\infty$ as $n \to \infty$.

(iii) $\sum_{n=1}^{\infty} \dfrac{1}{\sqrt{n(2n+1)}}$ diverges; compare it with $\sum b_n$ where

$$b_n = \dfrac{1}{\sqrt{n \cdot n}} = \dfrac{1}{n}. \text{ So, } \dfrac{a_n}{b_n} \to \dfrac{1}{\sqrt{2}} \text{ as } n \to \infty.$$

$\sum a_n$ and $\sum b_n$ will converge or diverge together but $\sum b_n$ diverges (Harmonic Series) and hence $\sum a_n$ diverges.

(iv) $\sum_{n=2}^{\infty} \dfrac{1}{\sqrt{n(n-1)}}$ diverges, compare it with $\sum b_n$ where $b_n = \dfrac{1}{\sqrt{n \cdot n}} = \dfrac{1}{n}$.

Example 6.6.7. *(Upper and Lower limits criterion)*

$$\text{Let } a_n = \begin{cases} \dfrac{1}{n^2}, & \text{when } n \text{ is not a squared integer} \\ \dfrac{1}{n^{2/3}}, & \text{when } n \text{ is a square.} \end{cases}$$

Then the series

$$\sum a_n = 1 + \dfrac{1}{2^2} + \dfrac{1}{3^2} + \dfrac{1}{4^{2/3}} + \dfrac{1}{5^2} + \dfrac{1}{6^2} + \dfrac{1}{7^2} + \dfrac{1}{8^2} + \dfrac{1}{9^{2/3}} + \cdots.$$

Take $b_n = 1/n$. Then $\underline{\lim} a_n/b_n = 0$ and $\overline{\lim} a_n/b_n = \infty$.

So that no conclusion from the **Comparison Test**. However $\sum a_n$ is convergent by **Integral Test** (given later).

Example 6.6.8. *Examine the convergence of a series whose nth term is*

$$(n^3 + 1)^{1/3} - n.$$

Solution. Let $\sum a_n$ be a positive series, where

$$\begin{aligned}
a_n &= (n^3 + 1)^{\frac{1}{3}} - n \\
&= n\left\{\left(1 + \frac{1}{n^3}\right)^{1/3} - 1\right\} \\
&= n\left\{1 + \frac{1}{3}\frac{1}{n^3} + \frac{\frac{1}{3}(\frac{1}{3} - 1)}{2!}\left(\frac{1}{n^3}\right)^2 + \cdots - 1\right\} \\
&= \frac{1}{3n^2} + \text{terms containing higher powers of } \frac{1}{n^2}.
\end{aligned}$$

We assume a positive series $\sum b_n$ where $b_n = \frac{1}{3n^2}$ and compare.

$$\lim_{n \to \infty} \frac{a_n}{b_n} = 1 \text{ (limit exists and equals to a positive constant)}.$$

$\therefore$ $\sum a_n$ and $\sum b_n$ will either both converge or will both diverge.

But $\sum b_n$ is convergent ($\because$ it is a p-series with $p > 1$).

Hence, by comparison test, $\sum a_n$ will converge.

Example 6.6.9. *Test the convergence of the following series:*

(i) $\sum_{n=1}^{\infty} \frac{1}{n^{1+\frac{1}{n}}}$; *Compare it with the series* $\sum_{n=1}^{\infty} \frac{1}{n}$.

(ii) $\sum_{n=1}^{\infty} \sin \frac{1}{n}$; *Compare it with the series* $\sum_{n=1}^{\infty} \frac{1}{n}$.

(iii) $\frac{1.2}{3^2.4^2} + \frac{3.4}{5^2.6^2} + \frac{5.6}{7^2.8^2} + \cdots$;

(*Compare it with the series* $\sum_{n=1}^{\infty} \frac{1}{n^2}$. *Hence* $a_n = \frac{(2n-1)(2n)}{(2n+1)^2(2n+2)^2}$, $b_n = \frac{n^2}{n^2.n^2} = \frac{1}{n^2}$)

(iv) $\sum_{n=2}^{\infty} \frac{1}{(\log n)^p} (p > 0)$

CHAPTER 6: INFINITE SERIES OF REAL CONSTANTS

$\left[\lim\limits_{n\to\infty} \dfrac{(\log n)^p}{n} = 0\right.$ and so $(\log n)^p < n$, $\forall n > 1$. Thus $\dfrac{1}{(\log n)^p} > \dfrac{1}{n}$, $\forall n > 1$. The given series can be compared with the series $\sum\limits_{n=1}^{\infty} \dfrac{1}{n}$ which is a divergent series and we conclude that the given series diverges.$\left.\right]$

(v) $\sum\limits_{n=1}^{\infty} \dfrac{1}{n!}$; Compare it with $\sum\limits_{n=1}^{\infty} \dfrac{1}{2^n}$ which is the convergent geometric series.

$[n! = n(n-1)(n-2)\cdots 3.2 > 2.2.2\cdots 2 = 2^{n-1}$, i.e., $\dfrac{1}{n!} < \dfrac{1}{2^{n-1}}$ and $\sum\limits_{n=1}^{\infty} \dfrac{1}{2^{n-1}}$ is a convergent geometric series]

ANSWERS (i) D; (ii) D; (iii) C; (iv) D; (v) C (C = Convergent; D = Divergent)

Example 6.6.10. Let

$$\sum_{n=1}^{\infty} a_n = \frac{1+2}{2^3} + \frac{1+2+3}{3^3} + \frac{1+2+3+4}{4^3} + \cdots + \frac{1+2+3+\cdots+(n+1)}{(n+1)^3} + \cdots$$

Here

$$a_n = \frac{1+2+3+\cdots+(n+1)}{(n+1)^3} = \frac{(n+1)(n+2)}{2(n+1)^3} = \frac{n+2}{2(n+1)^2}$$

Take $b_n = \dfrac{n}{2n^2} = \dfrac{1}{2n}$.

Compare the series $\sum\limits_{n=1}^{\infty} a_n$ and $\sum\limits_{n=1}^{\infty} b_n$:

$$\frac{a_n}{b_n} = \frac{\frac{n+2}{2(n+1)^2}}{\frac{1}{2n}} = \frac{1+2/n}{\left(1+\frac{1}{n}\right)^2} \to 1 \text{ as } n \to \infty$$

Thus $\lim\limits_{n\to\infty} \dfrac{a_n}{b_n} = 1$ and $\sum\limits_{n=1}^{\infty} b_n$ diverges and therefore the given series $\sum\limits_{n=1}^{\infty} a_n$ also diverges.

II. Cauchy's Root Test

STATEMENT. Let $\sum\limits_{n=1}^{\infty} a_n$ be a series of positive terms and let $\rho = \lim\limits_{n\to\infty} \sqrt[n]{a_n}$. Then

(i) If $\rho < 1$, then $\sum_{n=1}^{\infty} a_n$ Converges,

(ii) If $\rho > 1$, then $\sum_{n=1}^{\infty} a_n$ diverges,

(iii) If $\rho = 1$, then this test does not give a definite conclusion (We say that the test fails)

Proof:

Case 1. Let $\rho = \lim_{n \to \infty} a_n^{1/n} < 1$.

Choose $\epsilon > 0$ such that $\rho + \epsilon < 1$.

Let $\rho + \epsilon = r(< 1)$.

Since $\lim_{n \to \infty} a_n^{1/n} = \rho$, therefore $\exists$ a positive integer m such that

$$|a_n^{1/n} - \rho| < \epsilon, \quad \forall\, n \geq m$$
$$\Rightarrow \rho - \epsilon < a_n^{1/n} < \rho + \epsilon, \quad \forall n \geq m$$
$$\Rightarrow a_n < (\rho + \epsilon)^n \quad \text{i.e.,} \quad < r^n \;\; (\text{where } r < 1), \quad \forall n \geq m.$$

But since $\sum r^n$ is a convergent geometric series (with common ratio $r < 1$), therefore, by Comparison Test, the series Σa_n converges.

Case 2. Let $\rho > 1$.

Let us choose $\epsilon > 0$ such that $\rho - \epsilon > 1$. Suppose $\rho - \epsilon = \beta > 1$.

Since $\lim_{n \to \infty} a_n^{1/n} \rho$, therefore $\exists$ a positive integer m_1 such that

$$\rho - \epsilon < a_n^{1/n} < \rho + \epsilon, \quad \forall n \geq m_1$$
$$\Rightarrow (\rho - \epsilon) < a_n < (\rho + \epsilon)^n, \quad \forall n \geq m_1$$
$$\Rightarrow a_n > (\rho - \epsilon)^n, \quad \text{i.e.,} \quad a_n > \beta^n (\beta > 1) \;\; \forall\, n \geq m_1.$$

But since $\sum \beta^n$ is a divergent geometric (its common ratio being $\beta > 1$), therefore again by comparison test, the series $\sum a_n$ diverges.

Case 3. The test fails to give any definite information about convergence or divergence of $\sum a_n$ if $\rho = 1$.

Consider the two series

$$\sum \frac{1}{n} \quad \text{(harmonic series)}$$
$$\text{and} \quad \sum \frac{1}{n^2} \quad (p\text{-series with } p > 1)$$

Clearly $\sum \frac{1}{n}$ diverges while $\lim_{n \to \infty} \left(\frac{1}{n}\right)^{1/n} = 1$

CHAPTER 6: INFINITE SERIES OF REAL CONSTANTS

[Let $y = \left(\dfrac{1}{n}\right)^{1/n}$ $\therefore$ $\ln y = \dfrac{1}{n} \ln \dfrac{1}{n} = -\dfrac{\ln(n)}{n}$

$\therefore \lim\limits_{n \to \infty} \ln y = \lim\limits_{n \to \infty} -\dfrac{\ln(n)}{n} \left[\dfrac{\infty}{\infty}\right] = -\lim\limits_{n \to \infty} \dfrac{1}{n} = 0$

$\therefore \lim\limits_{n \to \infty} y = e^0 = 1.$]

and $\sum \dfrac{1}{n^2}$ converges while $\lim\limits_{n \to \infty} \left(\dfrac{1}{n^2}\right)^{1/n} = 1$.

So, when $\rho = 1$, the series may or may not converge.

As a simple application we give below an example:

Consider the series $\sum \left(1 + \dfrac{1}{\sqrt{n}}\right)^{-n\sqrt{n}}$

Here $a_n = \left(1 + \dfrac{1}{\sqrt{n}}\right)^{-n\sqrt{n}}$.

Then $\lim\limits_{n \to \infty} a_n^{1/n} = \lim\limits_{n \to \infty} \dfrac{1}{\left(1 + \dfrac{1}{\sqrt{n}}\right)^{\sqrt{n}}} = \dfrac{1}{e} < 1$.

Hence the series converges.

Cauchy's Root test may be stated in terms of upper limit of $\sqrt[n]{a_n}$.

Statement. Let $\sum a_n$ be a series of positive terms and let $\rho = \overline{\lim} \sqrt[n]{a_n}$.

(i) If $\rho < 1$, then Σa_n converges.

(ii) If $\rho > 1$, Σa_n diverges.

(iii) No definite conclusion can be drawn if $\rho = 1$.

The proof is similar. We give the details below:

Proof:

(a) When $\rho = \overline{\lim} \sqrt[n]{a_n} < 1$.

We choose a positive number ε so that $\rho + \varepsilon = r$ (say) < 1.

By the property of upper limit of a sequence $\exists$ a positive integer N such that
$$\forall n \geq N, \quad \sqrt[n]{a_n} < r, \text{ where } r < 1.$$

We shall prove that in this case $\sum a_n$ converges.

Since $\sqrt[n]{a_n} < r < 1$, for all $n \geq N$,

Therefore $a_n < r^n$, for all $n \geq N$.

But $\sum r^n$ is a convergent geometric series. ($\because$ common ratio $r < 1$)

$\therefore$ By Comparison Test the series $\sum a_n$ converges.

(b) When $\rho = \overline{\lim} \sqrt[n]{a_n} > 1$.

Let us choose a positive number ε (no matter how small) such that $\rho - \varepsilon > 1$.

Since $\overline{\lim} \sqrt[n]{a_n} = \rho$, $\exists$ infinite number of member of $\{\sqrt[n]{a_n}\} > \rho - \varepsilon > 1$.

In that case, $\lim\limits_{n \to \infty} \sqrt[n]{a_n}$ cannot tend to zero.

$\therefore$ the series $\sum a_n$ where $a_n = \sqrt[n]{a_n}$ **cannot** converge (since a necessary condition for the convergence of a series $\sum a_n$ is that $a_n \to 0$ as $n \to \infty$).

But $\sum a_n$ is a positive series. So it can either converge or diverge to $+\infty$, but it cannot oscillate. Now since $\sum a_n$ cannot converge, therefore it must diverge to ∞ when $\rho > 1$.

Observations:

1. What we have proved is that

    (a) *If, for a positive series $\sum a_n$, $\exists$ a positive integer N such that $\forall\, n \geq N$, $\sqrt[n]{a_n} \leq r < 1$, then $\sum a_n$ converges.*

    (b) *On the other hand, if for a positive series $\sum a_n$, $\sqrt[n]{a_n} \geq 1$, for infinitely many n, then $\sum a_n$ diverges.*

    The statements (a) and (b) can be remembered as the **statement of Cauchy's Root Test.**

2. In (a), unless $\sqrt[n]{a_n}$ ultimately remains less than *fixed number* $r < 1$, we cannot conclude that $\sum a_n$ converges. Thus simply $\sqrt[n]{a_n} < 1$ will not ensure convergence.

    e.g., in the **divergent series** $\sum \dfrac{1}{n}$, we find $\sqrt[n]{a_n} = \sqrt[n]{1/n} < 1$; here $\sqrt[n]{a_n} \to 1$ from below. But the **convergent series** $\sum (1/n^2)$ behaves in precisely the same way.

    Thus $\sqrt[n]{a_n} \to 1$ **from below,** no conclusion can be drawn from the root test. But if $\sqrt[n]{a_n} \to 1$ **from above,** (b) shows that $\sum a_n$ diverges.

3. In case if $\lim\limits_{n \to \infty} \sqrt[n]{a_n}$ exists $= l$ (say) then we state **the limit form of Cauchy's Root Test.**

    Thus, *a positive series $\sum a_n$ converges if $l < 1$, diverges if $l > 1$ and no conclusion if $l = 1$.*

Proof: (a) When $\lim_{n\to\infty} \sqrt[n]{a_n} = l < 1$, we choose a positive number ε such that $l + \varepsilon < 1$. Then $\exists$ a positive integer N such that

$$|\sqrt[n]{a_n} - l| < \varepsilon, \ \forall n \geq N$$

i.e., $\quad l - \varepsilon < \sqrt[n]{a_n} < l + \varepsilon, \ \forall n \geq N.$

Let $l + \varepsilon = r < 1$. Then

$$\sqrt[n]{a_n} < r, \ \forall n \geq N$$

or, $\quad a_n < r^n, \ \forall n \geq N.$

But $\sum r^n$ is a geometric series with common ratio $r < 1$ and hence convergent. Therefore by Comparison Test $\sum a_n$ converges.

(b) When $\lim_{n\to\infty} \sqrt[n]{a_n} = l > 1$, we choose a positive number ε such that $l - \varepsilon > 1$. Then $\exists$ a positive integer N such that $|\sqrt[n]{a_n} - l| < \varepsilon$ or $l - \varepsilon < \sqrt[n]{a_n} < l + \varepsilon, \ \forall n \geq N.$ Thus

$$\forall n \geq N, \quad \sqrt[n]{a_n} > l - \varepsilon, \text{ where } l - \varepsilon > 1.$$

i.e., $\quad \forall n \geq N, \quad a_n > (l - \varepsilon)^n > 1.$

$\therefore \ a_n$ cannot tend to zero as $n \to \infty$. Therefore the series $\sum a_n$ cannot converge. Being a positive series $\sum a_n$ cannot oscillate. Therefore $\sum a_n$ must be divergent.

Applications of Cauchy's Root test

Example 6.6.11. $\dfrac{1}{3} + \left(\dfrac{2}{5}\right)^2 + \left(\dfrac{3}{7}\right)^3 + \cdots + \left(\dfrac{n}{2n+1}\right)^n + \cdots$ *converges. Because the given series* $\sum a_n$ *where* $a_n = \left(\dfrac{n}{2n+1}\right)^n$ *is such that*

$$\sqrt[n]{a_n} = \dfrac{n}{2n+1} \to \dfrac{1}{2}(<1) \text{ as } n \to \infty.$$

$\therefore \ \sum a_n$ *converges by Cauchy's Root test.*

Example 6.6.12. *Investigate the convergence of the series:* $\sum\limits_{n=1}^{\infty} a_n.$

Solution. Here $a_n = 2^{-n-(-1)^n}$ converges. For, as $n \to \infty$,

$$a_n^{1/n} = 2^{-1 - \frac{(-1)^n}{n}} \longrightarrow 2^{-1} \ (<1).$$

$\therefore \ \sum a_n$ converges by Cauchy's Root test.

Check: The series $\sum 3^{-n+(-1)^n}$ is convergent.

Example 6.6.13. *Investigate the convergence of the series:* $\sum_{n=1}^{\infty} \dfrac{n^n}{n!}$.

Solution. Here $a_n = \dfrac{n^n}{n!}$ or $a_n^{1/n} = \dfrac{n}{\sqrt[n]{n!}} \to 0$ as $n \to \infty$ (limit is < 1) and hence by Cauchy's Root test it converges.

Example 6.6.14. *Investigate the convergence of the series:*

$$1 + \frac{1}{2^3} + \frac{1}{2^2} + \frac{1}{2^5} + \frac{1}{2^4} + \cdots.$$

Solution. Here $a_n = \dfrac{1}{2^{n+(-1)^n}}$ and $a_n^{1/n} = \dfrac{1}{2^{1+\frac{(-1)^n}{n}}}$

$\therefore a_n^{1/n} \to \frac{1}{2}$ as $n \to \infty$ (limit exists and limit is < 1).

$\therefore$ By Cauchy's root test the given series is convergent.

Our Third Test after Comparison Test and Cauchy Root Test will be a series of tests known as

III. Ratio Tests

A. D'Alembert's Ratio Test

(*D'Alembert (1717–1783) was a well-known French Mathematician*)

Statement: Let $\sum_{n=1}^{\infty} a_n$ (or $\sum a_n$) be a series of positive terms for which $\lim\limits_{n \to \infty} \dfrac{a_n}{a_{n+1}}$ exists and equal to l (say).

Then

(i) $\sum a_n$ converges if $l > 1$,

(ii) $\sum a_n$ diverges if $l < 1$,

(iii) $\sum a_n$ may or may not converge (i.e., the test fails to give a definite conclusion) if $l = 1$.

If case D'Alembert's test fails to give a definite conclusion, we shall use.

B. Raabe's Test

(*Raabe (1801–1859) was a German mathematician*)

Statement: Let $\sum a_n$ be a series of positive terms and let $\lim\limits_{n \to \infty} n\left(\dfrac{a_n}{a_{n+1}} - 1\right)$ exists and equals to l.

Then the series $\sum a_n$ is
(i) convergent if $l > 1$
(ii) divergent if $l < 1$
(iii) no definite conclusion if $l = 1$.

We often write: $R_n = n\left(\dfrac{a_n}{a_{n+1}} - 1\right)$ and call $\{R_n\}$ Raabe's sequence.

Remarks. (1) Raabe's Test will be certainly applicable if D'Alembert's Test is applicable.

(ii) But Raabe's Test may give a definite conclusion even if D'Alembert's test fails. So one can conclude:

Raabe's Test is stronger than D'Alembert's Test.

In case Raabe's Test also fails one may try test C given below:

C. De Morgan and Betrand's Test

Statement. Let Σa_n be a series of positive terms and let

$$B_n = \log n(R_n - 1) = \left\{n\left(\dfrac{a_n}{a_{n+1}} - 1\right) - 1\right\}\log n$$

Then (a) $\sum a_n$ converges if $\lim B_n > 1$.

(b) $\sum a_n$ diverges if $\lim B_n < 1$.

In each of the above statements we can make the statements **more precise** by using what we call **upper and lower limit criterion**.

SUMMARY: Ratio Tests

A. D'Alembert's Ratio Test

Let $\sum a_n$ be a series of positive terms.

Let $\bar{\rho} = \overline{\lim}\,\dfrac{a_n}{a_{n+1}}$ and let $\underline{\rho} = \underline{\lim}\,\dfrac{a_n}{a_{n+1}}$.

Then

(a) $\sum a_n$ converges if $\underline{\rho} > 1$

and (b) $\sum a_n$ diverges if $\bar{\rho} < 1$.

Special Case:

In case $\lim\limits_{n\to\infty}\dfrac{a_n}{a_{n+1}} = l$ then evidently $l = \bar{\rho} = \underline{\rho}$ and $\sum a_n$ converges if $l > 1$ and $\sum a_n$ diverges if $l < 1$.

B. Raabe's Test

Let $\sum a_n$ be a series of positive terms. Suppose
$$R_n = n\left(\frac{a_n}{a_{n+1}} - 1\right)$$

Then

(a) $\sum a_n$ converges if $\underline{\lim} R_n > 1$
(b) $\sum a_n$ diverges if $\overline{\lim} R_n < 1$

Special Case:

In case $\lim_{n\to\infty} R_n = l$ then $l = \underline{\lim} R_n = \overline{\lim} R_n$ and $\sum a_n$ converges if $l > 1$ and $\sum a_n$ diverges if $l < 1$.

C. De Morgan and Bertrand's Test

Let $\sum a_n$ be a series of positive terms. We define
$$B_n = \log n(R_n - 1) = \left\{n\left(\frac{a_n}{a_{n+1}} - 1\right) - 1\right\}\log n$$

Then

(a) $\sum a_n$ converges if $\underline{\lim} B_n > 1$
(b) $\sum a_n$ diverges if $\overline{\lim} B_n < 1$.

Special Case:

In case $\underline{\lim} B_n = \overline{\lim} B_n = \lim B_n = l$ (say)
Then $\sum a_n$ converges if $l > 1$ and Σa_n diverges if $l < 1$.

A Typical Example showing the applications of the three Ratio Tests to be used one after another:

Example 6.6.15. *Investigate the convergence of the series:*

$$\frac{1^2}{2^2} + \frac{1^2 \cdot 3^2}{2^2 \cdot 4^2}x + \frac{1^2 \cdot 3^2 \cdot 5^2}{2^2 \cdot 4^2 \cdot 6^2}x^2 + \cdots + \frac{1^2 \cdot 3^2 \cdot 5^2 \cdots (2n-1)^2}{2^2 \cdot 4^2 \cdot 6^2 \cdots (2n)^2}x^{n-1} + \cdots \quad (x > 0).$$

Solution.

Step 1. Let the series be $\sum a_n$ where
$$a_n = \frac{1^2 \cdot 3^2 \cdot 5^2 \cdots (2n-1)^2}{2^2 \cdot 4^2 \cdot 6^2 \cdots (2n)^2}x^{n-1}.$$

Since $x > 0$, each term of the series is positive i.e., $\sum a_n$ is a positive series.
Now
$$\frac{a_n}{a_{n+1}} = \left(\frac{1^2 \cdot 3^2 \cdot 5^2 \cdots (2n-1)^2}{2^2 \cdot 4^2 \cdot 6^2 \cdots (2n)^2} x^{n-1}\right) \cdot \left(\frac{2^2 \cdot 4^2 \cdot 6^2 \cdots (2n)^2 \cdot (2n+2)^2}{1^2 \cdot 3^2 \cdot 5^2 \cdots (2n-1)^2 \cdot (2n+1)^2} \frac{1}{x^n}\right)$$
$$= \frac{(2n+2)^2}{(2n+1)^2} \frac{1}{x}.$$
$$\therefore \lim_{n \to \infty} \frac{a_n}{a_{n+1}} = \lim_{n \to \infty} \frac{\left(1+\frac{1}{n}\right)^2}{\left(1+\frac{1}{2n}\right)^2} \frac{1}{x} = \frac{1}{x}.$$

$\therefore$ By D'Alembert's Ratio Test,

$\sum a_n$ converges, if $\frac{1}{x} > 1$ i.e., if $x < 1$ i.e., if $0 < x < 1$ ($\because x > 0$ given).

and $\sum a_n$ diverges, if $\frac{1}{x} < 1$ i.e., if $x > 1$.

No definite conclusion should be drawn if $x = 1$.

So when $x = 1$, we go to the next step (**Raabe's Test**):

Step 2.
$$R_n = n\left(\frac{a_n}{a_{n+1}} - 1\right) = n\left\{\left(\frac{2n+2}{2n+1}\right)^2 - 1\right\} = \frac{n(4n+3)}{(2n+1)^2}$$
$$\lim_{n \to \infty} R_n = \lim_{n \to \infty} \frac{4n^2 + 3n}{(2n+1)^2} = \lim_{n \to \infty} \frac{4 + 3/n}{\left(2+\frac{1}{n}\right)^2} = 1.$$

$\therefore$ Raabe's Test also does not lead to any definite conclusion, of the given series when $x = 1$.

We, therefore, go to the next step (**De Morgan and Bertrand's Test**):

Step 3.
$$B_n = \log n(R_n - 1) = \left(\frac{4n^2 + 3n}{(2n+1)^2} - 1\right) \log n$$
$$= \frac{\log n}{n}\left\{\frac{-(n+1)n}{(2n+1)^2}\right\} = \frac{\log n}{n}\left\{-\frac{\left(1+\frac{1}{n}\right)}{\left(2+\frac{1}{n}\right)^2}\right\}.$$
$$\therefore \lim_{n \to \infty} B_n = 0 \cdot \left(-\frac{1}{4}\right) = 0. \quad \left(\because \frac{\log n}{n} \to 0 \text{ as } n \to \infty\right)$$

Thus $\lim\limits_{n \to \infty} B_n$ exists and is less than 1. Hence by De Morgan and Bertrand's Test the given series $\sum a_n$ diverges for $x = 1$.

Conclusion. The given series $\sum a_n$ converges if $0 < x < 1$ and diverges if $x \geq 1$.

All this stage we give below the proofs of D'Alembert's Ratio Rest and Raabe's Test one after another:

Proof of D'Alembert's Ratio Test

STATEMENT. *For a series of real positive constants,* $\sum a_n$, *if*

$$\bar{\rho} = \overline{\lim} \frac{a_n}{a_{n+1}} \quad \text{and} \quad \underline{\rho} = \underline{\lim} \frac{a_n}{a_{n+1}},$$

then $\sum a_n$ *converges if* $\underline{\rho} > 1$ *and diverges if* $\bar{\rho} < 1$.

Proof:

1. If $\underline{\rho} > 1$, choose $\varepsilon > 0$ so that $\underline{\rho} - \varepsilon = k$ (say) > 1 then $\exists$ a positive integer m such that

$$\forall \ n \geq m, \frac{a_n}{a_{n+1}} > k \ \text{(where } k \text{ itself is greater than 1)}$$

or, $a_n - a_{n+1} > (k-1)a_{n+1}, \quad \forall \, n \geq m.$

Hence, taking $n \gg m$, we have

$$\left.\begin{array}{l} a_m \ -a_{m+1} > (k-1)\, a_{m+1} \\ a_{m+1} -a_{m+2} > (k-1)\, a_{m+2} \\ \cdots\cdots\cdots\cdots\cdots\cdots\cdots \\ a_{n-1} -a_n \ > (k-1)\, a_n. \end{array}\right\}$$

Adding we get $a_m - a_n > (k-1)[a_{m+1} + a_{m+2} + \cdots + a_n]$

i.e., $a_{m+1} + a_{m+2} + \cdots + a_n < \dfrac{1}{k-1} a_m - \dfrac{1}{k-1} a_n$

i.e., $< \dfrac{1}{k-1} a_m, \quad \forall \, n \gg m.$

Adding $(a_1 + a_2 + \cdots + a_m)$ to both sides we get

$$S_n = \sum_{r=1}^{n} a_r < \frac{1}{k-1} a_m + (a_1 + a_2 + \cdots + a_m).$$

The right hand side is a fixed positive number $= S$ (say).

The sequence $\{S_n\}$ of partial sums of the positive series $\sum a_n$ is always monotone increasing and now we have proved that it is bounded. Hence $\{S_n\}$ converges. Consequently, $\sum a_n$ converges.

2. **Next** consider $\bar{\rho} < 1$.

This implies that after a certain stage (say, for $n \geq m$), $\dfrac{a_n}{a_{n+1}}$ must always be less than 1.

Thus $\exists\, m$ such that
$$a_n < a_{n+1}, \quad \forall\, n \geq m.$$

Take n much larger than m.

Then, $a_m < a_{m+1}, a_{m+1} < a_{m+2}, \cdots, a_{n-1} < a_n$ or, $a_n > a_m$ if $n > m$.

So that $\lim\limits_{n \to \infty} a_n \neq 0$ and hence $\sum a_n$ **cannot converge**; further $\sum a_n$ being a positive series, **cannot oscillate**. The only possibility then is that $\sum a_n$ diverges.

Observations

1. In D'Alembert's Ratio Test we have actually proved.

   If in the positive series $\sum a_n$, the ratio a_n/a_{n+1} is greater than a positive constant $k < 1$, $\forall\, n \geq$ a fixed +ve integer m, then $\sum a_n$ converges; on the other hand if $a_n/a_{n+1} \leq 1$, $\forall\, n \geq m$, then $\sum a_n$ diverges.

   This is taken as the **alternative statement of D'Alembert's Ratio Test**.

   In practice, often we take the ratio $\dfrac{a_{n+1}}{a_n}$ instead of $\dfrac{a_n}{a_{n+1}}$ and make consequent changes in the statement.

   Then if $\dfrac{a_{n+1}}{a_n} \leq k < 1$, for $n \geq m$, then $\sum a_n$ converges.

   but if $\dfrac{a_{n+1}}{a_n} \geq 1$, for $n \geq m$, then $\sum a_n$ diverges

   Or, If $\underline{\lim}\dfrac{a_{n+1}}{a_n} = \underline{\rho}$ and $\overline{\lim}\dfrac{a_{n+1}}{a_n} = \bar{\rho}$, then $\begin{cases} \sum a_n \text{ converges if } \bar{\rho} < 1 \\ \sum a_n \text{ diverges if } \underline{\rho} > 1. \end{cases}$

2. D'Alembert's Ratio Test does not ensure the convergence of a series if only we know $\dfrac{a_n}{a_{n+1}} > 1$, $\forall\, n$; e.g., let the series be $\sum\limits_{n=1}^{\infty} a_n = \sum\limits_{n=1}^{\infty} \dfrac{1}{n}$. In this case $\dfrac{a_n}{a_{n+1}} = 1 + \dfrac{1}{n} > 1$, $\forall\, n$; yet the series is a divergent series. What is needed is that $a_n/a_{n+1} >$ a fixed constant k which itself is > 1, $\forall\, n > m$.

3. Again, it is not necessary for the convergence of the series $\sum a_n$ that a_n/a_{n+1} should have a definite limit. e.g., $\alpha+1+\alpha^3+\alpha^2+\alpha^5+\alpha^4+\cdots$ is a rearrangement of the geometric series $1+\alpha+\alpha^2+\alpha^3+\alpha^4+\cdots$ and so is convergent if $0<\alpha<1$. But in this series a_n/a_{n+1} is alternatively α and $1/\alpha^3$ and the ratio a_n/a_{n+1} does not tend to a finite limit.

4. In the language of Superior and Inferior limits

$$\underline{\lim} \frac{a_n}{a_{n+1}} > 1 \Rightarrow \sum a_n \text{ converges,}$$

and $\overline{\lim} \frac{a_n}{a_{n+1}} < 1 \Rightarrow \sum a_n \text{ diverges.}$

Now compare with Cauchy's Root Test, where

$$\overline{\lim} \sqrt[n]{a_n} < 1 \Rightarrow \sum a_n \text{ converges and } \overline{\lim} \sqrt[n]{a_n} > 1 \Rightarrow \sum a_n \text{ diverges.}$$

i.e., Only the *upper limit* can decide the convergence or divergence of the series.

But in Ratio Test both upper and lower limits of a_n/a_{n+1} are needed for deciding convergence and divergence of the series.

5. In general, Ratio Test is easier to apply than Cauchy's Root Test, but the latter is more general than the former. In many cases the Root Test gives positive information about the convergence or divergence of the series but Ratio Test fails to do so.

Take for example, a series $\sum a_n$ where $a_n = 2^{-n-(-1)^n}$.

Here $\sqrt[n]{a_n} = 2^{-1-\frac{(-1)^n}{n}}$ which tends to a limit 2^{-1} as $n \to \infty$ (and this limit is less than 1). Therefore, by Cauchy's root test, we conclude that $\sum a_n$ converges.

But see that,

$$\frac{a_n}{a_{n+1}} = \frac{2^{-n-(-1)^n}}{2^{-(n+1)-(-1)^{n+1}}} = 2 \cdot 2^{(-1)^{n+1}-(-1)^n}.$$

Hence, if n-is even $a_n/a_{n+1} = 2 \cdot 2^{-2} = \frac{1}{2}$ and, if n is odd $a_n/a_{n+1} = 2 \cdot 2^2 = 8$.

$$\therefore \overline{\lim} \frac{a_n}{a_{n+1}} = 8(>1) \text{ and } \underline{\lim} \frac{a_n}{a_{n+1}} = \frac{1}{2}(<1),$$

so that Ratio Test does not give any positive information.

But remember that the information about the series obtained from the root test can also be obtained by using the ratio test because of the fact: (Cauchy's Second Theorem on limit of a sequence, **art 5.13** Theorem **5.13.2**).

If $a_n > 0$ and if $\dfrac{a_{n+1}}{a_n} \to l$ as $n \to \infty$ then $a_n^{1/n}$ also tends to l as $n \to \infty$.

More generally, from Cauchy's Second theorem on limits of sequences we know that $\varlimsup \sqrt[n]{a_n}$ lies between the least and greatest limits of a_{n+1}/a_n. So D'Alembert's ratio test can be deduced from Cauchy's root test.

But since we only know $\varlimsup \sqrt[n]{a_n}$ falls between the extreme limits of a_{n+1}/a_n, we cannot deduce Cauchy's test in its full generality from D'Alembert's.

Proof of Raabe's Test

STATEMENT. *Let $\sum a_n$ be a series of positive terms.*
Suppose $R_n = n\left(\dfrac{a_n}{a_{n+1}} - 1\right)$.
Then
$$\begin{cases} \sum a_n \text{ converges, if } \varliminf R_n > 1, \\ \sum a_n \text{ diverges, if } \varlimsup R_n < 1. \end{cases}$$

Proof: $\varliminf R_n > 1 \Rightarrow \exists$ a positive constant $\alpha > 1$ such that for all n greater than a fixed positive integer m, we have $R_n > \alpha$

$$\text{or,} \quad \frac{a_n}{a_{n+1}} > 1 + \frac{\alpha}{n}, \quad \forall n \geq m.$$

Now let us take n much larger than m. Then $\forall n \geq m$

$$a_n > \left(1 + \frac{\alpha}{n}\right) a_{n+1}$$
$$\text{or,} \quad na_n > (n+1)a_{n+1} + (\alpha - 1)a_{n+1}$$
$$\text{or,} \quad na_n - (n+1)a_{n+1} > (\alpha - 1)a_{n+1} \quad (\alpha - 1 \text{ is positive}).$$

Putting successively $m, m+1, m+2, \cdots (n-1)$ in place of n and then adding $(n-m)$ inequalities we get

$$ma_m - na_n > (\alpha - 1)(a_{m+1} + a_{m+2} + \cdots + a_n)$$
$$\text{or,} \quad a_{m+1} + a_{m+2} + \cdots + a_n < \frac{1}{\alpha - 1}ma_m - \frac{1}{\alpha - 1}na_n$$
$$\text{i.e.,} \quad < \frac{1}{\alpha - 1}ma_m, \ \forall n$$
$$\text{or,} \quad (a_1 + a_2 + \cdots + a_m) + a_{m+1} + a_{m+2} + \cdots + a_n$$
$$< \frac{1}{\alpha - 1}ma_m + (a_1 + a_2 + \cdots + a_m)$$
$$\text{i.e.,} \quad S_n = \sum_{r=1}^{n} a_r < \text{a fixed positive number, for all } n.$$

The sequence of partial sums $\{S_n\}$ of the positive series $\sum a_n$ is obviously monotone increasing and now we have proved that it is bounded and hence $\{S_n\}$ converges and consequently $\sum a_n$ converges.

Next, consider $\overline{\lim} R_n < 1$ which implies that $\exists$ a positive integer m such that

$$\forall n \geq m, \quad R_n \leq 1$$

or, $\quad n\left(\dfrac{a_n}{a_{n+1}} - 1\right) \leq 1$

i.e., $\quad \dfrac{a_n}{a_{n+1}} \leq 1 + \dfrac{1}{n}, \quad \forall n \geq m.$

i.e., $\quad na_n \leq (n+1)a_{n+1}, \quad \forall n \geq m$

or, $\quad na_n \geq (n-1)a_{n-1} \geq (n-2)a_{n-2} \geq \cdots \geq ma_m.$

$\therefore \quad a_n \geq \dfrac{ma_m}{n}, \quad \forall n \geq m.$

i.e., $\quad a_n \geq k \cdot \dfrac{1}{n}, \quad \forall n \geq m$ (k is a fixed constant).

But $\sum(1/n)$ is the divergent harmonic series. Therefore, by the Comparison **Test** $\sum a_n$ diverges.

Observation: We have actually proved:

For a positive series $\sum a_n$, if $\exists$ a positive constant $\alpha > 1$ and a fixed positive integer m such that

$$\dfrac{a_n}{a_{n+1}} \geq 1 + \dfrac{\alpha}{n}, \quad \forall n \geq m,$$

then $\sum a_n$ converges. On the other hand, if $\dfrac{a_n}{a_{n+1}} \leq 1 + \dfrac{1}{n} \ \forall \ n \geq m$, then $\sum a_n$ diverges.

This statement is the **alternative statement** of Raabe's Test.

Applications on Ratio Tests

Example 6.6.16. *Prove that the series*

$$x + \dfrac{x^2}{2} + \dfrac{x^3}{3} + \dfrac{x^4}{4} + \cdots (x \geq 0)$$

is convergent if $0 \leq x < 1$ and divergent if $x \geq 1$.

Solution. Let the series $\sum_{n=1}^{\infty} a_n$ where $a_n = x^n/n$.

If $x = 0$, clearly the series $0 + 0 + 0 + \cdots$ converges.

CHAPTER 6: INFINITE SERIES OF REAL CONSTANTS

Next, consider the case $x \neq 0$.

$$a_n = \frac{x^n}{n} \quad \text{and} \quad a_{n+1} = \frac{x^{n+1}}{n+1}.$$

$\therefore$ if $x \neq 0$, $\dfrac{a_n}{a_{n+1}} = \dfrac{n+1}{n}\dfrac{1}{x} = \left(1 + \dfrac{1}{n}\right)\dfrac{1}{x} \to \dfrac{1}{x}$ as $n \to \infty$.

$\therefore$ By D'Alembert's Ratio Test.

If $0 < x < 1$, then $\dfrac{1}{x} > 1$ and the series $\sum a_n$ converges.

If $x > 1$, then $\dfrac{1}{x} < 1$ and series $\sum a_n$ diverges.

If $x = 1$, then Ratio Test does not give any definite conclusion.

Now,

$$\lim n\left(\frac{a_n}{a_{n+1}} - 1\right) = \lim n\left(\frac{n+1}{n} - 1\right), \text{ for } x = 1$$
$$= 1$$

$\therefore$ Raabe's test also fails to give an conclusion.

Now, consider

$$\lim \left\{n\left(\frac{a_n}{a_{n+1}} - 1\right) - 1\right\}\log n = \lim(1 - 1)\log n = 0 < 1$$

$\therefore$ The series is divergent when $x = 1$ by De Morgan and Bertrand's test.

$\therefore$ Given series converges for $0 \leq x < 1$ and diverges for $x \geq 1$.

Example 6.6.17. *The series* $\displaystyle\sum_{n=1}^{\infty} \frac{n}{2^n}$ *converges.*

Solution. Let $a_n = \dfrac{n}{2^n}$. Then $a_{n+1} = \dfrac{n+1}{2^{n+1}}$ so that

$$\frac{a_n}{a_{n+1}} = \frac{2^{n+1}}{2^n} \cdot \frac{n}{n+1} = 2 \cdot \frac{1}{1+\frac{1}{n}}.$$

$\therefore \displaystyle\lim_{n \to \infty} \frac{a_n}{a_{n+1}}$ exists $= 2 > 1$.

$\therefore$ By Ratio Test the series $\sum a_n$ converges.

Example 6.6.18. *Prove that the series*

$$1 + \frac{1}{2}\cdot\frac{1}{3} + \frac{1\cdot 3}{2\cdot 4}\cdot\frac{1}{5} + \frac{1\cdot 3\cdot 5}{2\cdot 4\cdot 6}\cdot\frac{1}{7} + \cdots$$

converges by using Raabe's Test.

Solution. Ignoring the first term we take
$$a_n = \frac{1 \cdot 3 \cdots (2n-1)}{2 \cdot 4 \cdots 2n} \cdot \frac{1}{2n+1}.$$
Then
$$a_{n+1} = \frac{1 \cdot 3 \cdot 5 \cdot 7 \cdots (2n-1)(2n+1)}{2 \cdot 4 \cdot 6 \cdot 8 \cdots 2n(2n+2)} \cdot \frac{1}{2n+3}.$$
$$\therefore \quad \frac{a_n}{a_{n+1}} = \frac{2n+2}{2n+1} \times \frac{2n+3}{2n+1} = \frac{1+1/n}{1+1/2n} \cdot \frac{1+3/2n}{1+1/2n} \to 1 \text{ as } n \to \infty.$$

Hence ratio test does not give any definite information. So we consider Raabe's sequence $\{R_n\}$ where
$$R_n = n\left(\frac{a_n}{a_{n+1}} - 1\right) = n\left(\frac{(2n+2)(2n+3)}{(2n+1)^2} - 1\right)$$
$$= n\left(\frac{4n^2 + 10n + 6 - 4n^2 - 4n - 1}{(2n+1)^2}\right)$$
i.e., $\quad R_n = \frac{n(6n+5)}{(2n+1)(2n+1)} = \frac{6+5/n}{(2+1/n)(2+1/n)}$

whence $R_n \to 3/2 (>1)$ as $n \to \infty.$

$\therefore$ By Raabe's Test $\sum a_n$ converges.

Example 6.6.19. *Test the series*
$$1 + \frac{1}{2} + \frac{1 \cdot 3}{2 \cdot 4} + \frac{1 \cdot 3 \cdot 5}{2 \cdot 4 \cdot 6} + \cdots + \frac{1 \cdot 3 \cdot 5 \cdot 7 \cdots (2n-1)}{2 \cdot 4 \cdot 6 \cdot 8 \cdots (2n)} + \cdots.$$

Solution. Take
$$a_n \doteq \frac{1 \cdot 3 \cdot 5 \cdot 7 \cdots (2n-1)}{2 \cdot 4 \cdot 6 \cdot 8 \cdots 2n} \quad \text{so that} \quad a_{n+1} = \frac{1 \cdot 3 \cdot 5 \cdot 7 \cdots (2n-1)(2n+1)}{2 \cdot 4 \cdot 6 \cdot 8 \cdots 2n(2n+2)}.$$
$$\therefore \quad \frac{a_n}{a_{n+1}} = \frac{2n+2}{2n+1} = \frac{2+2/n}{2+1/n} \to 1 \text{ as } n \to \infty.$$

$\therefore$ the ratio test fails to give any definite information. So we try Raabe's Test:
$$R_n = n\left(\frac{a_n}{a_{n+1}} - 1\right) = n\left(\frac{2n+2}{2n+1} - 1\right) = \frac{n}{2n+1} = \frac{1}{2+1/n} \to \frac{1}{2} \text{ as } n \to \infty.$$
i.e., $\lim_{n \to \infty} R_n < 1$ and hence by Raabe's Test the series $\sum a_n$ *diverges.*

Before we prove De Morgan and Bertrand's Test we consider a **general form** of various ratio tests, namely Kummer's Test:

CHAPTER 6: INFINITE SERIES OF REAL CONSTANTS

IV. Kummer's Test

STATEMENT. *Let $\sum a_n$ be a series of positive terms. Let $\{b_n\}$ be a sequence of positive real numbers. We now construct a sequence $\{w_n\}$ where*

$$w_n = \frac{a_n}{a_{n+1}} b_n - b_{n+1}.$$

Then

1. $\sum_{n=1}^{\infty} a_n$ *converges, if* $\underline{\lim} w_n > 0$;

2. $\sum_{n=1}^{\infty} a_n$ *diverges if* $\overline{\lim} w_n < 0$, *provided* $\sum_{n=1}^{\infty} \frac{1}{b_n}$ *is a divergent series.*

[For testing the convergence of the series we require $\underline{\lim} w_n > 0$ but then we do not require that $\sum \frac{1}{b_n}$ should be a divergent series; it is only when we test the divergence of $\sum a_n$ we require not only $\overline{\lim} w_n < 0$ but also that $\sum \frac{1}{b_n}$ should be a divergent series.

In practice, we choose a sequence $\{b_n\}$ such that $\sum \frac{1}{b_n}$ is a divergent series so that we can then use the same $\{b_n\}$ sequence for testing both convergence or divergence.]

Proof: 1. If $\underline{\rho} = \underline{\lim} w_n > 0$, we take $\varepsilon > 0$ such that $\rho - \varepsilon = h$ is positive, then $\exists$ a positive integer m such that

$$w_n = \frac{a_n}{a_{n+1}} b_n - b_{n+1} > h, \quad \forall n \geq m$$

i.e., $\quad a_n b_n - a_{n+1} b_{n+1} > h \, a_{n+1}, \quad \forall n \geq m.$

Replacing n by $m, m+1, m+2, \cdots, n-1$ (where $n \gg m$), and adding all the inequalities we obtain

$$a_m b_m - a_n b_n > h(a_{m+1} + a_{m+2} + \cdots + a_n).$$

Hence,

$$a_{m+1} + a_{m+2} + \cdots + a_n < \frac{a_m b_m}{h} - \frac{a_n b_n}{h}$$

i.e., $\quad < \frac{a_m b_m}{h}, \quad \forall n.$

or, $(a_1 + a_2 + \cdots + a_m) + a_{m+1} + a_{m+2} + \cdots + a_n < \frac{a_m b_m}{h} + (a_1 + a_2 + \cdots + a_m).$

We observe that the right hand side of the inequality does not involve n, so that $\sum_{r=1}^{n} a_r = S_n = n$th partial sum of the series $\sum a_n$ is bounded and $\{S_n\}$ is clearly monotone increasing. Hence $\{S_n\}$ converges and consequently $\sum a_n$ converges.

2. If $\bar{\rho} = \overline{\lim} w_n < 0$, all the expressions w_n must be negative after a certain stage i.e., $\exists$ a positive integer m such that

$$\frac{a_n}{a_{n+1}} b_n - b_{n+1} < 0, \quad \forall\, n \geq m$$
$$\text{or,} \quad a_n b_n < a_{n+1} b_{n+1}, \quad \forall\, n \geq m$$
$$\text{i.e.,} \quad a_n b_n > a_{n-1} b_{n-1} > a_{n-2} b_{n-2} > \cdots > a_m b_m,$$

provided we take $n \gg m$.

Thus $a_n > (a_m b_m) \left(\dfrac{1}{b_n}\right)$, if $n > m$ i.e., the terms of $\sum a_n$ are, after the mth term, greater than those of the divergent series $a_m b_m \sum (1/b_n)$. Hence, by Comparison Test, the series $\sum a_n$ diverges.

This completes the proof of Kummer's Test.

Special Cases

** In Case $\underline{\lim} w_n = \overline{\lim} w_n = \lim_{n \to \infty} w_n = l$ (say), then

(a) $\sum a_n$ converges if $l > 0$;

(b) $\sum a_n$ diverges if $l < 0$ (here $\sum (1/b_r)$ must be a divergent series).

** See that D'Alembert's Ratio Test is a particular case of Kummer's Test, where b_n is taken $= 1$, $\forall\, n$ so that $\sum b_n$ is clearly a divergent series. Then $w_n = \dfrac{a_n}{a_{n+1}} - 1$. So, by Kummer's Test

$$\sum a_n \text{ converges if } \underline{\lim} \left(\frac{a_n}{a_{n+1}} - 1\right) > 0 \quad \text{i.e., if } \underline{\lim} \frac{a_n}{a_{n+1}} > 1$$

$$\text{and } \sum a_n \text{ diverges if } \overline{\lim} \left(\frac{a_n}{a_{n+1}} - 1\right) < 0 \quad \text{i.e., if } \overline{\lim} \frac{a_n}{a_{n+1}} < 1.$$

This is exactly the statement of D'Alembert's Ratio Test (See Test **IV A.** given above).

** Raabe's Test is also a particular case of Kummer's Test where $b_n = n$ (so that

CHAPTER 6: INFINITE SERIES OF REAL CONSTANTS

$\sum \dfrac{1}{b_n} = \sum \dfrac{1}{n}$ is a divergent series). Now w_n becomes

$$= \dfrac{a_n}{a_{n+1}} \cdot n - (n+1)$$

$$= n\left(\dfrac{a_n}{a_{n+1}} - 1\right) - 1$$

$$= R_n - 1.$$

Then Kummer's Test gives.

$$\sum a_n \text{ converges if } \underline{\lim}(R_n - 1) > 0 \text{ i.e., if } \underline{\lim} R_n > 1$$

and $\sum a_n$ diverges if $\overline{\lim}(R_n - 1) < 0$ i.e., if $\overline{\lim} R_n < 1$.

(This is **Raabe's Test**)

Proof of De Morgan & Bertrand's Test

We now prove De Morgan and Bertrand's Test using Kummer's Test.

Take $b_n = n \log n$ so that $\sum \dfrac{1}{b_n}$ is a divergent series. Now

$$w_n = \dfrac{a_n}{a_{n+1}} b_n - b_{n+1} \text{ becomes}$$

$$w_n = \dfrac{a_n}{a_{n+1}} n \log n - (n+1) \log(n+1).$$

[In the statement of De Morgan and Bertrand's Test we assumed $B_n = \log n \left\{ n\left(\dfrac{a_n}{a_{n+1}} - 1\right) - 1 \right\}$. This gives $\dfrac{a_n}{a_{n+1}} = 1 + \dfrac{1}{n} + \dfrac{B_n}{n \log n}$].

Thus,

$$w_n = \left(1 + \dfrac{1}{n} + \dfrac{B_n}{n \log n}\right) n \log n - (n+1) \log(n+1)$$

$$= (n+1) \log \dfrac{n}{n+1} + B_n.$$

See that

$$(n+1) \log \dfrac{n}{n+1} = \log\left(\dfrac{n}{n+1}\right)^{n+1} = \log \dfrac{1}{\left(1 + \dfrac{1}{n}\right)^n \left(1 + \dfrac{1}{n}\right)}$$

which tends to $\log \dfrac{1}{e}$ as $n \to \infty$ i.e., it tends to -1 as $n \to \infty$.

∴ By Kummer's Test it follows:

$$\sum a_n \text{ converges if } \underline{\lim} \, w_n > 0 \quad \text{i.e., if } \underline{\lim} \, B_n > 1$$
$$\text{and} \quad \sum a_n \text{ diverges if } \overline{\lim} \, w_n < 0 \quad \text{i.e., if } \overline{\lim} \, B_n < 1.$$

An equivalent statement of De Morgan's and Bertrand's Test is the following: Let $\sum a_n$ be a series of positive terms. Suppose it is possible to express

$$\frac{a_n}{a_{n+1}} = 1 + \frac{1}{n} + \frac{B_n}{n \log n}$$

[in other words, we construct a sequence $\{B_n\}$ defined by
$B_n = \log n \left\{ n \left(\frac{a_n}{a_{n+1}} - 1 \right) - 1 \right\}.$] then

∴ By Kummer's Test it follows:

$$\sum a_n \text{ converges if } \underline{\lim} \, w_n > 0 \quad \text{i.e., if } \underline{\lim} \, B_n > 1$$
$$\text{and} \quad \sum a_n \text{ diverges if } \overline{\lim} \, w_n < 0 \quad \text{i.e., if } \overline{\lim} \, B_n < 1.$$

An equivalent statement of De Morgan's and Bertrand's Test is the following: Let $\sum a_n$ be a series of positive terms. Suppose it is possible to express

$$\frac{a_n}{a_{n+1}} = 1 + \frac{1}{n} + \frac{B_n}{n \log n}$$

[in other words, we construct a sequence $\{B_n\}$ defined by
$B_n = \log n \left\{ n \left(\frac{a_n}{a_{n+1}} - 1 \right) - 1 \right\}.$] then

(a) $\sum a_n$ converges if $\exists$ a positive constant k and a positive integer N such that $\forall \, n > N, \, B_n \geq 1 + k$; and

(b) $\sum a_n$ diverges if $B_n \leq 1$ for infinitely many n.

[This statement follows if we use the properties of lower limit and upper limit to be applied on $\underline{\lim} \, B_n > 1$ and $\overline{\lim} \, B_n < 1$.]

Using Kummer's Test we can also prove another Test called **Gauss Test** which gives the results of D'Alembert's Ratio Test and Raabe's Test in more compact form:

V. Gauss' Test

Let $\sum_{n=1}^{\infty} a_n$ be a series of positive real constants.

CHAPTER 6: INFINITE SERIES OF REAL CONSTANTS

Suppose we can express

$$\frac{a_n}{a_{n+1}} = 1 + \frac{a}{n} + \frac{\beta_n}{n^p} \quad (p > 1)$$

where $\{\beta_n\}$ is a bounded sequence (it will be sufficient if $\{\beta_n\}$ is a convergent sequence because then it will be necessarily bounded). Then

(a) $\sum a_n$ converges if $a > 1$; and

(b) $\sum a_n$ diverges if $a \leq 1$.

Note 6.6.1. *When D'Alembert's ratio test fails, one may at once try Gauss' Test without going through other tests. We find it more convenient to use Gauss' Test in Order-notation form (given later).*

The following example will illustrate the use of Gauss' Test:

Example 6.6.20. *Test the series*

$$\frac{1}{2} + \frac{1 \cdot 3}{2 \cdot 4} + \frac{1 \cdot 3 \cdot 5}{2 \cdot 4 \cdot 6} + \cdots + \frac{1 \cdot 3 \cdot 5 \cdots (2n-1)}{2 \cdot 4 \cdot 6 \cdots 2n} + \cdots .$$

Solution. Take $a_n = \dfrac{1 \cdot 3 \cdot 5 \cdots (2n-1)}{2 \cdot 4 \cdot 6 \cdots 2n}$. To Test the series $\sum a_n$

Ratio Test: $\dfrac{a_n}{a_{n+1}} = \dfrac{2n+2}{2n+1} \to 1$ as $n \to \infty$.

i.e., Ratio Test does not give a definite conclusion. We now try **Gauss' Test:**
Here

$$\frac{a_n}{a_{n+1}} = 1 + \frac{1}{2n+1} = 1 + \frac{2n}{2n(2n+1)}$$

$$= 1 + \frac{(2n+1) - 1}{2n(2n+1)} = 1 + \frac{1}{2n} - \frac{1}{2n(2n+1)}$$

$$= 1 + \frac{1/2}{n} - \frac{1}{n^2} \frac{n}{2(2n+1)}$$

$$= 1 + \frac{a}{n} + \frac{\beta_n}{n^2},$$

where $a = \dfrac{1}{2}$ and $\beta_n = -\dfrac{n}{2(2n+1)}$.

Note that $\beta_n = -\dfrac{1}{2(2 + 1/n)} \to -\dfrac{1}{4}$ as $n \to \infty$.

i.e., $\{\beta_n\}$ is convergent sequence and hence bounded.

∴ By Gauss' Test $\sum a_n$ diverges because, here $a = \frac{1}{2} < 1$.

Order Notations : A Short Note.

(i) The symbol O needs a little clarification. We call it O-notation (order-notation).

Suppose $\{a_n\}$ and $\{b_n\}$ are two sequences, such that $|a_n| < k|b_n|$ for all $n >$ some fixed positive integer n_0 and k is independent of n. Then we write

$$a_n = O(b_n).$$

In particular, if $b_n = 1$ for all n, then $a_n = O(1)$ implies that the sequence $\{a_n\}$ is bounded.

(ii) There is another notation—small letter o. When we write $a_n = o(b_n)$, it will mean $a_n/b_n \to o$ as $n \to \infty$.

In this case we say 'a_n is of lower order than b_n'.

(iii) The use of the symbols O, o was suggested by Landau and has proved to be of great help in various branches of mathematics (prime-number theory, in particular).

Examples

(a) $(n+2)^2 = O(n^2)$.

$\left[\text{Since } \left(\dfrac{n+2}{n}\right)^2 = 1 + \dfrac{4}{n} + \dfrac{4}{n^2} \text{ which converges to 1 as } n \to \infty \text{ and hence } \left(\dfrac{n+2}{n}\right)^2 \text{ is bounded.}\right]$

(b) $\dfrac{3n-4}{n+3} = O(1)$.

$\left[\because \dfrac{3n-4}{n+3} \to 3 \text{ as } n \to \infty \text{ and since a convergent sequence is bounded, the sequence } \left\{\dfrac{3n-4}{n+3}\right\} \text{ is bounded.}\right]$

(c) Let $a_n = \dfrac{1}{2}\left(\dfrac{n^2+2n+3}{n-2}\right)$.

See that $a_n = O(n)$; also $a_n = O(n^2)$.

For, $\dfrac{a_n}{n} = \dfrac{1}{2}\left(\dfrac{(n^2+2n+3)}{n(n-2)}\right) \to \dfrac{1}{2}$ as $n \to \infty$ and

$\dfrac{a_n}{n^2} = \dfrac{1}{2}\dfrac{n^2+2n+3}{n^2(n-2)} \to 0$ as $n \to \infty$.

(d) $(n^3+n)^{-\frac{1}{2}} = O\left(n^{-\frac{3}{2}}\right)$.

For, $\left|\dfrac{1}{\sqrt{(n^3+n)}}\right| < \dfrac{1}{\sqrt{n^3}} = n^{-\frac{3}{2}}$.

CHAPTER 6: INFINITE SERIES OF REAL CONSTANTS

(e) Sometimes the function a_n is comparable to some function other than a power of n. For example, the notations

$$a_n = O(e^{2n}), \quad a_n = O(n \log 2n)$$

mean respectively that

$$\left\{\frac{a_n}{e^{2n}}\right\}, \quad \left\{\frac{a_n}{n \log 2n}\right\} \quad (n = 1, 2, \cdots)$$

are bounded sequences.

(iv) Sometimes, n is restricted;

$$\text{e.g.,} \quad \frac{1}{n-1} = O\left(\frac{1}{n}\right), \quad \text{when } n \geq 2.$$

For $\dfrac{1}{n-1} \bigg/ \dfrac{1}{n} = \dfrac{n}{n-1}$, which takes the values $2, \frac{3}{2}, \cdots$ when $n = 2, 3$, and each of these values ≤ 2 : but we cannot admit the value $n = 1$.

Similarly,

$$\frac{1}{(n-a)(n-b)} = O\left(\frac{1}{n^2}\right) \quad \text{when } n \geq N > a, b.$$

(v) **The following techniques are very useful for practical work:**

If $f(n) = a_0 + \dfrac{a_1}{n} + \dfrac{a_2}{n^2} + \cdots$ $(a_0, a_1, a_2, \cdots$ are constants) be an absolutely convergent series, then we can write

$$f(n) = a_0 + O\left(\frac{1}{n}\right)$$

$$\text{or,} \quad f(n) = a_0 + \frac{a_1}{n} + O\left(\frac{1}{n^2}\right), \quad \text{etc.}$$

Again if $f(n) = 1 + \dfrac{a}{n} + O\left(\dfrac{1}{n^2}\right)$ and $\phi(n) = 1 + \dfrac{b}{n} + O\left(\dfrac{1}{n^2}\right)$ (either $\forall\, n$ or $\forall\, n > N$) then we can write

$$f(n)\phi(n) = 1 + \frac{a+b}{n} + O\left(\frac{1}{n^2}\right).$$

Check the following statements

1. $\left(1 + \dfrac{1}{n}\right)^k = 1 + \dfrac{k}{n} + O\left(\dfrac{1}{n^2}\right)$ when $n \geq 2$.

2. $\left(1 + \dfrac{a}{n}\right)^k = 1 + \dfrac{ka}{n} + O\left(\dfrac{1}{n^2}\right)$ when $n \geq N > |a|$.

3. $\log\left(1 + \dfrac{1}{n}\right) = \dfrac{1}{n} + O\left(\dfrac{1}{n^2}\right)$ when $n \geq 2$.

4. $e^{a/n} = 1 + \dfrac{a}{n} + O\left(\dfrac{1}{n^2}\right)$.

V(A). Gauss' Test in Order-Notation Form

Let $\sum a_n$ be a series of positive terms. If it is possible to express a_n/a_{n+1} in the form:

$$\dfrac{a_n}{a_{n+1}} = 1 + \dfrac{a}{n} + O\left(\dfrac{1}{n^p}\right),$$

where a is a constant and p is a positive integer > 1, (usually, for practical purposes we may take $p = 2$) then $\sum a_n$ converges if $a > 1$ and diverges if $a \leq 1$.

The most important practical application of Gauss' Test is the following:

Example 6.6.21. Hypergeometric Series:

$$1 + \dfrac{\alpha \cdot \beta}{1 \cdot \gamma} x + \dfrac{\alpha(\alpha+1)\beta(\beta+1)}{1 \cdot 2 \cdot \gamma(\gamma+1)} x^2 + \dfrac{\alpha(\alpha+1)(\alpha+2)\beta(\beta+1)(\beta+2)}{1 \cdot 2 \cdot 3 \cdot \gamma(\gamma+1)(\gamma+2)} x^3 + \cdots.$$

where α, β, γ are real and none of them is zero or a negative integer and let $x > 0$.

Solution. Let $a_n = \dfrac{\alpha(\alpha+1)\cdots(\alpha+n-1)\beta(\beta+1)\cdots(\beta+n-1)}{1 \cdot 2 \cdot 3 \cdots n \cdot \gamma(\gamma+1)\cdots(\gamma+n-1)} x^n$.

To Test the series $\sum a_n$.

Then

$$\dfrac{a_n}{a_{n+1}} = \dfrac{1}{(\alpha+n)(\beta+n)} \cdot (n+1)(\gamma+n) \cdot \dfrac{1}{x}$$

$$= \dfrac{(1+1/n)(1+\gamma/n)}{(1+\alpha/n)(1+\beta/n)} \cdot \dfrac{1}{x} \to \dfrac{1}{x} \text{ as } n \to \infty.$$

$\therefore$ By D'Alembert's Ratio test, $\sum a_n$ converges if $0 < x < 1$ and diverges if $x > 1$. If $x = 1$, consider

$$\dfrac{a_n}{a_{n+1}} = \dfrac{(n+1)(\gamma+n)}{(\alpha+n)(\beta+n)} = \dfrac{n^2 + (\gamma+1)n + \gamma}{n^2 + (\alpha+\beta)n + \alpha\beta}$$

$$= 1 + \dfrac{n(\gamma+1-\alpha-\beta) + \gamma - \alpha\beta}{n^2 + (\alpha+\beta)n + \alpha\beta}$$

$$= 1 + \dfrac{\gamma+1-\alpha-\beta}{n} + \dfrac{\beta_n}{n^2},$$

CHAPTER 6: INFINITE SERIES OF REAL CONSTANTS

where $\beta_n = \left(A + \dfrac{B}{n}\right) \Big/ (1+\alpha/n)(1+\beta/n)$, and

$$\left.\begin{array}{l} A = \gamma - \alpha\beta - (\alpha+\beta)(\gamma+1-\alpha-\beta) \\ B = -\alpha\beta(\gamma+1-\alpha-\beta) \end{array}\right\} \text{ are fixed numbers.}$$

Thus, $\beta_n \to A$ and as such $\{\beta_n\}$ is bounded.

Hence when $x = 1$ the hypergeometric series converges if $\gamma + 1 - \alpha - \beta > 1$ i.e., if $\gamma > \alpha + \beta$ and diverges if $\gamma \leq \alpha + \beta$.

Otherwise. See that

$$\dfrac{a_n}{a_{n+1}} = \left(1 + \dfrac{1}{n}\right)\left(1 + \dfrac{\gamma}{n}\right)\left(1 + \dfrac{\alpha}{n}\right)^{-1}\left(1 + \dfrac{\beta}{n}\right)^{-1}$$

$$= \left\{1 + \dfrac{\gamma+1}{n} + O\left(\dfrac{1}{n^2}\right)\right\}\left\{1 - \dfrac{\alpha}{n} + O\left(\dfrac{1}{n^2}\right)\right\}\left\{1 - \dfrac{\beta}{n} + O\left(\dfrac{1}{n^2}\right)\right\}$$

$$= \left\{1 + \dfrac{\gamma+1}{n} + O\left(\dfrac{1}{n^2}\right)\right\}\left\{1 - \dfrac{\alpha+\beta}{n} + O\left(\dfrac{1}{n^2}\right)\right\}$$

$$= 1 + \dfrac{\gamma+1-\alpha-\beta}{n} + O\left(\dfrac{1}{n^2}\right).$$

Hence

$$\begin{cases} \sum a_n \text{ converges if } \gamma + 1 - \alpha - \beta > 1 \text{ and} \\ \sum a_n \text{ diverges if } \gamma + 1 - \alpha - \beta \leq 1. \end{cases}$$

Example 6.6.22. Prove that the series $\sum \left\{\dfrac{2\cdot 4\cdot 6\cdot 8 \cdots 2n}{3\cdot 5\cdot 7\cdot 9 \cdots (2n+1)}\right\}^2$ diverges.

Solution.

$$\dfrac{a_n}{a_{n+1}} = \left(\dfrac{2n+3}{2n+2}\right)^2 = \left(1 + \dfrac{3}{2n}\right)^2 \left(1 + \dfrac{1}{n}\right)^{-2}$$

$$= \left\{1 + \dfrac{3}{n} + O\left(\dfrac{1}{n^2}\right)\right\}\left\{1 - \dfrac{2}{n} + O\left(\dfrac{1}{n^2}\right)\right\} = 1 + \dfrac{1}{n} + O\left(\dfrac{1}{n^2}\right).$$

$\therefore$ By Gauss' Test ($\because a = 1$) the series diverges.

Otherwise.

$$\frac{a_n}{a_{n+1}} = \left(\frac{2n+3}{2n+2}\right)^2 = \left(1 + \frac{1}{2n+2}\right)^2$$

$$= 1 + \frac{1}{n} + \frac{\beta_n}{n^2}, \text{ if } \beta_n = n^2\left\{\frac{1}{(2n+2)^2} + \frac{1}{n+1} - \frac{1}{n}\right\}$$

i.e., if $\beta_n = n^2\left\{\frac{1}{4(n+1)^2} - \frac{1}{n(n+1)}\right\} = n^2\left\{\frac{n-4(n+1)}{4n(n+1)^2}\right\}$

$$= \frac{n(-3n-4)}{4(n+1)^2} = \frac{-(3+4/n)}{4(1+1/n)^2} \to -\frac{3}{4}.$$

that is, $\{\beta_n\}$ converges and hence bounded; hence etc.

Example 6.6.23. Binomial Series: $1 + nx + \frac{n(n-1)}{2!}x^2 + \frac{n(n-1)(n-2)}{3!}x^3 + \cdots$ *converges if $n > 0$ and diverges if $n < 0$, if $x = -1$.*

Solution. When $x = -1$, after a certain stage all the terms have the same sign and hence we can use tests for positive series. Denoting the series by $\sum_{n=0}^{\infty} a_n$, we can show by division that

$$\frac{a_r}{a_{r+1}} = -\frac{r+1}{n-r} = 1 + \frac{n+1}{r} + \frac{\beta_r}{r^2}$$

where

$$\beta_r = n(n+1)/(1-n/r) \to n(n+1) \text{ as } r \to \infty.$$

Hence, by Gauss' Test, the series converges if $n + 1 > 1$, diverges if $n + 1 < 1$, that is according as $n >$ or < 0.

Note that the case $n = 0$, gives $1+0+0+\cdots$ and hence the series is then convergent.

Proof of Gauss' Test Using Kummer's Test

In Gauss' Test, we can express

$$\frac{a_n}{a_{n+1}} = 1 + \frac{a}{n} + \frac{\beta_n}{n^p} \quad (p > 1)$$

or, $n\left(\frac{a_n}{a_{n+1}} - 1\right) = a + \frac{\beta_n}{n^{p-1}} \to a \text{ as } n \to \infty.$

($\because \{\beta_n\}$ is bounded and $\frac{1}{n^{p-1}} \to 0$ as $n \to \infty$, $\because p-1 > 0$).

Therefore, by Raabe's Test $\sum a_n$ converges if $a > 1$ and diverges if $a < 1$.

CHAPTER 6: INFINITE SERIES OF REAL CONSTANTS

If, however, $a = 1$. We take $\{b_n\} = \{n \log n\}$ which is a positive sequence when $n > 1$ and $\sum \dfrac{1}{n \log n}$ is a divergent series (usually proved by Cauchy's Condensation Test).

Now w_n becomes

$$w_n = \dfrac{a_n}{a_{n+1}} n \log n - (n+1) \log(n+1)$$

$$= \left(1 + \dfrac{1}{n} + \dfrac{b_n}{n^p}\right) n \log n - (n+1) \log(n+1)$$

$$= (n+1) \log \dfrac{n}{n+1} + \dfrac{b_n}{n^{p-1}} \log n.$$

Observation:

$$\log \left(\dfrac{n}{n+1}\right)^{n+1} = \log \left\{\dfrac{1}{(1+\frac{1}{n})^n} \cdot \dfrac{1}{1+\frac{1}{n}}\right\} \to \log \dfrac{1}{e} = -1 \text{ as } n \to \infty.$$

Also $\{b_n\}$ being bounded and $\dfrac{\log n}{n^{p-1}} \to 0$ as $n \to \infty$ ($\because p - 1 > 0$), it follows

$$\dfrac{b_n}{n^{p-1}} \log n \to 0 \text{ as } n \to \infty.$$

Thus as $n \to \infty$, $w_n \to -1$ (negative).

$\therefore$ $\sum a_n$ diverges by Kummer's Test.

Thus
$$\begin{cases} \sum a_n \text{ converges, when } a > 1; \\ \sum a_n \text{ diverges, when } a \leq 1. \end{cases}$$

This completes the proof of **Gauss' Test**.

VI. Logarithmic Test

The series $\sum a_n$ of positive terms is convergent or divergent according as

$$\lim_{n \to \infty} n \log \left(\dfrac{a_n}{a_{n+1}}\right) > 1 \text{ or } < 1.$$

Note 6.6.2. *This test is an alternative to Raabe's Test and hence can be at once applied if D'Alembert's ratio test fails to give any conclusion.*

This test will be easier to apply if n occurs as a power of $\frac{a_n}{a_{n+1}}$.

Proof: Let $n \log \frac{a_n}{a_{n+1}} \to l$ as $n \to \infty$.

(i) Suppose, in the first instance, $l > 1$.

Choose any $\varepsilon > 0$ such that $l - \varepsilon = \alpha$ (say) > 1.

Since $\lim_{n \to \infty} \left(n \log \frac{a_n}{a_{n+1}} \right) = l$, $\exists$ a positive integer N such that $\forall n \geq N$

$$l - \varepsilon < n \log \frac{a_n}{a_{n+1}} < l + \varepsilon.$$

In particular, $n \log \frac{a_n}{a_{n+1}} > l - \varepsilon = \alpha$, $\forall n \geq N$.

$$\Rightarrow \quad \frac{a_n}{a_{n+1}} > e^{\alpha/n}, \quad \forall n \geq N. \quad (\alpha \text{ is greater than } 1)$$

Now since $\left\{ \left(1 + \frac{1}{n}\right)^n \right\}$ is m.i. and bounded above by the real number e so that

$$\left(1 + \frac{1}{n}\right)^n < e$$

or, $e^{\alpha/n} > \left(1 + \frac{1}{n}\right)^\alpha$, $\forall n \geq N$

$$\therefore \quad \frac{a_n}{a_{n+1}} > \left(1 + \frac{1}{n}\right)^\alpha \quad \text{i.e.,} \quad > \left(\frac{n+1}{n}\right)^\alpha, \quad \forall n \geq N.$$

Take $b_n = \frac{1}{n^\alpha}$ and so $\frac{b_n}{b_{n+1}} = \left(\frac{n+1}{n}\right)^\alpha$.

Thus,

$$\frac{a_n}{a_{n+1}} > \frac{b_n}{b_{n+1}} \quad \forall n \geq N.$$

Since $\sum \frac{1}{n^\alpha} (\alpha > 1)$ is convergent, by Comparison Test (See Test II) $\sum a_n$ also converges.

(ii) Next consider $0 \leq l \leq 1$, obtain in a similar manner

$$\frac{a_n}{a_{n+1}} < e^{k/n} \quad \forall n \geq N_1 \quad (k = l + \varepsilon < 1).$$

Now consider the sequence $\left\{ \left(1 + \frac{1}{n-1}\right)^n \right\}_{n=2}^{\infty}$ which is m.d. bounded below by e and hence

$$\left(1 + \frac{1}{n-1}\right)^n \geq e \quad \text{for all } n \geq 2$$

or, $\frac{a_n}{a_{n+1}} < \left(\frac{n}{n-1}\right)^k \quad \forall n \geq N_1 > 1.$

CHAPTER 6: INFINITE SERIES OF REAL CONSTANTS

Let $b_n = \frac{1}{(n-1)^k}$ $(k < 1)$. See that $\frac{a_n}{a_{n-1}} < \frac{b_n}{b_{n+1}}$ $\forall n \geq N_1 > 1$.

Since $\sum \frac{1}{(n-1)^k}$ is divergent, by Comparison Test $\sum a_n$ diverges.

(iii) We can also prove similarly that when $l < 0$, the series $\sum a_n$ diverges.

Example 6.6.24. *Test the convergence of the following series:*

a) $x + \frac{2^2 x^2}{2!} + \frac{3^3 x^3}{3!} + \frac{4^4 x^4}{4!} + \cdots$ $(x > 0)$; b) $\sum_{n=2}^{\infty} \frac{1}{(\log n)^p}$.

Solution. a)
$$\frac{u_n}{u_{n+1}} = \frac{n^n x^n}{n!} \div \frac{(n+1)^{n+1} x^{n+1}}{(n+1)!} = \left(\frac{n}{n+1}\right)^n \frac{1}{x}.$$
$$= \frac{1}{(1+1/n)^n} \frac{1}{x} \to \frac{1}{ex} \text{ as } n \to \infty.$$

$\therefore$ $\sum u_n$ converges if $1/ex > 1$ i.e., if $x < 1/e$ and diverges if $x > 1/e$ (by Ratio Test).

In case $x = 1/e$, Ratio Test does not give any definite conclusion.

Let us apply *Logarithmic Test* when $x = 1/e$.

$$\frac{u_n}{u_{n+1}} = \frac{e}{(1+1/n)^n}; \text{ hence } \log \frac{u_n}{u_{n+1}} = \log e - n \log\left(1 + \frac{1}{n}\right).$$

i.e., $\log \frac{u_n}{u_{n+1}} = 1 - n\left(\frac{1}{n} - \frac{1}{2n^2} + \frac{1}{3n^3} - \cdots\right)$
$$= \frac{1}{2n} - \frac{1}{3n^2} + \cdots$$

i.e., $n \log \frac{u_n}{u_{n+1}} \to \frac{1}{2}$ as $n \to \infty$.

Since this limit is < 1, by Logarithmic Test, the series is divergent.

[*Ans:* The given series converges if $0 < x < 1/e$ and diverges if $x \geq 1/e$.]

b)
$$u_n = \frac{1}{(\log n)^p}, \quad \frac{u_n}{u_{n+1}} = \left(\frac{\log(n+1)}{\log n}\right)^p.$$

$\therefore$ $\frac{u_n}{u_{n+1}} = \left\{\frac{\log n + \log(1 + 1/n)}{\log n}\right\}^p$
$$= \left\{\frac{\log n + \frac{1}{n} - \frac{1}{2}\cdot\frac{1}{n^2} + \frac{1}{3}\cdot\frac{1}{n^3} - \cdots}{\log n}\right\}^p$$
$$= \left\{1 + \frac{1}{n \log n} - \frac{1}{2n^2 \log n} + \frac{1}{3n^3 \log n} - \cdots\right\}^p.$$

Hence, $\lim\limits_{n\to\infty} \dfrac{u_n}{u_{n+1}} = 1$; D'Alembert's Ratio Test fails to give any definite conclusion. We turn to Logarithmic Test:

$$\log \frac{u_n}{u_{n+1}} = p \log \left\{ 1 + \frac{1}{n \log n} - \frac{1}{2n^2 \log n} + \cdots \right\}$$

$$= p \left\{ \left(\frac{1}{n \log n} - \frac{1}{2n^2 \log n} + \cdots \right) - \cdots \right\}.$$

$$\therefore \lim_{n\to\infty} n \log \frac{u_n}{u_{n+1}} = \lim_{n\to\infty} p \left\{ \frac{1}{\log n} - \frac{1}{2n \log n} + \cdots \right\}$$

$$= 0, \text{ for all values of } p.$$

Since the limit is less than 1, we conclude that the given series is divergent for all values of p.

VII. Another Logarithmic Test : Alternative to Bertrand's Test

The positive series $\sum u_n$ converges or diverges according as

$$\lim_{n\to\infty} \left\{ n \log \frac{u_n}{u_{n+1}} - 1 \right\} \log n > 1 \text{ or } < 1.$$

Example 6.6.25. *Test the convergence or divergence of the series:*

$$1^p + \left(\frac{1}{2}\right)^p + \left(\frac{1\cdot 3}{2\cdot 4}\right)^p + \left(\frac{1\cdot 3\cdot 5}{2\cdot 4\cdot 6}\right)^p + \cdots.$$

Solution. Let

$$u_n = \left\{ \frac{1\cdot 3\cdot 5 \cdots (2n-1)}{2\cdot 4\cdot 6 \cdots (2n)} \right\}^p.$$

Here

$$\frac{u_n}{u_{n+1}} = \left(\frac{2n+2}{2n+1}\right)^p = \left(\frac{1+1/n}{1+1/2n}\right)^p \to 1 \text{ as } n \to \infty.$$

$\therefore$ D'Alembert's Test does not give any conclusion.

CHAPTER 6: INFINITE SERIES OF REAL CONSTANTS

Now,

$$n \log \frac{u_n}{u_{n+1}} = n[p\log(1+1/n) - p\log(1+1/2n)]$$

$$= np\left[\left(\frac{1}{n} - \frac{1}{2}\cdot\frac{1}{n^2} + \frac{1}{3}\cdot\frac{1}{n^3} - \cdots\right)\right.$$

$$\left. - \left(\frac{1}{2n} - \frac{1}{2}\cdot\frac{1}{4n^2} + \frac{1}{3}\cdot\frac{1}{8n^3} - \cdots\right)\right]$$

$$= np\left[\frac{1}{2n} - \frac{3}{8n^2} + \frac{7}{24n^3} - \cdots\right]$$

$$= p\left[\frac{1}{2} - \frac{3}{8n} + \frac{7}{24n^2} - \cdots\right].$$

$\therefore \lim_{n\to\infty} n \log \frac{u_n}{u_{n+1}} = \frac{p}{2}.$

Hence, by Logarithmic Test, the series $\sum u_n$ converges if $\frac{p}{2} > 1$, and diverges if $\frac{p}{2} < 1$ (i.e., when $p > 2$ and $p < 2$).

When $\frac{p}{2} = 1$, i.e., when $p = 2$, this Logarithmic Test does not give any conclusion. We now turn to what we call alternative form of Bertrand's Test:

We find $\left(n \log \frac{u_n}{u_{n+1}} - 1\right) \log n$ (Taking $p = 2$)

$$= \left\{2\left(\frac{1}{2} - \frac{3}{8n} + \frac{7}{24n^2} - \cdots\right) - 1\right\} \log n$$

$$= n\left\{-\frac{3}{4n} + \frac{7}{12n^2} - \cdots\right\} \frac{\log n}{n}$$

$$= \left(-\frac{3}{4} + \frac{7}{12n} - \cdots\right) \frac{\log n}{n}$$

which tends to $-\frac{3}{4} \times 0 = 0$ as $n \to \infty$. $\left(\because \frac{\log n}{n} \to 0 \text{ as } n \to \infty\right).$

Thus, the limit is less than 1 and hence the series $\sum u_n$ diverges when $p = 2$.

[Ans: The given series converges if $p > 2$, diverges if $p \leq 2$.]

VIII. Cauchy's Condensation Test

STATEMENT. *If $\phi(n) > 0$, $\forall$ positive integers n and the sequence $\{\phi(n)\}$ is monotone decreasing, then the two series*

$$\sum_{n=1}^{\infty} \phi(n) \quad \text{and} \quad \sum_{n=1}^{\infty} h^n \phi(h^n) \quad (h \text{ being a positive integer } > 1) \quad \text{[C.H. 1999]}$$

are either both convergent or both divergent.

Proof: We group the terms of $\sum \phi(n)$ as follows:

$$\{\phi(1)\} + \{\phi(2) + \phi(3) + \cdots + \phi(h)\}$$
$$+ \{\phi(h+1) + \phi(h+2) + \cdots + \phi(h^2)\}$$
$$+ \{\phi(h^2+1) + \phi(h^2+2) + \cdots + \phi(h^3)\}$$
$$+ \cdots\cdots\cdots\cdots\cdots\cdots\cdots\cdots\cdots\cdots$$
$$+ \{\phi(h^{n-1}+1) + \phi(h^{n-1}+2) + \cdots + \phi(h^n)\}$$
$$+ \cdots\cdots\cdots\cdots\cdots\cdots\cdots\cdots\cdots\cdots.$$

We leave out the first group, namely $\{\phi(1)\}$ and call

$$v_1 = \phi(2) + \phi(3) + \cdots + \phi(h)$$
$$v_2 = \phi(h+1) + \phi(h+2) + \cdots + \phi(h^2)$$
$$\cdots\cdots\cdots\cdots\cdots\cdots\cdots\cdots\cdots\cdots$$
$$v_n = \phi(h^{n-1}+1) + \phi(h^{n-1}+2) + \cdots + \phi(h^n)$$
$$\cdots\cdots\cdots\cdots\cdots\cdots\cdots\cdots\cdots\cdots.$$

We discuss the convergence or divergence of the positive series $\sum v_n$.

Each term of v_n is $\leq \phi(h^{n-1}+1)$ and is $\geq \phi(h^n)$ because $\{\phi(n)\}$ is given to be monotone decreasing.

Moreover it is easy to see that there are, in all, $h^n - h^{n-1}$ terms in v_n.

$$\therefore \quad (h^n - h^{n-1})\phi(h^n) \leq v_n \leq (h^n - h^{n-1})\phi(h^{n-1})$$
$$\text{or,} \quad (h-1)h^{n-1}\phi(h^n) \leq v_n \leq (h-1)h^{n-1}\phi(h^{n-1})$$
$$\text{or,} \quad \frac{1}{h}(h-1)\{h^n \phi(h^n)\} \leq v_n \leq (h-1)\{h^{n-1}\phi(h^{n-1})\}. \quad (6.6.1)$$

We now see that

$$\forall\, n \geq 1, \quad \text{every } v_n \geq \frac{h-1}{h} h^n \phi(h^n) \quad \forall\, n \geq 1 \quad (6.6.2)$$
$$\text{or,} \quad h^n \phi(h^n) \leq \frac{h}{h-1} v_n, \quad \forall\, n \geq 1. \quad (6.6.3)$$

(h is a positive integer > 1 and so $\frac{h}{h-1}$ or $\frac{h-1}{h}$ is positive.)

CHAPTER 6: INFINITE SERIES OF REAL CONSTANTS

∴ By comparison Test

$\left.\begin{array}{l}\text{from (6.6.3), it follows that if } \sum v_n \text{ converges then } \sum h^n \phi(h^n) \\ \text{also converges;} \\ \text{and from (6.6.2) it follows that if } \sum h^n \phi(h^n) \text{ diverges} \\ \text{then } \sum v_n \text{ also diverges.}\end{array}\right\}$ (6.6.4)

Again the second inequality of (6.6.3) gives

$$v_n \leq (h-1)h^{n-1}\phi(h^{n-1}) \ \forall \ n \geq 2$$
$$\text{or,} \quad h^{n-1}\phi(h^{n-1}) \geq \frac{1}{h-1}v_n.$$

($h-1$ is a positive quantity since $h > 1$.)

Hence, as before, by Comparison Test, it follows that

$\left.\begin{array}{l}\text{if } \sum_{n=2}^{\infty} h^{n-1}\phi(h^{n-1}) \text{ is convergent then } \sum v_n \text{ converges} \\ \text{and if } \sum v_n \text{ diverges then } \sum_{n=2}^{\infty} h^{n-1}\phi(h^{n-1}) \text{ also diverges.}\end{array}\right\}$ (6.6.5)

Combining the results of (6.6.6) and (6.6.7) we get

$$\sum v_n \text{ and } \sum h^n \phi(h^n) \text{ converge or diverge together.}$$

But for a positive series $\sum v_n$ and $\sum \phi(n)$ converge or diverge together.

Hence finally $\sum \phi(n)$ and $\sum h^n \phi(h^n)$ converge or diverge together.

Applications

Example 6.6.26. *Test the convergence or divergence of the series* $\sum_{n=1}^{\infty} \frac{1}{n^p}$.

[C.H. 1981]

Solution. Let $\phi(n) = \frac{1}{n^p}$. The sequence $\left\{\frac{1}{n^p}\right\}$ is monotone decreasing.

Let h be a positive integer > 1; one may take $h = 2$.

Then $\phi(h^n) = \frac{1}{(h^n)^p}$ and so $h^n \phi(h^n) = \frac{1}{(h^{p-1})^n}$.

1. When $p > 1$, $\frac{1}{h^{p-1}}$ is positive and less than 1 i.e., $\sum_{n=1}^{\infty} \frac{1}{(h^{p-1})^n}$ is a geometric series with common ratio $\frac{1}{h^{p-1}} < 1$ and hence convergent.

∴ By Cauchy's Condensation Test

$$\sum \phi(n) = \sum \frac{1}{n^p} \quad \text{and} \quad \sum h^n \phi(h^n) = \sum \frac{1}{(h^{p-1})^n}$$

will either both converge or will both diverge.

But since $\sum \frac{1}{(h^{p-1})^n}$ is convergent, therefore, $\sum \frac{1}{n^p}$ converges when $p > 1$.

2. When $p < 1$, $\sum \frac{1}{(h^{p-1})^n}$ is a divergent geometric series because its common ratio h^{1-p} is > 1 (h being a positive integer > 1). Hence by Cauchys Condensation Test $\sum \frac{1}{n^p}$ ($p < 1$) diverges.

3. When $p = 1$, the given series is the harmonic series $\sum \frac{1}{n}$ and here $\sum \frac{1}{(h^{p-1})^n} = 1 + 1 + 1 + 1 + \cdots$ (clearly divergent).

∴ $\sum \frac{1}{n}$ also diverges by Cauchy's Condensation Test.

Example 6.6.27. (A very useful series) $\sum_{n=2}^{\infty} \frac{1}{n(\log n)^p}$.

To prove that this series converges when $p > 1$, diverges if $p \leq 1$. [C.H. 1999]

Solution. Let $\phi(n) = \frac{1}{n(\log n)^p}$ ($n \geq 2$).

The sequence $\{\phi(n)\}$ is a positive monotone decreasing sequence. Let h be a positive integer > 1.

Then

$$h^n \phi(h^n) = h^n \frac{1}{h^n (\log h^n)^p} = \frac{1}{n^p (\log h)^p}$$

$$= \frac{k}{n^p}, \quad \text{where } k = \frac{1}{(\log h)^p} = \text{constant}.$$

Hence by Cauchy's Condensation Test

$$\sum \frac{1}{n(\log n)^p} \quad \text{and} \quad k \sum \frac{1}{n^p} \quad \text{converge or diverge together}.$$

Since $k \sum \frac{1}{n^p}$ converges if $p > 1$ and diverges if $p \leq 1$ therefore, $\sum \frac{1}{n(\log n)^p}$ converges if $p > 1$ and diverges if $p \leq 1$.

CHAPTER 6: INFINITE SERIES OF REAL CONSTANTS

In particular, when $p = 1$ we have the **Abel's series**

$$\frac{1}{2\log 2} + \frac{1}{3\log 3} + \frac{1}{4\log 4} + \cdots + \frac{1}{n\log n} + \cdots.$$

The is a divergent series by Cauchys Condensation Test.
But since $\dfrac{1}{n\log n} < \dfrac{1}{n}$ $(n > 2)$ we note that Abel's series diverges more slowly than the harmonic series $\sum \frac{1}{n}$.

Example 6.6.28. *The series* $\displaystyle\sum_{n=2}^{\infty} \frac{1}{n\log n \log(\log n)}$ *diverges.*

Solution. Let h be a positive integer > 1. One may take $h = 2$.
The condensed series

$$\sum_{n=2}^{\infty} \frac{2^n}{2^n \log 2^n \log(\log 2^n)} = \sum_{n=2}^{\infty} \frac{1}{n \log 2 \log(n \log 2)}.$$

has larger terms than the Abel's series, since $\log 2 < 1$ and must therefore, diverge. The given series diverges even more slowly than Abel's series since

$$\frac{1}{n\log n \log(\log n)} < \frac{1}{n \log n} \quad \text{when} \ \ n > 15.$$

SUMMARY: TESTS FOR POSITIVE SERIES

Test	Statement & Proof	Illustrative Eamples
I. Comparison Test		
(a) Ratio Form **A** & **B**	P351	Ex.6.6.2–6.6.10
(b) Upper and Lower Limit form **C** & **D**	P352	
(c) Limit form **E**	P352	
(d) $\frac{a_n}{a_{n+1}} \geq \frac{b_n}{b_{n+1}}$ Form **F**	P354	
II. Cauchy's Root Test		
(a) Limit Form	P359	Ex.6.6.11–6.6.14
(b) in terms of upper limit: $\overline{\lim} \sqrt[n]{a_n}$	P362	
III. Ratio Tests		
(a) D'Alembert's Ratio Test	P364	Ex.6.6.15–6.6.19
(b) Raabe's Test	P364	
(c) De Morgan & Bertrand's Test	P365	
A General Case:		
IV. Kummer's Test	P375	
V. Gauss's Test	P378	Ex.6.6.20–6.6.23
Gauss's Test in Order-notation form	P382	
VI. Logarithmic Test	P385	Ex.6.6.24
Another Logarithmic Test:	P388	Ex.6.6.25
Alternative Bertrand's		
VII. Cauchy's Condensation Test	P389	Ex.6.6.26–Ex.6.6.28

Note: The students should be well acquainted with the working of Illustrative Examples (as indicate in col. 3).

6.7 Series of Positive and Negative Terms

We shall now consider *Mixed Series* with infinite number of positive and negative terms. Such a series may be *Absolutely convergent* or *Conditional convergent* or *Divergent*.

Definition 6.7.1. *An infinite series* $\sum_{n=1}^{\infty} a_n$ *of real constants is said to be* **absolutely convergent** *when the series* $\sum_{n=1}^{\infty} |a_n|$ *i.e., when* $|a_1|+|a_2|+\cdots+|a_n|+\cdots$ *is convergent.*

Obviously, a convergent series of real positive terms is absolutely convergent.

CHAPTER 6. INFINITE SERIES OF REAL CONSTANTS

Definition 6.7.2. *An infinite series* $\sum_{n=1}^{\infty} a_n$ *of real constants is said to be* **conditionally convergent** *if the series* $\sum a_n$ *converges but not absolutely.*

Example 6.7.1. *The series* $1 - \frac{1}{2} + \frac{1}{3} - \frac{1}{4} + \cdots$ *converges conditionally because it can be proved by Leibnitz's Test that the series converges but the positive series of its absolute values, namely, the Harmonic Series* $1 + \frac{1}{2} + \frac{1}{3} + \frac{1}{4} + \cdots$ *diverges.*

For the discussions that follow we shall recall the following:

(i) *An infinite series* $\sum a_n$ *of the real constants converges if the sequence* $\{s_n\}$ *of its partial sums converges, where* $s_n = a_1 + a_2 + a_3 + \cdots + a_n$.

(ii) *Further, a necessary and sufficient condition of convergence of* $\{s_n\}$ *is that given any* $\varepsilon > 0$, $\exists$ *a positive integer* n_0 *such that*

$$|s_{n+p} - s_n| < \varepsilon, \quad \forall\, n \geq n_0 \text{ and } \forall\, p = 1, 2, 3, \cdots.$$

[**Cauchy's General Principle of Convergence**]

Hence, a necessary as well as sufficient condition of convergence of the series $\sum a_n$ is that given any $\varepsilon > 0$, $\exists$ a positive integer n_0 such that

$$|a_{n+1} + a_{n+2} + \cdots + a_{n+p}| < \varepsilon, \quad \forall\, n \geq n_0 \text{ and } \forall\, p = 1, 2, 3, \cdots.$$

Theorem 6.7.1. *If the series* $\sum_{n=1}^{\infty} a_n$ *of real constants is absolutely convergent, then it converges, i.e., Absolute convergence of a series implies its convergence.* [C.H. 1999]

Proof: Given $\sum_{n=1}^{\infty} |a_n|$ converges.

To prove: $\sum_{n=1}^{\infty} a_n$ converges.

Since $\sum_{n=1}^{\infty} |a_n|$ converges, by what we have observed above.

Given any $\varepsilon > 0$, $\exists$ a positive integer n_0 such that

$$||a_{n+1}| + |a_{n+2}| + \cdots + |a_{n+p}|| < \varepsilon, \quad \forall\, n \geq n_0 \text{ and } \forall\, p = 1, 2, 3, \cdots$$

i.e., $\quad |a_{n+1}| + |a_{n+2}| + \cdots + |a_{n+p}| < \varepsilon, \quad \forall\, n \geq n_0 \text{ and } \forall\, p = 1, 2, 3, \cdots.$

Now for $n \geq n_0$ and $\forall\, p = 1, 2, 3, \cdots$ we also have

$$|a_{n+1} + a_{n+2} + \cdots + a_{n+p}| \leq |a_{n+1}| + |a_{n+2}| + \cdots + |a_{n+p}| < \varepsilon.$$

Hence, by sufficiency condition of Cauchy's General Principle, $\sum a_n$ converges.

Observations:

1. Given a mixed series, if we can prove its absolute convergence, we can at once conclude that the series itself converges.

2. For proving absolute convergence, we *usually* apply the known tests of positive series on $\sum_{n+1}^{\infty} |a_n|$.

   Thus $\sum a_n$ converges absolutely if $\lim_{n \to \infty} \sqrt[n]{|a_n|} < 1$ (**Root test**)

   or if $\lim_{n \to \infty} \left| \dfrac{a_{n+1}}{a_n} \right| < 1$ (**Ratio Test**).

3. Theorem 6.7.1 is **not true** for sequences.

   Thus $\{1, -1, 1, -1, \cdots\}$ does not converge, but $\{1, 1, 1, 1, \cdots\}$ does converge.

4. Clearly the series $\sum_{n=1}^{\infty} \dfrac{\sin n\theta}{n^p}$ or $\sum_{n=1}^{\infty} \dfrac{\cos n\theta}{n^p}$

   where $0 \leq \theta \leq 2\pi$ and $p > 1$ is absolutely convergent.

   Since, $\left| \dfrac{\sin n\theta}{n^p} \right| \leq \dfrac{1}{n^p}$ and $\left| \dfrac{\cos n\theta}{n^p} \right| \leq \dfrac{1}{n^p}$ and $\sum \dfrac{1}{n^p}$ converges if $p > 1$.

   (Use **Comparison Test**).

Alternating Series

There is one type of mixed series for which it is easy to decide whether it converges or not. The type is

$$\sum_{n=1}^{\infty} (-1)^{n-1} a_n = a_1 - a_2 + a_3 - a_4 + \cdots, \qquad (6.7.1)$$

where each a_r is positive. Such a series is called an **alternating series**, because the terms of the series are alternately positive and negative.

Theorem 6.7.2. Leibnitz's Test of Convergence for an alternating series

The alternating series (6.7.1) converges if

(a) the sequence $\{a_n\}$ is monotone decreasing (m.d.); and

(b) $a_n \to 0$ as $n \to \infty$.

CHAPTER 6: INFINITE SERIES OF REAL CONSTANTS

Proof: We first consider the sum of an even number of terms:

$$s_{2n} = a_1 - a_2 + a_3 - a_4 + \cdots + a_{2n-1} - a_{2n}.$$

Clearly, $\quad s_{2n+2} - s_{2n} = a_{2n+1} - a_{2n+2} \geq 0. \quad (\because \{a_n\} \text{ is m.d.})$.

Thus the sequence $\{s_2, s_4, s_6, \cdots, s_{2n}, s_{2n+2}, \cdots\}$ is monotone increasing (m.i.). Again we may write

$$s_{2n} = a_1 - (a_2 - a_3) - (a_4 - a_5) - \cdots - (a_{2n-2} - a_{2n-1}) - a_{2n}.$$

Therefore, $s_{2n} \leq a_1$, since the quantities $a_2 - a_3, a_4 - a_5$, etc. are all positive $\{a_n\}$ being m.d.

Thus the sequence $\{s_{2n}\}$ is m.i. and bounded above; hence

$$s_{2n} \to a \text{ finite limit } l \text{ (say) as } n \to \infty. \ (l \leq a_1).$$

Next we consider the sum of an odd number of terms of the alternating series (6.7.1):

$$\begin{aligned} s_{2n+1} &= a_1 - a_2 + a_3 - a_4 + \cdots - a_{2n} + a_{2n+1} \\ &= s_{2n} + a_{2n+1}. \end{aligned}$$

Since $s_{2n} \to l$ as $n \to \infty$ and also $a_{2n+1} \to 0$ as $n \to \infty$ by the given condition, (b) we conclude that $s_{2n+1} \to l$ as $n \to \infty$.

Thus $\{s_n\}$ tends to l, when $n \to \infty$ either through even or through odd positive integers.

In other words, the alternating series (6.7.1) converges.

Note 6.7.1. *We can easily verify that the intervals:*

$$(s_2, s_1), (s_4, s_3), (s_6, s_5), \cdots, (s_{2n}, s_{2n-1}), \cdots$$

*form a **nest** (i.e., each interval after the first is contained within the preceding and the lengths form a null sequence).*

Hence, by the Nested Interval Theorem, we have an unique real number l which is the common limit of $\{s_{2n}\}$ and $\{s_{2n+1}\}$.

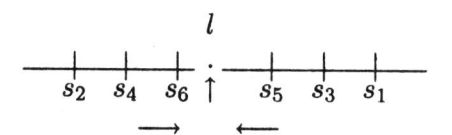

Since s_n and s_{n+1} lie on the opposite sides of l,

$$|r_n| = |\text{Remainder after } n \text{ terms}|.$$
$$= |l - s_n| < |s_{n+1} - s_n| \text{ i.e., } < |a_{n+1}|.$$

i.e., The remainder after n terms is always numerically less than the first of the terms neglected.

Properties of Mixed Series

Theorem 6.7.3. (a) *If a series is absolutely convergent, then the series formed by its positive terms alone is convergent, and the series formed by its negative terms alone is convergent.*

(b) *If a series is conditionally convergent (i.e., convergent but not absolutely), then the series formed by its positive (or negative) terms alone is divergent.*

Proof: Let us consider the series $\sum a_n$, where infinite number of terms are positive and also infinite number of terms are negative. Let $|a_n| = \alpha_n$.

Suppose,
$$p_n = \frac{1}{2}(a_n + \alpha_n) \quad \text{and} \quad q_n = \frac{1}{2}(a_n - \alpha_n).$$

Then
$$p_n = \begin{cases} a_n, & \text{where } a_n > 0 \\ 0, & \text{where } a_n < 0 \end{cases} \qquad q_n = \begin{cases} a_n, & \text{where } a_n < 0 \\ 0, & \text{where } a_n > 0. \end{cases}$$

Let
$$\left.\begin{array}{l} P_n = p_1 + p_2 + p_3 + \cdots + p_n \\ Q_n = q_1 + q_2 + q_3 + \cdots + q_n \end{array}\right\} \qquad \begin{array}{l} s_n = a_1 + a_2 + \cdots + a_n \\ \sigma_n = \alpha_1 + \alpha_2 + \cdots + \alpha_n. \end{array}$$

With these notations, we obtain
$$P_n = \frac{1}{2}(s_n + \sigma_n) \quad \text{and} \quad Q_n = \frac{1}{2}(s_n - \sigma_n). \tag{6.7.2}$$

To prove (a). Given: $\sum a_n$ converges absolutely; therefore, $\sum a_n$ is certainly convergent.

That is, $\sigma_n \to$ a finite limit σ' (say) and $s_n \to$ a finite limit s (say). Hence from (6.7.2),

$$P_n \longrightarrow \text{a finite limit } \frac{1}{2}(s + \sigma)$$
$$\text{and} \quad Q_n \longrightarrow \text{a finite limit } \frac{1}{2}(s - \sigma).$$

CHAPTER 6: INFINITE SERIES OF REAL CONSTANTS

This means that the series of *positive* (or *negative*) terms alone converges. This proves (a).

To prove (b). *Given:* $\sum a_n$ converges but $\sum \alpha_n$ diverges.

To prove: The series of *positive* (or *negative*) terms alone diverges.

Proof: If possible let the series of positive terms alone is convergent. Also according to the given condition $\sum a_n$ converges.

Therefore, $P_n \to$ a finite limit P and $s_n \to s$ as $n \to \infty$.

Hence from (6.7.2), $\sigma_n = 2P_n - s_n \to 2P - s$ (finite limit) as $n \to \infty$.

This goes against the given condition that $\sum \alpha_n$ diverges. Consequently, we cannot assume that the series of positive terms alone can converge (in fact, it *diverges* because a positive series cannot oscillate). Similarly, the series of negative terms alone diverges.

6.7.1 Power Series

A series of the form $\sum_{n=0}^{\infty} a_n x^n = a_0 + a_1 x + a_2 x^2 + \cdots$, where a_n and x are real, will be called a *Real Power Series*.

a) Every such power series converges at $x = 0$.

b) There are power series which converges only at $x = 0$ and at no other value of x,

[**Nowhere convergent**]

e.g., $\sum_{n=1}^{\infty} n! x^n$. For $x \neq 0$, nth term does not approach zero as $n \to \infty$; therefore, the series converges at $x = 0$ but at nowhere else.

c) There exist power series which converges for all real value of x. [**everywhere convergent**]

e.g., $\sum_{n=0}^{\infty} \dfrac{x^n}{n!}$. It converges for all real values of x.

[Justifications. (a) Let $x > 0$; then term ratio $\dfrac{a_{n+1}}{a_n} = \dfrac{x}{n+1}$, $\left(\text{where } a_n = \dfrac{x^n}{n!}\right)$ which tends to zero as $n \to \infty$ for any real positive value of x; hence by ratio test the series converges $\forall \, x > 0$.

(b) Let $x < 0$; put $x = -y$ ($y > 0$) then the series becomes $1 - y + \dfrac{y^2}{2!} - \dfrac{y^3}{3!} + \cdots$. From (a) it follows that the series is absolutely convergent and hence convergent; for all real $y > 0$ (i.e., for all real $x < 0$).

(c) Obviously for $x = 0$, the series converges. Hence $\sum x^n/n!$ converges for every real value of x.]

d) In general, however, a real power series converges within a certain interval with centre as origin. e.g., $\sum_{n=0}^{\infty} \dfrac{x^n}{n+1}$ converges absolutely (and hence converges) if $-1 < x < 1$.

When $x = 1$, the series is a harmonic series and diverges. When $x = -1$, the series is an alternating series and by Leibnitz's Test the series converges.

The interval $-1 \leq x < 1$ is then the *precise interval of convergence*.

6.8 Illustrative Examples

Example 6.8.1. *When we are given a power series such as*
$$1 + 2x + 3x^2 + 4x^3 + \cdots \quad (x \geq \text{ or } \leq 0).$$

Solution. We shall first consider the positive series
$$1 + 2|x| + 3|x|^2 + 4|x|^3 + \cdots$$

and use, say, ratio test (taking $a_n = (n+1)|x|^n$).

$$\lim_{n \to \infty} \dfrac{a_n}{a_{n+1}} = \lim_{n \to \infty} \dfrac{(n+1)|x|^n}{(n+2)|x|^{n+1}} = \lim_{n \to \infty} \dfrac{1+1/n}{1+2/n} \dfrac{1}{|x|} = \dfrac{1}{|x|}.$$

$\therefore$ The given series $\sum_{n=0}^{\infty}(n+1)x^n$ converges absolutely if $\dfrac{1}{|x|} > 1$. i.e., if $|x| < 1$.

Now when $|x| > 1$. a_n does not tend to zero as $n \to \infty$ and as such the series cannot converge.

If $x = 1$, the series becomes $1 + 2 + 3 + 4 + \cdots$ (*divergent*).

If $x = -1$, the series becomes $1 - 2 + 3 - 4 + \cdots$ (*oscillating*).

i.e., The power series $\sum_{n=0}^{\infty}(n+1)x^n$ converges only when $|x| < 1$ and the convergence is *absolute*.

Example 6.8.2. *If x and s are both real, discuss the convergence of the series*
$$\sum_{n=1}^{\infty} \dfrac{x^{n-1}}{n^s}.$$

CHAPTER 6: INFINITE SERIES OF REAL CONSTANTS

1. Absolute Convergence

We examine the series $\sum_{n=1}^{\infty} \dfrac{|x|^{n-1}}{n^s}$.

(a) $\left|\dfrac{a_{n+1}}{a_n}\right| = \left(\dfrac{n}{n+1}\right)^s |x|$, where $a_n = \dfrac{x^{n-1}}{n^s}$.

$\therefore \lim_{n \to \infty} \left|\dfrac{a_{n+1}}{a_n}\right| = |x|$.

Thus, $\sum |a_n|$ converges if $|x| < 1$ i.e., $\sum a_n$ converges absolutely if $|x| < 1$ and $\sum |a_n|$ diverges if $|x| > 1$ i.e., $\sum a_n$ does not converge absolutely if $|x| > 1$. In this case, since $\left|\dfrac{a_{n+1}}{a_n}\right| > 1$ for all $n >$ some fixed positive integer m, a_n does not tend to zero and so $\sum a_n$ cannot converge; also $\sum a_n$ cannot oscillate finitely because the terms increase.

Thus when $|x| > 1$, $\sum a_n$ either properly diverges or oscillates infinitely.

(b) Next we consider the case when $|x| = 1$. The series of moduli is then $\sum \dfrac{1}{n^s}$, which converges if $s > 1$ and diverges if $s \le 1$.

Conclusions. $\sum_{n=1}^{\infty} \dfrac{x^{n-1}}{n^s}$ *converges absolutely if* $|x| < 1$ *and also if* $|x| = 1$ *and* $s > 1$.

2. Non-Absolute Convergence

The only case, therefore, *where conditional convergence is possible is when* $|x| = 1$ *and* $s \le 1$ when the series of moduli diverges.

If $x = 1$, the case reduces to (b) above i.e., converges when $s > 1$, diverges if $s \le 1$.

When $x = -1$, the series $\sum_{n=1}^{\infty} \dfrac{(-1)^{n-1}}{n^s}$ is an alternating series; $\dfrac{1}{n^s} \to 0$ as $n \to \infty$ provided $s > 0$; also $\dfrac{1}{m^s} > \dfrac{1}{(m+1)^s}$ if $s > 0$.

Hence, by Leibnitz's Test, $\sum_{n=1}^{\infty} \dfrac{(-1)^{n-1}}{n^s}$ converges if $s > 0$.

3. Remaining Cases

(a) $|x| = 1$, and $s = 0$.

When $x = 1$, the series becomes $1 + 1 + 1 + \cdots$ (*obviously divergent*).

When $x = -1$, the series becomes $1 - 1 + 1 - \cdots$ (*oscillating finitely*).

(b) $|x| = 1$ and $s < 0$. We write $s = -p$ $(p > 0)$.

When $x = 1$, the series becomes $\sum n^p$ (*clearly diverges*).

When $x = -1$, the series is $\sum (-1)^{n-1} n^p$ (*oscillates infinitely*).

Conclusion. The series $\sum \dfrac{x^{n-1}}{n^s}$ is

a) Absolutely Convergent (and hence convergent) when $-1 < x < 1$ and s is any real number.

b) Absolutely convergent (and hence convergent) when $|x| = 1$ and $s > 1$.

c) Conditionally convergent if $x = -1$ and $0 < s \leq 1$.

d) In other cases, the series does not converge.

Example 6.8.3. The Use of Brackets in an infinite series (*Theorem* **6.5.3**)

a) The positive series $\left(1 - \frac{1}{2}\right) + \left(\frac{1}{3} - \frac{1}{4}\right) + \left(\frac{1}{5} - \frac{1}{6}\right) + \cdots$ *converges*, since its general term $\dfrac{1}{2n-1} - \dfrac{1}{2n} = \dfrac{1}{2n(2n-1)}$ is comparable with the general term of the convergent series $\sum \dfrac{1}{n^2}$.

What can we say about the following series where the brackets are removed?

$$1 - \frac{1}{2} + \frac{1}{3} - \frac{1}{4} + \frac{1}{5} - \frac{1}{6} + \cdots,$$

It does not at once follow that the series converges. It requires careful examination: $\{s_{2n}\}$ is m.i. and bounded, where

$$s_{2n} = 1 - \frac{1}{2} + \frac{1}{3} - \frac{1}{4} + \cdots + \frac{1}{2n-1} - \frac{1}{2n}.$$

Hence $s_{2n} \to$ a finite limit s (say).

Also $s_{2n+1} - s_{2n} = \dfrac{1}{2n+1} \to 0$ as $n \to \infty$.

$\therefore$ s_{2n+1} also tends to s as $n \to \infty$.

Thus, $\{s_n\} \to s$ as $n \to \infty$ whether n be odd or even.

i.e., the series converges even when the brackets are removed.

(b) $\left(2 - \frac{3}{2}\right) + \left(\frac{4}{3} - \frac{5}{4}\right) + \left(\frac{6}{5} - \frac{7}{6}\right) + \cdots$, is the same series as (a) and hence is convergent. But if we remove the brackets, we now get

$$2 - \frac{3}{2} + \frac{4}{3} - \frac{5}{4} + \frac{6}{5} - \frac{7}{6} + \cdots.$$

CHAPTER 6: INFINITE SERIES OF REAL CONSTANTS   403

Does it also converge?

We see that
$$s_{2n} = 2 - \frac{3}{2} + \frac{4}{3} - \frac{5}{4} + \cdots - \frac{2n+1}{2n}$$
$$= 1 - \frac{1}{2} + \frac{1}{3} - \frac{1}{4} + \cdots - \frac{1}{2n}.$$
$$s_{2n+1} = 2 - \frac{3}{2} + \frac{4}{3} - \frac{5}{4} + \cdots - \frac{2n+1}{2n} + \frac{2n+2}{2n+1}$$
$$= 1 + 1 - \frac{1}{2} + \frac{1}{3} - \frac{1}{4} + \cdots + \frac{1}{2n+1}.$$

Here $s_{2n+1} - s_{2n} \to 1$ if $n \to \infty$ and the series, therefore, does not converge.

Thus the series, without brackets, is **not convergent**.

6.9 Rearrangement of Series

For a finite number of terms, say,
$$a_1 + a_2 + a_3 + \cdots + a_n$$
the *algebraic sum* is not altered by writing the terms in a different order. This property cannot always be extended for an infinite series. We must first say what we mean by "writing the terms in a different order".

Definition 6.9.1. *The series*
$$b_1 + b_2 + b_3 + b_4 + \cdots + b_n + \cdots$$
is the series
$$a_1 + a_2 + a_3 + a_4 + \cdots + a_n + \cdots$$
with its terms in a different order it every b_r comes somewhere in the a series and every a_r comes somewhere in the b series; the series $\sum b_n$ is then said to be a Rearrangement of the series $\sum a_n$. e.g.,

(a) $1 - \frac{1}{2} + \frac{1}{3} - \frac{1}{4} + \frac{1}{5} - \frac{1}{6} + \cdots$ and $-\frac{1}{2} + 1 - \frac{1}{4} + \frac{1}{3} - \frac{1}{6} + \frac{1}{5} - \cdots$.

(b) $1 - \frac{1}{2} + \frac{1}{3} - \frac{1}{4} + \frac{1}{5} - \frac{1}{6} + \cdots$ and $1 + \frac{1}{3} - \frac{1}{2} + \frac{1}{5} + \frac{1}{7} - \frac{1}{4} + \frac{1}{9} + \frac{1}{11} - \frac{1}{6} + \cdots$.

In (a) and (b) the second series is a rearrangement of the first.

Since the sequence of partial sums of any infinite series and its rearranged series may be quite dissimilar their is no reason why they would behave in a similar way—one may tend to a finite limit, the other may not tend to the same limit (as a matter of fact, the other may not tend to any finite limit at all). We have an important general theorem:

Theorem 6.9.1. *If a series is absolutely convergent, then its sum remains unaltered by any change in the order of its terms.*

Proof:

Case 1. We first prove the theorem for a series of positive terms.

Let each a_n be positive and let

$$b_1 + b_2 + b_3 + \cdots + b_n + \cdots$$

be a rearrangement of the series

$$a_1 + a_2 + a_3 + \cdots + a_n + \cdots .$$

Let $s_n = a_1 + a_2 + a_3 + \cdots + a_n$ and $\sigma_n = b_1 + b_2 + \cdots + b_n$.

For a definite positive integer n, s_n contains n terms each of which comes, sooner or later, in the b-series and so $\exists$ a positive integer m such that σ_m contains all the terms of s_n (and possibly others not contained in s_n). Since each term is positive

$$\sigma_m \geq s_n.$$

Thus, Given n, $\exists$ a corresponding m such that $\sigma_m \geq s_n$.

Suppose now that the b-series is convergent. Then the sequence $\{\sigma_n\}$ has a finite upper bound σ (say).

Since $\sigma_m \geq s_n$, it follows $s_n \leq \sigma_m \leq \sigma$.

Hence the upper bound of $\{s_n\}$, say s, cannot exceed σ; hence the a-series is convergent and

$$s \leq \sigma. \tag{6.9.1}$$

Again, for any definite positive integer n, σ_n contains n terms each of which comes, sooner or later, in the a-series and so $\exists$ a positive integer M such that s_M contains all the terms of σ_n (and possibly others not contained in σ_n).

Since each term is positive, $s_M \geq \sigma_n$. Since $s_M \leq s$, it follows $\sigma_n \leq s_M \leq s$ and so σ, the upper bound of σ_n cannot exceed s; hence

$$\sigma \leq s. \tag{6.9.2}$$

By (6.9.1) and (6.9.2), $\sigma = s$ and the a-series has the same sum as b-series.

Next, suppose that the b-series is divergent. Then σ_n increases indefinitely and since we can find s_M to exceed any given σ_n, s_n must also increase indefinitely and hence the a-series also diverges.

Case 2. Suppose $\sum a_n$ is an absolutely convergent series.

We assume that $\sum a_n$ contains infinite number of positive terms and infinite number of negative terms.

Let P be the sum of its positive terms alone and Q be the sum of its negative terms alone.

Then, by Theorem **6.7.3(a)** of art. **6.7**, if s be the sum of the series $\sum a_n$,

$$s = P + Q \quad (Q \text{ is, of course, negative}).$$

Any change in the order of the terms of $\sum a_n$ gives a new series which, by **Case 1**, is such that
$$\begin{cases} \text{its positive terms alone converge to } P; \\ \text{its negative terms alone converge to } Q. \end{cases}$$

Hence the new series is absolutely convergent and its sum is $P+Q$, so that the new series has the same sum as the old one.

Example 6.9.1. $1 - \dfrac{1}{2^2} + \dfrac{1}{3^2} - \dfrac{1}{4^2} + \dfrac{1}{5^2} - \cdots$ is clearly absolutely convergent; let its sum be s. Therefore, the rearranged series $1 + \dfrac{1}{3^2} - \dfrac{1}{2^2} + \dfrac{1}{5^2} + \dfrac{1}{7^2} - \dfrac{1}{4^2} + \cdots$ is also absolutely convergent and its sum will be the same s.

Rearrangement of a non-absolutely convergent series

Let us now discuss the rearrangement of terms in a conditionally convergent series:

Example 6.9.2. The series $1 - \dfrac{1}{2} + \dfrac{1}{3} - \dfrac{1}{4} + \dfrac{1}{5} - \dfrac{1}{6} + \cdots$ and the rearranged series $1 + \dfrac{1}{3} - \dfrac{1}{2} + \dfrac{1}{5} + \dfrac{1}{7} - \dfrac{1}{4} + \cdots$ are both convergent and their sums are $\log 2, \dfrac{3}{2} \log 2$ respectively.

We shall assume the following result which we proved in the chapter on Sequences, art. **5.14**. Ex. **5.14.4**.

As $n \to \infty$, $\gamma_n = 1 + \dfrac{1}{2} + \dfrac{1}{3} + \cdots + \dfrac{1}{n} - \log n \to$ a constant γ [*Euler's Constant*].
Assuming this result, we may write

$$1 + \dfrac{1}{2} + \dfrac{1}{3} + \cdots + \dfrac{1}{n} = \log n + \gamma_n,$$

where $\gamma_n \to \gamma$ as $n \to \infty$.

406   AN INTRODUCTION TO ANALYSIS: DIFFERENTIAL CALCULUS

We have already proved that the series $1 - \frac{1}{2} + \frac{1}{3} - \frac{1}{4} + \frac{1}{5} - \frac{1}{6} + \cdots$ converges (alternating series; numerically terms steadily decrease and tend to zero).

In order to find the sum of this series we proceed thus:[Refer to **W.L. Ferrar—A Test Book of Convergence (O.U.P.)**].

Let s_n denote the algebraic sum of its first n terms; s the sum of the series. Then

$$s = \lim_{n \to \infty} s_n = \lim_{n \to \infty} s_{2n} = \lim_{n \to \infty} \left(1 - \frac{1}{2} + \frac{1}{3} - \frac{1}{4} + \cdots - \frac{1}{2n}\right)$$

$$= \lim_{n \to \infty} \left[\left(1 + \frac{1}{2} + \frac{1}{3} + \frac{1}{4} + \cdots + \frac{1}{2n}\right) - 2\left(\frac{1}{2} + \frac{1}{4} + \cdots + \frac{1}{2n}\right)\right]$$

$$= \lim_{n \to \infty} \left[\left(1 + \frac{1}{2} + \frac{1}{3} + \frac{1}{4} + \cdots + \frac{1}{2n}\right) - \left(1 + \frac{1}{2} + \frac{1}{3} + \cdots + \frac{1}{n}\right)\right]$$

$$= \lim_{n \to \infty} \left[(\log 2n + \gamma_{2n}) - (\log n + \gamma_n)\right]$$

$$= \lim_{n \to \infty} \left[\log 2 + \log n + \gamma_{2n} - \log n - \gamma_n\right]$$

$$= \lim_{n \to \infty} (\log 2 + \gamma_{2n} - \gamma_n) = \log 2. \quad (\because \gamma_{2n} - \gamma_n \to \gamma - \gamma = 0 \text{ as } n \to \infty.)$$

The series $1 + \frac{1}{3} - \frac{1}{2} + \frac{1}{5} + \frac{1}{7} - \frac{1}{4} + \cdots$ is not obviously convergent by any of the Standard tests. We begin by considering the sum of $3n$ terms : let σ_n denote the sum of first n terms of this series. Then,

$$\sigma_{3n} = \left(1 + \frac{1}{3} - \frac{1}{2}\right) + \left(\frac{1}{5} + \frac{1}{7} - \frac{1}{4}\right) + \cdots + \left(\frac{1}{4n-3} + \frac{1}{4n-1} - \frac{1}{2n}\right)$$

$$= \left(1 + \frac{1}{3} + \frac{1}{5} + \frac{1}{7} + \cdots + \frac{1}{4n-3} + \frac{1}{4n-1}\right) - \frac{1}{2}\left(1 + \frac{1}{2} + \frac{1}{3} + \cdots + \frac{1}{n}\right)$$

$$= \left(1 + \frac{1}{2} + \frac{1}{3} + \frac{1}{4} + \frac{1}{5} + \frac{1}{6} + \frac{1}{7} + \cdots + \frac{1}{4n-3} + \frac{1}{4n-2} + \frac{1}{4n-1} + \frac{1}{4n}\right)$$

$$\quad - \frac{1}{2} - \frac{1}{4} - \frac{1}{6} - \frac{1}{8} - \cdots - \frac{1}{4n-2} - \frac{1}{4n} - \frac{1}{2}\left(1 + \frac{1}{2} + \frac{1}{3} + \cdots + \frac{1}{n}\right)$$

$$= (\log 4n + \gamma_{4n}) - \frac{1}{2}\left(1 + \frac{1}{2} + \frac{1}{3} + \cdots + \frac{1}{2n}\right) - \frac{1}{2}(\log n + \gamma_n)$$

$$= (\log 4n + \gamma_{4n}) - \frac{1}{2}(\log 2n + \gamma_{2n}) - \frac{1}{2}(\log n + \gamma_n)$$

$$= \log 4 + \log n - \frac{1}{2}(\log 2 + \log n) - \frac{1}{2}\log n + \gamma_{4n} - \frac{1}{2}\gamma_{2n} - \frac{1}{2}\gamma_n$$

$$= \frac{3}{2}\log 2 + \gamma_{4n} - \frac{1}{2}\gamma_{2n} - \frac{1}{2}\gamma_n.$$

$\therefore \sigma_{3n} \to \frac{3}{2}\log 2$ as $n \to \infty$ $(\because \gamma_{4n} \to \gamma, \quad \gamma_{2n} \to \gamma, \quad \gamma_n \to \gamma$ as $n \to \infty)$.

Further,

$$\sigma_{3n+1} = \sigma_{3n} + \frac{1}{4n+1} \to \frac{3}{2}\log 2 \text{ as } n \to \infty.$$

$$\sigma_{3n+2} = \sigma_{3n} + \frac{1}{4n+1} + \frac{1}{4n+3} \to \frac{3}{2}\log 2 \text{ as } n \to \infty.$$

Hence not only the bracketed series

$$\left(1 + \frac{1}{3} - \frac{1}{2}\right) + \left(\frac{1}{5} + \frac{1}{7} - \frac{1}{4}\right) + \cdots$$

but also the series, without brackets,

$$1 + \frac{1}{3} - \frac{1}{2} + \frac{1}{5} + \frac{1}{7} - \frac{1}{4} + \cdots.$$

is convergent and that it has the sum $\frac{3}{2}\log 2$.

Riemann's Theorem

We have proved that an absolutely convergent series converges to the same sum in whatever order the terms are taken. In sharp contrast to this behaviour, the sum of a conditionally convergent series depends essentially on the order in which its terms are taken. Even more, a conditionally convergent series can be made a divergent series by a suitable rearrangement of its terms. This fact is stated in

Riemann's Theorem. *A conditionally convergent series can be made to converge to any arbitrary value, or even to diverge, by a suitable rearrangement of its terms.*

Justification. Let σ be the arbitrary sum to which the rearranged series shall converge. Let P_n be the sum of first n terms of the series of positive terms alone and $-Q_n$ be the sum of first n terms of the series of negative terms alone.

Since the series is conditionally convergent both $P_n \to \infty$, $Q_n \to \infty$. We can add just as many (possibly zero) positive terms as are needed to pass σ, then add just enough negative terms to repass σ again, and so on indefinitely.

Since the given series converges, its terms form a null sequence and we can approach σ by this process as closely as desired; for if σ_n be the sum of n terms of the series thus constructed, $\sigma - \sigma_n$ is always numerically less than that last term added.

In order to make the given series diverge by a rearrangement of terms, we carry out the same process when σ, instead of being a fixed constant, is assigned σ_n of a divergent sequence at the nth step. Thus, we may make the series diverge to infinity by choosing $\sigma_n = n$; or we may force s_n to have two limit points a and b $(a < b)$ by choosing $\sigma_1 = \sigma_3 = \sigma_5 = \cdots = b$ and $\sigma_2 = \sigma_4 = \sigma_6 = \cdots = a$.

6.10 Multiplication of two infinite series

If we multiply two power series

$$a_1 + a_2 x + a_3 x^2 + \cdots + a_n x^{n-1} + \cdots$$
$$\text{and} \quad b_1 + b_2 x + b_3 x^2 + \cdots + b_n x^{n-1} + \cdots,$$

the form of the answer is

$$a_1 b_1 + (a_1 b_2 + a_2 b_1)x + (a_1 b_3 + a_2 b_2 + a_3 b_1)x^2 + \cdots$$
$$+ \cdots + (a_1 b_n + a_2 b_{n-1} + \cdots + a_n b_1)x^{n-1} + \cdots.$$

We state our theorems about the multiplication of series in such a form that they can be used easily for power series. The coefficient of x^{n-1} in the previous work gives the reason for our choice of c_n in the following theorem of multiplication of two infinite series.

Theorem 6.10.1. *If* $\sum\limits_{n=1}^{\infty} a_n$ *and* $\sum\limits_{n=1}^{\infty} b_n$ *be two absolutely convergent series and if*

$$c_n = a_1 b_n + a_2 b_{n-1} + \cdots + a_n b_1$$

then $\sum\limits_{n=1}^{\infty} c_n$ *is absolutely convergent and*

$$\sum_{n=1}^{\infty} c_n = \sum_{n=1}^{\infty} a_n \cdot \sum_{n=1}^{\infty} b_n.$$

(We call $\sum c_n$. **Cauchy product** of $\sum a_n$ and $\sum b_n$)

Proof: Let

$$A_n = a_1 + a_2 + \cdots + a_n, \qquad B_n = b_1 + b_2 + \cdots + b_n,$$
$$A'_n = |a_1| + |a_2| + \cdots + |a_n|, \qquad B'_n = |b_1| + |b_2| + \cdots + |b_n|.$$

Let $A_n \to A$, $B_n \to B$, $A'_n \to A$, $B'_n \to B$ (as $n \to \infty$).
We first display the product $A_n B_n$ thus:

CHAPTER 6: INFINITE SERIES OF REAL CONSTANTS

$$\left.\begin{array}{ccccccc} a_1b_1 & a_1b_2 & a_1b_3 & a_1b_4 & \cdots & \cdots & a_1b_n \\ a_2b_1 & a_2b_2 & a_2b_3 & a_2b_4 & \cdots & \cdots & a_2b_n \\ a_3b_1 & a_3b_2 & a_3b_3 & a_3b_4 & \cdots & \cdots & a_3b_n \\ a_4b_1 & a_4b_2 & a_4b_3 & a_4b_4 & \cdots & \cdots & a_4b_n \\ \cdots & \cdots & \cdots & \cdots & \cdots & \cdots & \cdots \\ a_nb_1 & a_nb_2 & a_nb_3 & a_nb_4 & \cdots & \cdots & a_nb_n \end{array}\right\} \quad (S)$$

[If we *count by triangles*, as shown by arrows, we obtain a series of the form:

$$a_1b_1 + (a_1b_2 + a_2b_1) + (a_1b_3 + a_2b_2 + a_3b_1) + \cdots (Cauchy\ product).$$

If we *count by squares*, as shown by broken lines, we obtain a series of the form:

$$a_1b_1 + (a_2b_1 + a_2b_2 + a_1b_2) + (a_3b_1 + a_3b_2 + a_3b_3 + a_2b_3 + a_1b_3) + \cdots .]$$

Consider $A_n B_n$ set out in the form

$$\begin{aligned} A_n B_n &= A_1 B_1 + (A_2 B_2 - A_1 B_1) + (A_3 B_3 - A_2 B_2) + \cdots \\ &\quad + \cdots + (A_n B_n - A_{n-1} B_{n-1}) \\ &= a_1 b_1 + (a_2 b_1 + a_2 b_2 + a_1 b_2) \\ &\quad + (a_3 b_1 + a_3 b_2 + a_3 b_3 + a_2 b_3 + a_1 b_3) + \cdots \text{ to } n \text{ terms.} \end{aligned}$$

[Count by Squares] (6.10.1)

Now let us remove the brackets from (6.10.1) and consider the infinite series

$$a_1 b_1 + a_2 b_1 + a_2 b_2 + a_1 b_2 + a_3 b_1 + a_3 b_2 + a_3 b_3 + a_2 b_3 + a_1 b_3 + \cdots . \quad (6.10.2)$$

First Step. The sum of the moduli of any number, say m, of its terms $<$ some $A'_n B'_n$; for we can choose n big enough to ensure that all the m terms are in the square (S) displayed above and clearly the sum of the moduli of all terms in (S) is $A'_n B'_n$.

But A'_n is a m.i. sequence and so $A'_n < A'$ and similarly $B'_n < B'$.

Hence $A'_n B'_n < A' B'$.

Hence the series formed by taking the absolute values of the terms in (6.10.2) is convergent; that is (6.10.2) is absolutely convergent.

Second Step. The sum of the series (6.10.2), which has no brackets, is equal to the sum of the series with brackets:

i.e., $a_1 b_1 + (a_2 b_1 + a_2 b_2 + a_1 b_2) + \cdots + (a_n b_1 + a_n b_2 + \cdots + a_1 b_n) + \cdots$
$= \lim_{n \to \infty} A_n B_n,$

as we see by looking at (6.10.1) which is another way of writing $A_n B_n$.

But $A_n \to A$, $B_n \to B$. Hence (6.10.2) has the sum AB.

Third Step. Since (6.10.2) is absolutely convergent, any series got from it by rearranging its terms is also absolutely convergent and has the same sum. Hence

$$a_1 b_1 + a_2 b_1 + a_1 b_2 + a_3 b_1 + a_2 b_2 + a_1 b_3 + a_4 b_1 + \cdots \qquad (6.10.3)$$

is absolutely convergent and has the sum AB.

Finally the series obtained by putting brackets in (6.10.3), namely the series

$$a_1 b_1 + (a_2 b_1 + a_1 b_2) + (a_3 b_1 + a_2 b_2 + a_1 b_3) + \cdots$$
$$+ (a_n b_1 + a_{n-1} b_2 + \cdots + a_1 b_n) + \cdots \text{ (\textbf{Count by triangles})} \qquad (6.10.4)$$

i.e., $c_1 + c_2 + c_3 + \cdots + c_n + \cdots$

is absolutely convergent and has the sum AB.

Observation: The absolute convergence of **both** series $\sum a_n = A$, $\sum b_n = B$ is a *sufficient condition* for the convergence of their Cauchy product to AB, but is not necessary, as we see from

Merten's Theorem. *If one of the two convergent series $\sum a_n$, $\sum b_n$ converges absolutely, their Cauchy product will converge to AB.* (For proof one may refer to Knopp, *Theory and Application of Infinite Series*).

Abel also proved that, if the Cauchy product of two series converges, then it must converge to AB.

Abel's Theorem. *If the series $\sum a_n$, $\sum b_n$ and their Cauchy product $\sum c_n$ converge to A, B and C respectively, then $AB = C$.*

Example 6.10.1. *The Cauchy product of the two geometric series*

$$1 + x + x^2 + x^3 + \cdots = \frac{1}{1-x} \qquad (|x| < 1)$$
$$\text{and} \quad 1 + y + y^2 + y^3 + \cdots = \frac{1}{1-y} \qquad (|y| < 1)$$

is $1 + (x+y) + (x^2 + xy + y^2) + (x^3 + x^2 y + xy^2 + y^3) + \cdots = \dfrac{1}{(1-x)(1-y)}$.

In particular, if $x = y$,

$$1 + 2x + 3x^2 + 4x^3 + \cdots = \frac{1}{(1-x)^2} \qquad (|x| < 1).$$

Example 6.10.2. *The series for e is*

$$e = 1 + \frac{1}{\underline{|1}} + \frac{1}{\underline{|2}} + \frac{1}{\underline{|3}} + \cdots + \frac{1}{\underline{|n}} + \cdots.$$

Hence, on multiplying the series by itself,

$$e^2 = 1 + \frac{(1+1)}{\lfloor 1} + \frac{1+2+1}{\lfloor 2} + \frac{(1+3+3+1)}{\lfloor 3} + \cdots$$
$$= 1 + \frac{2}{\lfloor 1} + \frac{2^2}{\lfloor 2} + \frac{2^3}{\lfloor 3} + \cdots.$$

On multiplying the series for e and e^2, we get

$$e^3 = 1 + \frac{3}{\lfloor 1} + \frac{3^2}{\lfloor 2} + \frac{3^3}{\lfloor 3} + \cdots.$$

Section V: Sequence & Series: A Collection of Problems

MISCELLANEOUS EXERCISES

[A]

(Questions on Theory)

1. (a) Let $\sum u_n$ be a series of positive terms. Prove that if $\overline{\lim}\frac{u_{n+1}}{u_n} < 1$, then $\sum u_n$ converges but if $\underline{\lim}\frac{u_{n+1}}{u_n} > 1$, then $\sum u_n$ diverges. (*Ratio Test*)

    [C.H. 1994]

   (b) Prove that $\sum_{n=1}^{\infty} u_n$ and $\sum_{n=1}^{\infty} 2^n u_{2^n}$ either both converge or both diverge, provided $\{u_n\}$ is a monotone decreasing sequence of positive terms.
    (*Cauchy's Condensation Test*) [C.H. 1993, 96]
    Hence show that $\sum_{n=1}^{\infty} \frac{1}{n^p}$ converges if $p > 1$, diverges if $p \leq 1$.

   (c) Prove that a series of positive terms $\sum_{n=1}^{\infty} u_n$ converges if $\overline{\lim} \sqrt[n]{u_n} < 1$, diverges if $\overline{\lim} \sqrt[n]{u_n} > 1$. (*Root Test*)

   (d) Prove that if $\sum a_n$ is a convergent series of positive terms and $\{a_n\}$ is m.d., then (i) $a_n \to 0$ as $n \to \infty$ and also (ii) $na_n \to 0$ as $n \to \infty$.
    (*Abel's Theorem*) [C.H. 1992]

   (e) State and prove Leibnitz's Test on Alternating Series. [C.H. 1995]

   (f) State and prove Raabe's Test for convergence or divergence of a series of positive terms in terms of upper and lower limits. [C.H. 1991]

(g) If $s = \sum_{n=1}^{\infty}(-1)^n a_n$ ($a_n > 0\ \forall\ n$), where $\{a_n\}$ m.d. and $a_n \to 0$ as $n \to \infty$, then prove that $|s - s_n| \leq a_{n+1}$, $\forall\ n$ where s_n is the nth partial sum of the series.

(h) Define an absolutely convergent series and a conditionally convergent series. Give one example of each.

Prove that an absolutely convergent series is convergent and has the same sum even if it is rearranged.

With the help of examples show that a conditionally convergent series may have a different sum when it is rearranged.   [C.H. 1994]

(i) State and prove Gauss' Test for the convergence or divergence of a series of positive terms.

(j) If $\sum_{n=1}^{\infty} a_n$ converges, then prove that $\sum_{n=1}^{\infty} a_{2n}$ also converges.

[See P. 182. Theorem 5.9.1.]   [C.H. 1996]

[B]

(Maclaurin's Integral Test)

1. If $f(x)$ is a positive decreasing function of x when $x \geq 1$, then the series $\sum_{i=1}^{\infty} f(i)$ and the infinite integral $\int_1^{\infty} f(x)dx$ either both converge or both diverge and in case of convergence if $S = \sum_{i=1}^{\infty} f(i)$ and $I = \int_1^{\infty} f(x)dx$, then $I < S < I + f(1)$.

(*Integral Test*)

[*Proof:* When $i \geq 1$ $f(i+1) < f(x) < f(i)$ where $i < x < i+1$.

Taking integrals between the limits $x = i$ and $x = i+1$,

$$f(i+1) < \int_i^{i+1} f(x)dx < f(i).$$

If we add these inequalities for $i = 1, 2, 3, \cdots, n-1$, we have

$$f(2) + f(3) + \cdots + f(n) < \int_1^n f(x)dx < f(1) + f(2) + \cdots + f(n-1). \quad (6.10.5)$$

Now let $n \to \infty$; if the integral converges to I, then the series on the left converge and also the series on the right converges.

$$S - f(1) < I < S. \quad (6.10.6)$$

Conversely, if the series $\sum_{i=1}^{\infty} f(i)$ converges to S then the integral $\int_k^x f(x)dx$ is a positive increasing function of x which is bounded above by S; hence the integral converges. Therefore, the series and the integral either both converge or both diverge. Moreover $I < S < f(1) + I$ follows from (6.10.6).]

Deduce from the Integral Test:

If $f(x)$ is a positive decreasing function when $x \geq 1$, then the sequence $\{S_n\}$ where $S_n = f(1) + f(2) + f(3) + \cdots + f(n) - \int_1^n f(x)dx$ converges to a limit which lies between 0 and $f(1)$.

[From (6.10.5) of the above proof we have

$$0 < f(n) < S_n < f(1).$$

Moreover

$$S_{n+1} - S_n = f(n+1) - \int_n^{n+1} f(x)dx < 0$$

since $f(x) > f(n+1)$ in the interval $n \leq x < n+1$. Then $\{S_n\}$ is a positive decreasing sequence, must converge to a limit between 0 and $f(1)$.]

It at once follows (taking $f(x) = \frac{1}{x}$) that

$$\gamma_n = 1 + \frac{1}{2} + \frac{1}{3} + \cdots + \frac{1}{n} - \log n \to \gamma \text{ where } 0 < \gamma < 1.$$

In fact γ (called **Euler's Constant**) $= 0.5772156649\cdots$.

2. When $p \neq 1$, obtain $\int_1^x \frac{dx}{x^p} = \frac{1}{1-p}\left(\frac{1}{x^{p-1}} - 1\right)$; also $\int_1^x \frac{dx}{x} = \log x$. What can you say about $\int_1^\infty \frac{dx}{x^p}$ and about the series $\sum_{n=1}^{\infty} \frac{1}{n^p}$?

[*Ans:* The integral converges to $\frac{1}{p-1}$ when $p > 1$; diverges to ∞ if $p \leq 1$. The series, therefore, converges to a sum when $p > 1$, diverges to ∞ when $0 < p \leq 1$. In case of convergence we denote the sum $\zeta(p)$, **Riemann's Zeta Function**. When $p \leq 0$, the integral and the series both diverge.]

3. Prove by **integral test**, that the series $\sum_{n=2}^{\infty} \frac{1}{n(\log n)^p}$ converges if $p > 1$, diverges if $p \leq 1$.

4. Using Integral Test prove that the series $\sum_{n=1}^{\infty} \frac{1}{n^2+1}$ converges.

[Test $f(x) = \frac{1}{x^2+1}$; $\int_1^x \frac{dx}{x^2+1} = \tan^{-1} x - \tan^{-1} 1$; hence $\int_1^\infty \frac{dx}{1+x^2} = \pi/4$.]

5. Use integral test to examine the convergence or divergence of the series

(a) $\sum_{n=1}^{\infty} \frac{1}{n}$ (D).

(b) $\sum_{n=1}^{\infty} \frac{1}{n \log n}$ (D).

(c) $\sum_{n=1}^{\infty} \frac{n}{n^2+1}$ (D).

(d) $\sum_{n=1}^{\infty} \frac{n}{e^{n^2}}$.   (C).

[C]

(USE OF DEFINITION OF CONVERGENT/DIVERGENT SERIES, CAUCHY'S CRITERION AND COMPARISON TEST FOR CONVERGENT OR DIVERGENT SERIES)

[C = Convergent; D = Divergent]

1. For the following series write S_n, the sum of first n terms and then obtain the sum $\left(\text{i.e., } \lim_{n \to \infty} S_n\right)$ of the series:

(a) $\sum_{n=1}^{\infty} \frac{1}{n(n+1)(n+2)}$ (Sum = $\frac{1}{4}$).

(b) $\sum_{n=1}^{\infty} \frac{1}{(2n-1)(2n+1)}$ (Sum = $\frac{1}{2}$).

(c) $\sum_{n=1}^{\infty} \left(\frac{2}{3}\right)^n$ (Sum = 2).

(d) $\sum_{n=1}^{\infty} (-1)^{n-1} \frac{2n+1}{n(n+1)}$ (Sum = 1).

2. (a) Prove that the series $\sum_{n=1}^{\infty} \frac{n}{n+1}$ cannot converge.

(see art. 6.3. Theorem 6.3.2)

(b) Show that the series $\sum a_n$, whose nth term $a_n = \sqrt{n+1} - \sqrt{n}$ diverges although $\lim_{n \to \infty} a_n = 0$.

3. Apply Cauchy's convergence criterion (art. 6.3. Theorem 6.3.1) to prove that the series

(a) $1 + \frac{1}{2} + \frac{1}{3} + \cdots + \frac{1}{n} + \cdots$ diverges.

(b) $1 - \frac{1}{2} + \frac{1}{3} - \frac{1}{4} + \cdots + (-1)^{n+1}\frac{1}{n} + \cdots$ converges.

(c) $1 + \frac{1}{\underline{|1}} + \frac{1}{\underline{|2}} + \frac{1}{\underline{|3}} + \cdots + \frac{1}{\underline{|n}} + \cdots$ converges.

4. Investigate the convergence or divergence of the following series: [Use Comparison Test]

(a) $\sum_{n=1}^{\infty} \frac{1}{(2n-1)2n}$ (C).

(b) $\sum_{n=1}^{\infty} \frac{1}{\sqrt{n(2n+1)}}$ (D).

(c) $\sum_{n=1}^{\infty} (\sqrt[3]{n^3+1} - n)$ (C).

(d) $\sum_{n=2}^{\infty} \frac{1}{n \log n}$ (D).

(e) $\sum_{n=2}^{\infty} \frac{1}{\sqrt{n(n-1)}}$ (D).

(f) $\sum_{n=2}^{\infty} \frac{\log n}{2n^3 - 1}$ (C).

(g) $\sum_{n=1}^{\infty} \frac{4n^2 - n + 3}{n^3 + 2n}$ (D).

(h) $\sum_{n=1}^{\infty} \frac{n + \sqrt{n}}{2n^3 - 1}$ (C).

(i) $\sum_{n=1}^{\infty} \frac{\sin n\alpha}{n^2}$, where $\alpha > 0$; $\left(\frac{\sin n\alpha}{n^2} \leq \frac{1}{n^2}\right)$.   (C).   [C.H. 1983]

(j) $\sum_{n=1}^{\infty} \frac{2n+3}{2n(n+1)(n+3)}$,   (C).   [C.H. 1995]

5. Using Comparison Test prove that the series

$$\sum_{n=1}^{\infty} e^{-n^2} \quad \text{and} \quad \sum_{n=1}^{\infty} \sin^3\left(\frac{1}{n}\right) \quad \text{both converge.}$$

6. Test the convergence of the series

$$\frac{1}{1+x} + \frac{x}{1+x^2} + \frac{x^2}{1+x^3} + \cdots + \frac{x^{n-1}}{1+x^n} + \cdots (x > 0).$$

[*Hints:* When $0 < x < 1$, $\sum x^{n-1}$ is convergent. Compare : The given series converges.

When $x > 1$, $\frac{1}{x^n} \to 0$ and so $\frac{x^{n-1}}{1+x^n} = \frac{1}{x} \cdot \frac{1}{1/x^n+1} \to \frac{1}{x}$ (not zero); hence when $x > 1$, the given series diverges.

When $x = 1$, the series becomes $\frac{1}{2} + \frac{1}{2} + \frac{1}{2} + \cdots$ and it obviously diverges.]

7. (a) If $\sum a_n$ be a positive convergent series, then prove that $\sum a_n^2$ also converges.

   [C.H. 1999]

   [*Hints:* $\exists m$ s.t. $\forall n \geq m$, $a_n < 1$ because $a_n \to 0$; $\therefore a_n^2 < a_n$.]

   (b) If $\sum a_n^2$ is convergent, prove that $\sum (a_n/n)$ also converges $(a_n > 0 \ \forall \ n)$.

   $$\left[\frac{a_n}{n} \leq \frac{1}{2}\left(a_n^2 + \frac{1}{n^2}\right) \text{ using A.M.} \geq \text{G.M.}\right].$$

   (c) If $\sum a_n$ is a positive convergent series, prove that $\sum \dfrac{a_n}{1+a_n}$ also converges.

8. Let $\sum u_n$ and $\sum v_n$ be two positive series of real constants.

   (a) If $\forall n \geq m$, $\dfrac{v_{n+1}}{v_n} \leq \dfrac{u_{n+1}}{u_n}$ and $\sum u_n$ is convergent, then prove that $\sum v_n$ also converges.

   (b) If $\forall n \geq m$, $\dfrac{v_{n+1}}{v_n} \geq \dfrac{u_{n+1}}{u_n}$ and $\sum u_n$ diverges, then prove that $\sum v_n$ also diverges.

[D]

(Ratio Test; Root Test; Condensation Test and Abel's Theorem)

[C = Convergent; D = Divergent]

1. Using D'Alembert's Ratio Test examine the convergence of the following series:

   (a) $\sum_{n=1}^{\infty} \dfrac{n^2}{2^n} = \dfrac{1^2}{2} + \dfrac{2^2}{2^2} + \dfrac{3^2}{2^3} + \dfrac{4^2}{2^4} + \cdots$ (C).

   (b) $\sum_{n=1}^{\infty} \dfrac{n}{3^{n-1}}$ (C).

   (c) $\sum_{n=1}^{\infty} \dfrac{4^n}{n!}$ (C).

   (d) $\sum_{n=1}^{\infty} \dfrac{n^4}{e^{n^2}}$ (C).

CHAPTER 6: INFINITE SERIES OF REAL CONSTANTS

2. Verify that $\sum_{n=1}^{\infty} \frac{x^n}{n!}$ converges for all $x > 0$ but $\sum_{n=1}^{\infty} \frac{x^n}{n}$ converges if $0 < x < 1$ and diverges if $x \geq 1$.

3. Prove that the series
$$\frac{1}{x+1} + \frac{x}{x+2} + \frac{x^2}{x+3} + \frac{x^3}{x+4} + \cdots \quad (x > 0)$$
converges if $0 < x < 1$, diverges if $x \geq 1$.

Discuss the convergence and divergence of the series
$$\frac{1}{x+1} + \frac{1}{2x^2+1} + \frac{1}{3x^3+1} + \cdots \quad (x > 0).$$

[Ans: $0 < x \leq 1$ diverges; $x > 1$ converges.]

4. (a) Prove *Abel's Theorem*: If $\sum a_n$ is a convergent series of positive real constants and if $\{a_n\}$ is a monotone decreasing sequence, then $\lim_{n \to \infty} na_n = 0$.

   [C.H. 1983]

   Give an example to show that the converse is not true (i.e., if $\{a_n\}$ is m.d. and $\lim_{n \to \infty} na_n = 0$ still $\sum a_n$ may not be convergent).

   (b) If $a_n = n^{-1-1/n}$, prove that $\sum a_n$ is divergent.

   [$na_n \to 1$; $\sum a_n$ cannot converge though $\{a_n\}$ is m.d.
   Otherwise. $\because \log n < n$, $n < e^n$; $n^{1/n} < e$; thus $a_n = \frac{1}{n \cdot n^{1/n}} > \frac{1}{ne}$.
   But $\sum \frac{1}{n}$ diverges, therefore, by comparison test $\sum a_n$ diverges.]

5. Examine the convergence of $\sum a_n$ where
$$a_n = n^p \left\{ \frac{1}{\sqrt{n-1}} - \frac{1}{\sqrt{n}} \right\} \text{ for } n > 1.$$

[Ans: Converges if $p < \frac{1}{2}$, diverges if $p \geq \frac{1}{2}$.]

6. State and prove *Cauchy's Condensation Test*. Use this test to prove that the series

   (a) $\sum_{n=2}^{\infty} \frac{1}{n(\log n)^p}$ is convergent if $p > 1$ and divergent if $p \leq 1$.

   (b) $\sum \frac{1}{n^p}$ converges if $p > 1$, diverges if $p \leq 1$. [C.H. 1981, '83, '86]

(c) $\sum_{n=2}^{\infty} \frac{1}{n^2}(\log_e n)^2$ converges. [C.H. 1993]

7. Consider the series

$$1 + a + ab + a^2b + a^2b^2 + \cdots + a^n b^{n-1} + a^n b^n + \cdots \quad (0 < a < b).$$

(a) Use Ratio Test (in terms of $\overline{\lim}$ and $\underline{\lim}$) to prove that the series converges when $b < 1$ and diverges if $a > 1$.

(b) Will the ratio test give any conclusion of $a < 1 < b$?

(c) Prove by **Cauchy's Root Test** that the series converges when $ab < 1$, diverges when $ab > 1$. What happens if $ab = 1$? [C.H. 1995]

[**Conclusion.** The *ratio test* is less general than the *root test*.]

8. When $0 < a < b < 1$. Use the *root test* to prove that the series

$$a + b + a^2 + b^2 + a^3 + b^3 + \cdots \text{ converges.}$$

Does ratio test give this conclusion?

[*Ans:* No. Successive ratios are alternately > 1 and < 1;

in fact, $\dfrac{b^n}{a^n} \to \infty$ and $\dfrac{a^{n+1}}{b^n} = a\left(\dfrac{a}{b}\right)^n \to 0.$]

9. Show that the series

(a) $\sum_{n=0}^{\infty} \dfrac{n! x^n}{(2n)!}$ converges for all $x > 0$.

(b) $\sum_{n=2}^{\infty} \dfrac{1}{(\log n)^{\log n}}$ converges.

[For (b), note that $(\log n)^{\log n} = n^{\log(\log n)}$]

10. Explain why *to ensure convergence* the ratio $\dfrac{u_{n+1}}{u_n}$ (Ratio Test) or the root $\sqrt[n]{u_n}$ (Root Test) must eventually remain less than a *fixed number r which is less than* 1.

[*Hints:* When u_{n+1}/u_n or $u_n^{1/n} \to 1$ from below, we can draw no conclusions, thus the divergent series $\sum(1/n)$ and the convergent series $\sum(1/n^2)$ have this property.]

CHAPTER 6: INFINITE SERIES OF REAL CONSTANTS   419

[E]

(RAABE'S TEST, GAUSS' TEST)

1. State and prove Raabe's test for convergence or divergence of a series of positive terms in terms of upper and lower limits. [C.H. 1984, 91]

2. Examine the convergence or divergence of the following series:

    (a) $x + \dfrac{1}{2}\dfrac{x^3}{3} + \dfrac{1 \cdot 3}{2 \cdot 4}\dfrac{x^5}{5} + \cdots (x > 0)$. [C.H. 1983; '86; '89]
    [Conv. if $0 < x < 1$, diverges if $x > 1 \to$ use Ratio Test; if $x = 1$, use Raabe's Test and prove that the series converges.]

    (b) $1 + \dfrac{\alpha}{1} + \dfrac{\alpha(\alpha+1)}{2!} + \dfrac{\alpha(\alpha+1)(\alpha+2)}{3!} + \cdots$.
    [Note that $\dfrac{a_{n+1}}{a_n} = \dfrac{n+\alpha}{n+1} > 0$ when $n > -\alpha$. Hence the terms ultimately keep the same sign and so we can apply our tests—in this case apply Raabe's Test.]
    [Ans: Conv. if $\alpha < 0$, diverges if $\alpha > 0$.]

    (c) $\displaystyle\sum_{n=1}^{\infty} \dfrac{2 \cdot 4 \cdot 6 \cdots 2n}{1 \cdot 3 \cdot 5 \cdots (2n-1)} \cdot \dfrac{1}{2n+2}$ (D)

3. State and prove Gauss' Test of convergence of a series of positive terms. Use this test to prove that

    $\sum \dfrac{1}{n^p}$ converges if $p > 1$ and diverges if $0 < p \leq 1$. [C.H. 1978]

4. (a) Let $\sum a_n$ be a series of positive terms. If we can write

    $$R_n = n\left(1 - \dfrac{a_{n+1}}{a_n}\right) = h + \dfrac{f_n}{n^p} \quad (p > 0)$$

    where h is a constant and $\{f_n\}$ is bounded, then prove that $\sum a_n$ converges if $h > 1$, diverges when $h \leq 1$.
    [This is, in fact, Gauss' Test. Raabe's Test covers the cases $h \neq 1$. When $h = 1$, the series diverges by Bertrand's Test.]

    (b) Suppose we have a series $\sum a_n$ of positive terms and suppose

    $\dfrac{a_{n+1}}{a_n}$ = quotient of two polynomials in n having the same term of the highest degree

    $= \dfrac{n^k + bn^{k-1} + \cdots}{n^k + cn^{k-1} + \cdots}$.

Use Raabe's Test to prove that $\sum a_n$ converges when $c - b > 1$, diverges when $c - b < 1$. When $c - b = 1$, use Bertrand's Test to prove that the series diverges.

(c) **(Hypergeometric Series)** Show that the series

$$1 + \frac{\alpha \cdot \beta}{1 \cdot \gamma} + \frac{\alpha(\alpha + 1)}{1 \cdot 2} \frac{\beta(\beta + 1)}{\gamma(\gamma + 1)} + \cdots$$

(where α, β, γ are real, none of them zero or negative integer) converges if $\gamma > \alpha + \beta$ and diverges if $\gamma \leq \alpha + \beta$. [C.H. 1982]

5. Test the following series for convergence (Using Raabe's Test):

(a) $\sum_{n=1}^{\infty} \frac{n!}{(\alpha + 1)(\alpha + 2) \cdots (\alpha + n)}$ ($\alpha > 0$) [Conv. $\alpha > 1$, div. $\alpha \leq 1$].

(b) $\sum_{n=1}^{\infty} \frac{1 \cdot 3 \cdot 5 \cdots (2n - 1)}{2 \cdot 4 \cdot 6 \cdots 2n} \cdot \frac{1}{2n + 1}$ [Conv.]. [C.H. 1988]

(c) $\sum_{n=1}^{\infty} \left(\frac{1 \cdot 3 \cdot 5 \cdots (2n - 1)}{2 \cdot 4 \cdot 6 \cdots 2n} \right)^p$ when $p = 1, 2, 3$.

[*Hints:* When $p = 2$, prove that the series diverges by Gauss' Test also. Div. $p = 1$. Conv. $p = 3$.]

6. Show that the series

$$1 + \frac{\alpha \cdot \beta}{\gamma \cdot \delta} + \frac{\alpha(\alpha + 1)\beta(\beta + 1)}{\gamma(\gamma + 1)\delta(\delta + 1)} + \cdots$$

where $\alpha, \beta, \gamma, \delta$ are not zero or negative integers, converges or diverges according as

$$\gamma + \delta - (\alpha + \beta) > 1 \text{ or } \leq 1.$$

7. Use Gauss' Test to show that the series

$$\left(\frac{1}{2}\right)^2 + \left(\frac{1 \cdot 3}{2 \cdot 4}\right)^4 + \left(\frac{1 \cdot 3 \cdot 5}{2 \cdot 4 \cdot 6}\right)^2 + \cdots \text{ diverges.}$$

$$\left[\frac{a_n}{a_{n+1}} = \left(\frac{2n + 2}{2n + 1}\right)^2 = \left(1 + \frac{1}{n}\right)^2 \left(1 + \frac{1}{2n}\right)^{-2} \right.$$
$$\left. = \left(1 + \frac{2}{n}\right)\left(1 - \frac{1}{n}\right) + O\left(\frac{1}{n^2}\right) = 1 + \frac{1}{n} + O\left(\frac{1}{n^2}\right). \right]$$

CHAPTER 6: INFINITE SERIES OF REAL CONSTANTS

8. Use Raabe's Test to prove that the infinite series $\sum_{n=1}^{\infty} \dfrac{3 \cdot 6 \cdot 9 \cdots 3n}{7 \cdot 10 \cdot 13 \cdots (3n+4)}$ is convergent.

[C.H. 1995 (old)]

[F]

1. Prove that an alternating series $\sum_{n=0}^{\infty} (-1)^n a_n \, (a_n > 0)$ converges if $\{a_n\}$ is monotone decreasing and $\lim\limits_{n \to \infty} a_n = 0$. Show further that if s denotes the sum of the series and s_n its nth partial sum, then $|s - s_n| \leq a_{n+1}$, for all n.

2. (a) Define a conditionally convergent series and an absolutely convergent series with suitable examples.

   (b) If $\sum\limits_{k=1}^{\infty} u_k$ converges absolutely then prove that $\sum\limits_{k=1}^{\infty} u_k$ is convergent.

   (c) Show that the series $1 - \tfrac{1}{2} + \tfrac{1}{3} - \tfrac{1}{4} + \cdots$ converges but not absolutely.

   [C.H. 1980, 94]

   [To prove in details $\sum (-1)^{n-1} \tfrac{1}{n}$ converges and $\sum \dfrac{1}{n}$ diverges.]

3. Let $\log 2 = 1 - \tfrac{1}{2} + \tfrac{1}{3} - \tfrac{1}{4} + \cdots$. Prove that the following rearranged series give the indicated sum on the right:

   (a) $1 - \tfrac{1}{2} - \tfrac{1}{4} + \tfrac{1}{3} - \tfrac{1}{6} - \tfrac{1}{8} + \tfrac{1}{5} - \cdots$ $= \tfrac{1}{2} \log 2$ [C.H. 1990]

   (b) $1 + \tfrac{1}{3} - \tfrac{1}{2} + \tfrac{1}{5} + \tfrac{1}{7} - \tfrac{1}{4} + \tfrac{1}{9} + \tfrac{1}{11} - \tfrac{1}{6} + \cdots$ $= \tfrac{3}{2} \log 2$ [C.H. 1994]

   (c) $1 - \tfrac{1}{2} - \tfrac{1}{4} - \tfrac{1}{6} - \tfrac{1}{8} + \tfrac{1}{3} - \tfrac{1}{10} - \tfrac{1}{12} - \tfrac{1}{14} - \tfrac{1}{16} + \tfrac{1}{5} - \cdots$ $= 0$

   (d) $1 + \tfrac{1}{3} + \tfrac{1}{5} - \tfrac{1}{2} - \tfrac{1}{4} + \tfrac{1}{7} + \tfrac{1}{9} + \tfrac{1}{11} - \tfrac{1}{6} - \tfrac{1}{8} + \cdots$ $= \tfrac{1}{2} \log 6$

   (e) $1 + \tfrac{1}{2} - \tfrac{1}{3} + \tfrac{1}{4} + \tfrac{1}{5} - \tfrac{1}{6} + \cdots$ diverges.

4. (a) Let the sum of the series $1 - \tfrac{1}{2} - \tfrac{1}{4} + \tfrac{1}{3} - \tfrac{1}{6} - \cdots$ be $\tfrac{1}{2} \log 2$. If the series be rearranged so that the ratio of the number of positive terms to the ratio of negative terms is $k(> 0)$. Show that the sum is $\tfrac{1}{2} \log 4k$. How do you construct a deranged series so as to have an arbitrary sum S?

   [Let there be a series with $2n$ negative terms and $2kn$ positive terms and so on. Let S_r be the sum of first r terms of the given series and S'_r the sum of the first r terms of the deranged series. We then have:

   $$S'_{2kn+2n} = \left[1 - \frac{1}{2} - \frac{1}{4} + \frac{1}{3} - \frac{1}{6} - \frac{1}{8} + \frac{1}{5} - \cdots \text{ to } 3n \text{ terms}\right]$$
   $$+ \left[\frac{1}{2n+1} + \frac{1}{2n+3} + \cdots \text{ to } (2nk - n) \text{ terms}\right].$$

where the first bracket contains n positive and $2n$ negative terms and the second bracket contains $(2nk-n)$ positive terms. The last term in the second bracket is $\frac{1}{2n+\{2(2nk-2n)-1\}} = \frac{1}{4nk-1}$.

$$\therefore \lim_{n\to\infty} S'_{2kn+2n} = \lim_{n\to\infty} S_{3n} + \lim_{n\to\infty}\left[\frac{1}{2n+1} + \frac{1}{2n+3} + \cdots + \frac{1}{4nk-1}\right]$$

$$= \frac{1}{2}\log 2 + \int_0^{2k} \frac{dx}{2+2x} = \frac{1}{2}\log 2 + \frac{1}{2}\log 2k = \frac{1}{2}\log 4k.$$

$$\left(\because \lim_{n\to\infty} \sum_{r=1}^{2nk-1} \frac{1}{2n+2r-1} = \lim_{n\to\infty} \frac{1}{n}\sum_{r=1}^{2nk-1} \frac{1}{2+\frac{2r}{n}-\frac{1}{n}} = \int_0^{2k-1} \frac{dx}{2+2x}\right).$$

To construct a deranged series whose sum should be S we shall take k to be such that $\frac{1}{2}\log 4k = S$ or $K = \frac{1}{4}e^{2S}$].

(b) What deranged series of $1 - \frac{1}{2} + \frac{1}{3} - \frac{1}{4} + \frac{1}{5} + \frac{1}{6} + \cdots$ will give the sum zero? [Ans: $k = \frac{1}{4}$]

(c) Using (a) prove that the sum of the series

$$1 + \frac{1}{3} + \frac{1}{5} - \frac{1}{2} + \frac{1}{7} + \frac{1}{9} + \frac{1}{11} - \frac{1}{4} \cdots = \frac{1}{2}\log 12.$$

5. (a) Find the Cauchy product of

$$1 + x + x^2 + x^3 + \cdots \quad \text{and} \quad 1 - x + x^2 - x^3 + \cdots \quad (|x| < 1).$$

(b) Prove that $\sum_{n=0}^{\infty} \frac{x^n}{n!}$ is convergent for all real values of x.

If $E(x)$ denotes its sum prove that $E(x)\cdot E(y) = E(x+y)$.

[G]

[MISCELLANEOUS TYPES]

1. Prove that the series:

(a) $\left(\frac{2^2}{1^2} - \frac{2}{1}\right)^{-1} + \left(\frac{3^3}{2^3} - \frac{3}{2}\right)^{-2} + \left(\frac{4^4}{3^4} - \frac{4}{3}\right)^{-3} + \cdots$ is convergent.

(b) $2x + \frac{3x^2}{8} + \frac{4x^3}{27} + \cdots + \frac{(n+1)x^n}{n^3} + \cdots$ converges if $0 \leq x \leq 1$ and diverges if $x > 1$.

CHAPTER 6: INFINITE SERIES OF REAL CONSTANTS 423

(c) $1 + \dfrac{1}{2}\dfrac{x^2}{4} + \dfrac{1\cdot 3\cdot 5}{2\cdot 4\cdot 6}\dfrac{x^4}{8} + \dfrac{1\cdot 3\cdot 5\cdot 7\cdot 9}{2\cdot 4\cdot 6\cdot 8\cdot 10}\dfrac{x^6}{12} + \cdots$ converges if $x^2 \leq 1$ and diverges if $x^2 > 1$.

(d) $x + \dfrac{2^2 x^2}{2!} + \dfrac{3^3 x^3}{3!} + \dfrac{4^4 x^4}{4!} + \cdots (x > 0)$ converges if $x < \dfrac{1}{e}$, diverges if $x \geq \dfrac{1}{e}$.

[Use Logarithmic Test: Convergent or divergent acc. as $\lim \left(n \log \dfrac{a_n}{a_{n+1}}\right) >$ or < 1].

(e) $1^p + \left(\dfrac{1}{2}\right)^p + \left(\dfrac{1\cdot 3}{2\cdot 4}\right)^p + \left(\dfrac{1\cdot 3\cdot 5}{2\cdot 4\cdot 6}\right)^p + \cdots$ converges if $p > 2$, diverges if $p \leq 2$.

(f) $\dfrac{1}{1^p} - \dfrac{1}{2^p} + \dfrac{1}{3^p} - \dfrac{1}{4^p} + \cdots$ converges if $p > 0$.

(g) $\dfrac{1}{x+1} - \dfrac{1}{x+2} + \dfrac{1}{x+3} - \cdots$ ($x \neq$ any negative integer) converges.

2. (a) For what positive x, the series $\dfrac{x}{x+1} + \dfrac{x^2}{x+2} + \dfrac{x^3}{x+3} + \cdots$, where $x \neq$ any negative integer, is convergent?

[Ans: $x < 1$]

(b) Prove that the series $\dfrac{x}{1+x} - \dfrac{x^2}{1+x^2} + \dfrac{x^3}{1+x^3} - \cdots$ is convergent if $0 < x < 1$.

(c) Prove that the alternating series $1 - \dfrac{1}{2} + \dfrac{1\cdot 3}{2\cdot 4} - \dfrac{1\cdot 3\cdot 5}{2\cdot 4\cdot 6} + \cdots$ converges.

3. Prove that $1 - \dfrac{1}{\sqrt{2}} + \dfrac{1}{\sqrt{3}} - \dfrac{1}{\sqrt{4}} + \cdots$ is a convergent series. Also prove that $|S_{99} - S| < \dfrac{1}{10}$ where $S = 1 - \dfrac{1}{\sqrt{2}} + \dfrac{1}{\sqrt{3}} - \dfrac{1}{\sqrt{4}} + \cdots$ and S_{99} is the 99th partial sum of the series. [C.H. 1983]

4. Match the following with justifications:

A. $\displaystyle\sum_{n=1}^{\infty} \dfrac{x^n}{n!}$. I. Convergent but not absolutely convergent.

B. $\displaystyle\sum_{n=1}^{\infty} \dfrac{x^n}{n^n}$. II. Absolutely convergent.

C. $\displaystyle\sum_{n=1}^{\infty} \dfrac{(-1)^{n-1}}{\sqrt{n}}$. III. Divergent.

D. $\displaystyle\sum_{n=1}^{\infty} (-1)^{n-1} 3^n$.

E. $\displaystyle\sum \dfrac{(-1)^{n-1}}{x^2 + n}$.

[Ans: $A, B \to II$; $C, E \to I$; $D \to III$.]

5. Show that the series $\sum \dfrac{1 \cdot 3 \cdot 5 \cdots (2n-1)}{2 \cdot 4 \cdot 6 \cdots 2n}$; and $\sum \dfrac{1 \cdot 3 \cdot 5 \cdots (2n-1)}{2 \cdot 4 \cdot 6 \cdots 2n} \cdot \dfrac{4n+3}{2n+2}$ are both divergent.

6. Show that the series

$$\left(1 + \dfrac{1}{\sqrt{3}} - \dfrac{1}{\sqrt{2}}\right) + \left(\dfrac{1}{\sqrt{5}} + \dfrac{1}{\sqrt{7}} - \dfrac{1}{\sqrt{4}}\right) + \left(\dfrac{1}{\sqrt{9}} + \dfrac{1}{\sqrt{11}} - \dfrac{1}{\sqrt{6}}\right) + \cdots$$

formed by rearranging the terms of the convergent series

$$1 - \dfrac{1}{\sqrt{2}} + \dfrac{1}{\sqrt{3}} - \dfrac{1}{\sqrt{4}} + \dfrac{1}{\sqrt{5}} - \cdots \quad \text{is divergent.}$$

7. Investigate the convergence of

(a) $\sum\limits_{n=1}^{\infty} \dfrac{(-1)^{n-1} 2^n}{n^2}$ \hspace{2em} (D).

(b) $\sum\limits_{n=1}^{\infty} \dfrac{(-1)^{n-1} n}{n^2 + 1}$ \hspace{2em} (C).

(c) $\sum\limits_{n=2}^{\infty} \dfrac{(-1)^{n-1}}{n (\log n)^2}$ \hspace{2em} (Abs. Conv.).

(d) $\sum\limits_{n=1}^{\infty} \dfrac{\cos n\pi\alpha}{x^2 + n^2}$ \hspace{2em} (Abs. Conv.).

8. Verify the following statements: (D = divergent; C = Convergent).

a) $\sum \dfrac{1}{\sqrt{n}}$ \hspace{1em} (D). \hspace{1em} b) $\sum \dfrac{1}{n!}$ \hspace{1em} (C).

c) $\sum \dfrac{2}{n^n}$ \hspace{1em} (C). \hspace{1em} d) $\sum \dfrac{\sqrt{n}}{(n+1)^2}$ \hspace{1em} (C).

e) $\sum \dfrac{n+2}{2^n}$ \hspace{1em} (C). \hspace{1em} f) $\sum \dfrac{1}{n^2 + 1}$ \hspace{1em} (C).

CHAPTER 6: INFINITE SERIES OF REAL CONSTANTS

g) $\sum \dfrac{n}{(2n-1)(2n+1)}$ (D).   h) $\sum \dfrac{1}{n(n+1)^2}$ (C).

i) $\sum \dfrac{n(n+1)}{3^{n+1}}$ (C).   j) $\sum \dfrac{2n-1}{n!}$ (C).

k) $1 + \dfrac{2^2}{3^2} + \dfrac{2^2 \cdot 4^2}{3^2 \cdot 5^2} + \dfrac{2^2 \cdot 4^2 \cdot 6^2}{3^2 \cdot 5^2 \cdot 7^2} + \cdots$ (D).   [C.H. 1988]

l) $\dfrac{1 \cdot 2}{3^2 \cdot 4^2} + \dfrac{3 \cdot 4}{5^2 \cdot 6^2} + \dfrac{5 \cdot 6}{7^2 \cdot 8^2} + \cdots$ (C).

m) $\left(\dfrac{1}{2}\right)^2 + \left(\dfrac{1 \cdot 3}{2 \cdot 4}\right)^2 + \left(\dfrac{1 \cdot 3 \cdot 5}{2 \cdot 4 \cdot 6}\right)^2 + \cdots$ (D).   [C.H. 1991]

[Use De Morgan and Bertrand's Test]

n) $\dfrac{2}{9} + \dfrac{2 \cdot 5}{9 \cdot 12} + \dfrac{2 \cdot 5 \cdot 8}{9 \cdot 12 \cdot 15} + \cdots$ (C).

o) $\dfrac{2}{1^3} - \dfrac{3}{2^3} + \dfrac{4}{3^3} - \dfrac{5}{4^3} + \cdots$ (D).

p) $\dfrac{2}{1 \cdot 4 \cdot 5} + \dfrac{3}{2 \cdot 5 \cdot 6} + \dfrac{4}{3 \cdot 6 \cdot 7} + \cdots$ (C).

q) $\dfrac{1}{2 \cdot 3} + \dfrac{1 \cdot 3}{2 \cdot 4} \cdot \dfrac{1}{5} + \dfrac{1 \cdot 3 \cdot 5}{2 \cdot 4 \cdot 6} \cdot \dfrac{1}{7} + \cdots$ (C).   [C.H. 1988]

r) $\dfrac{2 \cdot 4}{3 \cdot 5} + \dfrac{2 \cdot 4 \cdot 6}{3 \cdot 5 \cdot 7} + \dfrac{2 \cdot 4 \cdot 6 \cdot 8}{3 \cdot 5 \cdot 7 \cdot 9} + \cdots$ (D).   [C.H. 1984]

(Raabe's Test)

AN INTRODUCTION TO
ANALYSIS

(Differential Calculus—Part I)

Chapter 7

Limit of a Function of a Single Real Variable

Chapter 7

Limit of a Function of a Single Real Variable

7.1 Introduction

In the present chapter we shall consider function of a real variable. The domain and range of the function that we shall consider are both subsets of R. Usually we shall take the independent variable x on a certain interval—open or closed: $x \in (a,b)$ or $x \in [a,b]$.

7.2 Meanings of the symbols: $x \to a$ and $x \to \pm\infty$

I. An independent variable x tends to a finite limit a

When we write $x \to a$ we shall assume the following:

(i) x is a real variable

(ii) a is a finite real number—its value remain fixed

(iii) x passes successively through a number of values so that if we consider any particular value taken by x, then we may discriminate the values that precede and the values that follow and none of the values of x can looked upon as the last.

Definition 7.2.1. $x \to a$ or $\lim x = a$ means.

Given any $\epsilon > 0$, the successive values of x will ultimately satisfy the inequality $0 < |x - a| < \epsilon$, i.e. ultimately x lies in the open interval $(a - \epsilon, a + \epsilon)(x \neq a)$, i.e. ultimately x satisfies $a - \epsilon < x < a$ and $a < x < a + \epsilon$.

When $a - \varepsilon < x < a$: we say that x tends to a *from the left* and we write $x \to a^-$

or $x \to a^{-0}$ but when $a < x < a + \varepsilon$; we say that x tends to a from the *right* and we write $x \to a^+$ or $x \to a^{+0}$.

> REMEMBER: $x \to a \Rightarrow x \to a^{-0} \wedge x \to a^{+0}$

Example 7.2.1. $x \to a$ and $x \to b$ $(a \neq b)$ *cannot be simultaneously true, since the successive values of $x - a$ and $x - b$ cannot simultaneously be less than* $\varepsilon = \frac{1}{2}|b - a|$.

The limit of x, when exists, is thus **unique**.

II. An independent variable x tending to $+\infty$

When we write $x \to +\infty$ or simply $x \to \infty$, we assume:

(i) x is a real variable which assumes successive values so that one can discriminate values that precede and the values that follow and *that no value is the last value attained by x.*

(ii) ∞ is not a number at all; it is a symbol whose meaning will be made clear in the following lines.

Now $x \to \infty$ means:

The successive values of x ultimately become and **remain more than** any arbitrary positive number G (no matter how large), given in advance.

III. An independent variable x tending to $-\infty$

$x \to -\infty$ is the same as $(-x) \to +\infty$.

$\therefore$ $x \to -\infty \Rightarrow$ the successive values of x ultimately become and remain less than $-G$, where G is any arbitrary positive number, given in advance.

Note 7.2.1. *There is no such number like* $+\infty$ *or* $-\infty$. *We use these symbols in the sense described above.*

7.3 Limit of a Function

1. The symbolism: $\lim_{x \to a} f(x) = l$ or '$f(x) \to l$ as $x \to a$' means the following:

For every positive number ε, $\exists$ a positive number δ such that

$$\text{whenever } 0 < |x - a| < \delta, \quad |f(x) - l| < \varepsilon.$$

CHAPTER 7: LIMIT OF A FUNCTION OF A SINGLE REAL VARIABLE

The definition requires that f must be defined in some **deleted neighbourhood** N of the point a. The value of f at $x = a$, is left out of the discussion. We are just not concerned with $f(a)$ (it may or may not exist).

Instead of $\lim_{x \to a} f(x) = l$, we may often write: $f(x) \to l$ as $x \to a$.

More compact notation: $f(x) \to l$ as $x \to a$

$$\Rightarrow \varepsilon > 0;\ \exists\ \delta > 0 \text{ s.t. } |f(x) - l| < \varepsilon,\text{ whenever } 0 < |x - a| < \delta.$$

[In this notation "$\varepsilon > 0$;" should be read as

"Given any arbitrary positive number ε, no matter how small."]

2. What is the negation of the statement: $\lim_{x \to a} f(x) = l$?

If $f(x)$ does not tend to l as $x \to a$, then *it is not true that* **for every** $\varepsilon > 0$, $\exists$ some $\delta > 0$ s.t.

$$|f(x) - l| < \varepsilon \quad \forall\ x \in 0 < |x - a| < \delta.$$

$\therefore$ If $f(x)$ does not tend to l as $x \to a$, then $\exists$ some $\varepsilon > 0$ s.t. **for every** $\delta > 0$ there is some x *satisfying* $0 < |x - a| < \delta$ *but not satisfying* $|f(x) - l| < \varepsilon$.

In short, if $f(x)$ does not tend to l as x tends to a, then we shall be able to obtain an $\varepsilon > 0$ s.t. for every positive number $\delta\ \exists$ some x for which

$$0 < |x - a| < \delta \text{ but } |f(x) - l| \geq \varepsilon.$$

Example 7.3.1. *Show that the function f, defined by $f(x) = \sin(1/x)$, when $x \neq 0$ and $f(0) = 0$, does not approach 0 as $x \to 0$.*

Solution: Take $\varepsilon = 1/2$. If δ be any positive number, then it is possible to find a positive integer n s.t.

$$\delta > \frac{2}{(4n+1)\pi} \quad \text{or,}\ 2n\pi + \pi/2 > \frac{1}{\delta}.$$

Then, clearly, $0 < |x - 0| < \delta$, when $x = \frac{1}{2n\pi + \pi/2}$ and

$$|f(x) - 0| = \left|\sin\frac{1}{x}\right| = |\sin(2n\pi + \pi/2)| = 1 > \varepsilon.$$

Hence $f(x)$ does not tend to zero as $x \to 0$; i.e., it is not true that

$$\lim_{x \to 0} \sin\frac{1}{x} = 0.$$

3. What does $\lim_{x \to a} f(x) = l$ mean from geometric point of view?

The existence of limit l of $f(x)$ as $x \to a$ is illustrated in Fig. 7.1.

l and ε, being known, one can draw the horizontal lines $y = l$, $y = l - \varepsilon$ and $y = l + \varepsilon$.

$y \to l$ as $x \to a$, if only a δ-nbd of the point $x = a$ can be found s.t. all points of the graph of $y = f(x)$ corresponding to points x of the nbd (excepting perhaps for the point $x = a$) lie within the strip bounded by the lines $y = l \pm \varepsilon$ and $x = a \pm \delta$.

The point $P(a, l)$ may or may not belong to the graph of $y = f(x)$. We are simply not concerned with it.

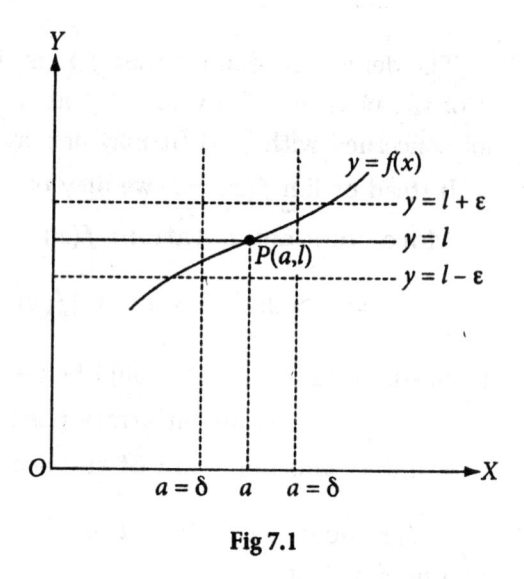

Fig 7.1

4. Right-hand and Left-hand limits

In the definition of the limit of a function f as x approaches a we consider the behaviour of $f(x)$ as x takes values near a. These values of x may be greater than a or may be less than a. In case we restrict ourselves to values greater than a, we say that x approaches a *from the right* (or *from above*). In the other case when x is restricted to take values less than a, we say that x approaches a *from the left* (or *from below*). The limits of f with these restrictions on the values of x, are called *right-hand* and *left-hand limits*.

Definition 7.3.1. (i) *A function f is said to approach l_1 as x approaches from the right, written as*

$$\lim_{x \to a+0} f(x) = l_1 \quad \text{or,} \quad f(a+0) = l_1$$

if corresponding to an arbitrary positive number ε, $\exists \, \delta > 0$ s.t.

$$|f(x) - l_1| < \varepsilon, \quad \text{whenever} \quad a < x < a + \delta.$$

(ii) *A function f is said to approach l_2 as x approaches a from the left, written as*

$$\lim_{x \to a-0} f(x) = l_2 \quad \text{or,} \quad f(a-0) = l_2$$

if corresponding to an arbitrary positive number ε, $\exists \, \delta' > 0$ s.t.

$$|f(x) - l_2| < \varepsilon, \quad \text{whenever} \quad a - \delta' < x < a.$$

CHAPTER 7: LIMIT OF A FUNCTION OF A SINGLE REAL VARIABLE

(iii) *Clearly* $\lim_{x \to a} f(x)$ *exists if and only if* $f(a+0)$ *and* $f(a-0)$ *both exist and are equal.*

(iv) **Infinite One-sided Limits:** *A function* f *is said to approach* $+\infty$ *as* x *approaches* a *from above, written as*

$$\lim_{x \to a+0} f(x) = +\infty$$

if, given any $G > 0$, $\exists\, \delta > 0$ *s.t.*

$$f(x) > G, \quad \text{whenever } a < x < a + \delta.$$

We now leave for the reader to write out similar definitions of

$$\lim_{x \to a-0} f(x) = +\infty; \quad \lim_{x \to a+0} f(x) = -\infty; \quad \lim_{x \to a-0} f(x) = -\infty.$$

5. A function f is said to tend to a finite limit l as x approaches $+\infty$, written as $\lim_{x \to \infty} f(x) = l$, if given any

$$\varepsilon > 0,\ \exists\, G > 0 \text{ s.t. } |f(x) - l| < \varepsilon, \quad \text{whenever } x > G.$$

Similar meanings may be attributed to

$$\lim_{x \to -\infty} f(x) = l \quad \text{or,} \quad \lim_{x \to \infty} f(x) = \infty.$$

Examples illustrating various definitions

Example 7.3.2. *Verify the validity of the statement:* $\lim_{x \to 2} 5x = 10$.

Solution: The statement is true if, for any given $\varepsilon > 0$, $\exists\, \delta > 0$ s.t. $|5x - 10| < \varepsilon$ whenever

$$0 < |x - 2| < \delta.$$

See that $|5x - 10| < \varepsilon \Rightarrow |x - 2| < \varepsilon/5$.

∴ Choosing $\delta = \varepsilon/5$, our definition criterion is satisfied; hence the statement $\lim_{x \to 2} 5x = 10$ is TRUE.

Example 7.3.3. $\lim_{x \to 3} \dfrac{1}{x} = \dfrac{1}{3}$. *Is the statement true?*

Solution: The statement is true if for any given $\varepsilon > 0$, $\exists\, \delta > 0$ s.t.

$$\left|\tfrac{1}{x} - \tfrac{1}{3}\right| < \varepsilon, \quad \text{whenever } 0 < |x - 3| < \delta.$$

See that $\left|\tfrac{1}{x} - \tfrac{1}{3}\right| < \varepsilon \;\Rightarrow\; \tfrac{1}{3} - \varepsilon < \tfrac{1}{x} < \tfrac{1}{3} + \varepsilon$

$$\Rightarrow \tfrac{1-3\varepsilon}{3} < \tfrac{1}{x} < \tfrac{1+3\varepsilon}{3}$$

$$\Rightarrow \tfrac{3}{1+3\varepsilon} < x < \tfrac{3}{1-3\varepsilon}$$

$$\Rightarrow \tfrac{3}{1+3\varepsilon} - 3 < x - 3 < \tfrac{3}{1-3\varepsilon} - 3$$

$$\Rightarrow \tfrac{-9\varepsilon}{1+3\varepsilon} < x - 3 < \tfrac{9\varepsilon}{1-3\varepsilon}.$$

Call $\delta_1 = \tfrac{9\varepsilon}{1+3\varepsilon}$, $\delta_2 = \tfrac{9\varepsilon}{1-3\varepsilon}$. Let $\delta = \min.\{\delta_1, \delta_2\} = \delta_1 = \tfrac{9\varepsilon}{1+3\varepsilon}$.

Then it is true that $\left|\tfrac{1}{x} - \tfrac{1}{3}\right| < \varepsilon$, whenever $|x - 3| < \delta = \tfrac{9\varepsilon}{1+3\varepsilon}$,

i.e., the statement $\lim\limits_{x \to 3} \tfrac{1}{x} = \tfrac{1}{3}$ is TRUE.

Example 7.3.4. Does $\lim\limits_{x \to 0} \dfrac{2x + x^2}{x}$ exists? If so, what is the limit?

Solution: Let $f(x) = \dfrac{x^2 + 2x}{x}$. When $x = 0$, f is undefined. But in the definition of limit of $f(x)$ as $x \to 0$ we are really not concerned with the value of f at $x = 0$. Hence we may consider $f(x) = x + 2$, for $x \neq 0$.

Here $\lim\limits_{x \to 0+0} f(x) = \lim\limits_{x \to 0-0} f(x) = 2$ (can be easily verified by ε-δ definition).

$\therefore\; \lim\limits_{x \to 0} f(x)$ exists and equals to 2.

Note 7.3.1. *See that here limit of $f(x)$ as $x \to 0$ exists but the functional value at $x = 0$ does not exist.*

Example 7.3.5. $f(x) = \tfrac{1}{x^2}$. $\lim\limits_{x \to 0} f(x) = +\infty$.

Solution: This statement is true if for any given $G > 0$ we can find a $\delta > 0$ s.t.

$f(x)$ i.e., $\tfrac{1}{x^2} > G$, whenever $0 < |x - 0| < \delta$.

See that $\tfrac{1}{x^2} > G \Rightarrow x^2 < \tfrac{1}{G} \Rightarrow -\tfrac{1}{\sqrt{G}} < x < \tfrac{1}{\sqrt{G}}$.

So choosing $\delta = 1/\sqrt{G}$, the definition criterion is satisfied and the statement

$$\lim\limits_{x \to 0} f(x) = +\infty \text{ is TRUE.}$$

CHAPTER 7: LIMIT OF A FUNCTION OF A SINGLE REAL VARIABLE   435

Example 7.3.6. $\lim_{x \to \infty} \frac{1}{x^2} = 0$ *is* TRUE.

Solution: For, given any $\varepsilon > 0$, $\left|\frac{1}{x^2} - 0\right| < \varepsilon \Rightarrow x^2 > 1/\varepsilon \Rightarrow x > 1/\sqrt{\varepsilon}$.

Choosing $G = 1/\sqrt{\varepsilon}$, we see that the definition criterion is satisfied.

Example 7.3.7. $\lim_{x \to \infty} \frac{[x]}{2x} = \frac{1}{2}$, *where* $[x]$ = *greatest integer in* x, *not exceeding* x. *For this statement to be* TRUE, *we are to find G in terms of a given* $\varepsilon > 0$, *(no matter how small) satisfying.*

Solution:

$$\left|\frac{[x]}{2x} - \frac{1}{2}\right| < \varepsilon \quad \text{for all } x > G, \quad \text{i.e.,} \quad \left|\frac{[x] - x}{2x}\right| < \varepsilon \quad \text{for all } x > G.$$

But $|[x] - x| < 1$ since $[x] - x$ always lies between 0 and -1, when $x > 0$.

$$\therefore \left|\frac{[x] - x}{2x}\right| < \frac{1}{2x} \text{ which is less than } \varepsilon, \text{ whenever } x > \frac{1}{2\varepsilon}.$$

Choosing $G = \frac{1}{2\varepsilon}$, we see that the definition-criterion is satisfied.

Example 7.3.8. $\lim_{x \to 0} \frac{1}{x}$ *does not exist.* (*For* $\lim_{x \to 0+0} \frac{1}{x} = +\infty$ *and* $\lim_{x \to 0-0} \frac{1}{x} = -\infty$.)

Example 7.3.9. $\lim_{x \to 0} \frac{1}{1 + e^{1/x}}$ *also does not exist.*

Solution: For, Right-hand limit $= \lim_{x \to 0+0} \frac{1}{1 + e^{1/x}} = 0$ (as $x \to 0+0$, $1/x \to +\infty$, $e^{1/x} \to \infty$)

but, Left-hand limit $= \lim_{x \to 0-0} \frac{1}{1 + e^{1/x}} = 1$ (as $x \to 0-0$, $1/x \to -\infty$, $e^{1/x} \to 0$)

and the two limits are not equal; hence the limit does not exist.

Example 7.3.10. *A very useful result:* $\lim_{x \to 0} x \sin \frac{1}{x} = 0$.

Solution: This statement is true, if given any $\varepsilon > 0$ we can find $\delta > 0$ s.t.

$$\left|x \sin \frac{1}{x} - 0\right| < \varepsilon, \; \forall \, x \text{ in } 0 < |x - 0| < \delta.$$

But see that $\left|x \sin \frac{1}{x} - 0\right| = \left|x \sin \frac{1}{x}\right| \leq |x|$.

[$\because |\sin 1/x| \leq 1$, except for $x = 0$, where it is undefined; in this case we are not concerned with the value of $x \sin 1/x$ at $x = 0$.]

Therefore,

$$\left| x \sin \frac{1}{x} - 0 \right| < \varepsilon \text{ is true if } |x| < \varepsilon \quad (x \neq 0)$$

i.e., if $0 < |x - 0| < \delta$, where $\delta = \varepsilon$.

Thus given any $\varepsilon > 0$, we can find $\delta (= \varepsilon)$ which will satisfy our definition test

$$\therefore \lim_{x \to 0} x \sin \frac{1}{x} = 0.$$

Similarly, the reader can prove:

$$\lim_{x \to 0} x^2 \sin \frac{1}{x} = 0; \quad \lim_{x \to 0} x \cos \frac{1}{x} = 0; \quad \lim_{x \to 0} x^2 \cos \frac{1}{x} = 0.$$

Example 7.3.11. *Prove that $\lim_{x \to a} f(x)$, when exists, is unique.*

Solution: If possible, let $\lim_{x \to a} f(x)$ has two distinct limits l_1 and l_2 where $l_1 \neq l_2$. Let $\varepsilon > 0$ be any given positive number.

We can choose $\delta > 0$ suitably such that

$$|f(x) - l_1| < \varepsilon/2, \quad \forall x \in 0 < |x - a| < \delta$$
$$|f(x) - l_2| < \varepsilon/2, \quad \forall x \in 0 < |x - a| < \delta.$$

Now
$$\begin{aligned} |l_1 - l_2| &= |l_1 - f(x) + f(x) - l_2| \\ &\leq |l_1 - f(x)| + |f(x) - l_2| \\ &< \varepsilon/2 + \varepsilon/2 = \varepsilon, \forall x \in 0 < |x - a| < \delta. \end{aligned}$$

Thus $|l_1 - l_2|$ is less than any positive number (no matter how small we take). This is only possible if $|l_1 - l_2| = 0$. i.e., if $l_1 = l_2$. Hence the result.

7.4 Theorems on Limits

Theorem 7.4.1. *Let a function f be defined in a certain interval containing 'a' as an interior point. Let $\lim_{x \to a} f(x) = l$. Then*

(i) $\exists$ *a positive number δ_1 for which*

$$|f(x)| < 1 + |l|, \text{ whenever } 0 < |x - a| < \delta_1$$

CHAPTER 7: LIMIT OF A FUNCTION OF A SINGLE REAL VARIABLE

and (ii) *if* $l \neq 0$, $\exists$ *a positive number* δ_2 *for which*

$$\frac{1}{2}|l| > |f(x)| > \frac{3}{2}|l| \text{ whenever } 0 < |x - a| < \delta_2.$$

[This is comparable to the theorem: *Every convergent sequence is bounded.*]

Proof: (i) Since $f(x) \to l$ as $x \to a$, hence, by definition, corresponding to $\varepsilon = 1$, $\exists \delta_1 > 0$ such that

$$|f(x) - l| < 1, \text{ whenever } 0 < |x - a| < \delta_1$$

and for such an x,

$$|f(x)| = |f(x) - l + l| \leq |f(x) - l| + |l|$$
$$\text{i.e., } < 1 + |l|.$$

(ii) Let $l \neq 0$. Then corresponding to $\varepsilon = \frac{1}{2}|l|$, $\exists \delta_2 > 0$ s.t.

$$|f(x) - l| < \frac{1}{2}|l|, \text{ whenever } 0 < |x - a| < \delta_2.$$

For all such x,

$$|f(x)| = |f(x) - l + l|$$
$$\leq |f(x) - l| + |l|$$
$$< \frac{3}{2}|l|$$

and $|f(x)| \geq |l| - |f(x) - l|$ (since $|a + b| \geq |b| - |a|$)
$$> \frac{1}{2}|l|.$$

Corollary 7.4.1. *With the hypothesis of the theorem it is easy to prove that k being a given positive number,*

$\exists \delta_3$ s.t. $|f(x)| < k + |l|$, whenever $0 < |x - a| < \delta_3$.

Moreover, if $0 < k < |l|$,

$\exists \delta_4$ s.t. $|l| - k < |f(x)| < |l| + k$, whenever $0 < |x - a| < \delta_4$.

Theorem 7.4.2. *Let f and g be two real-valued functions of x defined in an interval containing a point a in its interior.*

Suppose $\lim\limits_{x \to a} f(x) = l$ *and* $\lim\limits_{x \to a} g(x) = m$.

Then

(i) $\lim_{x \to a}\{f(x) \pm g(x)\} = l \pm m = \lim_{x \to a} f(x) \pm \lim_{x \to a} g(x).$

(ii) $\lim_{x \to a}\{f(x) \cdot g(x)\} = l \cdot m = \lim_{x \to a} f(x) \cdot \lim_{x \to a} g(x).$

(iii) $\lim_{x \to a}\left\{\dfrac{f(x)}{g(x)}\right\} = \dfrac{l}{m},$ provided $m \neq 0$

$\qquad\qquad\qquad\quad = \dfrac{\lim_{x \to a} f(x)}{\lim_{x \to a} g(x)},$ provided $\lim_{x \to a} g(x) \neq 0.$ [C.H. 1991, 95]

Proof: (i) *Given:* $\lim_{x \to a} f(x) = l$ and $\lim_{x \to a} g(x) = m.$

To prove: $\lim_{x \to a}[f(x) + g(x)] = l + m.$

Let ε be any given positive number, no matter how small. Corresponding to $\varepsilon/2$ ∃ a **suitable** positive number δ such that

$$|f(x) - l| < \frac{\varepsilon}{2}, \forall\, x \in 0 < |x - a| < \delta$$

$$|g(x) - m| < \frac{\varepsilon}{2}, \forall\, x \in 0 < |x - a| < \delta$$

$$\begin{aligned}|\{f(x) + g(x)\} - (l + m)| &= |\{f(x) - l\} + \{g(x) - m\}| \\ &\leq |f(x) - l| + |g(x) - m| \\ &< \frac{\varepsilon}{2} + \frac{\varepsilon}{2} = \varepsilon \quad \forall\, x \in 0 < |x - a| < \delta.\end{aligned}$$

Hence, by definition of limit

$$\lim_{x \to a}\{f(x) + g(x)\} = l + m. \text{ (proved)}$$

Exactly in a similar manner, we can prove

$$\lim_{x \to a}\{f(x) - g(x)\} = l - m.$$

(ii) *Given:* $\lim_{x \to a} f(x) = l$ and $\lim_{x \to a} g(x) = m.$

To prove: $\lim_{x \to a}\{f(x) \cdot g(x)\} = l \cdot m.$

$$\begin{aligned}|f(x) \cdot g(x) - lm| &= |f(x)\{g(x) - m\} + m\{f(x) - l\}| \\ &\leq |f(x)||g(x) - m| + |m||f(x) - l|.\end{aligned} \qquad (7.4.1)$$

CHAPTER 7: LIMIT OF A FUNCTION OF A SINGLE REAL VARIABLE

[We shall try to make $|f(x)g(x) - lm| <$ any given positive number ε within some deleted δ-nbd of a.]

Since $\lim_{x \to a} f(x) = l$, corresponding to $\varepsilon = 1$, $\exists\, \delta_1 > 0$ such that

$$|f(x) - l| < 1, \forall\, x \in 0 < |x - a| < \delta_1$$

whence,

$$|f(x)| = |f(x) - l + l| \le |f(x) - l| + |l| < 1 + |l| \quad \forall\, x \in 0 < |x - a| < \delta_1. \quad (7.4.2)$$

[We shall try to make $|f(x)||g(x) - m| < \varepsilon/2$.]

Let $\varepsilon > 0$ be given.

Since $\lim_{x \to a} g(x) = m$, corresponding to $\frac{\varepsilon/2}{1+|l|}$, $\exists$ a $\delta_2 > 0$ such that

$$|g(x) - m| < \frac{\varepsilon/2}{1 + |l|}, \quad \forall\, x \in 0 < |x - a| < \delta_2. \quad (7.4.3)$$

[See that from (7.4.2) and (7.4.3) $|f(x)||g(x) - m| < \varepsilon/2$ in some nbd of a.]

Again since $\lim_{x \to a} f(x) = l$, corresponding to a positive number $\frac{\varepsilon/2}{1+|m|}$, $\exists$ a positive number $\delta_3 > 0$ such that

$$|f(x) - l| < \frac{\varepsilon/2}{1 + |m|}, \quad \forall\, x \in 0 < |x - a| < \delta_3. \quad (7.4.4)$$

Let $\delta = \min.(\delta_1, \delta_2, \delta_3)$. Then using the results of **(7.4.1)**, **(7.4.2)**, **(7.4.3)** and **(7.4.4)** we get

$$\begin{aligned}|f(x)g(x) - lm| &< \{1 + |l|\}\frac{\varepsilon/2}{1 + |l|} + |m|\frac{\varepsilon/2}{1 + |m|} \\ &< \frac{\varepsilon}{2} + \frac{\varepsilon}{2} = \varepsilon \ \forall\, x \in 0 < |x - a| < \delta. \quad \left(\because \frac{|m|}{1 + |m|} < 1\right)\end{aligned}$$

Hence, by definition, $\lim_{x \to a} f(x)g(x) = l \cdot m$. (proved).

(iii) Given: $\lim_{x \to a} f(x) = l$ and $\lim_{x \to a} g(x) = m$ $(m \ne 0)$.

To prove: $\lim_{x \to a} \frac{f(x)}{g(x)} = \frac{l}{m}$.

$$\begin{aligned}\left|\frac{f(x)}{g(x)} - \frac{l}{m}\right| &= \left|\frac{mf(x) - lg(x)}{mg(x)}\right| = \left|\frac{m\{f(x) - l\} - l\{g(x) - m\}}{mg(x)}\right| \\ &\le \frac{|m||f(x) - l| + |l||g(x) - m|}{|m||g(x)|}. \quad (7.4.5)\end{aligned}$$

[Our efforts will be to make it $< \varepsilon/2$ in some deleted nbd of a. What we need here is to make $\frac{1}{|g(x)|} <$ certain fixed number.]

Since $\lim_{x \to a} g(x) = m (m \neq 0)$, corresponding to $\varepsilon = \frac{1}{2}|m| > 0 \; \exists$ a $\delta_1 > 0$ such that $|g(x) - m| < \frac{1}{2}|m|, \quad \forall \, x \in 0 < |x-a| < \delta_1$.

Now, $|m| = |m - g(x) + g(x)| \leq |g(x) - m| + |g(x)|$

i.e., $|m| < \frac{1}{2}|m| + |g(x)|, \quad \forall \, x \in 0 < |x-a| < \delta_1$.

In other words,
$$\frac{1}{|g(x)|} < \frac{2}{|m|}, \quad \forall \, x \in 0 < |x-a| < \delta_1. \tag{7.4.6}$$

Let $\varepsilon > 0$ be given.

Since $\lim_{x \to a} f(x) = l$ and $\lim_{x \to a} g(x) = m$.

Corresponding to $\frac{\varepsilon}{4}|m|, \; \exists \, \delta_2 > 0$ such that
$$|f(x) - l| < \frac{\varepsilon}{4}|m|, \; \forall \, x \in 0 < |x-a| < \delta_2. \tag{7.4.7}$$

Corresponding to $\frac{\varepsilon}{4} \frac{|m|^2}{|l|+1}, \; \exists \, \delta_3 > 0$ such that
$$|g(x) - m| < \frac{\varepsilon}{4} \frac{|m|^2}{|l|+1}, \; \forall \, x \in |x-a| < \delta_3. \tag{7.4.8}$$

Let $\delta = \min.(\delta_1, \delta_2, \delta_3)$ so that from (7.4.5), (7.4.6), (7.4.7) and (7.4.8) it follows

$$\left| \frac{f(x)}{g(x)} - \frac{l}{m} \right| < \left\{ |m| \cdot \frac{\varepsilon}{4}|m| + |l| \frac{\frac{\varepsilon}{4}|m|^2}{|l|+1} \right\} \frac{1}{|m|} \cdot \frac{2}{|m|} \quad \forall \, x \in 0 < |x-a| < \delta$$

$$< \left(\frac{\varepsilon}{4} + \frac{|l|}{|l|+1} \cdot \frac{\varepsilon}{4} \right) \cdot 2$$

$$< \frac{\varepsilon}{2} + \frac{|l|}{|l|+1} \cdot \frac{\varepsilon}{2}$$

$$< \frac{\varepsilon}{2} + \frac{\varepsilon}{2} = \varepsilon \quad \forall \, x \in 0 < |x-a| < \delta. \quad \left(\because \frac{|l|}{|l|+1} < 1 \right)$$

Hence, by definition, $\lim_{x \to a} \frac{f(x)}{g(x)} = \frac{l}{m} (m \neq 0)$.
[See that arguments do not hold if $m = 0$.]

Example 7.4.1. Try in a similar way and prove $\lim_{x \to a} \frac{1}{g(x)} = \frac{1}{m} (m \neq 0)$ if it is given $\lim_{x \to a} \tilde{g}(x) = m (\neq 0)$.

CHAPTER 7: LIMIT OF A FUNCTION OF A SINGLE REAL VARIABLE

Observation: The beginner may wonder how ε's have been chosen at different stages. We give below a line of reasoning which the students may find easily acceptable. It may, however, be noted that most of the text books use the methods shown above.

Suppose we are to prove (ii)

We start with two positive numbers ε and k of which ε is arbitrary but k is a fixed positive number which we shall specify later. In any case εk is an arbitrary positive number.

Since $\lim_{x \to a} f(x) = l$ and $\lim_{x \to a} g(x) = m$.

Corresponding to εk $\exists$ a suitable positive number δ_1 such that

$$\left. \begin{array}{c} |f(x) - l| < \varepsilon k \\ |g(x) - m| < \varepsilon k \end{array} \right\} \quad \forall\, x \in 0 < |x - a| < \delta_1. \tag{7.4.9}$$

Now, $|f(x)g(x) - lm| = |f(x)\{g(x) - m\} + m\{f(x) - l\}|$
$\leq |f(x)||g(x) - m| + |m||f(x) - l|$
$< \varepsilon k\{|f(x)| + |m|\} \quad \forall\, x \in 0 < |x - a| < \delta_1.$

Now since $\lim_{x \to a} f(x) = l$, $\exists\, \delta_2 > 0$ such that (see Theorem **7.4.1** above)

$$|f(x)| < 1 + |l| \quad \forall\, x \in 0 < |x - a| < \delta_2$$

we now take $\delta = \min.(\delta_1, \delta_2)$. Then

$$\forall\, x \in 0 < |x - a| < \delta \text{ we have}$$

$$|f(x)g(x) - lm| < \varepsilon k\{|l| + 1 + |m|\} \quad \forall\, x \in 0 < |x - a| < \delta.$$

[Our main objective is to make it $< \varepsilon$.]
If from the very beginning we would take $k = \frac{1}{|l|+|m|+1}$ or some thing less than $\frac{1}{|l|+|m|+1}$ then we would arrive at

$$|f(x)g(x) - lm| < \varepsilon \quad \forall\, x \in 0 < |x - a| < \delta$$

and by definition $\lim_{x \to a}\{f(x)g(x)\} = l \cdot m$.

We advise our students to try to prove (iv) in a similar way [i.e., taking εk to start with and then fix k later; you may require the result of Theorem **7.4.1**. (ii)].

Extension: For $r = 1, 2, 3, \cdots, n$, let $f_r(x) \to l_r$. Then

$$\lim_{x \to a}[f_1(x) \pm f_2(x) \pm \cdots \pm f_n(x)] = l_1 \pm l_2 \pm l_3 \pm \cdots \pm l_n$$

and $\lim_{x \to a}[f_1(x) \cdot f_2(x) \cdots f_n(x)] = l_1 \cdot l_2 \cdot l_3 \cdots l_n$.

Note that the extensions can be made for a finite number of functions. For infinite number of functions we shall require further conditions.

Theorem 7.4.3. Limit of a function of a function

Let $\lim_{x \to a} \phi(x) = l$ and let $\lim_{u \to l} f(u) = f(l)$. Then

$$\lim_{x \to a} f\{\phi(x)\} = f\{\lim_{x \to a} \phi(x)\} = f(l).$$

Proof: Since $f(u) \to f(l)$ as $u \to l$, we have
corresponding to any $\varepsilon > 0$, $\exists\, \delta_1 > 0$ such that

$$|f(u) - f(l)| < \varepsilon, \text{ whenever } 0 < |u - l| < \delta_1.$$

On writing $u = \phi(x)$ we have, corresponding to any $\varepsilon > 0$ $\exists\, \delta_1 > 0$ s.t.

$$|f(\phi(x)) - f(l)| < \varepsilon, \text{ whenever } 0 < |\phi(x) - l| < \delta_1.$$

Again, since $\lim_{x \to a} \phi(x) = l$, we have,
corresponding to the positive number δ_1 $\exists\, \delta > 0$ s.t.

$$|\phi(x) - l| < \delta_1, \text{ whenever } 0 < |x - a| < \delta.$$

Combining the two results:
Corresponding to any $\varepsilon > 0$ $\exists\, \delta > 0$ s.t.

$$|f(\phi(x)) - f(l)| < \varepsilon, \text{ whenever } 0 < |x - a| < \delta.$$

Hence $\lim_{x \to a} f(\phi(x)) = f(l)$.

Corollary 7.4.2. $\lim_{x \to a}\{f(x)\}^n = \left\{\lim_{x \to a} f(x)\right\}^n$; $\lim_{x \to a}\{\phi(x)\}^{f(x)} = \left\{\lim_{x \to a} \phi(x)\right\}^{\lim_{x \to a} f(x)}$ provided that $\phi(x) > 0, \forall\, x$.

Theorem 7.4.4. (Neighbourhood Property).
If $\lim_{x \to a} f(x) = l$ $(l \neq 0)$, then there exists some neighbourhood of a, at every point of which $f(x)$ will have the same sign as that of l.

Proof: Since $f(x) \to l$ as $x \to a$ and $l \neq 0$, corresponding to $\varepsilon = \frac{1}{2}|l|$, $\exists\, \delta > 0$ s.t.

$$|f(x) - l| < \varepsilon, \qquad \text{whenever } 0 < |x - a| < \delta$$
$$\text{i.e., } -\varepsilon < f(x) - l < \varepsilon, \text{ whenever } x \in \delta\text{-nbd of } a. \qquad (7.4.10)$$

We may take $\varepsilon = \frac{1}{2}l$, when $l > 0$ and $\varepsilon = -\frac{1}{2}l$, when $l < 0$.

If $l > 0$, then (**7.4.10**), gives a δ-nbd of a, where

$$-\frac{1}{2}l < f(x) - l < \frac{1}{2}l \qquad \left(\because \text{ here } \varepsilon = \frac{1}{2}l\right)$$

or, $\quad \dfrac{1}{2}l < f(x) < \dfrac{3}{2}l$

$\Rightarrow\ f(x)$ is positive $\forall\, x$ in the δ-nbd of a.

If $l < 0$, from (**7.4.10**), we get a δ-nbd of a, where

$$\frac{1}{2}l < f(x) - l < -\frac{1}{2}l \qquad \left(\because \text{ here } \varepsilon = -\frac{1}{2}l\right)$$

or, $\quad \dfrac{3}{2}l < f(x) < \dfrac{1}{2}l$ (l is negative)

$\Rightarrow\ f(x)$ is negative $\forall\, x$ in that δ-nbd of a.

Corollary 7.4.3. Let $\lim\limits_{x \to a} f(x) = l$. Suppose $f(x) > 0\ \forall\, x$ in a certain nbd of a.

Then, we shall conclude that l is **not negative** i.e., $l \geq 0$.

Similarly, when $f(x) < 0$, $\forall\, x$ in a certain nbd of a, we shall conclude $l \leq 0$.

See that even when $f(x) > 0$ for all x in a certain nbd of a, still $\lim\limits_{x \to a} f(x)$ may be zero e.g., $f(x) = x^2$; here $f(x) > 0\ \forall\, x \neq 0$ but $\lim\limits_{x \to 0} f(x) = 0$.

Se we stress upon the fact that $l \geq 0$.

Strong inequalities become weak after taking limit operations as will be proved in the next theorems.

Theorem 7.4.5. *If $f(x) < g(x)$ for $a - h < x < a + h$ and*

$$\lim_{x \to a} f(x) = l \quad \text{and} \quad \lim_{x \to a} g(x) = m$$

then

$$l \leq m.$$

Proof: Use the corollary of Theorem **7.4.4**.

Illustration. Let $f(x) = \frac{1}{2} - x^2$, $g(x) = 1 + x^2$. Then $f(x) < g(x)$ for all values of x and $f(x) \to \frac{1}{2}$ and $g(x) \to 1$ as $x \to 0$. Hence $l(=\frac{1}{2}) < m(=1)$. But if $f(x) = 1 - x^2$, $g(x) = 1 + x^2$, then both the functions tend to 1 as $x \to 0$ even though $f(x) < g(x)$ for all non-zero values of x.

Theorem 7.4.6. (Sandwich Theorem). *If, in a certain neighbourhood of a,*

$$f(x) \leq g(x) \leq h(x)$$

and $\quad \lim_{x \to a} f(x) = l = \lim_{x \to a} h(x)$

then $\lim_{x \to a} g(x)$ *exists and is equal to l.*

Proof: Suppose

$$f(x) \leq g(x) \leq h(x), \text{ when } 0 < |x - a| < \delta_1.$$

From **the** given limits, we have, for any $\varepsilon > 0$, $\exists\ \delta_2, \delta_3 > 0$ s.t.

$$l - \varepsilon < f(x) < l + \varepsilon, \text{ when } 0 < |x - a| < \delta_2.$$
and $\quad l - \varepsilon < h(x) < l + \varepsilon, \text{ when } 0 < |x - a| < \delta_3.$

If $\delta = \min.\{\delta_1, \delta_2, \delta_3\}$ then all the above inequalities hold when $0 < |x - a| < \delta$. Under these conditions

$$l - \varepsilon < f(x) \leq g(x) \text{ also } l + \varepsilon > h(x) \geq g(x)$$

i.e., $\quad l - \varepsilon < g(x) < l + \varepsilon, \text{ for } 0 < |x - a| < \delta$

i.e., $\quad \lim_{x \to a} g(x) = l.$

Corollary 7.4.4. *If $|f(x)| < |g(x)|$ in a certain neighbourhood of a and if $\lim_{x \to a} g(x) = 0$, then $\lim_{x \to a} f(x) = 0$.*

7.5 Illustrative Examples using Theorems on Limits

Example 7.5.1. *Use theorems on limits to establish*

$$\lim_{x \to 2} \frac{x^2 - 3x + 7}{4x^3 + x + 1} = \frac{1}{7}.$$

CHAPTER 7: LIMIT OF A FUNCTION OF A SINGLE REAL VARIABLE

Solution: Since $x \to 2$, we may put $x = 2 + h$, where $h \to 0$. Then,

$$\lim_{x \to 2} \frac{x^2 - 3x + 7}{4x^3 + x + 1} = \lim_{h \to 0} \frac{(2+h)^2 - 3(2+h) + 7}{4(2+h)^3 + (2+h) + 1}$$

$$= \lim_{h \to 0} \frac{5 + h + h^2}{35 + 49h + 24h^2 + 4h^3} = \frac{\lim_{h \to 0}(5 + h + h^2)}{\lim_{h \to 0}(35 + 49h + 24h^2 + 4h^3)}$$

$$= \frac{5 + \lim_{h \to 0} h + \lim_{h \to 0} h^2}{35 + \lim_{h \to 0} 49h + \lim_{h \to 0} 24h^2 + \lim_{h \to 0} 4h^3} = \frac{5 + 0 + 0}{35 + 0 + 0 + 0}$$

$$= \frac{5}{35} = \frac{1}{7}.$$

Example 7.5.2. $\lim_{x \to 3} \dfrac{x^2 + x - 12}{x - 3} = 7.$

Solution: Since $x \to 3$, put $x = 3 + h$ where $h \to 0$.

$$\lim_{x \to 3} \frac{x^2 + x - 12}{x - 3} = \lim_{h \to 0} \frac{(3+h)^2 + (3+h) - 12}{h}$$

$$= \lim_{h \to 0} \frac{h^2 + 7h}{h} = \lim_{h \to 0}(h + 7)$$

$$= \lim_{h \to 0} h + 7 = \mathbf{7}.$$

Example 7.5.3. $\lim_{x \to \infty} \dfrac{5x^{12} + 7x^9 + 12}{14x^{16} + 9x^2 - 3} = 0.$

Solution: Put $x = 1/y$. When $x \to \infty$, $y \to 0^+$.

$$\lim_{x \to \infty} \frac{5x^{12} + 7x^9 + 12}{14x^{16} + 9x^2 - 3} = \lim_{y \to 0^+} \frac{5/y^{12} + 7/y^9 + 12}{14/y^{16} + 9/y^2 - 3}$$

$$= \lim_{y \to 0^+} \frac{y^4(12y^{12} + 7y^3 + 5)}{(-3y^{16} + 9y^{14} + 14)}$$

$$= \frac{12 \lim_{y \to 0^+} y^{16} + 7 \lim_{y \to 0^+} y^7 + 5 \lim_{y \to 0^+} y^4}{-3 \lim_{y \to 0^+} y^{16} + 9 \lim_{y \to 0^+} y^{14} + 14}$$

$$= \frac{0 + 0 + 0}{0 + 0 + 14} = \frac{0}{14} = 0.$$

Similarly, check: $\lim_{x \to \infty} \dfrac{15x^7 + 12x + 17}{5x^7 + 9x^2 + 12} = 3.$

Example 7.5.4. $\lim_{x \to \infty} \dfrac{2x^8 - 7x^2 + 5}{9x^4 + 5x + 12} = \infty.$

Solution: As before, putting $x = 1/y$, $y \to 0^+$ we have

$$\text{Given limit} = \lim_{y \to 0^+} \frac{2/y^8 - 7/y^2 + 5}{9/y^4 + 5/y + 12} = \lim_{y \to 0^+} \frac{5y^8 - 7y^6 + 2}{12y^4 + 5y^3 + 9} \cdot \frac{1}{y^4} = \infty.$$

(Numr. $\to 2$ and Denr. $\to 0$; hence no finite limit exists here.)

7.6 Cauchy's Criterion for the existence of Limit of a Function

Theorem 7.6.1. *The necessary and sufficient condition that a function $f(x)$ may tend to a definite finite limit l as $x \to a$, is that, to any pre-assigned positive quantity ε, however small, there corresponds a positive number δ such that $|f(x_2) - f(x_1)| < \varepsilon$, for every pair x_1, x_2 of values of x satisfying*

$$0 < |x_1 - a| < \delta \quad \text{and} \quad 0 < |x_2 - a| < \delta.$$

[This theorem is known as **Cauchy's General Principle** as applied to limit of a function.]

Proof: The condition is necessary.

Suppose $f(x) \to l$ as $x \to a$. Then to $\frac{1}{2}\varepsilon$ (ε, small positive), there corresponds a positive number δ, such that

$$|f(x) - l| < \frac{1}{2}\varepsilon \text{ for } 0 < |x - a| < \delta.$$

Taking x_1, x_2 any two values of x satisfying

$$0 < |x_1 - a| < \delta, \quad 0 < |x_2 - a| < \delta,$$

we have

$$|f(x_1) - l| < \frac{1}{2}\varepsilon \quad \text{and} \quad |f(x_2) - l| < \frac{1}{2}\varepsilon$$

and accordingly it necessarily follows:

$$\begin{aligned} |f(x_2) - f(x_1)| &= |\{f(x_2) - l\} - \{f(x_1) - l\}| \\ &\leq |f(x_2) - l| + |f(x_1) - l| \\ &< \frac{1}{2}\varepsilon + \frac{1}{2}\varepsilon = \varepsilon. \end{aligned}$$

The condition is sufficient.

CHAPTER 7: LIMIT OF A FUNCTION OF A SINGLE REAL VARIABLE

Let the condition be satisfied. The proof that $f(x)$ then tends to a finite limit is effected in the following three steps:

Step 1. We shall prove that, it is not possible that there should be two sequences $\{x_n\}$ and $\{x'_n\}$ both tending to a as $n \to \infty$, for which the following statement is true:

$$f(x_n) \to l \quad \text{and} \quad f(x'_n) \to l' \text{ where } l \neq l'. \tag{7.6.1}$$

Suppose that it is possible. Then from (7.6.1), $|l - l'|$ is a positive number and there exists a positive integer N_1, for which

$$\left. \begin{array}{r} |f(x_n) - l| < \tfrac{1}{3}|l - l'| \\ \text{and } |f(x'_m) - l'| < \tfrac{1}{3}|l - l'| \end{array} \right\} \text{ for } n > N_1 \text{ and } m > N_1.$$

Again since the condition of the theorem is satisfied, there exists a positive number δ for which

$$|f(x_n) - f(x'_m)| < \frac{1}{3}|l - l'|, \text{ whenever } 0 < |x_n - a| < \delta \text{ and } 0 < |x'_m - a| < \delta.$$

Now since $x_n \to a$, $x'_m \to a$; it is possible to find a positive integer N_2 such that

$$0 < |x_n - a| < \delta \quad \text{and} \quad 0 < |x'_m - a| < \delta$$

whenever both n and m exceed N_2.

Thus, all the above results are true if n and m both exceed $\mu = \max.\{N_1, N_2\}$ and so for such n, m, we have

$$\begin{aligned} |l - l'| &= |l - f(x_n) + f(x_n) - f(x'_m) + f(x'_m) - l'| \\ &\leq |l - f(x_n)| + |f(x_n) - f(x'_m)| + |f(x'_m) - l'| \\ &< 3 \cdot \frac{1}{3}|l - l'|, \text{ which is clearly not true.} \end{aligned}$$

Hence (7.6.1) cannot hold. That means if $f(x_n) \to l$ and $f(x'_n) \to l'$ as $\{x_n\} \to a$ and $\{x'_n\} \to a$ respectively, we must conclude $l = l'$.

Step 2. Since the condition of the theorem holds for any positive number ε, there exists a positive number δ_1 for which $|f(x) - f(x_1)| < 1$, whenever

$$0 < |x - a| < \delta_1 \quad \text{and} \quad 0 < |x_1 - a| < \delta_1.$$

On taking a definite x_1 satisfying this, we get

$$|f(x)| < 1 + |f(x_1)|, \text{ whenever } 0 < |x - a| < \delta_1$$

and so $f(x)$ is bounded in $0 < |x - a| < \delta_1$.

Let I_n denote the closed interval $[a - \frac{1}{2^n}, a + \frac{1}{2^n}]$, the point $x = a$ being excluded. Then from and after some n, say from $n = N$,

$$2^{-n} < \delta_1$$

and $f(x)$ is bounded in I_n (for $n \geq N$).

Let G_n and g_n be respectively the exact upper and lower bounds respectively of $f(x)$ in I_n.

Since $I_{n+1} \subset I_n$, $G_{N+1} \leq G_N$. Also $G_{N+k} \geq g_n$ for all positive integers k. Hence the sequence

$$\{G_N, G_{N+1}, G_{N+2}, \cdots\}$$

is a monotone decreasing sequence bounded below and so as $n \to \infty$,

$$G_N \to \text{ a finite limit } G \text{ (say)}. \qquad (7.6.2)$$

Similarly, $\quad g_n \to$ a finite limit g (say). $\qquad (7.6.3)$

From (7.6.2), we may *prove* that there exists a sequence $\{x_n\}$ tending to a for which $f(x_n) \to G$. To prove this we observe that for all $n \geq N$, there exists x_n of I_n such that

$$G_n - 2^{-n} < f(x_n) \leq G_n.$$

From (7.6.2), as $n \to \infty$, $G_n - 2^{-n} \to G$ and $G_n \to G$ and so $f(x_n) \to G$.

Similarly we may prove that there is a sequence $\{x'_n\}$ tending to a for which $f(x'_n) \to g$.

Using the result of **Step 1**, we conclude $G = g$.

Step 3. Since both G_n and $g_n \to G$; for any $\varepsilon > 0$, $\exists$ a positive integer M s.t.

$$|G_n - G| < \varepsilon \quad \text{and} \quad |g_n - G| < \varepsilon \text{ for all } n > M.$$

Hence, when x lies in I_n and $n > M$,

$$|f(x) - G| \leq \max.\{|G_n - G|, |g_n - G|\} < \varepsilon$$

and so we have proved that, given any $\varepsilon > 0$, there exists a positive number $\delta = 1/2^M$ such that

$$|f(x) - G| < \varepsilon, \text{ whenever } 0 < |x - a| < \delta.$$

That is, $f(x) \to G$ as $x \to a$. Hence the theorem.

CHAPTER 7: LIMIT OF A FUNCTION OF A SINGLE REAL VARIABLE

Corollary 7.6.1. *We have supposed that $x \to a$ from both sides. The slight modification in the condition for convergence can easily be made when $x \to a$ from one side only.*

For limits when x → ∞

A necessary and sufficient condition that $f(x)$ may tend to a finite limit as $x \to \infty$, is that given any positive number ε, however, small, there exists a positive number X such that

$$|f(X_1) - f(X_2)| < \varepsilon, \text{ whenever } X_1 \text{ and } X_2 \text{ both exceed } X.$$

Proof: Put $y = a + 1/x$ ($x > 0$) and let $F(y) = f(x)$. The function $F(y)$ is, of course, defined only when $y > a$. As $x \to \infty$, $y \to a$. Now apply the above theorem, suitably modified to cover convergence from one side of a only, to the function $F(y)$.

Illustrative Examples using Cauchy's Principle

Example 7.6.1. *Examine whether $\sin(1/x)$ approaches any limit as $x \to 0$.*

[C.H. 1994]

Graphical approach. $\sin(1/x)$ is defined for all non-zero values of x. As $x \to 0$, the function $1/x$ increases without limit. And as $1/x$ increases indefinitely, the function itself oscillates endlessly between 1 and -1. At the points $x = \frac{2}{(4n+1)\pi}$, it will have the value $+1$ and at $x = \frac{2}{(4n-1)\pi}$, the function takes the value -1, n being an integer. Thus, no matter how close to zero we keep x, the function $\sin(1/x)$ swings backwards and forwards more and more rapidly between -1 and 1 and does not approach any definite value. Figure 7.2 suggests this behaviour.

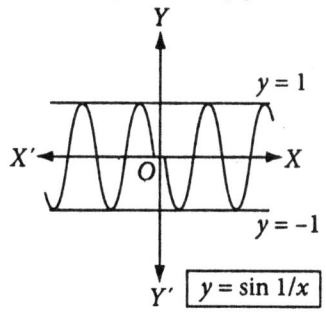

Fig 7.2

Analytic approach. (By Cauchy's principle of convergence.)

In order that the limit may exist, we have, by Cauchy's Principle to find a positive number δ corresponding to a pre-assigned positive quantity ε, however small, such that for $0 < |x_1| < \delta$ and $0 < |x_2| < \delta$

$$\left|\sin\frac{1}{x_2} - \sin\frac{1}{x_1}\right| < \varepsilon.$$

Now take $x_2 = \frac{2}{(4n+1)\pi}$ and $x_1 = \frac{1}{(4n-1)\pi}$, where n is an integer. Then for $0 < |x_1| < \delta$ and $0 < |x_2| < \delta$, where δ may be chosen by taking n sufficiently large, we have

$$\left|\sin\frac{1}{x_2} - \sin\frac{1}{x_1}\right| = \left|\sin\left(2n+\frac{1}{2}\right)\pi - \sin\left(2n-\frac{1}{2}\right)\pi\right| = 2 \not< \text{any } \varepsilon.$$

Hence $\lim_{x \to 0} \sin(1/x)$ does not exist.

Example 7.6.2. *Does* $\lim_{x \to 0} \dfrac{1}{1+e^{\frac{1}{x}}}$ *exist?*

Solution: Choose x_1 and x_2 two values of x very close to 0. Then writing $x_1 = -1/n$ and $x_2 = 1/n$ where n is arbitrarily large positive integer we can make x_1 and x_2 as close to 0 as we desire.

But then

$$|f(x_2) - f(x_1)| = \left|\frac{1}{1+e^n} - \frac{1}{1+e^{-n}}\right| = \frac{e^n - e^{-n}}{(1+e^n)(1+e^{-n})} = \frac{1-e^{-2n}}{(1+e^{-n})(1+e^{-n})}$$

which tends to 1 as $n \to \infty$. Therefore, $|f(x_2) - f(x_1)|$ cannot be made less than any chosen ε, however small. Hence the limit does not exist.

Note 7.6.1. Here $f(0+0) = 0$ and $f(0-0) = 1$ which also implies the non-existence of limit.

Theorem 7.6.2. *If* $\lim_{x \to a} f(x) = l$, *then* $\lim_{x \to a} |f(x)| = |l|$.

Proof: Since $\lim_{x \to a} f(x) = l$, given any $\varepsilon > 0$, $\exists\, \delta > 0$ s.t.

$$\forall x \text{ in } 0 < |x-a| < \delta, \text{ we have } |f(x) - l| < \varepsilon.$$

Now $\left||f(x)| - |l|\right| \leq \left|f(x) - l\right|$, $\forall x$ and hence $< \varepsilon$, $\forall x$ in $0 < |x-a| < \delta$.

This proves that $\lim_{x \to a} |f(x)| = |l|$.

Note 7.6.2. *The converse, however, is not necessarily true except when the limit is 0.*

7.7 Some Important Limits

(A) $\lim_{x \to 0} \dfrac{\sin x}{x} = 1$.

Let $0 < x < \frac{1}{2}\pi$, then it is evident from Fig. 7.3 that

CHAPTER 7: LIMIT OF A FUNCTION OF A SINGLE REAL VARIABLE

area of $\triangle OAB$ < area of sector OAB < area of $\triangle OAT$

i.e., $\quad \dfrac{1}{2} r \cdot r \cdot \sin x < \dfrac{1}{2} r^2 x < \dfrac{1}{2} r \cdot r \cdot \tan x$

i.e., $\quad \sin x < x < \tan x. \quad \left(\because \dfrac{1}{2} r^2 > 0 \right)$

or, $\quad 1 < \dfrac{x}{\sin x} < \dfrac{1}{\cos x}. \quad (\because \sin x > 0).$

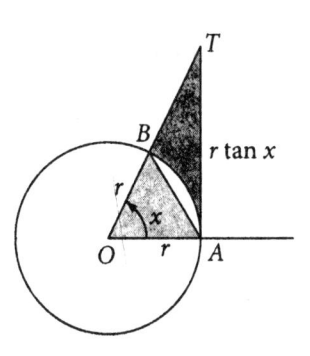

Fig 7.3

Let now $x \to 0+0$, then the first member remains 1 and $\cos x \to 1$ and accordingly $\dfrac{1}{\cos x} \to 1$. Hence by the *Sandwich Theorem*, $\lim\limits_{x \to 0+0} \dfrac{x}{\sin x} = 1$ and, therefore, $\lim\limits_{x \to 0+0} \dfrac{\sin x}{x} = 1.$

Also, $\lim\limits_{x \to 0-0} \dfrac{\sin x}{x} = \lim\limits_{y \to 0+0} \dfrac{\sin(-y)}{-y} = \lim\limits_{y \to 0+0} \dfrac{\sin y}{y} = 1.$

$\therefore \lim\limits_{x \to 0} \dfrac{\sin x}{x} = 1.$

Observation: In $\sin x$, x is taken in radians but in the denominator, x is a pure number so that the ratio $\sin x / x$ is a pure number.

Example 7.7.1. $\lim\limits_{x \to 0} \dfrac{\sin x (x \text{ in degrees})}{x} = \lim\limits_{x \to 0} \dfrac{\sin \frac{\pi x}{180}}{x} \quad \left(\because x° = \dfrac{\pi x}{180} \text{ radian} \right)$

$= \lim\limits_{x \to 0} \dfrac{\sin \frac{\pi x}{180}}{\frac{\pi x}{180}} \times \dfrac{\pi}{180} = \lim\limits_{\theta \to 0} \dfrac{\sin \theta}{\theta} \times \dfrac{\pi}{180} \quad \left(\text{writing } \theta = \dfrac{\pi x}{180} \right)$

$= 1 \times \dfrac{\pi}{180} = \dfrac{\pi}{180}.$

Example 7.7.2. $\lim\limits_{h \to 0} \dfrac{1 - \cos h}{h} = \lim\limits_{h \to 0} \dfrac{2 \sin^2 h/2}{h} = \lim\limits_{h \to 0} \dfrac{\sin h/2}{h/2} \times \dfrac{\sin h/2}{h/2} \times \dfrac{h}{2}$

$= \lim\limits_{\theta \to 0} \dfrac{\sin \theta}{\theta} \times \lim\limits_{\theta \to 0} \dfrac{\sin \theta}{\theta} \times \lim\limits_{h \to 0} h/2 \quad \left(\text{writing } \theta = \dfrac{h}{2} \right).$

$= 1 \times 1 \times 0 = 0.$

Example 7.7.3. $\lim\limits_{x \to 0} \dfrac{1 - \cos x}{x^2} = \lim\limits_{x \to 0} \dfrac{2 \sin^2 x/2}{x^2} = \lim\limits_{x \to 0} \dfrac{\sin x/2}{x/2} \times \dfrac{\sin x/2}{x/2} \times \dfrac{1}{2}$

$= \lim\limits_{\theta \to 0} \dfrac{\sin \theta}{\theta} \times \lim\limits_{\theta \to 0} \dfrac{\sin \theta}{\theta} \times \dfrac{1}{2} \quad (\text{writing } \theta = x/2).$

$= 1 \times 1 \times \dfrac{1}{2} = \dfrac{1}{2}.$

Example 7.7.4. $\lim\limits_{x \to 0} \dfrac{x^2 \sin(1/x)}{\sin x} = \lim\limits_{x \to 0} x \sin(1/x) \times \dfrac{x}{\sin x}$

$= \lim\limits_{x \to 0} x \sin(1/x) \times \lim\limits_{x \to 0} \dfrac{x}{\sin x} = 0 \times 1 = 0.$

$(\because \lim\limits_{x \to 0} x \sin(1/x) = 0,$ see **Ex. 7.3.10**).

Example 7.7.5. $\lim\limits_{x \to \infty} \dfrac{\sin x}{x + \cos x} = 0.$

$\lim\limits_{x \to \infty} \dfrac{\sin x}{x + \cos x} = \lim\limits_{y \to 0^+} \dfrac{\sin 1/y}{1/y + \cos 1/y} = \lim\limits_{y \to 0^+} \dfrac{y \sin 1/y}{1 + y \cos 1/y} = \dfrac{0}{1+0} = 0.$

Example 7.7.6. $\lim\limits_{x \to 0} \dfrac{\tan x}{x} = 1$ $\left(\text{since } \lim\limits_{x \to 0} \dfrac{\tan x}{x} = \lim\limits_{x \to 0} \dfrac{\sin x}{x} \times \lim\limits_{x \to 0} \dfrac{1}{\cos x} \right).$

(B) $\lim\limits_{x \to \pm\infty} \left(1 + \dfrac{1}{x}\right)^x = e.$

First, let $x \to \infty$ through positive real numbers only. For any $x > 0$, $\exists$ a positive integer n such that

$$n \le x < n+1 \quad \text{or,} \quad \dfrac{1}{n+1} < \dfrac{1}{x} \le \dfrac{1}{n}$$

or, $1 + \dfrac{1}{n+1} < 1 + \dfrac{1}{x} \le 1 + \dfrac{1}{n}.$

$\therefore \left(1 + \dfrac{1}{n+1}\right)^n < \left(1 + \dfrac{1}{x}\right)^x \le \left(1 + \dfrac{1}{n}\right)^{n+1}$

or, $\dfrac{\left(1 + \frac{1}{n+1}\right)^{n+1}}{1 + \frac{1}{n+1}} \le \left(1 + \dfrac{1}{x}\right)^x \le \left(1 + \dfrac{1}{n}\right)^n \cdot \left(1 + \dfrac{1}{n}\right).$

See that when $n \to \infty$ (through positive integers only),

$$\left(1 + \dfrac{1}{n}\right)^n \to e \quad \text{and} \quad \left(1 + \dfrac{1}{n+1}\right)^{n+1} \to e.$$

Also $\left(1 + \dfrac{1}{n+1}\right)$ or $\left(1 + \dfrac{1}{n}\right)$ tends to 1 as $n \to \infty$.

$\therefore$ Using Sandwich Theorem (remembering when $n \to \infty$, $x \to \infty$) we obtain
$\lim\limits_{x \to \infty} \left(1 + \dfrac{1}{x}\right)^x = e.$ ($x \to \infty$ through positive real numbers only.)

Next, we prove: $\lim\limits_{x \to -\infty} \left(1 + \dfrac{1}{x}\right)^x = e.$

CHAPTER 7: LIMIT OF A FUNCTION OF A SINGLE REAL VARIABLE

Here $\lim\limits_{x \to -\infty} \left(1 + \frac{1}{x}\right)^x = \lim\limits_{y \to \infty} \left(1 - \frac{1}{y}\right)^{-y}$ (putting $x = -y$)

$$= \lim_{y \to \infty} \left(\frac{y}{y-1}\right)^y = \lim_{y \to \infty} \left(1 + \frac{1}{y-1}\right)^{y-1} \cdot \left(1 + \frac{1}{y-1}\right)$$

$$= \lim_{t \to \infty} \left(1 + \frac{1}{t}\right)^t \times \lim_{t \to \infty} \left(1 + \frac{1}{t}\right) \quad \text{(putting } t = y - 1\text{)}$$

$$= e \times 1 = e.$$

$\therefore$ For all real values of x, $\lim\limits_{|x| \to \infty} \left(1 + \frac{1}{x}\right)^x = e$.

(C) $\lim\limits_{x \to 0} (1 + x)^{1/x} = e$. [put $x = 1/y$ in (B)]

(D) $\lim\limits_{x \to 0} \frac{a^x - 1}{x} = \log_e a$ $(a > 0)$.

Put $a^x - 1 = y$ so that $y \to 0$ as $x \to 0$. Also $x = \log(1 + y)/\log a$ (logarithms being taken to the base e).

$$\therefore \lim_{x \to 0} \frac{a^x - 1}{x} = \lim_{y \to 0} \frac{y}{\log_e(1 + y)/\log_e a} = \log_e a \cdot \lim_{y \to 0} \frac{1}{\log_e(1 + y)^{1/y}}.$$

$$= \log_e a \cdot \frac{1}{\log_e \left\{\lim\limits_{y \to 0} (1 + y)^{1/y}\right\}} = \frac{\log_e a}{\log_e e} = \log_e a.$$

(E) $\lim\limits_{x \to 0} \frac{e^x - 1}{x} = 1$.

Put $e^x - 1 = y$ and proceed as in (D).

(F) $\lim\limits_{x \to 0} \frac{\log(1+x)}{x} = 1$.

(G) $\lim\limits_{x \to a} \frac{x^n - a^n}{x - a} = n\, a^{n-1}$, for $a > 0$ and n, a rational number.

Case 1. When n is positive integer.

$$\frac{x^n - a^n}{x - a} = x^{n-1} + x^{n-2} a + x^{n-3} a^2 + \cdots + a^{n-1}.$$

$$\therefore \lim_{x \to a} \frac{x^n - a^n}{x - a} = a^{n-1} + a^{n-2} \cdot a + a^{n-3} \cdot a^2 + \cdots + a^{n-1} = n a^{n-1}.$$

Case 2. If n be a negative integer, put $n = -m$, where m is a positive integer.

$$\therefore \lim_{x \to a} \frac{x^n - a^n}{x - a} = \lim_{x \to a} \frac{x^{-m} - a^{-m}}{x - a} = \lim_{x \to a} \frac{a^m - x^m}{x^m a^m (x - a)}$$

$$= -\lim_{x \to a} \frac{1}{x^m a^m} \cdot \frac{x^m - a^m}{x - a} = -\frac{1}{a^m a^m} \cdot m a^{m-1} \quad \text{(by case 1)}$$

$$= -m a^{-m-1} = n a^{n-1}.$$

454 AN INTRODUCTION TO ANALYSIS: DIFFERENTIAL CALCULUS

Case 3. When n is a rational fraction; suppose $n = p/q$, where q is a positive integer but p is any integer, positive or negative.

Then $\lim_{x \to a} \dfrac{x^n - a^n}{x - a} = \lim_{x \to a} \dfrac{x^{\frac{p}{q}} - a^{\frac{p}{q}}}{x - a}$.

Next let $x^{\frac{1}{q}} = y$ and $a^{\frac{1}{q}} = b$; then as $x \to a$, $y \to b$.

$\therefore$ The limit $= \lim_{y \to b} \dfrac{y^p - b^p}{y^q - b^q} = \lim_{y \to b} \dfrac{(y^p - b^p)/(y - b)}{(y^q - b^q)/(y - b)}$

$= \dfrac{pb^{p-1}}{qb^{q-1}} = \dfrac{p}{q} b^{p-q} = \dfrac{p}{q} a^{\frac{p}{q}-1}$

$= na^{n-1}$. (by cases 1 and 2 and remembering $b = a^{\frac{1}{q}}$)

(H) $\lim_{x \to 0} \dfrac{(1+x)^n - 1}{x} = n$.

Put $(1+x)^n - 1 = y$ so that $y \to 0$ as $x \to 0$; and
$(1+x)^n = 1 + y$ or, $n \log(1+x) = \log(1+y)$..

$\therefore \lim_{x \to 0} \dfrac{(1+x)^n - 1}{x} = \lim_{x \to 0} \dfrac{y}{x} = \lim_{x \to 0} \dfrac{y}{\log(1+y)} \cdot \dfrac{\log(1+y)}{x}$

$= \lim_{x \to 0} \dfrac{y}{\log(1+y)} \cdot \dfrac{n \log(1+x)}{x}$

$= \lim_{y \to 0} \dfrac{y}{\log(1+y)} \cdot \lim_{x \to 0} n \dfrac{\log(1+x)}{x}$

$= \lim_{y \to 0} \dfrac{1}{\log(1+y)^{\frac{1}{y}}} \cdot n \cdot \lim_{x \to 0} \log(1+x)^{\frac{1}{x}}$

$= \dfrac{1}{\log e} \cdot n \cdot \log e = n$.

(I) $\lim_{x \to a} \dfrac{x^n - a^n}{x - a} = na^{n-1}$, for $a > 0$. (n, any real number)

Put $x = a(1 + y)$ so that $y \to 0$ as $x \to a$.

$\therefore \lim_{x \to a} \dfrac{x^n - a^n}{x - a} = \lim_{y \to 0} a^{n-1} \dfrac{(1+y)^n - 1}{(1+y) - 1} = na^{n-1}$. [using (H)]

A Few Applications

Example 7.7.7. $\lim_{x \to 0} (1 + 2x)^{\frac{x+3}{x}} = \lim_{x \to 0} (1 + 2x) \cdot (1 + 2x)^{\frac{6}{2x}}$

$= \lim_{x \to 0} (1 + 2x) \cdot \lim_{y \to 0} (1 + y)^{6/y}$ (putting $2x = y$)

$= 1 \cdot \left\{ \lim_{y \to 0} (1 + y)^{1/y} \right\}^6 = e^6$.

CHAPTER 7: LIMIT OF A FUNCTION OF A SINGLE REAL VARIABLE

Example 7.7.8. $\lim\limits_{x \to 0} \dfrac{x^{9/2} - a^{9/2}}{x^{3/2} - a^{3/2}}$ $(a > 0)$

$= \lim\limits_{x \to a} \dfrac{x^{9/2} - a^{9/2}}{x - a} \Big/ \dfrac{x^{3/2} - a^{3/2}}{x - a}$ ($\because$ when $x \to a$ we may ignore the value $x = a$)

$= \dfrac{\frac{9}{2} a^{7/2}}{\frac{3}{2} a^{1/2}} = 3a^3.$

7.8 Limit of a Function defined in the Language of Sequence

Let f be defined in some neighbourhood of a point a.

Definition 7.8.1. $\lim\limits_{x \to a} f(x) = l \Rightarrow$ Given any $\varepsilon > 0$, $\exists\, \delta > 0$ such that

$|f(x) - l| < \varepsilon$, for all $x \in 0 < |x - a| < \delta$.

Definition 7.8.2. Let $D \subseteq \mathbb{R}$ and $f : D \to \mathbb{R}$ be a function. Let 'a' be a limit point of D and l be a real number. Then $\lim\limits_{x \to a} f(x) = l$ iff for every sequence $\{x_n\}$ in $D - \{a\}$ and in some neighbourhood of a which converges to a, the sequence $\{f(x_n)\}$ converges to l.

We first prove that the two definitions are equivalent.

(a) DEFINITION **7.8.1** $\Longrightarrow$ DEFINITION **7.8.2**.

Let f have a limit l as $x \to a$ in the sense of Definition **7.8.1**.

i.e., for any $\varepsilon > 0$ $\exists\, \delta > 0$ such that $|f(x) - l| < \varepsilon$ $\forall\, x \in 0 < |x - a| < \delta$.

Let there be a sequences $\{x_n\}$ ($x_n \neq a$ for any n) converging to a so that corresponding to δ given above $\exists$ a positive integer n_0 such that $|x_n - a| < \delta$ for all $n \geq n_0$.

$\therefore\quad |f(x_n) - l| < \varepsilon \quad \forall\, n \geq n_0$

i.e., $\quad f(x_n) \to l$ as $n \to \infty.$

Thus for any sequence $\{x_n\}$ converging to a, the sequence $\{f(x_n)\} \to l$ converges to l

i.e., Definition **7.8.2** follows.

(b) **Definition 7.8.2** $\Longrightarrow$ **Definition 7.8.1**.

Let f have a limit l as $x \to a$ in the sense of Definition **7.8.2**. i.e., $f(x_n) \to l$ for every sequence $\{x_n\}$ converging to a as $n \to \infty$.

If possible, let f have no limit in the sense of Definition **7.8.1**.

Then $\exists$ an $\varepsilon > 0$ (say ε_0), for which no suitable δ can be found to satisfy the definition of limit in the sense of Definition **7.8.1**. Thus whatever positive integer n may be, $1/n$ is not a suitable δ corresponding to $\varepsilon = \varepsilon_0$. This means that $\exists$ at least one x in $0 < |x - a| < \frac{1}{n}$ for which it is NOT TRUE that $|f(x) - l| < \varepsilon_0$. Let x_n be such an x.

Then $0 < |x_n - a| < \frac{1}{n}$, and $|f(x_n) - l| \not< \varepsilon_0$ i.e., $|f(x_n) - l| \geq \varepsilon_0$. The first inequality shows $x_n \to a$ when $n \to \infty$ and the second inequality shows $f(x_n) \not\to l$ as $n \to \infty$. This contradicts the second definition of limit which we have assumed to be true. Thus our supposition that Definition **7.8.1** is not true was wrong. i.e., Definition **7.8.1** must be true if Definition **7.8.2** is true.

Example 7.8.1. Let

$$f(x) = \begin{cases} 1, & \text{if } x \text{ is rational} \\ 0, & \text{if } x \text{ is irrational.} \end{cases}$$

Prove that $\lim_{x \to a} f(x)$ does not exist for any real number a.

Solution: Let $\{x_n\}$ be a sequence of rational numbers such that $x_n \to a$ and $x_n \neq a$, $n = 1, 2, 3, \cdots$; let $\{y_n\}$ be a sequence of irrational numbers such that $y_n \to a$ and $y_n \neq a$ for $n = 1, 2, 3, \cdots$. Clearly $f(x_n) = 1$ and $f(y_n) = 0$ for all $n = 1, 2, 3, \cdots$.

Therefore $f(x_n) \to 1$ and $f(y_n) \to 0$ as $n \to \infty$ and $\lim_{x \to a} f(x)$ does not exist. Since a was arbitrary, the limit fails to exist at every real number.

Example 7.8.2. Let

$$f(x) = \begin{cases} x, & \text{if } x \text{ is rational} \\ 0, & \text{if } x \text{ is irrational.} \end{cases}$$

Prove that $\lim_{x \to a} f(x)$ exists only if $a = 0$.

Solution: (i) Suppose $a \neq 0$. We can choose, (as in **Ex. 7.8.1.** above) sequences $\{x_n\}$ and $\{y_n\}$ each converging to $x = a$ such that x_n is rational and $\neq a$ and y_n is irrational and not equal to a for $n = 1, 2, 3, \cdots$. Then $f(x_n) = x_n \to a$ and $f(y_n) = 0 \to 0$ and since $a \neq 0$ it follows that $\lim_{x \to a} f(x)$ does not exist. a is arbitrary and so the limit fails to exist at every non-zero real number.

Now for any sequence $\{x_n\}$, $f(x_n) = x_n$ or 0 and so if $x_n \to 0$ then clearly $f(x_n) \to 0$. Therefore, by Definition **7.8.2**, $\lim_{x \to 0} f(x) = 0$.

Main Points : A Recapitulation

---

1. $x \to a \Rightarrow 0 < |x - a| < \varepsilon$, where ε is any given positive number.
   $\Rightarrow$ (i) $a - \varepsilon < x < a$ (then $x \to a^-$ or $x \to a^{-0}$)
   and (ii) $a < x < a + \varepsilon$ (then $x \to a^+$ or $x \to a^{+0}$).

2. $x \to \infty \Rightarrow x$ ultimately exceed any $G > 0$ and remains $> G$;
   $$x \to -\infty \Rightarrow -x \to \infty.$$

3. $\lim_{x \to a} f(x) = l \Rightarrow$ Given any $\varepsilon > 0$ we can find $\delta > 0$ s.t.
   $\forall\, x$ in $0 < |x - a| < \delta$, we have $|f(x) - l| < \varepsilon$.
   $\lim_{x \to a+0} f(x) = l_1 \Rightarrow$ Given any $\varepsilon > 0$, we can find $\delta_1 > 0$ s.t.
   $\forall\, x$ in $a < x < a + \delta_1$, $|f(x) - l_1| < \varepsilon$.
   Similar definition for $\lim_{x \to a-0} f(x) = l_2$.
   $\lim_{x \to a} f(x)$ exists iff $\lim_{x \to a+0} f(x)$ and $\lim_{x \to a-0} f(x)$ both exist and are equal.

4. $\lim_{x \to a} f(x) = \infty \Rightarrow$ Given any $G > 0$, $\exists\, \delta > 0$, s.t. $\forall\, x$ in $0 < |x - a| < \delta$, $f(x) > G$.
   $\lim_{x \to \infty} f(x) = l \Rightarrow$ Given any $\varepsilon > 0$, $\exists\, G > 0$, s.t. $\forall\, x > G, |f(x) - l| < \varepsilon$.
   Similar definitions with $-\infty$.

5. When does $\lim_{x \to a} f(x)$ exist finitely? We have the **Cauchy's criterion** for this purpose: $\lim_{x \to a} f(x) = $ a finite quantity iff given any $\varepsilon > 0$, $\exists\, \delta > 0$ s.t. for any two points $x_1, x_2 \in 0 < |x - a| < \delta$, we have $|f(x_1) - f(x_2)| < \varepsilon$.

6. We have several theorems on limits: Useful results are the following:

   (a) If $f(x) \to l$ and $g(x) \to m$ as $x \to a$, then $f(x) \pm g(x) \to l \pm m$, $f(x) \cdot g(x) \to lm$ and $f(x)/g(x) \to l/m$ as $x \to a$ provided $m \neq 0$ in the last case.

   (b) Let $\phi(x) \to l$ as $x \to a$ and let $f(u) \to f(l)$ when $u \to l$. Then
   $$\lim_{x \to a} f(\phi(x)) = f\{\lim_{x \to a} \phi(x)\} = f(l).$$

   (c) If $\lim_{x \to a} f(a) = l$ where $l \neq 0$, then $\exists$ a δ-nbd of a where, for every x, $f(x)$ has the same sign as l. (Neighbourhood property.)

   (d) If in a certain *nbd* of a, $f(x) \leq g(x) \leq h(x)$ and if $f(x)$ and $h(x)$ both tend to l as $x \to a$, then
   $$\lim_{x \to a} g(x) = l \quad (\text{Sandwich Theorem}).$$

(e) $f(x) \to l$ as $x \to a \Rightarrow |f(x)| \to |l|$ as $x \to a$ but the converse is not always true, except when the limit is zero.

7. We have the following important limits:

(a) $\lim_{x \to 0} \sin 1/x$ does not exist but $\lim_{x \to 0} x \sin 1/x = 0$ and $\lim_{x \to 0} x^2 \sin 1/x = 0$.

(b) $\lim_{x \to 0} \dfrac{\sin x}{x} = 1$ and $\lim_{x \to 0} \cos x = 1$; $\lim_{x \to 0} \dfrac{\tan x}{x} = 1$.

(c) $\lim_{x \to 0} \lim (1+x)^{1/x} = e$; $\lim_{x \to 0} \dfrac{\log(1+x)}{x} = 1$; $\lim_{x \to 0} \dfrac{a^x - 1}{x} = \log_e a$; $\lim_{x \to 0} \dfrac{e^x - 1}{x} = 1$.

(d) $\lim_{x \to 0} \dfrac{(1+x)^n - 1}{x} = n$; $\lim_{x \to a} \dfrac{x^n - a^n}{x - a} = na^{n-1}$

(n, any real number and $a > 0$).

Section VI: Chapter 7

Exercises VII

1. Use the definition of limit of establish

(a) $\lim_{x \to 1} \sqrt{x} = 1$;

(b) $\lim_{x \to 3} \dfrac{x^2 - 9}{x - 3} = 6$;

(c) $\lim_{x \to 1} \dfrac{x^2 + x + 1}{x^2 + 2x + 1} = \dfrac{3}{4}$;

(d) $\lim_{x \to \infty} \dfrac{x}{1 + x} = 1$;

(e) $\lim_{x \to 1} \dfrac{1}{(x-1)^2} = \infty$;

(f) $\lim_{x \to 1+0} \dfrac{x}{x - 1} = \infty$;

(g) $\lim_{x \to 3} \dfrac{1}{x} = \dfrac{1}{3}$;

(h) $\lim_{x \to \infty} \dfrac{x^2}{1 + x} = \infty$;

(i) $\lim_{x \to \infty} \dfrac{x - [x]}{x} = 0$;

(j) $\lim_{x \to \infty} \dfrac{[x] + x}{x^2} = 0$;

(k) $\lim_{x \to \infty} \dfrac{2x^2 + x + 1}{x^2 - 3x + 1} = 2$;

(l) $\lim_{x \to 0} x \cos \dfrac{1}{x} = 0$;

(m) $\lim_{x \to 0} x^2 \sin \dfrac{1}{x} = 0$.

2. Find the right-hand and left-hand limits and discuss the existence of limit as x tends to the indicated value:

(a) $f(x) = \dfrac{|x|}{x}$, $x \to 0$;

(b) $f(x) = \dfrac{x}{1 + e^{-1/x}}$ as $x \to 0$;

(c) $f(x) = x^2$, when $x > 1$; $= 2$, when $x = 1$ and $= x$ when $x < 1$, as $x \to 1$.

[Ans: (a) R.H. limit $= 1$, L.H. limit $= -1$; no. (b) Yes, 0; (c) Yes, 1.]

3. Suppose $f(x) \to l$ as $x \to \infty$. Let $\{x_n\}$ be a sequence s.t. $x_n \to \infty$ as $n \to \infty$, all x_n's belong to the domain of definition of $f(x)$. Prove that $f(x_n) \to l$ as $n \to \infty$.

4. Evaluate (using the theorems on limits) the following limits:

(a) $\lim\limits_{x \to \infty} \dfrac{2x^2 - x - 6}{3x^2 + 2x + 1}$;

(b) $\lim\limits_{x \to 3} \dfrac{x - 3}{\sqrt{x - 2} - \sqrt{4 - x}}$;

(c) $\lim\limits_{x \to 0} \dfrac{\frac{1}{x-2} - \frac{1}{x}}{\frac{1}{x-3} + \frac{1}{x}}$;

(d) $\lim\limits_{x \to \infty} (\sqrt[3]{x+1} - \sqrt[3]{x})$;

(e) $\lim\limits_{x \to 1} \dfrac{x^7 - 2x^5 + 1}{x^3 - 3x^2 + 2}$;

(f) $\lim\limits_{x \to 0} \dfrac{\sqrt{1+x} - \sqrt{1-x}}{x}$;

(g) $\lim\limits_{x \to 0} \dfrac{\sqrt{1+x} - \sqrt{1+x^2}}{\sqrt{1-x^2} - \sqrt{1-x}}$;

(h) $\lim\limits_{x \to \infty} \sqrt{x}(\sqrt{x+a} - \sqrt{x})$;

(i) $\lim\limits_{x \to 0} \dfrac{a_0 x^m + a_1 x^{m-1} + a_2 x^{m-2} + \cdots + a_{m-1} x + a_m}{b_0 x^n + b_1 x^{n-1} + b_2 x^{n-2} + \cdots + b_{n-1} x + b_n} \cdot (b_n \neq 0)$.

[Ans: (a) 2/3; (b) 1; (c) -1; (d) 0; (e) 1; (f) 1; (g) 1; (h) $\frac{1}{2}a$; (i) a_m/b_n.]

5. Use Cauchy's General Principle to prove that none of the following limits exist:

(a) $\lim\limits_{x \to 0} \cos 1/x$;

(b) $\lim\limits_{x \to \infty} \cos x$;

(c) $\lim\limits_{x \to 0} \dfrac{1}{2 + e^{1/x}}$;

(d) $\lim\limits_{x \to 0} \left(\sin \dfrac{1}{x} + x \sin \dfrac{1}{x} \right)$. [C.H. 1989]

6. Using standard results [$\lim\limits_{x \to 0}(\sin x/x) = 1$ and $\lim\limits_{x \to 0} \cos x = 1$], show that

(a) $\lim\limits_{x \to 0} \dfrac{3x}{\sin x} = 3$;

(b) $\lim\limits_{x \to 0} \dfrac{1 - \cos 2x}{4x^2} = \dfrac{1}{2}$;

(c) $\lim\limits_{h \to 0} \dfrac{\sinh -h}{h} = 0$;

(d) $\lim\limits_{x \to 0} \dfrac{1 - \sec 2x}{x^2} = -2$;

(e) $\lim\limits_{x \to 0} \dfrac{\sin^2 2x}{x \sin 3x} = \dfrac{4}{3}$;

(f) $\lim\limits_{x \to 0} \dfrac{\tan ax}{\sin bx} = \dfrac{a}{b}$;

(g) $\lim_{x \to 0} \dfrac{\cos 7x - \cos 9x}{\cos x - \cos 5x} = \dfrac{4}{3}$; (h) $\lim_{x \to \pi} \dfrac{1 + \cos x}{\pi - x} = 0$;

(i) $\lim_{x \to 0} \dfrac{x^2 \sin 1/x}{\sin x} = 0$; (j) $\lim_{x \to 0} \dfrac{\sin^{-1} x}{x} = 1$;

(k) $\lim_{x \to \infty} \dfrac{\sin x}{x + \cos x} = 0$; (l) $\lim_{x \to 0} (\operatorname{cosec} x - \cot x) = 0$;

(m) $\lim_{x \to 0} \dfrac{1 - \cos x}{\sin^2 x} = \dfrac{1}{2}$; (n) $\lim_{x \to 1} \dfrac{\sin(1-x)}{1 - x^2} = \dfrac{1}{2}$.

7. Show that [you may assume that $\lim_{x \to 1}(1+x)^{1/x} = e$]:

(a) $\lim_{x \to 0}(1 + ax)^{1/x} = e^a$; (b) $\lim_{x \to 0}(1 - x^2)^{1/x} = 1$;

(c) $\lim_{x \to a} \dfrac{x^4 - a^4}{x^3 - a^3} = \dfrac{4}{3}a$; (d) $\lim_{x \to a} \dfrac{x^m - a^m}{x^n - a^n} = \dfrac{m}{n} a^{m-n} \; (m > n)$;

(e) $\lim_{x \to 1} x^{1/(1-x)} = e$; (f) $\lim_{x \to \infty} \left(1 + \dfrac{2}{x}\right)^x = e^2$.

8. Let $f(x)$ be defined by $\lim_{n \to \infty} \dfrac{1}{1 + n \sin^2 \pi x}$.

   Prove that $f(x) = 1$ or 0 according as x is an integer or not.

9. Show that the signum function sgn x can be written as $\lim_{n \to \infty} \left(\dfrac{2}{\pi} \tan^{-1} nx\right)$ where sgn $x = 1, 0$ or -1 according as $x > 0, = 0$ or < 0.

10. If $f(x) = \lim_{n \to \infty} \dfrac{1}{1 + x^{2n}}$, show that

$$f(x) = \begin{cases} 1, & \text{if } |x| < 1; \\ \tfrac{1}{2}, & \text{if } x = 1 \\ 0, & \text{if } |x| > 1. \end{cases}$$

11. If $\phi(x) = \lim_{n \to \infty} \dfrac{x^n f(x) + g(x)}{x^n + 1}$, prove that $\phi(x) = f(x)$, if $|x| > 1; = g(x)$, if $|x| < 1; = \tfrac{1}{2}\{f(x) + g(x)\}$ if $x = 1$ but $\phi(x)$ is not defined at $x = -1$.

12. If $\phi(x) = \lim_{n \to \infty} \dfrac{f(x) + \cos^{2n} \pi x \{2g(x) - f(x)\}}{1 + \cos^{2n} \pi x}$, prove that $\phi(x) = g(x)$ or $f(x)$ according as x is an integer or not.

13. If $f(x) = \lim_{n \to \infty} \sin n\pi x$, verify that $f(x) = 0$ if x is an integer but $f(x)$ is not defined if x is not an integer.

14. Show that $\lim\limits_{x \to 0} \dfrac{\sin x}{\sin x} = 1$, but $\lim\limits_{x \to 0} \dfrac{\sin(1/x)}{\sin(1/x)}$ does not exist.

15. (a) Prove, as a consequence of the definitions on limits, that if $f(x) \to \infty$, then $1/f(x) \to 0$. But the converse is not *necessarily* true.

    [Hints: When $x \to 0$, $1/x^2 \to \infty$ but $1/x$ oscillates infinitely.]

    (b) If $f(x) \to l (\neq 0)$, then show that $f(x) \cos \pi x$ and $f(x) \sin \pi x$ oscillate finitely. If however, $f(x) \to \infty$ or $f(x) \to -\infty$, then they oscillate infinitely. Show also that the graph of either function is a wavy curve oscillating between the curves $y = f(x)$ and $y = -f(x)$.

    [Hints: $|\cos \pi x| \leq 1$ and $\sin \pi x| \leq 1$.]

16. (a) When $a_0, \cdots, a_k$ and $b_0, \cdots, b_l$ are given constants, neither a_0 nor b_0 is zero, and $x \to 0$, show that

$$\frac{a_0 x^m + a_1 x^{m+1} + \cdots + a_k x^{m+k}}{b_0 x^n + b_1 x^{n+1} + \cdots + b_l x^{n+l}}$$

$\to 0$ if $m > n$; $\to a_0/b_0$ if $m = n$; $\to +\infty$ or $-\infty$ according as $a_0/b_0 > 0$ or $a_0/b_0 < 0$ if $m < n$ and $n - m$ is even.

(b) Verify that

$$\lim_{x \to \infty} \frac{a_0 x^m + a_1 x^{m-1} + \cdots + a_m}{b_0 x^n + b_1 x^{n-1} + \cdots + b_n} \quad (a_0 \neq 0, b_0 \neq 0)$$

is 0 if $m < n$, $= \infty$ if $m > n$ and $= a_0/b_0$ if $m = n$.

17. (a) Prove that, if $a > 0$, then $\lim\limits_{x \to 0+0} \dfrac{x}{a}\left[\dfrac{b}{x}\right] = \dfrac{b}{a}$ and $\lim\limits_{x \to 0+0} \left[\dfrac{x}{a}\right]\dfrac{b}{x} = 0$, where $[x]$ is the greatest integer in x but not greater than x. Discuss the left-hand limits of these functions.

(b) Prove that if m be any integer, then

$[x] \to m$ and $x - [x] \to 0$ as $x \to m^{+0}$ and $[x] \to m-1$, $x - [x] \to 1$ as $x \to m^{-0}$.

18. Discuss about the existence of the following limits:

(a) $\lim\limits_{x \to \infty} \cos\left[1 + (-1)^{[x]} \sin \frac{1}{x^2}\right] \frac{\pi}{2}$; (b) $\lim\limits_{x \to \infty} \dfrac{\sin x}{x}$; (c) $\lim\limits_{x \to 1} \sin \frac{1}{2}\pi[x]$.

19. Show that $\lim\limits_{x \to \infty} a^x \sin \dfrac{b}{a^x} = 0$, if $0 < a < 1$; $= b$ if $a > 1$

    [When $0 < a < 1$, put $a^x = yb$, limit becomes $\lim\limits_{y \to 0} y \sin \frac{1}{y} = 0$;

    When $a > 1$, put $a^x = \frac{1}{y}$, limit becomes $\lim\limits_{y \to 0} b\frac{\sin by}{by} = b$.] [C.H. 1999]

AN INTRODUCTION TO ANALYSIS

(Differential Calculus—Part I)

Chapter 8
Continuity of Functions of a Single Real Variable

Chapter 8

Continuity of Functions of a Single Real Variable

8.1 Introduction

We now make use of the concept of limit of a function (Section VI: Chapter 7) to give a mathematical shape to our intuitive concept of continuity.

In defining the symbol $\lim_{x \to c} f(x) = l$ (Chapter 7) we did not make use of the value of f at $x = c$ (in fact, such a value may not be defined). Even if f is defined at $x = c$, the value of $f(c)$ need not always be equal to the limiting value l. When we do have $f(c) = l$ i.e., when $\lim_{x \to c} f(x) = f(c)$, we say that f is *continuous at* $x = c$. In the present chapter we shall study some deeper properties of continuous functions defined over a certain interval I (open or closed).

In Part II of the present treatise we have discussed a more general approach, namely, continuity defined on a subset S of real numbers not necessarily over an interval.

8.2 Continuity at a point and on an interval I

Continuity at a point

Let f be defined on an interval $I \subseteq \mathbb{R}$.

Suppose c is an interior point of I. The continuity of f at $x = c$ may be defined in the following way:

Definition 8.2.1. *f is said to be continuous at $x = c$ if for any arbitrary positive number ε (i.e., $\varepsilon > 0$), no matter how small, $\exists$ a positive number $\delta (\delta > 0)$ such that* for all points $x \in |x - c| < \delta \cap I$ we have

$$|f(x) - f(c)| < \varepsilon.$$

It means that if f is continuous at c then $\lim_{x \to c} f(x)$ exists and is equal to $f(c)$.

This means that f is continuous at $x = c$ if the functional values $f(x)$ are close to $f(c)$ when x is close to c. Thus for continuity of f at $x = c$, f must be defined at $x = c$ and also in some neighbourhood around c; otherwise we can not find $\lim_{x \to c} f(x)$ which must exist and equals to $f(c)$. Hence for continuity of f at a point $x = c$, we must have

1. f is defined at c i.e., $f(c)$ exists.

2. $\lim_{x \to c} f(x)$ exists i.e., $\lim_{x \to c^+} f(x) = \lim_{x \to c^-} f(x)$, or, $f(c+0) = f(c-0)$.

3. $f(c)$ and $\lim_{x \to c} f(x)$ must be equal.

If any of these conditions fail, then f is NOT continuous at c or f is **discontinuous** at $x = c$.

Sometimes a function f may be continuous on one side of c. We call such a continuity as **one-sided continuity**. More explicitly,

Definition 8.2.2. 1. f is said to be *right-continuous (or continuous from the right)* at a point $x = a$, if $\lim_{x \to a+0} f(x) = f(a)$, i.e., given any $\varepsilon > 0$, $\exists \delta > 0$ such that

$$\forall x \in (a, a+\delta) \text{ we have } |f(x) - f(a)| < \varepsilon.$$

[Here f must be defined in some right neighbourhood of a.]

2. f is said to be *left-continuous (or, continuous from the left)* at a point $x = b$, if $\lim_{x \to b^-} f(x) = f(b)$, i.e., given any $\varepsilon > 0$, $\exists \delta > 0$ such that

$$\forall x \in (b-\delta, b) \text{ we have } |f(x) - f(b)| < \varepsilon.$$

[Here f must be defined in some left-neighbourhood of b.]

We next define **continuity in an Interval**

Continuity of f in an open interval $(a, b) \Rightarrow f$ is continuous at every point c, where $a < c < b$, in the sense of Definition **8.2.1**.

Continuity of f in a closed interval $[a, b] \Rightarrow$

1. If c is an interior point of $[a, b]$, then f is continuous at $x = c$ in the sense of Definition **8.2.1**.

2. At the left end point a, f is *right-continuous*.

CHAPTER 8: CONTINUITY OF FUNCTIONS OF A SINGLE REAL VARIABLE

and 3. At the right end point b, f is *left-continuous*.

The definition of continuity at a point can also be defined in the language of sequence:

Definition 8.2.3. *f is said to be continuous at $x = c$ if every convergent sequence $\{x_n\}$ in I having the limit c, the sequence $\{f(x_n)\}$ is convergent and $\lim_{n \to \infty} f(x_n) = f(c)$.*

We first prove that the two definitions are equivalent.

(a) *To prove*: DEFINITION **8.2.1** $\Longrightarrow$ DEFINITION **8.2.3**.
If f is continuous according to Definition **8.2.1**, then for each $\varepsilon > 0$ $\exists$ a $\delta > 0$ such that
$$|f(x) - f(c)| < \varepsilon, \text{ whenever } x \in |x - c| < \delta \cap I.$$
Let an arbitrary sequence $\{x_n\} \in I$ converges to c.
Then, corresponding to this $\delta > 0$, $\exists$ a natural number n_0 such that
$$|x_n - c| < \delta, \quad \forall n \geq n_0.$$
But $x_n \in |x_n - c| < \delta \Rightarrow |f(x_n) - f(c)| < \varepsilon, \forall n \geq n_0$.
This last line proves that $\{f(x_n)\}$ converges to $f(c)$.

(b) *To prove*: DEFINITION **8.2.3** $\Longrightarrow$ DEFINITION **8.2.1**.
We shall use here indirect method. Suppose $\exists$ a point $c \in I$ such that f is continuous according to Definition **8.2.3**, but not according to Definition **8.2.1**. This means that f is continuous at $x = c$ by Definition **8.2.3** but $\exists$ an $\varepsilon_0 > 0$ such that for *every* $\delta > 0$, $|f(x) - f(c)| \geq \varepsilon_0$, for at least one point x in $|x - c| < \delta \cap I$.
Let $\delta_0 = \frac{1}{2}$. We can find $x_1 \in |x - c| < \delta_0 \cap I$ such that
$$|f(x_1) - f(c)| \geq \varepsilon_0.$$
Notice $x_1 \neq c$. Let $\delta_1 = \min.\left\{\frac{1}{2}, |x_1 - c|\right\}$.
We can find $x_2 \in \delta_1$-*nbd* of c, such that $|f(x_2) - f(c)| \geq \varepsilon_0$.
Again $x_2 \neq c$ and $x_2 \neq x_1$ for x_1 does not lie in the δ_1-*nbd* of c. It must be observed that $\delta_1 \leq \frac{1}{2}\delta_0$.
Proceeding in this manner, we can construct a sequence $\{x_n\}$ such that $x_n \in \delta_n$-*nbd* of c and $\delta_n \leq \frac{1}{2}\delta_{n-1}$, and $|f(x_n) - f(c)| \geq \varepsilon$ for every n.
It is easy to see that $\{x_n\}$ converges to c but $\{f(x_n)\}$ does not converge to $f(c)$, which contradicts our hypothesis (i.e., Definition **8.2.3**). Hence we must conclude that whenever Definition **8.2.3** holds Definition **8.2.1** cannot but be true.

Discontinuity Criterion of f at x = c by Definition 8.2.3:

f is *discontinuous* at c iff there exists a sequence $\{x_n\} \in I$ such that $\{x_n\}$ converges to c but the corresponding sequence $\{f(x_n)\}$ does not converge to $f(c)$.

Note 8.2.1. *Let $A \subseteq B \subseteq \mathbb{R}$, let $f : B \to \mathbb{R}$ and let g be the restriction of f to A (i.e., $g(x) = f(x) \ \forall \ x \in A$). Then it is easy to see that*

(a) *If f is continuous at $c \in A$, then g is also continuous at c.*

(b) *The following example shows that if g is continuous at c, it need not follow that f is continuous at c.*

Example 8.2.1. *Let $f(x) = 1, \ \forall \ x \geq 0$, $f(x) = -1$ for $x < 0$, $A = [0,1]$ and $B = [-1, 1]$.*

Graphical Considerations

(i) $f(x)$ is continuous at $x_0 \Rightarrow$ There is no break in the graph of $y = f(x)$ at the point where the abscissa is x_0 and in its immediate neighbourhood.

(ii) $f(x)$ is continuous in $[a,b] \Rightarrow$ The graph of $y = f(x)$ is **unbroken** from the point where the abscissa is a to the point where the abscissa is b.

The term *unbroken* used above has the following significance:

Small changes in the values of x cause small changes in the ordinates of the graph. i.e., when $x - a$ is small, $f(x) - f(a)$ is also small.

Examples:

Example 8.2.2. *Show that the function f defined by $f(x) = 5x + 3$ is continuous at $x = 1$.*

Solution: Here the domain D of definition of the function f is $\mathbb{R}$. f is continuous at $x = 1$ if $\lim_{x \to 1} f(x) = f(1)$ i.e., if $\lim_{x \to 1}(5x + 3) = 5.1 + 3 = 8$.

This statement is true if, for **any** $\varepsilon > 0$, we can find $\delta > 0$ such that $\forall \ x \in \mathbb{R} \cap |x - 1| < \delta$, we have $|5x + 3 - 8| < \varepsilon$.

But $|5x + 3 - 8| = 5|x - 1|$. Hence $|5x + 3 - 8| < \varepsilon$ whenever $|x - 1| < \varepsilon/5$.

Taking $\delta = \varepsilon/5$, our requirement for the definition of limit is met.

$\therefore \ f$ is continuous at $x = 1$.

Example 8.2.3. *Show that $f(x) = \sin x$ is continuous for all values of x. What about the continuity of $\cos x$ for all real values of x?*

Solution: Here also the domain of $f(x) = \sin x$ is $\mathbb{R}$, the set of all real numbers.

Take any value a of x, where $a \in \mathbb{R}$.

$\sin x$ will be continuous for $x = a$ if for any $\varepsilon > 0$, we can find a value of $\delta (> 0)$ such that
$$|\sin x - \sin a| < \varepsilon, \text{ whenever } |x - a| < \delta.$$
Now, $|\sin x - \sin a| = \left|2 \sin \frac{x-a}{2} \cos \frac{x+a}{2}\right| = 2 \left|\sin \frac{x-a}{2}\right| \left|\cos \frac{x+a}{2}\right|$.

Observe that $\left|\cos \frac{x+a}{2}\right| \leq 1$ for every value of $x \in \mathbb{R}$.

and $\left|\sin \frac{x-a}{2}\right| < \left|\frac{x-a}{2}\right|$, for $0 < \left|\frac{x-a}{2}\right| < \frac{\pi}{2}$

$(\because |\sin x| < |x| < |\tan x|$ for $0 < x < \pi/2.)$

$\therefore \quad |\sin x - \sin a| = 2 \left|\sin \frac{x-a}{2}\right| \left|\cos \frac{x+a}{2}\right| < 2 \left|\frac{x-a}{2}\right| = |x-a|.$

$\therefore \quad |\sin x - \sin a| < \varepsilon$ for all x for which $|x - a| < \delta$, where $\delta = \varepsilon$.

i.e., $\sin x$ is continuous at any arbitrary point $a \in \mathbb{R}$. Thus $\sin x$ is continuous for all real values of x.

We can similarly prove the continuity of $\cos x$ for all real values of x.

Example 8.2.4. *Show that $f(x)$ is continuous at $x = 0$ and $x = 1$ where*

$$f(x) = \begin{cases} -x, & x \leq 0 \\ x, & 0 < x < 1 \\ 2 - x, & x \geq 1. \end{cases}$$

Solution: (i) when $x = 0$, $f(0) = 0$, $f(0 - 0) = \lim_{x \to 0-0} f(x) = \lim_{x \to 0-0} (-x) = 0$.

and $f(0 + 0) = \lim_{x \to 0+0} f(x) = \lim_{x \to 0+0} x = 0$.

$\therefore \quad f(0 + 0) = f(0 - 0) = 0$.

Hence f is continuous at $x = 0$.

(ii) when $x = 1$, $f(1) = 1$. Now see that

$$f(1 + 0) = \lim_{x \to 1+} f(x) = \lim_{x \to 1+} (2 - x) = \lim_{h \to 0+} (2 - \overline{1 + h}) = 1.$$

$$f(1 - 0) = \lim_{x \to 1-} f(x) = \lim_{x \to 1-} x = \lim_{h \to 0-} (1 + h) = 1.$$

$\therefore \quad f(1 + 0) = f(1 - 0) = 1$ and hence f is continuous at $x = 1$.

Example 8.2.5. *Investigate the continuity of the function:*

$$f(x) = \begin{cases} |x|/x, & \text{when } x \neq 0 \\ -1, & \text{when } x = 0. \end{cases}$$

Solution: The function has a left-hand continuity at $x = 0$; since
$$f(0-0) = \lim_{x \to 0-0} f(x) = \lim_{x \to 0-} -x/x = -1 = f(0).$$
But see that $f(0+0) = \lim_{x \to 0+0} f(x) = \lim_{x \to 0+0} x/x = +1 \neq f(0).$

$\therefore$ $f(0+0) \neq f(0-0)$ and hence f is **not continuous** at $x = 0$.

Example 8.2.6.
$$f(x) = \begin{cases} \frac{\sin x}{x}, & \text{when } x \neq 0 \\ 1, & \text{when } x = 0. \end{cases}$$

This function is continuous at $x = 0$. $\left[\because \lim_{x \to 0} \frac{\sin x}{x} = 1 = f(0). \right]$

Example 8.2.7. *Prove that the following function $f(x)$ is continuous at $x = 0$,*
$$f(x) = \begin{cases} x \sin \frac{1}{x}, & \text{when } x \neq 0 \\ 0, & \text{when } x = 0. \end{cases}$$

We can easily prove $\lim_{x \to 0} x \sin(1/x) = 0 = f(0)$ and hence f is continuous at $x = 0$. The validity of the existence of $\lim_{x \to 0} x \sin(1/x) = 0$ has been shown in Chapter 7 (see worked Ex. **7.3.10**).

Example 8.2.8. *Prove that x^2 is continuous in the closed interval $[-a, a]$ $(a > 0)$.*

Solution: x^2 will be continuous in $[-a, a]$ if we can prove the following three statements:

(i) x^2 is continuous in $-a < x < a$.

(ii) x^2 is left-continuous at a.

(iii) x^2 is right-continuous at $-a$.

(i) **First**, let c be any point in $0 \leq x < a$; then $c \geq 0$. To prove x^2 is continuous at c is the same as to prove
$$\lim_{x \to c} x^2 = c^2.$$

This limit-statement is TRUE if, for any $\varepsilon > 0$, we can find $\delta > 0$ such that $|x^2 - c^2| < \varepsilon$ for all $x \in |x - c| < \delta$.

Now $|x^2 - c^2| < \varepsilon \Rightarrow c^2 - \varepsilon < x^2 < c^2 + \varepsilon$ and hence clearly,
$$0 \leq x^2 < c^2 + \varepsilon$$
$$\text{or,} \quad -\sqrt{c^2 + \varepsilon} < x < \sqrt{c^2 + \varepsilon}$$
$$\text{or,} \quad -(c + \sqrt{c^2 + \varepsilon}) < x - c < \sqrt{c^2 + \varepsilon} - c.$$

Since $\sqrt{c^2+\varepsilon}-c < \sqrt{c^2+\varepsilon}+c$, we choose $\delta =$ the lesser of the two values $c+\sqrt{c^2+\varepsilon}$ and $\sqrt{c^2+\varepsilon}-c = \sqrt{c^2+\varepsilon}-c$.

Thus $|x^2-c^2|<\varepsilon$, whenever $x \in |x-c|<\delta$ where $\delta = \sqrt{c^2+\varepsilon}-c$.

The continuity of x^2 at any point $c \in 0 \leq x < a$ implies that x^2 is continuous in $0 \leq x < a$.

Next, let $(-c)$ be any point in $-a < x \leq 0;\ c \geq 0$.

To prove: x^2 is continuous at $-c$ i.e., to prove $\lim_{x \to -c} x^2 = (-c)^2 = c^2$.

i.e., for any $\varepsilon > 0$, we are to find $\delta > 0$ such that

$$|x^2 - c^2| < \varepsilon,\quad \forall\, x \in |x+c| < \delta.$$

As before, see that $|x^2-c^2|<\varepsilon$, whenever $x \in |x+c|<\delta$, where δ is the lesser of the two values $\sqrt{c^2+\varepsilon}-c$ and $\sqrt{c^2+\varepsilon}+c = \sqrt{c^2+\varepsilon}-c$.

The continuity of x^2 at any point of $-a<x\leq 0$ is established.

Thus x^2 is continuous at any point of $-a < x < a$.

(ii) x^2 is *left-continuous* at a, if $\lim_{x \to a-0} x^2 = a^2$.

i.e., if, for any $\varepsilon > 0$, $\exists\, \delta > 0$ such that $|x^2 - a^2| < \varepsilon$, $\forall\, x \in a-\delta < x \leq a$. i.e., if $|x^2-a^2|<\varepsilon$, whenever $0 \leq a-x < \delta$.

As before,

$$\begin{aligned}
|x^2-a^2|<\varepsilon &\Rightarrow 0 < x^2 < a^2+\varepsilon \\
&\Rightarrow -\sqrt{a^2+\varepsilon} < x < \sqrt{a^2+\varepsilon} \\
&\Rightarrow -\sqrt{a^2+\varepsilon} < -x < \sqrt{a^2+\varepsilon} \\
&\Rightarrow a-\sqrt{a^2+\varepsilon} < a-x < \sqrt{a^2+\varepsilon}+a. \\
&\Rightarrow 0 \leq a-x < a+\sqrt{a^2+\varepsilon} = \delta.
\end{aligned}$$

(iii) In a similar manner we can prove that x^2 is *right-continuous* at $-a$.

Note: As a is arbitrary and so x^2 is continuous in $(-\infty,\infty)$, i.e., in $\mathbb{R}$. Similarly, $x^n (n \in \mathbb{N} \cup \{0\})$ is continuous in $\mathbb{R}$. $a_0 x^n (a_0 \in \mathbb{R})$ is also continuous in $\mathbb{R}$. As sum of finite number of continuous functions is continuous, therefore the polynomial function:

$$f(x) = a_0 x^n + a_1 x^{n-1} + a_2 x^{n-2} + \cdots + a_{n-1} x + a_n \quad (a_i \in \mathbb{R})$$

is continuous throughout the real line $\mathbb{R}$.

Example 8.2.9. Let f be defined on $\mathbb{R}$ by

$$f(x) = \begin{cases} 1, & \text{if } x \text{ is rational} \\ 0, & \text{if } x \text{ is irrational}. \end{cases}$$

Prove that f is NOT *continuous at any point of* $\mathbb{R}$.

(This function was first introduced by P.G.L. Dirichlet in 1829.)

[Compare Ex. 2. Exercises VIII Section [B]]

Solution: **Case 1.** Let c be any arbitrary rational number. Let $\{x_n\}$ be a sequence of irrational numbers that converges to c.

[Density property i.e., $\exists$ an irrational number between any two real numbers, assures us that such a sequence does exist.]

Since $f(x_n) = 0$, $\forall\, n \in \mathbb{N}$, we have $\lim_{n \to \infty} f(x_n) = 0$ while $f(c) = 1$. Therefore f is not continuous at c.

Case 2. Let b be any arbitrary irrational number. Let $\{y_n\}$ be a sequence of rational numbers that converge to b.

[Density property i.e., $\exists$ a rational number between any two real numbers, assures us that such a sequence does exist.]

Since $f(y_n) = 1$, $\forall\, n \in \mathbb{N}$, we have $\lim_{n \to \infty} f(y_n) = 1$ while $f(b) = 0$. Therefore, f is not continuous at b. Since every point of $\mathbb{R}$ is either rational or irrational.

We conclude that f is not continuous at any point of $\mathbb{R}$.

Note 8.2.2. Let $f(x) = 0$, *if x is rational and* $f(x) = 1$, *if x is irrational. It can be similarly shown that f is everywhere* **discontinuous** *in* $\mathbb{R}$.

Example 8.2.10. *Let f be defined on $\mathbb{R}^+$ by*

$$f(x) = \begin{cases} 0, & \text{if } x \text{ is irrational} \\ \frac{1}{q}, & \text{if } x \text{ is rational} \end{cases}$$

Prove that f is continuous at every irrational number in $\mathbb{R}^+$ and discontinuous at every rational number in $\mathbb{R}^+$.

(This function was introduced in 1875 by K.J. Thomac.)

Solution: **Case 1.** Suppose a is any positive rational number.

Let $\{x_n\}$ be a sequence of irrational numbers in $\mathbb{R}^+$ that coverges to a. Then $\lim_{x \to \infty} f(x_n) = 0$ while $f(a) = a$ positive number of the form $1/q$ where q is a positive integer. Hence f is discontinuous at a.

Case 2. Let b be an irrational number in $\mathbb{R}^+$.

Let $\varepsilon > 0$. Then by the Archimedean property, $\exists$ a positive integer n_0 such that $n_0 \varepsilon > 1$ or $1/n_0 < \varepsilon$. There are only a finite number of rational numbers with denominator less than n_0 in the open interval $(b-1, b+1)$.

[For further explanations the students are advised to see the solution of **Ex. 3.** of Exercises VIII Section [B].]

Hence $\delta > 0$ can be chosen so small that the neighbourhood $(b - \delta, b + \delta)$ contains no rational numbers with denominators less than n_0.

It then follows that for $|x - b| < \delta$ and $x \in \mathbb{R}^+$ we have
$$|f(x) - f(b)| = |f(x)| \leq \tfrac{1}{n_0} < \varepsilon.$$

Thus f is continuous at the irrational number b.

Consequently Thomae's function f is continuous precisely at the irrational points in $\mathbb{R}^+$.

Example 8.2.11. *Let $f(x) = x$, if x is rational but $f(x) = 1 - x$, if x is irrational. Prove that f is continuous at $x = 1/2$ but not at any other point.*

Solution: At $x = \tfrac{1}{2}$, $f\left(\tfrac{1}{2}\right) = \tfrac{1}{2}$.

$$\left|f(x) - f\left(\tfrac{1}{2}\right)\right| = \left|x - \tfrac{1}{2}\right|, \text{ if } x \text{ is rational}$$
$$= \left|(1-x) - \left(1 - \tfrac{1}{2}\right)\right|, \text{ if } x \text{ is irrational}$$
$$\text{i.e., } = \left|x - \tfrac{1}{2}\right|, \text{ even if } x \text{ is irrational}.$$

$\therefore$ Given any $\varepsilon > 0$, we can make $\left|f(x) - f\left(\tfrac{1}{2}\right)\right| < \varepsilon \; \forall \, x$ (rational or irrational) $\in \left|x - \tfrac{1}{2}\right| < \delta$ where $\delta = \varepsilon$, i.e., f is continuous at $x = \tfrac{1}{2}$.

At other points f is clearly discontinuous (why?).

Observation: The previous examples give answers to some intriguing questions: Can a function fail to be continuous everywhere? If a function is continuous at a point, does it mean that it should be continuous in some neighbourhood of that point or is it possible that a function may be continuous at merely one point? In the next article we shall see that how discontinuities can arise in various ways.

8.3 Piecewise Continuous Functions: Discontinuous Functions

Piecewise Continuous Functions

It is sometimes useful to consider functions which, while not continuous in a whole interval, are nevertheless made up of a finite number of continuous pieces. They are called *piecewise continuous functions*. The formal definition is as follows:

Definition 8.3.1. *We say that f, defined in $[a, b]$, is piecewise continuous if*

(i) $\lim_{x \to a+0} f(x)$ *and* $\lim_{x \to b-0} f(x)$ *both exist;*

(ii) *f is continuous at all but a finite number of points in (a, b);*

(iii) *at all points of (a, b) both left and right-hand limits exist.*

Discontinuities

A point at which a function f is not continuous is called a *point of discontinuity* of f.

1. A point x_0 will be obviously a point of discontinuity of f if f is not defined at x_0.

2. Suppose that f is defined at x_0, then also x_0 can be a point of discontinuity under any one of the following conditions:

(i) If either $f(x_0 + 0)$ or $f(x_0 - 0)$ does not exist. (**Discontinuity of the Second kind.**)

(ii) If both $f(x_0 + 0)$ and $f(x_0 - 0)$ exist but have different values (**Non-removable Discontinuity of the First kind**).

(iii) If $f(x_0 + 0)$ and $f(x_0 - 0)$ both exist and are equal but this equal value is not equal to $f(x_0)$. (**Removable Discontinuity of First kind**)

3. If either $f(x_0 + 0)$ or $f(x_0 - 0)$ or both are infinite ($-\infty$ or $+\infty$), then f is said to have an **infinite discontinuity** at x_0.

4. Any point of discontinuity *which is not* a point of discontinuity of first kind (removable or not) **nor** a point of discontinuity of second kind, **nor also** a point of infinite discontinuity is called a point of **Oscillatory Discontinuity**.

5. **Jump Discontinuity**

Let f be a real-valued function defined on a closed interval $[a, b]$.

(i) Let c be *an interior point* of $[a, b]$.

If $f(c+0) = \lim_{x \to c+0} f(x)$ and $f(c-0) = \lim_{x \to c-0} f(x)$ both exist and are finite and **at least two** of the three numbers $f(c), f(c-0)$ and $f(c+0)$ are different, then f is said to have a **jump discontinuity** at c. Moreover, $f(c+0) - f(c-0)$ is called the *height of the jump* at c. (also called *Saltus of $f(x)$ at $x = c$*.)

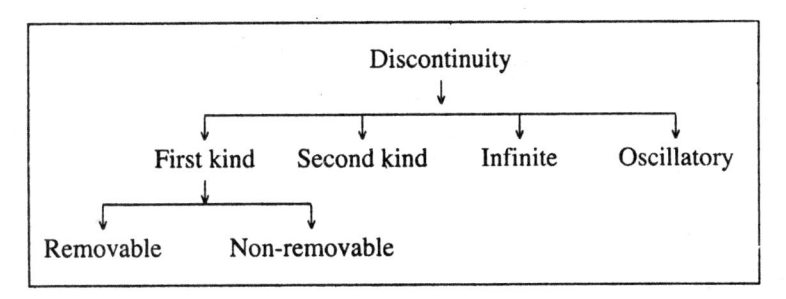

```
                        Discontinuity
                             ↓
         ↓              ↓              ↓              ↓
    First kind     Second kind      Infinite      Oscillatory
         ↓
    ↓         ↓
Removable  Non-removable
```

e.g., $f(x) = |x|/x$, for $x \neq 0$ and $f(0) = A$ (whatever A may be). For this function $f(0-0) = -1$ $f(0) = A$; $f(0+0) = 1$. Thus f has a jump discontinuity at $x = 0$, the height of the jump being 2.

(ii) *At the point $x = a$*: If $f(a+0)$ exists finitely and is different from $f(a)$, then we say that f has a right-hand jump at a, the height of the jump being $f(a+0) - f(a)$.

(iii) *At the point $x = b$*: If $f(b-0)$ exists finitely and is different from $f(b)$ then we say that f has a left-hand jump at b, the height in this being $f(b) - f(b-0)$.

An Important Example of Jump discontinuity is the following:

Example 8.3.1. $f(x) = [x]$, *where* $[x] =$ *integral part of x but not greater than x, has a jump-discontinuity at each integral value of x, the height being* 1.

Solution: Let $n > 0$ be a positive integer. We know

$$\left.\begin{array}{l} \text{for } n-1 \leq x < n, \quad f(x) = n-1 \\ \text{for } n \leq x < n+1, \quad f(x) = n. \end{array}\right\}$$

Hence $f(n+0) = n$ and $f(n-0) = n-1$ and $f(n) = n$.

∴ f has a jump discontinuity at $x = n$, the height of the jump being

$$f(n+0) - f(n-0) = n - (n-1) = 1.$$

Now take n, a negative integer or zero. Arguing in a similar manner we may arrive at the conclusion: for every integer n, f has a jump discontinuity of height 1.

Another example of Jump discontinuity:

$f(x) = x - [x]$. It can be shown that

$$f(1+0) = \lim_{x \to 1+0}(x - [x]) = \lim_{h \to 0+}(1+h) - 1 = 0;$$

$$f(1-0) = \lim_{x \to 1-0}(x - [x]) = \lim_{h \to 0-}(1+h) - 0 = 1 \text{ and also } f(1) = 0.$$

Hence f has a jump discontinuity at $x = 1$ of height 1. [C.H. 1981]

We now give below a few illustrations of various types of Discontinuities:

Discontinuity of FIRST and SECOND kind:

Example 8.3.2. (*Removable*): *Investigate the continuity of the function:*

$$f(x) = x^2/x;$$

For $x \neq 0$, $f(x) = x$ but $f(0)$ is not defined.

Solution: Here $f(0+0) = 0$ and $f(0-0) = 0$ i.e., $f(0+0), f(0-0)$ both exist and are equal; but $f(0)$ is not defined. Hence f has a discontinuity of first kind of Removable type (*Removable*, because if we would have defined $f(0) = 0$, then f would be continuous at $x = 0$ i.e., the discontinuity would be removed).

Example 8.3.3. (*Removable*) *Investigate the continuity of the function:*

$$f(x) = \begin{cases} \dfrac{1-\cos x}{x^2}, & \text{for } x \neq 0 \\ 1, & \text{for } x = 0 \end{cases}$$

at $x = 0$.

Solution: Here $f(0+0) = f(0-0) = 1/2$

In fact, $\lim\limits_{x \to 0} \dfrac{1-\cos x}{x^2} = \lim\limits_{x \to 0} \dfrac{2\sin^2 x/2}{x^2} = \lim\limits_{x/2 \to 0} \dfrac{\sin x/2}{x/2} \cdot \dfrac{\sin x/2}{x/2} \cdot \dfrac{1}{2} = 1 \cdot 1 \cdot \dfrac{1}{2} = \dfrac{1}{2}$

but $f(0) = 1$.

$\therefore$ at $x = 0$, f has a discontinuity of first kind (removable type).

The discontinuity will be removed if we define $f(0) = 1/2$.

Example 8.3.4. $f(x) = [x]$ has a discontinuity of first kind at $x = 0$ **but not removable.** $f(0+0) = 0$, $f(0-0) = -1$, $f(0) = 0$. Here $f(0+0) \neq f(0-0)$. So in whatever we may define $f(0)$ the discontinuity cannot be removed.

Example 8.3.5. $f(x) = \dfrac{x-3}{\sqrt{x-3}}$ *has a discontinuity of second kind at $x = 3$.*

Solution: Here $f(3+0) = 0$; $f(3)$ is not defined and $f(3-0)$ does not exist. Since $f(3-0)$ does not exist, we conclude that the given function has a discontinuity at $x = 3$. This is a discontinuity of second kind.

Infinite Discontinuity

Example 8.3.6. $f(x) = 1/x^2$ has an infinite discontinuity at $x = 0$.

See that here $f(0+0) = f(0-0) = \infty$.

Oscillatory Discontinuity

Example 8.3.7.
$$f(x) = \begin{cases} \sin(1/x), & \text{for } x \neq 0 \\ 0, & \text{for } x = 0 \end{cases}$$
has an oscillatory discontinuity at $x = 0$.

$\sin 1/x$ assumes all values between -1 and $+1$ infinitely often i.e., it oscillates infinitely often between -1 and $+1$ in any *nbd* of $x = 0$, no matter however small that *nbd* may be.

Example 8.3.8. *Give an example of a function f and sketch its graph showing that:*

(i) *f has a jump discontinuity at a point in its domain of definition;*

(ii) *f has a removable discontinuity at a certain point in its domain;*

(iii) *f has an infinite discontinuity in its domain of definition;*

(iv) *f has a point c in its domain of definition where neither $f(c+0)$ nor $f(c-0)$ exists.*

Solution: (i) The function f defined by $f(x) = x/|x|$, if $x \neq 0$; $= k$ if $x = 0$ has a jump discontinuity at $x = 0$, whatever may be the value of k. (See Fig. 8.1(a))

(ii) The function f defined by $f(x) = x \sin(1/x)$ if $x \neq 0$, $f(0) = 1$ has a removable discontinuity at $x = 0$ (see that $f(0+0) = f(0-0) = 0$). (See Fig. 8.1(b))

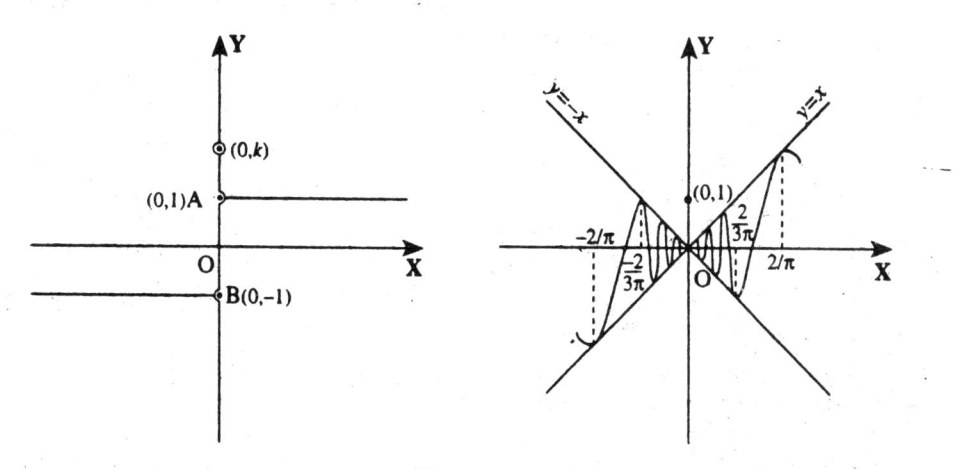

Fig 8.1(a) Jump discontinuity at $x = 0$. The points $(0, 1)$ and $(0, -1)$ are not included in the graph.

Fig. 8.1(b) Removable discontinuity at $x = 0$.

(iii) The function f defined by $f(x) = 1/x$ if $x \neq 0$, $f(0) = k$ has an infinite discontinuity at $x = 0$. (See Fig. 8.1(c))

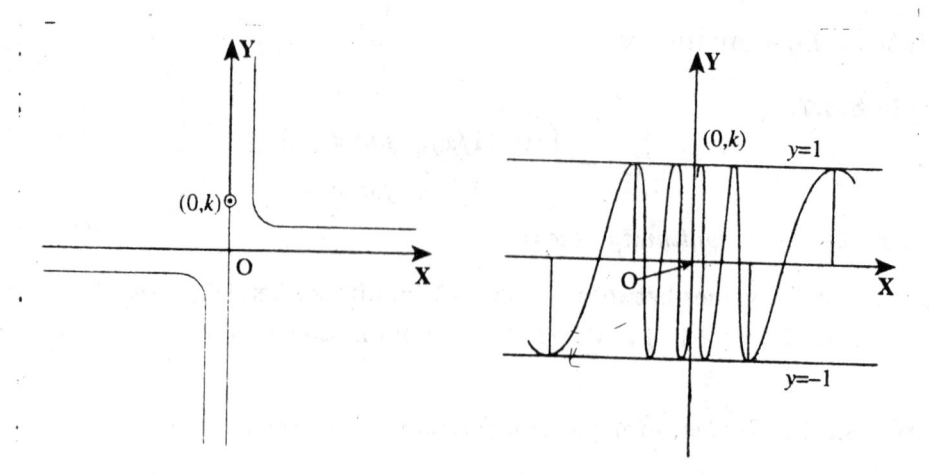

Fig. 8.1(c) Infinite discontinuity at $x = 0$.

Fig. 8.1(d) Neither $f(0+0)$ nor $f(0-0)$ exists.

(iv) The function f defined by $f(x) = \sin(1/x)$ if $x \neq 0$, $f(0) = k$ has a discontinuity at $x = 0$, whatever may be the value of k. In this case neither $f(0+0)$ nor $f(0-0)$ exists. (See Fig. 8.1(d)).

8.4 Elementary Properties of Continuous Functions

Theorem 8.4.1. *Let f and g be two functions of x defined in some neighbourhood of c. If f and g are both continuous at c. Then*

(i) $f + g, f - g$ *and* $f.g$ *are all continuous at c.*

(ii) f/g *is also continuous at c, provided that $g(c) \neq 0$.*

(iii) $|f|$ *or* $|g|$ *is continuous at c.*

(iv) $kf(x)$ *is continuous where k is a constant.*

Proof: All these results follow as immediate consequences of theorems on limits of Chapter 7 (see Theorem **7.4.2**).

For example, we wish to prove that $f + g$ is continuous at c if f and g are both continuous there.

Since f is continuous at c, we have, $\lim\limits_{x \to c} f(x) = f(c)$.

Also since g is continuous at c, we have, $\lim\limits_{x \to c} g(x) = g(c)$.

By the theorem on limits,

$$\lim_{x \to c}\{f(x) + g(x)\} = \lim_{x \to c} f(x) + \lim_{x \to c} g(x) = f(c) + g(c).$$

Hence, $f + g$ is continuous at c.

Similarly, we may discuss the cases for $f - g$, $f \cdot g$ and f/g ($g(c) \neq 0$).

Again, from the result: $\lim_{x \to c} |f(x)| = \left|\lim_{x \to c} f(x)\right|$ we may obtain part (iii).

(iv) $\lim_{x \to c} kf(x) = k \lim_{x \to c} f(x) = kf(c)$. $\therefore kf(x)$ is continuous at c.

Composite Function

Let $f : A \to B$ and $g : C \to D$ be two functions and $B \subseteq C$. Define a function $h : A \to D$ such that for each $x \in A$, $h(x) = g(f(x))$. The function h is called the composite function of f and g. The composite functions $g(f(x))$ and $f(g(x))$ are in general different functions; even the existence of one does not imply the existence of the other.

Theorem 8.4.2. *Let f and g satisfy the conditions of the previous paragraph. If f is continuous at a point $c \in A$ and g is continuous at $f(c)$ then the composite function $h \equiv g(f)$ is also continuous at c.*

Proof: Let $u_0 = f(c)$, $u_0 \in B$, i.e., $\in C$. Let ε be any given arbitrary positive number.

Since g is continuous at $u_0 = (= f(c))$, corresponding to the given ε, $\exists$ a positive number δ_1 such that
$$|g(u) - g(u_0)| < \varepsilon, \forall u \text{ in } |u - u_0| < \delta_1.$$

Again, since f is continuous at $x = c$, corresponding to **that determined positive number** δ_1, $\exists$ a positive number δ_2 s.t.
$$|f(x) - f(c)| < \delta_1 \quad \forall x \in |x - c| < \delta_2.$$
i.e., $|u - u_0| < \delta_1 \qquad \forall x \in |x - c| < \delta_2.$

$\therefore$ Given $\varepsilon > 0$, $\exists \delta_1(\varepsilon)$ whence we can find $\delta_2(\delta_1, \varepsilon)$ such that $|g(u) - g(u_0)| < \varepsilon$ $\forall x \in |x - c| < \delta_2$

i.e., $|g\{f(x)\} - g\{f(c)\}| < \varepsilon, \quad \forall x \in |x - c| < \delta_2.$

or, $|h(x) - h(c)| < \varepsilon, \qquad \forall x \in |x - c| < \delta_2.$

i.e., h is continuous at $x = c \in A$.

The theorem is thus proved.

> REMEMBER: A continuous function of a continuous function is also a continuous function.

Corollary 8.4.1. *If a function f has its values ≥ 0, and if f is continuous at a point $x = c$, then $\sqrt{f}$ is also continuous at c.*

Neighbourhood Property

Theorem 8.4.3. *If f is continuous at a certain point c and if $f(c) \neq 0$, then there exists a certain neighbourhood of c such that $\forall\, x$ in that neighbourhood $f(x)$ has the same sign as that of $f(c)$.*

Proof: (i) First, let c be an interior point of $[a, b]$, where f is defined. Since it is given that f is continuous at c corresponding to $\varepsilon = \frac{1}{2}|f(c)|$. There exists a positive number δ such that

$$|f(x) - f(c)| < \frac{1}{2}|f(c)|,\ \forall\, x \in |x - c| < \delta.$$

i.e., $\quad f(c) - \frac{1}{2}|f(c)| < f(x) < f(c) + \frac{1}{2}|f(c)|,\ \forall\, x \in |x - c| < \delta. \qquad (8.4.1)$

(a) If $f(c) > 0$, (8.4.1) gives $\frac{1}{2}f(c) < f(x) < \frac{3}{2}f(c),\ \forall\, x \in |x - c| < \delta$. Since $\frac{1}{2}f(c)$ and $\frac{3}{2}f(c)$ are both positive, we conclude that

$$\forall\, x \in |x - c| < \delta,\, f(x)\ \text{is positive}.$$

(b) If $f(c) < 0$, (8.4.1) gives $\frac{3}{2}f(c) < f(x) < \frac{1}{2}f(c),\ \forall\, x \in |x - c| < \delta$. Since $\frac{3}{2}f(c)$ and $\frac{1}{2}f(c)$ are now both negative, we conclude that

$$\forall\, x \in |x - c| < \delta,\ \text{every}\ f(x)\ \text{is negative}.$$

(ii) Next let c coincide with the left-end a of $[a, b]$ i.e., f is continuous at a and $f(a) \neq 0$. We know that $f(x)$ is continuous at $a \Rightarrow f(x)$ has a right-continuity at a.

From the definition of right-continuity it will follow, as before, $f(x)$ has the same sign as that of $f(a)$ in a certain nbd like $(a, a + \delta)$.

(iii) If c coincides with b i.e., if f is continuous at b and $f(b) \neq 0$ then $\exists$ a nbd $(b - \delta, b)$ where every $f(x)$ has the same sign as that of $f(b)$.

Applications of the previous theorems

Example 8.4.1. *A constant function is continuous at any point in its domain.*

Let $f(x) = c\ \forall\, x \in \mathbb{R}$. At any point $x_0 \in \mathbb{R}$.

$$|f(x) - f(x_0)| = |c - c| = 0 < \text{any given}\ \varepsilon(> 0);\ \text{hence etc.}.$$

Example 8.4.2. *Identity function:* Let $f(x) = x,\ \forall\, x \in \mathbb{R}$. Then f is continuous at any point x_0 in its domain.

$$|f(x) - f(c)| = |x - c| < \varepsilon,\ \forall\, x \in |x - c| < \delta\ \text{where}\ \delta = \varepsilon;\ \text{hence etc.}$$

CHAPTER 8: CONTINUITY OF FUNCTIONS OF A SINGLE REAL VARIABLE

Example 8.4.3. $f(x) = x^n$, (where n is a positive integer) is continuous for all values of $x \in \mathbb{R}$. For any arbitrary $c \in \mathbb{R}$, we have

$$\lim_{x \to c} f(x) = \lim_{x \to c} x^n = \lim_{x \to c} x \cdot \lim_{x \to c} x \cdot \lim_{x \to c} x \cdots \lim_{x \to c} x = c \cdot c \cdot c \cdots c = c^n = f(c).$$

Hence x^n is continuous at c and c being any arbitrary point of $\mathbb{R}$, it is continuous for all $x \in \mathbb{R}$.

Polynomials

Example 8.4.4. Let $P(x) = a_0 x^n + a_1 x^{n-1} + \cdots + a_{n-1} x + a_n$ be a **polynomial in x of degree n**. Then $P(x)$ is continuous for all values of $x \in \mathbb{R}$.

Solution: This statement is true because for any real c, we have

$$\lim_{x \to c} P(x) = a_0 \lim_{x \to c} x^n + a_1 \lim_{x \to c} x^{n-1} + \cdots + a_{n-1} \lim_{x \to c} x + \lim_{x \to c} a_n$$
$$= a_0 c^n + a_1 c^{n-1} + \cdots + a_{n-1} c + a_n = P(c); \text{ hence etc.}$$

Note: $P(x)$ is continuous is due to the fact that sum and product of continuous functions in continuous. (Here $v(x) = a_i, i = 0(1)n$ are all constant functions and hence continuous).

Rational Functions

Example 8.4.5. $f(x) = x^n$, where n is a negative integer, is continuous for all x except for $x = 0$. More generally, a rational function $P(x)/Q(x)$ is continuous at every point of $\mathbb{R}$ which is not a zero of $Q(x)$ (i.e., where $Q(x) \neq 0$).

$$\left[\text{See that } x^n = x^{-m} = \frac{1}{x^m}; \text{ where } n = -m (m \text{ being a positive integer}). \right]$$

Trigonometric Functions

Example 8.4.6. $\sin x$ and $\cos x$ are continuous at any arbitrary point $x = c$; hence we say they are continuous for all real values of x.

$$\left[\text{See that } |\sin x - \sin c| = \left| 2 \cos \frac{x+c}{2} \sin \frac{x-c}{2} \right| \right.$$
$$\leq 2 \cdot 1 \cdot \left| \frac{x-c}{2} \right| < |x - c| < \text{ any } \varepsilon (> 0)$$
$$\left. \text{whenever } |x - c| < \delta, \text{ choosing } \delta = \varepsilon. \right]$$

Example 8.4.7. $\tan x$ and $\sec x$ are continuous for all values of x except when $x = n\pi + \pi/2$; $\cot x$ and $\operatorname{cosec} x$ are continuous for all values of x except when $x = n\pi$ (n being 0 or any positive or negative integer).

Transcendental Functions

Example 8.4.8. *The exponential function e^x is continuous everywhere and its inverse function $\log_e x$ is continuous for all values of $x > 0$.*

[See **Ex. 8.5.11** for detailed discussions.]

8.5 Deeper Properties of Continuous Functions

We shall now discuss (with proofs) the following more deep properties of continuous functions which form an important part of Real Analysis:

1. Intermediate Value Property.
2. Boundedness Property.
3. Uniform Continuity Property.
4. Compactness and Continuity.
5. Continuity and Existence of Inverse functions.
6. Continuity and Monotoncity.

1. Intermediate Value Property

Theorem 8.5.1. (Bolzano's Theorem on Continuity). *Let f be a real-valued continuous function in a closed interval $[a, b]$. Suppose that $f(a)$ and $f(b)$ are of opposite signs (i.e., let $f(a) \cdot f(b) < 0$). Then there exists at least one point c where $a < c < b$ such that $f(c) = 0$.*

[The truth of this theorem seems obvious and indeed it is evident geometrically: If the graph of a continuous function lies above the X-axis at one end of an interval $[a, b]$ and below the X-axis at another end, then it must cross (at least once) the X-axis somewhere in-between. However, the point is to give a proof based on the arithmetical properties of real numbers rather than on a geometrical picture. Remember that there are continuous functions for which no geometrical picture can be drawn].

Proof: For the sake of definiteness, assume $f(a) > 0$ and $f(b) < 0$.

Since $f(a) > 0$, by neighbourhood property $\exists$ some right-neighbourhood of a where, for every x, $f(x) > 0$. Let

$$A = \{x : x \in [a, b] \text{ and } f(x) > 0\}.$$

Clearly, A is non-empty subset of $[a, b]$ ($\because a \in A$); A is obviously bounded. Therefore, A has a supremum $= c$ (say).

Since $\exists$ a right-nbd of a, where $f(x) > 0$, $c \neq a$.
Since $\exists$ a left-nbd of b, where $f(x) < 0$, $c \neq b$.

$$\therefore \quad a < c < b.$$

We shall establish that $f(c) = 0$.

If $f(c) \neq 0$, then either $f(c) > 0$ or $f(c) < 0$.

(i) Let, if possible, $f(c) > 0$. Then, by the neighbourhood property, $\exists \delta > 0$ such that $f(x) > 0$ for all $x \in [c - \delta, c + \delta]$.

$\therefore$ For some $x > c$, $f(x) > 0$. This contradicts the fact that $c = \sup A$. Hence $f(c)$ *cannot be positive.*

(ii) Next suppose $f(c) < 0$. Then $\exists \delta > 0$ such that

$$f(x) < 0 \text{ when } x \in [c - \delta, c].$$

Since $c = \sup A$, by definition of Supremum it follows that $\exists \eta \in A$ such that

$$c - \delta < \eta \leq c.$$

Now we see that $\eta \in A \Rightarrow f(\eta) > 0$ and $\eta \in (c - \delta, c] \Rightarrow f(\eta) < 0$.

Thus, we arrive at a contradiction and as such $f(c)$ *is not negative.*

$\therefore$ From (i) and (ii) we arrive at the conclusion that $f(c) = 0$.

Alternative Proof

Proof of Bolzano's Theorem by the method of Bisection

Take $\alpha = \frac{1}{2}(a + b)$, the mid-point of $[a, b]$.

In case, we find $f(\alpha) = 0$ the proof is complete.

If, however, $f(\alpha) \neq 0$ then either $f(a), f(\alpha)$ or $f(\alpha), f(b)$ must have opposite signs. Of the two intervals $[a, \alpha]$ and $[\alpha, b]$ the one at the ends of which $f(x)$ has opposite signs, we re-name as $[a_1, b_1]$. Hence in either case, we get

$$a \leq a_1 < b_1 \leq b;$$
$$b_1 - a_1 = \frac{1}{2}(b - a);$$

and $f(a_1), f(b_1)$ have opposite signs.

We now bisect $[a_1, b_1]$ and proceed as above. We shall see that either $f(x) = 0$ at the mid-point $c_1 = \frac{1}{2}(a_1 + b_1)$ or otherwise we obtain an interval $[a_2, b_2]$ such that

$$a_1 \leq a_2 < b_2 \leq b_1;$$
$$b_2 - a_2 = \frac{1}{2}(b_1 - a_1) = \left(\frac{1}{2}\right)^2 (b - a);$$

and $f(a_2)$, $f(b_2)$ have opposite signs.

Proceeding as above we see that, either after a finite number of steps we shall obtain a point at which the function vanishes or we shall obtain an infinite sequence of intervals.

$$[a_1, b_1], [a_2, b_2], \cdots, [a_n, b_n], \cdots$$

such that

$$a \leq a_1 \leq a_2 \leq \cdots \leq a_n < b_n \leq b_{n-1} \leq \cdots \leq b_2 \leq b_1 \leq b \qquad (8.5.1)$$

$$b_n - a_n = \left(\frac{1}{2}\right)^n (b - a) \qquad (8.5.2)$$

and $f(a_n)$, $f(b_n)$ are of opposite signs.

$$(8.5.3)$$

From (8.5.1) $\{a_n\}$ is $m.i.$, bounded above and hence convergent and $\{b_n\}$ is $m.d.$, bounded below and hence convergent.

From (8.5.2)

$$\lim_{n \to \infty}(b_n - a_n) = \lim_{n \to \infty} \left(\frac{1}{2}\right)^n \cdot (b - a) = 0.$$

$$\therefore \quad \lim_{n \to \infty} a_n = \lim_{n \to \infty} b_n = l \text{ (say)}.$$

The point l may either be an interior point of $[a, b]$ or it may be an end-point of $[a, b]$. We shall prove that $f(l) = 0$.

If possible, let $f(l) \neq 0$.

Case I. Let l be an interior point of $[a, b]$. Since $f(l) \neq 0$ it may be $f(l) > 0$ or $f(l) < 0$. By neighbourhood property there exists an interval $[l - \delta, l + \delta]$ such that for every point x of this interval $f(x)$ and $f(l)$ have the same sign.

Since $a_n \to l$ from below and $b_n \to l$ from above, there exists a positive integer m such that $[a_m, b_m]$ lies within $[l - \delta, l + \delta]$ and accordingly $f(a_m)$, $f(b_m)$ have the same sign. Thus, we arrive at a contradiction of (8.5.3). Hence the only conclusion is $f(l) = 0$.

Case II. Let l coincide with a. In this case $a_n = a$, $\forall\, n$.

Since $f(a) \neq 0$, $\exists$ an interval $[a, a+\delta]$ such that for every point x of this interval $f(x)$ and $f(a)$ have the same sign. Since $b_n \to l = a$, $\exists$ a positive integer m such that
$$a < b_m < a + \delta.$$
Accordingly, $f(a_m) = f(a)$ and $f(b_m)$ have the same sign so that we again have a contradiction. Hence l cannot coincide with a.

Similarly, it may be shown that l cannot coincide with b.

Thus, we have proved that $\exists$ an interior point l of $[a,b]$ where $f(l) = 0$. Hence the theorem.

An immediate consequence of Bolzano's Theorem is the following:

Theorem 8.5.2. (Intermediate Value Theorem). *Let f be a real-valued continuous function in a closed interval $[a, b]$. Suppose that $f(a) \neq f(b)$.*

Then f assumes every value between $f(a)$ and $f(b)$ at least once.

Proof: Let μ be a number between $f(a)$ and $f(b)$.

We construct a function ϕ such that $\phi(x) = f(x) - \mu$.

ϕ is obviously continuous in $[a, b]$ since f is so and also
$$\phi(a) = f(a) - \mu, \quad \phi(b) = f(b) - \mu$$
are of opposite signs.

Hence $\exists$ at least one point c, where $a < c < b$ such that $\phi(c) = 0$.

But $\phi(c) = 0 \Rightarrow f(c) = \mu$. Hence the theorem.

As an application of *Intermediate Value Theorem* we prove the following theorem:

Theorem (Fixed-Point Theorem). *If f is continuous on $[a, b]$ and $f(x) \in [a, b]$ for every $x \in [a, b]$, [in other words, $f : [a, b] \to [a, b]$] then f has a fixed point i.e., $\exists\, c \in [a, b]$ whose $f(c) = c$.* [C.H. 1999]

Proof. Suppose f is continuous $[a, b]$ and $f(x) \in [a, b]$ for every $x \in [a, b]$. If $f(a) = a$ or $f(b) = b$ then the theorem is true ($\because c = a$ or $c = b$).

So assume that $f(a) > a$ and $f(b) < b$. Construct a function $\phi(x) = f(x) - x$ for every $x \in [a, b]$.

Clearly $\phi(a) > 0$ and $\phi(b) < 0$ and ϕ is continuous $[a, b]$. Now 0 (zero) is an intermediate value for ϕ on $[a, b]$.

Hence by intermediate value theorem, there exists some $c \in [a, b]$ such that $\phi(c) = 0$ or $f(c) - c = 0$ or $f(c) = c$.

Observation: The theorem asserts the existence of at least one point c for which $f(x) = \mu$. Clearly there may be many such points.

But in one important situation we can assert that there is **exactly one point c** for which $f(c) = \mu$. We give the result in the form of the following theorem:

Theorem 8.5.3. *Let f be a real-valued continuous function in the closed interval $[a, b]$. Suppose that $f(a) \neq f(b)$. Let μ be a number between $f(a)$ and $f(b)$. Moreover, assume that f is **strictly monotone** in the interval $[a, b]$. Then $\exists$ exactly one value of $x = c$ (say) where $a < c < b$ such that $f(c) = \mu$.*

Proof. For definiteness, suppose f is monotonically increasing in $[a, b]$. If possible, let there be two points c and $c' \in (a, b)$, with $c > c'$, for which $f(c) = \mu$ and $f(c') = \mu$.

But being *strictly m.i.* $c > c' \Rightarrow f(c) > f(c')$ i.e., $\mu > \mu$, which is clearly not possible. Therefore $\mu = f(c) = f(c') \Rightarrow c = c'$. Hence the theorem.

Note 8.5.1. *Now compare the results of Theorems 8.5.11 and 8.5.12 given towards the end of this chapter.*

> REMEMBER: The converse of Theorem **8.5.2** is not necessarily true.
> i.e., If in $[a, b]$, f assumes at least once every value between $f(a)$ and $f(b)$ with $f(a) \neq f(b)$, then f need not necessarily be continuous in $[a, b]$.

We give an example in support of this assertion:

Example 8.5.1. *Let $f : [0, 1] \to [0, 1]$ be a function defined by*

$$f(0) = 0, \quad f(1) = 1, \quad f\left(\tfrac{1}{2}\right) = \tfrac{1}{2} \text{ and}$$

$$f(x) = \begin{cases} \tfrac{1}{2} - x, & \text{when } 0 < x < \tfrac{1}{2} \\ \tfrac{3}{2} - x, & \text{when } \tfrac{1}{2} < x < 1 \end{cases}$$

Graphical Approach: We refer to Fig. 8.2. The graph consists of (i) the points $(0, 0)$, $\left(\tfrac{1}{2}, \tfrac{1}{2}\right)$ and $(1, 1)$; (ii) the segment AB & CD; the points $A\left(0, \tfrac{1}{2}\right)$, $B\left(\tfrac{1}{2}, 0\right)$, $C\left(\tfrac{1}{2}, 1\right)$, $D\left(1, \tfrac{1}{2}\right)$ are not included in the graph.

Clearly, there are discontinuities at $x = 0$, $x = 1/2, x = 1$.

But every value between 0 and 1 is assumed by f. Take a line parallel to X-axis between $y = 0$ and $y = 1$. [e.g., a dotted line shown in the figure, $y = d (0 \leq d \leq 1)$. $\exists$ a corresponding

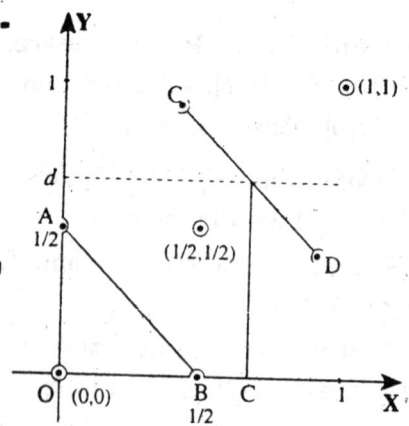

Fig. 8.2 Graph of a Discontinuous Function whose values are all attained.

CHAPTER 8: CONTINUITY OF FUNCTIONS OF A SINGLE REAL VARIABLE 485

$x = c$ for which $f(c) = d$]. Thus f assumes every value between $f(0)$ and $f(1)$ but f is not continuous in $0 \le x \le 1$.

Analytical approach: Let us take any value of y between 0 and 1, say $y = d$.

If $0 < d < 1/2$, we have $d = 1/2 - x$ or, $x = 1/2 - d$ for which $f(x) = d$.

If $\frac{1}{2} < d < 1$, we have $d = 3/2 - x$ or, $x = 3/2 - d$ for which $f(x) = d$.

If $d = 0$, then $x = 0$ for which $f(0) = d$. If $d = 1/2, x = 1/2$ for which $f(x) = d$.

If $d = 1$, then $x = 1$ for which $f(x) = d$.

Thus every value between 0 and 1 is assumed by $f(x) \in [0, 1]$.

But it is easy to see that f is discontinuous at $x = \frac{1}{2}$:

$$f\left(\frac{1}{2}+0\right) = \lim_{x \to \frac{1}{2}+0}\left(\frac{3}{2}-x\right) = 1; \quad f\left(\frac{1}{2}-0\right) = \lim_{x \to \frac{1}{2}-0}\left(\frac{1}{2}-x\right) = 0;$$

and $f\left(\frac{1}{2}-0\right) \ne f\left(\frac{1}{2}+0\right)$; hence f is not continuous at $x = 1/2$ i.e., f is not continuous in $[0, 1]$.

2. Boundedness Property of Continuous Functions

Theorem 8.5.4. *If a real-valued function f is continuous on a closed interval $I = [a, b] = a \le x \le b$, then it is bounded there.*

Proof: We have to show that the set A of values of $f(x)$, for $x \in I$, is a bounded set.

If this is not true then at least we can construct a set S with the following properties:

$$S = \{c : c \in [a, b] \text{ and } f \text{ is bounded in } a \le x \le c\}.$$

The set S is non-empty. For, by continuity of f at $x = a$, corresponding to $\varepsilon = 1$, we have a positive number δ_1 such that

$$|f(x) - f(a)| < 1, \ \forall \, x \in a \le x \le a + \delta_1.$$

and hence

$$|f(x)| = |f(x) - f(a) + f(a)|$$
$$\le |f(x) - f(a)| + |f(a)|$$
$$\text{i.e.,} \quad < 1 + |f(a)|, \ \forall \, x \in a \le x \le a + \delta_1$$

i.e., $\exists \, \delta_1 > 0$ for which $a + \delta_1 \in S$. In other words S is not empty.

The set S is clearly bounded. (Since every $c \in [a, b]$).

Since S is non-empty and bounded, by lub-property, $\exists$ a supremum of S; call it c_0 i.e.,
$$c_0 = \sup_S c.$$

We now wish to prove that $c_0 = b$.

Suppose that this is not so. Then, by the same sort of continuity argument for the point c_0 $\exists$ **an interval** $c_0 - \delta \leq x \leq c_0 + \delta$ in which

$$|f(x)| < 1 + |f(c_0)| = M_1 \text{ (say)}.$$

Since $c_0 - \delta < c_0$, we have f bounded in $a \leq x \leq c_0 - \delta$ say by M_2. Then for $M = \max. [M_1, M_2]$ we have

$$|f(x)| \leq M \in a \leq x \leq c_0 + \delta.$$

Since $c_0 + \delta > c_0$, this contradicts the definition of c_0 as $\sup_S c$.

Therefore, $c_0 = b$. This proves that f is bounded in $a \leq x \leq b$.

$\therefore$ The statement that "the set of $f(x)$'s is not bounded on I" is not true.

This proves our theorem of boundedness of a continuous function in a closed interval.

Observation: The result of this theorem is extremely important in Real Analysis. There is a more general theorem which we shall prove in Part II of this treatise:

The range of a continuous function with compact domain is compact.

The present theorem will then follow as a corollary. But we advise our students to learn this independent proof. We give below two other proofs of this important theorem:

Alternative Proof of Theorem 8.5.4

Given: f is continuous in $[a, b]$.

To prove: f is bounded in this interval.

i.e., the set $\{f(x) : x \in [a, b]\}$ is a bounded set.

Proof: If possible, let f be not bounded (*rather*, not bounded above), so that for each positive integer n $\exists$ a point $x_n \in [a, b]$ such that $|f(x_n)| > n$.

We have thus constructed a sequence $\{x_n\}$ whose members all lie in $[a, b]$ and hence it is a bounded sequence.

By Bolzano Weierstrass Theorem the sequence $\{x_n\}$ has, therefore, at least one cluster point $\xi \in [a, b]$.

Since ξ is a cluster point of $\{x_n\}$, therefore $\exists$ a subsequence $\{x_{n_k}\}$ of the sequence $\{x_n\}$ such that $x_{n_k} \to \xi$ as $k \to \infty$.

Also since, by construction, $|f(x_{n_k})| > n_k \ \forall \ k$, therefore the sequence $\{f(x_{n_k})\}$ diverges to ∞.

Thus $\exists$ a point ξ of $[a, b]$ such that $\{x_n\}$ in $[a, b]$ converges to ξ but $\lim\limits_{k \to \infty} f(x_{n_k}) \neq f(\xi)$.

Thus f is not continuous at $\xi \in [a, b]$, which contradicts our given condition and hence the function f must be bounded above. By considering the function $-f$, it can be shown that f is also bounded below. Hence the function is bounded in $[a, b]$.

Another Proof of this theorem: using Nested-Interval Property

Let f be not bounded in $I = [a, b]$. If $c = \frac{1}{2}(a + b)$, then f must not be bounded above in at least one of the closed intervals $[a, c]$ and $[c, b]$. Let us call it I_1 (if f is unbounded above on both, let us take $[a, c]$ to be I_1), and re-label it as $[a_1, b_1]$. If $c_1 = \frac{1}{2}(a_1 + b_1)$, then f must be unbounded above at least in one of the intervals $[a_1, c_1]$ and $[c_1, b_1]$. Let us call it I_2 (if it is unbounded above on both, then let us call $[a_1, c_1]$ to be I_2) and re-label it as $[a_2, b_2]$.

Proceeding in this manner we construct a sequence of closed intervals $\{I_1, I_2, I_3, \cdots, I_n, \cdots\}$ with the following properties:

$$I_1 \supset I_2 \supset I_3 \supset \cdots \supset I_n \supset I_{n+1} \supset \cdots$$

and length of $I_n = d_n = \frac{1}{2^n}(b - a) (d_n \to 0$ as $n \to \infty)$ and f is unbounded above on each I_n.

$\therefore$ By Nested-Interval property we have exactly one point x_0 common to all the intervals I_n, where $x_0 \in I$. Since f is continuous at x_0, therefore, given any $\varepsilon > 0$, say $\varepsilon = 1$, $\exists$ a positive number δ such that

$$\forall \ x \in |x - x_0| < \delta, \text{ we have } |f(x) - f(x_0)| < 1.$$

Since $d_n \to 0$ as $n \to \infty$, we can find a positive integer m such that

$$I_m \subset (x_0 - \delta, x_0 + \delta).$$

If $x \in I_m$, then $|f(x)| = |f(x) - f(x_0) + f(x_0)| \leq 1 + |f(x_0)|$ so that $f(x)$ is bounded in I_m, **contradicting** our assumption that on each I_m, f is unbounded above.

Hence we cannot accept that f is not bounded in I i.e., f must be bounded in I and the theorem is proved.

Note 8.5.2. *The theorem requires that the interval should be closed. A function which is continuous in only an open interval, say $]a, b]$ or $]a, b[$ or $[a, b[$, may not be bounded there.*

For example: Let f be defined by $f(x) = 1/x \ \forall \ x \in \ ]0,1[$. This function f is certainly continuous in $]0,1[$ but is not bounded there (see that $f(x) \to +\infty$ as $x \to 0^+$). Continuity at the end points would prevent the function from evading the theorem in this way.

Note 8.5.3. The converse of this theorem is not true, *i.e., a function which is bounded in an interval may not be continuous throughout the interval.*

For example: Let f be defined by $f(x) = x$, when $0 < x < 1$; $= \frac{1}{2}$, when $x = 0$ and when $x = 1$. The set of functional values is bounded between 0 and 1 but certainly it is not continuous in $0 \leq x \leq 1$ (it is discontinuous at $x = 0$ and at $x = 1$).

Attainment of Bounds of a continuous function

Theorem 8.5.5. *If f be continuous on a closed interval, then it assumes its least upper bound and its greatest lower bound in that interval.*

Proof: Let f be continuous on the closed interval $I : \{a \leq x \leq b\}$.

Then by the previous theorem, f is bounded on I.

Hence $M = \sup_I f(x)$ and $m = \inf_I f(x)$ both exist.

That is, $m \leq f(x) \leq M, \quad \forall \, x \in I$.

What we are to prove is that there exists a point, say $x = \alpha \in I$ and a **point** $x = \beta \in I$ such that $f(\alpha) = M$ and $f(\beta) = m$.

(i) Suppose there is no point in I for which $f(x) = M$. Then $M - f(x)$ is everywhere positive in I and is, of course, continuous there. This implies that $1/(M - f(x))$ is continuous in I.

Hence, by the last theorem, this function must be bounded on I; i.e., $\exists \ B$ s.t.

$$\frac{1}{M - f(x)} \leq B, \quad \forall \, x \in I$$

which implies that

$$M - f(x) \geq \frac{1}{B}, \quad \forall \, x \in I$$

$$\text{or,} \quad f(x) \leq M - \frac{1}{B}, \quad \forall \, x \in I.$$

i.e., $M - \frac{1}{B}$ is an upper bound of $f(x) \in I$.

This shows that a number $< M$ can be the supremum of $f(x) \in I$, contradicting the assumption that $\sup_I f(x) = M$.

This contradiction proves that $\exists$ a point $x = \alpha$ (say) $\in I$ where $M = f(\alpha)$.

(ii) Similarly suppose there does not exist any $x \in I$ such that $f(x) = m$. In that case, $f(x) > m, \forall x \in I$ and $\frac{1}{f(x)-m}$ is continuous in I and consequently it should be bounded in I. Let it be bounded above by k. Then

$$\frac{1}{f(x) - m} \leq k, \quad \forall x \in I$$
$$\Rightarrow f(x) - m \geq \frac{1}{k}, \quad \forall x \in I$$
$$\Rightarrow f(x) \geq m + \frac{1}{k}, \quad \forall x \in I.$$

Thus a number greater than m could be the $\inf_I f(x)$.

This contradicts the fact that $m =$ infimum of $f(x) \in I$. Hence $\exists$ some x, say $x = \beta \in I$ such that $f(\beta) = m$.

Note 8.5.4. *In a bounded set of real numbers if the lub and glb are actually members of the set, then they are respectively called the **greatest value** and the **least value** of the set. So we could state this theorem as follows:*

A function f continuous on a closed interval $[a, b] = I$ has its greatest and least values in I:

In fact, Greatest value of $f(x) = \sup_I f(x)$ and Least value of $f(x) = \inf_I f(x)$.

Note 8.5.5. *This theorem also requires that the interval in which f is continuous must be closed, otherwise the theorem may not be true. For example, $f(x) = x^2$ is continuous (and even bounded) $\forall x \in 0 < x < 1$.*

Here $m = 0$ and $M = 1$. See that f does not assume either the value 0 or the value 1 anywhere in $(0, 1)$; f has no greatest or least value in $(0, 1)$.

REMEMBER: A discontinuous function may not possess greatest or least values:

e.g., $f(x) = \begin{cases} 1 - x, & \text{when } 0 < x \leq 1; \\ \frac{1}{2}, & \text{when } x = 0. \end{cases}$

At $x = 0$, there is a point of discontinuity i.e., f is not continuous in $[0, 1]$ and f has no greatest value in $[0, 1]$.

Theorem 8.5.6. *If a function f is continuous on a closed interval $I = \{a \leq x \leq b\}$, then f assumes all the values between its exact upper and lower bounds.*

Proof: Let f be continuous in a closed interval $I = [a, b]$.

Then, by Theorem **8.5.4**, f must be bounded in $[a,b]$ and by Theorem **8.5.5**, the glb M and the lub m of f are both attained in I. So there exist points $\alpha, \beta \in I$ such that

$$f(\alpha) = M \quad \text{and} \quad f(\beta) = m.$$

Now, by the Intermediate value Theorem, if $m \neq M$, then f must assume every value between $f(\alpha)$ and $f(\beta)$. Hence the theorem.

If $m = M$, f is constant and the theorem is trivial.

3. Uniform Continuity Property

Let us recall here the definition of continuity of a function f at any interior point x_0 of its domain of definition:

*f is continuous at x_0 if, given **any** $\varepsilon > 0$, $\exists$ a positive number δ such that*

$$\forall\, x \in |x - x_0| < \delta \text{ we have } |f(x) - f(x_0)| < \varepsilon.$$

Clearly δ depends on ε, a fact which we often denote by $\delta(\varepsilon)$.

But δ depends not only on ε but also it depends upon the particular point x_0 at which the continuity of f is considered.

For example, consider the function f defined by $f(x) = x^2$, $\forall\, x \in \mathbb{R}$.

One can easily verify that f is continuous both at $x_0 = 0$ and at $x_0 = 1$. If $\varepsilon = 1/4$, at $x_0 = 0$, suitable value of δ is $\delta = 1/2$. But with same $\varepsilon = 1/4$, at $x_0 = 1$, $1/2$ will not be a suitable δ.

[Taking $x = 1.4$, $|x - x_0| = |1.4 - 1| = 0.4 < \frac{1}{2}$, but $|f(x) - f(x_0)| = 0.96 > 1/4$.]

The fact that δ depends on the particular value of x in the domain gives the idea that f may change its values rapidly near certain points and slowly near other points; see the graph of $f(x) = \sin(1/x)$.

We shall now define *Uniform continuity* by removing this dependence of δ on the particular point.

It must be noted that *continuity* of a function is basically a **local concept** (i.e., it describes what is happening to a function in the neighbourhood of a point). On the other hand, the concept of uniform continuity is *not a local* but **a global concept** (i.e., it describes what is happening to a function *in a certain domain*).

Definition 8.5.1. *A function f defined in a domain D is said to be uniformly continuous on a set S (where $S \subseteq D$) if, for any $\varepsilon > 0$, $\exists$ a $\delta(\varepsilon) > 0$, such that $|f(x_1) - f(x_2)| < \varepsilon$ for any two points x_1 and $x_2 \in S$ with $|x_1 - x_2| < \delta$.*

CHAPTER 8: CONTINUITY OF FUNCTIONS OF A SINGLE REAL VARIABLE   491

Note 8.5.6. *In this definition δ depends upon ε but is independent of the point chosen.*

Non-uniform Continuity Criteria: f is not uniformly continuous on $D \equiv \exists$ some $\varepsilon > 0$ such that for any $\delta > 0$, $\exists$ points $x_1, x_2 \in D$ with $|x_1 - x_2| < \delta$ but $|f(x_1) - f(x_2)| \geq \varepsilon$. In the language of sequences we can give the following criterion for non-uniform continuity: f is not uniformly continuous $\equiv \exists$ some $\varepsilon_0 > 0$ and two sequences $\{x_n\}$ and $\{y_n\} \in D$ such that $\lim_{n \to \infty}(x_n - y_n) = 0$ and $|f(x_n) - f(y_n)| \geq \varepsilon_0$ for all $n \in \mathbb{N}$.

[For justification, see the first part of Theorem **8.5.8** given below].

Example 8.5.2. $f(x) = x^2$, $\forall\, x \in -a \leq x \leq a$. We prove that f is uniformly continuous in $[-a, a]$. For any two points $x_1, x_2 \in [-a, a]$, $|f(x_1) - f(x_2)| = |x_1^2 - x_2^2| = |x_1 - x_2||x_1 + x_2| \leq 2a|x_1 - x_2|$.

$\therefore |f(x_1) - f(x_2)| <$ any given ε, provided $|x_1 - x_2| \leq \delta$ where $\delta = \varepsilon/2a$.

Hence f is uniformly continuous in $[-a, a]$.

Example 8.5.3. We note here that f defined by $f(x) = x^2$, $\forall\, x \in \mathbb{R}$ is not uniformly continuous in $\mathbb{R}$ (through it is continuous at every given point). For if we choose $x_2(> 0)$ s.t. x_2 is less than a fixed positive number N then

$$|f(x_1) - f(x_2)| = |x_1^2 - x_2^2| = |x_1 - x_2| \cdot |x_1 + x_2| < 2N|x_1 - x_2|$$

and hence $|f(x_1) - f(x_2)| < \varepsilon$ if $|x_1 - x_2| < \delta$ when $\delta = \varepsilon/2N$.

This δ depends on N, which again depends on the choice of x_2.

Thus, the definition of uniform continuity is not satisfied throughout $\mathbb{R}$.

Example 8.5.4. Again, $f(x) = \frac{1}{x}(x > 0)$ is not uniformly continuous in $[0, 1]$.

If possible let $\exists\, \delta$ s.t. $|x_1 - x_2| < \delta$ for which $|f(x_1) - f(x_2)| < \varepsilon$.

Take $x_1 = \delta$; $x_2 = \dfrac{\delta}{\varepsilon + 1}$, so that $|x_1 - x_2| = \dfrac{\varepsilon \delta}{\varepsilon + 1} < \delta$ whereas,

$$|f(x_1) - f(x_2)| = \left|\frac{1}{x_1} - \frac{1}{x_2}\right| = \left|\frac{1}{\delta} - \frac{\varepsilon + 1}{\delta}\right| = \frac{\varepsilon}{\delta} > \varepsilon.$$

Alternatively: $x_n = \dfrac{1}{n}$, $y_n = \dfrac{1}{n+1}$ then we have $\lim_{n \to \infty}(x_n - y_n) = 0$ but $|f(x_n) - f(y_n)| = 1$ for all $n \in \mathbb{N}$.

Example 8.5.5. *A function continuous in a certain domain but not uniformly continuous there.*

Let us consider a function f defined by $f(x) = \sin(1/x)$ for all $x > 0$. (domain $= \mathbb{R}^+$) where $\mathbb{R}^+$ represents the set of all **positive** real numbers.   [C.H. 1980]

We can see that f is clearly continuous on $\mathbb{R}^+$.

To show that this function f is not uniformly continuous on $\mathbb{R}^+$.

For this we are to obtain an $\varepsilon > 0$ for which no δ works in the definition of continuity, i.e., whatever $\delta > 0$ we may choose, we can find two points x_1 and x_2 in $\mathbb{R}^+$ such that

$$|x_1 - x_2| < \delta \text{ but } |f(x_1) - f(x_2)| \text{ is not less than } \varepsilon \text{ (or } \geq \varepsilon).$$

Out problem will be solved if we can show that no δ works for $\varepsilon = 1$. Let δ be any positive number, and let

$$x_1 = \frac{1}{n\pi} \text{ and } x_2 = \frac{2}{(2n+1)\pi}, \text{ where } n \text{ is a positive integer such that}$$

$$|x_1 - x_2| = \left| \frac{1}{n\pi} - \frac{2}{(2n+1)\pi} \right| < \delta. \quad \text{[such a positive integer definitely exists for each positive } \delta.\text{]}$$

Then $|x_1 - x_2| < \delta$ but $|f(x_1) - f(x_2)| = \left|\sin n\pi - \sin(2n+1)\frac{\pi}{2}\right| = 1$ (not less than ε).

Example 8.5.6. *Another interesting example of a function which is continuous but not uniformly continuous.*

Let f be defined on $\mathbb{Q}$ (the set of all rational numbers) in the following manner:

$$f(x) = \begin{cases} 1 & \text{if } x < \pi; \\ 2 & \text{if } x > \pi; \end{cases} \quad x \in \mathbb{Q}.$$

Here f is continuous on $\mathbb{Q}$, for if $\varepsilon > 0$ and if $x_0 < \pi$, then with $\delta < (\pi - x_0)$, we have $|f(x) - f(x_0)| = 1 - 1 = 0 < \varepsilon$, whenever $|x - x_0| < \delta$.

So f is continuous for any $x_0 \in \mathbb{Q}$ but $< \pi$.

Similarly one can show that f is continuous for any $x_0 \in \mathbb{R}$ but $> \pi$.

However, f is not uniformly continuous on $\mathbb{Q}$.

Consider $0 < \varepsilon < 1$. For any $\varepsilon > 0$ we can always find $x < \pi$ and $y > \pi$ such that $y - x < \delta$. Now $|f(x) - f(y)| = |1 - 2| = 1 > \varepsilon$.

REMEMBER: Whenever we are given a function continuous on a set D, then for $x_0 \in D$ and $\varepsilon > 0$ we can find $\delta(\varepsilon, x_0) > 0$ such that
$$|f(x) - f(x_0)| < \varepsilon \text{ for } x \in \,]x_0 - \delta, x_0 + \delta[.$$

The set $S = \{\delta(\varepsilon, x_0) : x_0 \in D\}$ is always bounded below.

If S has a **positive greatest lower bound**, then the function must be uniformly continuous on D and not otherwise.

CHAPTER 8: CONTINUITY OF FUNCTIONS OF A SINGLE REAL VARIABLE

We next establish a relation between continuity and uniform continuity through the following two theorems:

Theorem 8.5.7. *If a function f is uniformly continuous in a certain interval I, then it is necessarily continuous on I.*

Proof: (follows from the definition of Uniform Continuity).

Let f be uniformly continuous on I. Let $x_0 \in I$ and let ε be any given positive number.

Since f is uniformly continuous on I, therefore, by definition $\exists$ a positive number δ s.t. if x and $y \in I$ and $|x - y| < \delta$ then

$$|f(x) - f(y)| < \varepsilon.$$

We put $y = x_0$ and thus obtain:

if $x \in I$ and $|x - x_0| < \delta$, then $|f(x) - f(x_0)| < \varepsilon$.

This, of course, means that f is continuous at x_0.

Since x_0 is any point of I, it follows f is continuous on I.

> REMEMBER: Uniform continuity $\Rightarrow$ Continuity.

Theorem 8.5.8. *If a function f is continuous on a closed interval $I = \{a \leq x \leq b\}$, then f is uniformly continuous on I.*

Proof: (By contradiction).

Suppose that the theorem were not true: This would mean that there is an exceptional ε, say ε_0, such that for every $\delta > 0$ there exists a pair of points x and $x' \in I$ with $|x - x'| < \delta$ for which $|f(x) - f(x')|$ is not less than ε_0 (i.e., $|f(x) - f(x')| \geq \varepsilon_0$). Take δ successively to be $1, 1/2, 1/3, 1/4, 1/5, \cdots$. Thus there are

$x_1, x_1' \in I$ with $|x_1 - x_1'| < 1 \quad$ for which $|f(x_1) - f(x_1')| \geq \varepsilon_0$;

$x_2, x_2' \in I$ with $|x_2 - x_2'| < 1/2 \quad$ for which $|f(x_2) - f(x_2')| \geq \varepsilon_0$;

$\cdots\cdots\cdots\cdots\cdots\cdots\cdots\cdots\cdots\cdots\cdots\cdots\cdots\cdots\cdots\cdots\cdots\cdots\cdots$

$x_n, x_n' \in I$ with $|x_n - x_n'| < 1/n$ for which $|f(x_n) - f(x_n')| \geq \varepsilon_0$.

Hence we have two sequences $\{x_n\}$ and $\{x_n'\} \in I$ for which

$$|x_n - x_n'| < 1/n \quad \text{and} \quad |f(x_n) - f(x_n')| \geq \varepsilon_0.$$

Now $\{x_n\}$ is bounded, for $a \leq x_n \leq b$ **for every** n and hence $\{x_n\}$ has a **cluster point** which we denote by β, i.e., there is a subsequence $[x_{n_k}]$ converging to β. Thus

$$\left. \begin{array}{l} \beta = \lim_{k \to \infty} x_{n_k} \geq a, \text{ since } x_{n_k} \geq a \\ \text{and } \beta = \lim_{k \to \infty} x_{n_k} \leq b, \text{ since } x_{n_k} \leq b \end{array} \right\}.$$

This means that β is in I and therefore f is continuous at β. [That $\beta \in I$ follows from the fact that I is closed—a closed interval is a closed set.]

Also

$$\begin{aligned} |\beta - x'_{n_k}| &= |\beta - x_{n_k} + x_{n_k} - x'_{n_k}| \\ &\leq |\beta - x_{n_k}| + |x_{n_k} - x'_{n_k}| \\ &\leq |\beta - x_{n_k}| + \frac{1}{n_k}. \end{aligned}$$

As $k \to \infty$, $n_k \to \infty$ and $x_{n_k} \to \beta$ so that clearly we get

$$|\beta - x'_{n_k}| \to 0$$

i.e., $x'_{n_k} \to \beta$ as $k \to \infty$.

Then by the continuity of f at β,

$$|f(x_{n_k}) - f(x'_{n_k})| \to |f(\beta) - f(\beta)| = 0.$$

[see that $\lim_{k \to \infty} f(x_{n_k}) = f\left\{\lim_{k \to \infty} x_{n_k}\right\} = f(\beta).$]
which contradicts the construction that

$$|f(x_{n_k}) - f(x'_{n_k})| \geq \varepsilon_0.$$

The contradiction leads to the conclusion that f is uniformly continuous on I.

> REMEMBER: Continuity in a closed interval $\Rightarrow$ Uniform continuity.

4. Continuity of Inverse Functions

In Chapter 3 we introduced the concepts of monotone functions and inverse functions. It was proved that if f is strictly increasing (or *strictly decreasing*), then f is one-to-one and hence the inverse of f, i.e., f^{-1} exists and the inverse function f^{-1} is also strictly increasing (or *strictly decreasing*). What we shall now prove is that in case f is both strictly monotone and continuous, then f^{-1} exists, strictly monotone and is continuous in its own domain.

CHAPTER 8: CONTINUITY OF FUNCTIONS OF A SINGLE REAL VARIABLE   495

Theorem 8.5.9. *Let f be a strictly increasing function on $a \leq x \leq b$. Then*

(i) *The inverse function f^{-1} exists and is strictly monotone increasing in its domain of definition.*

(ii) *If, further, f is continuous on $a \leq x \leq b$, then f^{-1} is also continuous on $[\alpha, \beta]$ where $\alpha = f(a)$ and $\beta = f(b)$.*

Proof: (i) To prove the existence of f^{-1} and its strict monotonicity we require to assume only that f is *strictly* increasing in *its domain* (but f need not be continuous).
But suppose it is given that f is both continuous and strictly increasing in $a \leq x \leq b$.
Let $y = f(x)$ be *continuous* and *strictly increasing* in $[a, b]$.
Let x_1 and x_2 be any two points in the interval $[a, b]$.
Since f is strictly increasing,

$$x_2 > x_1 \Rightarrow f(x_2) > f(x_1) \text{ and } \mathbf{conversely.} \tag{8.5.4}$$

[if $f(x_2) > f(x_1)$ one cannot conclude $x_2 < x_1$ or $x_2 = x_1$; because then $f(x_2)$ would be less than $f(x_1)$ or $f(x_2) = f(x_1)$ since f is strictly increasing.]
Let $f(a) = \alpha$ and $f(b) = \beta$. Let y_1 be a number between α and β.
Since f is continuous in $[a, b]$, by intermediate-value theorem we can find **some value** $x = x_1$ between a and b for which $f(x_1) = y_1$.
Moreover this value x_1 is UNIQUE; for

$$f(x_1) = f(x_2) = y_1 \Rightarrow x_1 = x_2 \text{ (since (\mathbf{8.5.4}) excludes the possibilities } x_1 \lessgtr x_2).$$

Thus the equation $y = f(x)$ sets up the one-to-one correspondence $x \longleftrightarrow y$ over the intervals $a \leq x \leq b$ and $\alpha \leq y \leq \beta$.
Hence x is a function of y over $\alpha \leq y \leq \beta$. We denote the **inverse function** by $x = f^{-1}(y)$ or $x = \phi(y)$, where $\alpha \leq y \leq \beta$.
Now we observe that if $x_1 = \phi(y_1)$, $x_2 = \phi(y_2)$, the converse case of (**8.5.4**) gives

$$y_2 > y_1 \Rightarrow \phi(y_2) > \phi(y_1)$$

i.e., $\phi(y)$ is also strictly increasing in the interval $\alpha \leq y \leq \beta$.

(ii) To prove the continuity of the inverse function of f.
Let y_0 satisfy $\alpha \leq y_0 \leq \beta$. We have to show that given any $\varepsilon > 0$, $\exists\, \delta > 0$ such that

$$|f^{-1}(y) - f^{-1}(y_0)| < \varepsilon, \text{ for all } y \in |y - y_0| < \delta. \tag{8.5.5}$$

Consider y_0, an intermediate point of $[\alpha, \beta]$ i.e., $\alpha < y_0 < \beta$.

[The other possibility that y_0 is an end-point is rather similar with a slight modification which can be easily made by the readers.]

Figure 8.3 may be helpful to appreciate the arguments.

Let $x_0 = f^{-1}(y_0)$. Then $a < x_0 < b$

($\because f$ is strictly increasing).

We can choose two points x_1 and $x_2(x_1 < x_2) \in [a, b]$ on **either side** of x_0 and within ε-neighbourhood of x_0 (x_1 is on the left of x_0 and x_2 on the right of x_0).

Let $y_1 = f(x_1)$, $y_2 = f(x_2)$. Then $y_1 < y_0 < y_2$.

Now if y is any number between y_1 and y_2 it follows that

$$f^{-1}(y_1) < f^{-1}(y) < f^{-1}(y_2)$$

i.e., $x_1 < f^{-1}(y) < x_2$.

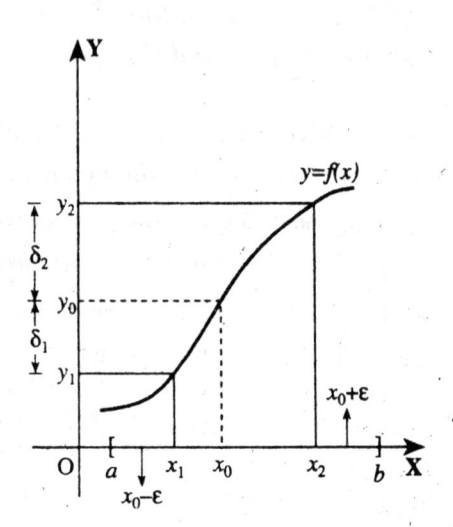

Fig. 8.3

Hence $|f^{-1}(y) - f^{-1}(y_0)| < \varepsilon$ for such y, and the smaller of $\delta_1 = y_0 - y_1$ and $\delta_2 = y_2 - y_0$ will be a suitable δ.

This is what we wished to obtain in (8.5.5).

Alternative Proof. To prove the continuity of $f^{-1}(y)$ or $\phi(y)$ in $\alpha \leq y \leq \beta$. We may consider the **positive function**

$$F(x) = f(x + \varepsilon) - f(x), \text{ where } \varepsilon \text{ is any arbitrary positive number.}$$

The domain of definition of $F(x)$ is $[a, b - \varepsilon]$. Since $F(x)$ is continuous in this closed interval $[a, b - \varepsilon]$ it must assume its minimum value δ (say) at some point of this interval. Hence if x_1, x_2 are two points for which $|x_1 - x_2| \geq \varepsilon$, then, since f is **strictly** increasing,

$$|f(x_1) - f(x_2)| = |y_1 - y_2| \geq \delta. \quad \text{(taking } x_1 = x + \varepsilon, x_2 = x.\text{)}$$

The contrapositive of this statement is: When $|y_1 - y_2| < \delta$, then

$$|x_1 - x_2| = |f^{-1}(y_1) - f^{-1}(y_2)| < \varepsilon$$
$$\text{or,} \quad |\phi(y_1) - \phi(y_2)| < \varepsilon$$

i.e., $x = f^{-1}(y)$ or $x = \phi(y)$ is uniformly continuous in $[a, b]$ and consequently continuous on $[a, b]$.

The corresponding statement for a strictly decreasing function is the following:

CHAPTER 8: CONTINUITY OF FUNCTIONS OF A SINGLE REAL VARIABLE

If the continuous function f is strictly decreasing in $a \leq x \leq b$, then $y = f(x)$ defines an inverse function $x = f^{-1}(y)$ or $x = \phi(y)$, where ϕ is also continuous and strictly decreasing in $f(b) \leq y \leq f(a)$.

Observation: If $f(x)$ is strictly decreasing we may apply the theorem of increasing function to the function $-f(x)$ (which is strictly increasing).

Then $-\phi(y)$ is increasing in the interval $-f(a) \leq y \leq -f(b)$ and hence $\phi(y)$ is decreasing in the interval $f(b) \leq y \leq f(a)$.

Illustrative Examples on Inverse Functions

Example 8.5.7. *The increasing continuous function $y = x^2$ $(0 \leq x < \infty)$ has the inverse $x = \sqrt{y}$ $(0 \leq y < \infty)$ where $\sqrt{y}$ is the positive root of y. The inverse function $x = \sqrt{y}$ can also be written as $y = \sqrt{x}$ $(0 \leq x < \infty)$ (taking x as the independent variable for the function as well as for its inverse).*

Then the two semi-parabolas $y = x^2$ $(0 \leq x < \infty)$ and $y = \sqrt{x}$ $(x \geq 0)$ are symmetric to the line $y = x$. [see the Figure 3.14.]

Example 8.5.8. *Show that $y = \sin x$ is a continuous function of x, for all real values of x, the range set being $-1 \leq y \leq 1$.*

Solution: Now see that for each $y \in [-1, 1]$ there are infinite number of values of x for which $\sin x$ is equal to that given y i.e., $x = \sin^{-1} y$ is a *multiple-valued function of y.*

We, however, take a restricted domain of x, say $-\pi/2 \leq x \leq \pi/2$.

In this interval $y = \sin x$ is a *continuous, strictly increasing function*, the range set being $-1 \leq y \leq 1$.

Hence $\exists$ an inverse function $x = \sin^{-1} y$ $(-1 \leq y \leq 1$ and $-\pi/2 \leq x \leq \pi/2)$.

We may take x as independent and y as dependent variable and thus obtain the inverse function

$$y = \sin^{-1} x, \quad (-1 \leq x \leq 1 \text{ and } -\pi/2 \leq y \leq \pi/2).$$

This is what we mean by the **principal branch** of the inverse of sine.

Observe that instead of taking $-\pi/2 \leq x \leq \pi/2$, we could take intervals

$$\pi/2 \leq x \leq 3\pi/2, \quad 3\pi/2 \leq x \leq 5\pi/2, \cdots, -3\pi/2 \leq x \leq -\pi/2, \cdots.$$

On any one of these intervals $y = \sin x$ will define a continuous, inverse function of sine.

Thus inverse function of sine is basically a multiple-valued function but it can be regarded as composed of an *enumerable set of single-valued, continuous, monotonic* functions—the *principal branch* is obtained when we consider $y = \sin x$ for $-\pi/2 \leq x \leq \pi/2$.

Example 8.5.9. *The strictly increasing function* $y = \tan x \left(-\frac{\pi}{2} < x < \frac{\pi}{2}\right)$ *has the inverse* $x = \tan^{-1} y$ $(-\infty < y < \infty)$ *where* $\tan^{-1} y$ *denotes the principal branch of the inverse tangent; its values are restricted to the interval* $(-\pi/2, \pi/2)$.

Example 8.5.10. *The strictly decreasing function* $y = \cos x (0 \leq x \leq \pi)$ *has inverse* $x = \cos^{-1} y (-1 \leq y \leq 1)$, *where* $\cos^{-1} y$ *denotes the principal branch of the inverse cosine; its values are restricted to the interval* $[0, \pi]$. *Note that* $\cos^{-1} y$ *is strictly decreasing in* $[-1, 1]$.

Example 8.5.11. Exponential and Logarithm:

(i) When x is a positive integer $= n$ (say), then we define

$$a^n = a \times a \times a \times \cdots \text{ to } n \text{ such factors.}$$

For positive integral exponents we can prove the *laws of exponents*:

$$a^m \cdot a^n = a^{m+n} \quad \text{and} \quad (a^m)^n = a^{mn}.$$

(ii) In order to maintain the validity of these laws for ALL rational exponents, we define

$$a^0 = 1, \quad a^{p/q} = \sqrt[q]{a^p}, \quad a^{-r} = \frac{1}{a^r}.$$

(iii) We assume $a > 0$. When ρ is an irrational number. We can always find a monotonic sequence of rationals $\{r_n\}$ having ρ as its limit. Then $\{a^{r_n}\}$ is also a bounded monotonic sequence which has a limit. The limit we define as a^ρ.

Thus $a^x (a > 0)$ is defined for all real exponents x.
The function a^x is positive and monotone in $-\infty < x < \infty$

$$\begin{cases} \text{if } a > 1, & a^x \text{ increases from 0 to } \infty. \\ \text{if } 0 < a < 1, & a^x \text{ decreases from } \infty \text{ to 0.} \end{cases}$$

Finally a^x is everywhere continuous. Continuity at $x = 0$ means that

$$\lim_{h \to 0} a^h = 1 = a^0.$$

When $h \to 0^+$, this follows from the fact $\{r^{1/n} \to 1 \text{ as } n \to \infty, r > 0\}$.

Since $a^{-h} = 1/a^h$, we have the same limit when $h \to 0^-$.

Therefore,
$$\lim_{h \to 0} a^{x+h} = \lim_{h \to 0} a^x \cdot a^h = a^x,$$

so that a^x is continuous for all real values of x.

When $a > 1$, $y = a^x$ is continuous and strictly increasing in $-\infty < x < \infty$; hence the inverse function $x = \log_a y$ $(0 < y < \infty)$, *the logarithm of y to the base a*, is also continuous and strictly increasing in $0 < y < \infty$.

In particular, when $a = e = 2.71825\cdots$ we have the exponential function e^x (or exp x) and its inverse, the **natural logarithm** $\log x$ (the base e is usually omitted) or $\ln x$ for which
$$e^{\log y} = y \quad \text{and} \quad \log e^x = x.$$

See Fig. 3.16: From the graph of $y = e^x$ we may obtain the graph of $y = \log x$ by a reflection in the line $y = x$.

6. Continuity and Monotonicity

We advise our students to study the two theorems given below in the light of the results proved under Theorem **8.5.3**.

Theorem 8.5.10. *If $f(x)$ be continuous in the interval $a \leq x \leq b$ and assumes each values between $f(a)$ and $f(b)$ just once, it is strictly monotonic in this interval.*

(Compare, Theorem 8.5.3)

Proof: Suppose $f(a) < f(b)$. Then if $a < x_1 < b$, we are to prove
$$f(a) < f(x_1) < f(b).$$

For this we show that the hypothesis rules out all other possibilities;

(i) $f(x_1) = f(a)$ or $f(x_1) = f(b)$;

(ii) $f(x_1) < f(a)$;   (iii) $f(x_1) > f(b)$.

Case (i) is obviously impossible. In case (ii), $f(x)$ must be equal to $f(a)$ for some value of x between x_1 and b by Theorem **8.5.2**, which again contradicts the condition of the theorem. In case (iii) $f(x) = f(b)$ for some x between a and x_1, hence impossible.

Now, if $a < x_1 < x_2 < b$, we have just seen that $f(x_1) < f(b)$ and consequently $f(x_1) < f(x_2)$. Thus $f(x)$ is strictly monotone increasing.

If $f(a) > f(b)$, we can show in the same way that $f(x)$ is strictly monotone decreasing.

Hence the theorem.

Theorem 8.5.11. *Let f be monotonic increasing in $[a,b]$. Then if c be s.t. $a < c < b$ then $f(c-0)$ and $f(c+0)$ both exist and*

$$\sup_{a<x<c} f = f(c-0) \quad \text{and} \quad \inf_{c<x<b} f = f(c+0)$$

Also, $f(c-0) \leq f(c) \leq f(c+0)$.

Proof: The set $A = \{f(x) : a < x < c\}$ is bounded above. In fact, $f(c)$ is an upper bound of A.

Let $\sup A = \xi$. Then $\xi \leq f(c)$ since ξ is the lub of A.

Let $\varepsilon > 0$ be given. We have

$$f(x) \leq \xi, \quad \forall\, x \in (a,c)$$
$$\text{hence,} \quad f(x) < \xi + \varepsilon, \quad \forall\, x \in (a,c).$$

Also $\exists\, \delta > 0$ s.t. $a < c - \delta < c$ and $f(c-\delta) > \xi - \varepsilon$.
Since f is m.i. $f(x) > f(c-\delta) > \xi - \varepsilon$, whenever $x > c - \delta$.
$\therefore\ \exists\, \delta > 0$ s.t. $\xi - \varepsilon < f(x) < \xi + \varepsilon$, whenever $c - \delta < x < c$.
This implies $f(c-0) = \lim_{x \to c-0} f(x) = \xi = \sup_{a<x<c} f(x)$.

We may similarly prove that $f(c+0) = \inf_{c<x<b} f(x)$.

In case of m.d. function $f \in [a,b]$, we have the result

$$f(c-0) = \inf_{a<x<c} f(x), \quad f(c+0) = \sup_{c<x<b} f(x).$$

Also, $f(c-0) \geq f(c) \geq f(c+0)$.

Main Points : A Recapitulation

1. Continuity at a point x_0. Let f be defined on an interval I and x_0 is an interior point of I.

f is *continuous at x_0 iff* for each $\varepsilon > 0$, $\exists\, \delta > 0$ s.t. $|f(x) - f(x_0)| < \varepsilon$, $\forall\, x \in |x - x_0| < \delta \cap I$.

The definition implies $f(x_0 + 0) = f(x_0 - 0) = f(x_0)$. At the end of point of an interval I we shall take one-sided continuity.

Discontinuity (not continuous) may arise in more than one way:

(i) *First kind* or *Simple discontinuity* if $f(x_0 + 0) = f(x_0 - 0) \neq f(x_0)$ (*removable*).

(ii) *Second kind*, if $\lim f(x)$ as $x \to x_0$ does not exist.

(iii) We have also *infinite discontinuity* and *oscillatory discontinuity*.

(iv) In case, $f(x_0 + 0)$ and $f(x_0 - 0)$ both exist, then their difference is called the *jump of f* at x_0. If either this difference or $f(x_0 + 0) - f(x_0)$ or $f(x_0) - f(x_0 - 0)$ be different from zero, then f is said to have a *jump discontinuity* at x_0.

2. Continuity in a closed interval $[a,b]$: Continuity at interior points and *one-sided continuity* at a and b.

3. f continuous at $x_0 \in I \Leftrightarrow$ For every convergent sequence $\{x_n\} \in I$ which has x_0 as limit, $\{f(x_n)\} \to f(x_0)$ as $n \to \infty$.

4. Sum, difference, product and quotient of two continuous functions is continuous. Also *composite function Fof* is continuous at x_0, if f is continuous at x_0 and $F(t)$ is continuous at $t = f(x_0)$.

5. Important Properties: Neighbourhood Property, Intermediate value property (Bolzano's Theorem) and property of Boundedness and Attainment of Bounds: f continuous in $[a,b] \Rightarrow f$ attains every value between its *infimum* and *supremum*.

6. Uniform Continuity: A function f defined over $D \subseteq R$ is uniformly continuous on a set $S \subseteq D$, if for each $\varepsilon > 0$, $\exists$ a δ s.t.

$|f(x) - f(y)| < \varepsilon$ for all x and y of S satisfying $|x - y| < \delta$.

If f is continuous in a closed interval $[a,b]$, then f is uniformly continuous there.

> **7. Monotonic functions:** (i) f continuous in $[a,b]$ and assumes each value between $f(a)$ and $f(b)$ just once, then f is strictly monotonic.
>
> (ii) If f is m.i. $\in [a,b]$, then $f(x_0+0)$ and $f(x_0-0)$ both exist for every $x_0 \in (a,b)$, and $f(x_0-0) \leq f(x_0) \leq f(x_0+0)$. At end points $f(a) \leq f(a+0)$ and $f(b-0) \leq f(b)$. Similar results for f m.d.
>
> **8. Inverse Functions:** Let f be strictly increasing on $[a,b]$. Then
>
> (i) f^{-1} exists and f^{-1} is strictly increasing in its own domain of definition.
>
> (ii) If f is continuous on $[a,b]$, then f^{-1} also is continuous in $[f(a), f(b)]$. Similar results for strictly decreasing function f on $[a,b]$.

Exercises VIII
(Theoretical Questions on Continuity)

1. Define continuity of a function f at a point in its domain of definition and also in a certain closed interval in which f is defined.

2. Give one example of (i) a removable discontinuity, (ii) a discontinuity of the first kind (non-removable), (iii) a discontinuity of second kind, (iv) an Infinite discontinuity and (v) an Oscillatory discontinuity.

3. Define jump discontinuity of a function at a point and the height of the jump at such a point of discontinuity. Give an example. **[C.H. 1982]**

4. Prove that the necessary and sufficient condition for a function defined on an interval I to be continuous at a point $a \in I$ is that for each sequence $\{a_n\}$ which converges to a, we have
$$\lim_{n \to \infty} f(a_n) = f(a).$$

   Use this theorem to show that the **Dirichlet's function** f defined on $\mathbb{R}$ by $f(x) = 1$, *when x is rational;* $= -1$, *when x is irrational* is discontinuous for every $x \in \mathbb{R}$.

5. Let f and g be two real-valued functions. If f is continuous at a, and g is continuous at $f(a)$, then prove that the composite function gof is also continuous at a.

6. If f is continuous in $[a, b]$ then prove that $f(x)$ is bounded there and that the exact bounds are attained in $[a, b]$. [C.H. 1994]

7. Let f be continuous in $[a, b]$ and let $f(c) > 0$, where $a < c < b$. Show that there exists a sub-interval $(c - \delta, c + \delta)$ of $[a, b]$ in which every $f(x)$ is positive.

[C.H. 1992, 93, 96]

8. **(Bolzano's Theorem)**. If f is continuous in $[a, b]$ and $f(a) \cdot f(b) < 0$, then prove that there exists a point $c \in [a, b]$ such that $f(c) = 0$. [C.H. 1982, 91, 95]

9. (i) **(Intermediate Value Theorem)**. If f is continuous in $[a, b]$ then prove that $f(x)$ must assume every value between $f(a)$ and $f(b)$ at least once in that interval. [C.H. 1980]

   (ii) Let f be continuous in $[a, b]$ and let M and m be respectively the lub and glb of $f(x)$'s in $[a, b]$ and let η be a number between m and $M (m \leq \eta \leq M)$. Then show that $\exists$ a point $\xi \in [a, b]$ such that $f(\xi) = \eta$.

10. (i) Define uniform continuity of a function in an interval I.

    (ii) Show that $\sin(1/x)$ is not uniformly continuous in $(0, 1)$ though it is continuous at every point of $(0, 1)$.

    (iii) Prove that a function f which is continuous in a closed interval $[a, b]$ is uniformly continuous there.

11. (i) Let $f : \mathbb{R} \to \mathbb{R}$ be *strictly increasing* and continuous and let $S = f(\mathbb{R})$. Then $f^{-1} : S \to \mathbb{R}$ is also strictly increasing and continuous.

    (ii) Let f be continuous and one-to-one on an interval I. Then prove that f is strictly monotonic in I.

12. Let f be a real-valued function whose domain of definition is a point set E and whose range is X. If E is a compact set and f is continuous on E, then prove that:

    (i) X is a compact set,

    (ii) f is bounded on E and lub and glb of f on E are attained on E.

[C.H. 1981]

13. (i) Let f be monotonically increasing in $[a, b]$. If c is any point such that $a < c < b$, then prove that $f(c - 0)$ and $f(c + 0)$ both exist and

$$f(c - 0) = \sup_{a < x < c} f(x), \quad f(c + 0) = \inf_{c < x < b} f(x).$$

Also prove that $f(c-0) \le f(c) \le f(c+0)$.

How will you modify the results if f be monotonically decreasing in $[a,b]$?

(ii) If f be a continuous function in $[a,b]$ and if f assumes every value between $f(a)$ and $f(b)$ just once, then prove that it is strictly monotonic in this interval.

[C.H. 1990]

14. Let f be continuous on $[a,b]$ and suppose that a function g is defined thus: $g(a) = f(a)$ and $\forall x \in a < x \le b, g(x)$ is the maximum value of $f \in [a,x]$. Show that g is continuous on $[a,b]$.

15. If $f : [a,b] \to \mathbb{R}$, $g : [a,b] \to \mathbb{R}$ be continuous on $[a,b]$. Prove that the function $\phi(x)$, where
$$\phi(x) = \max.\{f(x), g(x)\} : [a,b] \to \mathbb{R}$$
is continuous on $[a,b]$.

[A]

1. Show that the following functions are continuous at the points indicated:

   (a) $f(x) = \begin{cases} x^2, & \text{when } 0 < x < 1; \\ x, & \text{when } 1 \le x \le 2; \\ \frac{1}{4}x^3, & \text{when } 2 \le x < 3. \end{cases}$

   (at the points, $x = 1$ and $x = 2$).

   (b) $f(x) = \begin{cases} (1+2x)^{1/x}, & x \ne 0; \\ e^2, & x = 0. \end{cases}$

   (at the point $x = 0$).

   (c) $f(x) = \begin{cases} x^2 \sin(1/x), & (x \ne 0); \\ 0, & \text{when } x = 0. \end{cases}$

   (at the point $x = 0$).

2. Discuss the continuity of $f(x)$ for $x \ge 0$, where
$$f(x) = \begin{cases} 0, & \text{when } x = 0; \\ 1/x, & \text{when } 0 < x < 1; \\ \frac{1}{x^2} \sin \frac{\pi x}{2}, & 1 \le x \le 2; \\ 1 - e^{x-2}, & \text{for } x > 2. \end{cases}$$

[C.H. 1980]

3. f is defined in $(0,2)$ by $f(x) = x - [x]$. Prove that f is not continuous at $x = 1$.

[C.H. 1981]

4. Discuss the right-continuity, the left continuity and then continuity of the following functions at the indicated points:

(a) $f(x) = \begin{cases} (x^2 - 1)/(x - 1), & \text{for } x < 1 \\ 2, & \text{for } x = 1 \\ x + 2, & \text{for } x > 1 \end{cases}$ at $x = 1$.

(b) $f(x) = \begin{cases} 1 - x, & x > 0 \\ 2 + x, & x < 0 \\ 1, & x = 0 \end{cases}$ at $x = 0$.

(c) $f(x) = \begin{cases} x, & 0 \leq x \leq 1 \\ x - 2, & 1 < x \leq 2 \end{cases}$ at $x = 1$. [C.H. 1963]

(d) $f(x) = \begin{cases} e^{-1/x^2}, & x \neq 0 \\ 1, & x = 0 \end{cases}$ at $x = 0$.

(e) $f(x) = \begin{cases} \dfrac{xe^{1/x}}{1 + e^{1/x}}, & x \neq 0 \\ 0, & x = 0 \end{cases}$ at $x = 0$. [C.H. 1967]

(f) $f(x) = \begin{cases} 1 + x, & x \leq 0 \\ x, & 0 < x < 1 \\ 2 - x, & 1 \leq x \leq 2 \\ 3x - x^2, & x > 2 \end{cases}$ at $x = 0$, at $x = 1$, at $x = 2$. [C.H. 1966]

[Ans: (a) left; (b) right; (c) left; (d) neither; (e) all; (f) at $x = 0$ left, at $x = 1$ cont., at $x = 2$ left.]

5. What is the nature of discontinuity at the indicated points of each of the following functions?

(a) $[x] + [-x]$ at $x = 0$;

(b) $e^{1/(x-a)}$ at $x = a$;

(c) $\dfrac{1}{(x-a)^{2m}}$ at $x = a$ (m is a positive integer);

(d) $\dfrac{1}{(x-a)^{2m+1}}$ at $x = a$ (m is a positive integer);

(e) $\dfrac{1}{x-a} \sin \dfrac{1}{x-a}$ at $x = a$;

(f) $e^{-1/(x-1)}$ at $x = 1$;

(g) $\dfrac{x^2 - 4}{x - 2}$ at $x = 2$.

[*Ans:* (a) simple, non-removable; (b)–(d) infinite discontinuities; (e) oscillatory; (f) infinite; (g) simple, removable if $f(2) = 4$.]

6. Show that $f(x) = \dfrac{x^4 + 3x^3 + 5x^2}{\sin x}$, for $x \neq 0$; $= 0$, for $x = 0$, is continuous at $x = 0$, but $\phi(x) = \dfrac{x^4 + 3x^3 + 5x^2 + 4x}{\sin x}$, for $x \neq 0$; $= 0$ for $x = 0$, is discontinuous at $x = 0$.

7. Discuss the continuity of $f(x)$ at $x = 0$ and $\phi(x)$ at $x = \pm 1$, where

   (a) $f(x) = \dfrac{1}{1 - e^{1/x}}$, for $x \neq 0$; $= 0$ for $x = 0$. [C.H. 1976]

   (b) $\phi(x) = 0$, for $x^2 > 1$; $= 1$ for $x^2 < 1$; $= \frac{1}{2}$ for $x^2 = 1$.

8. (a) Show that $\sin(1/x)$ is discontinuous at $x = 0$, no matter how $f(0)$ is defined. [C.H. 1977]

   (b) Show that $\dfrac{x}{1 + e^{1/x}}$ is continuous everywhere except at $x = 0$ where there is a *removable* discontinuity. How do you define the function at $x = 0$ in order to remove the discontinuity?

   (c) Show that $\dfrac{x}{x - 1}$ is continuous everywhere except at $x = 1$ where there is an infinite discontinuity.

9. Show that the function f defined by $f(x) = 0$, for $x = 0$, $= \frac{1}{2} - x$, for $0 < x < 1/2$; $= 1/2$ for $x = \frac{1}{2}$; $= 3/2 - x$, for $\frac{1}{2} < x < 1$; $= 1$ for $x = 1$, has three points of discontinuity; find them. Also draw the graph of this function. [C.H. 1965]

10. Investigate the points of continuity or discontinuity of the function ϕ defined by $\phi(x) = (x^2/a) - a$, for $0 < x < a$; $= 0$ for $x = 0$; $= a - a^3/x^2$, when $x > a$.

    [*Ans:* Continuous for all real x.]

[B]

[Solutions for the problems of this section are given after [C]]

1. $f(x) = x$, when x is rational; $= 1 - x$, when x is irrational.

   Show that $f(x)$ is continuous at $x = 1/2$. [C.H. 1974, 1984]

CHAPTER 8: CONTINUITY OF FUNCTIONS OF A SINGLE REAL VARIABLE

2. $f(x) = x$, when x is rational; $= 0$, when x is irrational.

   Show that $f(x)$ is continuous only at $x = 0$, but discontinuous at all other points.

3. f is defined for positive x in the following way:

   $f(x) = 0$, when x is irrational; $= 1/q$ when $x = p/q$, where p and q are integers in their lowest terms.

   Show that $f(x)$ is continuous for any irrational value of x, but discontinuous for all rational values of x.

4. Verify that the function defined in Ex. 1 above assumes every value between 0 and 1 once, and only once as x increases from 0 to 1 but the function is discontinuous for every value of x except at $x = 1/2$.

5. A function f assumes only rational values in $0 \leq x \leq 1$ but continuous in this interval. If $f(1/2) = \frac{1}{2}$, prove that $f(x) = \frac{1}{2}$ everywhere in $[0, 1]$.

   [C.H. 1991, 96]

6. Let f be continuous on $[a, b]$ and let $f(x) = 0$ when x is rational. Show that $f(x) = 0$ for every $x \in [a, b]$.   [C.H. 1978]

7. $f(x) = \lim\limits_{n \to \infty} \dfrac{\log(2 + x) - x^{2n} \sin x}{1 + x^{2n}}.$

   (a) Discuss the nature of discontinuity of this function at $x = 1$.

   (b) Show that $f(0)$ and $f(\pi/2)$ differ in sign.

   (c) Explain why this function does not vanish anywhere in $[0, \pi/2]$ although $f(0)$ and $f(\pi/2)$ differ in sign.   [C.H. 1969, 90]

8. Prove that any function f, which is continuous in an interval (no matter how small) and which satisfies for all values of x and y, the relation

   $$f(x + y) = f(x) + f(y),$$

   is of the form $f(x) = ax$, where a is a constant.

9. Prove that $x^n = 2$ has one and only one positive root (n is a positive integer).

10. (a) Prove that if $f(x)$ is continuous at $x = a$ and for every $\delta > 0$ there is a point in $|x - a| < \delta$, where $f(x) = 0$, then $f(a) = 0$.   [C.H. 1963]

    (b) Let f be continuous in an interval and suppose it does not vanish for any x in this interval. Show that $f(x)$ has a constant sign in this interval.

[C]
(Further Miscellaneous Exercises)

1. Define uniform continuity of a function on a set.

   (a) Show that $\sin(1/x)$ or $\cos(1/x)$ is *not uniformly continuous* in $(0,1)$.
   [C.H. 1980, 83, 91]

   (b) $f(x) = x^2$ is uniformly continuous on any bounded set of real numbers, but not uniformly continuous on $\mathbb{R}$. Prove this statement.

   (c) Prove that $\sqrt{x}$ and $\sin x$ are uniformly continuous on $\mathbb{R}$ and that $\tan x$ is uniformly continuous over $[a,b]$ where $-\frac{\pi}{2} < a < b < \frac{\pi}{2}$. [C.H. 1995]

2. Prove that $\frac{5}{x-1} + \frac{7}{x-2} + \frac{16}{x-3} = 0$ has one solution between 1 and 2 and another solution between 2 and 3.

3. Suppose that $f(x)$ is continuous in an interval and does not vanish anywhere in the interval. Show that $f(x)$ has one constant sign on this interval.

4. Show that for each non-negative number y, $\exists$ a unique $x \geq 0$ for which $x^2 = y$.

5. Show that $f(x) = [x]\sin^2 \pi x$ is continuous at every positive integral value of x. Show also that $[x]\cos^2 \pi x$ is not continuous for positive integral values of x.

6. If $f(x) = x - [x]$, show that

   (a) $f(x)$ is discontinuous for all integral values of x and is continuous for all other values. [Banaras M.Sc.]

   (b) In every interval which includes an integer it is bounded between 0 and 1.

   (c) The lower bound 0 is attained but not the upper bound 1 in any interval including an integer.

7. Investigate the inverse of the following functions (use Theorem 8.5.10), stating in each case the domain of the inverses:

   (a) $y = x^2$; (b) $y = x^3$; (c) $y = \sin x$ $(-\pi/2 \leq x \leq \pi/2)$;
   (d) $y = \cos x$ $(0 \leq x \leq \pi)$; (e) $y = \tan x$ $(-\pi/2 < x < \pi/2)$; (f) $y = a^x$;
   (g) $y = e^x$.

   [Ans: (a) $x = +\sqrt{y}$ $(0 \leq y < \infty)$; (b) $x = \sqrt[3]{y}$ $(-\infty < y < \infty)$; (c) $x = \sin^{-1} y$ $(-1 \leq y \leq 1)$; (d) $x = \cos^{-1} y$ $(-1 \leq y \leq 1)$; (e) $x = \tan^{-1} y$ $(-\infty < y < \infty)$; (f) $a > 1; x = \log_a y$ $(0 < y < \infty)$; (g) $x = \log y$ $(0 < y < \infty)$.]

CHAPTER 8: CONTINUITY OF FUNCTIONS OF A SINGLE REAL VARIABLE

8. Draw the graph of the function y of x defined by y *is the smallest positive number that makes $x + y$ an integer.*

9. (*Harder*) Show that the function f defined by

$$f(x) = \frac{1}{1 + e^{1/(\sin n!\pi x)}}$$

can be made discontinuous at any rational point in $[0, 1]$ by a proper choice of n.

10. (*Harder*) Show that the function f defined by

$$f(x) = \lim_{n \to \infty} \left[\lim_{t \to 0} \frac{\sin^2(n!\pi x)}{\sin^2(n!\pi x) + t^2} \right]$$

is equal to 0, when x is rational and is equal to 1 when x is irrational.

Now show that f is discontinuous for every real value of x (Totally Discontinuous).

11. Determine the discontinuties of the function f defined by

$$f(x) = \lim_{n \to \infty} \frac{(1 + \sin \pi/x)^n - 1}{(1 + \sin \pi/x)^n + 1} \quad (0 < x < 1).$$

[*Ans: f has discontinuity of second kind at $x = 0$, and simple discontinuities at $x = 1, 1/2, 1/3, \cdots$.*]

12. If f is continuous on a closed set F, show that the set of all those points $x \in F$ for which $f(x) = 0$ is closed.

13. Let $\phi(x) = \frac{x^{2n+2} - \cos x}{x^{2n} + 1}$. Show that $\phi(0)$ and $\phi(2)$ differ in sign but $\phi(x)$ does not vanish for any $x \in [0, 2]$. Explain the reason.

14. If a continuous function f satisfies the relation $f(x + y) = f(x) \cdot f(y)$, then prove that either $f(x) = 0$ or else $f(x) = e^{ax}$, $\forall x$ (a is a constant).

15. Give an example of

    (a) a continuous bounded function on $\mathbb{R}$ may attain its supremum but not its infimum;

    (b) a continuous bounded function on $\mathbb{R}$ may attain its infimum but not its supremum;

    (c) a continuous bounded function defined on an interval may attain neither its supremum nor its infimum.

[Ans: (a) $f(x) = 1/x^2 + 1$, $\forall x \in \mathbb{R}$; (b) $f(x) = -1/x^2 + 1$, $\forall x \in \mathbb{R}$; (c) $f(x) = x$, $\forall x \in]0, 1[$.]

[We give below Solutions of some of the problems under [B]. These problems are extremely important.]

1. $f(1/2) = 1/2$. $|f(x) - f(1/2)| = \begin{cases} |x - 1/2|, & \text{when } x \text{ is rational;} \\ |1 - x - \frac{1}{2}| = |\frac{1}{2} - x| = |x - \frac{1}{2}|, & \text{when } x \text{ is irrational} \end{cases}$

$\therefore |f(x) - f(1/2)| <$ any given ε (positive) for all x (rational and irrational) in $|x - 1/2| < \delta$, where $\delta = \varepsilon$; hence f is continuous at $x = 1/2$.

2. **Case 1.** At x = 0. $f(0) = 0$. Now

$$|f(x) - f(0)| = \begin{cases} |x - 0| = |x|, & \text{when } x \text{ is rational;} \\ |0 - 0| = 0, & \text{when } x \text{ is irrational.} \end{cases}$$

$\therefore |f(x) - f(0)| \leq |x|$ (equality, when x is rational),

$< \varepsilon$, whenever $|x - 0| < \delta$, where $\delta = \varepsilon$.

$\therefore \lim_{x \to 0} f(x_0) = f(0)$ and f is continuous at $x = 0$.

Case 2. At $x = a$, where $a \neq 0$ but a is a rational number so that $f(a) = a$. Take any *nbd* N of a. Consider *only irrational values* of $x \in N$. We then have $|f(x) - f(a)| = |0 - a| = |a|$.

If we choose $\varepsilon > 0$ but $\leq |a|$ then $|f(x) - f(a)|$ cannot be made less than ε for all $x \in N$ (at least not for irrational x of N).

$\therefore f$ is not continuous at $x = a$, where a is a *non-zero rational number*.

Case 3. At $x = a$, where a is irrational so that $f(a) = 0$.

Considering **any** *nbd* N_1 of a, we now take *only those rational values* of x for which $|x| > |a|$. For such values of x, $|f(x) - f(a)| = |x - 0| = |x|$.

$\therefore |f(x) - f(a)| > |a|$ for such x's of N_1.

Hence if we choose $\varepsilon < |a|$, then $|f(x) - f(a)|$ will not be less than ε for all values of x in any *nbd* N_1 of a.

$\therefore f$ cannot be continuous at $x = a$, where a is *irrational*.

3. **Case 1.** Let $x = p/q$ ($q \neq 0$), p and q are positive integers in their lowest terms. Then $f(p/q) = 1/q$. We **know** that in any *nbd* N of p/q $\exists$ infinite number of

rational as well as irrational numbers. We consider only irrational values of x in N. Then we have
$$|f(x) - f(p/q)| = |0 - 1/q| = 1/q,$$
which is not less than any prescribed ε (we may choose $\varepsilon < 1/q$).

Thus there cannot exist a nbd of p/q for all points of which $|f(x) - f(p/q)|$ can be made less than any given ε.

i.e., f is not continuous at $x = p/q$.

Case 2. Let us now consider the case $x = c$, where c is any positive irrational number. Then $f(c) = 0$. Let the integral part of c be m. Then $m < c < m+1$.

We can arrange the rational numbers between m and $(m+1)$ in the order of increasing denominators. Let ε be any given positive number.

Choose a positive integer $n > 1/\varepsilon$. We find all positive fractions between m and $(m+1)$. There can be only a finite number of such fractions whose denominators $\leq n$. Take that fraction as x' which is nearest to c. Choose $\delta = |x' - c|$.

Now in the δ-nbd of c, $\exists$ infinite number of irrational numbers and also an infinite number of fractions like p/q whose $q > n$.

In this δ-nbd of c, for all irrational x, $|f(x) - f(c)| = |0 - 0| = 0 < \varepsilon$ and, for all rational x,

$$|f(x) - f(c)| = |f(p/q) - 0| = \left|\frac{1}{q} - 0\right| = \frac{1}{q} < \frac{1}{n} < \varepsilon.$$

$\therefore$ Given any $\varepsilon > 0$, we can find δ s.t. $|f(x) - f(c)| < \varepsilon$, whenever $|x - c| < \delta$.

5. Take any point $\alpha \in [0, 1]$, other than $1/2$. Since f is continuous in the closed interval $[0, 1]$, $f(\alpha)$ exists finitely. If $f(\alpha) \neq 1/2$, f must assume every value between $f(1/2) = \frac{1}{2}$ and $f(\alpha)$ at least once for values of x between $1/2$ and α.

Between $f(1/2)$ and $f(\alpha)$, $\exists$ infinite number of rational as well as irrationals.

But by the given condition $f(x)$ assumes only rational values. Hence we must have $f(\alpha) = 1/2$. Since α was taken arbitrarily, hence $f(x) = 1/2$ everywhere in $0 \leq x \leq 1$.

7. We know: $\lim\limits_{n \to \infty} x^{2n} = \infty$, if $x^2 > 1$; $= 1$, if $x^2 = 1$; $= 0$, if $x^2 < 1$ ($\because (x^2)^n = x^{2n}$).

or, $\lim\limits_{n \to \infty} 1/x^{2n} = 0$, when $x > 1$ and $\lim\limits_{n \to \infty} x^{2n} = 0$, when $-1 < x < 1$.

$\therefore$ For $x > 1$,

$$f(x) = \lim_{n \to \infty} \frac{(1/x^{2n}) \log(2 + x) - \sin x}{(1/x^{2n}) + 1} = -\sin x. \tag{1}$$

For $-1 < x < 1$,
$$f(x) = \lim_{n \to \infty} \frac{\log(2+x) - x^{2n} \cdot \sin x}{1 + x^{2n}} = \log(2+x). \tag{2}$$
But clearly for $x = 1$, $f(x) = f(1) = \frac{\log 3 - \sin 1}{2}$.

$\therefore \lim_{x \to 1+0} f(x) = -\sin 1$ and $\lim_{x \to 1-0} f(x) = \log 3$. Certainly $-\sin 1$ and $\log 3$ are of opposite signs and cannot be equal, we see that $\lim_{x \to 1} f(x)$ does not exist. Thus, f is discontinuous at $x = 1$ and this discontinuity is of *second kind* (non-removable).

Since $\pi/2 >, 1$ we have, from (1), $f(\pi/2) = -1 < 0$ and since $-1 < 0 < 1$ we have from (2), $f(0) = \log 2 > 0$.

i.e., $f(0)$ and $f(\pi/2)$ are of opposite signs.

We see that f is **not continuous** in $[0, \pi/2]$ ($x = 1$ is a point of discontinuity) and hence f may not vanish anywhere in $[0, \pi/2]$ although $f(0)$ and $f(\pi/2)$ are of opposite signs. It can be easily checked that in $[0, \pi/2]$, the function f is defined as follows:
$$f(x) = \begin{cases} \log(2+x), & 0 \le x < 1; \\ \frac{1}{2}(\log 3 - \sin 1), & \text{if } x = 1, \\ -\sin x, & 1 < x \le \pi/2. \end{cases}$$
so that f is no-where zero in $[0, \pi/2]$.

8. Take $x = 0, y = 0$; then $f(0) = 0$. Take $y = -x$; $f(0) = f(x) + f(-x)$ so that $f(x) = -f(-x)$.

    (a) Let x be any given positive integer. Then
    $$f(x) = f(1 + 1 + \cdots + 1) = f(1) + f(1) + \cdots = xf(1) = ax, \text{ where } a = f(1).$$

    (b) Let x now be a negative integer. Write $x = -y (y > 0)$ then
    $$f(x) = f(-y) = -f(y) = -ay = ax.$$

    (c) Let $x = p/q (q > 0)$. We have ($\because$ q is a positive integer)
    $$f(p) = f\left(\frac{p}{q} \cdot q\right) = f\left(\frac{p}{q} + \frac{p}{q} + \cdots q \text{ terms}\right)$$
    $$= f\left(\frac{p}{q}\right) + f\left(\frac{p}{q}\right) + f\left(\frac{p}{q}\right) + \cdots q \text{ terms} = qf(p/q).$$

    Thus, $f(p) = qf(p/q)$.
    $\therefore ap = qf(p/q) \Rightarrow f(x) = ap/q$ i.e., $f(x) = ax$ ($x =$ rational).

(d) Suppose that x is any real number. Let $\{x_n\}$ be any sequence of rational numbers converging to x. We have x_n, being rational, $f(x_n) = ax_n$. Since f is continuous, we now obtain, as $n \to \infty$.

L.H.S. $= \lim_{n\to\infty} f(x_n) = f(\lim_{n\to\infty} x_n) = f(x)$; R.H.S. $= a \lim_{n\to\infty} x_n = ax$.

$\therefore f(x) = ax, \forall\, x \in \mathbb{R}$.

9. $x^n = 0$, if $x = 0$; $x^n = b^n > 2$, if we take $x = b > 2$.

Therefore, as x varies from 0 to b, the *continuous function* $f(x) = x^n$ must, for at least one value of x, be equal to 2.

Further, if $x > 0$ and $y > 0 (x \neq y)$, x^n and y^n are unequal, so that there is only one positive value of x that makes $x^n = 2$.

$\therefore x^n = 2$ has one and only one positive root.

10. (a) Use neighbourhood property. If possible, let $f(a) \neq 0$, then $\exists$ a δ-*nbd* of a where in for every x, $f(x)$ has the same sign as $f(a)$ but this will contradict the given condition; hence etc.

AN INTRODUCTION TO ANALYSIS

(Differential Calculus—Part I)

Chapter 9

Differentiation

Chapter 9

Differentiation

9.0 Definition: Let $f : \mathbb{R} \to \mathbb{R}$ be a function and $x \in \mathbb{R}$. Then the derivative $f'(x)$ of the function f at any point x of $\mathbb{R}$ is defined as

$$f'(x) = \lim_{t \to x} \frac{f(t) - f(x)}{t - x}$$

provided the limit exists.

Taking $t - x = h \in \mathbb{R}$ we get $h \to 0$ as $t \to x$.

So, equivalently

$$f'(x) = \lim_{h \to 0} \frac{f(x+h) - f(x)}{h}$$

provided the limit exists.

The process of finding derivative is called *differentiation*. In particular, the derivative of a function f at a point $c (\in \mathbb{R})$ is denoted by $f'(c)$ and is defined as

$$f'(c) = \lim_{x \to c} \frac{f(x) - f(c)}{x - c}$$

or, equivalently

$$f'(c) = \lim_{x \to c} \frac{f(c+h) - f(c)}{h}$$

provided the limits exist.

Right-hand derivative

Right-hand derivative of a function f at a point c is denoted by $Rf'(c)$ and defined as

$$Rf'(c) = \lim_{h \to 0+} \frac{f(c+h) - f(c)}{h}$$

provided the limit exists.

Similarly left-hand derivative

$$Lf'(c) = \lim_{h \to 0-} \frac{f(c+h) - f(c)}{h}$$

provided the limit exists.

Note that $f'(c)$ exists iff $Lf'(c) = Rf'(c)$.

Derivative at an end point

Suppose $f : [a, b] \to \mathbb{R}$ be a function and c is an interior point of $[a, b]$, i.e., $\exists \delta > 0$ such that $(c - \delta, c + \delta) \subset [a, b]$ then

$$f'(c) = \lim_{h \to 0} \frac{f(c+h) - f(c)}{h}$$

as defined earlier.

At the left end point a, $f'(a)$ is defined as

$$f'(a) = \lim_{h \to 0+} \frac{f(a+h) - f(a)}{h} \quad \text{i.e.} \quad f'(a) = Rf'(a)$$

Similarly at the right end point b,

$$f'(b) = Lf'(b) = \lim_{h \to 0-} \frac{f(b+h) - f(b)}{h}$$

$$\text{or} \quad f'(b) = Lf'(b) = \lim_{h \to 0+} \frac{f(b) - f(b-h)}{h}$$

provided the limits exist.

If f is differentiable at each point of $[a, b]$ then f is said to be differentiable on $[a, b]$.

Note: (i) If $\lim_{h \to 0} \frac{f(c+h) - f(c)}{h} = \infty (\text{or} - \infty)$, we say that $f'(c) = \infty (\text{or} - \infty)$. However f is said to be differentiable at c if $f'(c)$ is finite.

(ii) $Lf'(c)$ is also denoted by $f'(c - 0)$
and $Rf'(c)$ is also denoted by $f'(c + 0)$.

Illustrative Examples

Example 9.0.1. *Is the function f defined by $f(x) = |x|, x \in \mathbb{R}$ (i.e., $-\infty < x < \infty$) derivable at $x = 0$?*

CHAPTER 9: DIFFERENTIATION

Solution:

$$\lim_{h \to 0^+} \frac{f(0+h) - f(0)}{h} = \lim_{h \to 0^+} \frac{h - 0}{h} = 1 \quad [\text{see that when } h \to 0^+, h > 0; \text{ then } f(h) = h].$$

$$\lim_{h \to 0^-} \frac{f(0+h) - f(0)}{h} = \lim_{h \to 0^-} \frac{-h - 0}{h} = -1 \quad [\text{when } h \to 0^-, h < 0; \text{ then } f(h) = -h].$$

$\therefore f'(0+0) = 1$ but $f'(0-0) = -1$ and consequently $f'(0)$ does not exist i.e., **f is not derivable at x = 0**.

Example 9.0.2. Let f be defined in $1 \leq x < \infty$ by

$$f(x) = \begin{cases} 2 - x, & \text{when } 1 \leq x \leq 2; \\ 3x - x^2, & \text{when } x > 2. \end{cases}$$

Is the function f derivable at $x = 1$? at $x = 2$?

Solution: Note, derivability of f at $x = 1$ requires the existence of right-derivative at $x = 1$.

$$f'(1+0) = \lim_{h \to 0^+} \frac{f(1+h) - f(1)}{h} = \lim_{h \to 0^+} \frac{2 - (1+h) - (2-1)}{h} = \lim_{h \to 0^+} \frac{-h}{h} = -1.$$

i.e., f is derivable at the **left-end** $x = 1$ of its domain and the derivative there is -1.

At x = 2,

$$\lim_{h \to 0^+} \frac{f(2+h) - f(2)}{h} = \lim_{h \to 0^+} \frac{3(2+h) - (2+h)^2 - (2-2)}{h}$$

$$= \lim_{h \to 0^+} \frac{2 - h - h^2}{h},$$

which does not have a finite limit. But

$$\lim_{h \to 0^-} \frac{f(2+h) - f(2)}{h} = \lim_{h \to 0^-} \frac{2 - (2+h) - (2-2)}{h} = \lim_{h \to 0^-} \frac{-h}{h} = -1$$

i.e., $f'(2+0)$ does not exist, $f'(2-0) = -1$ but in any case $f'(2)$ does not exist.

Thus, f is not derivable at the right end point $x = 2$ of the domain of definition of f.

Example 9.0.3. f is a real-valued function defined on $(-1, 2)$ by

$$f(x) = |x| + |x - 1|.$$

Is f derivable at 0 and 1? Find the derived function f' and its domain.

Solution: The function may be defined as:

$$f(x) = \begin{cases} -x + (1-x) = 1 - 2x, & \text{if } -1 < x \leq 0; \\ x + 1 - x = 1, & \text{if } 0 < x \leq 1; \\ x + x - 1 = 2x - 1, & \text{if } 1 < x < 2. \end{cases}$$

See that (i) $f'(x) = -2$, if $-1 < x < 0$

(ii)
$$Rf'(0) = \lim_{x \to 0^+} \frac{f(x) - f(0)}{x - 0} = \lim_{x \to 0^+} \frac{1 - 1}{x} = 0$$
$$Lf'(0) = \lim_{x \to 0^-} \frac{f(x) - f(0)}{x - 0} = \lim_{x \to 0^-} \frac{1 - 2x - 1}{x} = -2.$$

∴ $f'(0)$ does not exist.

(iii) $f'(x) = 0$, if $0 < x < 1$.

(iv)
$$Rf'(1) = \lim_{x \to 1^+} \frac{f(x) - f(1)}{x - 1} = \lim_{x \to 1} \frac{2x - 1 - 1}{x - 1} = 2.$$
$$Lf'(1) = \lim_{x \to 1^-} \frac{f(x) - f(1)}{x - 1} = \lim_{x \to 1^-} \frac{1 - 1}{x - 1} = 0.$$

∴ $f'(1)$ does not exist.

(v) $f'(x) = 2$, if $1 < x < 2$.

[*Ans:* Thus the domain of f' is $\{-1 < x < 0\} \cup \{0 < x < 1\} \cup \{1 < x < 2\}$ defined by

$$\begin{aligned} f'(x) &= -2, \text{ if } -1 < x < 0 \\ &= 0, \text{ if } 0 < x < 1 \\ &= 2, \text{ if } 1 < x < 2. \end{aligned}$$

f' is not defined at $x = 0$ or at $x = 1$.]

Infinite derivatives: Some authors prefer to introduce the concept of *Infinite derivatives* so that the usual geometric interpretation of a derivative as the slope of a tangent line will still be valid in case the tangent line happens to be vertical.

Let f be defined on $I \subseteq \mathbb{R}$. At an interior point $x \in I$, f has an infinite derivative $+\infty$, if both the right-and the left-derivatives at that be point be $+\infty$, i.e., if $f'(x+0) = +\infty$, $f'(x-0) = +\infty$, we then write $f'(x) = +\infty$.

The derivative $f'(x) = -\infty$ is similarly defined.

If x is the left-end point of interval, then $f'(x) = +\infty$ if only $f'(x+0) = +\infty$.

Similarly, if x is the right-end point, then $f'(x) = -\infty$, if only $f'(x-0) = -\infty$.

CHAPTER 9: DIFFERENTIATION

> REMEMBER: Unless explicitly stated, existence of derivative will always mean existence of a finite derivative.

9.1 Relation between Continuity and Derivability

Theorem 9.1.1. *If a function f has a finite derivative at $x = c$, then it is continuous at $x = c$; i.e, derivability ensures continuity.*

Proof: Let $f(x)$ have a finite derivative $f'(c)$ at the point $x = c$.
Then,
$$f(c+h) - f(c) = \frac{f(c+h) - f(c)}{h} \cdot h.$$
$$\therefore \lim_{h \to 0}\{f(c+h) - f(c)\} = \lim_{h \to 0} \frac{f(c+h) - f(c)}{h} \times \lim_{h \to 0} h = f'(c) \times 0 = 0.$$

i.e., for any $\varepsilon > 0$, $\exists\, \delta(\varepsilon) > 0$ s.t.

$$|f(c+h) - f(c)| < \varepsilon,\ \forall\, h \in |h| < \delta$$

or, $|f(x) - f(c)| < \varepsilon,\ \forall\, x \in |x - c| < \delta.$ (writing x for $c + h$ and hence $x - c$ for h)

This proves that $f(x)$ be continuous at $x = c$.

Corollary 9.1.1. *If a function f is* NOT CONTINUOUS *at a certain point $x = c$, then the function f can* NOT BE DERIVABLE *there. (contrapositive statement)*

[Cf. In Logic, $p \Rightarrow q$ and $\sim q \Rightarrow \sim p$ are logically equivalent.]

Observation: The converse of this theorem is not true in all cases: there are functions which are continuous but do not have a derivative.

An example: $f(x) = |x|$ is continuous at $x = 0$, for $|f(x) - f(0)| = |x| <$ any given positive number ε whenever $|x| < \delta$, if one chooses $\delta = \varepsilon$, i.e., ε-δ condition for continuity is satisfied.

Otherwise: $\lim_{x \to 0+} f(x) = \lim_{x \to 0+} x = 0;\ \lim_{x \to 0-} f(x) = \lim_{x \to 0-} -x = 0$ and hence $\lim_{x \to 0} f(x)$ exists and $= 0 = f(0)$. So $f(x)$ is continuous at $x = 0$.

But it has already been shown in Ex. **9.0.1** that $|x|$ is not derivable at $x = 0$.

Example 9.1.1.

$$f(x) = \begin{cases} x \sin \frac{1}{x}, & \text{if } x \neq 0; \\ 0, & \text{if } x = 0. \end{cases} \quad \textit{To prove that } f \textit{ is continuous but not derivable at } x = 0.$$

Solution: At $x = 0$, this function is continuous.

$|f(x) - f(0)| = |x\sin(1/x)| \le |x| <$ any given positive ε, whenever $|x| < \delta$, where $\delta = \varepsilon$.

But observe that

$$\frac{f(x) - f(0)}{x - 0} = \frac{x\sin(1/x)}{x} = \sin(1/x) \text{ (for } x \ne 0).$$

We can, by using Cauchy's General Principle, prove that $\lim_{x \to 0} \sin(1/x)$ does **not** exist.

Then $\lim_{x \to 0} \dfrac{f(x) - f(0)}{x - 0}$ does not exist, i.e., **f(x) is not derivable at x = 0**.

[To prove that $\lim_{x \to 0} \sin(1/x)$ does not exist, we choose two points

$$x_1 = \frac{2}{(4n+1)\pi} \quad \text{and} \quad x_2 = \frac{2}{(4n-1)\pi}.$$

But taking suitable large n, x_1 and x_2 can be taken in any nbd of $x = 0$. For such two points,

$$|f(x_1) - f(x_2)| = \left|\sin\left(2n + \frac{1}{2}\right)\pi - \sin\left(2n - \frac{1}{2}\right)\pi\right| = |\sin \pi/2 - \sin(-\pi/2)| = 2,$$

which is not less than any ε we choose.

i.e., Cauchy's condition for the existence of limit is not satisfied and as such $\lim_{x \to 0} \sin(1/x)$ does not exist.]

Example 9.1.2. $f(x) = x^2 \sin(1/x), x \ne 0; = 0, x = 0$.

To prove f is derivable at $x = 0$, but the derived function f' is NOT CONTINUOUS at $x = 0$. [C.H. 1993]

Solution:

$$\lim_{x \to 0} \frac{f(x) - f(0)}{x - 0} = \lim_{x \to 0} \frac{x^2 \sin(1/x)}{x} = \lim_{x \to 0} x \sin\frac{1}{x} = 0 \text{ (see Ex. \textbf{9.1.1} above)}.$$

i.e., f has a derivative zero at $x = 0$.

Assuming that the reader is acquainted with the technique of differentiation we may obtain the *derived function* f' thus:

$$f'(x) = \begin{cases} 2x\sin(1/x) - \cos(1/x), & \text{when } x \ne 0 \\ 0, & \text{when } x = 0. \end{cases}$$

Clearly, $\lim_{x \to 0} f'(x)$ does not exist ($\because \lim \cos(1/x)$ does not exist) i.e., the derived function f' cannot be continuous at $x = 0$.

[To prove that $\lim_{x \to 0} \cos(1/x)$ does not exist, take two points $x_1 = 1/2n\pi$ & $x_2 = 1/(2n+1)\pi$ and see that Cauchy's condition is not satisfied; hence etc.]

Example 9.1.3. $f(x) = x + 1$, when $0 \leq x \leq 1$; $= 3 - x$ when $1 \leq x \leq 2$.

To prove: (i) $f(x)$ is continuous at $x = 1$, but not derivable there.

[Check: $\lim_{x \to 1+} f(x) = \lim_{x \to 1-} f(x) = f(1)$; hence continuous.

but $\lim_{x \to 1+} \dfrac{f(x) - f(1)}{x - 1} = -1$, $\lim_{x \to 1-} \dfrac{f(x) - f(1)}{x - 1} = +1$; hence $f'(1)$ does not exist.]

(ii) *Between 0 and 1, the function* $f(x) = x + 1$, $f'(x) = 1$.

(iii) *Between 1 and 2,* $f(x) = 3 - x$ *and* $f'(x) = -1$.

We leave the details for the students to complete.

9.2 Standard Formulae and General Rules of Differentiation

Using the definition of derivative and applying some standard limits like $\lim_{x \to 0} \dfrac{\sin x}{x} = 1$, $\lim_{x \to 0} (1+x)^{1/x} = e$ the following results can be deduced:

I. $\dfrac{d}{dx}(x^n) = nx^{n-1}$, where n is a positive integer; $\dfrac{d}{dx}$ (any const. function) $= 0$.

II. $\dfrac{d}{dx}(\sin x) = \cos x$; $\dfrac{d}{dx}(\cos x) = -\sin x$; $\dfrac{d}{dx}(\tan x) = \sec^2 x$.

$\dfrac{d}{dx}(\cot x) = -\csc^2 x$; $\dfrac{d}{dx}(\sec x) = \sec x \tan x$; $\dfrac{d}{dx}(\csc x) = -\csc x \cot x$.

III. $\dfrac{d}{dx}(e^x) = e^x$; $\dfrac{d}{dx}(a^x) = a^x \log_e a (a > 0)$; $\dfrac{d}{dx}(\log_e x) = \dfrac{1}{x} (x > 0)$.

IV. $\dfrac{d}{dx}(x^m) = mx^{m-1}$, where m is any real number and the function is defined at and near the point x, where derivative is found.

In particular, $\dfrac{d}{dx}(\sqrt{x}) = \dfrac{1}{2\sqrt{x}}$, $x \neq 0$.

A Few General Rules

A. Addition Rule:

$$\frac{d}{dx}(u \pm v \pm w \pm \text{etc.}) = \frac{du}{dx} \pm \frac{dv}{dx} \pm \frac{dw}{dx} \pm \text{etc.}$$

[number of functions being finite]

B. Product Rule:

$$\frac{d}{dx}(uv) = u\frac{dv}{dx} + v\frac{du}{dx};$$

$$\frac{d}{dx}(uvw) = uv\frac{dw}{dx} + uw\frac{dv}{dx} + vw\frac{du}{dx};$$

$$\frac{d}{dx}(cu) = c\frac{du}{dx}.$$

C. Quotient Rule:

$$\frac{d}{dx}\left(\frac{u}{v}\right) = \frac{v\dfrac{du}{dx} - u\dfrac{dv}{dx}}{v^2} \quad (v \neq 0).$$

D. Derivation of Composite Functions: Chain Rule

(i) $y = f(u)$ and $u = \phi(x) \Rightarrow \dfrac{dy}{dx} = \dfrac{dy}{du} \cdot \dfrac{du}{dx} = f'(u) \cdot \phi'(x).$

(ii) $y = f(u)$, $u = \phi(v)$ and $v = \psi(x) \Rightarrow \dfrac{dy}{dx} = \dfrac{dy}{du} \cdot \dfrac{du}{dv} \cdot \dfrac{dv}{dx} = f'(u) \cdot \phi'(v) \cdot \psi'(x).$

E. If $y = f(x)$ admits of an inverse function $x = \phi(y)$ and if $f(x)$ has a finite non-zero derivative at x, then $\phi(y)$ also possesses a finite non-zero derivative at the corresponding y s.t.

$$\frac{dx}{dy} = 1 \bigg/ \frac{dy}{dx} \quad \left(\frac{dy}{dx} \neq 0\right).$$

Note 9.2.1. *This rule is used for finding the derivative of an inverse function.*

CHAPTER 9: DIFFERENTIATION

V. Some more Standard Results

$$\frac{d}{dx}(\sin^{-1} x) = \frac{1}{+\sqrt{1-x^2}} \quad (-1 < x < 1). \quad [-\pi/2 < \sin^{-1} x < \pi/2]$$

$$\frac{d}{dx}(\cos^{-1} x) = \frac{-1}{\sqrt{1-x^2}} \quad (-1 < x < 1). \quad [0 < \cos^{-1} x < \pi]$$

$$\frac{d}{dx}(\tan^{-1} x) = \frac{1}{1+x^2} \quad (-\infty < x < \infty). \quad [-\pi/2 < \tan^{-1} x < \pi/2]$$

$$\frac{d}{dx}(\cot^{-1} x) = \frac{-1}{1+x^2} \quad (-\infty < x < \infty). \quad [0 < \cot^{-1} x < \pi]$$

$$\frac{d}{dx}(\sec^{-1} x) = \frac{1}{x\sqrt{x^2-1}} \quad (|x| > 1). \quad [0 < \sec^{-1} x < \pi]$$

$$\frac{d}{dx}(\operatorname{cosec}^{-1} x) = \frac{-1}{x\sqrt{x^2-1}} \quad (|x| > 1). \quad \left[-\pi/2 < \operatorname{cosec}^{-1} x < \frac{\pi}{2}\right]$$

Notations: We often write **arc sin** x for $\sin^{-1} x$, **arc cos** x for $\cos^{-1} x$, etc.

VI. On Hyperbolic Functions

Definition 9.2.1. $\cosh x = \dfrac{e^x + e^{-x}}{2}, \quad \sinh x = \dfrac{e^x - e^{-x}}{2},$

$$\tanh x = \frac{\sinh x}{\cosh x} = \frac{e^x - e^{-x}}{e^x + e^{-x}}$$

$\coth x = 1/\tanh x, \quad \operatorname{sech} x = 1/\cosh x, \quad \operatorname{cosech} x = 1/\sinh x.$

> REMEMBER: $\cosh^2 x - \sinh^2 x = 1.$

General Rule: Take any standard formula of trigonometric functions. Replace each trigonometric function by the corresponding hyperbolic functions and *change the sign before any product of two sines.*

Trigonometric Formula	Hyperbolic Formula
$\cos(A+B) = \cos A \cos B - \sin A \sin B$	$\cosh(x+y) = \cosh x \cosh y + \sinh x \sinh y$
$\cos^2 A + \sin^2 A = 1.$	$\cosh^2 x - \sinh^2 x = 1.$

(i) $\dfrac{d}{dx}(\sinh x) = \dfrac{d}{dx}\left(\dfrac{e^x - e^{-x}}{2}\right) = \dfrac{e^x + e^{-x}}{2} = \cosh x.$

(ii) $\dfrac{d}{dx}(\cosh x) = \dfrac{d}{dx}\left(\dfrac{e^x + e^{-x}}{2}\right) = \dfrac{e^x - e^{-x}}{2} = \sinh x.$

Inverse hyperbolic functions

$\dfrac{d}{dx}(\sinh^{-1} x) = \dfrac{1}{+\sqrt{1+x^2}};$ $\qquad \dfrac{d}{dx}(\cosh^{-1} x) = \dfrac{1}{+\sqrt{x^2-1}};$

$\dfrac{d}{dx}(\tanh^{-1} x) = \dfrac{1}{1-x^2}(-1 < x < 1);$ $\qquad \dfrac{d}{dx}(\coth^{-1} x) = \dfrac{-1}{x^2-1}(|x| > 1).$

$\dfrac{d}{dx}(\operatorname{sech}^{-1} x) = \dfrac{-1}{x\sqrt{1-x^2}}(0 < x < 1);$ $\qquad \dfrac{d}{dx}(\operatorname{cosech}^{-1} x) = \dfrac{-1}{x\sqrt{x^2+1}}, x > 0$

$$= \dfrac{1}{x\sqrt{x^2+1}}, x < 0.$$

REMEMBER: Inverse hyperbolic functions can be expressed as logarithms.

$\sinh^{-1} x = \log_e(x + \sqrt{x^2+1});$ $\qquad \cosh^{-1} x = \log_e(x + \sqrt{x^2-1})(x > 1);$

$\tanh^{-1} x = \dfrac{1}{2}\log\dfrac{1+x}{1-x}(|x| < 1);$ $\qquad \coth^{-1} x = \dfrac{1}{2}\log\dfrac{x+1}{x-1}$ if $|x| > 1;$

$\operatorname{sech}^{-1} x = \log\left(\dfrac{1+\sqrt{1-x^2}}{x}\right)(0 < x < 1);$ $\qquad \operatorname{cosech}^{-1} x = \log\left(\dfrac{1 \pm \sqrt{1+x^2}}{x}\right).$

[+ve sign, if $x > 0$

−ve sign, if $x < 0$]

CHAPTER 9: DIFFERENTIATION

Applications

Example 9.2.1. *Differentiate* $\dfrac{(1-x)^4}{(1+x)^5}$ *w.r.t. 'x'.*

Solution:

$$\frac{d}{dx}\left\{\frac{(1-x)^4}{(1+x)^5}\right\} = \frac{(1+x)^5 \frac{d}{dx}(1-x)^4 - (1-x)^4 \frac{d}{dx}(1+x)^5}{(1+x)^{10}} \quad \text{(Quotient Rule)}$$

$$= \frac{-4(1+x)^5(1-x)^3 - 5(1-x)^4(1+x)^4}{(1+x)^{10}}$$

$$= \frac{\{-4(1+x) - 5(1-x)\}(1-x)^3}{(1+x)^6}$$

$$= -\frac{(x-9)(1-x)^3}{(1+x)^6}.$$

Example 9.2.2. *Differentiate* $\left\{\sin\left(\tan^{-1}\dfrac{1}{1-x^2}\right)\right\}$ *w.r.t. 'x'.*

Solution:

$$\frac{d}{dx}\left\{\sin\left(\tan^{-1}\frac{1}{1-x^2}\right)\right\}$$

$$= \cos\left(\tan^{-1}\frac{1}{1-x^2}\right)\frac{d}{dx}\left(\tan^{-1}\frac{1}{1-x^2}\right)$$

$$= \cos\left(\tan^{-1}\frac{1}{1-x^2}\right) \cdot \frac{1}{1+(\frac{1}{1-x^2})^2}\frac{d}{dx}\left(\frac{1}{1-x^2}\right)$$

$$= \cos\left(\tan^{-1}\frac{1}{1-x^2}\right) \cdot \frac{(1-x^2)^2}{(1-x^2)^2+1}\left(\frac{-1}{(1-x^2)^2}\right)\frac{d}{dx}(1-x^2)$$

$$= \cos\left(\tan^{-1}\frac{1}{1-x^2}\right) \cdot \frac{2x}{(1-x^2)^2+1}.$$

Example 9.2.3. *To find* $\dfrac{d}{dx}\left\{\sqrt{e^{\sin(\log(x^2+7)^5)}}\right\}.$

Solution: Let $y = u^{1/2}$, where $u = e^v$, $v = \sin w$, $w = \log t$, $t = s^5$, $s = x^2 + 7$.

Then
$$\frac{dy}{dx} = \frac{dy}{du} \cdot \frac{du}{dv} \cdot \frac{dv}{dw} \cdot \frac{dw}{dt} \cdot \frac{dt}{ds} \cdot \frac{ds}{dx} = \frac{1}{2\sqrt{u}} \cdot e^v \cdot \cos w \cdot \frac{1}{t} \cdot 5s^4 \cdot 2x$$
$$= \frac{1}{2\sqrt{e^{\sin(\log(x^2+7)^5)}}} \cdot e^{\sin(\log(x^2+7)^5)} \cdot \cos(\log(x^2+7)^5) \times$$
$$\frac{1}{(x^2+7)^5} \cdot 5(x^2+7)^4 \cdot 2x.$$

REMEMBER: Let $z = f(y)$ where it is known $y = \phi(x)$, not explicitly stated, then
$$\frac{dz}{dx} = \frac{dz}{dy}\frac{dy}{dx}; \text{ e.g., } \frac{d}{dx}(\sin y) = \cos y \frac{dy}{dx}, \quad \frac{d}{dx}(y^5) = 5y^4 \frac{dy}{dx} \text{ etc.}$$

Logarithmic Differentiation

Example 9.2.4. $y = x^{\tan x} + (\sin x)^{\cos x} + 10x^2$. To find $\frac{dy}{dx}$.

Solution: Let $y = u + v + w$, where $u = x^{\tan x}$, $v = (\sin x)^{\cos x}$, $w = 10x^2$.

So that $\frac{dy}{dx} = \frac{du}{dx} + \frac{dv}{dx} + \frac{dw}{dx}$.

To find $\frac{du}{dx}$, we take logarithms w.r.t the base e: $\log u = \tan x \log x$ and then

$$\frac{d}{dx}(\log u) = \frac{d}{dx}(\tan x) \cdot \log x + \frac{d}{dx}(\log x) \cdot \tan x = \sec^2 x \log x + \frac{\tan x}{x}.$$

$$\therefore \quad \frac{1}{u}\frac{du}{dx} = \sec^2 x \log x + \frac{\tan x}{x} \quad \text{and} \quad \frac{du}{dx} = x^{\tan x}\left(\sec^2 x \log x + \frac{\tan x}{x}\right).$$

To find $\frac{dv}{dx}$, we take logarithms again on $v = (\sin x)^{\cos x}$ $\therefore$ $\log v = \cos x \log \sin x$.

$$\therefore \quad \frac{d}{dx}(\log v) = \frac{d}{dx}(\cos x) \cdot \log \sin x + \cos x \frac{d}{dx} \log \sin x$$
$$= -\sin x \log \sin x + \cos x \frac{d}{dx} \log t \cdot \text{ if } t = \sin x$$
$$= -\sin x \log \sin x + \cos x \cdot \frac{1}{t}\frac{dt}{dx}$$

i.e., $\frac{1}{v} \cdot \frac{dv}{dx} = -\sin x \log \sin x + \frac{\cos x}{\sin x} \cdot \cos x$

or, $\frac{dv}{dx} = (\sin x)^{\cos x}\left\{-\sin x \log \sin x + \frac{\cos^2 x}{\sin x}\right\}.$

CHAPTER 9: DIFFERENTIATION

Finally, $\dfrac{dw}{dx} = \dfrac{d}{dx}(10x^2) = 20x$.

$\therefore \quad \dfrac{dy}{dx} = \dfrac{du}{dx} + \dfrac{dv}{dx} + \dfrac{dw}{dx} = x^{\tan x}\left(\sec^2 x \log x + \dfrac{\tan x}{x}\right)$

$\qquad\qquad\qquad\qquad\qquad + (\sin x)^{\cos x}\left\{-\sin x \log \sin x + \dfrac{\cos^2 x}{\sin x}\right\} + 20x.$

9.3 Implicit Functions

For an implicit relation between x and y (i.e., where y is not directly or explicitly stated in terms of x) we differentiate each term w.r.t. x and then solve for dy/dx.

Example 9.3.1. $(x+y)^{m+n} = x^m y^n$; prove that $\dfrac{dy}{dx}$ is independent of m and n.

Solution: On taking logarithms; we obtain the implicit relation as

$$(m+n)\log(x+y) = m \log x + n \log y.$$

Differentiating each term w.r.t. 'x', we get,

$$(m+n)\dfrac{d}{dx}\{\log(x+y)\} = m\dfrac{d}{dx}(\log x) + n\dfrac{d}{dx}(\log y)$$

or, $\quad (m+n)\dfrac{1}{x+y}\dfrac{d}{dx}(x+y) = \dfrac{m}{x} + n\dfrac{1}{y}\dfrac{dy}{dx}$

or, $\quad (m+n)\dfrac{1}{x+y}\left(1+\dfrac{dy}{dx}\right) = \dfrac{m}{x} + \dfrac{n}{y}\dfrac{dy}{dx}$

or, $\quad \dfrac{dy}{dx}\left\{\dfrac{m+n}{x+y} - \dfrac{n}{y}\right\} = \dfrac{m}{x} - \dfrac{m+n}{x+y} = \dfrac{mx+my-mx-nx}{x(x+y)}$

or, $\quad \dfrac{dy}{dx}\left\{\dfrac{my+ny-nx-ny}{y(x+y)}\right\} = \dfrac{my-nx}{x(x+y)},$

which gives $\dfrac{dy}{dx} = \dfrac{y}{x}$ (independent of m and n).

9.4 Parametric Functions

If $x = f(t)$, $y = \phi(t)$ be two derivable functions of t and if $x = f(t)$ admits of an inverse function $t = \psi(x)$ then we may write

$$y = \phi(t), \quad t = \psi(x)$$

and hence $\dfrac{dy}{dx} = \dfrac{dy}{dt} \cdot \dfrac{dt}{dx} = \dfrac{dy}{dt} \Big/ \dfrac{dx}{dt} = \dfrac{\phi'(t)}{f'(t)}. \quad (f'(t) \neq 0).$

Example 9.4.1. $y = a(\cos t + t \sin t)$, $x = a(\sin t - t \cos t)$: *To find dy/dx.*

Solution:

$$\frac{dy}{dx} = \frac{dy}{dt} \Big/ \frac{dx}{dt} = \frac{a(-\sin t + 1 \cdot \sin t + t \cdot \cos t)}{a\{\cos t - 1 \cdot \cos t - t(-\sin t)\}} = \frac{\cos t}{\sin t} = \cot t.$$

Miscellaneous Problems

Example 9.4.2. Trigonometric Simplification. *To find* $\dfrac{d}{dx}\left\{\tan^{-1}\dfrac{\cos x}{1+\sin x}\right\}$.

Solution:

$$\frac{\cos x}{1+\sin x} = \frac{\cos^2 x/2 - \sin^2 x/2}{(\cos x/2 + \sin x/2)^2} = \frac{\cos x/2 - \sin x/2}{\cos x/2 + \sin x/2}$$

$$= \frac{1 - \tan x/2}{1 + \tan x/2} = \tan(\pi/4 - x/2).$$

Hence

$$\frac{d}{dx}\tan^{-1}\left(\frac{\cos x}{1+\sin x}\right) = \frac{d}{dx}\left\{\tan^{-1}\tan\left(\frac{\pi}{4} - \frac{x}{2}\right)\right\} = \frac{d}{dx}\left\{\frac{\pi}{4} - \frac{x}{2}\right\} = -\frac{1}{2}.$$

Example 9.4.3. *Differentiate* $x^{\sin^{-1} x}$ *with respect to* $\sin^{-1} x$.

Let $y = x^{\sin^{-1} x}$, $z = \sin^{-1} x$. To find $\dfrac{dy}{dz}$.

We have

$$\frac{dy}{dz} = \frac{dy}{dx}\Big/\frac{dz}{dx}.$$

On taking logarithms of $y = x^{\sin^{-1} x}$, we get

$$\log y = \sin^{-1} x \log x.$$

$$\therefore \quad \frac{d}{dx}(\log y) = \sin^{-1} x \frac{d}{dx}(\log x) + \log x \frac{d}{dx}(\sin^{-1} x),$$

or, $\quad \dfrac{1}{y}\dfrac{dy}{dx} = \dfrac{\sin^{-1} x}{x} + \dfrac{\log x}{\sqrt{1-x^2}} \quad$ or $\quad \dfrac{dy}{dx} = x^{\sin^{-1} x}\left\{\dfrac{\sin^{-1} x}{x} + \dfrac{\log x}{\sqrt{1-x^2}}\right\}.$

Again, $z = \sin^{-1} x$ and $\dfrac{dz}{dx} = \dfrac{1}{\sqrt{1-x^2}}.$

$$\therefore \quad \frac{dy}{dz} = \frac{dy}{dx}\Big/\frac{dz}{dx} = x^{\sin^{-1} x}\left\{\frac{\sin^{-1} x}{x} + \frac{\log x}{\sqrt{1-x^2}}\right\}\Big/\frac{1}{\sqrt{1-x^2}}$$

$$= x^{\sin^{-1} x}\left\{\frac{x \log x + \sqrt{1-x^2}\sin^{-1} x}{x}\right\}.$$

CHAPTER 9: DIFFERENTIATION

Example 9.4.4. *Prove that*

$$\frac{d}{dx}\left\{\frac{1}{4\sqrt{2}}\log\frac{x^2+x\sqrt{2}+1}{x^2-x\sqrt{2}+1}+\frac{1}{2\sqrt{2}}\tan^{-1}\frac{x\sqrt{2}}{1-x^2}\right\}=\frac{1}{x^4+1}.$$

Solution: Required derivative

$$=\frac{d}{dx}\left\{\frac{1}{4\sqrt{2}}[\log(x^2+x\sqrt{2}+1)-\log(x^2-x\sqrt{2}+1)]+\frac{1}{2\sqrt{2}}\tan^{-1}\frac{x\sqrt{2}}{1-x^2}\right\}$$

$$=\frac{1}{4\sqrt{2}}\frac{1}{x^2+x\sqrt{2}+1}(2x+\sqrt{2})-\frac{1}{4\sqrt{2}}\frac{1}{x^2-x\sqrt{2}+1}(2x-\sqrt{2})$$

$$+\frac{1}{2\sqrt{2}}\frac{1}{1+\frac{2x^2}{(1-x^2)^2}}\frac{d}{dx}\left(\frac{x\sqrt{2}}{1-x^2}\right)$$

$$=\frac{\sqrt{2}x+1}{4(x^2+x\sqrt{2}+1)}-\frac{\sqrt{2}x-1}{4(x^2-x\sqrt{2}+1)}+\frac{1}{2\sqrt{2}}\frac{(1-x^2)^2}{1+x^4}\cdot\frac{\sqrt{2}(1-x^2)-x\sqrt{2}(-2x)}{(1-x^2)^2}$$

$$=\frac{1}{4}\frac{(\sqrt{2}x+1)(x^2-x\sqrt{2}+1)-(\sqrt{2}x-1)(x^2+x\sqrt{2}+1)}{(x^2+x\sqrt{2}+1)(x^2-x\sqrt{2}+1)}+\frac{1}{2\sqrt{2}}\frac{\sqrt{2}(1-x^2+2x^2)}{1+x^4}$$

$$=\frac{1}{4}\frac{\sqrt{2}x(x^2-x\sqrt{2}+1-x^2-x\sqrt{2}-1)+x^2-x\sqrt{2}+1+x^2+x\sqrt{2}+1}{(x^2+1)^2-(x\sqrt{2})^2}+\frac{1}{2}\frac{1+x^2}{1+x^4}$$

$$=\frac{1}{4}\frac{-4x^2+2x^2+2}{1+x^4}+\frac{1}{2}\frac{1+x^2}{1+x^4}=\frac{1}{2}\frac{1-x^2}{1+x^4}+\frac{1}{2}\frac{1+x^2}{1+x^4}=\frac{(1-x^2)+(1+x^2)}{2(1+x^4)}$$

$$=\frac{1}{1+x^4}.$$

Example 9.4.5. *Trigonometric substitution.*

If $\sqrt{1-x^2}+\sqrt{1-y^2}=a(x-y)$, *prove that one of the values of* $\frac{dy}{dx}=\sqrt{\frac{1-y^2}{1-x^2}}$.

Solution: Let $x=\sin\theta, y=\sin\phi$. Then the given relation reduces to

$$\cos\theta+\cos\phi=a(\sin\theta-\sin\phi)$$

or, $\quad 2\cos\dfrac{\theta+\phi}{2}\cos\dfrac{\theta-\phi}{2}=2a\cos\dfrac{\theta+\phi}{2}\sin\dfrac{\theta-\phi}{2}$

i.e., $\quad\tan\dfrac{\theta-\phi}{2}=\dfrac{1}{a}$, unless $\cos\dfrac{\theta+\phi}{2}=0$

or, $\quad\theta-\phi=2\tan^{-1}\dfrac{1}{a}$, or, $\sin^{-1}x-\sin^{-1}y=2\tan^{-1}(1/a)$.

∴ Differentiating w.r.t. x, we get $\dfrac{1}{\sqrt{1-x^2}} - \dfrac{1}{\sqrt{1-y^2}}\dfrac{dy}{dx} = 0$.

($\because 2\tan^{-1}(1/a)$ is a constant).

Hence $\dfrac{dy}{dx} = \sqrt{\dfrac{1-y^2}{1-x^2}}$.

In case, we assume $\cos\dfrac{\theta+\phi}{2} = 0$ i.e., $\theta + \phi = \pi$ i.e., $\sin^{-1}x + \sin^{-1}y = \pi$.

We would get $\dfrac{dy}{dx} = -\sqrt{\dfrac{1-y^2}{1-x^2}}$.

Example 9.4.6. *If* $y = \dfrac{1}{\sqrt{(a^2-b^2-c^2)}}\cos^{-1}\left(\dfrac{a\theta - a^2 + b^2 + c^2}{\theta\sqrt{b^2+c^2}}\right)$ *and*
$\theta = a + b\cos x + c\sin x$, *prove that* $\dfrac{dy}{dx} = \dfrac{1}{\theta}$.

Solution: Let us put $b^2 + c^2 = h^2$ and $a^2 - h^2 = k^2$. Then the given condition gives,

$$\cos ky = \dfrac{a\theta - k^2}{\theta h} = \dfrac{a}{h} - \dfrac{k^2}{h}\dfrac{1}{\theta}$$

∴ $-\sin ky \cdot k\dfrac{dy}{dx} = \dfrac{k^2}{h}\dfrac{1}{\theta^2}\dfrac{d\theta}{dx}$ (differentiating w.r.t. x),

∴ $\dfrac{dy}{dx} = -\dfrac{k}{\theta^2 h\sqrt{1-\cos^2 ky}}\dfrac{d\theta}{dx}$

$= -\dfrac{k}{\theta^2 h\sqrt{1-\left(\dfrac{a\theta-k^2}{\theta h}\right)^2}} \cdot (-b\sin x + c\cos x)$

$= \dfrac{k}{\theta\sqrt{\theta^2 h^2 - (a\theta - k^2)^2}}(-c\cos x + b\sin x)$.

$[\theta^2 h^2 - (a\theta - k^2)^2 = \theta^2(h^2 - a^2) - k^4 + 2a\theta k^2 = -\theta^2 k^2 - k^4 + 2a\theta k^2$
$= k^2\{2a\theta - k^2 - \theta^2\} = k^2\{-(\theta - a)^2 - k^2 + a^2\}$
$= k^2\{-(b\cos x + c\sin x)^2 + h^2\}$
$= k^2\{-b^2\cos^2 x - c^2\sin^2 x - 2bc\sin x\cos x + b^2 + c^2\}$
$= k^2\{b^2\sin^2 x + c^2\cos^2 x - 2bc\sin x\cos x\}$
$= k^2(b\sin x - c\cos x)^2.]$

∴ $\dfrac{dy}{dx} = \dfrac{k}{\theta\cdot k(b\sin x - c\cos x)}(b\sin x - c\cos x) = \dfrac{1}{\theta}$. **(Proved)**.

Example 9.4.7. Differentiation of a determinant whose elements are functions of x.

CHAPTER 9: DIFFERENTIATION

Let u_1, v_1, w_1 etc. are all derivable functions of x.
Then

$$\frac{d}{dx}\begin{vmatrix} u_1 & v_1 & w_1 \\ u_2 & v_2 & w_2 \\ u_3 & v_3 & w_3 \end{vmatrix} = \begin{vmatrix} \frac{du_1}{dx} & \frac{dv_1}{dx} & \frac{dw_1}{dx} \\ u_2 & v_2 & w_2 \\ u_3 & v_3 & w_3 \end{vmatrix} + \begin{vmatrix} u_1 & v_1 & w_1 \\ \frac{du_2}{dx} & \frac{dv_2}{dx} & \frac{dw_2}{dx} \\ u_3 & v_3 & w_3 \end{vmatrix} + \begin{vmatrix} u_1 & v_1 & w_1 \\ u_2 & v_2 & w_2 \\ \frac{du_3}{dx} & \frac{dv_3}{dx} & \frac{dw_3}{dx} \end{vmatrix}.$$

This rule can be extended to derminants of order n (say): For examples,

$$\frac{d}{dx}\begin{vmatrix} 1 & x & x^2 & x^3 \\ 1 & \sin x & e^x & \cos x \\ 1 & a^x & \sinh x & \tan^{-1} x \\ 1 & \sinh^{-1} x & \tan x & 2^x \end{vmatrix} = \begin{vmatrix} 0 & 1 & 2x & 3x^2 \\ 1 & \sin x & e^x & \cos x \\ 1 & a^x & \sinh x & \tan^{-1} x \\ 1 & \sinh^{-1} x & \tan x & 2^x \end{vmatrix}$$

$$+ \begin{vmatrix} 1 & x & x^2 & x^3 \\ 0 & \cos x & e^x & -\sin x \\ 1 & a^x & \sinh x & \tan^{-1} x \\ 1 & \sinh^{-1} x & \tan x & 2^x \end{vmatrix} + \begin{vmatrix} 1 & x & x^2 & x^3 \\ 1 & \sin x & e^x & \cos x \\ 0 & a^x \log_e a & \cosh x & 1/(1+x^2) \\ 1 & \sinh^{-1} x & \tan x & 2^x \end{vmatrix}$$

$$+ \begin{vmatrix} 1 & x & x^2 & x^3 \\ 1 & \sin x & e^x & \cos x \\ 1 & a^x & \sinh x & \tan^{-1} x \\ 1 & \frac{1}{\sqrt{1+x^2}} & \sec^2 x & 2^x \log_e a \end{vmatrix}.$$

Exercises IX (I)

[A]

(Derivatives from definition)

1. Use the definition of derivative to find $\frac{d}{dx}(e^{\sin \sqrt{x}})(x > 0)$ and $\frac{d}{dx}(\log \tan^{-1} x)$.

2. Find, from *first principles*, the derivative of $\sqrt{x}, \sin^{-1} x, xe^{-x}, \sqrt{\tan x}$, $\log \sin(x/a), \log_{10} x$ (at the point x, where the function is defined at and near x).

3. (a) If $f(x) = x^{1/3}$, verify that at $x = 0$, right-derivative and left-derivative each equals to $+\infty$.

   (b) Show that $f(x) = x^{2/3}$ is not derivable at $x = 0$.

4. If $f(x) = 0 (x \le 0)$ and $= 1 (x > 0)$, find $f'(0), f'(0+0), f'(0-0)$, in case they exist.

[*Continuity and Derivability:*
For further problems of this type see **Ex. IX (II)** Section [A]]

5. $f(x) = x + 1 (x < 0), f(x) = e^x (x \ge 0)$. Does $f'(0)$ exist? If $f'(x)$ continuous at $x = 0$?

6. Let $f(x) = x\cos(1/x), x \ne 0$ and $f(0) = 0$. Show that f is continuous at $x = 0$ but is not derivable there.

7. Let $\phi(x) = x^2 \cos(1/x), x \ne 0$ and $\phi(0) = 0$. Show that f is derivable at $x = 0$. Is it continuous at $x = 0$?

8. Show that the function f defined by
$$f(x) = \begin{cases} 3 + 2x, & \text{for } -3/2 < x \le 0; \\ 3 - 2x, & \text{for } 0 < x < 3/2; \end{cases}$$
is continuous but not derivable at $x = 0$.

9. A function ϕ is defined by
$$\phi(x) = x, 0 < x < 1; \quad = 2 - x, 1 \le x \le 2; \quad = x - \frac{1}{2}x^2, x > 2.$$
Show ϕ is continuous at $x = 1$ and $x = 2$ and that $\phi'(2)$ exists but $\phi'(1)$ does not exist.

10. Let
$$\phi(x) = \begin{cases} \frac{1}{2}(b^2 - a^2), & 0 \le x \le a \\ \frac{1}{2}b^2 - \frac{1}{6}x^2 - \frac{1}{3}\frac{a^3}{x}, & a < x \le b \\ \frac{1}{3}\frac{(b^3 - a^3)}{x}, & x > b. \end{cases}$$
Show that the derived function ϕ' is continuous $\forall x > 0$.

11. Examine the differentiability of the function f at $x = 1$ where f is defined by $f(x) = \sin|x - 1|$.

CHAPTER 9: DIFFERENTIATION

[B]
Calculation of derivatives
REVISION OF PREVIOUS KNOWLEDGE

1. Find the derivatives of the following functions with respect to the independent variable appearing in the function:

a) $\dfrac{a + bx^{3/2}}{c\sqrt[4]{x^5}}$;

b) $\sqrt{\sin\sqrt{x}}$;

c) $(\log \cot x) \cdot \tan^{-1}(e^x)$;

d) $\log \cosh x$;

e) $\tan^{-1}(\tanh x)$;

f) $\tan^{-1}\dfrac{\sqrt{x} - x}{1 + x^{3/2}}$;

g) $\sin(e^x \log x) \cdot \sqrt{1 - (\log x)^2}$;

h) $\sin^{-1}\dfrac{a + b\cos x}{b + a\cos x}$;

i) $\sec(\log_a \sqrt{a^2 + x^2})$;

j) $\tanh^{-1}\dfrac{3x + x^3}{1 + 3x^2} + \tan^{-1}\dfrac{3x - x^3}{1 - 3x^2}$;

k) $\dfrac{1}{\sqrt{b^2 - a^2}} \log \dfrac{\sqrt{b+a} + \sqrt{b-a}\tan x/2}{\sqrt{b+a} - \sqrt{b-a}\tan x/2}$;

l) $\log\left\{e^x \sqrt[4]{\left(\dfrac{x-2}{x+2}\right)^3}\right\}$;

m) x^{x^x};

n) $x^x + x^{1/x}$;

o) $(\sin x)^{\cos x} + (\cos x)^{\sin x}$;

p) $\tan^{-1}(a^{cx} x^{\sin x}) \cdot \dfrac{\sqrt{x}}{1 + x^{3/2}}$;

q) $\tan^{-1}\sqrt{\sqrt{x} + \cos^{-1} x}$;

r) $\left(\dfrac{1 + \sqrt{x}}{1 + 2\sqrt{x}}\right)^{\sin e^{x^2}}$;

s) $\left(1 + \dfrac{1}{x}\right)^x + x^{1 + \frac{1}{x}}$;

t) $\tan^{-1}(e^{\cos^2 x} \sin x)$;

u) $\tan^{-1}\dfrac{\cos x - \sin x}{\cos x + \sin x}$;

v) $\tan^{-1}\sqrt{\dfrac{1-x}{1+x}}$.

2. Find $\dfrac{dy}{dx}$ in the following cases:

a) $y = \dfrac{\sin x - \cos x}{\sin x + \cos x}$;

b) $y = \sqrt{x \sin x}$;

c) $y = \log \sin x° \ (x \text{ degrees})$;

d) $y = x^{\cos^{-1} x}$;

e) $y = (\tan x)^{\cot x} + x^{\cos^{-1} x} + (\cot x)^{\tan x}$;

f) $y = 10^{\log \sin x}$;

g) $x^3 + y^3 = 3axy$;

h) $(\cos x)^y = (\sin y)^x$;

i) $y = x^{y^x}$;

j) $x^m y^n = (x+y)^{m+n}$;

k) $y = e^{\tan^{-1} y} \log \sec^2 x^3$;

l) $x = \text{arc sec } \dfrac{1}{2t^2 - 1}$,
$y = \text{arc tan } \dfrac{t}{\sqrt{1-t^2}}$;

m) $x = a(\cos t + t \sin t)$,
$y = a(\sin t - t \cot t)$;

n) $x = a(\theta - \sin \theta)$,
$y = a(1 - \cos \theta)$;

o) $y = \log_{\sin x} x$;

p) $y = \dfrac{\sin^m mx}{\cos^n nx}$;

q) $y = \sqrt{\dfrac{(x-a)(x-b)}{(x-c)(x-d)}}$;

r) $y = \left(x^{\frac{l+m}{m-n}}\right)^{\frac{1}{n-l}} \cdot \left(x^{\frac{m+n}{n-l}}\right)^{\frac{1}{l-m}} \cdot \left(x^{\frac{n+l}{l-m}}\right)^{\frac{1}{m-n}}$;

s) $y = \dfrac{x\sqrt{a^2 - x^2}}{2} + \dfrac{a^2}{2} \sin^{-1} \dfrac{x}{a}$;

t) $y = \tan^{-1} \dfrac{\sqrt{1+x^2} + \sqrt{1-x^2}}{\sqrt{1+x^2} - \sqrt{1-x^2}}$.

3. Differentiate:

(a) $(x^2 + ax + a^2)^n \log \cot(x/2)$ with respect to (w.r.t.) $\tan^{-1}(a \cos bx)$.

(b) $\log \dfrac{a + b \tan(x/2)}{a - b \tan(x/2)}$ w.r.t. $\dfrac{1}{a^2 \cos^2(x/2) - b^2 \sin^2(x/2)}$.

(c) $x^{\sin^{-1} x}$ w.r.t. $\sin^{-1} x$; $x^{\sin x}$ w.r.t. $(\sin x)^x$.

(d) $\dfrac{\sqrt{1+x^2} + \sqrt{1-x^2}}{\sqrt{1+x^2} - \sqrt{1-x^2}}$ w.r.t. $\sqrt{1-x^4}$.

(e) $\tan^{-1} \dfrac{x}{\sqrt{1-x^2}}$ w.r.t. $\sec^{-1} \dfrac{1}{2x^2 - 1}$.

(f) $x^n \log \tan^{-1} x$ w.r.t. $\dfrac{\sin \sqrt{x}}{x^{3/2}}$.

(g) $\arccos \sqrt{\dfrac{\cos 3\theta}{\cos^3 \theta}}$ w.r.t. θ.

(h) $\log_e \cos \sqrt{\arcsin 3^{-2x}}$ w.r.t. $\sqrt[3]{\arctan \sqrt[5]{\cos(\log x)^3}}$.

(i) $\sin \dfrac{x}{\sqrt{1-x^2}}$ w.r.t. $\arcsin \dfrac{x}{\sqrt{1+x^2}}$.

(j) $\tan^{-1} \dfrac{\sqrt{1+x^2}-1}{x}$ w.r.t. $\tan^{-1} x$.

4. If S_n = the sum of a geometric progression to n terms of which r is the common ratio, prove that
$$(r-1)\dfrac{dS_n}{dr} = (n-1)S_n - nS_{n-1}.$$

5. If $y = \sec 4x$ and $x = \tan^{-1} t$, prove that
$$\dfrac{dy}{dt} = \dfrac{16t(1-t^4)}{(1-6t^2+t^4)^2}.$$

6. If $C = 1 + r\cos\theta + \dfrac{r^2}{2!}\cos 2\theta + \dfrac{r^3}{3!}\cos 3\theta + \cdots\cdots$

$S = r\sin\theta + \dfrac{r^2}{2!}\sin 2\theta + \dfrac{r^3}{3!}\sin 3\theta + \cdots$.

Prove that
$$\left.\begin{array}{l} C\dfrac{dC}{dr} + S\dfrac{dS}{dr} = R^2 \cos\theta \\[4pt] C\dfrac{dS}{dr} - S\dfrac{dC}{dr} = R^2 \sin\theta \end{array}\right\} \quad R^2 = C^2 + S^2.$$

7. If $y = e^{-xt} \sec^{-1}(x\sqrt{t})$ and $t^4 + x^2 t = x^5$, find dy/dx in terms of x and t.

8. Prove that if $|x| < 1$, then
$$\dfrac{1}{1+x} + \dfrac{2x}{1+x^2} + \dfrac{4x^3}{1+x^4} + \dfrac{8x^7}{1+x^8} + \cdots \infty = \dfrac{1}{1-x}.$$

9. Given Euler's Theorem
$$\lim_{n\to\infty}\left(\cos\dfrac{x}{2}\cos\dfrac{x}{2^2}\cos\dfrac{x}{2^3}\cdots\cos\dfrac{x}{2^n}\right) = \dfrac{\sin x}{x},$$
prove that

(a) $\dfrac{1}{2}\tan\dfrac{x}{2} + \dfrac{1}{2^2}\tan\dfrac{x}{2^2} + \dfrac{1}{2^3}\tan\dfrac{x}{2^3} + \cdots \infty = \dfrac{1}{x} - \cot x.$

(b) $\dfrac{1}{2^2}\sec^2\dfrac{x}{2}+\dfrac{1}{2^4}\sec^2\dfrac{x}{2^2}+\dfrac{1}{2^6}\sec^2\dfrac{x}{2^3}+\cdots\infty=\operatorname{cosec}^2 x-\dfrac{1}{x^2}$.

10. (a) If $y=x^{x^{x^{\cdot^{\cdot^{\cdot \text{to}\infty}}}}}$, prove that $x\dfrac{dy}{dx}=\dfrac{y^2}{1-y\log x}$.

(b) If $y=\sqrt{\sin x+\sqrt{\sin x+\sqrt{\sin x+\cdots\infty}}}$, prove that $\dfrac{dy}{dx}=\dfrac{\cos x}{2y-1}$.

(c) If $y=\log^n(x)$, where $\log^n$ means $\log\log\log\cdots$ (repeated n times), prove that $\dfrac{dy}{dx}=\dfrac{1}{x\log x\log^2(x)\log^3(x)\cdots\log^{n-1}(x)}$.

11. If $y=1+\dfrac{a_1}{x-a_1}+\dfrac{a_2 x}{(x-a_1)(x-a_2)}+\dfrac{a_3 x^2}{(x-a_1)(x-a_2)(x-a_3)}$

$+\dfrac{a_4 x^3}{(x-a_1)(x-a_2)(x-a_3)(x-a_4)}$, then prove that

$\dfrac{dy}{dx}=\dfrac{y}{x}\left(\dfrac{a_1}{a_1-x}+\dfrac{a_2}{a_2-x}+\dfrac{a_3}{a_3-x}+\dfrac{a_4}{a_4-x}\right)$.

12. Let $X=a^2\cos^2 x-b^2\sin^2 x$, $\phi(x)=\log\dfrac{a+b\tan x}{a-b\tan x}\cdot\sqrt{\dfrac{a\cos x-b\sin x}{a\cos x+b\sin x}}$.
Prove that $\phi'(x)=ab/X$.

13. If $y=\dfrac{1}{2}(x-a)\theta+\dfrac{1}{2}a^2\sin^{-1}\dfrac{x-a}{a}$, where $\theta^2=2ax-x^2$, prove that $\dfrac{dy}{dx}=\theta$.

14. (a) If $\sin y=x\sin(a+y)$, prove that $\dfrac{dy}{dx}=\dfrac{\sin^2(a+y)}{\sin a}$.

(b) If $x^y=e^{x-y}$, prove that $\dfrac{dy}{dx}=\dfrac{\log x}{(\log ex)^2}$.

(c) If $f(x)=\left(\dfrac{a+x}{b+x}\right)^{a+b+2x}$, prove that $f'(0)=\left(2\log\dfrac{a}{b}+\dfrac{b^2-a^2}{ab}\right)\left(\dfrac{a}{b}\right)^{a+b}$.

15. Find the derivatives of the following functions w.r.t. x in simplest forms:

(a) $\dfrac{1}{2}\log(x+\sqrt[3]{1-x^3})-\dfrac{1}{\sqrt{3}}\tan^{-1}\dfrac{2\sqrt[3]{1-x^3}-x}{\sqrt{3}\cdot x}$.

(b) $\dfrac{1}{3}\log\dfrac{x-1}{\sqrt{x^2+x+1}}+\dfrac{1}{\sqrt{3}}\tan^{-1}\dfrac{2x+1}{\sqrt{3}}$.

16. If $x=\dfrac{\sin^3 t}{\sqrt{\cos 2t}}$ and $y=\dfrac{\cos^3 t}{\sqrt{\cos 2t}}$, find $\dfrac{dy}{dx}$ at $t=\dfrac{\pi}{6}$.

17. $y=\dfrac{(3+x)(2-x)^2}{(4-3x)^3}$, find the values of x for which $\dfrac{dy}{dx}=0$.

CHAPTER 9: DIFFERENTIATION

18. (a) If $\phi(x) = \begin{vmatrix} 1 & 1 & 1 & 1 \\ 1 & x & 1 & 1 \\ 1 & 1 & x & 1 \\ 1 & 1 & 1 & x \end{vmatrix}$, prove that $\phi'(x) = 3(x-1)^2$.

(b) If $D_4 = \begin{vmatrix} x & a & a & a \\ a & x & a & a \\ a & a & x & a \\ a & a & a & x \end{vmatrix}$, prove that $\dfrac{d}{dx}(D_4) = 4D_3$.

(c) If $f(x) = \begin{vmatrix} \sin x & \cos x & \sin x \\ \cos x & -\sin x & \cos x \\ x & 1 & 1 \end{vmatrix}$, prove that $f'(x) = 1$.

19. If $a^2 > b^2$, prove that the derivative of either

$$\frac{1}{\sqrt{a^2-b^2}} \cos^{-1} \frac{a\cos x + b}{a + b\cos x} \quad \text{or} \quad \frac{2}{\sqrt{a^2-b^2}} \tan^{-1}\left(\sqrt{\frac{a-b}{a+b}} \tan \frac{x}{2}\right)$$

w.r.t. x will be $\dfrac{1}{a + b\cos x}$.

20. If $\sin x \sin\left(\dfrac{\pi}{n} + x\right) \sin\left(\dfrac{2\pi}{n} + x\right) \cdots \sin\left((n-1)\dfrac{\pi}{n} + x\right) = \dfrac{\sin nx}{2^{n-1}}$, then prove that

$$\cot x + \cot(\pi/n + x) + \cot(2\pi/n + x) + \cdots + \cot\{(n-1)\pi/n + x\} = n\cot nx.$$

21. (a) Prove that $\dfrac{d}{dx}\left(\dfrac{2}{3} \tan^{-1} \dfrac{1}{3}(5\tan\dfrac{x}{2} + 4)\right) = \dfrac{1}{5 + 4\sin x}$.

(b) Prove that if $|x| < 1$,

$$\frac{d}{dx}\left\{\frac{x}{1 + \sqrt{1-x^2}}\right\}^n = -ny\frac{d}{dx}(\text{sech}^{-1}x), \quad \text{where } y = \left(\frac{x}{1 + \sqrt{1-x}}\right)^n$$

22. (a) If $y = \log\dfrac{1+x}{1-x} + \dfrac{1}{2}\log\dfrac{1+x+x^2}{1-x+x^2} + \sqrt{3}\tan^{-1}\dfrac{x\sqrt{3}}{1-x^2}$, prove that

$$\frac{dy}{dx} = \frac{6}{1-x^6}.$$

(b) If $y = \dfrac{2^x}{\sqrt{1 + (\log 2)^2}} \cos\{x - \cot^{-1}(\log 2)\}$, prove that $\dfrac{dy}{dx} = 2^x \cos x$.

(c) If $\sqrt{1-x^2} + \sqrt{1-y^2} = a(x-y)$, prove that $\dfrac{dy}{dx} = \sqrt{\dfrac{1-y^2}{1-x^2}}$.

(d) Prove that $\dfrac{dy}{dx} = \dfrac{1}{\cos x + \cos \alpha}$, if $y \sin \alpha = \log \dfrac{\cos \frac{1}{2}(x-\alpha)}{\cos \frac{1}{2}(x+\alpha)}$.

(e) Prove that $\dfrac{dy}{dx} = \sec x + \operatorname{cosec} x$, if $y = \log \sqrt{\dfrac{(1+\sin x)(1-\cos x)}{(1-\sin x)(1+\cos x)}}$.

23. $f(x) = \log \dfrac{\sqrt{a+bx}-\sqrt{a-bx}}{\sqrt{a+bx}+\sqrt{a-bx}}$, find for what value of x, $\dfrac{1}{f'(x)} = 0$.

24. (a) Differentiate $\dfrac{\sin x - \cos x}{\sin x + \cos x}$ w.r.t. x (w.r.t. means 'with respect to').

(b) Differentiate $2 \tan^{-1} \sqrt{\dfrac{x-a}{b-x}}$ w.r.t. x.

(c) Differentiate $\tan^{-1} \dfrac{2t}{1-t^2}$ w.r.t. $\sin^{-1} \dfrac{2t}{1+t^2}$.

(d) Differentiate $\dfrac{3at^2}{1+t^3}$ w.r.t. $\dfrac{3at}{1+t^3}$.

Answers

1. a) $\dfrac{bx^{3/2}-5a}{4cx^{9/4}}$; b) $\dfrac{\cos\sqrt{x}}{4\sqrt{x \sin \sqrt{x}}}$; c) $\dfrac{\log\sqrt{\cot x}}{\cosh x} - \dfrac{2\tan^{-1} e^x}{\sin 2x}$;

d) $\tanh x$; e) $\operatorname{sech} 2x$; f) $\dfrac{x^2 - 2x^{3/2} - 2x^{1/2} + 1}{2x^{1/2}(1+x)(1+x^2)}$;

g) $\cos(e^x \log x) \cdot e^x \log(xe^{1/x}) \cdot \sqrt{1-(\log x)^2} - \sin(e^x \log x) \cdot \dfrac{\log x^{1/x}}{\sqrt{1-(\log x)^2}}$;

h) $-\dfrac{\sqrt{b^2-a^2}}{b+a\cos x}$; i) $\{x \log_a e \cdot \sin(\log_a \sqrt{a^2+x^2})\}/(a^2+x^2)\cos^2(\log_a \sqrt{a^2+x^2})$;

j) $\dfrac{6}{1-x^4}$; k) $\dfrac{1}{a+b\cos x}$; l) $\dfrac{x^2-1}{x^2-4}$;

m) $x^{x^x} \cdot x^x \left\{(\log x)^2 + \log x + \dfrac{1}{x}\right\}$; n) $x^x \log(ex) - x^{1/x-2} \log \dfrac{x}{e}$;

o) $(\sin x)^{\cos x}\left\{\dfrac{\cos^2 x}{\sin x} - \sin x \log \sin x\right\} - (\cos x)^{\sin x}\left\{\dfrac{\sin^2 x}{\cos x} - \cos x \log \cos x\right\}$;

CHAPTER 9: DIFFERENTIATION 541

p) $\dfrac{\sqrt{x}}{1+x^{3/2}} \cdot \dfrac{a^{cx} \cdot x^{\sin x}}{1+a^{2cx} \cdot x^{2\sin x}} \log\left\{ a^c \cdot x^{\cos x} \cdot e^{\frac{\sin x}{x}} \right\}$

$+ \tan^{-1}\left(a^{cx} \cdot x^{\sin x}\right) \dfrac{1 - 2x^{3/2}}{2x^{1/2}(1+x^{3/2})^2};$

q) $\dfrac{\sqrt{1-x^2} - 2\sqrt{x}}{4\sqrt{x}\sqrt{1-x^2}\sqrt{\sqrt{x}+\cos^{-1}x}\,(1+\sqrt{x}+\cos^{-1}x)};$

r) $y\left\{ 2xe^{x^2}\cos e^{x^2}\log\dfrac{1+\sqrt{x}}{1+2\sqrt{x}} - \dfrac{\sin e^{x^2}}{2\sqrt{x}(1+\sqrt{x})(1+2\sqrt{x})}\right\},$ where $y =$ given function.

s) $\left(1+\dfrac{1}{x}\right)^x\left\{\log\dfrac{x+1}{x} - \dfrac{1}{x+1}\right\} + x^{1/x-1}\{x+1-\log x\};$

t) $\cos x \cdot \cos 2x \cdot \cos^2 y \cdot e^{\cos^2 x}$, where $y = \tan^{-1}(e^{\cos^2 x}\sin x);$

u) $-1;$ v) $-\dfrac{1}{2\sqrt{1-x^2}}.$

2. a) $\dfrac{2}{(\sin x + \cos x)^2};$ b) $\dfrac{x\cos x + \sin x}{2\sqrt{x\sin x}};$ c) $\dfrac{\pi}{180}\cot x°;$

d) $x^{\cos^{-1}x}\left\{\dfrac{1}{x}\cos^{-1}x - \dfrac{\log x}{\sqrt{1-x^2}}\right\};$

e) $x^{\cos^{-1}x}\left\{\dfrac{1}{x}\cos^{-1}x - \dfrac{\log x}{\sqrt{1-x^2}}\right\} + (\tan x)^{\cot x}\cdot\{\operatorname{cosec}^2 x(1-\log\tan x)\}$

$+ (\cot x)^{\tan x}\{\sec^2 x(\log \cot x - 1)\};$

f) $y\cot x\log 10;$ g) $-\dfrac{x^2 - ay}{y^2 - ax};$ h) $\dfrac{y\tan x + \log\sin y}{\log\cos x - x\cot y};$

i) $\dfrac{y\log y}{x\log x} \times \dfrac{1+x\log x\log y}{1 - x\log y};$ j) $\dfrac{y}{x};$ k) $\dfrac{6x^2(1+y^2)\tan x^3 \cdot e^{\tan^{-1}y}}{1 + y^2 - \log\sec^2 x^3 \cdot e^{\tan^{-1}y}};$

l) $-\dfrac{1}{2};$ m) $\tan t;$ n) $\cot\theta/2;$

o) $\dfrac{x^{-1}\log\sin x - \cot x \log x}{(\log\sin x)^2};$

p) $\{\sin^{m-1}mx\cos^{n-1}nx(m^2\cos mx\cos nx + n^2\sin mx\sin nx)\}/\cos^{2n}nx;$

q) $\dfrac{1}{2(x-c)(x-d)}\left[\sqrt{\dfrac{(x-c)(x-d)}{(x-a)(x-b)}}(2x - a - b) - \sqrt{\dfrac{(x-a)(x-b)}{(x-c)(x-d)}}(2x - c - d)\right];$

A.I.T.D.C.[P-I]—35

r) 0;   s) $\sqrt{a^2 - x^2}$;   t) $\dfrac{-x}{\sqrt{1-x^4}}$.

3. a) $\dfrac{(1+a^2\cos^2 bx)(x^2+ax+a^2)^{n-1}\left[n(2x+a)\log\cot\dfrac{x}{2} - \csc x\cdot(x^2+ax+a^2)\right]}{-ab\sin bx}$;

b) $\dfrac{ab}{a^2+b^2}\left(a^2\cot\dfrac{x}{2} - b^2\tan\dfrac{x}{2}\right)$;   c) $x^{\sin^{-1} x}\left(\log x + \dfrac{\sqrt{1-x^2}}{x}\sin^{-1} x\right)$;

$\dfrac{x^{\sin x-1}(x\cos x\log x + \sin x)}{(\sin x)^x(\log \sin x + x\cot x)}$;   d) $\dfrac{1}{x^4}\cdot\dfrac{\sqrt{1+x^2}+\sqrt{1-x^2}}{\sqrt{1+x^2}-\sqrt{1-x^2}}$;

e) $-\dfrac{1}{2}$;   f) $2\dfrac{n(1+x^2)\tan^{-1} x\log\tan^{-1} x + x}{(1+x^2)\tan^{-1} x(\sqrt{x}\cos\sqrt{x} - 3\sin\sqrt{x})}\cdot x^{\frac{1}{2}(2n+3)}$;

g) $\sqrt{\left(\dfrac{3}{\cos\theta\cos 3\theta}\right)}$;

h) $\left\{\dfrac{\tan\sqrt{\arcsin 3^{-2x}}}{\sqrt{\arcsin 3^{-2x}}} \times \dfrac{1}{\sqrt{1-3^{-4x}}} \times \dfrac{\log 3}{3^{2x}}\right\} \Big/ \left[-\dfrac{1}{\{\arctan\sqrt[5]{\cos(\log x)^3}\}^{2/3}}\right.$

$\left.\times \dfrac{1}{1+\{\cos(\log x)^3\}^{2/5}} \times \dfrac{\sin(\log x)^3 \times (\log x)^2}{x}\right]$;

i) $\left\{\cos\dfrac{x}{\sqrt{1-x^2}} \times \dfrac{1}{(1-x^2)^{3/2}}\right\} \Big/ \dfrac{1}{1+x^2}$;   j) $\dfrac{1}{2}$;

7. $\dfrac{e^{-xt}}{4t^3+x^2}\left[\dfrac{8t^4+5x^5}{2xt\sqrt{x^2 t-1}} - (5t^4+4x^5)\sec^{-1} x\sqrt{t}\right]$.   15. a) $\dfrac{1}{\sqrt[3]{1-x^3}}$;   b) $\dfrac{x}{x^3-1}$.

16. 0.   17. $x = 2$ or $x = 38/9$.   23. $x = 0$ or $x = \pm a/b$.

24. a) $\dfrac{2}{(\sin x + \cos x)^2}$;   b) $\dfrac{1}{\sqrt{(x-a)(b-x)}}$;   c) 1;   d) $\dfrac{2t-t^4}{1-2t^3}$.

9.5 Concepts of Differentiability and Differentials

Let f be a real-valued function defined on a closed interval $[a,b]$. The derivative of a function f at a point $x \in [a,b]$ is defined in the following manner:

(a) We construct the quotient $\dfrac{f(t)-f(x)}{t-x}$ $(a < t < b, t \neq x)$.

(b) We denote this quotient by $\phi(t)$: it is certainly a function of t.

CHAPTER 9: DIFFERENTIATION

(c) If $\lim_{t \to x} \phi(t)$ exists, then f is said to be *derivable* at x.

(d) We then denote this limits by $f'(x)$ and call it is the **derivative** of f at the point x.

We thus associate with the function f another function f' **whose domain** is the **set of points x at which** $\lim_{t \to x} \phi(t)$ **exists.**

This function f' is then called the **derived function** of f and its domain is a **subset** of the domain of f.

According to some authors, if f' is defined at a point x, then f is **differentiable at x**. If f' is defined at every point of a set $E \subset [a,b]$, then f is said to be **differentiable on E**.

Keeping in mind the use of the term differentiability in case of functions of more than one independent variable we shall, however, give the following definition of differentiability of a function of a single variable:

Definition 9.5.1. *Let f be a real-valued function of a single variable x, defined over a set of real numbers, say over $[a,b]$. Let $x \in [a,b]$.*

We take another point $x + \Delta x \in [a,b]$. Then f is said to be differentiable at the point x if we can express

$$f(x + \Delta x) - f(x) = A\Delta x + \varepsilon \Delta x,$$

where A is independent of Δx and ε is a function of Δx which tends to zero as $\Delta x \to 0$.

Definition 9.5.2. *The linear part $A\Delta x$ is called the **Differential** of f at the point x, and is denoted by df (thus, $df = A\Delta x$). The other part $\varepsilon \Delta x$ is called the **Error** if we accept $A\Delta x$ for the increment of f, namely $\Delta f = f(x + \Delta x) - f(x)$.*

> Derivability of f at a point $x \iff$ Differentiability of f at the point x.

We shall now prove the following theorem:

Theorem 9.5.1. (Condition for differentiability). *The necessary and sufficient condition that a function $f(x)$ be differentiable at a given point is that it possessess a finite derivative at that point.*

Proof: Let $f(x)$ be differentiable at x, then from the definition of differentiability

$$\frac{f(x + \Delta x) - f(x)}{\Delta x} = A + \varepsilon,$$

where A is independent of Δx, but $\varepsilon \to 0$ as $\Delta x \to 0$.

Letting $\Delta x \to 0$, we see that $f'(x) = A$, i.e., $f(x)$ is finitely derivable at x, the derivative $f'(x)$ being equal to A.

Conversely, let $f(x)$ possess a derivative $f'(x)$ at x, then by the definition of derivability,

$$\lim_{\Delta x \to 0} \frac{f(x+\Delta x)-f(x)}{\Delta x} = f'(x).$$

i.e., $\quad \dfrac{f(x+\Delta x)-f(x)}{\Delta x} = f'(x) + \varepsilon$, when $\varepsilon \to 0$ as $\Delta x \to 0$,

or, $\quad f(x+\Delta x) - f(x) = f'(x)\Delta x + \varepsilon \Delta x, \varepsilon \to 0$ as $\Delta x \to 0$.

Thus, $f(x)$ is differentiable at x. Hence the theorem.

Observation: If $y = f(x)$, we denote its differential by dy and write

$$dy = A\Delta x = f'(x)\Delta x.$$

It is customary to replace Δx by dx and *to define the differential of the independent variable x by the relation $dx = \Delta x$*. This is consistent with the fact that if we take $y = f(x) = x$ in the definition, $f'(x) = 1$ and we have $dx = 1 \cdot \Delta x = \Delta x$.

Finally, whether x be the independent variable or not, the differential of $y = f(x)$ is defined by

$$\boxed{dy = f'(x)\, dx.}$$

The symbol dy/dx was defined as the limit of $\Delta y/\Delta x$ when $\Delta x \to 0$, but the dy and dx were not given independent meanings there. We may now think of

$$dy \div dx = f'(x) = \text{(differential of } y\text{)} \div \text{(differential of } x\text{)}$$

Example 9.5.1. *If $y = x^4$, then $dy = 4x^3 dx$.*

If $f(x) = \log x$, then $df(x) = f'(x)dx = (1/x)dx$.

Note 9.5.1. *See that dy not only depends on the value of x but also on the choice of the increment Δx; i.e., dy is a function of two variables x and Δx.*

Note 9.5.2. *Since the expression for the differential of a function, the coefficient of the differential of the independent variable is always the derivative of the function, the derivative is often called the differential coefficient.*

CHAPTER 9: DIFFERENTIATION   545

9.6 Geometrical Interpretation of the Derivative and Differential

In many applications we shall use the fact that the problem of finding the direction of a curve is the same as the problem of finding the derivative of the function which represents the curve. We recall here the definition of the tangent line at a point of a curve:

Definition 9.6.1. *Let P be an arbitrary point of a given curve C. We shall assume that the curve is continuous at and near P. We next take a second point Q (near P) on the curve. Suppose that Q can move along the curve so as to approach to P from either side.*

As Q moves along the curve C and approaches P, the secant PQ rotates about P as a **pivot** and **sometimes** (may not be *always*) approaches a definite limiting position PT which we call **the tangent line at P.**

In Analytic Geometry we defined the *slope* or *gradient* of a line, not parallel to Y-axis as the trigonometric tangent of the angle θ which the line makes with the positive axis of $x : OX$.

Thus, in Fig. 9.1, the gradient of the secant PQ is $\tan\theta$ and the gradient of the tangent line PT is $\tan\psi$. The existence of the limiting position of the secant PQ implies that the limit of the angle θ is ψ when $Q \to P$.

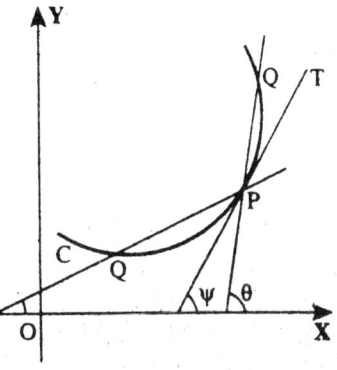

The direction of the curve at P is determined by the gradient of the tangent line at P.

∴ slope or gradient of the *curve* at $P = \tan\psi$.

Fig. 9.1

Relation between gradient of the tangent-line and derivative

Let $P : (x, y)$ be a point on a *continuous curve* whose equation is $y = f(x)$, the system of axes being supposed to be rectangular.

We assume here that the *curve possesses a definite tangent PT at P and that this tangent is not parallel to the Y-axis.*

Let Q be the point on the curve whose co-ordinates are $(x + \Delta x, y + \Delta y)$. Draw PL, QM perpendiculars on OX and draw PN parallel to OX to meet QM in N (Fig. 9.2).

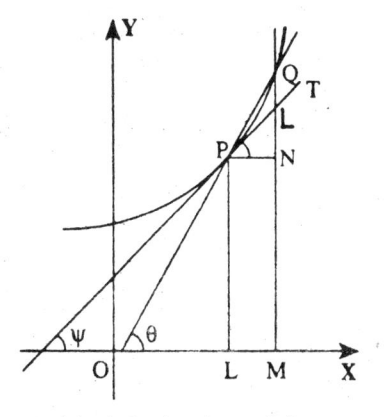

$P \equiv (x, y), Q \equiv (x + \Delta x, y + \Delta y)$
$PN = \Delta x, NQ = \Delta y, NL = dy, PN = dx$

Fig. 9.2

Then
$$PN = LM = OM - OL$$
$$= (x + \Delta x) - x = \Delta x,$$
$$NQ = MQ - MN = MQ - LP$$
$$= (y + \Delta y) - y = \Delta y,$$

and since P and Q are points on the curve, we have
$$y = f(x) \quad \text{and} \quad y + \Delta y = f(x + \Delta x)$$
so that $\Delta y = f(x + \Delta x) - f(x)$.

∴ gradient of secant $PQ = \tan \theta = \tan QPN = \frac{NQ}{PN} = \frac{\Delta y}{\Delta x}$.

Now as Q tends to P, PQ tends to a definite limiting position PT and Δx tends to zero.

∴ gradient of $PT = \tan \psi = \lim_{\Delta x \to 0} \frac{\Delta y}{\Delta x} = \frac{dy}{dx}$ or $f'(x)$.

Hence $f'(x)$ is the gradient of the tangent to the curve $y = f(x)$. This is also the gradient or slope of the curve at (x, y).

In particular, $f'(x) = 0$ implies that the tangent at the point (x, y) is parallel to the X-axis.

Note 9.6.1. *The equation of the tangent line at the point $(a, f(a))$ of the curve $y = f(x)$ is given by:*
$$y - f(a) = f'(a)(x - a), \quad \text{provided } f'(a) \text{ exists}.$$

At a point $(b, f(b))$ the tangent is given by $x = b$, where $f'(b)$ does not exist.

In this case the tangent is the straight line through $(b, f(b))$ and is parallel to y-axis.

Example 9.6.1. *Find the slope of the parabola $y = x^2$ at the vertex and at the point $\left(\frac{1}{2}, \frac{1}{4}\right)$. Obtain the equation of the tangent at $\left(\frac{1}{2}, \frac{1}{4}\right)$.*

Solution: Differentiating $y = x^2$ with respect to x, we obtain
$$\frac{dy}{dx} = 2x = \text{slope of the tangent line at any point } (x, y) \text{ of the curve}$$
$$= \text{slope of the curve at } (x, y). \tag{9.6.1}$$

At the vertex $(0, 0)$, slope $= 0$ [substitute $x = 0$ in (9.6.1)].

Therefore, the tangent at the vertex is parallel to the X-axis and in this case coincides with it.

CHAPTER 9: DIFFERENTIATION 547

At the point $\left(\frac{1}{2}, \frac{1}{4}\right)$, slope $= 1 = \tan 45°$ i.e., the tangent at this point makes an angle of 45° with the X-axis. Its equation will be:

$$y - \frac{1}{4} = 1\left(x - \frac{1}{2}\right) \text{ or, } 4y - 4x + 1 = 0.$$

> **REMEMBER:**
> 1. If $f'(x) > 0$, the tangent line makes an acute angle with the X-axis.
> 2. If $f'(x) < 0$, the tangent line makes an obtuse angle with the X-axis.
> 3. If $f'(x) = 0$, the tangent line is parallel to the X-axis. *In particular*, if at $(0, 0)$ the derivative vanishes, the tangent line coincides with the X-axis.
> 4. We left out the exceptional case where $\psi = \frac{\pi}{2}$. Here $f'(x)$ does not exist. We may say that the gradient becomes undefined and the tangent is perpendicular to OX.
>
> Fig. 9.3 illustrates different situations.
>
> Fig. 9.3

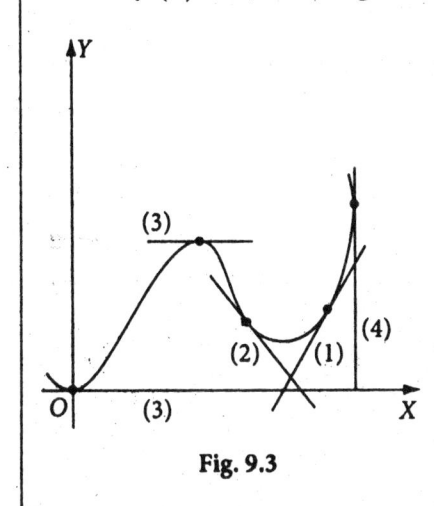

Geometric interpretation of differential

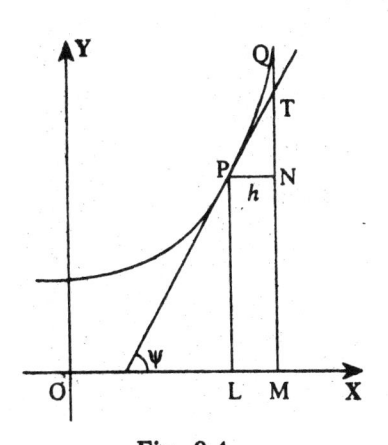

We now give a geometrical interpretation of $d\{f(x)\}$. We take a point $P(x, y)$ on the curve given by $y = f(x)$. Then, from Fig. 9.4.

$$f'(x) = \text{derivative of } f(x) \text{ at } P$$
$$= \tan \psi.$$

Fig. 9.4

Take an increment of $x = h = dx = PN$, then

$$dy = hf'(x) = \tan\psi . PN = \frac{NT}{PN}. PN = NT.$$

Therefore, $df(x)$ or dy is the increment NT of the ordinate of the tangent at $P(x, y)$ corresponding to the increment $h = dx$ of x.

Note again, that NQ is the increment of the function and NT is the differential. So they are not, in general, identical. But in making approximate calculations we may consider:

increment of the function = differential of the function, provided the increment in independent variable is sufficiently small.

Example 9.6.2. *Obtain* $dx, \Delta y, dy, \Delta y - dy$, *given*

$$y = \frac{x^2}{2} + 3x, \quad x = 2 \quad \text{and} \quad \Delta x = 0.5.$$

Solution:

$$dx = \Delta x = 0.5; \quad \Delta y = \left\{\frac{(2.5)^2}{2} + 3 \times 2.5\right\} - \left\{\frac{2^2}{2} + 3 \times 2\right\} = 2.625.$$

$$dy = \frac{dy}{dx} dx = (x+3)dx = (2+3) \times 0.5 = 2.5$$

whence $\Delta y - dy = 2.625 - 2.5 = 0.125$.

Example 9.6.3. *Find approximately the volume of the material portion of a spherical shell of external diameter 10 cm. and thickness* $\frac{1}{16}$ *cm.*

Solution: Since $V = \frac{4}{3}\pi r^3$ ($r=$ radius; $V =$ volume).

The *exact* volume is $\Delta V =$ volume of the shell of radius 5 cm. $-$ volume of the shell of radius $4\frac{15}{16}$ cm.

Since we require an approximate value of ΔV, we proceed to find dV.

$$dV = \frac{dV}{dr} \cdot dr = 4\pi r^2 dr.$$

Substitute $r = 5$ and $dr = \frac{1}{16}$ we obtain

$$dV = 19.625 \text{ c.c.}$$

Observe that the exact volume $\Delta V = 19.4$ c.c.

Absolute error $= -0.225$, Relative error $= -0.0116$, Percentage error $= -1.16$.

Use of differentials: Small Errors

It is occasionally convenient to approximate small errors by use of differential. Small errors in the value of the function arise due to a small error in the independent variable.

CHAPTER 9: DIFFERENTIATION

Example 9.6.4. *A right circular cone has its altitude equal to the radius of its base. The altitude is measured as 10 cm. with a possible error in the measurement of 0.01 cm. Find approximately the greatest possible error in the calculated volume which this possible error might produce.*

Solution: Given $h = r$ and we know

$$V = \frac{1}{3}\pi r^2 h = \frac{1}{3}\pi h^3. \tag{9.6.2}$$

(V = volume; r = radius of the base; h = altitude)

The exact maximum error in V will be the increment ΔV in its value obtained from (9.6.2) when h changes from 10 cm to 10.01 cm. The approximate error is the value of dV.

Hence $dV = \pi h^2 dh = \pi \cdot 10^2 \cdot 0.01 = \pi$ cu.cm. This means that the calculated volume $\frac{1000\pi}{3}$ cu.cm. will differ by π cu.cm. approximately; the computed volume may be too great or too small according as we substitute $df = \pm 0.01$.

Use of differentials: Relative and Percentage Errors

Definition 9.6.2. *If dx is the error in x, then dx is called the absolute error, dx/x is called the relative error, and $100 \times (dx/x)$ is called the percentage error.*

Thus, if an error of 0.02 cm. be made in measuring a length of 10 cm., then the absolute error is 0.02, the relative error is 0.002 and the percentage error is 0.2%.

Example 9.6.5. *The time T of a complete oscillation of a simple pendulum of length l is given by*

$$T = 2\pi\sqrt{\frac{l}{g}},$$

where g is a constant. Find the approximate error in the calculated value of T corresponding to an error of 2 per cent in the value of l.

Solution: Logarithmic differentiation gives

$$\frac{dT}{T} = \frac{1}{2}\frac{dl}{l} = \frac{1}{2} \cdot \frac{2}{100} = \frac{1}{100}. \quad \left(\because \frac{dl}{l} = \frac{2}{100}\right)$$

$\therefore$ Error in T is 1 per cent corresponding to an error of 2 per cent in l. (g is the acceleration due to gravity).

Example 9.6.6. *Given that $\log_{10} e = 0.4343$, find $\log_{10} 10.1$.*

Solution: $d(\log_{10} x) = d(\log_e x \times \log_{10} e) = (1/x)\log_{10} e\, dx$.
Put $x = 10, \log_{10} e = 0.4343, dx = 0.1$ then $d(\log_{10} x) = \frac{0.4343 \times 0.1}{10} = 0.004343$.
$\therefore \log_{10} 10.1 = \log_{10} 10 + 0.004343 = 1.004343$.

Example 9.6.7. *Using differentials, approximate* $\sin 60° 1'$.

Solution: For $x = \pi/3$ and $dx = 1' = 0.0003$ radian

$$y = \sin x = 0.86603.$$

Again $dy = \cos x\, dx = 0.5(0.0003) = 0.00015$.
$\therefore \sin 60° 1' = \sin 60° + 0.00015 = 0.86618$ (approx).

9.7 Meaning of the sign of the Derivative at a point: Geometrical considerations

We shall first give geometrical explanations of the following important result:

A function is increasing at a point when its derivative at the point is positive and decreasing when its derivative there is negative.

> For an Analytical proof on the sign of Derivatives.
> See art 9.8 Theorem 9.8.4

Justifications: Geometrical considerations

If $f'(a)$, the derivative of $f(x)$ at the point $x = a$, is positive, the ratio of the increments $\Delta y/\Delta x$ must have a positive limiting value, so that Δy and Δx must have ultimately the same sign-either both positive or both negative, where $|\Delta x|$ is sufficiently small. Represented graphically (Fig. 9.2) if we take $OL = a$ and if $f'(a)$ is positive then there is a certain interval to the right of L for every point of which the value of the function $f(x)$ is greater than its value at L, and a certain interval to the left of L at every point of which the value of the function is less than its value at L. Thus $f(x)$ increases as x increases in a small interval round the point $x = a$. In other words, $f(x)$ is increasing at $x = a$.

1. If $f'(a)$ is negative, we can similarly show that $f(x)$ is decreasing at $x = a$.
2. It now follows that if $f'(x)$ be positive over *any finite range* the value of $f(x)$ steadily increases with x throughout this range.
3. Similarly, if $f'(x)$ be negative *over any finite range*, then as x increases $f(x)$ steadily decreases throughout this range.

4. If $f'(a) = 0$ then $f(x)$ is neither increasing nor decreasing at the point $x = a$. We say that the function is *stationary* there.

5. The converse statements that "if $f(x)$ steadily increases as x increases in a sufficiently small interval round the point $x = a$, then $f'(a)$ **cannot be negative** and that if $f(x)$ steadily decreases as x increases in a small interval round $x = a$, then $f'(x)$ **cannot be positive**" follow immediately from the definition of $f'(a)$.

Example 9.7.1. *Find the range of values of x for which the function $f(x) = x^3 - 6x^2 - 36x + 7$ increases with x.*

Solution: Since, $f'(x) = 3(x^2 - 4x - 12) = 3(x-6)(x+2)$, $f'(x)$ *is positive*: when both the factors $x - 6$ and $x + 2$ are positive i.e., when $x > 6$ and $x > -2$.

OR, when both the factors are negative i.e., when $x < 6$ *as well as* $x < -2$.

Therefore, the required range is $x > 6$ or $x < -2$.

Example 9.7.2. *Show that*

$$\frac{\sin \theta}{\theta} \text{ continually decreases as } \theta \text{ continually increases, in } 0 < \theta < \frac{\pi}{2}.$$

Solution: Let $f(\theta) = \frac{\sin \theta}{\theta}$. Then $f'(\theta) = \frac{\theta \cos \theta - \sin \theta}{\theta^2}$.

Hence $f'(\theta)$ is negative if $\theta \cos \theta - \sin \theta < 0$ i.e., if $\theta < \tan \theta$.

We know that if $0 < \theta < \pi/2$, then $\sin \theta < \theta < \tan \theta$.

Hence under the given condition $f'(\theta)$ is negative i.e., $\frac{\sin \theta}{\theta}$ decreases as θ increases.

Example 9.7.3. *The rate of change of total cost (y) of a commodity per unit change of output (x) is called the marginal cost of the commodity.*

Now if a relation of the form $y = 3x \cdot \frac{x+a}{x+b} + 5 (a > b)$ exists (a and b are positive constants), then show that the marginal cost falls continuously as the output increases.

Solution:

$$\frac{dy}{dx} = \text{marginal cost} = 3\frac{x^2 + 2bx + ab}{(x+b)^2}$$

$$= 3\left\{\frac{(x+b)^2 + ab - b^2}{(x+b)^2}\right\} = 3\left\{1 + \frac{b(a-b)}{(x+b)^2}\right\}.$$

The expression for dy/dx suggests that as x increases, dy/dx decreases. In other words, the marginal cost decreases as output increases.

9.8 Important Theorems on Derivatives

Theorem 9.8.1. *If f has a derivative at an interior point c of (a,b), then $\exists$ a δ-nbd of c and a positive number M such that*

$$\forall x \in 0 < |x - c| < \delta, \ |f(x) - f(c)| < M|x - c|,$$

whence it will follow that f is continuous at c. [C.H. 1981]

Proof: Since f has a derivative at c,

$$\lim_{x \to c} \frac{f(x) - f(c)}{x - c} \text{ exists, equals to } f'(c).$$

$\therefore$ Given any $\varepsilon > 0$, $\exists$ a δ-neighbourhood of c which lies in (a,b) such that

$$\forall x \in 0 < |x - c| < \delta, \ \left|\frac{f(x) - f(c)}{x - c} - f'(c)\right| < \varepsilon.$$

In particular, we take the δ corresponding to $\varepsilon = 1$. Then

$$\left|\frac{f(x) - f(c)}{x - c}\right| = \left|\frac{f(x) - f(c)}{x - c} - f'(c) + f'(c)\right|$$

$$\leq \left|\frac{f(x) - f(c)}{x - c} - f'(c)\right| + |f'(c)| < 1 + |f'(c)|$$

whenever x belongs to that δ-nbd of c ($x \neq c$).

The theorem follows by taking $M = |f'(c)| + 1$.

Now if we choose $\delta = \varepsilon/M$, then

$$\forall x \in 0 < |x - c| < \varepsilon/M, \ |f(x) - f(c)| < \varepsilon,$$

whence it follows f is continuous at c.

Observation: Functions for which the conclusion of this theorem holds are said to satisfy **Lipschitz Condition** at c. Geometrically, Lipschitz condition means that the graph of the function must lie between the two lines passing through $(c, f(c))$ having slopes M and $-M$ whenever $x \in$ to that specified δ-nbd of c.

Functions which satisfy Lipschitz condition at c are automatically continuous at c, but the converse is not always true.

Theorem 9.8.2. (Chain Rule). *Suppose f is continuous in $[a,b]$ and suppose that $f'(x)$ exists at some point $x \in [a,b]$. Let g be defined on an interval I which contains the range of f and let g be differentiable at the point $f(x) \in I$. If*

$$h(t) = g\{f(t)\}, (a \leq t \leq b)$$

CHAPTER 9: DIFFERENTIATION

then g is differentiable at x and

$$h'(x) = g'\{f(x)\}.f'(x).$$

Proof: Let $y = f(x)$. By definition of derivative at the point $x \in [a, b]$ we have

$$\lim_{t \to x} \frac{f(t) - f(x)}{t - x} = f'(x)$$

or, $\quad f(t) - f(x) = (t - x)f'(x) + \varepsilon_1(t - x),$ \hfill (9.8.1)

where $t \in [a, b], \varepsilon_1$ is a function of t and $\varepsilon_1 \to 0$ as $t \to x$.

Also, since g is differentiable at the point y, we have

$$g(s) - g(y) = (s - y)g'(y) + \varepsilon_2(s - y), \quad (9.8.2)$$

where $s \in I$, ε_2 is a function of s and $\varepsilon_2 \to 0$ as $s \to y$.

Let $s = f(t)$. From (9.8.1) and (9.8.2) we obtain,

$$h(t) - h(x) = g\{f(t)\} - g\{f(x)\}$$
$$= \{f(t) - f(x)\} \cdot g'\{f(x)\} + \varepsilon_2\{f(t) - f(x)\}, \quad \text{[using (9.8.2)]}$$
$$= (t - x)[f'(x) + \varepsilon_1] \cdot [g'\{f(x)\} + \varepsilon_2] \quad \text{[using (9.8.1)]}$$

or, if $t \neq x$, then

$$\frac{h(t) - h(x)}{t - x} = [f'(x) + \varepsilon_1][g'\{f(x)\} + \varepsilon]. \quad (9.8.3)$$

Letting $t \to x$, we see that $\lim s = \lim f(t) = f(x) = y$.

$(\because f$ is continuous for every x in $[a, b]$.)

Hence from (9.8.3)

$$\lim_{t \to x} \frac{h(t) - h(x)}{t - x} = \lim_{t \to x}[f'(x) + \varepsilon_1] \cdot \lim_{s \to y = f(x)}[g'\{f(x)\} + \varepsilon_2]$$

or, $\quad h'(x) = f'(x) \cdot g'\{f(x)\}$

[In applications, we may use: Let $y = f(u)$, where $u = \phi(x)$. Then, $\dfrac{dy}{dx} = \dfrac{dy}{du} \cdot \dfrac{du}{dx}$.]

Note 9.8.1. *If y is compounded of three functions:*

$$y = f(u); \quad u = g(v); \quad v = h(x),$$

then y is ultimately a function of x through intermediary functions u and v.

Then, if we regard $u = g(v) = g\{h(x)\}$, as a function of x, we have, by Chain Rule:

$$\frac{dy}{dx} = \frac{dy}{du} \cdot \frac{du}{dx}; \quad \frac{du}{dx} = \frac{du}{dv} \cdot \frac{dv}{dx}$$

and hence
$$\frac{dy}{dx} = \frac{dy}{du} \cdot \frac{du}{dv} \cdot \frac{dv}{dx},$$

in which the chain has three links. Further generalisations are obvious.

Theorem 9.8.3. Derivative of an inverse function.

Let f be a derivable (hence continuous) strictly monotonic function with domain $[a, b]$. The inverse of f certainly exists (because f is continuous as well as strictly monotonic).

We shall denote the inverse of f by ϕ.

Now, let $x \in [a, b]$. Then f is derivable at x. Suppose $f'(x) \neq 0$. Then we prove that the inverse function ϕ is also derivable at $y = f(x)$ and that

$$\phi'(y) = \frac{1}{f'(x)}.$$

Proof: Let us write $t = f(u)$ so that the domain of f is $a \leq u \leq b$ and $u = x$ is a point of this domain.

Next write $u = \phi(t)$ (the inverse of f is ϕ) so that the domain of ϕ ranges between $f(a)$ and $f(b)$. When $u = x$, $t = f(x) = y$ (say), i.e., $\phi(y) = x$.

We consider,

$$\frac{\phi(t) - \phi(y)}{t - y} = \frac{u - x}{f(u) - f(x)} = 1 \div \frac{f(u) - f(x)}{u - x}. \tag{9.8.4}$$

Since f is continuous on $[a, b]$, ϕ is also continuous on $[f(a), f(b)]$ so that

$$\lim_{t \to y} \phi(t) = \phi\left\{\lim_{t \to y} t\right\} = \phi(y) \Leftrightarrow \lim_{t \to y} u = x.$$

On taking limit in (9.8.4), when $t \to u$, we obtain

$$\phi'(y) = \frac{1}{f'(x)}.$$

In more familiar form,

$$\frac{dx}{dy} = 1 \bigg/ \frac{dy}{dx}.$$

Theorem 9.8.4. (Meaning of sign of derivative at a point).

CHAPTER 9: DIFFERENTIATION

We first give the following precise definitions:

Definition 9.8.1. *Let f be defined on a closed interval $[a,b] = I$ (say).*

*1. Let $c \in (a,b)$. Then f is said to be **increasing** at c, if $\exists$ positive numbers δ_1 and δ_2 such that*

$$\left. \begin{array}{l} a) \ \forall \ x \in (c < x < c+\delta_1), \ f(x) > f(c) \\[6pt] and \ b) \ \forall \ x \in (c-\delta_2 < x < c), \ f(x) < f(c) \end{array} \right\}.$$

2. In case f is decreasing at c, the inequalities are reversed.

3. At the end points a, b of I suitable modifications are to be made such that inequalities occur within the domain I.

4. f is said to the increasing (or decreasing) in $[a,b]$ if at each point of the interval f increases (or decreases) in the sense defined above.

Statement of the Theorem: *Let f be defined in the closed interval $[a,b] = I$ (say). Let c be an interior point of I. Let $f'(c)$ be either a finite positive number or $f'(c) = +\infty$. Then f increases at c.*

Proof:

Case 1. Suppose $f'(c)$ is a finite positive number l.

By definition, $\lim\limits_{h \to 0} \dfrac{f(c+h) - f(c)}{h} = f'(c) = l$ (where $l > 0$).

$\therefore \ \exists \ \delta > 0$ s.t. when $0 < |h| < \delta$

$$\left| \frac{f(c+h) - f(c)}{h} - l \right| < \frac{1}{2}l \ \left(\text{choosing } \varepsilon = \frac{1}{2}l \right).$$

It, therefore, follows $l - \dfrac{1}{2}l < \dfrac{f(c+h) - f(c)}{h} < l + \dfrac{1}{2}l, \ \forall \ h \in 0 < |h| < \delta$ i.e.,

$$\frac{f(c+h) - f(c)}{h} > \frac{1}{2}l, \ \forall \ h \in 0 < |h| < \delta.$$

Therefore, when $\left. \begin{array}{l} 0 < h < \delta, \ f(c+h) - f(c) > 0 \\[6pt] and \ when \quad -\delta < h < 0, \ f(c+h) - f(c) < 0. \end{array} \right\}$

In other words, when $c < c+h < c+\delta$, $f(c+h) > f(c)$

and, when $c-\delta < c+h < c$, $f(c+h) < f(c)$

i.e., for $\quad c < x < c+\delta$, $f(x) > f(c)$

and $\quad c-\delta < x < c$, $f(x) < f(c)$

i.e., f is increasing at c (by definition).

Case 2. Suppose $f'(c) = +\infty$. Then for a given G (where G is any positive number, no matter how large),

$$\exists \delta \text{ such that } \frac{f(c+h) - f(c)}{h} > G \quad \forall h \in 0 < |h| < \delta.$$

Now the arguments will be exactly same as in Case 1.

$\therefore$ f increases at c will then follow.

Corollary 9.8.1. *At the end-points of $[a, b]$ modifications of intervals around them are to be made. The theorem is, however true, even at end points of I.*

In exactly similar way we can establish:

Let f be defined in $I = [a, b]$. If at some point c (including end points) of I, $f'(c)$ is either a negative finite number or $f'(c) = -\infty$, then f decreases at c.

Observations:

1. This theorem proves that if $f'(x)$ is positive at a single point $x = c$ then $f(x_2) > f(x_1)$ where x_1 and x_2 are two values of x, sufficiently near to c, and $x_1 < c < x_2$. [For $f(x_1) < f(c)$ and $f(c) < f(x_2)$.]

   But this does not prove that there is an interval including c throughout which f is a steadily increasing function, for x_1 and x_2 to lie on the opposite sides of c is essential for our conclusion.

   [If $c < x_1 < x_2$, we cannot necessarily conclude $f(x_2) > f(x_1)$.]

   EXAMPLE 1. *Let f be defined in $\mathbb{R}$ by*

   $$f(x) = \begin{cases} x^2 \sin \frac{1}{x} + kx & (0 < k < 1), \text{ when } x \neq 0, \\ 0, & \text{when } x = 0. \end{cases}$$

CHAPTER 9: DIFFERENTIATION

Verify that here $f'(0) = k$ (positive).

SOLUTION: Now if $x \neq 0$, then $f'(x) = 2x \sin \frac{1}{x} - \cos \frac{1}{x} + k$, which oscillates between $k-1$ and $k+1$ as $x \to 0$.

Since $k - 1 < 0$, we can find values of x, as near to 0 as we wish, for which $f'(x) < 0$ and it is, therefore, impossible to find an interval including $x = 0$, throughout which f is a steadily increasing function of x.

2. $f'(c) > 0$ **is sufficient but not a necessary condition for $f(x)$ to be increasing at $x = c$.**

EXAMPLE 2.

$f(x) = x^3$ increases at $x = 0$; here $f'(0) = 0$ (NOT POSITIVE).

$f(x) = \sqrt[3]{x}$ increases at $x = 0$; here $f'(0)$ does not exist.

3. It follows from the theorem that if the derivative does not vanish at the point x, the function $f(x)$ has always, in a suitable neighbourhood of the point, values which are greater and lesser than those that it has at the point itself. For instance, greater values will occur to the right of the point if the function is increasing and to the left if it is decreasing. Hence the function cannot attain its greatest nor its least value at the point x. This simple remark is the base of an important theorem of Calculus known as *Rolle's Theorem* although Rolle had originally used it only for a polynomial function.

Theorem 9.8.5. (Darboux's Theorem). *Assume that f is defined on the closed interval $[a, b]$ and that f has a finite derivative at each interior point. Assume also that f has a right-hand derivative $f'(a)$ at a and a left-hand derivative $f'(b)$ at b where $f'(a) \cdot f'(b) < 0$. Then there exists at least one point c of the interval $[a, b]$ where $f'(c) = 0$.* [C.H. 1996]

Proof: For the sake of definiteness let us suppose that $f'(a) > 0$ and $f'(b) < 0$.

$\therefore \exists$ intervals $(a, a+h]$ and $[b-h, b)$ (h is a suitable positive number) such that

$$\forall x \in a < x \leq a+h, \quad f(x) > f(a) \qquad (9.8.5)$$
$$\forall x \in b - h \leq x < b, \quad f(x) > f(b). \qquad (9.8.6)$$

Since f is derivable in $[a, b]$, f is continuous on $[a, b]$.

Therefore, by the property of continuous functions on a closed interval, f is bounded in $[a, b]$ and the bounds (sup and inf) *are attained there*.

Thus if M be the Supremum of f in $[a, b]$, $\exists$ a point $c \in [a, b]$ such that

$$f(c) = M.$$

The inequalities of (9.8.5) and (9.8.6) indicate that the Supremum **is not attained** at the end points a and b for otherwise there would exist points where the functional values $f(x)$ would be $> f(c)$. Hence c must be an interior point of $[a, b]$. We shall prove that $f'(c) = 0$.

(a) If $f'(c)$ were positive, then there would exist an interval $(c, c + \eta], (\eta > 0)$ such that for every point x of this interval

$$f(x) > f(c) = M.$$

This would contradict the very definition of the term Supremum.

(b) If $f'(c)$ were negative, there would exist an interval $[c - \eta, c)(\eta > 0)$ such that for every point x of this interval

$$f(x) > f(c) = M$$

and this is again a contradiction.

Hence, from (a) and (b) we conclude that $f'(c) = 0$.

This proves the theorem.

Note 9.8.2. *If $f'(a) < 0$ and $f'(b) > 0$ then with similar arguments it can be shown that $\exists$ an interior pt. $d \in [a, b]$ where $f'(d) = 0$. In this case d is the point where f attains the Infimum.*

Intermediate-Value Property for Derivatives

Corollary 9.8.2. *Let f be derivable in $[a, b]$ and $f'(a) \neq f'(b)$. Let γ be **any** number between $f'(a)$ and $f'(b)$. Then there exists at least one point $c \in [a, b]$ such that $f'(c) = \gamma$.*

Proof: We construct $F(x) = f(x) - \gamma x$.

Then F is derivable in $[a, b]$ (since f is so) and

$$F'(a) = f'(a) - \gamma, \quad F'(b) = f'(b) - \gamma.$$

Since γ lies between $f'(a)$ and $f'(b)$, it follows that $F'(a)$ and $F'(b)$ are of opposite signs. Therefore, according to Theorem **9.8.5** (just proved), there exists at least one point c of $[a, b]$ such that

$$F'(c) = 0 \quad \text{or} \quad f'(c) = \gamma.$$

Observation: In the Chapter on continuity we proved that if $f(x)$ be continuous in $[a, b]$ and $f(a) \neq f(b)$, then $f(x)$ assumes at least once every value between $f(a)$

CHAPTER 9: DIFFERENTIATION

and $f(b)$ at some interior point of $[a,b]$. This is what is known as the Intermediate Value property of Continuous Functions on a closed interval. This property does not characterise continuous functions, because we have just proved that this property is also shared by the class of derivatives $f'(x)$, **whether continuous or not**.

Use of Derivatives: Practical Problems

9.9 Derivative as a Rate-measurer

In applied problems we use the fact that the derivative gives us a measure of the rate of change of a function with respect to its independent variable.

Consider the functional relation $y = f(x)$. Choose a *fixed value* x_1 for x. Suppose the corresponding value of y is $y_1 = f(x_1)$.

For another value of $x = x_2$ (say), let $y = y_2 = f(x_2)$.

The increment ratio

$$\frac{\Delta y}{\Delta x} = \frac{y_2 - y_1}{x_2 - x_1} = \frac{f(x_2) - f(x_1)}{x_2 - x_1}$$

is called the *average rate of change of y w.r.t. x in the interval* $[x_1, x_2]$. We denote this average rate by the symbol

$$\left(\frac{\Delta y}{\Delta x}\right)_{x_1, x_2}$$

Now as $\Delta x \to 0$, if

$$\lim_{\Delta x \to 0} \frac{\Delta y}{\Delta x} = \lim_{x_2 \to x_1} \frac{f(x_2) - f(x_1)}{x_2 - x_1} \text{ exists, } = f'(x_1) = \left(\frac{dy}{dx}\right)_{x=x_1}$$

then the limiting value is called the *instantaneous rate of change of y* with respect to x for the value $x = x_1$. We simply say:

derivative $\dfrac{dy}{dx}$ *at $x = x_1$ is the rate of change of y with respect to x for the value $x = x_1$.*

Example 9.9.1. *The population of a certain city was as given in the following table.*

Date of Census (x)	1920	1930	1940	1950	1960
Population (y)	12,142	14,620	15,721	20,142	25,721

Find the average rate of change of population over the periods 1920-1930, 1930-1940, 1940-1950, 1950-1960.

Solution: With our usual notation

$$\left(\frac{\Delta y}{\Delta x}\right)_{1920-1930} = \frac{14620 - 12142}{1930 - 1920} = \frac{2478}{10} = 247.8.$$

Similarly we may obtain the other average rates: 110.1, 442.1, 557.9.

Example 9.9.2. *The volume of a certain gas varies with pressure according to the law $v = 600/p$. Find the average rate of change of volume with respect to pressure when pressure increases from $p = 30$ to $p = 30.5$ and also obtain the instantaneous rate of change of volume at $p = 30$.*

Solution:

$$\left(\frac{\Delta v}{\Delta p}\right)_{30,30.5} = \frac{19.67 - 20}{30.5 - 30}$$

p	30	30.5
v	20	19.67

$$= -0.66$$

Since, $\dfrac{dv}{dp} = -\dfrac{600}{p^2}$, instantaneous rate at $p = 30$ is $-.667$. The negative sign suggests the *diminution* of volume with increase of pressure.

Time-rate of change: Rectilinear Motion

Important applications arise when the independent variable in a rate is the *time*. We call it *time-rate of change*. To illustrate this, we first consider the motion of a particle along a straight line (*Rectilinear motion*).

Velocity: Suppose a particle moves along a straight line AB (Fig. 9.5). Let s be the displacement measured from some fixed origin O in the line to some position P of the moving particle and let t be the corresponding time elapsed. To each value of t there corresponds a point P in the line AB and therefore, a displacement s. Hence s is a function of t and we may write: $s = f(t)$.

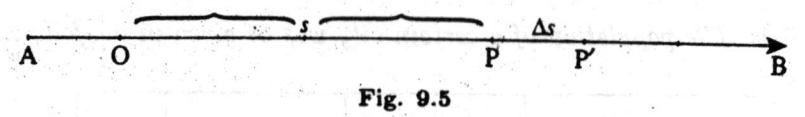

Fig. 9.5

Suppose at time $t + \Delta t$ the moving particle occupies the position P' so that $OP' = s + \Delta s$. Hence

increment in displacement $= \Delta s$; increment in time $= \Delta t$.

CHAPTER 9: DIFFERENTIATION

The ratio $\Delta s/\Delta t$ is then *the average velocity* during the time interval Δt. When $\Delta t \to 0$, the limiting value to which this average velocity tends is called the *velocity v at the point P* or *the velocity v at the instant t*. Thus,

$$v = \lim_{\Delta t \to 0} \frac{\Delta s}{\Delta t} = \frac{ds}{dt} \text{ at time } t.$$

Thus in the notation of Calculus *velocity at any instant is the derivative of the displacement with respect to time* or *time-rate of change of displacement*.

Note 9.9.1. *When ds/dt is positive, the point P moves in the direction AB i.e., s increases as t increases. But ds/dt is negative when the point P moves in the direction BA i.e., s decreases as t increases. According as ds/dt is positive or negative, s is an increasing or a decreasing function of t.*

Acceleration: Again the velocity v is itself a function of t. Considerations similar to the above will now lead us to the following definitions. If Δv be the increment of velocity in the interval Δt then $\Delta v/\Delta t$ will be called the *average rate of change of velocity* or *average acceleration* in this interval. The limiting value to which this average acceleration tends when $\Delta t \to 0$ will be called the *acceleration a at the instant t*. Thus

$$a = \lim_{\Delta t \to 0} \frac{\Delta v}{\Delta t} = \frac{dv}{dt} = \frac{d}{dt}\left(\frac{ds}{dt}\right) = \frac{d^2s}{dt^2} \text{ at the instant } t.$$

$\left(\text{we could also write: } a = \frac{dv}{dt} = \frac{dv}{ds}\frac{ds}{dt} = v\frac{dv}{ds}.\right)$

Thus *acceleration at any instant is the derivative of the velocity with respect to time or time-rate of change of velocity*.

Note 9.9.2. *If $a > 0$, then v is increasing and if $a < 0$, then v is decreasing.*

In the case of a rigid body revolving about a fixed axis, suppose θ be the angle through which the body has revolved from some standard position and let t be the corresponding time elapsed. Then if $\theta + \Delta\theta$ be the angle through which the body has revolved after time $t + \Delta t$, then

$\Delta\theta/\Delta t$ = average angular velocity in the time interval Δt and

$$\lim_{\Delta t \to 0} \frac{\Delta\theta}{\Delta t} = \frac{d\theta}{dt} = \text{angular velocity at the instant } t.$$

If, further, we denote this angular velocity by ω, then

$\Delta\omega/\Delta t$ = average angular acceleration in the time interval Δt and $d\omega/dt$ = angular acceleration at the instant t.

Example 9.9.3. *A point moves in a line so that its distances in cm. measured from a fixed point O on the line at time t seconds reckoned from some fixed epoch is given by*

$$s = t^3 - 6t^2 - 15t$$

find (a) *velocity and acceleration at any instant t; at the end of first second;*

(b) *the average velocity while t changes from 1 to 6;*

(c) *when and where the body stops;*

(d) *the time interval during which the velocity is negative but the acceleration is positive.*

Solution: (a) Velocity v at any instant $t = \dfrac{ds}{dt} = 3t^2 - 12t - 15 = 3(t^2 - 4t - 5)$

$$\text{acceleration at any instant } t \quad = \frac{dv}{dt} = 6(t - 2)$$

at the end of *first* second, velocity $= -24$ cm./sec.

$$\text{acceleration} \quad = -6 \text{ cm./sec}^2.$$

(b)

t	1	9
s	-20	-90

$$\left(\frac{\Delta s}{\Delta t}\right)_{1,6} = \frac{-90 - (-20)}{6 - 1} = -14 \text{ cm./sec.}$$

(c) The body stops when $\dfrac{ds}{dt} = 0$.

i.e., when $3(t^2 - 4t - 5) = 0$ which gives $t = -1$, $t = 5$.

Again, if $t = -1$, $s = 8$ and if $t = 5$, $s = -100$.

Conclusion. One second before the fixed epoch from which time is measured the particle was at rest being at a distance 8 cm. from O. Five seconds after the time-epoch the particle is again at rest being at a distance 100 cm. from O but this time it is on the *other side* of O.

(d) Velocity is negative when $t^2 - 4t - 5 < 0$

i.e., when $(t + 1)(t - 5) < 0$.

i.e., when $-1 < t < 5$

acceleration is positive when $t - 2 > 0$ i.e., when $t > 2$.

CHAPTER 9: DIFFERENTIATION

Thus, velocity is negative and acceleration is positive during the interval

$$2 < t < 5.$$

Related Rates

We shall deal with problems where two or more variables are involved each of which being a function of t (time). The conditions of the problem will suggest a relation between these variables. The relation between their time-rates of change may then be obtained by differentiation with respect to t.

Example 9.9.4. *Water is running into a conical reservoir, 10 cm. deep and 5 cm. in radius at the rate of 1.5 c.c. per minute.*

(a) *At what rate is the water-level rising when the water is 4 cm. deep?*

(b) *At what rate is the area of water-surface increasing when the water is 6 cm. deep?*

(c) *At what rate is the wetted surface of the reservoir increasing when water is 8 cm. deep?*

Solution: Let r be the radius and h the height of the water-level at time t.

Then, from similar triangles (Fig. 9.6), we obtain,

$$\frac{r}{5} = \frac{h}{10} \text{ i.e., } r = \frac{1}{2}h.$$

Volume V of the cone $= \frac{1}{3}\pi r^2 h = \frac{1}{12} \cdot \pi h^3$.

$$\therefore \frac{dV}{dt} = \frac{\pi}{4}h^2 \frac{dh}{dt}.$$

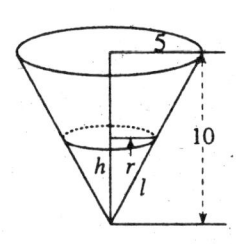

Fig. 9.6

Given that $\dfrac{dV}{dt} = \dfrac{3}{2}$. $\therefore \dfrac{dh}{dt} = \dfrac{6}{\pi h^2}$.

(a) when $h = 4$, $\dfrac{dh}{dt} = \dfrac{6}{\pi \cdot 16} = \dfrac{3}{8\pi}$ cm/min.

(b) Area of water-surface $= A = \pi r^2 = \dfrac{\pi}{4}h^2$.

$$\therefore \frac{dA}{dt} = \frac{\pi}{2}h\frac{dh}{dt} = \frac{\pi}{2}h \cdot \frac{6}{\pi h^2} = \frac{3}{h};$$

when $h = 6$, $\dfrac{dA}{dt} = \dfrac{1}{2}$ cm^2/min.

(c) Wetted surface is a cone whose lateral area

$$S = \pi(\text{radius}) \cdot (\text{slant height}) = \pi r \sqrt{r^2 + h^2}.$$

[∵ from the right-angled triangle (see Fig. 9.6), $l^2 = h^2 + r^2$.]

$$\therefore S = \pi \frac{h}{2} \sqrt{\frac{h^2}{4} + h^2} = \frac{\sqrt{5}}{4} \pi h^2.$$

Hence $\dfrac{ds}{dt} = \dfrac{\sqrt{5}}{2} \pi h$. When $h = 8$, $\dfrac{ds}{dt} = 4\sqrt{5} \cdot \pi$ cm²/min.

Example 9.9.5. *One ship A was sailing due south at the rate of 6 miles per hour; another ship B was sailing due east at the rate of 8 miles per hour. At 4 P.M. the second ship B crossed the position where first ship A was 2 hours before.*

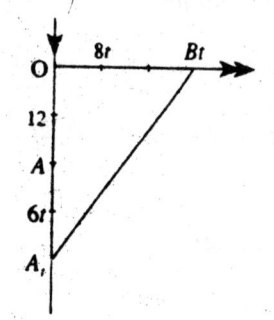

(a) *At what rate were they approaching or separating at 3 P.M.?*

(b) *At what rate were they changing the distance between them at 5 P.M.?*

(e) *When was the distance between them not changing?*

Fig. 9.7

Solution: At 4 P.M. suppose the position of B was at O (Fig. 9.7). The corresponding position of the first ship should be at A where $OA = 12$ miles. Suppose now B_t, A_t be the positions of the first and second ship respectively, t hours later. Let D be the distance between them at this position.

Then clearly, $D^2 = (12 + 6t)^2 + (8t)^2$

and $\dfrac{dD}{dt} = \dfrac{100t + 72}{D}.$

(a) At 3 P.M., $t = -1$ [fixed epoch from which time is measured being 4 P.M.] Then D becomes 10 so that

$$\frac{dD}{dt} = -2.8 \text{ miles per hour}$$

i.e., the distance between them was *decreasing* at the rate of 2.8 miles per hour.

(b) At 5 P.M., $t = 1$, $D = \sqrt{388}$ so that

$$\frac{dD}{dt} = +8.73 \text{ miles per hour (increasing at this rate)}.$$

(c) When the distance between them was not changing,

$$\frac{dD}{dt} = 0 \text{ which happened when } t = -\frac{18}{25} \text{ hrs.} = -43 \text{ min. (approx.)}$$

∴ At 3.17 P.M., the distance was not changing.

Example 9.9.6. *The top of a ladder 30 feet long leans against a vertical wall and the lower end rests on a level pavement. The ladder begins to slide outwards so that the lower end is being moved away from the wall at a rate (per second) equal to twice its distance from the wall. How fast is the top sliding downwards at the instant when the lower end is 10 feet away from the wall? How far is the lower end from the wall when it and the top are moving at the same rate?*

Solution: Let x ft. be the distance of the lower end of the ladder from the vertical wall and y be the height of the top from the pavement at any instant t. Then clearly (Fig. 9.8),

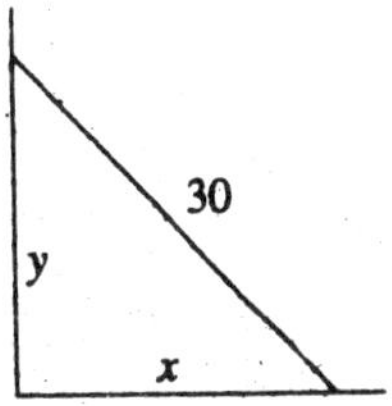

$$900 = x^2 + y^2$$

or, $\quad x\dfrac{dx}{dt} + y\dfrac{dy}{dt} = 0.$ \hfill (9.9.1)

Fig. 9.8

But as given in the problem, $\dfrac{dx}{dt} = 2x$. Hence,

$$\frac{dy}{dt} = -\frac{2x^2}{y} = \frac{-2x^2}{\sqrt{900 - x^2}} \text{ ft./sec.}$$

If $x = 10$, $\dfrac{dy}{dt} = -5\sqrt{2}$ ft/sec.

The negative sign indicates that y decreases as t increases i.e., the top is sliding downwards at the rate of $5\sqrt{2}$ ft/sec.

Second part. If $\dfrac{dx}{dt}$ and $\dfrac{dy}{dt}$ are equal numerically, then (9.9.1) gives that x must be equal to y (numerically). Hence

$$x^2 + x^2 = 900$$

or, $\quad x = 15\sqrt{2}$ feet, which gives the required distance. •

Main Points : A Recapitulation

1. Derivative : Definition. Let f be defined on a set $D \subseteq \mathbb{R}$. Let $x \in D$. If $\lim\limits_{t \to x} \dfrac{f(t) - f(x)}{t - x}$ exists (and is finite), then we say that f has a derivative at the point $t = x$.

NOTATIONS: $f'(x)$ or $\dfrac{df}{dt}$ at $t = x$, or $\dfrac{dy}{dt}$ at $t = x$ where $y = f(t)$.

2. One sided derivative:

$$\text{right derivative } f'(x+0), \quad \text{if } \lim_{t \to x^+} \frac{f(t) - f(x)}{t - x} \text{ exists,}$$

$$\text{left derivative } f'(x-0), \quad \text{if } \lim_{t \to x^-} \frac{f(t) - f(x)}{t - x} \text{ exists.}$$

3. f has a derivative iff $f'(x+0) = f'(x-0) =$ a finite quantity.

f is derivable in $[a,b]$ if (i) at every interior point x, $f'(x)$ exists; and (ii) $f'(a+0)$ and $f'(b-0)$ also exist.

4. *Infinite Derivative:* $f'(x+0) = f'(x-0) = +\infty$; $f'(x+0) = f'(x-0) = -\infty$.

5. Relation with continuity: If f has a finite derivative at the point x, then it is continuous there; but the converse is not always true (there are examples in support of this statement).

6. Standard Rules and Formulas: For calculation of derivatives: Supposed to be familiar with the readers of this treatise.

In particular, one should be familiar with the *Chain Rule* and derivation of *parametric functions, inverse functions* and *logarithmic differentiation*.

7. Differentiability: f is differentiable at x if the increament $f(x + \Delta x) - f(x)$ can be expressed as $A\Delta x + \varepsilon \cdot \Delta x$, where A is independent of Δx and $\varepsilon \to 0$ as $\Delta x \to 0$. $A\Delta x$ is called the *differential* of f denoted by df and $\varepsilon \Delta x$ is called the *error*.

When f is differentiable at x, it is derivable there and conversely.

When f is differentiable at x, $A = f'(x)$ and the differential of $f = df = f'(x) dx$.

8. Geometric Interpretation: $f'(x)$ is the gradient of the tangent to the curve given by $y = f(t)$ at $t = x$.

9. Sign of derivative: Increasing at the point where the derivative is positive but decreasing there if the derivative is negative: *Stationary*, where the derivative vanishes.

CHAPTER 9: DIFFERENTIATION

> **10.** If f has a derivative at an interior point c of (a,b) then $\exists$ a δ-neighbourhood of c and a positive number M such that
>
> $$\forall x \in \; 0 < |x - c| < \delta, \quad |f(x) - f(c)| < M|x - c| \quad (Lipschitz\ condition)$$
>
> whence it follows f is continuous at c.
>
> **11. Darboux's Theorem:** Intermediate value property is not only a property of a continuous function in a closed interval but it also holds for a derived function, (*continuous or not*) over an interval: *Let f be derivable in $[a,b]$, where $f'(a) \neq f'(b)$. Any number γ between $f'(a)$ and $f'(b)$ is obtained somewhere in $[a,b]$.*
>
> **12. Applications:** (i) We use differentials in calculating small errors whence Relative and Percentage errors are also computed. If dx is the error in x, then dx/x is the relative error and $100 \times (dx/x)$ is the percentage error.
>
> (ii) In all physical problems, we use derivative as a rate-measurer: In rectilinear motion, for instance, $v = ds/dt$ is the velocity at time t and dv/dt is the acceleration at the instant t.
>
> (iii) When two or more variables are involved each of the variables being a function of t we may obtain the relation between their time rates of change.

Exercises IX (II)

[A]

[The problems given under this section need **special attention**]

1. A function f is defined in $[-1, 2]$ by

$$f(x) = \begin{cases} 1 - x^2, & \text{for } -1 \leq x < 0 \\ x^2 + x + 1, & \text{for } 0 \leq x < 1 \\ x^3 + 2, & \text{for } 1 \leq x \leq 2. \end{cases}$$

Examine if $f'(x)$ exists for $x = 0$ and $x = 1$. [C.H. 1983]

2. (a) f is a real-valued function defined in $(-1, 2)$ by

$$f(x) = |x| + |x - 1|.$$

Discuss the continuity of f at $x = 0$ and at $x = 1$.

Is f differentiable at $x = 0$ and at $x = 1$? Find the derived function f' and its domain of definition and show that f' is continuous in its domain.

[C.H. 1982]

(b) A function f defined on $(0, 2)$ by
$$f(x) = x - [x].$$
where $[x]$ denotes the integral part of x. Is f continuous at $x = 1$? Compute the derived function f' and its domain. [C.H. 1987]

3. A function f is defined in $(-1, 1)$ by
$$f(x) = \begin{cases} x^\alpha \sin \dfrac{1}{x^\beta}, & \text{for } x \neq 0; \\ 0, & \text{for } x = 0. \end{cases}$$

Prove the following statements:

(a) If $\alpha = 1$, $\beta > 0$, then $f'(0)$ does not exist; but if $\alpha > 1$ and $\beta > 0$, then $f'(0)$ does exist.

(b) If $0 < \beta < \alpha - 1$, then f' is continuous at $x = 0$.

(c) If $0 < \alpha - 1 \leq \beta$, then f' is discontinuous at $x = 0$. [C.H. 1983]

4. Let
$$f(x) = \begin{cases} x \tan^{-1}(1/x), & \text{for } x \neq 0; \\ 0, & \text{for } x = 0. \end{cases}$$

Find the right- and left-derivatives of $f(x)$ at $x = 0$ and then examine if $f'(0)$ exists.

[C.H. 1989 Old]

5. A function is defined in $(0, \infty)$ by
$$f(x) = \begin{cases} 1 - x^2, & \text{in } 0 < x \leq 1; \\ \log x, & \text{in } 1 < x \leq 2; \\ \log 2 - 1 + \tfrac{1}{2}x, & \text{in } 2 < x < \infty. \end{cases}$$

Obtain the derived function f' and its domain. [C.H. 1979, 1986]

6. Let
$$f(x) = \begin{cases} x^\alpha \sin(1/x), & \text{when } x \neq 0; \\ 0, & \text{when } x = 0. \end{cases}$$

Prove that if $\alpha = 1$, $f'(0)$ does not exist and if $\alpha = 2$, $f'(x)$ exists everywhere but $f'(x)$ is not continuous at $x = 0$. [C.H. 1993]

CHAPTER 9: DIFFERENTIATION

7. If
$$f(x) = \begin{cases} x, & \text{for } 0 \leq x < 1; \\ 2 - x, & \text{for } 1 \leq x \leq 2; \\ 3x - x^2 - 2, & \text{for } x > 2; \end{cases}$$
then prove that $f'(x)$ does not exist at $x = 1$ but $f'(2)$ exists and $= -1$.

8. Let $f(x)$ be continuous in $[a, b]$ and differentiable in (a, b).

   Prove that $f(x)$ is m.i. in $[a, b]$ if $f'(x) \geq 0$ in (a, b).

   Prove that $f(x)$ is m.d. in $[a, b]$ if $f'(x) \leq 0$ in (a, b).

   Prove that $f(x)$ is constant $\forall x$ in $[a, b]$ if $f'(x) = 0$ in (a, b).

   Prove that $\dfrac{2x}{\pi} \leq \sin x \leq x$, when $0 \leq x \leq \pi/2$. [C.H. 1984, 96]

9. Give illustrations in support of the following statements:

   (a) if a function $f(x)$ be continuous at $x = c$ but $f'(c)$ does not exist, then $f'(x)$ cannot be continuous at $x = c$.

   (b) A function $f(x)$ which is continuous and possesses a derivative at every point may have a derived function $f'(x)$ which is not continuous at some point.

10. (a) Let $y = f(x) = \dfrac{ax + b}{cx + d}$. Obtain $f'(x)$ and then show that $f(x)$ is an increasing or decreasing function in any interval not containing $-d/c$ according as $ad - bc > 0$ or < 0. Discuss the case when $ad - bc = 0$.

    (b) Find the inverse function $x = \phi$ in (a) and its derivative $\phi'(y)$. When is $\phi(y)$ an increasing function?

11. If $f(x) = 1/(1 + e^{1/x})$, show that $f(x)$ and $f'(x)$ are continuous when $x \neq 0$; and that $f(0 + 0) = 0$, $f(0 - 0) = 1$.

    Obtain $\lim\limits_{x \to 0+} f'(x)$ and $\lim\limits_{x \to 0-} f'(x)$ and verify that they are equal. Does the derivative of f at $x = 0$ exist? Is the function
    $$g(x) = \begin{cases} f'(x), & x \neq 0 \\ 0, & x = 0 \end{cases} \text{ continuous at } x = 0?$$

    Describe the behaviour of $f(x)$ and $f'(x)$ as $x \to \pm\infty$.

12. Show that f defined by
    $$f(x) = |x| + |x - 1| + |x - 2|$$
    is continuous but not derivable at $x = 0, x = 1, x = 2$.

13. If $f(x) = x^2 \sin(1/x)$, when $x \neq 0; = 0$, when $x = 0$ and $g(x) = x$, show that
$$\lim_{x \to 0} \frac{f'(x)}{g'(x)} \text{ does not exist}$$
but $\lim_{x \to 0} \frac{f(x)}{g(x)}$ exists and $= \frac{f'(0)}{g'(0)}$.

14. If $f(x) = |x|$, $g(x) = 2|x|$, show that $f'(0)$ and $g'(0)$ do not exist but $\lim_{x \to 0} f(x)/g(x)$ exists and is equal to $\lim_{x \to 0} f'(x)/g'(x)$.

15. If
$$f(x) = \begin{cases} \sqrt{x}(1 + x \sin 1/x), & x > 0; \\ -\sqrt{x}(1 + x \sin 1/x), & x < 0; \\ 0, & x = 0. \end{cases}$$

Show that f' exists everywhere and is finite except at $x = 0$ in the neighbourhood of which it oscillates between $-\infty$ and $+\infty$.

16. $f(x) = x\left(1 + \frac{1}{3}\sin(\log x^2)\right)$, when $x \neq 0$, but $f(0) = 0$. Show that $f(x)$ is everywhere continuous but $f'(0)$ does not exist.

17. $f(x) = \begin{cases} \sin x \sin(1/\sin x), & \text{when } 0 < x < \pi \text{ and } \pi < x < 2\pi; \\ 0, & \text{when } x = 0, \pi, 2\pi. \end{cases}$

Show that f is continuous but not derivable at $x = 0, \pi, 2\pi$. [C.H. 1989]

18. Let f be continuous on $[0, 1]$, $f(0) = 0$, $f'(x)$ finite for each x in $(0, 1)$. Prove that if f' is an increasing function on $(0, 1)$, then so too is the function g defined by $g(x) = f(x)/x$.

19. Assume that f has a finite derivative at each point of open interval (a, b). Assume also that $\lim_{x \to x_0} f'(x)$ exists and is finite for some interior point x_0. Prove that the value of this limit is $f'(x_0)$.

20. Let f be continuous on (a, b) with a finite derivative f' everywhere in (a, b) except possibly at x_0. If $\lim_{x \to x_0} f'(x)$ exists and has the value A show that $f'(x_0)$ must exist and have the value A.

[B]

Applications : Derivative as a Rate measurer

1. A balloon, which always remains spherical, has a variable radius. Find the rate at which its volume is increasing with the radius, when it is 10 cm.

CHAPTER 9: DIFFERENTIATION

2. From the Table given below, find the average rate of change of Q with respect to t:

   (a) for the first thousand years;
   (b) for the thousand years following the four thousandth year;
   (c) for the total period of seven thousand years.

t	0	1000	2000	3000	4000	5000	7000
Q	1000	681	463	315	214	146	68

3. If the rate of change of y with respect to x is 5 and x is changing at 3 units per second, how fast is y changing?

4. The area A of a circle of radius r is given by $A = \pi r^2$. Find the instantaneous rate of change of area with respect to radius when $r = R$ and also find the average rate if r changes from R to $R + \Delta R$.

5. A kite 80 feet high with 100 feet of cord starts moving away horizontally at the rate of 4 miles per hour. How fast is the cord being payed out?

6. If $y = x^3$ and x is increasing at the rate of 10 units per minute when $x = 3$, find how fast y is changing.

7. A man is walking at the rate of 5 miles per hour towards the foot of a tower 60 feet high. At what rate is he *approaching* the top when he is 80 feet from the foot of the tower?

8. A point moves along the curve $y^2 = 12x$ in such a way that its abscissa increases uniformly at the rate of 2 cm. per second. At what points do the abscissa and ordinate increase at the same rate?

9. The side of an equilateral triangle is 5 cm. and is increasing at the rate of $\sqrt{3}$ cm. per sec. How fast is the area increasing?

10. Prove the relation $2\dfrac{dV}{dt} = r\dfrac{dS}{dt}$ in a sphere of radius r, S being its surface and V its volume.

11. Prove that if a particle moves along a line so that the distance described is proportional to the square of the time of description, the velocity will be proportional to the time and the rate of increase of the velocity will be constant.

12. A man six feet long walks away from the foot of a lamp post 10 feet high, along a line and moves at the rate of 2 feet per sec. Find the rate at which the shadow is increasing.

13. A 20-foot ladder leans against a building while its base is drawn away from the wall at the rate of 2 ft./sec. How fast is the top of the ladder descending when the ladder is inclined at an angle 60° to the horizontal.

14. A ship sailing east at 15 m.p.h. passed a point A at 11 a.m. A second ship sailing south at 10 m.p.h., passed A at 9 a.m. How fast were the ships separating at 1 p.m.?

15. A boat with its anchor fast on the bottom at a depth of 40 feet is drifting at 4 ft./sec., while the cable attached to the anchor slips from the boat at water-level. At what rate is the cable leaving the boat at the instant when 50 feet are out? (assume that the cable from the boat to anchor is straight and remains in a fixed vertical plane.)

16. A vertical cylindrical tank of radius 10 inches has a hole of radius 1 inch in its base. The velocity v with which the water contained runs out of the tank is given by $v^2 = 2gh$, where h is the depth of the water and g is the acceleration due to gravity. How rapidly is the velocity changing?

17. Sand is pouring on the ground from the orifice of an elevated pipe, and forms a pile which has always the shape of a right circular cone whose height is equal to the radius of the base. If sand is falling at the rate of 10 cubic feet per second, how fast is the height of the pile increasing when the height is 20 feet.

18. A rod AB, 10 feet long, moves with its ends A and B on two perpendicular lines OX and OY respectively. If A is 8 feet from O and is moving away at the rate of 2 feet/sec., find at what rate the end B is moving.

19. A vessel is in the form of an inverted right circular cone, the vertical angle of which is 60°. Water leaks from the vessel at the rate of 0.05 cu. in per minute. At what rate is the inner surface of the vessel being exposed when the water is 6 in. deep?

20. The pressure p and the volume v of a gas at constant temperature obey the law $pv = c$. When pressure is 10 lb/ft^2 the volume is 3 cu. ft. and is changing at 0.3 cu ft/sec. How fast is p changing at that instant? Replace the law $pv = c$ by the adiabatic law $pv^{1.4} = c$ and solve the problem.

CHAPTER 9: DIFFERENTIATION

21. A spherical ice-ball is melting the radius decreasing at a constant rate of 0.1 cm. per sec. Find the amount of water formed in one second when the radius of the sphere is 7 cm. [Given $\pi = \frac{22}{7}$, sp. gr. of ice = 0.9]

22. If the area of a circle increases at a uniform rate, prove that the rate of increase of the perimeter varies inversely as the radius.

23. A circular plate of metal expands by heat so that its radius increases at the rate of a inches per second. At what rate is the surface increasing when the radius is b in.?

24. Two roads intersect at an angle of 60°. A car is 500 feet from the crossing and moving away from it at the rate of 60 miles per hour. Another car is 500 feet from the crossing on the other road and moving towards the intersecting point at the rate of 30 miles per hour. What is the rate of change of distance between them?

[C]

(Applications : Geometrical Problems)

1. For the curve $y = 8 - 2x - x^2$, find the average rate of change of y with respect to x as x changes from (i) 0 to 1; (ii) −3 to 1; (iii) −5 to −3. Plot the curve for values of x ranging from −5 to −3. Draw suitable secant lines and satisfy yourself that their slopes are the average rates required in (i), (ii) and (iii). Differentiate y with respect to x and obtain the instantaneous rates of y for which $x = -4$, $-2, -1, 2$. Draw tangent lines at these points of the curve and satisfy yourself that the slopes of these tangent lines give the desired instantaneous rates.

   (*This problem gives geometric interpretations of average and instantaneous rates.*)

2. Find the slope of the tangent line at the point $(0, 2)$ of the curve $8y = x^3 - 12x + 16$. Hence write down its equation. Obtain the equation of the normal at this point. At what *points* on this curve the slope of the tangent is equal to $\frac{9}{2}$ and at what *points* are tangents parallel to the X-axis?

3. At what points do the curves $y = x^2 - 1$ and $y = -2x^2 + 2$ intersect? Which curve is steeper at these points of intersection?

4. Find the area of the triangle formed by the X-axis and the tangent and normal to the curve $y = 6x - x^2$ at the point $(5, 5)$.

5. Show that the function $y = x^3$ steadily increases from $x = -\infty$ to $x = +\infty$ but $y = x^4$ decreases from $x = -\infty$ to $x = 0$ and then increases.

6. Consider the function

$$f(x) = \begin{cases} x, & -1 \leq x < 1; \\ x - 2, & 1 \leq x < 2. \end{cases}$$

Is $f(x)$ an increasing function when $x = 0, 1.5$? Is $f(x)$ increasing or decreasing when $x = 1$; when $x = 2$?

7. Prove that as x increases

$$\frac{a \sin x + b \cos x}{c \sin x + d \cos x} \quad (a, b, c, d \text{ are constants})$$

either increases for all values of x or decreases for all values of x.

8. If $f(x) = 2x - \tan^{-1} x - \log(x + \sqrt{1 + x^2})$, show that y increases as x changes from $x = 0$ to $x \to +\infty$.

9. If $f(x) = (x - 1)e^x + 1$, show that $f(x)$ is positive for all positive values of x.

10. If $0 < x < \frac{1}{2}\pi$, then prove that

(a) $1 - \dfrac{x^2}{2} < \cos x < 1 - \dfrac{x^2}{2} + \dfrac{x^4}{4!}$;

(b) $x - \dfrac{x^3}{3!} < \sin x < x - \dfrac{x^3}{3!} + \dfrac{x^5}{5!}$.

11. Find the range of values of x for which the function

$$x^3 - 6x^2 - 36x + 7$$

increases with x.

12. Separate the intervals in which the function $2x^3 - 15x^2 + 36x + 1$ is increasing or decreasing. Draw the graph of the function.

13. Show that

(a) $\dfrac{x}{1+x} < \log(1 + x) < x; \quad (x > 0)$ [C.H. 1992]

(b) $x > \log(1 + x) > x - \frac{1}{2}x^2; \quad (x > 0)$

(c) $x - \dfrac{x^2}{2} + \dfrac{x^3}{3(1+x)} < \log(1 + x) < x - \dfrac{1}{2}x^2 + \dfrac{x^3}{3}; \quad (x > 0)$.

14. (a) Show that $x/\sin x$ increases steadily from $x = 0$ to $x = \pi/2$.

CHAPTER 9: DIFFERENTIATION

(b) Show that $\sin x/x$ decreases steadily from 1 to 0 as x increases from 0 to π.

[D]

(Approximate Calculations: Small errors)

1. Find $d\{f(x)\}$ for each of the following functions:

   (a) $f(x) = x^3 + \log x + \sin^3 \pi x$.

   (b) $f(x) = x \sin^{-1} 3x$.

   (c) $f(x) = 8e^{-\frac{1}{2}x} \tan^{-1} 2x$.

2. Show by means of differentials that

$$\frac{1}{x+dx} = \frac{1}{x} - \frac{dx}{x^2} \text{ (approximately)}.$$

3. find Δy and dy for $x = 2$ and $\Delta x = 0.1$ if

$$y = 2x^3 - 2x^2 + 3x - 5.$$

4. If A be the area of a square of side x, what are $\Delta A, dA$ and $\Delta A - dA$ for $x = 5$ and $\Delta x = 0.0002$ in.?

5. A man makes a cubical box to hold a cubic yard of sand. It is found to hold only 98.2 per cent of a cubic yard. What actual error and percentage error were made in the length of an edge?

6. Find approximate values of the following (using differentials):

   (a) $\sqrt[5]{33}$,

   (b) $\sqrt[3]{101}$.

[Hints: $f(x + \Delta x) = f(x) + f'(x)dx$; Use in (a) $f(x) = x^{1/5}$; put $x = 32$ and $\Delta x = 1$.]

7. Find the approximate values of the following by the method of differentials:

   (a) $\log_{10} 404$. (Given $\log_{10} 4 = 0.6021$, $\log_{10} e = 0.4343$);

   (b) $\log_e 10.1$. (Given $\log_e 10 = 2.303$);

   (c) $\tan 46°$. (Given $\tan 45° = 1$, $\sec 45° = \sqrt{2}$, $1° = 0.01745^c$);

   (d) $\sin 62°$, $\cos 61°$, $\sin 59°$, $\cos 58°$.

(Given $\sin 60° = 0.86603$, $\cos 60° = 0.5$ and $1° = 0.01745$ radians).

8. (a) For a simple pendulum or length l feet, the time T seconds of a beat is given by the formula
$$T = \frac{\pi}{\sqrt{32.2}} \cdot \sqrt{l}.$$
If l be measured as 3 feet, and this value instead of the true one 3.05 feet, is used to compute T, find the relative error and the percentage error in the computed value of T.

   (b) How much a clock with this error would lose or gain in a day?

   (c) A clock gains 2 minutes a day. The length of its beat is supposed to be one second. How much should the pendulum be lengthened so that the clock will keep correct time?

9. Show that the percentage error in the nth root of a number is approximately $1/n$ times the percentage error in the number.

10. The angle of elevation of the top of a tower as observed from a distance of 500 feet from the foot of the tower is found to be $30°$. If the angle of elevation was actually $30°12'$, what percentage error has crept into the calculated height of the tower?

    (Given $12' = 0.0035$ radian)

11. If the three sides a, b, c of a triangle are measured, the error in the angle A due to given small errors in the sides is
$$dA = \frac{\sin A}{\sin B \sin C} \frac{dA}{a} - \cot C \frac{db}{b} - \cot B \frac{dc}{c}.$$

12. If a triangle ABC be slightly varied, so as to remain inscribed in the *same* circle, prove that
$$\frac{da}{\cos A} + \frac{db}{\cos B} + \frac{dc}{\cos C} = 0. \qquad \text{[C.H. 1967]}$$
[Hints: $\dfrac{a}{\sin A} = \dfrac{b}{\sin B} = \dfrac{c}{\sin C} = 2R$; $a = 2R \sin A$ gives $da = 2R \cos A dA$;

L.H.S. $= 2R.d(A + B + C) = 2R.d(\pi) = 0$.]

13. The side a and the opposite angle A of a $\triangle ABC$ remains constant; show that when the other sides and angles are slightly varied,
$$\frac{\delta b}{\cos B} + \frac{\delta c}{\cos C} = 0. \qquad \text{[C.H. 1969]}$$

CHAPTER 9: DIFFERENTIATION

14. If A be the area of a triangle, prove that the error in A resulting from small error in the measurement of c is given by

$$dA = \frac{1}{4}A\left\{\frac{1}{s} + \frac{1}{s-a} + \frac{1}{s-b} - \frac{1}{s-c}\right\}dc, \text{ where } 2s = a+b+c.$$

15. If $ABCD$ be a deformable plane quadrilateral of jointed rods, and if x, y be the lengths of the diagonals AC, BD the infinitesimal variations of these lengths are connected by the relation

$$\sin A \sin C \cdot x\, dx + \sin B \sin D \cdot y\, dy = 0.$$

Answers

[A]

(1) $f'(0)$ does not exist but $f'(1) = 3$. (2) Cont. but not differentiable at $x = 1, x = 0$. (4) $Rf'(0) = \pi/2$, $Lf'(0) = -\pi/2$, $f'(0)$ does not exist. (5) $f'(x) = -2x, 0 < x < 1$; undefined at $x = 1$; $= 1/x, 1 < x < 2$; $= \frac{1}{2}, x \geq 2$. (9) See Ex. 6; (a) $\alpha = 1$; (b) $\alpha = 2$. (10) (a) $f'(x) = \dfrac{ad - bc}{(cx+d)^2}$, where $x \neq -\frac{d}{c}$. (b) $\phi'(y) = \dfrac{ad - bc}{(cy - a)^2}$. When $ad - bc > 0$, $y \neq a/c$. (11) Each limit $= 0$; $f'(0)$ does not exist; $g(x)$ is cont. at $x = 0$; limits are $1/2$ and 0.

[B]

(1) 1257.14 cu. cm/unit increase in radius. (2) (a) -0.32. (b) -0.068. (c) -0.133. (3) 15 units/sec. (4) $2\pi R$; $\pi(2R + \Delta R)$. (5) 2.4 m.p.h. (6) 270 units/min. (7) 4 m.p.h. (8) $(3, 6)$. (9) 7.5 sq. cm./sec. (12) 3 ft/sec. (13) $\frac{2}{3}\sqrt{3}$ ft/sec. (14) 17 m.p.h. (15) 2.4 ft/sec. (16) decreasing $g/100$ ft/sec^2. (17) $1/40\pi$ ft/sec. (18) $2\frac{2}{3}$ ft/sec. (19) $\frac{1}{30}$ in.^{2}/min. (20) 1 lb/ft^2. sec; 1.4 lb/ft^2.sec. (21) 55.44 gms. (23) $2\pi ab$ in.^{2}/sec. (24) increasing 15 m.p.h. or $15\sqrt{3}$ m.p.h.

[C]

(2) $-3/2$; $2y + 3x = 4$; $2x - 3y + 6 = 0$; $(-4, 0)$ and $(4, 4)$; $(2, 0)$ and $(-2, 4)$. (3) $(1, 0)$ and $(-1, 0)$; second is steeper. (4) $53\frac{1}{8}$ sq. units. (6) increasing; increasing; neither; no definite answer. (11) $x < -2$ and $x > 6$. (12) increasing in $-\infty < x < 2$ and $3 < x < \infty$, decreasing in $2 < x < 3$.

[D]

(1) (a) $\left(3x^2 + \frac{1}{x} + 3\pi \sin^2 \pi x \cos \pi x\right) dx$; (b) $\left(\sin^{-1} 3x + \frac{3x}{\sqrt{1-9x^2}}\right) dx$; (c) $4e^{-\frac{1}{2}x} \times \left(-\tan^{-1} 2x + \frac{4}{1+4x^2}\right) dx$. (3) $\Delta y = 2.002, dy = 1.9$. (4) $\Delta A = 0.00200004, dA = 0.002$. (5) $-0.006, -0.6\%$. (6) (a) 2.01235. (b) 4.6570. (7) (a) .6064. (b) 2.313. (c) 1.0350. (d) 0.8835, 0.4849, 0.8578, 0.5302. (8) (c) 0.00905 ft. (10) 0.808 per cent.

AN INTRODUCTION TO ANALYSIS

(Differential Calculus—Part I)

Chapter 10

**Repeated Differentiation:
Second and Higher Order Derivatives**

Chapter 10

Repeated Differentiation: Second and Higher Order Derivatives

10.1 Introduction

The present chapter will be mainly devoted to *calculations of higher order derivatives*.

If f has a derivative f' in a given interval and if f' is itself derivable, we denote the derivative of f' by f'' and call f'', the *second order derivative* of f. Continuing in this manner, we obtain functions $f, f', f'', f''', f^{iv}, \cdots, f^{(n)}$, each of which is the derivative of the previous one. We call $f^{(n)}$, *the nth derivative* of f or $f^{(n)}$ is the *derivative of order n* of the function f. When we write $y = f(x)$, we use notations

$$\frac{dy}{dx} \text{ for } f'(x), \quad \frac{d^2y}{dx^2} \text{ for } f''(x), \quad \cdots, \quad \frac{d^ny}{dx^n} \text{ for } f^n(x).$$

Alternative Notations:

$$y_1 \text{ for } \frac{dy}{dx}, \quad y_2 \text{ for } \frac{d^2y}{dx^2}, \quad \cdots\cdots, \quad y_n \text{ for } \frac{d^ny}{dx^n}.$$

$$\text{or} \quad Dy \text{ for } \frac{dy}{dx}, \quad D^2y \text{ for } \frac{d^2y}{dx^2}, \quad \cdots\cdots, \quad D^ny \text{ for } \frac{d^ny}{dx^n}.$$

It is to be observed that $\dfrac{d^ny}{dx^n} = \dfrac{d}{dx}\left(\dfrac{d^{n-1}y}{dx^{n-1}}\right) = \dfrac{d^2}{dx^2}\left(\dfrac{d^{n-2}y}{dx^{n-2}}\right) = $ etc.

Observation: In order that $f^n(x)$ may exist at a certain point x, it is clear that $f^{(n-1)}$ must exist in a certain neighbourhood of x and $f^{(n-1)}$ must be derivable at x. Since $f^{(n-1)}$ must exist in a neighbourhood of x, $f^{(n-2)}$ must be derivable in that neighbourhood and so on.

10.2 A Standard Result: nth Derivative of x^k

Let us begin with $y = x^k$, where k is any real number.
We shall assume that derivatives obtained here all exist.

$$y_1 = Dy = D(x^k) = kx^{k-1}, \quad y_2 = D^2y = D(Dy) = D(kx^{k-1}) = k(k-1)x^{k-2}.$$

We may infer that

$$\boxed{\begin{aligned} y_n = D^n y &= \frac{d^n y}{dx^n} = \frac{d^n}{dx^n}(x^k) \\ &= k(k-1)(k-2)\cdots(k-n+1)x^{k-n}, \; \forall \text{ positive integer } n. \end{aligned}}$$

[**Justification by Mathematical Induction**: The result is clearly true for $n = 1$. Let it be true for $n = a$ certain positive integer m, i.e., let $y_m = k(k-1)(k-2)\cdots(k-m+1)x^{k-m}$. Differentiating once more,

$$y_{m+1} = k(k-1)(k-2)\cdots(k-m+1)(k-m)x^{k-m-1}$$

whence the result is seen to be true for $n = m+1$ if it is accepted to be true for $n = m$. But since the result is true for $n = 1$, it is true for $n = 1+1 = 2$ and hence for $n = 3, 4, \cdots$, i.e., the result is true for all positive integers n.]

We now give below some very useful **special cases of**: $y = x^k$.

1. Let k be a positive integer. Then,

$$y_k = k(k-1)(k-2)\cdots(k-k+1)x^0; \quad \text{i.e., } \mathbf{y_k = \lfloor k}.$$

2. Let k be a positive integer, but n is a positive integer $> k$. Then $\mathbf{y_n = 0}$.

3. Let k be a positive real number so that $-k$ is a negative real number.
Then if $y = x^{-k}$,

$$\begin{aligned} y_n &= -k(-k-1)(-k-2)\cdots(-k-n+1)x^{-k-n} \\ &= (-1)^n \frac{k(k+1)(k+2)\cdots(k+n-1)}{x^{n+k}}. \end{aligned}$$

If k be a positive integer, then $y_n = \boxed{(-1)^n \frac{\lfloor k+n-1}{\lfloor k-1} \frac{1}{x^{n+k}}}.$

CHAPTER 10: REPEATED DIFFERENTIATION

e.g., $y = x^{-1}$, $y_n = \dfrac{(-1)^n \lfloor n}{x^{n+1}}$ (taking $k = 1$);

$y = x^{-2}$, $y_n = \dfrac{(-1)^n \lfloor n+1}{\lfloor 1 \, x^{n+2}}$ (taking $k = 2$), etc.

4. $y = \log x$ $(x > 0)$; $y_1 = \dfrac{1}{x}$. Hence

$$y_n = n\text{th derivative of } \log x$$
$$= (n-1)\text{th derivative of } \dfrac{1}{x} = \boxed{\dfrac{(-1)^{n-1}\lfloor n-1}{x^n}}.$$

5. $y = \dfrac{1}{x-a}$, $y_n = \dfrac{(-1)^n \lfloor n}{(x-a)^{n+1}}$

$y = \dfrac{1}{ax+b}$, $y_n = \boxed{\dfrac{(-1)^n \lfloor n}{(ax+b)^{n+1}} \cdot a^n}$

(Justify by using Mathematical Induction).

Applications

Example 10.2.1. *If* $y = \dfrac{1}{x^2 - a^2}$, *then find* y_n.

Solution: $y = \dfrac{1}{x^2 - a^2} = \dfrac{1}{(x-a)(x+a)} = \dfrac{1}{2a}\left[\dfrac{1}{x-a} - \dfrac{1}{x+a}\right].$

$\therefore y_n = \dfrac{1}{2a} \cdot (-1)^n \cdot \lfloor n \left[\dfrac{1}{(x-a)^{n+1}} - \dfrac{1}{(x+a)^{n+1}}\right].$

When $x > a > 0$, we write $x = r \cosh \theta$, $a = r \sinh \theta$ so that

$$r = +\sqrt{x^2 - a^2}, \quad \theta = \sinh^{-1}(a/r).$$

We have now,

$$x - a = r(\cosh \theta - \sinh \theta) = \dfrac{r}{2}\{(e^\theta + e^{-\theta}) - (e^\theta - e^{-\theta})\} = re^{-\theta}$$

and $x + a = r(\cosh \theta + \sinh \theta) = \dfrac{r}{2}\{(e^\theta + e^{-\theta}) + (e^\theta - e^{-\theta})\} = re^\theta.$

$\therefore y_n = \dfrac{(-1)^n \lfloor n}{2a \cdot r^{n+1}}[e^{(n+1)\theta} - e^{-(n+1)\theta}] = \dfrac{(-1)^n \lfloor n}{r^{n+1} \cdot a} \sinh(n+1)\theta.$

i.e., $\dfrac{d^n}{dx^n}\left(\dfrac{1}{x^2 - a^2}\right) = \dfrac{(-1)^n \lfloor n \, \sinh^{n+1}\theta \cdot \sinh(n+1)\theta}{a^{n+2}}$, where $\theta = \sinh^{-1}(a/r)$.

Example 10.2.2. $y = \dfrac{1}{x^2 + a^2}$, find y_n.

Solution: $y = \dfrac{1}{x^2 + a^2} = \dfrac{1}{(x+ia)(x-ia)} = \dfrac{1}{2ia}\left[\dfrac{1}{x-ia} - \dfrac{1}{x+ia}\right].$

$\therefore y_n = \dfrac{(-1)^n \lfloor n}{2ia}\left[\dfrac{1}{(x-ia)^{n+1}} - \dfrac{1}{(x+ia)^{n+1}}\right].$

We now put $x = r\cos\theta, a = r\sin\theta$; then $r = +\sqrt{x^2 + a^2}$ and $\theta = \sin^{-1}(a/r)$ $(0 < \theta < \pi)$.

We have now
$$x - ia = r(\cos\theta - i\sin\theta) = re^{-i\theta},$$
$$x + ia = r(\cos\theta + i\sin\theta) = re^{i\theta}.$$

$\therefore y_n = \dfrac{(-1)^n \lfloor n}{2ia \cdot r^{n+1}}[e^{i(n+1)\theta} - e^{-i(n+1)\theta}] = \dfrac{(-1)^n \lfloor n}{a \cdot r^{n+1}} \cdot \sin(n+1)\theta$

$= \dfrac{(-1)^n \lfloor n}{a^{n+2}} \sin^{n+1}\theta \sin(n+1)\theta$, where $\theta = \sin^{-1}(a/r)(0 < \theta < \pi)$.

Thus, $\dfrac{d^n}{dx^n}\left(\dfrac{1}{x^2+a^2}\right) = \dfrac{(-1)^n \lfloor n}{a^{n+2}} \sin^{n+1}\theta \cdot \sin(n+1)\theta.$ [C.H. 1991]

Corollary 10.2.1. If $y = \tan^{-1} x/a$, then

$$y_n = \dfrac{(-1)^{n-1}\lfloor n-1}{a^n} \sin^n\theta \sin n\theta, \text{ where } \theta = \tan^{-1}(a/x).$$

Example 10.2.3. If $ac > b^2$, then prove that

$$D^n\left(\dfrac{b+cx}{a + 2bx + cx^2}\right) = (-1)^n \lfloor n \left(\dfrac{c}{a+2bx+cx^2}\right)^{\frac{n+1}{2}} \times \cos\left\{(n+1)\tan^{-1}\dfrac{\sqrt{ac-b^2}}{b+cx}\right\}.$$

Solution: Let
$$y = \dfrac{b+cx}{cx^2 + 2bx + a} = \dfrac{x + b/c}{x^2 + 2\frac{b}{c} \cdot x + a/c} = \dfrac{x+b/c}{(x+b/c)^2 + (a/c - b^2/c^2)}$$

$$= \dfrac{x+b/c}{(x+b/c)^2 + k^2}, \text{ where } k^2 = \dfrac{ac - b^2}{c^2} \text{ (given } ac - b^2 > 0)$$

$$= \dfrac{x+b/c}{(x+b/c+ik)(x+b/c-ik)} = \dfrac{1}{2}\left[\dfrac{1}{x+b/c+ik} + \dfrac{1}{x+b/c-ik}\right].$$

$\therefore y_n = \dfrac{1}{2}(-1)^n \lfloor n \left[\dfrac{1}{(x+b/c+ik)^{n+1}} + \dfrac{1}{(x+b/c-ik)^{n+1}}\right].$

CHAPTER 10: REPEATED DIFFERENTIATION

Put $x + b/c = r\cos\theta, k = r\sin\theta$ so that

$$r^2 = \left(x + \frac{b}{c}\right)^2 + k^2 = \left(x + \frac{b}{c}\right)^2 + \frac{ac - b^2}{c^2}$$

$$= x^2 + \frac{2b}{c}x + \frac{b^2}{c^2} + \frac{a}{c} - \frac{b^2}{c^2} = \frac{cx^2 + 2bx + a}{c}$$

and $\tan\theta = \dfrac{k}{x + b/c} = \dfrac{\sqrt{ac - b^2}}{cx + b}$ or, $\theta = \tan^{-1}\dfrac{\sqrt{ac - b^2}}{cx + b}$.

We have now

$$x + b/c + ik = r(\cos\theta + i\sin\theta) = re^{i\theta}$$

$$x + b/c - ik = r(\cos\theta - i\sin\theta) = re^{-i\theta}.$$

$$\therefore\; y_n = \frac{1}{2}(-1)^n\underline{\lfloor n}\,\frac{1}{r^{n+1}}\left[e^{-i(n+1)\theta} + e^{i(n+1)\theta}\right]$$

$$= \frac{(-1)^n\underline{\lfloor n}}{r^{n+1}}\cos(n+1)\theta$$

$$= \frac{(-1)^n\underline{\lfloor n}}{\left(\frac{a + 2bx + cx^2}{c}\right)^{\frac{1}{2}(n+1)}}\cos\left\{(n+1)\tan^{-1}\frac{\sqrt{ac - b^2}}{b + cx}\right\}$$

$$= (-1)^n\underline{\lfloor n}\left(\frac{c}{a + 2bx + cx^2}\right)^{\frac{1}{2}(n+1)}\cos\left\{(n+1)\tan^{-1}\frac{\sqrt{ac - b^2}}{b + cx}\right\}.$$

10.3 Other Standard Results of nth Order Derivatives

1. $y = e^{ax}, D^n y = a^n e^{ax}$ (*Justify by Mathematical Induction*).

2. $y = \sin ax, D^n y = a^n \sin\left(ax + \frac{n\pi}{2}\right)$.

$$\boxed{D^{2n}(\sin\mathbf{ax}) = (-\mathbf{a}^2)^n \sin\mathbf{ax}.}$$

3. $y = \cos ax, D^n y = a^n \cos\left(ax + \frac{n\pi}{2}\right)$

$$\boxed{D^{2n}(\cos\mathbf{ax}) = (-\mathbf{a}^2)^n \cos\mathbf{ax}.}$$

4. $y = e^{ax}\cos(bx + c)(a > 0), \mathbf{y}_n = \mathbf{r}^n e^{ax}\cos(\mathbf{bx + c} + n\phi)$,
   where $r = +\sqrt{a^2 + b^2}, a = r\cos\phi, b = r\sin\phi, -\pi/2 < \phi < \pi/2$.

5. $D^n\{e^{-ax}\cos(bx+c)\} = \mathbf{r}^n e^{-ax}\cos(\mathbf{bx+c+n\phi})$,

   where $r = +\sqrt{a^2+b^2}, -a = r\cos\phi, b = r\sin\phi$.

   [One value of ϕ lies between $\pi/2$ and $3\pi/2$ and is equal to $\pi - \sin^{-1}(b/r)$.]

6. $y = e^{ax}\sin(bx+c)(a>0)$, $y_n = r^n e^{ax}\sin(bx+c+n\phi)$,

   where $r = +\sqrt{a^2+b^2}, a = r\cos\phi, b = r\sin\phi; -\pi/2 < \phi < \pi/2$.

   [write down the corresponding result when $a < 0$.]

We give below proofs for **2** and **4** and leave others as exercises for the students:

Proof: **2.** $y = \sin ax$, $y_1 = a\cos ax = a\sin(ax+\pi/2)$,

$$y_2 = a^2\sin\left(ax+2\frac{\pi}{2}\right), \quad y_3 = a^3\sin\left(ax+\frac{3\pi}{2}\right).$$

We infer $D^n(\sin ax) = a^n\sin\left(ax+\frac{n\pi}{2}\right)$.

We can now *justify this assertion by Mathematical Induction.*

See that

$$\begin{aligned}D^{2n}(\sin ax) &= a^{2n}\sin\left(ax+2n\cdot\frac{\pi}{2}\right)\\ &= a^{2n}[\sin ax\cos n\pi + \cos ax\sin n\pi]\\ &= a^{2n}(-1)^n\sin ax \quad (\because \cos n\pi = (-1)^n, \sin n\pi = 0)\\ &= (-a^2)^n\sin ax.\end{aligned}$$

Note 10.3.1. *This technique is used in finding Particular Integrals of* $\sin ax$ *or* $\cos ax$ *in solving linear differential equations of higher orders with constant coefficients. See Authors'* **An Introduction to Differential Equations.**

Proof: **4.** $y = e^{ax}\cos(bx+c)(a>0)$. Then

$$y_1 = ae^{ax}\cos(bx+c) - be^{ax}\sin(bx+c) = e^{ax}[a\cos(bx+c) - b\sin(bx+c)]$$
$$= e^{ax}r[\cos\phi\cos(bx+c) - \sin\phi\sin(bx+c)], \text{ putting } a = r\cos\phi, b = r\sin\phi.$$

$\therefore D\{e^{ax}\cos(bx+c)\} = e^{ax}r\cos(bx+c+\phi)$.

We can now justify by induction: $D^n\{e^{ax}\cos(bx+c)\} = e^{ax}\cdot r^n\cdot\cos(bx+c+n\phi)$, where $r = +\sqrt{a^2+b^2}, -\pi/2 < \phi < \pi/2$ ($\because a$ is positive).

CHAPTER 10: REPEATED DIFFERENTIATION

Example 10.3.1. If $y = \sin kx + \cos kx$, prove that $y_n = k^n \{1 + (-1)^n \sin 2kx\}^{1/2}$.

Solution:

$$y_n = k^n \left[\sin\left(kx + \frac{n\pi}{2}\right) + \cos\left(kx + \frac{n\pi}{2}\right)\right]$$

$$= k^n \left[\left\{\sin\left(kx + \frac{n\pi}{2}\right) + \cos\left(kx + \frac{n\pi}{2}\right)\right\}^2\right]^{1/2}$$

$$= k^n \left[\sin^2\left(kx + \frac{n\pi}{2}\right) + \cos^2\left(kx + \frac{n\pi}{2}\right) + 2\sin\left(kx + \frac{n\pi}{2}\right)\cos\left(kx + \frac{n\pi}{2}\right)\right]^{1/2}$$

$$= k^n[1 + \sin(2kx + n\pi)]^{1/2} = k^n[1 + \sin 2kx \cos n\pi + \cos 2kx \sin n\pi]^{1/2}$$

$$= k^n[1 + (-1)^n \sin 2kx]^{1/2}. \text{ (since } \cos n\pi = (-1)^n \text{ and } \sin n\pi = 0\text{)}$$

Example 10.3.2. If $y = e^{ax} \cos^2 bx$, find $y_n (a, b > 0)$.

Solution: $y = \frac{1}{2}e^{ax}(1 + \cos 2bx) = \frac{1}{2}e^{ax} + \frac{1}{2}e^{ax}\cos 2bx$.

$$\therefore y_n = \frac{1}{2}a^n e^{ax} + \frac{1}{2}e^{ax} \cdot r^n \cdot [\cos(2bx + n\phi)]$$

$$= \frac{1}{2}e^{ax}[a^n + (a^2 + 4b^2)^{n/2} \cos(2bx + n\tan^{-1} 2b/a)].$$

(putting $a = r\cos\phi$, $2b = r\sin\phi$ and hence $r = +\sqrt{a^2 + 4b^2}$, $\tan\phi = 2b/a]$.

10.4 Second-Order Derivatives : Typical Problems

Example 10.4.1. Find $\frac{d^2y}{dx^2}$, if $y = \tan e^{3x}$.

Solution: $Dy = \sec^2 e^{3x} \cdot e^{3x} \cdot 3$
$= 3e^{3x} \sec^2 e^{3x}$.

$$\therefore D^2 y = 3[3e^{3x} \sec^2 e^{3x} + e^{3x} \cdot 2\sec^2 e^{3x} \tan e^{3x} \cdot e^{3x} \cdot 3]$$

$$= 9e^{3x} \sec^2 e^{3x} \{1 + 2e^{3x} \tan e^{3x}\}.$$

Example 10.4.2. If $v = \frac{dx}{dt}$, show that $\frac{dv}{dt} = v\frac{dv}{dx} = \frac{d^2x}{dt^2}$.

Solution: $\dfrac{dv}{dt} = \dfrac{d}{dt}\left(\dfrac{dx}{dt}\right) = \dfrac{d^2x}{dt^2}.$

Again $\dfrac{dv}{dt} = \dfrac{dv}{dx} \cdot \dfrac{dx}{dt} = v \cdot \dfrac{dv}{dx}$ $\left(\because \dfrac{dx}{dt} = v\right).$

Implicit Functions

Example 10.4.3. If $ax^2 + 2hxy + by^2 + 2gx + 2fy + c = 0$, prove that

$$\frac{d^2y}{dx^2} = \frac{abc + 2fgh - af^2 - bg^2 - ch^2}{(hx + by + f)^3}.$$

Solution: Differentiating each term w.r.t. x we get,

$$2ax + 2h\left[y + x\frac{dy}{dx}\right] + 2by\frac{dy}{dx} + 2g + 2f\frac{dy}{dx} = 0,$$

whence, $\dfrac{dy}{dx} = -\dfrac{ax + hy + g}{hx + by + f}.$

Hence,

$$\frac{d^2y}{dx^2} = -\frac{(a + h\,dy/dx)(hx + by + f) - (h + b\,dy/dx)(ax + hy + g)}{(hx + by + f)^2}$$

$$= -\frac{\left\{a - \frac{h(ax+hy+g)}{hx+by+f}\right\}(hx + by + f) - \left\{h - \frac{b(ax+hy+g)}{hx+by+f}\right\}(ax + hy + g)}{(hx + by + f)^2}$$

$$= -\frac{(aby - h^2y + af - gh)(hx + by + f) - \{h^2x - abx + fh - bg\} \times (ax + hy + g)}{(hx + by + f)^3}$$

$$= \frac{(abc + 2fgh - af^2 - bg^2 - ch^2)}{(hx + by + f)^3}.$$

CHAPTER 10: REPEATED DIFFERENTIATION

[∵ Numerator

$= -\{(ab - h^2)y + (af - gh)\}(hx + by + f) + \{(h^2 - ab)x + (fh - bg)\}(ax + hy + g)$

$= \{(h^2 - ab)y + (gh - af)\}(hx + by + f) + \{(h^2 - ab)x + (fh - bg)\}(ax + hy + g)$

$= (h^2 - ab)hxy + (h^2 - ab)by^2 + (h^2 - ab)fy + (gh - af)hx + (gh - af)by$
$\quad + (gh - af)f + (h^2 - ab)ax^2 + (h^2 - ab)hxy + (h^2 - ab)gx + (fh - bg)ax$
$\quad + (fh - bg)hy + (fh - bg)g$

$= (h^2 - ab)(ax^2 + 2hxy + by^2) + y(h^2f - abf + gbh - afb + fh^2 - gbh)$
$\quad + x(gh^2 - afh + h^2g - abg + afh - abg) + 2fgh - af^2 - bg^2$

$= (h^2 - ab)(ax^2 + 2hxy + by^2 + 2fy + 2gx) + 2fgh - af^2 - bg^2$

$= (h^2 - ab)(-c) + 2fgh - af^2 - bg^2$

$= abc + 2fgh - af^2 - bg^2 - ch^2.]$

Parametric Functions

Example 10.4.4. *Let $x = \phi(t)$ and $y = \psi(t)$, show that*

$$\frac{d^2y}{dx^2} = \frac{x_1 y_2 - x_2 y_1}{x_1^3},$$

where the suffixes denote the order of differentiations with respect to t. Hence obtain

$$\frac{d^2y}{dx^2}, \text{ if } x = a(\cos\theta + \theta\sin\theta), \text{ and } y = a(\sin\theta - \theta\cos\theta).$$

Solution: First part. $x_1 = \phi'(t), y_1 = \psi'(t)$ so that $\dfrac{dy}{dx} = \dfrac{dy}{dt} \Big/ \dfrac{dx}{dt} = \dfrac{y_1}{x_1}$.

$\therefore \dfrac{d^2y}{dx^2} = \dfrac{d}{dx}\left(\dfrac{dy}{dx}\right) = \dfrac{d}{dt}\left(\dfrac{y_1}{x_1}\right) \times \dfrac{dt}{dx}$ [since y_1 and x_1 are functions of t only]

$= \dfrac{\frac{d}{dt}(y_1) \cdot x_1 - \frac{d}{dt}(x_1) \cdot y_1}{x_1^2} \cdot \dfrac{1}{x_1}$

$= \dfrac{x_1 y_2 - x_2 y_1}{x_1^3}.$

Second part. $x_1 = \dfrac{dx}{d\theta} = a(-\sin\theta + \sin\theta + \theta\cos\theta) = a\theta\cos\theta.$

$\therefore x_2 = \dfrac{d^2x}{d\theta^2} = a(\cos\theta - \theta\sin\theta).$

Again, $y_1 = \dfrac{dy}{d\theta} = a(\cos\theta - \cos\theta + \theta\sin\theta)$
$= a\theta\sin\theta.$

$$\therefore y_2 = \dfrac{d^2y}{d\theta^2} = a(\sin\theta + \theta\cos\theta).$$

Hence

$$\dfrac{d^2y}{dx^2} = \dfrac{x_1 y_2 - x_2 y_1}{x_1^3}$$

$$= \dfrac{a\theta\cos\theta \cdot a(\sin\theta + \theta\cos\theta) - a(\cos\theta - \theta\sin\theta) \cdot a\theta\sin\theta}{(a\theta\cos\theta)^3}$$

$$= \dfrac{a^2\theta\sin\theta\cos\theta + a^2\theta^2\cos^2\theta - a^2\theta\sin\theta\cos\theta + a^2\theta^2\sin^2\theta}{a^3\theta^3\cos^3\theta}$$

$$= \dfrac{a^2\theta^2}{a^3\theta^3\cos^3\theta} = \dfrac{1}{a\theta\cos^3\theta} = \dfrac{\sec^3\theta}{a\theta}.$$

Function of a Function

Example 10.4.5. *If $y = f(u)$ and $u = g(x)$ are derivable functions such that $dy/dx = dy/du \cdot du/dx = f'(u) \cdot g'(x)$, prove that*

$$\dfrac{d^2y}{dx^2} = \dfrac{d^2u}{dx^2} \cdot \dfrac{dy}{du} + \left(\dfrac{du}{dx}\right)^2 \cdot \dfrac{d^2y}{du^2}.$$

Solution: $\dfrac{dy}{dx} = \dfrac{dy}{du} \cdot \dfrac{du}{dx};$
hence

$$\dfrac{d^2y}{dx^2} = \dfrac{d}{dx}\left(\dfrac{dy}{du} \cdot \dfrac{du}{dx}\right) = \dfrac{d}{dx}\left(\dfrac{dy}{du}\right) \cdot \dfrac{du}{dx} + \dfrac{d}{dx}\left(\dfrac{du}{dx}\right) \cdot \dfrac{dy}{du}$$

$$= \dfrac{d}{du}\left(\dfrac{dy}{du}\right) \cdot \dfrac{du}{dx} \cdot \dfrac{du}{dx} + \dfrac{d^2u}{dx^2} \cdot \dfrac{dy}{du}$$

$$= \dfrac{d^2y}{du^2} \cdot \left(\dfrac{du}{dx}\right)^2 + \dfrac{d^2u}{dx^2} \cdot \dfrac{dy}{du}. \text{ Proved.}$$

Example 10.4.6. *Starting from $\dfrac{dx}{dy} = 1 \Big/ \dfrac{dy}{dx}$, obtain*

$$\dfrac{d^2x}{dy^2} = -\dfrac{\dfrac{d^2y}{dx^2}}{\left(\dfrac{dy}{dx}\right)^3}; \qquad \dfrac{d^3x}{dy^3} = -\dfrac{\dfrac{d^3y}{dx^3} \cdot \dfrac{dy}{dx} - 3\left(\dfrac{d^2y}{dx^2}\right)^2}{\left(\dfrac{dy}{dx}\right)^5}.$$

[C.H. 1968]

CHAPTER 10: REPEATED DIFFERENTIATION

Solution:

$$\frac{d^2x}{dy^2} = \frac{d}{dy}\left(1 \Big/ \frac{dy}{dx}\right) = -\frac{1}{(dy/dx)^2} \cdot \frac{d}{dy}\left(\frac{dy}{dx}\right) = -\frac{1}{(dy/dx)^2} \cdot \frac{d}{dx}\left(\frac{dy}{dx}\right) \cdot \frac{dx}{dy}$$

$$= -\frac{d^2y}{dx^2} \Big/ \left(\frac{dy}{dx}\right)^3.$$

Now

$$\frac{d^3x}{dy^3} = \frac{d}{dy}\left(\frac{d^2x}{dy^2}\right) = \frac{d}{dy}\left\{-\frac{d^2y/dx^2}{(dy/dx)^3}\right\} = -\frac{\frac{d}{dy}\left(\frac{d^2y}{dx^2}\right)\left(\frac{dy}{dx}\right)^3 - \frac{d^2y}{dx^2} \cdot \frac{d}{dy}\left(\frac{dy}{dx}\right)^3}{(dy/dx)^6}$$

$$= -\frac{\left\{\frac{d^3y}{dx^3}\cdot\left(\frac{dy}{dx}\right)^3 - \frac{d^2y}{dx^2}\frac{d}{dx}\left(\frac{dy}{dx}\right)^3\right\}\frac{dx}{dy}}{\left(\frac{dy}{dx}\right)^6}$$

$$= -\frac{\left\{\frac{d^3y}{dx^3}\cdot\left(\frac{dy}{dx}\right)^3 - \frac{d^2y}{dx^2}\cdot 3\left(\frac{dy}{dx}\right)^2\frac{d^2y}{dx^2}\right\}}{\left(\frac{dy}{dx}\right)^7} = -\frac{\frac{d^3y}{dx^3}\cdot\left(\frac{dy}{dx}\right) - 3\left(\frac{d^2y}{dx^2}\right)^2}{\left(\frac{dy}{dx}\right)^5}.$$

Example 10.4.7. *If $p^2 = a^2\cos^2\theta + b^2\sin^2\theta$, prove that $p + \dfrac{d^2p}{d\theta^2} = \dfrac{a^2b^2}{p^3}$.*

Solution: Given $p^2 = a^2\cos^2\theta + b^2\sin^2\theta$.

$$\therefore\ 2p\frac{dp}{d\theta} = (b^2 - a^2)\sin 2\theta;\ \text{hence } 2\left(\frac{dp}{d\theta}\right)^2 + 2p\frac{d^2p}{d\theta^2} = 2(b^2 - a^2)\cos 2\theta.$$

i.e., $\quad p\dfrac{d^2p}{d\theta^2} = (b^2 - a^2)\cos 2\theta - \left(\dfrac{dp}{d\theta}\right)^2$

$$= (b^2 - a^2)\cos 2\theta - \frac{(b^2 - a^2)^2 \sin^2 2\theta}{4p^2}$$

or, $\quad p^3\dfrac{d^2p}{d\theta^2} = (b^2 - a^2)p^2\cos 2\theta - \dfrac{(b^2 - a^2)^2 \sin^2 2\theta}{4}.$

$$\therefore\ p^4 + p^3\frac{d^2p}{d\theta^2}$$

$$= (a^2\cos^2\theta + b^2\sin^2\theta)^2 + (b^2 - a^2)(a^2\cos^2\theta + b^2\sin^2\theta)(\cos^2\theta - \sin^2\theta)$$
$$-(b^2 - a^2)^2\sin^2\theta\cos^2\theta \quad \text{(using, } p^2 = a^2\cos^2\theta + b^2\sin^2\theta,$$
$$\cos 2\theta = \cos^2\theta - \sin^2\theta \quad \text{and} \quad \sin 2\theta = 2\sin\theta\cos\theta\text{)}$$
$$= a^4\cos^4\theta + 2a^2b^2\sin^2\theta\cos^2\theta + b^4\sin^4\theta + (b^2 - a^2)(a^2\cos^4\theta + b^2\sin^2\theta\cos^2\theta$$
$$-a^2\sin^2\theta\cos^2\theta - b^2\sin^4\theta) - (b^4 - 2a^2b^2 + a^4)\sin^2\theta\cos^2\theta$$
$$= a^2b^2(\cos^4\theta + \sin^4\theta + 2\sin^2\theta\cos^2\theta) = a^2b^2(\sin^2\theta + \cos^2\theta)^2 = a^2b^2.$$

We have thus obtained, $p^4 + p^3\dfrac{d^2p}{d\theta^2} = a^2b^2$.

$$\therefore\ p + \frac{d^2p}{d\theta^2} = \frac{a^2b^2}{p^3}.$$

10.5 Use of Partial Fractions in finding nth Derivatives

We are familiar with the process of combining fractions by addition or subtraction e.g.,

$$\frac{3}{x-1} - \frac{4}{x+1} + \frac{x+2}{x^2+2} = \frac{5x^2 - 3x + 16}{(x^2-1)(x^2+2)}.$$

It is sometimes necessary (particularly *in Integration* and *in finding nth derivative* of a rational function) to reverse the process, i.e., when a rational function (quotient of two polynomials) such as

$$\frac{5x^2 - 3x + 16}{(x^2-1)(x^2+2)}$$

is given, we may have to break it up into simpler component fractions called **partial fractions**. [See Authors' book on Integral Calculus]

We shall show the techniques of breaking into partial fractions, rational functions like $\dfrac{f(x)}{\phi(x)}$, (where both $f(x)$ and $\phi(x)$ are polynomials in x).

If the numerator is of degree **equal** to or **higher** than the degree of the denominator, $f(x)/\phi(x)$ is reduced to a mixed expression **by division**, e.g.,

$$\frac{x^4 + 7x^3 + 21x^2 + 33x + 20}{x^3 + 6x^2 + 11x + 6} = x + 1 + \frac{4x^2 + 16x + 14}{x^3 + 6x^2 + 11x + 6}.$$

The fractional part is then separated into partial fractions.

CHAPTER 10: REPEATED DIFFERENTIATION

Next step is to factorise the denominator of the fractional expression into **simple real factors**: e.g., $x^3 + 6x^2 + 11x + 6 = (x+1)(x+2)(x+3)$.

The following four cases may arise:

Case 1. When the factors of the denominator are linear but not repeated

$$\text{e.g., } (ax+b)(px+q)(lx+m)\cdots.$$

For each factor of the form $(ax+b)$ there should be a fraction of the type $\dfrac{A}{ax+b}$, where A is a certain suitable constant.

Example 10.5.1. $\dfrac{x^2-6}{x^3-x^2-2x} = \dfrac{x^2-6}{x(x+1)(x-2)} = \dfrac{A}{x} + \dfrac{B}{x+1} + \dfrac{C}{x-2}$ where A, B, C are constants to be determined so that the equality may hold identically.

Solution: **Clearly**, we get the **identity**

$$A(x+1)(x-2) + Bx(x-2) + Cx(x+1) = x^2 - 6. \qquad (10.5.1)$$

We equate the coefficients of like powers of x:

$A+B+C = 1$ (coeff. of x^2); $-A-2B+C = 0$ (coeff. of x); $-2A = -6$ (constant term).

Solving we get, $A = 3$, $B = -5/3$, $C = -1/3$.

Another convenient method for determining the constants A, B, C.

In the **identity** (10.5.1) we assign suitable values for x on both sides.

Putting $x = 0$ on both sides of (10.5.1), we get $-2A = -6$ or $A = 3$.
Putting $x = -1$ on both sides of (10.5.1), we get $3B = -5$ or $B = -5/3$.
Putting $x = 2$ on both sides of (10.5.1), we get $6C = -2$ or $C = -1/3$.

Next suppose the problem is:

$$\text{if } y = \dfrac{x^2-6}{x^3-x^2-2x}, \text{ find } y_n.$$

We first break up y into partial fractions:

$$y = \dfrac{x^2-6}{x^3-x^2-2x} = \dfrac{x^2-6}{x(x+1)(x-2)} = \dfrac{A}{x} + \dfrac{B}{x+1} + \dfrac{C}{x-2},$$

where $A = 3$, $B = -5/3$, $C = -1/3$.

$$\therefore y_n = (-1)^n \underline{|n} \left[\dfrac{3}{x^{n+1}} - \dfrac{5/3}{(x+1)^{n+1}} - \dfrac{1/3}{(x-2)^{n+1}} \right].$$

Try in a similar manner:
$$D^n \left\{ \frac{x^4 + 7x^3 + 21x^2 + 33x + 20}{x^3 + 6x^2 + 11x + 6} \right\}$$
$$= (-1)^n \underline{\lfloor n} \left[\frac{1}{(x+1)^{n+1}} + \frac{2}{(x+2)^{n+1}} + \frac{3}{(x+3)^{n+1}} \right] (n \geq 1).$$

Case 2. When the factors of the denominator are linear but some or all of them are repeated

$$\text{e.g., } (ax+b)^k (px+q)^l \cdots .$$

For each factor of the form $(ax+b)^k$ there will be k fractions, namely

$$\frac{A_1}{(ax+b)^k} + \frac{A_2}{(ax+b)^{k-1}} + \frac{A_3}{(ax+b)^{k-2}} + \cdots + \frac{A_k}{ax+b},$$

where $A_1, A_2, A_3, \cdots, A_k$ are constants.

Example 10.5.2. $\dfrac{x-2}{x^3 - x^2 - 5x - 3} = \dfrac{x-2}{(x+1)^2(x-3)} = \dfrac{A_1}{(x+1)^2} + \dfrac{A_2}{x+1} + \dfrac{A_3}{x-3}$,
where A_1, A_2, A_3 are constants.

Solution: We obtain the identity

$$A_1(x-3) + A_2(x+1)(x-3) + A_3(x+1)^2 = x - 2.$$

Equating the coefficients of like powers of x:

$$A_2 + A_3 = 0; \quad A_1 - .2A_2 + 2A_3 = 1; \quad -3A_1 - 3A_2 + A_3 = -2.$$

Solving we get $A_1 = 3/4$, $A_2 = -1/16$, $A_3 = 1/16$.

$\therefore$ Given fraction $= \dfrac{3/4}{(x+1)^2} + \dfrac{-1/16}{x+1} + \dfrac{1/16}{x-3}.$

Now, if the problem is to find y_n where $y = \dfrac{x-2}{x^3 - x^2 - 5x - 3}$, then we first break up y into partial fractions as shown above. Then

$$y_n = \frac{3}{4} \cdot \frac{(-1)^n \underline{\lfloor n+1}}{(x+1)^{n+2}} - \frac{1}{16} \frac{(-1)^n \underline{\lfloor n}}{(x+1)^{n+1}} + \frac{1}{16} \frac{(-1)^n \underline{\lfloor n}}{(x-3)^{n+1}}.$$

Case 3. When the denominator contains factors of second degree (not resolvable into further real simple factors) but not repeated.

For a factor of the type $ax^2 + bx + c$, the fraction should be

$$\frac{Ax+B}{ax^2+bx+c} \quad (A, B \text{ are constants}).$$

CHAPTER 10: REPEATED DIFFERENTIATION

Example 10.5.3. $\dfrac{13x}{(x-3)(x^2+x+1)} = \dfrac{A}{x-3} + \dfrac{Bx+C}{x^2+x+1}$ $(A, B, C$ are constants to be determined).

Solution: We have then the identity

$$A(x^2 + x + 1) + (Bx + C)(x - 3) = 13x.$$

Equating coefficients of like powers of x we get

$$A + B = 0; \quad A - 3B + C = 13; \quad A - 3C = 0.$$

Solving, we get $A = 3$, $B = -3$ and $C = 1$.

$$\therefore \dfrac{13x}{(x-3)(x^2+x+1)} = \dfrac{3}{x-3} - \dfrac{3x-1}{x^2+x+1}.$$

To find its nth derivative, follow for the second fraction the method of worked **Ex. 10.2.3**.

Case 4. When the denominator contains factors of second degree but repeated.

Example 10.5.4. $\dfrac{x^4 + 4x^2 + 2x + 9}{(x-1)(x^2+x+2)^2} = \dfrac{A}{x-1} + \dfrac{Bx+C}{(x^2+x+2)^2} + \dfrac{Dx+E}{(x^2+x+2)}$, where A, B, C, D, E are constants to be determined.

Solution: We have the identity

$$x^4 + 4x^2 + 2x + 9 = A(x^2 + x + 2)^2 + (Bx + C)(x - 1) + (Dx + E)(x^2 + x + 2)(x - 1).$$

Equating coefficients of like powers of x we get

$$A + D = 1, \qquad 2A + E = 0, \qquad 5A + B + D = 4,$$
$$4A - B + C - 2D + E = 2, \quad 4A - C - 2E = 9.$$

Solving, we get $A = 1$, $B = -1$, $C = -1$, $D = 0$, $E = -2$ and we have

$$\dfrac{x^4 + 4x^2 + 2x + 9}{(x-1)(x^2+x+2)^2} = \dfrac{1}{x-1} - \dfrac{2}{x^2+x+2} - \dfrac{x+1}{(x^2+x+2)^2}.$$

10.6 Leibnitz's Theorem on Successive Derivatives

Theorem 10.6.1. *If u and v be two functions of x, both derivable at least upto n times, then $y = uv$ is derivable n times and the nth derivatives of $y = y_n = (uv)_n$ is* given by the formula:

$$(uv)_n = \sum_{r=0}^{n} {}^nC_r u_{n-r} v_r = {}^nC_0 u_n v + {}^nC_1 u_{n-1} v_1 + \cdots + {}^nC_n u v_n.$$

$$= u_n v + n u_{n-1} v_1 + \frac{n(n-1)}{2!} u_{n-2} v_2 + \cdots + u v_n,$$

where the suffixes denote the order of differentiations.

This theorem has many applications: Its proof may be established by Mathematical Induction. The students are advised to get acquainted with the technique of using this formula. So, before giving the proof we show below a number of applications of this very important theorem.

Example 10.6.1. Find y_n, when $y = x^3 \log x$.

Solution: Let $v = x^3$. Then $v_k = 0$ for $k \geq 4$.

If we assume $u = \log x$, then

$$u_k = \frac{(-1)^{k-1} \lfloor k-1}{x^k}.$$

$\therefore$ Leibnitz's formula gives

$$y_n = (x^3 \log x)_n = (uv)_n = u_n v + n u_{n-1} v_1 + \frac{n(n-1)}{2!} u_{n-2} v_2$$
$$+ \frac{n(n-1)(n-2)}{3!} u_{n-3} v_3$$

(other terms will all be zero, since $v_k = 0$ for $k \geq 4$.)

$$= \frac{(-1)^{n-1} \lfloor n-1}{x^n} \cdot x^3 + n \frac{(-1)^{n-2} \lfloor n-2}{x^{n-1}} \cdot 3x^2 + \frac{n(n-1)}{2!} \frac{(-1)^{n-3} \lfloor n-3}{x^{n-2}} \cdot 6x$$
$$+ \frac{n(n-1)(n-2)}{3!} \frac{(-1)^{n-4} \lfloor n-4}{x^{n-3}} \cdot 6.$$

$$= (-1)^n \frac{6 \lfloor n-4}{x^{n-3}} \quad \text{(on simplification)}.$$

Example 10.6.2. Differentiate n times the equation

$$(1-x^2) y_2 - x y_1 + m^2 y = 0.$$

CHAPTER 10: REPEATED DIFFERENTIATION 597

Solution: We use Leibnitz's formula and differentiate each term n times:

$$\{(1-x^2)y_2\}_n = (uv)_n, \text{ where } u = y_2 \text{ and } v = 1-x^2.$$
$$= u_n v + n u_{n-1} v_1 + \frac{n(n-1)}{2!} u_{n-2} v_2$$

(other terms are all zero since $v_k = 0$ for $k \geq 3$.)

$$= y_{n+2}(1-x^2) + n y_{n+1}(-2x) + \frac{n(n-1)}{2} y_n(-2).$$
$$= (1-x^2) y_{n+2} - 2nx y_{n+1} - n(n-1) y_n.$$

$$\{xy_1\}_n = (uv)_n, \text{ where } u = y_1 \text{ and } v = x$$
$$= u_n v + n u_{n-1} v_1 \text{ (here } v_k = 0 \text{ for } k \geq 2)$$
$$= y_{n+1} x + n y_n(1) = x y_{n+1} + n y_n.$$

$$(m^2 y)_n = m^2 y_n (m^2 \text{ being a constant}).$$

Hence from the given equation (differentiating each term n times by Leibnitz's formula) we get,

$$[(1-x^2) y_{n+2} - 2nx y_{n+1} - n(n-1) y_n] - [x y_{n+1} + n y_n] + m^2 y_n = 0$$
or, $(1-x^2) y_{n+2} - (2n+1) x y_{n+1} - (n^2 - n + n - m^2) y_n = 0$
or, $(1-x^2) y_{n+2} - (2n+1) x y_{n+1} + (m^2 - n^2) y_n = 0.$

Note 10.6.1. *At $x = 0$, this last relation gives a recurrence relation*

$$(y_{n+2})_0 = -(m^2 - n^2)(y_n)_0$$

where $(y_k)_0$ is a notation for the value of y_k when $x = 0$.

The following example should be very carefully observed:

Example 10.6.3. (a) *If $y = \cosh(\sin^{-1} x)$, prove that*
(i) $(1-x^2) y_2 - x y_1 - y = 0$; (ii) $(1-x^2) y_{n+2} - (2n+1) x y_{n+1} = (n^2 + 1) y_n.$
Now find the value of y_n when $x = 0$.
(b) *If we assume that $\cosh(\sin^{-1} x)$ can be expanded in an infinite series of the form*

$$a_0 + a_1 x + a_2 x^2 + \cdots + a_n x^n + \cdots,$$

prove that $a_{2n+1} = 0$ and $a_{2n} = \frac{4n^2 - 8n + 5}{2n(2n-1)} a_{2n-2}$ and hence write down the expansion of $\cosh(\sin^{-1} x)$ as far as x^6.

Solution: (a) (i) $y = \cosh(\sin^{-1} x)$, $y_1 = \sinh(\sin^{-1} x) \cdot \dfrac{1}{\sqrt{1-x^2}}$.

or, $\sqrt{1-x^2}\, y_1 = \sinh(\sin^{-1} x)$.

Differentiating once more, we get

$$\sqrt{1-x^2}\, y_2 - \dfrac{x}{\sqrt{1-x^2}} y_1 = \cosh(\sin^{-1} x) \cdot \dfrac{1}{\sqrt{1-x^2}}$$

or, $(1-x^2) y_2 - x y_1 - y = 0$.

(ii) We differentiate each term n times (using Leibnitz's formula):

$$\{(1-x^2) y_2\}_n - \{x y_1\}_n - \{y\}_n = 0$$

i.e., $\left[(1-x^2) y_{n+2} + n(-2x) y_{n+1} + \dfrac{n(n-1)}{2}(-2) y_n \right]$

$\qquad - [x y_{n+1} + n \cdot (1) \cdot y_n] - y_n = 0$

or, $(1-x^2) y_{n+2} - (2n+1) x y_{n+1} - (n^2 - n + n + 1) y_n = 0$.

$\therefore\ (1-x^2) y_{n+2} - (2n+1) x y_{n+1} = (n^2 + 1) y_n$.

Determination of $(y_n)_0$: See that we have obtained four results:

$y = \cosh(\sin^{-1} x)$; $\qquad y_1 = \dfrac{\sinh(\sin^{-1} x)}{\sqrt{1-x^2}}$;

$(1-x^2) y_2 - x y_1 - y = 0$; and $(1-x^2) y_{n+2} - (2n+1) x y_{n+1} = (n^2+1) y_n$.

Put $x = 0$ in each of the above four relations:

$(y)_0 = \cosh(\sin^{-1} 0) = \cosh 0 = 1$; $\quad (y_1)_0 = \dfrac{\sinh(\sin^{-1} 0)}{\sqrt{1-0}} = 0$.

$(y_2)_0 = (y)_0 = 1$; finally $(y_{n+2})_0 = (n^2+1)(y_n)_0$.

In the last relation, put $n = 1, 3, 5, 7, \cdots$;

$(y_3)_0 = (1^2 + 1)(y_1)_0 = 0$; $\quad (y_5)_0 = (3^2 + 1)(y_3)_0 = 0$; etc.

We infer $(y_{2n+1})_0 = 0$, for all positive integral values of n.

Next we put $n = 2$, we get

$$(y_4)_0 = (2^2 + 1)(y_2)_0 = (2^2 + 1) \cdot 1 \ (\because (y_2)_0 = 1).$$

Putting $n = 4$, we get

$$(y_6)_0 = (4^2 + 1)(y_4)_0 = (4^2 + 1)(2^2 + 1) \cdot 1.\ \text{etc.}$$

CHAPTER 10: REPEATED DIFFERENTIATION

Thus
$$(y_{2n})_0 = \{(2n-2)^2+1\}\{(2n-4)^2+1\}\cdots\{4^2+1\}\{2^2+1\}\cdot 1.$$

We conclude
$$(y_n)_0 = \{(n-2)^2+1\}\{(n-4)^2+1\}\cdots(4^2+1)(2^2+1)\cdot 1, \text{ if } n \text{ be even.}$$
$$= 0, \text{ if } n \text{ be odd.}$$

(b) Given
$$y = \cosh(\sin^{-1} x) = a_0 + a_1 x + a_2 x^2 + \cdots + a_n x^n + \cdots$$
$$y_1 = \frac{\sinh(\sin^{-1} x)}{\sqrt{1-x^2}} = a_1 + 2a_2 x + \cdots + na_n x^{n-1} + \cdots$$
(assuming that term-by-term differentiation is permissible)
$$y_2 = 2a_2 + 3\cdot 2a_3 x + \cdots + n(n-1)a_n x^{n-2} + \cdots.$$

Putting $x = 0$, we get $(y)_0 = 1 = a_0$, $(y_1)_0 = 0 = a_1$.

We have obtained before
$$(1-x^2)y_2 - xy_1 - y = 0$$
or, $$(1-x^2)(\Sigma n(n-1)a_n x^{n-2}) = x\Sigma na_n x^{n-1} + \Sigma a_n x^n.$$

In this identity we equate the coefficient of x^n:
$$(n+2)(n+1)a_{n+2} - n(n-1)a_n = na_n + a_n$$
or, $$(n+1)(n+2)a_{n+2} = (n^2+1)a_n \quad (10.6.1)$$
i.e., $$a_{n+2} = \frac{n^2+1}{(n+1)(n+2)} a_n.$$

Putting $n = 0, 1, 2, 3, 4, \cdots$ and using $a_0 = 1, a_1 = 0$ (already obtained) we get $a_2, a_3, a_4, \cdots$. Thus
$$a_2 = \frac{1}{2}, \quad a_3 = 0, \quad a_4 = \frac{5}{24}, \quad a_5 = 0, \quad a_6 = \frac{17}{144} \text{ etc.}$$

In general, $a_{2n+1} = 0$, and $(2n-1)(2n)a_{2n} = \{(2n-2)^2+1\}a_{2n-2}$ (replacing n by $2n-2$).
$$\therefore a_{2n} = \frac{4n^2 - 8n + 5}{4n^2 - 2n} a_{2n-2}.$$

Finally, the expansion of $\cosh(\sin^{-1} x)$ is given by

$$\cosh(\sin^{-1} x) = 1 + \frac{1}{2}x^2 + \frac{5}{24}x^4 + \frac{17}{144}x^6 + \cdots.$$

Example 10.6.4. *Prove that if $y = (x^2 - 1)^n$, then $(x^2 - 1) y_{n+2} + 2x\, y_{n+1} - n(n+1) y_n = 0$. Hence show that if $z = D^n(x^2 - 1)^n$, then z satisfies the following second order differential equation (known as Legendre's Equation):*

$$(1 - x^2)\frac{d^2 z}{dx^2} - 2x\frac{dz}{dx} + n(n+1)z = 0. \qquad \text{[C.H. 1974, 1981]}$$

Solution:

$$y = (x^2 - 1)^n, \quad y_1 = n(x^2 - 1)^{n-1} \cdot (2x) = \frac{2nx(x^2-1)^n}{x^2-1}$$

or, $(x^2 - 1)y_1 = 2nxy.$

Differentiate each term of this equation $(n+1)$ times (using Leibnitz's formula):

$$\{(x^2 - 1)y_1\}_{n+1} = 2n(xy)_{n+1}$$

or, $(x^2 - 1)y_{n+2} + (n+1)y_{n+1}(2x) + \dfrac{n(n+1)}{\lfloor 2}y_n \cdot (2) = 2n[xy_{n+1} + (n+1)y_n]$

or, $(x^2 - 1)y_{n+2} + 2xy_{n+1} + \{(n+1)n - 2n(n+1)\}y_n = 0$

or, $(x^2 - 1)y_{n+2} + 2xy_{n+1} - n(n+1)y_n = 0.$ (first part proved)

Second Part. The last relation can be written as

$$[(1 - x^2)y_{n+2} - 2xy_{n+1}] + n(n+1)y_n = 0$$

or, $\dfrac{d}{dx}\{(1 - x^2)y_{n+1}\} + n(n+1)y_n = 0$

or, $\dfrac{d}{dx}\left\{(1 - x^2)\dfrac{dz}{dx}\right\} + n(n+1)z = 0 \quad (\because z = D^n y).$

Thus z satisfies: $(1 - x^2)\dfrac{d^2 z}{dx^2} - 2x\dfrac{dz}{dx} + n(n+1)z = 0.$

We conclude this article by giving the proof of Leibnitz's Theorem (*Statement given earlier*).

Proof of Leibnitz's Theorem.

Step 1. By direct differentiation, we get

$$(uv)_1 = u_1 v + u v_1$$

CHAPTER 10: REPEATED DIFFERENTIATION

Thus the theorem is true for $n = 1$.

Step 2. Let us assume that the theorem is true for a certain positive integer m (say) $(m < n)$ i.e., we assume the following result to be true:

$$(uv)_m = u_m v + {}^mC_1 u_{m-1} v_1 + {}^mC_2 u_{m-2} v_2 + \cdots + {}^mC_r u_{m-r} v_r + \cdots + {}^mC_m u v_m.$$

Differentiating both sides once more we get

$$(uv)_{m+1} = \{u_{m+1}v + u_m v_1\} + {}^mC_1\{u_m v_1 + u_{m-1} v_2\} + \cdots$$
$$\cdots + {}^mC_r\{u_{m-r+1} v_r + u_{m-r} v_{r+1}\} + \cdots + {}^mC_m\{u_1 v_m + u v_{m+1}\}$$
$$= u_{m+1} v + (1 + {}^mC_1) u_m v_1 + ({}^mC_1 + {}^mC_2) u_{m-1} v_2 + \cdots$$
$$\cdots + ({}^mC_{r-1} + {}^mC_r) u_{m-r+1} v_r + \cdots + ({}^mC_{m-1} + {}^mC_m) u_1 v_m + u v_{m+1}.$$
$$= u_{m+1} v + {}^{m+1}C_1 u_m v_1 + {}^{m+1}C_2 u_{m-1} v_2 + \cdots + {}^{m+1}C_r u_{m-r+1} v_r + \cdots$$
$$\cdots + {}^{m+1}C_m u_1 v_m + u v_{m+1}.$$

(using the known formula of combinations: ${}^mC_r + {}^mC_{r-1} = {}^{m+1}C_r$)

Thus the theorem is found to be true for $n = m+1$ if it be assumed to be true for $n = m$.

[Conclusion. In step **1**, we observed that the theorem is true for $n = 1$ or $n = 2$.

∴ From what we proved in step **2**, the theorem is true for $n = 2 + 1 = 3$.

If it is true for $n = 3$, it should be true for $n = 3 + 1 = 4$ and so on.]

∴ By the principle of Mathematical Induction the theorem is true for all positive integers n.

Alternative method

One may give the justification of Leibnitz's theorem in the following way:

For the product of two functions u and v we have

$$(uv)_1 = u_1 v + u v_1, \quad (uv)_2 = u_2 v + 2 u_1 v_1 + u v_2 \text{ etc.}$$

and in general, we may infer that

$$(uv)_n = u_n v + C_1 u_{n-1} v_1 + C_2 u_{n-2} v_2 + \cdots \text{ to } (n+1) \text{ terms.} \quad (10.6.2)$$

Since the coefficients C_k are constants independent of u and v we may take special functions to determine them; thus if

$$u = e^{rx}, \quad v = e^x, \text{ then } uv = e^{(1+r)x}.$$
$$(uv)_n = (1+r)^n e^{(1+r)x} \quad \text{and} \quad u_k v_{n-k} = r^k e^{rx} \cdot e^x = r^k e^{(1+r)x}.$$

On substituting these values in (10.6.2) and dividing by the common factor, namely $e^{(1+r)x}$, we get

$$(1+r)^n = 1 + C_1 r + C_2 r^2 + \cdots + C_n r^n.$$

$\therefore$ The coefficients C_k in (10.6.2) are the binomial coefficients ${}^n C_k$.

This is what is to be proved in **Leibnitz's Theorem**.

D as an operator

If D denotes the operator d/dx while D_1 and D_2 operate only on the first or second of the two factors u and v in $y = uv$, then Leibnitz's theorem can be put in the symbolic form:

$$D^n(uv) = (D_1 + D_2)^n uv. \tag{10.6.3}$$

In particular let $u = e^{ax}$, then $D_1 e^{ax} = ae^{ax}$. We may replace D_1 by a in (10.6.3) and obtain a very useful technique used in solving Differential Equations:

$$D^n(e^{ax}v) = (a + D_2)^n e^{ax} v = e^{ax}(D+a)^n v \text{ (A shift formula shifting } D \text{ to } D+a\text{)}.$$

In fact if $P(D)$ is a polynomial in D, we can deduce the shift formula

$$P(D)e^{ax}v = e^{ax} P(D+a)v.$$

[See **Authors' book on Differential Equations**]

10.7 A Few Important Problems of Harder Type

Example 10.7.1. Prove, by Mathematical Induction, that

$$\frac{d^n}{dx^n}(x^{n-1} e^{1/x}) = (-1)^n \frac{e^{1/x}}{x^{n+1}}. \tag{10.7.1}$$

Solution: For $n = 1$, $\dfrac{d}{dx}(x^0 e^{1/x}) = e^{1/x}\left(\dfrac{-1}{x^2}\right) = \dfrac{(-1)^1 e^{1/x}}{x^2}$ i.e., the result is true for $n = 1$.

Let us assume that the result (10.7.1) is true for some positive integer $k < n$ i.e., assume that

$$D^k(x^{k-1} e^{1/x}) = \frac{d^k}{dx^k}(x^{k-1} e^{1/x}) = \frac{(-1)^k e^{1/x}}{x^{k+1}}. \tag{10.7.2}$$

CHAPTER 10: REPEATED DIFFERENTIATION

Now

$$D^{k+1}(x^k e^{1/x}) = \frac{d}{dx}\{(D^k(x^k e^{1/x}))\} = \frac{d}{dx}\left\{D^k\left(\underbrace{x}_{v}\cdot\underbrace{x^{k-1}e^{1/x}}_{u}\right)\right\}$$

$$= \frac{d}{dx}\left\{D^k(x^{k-1}e^{1/x})\cdot x + {}^kC_1 D^{k-1}(x^{k-1}e^{1/x})\cdot 1\right\}$$
(using Leibnitz's formula)

$$= \frac{d}{dx}\left\{\frac{(-1)^k e^{1/x}}{x^{k+1}}\cdot x + k\, D^{k-1}(x^{k-1}e^{1/x})\right\} \quad \text{(using (10.7.2) above)}$$

$$= \frac{d}{dx}\left\{\frac{(-1)^k e^{1/x}}{x^k}\right\} + k\, D^k(x^{k-1}e^{1/x})$$

$$= (-1)^k \frac{e^{1/x}}{x^k}\left(\frac{-1}{x^2}\right) + \frac{(-1)^k e^{1/x}(-k)}{x^{k+1}} + k\frac{(-1)^k e^{1/x}}{x^{k+1}}$$
(using (10.7.2) again)

$$= (-1)^{k+1}\frac{e^{1/x}}{x^{k+2}} - k\frac{(-1)^k e^{1/x}}{x^{k+1}} + k\frac{(-1)^k e^{1/x}}{x^{k+1}} = (-1)^{k+1}\frac{e^{1/x}}{x^{k+2}}.$$

Thus the result (10.7.1) is true for $n = k + 1$, if it is assumed to be true for $n = k$. [But it is true for $n = 1$ (proved by direct differentiation).

∴ It must be true for $n = 1 + 1 = 2$; since it is true for $n = 2$, it must be true for $n = 3$ and so on.]

The result (10.7.1) is thus true for all positive integers n by the method of induction.

Example 10.7.2. Let $P_n = D^n(x^n \log x)$. Prove the recurrence relation

$$P_n = n\cdot P_{n-1} + \lfloor n-1.$$

Hence show that

$$P_n = n!\left(\log x + 1 + \frac{1}{2} + \frac{1}{3} + \cdots + \frac{1}{n}\right).$$

Solution: Given

$$P_n = D^n(x^n \log x)$$

$$= D^{n-1}\left\{\frac{d}{dx}(x^n \log x)\right\}$$

$$= D^{n-1}\left\{x^n \cdot \frac{1}{x} + \log x \cdot n\, x^{n-1}\right\} = D^{n-1}\{x^{n-1} + n\, x^{n-1}\log x\}$$

$$= D^{n-1}(x^{n-1}) + n\cdot D^{n-1}(x^{n-1}\log x).$$

∴ $P_n = \lfloor n-1 + n\cdot P_{n-1}$ (First part proved).

Next part.

$$\frac{P_n}{\lfloor n} = \frac{\lfloor n-1}{\lfloor n} + \frac{n}{\lfloor n} P_{n-1} \quad \text{(dividing both sides by } \lfloor n \text{)}$$

or, $\quad \dfrac{P_n}{\lfloor n} - \dfrac{P_{n-1}}{\lfloor n-1} = \dfrac{1}{n}.$

Replacing successively n by $n-1, n-2, n-3, \cdots, 3, 2$ we get

$$\frac{P_{n-1}}{\lfloor n-1} - \frac{P_{n-2}}{\lfloor n-2} = \frac{1}{n-1}$$

$$\frac{P_{n-2}}{\lfloor n-2} - \frac{P_{n-3}}{\lfloor n-3} = \frac{1}{n-2}$$

$$\cdots\cdots \qquad \cdots\cdots \qquad \cdots\cdots$$

$$\frac{P_3}{\lfloor 3} - \frac{P_2}{\lfloor 2} = \frac{1}{3}$$

$$\frac{P_2}{\lfloor 2} - \frac{P_1}{\lfloor 1} = \frac{1}{2}.$$

Adding these results we get

$$\frac{P_n}{\lfloor n} - \frac{P_1}{\lfloor 1} = \frac{1}{2} + \frac{1}{3} + \frac{1}{4} + \cdots + \frac{1}{n-1} + \frac{1}{n},$$

or, $\quad \dfrac{P_n}{\lfloor n} = P_1 + \dfrac{1}{2} + \dfrac{1}{3} + \dfrac{1}{4} + \cdots + \dfrac{1}{n}$

$$= \log x + 1 + \frac{1}{2} + \frac{1}{3} + \cdots + \frac{1}{n}$$

$$[\because P_1 = D(x \log x) = \log x + x\frac{1}{x} = 1 + \log x]$$

i.e., $P_n = \lfloor n \left(\log x + 1 + \dfrac{1}{2} + \dfrac{1}{3} + \cdots + \dfrac{1}{n} \right).$

Example 10.7.3. *Given that y is a function of x and $z = 1/y$, $Y_r \cdot \lfloor r = \dfrac{d^r y}{dx^r}$, $Z_r \cdot \lfloor r = \dfrac{d^r z}{dx^r}$ when $r \geq 1$ while $Y = y$ and $Z = z$. Prove that*

$$Y^2 \begin{vmatrix} Z & Z_1 & Z_2 \\ Z_1 & Z_2 & Z_3 \\ Z_2 & Z_3 & Z_4 \end{vmatrix} = Z^3 \begin{vmatrix} Y_2 & Y_3 \\ Y_3 & Y_4 \end{vmatrix}.$$

CHAPTER 10: REPEATED DIFFERENTIATION

Solution: Given $yz = 1$. Hence, for $n \geq 1$, taking nth derivative by using Leibnitz's formula we obtain

$$\sum_{r=0}^{n} {}^nC_r \cdot y_r z_{n-r} = 0.$$

i.e., $\quad \displaystyle\sum_{r=0}^{n} \frac{\lfloor n}{\lfloor n-r \rfloor \lfloor r} y_r z_{n-r} = 0$

or, $\quad \displaystyle\sum_{r=0}^{n} Y_r Z_{n-r} = 0 \ (\because y_r/\lfloor r = Y_r, \ z_{n-r}/\lfloor n-r = Z_{n-r} \text{ and } \lfloor n \neq 0).$

For $n = 1, 2, 3, 4$ this gives

$$YZ_1 + Y_1 Z = 0 \qquad (10.7.3)$$
$$YZ_2 + Y_1 Z_1 + Y_2 Z = 0 \qquad (10.7.4)$$
$$YZ_3 + Y_1 Z_2 + Y_2 Z_1 = -Y_3 Z \qquad (10.7.5)$$
$$YZ_4 + Y_1 Z_3 + Y_2 Z_2 = -Y_3 Z_1 - Y_4 Z. \qquad (10.7.6)$$

From the last three linear equations in Y, Y_1, Y_2 we solve for Y by using Crammer's Rule:

$$Y \begin{vmatrix} Z_2 & Z_1 & Z \\ Z_3 & Z_2 & Z_1 \\ Z_4 & Z_3 & Z_2 \end{vmatrix} = \begin{vmatrix} 0 & Z_1 & Z \\ -Y_3 Z & Z_2 & Z_1 \\ -Y_3 Z_1 - Y_4 Z & Z_3 & Z_2 \end{vmatrix}. \qquad (10.7.7)$$

R.H.S. of (10.7.7) $= \dfrac{1}{Y} \begin{vmatrix} 0 & YZ_1 & Z \\ -Y_3 Z & YZ_2 & Z_1 \\ -Y_3 Z_1 - Y_4 Z & YZ_3 & Z_2 \end{vmatrix}$ (multiply Col. 2 by Y/Y).

$= \dfrac{1}{Y} \begin{vmatrix} 0 & YZ_1 + Y_1 Z & Z \\ -Y_3 Z & YZ_2 + Y_1 Z_1 & Z_1 \\ -Y_3 Z_1 - Y_4 Z & YZ_3 + Y_1 Z_2 & Z_2 \end{vmatrix}$ (multiply Col. 3 by Y_1 and add with Col. 2).

$= \dfrac{1}{Y} \begin{vmatrix} 0 & 0 & Z \\ -Y_3 Z & -Y_2 Z & Z_1 \\ -Y_3 Z_1 - Y_4 Z & -Y_3 Z - Y_2 Z_1 & Z_2 \end{vmatrix}$ [using (10.7.3), (10.7.4), (10.7.5)].

$$= \frac{Z}{Y} \begin{vmatrix} Y_3Z & Y_2Z \\ Y_3Z_1 + Y_4Z & Y_3Z + Y_2Z_1 \end{vmatrix}$$

$$= \frac{Z^2}{Y} \begin{vmatrix} Y_3 & Y_2 \\ Y_3Z_1 + Y_4Z & Y_3Z + Y_2Z_1 \end{vmatrix}$$

$$= \frac{Z^2}{Y} \{Y_3^2 Z + Y_3Y_2Z_1 - Y_2Y_3Z_1 - Y_2Y_4Z\}$$

$$= \frac{Z^3}{Y}(Y_3^2 - Y_2Y_4) = -\frac{Z^3}{Y} \begin{vmatrix} Y_2 & Y_3 \\ Y_3 & Y_4 \end{vmatrix}.$$

$\therefore$ From (10...); we get

$$-Y \begin{vmatrix} Z & Z_1 & Z_2 \\ Z_1 & Z_2 & Z_3 \\ Z_2 & Z_3 & Z_4 \end{vmatrix} = -\frac{Z^3}{Y} \begin{vmatrix} Y_2 & Y_3 \\ Y_3 & Y_4 \end{vmatrix}$$

or, $$Y^2 \begin{vmatrix} Z & Z_1 & Z_2 \\ Z_1 & Z_2 & Z_3 \\ Z_2 & Z_3 & Z_4 \end{vmatrix} = Z^3 \begin{vmatrix} Y_2 & Y_3 \\ Y_3 & Y_4 \end{vmatrix}.$$

Main Points : A Recapitulation

1. If the derived function $f'(x)$ of the function $f(x)$ is again derivable we may obtain the second derivative of $f(x)$, denoted by $f''(x)$; third derivative, fourth-derivative etc., nth derivative $f^n(x)$ can be obtained successively. If $y = f(x)$, we may write the nth derivative by $\dfrac{d^n y}{dx^n}$ or simply y_n or $D^n y$ where $D \equiv \dfrac{d}{dx}$.

2. Some standard nth derivatives are the following:

(a) $y = x^k$ (k, any real number), $y_n = k(k-1)\cdots(k-n+1)x^{k-n}$.

(b) $y = \frac{1}{ax+b}$, $y_n = \frac{(-1)^n n!}{(ax+b)^{n+1}} a^n$.

CHAPTER 10: REPEATED DIFFERENTIATION

In particular, $y = \frac{1}{x+b}$, $y_n = \frac{(-1)^n n!}{(x+b)^{n+1}}$.

(c) $y = \log x$, $y_1 = \frac{1}{x}$, $y_n = (n-1)$th derivative of $y_1 = \frac{(-1)^{n-1}(n-1)!}{x^n}$.

(d) $y = e^{ax}$, $y_n = a^n e^{ax}$.

(e) $y = \cos ax$ (or $\sin ax$); $y_n = a^n \cos\left(ax + \frac{n\pi}{2}\right)$ [or, $a^n \sin\left(ax + \frac{n\pi}{2}\right)$].

3. For the nth derivative of $y = \frac{1}{x^2+a^2}$, write
$$y = \frac{1}{2ai}\left[\frac{1}{x-ai} - \frac{1}{x+ai}\right]; \text{ then use 2. (b)}.$$
Putting $x = r\cos\theta$, $a = r\sin\theta$ and using De Moivre's Theorem we obtain
$$y_n = \frac{(-1)^n n!}{ar^{n+1}} \sin(n+1)\theta = \frac{(-1)^n n!}{a^{n+2}} \sin^{n+1}\theta \sin(n+1)\theta.$$

4. Leibnitz's Theorem: $(uv)_n = \sum_{r=0}^{n} {}^nC_r \cdot (u_{n-r} \cdot v_r)$ the suffixes denoting the order of derivations.

5. We often use **Leibnitz's Theorem** in forming a differential equation of say $(n+2)$th order; then putting $x = 0$, a recurrence relation is obtained connecting $(y_{n+2})_{x=0}$ and $(y_n)_{x=0}$, when all values like $(y_1)_0$, $(y_2)_0$, $(y_3)_0$, $(y_4)_0$ etc. are computed.

6. A few Important Deductions are:

(a) Expressions for $\frac{dx}{dy}$, $\frac{d^2x}{dy^2}$, $\frac{d^3x}{d^3y}$ etc. in terms of $\frac{dy}{dx}$, $\frac{d^2y}{dx^2}$, ...

(b) y is a function of x, $x = \sin t$; to find $\frac{d^2y}{dx^2}$ in terms of $\frac{dy}{dt}$ and $\frac{d^2y}{dt^2}$.

(c) y is a function of x, $t = \sin x$, to find $\frac{d^2y}{dx^2}$ in terms of $\frac{dy}{dt}$ and $\frac{d^2y}{dt^2}$.

[These results help in solving second order differential equations by Transformations.]

Exercises X
[A]

1. Find d^2y/dx^2 if y is given by
   a) $x^4 e^{5x}$.
   b) $\sin(\cos x)$.
   c) $1/(x^2 - x - 2)$.
   d) $\dfrac{1+x}{1-x}$.
   e) $\sin^2 x$.
   f) $x^2 \sin x$.
   g) $\sin^3 x \cos x$.
   h) $\sin^{-1} x$.
   i) $\tan e^{5x}$.
   j) $x^2 \log(x/a)$.
   k) $x\sqrt{a^2 - x^2}$.
   l) $\dfrac{\log x}{x}$.
   m) $e^{\sin x}$.
   n) $x = t - t^4, y = t^2 + t^3$.
   o) $\dfrac{x^2}{a^2} + \dfrac{y^2}{b^2} = 1$.
   p) $y = \tan^{-1} x^2$.
   q) $ax^2 + 2hxy + by^2 + 2gx + 2fy + c = 0$.

2. Find the second order derivatives of the following functions at the particular value of x given below:
   a) $y = x\sqrt{x^2 + 9}$ at $x = 4$.
   b) $y = (1/x)^x$ at $x = 1$.
   c) $x^2 + 4xy + y^2 + 3 = 0$ at $x = 2, y = -1$.
   d) $\left.\begin{array}{l} x = 2\cos\theta - \cos 2\theta \\ y = 2\sin\theta - \sin 2\theta \end{array}\right\}$ at $\theta = \pi/2$.

3. If $y = \sin pt, x = \sin t$, prove that
$$(1 - x^2)\frac{d^2y}{dx^2} - x\frac{dy}{dx} + p^2 y = 0.$$

4. If $x = Ae^{-kt/2} \cos(pt + \varepsilon)$, prove that
$$\frac{d^2x}{dt^2} + k\frac{dx}{dt} + n^2 x = 0, \text{ where } n^2 = p^2 + \frac{1}{4}k^2.$$

5. A point describing rectilinear harmonic motion of frequency $w/2\pi$ per second has displacement x at time t given by
$$x = A\sin wt + B\cos wt, \text{ or, } x = R\sin(wt + \phi).$$

Show that x satisfies, in either case, the second order differential equation
$$\frac{d^2x}{dt^2} + w^2 x = 0.$$

CHAPTER 10: REPEATED DIFFERENTIATION

[B]

1. Find the nth derivative of
   a) $\sqrt{x}$.
   b) $x^{-1/2}$.
   c) $x^{-7/3}$.
   d) $(ax+b)^{-4/15}$.
   e) $\log \dfrac{a+x}{a-x}$.
   f) a^x.
   g) $\sin^2 x$.
   h) $\cos^2 x$.
   i) $\sin^3 x$.
   j) $\cos 5x \cos 2x$.
   k) $\sin x \sin 2x \sin 3x$.
   l) $e^x \cos x$.
   m) $e^{3x} \sin 4x$.
   n) $e^{3x} \cos^2 5x$.
   o) $e^x \sin x \sin 2x$.

2. Find the nth derivatives of
   a) $\dfrac{x}{x-1}$.
   b) $\dfrac{x^2}{x-1}$.
   c) $\dfrac{1}{x^2-16}$.
   d) $\dfrac{1}{x^2+16}$.
   e) $\dfrac{x^2+1}{(x-1)(x-2)(x-3)}$.
   f) $\dfrac{4x}{(x-1)^2(x+1)}$.
   g) $\dfrac{1}{x^2+x+1}$.
   h) $\tan^{-1} \dfrac{x}{a}$.
   i) $\tan^{-1} \dfrac{1+x}{1-x}$.
   j) $\cos^{-1} \dfrac{1-x^2}{1+x^2}$.
   k) $\dfrac{x^4}{(x-1)(x-2)}$.
   l) $\dfrac{1}{6x^2-5x+1}$.
   m) $\dfrac{1}{(x-1)^3(x-2)}$.
   n) $\dfrac{x}{2x^2+3x+1}$.
   o) $\dfrac{x^2}{(x-a)(x-b)}$.
   p) $\dfrac{1}{x^4-a^4}$.

3. If $\theta = \arctan(a/x)$, then prove that
$$D^n \left(\dfrac{a}{x^2+a^2} \right) = (-1)^n n! (x^2+a^2)^{-\frac{1}{2}(n+1)} \sin(n+1)\theta$$
and $$D^n \left(\dfrac{x}{x^2+a^2} \right) = (-1)^n n! (x^2+a^2)^{-\frac{1}{2}(n+1)} \cos(n+1)\theta.$$

4. If $u = \dfrac{x^2-x+1}{x^2+x+1}$ and $\theta = \tan^{-1} \dfrac{\sqrt{3}}{2x+1}$, then prove that
$$D^n u = \dfrac{4}{\sqrt{3}} \cdot \dfrac{(-1)^n \underline{|n}\, \sin[(n+1)\theta - \pi/3]}{(x^2+x+1)^{\frac{1}{2}(n+1)}}.$$

5. Show that the nth derivatives of $\dfrac{\cos x}{x}$ and $\dfrac{\sin x}{x}$ are respectively of the forms:
$$\left\{ P_n \cos\left(x + \dfrac{1}{2}n\pi\right) - Q_n \cos\left(x + \dfrac{1}{2}n\pi\right) \right\} / x^{n+1}$$
and $$\left\{ P_n \sin\left(x + \dfrac{1}{2}n\pi\right) + Q_n \cos\left(x + \dfrac{1}{2}n\pi\right) \right\} / x^{n+1}$$

where
$$P_n = x^n - n(n-1)x^{n-2} + n(n-1)(n-2)(n-3)x^{n-4} - \cdots$$
and $$Q_n = nx^{n-1} - n(n-1)(n-2)x^{n-3}$$
$$+ n(n-1)(n-2)(n-3)(n-4)x^{n-5} - \cdots.$$

[C]

1. Use Leibnitz's formula to find the nth derivatives of
   a) $x^3 e^{ax}$. b) $x^3 \sin x$. c) $e^{ax} \cos bx$.
   d) $e^x \log x$. e) $x^2 \tan^{-1} x$. f) $x^2 e^x \cos x$.

2. Obtain the differential equations (given on the right side) by eliminating the arbitrary constants A, B, C in the expressions for y:

   (a) $y = Ae^{2x} + Be^{-2x}$;
   $$\frac{d^2y}{dx^2} = 4y.$$

   (b) $y = e^{-x}(A \cos x + B \sin x)$;
   $$\frac{d^2y}{dx^2} + 2\frac{dy}{dx} + 2y = 0.$$

   (c) $y = A \cos 3x + B \sin 3x + \frac{1}{2}x \sin 3x$;
   $$y_2 + 9y = 3 \cos 3x.$$

   (d) $y = Ae^x + Be^{2x} + Ce^{3x}$;
   $$\frac{d^3y}{dx^3} - 6\frac{d^2y}{dx^2} + 11\frac{dy}{dx} - 6y = 0.$$

3. Show that a family of *confocal conics*
   $$\frac{x^2}{a^2 + \lambda} + \frac{y^2}{b^2 + \lambda} = 1 \quad (a, b \text{ are fixed constants}; \lambda \text{ is the only variable parameter})$$
   gives the first-order differential equation
   $$\frac{dy}{dx}(a^2 - b^2) = \left(x + y\frac{dy}{dx}\right)\left(x\frac{dy}{dx} - y\right),$$
   by the elimination of the parameter λ.

   [*Hints:* Differentiate once w.r.t. x and then eliminate λ].

CHAPTER 10: REPEATED DIFFERENTIATION

4. (a) Given $y = \log(x + \sqrt{a^2 + x^2})$, deduce $(x^2 + a^2)y_2 + xy_1 = 0$.

   (b) Given $y = (x + \sqrt{1 + x^2})^m$, deduce $(1 + x^2)y_2 + xy_1 - m^2 y = 0$.

5. If $y = a \cosh nx + b \sinh nx$ (a, b are two parameters), then prove that
$$\frac{d^2 y}{dx^2} = n^2 y.$$

[D]

[Use of **Leibnitz's Theorem**: If f and g are two real-valued functions possessing derivatives of all orders at a given point $x = c$ and if $h(x) = f(x) \cdot g(x)$ then for any positive integer n, $h^n(c) = \sum_{k=0}^{n} \frac{n!}{k!(n-k)!} f^n(c) g^n(c)$.]

1. If $y = \sin(m \sin^{-1} x)$, show that

   (a) $(1 - x^2)y_2 - xy_1 + m^2 y = 0$.

   (b) $(1 - x^2)y_{n+2} - (2n + 1)xy_{n+1} + (m^2 - n^2)y_n = 0$. [C.H. 1976]

2. If $y = \tan^{-1} x$, deduce that

   (a) $(1 + x^2)y_{n+1} + 2nxy_n + n(n - 1)y_n = 0$.

   (b) $(1 + x^2)y_{n+2} + 2(n + 1)xy_{n+1} + n(n + 1)y_n = 0$.

   Also prove that
$$(y_n)_0 = \begin{cases} 0, & \text{if } n \text{ is even}; \\ (-1)^{\frac{1}{2}(n-1)} \lfloor n - 1, & \text{if } n \text{ is odd}. \end{cases}$$

3. If $y = (\sin^{-1} x)^2$, prove that $(1 - x^2)y_{n+2} - (2n + 1)xy_{n+1} - n^2 y_n = 0$.

4. If $y = a \cos(\log x) + b \sin(\log x)$, show that
$$x^2 y_{n+2} + (2n + 1)xy_{n+1} + (n^2 + 1)y_n = 0.$$ [C.H. 1990]

5. If $y = \cos(m \sin^{-1} x)$, then prove that
$$(1 - x^2)y_{n+2} - (2n + 1)xy_{n+1} + (m^2 - n^2)y_n = 0.$$

Now find y_n for $x = 0$.

6. If $y = e^{m \sin^{-1} x}$, show that

$$(1 - x^2)y_{n+2} - (2n + 1)xy_{n+1} - (n^2 + m^2)y_n = 0.$$

Also find y_n when $x = 0$.

7. If $y = \dfrac{x^3}{x^2 - 1}$, then prove that

$$(y_n)_0 = \begin{cases} 0, & \text{if } n \text{ is even;} \\ -\underline{\lfloor n \rfloor}, & \text{if } n \text{ is odd;} \end{cases} \quad n > 1.$$

[Hints: $y = x + \dfrac{1}{2}\left(\dfrac{1}{x-1} + \dfrac{1}{x+1}\right)$, $y_n = \dfrac{1}{2}(-1)^n \underline{\lfloor n \rfloor}\left[\dfrac{1}{(x-1)^{n+1}} + \dfrac{1}{(x+1)^{n+1}}\right]$.
Take $n = 2k$, $n = 2k + 1$.] [C.H. 1989, 99]

8. If $y^{1/m} + y^{-1/m} = 2x$, prove that

$$(x^2 - 1)y_{n+2} + (2n + 1)xy_{n+1} + (n^2 - m^2)y_n = 0.$$ [C.H. 1994]

9. If $x + y = 1$, prove that the nth derivative of $x^n y^n$ is

$$\underline{\lfloor n \rfloor}\{y^n - (^nC_1)^2 y^{n-1}x + (^nC_2)^2 y^{n-2}x^2 - (^nC_3)^2 y^{n-3}x^3 + \cdots + (-1)^n x^n\}.$$

10. If $\cos^{-1}\left(\dfrac{y}{b}\right) = \log\left(\dfrac{x}{n}\right)^n$, prove that $x^2 y_{n+2} + (2n+1)xy_{n+1} + 2n^2 y_n = 0$.

11. If $y = (\sinh^{-1} x)^2$, prove that

$$(1 + x^2)y_{n+2} + (2n + 1)xy_{n+1} + n^2 y_n = 0.$$

Obtain y_n for $x = 0$.

12. Prove that if $y = x^{n-1} e^{1/x}$, then $D^n y = \{(-1)^n e^{1/x}\}/x^{n+1}$.

13. If $y = e^{\tan^{-1} x} = a_0 + a_1 x + a_2 x^2 + \cdots + a_n x^n + \cdots$, prove that

(a) $(1 + x^2)y_2 + (2x - 1)y_1 = 0$;
(b) $(1 + x^2)y_{n+2} + \{2(n+1)x - 1\}y_{n+1} + n(n+1)y_n = 0$;
(c) $a_{n+2} = \dfrac{a_{n+1} - na_n}{n + 2}$;
(d) $e^{\tan^{-1} x} = 1 + x + \dfrac{1}{2}x^2 - \dfrac{1}{6}x^3 - \dfrac{7}{24}x^4 - \cdots$. [C.H. 1988, Old]

14. If $y = e^{a \sin^{-1} x} = a_0 + a_1 x + a_2 x^2 + \cdots + a_n x^n + \cdots$, prove that

$$(n + 1)(n + 2)a_{n+2} = (n^2 + a^2)a_n.$$

CHAPTER 10: REPEATED DIFFERENTIATION

15. If $y = x^{n-1} \log x$, then prove that $y_n = (n-1)!/x$.

16. Show that the nth derivative of $\log x/x$ is

$$\frac{(-1)^n n!}{x^{n+1}}\left[\log x - 1 - \frac{1}{2} - \frac{1}{3} - \frac{1}{4} - \cdots - \frac{1}{n}\right].$$

17. If $x = \tan(\log y)$, prove that

$$(1+x^2)y_{n+1} + (2nx-1)y_n + n(n-1)y_{n-1} = 0.$$

18. If $y = (\tan^{-1} x)^2$, prove that

(a) $(x^2+1)^2 y_2 + 2x(x^2+1)y_1 = 2$;

(b) $(x^2+1)^2 y_{n+2} + (4n+2)x(x^2+1)y_{n+1} + 2n^2(3x^2+1)y_n$
$\quad + 2n(n-1)(2n-1)x y_{n-1} + n(n-1)^2(n-2)y_{n-2} = 0.$

19. If $y = x(x+1)\log(x+1)^3$, prove that

$$y_n = \frac{3(-1)^{n-1} \lfloor n-3 \rfloor (2x+n)}{(x+1)^{n-1}} \quad (n \geq 3).$$

20. If $y = \sin(m \cos^{-1} \sqrt{x})$, then prove that

$$\lim_{x \to 0} \frac{y_{n+1}}{y_n} = \frac{4n^2 - m^2}{4n+2}.$$

21. If $y = x \log \dfrac{x-1}{x+1}$, prove that

$$y_n = (-1)^n \lfloor n-2 \left[\frac{x-n}{(x-1)^n} - \frac{x+n}{(x+1)^n}\right].$$

[E]
(Miscellaneous—Harder)

1. Change the independent variable from x to θ in the equation

(a) $\dfrac{d^2y}{dx^2} + \dfrac{2x}{1+x^2}\dfrac{dy}{dx} + \dfrac{y}{(1+x^2)^2} = 0$, using the transformation $x = \tan \theta$.

(b) $\dfrac{d^2y}{dx^2} + \cos x \dfrac{dy}{dx} + 4y \operatorname{cosec}^2 x = 0$, using $\theta = \log \tan(x/2)$.

(c) $\sin x \cos^2 x \dfrac{d^2y}{dx^2} - \cos x \dfrac{dy}{dx} + y \sin^3 x = 0$, using $\theta = \log \cos x$.

[Hints: (a) $\frac{dy}{dx} = \frac{dy}{d\theta} \cdot \frac{d\theta}{dx} = \frac{dy}{d\theta}/(1+x^2)$; i.e., $(1+x^2)\frac{dy}{dx} = \frac{dy}{d\theta}$. Now differentiate both sides w.r.t. x and obtain $(1+x^2)^2\frac{d^2y}{dx^2} + 2x(1+x^2)\frac{dy}{dx} = \frac{d^2y}{d\theta^2}$; ∴ the required transformed eqn. is $\frac{d^2y}{d\theta^2} + y = 0$.]

2. Show that the nth derivative of $(1 + x + x^2 + x^3)^{-1}$ is

$$\frac{1}{2}(-1)^n \lfloor n\, \sin^{n+1}\theta [\sin(n+1)\theta - \cos(n+1)\theta + (\sin\theta + \cos\theta)^{-n-1}],\ (\theta = \cot^{-1} x).$$

3. Let P_n denote the nth derivative of $(lx + m)/(x^2 - 2bx + c)$. Prove that

$$\frac{x^2 - 2bx + c}{(n+1)(n+2)}P_{n+2} + \frac{2(x-b)}{n+1}P_{n+1} + P_n = 0.$$

4. Prove that

$$\left(\sin^2\theta \frac{d}{d\theta}\right)^n \sin^2\theta = n!\sin^{n+1}\theta \sin(n+1)\theta.$$

5. If $y = \dfrac{1}{x^3 - a^3}$, prove that

$$y_n = \frac{(-1)^n n!}{3a^2} \cdot \frac{1}{(x-a)^{n+1}} + \frac{(-1)^n n!}{3a^2} \cdot \frac{2\cos\left[(n+1)\theta + \frac{2}{3}\pi\right]}{(x^2 + ax + a^2)^{\frac{1}{2}(n+1)}},$$

where $\tan\theta = a\sqrt{3}/(2x + a)$.

6. Prove, by induction or otherwise, that when u and v are functions of x,

$$vD^n(u) = D^n(uv) - nD^{n-1}(uDv) + \frac{1}{\lfloor 2}n(n-1)D^{n-2}(uD^2v) - \cdots + (-1)^n D^n v.$$

7. (a) Differentiate the identity $x^m \cdot x^n = x^{m+n}$ using Leibnitz's formula and hence deduce the following result (known as *Vandermonde's Theorem*):

$$[m+n]_r = [m]_r + \frac{r}{\lfloor 1}[m]_{r-1}[n]_1 + \frac{r(r-1)}{\lfloor 2}[m]_{r-2}[n]_2 + \cdots + [n]_r,$$

where $[m]_r = m(m-1)(m-2)\cdots(m-r+1)$.

(b) Put $m = n$ and obtain by taking nth derivative of the identity $x^{2n} = x^n \cdot x^n$,

$$\frac{\lfloor 2n}{\lfloor n \lfloor n} = 1 + \frac{n^2}{1^2} + \frac{n^2(n-1)^2}{1^2 \cdot 2^2} + \frac{n^2(n-1)^2(n-2)^2}{1^2 \cdot 2^2 \cdot 3^2} + \cdots.$$

8. Show that, when $y = x^2 e^x$,

$$y_n = \frac{1}{2}n(n-1)y_2 - n(n-2)y_1 + \frac{1}{2}(n-1)(n-2)y.$$

CHAPTER 10: REPEATED DIFFERENTIATION

9. Prove that
$$e^{-x} D^n(x^n e^x) = \sum_{r=0}^{n} \left\{ \frac{n!}{(n-r)!} \right\}^2 \frac{x^{n-r}}{r!}.$$

10. When $y = x^k e^x$, prove that $(y_1 - y)x = ky$ and that
$$xy_{n+2} + (n+1-k-x)y_{n+1} - (n+1)y_n = 0.$$

Show that if $z = D^n(x^{n+1} e^x)$, then $xz_2 - xz_1 = (n+1)z$.

11. Prove that
$$\frac{d^n}{dx^n} \left(\frac{\log x}{1-x} \right) = \frac{n!}{(1-x)^{n+1}} \left\{ \log x + \sum_{r=1}^{n} \frac{(-1)^{r-1}}{r} \left(\frac{1-x}{x} \right)^r \right\}.$$

12. Verify that the nth derivative of $e^x/(x^2+1)$, when $x = 1$ is
$$n! \, e \sum_{r=0}^{n} \frac{(-1)^r \sin \frac{1}{4}(r+1)\pi}{(\sqrt{2})^{r+1} \underline{|n-r|}}.$$

13. Show that
$$\frac{d^r}{dx^r}(1+x)^{n-\frac{1}{2}} = \left(n - \frac{1}{2} \right)\left(n - \frac{3}{2} \right) \cdots \left(n - r + \frac{1}{2} \right)(1+x)^{n-r-\frac{1}{2}}$$

and that the product of this and $(2n-r)$th derivative of $(1-x)^{n-\frac{1}{2}}$ is equal to
$$(-1)^n \left\{ \frac{1 \cdot 3 \cdot 5 \cdots (2n-1)}{2 \cdot 2 \cdot 2 \cdots 2} \right\}^2 \frac{(1+x)^{n-r-\frac{1}{2}}}{(1-x)^{n-r+\frac{1}{2}}}.$$

Differentiate $2n$ times the product $(1+x)^{n-\frac{1}{2}}(1-x)^{n-\frac{1}{2}}$ to show that when $x = \cos\theta$,
$$\sin^{2n+1}\theta \frac{d^{2n}}{dx^{2n}}(\sin^{2n-1}\theta) = (-1)^n \cdot 1^2 \cdot 3^2 \cdot 5^2 \cdots (2n-1)^2.$$

14. When $y = \sqrt{\dfrac{1+x}{1-x}}$, prove that $(1-x^2)y_1 = y$ and deduce by Leibnitz's formula that
$$(1-x^2)y_n - \{2(n-1)x + 1\}y_{n-1} - (n-1)(n-2)y_{n-2} = 0.$$

Prove that if the function $P_n(x)$ is defined by $(1-x^2)^n y_n = y \, P_n(x)$, then $P_n(x)$ is a polynomial in x of degree $(n-1)$.

15. Prove that $D^n(e^{-x^2}) = -2\{xD^{n-1}(e^{-x^2}) + (n-1)D^{n-2}(e^{-x^2})\}$ and that $P_n(x) = e^{x^2}D^n(e^{-x^2})$ is a polynomial of degree n in x. [This polynomial is known as *Hermite polynomial*.]

    Prove that $DP_n(x) = e^{x^2}\{2xD^n(e^{-x^2}) + D^{n+1}(e^{-x^2})\} = -2nP_{n-1}(x)$.

16. If y and z are two functions of x such that $yz = 1$, then prove that

    $$\frac{1}{z^3}\begin{vmatrix} z & z_1 & z_2 \\ z_1 & z_2 & z_3 \\ z_2 & z_3 & z_4 \end{vmatrix} = \frac{1}{y^2}\begin{vmatrix} y_2 & y_3 \\ y_3 & y_4 \end{vmatrix}$$

    where $y_r = \dfrac{1}{r!}\dfrac{d^r y}{dx^r}$ and $z_r = \dfrac{1}{r!}\dfrac{d^r z}{dx^r}$.

17. Show that if $y = 1/(1+x^2)$, then $y_n = f_n(x)/(1+x^2)^{n+1}$, where $f_n(x)$ is a polynomial in x of degree n. Show also that

    (a) $f_{n+1} = (1+x^2)f_n' - 2(n+1)xf_n$.
    (b) $f_{n+2} + 2(n+2)xf_{n+1} + (n+2)(n+1)(1+x^2)f_n = 0$.
    (c) $(1+x^2)f_n'' - 2nxf_n' + n(n+1)f_n = 0$.
    (d) $f_n = (-1)^n n!\left\{(n+1)x^n - \dfrac{(n+1)n(n-1)}{3!}x^{n-2} + \cdots\right\}$.

    [See that $f_n = f_n(x)$, $f_n' = f_n'(x)$ etc.]

18. Show that the nth derivative of $x^n e^{-x}$ is of the form $L_n(x)e^{-x}$ when $L_n(x)$ is a polynomial in x of degree n. [This polynomial is known as *Laguerre polynomial*.]

19. Given that $y = f(x) = \dfrac{1}{\sqrt{1+2x}}$, prove that

    $$(1+2x)y_{n+1} + (2n+1)y_n = 0.$$

    Find y_n for $x = 0$, where n is a positive integer (odd or even). [C.H. 1983]

20. Show that if $x = \cot y$,

    $$\frac{d^n}{dx^n}\left(\frac{x^n}{1+x^2}\right) = n!\sin y\{\sin y - {}^nC_1 \cos y \sin 2y + {}^nC_2 \cos^2 y \sin 3y - \cdots\}.$$

    [C.H. 1991]

CHAPTER 10: REPEATED DIFFERENTIATION   617

Answers

[A]

1. (a) $e^{5x}(25x^4 + 40x^3 + 12x^2)$. (b) $-\sin(\cos x)\sin^2 x - \cos(\cos x)\cos x$.

(c) $\dfrac{2}{3}\left\{\dfrac{1}{(x-2)^3} - \dfrac{1}{(x+1)^3}\right\}$. (d) $\dfrac{4}{(1-x)^3}$. (e) $2\cos 2x$. (f) $4x\cos x - (x^2 - 2)\sin x$.

(g) $\sin x \cos x\, [6\cos 2x - 2\sin 2x \sin x]$. (h) $\dfrac{x}{(1-x^2)^{\frac{3}{2}}}$.

(i) $25e^{5x}\sec^2(e^{5x})\{1 + 2e^{5x}\tan e^{5x}\}$.

(j) $2\log\dfrac{x}{a} + 3$. (k) $\dfrac{2x^3 - 3a^2 x}{(a^2 - x^2)^{\frac{3}{2}}}$. (l) $\dfrac{2\log x - 3}{x^2}$. (m) $e^{\sin x}(\cos^2 x - \sin x)$.

(n) $\dfrac{12t^4 + 16t^3 + 6t + 2}{(1 - 4t^3)^3}$. (o) $-\dfrac{b^4}{a^2 y^3}$.

(p) $\dfrac{2(1 - 3x^4)}{(1 + x^4)^2}$. (q) $\dfrac{abc + 2fgh - af^2 - bg^2 - ch^2}{(hx + by + f)^3}$.

2. (a) $236/125$. (b) 0. (c) $-1/3$. d) $-3/2$.

[B]

1. (a) $(-1)^{n-1}\cdot\dfrac{1\cdot 3\cdot 5\cdots(2n-3)}{2^n\cdot x^{n-1/2}}$. (b) $(-1)^n\dfrac{1\cdot 3\cdot 5\cdots(2n-1)}{2^n x^{n+\frac{1}{2}}}$.

(c) $(-1)^n\dfrac{7\cdot 10\cdot 13\cdots(3n+4)}{3^n x^{n+\frac{7}{3}}}$.

(d) $(-1)^n\dfrac{4\cdot 19\cdots(15n-11)}{15^n(ax+b)^{n+\frac{4}{15}}}a^n$. (e) $(n-1)!\left\{\dfrac{(-1)^{n-1}}{(a+x)^n} + \dfrac{1}{(a-x)^n}\right\}$. (f) $(\log a)^n \cdot a^x$.

(g) $-2^{n-1}\cos\left(\dfrac{1}{2}n\pi + 2x\right)$. (h) $2^{n-1}\cos\left(\dfrac{1}{2}nx + 2x\right)$.

(i) $\dfrac{1}{4}\left\{3\sin\left(\dfrac{1}{2}n\pi + x\right) - 3^n \sin\left(\dfrac{1}{2}n\pi + 3x\right)\right\}$.

(j) $\dfrac{1}{2}\left\{7^n \cos\left(\dfrac{1}{2}n\pi + 7x\right) + 3^n \cos\left(\dfrac{1}{2}n\pi + 3x\right)\right\}$.

(k) $\dfrac{1}{4}\left\{4^n \sin\left(\dfrac{1}{2}n\pi + 4x\right) + 2^n \sin\left(\dfrac{1}{2}n\pi + 2x\right) - 6^n \sin\left(\dfrac{1}{2}n\pi + 6x\right)\right\}$.

(l) $2^{\frac{n}{2}} e^x \cos\left(x + \frac{1}{4}n\pi\right)$. (m) $5^n e^{3x} \sin\left(4x + n\tan^{-1}\frac{4}{3}\right)$.

(n) $\frac{1}{2} e^{3x} \left\{ 3^n + 109^{n/2} \cos\left(10x + n\tan^{-1}\frac{10}{3}\right) \right\}$.

(o) $\frac{1}{2} e^x \left\{ 2^{\frac{1}{2}n} \cos\left(x + \frac{1}{4}n\pi\right) - 10^{\frac{1}{2}n} \cos(3x + n\tan^{-1} 3) \right\}$.

2. (a) and (b) $(-1)^n \cdot \dfrac{n!}{(x-1)^{n+1}}$. (c) $\dfrac{(-1)^n n!}{8} \left\{ \dfrac{1}{(x-4)^{n+1}} - \dfrac{1}{(x+4)^{n+1}} \right\}$.

(d) $\dfrac{(-1)^n n! \sin(n+1)\theta}{4 \cdot r^{n+1}}$, where $x = r\cos\theta, 4 = r\sin\theta$.

(e) $(-1)^n n! \left\{ \dfrac{1}{(x-1)^{n+1}} - \dfrac{5}{(x-2)^{n+1}} + \dfrac{5}{(x-3)^{n+1}} \right\}$.

(f) $(-1)^n n! \left\{ \dfrac{1}{(x-1)^{n+1}} - \dfrac{1}{(x+1)^{n+1}} + \dfrac{2(n+1)}{(x-1)^{n+2}} \right\}$.

(g) $\dfrac{(-1)^n \cdot 2^{n+2} \cdot n!}{3^{\frac{n}{2}+1}} \sin^{n+1}\theta \cdot \sin(n+1)\theta$, where $\theta = \tan^{-1}\{\sqrt{3}/(2x+1)\}$.

(h) $\dfrac{(-1)^{n-1}(n-1)! \sin^n \theta \sin n\theta}{a^n}$; $\theta = \tan^{-1}(a/x)$.

(i) $(-1)^{n-1} (n-1)! \sin^n \theta \sin n\theta$; $\theta = \cot^{-1} x$.

(j) $2(-1)^{n-1}(n-1)! \sin^n \theta \sin n\theta$; $\theta = \cot^{-1} x$.

(k) $(-1)^n (n!)\{16(x-2)^{-n-1} - (x-1)^{-n-1}\}, n > 2$.

(l) $(n!)\{3^{n+1}(1-3x)^{-n-1} - 2^{n+1}(1-2x)^{-n-1}\}$.

(m) $(-1)^{n-1}(n!) \left\{ \dfrac{(n+2)(n+1)}{2(x-1)^{n+3}} + \dfrac{n+1}{(x-1)^{n+2}} + \dfrac{1}{(x-1)^{n+1}} - \dfrac{1}{(x-2)^{n+1}} \right\}$.

(n) $(-1)^n (n!) \left\{ \dfrac{1}{(x+1)^{n+1}} - \dfrac{2^n}{(2x+1)^{n+1}} \right\}$.

(o) $\dfrac{(-1)^n (n!)}{a-b} \left\{ \dfrac{a^2}{(x-a)^{n+1}} - \dfrac{b^2}{(x-b)^{n+1}} \right\}$.

(p) $\dfrac{(-1)^n (n!)}{4a^3} \left\{ \dfrac{1}{(x-a)^{n+1}} - \dfrac{1}{(x+a)^{n+1}} - \dfrac{2\sin[(n+1)\cot^{-1}(x/a)]}{(a^2+x^2)^{\frac{n+1}{2}}} \right\}$.

CHAPTER 10: REPEATED DIFFERENTIATION

[C]

1. (a) $e^{ax}a^{n-3}\{a^3x^3 + 3na^2x^2 + 3n(n-1)ax + n(n-1)(n-2)\}$.

   (b) $x^3 \sin\left(\dfrac{1}{2}n\pi + x\right) + 3nx^2 \sin\left[\dfrac{1}{2}(n-1)\pi + x\right] + 3n(n-1)x \cdot \sin\left\{\dfrac{1}{2}(n-2)\pi + x\right\}$
   $+ n(n-1)(n-2)\sin\left[\dfrac{1}{2}(n-3)\pi + x\right]$.

   (c) $e^{ax}\left\{a^n \cos bx + {}^nC_1 a^{n-1} b \cos\left(bx + \dfrac{1}{2}\pi\right) + \cdots + b^n \cos\left(bx + n \cdot \dfrac{1}{2}\pi\right)\right\}$.

   (d) $e^x\left\{\log x + nx^{-1} - {}^nC_2 x^{-2} + \cdots + (-1)^{n-1}(n-1)! x^{-n}\right\}$.

   (e) $(-1)^{n-1}(n-3)! \sin^{n-2}\theta \left\{(n-1)(n-2)\sin n\theta \cos^2\theta - 2n(n-2)\sin(n-1)\theta \cos\theta + n(n-1)\sin(n-2)\theta\right\}$, where $\cot\theta = x$.

   (f) $2^{\frac{1}{2}(n-2)} e^x \left[2x^2 \cos\left(x + \dfrac{n\pi}{4}\right) + 2^{3/2} \cdot nx \cos\left(x + (n-1)\dfrac{\pi}{4}\right)\right.$
   $\left. + n(n-1)\cos\left(x + (n-2)\dfrac{\pi}{4}\right)\right]$.

[D]

5. $(y_n)_0 = \begin{cases} 0, & \text{when } n \text{ is odd;} \\ -m^2(2^2 - m^2)(4^2 - m^2)\cdots\{(n-2)^2 - m^2\}, & \text{when } n \text{ is even.} \end{cases}$

6. $(y_n)_0 = \begin{cases} m(m^2 + 1^2)(m^2 + 3^2)\cdots\{m^2 + (n-2)^2\}, & \text{when } n \text{ is odd} \\ m^2(m^2 + 2^2)(m^2 + 4^2)\cdots(m^2 + (n-2)^2\}, & \text{when } n \text{ is even.} \end{cases}$

11. $(y_{2n+1})_0 = 0;\quad (y_{2n})_0 = \begin{cases} (-1)^{n-1} \cdot 2 \cdot 2^2 \cdot 4^2 \cdot 6^2 \cdots (2n-2)^2; \\ (-1)^{n-1} 2^{2n-1}\{(n-1)!\}^2. \end{cases}$

AN INTRODUCTION TO ANALYSIS

(Differential Calculus—Part I)

Chapter 11

**Rolle's Theorem:
Mean Value Theorem: Lagrange's Form
and Cauchy's Form**

Chapter 11

Rolle's Theorem: Mean Value Theorem: Lagrange's Form and Cauchy's Form

11.1 Introduction

It is geometrically evident that a sufficiently smooth curve which cross the x-axis at both ends of an interval $[a, b]$ must have a turning point somewhere between a and b. This is exactly what we mean by the well-known Rolle's Theorem from which Mean Value Theorems follow as Corollaries.

11.2 Rolle's Theorem

STATEMENT. *Let a real-valued function f be defined on a closed interval $[a, b]$. Suppose, further, that*

1. *f is continuous on the closed interval $[a, b]$;*
2. *f is derivable in the open interval (a, b) and*
3. *$f(a) = f(b)$.*

Then, Rolle's theorem states that $\exists$ at least one value α, when $a < \alpha < b$, such that $f'(\alpha) = 0$.

Proof. Since f is continous on a closed interval $[a, b]$ it must be bounded there and the bounds $M(\text{lub})$ and $m(\text{glb})$ must be attained in $[a, b]$ i.e. $\exists$ two points α and β of $[a, b]$ such that

$$f(\alpha) = M \quad \text{and} \quad f(\beta) = m.$$

Now either $M = m$ or $M \neq m$.

1. The case $M = m$ is trivial. Four, in, that case,
$$f(x) = M, \quad \forall x \text{ in } [a, b].$$

Then $f(x)$ must be constant function in $[a, b]$ and hence $f'(x) = 0$ for each point of $[a, b]$ and the truth of the theorem in assured.

2. Next, suppose $M \neq m$.

Then since $f(a) = f(b)$ and $m \neq M$, either M or m, if not both, must be distinct from $f(a)$ or $f(b)$. Suppose $M \neq f(a)$.

Then
$$\left. \begin{array}{l} f(\alpha) = M \neq f(a) \Rightarrow \alpha \neq a. \\ \text{and } f(\alpha) = M \neq f(b) \Rightarrow \alpha \neq b. \end{array} \right\} \Rightarrow \alpha \text{ is neither } a \text{ nor } b.$$

$$\therefore a < \alpha < b.$$

By hypothesis $f(x)$ is derivable in the open interval (a, b). So, $f'(\alpha)$ exists. We shall prove: $f'(\alpha) = 0$.

If not, then

either $f'(\alpha)$ is a finite positive number

or, $f'(\alpha)$ is a finite negative number.

If $f'(\alpha)$ were a finite positive number then $f(x)$ would be increasing then there would exist an open interval $(\alpha, \alpha + \delta)$ at every point x of which
$$f(x) > f(\alpha) = M,$$
and so there would be values of x in $a < x < b$ for which
$$f(x) > M.$$

This would then contradict the definition of M.

Hence we cannot assume $f'(\alpha) =$ a finite positive number.

On the other hand, if $f'(\alpha)$ were a finite negative number, there would be an open interval $(\alpha - \delta_1, \alpha)$ at every point x of which $f(x) > f(\alpha)$. Exactly as before, we would face a contradiction. *Hence $f'(\alpha)$ is not a finite negative number.*

$\therefore$ the only conclusion is $f'(\alpha) = 0$.

Thus, $\exists$ at least one $x = \alpha$ $(a < \alpha < b)$ s.t. $f'(\alpha) = 0$.

CHAPTER 11: ROLLE'S THEOREM: MEAN VALUE THEOREM

Note 11.2.1. *The statement "f is derivable in (a,b)" means that at any point $x \in (a,b)$, **right-hand** and **left-hand** derivatives exist and that they are equal. It may be that at some $x \in (a,b)$,*

$$f'(x+0) = f'(x-0) = +\infty$$

or it may be $f'(x+0) = f'(x-0) = -\infty$, still the theorem holds.

Alternative Proof of $f'(\alpha) = 0$.

We shall use the fact

$$f'(\alpha) \text{ exists} \Rightarrow Rf'(\alpha) = Lf'(\alpha) = \text{a finite number or } \pm\infty.$$

Since $f(\alpha) = M$ (the greatest value of $f(x)$ in $[a,b]$).

$$\therefore \ f(\alpha + h) - f(\alpha) \leq 0, \ \forall \ (\alpha + h) \in [a,b].$$

If $h > 0$, then

$$\frac{f(\alpha + h) - f(\alpha)}{h} \leq 0$$

but if $h < 0$, then

$$\frac{f(\alpha + h) - f(\alpha)}{h} \geq 0.$$

Taking limits as $h \to 0^+$ and $h \to 0^-$ in the two cases we get

$$Rf'(\alpha) \leq 0 \quad \text{and} \quad Lf'(\alpha) \geq 0.$$

But $Rf'(\alpha) = Lf'(\alpha) = f'(\alpha)$. Hence the only possibility is $f'(\alpha) = 0$.

Corollary 11.2.1. *If a, b are two roots of the equation $f(x) = 0$, then the equation $f'(x) = 0$ will have at least one root between a and b, provided*

1. *$f(x)$ is continuous in $a \leq x \leq b$; and*
2. *$f'(x)$ exists in $a < x < b$.*

If $f(x)$ be a *polynomial*, the conditions (1) and (2) are evidently satisfied. Hence:

Between any two zeros of a polynomial $f(x)$ there lies at least one zero of the polynomial $f'(x)$. (Rolle's Theorem for Polynomials)

Note: $x = a$ is a zero of a polynomial $f(x)$ if $f(a) = 0$, i.e., a is a root of the polynomial equation $f(x) = 0$.

Geometric Interpretation

If the graph of $y = f(x)$ has the ordinates at two points A, B equal, and if the graph be continuous throughout the interval from A to B and if the curve has a tangent at every point on it from A to B except possibly at the two extreme points A and B, then there must exist at least one point on the curve intervening between A and B where the tangent is parallel to the x-axis.

In this geometrical language, the theorem is almost self-evident. Compare the Figs. 11.1 and 11.2.

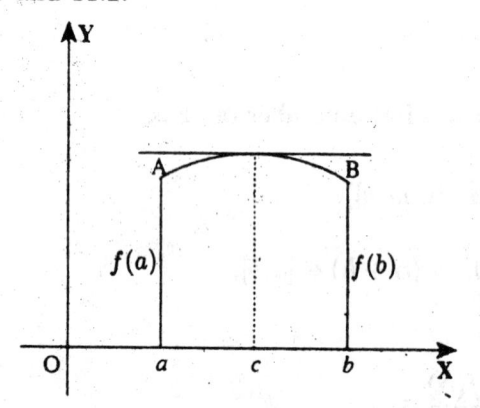

Fig. 11.1

Fig. 11.2

Observations:

1. The three conditions of Rolle's theorem are a set of *sufficient conditions*. They are by no means *necessary*. The following illustrations will make the point clear.

   EXAMPLE 1. $f(x) = x\sqrt{a^2 - x^2}$ in $[0, a]$.
   Here (a) $f(x)$ is continuous in $0 \leq x \leq a$;

   (b) $f'(x) = \dfrac{a^2 - 2x^2}{\sqrt{a^2 - x^2}}$ exists in $0 < x < a$;

   (c) $f(0) = f(a) = 0$.

   All the conditions of Rolle's theorem are satisfied and as such there exists c where $f'(c) = 0$ namely, $c = a/\sqrt{2}$ where $0 < c < a$.

   EXAMPLE 2. $f(x) = |x|$ in $[-1, 1]$.
   Here (a) $f(x)$ is continuous in $-1 \leq x \leq 1$;

   (b) $f'(x) = 1$ in $0 < x \leq 1$; $= -1$ in $-1 \leq x < 0$; $f'(0)$ does not exist.

   $\therefore f'(x)$ does not exist for any $x \in (-1, 1)$.

   (c) $f(1) = f(-1) = 1$.

   Note that $f'(x)$ nowhere vanishes in $[-1, 1]$ and as such Rolle's theorem fails. The failure is explained by the fact that $|x|$ is not derivable in $-1 < x < 1$, all other conditions being satisfied.

EXAMPLE 3. $f(x) = \sin(1/x)$ is discontinuous at $x = 0$ in the interval $[-1, 2]$ **but** still $f'(x) = 0$ for an infinite number of points in $[-1, 2]$, namely, $x = 2/(2n+1)\pi$. where $\cos(1/x) = 0$.

EXAMPLE 4. $f(x) = \dfrac{1}{x} + \dfrac{1}{1-x}$ in $[0, 1]$.

Here (a) $f(x)$ is continuous in $0 < x < 1$ (not in $0 \leq x \leq 1$);

(b) $f'(x) = \dfrac{1}{(1-x)^2} - \dfrac{1}{x^2}$ exists in $0 < x < 1$.

(c) $f(0) \neq f(1)$, both being undefined.

Thus, all the conditions of Rolle's theorem do not hold. But yet there exists a c where $f'(c) = 0$ (namely $c = \frac{1}{2}$) where $0 < c < 1$.

The above illustrations lead to the following conclusion:

> If $f(x)$ satisfies all the conditions of Rolle's theorem in $[a, b]$ then the conclusion $f'(c) = 0$ where $a < c < b$ is assured, but if any of the conditions are violated then Rolle's theorem will not be necessarily true; it may still be true but the truth is not ensured.

Note: The conditions of Rolle's theorem are sufficient but not necessary.

EXAMPLE 5. $f(x) = [x], x \in [0, 2]$.

No conditions of Rolle's theorem is satisfied viz. at $x = 1$, $f(x)$ is not continuous, not differentiable so that $f(x)$ is not continuous in $[0, 2]$ and not differentiable in $(0, 2)$ and last of all $f(0) \neq f(2)$ as $f(0) = 0, f(2) = 2$. But $f'(x) = 0$ for infinitely many points in $(0, 2)$ viz. $f'(x) = 0$ for $x \in (0, 1) \cup (1, 2)$ i.e., $f'(x) = 0$ at all points of $(0, 2)$ except $x = 1$.

The following fig. may be helpful to understand visually.

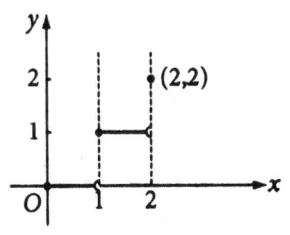

Fig. 11.3

2. The theorem asserts the existence of **at least one value** c, where $f'(c) = 0$ and $a < c < b$. There may be several such values where the derivatives will vanish. e.g.,

(a) $f(x) = \sin x$; $a = \frac{1}{2}\pi$, $b = (100)\frac{1}{2}\pi$;

(b) $f(x) = 0$; $a \leq x \leq b$.

In each case $f'(x) = 0$ for several values in (a, b).

Again, according to the theorem, *c is strictly within a and b*. In special cases (e.g., (b) above) a or b themselves may be points where $f'(x) = 0$ but then also there will be other points between a and b where $f'(x)$ should vanish, if Rolle's theorem is to remain valid.

11.3 Mean-Value Theorem: Lagrange's Form[1]

STATEMENT. *If a function f is*

(a) *continuous in the closed interval* $[a,b]$;

(b) *derivable in the open interval* (a,b);

then there exists at least one value of x, say c in (a,b) such that

$$\frac{f(b) - f(a)}{b - a} = f'(c).$$

Proof. If $f(a) = f(b)$, the theorem reduces to Rolle's theorem. We, therefore, assume $f(a) \neq f(b)$ and construct a function ϕ s.t.

$$\phi(x) = f(x) + Ax \qquad (11.3.1)$$

and choose the constant A in such a manner that $\phi(a) = \phi(b)$. Obviously then

$$f(a) + Aa = f(b) + Ab$$
$$\text{or,} \quad A = -\frac{f(b) - f(a)}{b - a}.$$

Since $f(x)$ is continuous in $[a,b]$, derivable in (a,b) and Ax is always continuous and derivable we conclude that $\phi(x)$ is continuous in $[a,b]$, derivable in (a,b) and $\phi(a) = \phi(b)$.

Thus $\phi(x)$ satisfies *all the three* conditions of Rolle's theorem in $[a,b]$ and hence ∃ at least one value c, in (a,b) where

$$\phi'(c) = 0.$$

From (11.3.1), differentiating w.r.t. x we get

$$\phi'(x) = f'(x) + A. \qquad (11.3.2)$$

---

[1] Also known as '*Formula of finite increment of Lagrange or Average Value Theorem*' or *The Law of Mean*.

Then from (11.3.2),

$$\phi'(c) = f'(c) + A = 0 \quad \text{for } a < c < b.$$
i.e., $\quad -A = f'(c) \quad \text{for } a < c < b.$
i.e., $\quad \dfrac{f(b) - f(a)}{b - a} = f'(c) \quad \text{for } a < c < b.$

Hence the theorem.

Observations:

1. The whole point of the proof is to construct a function $\phi(x)$ which will satisfy the hypothesis of Rolle's Theorem. One could proceed by constructing

$$\phi(x) = f(x) - f(a) - A(x - a)$$

where A is a constant so chosen that $\phi(b) = 0$. See that $\phi(a)$ is clearly $= 0$. The proof of the Mean-Value Theorem (MVT) can then be carried out as above.

2. **h-θ form of MVT**: When $b = a + h$, a number lying between a and b may be written as $a + \theta h$ $(0 < \theta < 1)$. Then the theorem takes the following form:

   Let f be continuous in the closed interval $[a, a + h]$ and have a derivative $f'(x)$ whenever $a < x < a + h$. Then $\exists$ a number θ, between 0 and 1 for which

$$f(a + h) - f(a) = h f'(a + \theta h) \quad (0 < \theta < 1).$$

   The students are advised to construct a function

$$\phi(x) = f(a + x) - f(a) - Ax,$$

   where A is a constant so chosen that $\phi(h) = 0$ (then, of course $\phi(0) = \phi(h) = 0$) and the proof can be carried out independently.

   REMEMBER: h-θ form of MVT is usually referred to as **Lagrange's MVT.**

3. Putting $h = x$, $a = 0$, we have the MVT in the form:

$$f(x) - f(0) = x f'(\theta x) \quad (0 < \theta < 1).$$

   This form is due to **Maclaurin** (*Maclaurin's formula* as it is often called)

4. The two hypothesis of MVT are only a set of **sufficient conditions** and by no means necessary for the conclusion of the theorem.

   Take, for instance, $f(x) = 1/|x|$, $a = -1$, $b \geq 1$.

   The conditions of the theorem are not satisfied in $[-1, 1]$ (Why?). But see that the conclusion is true iff $b > 1 + \sqrt{2}$. [see worked Example No. 11.5.9 of art. 11.5]

Geometric Interpretation of MVT

Let $y = f(x)$ be represented by the curve AB, in Fig. 11.3, then

$$\frac{f(b) - f(a)}{b - a} = \frac{BL}{AL} = \tan BAL.$$

Again from art. 9.6, it is clear that $f'(\xi)$ is the trigonometrical tangent of the angle, which the tangent at $P = (\xi, f(\xi))$ makes with OX.

The theorem $\frac{f(b) - f(a)}{b - a} = f'(\xi)$, therefore states that at some point $P(x = \xi)$ of the curve between $A(x = a)$ and $B(x = b)$, the tangent to the curve is parallel to the chord AB.

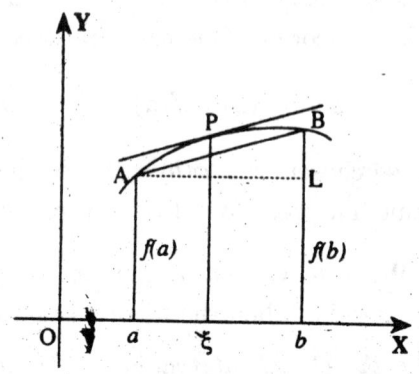

Fig. 11.4

REMEMBER: The fraction $\frac{f(b) - f(a)}{b - a}$ measures the mean (or average) rate of increase of the function in the interval of length $b - a$. Hence the theorem expresses the fact that, under the conditions stated, the mean rate of increase in any interval is equal to the actual rate of increase at some point within the interval. For instance, the mean velocity of a moving point in any interval of time is equal to the actual velocity at some instant within the interval. This justifies the name **Mean-Value Theorem**.

Important Deductions from MVT

(a) *Suppose that f is continuous on the closed interval $[a, b]$, that f is derivable in the open interval (a, b) and that $f'(x) = 0$ for every $x \in (a, b)$. Then f is constant on $[a, b]$.* In fact, we shall prove that $f(x) = f(a)$, $\forall x \in [a, b]$.

Proof: Let x_1 be any number such that $a < x_1 \leq b$.

Then $f(x)$ is continuous throughout the interval $[a, x_1]$.

Also $f'(x)$ exists $\forall x$ in $a < x < x_1$.

CHAPTER 11: ROLLE'S THEOREM: MEAN VALUE THEOREM

Thus f satisfies all the conditions of the MVT in the interval $[a, x_1]$ and hence

$$f(x_1) - f(a) = (x_1 - a)f'(c),$$

for some c satisfying $a < c < x_1$.

But, by hypothesis $f'(c) = 0$; therefore $f(x_1) = f(a)$ and since $a < x_1 \leq b$, we see that $f(x) = f(a)$ for every x in $a \leq x \leq b$.

(b) *If F and G are both continuous on $[a, b]$, that they are both derivable on (a, b) and that $F'(x) = G'(x)$ $\forall\, x \in (a, b)$ then $F(x)$ and $G(x)$ differ by a constant in this interval i.e., $F(x) = G(x) + C$ on $[a, b]$.*

Proof: The result follows from above because according to the given condition, $\forall\, x$ in (a, b)

$$\frac{d}{dx}\{F(x) - G(x)\} = \frac{d}{dx}F(x) - \frac{d}{dx}G(x) = 0$$

$\therefore F(x) - G(x) = F(a) - G(a) = $ constant, for every $x \in [a, b]$.

Note 11.3.1. *This result is familiar in the symbols of integration:*

If $\dfrac{d}{dx}F(x) = f(x)$, then $\int f(x)\,dx = F(x) + C$.

i.e., if $F(x)$ is an integral of $f(x)$, then another integral (in fact, the only other integral) is $F(x) + C$.

(c) *MVT is very useful in estimating the size of functions and in making numerical approximations:*

EXAMPLE. Let us estimate $\sqrt[3]{28}$.

Take $f(x) = \sqrt[3]{x}$, so that $f'(x) = \dfrac{1}{3}(x)^{-2/3}$ in $[27, 28]$

By MVT, $f(28) = f(27) + (28 - 27)f'(x_0)$, where $27 < x_0 < 28$

i.e., $\sqrt[3]{28} = \sqrt[3]{27} + \dfrac{1}{3}(x_0)^{-2/3} < 3 + \dfrac{1}{3(27)^{2/3}} = 3 + \dfrac{1}{27}$

i.e., $\sqrt[3]{28} < 3\dfrac{1}{27}$.

As another example, let us consider $f(x) = \text{arc tan } x$ or $\tan^{-1} x$ near $x = 0$. By the MVT in $[0, x]$,

$$\text{arc tan } x = \text{arc tan } 0 + x \cdot \frac{1}{1 + x_0^2} = \frac{x}{1 + x_0^2},$$

where x_0 is between 0 and x. Hence for $x > 0$,

$$\frac{x}{1 + x^2} < \text{arc tan } x < x, \quad \text{i.e.,} \quad \frac{x}{1 + x^2} < \tan^{-1} x < x,$$

where x_0 is replaced by 0 on the right and by x on the left.

(d) *We recall the following definitions:*
Let a functions $f : I \to \mathbb{R}$ be defined over an interval I.

(i) f *is said to be monotone increasing or simply, increasing on the interval* I *if whenever* $x_1, x_2 \in I$ *satisfy* $x_1 < x_2$, *then* $f(x_1) \leq f(x_2)$; f *is strictly increasing on* I *when* $x_1 < x_2 \Rightarrow f(x_1) < f(x_2)$.

(ii) f *is said to be increasing at a given point* $c \in I$ *if* $\exists$ *a neighbourhood (nbd) of* c *where* f *is increasing on this interval.*

(iii) f *is said to be decreasing, strictly decreasing on an interval or* f *is decreasing at a point* $c \in I$, *if* $-f$ *is increasing, strictly increasing on the interval* I *or* $-f$ *is increasing at* $c \in I$.

We have then the following three **theorems: 11.3.1–11.3.3**.

Theorem 11.3.1. *Let* $f : I \to \mathbb{R}$ *be a derivable function on the interval* I. *Then* (i) f *is increasing on* I *if and only if* $f'(x) \geq 0$, $\forall x \in I$, (ii) f *is decreasing on* I *if and only if* $f'(x) \leq 0$, $\forall x \in I$.

Proof. (i) Suppose $f'(x) \geq 0$, $\forall x \in I$.

If $x_1, x_2 \in I$ where $x_1 < x_2$ we can apply MVT in $[x_1, x_2]$ and obtain a point $c \in (x_1, x_2)$ such that
$$f(x_2) - f(x_1) = (x_2 - x_1)f'(c)$$
since $f'(c) \geq 0$ and $x_2 - x_1 > 0$, it follows that $f(x_2) - f(x_1) \geq 0$ i.e., $f(x_1) \leq f(x_2)$ and since x_1, x_2 are two arbitrary points in I, we conclude that f is *increasing* on I.

Conversely, suppose that f is derivable and increasing on I.

For any point c on I, if either $x > c$ or $x < c$ for $x \in I$, $\dfrac{f(x) - f(c)}{x - c} \geq 0$. Taking limit as $x \to c$, $f'(c) \geq 0$.

The proof of (ii) is exactly similar.

Theorem 11.3.2. *If* f *is continuous on* $[a, b]$ *and* $f'(x) > 0$ *in* (a, b) *then* f *is strictly increasing in* $[a, b]$.

Proof. Let x_1 and x_2 be two points so chosen that
$$a \leq x_1 < x_2 \leq b.$$

Since $f'(x)$ exists in (a, b) and $f(x)$ is continuous in $[a, b]$, we can apply MVT in $[x_1, x_2]$ and obtain a point c where $x_1 < c < x_2$ such that $f(x_2) - f(x_1) = (x_2 - x_1)f'(c)$.

Since $f'(c) > 0$ by hypothesis and $x_2 - x_1 > 0$ we see that $f(x_2) - f(x_1) > 0$ i.e., $f(x_1) < f(x_2)$ and the function is **strictly increasing** in $[x_1, x_2]$ and hence in $[a, b]$.

See that we may take $x_1 = a$, $x_2 = b$ but $x_1 < x_2$.

A similar result is:

Theorem 11.3.3. *If f is continuous in $a \leq x \leq b$ and $f'(x) < 0$ in $a < x < b$ then f is strictly decreasing in $a \leq x \leq b$.*

(*Proof is similar and is left out*)

What about the converses of Theorems 11.3.2 and 11.3.3?

The converses may not be true. A strictly increasing (or decreasing) derivable function may have a derivative that vanishes at certain points [i.e., derivatives may not be strictly positive (or negative)].

Let $f(x) = x^3$. This function is strictly increasing on any interval of $\mathbb{R}$ but $\exists\, x = 0$, where $f'(0) = 0$.

Observations:

1. The function f is such that the derivative is **strictly positive at a point** c of an interval I where f is defined. The function may not be **increasing at the point** c

$$\text{e.g., } f(x) = \begin{cases} x + 2x^2 \sin \dfrac{1}{x}, & x \neq 0; \\ 0, & x = 0. \end{cases}$$

See that $f'(0) = 1$ strictly positive but it can be shown that f is not increasing in any *nbd* of $x = 0$.

(In every *nbd* of 0 $f'(x)$ takes on both positive and negative values and f is not monotonic in any *nbd* of 0).

2. If $f'(x) > 0$ in (a, b) and $f(a) \geq 0$ then $f(x)$ is positive throughout the open interval $a < x < b$ and if $f'(x) < 0$ in (a, b) and $f(a) < 0$ then $f(x)$ is negative throughout the open interval (a, b).

(e) *If $f'(x)$ exists for $a < x \leq b$ and $|f(x)| \to \infty$ as $x \to a^{+0}$, then*

$$|f'(x)| \to \infty \text{ as } x \to a^{+0}.$$

Proof: From the MVT

$$|f(b) - f(x)| = (b - x)|f'(c)|, \quad a < x < c < b.$$

If $f'(x)$ were bounded in $a < x \leq b$, say $|f'(x)| < M$, we would have

$$|f(b) - f(x)| < (b-a)M,$$

and $f(x)$ would remain bounded as $x \to a^{+0}$. But this contradicts the hypothesis; hence the result.

(f) An Important Result: *If f is continuous at c, and if $\lim_{x \to c} f'(x) = l$ then f is derivable at c and l is the derivative there*, i.e., $f'(c) = l$.

Proof: The condition that $f'(x) \to l$ as $x \to c^{+0}$ implies that $\exists$ an interval $c < x \leq c + h$ ($h > 0$) for every point x of which f' exists and consequently f is continuous there. Since it is given that f is continuous at c also, we see that f is continuous in the closed interval $[c, c+h]$.

It x be a point of this interval, we have

$$f(x) - f(c) = (x-c)f'(t), \quad (c < t < x)$$

$$\Rightarrow \frac{f(x) - f(c)}{x - c} = f'(t).$$

Let $x \to c^{+0}$. Then we have

$$\lim_{x \to c^{+0}} \frac{f(x) - f(c)}{x - c} = \lim_{x \to c^{+0}} f'(t) = \lim_{x \to c^{+0}} f'(x) = l.$$

Similarly, $\lim_{x \to c^{-0}} \frac{f(x) - f(c)}{x - c} = l.$

i.e., f is derivable at c and the derivative is l.

Note 11.3.2. *This theorem shows that the derived function f' cannot have a* **discontinuity of the first kind.**

11.4 Mean-Value Theorem: Cauchy's Form

If we apply Lagrange's mean-value theorem to two functions $f(x)$ and $g(x)$, each of which satisfies the conditions of the theorem in the interval $[a, b]$, we would obtain

$$f(b) - f(a) = (b-a)f'(c_1), \quad a < c_1 < b.$$

and $\quad g(b) - g(a) = (b-a)g'(c_2), \quad a < c_2 < b.$

Dividing one by the other

$$\frac{f(b) - f(a)}{g(b) - g(a)} = \frac{f'(c_1)}{g'(c_2)}, \quad c_1, c_2 \text{ being, in general, different.}$$

CHAPTER 11: ROLLE'S THEOREM: MEAN VALUE THEOREM

Cauchy takes a step further to make $c_1 = c_2$ and establishes a theorem, which we are going to study now:

Cauchy's Mean-Value Theorem. *It two functions f and g*

(a) *be both continuous in $a \leq x \leq b$,*

(b) *are both derivable in $a < x < b$,*

(c) *$g'(x)$ does not vanish at any value of x in $a < x < b$,*

then there exists at least one value of x, say c, such that

$$\frac{f(b) - f(a)}{g(b) - g(a)} = \frac{f'(c)}{g'(c)} \quad \text{for } a < c < b.$$

Proof: Construct the function

$$\phi(x) = f(x) + Ag(x)$$

and choose A in such a manner that $\phi(a) = \phi(b)$. Obviously then

$$A = -\frac{f(b) - f(a)}{g(b) - g(a)} \quad [g(b) \neq g(a)].$$

We observe that A has a definite finite value, in view of $\{g(b) - g(a)\} \neq 0$, since if it were zero, $g(x)$ would satisfy all the conditions of Rolle's theorem and its derivative would, therefore, vanish at least once in $a < x < b$, contradicting given hypothesis (iii), that $g'(x) \neq 0$ there.

We now see that

(a) ϕ is continuous in $[a, b]$ ($\because$ f and g are both continuous in $[a, b]$.)

(b) ϕ' exists for every x in (a, b) [$\because$ f and g are both derivable in (a, b).]

(c) $\phi(a) = \phi(b)$, by construction.

$\therefore$ ϕ satisfies all the conditions of Rolle's Theorem and hence $\exists$ at least one c, where $a < c < b$ such that $\phi'(c) = 0$,

i.e., $f'(c) + Ag'(c) = 0$

i.e., $\dfrac{f'(c)}{g'(c)} = -A = \dfrac{f(b) - f(a)}{g(b) - g(a)}.$

Note 11.4.1. *Condition (c) given in the statement of the theorem may be replaced by*

(a) $f'(x)$ and $g'(x)$ do not both vanish at any value of x in the open interval (a, b); and

(b) $g(a) \neq g(b)$.

The students should verify that these two conditions and $g'(x) \neq 0$ for any x in (a, b) are equivalent.

h-θ Form: If $b = a + h$, then $\exists$ at least one θ such that

$$\frac{f(a+h) - f(a)}{g(a+h) - g(a)} = \frac{f'(a+\theta h)}{g'(a+\theta h)}, \quad \text{where } 0 < \theta < 1.$$

Geometric interpretation of Cauchy's MVT:

The functions f and g can be considered as determining a curve in the plane by means of parametric equations $x = f(t)$, $y = g(t)$, where $a \leq t \leq b$. Cauchy's MVT concludes that $\exists$ a point $(f(c), g(c))$ of the curve for some $t = c$ in (a, b) such that the slope of the line segment joining the end points of the curve = slope $\dfrac{g'(c)}{f'(c)}$ of the tangent to the curve at $t = c$.

Corollary 11.4.1. If $f(a) = g(a) = 0$, Cauchy's result would reduce to

$$\frac{f(b)}{g(b)} = \frac{f'(c)}{g'(c)}, \quad a < c < b,$$

which may be put in this form.

If $f(x) = 0$, $g(x) = 0$ have a common root $x = a$, then

$$\frac{f(x)}{g(x)} = \frac{f'(c)}{g'(c)} \text{ for some } c \in (a, x).$$

Note 11.4.2. *When $g(x) = x$, Lagrange's Mean-Value theorem becomes a particular case of Cauchy's Mean-Value theorem.*

It is to be observed that Rolle's Theorem can be obtained from Cauchy's MVT by letting $g(x) = x$ and $f(a) = f(b)$. However, we could not easily prove this theorem without using Rolle's Theorem. Here is a case where we have to use a special case to prove a general result.

We remind our readers that the conditions of the present theorem are also *sufficient* but *not necessary*.

11.5 Illustrative Examples

Example 11.5.1. *Are the conditions of Rolle's Theorem satisfied in the case of the following functions defined over given intervals:*

(a) $f(x) = |x|$ in $[-1, 1]$; (b) $g(x) = x^2$ in $2 \leq x \leq 3$;
(c) $h(x) = \cos \dfrac{1}{x}$ in $[-1, 1]$; (d) $\phi(x) = \tan x$ in $0 \leq x \leq \pi$.

Solution: (a) $f(-1) = f(1) = 1$. The function is continuous throughout the closed interval $[-1, 1]$ but it is not derivable at $x = 0$ which is an interior point of $[-1, 1]$.

See that $\lim\limits_{x \to 0^-} f(x) = \lim -x = 0$ and $\lim\limits_{x \to 0^+} f(x) = \lim\limits_{x \to 0^+} x = 0$ and $f(0) = 0$; hence $f(x)$ is continuous at $x = 0$ but f is not derivable at $x = 0$; for

$$\left. \begin{array}{l} \lim\limits_{x \to 0^-} \dfrac{f(x) - f(0)}{x - 0} = \lim\limits_{x \to 0^-} \dfrac{-x}{x} = -1 \\ \text{and } \lim\limits_{x \to 0^+} \dfrac{f(x) - f(0)}{x - 0} = \lim\limits_{x \to 0^+} \dfrac{x}{x} = +1 \end{array} \right\} Rf'(0) = 1 \neq Lf'(0) = -1.$$

Hence condition (b) of Rolle's Theorem is not satisfied.

(b) $g(x) = x^2$ is always continuous and derivable. In particular it is continuous and derivable in $[2, 3]$. i.e., the first two conditions of Rolle's theorem are satisfied. But the third condition is not satisfied: since $f(2) = 4$ and $f(3) = 9$ so that $f(2) \neq f(3)$.

(c) Here $h(-1) = h(1) = \cos 1$.

But the function $h(x) = \cos(1/x)$ is discontinuous at $x = 0$ $\left[\lim\limits_{x \to 0} \cos(1/x) \text{ does not exist} \right]$ and as such it cannot be derivable at $x = 0$. Hence first two conditions of Rolle's theorem are not satisfied whereas the third condition is satisfied.

(d) Here $\phi(0) = \tan 0 = 0$; $\phi(\pi) = \tan \pi = 0$ so that $f(0) = f(\pi)$. But the function is not continuous at $x = \pi/2$ and consequently not derivable there. Hence the first two conditions are not satisfied in the interval $[0, \pi]$.

Example 11.5.2. *Verify Rolle's theorem for $f(x) = 2x^3 + x^2 - 4x - 2$.*

Solution: Since every polynomial in x is continuous and derivable for every real value of x. Hence first two conditions of Rolle's theorem are satisfied in any interval.

Now we see that $f(x) = 0$ gives $2x^3 + x^2 - 4x - 2 = 0$

or, $(x^2 - 2)(2x + 1) = 0$ or, $x = -\dfrac{1}{2}, -\sqrt{2}, +\sqrt{2}$.

We take the interval $[-\sqrt{2}, \sqrt{2}]$ so that in this interval all the three conditions of Rolle's theorem are satisfied.

Now $f'(x) = 6x^2 + 2x - 4 = 0$ if $(3x - 2)(x + 1) = 0$

i.e., if $x = -1$ or if $x = \dfrac{2}{3}$ i.e., $f'(-1) = f'\left(\dfrac{2}{3}\right) = 0$.

Since both the points $x = -1$ and $x = 2/3$ lie in the open interval $]-\sqrt{2}, \sqrt{2}[$, Rolle's theorem is verified.

Example 11.5.3. (Rolle's theorem for polynomials). *If $f(x)$ be any polynomial then between any pair of roots of $f(x) = 0$ lies a root of $f'(x) = 0$.*

Solution: In art. **11.2** we have proved Rolle's theorem for general type of functions. We give here an algebraical proof of this important theorem valid for polynomials only.

Let α and β be two consecutive roots, repeated respectively m and n times, so that

$$f(x) = (x - \alpha)^m (x - \beta)^n \phi(x),$$

where $\phi(x)$ is a polynomial which has the same sign, *say the positive sign*, for $\alpha \leq x \leq \beta$ (since α and β are two consecutive roots of $f(x) = 0$, there cannot be any root of $\phi(x) = 0$ between them).

Now

$$\begin{aligned} f'(x) &= (x-\alpha)^m (x-\beta)^n \phi'(x) \\ &\quad + \{m(x-\alpha)^{m-1}(x-\beta)^n + n(x-\beta)^{n-1}(x-\alpha)^m\}\phi(x) \\ &= (x-\alpha)^{m-1}(x-\beta)^{n-1}\psi(x), \end{aligned}$$

where $\psi(x) = m\phi(x)(x-\beta) + n(x-\alpha)\phi(x) + (x-\alpha)(x-\beta)\phi'(x)$.

Observe that $\psi(\alpha) = m(\alpha - \beta)\phi(\alpha)$ and $\psi(\beta) = -n(\alpha - \beta)\phi(\beta)$ have opposite signs. Hence $\psi(x)$ must vanish for some value of x between α and β.

$\therefore$ $f'(x)$ must vanish for some value of x between α and β.

Hence the theorem.

Example 11.5.4. If $f(x) = \tan x$, then $f(x)$ vanishes for $x = 0$ and $x = \pi$. Is Rolle's theorem applicable to the function $f(x)$ in $[0, \pi]$? Give justifications of your answer.

SOLUTION. Two consecutive roots of $f(x) = 0$ are $x = 0$ and $x = \pi$. But see that $f'(x) = \sec^2 x$ does not vanish for any value of x between 0 and π. Hence Rolle's theorem is not applicable.

Note that $f'(x) = \sec^2 x$ exists for every $x \in [0, \pi]$ excepting at $x = \pi/2$ and further $f(x)$ is continuous in $[0, \pi]$ excepting at $x = \pi/2$. Hence the conditions under which Rolle's theorem is valid do not hold in the present case and this explains the failure of the theorem.

CHAPTER 11: ROLLE'S THEOREM: MEAN VALUE THEOREM

Example 11.5.5. *If $f(x) = (x-a)^m(x-b)^n$ where m and n are positive integers, show that c in Rolle's theorem divides the segment $a \le x \le b$ in the ratio $m:n$.*

Solution: $f(x) = (x-a)^m(x-b)^n$ is continuous in $a \le x \le b$;

$$f'(x) = (x-a)^{m-1}(x-b)^{n-1}\{m(x-b) + n(x-a)\} \text{ exists in } a < x < b.$$

Also $f(a) = f(b) = 0$. Hence by Rolle's theorem, $\exists\, c$ s.t.

$$f'(c) = (c-a)^{m-1}(c-b)^{n-1}\{m(c-b) + n(c-a)\} = 0 \text{ for } a < c < b.$$

Hence,
$$m(c-b) = n(a-c),$$
$$\text{or,} \quad c = (bm + an)/(m+n),$$

i.e., c divides the segment $a \le x \le b$ in the ratio $m:n$.

Example 11.5.6. *Is mean-value theorem valid for $f(x) = x^2 + 3x + 2$ in $1 \le x \le 2$? Find c, if the theorem be applicable.*

Solution: Here $f'(x) = 2x + 3$ exists in $1 < x < 2$; $f(x)$ is continuous in $1 \le x \le 2$.

Hence MVT is applicable. Using the theorem,

$$f'(c) = \frac{f(2) - f(1)}{2 - 1} = 12 - 6 = 6, \text{ i.e., } 2c + 3 = 6 \text{ or, } c = \frac{3}{2}.$$

Example 11.5.7. *If in the Cauchy's MVT we take $f(x) = e^x$ and $g(x) = e^{-x}$, then prove that c is the arithmetic mean between a and b. (Refer to the statement of Cauchy's MVT).*

Solution:

$$\frac{f(b) - f(a)}{g(b) - g(a)} = \frac{e^b - e^a}{e^{-b} - e^{-a}} = \frac{e^b - e^a}{e^a - e^b} \times e^b e^a = -e^{a+b}.$$

$$\frac{f'(c)}{g'(c)} = \frac{e^c}{-e^{-c}} = -e^{2c}.$$

∴ From Cauchy's MVT it follows:

$$-e^{a+b} = -e^{2c}$$
$$\text{or,} \quad \frac{a+b}{2} = c; \text{ hence etc.}$$

Note 11.5.1. *Similarly, it is easy to check:*

(a) *if $f(x) = \sqrt{x}$ and $g(x) = 1/\sqrt{x}$, then $c =$ geometric mean between a and b.*

(b) *if $f(x) = 1/x^2$ and $g(x) = 1/x$, then $c =$ harmonic mean between a and b.*

Example 11.5.8. *Let f be a real-valued function defined over $[-1, 1]$ such that*

$$f(x) = \begin{cases} x \sin \dfrac{1}{x}, & \text{when } x \neq 0; \\ 0, & \text{when } x = 0. \end{cases}$$

Does the Mean-value Theorem hold for f in $[-1, 1]$?

Solution: The function f is continuous in $[-1, 1]$ but not derivable in $(-1, 1)$ (in fact, f has no derivative at $x = 0$).

∴ MVT is not applicable for f in $[-1, 1]$.

[The students should justify that f is continuous in $[-1, 1]$ but that f is not derivable at $x = 0$].

Example 11.5.9. *If $a = -1, b \geq 1$ and $f(x) = 1/|x|$, prove that Lagrange's MVT is not applicable for f in $[a, b]$. But check that the conclusion of the theorem is TRUE if $b > 1 + \sqrt{2}$.*

Solution: The function is not defined at $x = 0$. Even if we define $f(0) = A$, where A is some finite quantity, then also

$$f'(0+0) = \lim_{h \to 0+} \frac{1/|h| - A}{h} = \lim_{h \to 0+} \frac{1}{h} \left\{ \frac{1}{|h|} - A \right\} = \infty$$

and $$f'(0-0) = \lim_{h \to 0-} \frac{1/|h| - A}{h} = \lim_{h \to 0-} \frac{1}{h} \left\{ \frac{1}{|h|} - A \right\} = -\infty$$

i.e., f is not derivable at $x = 0$.

∴ The conditions of MVT are not satisfied in $[a, b]$ which always includes the origin. Now, the conclusion of the MVT is

$$\frac{f(b) - f(a)}{b - a} = f'(c), \quad a < c < b.$$

This will be true iff

$$\left(\frac{1}{|b|} - \frac{1}{|a|} \right) \Big/ (b - a) = \frac{d}{dx} \left\{ \frac{1}{|x|} \right\} \text{ at } x = c$$

$$= -\frac{1}{|c|^2}$$

CHAPTER 11: ROLLE'S THEOREM: MEAN VALUE THEOREM

or, $\dfrac{1}{b} - 1 = (b+1) \times -\dfrac{1}{|c|^2} = -\dfrac{(b+1)}{c^2}$ ($\because$ $b \geq 1$ and $a = -1$)

or, $\dfrac{1-b}{b} = -\dfrac{(b+1)}{c^2}$ i.e., $c^2 = \dfrac{b^2+b}{b-1}$; since $c^2 < b^2$, therefore $\dfrac{b^2+b}{b-1} < b^2$.

or, $b+1 < b^2 - b$ i.e., $b^2 - 2b - 1 > 0$

i.e., $(b-1)^2 > 2$, i.e., $b > 1 + \sqrt{2}$.

Under this condition ($b > 1 + \sqrt{2}$) the conclusion of MVT is true but the conditions for the validity of the theorem are not satisfied.

Example 11.5.10. *Show that* $x > \sin x$ *for* $0 < x < \dfrac{\pi}{2}$.

SOLUTION. Consider $f(x) = x - \sin x$. Then $f'(x) = 1 - \cos x > 0$ for $0 < x < \dfrac{\pi}{2}$. Also at $x = 0$, $f(x) = 0$. Thus $f(x) = x - \sin x$, being an increasing function, is positive. Hence $x - \sin x > 0$ or $x > \sin x$ in $0 < x < \dfrac{\pi}{2}$.

Example 11.5.11. *Show that*

$$\dfrac{x}{1+x} < \log(1+x) < x, \text{ for } \forall \, x > 0.$$ [C.H. 1992]

Solution: **First Method.**

Let $f(x) = \log(1+x) - \dfrac{x}{1+x}$.

$\therefore f'(x) = \dfrac{1}{1+x} - \dfrac{1}{1+x} + \dfrac{x}{(1+x)^2} = \dfrac{x}{(1+x)^2}$.

$\therefore f'(x) > 0$, for $x > 0$ and $f'(x) = 0$, for $x = 0$.

i.e., f is monotone increasing in $[0, \infty[$.

Again $f(0) = 0$; therefore; $f(x) > f(0) = 0$, when $x > 0$ i.e., $\log(1+x) - \dfrac{x}{1+x} > 0$, for $x > 0$.

Thus the first part of the inequality is proved.

Taking $\phi(x) = x - \log(1+x)$ we can prove the second part in a similar manner. Hence the result.

Second Method. (Using MVT).

By MVT applied on $f(x)$ is $[0, x]$ we have

$$f(x) = f(0) + x f'(\theta x), \quad 0 < \theta < 1$$

Take $f(x) = \log(1+x)$. Then $f(0) = 0$ and $f'(\theta x) = \dfrac{1}{1+\theta x}$.

Hence $\log(1+x) = \dfrac{x}{1+\theta x}$, $0 < \theta < 1$. Since $1 + \theta x > 1$, $\dfrac{x}{1+\theta x} < x$.

Since $1 + \theta x < 1 + x$, $\dfrac{x}{1+\theta x} > \dfrac{x}{1+x}$.

$$\therefore \quad \dfrac{x}{1+x} < \dfrac{x}{1+\theta x} = \log(1+x) < x.$$

Hence etc.

Example 11.5.12. An Important Illustration. *If $f(x)$ and $\phi(x)$ be continuous in $a \le x \le b$ and derivable in $a < x < b$, then*

$$\begin{vmatrix} f(a) & f(b) \\ \phi(a) & \phi(b) \end{vmatrix} = (b-a) \begin{vmatrix} f(a) & f'(c) \\ \phi(a) & \phi'(c) \end{vmatrix}, \quad a < c < b. \qquad \text{[C.H. 1968]}$$

Solution: Let us construct the function

$$F(x) = \begin{vmatrix} f(a) & f(x) \\ \phi(a) & \phi(x) \end{vmatrix}; \quad \text{then } F'(x) = \begin{vmatrix} f(a) & f'(x) \\ \phi(a) & \phi'(x) \end{vmatrix}.$$

We observe that $F(x)$ is continuous in $a \le x \le b$ and $F'(x)$ exists in $a < x < b$. Hence using MVT, we get

$$F(b) - F(a) = (b-a)F'(c) \text{ where } a < c < b.$$

$$\begin{vmatrix} f(a) & f(b) \\ \phi(a) & \phi(b) \end{vmatrix} = (b-a) \begin{vmatrix} f(a) & f'(c) \\ \phi(a) & \phi'(c) \end{vmatrix}. \quad (\because F(a) = 0)$$

Hence the result.

Note 11.5.2. *If $\phi(x) = 1$, we get $f(b) - f(a) = (b-a)f'(c)$: Lagrange's MVT.*

In a more general way, we have the following result.

Example 11.5.13. *If $f(x), \phi(x)$ and $\psi(x)$ satisfy the first two conditions of Rolle's theorem, viz., continuity in $a \le x \le b$ and derivability in $a < x < b$, then there is a value c between a and b, such that*

$$\begin{vmatrix} f(a) & \phi(a) & \psi(a) \\ f(b) & \phi(b) & \psi(b) \\ f'(c) & \phi'(c) & \psi'(c) \end{vmatrix} = 0.$$

Solution: Construct the function $F(x) = \begin{vmatrix} f(a) & \phi(a) & \psi(a) \\ f(b) & \phi(b) & \psi(b) \\ f(x) & \phi(x) & \psi(x) \end{vmatrix}$ and proceed as in Ex. 11.5.12. Here notice $F(a) = 0 = F(b)$ and so we can at once apply Rolle's theorem.

CHAPTER 11: ROLLE'S THEOREM: MEAN VALUE THEOREM

Note 11.5.3. 1. Put $\phi(x) = $ a constant k; we have the Cauchy's MVT.

2. Put $\psi(x) = x$, we get Lagrange's MVT.

3. Put $f(a) = f(b)$, the conclusion of Rolle's theorem is obtained.

Example 11.5.14. (An Important Theorem). *In any interval in which the functions $u(x), v(x), u'(x), v'(x)$ are continuous and $uv' - u'v \neq 0$, the roots of $u(x)$ and $v(x)$ separate each other. Verify this theorem when $u = \sin x, v = \cos x$.* [C.H. 1992]

Proof: Let α and β be two consecutive roots of $u(x) = 0$ ($\alpha < \beta$). We shall prove that $\exists$ one and only one root of $v(x) = 0$ in $[\alpha, \beta]$. If possible, let $v(x) \neq 0$ when $\alpha < x < \beta$, then the function (u/v) is continuous in $[\alpha, \beta]$. Also,

$$\left(\frac{u}{v}\right)' = \frac{u'v - uv'}{(v)^2} = -\frac{uv' - u'v}{(v)^2}.$$

Thus, (u/v) is derivable in (α, β) and $u/v = 0$ for $x = \alpha$ and also for $x = \beta$.

So, by Rolle's Theorem applied to (u/v) in $[\alpha, \beta]$, $\exists$ at least one value γ where $\alpha < \gamma < \beta$ s.t. $(u/v)' = 0$. But, by hypothesis, $u'v - uv' \neq 0$ for any x in (α, β). Thus, we have a contradiction.

$\therefore v(x) = 0$ for at least one x in (α, β).

Moreover, by similar arguments, $\exists$ at least one root of $u(x) = 0$ between two consecutive roots of $v(x) = 0$.

Therefore, $v(x)$ cannot vanish *twice* in (α, β) for then $u(x)$ would have a root between α and β (*not possible*, since α and β were assumed to be two consecutive roots of $u(x) = 0$).

Hence the roots of u and v separate each other.

It is easy to verify that between any two consecutive roots of $\sin x = 0$ (say $x = 0$ and $x = \pi$) there exists only one root of $\cos x = 0$ (namely $x = \pi/2$).

Example 11.5.15. *A twice differentiable function $f(x)$ on a closed interval $[a, b]$ is such that $f(a) = f(b) = 0$ and $f(x_0) > 0$ where $a < x_0 < b$. Prove that $\exists$ at least one value of $x = c$ (say) between a and b for which $f''(c) < 0$.*

Solution: Since $f''(x)$ exists in $[a, b]$, we conclude that $f(x)$ and $f'(x)$ both exist and are continuous on $[a, b]$.

Since $a < x_0 < b$ we can apply Lagrange's MVT to $f(x)$ on two intervals $[a, x_0]$ and $[x_0, b]$ and obtain

$$f(x_0) = f(a) + (x_0 - a)f'(c_1) \quad \text{when} \quad a < c_1 < x_0$$

and $\quad f(b) = f(x_0) + (b - x_0)f'(c_2) \quad \text{when} \quad x_0 < c_2 < b.$

But it is given that $f(a) = f(b) = 0$.

$$\therefore \quad f'(c_1) = \frac{f(x_0)}{x_0 - a} \quad \text{and} \quad f'(c_2) = -\frac{f(x_0)}{b - x_0} \tag{11.5.1}$$

where $a < c_1 < x_0 < c_2 < b$.

Now since $f'(x)$ is continuous and derivable on $[c_1, c_2]$ we can again apply Lagrange's MVT and obtain

$$\frac{f'(c_2) - f'(c_1)}{c_2 - c_1} = f''(c) \text{ where } c_1 < c < c_2.$$

Using (11.5.1) we see that

$$f''(c) = \frac{-(b-a)f(x_0)}{(c_2 - c_1)(b - x_0)(x_0 - a)} < 0.$$

Now see No. **6**. Exercises **XI(I)**: Similar problem.

Example 11.5.16. *Show that* $\dfrac{\tan x}{x} > \dfrac{x}{\sin x}$ *when* $0 < x < \dfrac{\pi}{2}$.

Solution: To prove: $\dfrac{\tan x}{x} > \dfrac{x}{\sin x}$, if $0 < x < \dfrac{\pi}{2}$.

i.e., To prove: $\dfrac{\sin x \tan x - x^2}{x \sin x} > 0$ for $0 < x < \dfrac{\pi}{2}$.

Since the Denominator $x \sin x > 0$ for $0 < x < \dfrac{\pi}{2}$ it will suffice to prove that $f(x) = \sin x \tan x - x^2 > 0$ if $0 < x < \dfrac{\pi}{2}$.

Now $f'(x) = \sin x \sec^2 x + \cos x \tan x - 2x = \sin x + \sin x \sec^2 x - 2x$.

The function $f'(x)$ is continuous and derivable in $0 < x < \dfrac{\pi}{2}$.

$$\begin{aligned}
\therefore \quad f''(x) &= \cos x + \cos x \sec^2 x + 2 \sin x \sec^2 x \tan x - 2 \\
&= \{\cos x + \sec x - 2\} + 2 \sec x \tan^2 x \\
&= (\sqrt{\cos x} - \sqrt{\sec x})^2 + 2 \tan^2 x \sec x.
\end{aligned}$$

Clearly $f''(x) > 0$ for $0 < x < \pi/2$.

$\therefore$ $f'(x)$ is an increasing function in that interval. Moreover $f'(0) = 0$. Therefore, $f'(x) > 0$ for $0 < x < \dfrac{\pi}{2}$.

Consequently, $f(x)$ is an increasing function and $f(0) = 0$.

$$\therefore f(x) > 0 \text{ in } 0 < x < \dfrac{\pi}{2}.$$

$\therefore$ it follows from the previous arguments

$$\frac{\tan x}{x} > \frac{x}{\sin x} \text{ for } 0 < x < \frac{\pi}{2}.$$

CHAPTER 11: ROLLE'S THEOREM: MEAN VALUE THEOREM

Example 11.5.17. *Assuming $f''(x)$ to be continuous on $[a,b]$, prove that*

$$f(c) - f(a)\frac{b-c}{b-a} - \frac{c-a}{b-a}f(b) = \frac{1}{2}(c-a)(c-b)f''(\xi)$$

where c and ξ both lies in $[a,b]$.

Solution: To prove: $(b-a)f(c) - (b-c)f(a) - (c-a)f(b)$
$$= \frac{1}{2}(b-a)(c-a)(c-b)f''(\xi).$$

Choose a function

$$\phi(x) = (b-a)f(x) - (b-x)f(a) - (x-a)f(b) - (b-a)(x-a)(x-b)A$$

where A is a constant so determined that $\phi(c) = 0$.

i.e., $(b-a)f(c) - (b-c)f(a) - (c-a)f(b) - (b-a)(c-a)(c-b)A = 0$

or, $\quad A = \dfrac{(b-a)f(c) - (b-c)f(a) - (c-a)f(b)}{(b-a)(c-a)(c-b)}.$ (11.5.2)

Clearly, $\phi(a) = \phi(b) = 0$ and ϕ satisfies all the conditions of Rolle's Theorem in each of the two intervals $[a,c]$ and $[c,b]$ and hence $\exists\ \xi_1$ and ξ_2 in $]a,c[$ and $]c,b[$ respectively so that $\phi'(\xi_1) = 0$ and $\phi'(\xi_2) = 0$.

Again, $\phi'(x) = (b-a)f'(x) + f(a) - f(b) - (b-a)\{2x - (a+b)\}A$ which is continuous on $[a,b]$ and derivable on $]a,b[$ and in particular on $[\xi_1,\xi_2]$. Also $\phi'(\xi_1) = \phi'(\xi_2) = 0$.

Therefore by Rolle's Theorem $\exists\ \xi \in\ ]\xi_1,\xi_2[$ such that $\phi''(\xi) = 0$.

But $\phi''(x) = (b-a)f''(x) - (b-a)2A$ so that

$$f''(\xi) = 2A \quad (\because b \neq a)$$
$$\text{or,} \quad A = \frac{1}{2}f''(\xi) \text{ when } a < \xi_1 < \xi < \xi_2 < b. \quad (11.5.3)$$

From (11.5.2) and (11.5.3) the result follows.

Exercises XI(I)

1. Verify Rolle's Theorem in each of the following cases:

    (a) $f(x) = x^2$ in $[-1,1]$;
    (b) $f(x) = x^3 - 6x^2 + 11x - 6$ in $[1,3]$;
    (c) $f(x) = 1 - x^{2/3}$ in $[-1,1]$;
    (d) $f(x) = \sin x \cos x$ in $[0, \pi/2]$;

(e) $f(x) = x(x+3)e^{-\frac{1}{2}x}$ in $[-3, 0]$;

(f) $f(x) = \begin{cases} 1 - x^2, & \text{when } x \leq 0, \\ \cos x, & \text{when } x > 0; \end{cases}$ in $[-1, \pi/2]$.

N.B. In each case check the conditions of validity of Rolle's Theorem in the specified interval and then find c where $f'(c) = 0$, in case Rolle's Theorem is applicable. See the Answers given below.

2. Examine the validity of the hypothesis and conclusion of Lagrange's Mean-Value Theorem in each of the following cases:

    (a) $f(x) = x^2 + 3x + 2$ in $[1, 2]$;

    (b) $f(x) = x(x-1)(x-2)$ in $[0, \frac{1}{2}]$; [C.H.]

    (c) $f(x) = |x|$ in $[-1, 1]$ and also in $[0, 1]$;

    (d) $f(x) = \begin{cases} x \cos \frac{1}{x}, & x \neq 0, \\ 0, & x = 0; \end{cases}$ in $[-1, 1]$;

    (e) $f(x) = 1 + x^{2/3}$ in $[-8, 1]$;

    (f) $f(x) = 1/x$ in $[-1, 1]$. [See the Answers]

3. Verify that on the curve $y = x^2 + 2ax + b$, the chord joining the points $x = \alpha$ and $x = \beta$ is parallel to the tangent at the point $x = \frac{1}{2}(\alpha + \beta)$.

    [Geometrical interpretation of MVT]

4. (a) In the MVT applied to $f(x)$ in $[0, h]$ i.e., in

$$f(h) = f(0) + hf'(\theta h), \quad 0 < \theta < 1$$

prove that $\lim_{h \to 0^+} \theta = \frac{1}{2}$ when $f(x) = \cos x$.

(b) Show that θ, which occures in Lagrange's MVT, approaches the limit $\frac{1}{2}$ as $h \to 0$, provided that $f''(a) \neq 0$ (assume that $f''(x)$ is continuous at $x = a$).

[See the Answers]

5. Apply MVT to prove that if $\phi(x) = F\{f(x)\}$, then

$$\phi'(x) = F'\{f(x)\} \cdot f'(x),$$ [C.H. 1990]

assuming that the derivatives that occur are continuous. [Fully solved in the Answers]

CHAPTER 11: ROLLE'S THEOREM: MEAN VALUE THEOREM

6. The function $f(x)$ is twice differentiable on $[a,b]$ and is such that $f(a) = 0$; $f(b) = 0$, while $f(x_0) < 0$ for some x_0 in (a,b). Prove that $\exists$ at least one point c of the open interval (a,b) such that $f''(c) > 0$.   [Compare worked example No. 11.5.15 art. 11.5]

7. The functions $f(x), \phi(x)$ and their derivatives $f'(x), \phi'(x)$ are all continuous in $[a,b]$ and $f(x) \cdot \phi'(x) - f'(x)\phi(x) \neq 0$ at any point of the interval. Show that between any two roots of $f(x) = 0$ in the interval lies one root of $\phi(x) = 0$ and conversely. Verify the result when $f(x) = \cos x$, $\phi(x) = \sin x$.   [C.H. 1967, 92]

[See Ex. 11.5.14 worked out in art 11.5]

8. The function $f(x)$ is derivable in I $(a < x < b)$.

   Show that $f(x)$ is constant in I iff $f'(x) = 0$.

   Further if $f'(x) > 0$ at some point in I, prove that the set of points in I for which $f'(x) > 0$ is infinite.   [Consider points sufficiently near d, where $f'(d) > 0$. Use MVT]

9. Use MVT to prove that

   (a) $\sqrt{101}$ lies between 10 and 10.05.

   [Use MVT on $f(x) = \sqrt{x}$ in $[100, 101]$ i.e.,

   $$f(101) = f(100) + (101 - 100)f'(c) \text{ where } 100 < c < 101$$

   i.e., $\sqrt{101} = \sqrt{100} + \dfrac{1}{2\sqrt{c}}$ where $100 < c < 101$

   or, $\sqrt{101} = 10 + \dfrac{1}{2\sqrt{c}} < 10 + \dfrac{1}{2.10} = 10.05 \quad \left(\because c > 100; \ \therefore \ \dfrac{1}{2\sqrt{c}} < \dfrac{1}{2.10}\right).$]

   (b) $\sin 46°$ is approximately equal to $\dfrac{1}{2}\sqrt{2}\left(1 + \dfrac{\pi}{180}\right)$.

10. Verify that

    (a) $f(x) = x^2$ and $\phi(x) = \sqrt{x}$ are both increasing functions for $x > 0$.

    (b) $f(x) = \cot x$ is a decreasing function in $0 < x < \dfrac{\pi}{2}$.

    (c) $\dfrac{\tan x}{x} > \dfrac{x}{\sin x}$, for $0 < x < \dfrac{\pi}{2}$.

    (d) $\cos x < \left(\dfrac{\sin x}{x}\right)^3$, for $0 < x < \dfrac{\pi}{2}$.   [take $f(x) = x - \sin x(\cos x)^{-1/3}$]

11. Use MVT to prove the following inequalities:

   (a) $\dfrac{x}{\sqrt{1-x^2}} \leq \sin^{-1} x < x$, if $0 \leq x < 1$. When does the equality hold?

   (b) $\dfrac{x}{1+x} < \log_e(1+x) < x$; $\dfrac{x}{r+x} < \log_e(r+x) < x$, $x > 0$, $r \geq 1$.

   [C.H. 1992]

   (c) $0 < \dfrac{1}{\log_e(1+x)} - \dfrac{1}{x} < 1$.   [C.H. 1981]

   (d) $0 < \dfrac{1}{x} \log_e \dfrac{e^x - 1}{x} < 1$.   [C.H. 1983]

   [Hints: (c) Let $f(x) = \log_e(1+x)$ $(x > -1)$. Use $f(x) = f(0) + xf'(\theta x), 0 < \theta < 1$

   i.e., $\log_e(1+x) = \dfrac{x}{1+\theta x}$ or, $\dfrac{1+\theta x}{x} = \dfrac{1}{\log(1+x)}$ or, $\theta = \dfrac{1}{\log(1+x)} - \dfrac{1}{x}$

   But we know, $0 < \theta < 1$

   $\therefore\ 0 < \dfrac{1}{\log(1+x)} - \dfrac{1}{x} < 1$.

   (d) Let $f(x) = e^x$. Then proceed as in (c).]

12. Use the sign of derivative to prove that

   (a) $x < \log \dfrac{1}{1-x} < \dfrac{x}{1-x}$, if $0 < x < 1$.   [C.H. 1967]

   (b) $\dfrac{2x}{1-x^2} > \log \dfrac{1+x}{1-x} > 2x$, if $0 < x < 1$.

   (c) $\dfrac{2x}{\pi} \leq \sin x \leq x$, when $0 \leq x \leq \dfrac{\pi}{2}$ i.e., $\dfrac{2}{\pi} \leq \dfrac{\sin x}{x} < 1$ in $\left(0, \dfrac{\pi}{2}\right)$.

   [C.H. 1996]

   (d) $x < \log \dfrac{1}{1-x} k \dfrac{1}{1-x}$, for $0 < x < 1$.   [C.H. 1980]

   (e) $x^2 > (1+x)[\log(1+x)]^2$, when $x > 0$.

   [All these problems can be solved by using $f'(x) > 0 \Rightarrow f(x) > 0$. See (b) worked out in the **Answers**.]

13. If $\phi(x)$ be a polynomial in x and λ is a real number, then prove that $\exists$ a root of $\phi'(x) + \lambda\phi(x) = 0$ between any pair of roots of $\phi(x) = 0$.

   [use Rolle's Theorem on $F(x) = e^{\lambda x}\phi(x)$. Details worked out in the **Answers**.]

14. Prove that $\left(1 + \dfrac{1}{x}\right)^x$ and $\left(1 - \dfrac{1}{x}\right)^x$ $(x > 0)$ are both increasing functions.

   [See **Answers**]

15. If $f(x) = \begin{vmatrix} \sin x & \sin \alpha & \sin \beta \\ \cos x & \cos \alpha & \cos \beta \\ \tan x & \tan \alpha & \tan \beta \end{vmatrix}, 0 < \alpha < \beta < \dfrac{\pi}{2}$, show that $f'(x) = 0$ for some $x = c$, where $\alpha < c < \beta$. (apply Rolle's Theorem)

16. Let $f(x)$ be a *non-decreasing* function of x in a certain interval. If $f(x)$ be derivable at a point x_0 of the interval, then prove that $f'(x_0) \geq 0$.

    If $f(x)$ is *strictly increasing* at x_0, is it always true that $f'(x_0) > 0$? Give an example.

17. If $\phi(x)$ and $\psi(x)$ are both continuous in $[a, b]$ and are both derivable in (a, b) and if $\phi'(x) \neq 0, \psi'(x) \neq 0$ for any x in this interval, then prove that

    $$\frac{\phi(c) - \phi(a)}{\psi(b) - \psi(c)} = \frac{\phi'(c)}{\psi'(c)}, \quad a < c < b.$$

    [apply Rolle's Theorem to the function $\{\phi(x) - \phi(a)\} \cdot \{\psi(b) - \psi(x)\}$]

18. Show that $(x - \sin x)$ is an increasing function throughout any interval of values of x, and also prove that $(\tan x - x)$ increases as x increases from $-\frac{1}{2}\pi$ to $+\frac{1}{2}\pi$. For what values of a, $(ax - \sin x)$ is steadily increasing or decreasing function of x? [C.H. 1966]

19. Show that $(\tan x - x)$ increases from $x = \frac{1}{2}\pi$ to $x = \frac{3}{2}\pi$, from $x = \frac{3}{2}\pi$ to $x = \frac{5}{2}\pi$, and so on, and deduce that there is one and only one root of the equation $x = \tan x$ in each of these intervals.

20. Use Mean-Value Theorem to prove the following inequality:

    $$0 < \frac{1}{x} \log \frac{e^x - 1}{x} < 1. \qquad \text{[C.H. 1975]}$$

    [*Hints*: take $f(x) = e^x$ MVT in $[0, x]$ gives

    $$f(x) = f(0) + xf'(\theta x), 0 < \theta < 1$$

    i.e., $e^x = 1 + xe^{\theta x}$ i.e., $\dfrac{e^x - 1}{x} = e^{\theta x}$

    i.e., $\theta = \dfrac{1}{x} \log \dfrac{e^x - 1}{x}$ and $0 < \theta < 1$.]

21. Show that the function f where

    $$f(x) = \begin{cases} x \left[1 + \dfrac{1}{3} \sin(\log x^2)\right], & x \neq 0, \\ 0, & x = 0; \end{cases}$$

    is everywhere continuous and monotonic but has no differential coefficient at $x = 0$.

22. If the functions f, g, h are continuous on $[a, b]$ and twice differentiable on (a, b), prove that there exists ξ and $\eta \in (a, b)$ such that

$$\begin{vmatrix} f(a) & f(b) & f(c) \\ g(a) & g(b) & g(c) \\ h(a) & h(b) & h(c) \end{vmatrix} = \frac{1}{2}(b-c)(c-a)(a-b) \begin{vmatrix} f(a) & f'(\xi) & f''(\eta) \\ g(a) & g'(\xi) & g''(\eta) \\ h(a) & h'(\xi) & h''(\eta) \end{vmatrix}$$

where $a < c < b$. [See **Hints and Answers**]

23. Let f be continuous on $[a-h, a+h]$ and derivable on $(a-h, a+h)$. Prove that there exists a real number $\theta (0 < \theta < 1)$ for which

$$f(a+h) - 2f(a) + f(a-h) = h[f'(a+\theta h) - f'(a-\theta h)].$$

[*Hints*: Let $F(t) = f(a+ht) + f(a-ht)$, for all $t \in [0, 1]$. F is continuous on $[0, 1]$ and derivable on $(0, 1)$. By MVT $\exists\, \theta$ between 0 and 1 such that

$$F(1) - F(0) = (1-0)F'(\theta) \quad (0 < \theta < 1)$$

write down the values of $F(1)$, $F(0)$ and $F'(\theta)$; the result follows.]

24. Show that the derivative of $(x^2 - 1)e^{kx}$ has only one zero between -1 and $+1$ and that it lies between 0 and 1 when $k > 0$ and between -1 and 0 when $k < 0$.

25. Assuming that $f''(x)$ exists for all in $[a, b]$, show that

$$f(c) - f(a)\frac{b-c}{b-a} - f(b)\frac{c-a}{b-a} - \frac{1}{2}(c-a)(c-b)f''(\xi) = 0$$

where c and ξ both lie in (a, b).

[*Hints*: An equivalent form is

$$\frac{f(a)}{(a-b)(a-c)} + \frac{f(b)}{(b-c)(b-a)} + \frac{f(c)}{(c-a)(c-b)} = \frac{1}{2}f''(\xi).]$$

[see that if the result to be proved be multiplied by $(b-a)$ and the terms be re-arranged, then it can be shown to be equivalent to

$$\begin{vmatrix} 1 & 1 & 1 \\ a & b & c \\ f(a) & f(b) & f(c) \end{vmatrix} - \frac{1}{2}f''(\xi) \begin{vmatrix} 1 & 1 & 1 \\ a & b & c \\ a^2 & b^2 & c^2 \end{vmatrix} = 0.$$

We therefore, consider the function F defined by

$$F(x) = \begin{vmatrix} 1 & 1 & 1 \\ a & b & x \\ f(a) & f(b) & f(x) \end{vmatrix} - A \begin{vmatrix} 1 & 1 & 1 \\ a & b & x \\ a^2 & b^2 & x^2 \end{vmatrix}$$

where A is a constant so chosen that
$$F(c) = 0.$$

Since $F(a) = F(b) = 0$ and since $F'(x)$ exists in (a, b), therefore F satisfies all the conditions of Rolle's Theorem in each of the intervals $[a, c]$ and $[c, b]$. Consequently $\exists$ real numbers x_1 and x_2 such that $a < x_1 < c$ and $c < x_2 < b$ and $F'(x_1) = 0$ and $F'(x_2) = 0$.

Now
$$F'(x) = \begin{vmatrix} 1 & 1 & 0 \\ a & b & 1 \\ f(a) & f(b) & f'(x) \end{vmatrix} - A \begin{vmatrix} 1 & 1 & 0 \\ a & b & 1 \\ a^2 & b^2 & 2x \end{vmatrix}$$

$F'(x)$ satisfies all the conditions of Rolle's Theorem in $[x_1, x_2]$ and $\exists\,\xi$ such that $F''(\xi) = 0$ which yields $A = \frac{1}{2} f''(\xi)$. Since $F(c) = 0$, putting $A = \frac{1}{2} f''(\xi)$ the result follows.]

26. (a) If $f(0) = f'(0) = 0$ and $f'(x)$ exists in $0 \leq x \leq h$, prove that
$$f(h) = \frac{1}{2} h^2 f''(c), \quad 0 < c < h.$$

(b) If $f''(x)$ exists in $a - h \leq x \leq a + h$, prove that
$$f(a+h) - 2f(a) + f(a-h) = h^2 f''(c), \quad (a - h < c < a + h).$$

27. (a) Show that $a^x > x^a$ if $x > a \geq e$ [Take $f(x) = x \log a - a \log x$]. [C.H. 1990]
    (b) Prove that if $0 \leq x \leq 1$
$$\left| \log(1 + x) - x + \frac{1}{2} x^2 \right| \leq \frac{1}{3} x^3.$$

28. If f' exists and is bounded on some internal I, then prove that f is uniformly continuous on I.

[Hints: $|f(x_2) - f(x_1)| < M|x_2 - x_1|$. Choose $\delta = \varepsilon/M$. Then with $|x_1 - x_2| < \delta$, $|f(x_2) - f(x_1)| < M\delta = \varepsilon$.]   [C.H. 1999]

Hints & Answers [Exercises XI (I)]

1. (a) $f(x) = x^2$ is continuous in $[-1, 1]$, derivable in $(-1, 1)$ and $f(1) = f(-1) = 1$. Hence all the conditions of Rolle's Theorem are satisfied. *Conclusion*: $\exists$ a point c, namely $c = 0$ where $f'(x) = 2x = 0$.

(b) All the hypothesis of Rolle's Theorem are valid. *Conclusion:* $c = 2 \pm 1/\sqrt{3}$.

(c) Hypothesis not all valid; hence no c can be obtained.

(d) All hypothesis are valid and conclusion is $c = \pi/4$.

(e) Hypothesis valid, $c = -2$.

(f) Hypothesis valid, $c = 0$.

$$\lim_{x \to 0^-} -\frac{f(x) - f(0)}{x} = \lim_{x \to 0^-} \frac{1 - x^2 - 1}{x} = 0$$

$$\lim_{x \to 0^+} \frac{f(x) - f(0)}{x} = \lim_{x \to 0^+} \frac{\cos x - 1}{x} = \lim_{x \to 0} -\frac{2 \sin^2 x/2}{x} = 0.$$

2. (a) $c = 3/2$.

(b) $c = \frac{1}{6}(6 - \sqrt{21})$.

(c) $f(x)$ not derivable at $x = 0$ i.e., not derivable in $(-1, 1)$ but in the second case hypothesis of Lagrange's MVT are valid; here $c =$ any number between 0 and 1.

(d) and (e) Hypothesis not valid.

f) Hypothesis not valid.

4. (a) Applying MVT ($f(x) = \cos x$ is continuous and derivable at all points) in $[0, h]$

$$\cos h = \cos 0 + h(-\sin \theta h)$$

$$\text{or,} \quad \frac{\sin \theta h}{\theta h} \cdot \frac{\theta h}{h} = \frac{1 - \cos h}{h^2}.$$

Taking limits as $h \to 0^+$ we get

$$\lim_{h \to 0^+} \frac{\sin \theta h}{\theta h} \cdot \lim_{h \to 0^+} \frac{\theta h}{h} = \lim_{h \to 0^+} \frac{1 - \cos h}{h^2}$$

$$\text{or,} \quad \lim_{h \to 0^+} \theta = \lim_{h \to 0^+} \frac{2 \sin^2 h/2}{h^2} = \lim_{h \to 0^+} \frac{\sin h/2}{h/2} \cdot \frac{\sin h/2}{h/2} \cdot \frac{1}{2}$$

i.e., $\lim_{h \to 0^+} \theta = \frac{1}{2}$.

(b) Since $f''(x)$ is continuous at $x = a$, $f''(a)$ exists and given that $f''(a) \neq 0$. By MVT of second order

$$f(a + h) = f(a) + hf'(a) + \frac{h^2}{2!}f''(a + \theta' h), \quad 0 < \theta' < 1.$$

CHAPTER 11: ROLLE'S THEOREM: MEAN VALUE THEOREM

Also by MVT of 1st order,

$$f(a+h) = f(a) + hf'(a+\theta h), \quad 0 < \theta < 1.$$

It therefore follows that

$$0 = hf'(a) + \frac{h^2}{2!}f''(a+\theta'h) - hf'(a+\theta h)$$

or, $\quad f'(a+\theta h) = f(a) + \frac{h}{2}f''(a+\theta'h).$

Since f' is continuous and derivable, we have by MVT

$$f'(a+\theta h) = f'(a) + \theta h f''(a+\theta''\theta h).$$

$$\therefore \quad \frac{h}{2}f''(a+\theta'h) = \theta h f''(a+\theta''\theta h)$$

i.e., $\quad \theta = \frac{1}{2}\frac{f''(a+\theta'h)}{f''(a+\theta''\theta h)}.$

Hence

$$\lim_{h \to 0} \theta = \frac{1}{2}\lim_{h \to 0}\frac{f''(a+\theta'h)}{f''(a+\theta''\theta h)} = \frac{1}{2}\frac{f''(a)}{f''(a)}$$

$$(\because f''(x) \text{ is continuous at } x = a)$$

$$= \frac{1}{2} \quad (\because f''(a) \neq 0).$$

5. Let $f(x) = t$ so that $\phi(x) = F(t)$. Now

$$\phi'(x) = \lim_{h \to 0}\frac{\phi(x+h) - \phi(x)}{h}$$

$$= \lim_{h \to 0}\frac{F[f(x+h)] - F[f(x)]}{h}$$

$$= \lim_{h \to 0}\frac{F[f(x) + hf'(x+\theta_1 h)] - F[f(x)]}{h} \quad (0 < \theta_1 < 1)$$

[$\because$ using MVT on $f(x)$ in $[x, x+h]$ we get

$f(x+h) = f(x) + hf'(x+\theta_1 h), \quad 0 < \theta_1 < 1$]

$$= \lim_{h \to 0}\frac{F(t+H) - F(t)}{h} \text{ where } H = hf'(x+\theta_1 h)$$

$$= \lim \frac{F(t) + HF'(t+\theta_2 H) - F(t)}{h}, \quad 0 < \theta_2 < 1$$

($\because$ By MVT on $F(t)$ in $[t, t+H]$ we obtain

$F(t+H) = F(t) + HF'(t+\theta_2 H), \quad 0 < \theta_2 < 1$)

$$
\begin{aligned}
&= \lim_{h\to 0} \frac{HF'(t+\theta_2 H)}{h} \\
&= \lim_{h\to 0} \frac{hf'(x+\theta_1 h)F'\{f(x)+\theta_2 hf'(x+\theta_1 h)\}}{h} \\
&= \lim_{h\to 0} f'(x+\theta_1 h)F'[f(x)+\theta_2 hf'(\theta+\theta_1 h)] \\
&= f'(x)F'[f(x)].
\end{aligned}
$$

[Last line follows from the assumed continuity of $f'(x)$ and $F'(t)$.]

12. b) Let $f(x) = \dfrac{2x}{1-x^2} - \log\dfrac{1+x}{1-x}$. Then $f'(x) = \dfrac{4x^2}{(-x^2)^2} > 0$ in $0 < x < 1$. Also $f(0) = 0$. Hence $f(x) > 0$ when $0 < x < 1$ (1).

Again consider $\phi(x) = \log\dfrac{1+x}{1-x} - 2x$. Then $\phi'(x) = \dfrac{2x^2}{1-x^2} > 0$ in $0 < x < 1$.

Also $\phi(0) = 0$. Hence $\phi(x) > 0$ for $0 < x < 1$ (2). The results (1) and (2) together give the desired inequality.

13. **Let $F(x) = e^{\lambda x}\phi(x)$** [$\phi$ is a polynomial; hence always continuous and derivable; so it is the exponential function $e^{\lambda x}$.]

Let α, β be two consecutive roots of $\phi(x) = 0$ i.e., let $\phi(\alpha) = \phi(\beta) = 0$.

Then $F(x)$ is continuous and derivable in any interval, say in $[\alpha, \beta]$.

Further $F(\alpha) = 0$ and $F(\beta) = 0$.

$\therefore$ By Rolle's Theorem $\exists\, \gamma$ between α and β such that

$$
\begin{aligned}
F'(\gamma) &= 0 \\
\text{i.e., } e^{\lambda\gamma}[\lambda\phi(\gamma) + \phi'(\gamma)] &= 0
\end{aligned}
$$

i.e., $\exists$ a root of $\phi'(x) + \lambda\phi(x) = 0$ between a pair of roots of $\phi(x) = 0$.

14. Take logarithms and show that $f(x) = x\log\left(1+\frac{1}{x}\right)$ and $g(x) = x\log\left(1-\frac{1}{x}\right)$ have positive derivatives by using MVT.

21. To prove *monotonicity* see that $f'(x) \geq 0$ for all x other than zero. Let x_1, x_2 be two values of x other than zero such that $x_2 > x$. Then by the MVT,

$$
\begin{aligned}
f(x_2) - f(x_1) &= (x_2 - x_1)f'(c) \quad (x_1 < c < x_2) \\
&\geq 0. \quad (\because f'(c) \geq 0 \text{ and } x_2 > x_1)
\end{aligned}
$$

Hence $f(x_2) \geq f(x_1)$ i.e., f is m.i.

CHAPTER 11: ROLLE'S THEOREM: MEAN VALUE THEOREM

At $x = 0$ $f(x)$ has no derivative: See that

$$\lim_{x \to 0} \frac{f(x) - f(0)}{x} = \lim_{x \to 0} \frac{x + \frac{1}{3}x \sin(\log x^2) - 0}{x}$$

$$= \lim_{x \to 0} \left\{ 1 + \frac{1}{3} \sin(\log x^2) \right\} = \text{No limit.}$$

$$\left(\because \lim_{x \to 0} \log x^2 \text{ does not exist.} \right)$$

22. Consider the function F defined by

$$F(x) = \begin{vmatrix} f(a) & f(b) & f(x) \\ g(a) & g(b) & g(x) \\ h(a) & h(b) & h(x) \end{vmatrix} - \frac{(x-a)(x-b)}{(c-a)(c-b)} \begin{vmatrix} f(a) & f(b) & f(c) \\ g(a) & g(b) & g(c) \\ h(a) & h(b) & h(c) \end{vmatrix}.$$

Then $F(a) = F(b) = F(c) = 0$. Applying Rolle's Theorem on $[a, c]$ and $[c, b]$ we get

$$F'(x_1) = 0 \quad \text{and} \quad F'(x_2) = 0$$
$$\text{where} \quad a < x_1 < c \quad \text{and} \quad c < x_2 < b.$$

Therefore a second application of Rolle's Theorem on $[x_1, x_2]$ gives $F''(\eta) = 0$ where $x_1 < \eta < x_2$ but

$$F''(\eta) = 0 \Rightarrow \begin{vmatrix} f(a) & f(b) & f''(\eta) \\ g(a) & g(b) & g''(\eta) \\ h(a) & h(b) & h''(\eta) \end{vmatrix} = \frac{2}{(c-a)(c-b)} \begin{vmatrix} f(a) & f(b) & f(c) \\ g(a) & g(b) & g(c) \\ h(a) & h(b) & h(c) \end{vmatrix}. \quad (1)$$

Again consider the function ϕ defined by

$$\phi(x) = \begin{vmatrix} f(a) & f(x) & f''(\eta) \\ g(a) & g(x) & g''(\eta) \\ h(a) & h(x) & h''(\eta) \end{vmatrix} - \frac{2(x-a)}{(c-a)(c-b)(b-a)} \begin{vmatrix} f(a) & f(b) & f(c) \\ g(a) & g(b) & g(c) \\ h(a) & h(b) & h(c) \end{vmatrix}.$$

Now $\phi(a) = 0$ and using (1) we see that $\phi(b) = 0$.

Hence applying Rolle's Theorem on ϕ in $[a, b]$ we get

$$\begin{vmatrix} f(a) & f'(\xi) & f''(\eta) \\ g(a) & g'(\xi) & g''(\eta) \\ h(a) & h'(\xi) & h''(\eta) \end{vmatrix} = \frac{2}{(c-a)(c-b)(b-a)} \begin{vmatrix} f(a) & f(b) & f(c) \\ g(a) & g(b) & g(c) \\ h(a) & h(b) & h(c) \end{vmatrix}.$$

Hence

$$\begin{vmatrix} f(a) & f(b) & f(c) \\ g(a) & g(b) & g(c) \\ h(a) & h(b) & h(c) \end{vmatrix} = \frac{1}{2}(b-c)(c-a)(a-b) \begin{vmatrix} f(a) & f'(\xi) & f''(\eta) \\ g(a) & g'(\xi) & g''(\eta) \\ h(a) & h'(\xi) & h''(\eta) \end{vmatrix}.$$

CHAPTER 11B

11.6 Generalised Mean-Value Theorem: Taylor's Theorem

I. Taylor's Theorem with Lagrange's form of Remainder

STATEMENT. *If a real-valued function f of a real variable x defined in $[a, a+h]$ be such that*

1. *the $(n-1)$th derivative f^{n-1} is continuous in $[a, a+h]$,*

2. *the nth derivative f^n exists in the open interval $(a, a+h)$, then $\exists$ at least one number θ, where $0 < \theta < 1$, such that*

$$f(a+h) = f(a) + hf'(a) + \frac{h^2}{2!}f''(a) + \cdots + \frac{h^{n-1}}{\underline{|n-1}}f^{n-1}(a) + \frac{h^n}{\underline{|n}}f^n(a+\theta h).$$

Proof: The hypothesis (1) implies the existence and continuity of

$$f, f', f'', f''', \cdots, f^{n-2} \text{ in } [a, a+h].$$

Let us construct a function ϕ defined by

$$\phi(x) = f(x) + (a+h-x)f'(x) + \frac{(a+h-x)^2}{2!}f''(x) + \cdots$$
$$\cdots + \frac{(a+h-x)^{n-1}}{(n-1)!}f^{n-1}(x) + (a+h-x)^n A, \quad (11.6.1)$$

where A is a constant so chosen that $\phi(a) = \phi(a+h)$.

Then A is defined by

$$f(a) + hf'(a) + \frac{h^2}{2!}f''(a) + \cdots + \frac{h^{n-1}}{(n-1)!}f^{n-1}(a) + h^n A = f(a+h). \quad (11.6.2)$$

(a) ϕ is continuous in $[a, a+h]$.

[Evident from (11.6.1), since $f, f', \cdots, f^{n-1}$ are all continuous in $[a, a+h]$ and $(a+h-x), (a+h-x)^2, \cdots (a+h-x)^n$ are everywhere continuous.]

(b) ϕ is derivable in $(a, a+h)$.

[∵ the functions $f, f', \cdots, f^{n-1}$ are all derivable in $(a, a+h)$ and the terms $a+h-x$, $(a+h-x)^2, \cdots, (a+h-x)^{n-1}$ and $(a+h-x)^n$ are always derivable hence $\phi(x)$ must be derivable in $(a, a+h)$.]

(c) $\phi(a) = \phi(a+h)$, by construction i.e., ϕ satisfies all the conditions of Rolle's Theorem in $[a, a+h]$.

∴ By Rolle's Theorem, ∃ at least one number θ, where $0 < \theta < 1$ such that

$$\phi'(a + \theta h) = 0.$$

But from (11.6.1) we see that

$$\phi'(x) = f'(x) - f'(x) + (a+h-x)f''(x) - (a+h-x)f''(x) + \cdots$$
$$\cdots + \frac{(a+h-x)^{n-2}}{(n-2)!}f^{n-1}(x) - \frac{(a+h-x)^{n-2}}{(n-2)!}f^{n-1}(x)$$
$$+ \frac{(a+h-x)^{n-1}}{(n-1)!}f^n(x) - n(a+h-x)^{n-1}A$$

i.e., $\phi'(x) = \dfrac{(a+h-x)^{n-1}}{(n-1)!}f^n(x) - n(a+h-x)^{n-1}A.$ (11.6.3)

Hence, from (11.6.3),

$$\{(a+h) - (a+\theta h)\}^{n-1}\left\{\frac{f^n(a+\theta h)}{\lfloor n-1}- nA\right\} = 0$$

i.e., $A = \dfrac{f^n(a+\theta h)}{n\lfloor n-1} = \dfrac{f^n(a+\theta h)}{\lfloor n}.$

Putting this value of A in (11.6.2) we get the desired result of Taylor's Theorem. We call the $(n+1)$th term, namely,

$$\frac{h^n}{\lfloor n} f^n(a + \theta h), \quad 0 < \theta < 1.$$

the *Lagrange's Remainder after n terms*, and denote it by R_n.

II. Taylor's Theorem with Cauchy's form of Remainder [C.H. 1977, 1980]

STATEMENT. *If a real-valued function $f(x)$ be such that*

1. $f^{n-1}(x)$ *be continuous in* $[a, a+h]$,

2. $f^n(x)$ exists in $(a, a+h)$

then $\exists$ at least one number θ, where $0 < \theta < 1$, such that

$$f(a+h) = f(a) + hf'(a) + \frac{h^2}{2!}f''(a) + \cdots + \frac{h^{n-1}}{\lfloor n-1}f^{n-1}(a) + \frac{h^n(1-\theta)^{n-1}}{\lfloor n-1}f^n(a+\theta h).$$

In this case, $R_n = \dfrac{h^n(1-\theta)^{n-1}}{\lfloor n-1}f^n(a+\theta h)$ $(0 < \theta < 1)$ is called *Cauchy's Remainder after n terms*.

Proof: The proof is exactly the same as the previous one; only ϕ is to be defined as follows:

$$\phi(x) = f(x) + (a+h-x)f'(x) + \cdots + \frac{(a+h-x)^{n-1}}{\lfloor n-1}f^{n-1}(x) + (a+h-x)A,$$

where A is so chosen that $\phi(a) = \phi(a+h)$.

We leave the details to the students.

III. Taylor's Theorem with Generalised form of Remainder

STATEMENT. *If a function $f(x)$ be such that*

1. $f^{n-1}(x)$ be continuous in $a \leq x \leq a+h$,

2. $f^n(x)$ exists in $(a, a+h)$,

3. p is any given positive integer,

then $\exists$ at least one number θ, where $0 < \theta < 1$ such that

$$f(a+h) = f(a) + hf'(a) + \frac{h^2}{2!}f''(a) + \cdots + \frac{h^{n-1}}{\lfloor n-1}f^{n-1}(a) + \frac{h^n(1-\theta)^{n-p}}{p\lfloor n-1}f^n(a+\theta h).$$

In this case,

$$R_n = \frac{h^n(1-\theta)^{n-p}}{p\lfloor n-1}f^n(a+\theta h),\ 0 < \theta < 1$$

is called the **Schlömilch-Röche's** form of remainder after n-terms.

The proof is similar to **I**. Only we are to construct a function ϕ such that

$$\phi(x) = f(x) + (a+h-x)f'(x) + \cdots + \frac{(a+h-x)^{n-1}}{\lfloor n-1}f^{n-1}(x) + (a+h-x)^p A$$

where A is so chosen that $\phi(a) = \phi(a+h)$.

Note 11.6.1. *In Schlömilch-Röche's remainder, put $p = n$; then Lagrange's form of Remainder will be obtained. Similarly putting $p = 1$, one may obtain Cauchy's form of remainder.*

IV. An Important Deduction (Taylor's Theorem from Cauchy's MVT)

To obtain Lagrange's form of remainder after n terms by using Cauchy's Mean-Value Theorem:

Put
$$f(a+h) = f(a) + hf'(a) + \cdots + \frac{h^{n-1}}{\lfloor n-1} f^{n-1}(a) + \phi(h)$$

and $\psi(h) = \dfrac{h^n}{n!}.$

Both $\phi(h)$ and $\psi(h)$ and their first $(n-1)$ derivatives vanish for $h = 0$. (The students should check that the assertion is valid).

Using Cauchy's MVT successively,

$$\frac{\phi(h)}{\psi(h)} = \frac{\phi(h) - \phi(0)}{\psi(h) - \psi(0)} = \frac{\phi'(h_1)}{\psi'(h_1)} \quad (0 < h_1 < h).$$

Again, $\dfrac{\phi'(h_1)}{\psi'(h_1)} = \dfrac{\phi'(h_1) - \phi'(0)}{\psi'(h_1) - \psi'(0)} = \dfrac{\phi'(h_2)}{\psi'(h_2)} \quad (0 < h_2 < h_1 < h)$

$$\frac{\phi^{n-1}(h_{n-1})}{\psi^{n-1}(h_{n-1})} = \frac{\phi^{n-1}(h_{n-1}) - \phi^{n-1}(0)}{\psi^{n-1}(h_{n-1}) - \psi^{n-1}(0)} = \frac{\psi^n(h_n)}{\psi^n(h_n)} \quad (0 < h_n < h_{n-1}).$$

$$\therefore \quad \frac{\phi(h)}{\psi(h)} = \frac{\phi^n(\theta h)}{\psi^n(\theta h)}, \quad 0 < \theta < 1.$$

Thus,
$$\phi(h) = \psi(h) \frac{\phi^n(\theta h)}{\psi^n(\theta h)} = \frac{h^n}{\lfloor n} \frac{f^n(a+\theta h)}{1} \quad (0 < \theta < 1).$$

i.e., $R_n = \phi(h) = \dfrac{h^n}{\lfloor n} f^n(a+\theta h) \quad (0 < \theta < 1).$

V. Expressing $f(x)$ as a power series of x about a
(Another form of Taylor's Theorem)

Suppose f satisfies the two conditions of Taylor's Theorem:

1. $f^{n-1}(x)$ *is continuous in* $a \leq x \leq b = \mathbf{I}$ *(say) and*

2. $f^n(x)$ *exists in* $a < x < b$.

Then if $x \in I$ we can apply Taylor's Theorem in $[a, x]$ and obtain

$$f(x) = f(a) + (x-a)f'(a) + \frac{(x-a)^2}{2!}f''(a) + \cdots$$
$$+ \frac{(x-a)^{n-1}}{\lfloor n-1}f^{n-1}(a) + R_n,$$

where $R_n = \dfrac{(x-a)^n}{p\lfloor n-1}(1-\theta)^{n-p}f^n[a+\theta(x-a)].$

(*Schlömilch* and *Röche's Remainder* after n **terms**)

[putting $h = n$ we get *Lagrange's form* of remainder and putting $p = 1$ we obtain *Cauchy's form* of remainder.]

VI. Maclaurin's Theorem

In case the interval is $[0, x]$ instead of $[a, b]$, the corresponding results will be named after Maclaurin. Thus Maclaurin's theorem with generalised form of remainder can be written as:

$$f(x) = f(0) + xf'(0) + \frac{x^2}{\lfloor 2}f''(0) + \cdots + \frac{x^{n-1}}{\lfloor n-1}f^{n-1}(0)$$
$$+ \frac{x^n(1-\theta)^{n-p}}{p\lfloor n-1}f^n(\theta x) \quad (0 < \theta < 1).$$

Here again the remainder after n terms

$$R_n = \frac{x^n(1-\theta)^{n-p}}{p\lfloor n-1}f^n(\theta x) \quad (0 < \theta < 1) \text{ [Schlömilch-Röche's form]}$$

$$= \frac{x^n}{\lfloor n}f^n(\theta x) \quad (0 < \theta < 1) \text{ [Lagrange's form]}$$

$$= \frac{x^n(1-\theta)^{n-1}}{\lfloor n-1}f^n(\theta x) \quad (0 < \theta < 1). \text{ [Cauchy's form]}$$

To give an independent proof of Maclaurin's Theorem construct a function ϕ defined by

$$\phi(t) = f(t) + (x-t)f'(t) + \frac{(x-t)^2}{2!}f''(t) + \cdots$$
$$\cdots + \frac{(x-t)^{n-1}}{\lfloor n-1}f^{n-1}(t) + (x-t)^p \cdot A$$

where A is so chosen that $\phi(0) = \phi(x)$.

CHAPTER 11: ROLLE'S THEOREM: MEAN VALUE THEOREM

VII. Young's form of Taylor's Theorem

STATEMENT. *If a function f be such that $f^n(a)$ exists and M is defined by the equation*

$$f(a+h) = f(a) + hf'(a) + \frac{h^2}{2!}f''(a) + \cdots + \frac{h^{n-1}}{(n-1)!}f^{n-1}(a) + \frac{h^n}{n!}M$$

then $M \to f^n(a)$ as $h \to 0$.

Proof: The fact that $f^n(a)$ exists implies the existence of

$$f, f', f'', f''', \cdots, f^{n-1}$$

in a certain neighbourhood $(a - \delta, a + \delta)$ of $a (\delta > 0)$.

Let ε be any arbitrary positive number.

Case 1. First suppose $h \geq 0$. We construct a function ϕ of h such that

$$\phi(h) = f(a) + hf'(a) + \frac{h^2}{2!}f''(a) + \cdots + \frac{h^{n-1}}{(n-1)!}f^{n-1}(a)$$
$$+ \frac{h^n}{n!}[f^n(a) + \varepsilon] - f(a+h).$$

Then we see that

$$\phi(0) = 0, \quad \phi'(0) = 0, \cdots, \phi^{n-1}(0) = 0 \text{ but } \phi^n(0) = \varepsilon(> 0).$$

Now since $\phi^n(0) > 0$ and $\phi^{n-1}(0) = 0$, $\exists$ an interval $(0, \delta_1]$ such that for every point h of this semi-open interval $\phi^{n-1}(h)$ is positive.

Since $\phi^{n-1}(h) > 0$ when $h \in (0, \delta_1]$ and $\phi^{n-2}(0) = 0$; hence $\phi^{n-2}(h)$ is positive in $0 < h \leq \delta_1$.

We successively apply the result and finally deduce that $\phi(h)$ is positive when $0 < h \leq \delta_1$ i.e., $\exists\ \delta_1 > 0$ such that for

$$0 < h < \delta_1$$

$$f(a) + hf'(a) + \frac{h^2}{2!}f''(a) + \cdots + \frac{h^{n-1}}{\underline{|n-1}}f^{n-1}(a) + \frac{h^n}{\underline{|n}}[f^n(a) + \varepsilon] - f(a+h) > 0. \quad (11.6.4)$$

Similarly, we can prove that $\exists$ a positive number δ_2 such that for

$$0 < h < \delta_2$$

$$f(a) + hf'(a) + \frac{h^2}{2!}f''(a) + \cdots + \frac{h^{n-1}}{\underline{|n-1}}f^{n-1}(a) + \frac{h^n}{\underline{|n}}[f^n(a) - \varepsilon] - f(a+h) < 0. \quad (11.6.5)$$

But, by hypothesis.

$$f(a) + hf'(a) + \frac{h^2}{2!}f''(a) + \cdots + \frac{h^{n-1}}{\underline{|n-1}}f^{n-1}(a) + \frac{h^n}{\underline{|n}}M - f(a+h) = 0. \quad (11.6.6)$$

Let $\delta' = \min. \{\delta_1, \delta_2\}$.

Then, from above, it follows: corresponding to every positive number ε, $\exists\ \delta' > 0$, such that when $0 < h < \delta'$, from (11.6.4) and (11.6.6),

$$\frac{h^n}{\underline{|n}}[f^n(a) + \varepsilon] - \frac{h^n}{\underline{|n}}M > 0.$$

From (11.6.5) and (11.6.6),

$$\frac{h^n}{\underline{|n}}[f^n(a) - \varepsilon] - \frac{h^n}{\underline{|n}}M < 0,$$

i.e., when $0 < h < \delta'$, we have

$$f^n(a) - \varepsilon < M < f^n(a) + \varepsilon \quad \text{i.e.,} \quad \lim_{h \to 0+0} M = f^n(a).$$

Case 2. Now take the case when $h < 0$. We may exactly in a similar manner prove that

$$\lim_{h \to 0-0} M = f^n(a).$$

Hence from Cases 1 and 2, it follows

$$\lim_{h \to 0} M = f^n(a).$$

11.7 Illustrative Examples

Example 11.7.1. *Expand $\sin x$ in a finite series in powers of x, with remainder in Lagrange's form.*

Solution: Let $f(x) = \sin x$, then $f^n(x) = \sin\left(\frac{1}{2}n\pi + x\right)$, $f^n(0) = \sin\frac{1}{2}n\pi$ so that $f(x)$ possesses derivatives of every order for every value of x.

By Maclaurin's theorem with Lagrange's Remainder after n terms,

$$f(x) = f(0) + xf'(0) + \frac{x^2}{2!}f''(0) + \cdots + \frac{x^{n-1}}{(n-1)!}f^{n-1}(0) + \frac{x^n}{n!}f^n(\theta x), \quad (0 < \theta < 1).$$

i.e., $\sin x = \sin 0 + x\sin\frac{1}{2}\pi + \frac{x^2}{2!}\sin\pi + \cdots + \frac{x^{n-1}}{(n-1)!}\sin\frac{1}{2}(n-1)\pi$
$\qquad + \frac{x^n}{n!}\sin\left(\frac{1}{2}n\pi + \theta x\right), \quad (0 < \theta < 1)$

$\qquad = x - \frac{x^3}{3!} + \cdots + \frac{x^{n-1}}{(n-1)!}\sin\frac{1}{2}(n-1)\pi$
$\qquad + \frac{x^n}{n!}\sin\left(\frac{1}{2}n\pi + \theta x\right), \quad (0 < \theta < 1).$

Example 11.7.2. An Important Problem:

Show that $\lim\limits_{h\to 0}\theta = \dfrac{1}{n+1}$, *where θ is given by*

$$f(a+h) = f(a) + hf'(a) + \frac{h^2}{2!}f''(a) + \cdots + \frac{h^{n-1}}{(n-1)!}f^{n-1}(a) + \frac{h^n}{n!}f^n(a+\theta h)$$

provided that $f^{n+1}(x)$ is continuous at a, $f^{n+1}(a) \neq 0$. [C.H. 1992]

Solution: Since $f^{n+1}(x)$ is continuous at a, then $\exists\, \delta > 0$, s.t. in $[a-\delta, a+\delta]$, $f^{n+1}(x)$ exists.

Also then $f(x), f'(x), \cdots, f^n(x)$ are all continuous in $[a-\delta, a+\delta]$. Taking $a+h$, a point of this interval,

$$f(a+h) = f(a) + hf'(a) + \frac{h^2}{2!}f''(a) + \cdots + \frac{h^{n-1}}{(n-1)!}f^{n-1}(a) + \frac{h^n}{n!}f''(a+\theta h),\ (0<\theta<1)$$

(applying Taylor's Theorem with remainder after n terms) and

$$f(a+h) = f(a) + hf'(a) + \frac{h^2}{2!}f''(a) + \cdots + \frac{h^{n-1}}{(n-1)!}f^{n-1}(a) + \frac{h^n}{n!}f^n(a)$$
$$+ \frac{h^{n+1}}{(n+1)!}f^{n+1}(a+\theta' h),\ (0<\theta'<1).$$

(applying Taylor's Theorem with remainder after $n+1$ terms.)

Then, $f^n(a+\theta h) = f^n(a) + \dfrac{h}{n+1}f^{n+1}(a+\theta' h)$.

Using Lagrange's MVT on $f^n(x)$ in $[a, a+\theta h]$ we get

$$f^n(a) + \theta h f^{n+1}(a+\theta\theta'' h) = f^n(a) + \frac{h}{n+1}f^{n+1}(a+\theta' h),\ (0<\theta''<1),$$

i.e., $\theta \cdot f^{n+1}(a+\theta\theta'' h) = \dfrac{1}{n+1}f^{n+1}(a+\theta' h)$

or, $\lim\limits_{h\to 0}\theta\cdot f^{n+1}(a) = \dfrac{1}{n+1}\cdot f^{n+1}(a)$, since $f^{n+1}(x)$ is continuous

or, $\lim\limits_{h\to 0}\theta = \dfrac{1}{n+1}$, since $f^{n+1}(a) \neq 0$.

[See No. 4(a). Exercises XI(I)]

Example 11.7.3. *Show that (Using Maclaurin's development in the finite form):*

$$\log_e(1+x) = x - \frac{x^2}{2} + \frac{x^3}{3} - \frac{x^4}{4} + \cdots + (-1)^{n-2}\frac{x^{n-1}}{n-1} + R_n$$

where, $R_n = (-1)^{n-1}(1-\theta)^{n-1}\dfrac{x^n}{(1+\theta x)^n}$, $(0 < \theta < 1)$ *(Cauchy's form of remainder)*

or, $R_n = \dfrac{(-1)^{n-1}}{n}\left(\dfrac{x}{1+\theta x}\right)^n$, $(0 < \theta < 1)$, *(Lagrange's form of remainder)*.

Solution: Let $f(x) = \log(1+x)$. This function is defined and its derivatives of every order exist for every $x > -1$.

$$f^n(x) = (-1)^{n-1}\frac{\lfloor n-1}{(1+x)^n};\quad f^n(0) = (-1)^{n-1}\lfloor n-1;$$

$$f^n(\theta x) = (-1)^{n-1}\frac{\lfloor n-1}{(1+\theta x)^n}.$$

$$R_n \text{ (due to Lagrange)} = \frac{x^n}{n!}f^n(\theta x),\ (0 < \theta < 1);$$

$$= \frac{x^n}{n!}(-1)^{n-1}\frac{\lfloor n-1}{(1+\theta x)^n},$$

$$= \frac{(-1)^{n-1}}{n}\left(\frac{x}{1+\theta x}\right)^n. \qquad (11.7.1)$$

$$R_n \text{ (due to Cauchy)} = \frac{x^n}{\lfloor n-1}(1-\theta)^{n-1}f^n(\theta x),\ (0 < \theta < 1)$$

$$= \frac{x^n}{\lfloor n-1}(1-\theta)^{n-1}\cdot(-1)^{n-1}\frac{\lfloor n-1}{(1+\theta x)^n}.$$

$$= (-1)^{n-1}(1-\theta)^{n-1}\left(\frac{x}{1+\theta x}\right)^n. \qquad (11.7.2)$$

Now Maclaurin's Theorem in the finite form states

$$f(x) = f(0) + xf'(0) + \frac{x^2}{2!}f''(0) + \cdots + \frac{x^{n-1}}{(n-1)!}f^{n-1}(0) + R_n,$$

where $R_n =$ remainder after n terms (either due to Lagrange or due to Cauchy).

Putting the results for $f^n(0)$ and R_n we get

$$\log(1+x) = x - \frac{x^2}{2} + \frac{x^3}{3} - \cdots + (-1)^{n-2}\frac{x^{n-1}}{n-1} + R_n,$$

where R_n due to Lagrange and due to Cauchy are given above by (11.7.1) and (11.7.2) respectively.

CHAPTER 11: ROLLE'S THEOREM: MEAN VALUE THEOREM

Another useful result:

$$\log(x+h) = \log x + \frac{h}{x} - \frac{h^2}{2x^2} + \frac{h^3}{3x^3} - \cdots + (-1)^{n-2}\frac{h^{n-1}}{(n-1)x^{n-1}} + R_n,$$

where R_n (due to Lagrange) $= (-1)^{n-1}\frac{h^n}{n}\frac{1}{(x+\theta h)^n}$, $(0 < \theta < 1)$ and

$$R_n \text{ (due to Cauchy)} = (-1)^{n-1}h^n(1-\theta)^{n-1}\frac{1}{(x+\theta h)^n}, \quad (0 < \theta < 1).$$

Here take $f(x) = \log x\, (x > 0)$. Take h(+ve or −ve) such that $x + h > 0$. Then in $[x, x+h]$ apply *Taylor's theorem in the finite form*:

$$f(x+h) = f(x) + hf'(x) + \frac{h^2}{2!}f''(x) + \cdots + \frac{h^{n-1}}{\underline{n-1}}f^{n-1}(x) + R_n$$

and the result will follow.

Example 11.7.4. *Given:* $y = f(x) = \dfrac{1}{\sqrt{1+2x}},$

1. *Prove that* $(1+2x)y_{n+1} + (2n+1)y_n = 0.$

2. *Expand $f(x)$ by Maclaurin's Theorem with remainder after n terms. Write the remainders both in Lagrange's and Cauchy's forms.* [C.H. 1983]

Solution: (1) Given $y = \dfrac{1}{\sqrt{1+2x}}$, or, $\sqrt{1+2x}\cdot y = 1$.

Differentiating once we get, $\sqrt{1+2x}\cdot y_1 + \dfrac{1}{\sqrt{1+2x}}\cdot y = 0$,

i.e., $\sqrt{1+2x}\cdot\sqrt{1+2x}\cdot y_1 + y = 0$ or, $(1+2x)y_1 + y = 0.$

Differentiating n times by Leibnitz's Theorem we obtain

$$(1+2x)y_{n+1} + {}^nC_1\cdot y_n\cdot(2) + y_n = 0$$

i.e., $(1+2x)y_{n+1} + (2n+1)y_n = 0.$ (proved)

(2) Also if $y = u^\alpha$ (α, any real number) then

$$y_n = \alpha(\alpha-1)(\alpha-2)\cdots(\alpha-n+1)u^{\alpha-n}.$$

∴ When $y = (1+2x)^{-1/2}$, we have

$$y_n = -\frac{1}{2}\left(-\frac{1}{2}-1\right)\cdots\left(-\frac{1}{2}-n+1\right)(1+2x)^{-\frac{1}{2}-n}\cdot 2^n$$

$$= (-1)^n \frac{1\cdot 3\cdot 5\cdots(2n-1)}{2^n}(1+2x)^{-\frac{1}{2}-n}\cdot 2^n$$

i.e., $f^n(x) = (-1)^n \dfrac{\{1\cdot 3\cdot 5\cdots(2n-1)\}}{(1+2x)^{(2n+1)/2}}$

i.e., $f^n(0) = (-1)^n\{1\cdot 3\cdot 5\cdots(2n-1)\}$

and $f^n(\theta x) = (-1)^n \dfrac{\{1\cdot 3\cdot 5\cdots(2n-1)\}}{(1+2\theta x)^{(2n+1)/2}}$.

By Maclaurin's Theorem in $[0, x]$

$$f(x) = f(0) + xf'(0) + \cdots + \frac{x^{n-1}}{\lfloor n-1}f^{n-1}(0) + R_n$$

$$= 1 - x + \frac{1\cdot 3}{\lfloor 2}x^2 - \frac{1\cdot 3\cdot 5}{\lfloor 3}x^3 + \cdots + (-1)^{n-1}\frac{1\cdot 3\cdot 5\cdots(2n-3)}{\lfloor n-1}x^{n-1} + R_n$$

where

$$R_n = \text{remainder after } n \text{ terms (Lagrange's)}$$

$$= \frac{x^n}{\lfloor n}f^n(\theta x), \quad (0 < \theta \not< 1)$$

$$= \frac{x^n}{\lfloor n}(-1)^n \frac{1\cdot 3\cdot 5\cdots(2n-1)}{(1+2\theta x)^{(2n+1)/2}}, \quad (0 < \theta < 1)$$

and

$$R_n \text{ (due to Cauchy)} = \frac{x^n}{\lfloor n-1}(1-\theta)^{n-1}\cdot f^n(\theta x) \quad (0 < \theta < 1)$$

$$= \frac{x^n}{\lfloor n-1}(1-\theta)^{n-1}\cdot(-1)^n\frac{1\cdot 3\cdot 5\cdots(2n-1)}{(1+2\theta x)^{(2n+1)/2}}.$$

Exercises XI(II)

1. Obtain the expansions of the following functions with the remainder in Lagrange's form: $[0 < \theta < 1]$.

   (a) $e^x = 1 + x + \dfrac{1}{2!}x^2 + \dfrac{1}{3!}x^3 + \dfrac{1}{4!}x^4 e^{\theta x}$.

   (b) $\sin x + \cos x = 1 + x - \dfrac{1}{2!}x^2 - \dfrac{1}{3!}x^3 + \dfrac{1}{4!}x^4(\sin\theta x + \cos\theta x)$.

   (c) $(1+x)^{1/5} = 1 + \dfrac{1}{5}x - \dfrac{2}{25}x^2 + \dfrac{6}{125}x^3 - \dfrac{42x^4}{1250}(1+\theta x)^{-19/5}$.

(d) $(x+h)^{3/2} = x^{3/2} + \dfrac{3}{2}x^{1/2}h + \dfrac{3\cdot 1}{2\cdot 2}\dfrac{h^2}{2!}\dfrac{1}{\sqrt{(x+\theta h)}}.$

(e) $\log(x+h) = \log x + \dfrac{h}{x} - \dfrac{h^2}{2x^2} + \cdots + (-1)^{n-1}\dfrac{h^n}{n!(x+\theta h)^n}.$

2. Apply mean-value theorem (of appropriate order) to prove that

   (a) $x > \log(1+x) > x - \dfrac{1}{2}x^2;\ x > 0.$ [C.H. 1968, 85]

   (b) $x > \sin x > x - \dfrac{1}{6}x^3;\ 0 < x < \dfrac{1}{2}x.$

   (c) $0 < 1/\log(1+x) - 1/x < 1.$ [C.H. 1974, 81]

   (d) $0 < \dfrac{1}{x}\log\dfrac{e^x - 1}{x} < 1.$ [C.H. Spl. 1983]

3. If $f(h) = f(0) + hf'(0) + \dfrac{h^2}{2!}f''(\theta h),\ 0 < \theta < 1$, then prove that $\theta = \dfrac{9}{25}$, when $h = 1$ and $f(x) = (1-x)^{5/2}.$

4. (a) Find *Lagrange's Remainder* after n terms in the expansion of $e^{ax}\cos bx$, $e^{ax}\sin bx$ and $(1+x)^m$ in powers of x.

   (b) Find *Cauchy's Remainder* after n terms in the expansion of $(1+x)^m$ and $\log(1+x)$ in powers of x.

5. (a) In the equation
   $$f(a+h) = f(a) + hf'(a + \theta h),\ (0 < \theta < 1),$$
   prove that $\lim\limits_{h\to 0} \theta = \dfrac{1}{2}$, provided $f''(x)$ is continuous in $a \le x \le a+h$ and $f''(a) \ne 0$.
   Verify this limit taking $f(x) = \sin x$.

   (b) Show that
   $$f(a+h) = f(a) + hf'(a) + \dfrac{h^2}{2}f''(a+\theta h),\ (0 < \theta < 1),$$
   and prove that $\lim\limits_{h\to 0}\theta = \dfrac{1}{3}$, specifying the necessary conditions.

6. (a) If $f''(x)$ is continuous at $x = a$, then show that
   $$\lim_{h\to 0}\dfrac{f(a+h) - 2f(a) + f(a-h)}{h^2} = f''(a).$$

(b) Prove that $\lim\limits_{h \to 0} \dfrac{f(a+h) - f(a-h)}{2h} = f'(a)$, if f is derivable at $x = a$.

(c) Show that the result in (a) can be obtained if we simply assume that $f''(a)$ exists.

7. If $a < c < b$ and $f''(x)$ exists throughout the interval $[a, b]$, prove that
$$\frac{f(a)}{(a-b)(a-c)} + \frac{f(b)}{(b-c)(b-a)} + \frac{f(c)}{(c-a)(c-b)} = \frac{1}{2}f''(\xi),$$
where $a < \xi < b$ and now deduce that
$$\frac{f(a+h) - 2f(a) + f(a-h)}{h^2} = f''(a + \theta h), \quad (1 < \theta < 1).$$

Answers

4. (a) $\dfrac{(a^2 + b^2)^{n/2}}{n!} x^n e^{a\theta x} \cos\left(b\theta x + n \tan^{-1} \dfrac{b}{a}\right);$

$\dfrac{(a^2 + b^2)^{n/2}}{n!} x^n e^{a\theta x} \sin\left(b\theta x + n \tan^{-1} \dfrac{b}{a}\right);$

$\dfrac{m(m-1)(m-2)\cdots(m-n+1)}{n!} x^n (1 + \theta x)^{m-n}.$

(b) $\dfrac{m(m-1)(m-2)\cdots(m-n+1)}{(n-1)!} x^n (1-\theta)^{n-1}(1+\theta x)^{m-n}; \quad (-1)^{n-1} x^n \dfrac{(1-\theta)^{n-1}}{(1+\theta x)^n}.$

11.8 Taylor's Infinite Series[2]

I. Introduction

Let a function $f(x)$ be defined at $x = a$ and have derivatives there upto the nth order. Then the polynomial of degree n,
$$f(a) + (x-a)f'(a) + \frac{(x-a)^2}{2!}f''(a) + \cdots + \frac{(x-a)^n}{n!}f^n(a)$$
is such that at $x = a$, its value and its first n derivatives coincide in value with $f(x)$ and its first n derivatives. In this sense, this polynomial is an *approximating polynomial to* $f(x)$ at $x = a$.

e.g., the *approximating polynomial* to $f(x) = e^x$ at $x = 0$ is
$$f(0) + xf'(0) + \frac{x^2}{2!}f''(0) + \cdots + \frac{x^n}{n!}f^n(0) = 1 + x + \frac{x^2}{2!} + \cdots + \frac{x^n}{n!}.$$

[Since, $f(0) = e^0 = 1, f'(0) = e^x = 1$ at $x = 0$, etc.]

---
[2]For the purpose of this chapter we assume that our students are acquainted with *the definition* of infinite series and the *meaning of its convergence.*

II. Series Expansion of a Function

Next we wish to examine if we can express $f(x)$ as an infinite series* of the form:

$$f(a) + (x-a)f'(a) + \frac{(x-a)^2}{2!}f''(a) + \cdots + \frac{(x-a)^n}{n!}f^n(a) + \cdots . \qquad (11.8.1)$$

The following question arise in this connection:

1. Given any function $f(x)$, can we always construct a series of the form (11.8.1)?

    Ans: Such an infinite series can be constructed

    iff $f(x)$ has derivatives of all orders at $x = a$

    i.e., iff $f^n(a)$ exists for each positive integer n.]

2. Assuming that $f^n(a)$ exists for every positive integer n, then the question remains, will the series (11.8.1) converge?

    Ans: Let $S_n = f(a) + (x-a)f'(a) + \frac{(x-a)^2}{2!}f''(a) + \cdots + \frac{(x-a)^{n-1}}{(n-1)!}f^{n-1}(a)$.

    If $\lim_{n\to\infty} S_n$ exists and finite $= S$ (say), then the series (11.8.1) converges and its sum is S.]

3. Assume that f satisfies all the conditions of the Generalised Taylor's Theorem with remainder after n terms in an interval $[a-h, a+h]$ so that for each $x \in [a-h, a+h]$.

    $$f(x) = f(a) + (x-a)f'(a) + \frac{(x-a)^2}{2!}f''(a) + \cdots + \frac{(x-a)^{n-1}}{\lfloor n-1}f^{n-1}(a) + R_n,$$

    where R_n is the remainder after n terms (Lagrange or Cauchy or any other remainder). Under these conditions will the series (11.8.1) converge? If so, to what it converges?

    Ans: Since, we can now write

    $$f(x) = S_n + R_n \text{ i.e., } S_n = f(x) - R_n.$$

    $\lim_{n\to\infty} S_n$ exists iff $\lim_{n\to\infty} R_n$ exists finitely.

    $\therefore$ the infinite series (11.8.1) converges iff $\lim_{n\to\infty} R_n$ exists finitely.

    Moreover, we see that $\lim_{n\to\infty} S_n = f(x)$ if $\lim_{n\to\infty} R_n = 0$.

    i.e., The infinite series (11.8.1) converges to $f(x)$.

    i,e., $f(x) = f(a) + (x-a)f'(a) + \cdots + \frac{(x-a)^n}{\lfloor n}f^n(a) + \cdots$

    iff $\lim_{n\to\infty} R_n = 0$ (and, of course, derivatives of all orders must exist).]

We thus come to the following conclusion:

Taylor's Infinite Series. If f possesses derivatives of all orders at any $x \in [a-h, a+h]$ and if Taylor's Theorem holds in $[a-h, a+h]$ with its remainder $R_n \to 0$ as $n \to \infty$, then the infinite series

$$f(a) + (x-a)f'(a) + \frac{(x-a)^2}{2!}f''(a) + \cdots + \frac{(x-a)^n}{\underline{n}}f^n(a) + \cdots$$

converges to $f(x)$ for every $x \in [a-h, a+h]$ and we can then write

$$f(x) = f(a) + (x-a)f'(a) + \frac{(x-a)^2}{2!}f''(a) + \cdots + \frac{(x-a)^n}{\underline{n}}f^n(a) + \cdots .$$

We say: *the right-hand side series is the Taylor's expansion of $f(x)$ about $x = a$.*

Example 11.8.1. *If $f(x) = \sin x$, expand $f(x)$ about $x = \pi/2$.*

Note 11.8.1. *Changing x to $a + h$ we obtain under the stated conditions,*

$$f(a+h) = f(a) + hf'(a) + \frac{h^2}{2!}f''(a) + \cdots + \frac{h^n}{\underline{n}}f^n(a) + \cdots .$$

Maclaurin's Infinite Series. Putting $a = 0$, we have the following result: If

1. f be defined in $[-h, h]$;

2. for each positive integer n, $f^n(x)$ exists for $x \in [-h, h]$;

3. $\lim_{n \to \infty} R_n = 0$ for each x in $[-h, h]$, where R_n is the Maclaurin's remainder after n terms,

then for each x in $[-h, h]$

$$f(x) = f(0) + xf'(0) + \frac{x^2}{2!}f''(0) + \cdots + \frac{x^n}{n!}f^n(0) + \cdots . \qquad (11.8.2)$$

The series (11.8.2) is then called *Maclaurin's infinite series expansion of $f(x)$ about $x = 0$.*

REMEMBER: The series $f(0) + xf'(0) + \frac{x^2}{2!}f''(0) + \cdots + \frac{x^n}{n!}f^n(0) + \cdots$, converges if $\lim_{n \to \infty} S_n$ exists finitely, where

$$S_n = f(0) + xf'(0) + \frac{x^2}{2!}f''(0) + \cdots + \frac{x^{n-1}}{(n-1)!}f^n(0).$$

But the mere convergence of the series does not always imply that the series has a sum $f(x)$—it may have a different sum.

CHAPTER 11: ROLLE'S THEOREM: MEAN VALUE THEOREM

Take, for instance, the following example:

Example 11.8.2. (Cauchy Function). $f(x) = e^{-1/x^2}$, $x \neq 0$, $f(0) = 0$.

Since, if we put $u = 1/x$, $u_1 = -1/x^2 = -u^2$ and $f(x) = e^{-u^2}$.

$$f'(0) = \lim_{x \to 0} \frac{f(x) - f(0)}{x - 0} = \lim_{x \to 0} \frac{e^{-1/x^2}}{x} = \lim_{u \to \infty} \frac{e^{-u^2}}{1/u} = \lim_{u \to \infty} \frac{u}{e^{u^2}} = 0$$

(Using L'Hospital's Rule for the indeterminate form ∞/∞).

Also, for $x \neq 0$, $f'(x) = e^{-u^2} \cdot (-2u) \cdot u_1 = 2u^3 e^{-u^2}$.

$$\therefore f''(0) = \lim_{x \to 0} \frac{f'(x) - f'(0)}{x - 0} = \lim_{u \to \infty} \frac{2u^3}{e^{u^2}} \times u = \lim_{u \to \infty} \frac{2u^4}{e^{u^2}} = 0.$$

But for $x \neq 0$, $f''(x) = e^{-u^2}(6u^2 u_1 - 4u^4 u_1) = e^{-u^2}(4u^6 - 6u^4)$

and $f'''(0) = \lim_{x \to 0} \dfrac{f''(x) - f''(0)}{x - 0} = \lim_{u \to \infty} \dfrac{4u^7 - 6u^5}{e^{u^2}} = 0$

and so on.

Similarly, for every n,

$$f^n(0) = \lim_{x \to 0} \frac{f^{n-1}(x) - f^{n-1}(0)}{x - 0} = \lim_{u \to \infty} \frac{P_n(u)}{e^{u^2}} = 0.$$

where $P_n(u)$ is a polynomial in u. (Use L'Hospital's Rule n times.)

[This $P_n(u)$ may be determined from:

if $f^n(x) = e^{-u^2} P_n(u)$, then $P_{n+1}(u) = 2u^3 P_n(u) - u^2 P'_n(u)$ and since $P_0(u) = 1$, we have $P_1(u) = 2u^3, P_2(u) = 4u^6 - 6u^4$, etc.]

Thus, the Maclaurin's infinite series corresponding to $f(x)$ namely,

$$f(0) + xf'(0) + \frac{x^2}{2!}f''(0) + \cdots + \frac{x^n}{n!}f^n(0) + \cdots$$

becomes, in this case,

$$0 + x \cdot 0 + \frac{x^2}{2!} \cdot 0 + \cdots + \frac{x^n}{n!} \cdot 0 + \cdots = 0 + 0 + 0 + \cdots + 0 + \cdots$$

which obviously converges to **zero** for every x (not to e^{-1/x^2}).

i.e., Maclaurin's infinite series corresponding to $f(x) = e^{-1/x^2}$ converges, but it does not converge to $f(x)$.

Remember

1. $R_n \to 0$ is both *necessary* and *sufficient condition* for the **equality** of the infinite series with the function to which it corresponds.

2. $R_n \to 0$ is *sufficient but not necessary* for the mere **convergence** of the Maclaurin's infinite series.

III. Power series expansions of some standard functions

Definition 11.8.1. *A series of the form* $\sum_{n=0}^{\infty} a_n x^n = a_0 + a_1 x + a_2 x^2 + \cdots + a_n x^n + \cdots$ *where x is real variable, $a_0, a_1, a_2, \cdots, a_n, \cdots$ are real constants is called a real power series. Every such power series converges at $x = 0$. In general, $\exists\, R > 0$, s.t. the power series converges $\forall\, x$ in $-R < x < R$ but at the end points the behaviour is in doubt. Outside this interval the series diverges. We call the interval, interval of convergence or range of validity and R, the radius of convergence. Some power series converges for every real value of x. We then say $R = \infty$.*

(A) $f(x) = e^x$.

For every positive integer n, $f^n(x) = e^x$.

Thus f has derivatives of all orders in any interval $[-h, h]$, whatever positive real number h may be. We shall take $x \in [-h, h]$.

$$f(0) = 1, \quad f^n(0) = 1 \quad \forall\, n \in \mathbb{N}.$$

Lagrange's Remainder R_n after n terms in Maclaurin's finite development

$$= R_n = \frac{x^n}{\underline{|n}} f^n(\theta x) = \frac{x^n}{\underline{|n}} e^{\theta x}, \quad (0 < \theta < 1).$$

The Maclaurin's infinite series corresponding to $f(x) = e^x$ is

$$f(0) + x f'(0) + \frac{x^2}{\underline{|2}} f''(0) + \cdots + \frac{x^n}{\underline{|n}} f^n(0) + \cdots = 1 + x + \frac{x^2}{\underline{|2}} + \cdots + \frac{x^n}{\underline{|n}} + \cdots$$

This infinite series is *convergent* and *converges to* e^x iff $\lim_{n \to \infty} R_n = 0$.

We know that $\lim_{n \to \infty} \frac{x^n}{\underline{|n}} = 0, \forall\, x$.

REMEMBER: Writing $a_n = \frac{x^n}{\underline{|n}}$, $a_{n+1} = \frac{x^{n+1}}{\underline{|n+1}}$, $\lim_{n \to \infty} \left| \frac{a_{n+1}}{a_n} \right| = 0$ and hence $a_n \to 0$ as $n \to \infty$. (See art. 5.6, worked out Ex. 5.6.2).

CHAPTER 11: ROLLE'S THEOREM: MEAN VALUE THEOREM

$$\therefore \lim_{n \to \infty} R_n = \lim_{n \to \infty} \frac{x^n}{\lfloor n} \cdot e^{\theta x}, \quad (0 < \theta < 1)$$

$$= \lim_{n \to \infty} \frac{x^n}{\lfloor n} \cdot e^{\theta x} \quad (e^{\theta x} \text{ lies between } e^0 \text{ and } e^x \; \forall \, x \text{ and hence it is bounded})$$

$$= 0, \quad \forall \, x \in [-h, h], \text{ whatever } h \text{ may be.}$$

$\therefore$ We can write

$$e^x = 1 + \frac{x}{\lfloor 1} + \frac{x^2}{\lfloor 2} + \cdots + \frac{x^n}{\lfloor n} + \cdots \; (\forall \, x \in \mathbb{R}).$$

This is what we call power series expansion of e^x.

Now see that $\forall \, a > 0 \; (a \neq 1)$ and $\forall \, x \in \mathbb{R}$

$$a^x = e^{x \log a} = 1 + \frac{x}{\lfloor 1} \log a + \frac{x^2}{\lfloor 2}(\log a)^2 + \cdots + \frac{x^n}{\lfloor n}(\log a)^n + \cdots. \quad \text{[C.H. 1982]}$$

The expansions of e^x or a^x are called *Exponential Series*.

Two deductions

(i) $e^x = 1 + x + \dfrac{x^2}{\lfloor 2} + \dfrac{x^3}{\lfloor 3} + \cdots + \dfrac{x^n}{\lfloor n} + \dfrac{x^{n+1}}{\lfloor n+1} e^{\theta x}, \quad (0 < \theta < 1)$

(Lagrange's Remainder taken after $(n+1)$ terms).

For $x = 1$,

$$e = 1 + 1 + \frac{1}{\lfloor 2} + \frac{1}{\lfloor 3} + \cdots + \frac{1}{\lfloor n} + \frac{1}{\lfloor n+1} e^\theta \quad (0 < \theta < 1) \quad (11.8.3)$$

where

$$R_n = \frac{1}{(n+1)!} e^\theta < \frac{3}{(n+1)!}, \quad (\because e < 3).$$

If we wish to calculate e with an error of at most 10^{-5} we need only to choose n so large that

$$R_n < 1/100000.$$

Choosing $n = 9$, check that $R_n < \dfrac{3}{(9+1)!} = \dfrac{3}{3628800} < 0.000001.$

The value of e is then given correctly upto the sixth decimal place by

$$1 + \frac{1}{1!} + \frac{1}{2!} + \frac{1}{3!} + \cdots + \frac{1}{9!} = 2.718281.$$

(ii) From the result (11.8.3) we can also deduce that e is an irrational number. For if $e = p/q$ (ratio of two integers p and q) we can certainly choose $n > q$. Then multiplying (11.8.3) by $n!$ we have

$$n!e = 2 \cdot \lfloor n + \frac{\lfloor n}{\lfloor 2} + \frac{\lfloor n}{\lfloor 3} + \cdots + \frac{\lfloor n}{\lfloor n} + \frac{\lfloor n}{\lfloor n+1} e^\theta, \quad (0 < \theta < 1).$$

Since e has been assumed to be p/q and $q < n$, the left side must be an integer, whereas the right-hand side = an integer $+ \dfrac{e^\theta}{n+1}$.

But $e^\theta < e < 3$; hence $\dfrac{e^\theta}{n+1}$ is positive but < 1, for sufficiently large n.

Thus the right-hand side = an integer + a number between 0 and 1 i.e., $\neq$ an integer.

That left-side = an integer and right-side $\neq$ an integer, is not possible. So we cannot assume e in the form p/q.

In other words, *e is not a rational number*.

(B) Sine and Cosine Series. $f(x) = \sin x$ (say). [C.H. 1993]

Then $f^n(x) = \sin\left(\dfrac{1}{2}n\pi + x\right)$ so that $\sin x$ possesses derivatives of every order for all real values of x.

$$\therefore \quad f^n(0) = \sin\left(\dfrac{1}{2}n\pi\right) \text{ and as such}$$
$$f'(0) = 1, \quad f''(0) = 0, \quad f'''(0) = -1, \quad f^{iv}(0) = 0,$$
$$f^v(0) = 1, \quad f^{vi}(0) = 0, \quad f^{vii}(0) = -1, \quad f^{viii}(0) = 0, \text{ etc.}$$

Also R_n [Lagrange's form of remainder after n terms in Maclaurin's finite development of $f(x)$]

$$= \frac{x^n}{\lfloor n} f^n(\theta x) = \frac{x^n}{\lfloor n} \sin\left(\frac{1}{2}n\pi + \theta x\right).$$

We first construct Maclaurin's infinite series corresponding to $f(x) = \sin x$,

$$f(0) + xf'(0) + \frac{x^2}{2!}f''(0) + \cdots + \frac{x^n}{\lfloor n}f^n(0) + \cdots = x - \frac{x^3}{3!} + \frac{x^5}{5!} - \cdots + (-1)^n \frac{x^{2n+1}}{\lfloor 2n+1} + \cdots.$$

This series converges to $\sin x \ \forall\ x$ iff $\lim\limits_{n\to\infty} R_n = 0, \ \forall\ x$.

Here $\lim\limits_{n\to\infty} R_n = \lim\limits_{n\to\infty} \dfrac{x^n}{\lfloor n} \sin\left(\dfrac{1}{2}n\pi + \theta x\right) = 0$

$\left(\because \lim\limits_{n\to\infty} \dfrac{x^n}{\lfloor n} = 0, \forall\ x \text{ and } \left|\sin\left(\dfrac{1}{2}n\pi + \theta x\right)\right| \leq 1\right).$

$$\therefore \ \sin x = x - \frac{x^3}{3!} + \frac{x^5}{5!} - \frac{x^7}{7!} + \cdots + (-1)^n \frac{x^{2n+1}}{\lfloor 2n-1} + \cdots (\forall\ x \in \mathbb{R}).$$

CHAPTER 11: ROLLE'S THEOREM: MEAN VALUE THEOREM

Similarly,

$$\cos x = 1 - \frac{x^2}{2!} + \frac{x^4}{4!} - \frac{x^6}{6!} + \cdots + (-1)^n \frac{x^{2n}}{\underline{|2n}} + \cdots (\forall x \in \mathbb{R}).$$

Example 11.8.3. $f(x) = \sin x$. Expansion of $f(x)$ about $x = \pi/2$ is

$$f(\pi/2) - (x - \pi/2)f'(\pi/2) + \frac{(x - \pi/2)^2}{\underline{|2}} f''(\pi/2) + \cdots.$$

Substitute now the values of $f(\pi/2), f'(\pi/2)$, etc. [C.H. 1979]

(C) Logarithmic Series. $f(x) = \log(1+x)$. [C.H. 1974, 76, 88, 95, 96]

$\log(1+x)$ is defined when $1+x > 0$ i.e., when $x > -1$.

Moreover, $\log(1+x)$ possesses continuous derivatives of all orders for every value of x when $1+x > 0$ (or $x > -1$).

In fact,

$$f^n(x) = \frac{(-1)^{n-1}(n-1)!}{(1+x)^n}, \text{ where } x > -1$$

i.e., $f^n(0) = (-1)^{n-1}(n-1)!$.

Maclaurin's infinite series corresponding to $\log(1+x)$ is

$$f(0) + xf'(0) + \frac{x^2}{2!}f''(0) + \frac{x^3}{3!}f'''(0) + \cdots + \frac{x^n}{\underline{|n}}f^n(0) + \cdots$$

$$= x - \frac{x^2}{2} + \frac{x^3}{3} - \frac{x^4}{4} + \cdots + (-1)^{n-1}\frac{x^n}{n} + \cdots, \quad (-1 < x \leq 1).$$

We shall consider the following cases:

(i) Let $0 \leq x \leq 1$. We consider

Lagrange's remainder R_n after n terms $= \dfrac{x^n}{\underline{|n}} f^n(\theta x), \ (0 < \theta < 1)$.

$$\therefore \quad R_n = \frac{x^n}{\underline{|n}} \cdot \frac{(-1)^{n-1}(n-1)!}{(1+\theta x)^n}$$

$$= \frac{(-1)^{n-1}}{n} \cdot \left(\frac{x}{1+\theta x}\right)^n.$$

Since, $0 \leq x \leq 1$ and $0 < \theta < 1$, it follows

$$0 < \frac{x}{1+\theta x} < 1.$$

∴ $|R_n| < 1/n$ and hence can be made $<$ any positive ε for sufficiently large n, i.e., $R_n \to 0$ as $n \to \infty$.

(ii) Let $-1 < x < 0$.

In this case $\left|\dfrac{x}{1+\theta x}\right|$ need not necessarily be less than 1.

∴ We may not be able to prove easily that R_n (due to Lagrange) $\to 0$ as $n \to \infty$.

But if we take Cauchy's form of remainder then it will be easy to show that $R_n \to 0$ as $n \to \infty$.

$$\begin{aligned}
R_n \text{ (due to Cauchy)} &= \frac{x^n}{\lfloor n-1}(1-\theta)^{n-1} f^n(\theta x), \quad (0 < \theta < 1) \\
&= \frac{x^n}{\lfloor n-1}(1-\theta)^{n-1} \cdot \frac{(-1)^{n-1}(n-1)!}{(1+\theta x)^n} \\
&= (-1)^{n-1}\left(\frac{1-\theta}{1+\theta x}\right)^{n-1} \cdot x^n \cdot \frac{1}{1+\theta x}.
\end{aligned}$$

We remark here that our reasoning will be true if we take $|x| < 1$ and hence for $-1 < x < 0$. *Case* (i) *above was needed to cover the case* $x = 1$. See that

If $|x| < 1$, then

$$0 < \left|\frac{1-\theta}{1+\theta x}\right| < 1 \text{ so that } 0 < \left|\frac{1-\theta}{1+\theta x}\right|^{n-1} < 1$$

and $\left|\dfrac{1}{1+\theta x}\right| < \dfrac{1}{1-|x|}.$

∴ $|R_n \text{ (due to Cauchy)}| < \dfrac{|x|^n}{1-|x|}.$

This leads to the conclusion

$$\lim_{n\to\infty} R_n = 0 \text{ for } |x| < 1.$$

$(\because \lim_{n\to\infty} x^n = 0 \text{ for } |x| < 1 \text{ and } \dfrac{1}{1-|x|} \text{ is bounded})$

We now conclude:

If $-1 < x \leq 1$, some remainder of Maclaurin's finite development tends to zero as $n \to \infty$.

∴ When $-1 < x \leq 1$.

$$f(x) = \log(1+x) = x - \frac{x^2}{2} + \frac{x^3}{3} - \frac{x^4}{4} + \cdots + (-1)^{n-1}\frac{x^n}{n} + \cdots.$$

We can prove that the series does not converge if $x > 1$ or if $x \leq -1$. At $x = 1$, we have
$$\log 2 = 1 - \frac{1}{2} + \frac{1}{3} - \frac{1}{4} + \cdots .$$

Example 11.8.4. *Expand $\log_e(1+ax)$ in powers of x and indicate the range of validity of the expansion where $a \neq 0$.*
[C.H. 1980]

[*Hints:* One may establish the expansion of $\log(1+t)$ in $-1 < t \leq 1$. Then put $t = ax$. The range of validity in the expansion of $\log_e(1+ax)$ will be $-1 < ax \leq 1$ or $-1/a < x \leq 1/a$, a being $\neq 0$.]

(D) Binomial Series: $f(x) = (1+x)^m$, where m is any real number.
[C.H. 1978, 84, 90]

Here $f^k(x) = m(m-1)(m-2)\cdots(m-k+1)(1+x)^{m-k}$
and so, $f^k(0) = m(m-1)(m-2)\cdots(m-k+1)$.

Case 1. Let m be a positive integer.

Then $f^k(x)$ exists for all $x \in \mathbb{R}$ and $f^k(x) = 0$, $\forall k > m$.
$\therefore R_n = 0$, whenever $n > m$ and consequently
$$\lim_{n \to \infty} R_n = 0, \quad \forall x \in \mathbb{R}.$$

We have, then
$$f(x) = f(0) + xf'(0) + \frac{x^2}{2!}f''(0) + \cdots + \frac{x^m}{m!}f^m(0)$$

since all other terms vanish and hence
$$(1+x)^m = 1 + mx + \frac{m(m-1)}{\lfloor 2}x^2 + \cdots + x^m, \quad \forall x \in \mathbb{R}.$$

Case 2. Let m be any real number not necessarily a positive integer.

In this case $f(x) = (1+x)^m$ possesses continuous derivatives of every order for values of x such that $1+x$ is > 0 or $x > -1$, and $f^n(x) = m(m-1)\cdots(m-n+1)(1+x)^{m-n}$.

We consider Cauchy's Remainder after n terms:

$$\begin{aligned}
R_n \text{ (due to Cauchy)} &= \frac{x^n(1-\theta)^{n-1}}{(n-1)!} f^n(\theta x), 0 < \theta < 1. \\
&= \frac{x^n(1-\theta)^{n-1}}{(n-1)!} m(m-1)(m-2)\cdots(m-n+1)(1+\theta x)^{m-n} \\
&= \frac{m(m-1)(m-2)\cdots(m-n+1)}{(n-1)!} \cdot x^n \times \\
&\quad \left(\frac{1-\theta}{1+\theta x}\right)^{n-1} \cdot (1+\theta x)^{m-1}.
\end{aligned}$$

We now see that

(a) $\lim_{n\to\infty} \dfrac{m(m-1)(m-2)\cdots(m-n+1)}{(n-1)!} x^n = 0$, if $|x| < 1$.

[Write $a_n = \dfrac{m(m-1)(m-2)\cdots(m-n+1)}{(n-1)!} x^n$, then

$$a_{n+1} = \dfrac{m(m-1)(m-2)\cdots(m-n)}{n!} x^{n+1}$$

$$\therefore \left|\dfrac{a_{n+1}}{a_n}\right| = \left|\dfrac{m-n}{n}\right| |x| \to |x| \text{ as } x \to \infty.$$

$\therefore$ if $|x| < 1$, $\lim_{n\to\infty} a_n = 0$ (See art 5.6. worked out Ex. 5.6.4).]

(b) $\lim_{n\to\infty} \left(\dfrac{1-\theta}{1+\theta x}\right)^{n-1} = 0.$

In fact, since $0 < \theta < 1$ and $-1 < x < 1$, therefore

$$0 < \dfrac{1-\theta}{1+\theta x} < 1,$$

and hence

$$\lim_{n\to\infty} \left(\dfrac{1-\theta}{1+\theta x}\right)^{n-1} = 0.$$

(c) If $m < 1$, then $(1+\theta x)^{m-1} < (1-|x|)^{m-1}$.

If $m > 1$ then $(1+\theta x)^{m-1} < (1+1)^{m-1} = 2^{m-1}$. In any case, $(1+\theta x)^{m-1}$ is finite.

$\therefore$ From (a), (b), (c) we observe that $\forall\, x \in\,]-1, 1[\cdot \lim_{n\to\infty} R_n = 0.$

Hence the conditions for Maclaurin's infinite series expansion are satisfied if $|x| < 1$.

i.e., $f(x) = f(0) + xf'(0) + \dfrac{x^2}{\underline{|2}} f''(0) + \cdots + \dfrac{x^n}{\underline{|n}} f^n(0) + \cdots\; (\forall\, x \in\,]-1, 1[)$

or, $(1+x)^m = 1 + mx + \dfrac{m(m-1)}{2!} x^2 + \cdots + \dfrac{m(m-1)\cdots(m-n+1)}{\underline{|n}} x^n + \cdots,$

whenever $-1 < x < 1$.

This is called the **Binomial Series**.

Note 11.8.2. When $m \neq$ a positive integer the expansion is not possible if $|x| > 1$ (i.e., if $x > 1$ or $x < -1$). For then as $n \to \infty$ $\dfrac{m(m-1)\cdots(m-n+1)}{\underline{|n-1}} x^n$ and so R_n does not tend to zero.

Note 11.8.3. If we adopt Lagrange's form of remainder

$$R_n = \dfrac{m(m-1)\cdots(m-n+1)}{n!} x^n (1+\theta x)^{m-n}.$$

We cannot establish the validity of the expansion in $-1 < x < 0$. For, in that case, $1 + \theta x < 1$ and $(1 + \theta x)^{m-n} \to \infty$.

$\therefore$ Although $\dfrac{m(m-1)\cdots(m-n+1)}{n!} x^n \to 0$, we cannot conclude R_n (due to Lagrange) $\to 0$ as $n \to \infty$.

IV. Use of Power Series corresponding to standard functions

We shall prove later that a *function has no more than one series representation in powers of x in a given interval of convergence*. Thus a function representable by Maclaurin's infinite series in a certain interval is represented in one way only, i.e., the power series is itself the Maclaurin's infinite series of the function which it represents.

We shall give below some illustrations which involve power series representation of standard functions like e^x, $\sin x$ or $\cos x$, $\log(1+x)$, $(1+x)^m$.

Example 11.8.5. *Let $f(x) = e^x \sin x$.*

Solution: Here we could, though laborious, obtain the coefficients of the required Maclaurin's expansion by evaluating the various derivatives of $f(x) = e^x \sin x$ at $x = 0$. A much simpler way is to multiply together the known expansions of e^x and $\sin x$:

$$e^x \sin x = \left(1 + x + \frac{x^2}{2!} + \cdots\right)\left(x - \frac{x^3}{3!} + \frac{x^5}{5!} - \cdots\right)$$

$$= x + x^2 + \frac{x^3}{3} - \frac{x^5}{30} + \cdots,$$

which is valid for every value of x, since e^x and $\sin x$ separately converges for every value of x.

Example 11.8.6. *Obtain the series for $(1+x)^{-2}$ by differentiation.*

Solution: Since $(1+x)^{-1} = \dfrac{1}{1+x} = \sum_{n=0}^{\infty}(-1)^n x^n$ is valid for $|x| < 1$ differentiating term by term (which is permissible here),

$$-(1+x)^{-2} = \sum_{n=1}^{\infty}(-1)^n n x^{n-1} \text{ for } |x| < 1;$$

or on replacing n by $(n+1)$,

$$(1+x)^{-2} = \frac{1}{(1+x)^2} = \sum_{n=0}^{\infty}(-1)^n (n+1) x^n \text{ for } |x| < 1.$$

Example 11.8.7. (Method of undetermined co-efficients). *To find the expansion of* $\sec x$.

Solution: If $f(x) = \sec x$, then assuming that it is representable by a power series, we have
$$\sec x = \frac{1}{\cos x} = \frac{1}{1 - \frac{x^2}{2!} + \frac{x^4}{4!} - \cdots} = a_0 + a_1 x + a_2 x^2 + \cdots,$$

where, $a_0, a_1, a_2, \cdots$ are constants to be determined (i.e., *undetermined constants*). Then, cross-multiplying

$$\begin{aligned} 1 &= \left(1 - \frac{x^2}{2!} + \frac{x^4}{4!} - \cdots\right)(a_0 + a_1 x + a_2 x^2 + a_3 x^3 + a_4 x^4 + \cdots) \\ &= a_0 + a_1 x + \left(a_2 - \frac{a_0}{2}\right) x^2 + \left(a_3 - \frac{a_1}{2}\right) x^3 + \left(a_4 - \frac{a_2}{2} + \frac{a_0}{24}\right) x^4 + \cdots \end{aligned}$$

and equating like powers of x,

$$a_0 = 1, \quad a_1 = 0, \quad a_2 - \frac{a_0}{2} = 0, \quad a_3 - \frac{a_1}{2} = 0, \quad a_4 - \frac{a_2}{2} + \frac{a_0}{24} = 0, \cdots,$$

which gives

$$a_0 = 1, \quad a_1 = 0, \quad a_2 = \frac{1}{2}, \quad a_3 = 0, \quad a_4 = \frac{-5}{24}, \cdots,$$

and hence

$$\sec x = 1 + \frac{x^2}{2} + \frac{5x^4}{24} + \cdots.$$

Method of formation of Differential Equations

Example 11.8.8. *If* $y = e^{ax} \cos bx$, *prove that* $y_2 - 2ay_1 + (a^2 + b^2)y = 0$ *and hence obtain the expansion of* $e^{ax} \cos bx$. *Deduce the expansion of* e^{ax} *and* $\cos bx$.

Solution: Let

$$\begin{aligned} y &= e^{ax} \cos bx = a_0 + a_1 x + a_2 x^2 + a_3 x^3 + \cdots. & (11.8.4) \\ \therefore \quad y_1 &= ae^{ax} \cos bx - be^{ax} \sin bx = a_1 + 2a_2 x + 3a_3 x^2 + \cdots & (11.8.5) \\ y_2 &= ae^{ax}(a \cos bx - b \sin bx) - be^{ax}(a \sin bx + b \cos bx) \\ &= 1 \cdot 2 a_2 + 2 \cdot 3 a_3 x + 3 \cdot 4 a_4 x^2 + \cdots. & (11.8.6) \end{aligned}$$

Clearly from (11.8.5) and (11.8.6),

$$y_2 - 2ay_1 = -(a^2 + b^2)e^{ax} \cos bx$$
$$\text{or,} \quad y_2 - 2ay_1 + (a^2 + b^2)y = 0. \qquad (11.8.7)$$

Put $x = 0$ in (11.8.4), (11.8.5), (11.8.6) and obtain $a_0 = 1; a_1 = a; a_2 = \frac{1}{2}(a^2 - b^2)$.
Now substituting the series for y_2, y_1 and y in (11.8.7) and equating the coefficients of x, x^2, etc. we may obtain a_3, a_4 etc.

Hence,
$$e^{ax} \cos bx = 1 + ax + \frac{a^2 - b^2}{2!}x^2 + \frac{a(a^2 - 3b^2)}{3!}x^3 + \cdots . \qquad (11.8.8)$$

Put $b = 0$ and $a = 0$ successively in (11.8.8) and obtain

$$e^{ax} = 1 + ax + \frac{a^2 x^2}{2!} + \cdots \quad \text{and} \quad \cos bx = 1 - \frac{b^2 x^2}{2!} + \frac{b^4 x^4}{4!} - \cdots .$$

Chapters 11A & 11B: Main Points

1. **Rolle's theorem** states that if a function f is
a) continuous in $[a, b]$,
b) derivable in (a, b) and
c) $f(a) = f(b)$,
then $\exists$ at least one point c where $a < c < b$ such that $f'(c) = 0$.

The conditions are *sufficient* but *not necessary* to obtain a c for which $f'(c) = 0$.

2. **Mean-Value Theorem.** When the first two conditions of Rolle's Theorem hold [i.e., when $f(a)$ need not be equal to $f(b)$] we get at least one c, where $a < c < b$ such that

$$f(b) - f(a) = (b - a)f'(c).$$

h–θ form : $\quad f(a + h) = f(a) + hf'(a + \theta h), \quad 0 < \theta < 1.$ (Lagrange's).

Useful deductions:

a) If f is continuous in $[a, b]$ and $f'(x) > 0$ in (a, b), then f is strictly increasing in $[a, b]$. Similar result for strictly decreasing functions.

b) If f is continuous at c and if $\lim_{x \to c} f'(x) = l$, then f is derivable at the point c and $f'(c) = l$.

3. Cauchy's Mean-Value Theorem. If the functions f and g be

a) both continuous in $[a, b]$,

b) both derivable in (a, b), and

c) $g'(x) \neq 0$ for any x in (a, b),

then $\exists$ at least one point c where $a < c < b$ such that

$$\frac{f(b) - f(a)}{g(b) - g(a)} = \frac{f'(c)}{g'(c)} \quad (a < c < b).$$

h–θ form: $\quad \dfrac{f(a+h) - f(a)}{g(a+h) - g(a)} = \dfrac{f'(a+\theta h)}{g'(a+\theta h)} \quad (0 < \theta < 1).$

Theorem of Cauchy reduces to the *Mean-Value Theorem* when $g(x) = x$. This, in turn, becomes *Rolle's Theorem* when $f(a) = f(b)$.

4. Taylor's Theorem.

a) *with Lagrange's form of remainder*: If f^{n-1} is continuous in $[a, b]$ and f^n exists in $(a, a+h)$, then $\exists$ at least one number $\theta, 0 < \theta < 1$, such that

$$f(a+h) = f(a) + hf'(a) + \frac{h^2}{2!}f''(a) + \cdots + \frac{h^{n-1}}{\underline{|n-1}}f^{n-1}(a)$$
$$+ \frac{h^n}{\underline{|n}}f^n(a+\theta h), \quad (0 < \theta < 1).$$

We call $R_n = \dfrac{h^n}{\underline{|n}} f^n(a + \theta h)$, Lagrange's Remainder after n terms.

b) **With Cauchy's form of remainder.**

$$f(a+h) = f(a) + hf'(a) + \frac{h^2}{2!}f''(a) + \cdots + \frac{h^{n-1}}{\underline{|n-1}}f^{n-1}(a)$$
$$+ \frac{h^n(1-\theta)^{n-1}}{\underline{|n-1}}f^n(a+\theta h), \quad (0 < \theta < 1).$$

We call $R_n = \dfrac{h^n(1-\theta)^{n-1}}{\underline{|n-1}} f^n(a+\theta h)$, Cauchy's form of Remainder after n terms.

5. Maclaurin Expansion of $f(x)$: Using Taylor's Theorem in $[0, x]$.

$$f(x) = f(0) + xf'(0) + \frac{x^2}{2!}f''(0) + \cdots + \frac{x^{n-1}}{\lfloor n-1}f^{n-1}(0) + R_n$$

where

$$\left. \begin{array}{l} R_n = \dfrac{x^n}{\lfloor n} f^n(\theta x) \text{ (due to } Lagrange\text{)} \\ = \dfrac{x^n(1-\theta)^{n-1}}{\lfloor n-1} f^n(\theta x) \text{ (due to } Cauchy\text{)} \end{array} \right\} \quad 0 < \theta < 1.$$

6. Young's form of Taylor's Theorem. If a function f be such that its nth derivative exists at $x = a$ and if M is defined by

$$f(a+h) = f(a) + hf'(a) + \frac{h^2}{2!}f''(a) + \cdots + \frac{h^{n-1}}{\lfloor n-1}f^{n-1}(a) + \frac{h^n}{\lfloor h}M,$$

then $M \to f^n(a)$ as $h \to 0$.

7. Taylor's Infinite Series. If f possesses derivatives of all orders at any point x in $(a-h, a+h)$ and if the remainder R_n (any form) after n terms of Taylor's Theorem tends to zero as $n \to \infty$, then

$$f(x) = f(a) + (x-a)f'(a) + \frac{(x-a)^2}{2!}f''(a) + \cdots + \frac{(x-a)^n}{\lfloor n}f^n(a) + \cdots.$$

Putting $a = 0$, we have the Maclaurin's Infinite series for $f(x)$ about the origin

$$f(x) = f(0) + xf'(0) + \frac{x^2}{2!}f''(0) + \cdots + \frac{x^n}{\lfloor n}f^n(0) + \cdots.$$

To remember: The mere convergence of an infinite series like the right side above does not always imply that the series has the sum $f(x)$—it may have a different sum. Example: *Cauchy Function*: $f(x) = e^{-1/x^2}$, $x \neq 0$ and $f(0) = 0$.

8. Useful Power Series. (Valid for all real x)

$$e^x = 1 + x + \frac{x^2}{2!} + \frac{x^3}{3!} + \cdots;$$

$$\sin x = x - \frac{x^3}{3!} + \frac{x^5}{5!} - \frac{x^7}{7!} + \cdots \qquad \cos x = 1 - \frac{x^2}{2!} + \frac{x^4}{4!} - \frac{x^6}{6!} + \cdots;$$

$$\sinh x = x + \frac{x^3}{3!} + \frac{x^5}{5!} + \frac{x^7}{7!} + \cdots \qquad \cosh x = 1 + \frac{x^2}{2!} + \frac{x^4}{4!} + \frac{x^6}{6!} + \cdots.$$

9. Power Series Valid in Restricted Intervals.

The *Logarithmic series*: $\log(1+x) = x - \dfrac{x^2}{2} + \dfrac{x^3}{3} - \dfrac{x^4}{4} + \cdots$, $(-1 < x \leq 1)$

The *Binomial series*:

$$(1+x)^m = 1 + mx + \frac{m(m-1)}{2!}x^2$$
$$+ \frac{m(m-1)(m-2)}{3!}x^3 + \cdots \; (-1 < x < 1).$$

Both diverge when $|x| > 1$. The Binomial Series, however, converges at $x = 1$ only when the index $m > -1$, at $x = -1$ when $m > 0$.

Exercises XI(III)

1. Verifying Maclaurin's infinite series corresponding to the following functions in the respective intervals in which the series converge to the functions:

(a) $\log(1 + 2x) = \sum\limits_{r=1}^{\infty}(-1)^{r-1}(2x)^r/r$; $\quad -\dfrac{1}{2} < x \leq \dfrac{1}{2}$.

(b) $(1-x)^{-1/2} = \sum\limits_{r=0}^{\infty} 1 \cdot 3 \cdots (2r-1)x^r/(r!2^r)$; $\quad |x| < 1$.

(c) $\tan^{-1} x = \sum\limits_{r=1}^{\infty}(-1)^{r+1}\dfrac{1}{2r-1}x^{2r-1}$; $\quad |x| \leq 1$.

(d) $e^x \sin x = \sum\limits_{r=1}^{\infty}\dfrac{x^r}{r!}2^{r/2}\sin\left(\dfrac{1}{4}r\pi\right)$; for all x.

2. Show that

(a) $\log x = (x-1) - \dfrac{1}{2}(x-1)^2 + \dfrac{1}{3}(x-1)^3 - \cdots$; $\quad 0 < x \leq 2$.

(b) $\dfrac{1}{x} = \dfrac{1}{2} - \dfrac{1}{2^2}(x-2) + \dfrac{1}{2^3}(x-2)^2 - \cdots$; $\quad 0 < x < 4$.

3. Assuming the possibility of expansion, verify the following:

(a) $\tan x = x + \dfrac{1}{3}x^3 + \dfrac{2}{15}x^5 + \cdots$.

(b) $\cos^2 x = 1 - x^2 + \dfrac{1}{3}x^4 - \cdots$.

(c) $e^{\sin x} = 1 + x + \dfrac{1}{2}x^2 - \dfrac{1}{8}x^4 + \cdots$.

(d) $e^x \cdot \sin x = x + x^2 + \dfrac{1}{3}x^3 - \cdots$.

(e) $\log(1 + \sin x) = x - \dfrac{1}{2}x^2 + \dfrac{1}{6}x^3 - \cdots$.

(f) $\log \sec x = \tfrac{1}{2}x^2 + \dfrac{1}{12}x^4 + \dfrac{1}{45}x^6 + \cdots$.

(g) $e^x = e^2 \left\{ 1 + (x-2) + \dfrac{1}{2!}(x-2)^2 + \dfrac{1}{3!}(x-2)^3 + \cdots \right\}$.

(h) $\sin^{-1} x = x + \dfrac{1}{2}\dfrac{x^3}{3} + \dfrac{1 \cdot 3}{2 \cdot 4}\dfrac{x^5}{5} + \dfrac{1 \cdot 3 \cdot 5}{2 \cdot 4 \cdot 6}\dfrac{x^7}{7} + \cdots$.

4. Show that $f(x) = x^{3/2}$ cannot be expanded in Maclaurin's infinite series or that for this function the expansion of $f(x+h)$ fails when $x = 0$, but that there exists a proper fraction θ such that

$$f(x+h) = f(x) + hf'(x) + \frac{1}{2}h^2 f''(x + \theta h) \text{ holds when } x = 0.$$

5. Show by the addition of power series that

$$\frac{1}{1-x} + \cos x = 1 + 1 + x - \frac{x^2}{2!} + x^2 + \frac{x^4}{4!} + \cdots, \text{ for } -1 < x < 1.$$

6. Show by squaring the series that, for $-1 < x < 1$,

$$(1 + x + x^2 + \cdots + x^n + \cdots)^2 = 1 + 2x + 3x^2 + \cdots + (n+1)x^n + \cdots = \frac{1}{(1-x)^2}.$$

7. If $f(x) = 1 + x + \dfrac{x^2}{2!} + \dfrac{x^3}{3!} + \cdots$, prove by multiplication of power series that

$$f(x) \cdot f(x') = f(x + x').$$

8. By power series multiplication, compute the first few terms in the expansions about $x = 0$ of:
   a) $\sin x \cos x$; b) $e^x \sin x$; c) $\sin x \cdot \arcsin x$;
   d) $e^x \cos x$; e) $(e^x)^2$; f) $(e^x)^3$.

9. By division of power series, compute the first few terms in the expansions about $x = 0$ of:
   a) $\dfrac{1}{\cos x}$; b) $\dfrac{\sin x}{\cos x}$; c) $\dfrac{1}{\sin x}$; d) $\dfrac{1}{e^x + 1}$.

10. Prove by repeated differentiation of the identity
$$(1-x)^{-1} = 1 + x + x^2 + x^3 + \cdots, \text{ where } |x| < 1$$
that if m be a positive integer
$$(1-x)^{-m} = 1 + mx + \frac{m(m+1)}{1 \cdot 2}x^2 + \frac{m(m+1)(m+2)}{1 \cdot 2 \cdot 3}x^3 + \cdots.$$

11. Obtain the series for $\cos x, \sec^2 x, \sec x \tan x, \sec x - \tan x$ by differentiating the respective series of $\sin x, \tan x, \sec x, \log(1 + \sin x)$.

12. If $y = e^{a \sin^{-1} x} = a_0 = a_1 x + a_2 x^2 + \cdots + a_n x^n + \cdots$, show that

   (a) $(1 - x^2)y_2 = xy_1 + a^2 y$;

   (b) $(n+1)(n+2)a_{n+2} = (n^2 + a^2)a_n$;

and hence obtain the expansion of $e^{a \sin^{-1} x}$. Find the general term. Deduce the expansion of $\sin^{-1} x$.

13. If $y = e^{\tan^{-1} x} = \sum_{n=0}^{\infty} a_n x^n$, show that
$$(n+2)a_{n+2} + n a_n = a_{n+1},$$
and hence obtain the expansion of $e^{\tan^{-1} x}$ after the determination of $a_0, a_1, a_2, a_3, \cdots$.

14. If $y = \sin(m \sin^{-1} x)$, show that
$$(1 - x^2)y_{n+2} - (2n+1)xy_{n+1} + (m^2 - n^2)y_n = 0,$$
and hence obtain the expansion of $\sin(m \sin^{-1} x)$ in powers of x.

15. Assuming that $\sin^{-1} x / \sqrt{1 - x^2}$ can be expanded by Maclaurin's theorem in the form
$$a_0 + a_1 x + a_2 x^2 + \cdots,$$
find a_{2n} and a_{2n+1}.

16. Apply Maclaurin's Theorem with Lagrange's remainder to find the power series expansion of $e^{\alpha x + \beta}$, where α, β are constants and $\beta \neq 0$. Indicate the range of validity.   [C.H. 1980 old]

Answers

12. $e^{a\sin^{-1}x} = 1 + ax + \dfrac{a^2x^2}{2!} + \dfrac{a(a^2+1^2)}{3!}x^3 + \dfrac{a^2(a^2+2^2)}{4!} + \cdots$.

General term $= \dfrac{a(a^2+1^2)(a^2+3^2)\cdots\{a^2+(n-2)^2\}}{n!}x^n$, when n is odd

$\phantom{\text{General term }} = \dfrac{a^2(a^2+2^2)(a^2+4^2)\cdots\{a^2+(n-2)^2\}}{n!}x^n$, when n is even.

$\sin^{-1}x = x + \dfrac{1}{2}\cdot\dfrac{x^3}{3} + \dfrac{1\cdot 3}{2\cdot 4}\cdot\dfrac{x^5}{5} + \dfrac{1\cdot 3\cdot 5}{2\cdot 4\cdot 6}\dfrac{x^7}{7} + \cdots$.

13. $1 + x + \dfrac{1}{2}x^2 - \dfrac{1}{6}x^3 - \dfrac{7}{24}x^4 - \cdots$.

14. $mx - \dfrac{m(m^2-1^2)}{3!}x^3 + \dfrac{m(m^2-1^2)(m^2-3^2)}{3!}x^5 - \cdots$.

15. $a_{2n} = 0$ and $a_{2n+1} = \dfrac{2n(2n-2)\cdots 2}{(2n+1)(2n-1)\cdots 3}$.

AN INTRODUCTION TO ANALYSIS

(Differential Calculus—Part I)

Chapter 12

Indeterminate Forms and Maxima & Minima

Chapter 12

Indeterminate Forms and Maxima & Minima

Problems of Indeterminate Forms

12.1 Introduction

In the present chapter we shall discuss two important applications of Mean Value Theorems:

1. Evaluation of certain limits which are called Indeterminate forms by L'Hospital's Rule.

2. Problems of extreme values of a function of a single real variable.

1. In general, $\lim_{x \to a} \dfrac{\phi(x)}{\psi(x)} = \dfrac{\lim_{x \to a} \phi(x)}{\lim_{x \to a} \psi(x)}$, in case both the limits—$\lim_{x \to a} \phi(x)$ and $\lim_{x \to a} \psi(x)$ exist and the latter limit $\neq 0$. But if both these two limits are zero then we face with the problem like 0/0 which is meaningless. Such a case is known as **Indeterminate form.**

Other forms of indeterminate forms are $\infty/\infty, 0 \times \infty, \infty - \infty, 0^0, 1^\infty$ and ∞^0. For evaluation of indeterminate forms: ∞/∞ or $0/0$ we shall use a particular device known as L'Hospital's Rule.

2. The other application namely, determination of extreme values, is one of the most useful applications of differential calculus.

** f is said to have a *maximum value* at $x = c$ if $f(c)$ is the greatest value of the function in some small neighbourhood $]c - \delta, c + \delta[$, $\delta > 0$ of c i.e., $f(c) > f(x)$ for all $x \in]c - \delta, c + \delta[$, $x \neq c$.

** f is said to have a *minimum value* at $x = c$ if $f(c)$ is the least value of the function in some small neighbourhood $]c - \delta, c + \delta[$, $\delta > 0$ of c i.e.,

$$f(c) < f(x) \ \forall \ x \in]c - \delta, c + \delta[, \ x \neq c.$$

** The function f is said to have an extreme value at $x = c$ if $f(c)$ is either a maximum or a minimum value of f. i.e., *at an extreme point* c, $f(x) - f(c)$ keeps a constant sign, for all values of x in some deleted neighbourhood of c.

12.2 Indeterminate Forms : L'Hospital's Rule

(A) Form 0/0

The ratio $f(x)/\phi(x)$ of two functions $f(x)$ and $\phi(x)$ is *undefined* at $x = a$, if $\phi(a) = 0$. What happens if both $f(a) = 0$ and $\phi(a) = 0$? In this case obviously the ratio is of the form 0/0; we call it an *Indeterminate Form*. But if $\lim_{x \to a} f(x)/\phi(x)$ has a definite real value, we call that limit as the *true* or *appropriate* value of $f(x)/\phi(x)$ at $x = a$. e.g., whenever we compute a derivative

$$\lim_{x \to a} \frac{\phi(x) - \phi(a)}{x - a} = \phi'(a),$$

we have a problem of this sort; for $\phi(x) - \phi(a)$ and $x - a$ are functions of x which vanish for $x = a$. In fact the knowledge of derivatives helps us to evaluate the *True Value* of an indeterminate form:

A basic theorem in this connection is given by:

L'Hospital's Rule: *If $f(x)$ and $g(x)$ are*

(a) *continuous in the closed interval $a \leq x \leq a + h$,*

(b) *derivable in the open interval $a < x < a + h$, and*

(c) $f(a) = g(a) = 0$

then, $\lim_{x \to a+} \frac{f(x)}{g(x)} = A$, *provided* $\lim_{x \to a+} \frac{f'(x)}{g'(x)}$ *exists and is equal to A.*

Proof: Suppose that $\lim_{x \to a+} \frac{f'(x)}{g'(x)}$ exists and equals a finite number A.

Then $g'(x)$ cannot vanish $\forall \ x$ in *some suitable right neighbourhood* of a; let the neighbourhood be $a < x \leq a + h$, where $h > 0$ is chosen suitably small.

CHAPTER 12: INDETERMINATE FORMS AND MAXIMA & MINIMA

For, otherwise $f'(x)/g'(x)$ would not be defined for an infinite number of values of x between a and $a+h$ (no matter how small h is chosen) and thereby the limit A would not exist.

We now see that *all the conditions* of Cauchy's MVT are met in the interval $[a, a+h]$.

[Continuity of f and g in $[a, a+h]$; derivability of f and g in $(a, a+h)$ and $g'(x) \neq 0$ for any x in $(a, a+h)$].

Now,

$$\frac{f(x)}{g(x)} = \frac{f(x) - f(a)}{g(x) - g(a)} \quad (\because f(a) = g(a) = 0, \text{ given})$$

$$= \frac{f'(c)}{g'(c)} \text{ for } a < c < b \text{ (using Cauchy's MVT)}.$$

$$\therefore \lim_{x \to a^+} \frac{f(x)}{g(x)} = \lim_{c \to a^+} \frac{f'(c)}{g'(c)} = \lim_{x \to a^+} \frac{f'(x)}{g'(x)} = A.$$

We note here that, no matter how $x \to a^+$, $c \to a^+$ in some way (in general, unknown) and hence $f'(c)/g'(c) \to A$ as $x \to a^+$.

Observations:

1. If the ratio $f'(x)/g'(x)$ has no determined limit as $x \to a^+$, then we *should not* conclude that $f(x)/g(x)$ also has no limit as $x \to a^+$.

   For all sequences $\{x_n\} \to a^+$ may correspond to only **special sequences** $\{c_n\}$ such that $c_n \to a^+$, and for these $f'(c)/g'(c)$ may have a limit whereas, for sequences other than these special ones, $f'(c)/g'(c)$ may not have a limit. Consider the following:

   Example: Let $f(x) = x^2 \sin(1/x)$, $x \neq 0$; $= 0$, $x = 0$ and $g(x) = \sin x$.

   Then as $x \to 0^+$, $\dfrac{f(x)}{g(x)} = \dfrac{x}{\sin x} \cdot x \sin \dfrac{1}{x} \to 0$ but $\dfrac{f'(x)}{g'(x)} = \dfrac{2x \sin(1/x) - \cos(1/x)}{\cos x}$ does not possess a limit as $x \to 0^+$.

   This example shows that the conditions of the theorem are *sufficient* but *not necessary* for the existence of the *true value* of the indeterminate form $0/0$.

2. If the conditions of the theorem hold for the interval $[a-h, a]$ or $[a-h, a+h]$, then the theorem holds for the limits $x \to a^-$ or $x \to a$, respectively.

3. The proof given above can be carried out without any essential change if as $x \to a^+$, $A = +\infty$ or $A = -\infty$.

4. *If conditions* (a) *and* (b) *of L'Hospital's rule hold when x is sufficiently large (say $x > k$) and* (c) $\lim_{x \to \infty} f(x) = 0 = \lim_{x \to \infty} g(x)$, *then* $\lim_{x \to \infty} \dfrac{f(x)}{g(x)} = \lim_{x \to \infty} \dfrac{f'(x)}{g'(x)}$, *when the latter limit exists.*

Proof: Put $x = 1/t$, then

$$\lim_{x \to \infty} \frac{f(x)}{g(x)} = \lim_{t \to 0+} \frac{f(1/t)}{g(1/t)} = \lim_{t \to 0+} \frac{-t^{-2}f'(1/t)}{-t^{-2}g'(1/t)} = \lim_{x \to \infty} \frac{f'(x)}{g'(x)}.$$

Thus L'Hospital's rule holds when $a = \infty$.

Extension of the Rule.

If also $\lim_{x \to a} f'(x) = 0 = \lim_{x \to a} g'(x)$, then

$$\lim_{x \to a} \frac{f(x)}{g(x)} = \lim_{x \to a} \frac{f'(x)}{g'(x)} = \lim_{x \to a} \frac{f''(x)}{g''(x)},$$

if this last limit exists.

We may continue in this fashion until at least one derivative in the ratio is not zero when $x \to a$. In the ultimate stage, the numerator, say $f^r(a) \neq 0$ and $g^r(a) = 0$, the limit does not exist, but if $f^r(a) = 0$ and $g^r(a) \neq 0$ the limit is obviously 0.

Applications

Example 12.2.1. $\lim_{x \to 0} \dfrac{\log(1+x)}{x}$ $\left(\dfrac{0}{0} \text{ form}\right)$.

Solution: Here $f(x) = \log(1+x)$ and $g(x) = x$.

$$\frac{f'(x)}{g'(x)} = \frac{1}{1+x} \bigg/ 1 = \frac{1}{1+x} \to 1 \text{ as } x \to 0.$$

$\therefore$ By L'Hospital's rule, given limit i.e., $\lim_{x \to 0} \dfrac{\log(1+x)}{x} = 1$.

Example 12.2.2. $\lim_{x \to 0} \dfrac{\tan x - x}{x - \sin x}$ $\left(\dfrac{0}{0} \text{ form}\right)$.

Solution: Here $f(x) = \tan x - x$ and $g(x) = x - \sin x$.

$$\frac{f'(x)}{g'(x)} = \frac{\sec^2 x - 1}{1 - \cos x} \quad \left(\frac{0}{0} \text{ form for } x = 0\right).$$

Now,
$$\frac{f''(x)}{g''(x)} = \frac{2\sec^2 x \cdot \tan x}{\sin x} = 2\sec^3 x, \text{ when } x \neq 0.$$

$$\therefore \lim_{x \to 0} \frac{f''(x)}{g''(x)} = \lim_{x \to 0} 2\sec^3 x = 2.$$

∴ Using the extended rule of L'Hospital's
$$\lim_{x \to 0} \frac{f(x)}{g(x)} = \lim_{x \to 0} \frac{f''(x)}{g''(x)} = 2.$$

Example 12.2.3. (a) If $f'(x)$ exists in the neighbourhood of a,
$$\lim_{h \to 0} \frac{f(a+h) - f(a-h)}{2h} = \lim_{h \to 0} \frac{f'(a+h) + f'(a-h)}{2} = f'(a).$$

(b) If $f''(x)$ exists in the neighbourhood of a,
$$\lim_{h \to 0} \frac{f(a+h) + f(a-h) - 2f(a)}{h^2} = \lim_{h \to 0} \frac{f'(a+h) - f'(a-h)}{2h} = f''(a).$$

Example 12.2.4. $\lim_{x \to \infty} \frac{e^{-2x}(\cos x + 2\sin x)}{e^{-x}(\cos x + \sin x)} \quad \left(\frac{0}{0} \text{ form}\right).$

Solution: Here $f(x) = e^{-2x}(\cos x + 2\sin x); \quad g(x) = e^{-x}(\cos x + \sin x).$

Then,
$$\frac{f'(x)}{g'(x)} = \frac{-5e^{-2x} \cdot (\sin x)}{-2e^{-x} \cdot (\sin x)}.$$

When $\sin x \neq 0$, $f'(x)/g'(x) = \frac{5}{2}e^{-x} \to 0$ as $x \to \infty$.

But certainly f'/g' here has no limit as $x \to \infty$, for as $x \to \infty$ along the sequence $\{n\pi\}$, f'/g' is not defined.

What is to be stressed upon is that
$$\lim_{x \to \infty} \frac{f'(x)}{g'(x)} = A \text{ is true when } \lim_{x \to \infty} \frac{f'(x)}{g'(x)} = A$$

for all sequences $\{x_n\}$ such that $x_n \to \infty$.

If we cancel $\sin x$ in f'/g' and then take $\lim_{x \to \infty} f'/g' = 0$, we arrive at the false conclusion that $\lim_{x \to \infty} f/g = 0$; see that as $x \to \infty$
$$\frac{f(x)}{g(x)} = e^{-x} \frac{1 + 2\tan x}{1 + \tan x}$$

approaches no limit whatever, since it takes on all real values in every neighbourhood of ∞.

Example 12.2.5. $\lim\limits_{x \to 0} \dfrac{(1+x)^{1/x} - e}{x}$.

Solution: Here $f(x) = (1+x)^{1/x} - e \to 0$ as $x \to 0$
$g(x) = x \to 0$ as $x \to 0$.

Therefore, it is of the form $\dfrac{0}{0}$.

Using L'Hospital's Rule

$$\begin{aligned}
\text{Given limit} &= \lim_{x \to 0} \frac{\frac{d}{dx}(1+x)^{1/x}}{1} = \lim_{x \to 0} \frac{(1+x)^{1/x}\{x - (1+x)\log(1+x)\}}{x^2(1+x)} \\
&= e \cdot \lim_{x \to 0} \frac{x - (1+x)\log(1+x)}{x^2(1+x)} \quad \left(\frac{0}{0} \text{ form}\right) \\
&= -e \cdot \lim_{x \to 0} \frac{\log(1+x)}{2x + 3x^2} \quad \left(\frac{0}{0} \text{ form}\right) \\
&= -e \cdot \lim_{x \to 0} \frac{1}{(1+x)(2+6x)} = -e/2.
\end{aligned}$$

(B) Another Useful form $\dfrac{\infty}{\infty}$

Suppose $f(x) \to \infty$ and $g(x) \to \infty$ as $x \to a$. To evaluate

$$\lim_{x \to a} \frac{f(x)}{g(x)}.$$

We shall call it the *indeterminate form* $\dfrac{\infty}{\infty}$.

L'Hospital's rule can be applied in this case also but the proof of theorem is more difficult. If we try to reduce this case to the form $\dfrac{0}{0}$, we have

$$\lim_{x \to a} \frac{f(x)}{g(x)} = \lim_{x \to a} \frac{1/g(x)}{1/f(x)} = \lim_{x \to a} \frac{-g'(x)/\{g(x)\}^2}{-f'(x)/\{f(x)\}^2} = \lim_{x \to a} \left\{\frac{f(x)}{g(x)}\right\}^2 \cdot \frac{g'(x)}{f'(x)}.$$

Hence if it is given in advance that

$$\lim_{x \to a} \frac{f(x)}{g(x)} = L \neq 0, \quad \text{and} \quad \lim_{x \to a} \frac{f'(x)}{g'(x)} = A \neq 0$$

then we have $L = L^2/A$ and we get $L = A$.

But since, the existence of L is generally not known in advance, these arguments will not be applicable. What we need is a rule where A exists (given), to prove L exists and $L = A$. We thus give the following theorem:

CHAPTER 12: INDETERMINATE FORMS AND MAXIMA & MINIMA

Theorem 12.2.1. *If the functions $f(x)$ and $\phi(x)$ are*

1. *continuous in $a \leq x \leq a + h$,*

2. *derivable in $a < x \leq a + h$,*

3. $\lim\limits_{x \to a+} f(x) = \infty = \lim\limits_{x \to a+} \phi(x)$, *and*

4. $\lim\limits_{x \to a+} \dfrac{f'(x)}{\phi'(x)} = l \ (or \ \pm \infty)$,

$$\text{then,} \quad \lim_{x \to a+} \frac{f(x)}{\phi(x)} = l \ (or \ \pm \infty).$$

Proof: Since the limit in (4) exists, $\phi'(x) \neq 0$ in some interval $a < x \leq a+h$. Hence by applying Cauchy's MVT in $[a, a+h]$ and if $a < x < b < a+h$,

$$\frac{f(x) - f(b)}{\phi(x) - \phi(b)} = \frac{f(x)}{\phi(x)} \cdot \frac{1 - f(b)/f(x)}{1 - \phi(b)/\phi(x)} = \frac{f'(c)}{\phi'(c)}$$

where $a < x < c < b$. Hence

$$\frac{f(x)}{\phi(x)} = \frac{f'(c)}{\phi'(c)} \cdot \frac{1 - \phi(b)/\phi(x)}{1 - f(b)/f(x)}. \tag{12.2.1}$$

Case 1. When l is finite.

Choose ε, arbitrarily small between 0 and 1 and take b (and hence c) close enough to a so that by virtue of (4), we can make $f'(c)/\phi'(c)$ differ from l by less than ε,

$$\text{i.e.,} \quad \frac{f'(c)}{\phi'(c)} = l + \eta_1, \ |\eta_1| < \varepsilon < 1.$$

When b is fixed and $x \to a$, from (3) it is clear that $\dfrac{1 - \phi(b)/\phi(x)}{1 - f(b)/f(x)} \to 1$. Hence taking x close enough to a, we have

$$\frac{1 - \phi(b)/\phi(x)}{1 - f(b)/f(x)} = 1 + \eta_2 \ \text{where} \ |\eta_2| < \begin{cases} \varepsilon & \text{if } |l| < 1; \\ \varepsilon/|l| & \text{if } |l| > 1. \end{cases}$$

Then, $\dfrac{f(x)}{\phi(x)} = (l + \eta_1)(1 + \eta_2) = l + \eta_1 + l\eta_2 + \eta_1\eta_2$

$$\text{i.e.,} \quad \left|\frac{f(x)}{\phi(x)} - l\right| \leq |\eta_1| + |l||\eta_2| + |\eta_1||\eta_2| < 3\varepsilon = \varepsilon' \ (\text{say}).$$

Hence $\lim_{x \to a^+} \{f(x)/\phi(x)\} = l$.

Case 2. $l = +\infty$. Choose b close enough to a so that

$$\frac{f'(x)}{\phi'(x)} > M, \quad M \text{ being aribitrarily large.}$$

Keeping b fixed, choose x so near to a that

$$\frac{1 - \phi(b)/\phi(x)}{1 - f(b)/f(x)} > 1 - \frac{1}{M}$$

then from (12.2.1), $\dfrac{f(x)}{\phi(x)} > M\left(1 - \dfrac{1}{M}\right) = M - 1$ and hence $f(x)/\phi(x) \to \infty$.

Case 3. $l = -\infty$. Since $-f'(x)/\phi'(x) \to \infty$, by Case 2, $-f(x)/\phi(x) \to \infty$.

Hence the theorem. Similar things happen when $x \to a^-$.

Corollary 12.2.1. *If conditions 1 and 2 hold when x is sufficiently large (say $x > k$) and*

$$\lim_{x \to \infty} f(x) = \infty = \lim_{x \to \infty} \phi(x),$$

then $\lim_{x \to \infty} \{f(x)/\phi(x)\} = \lim_{x \to \infty} \{f'(x)/\phi'(x)\}$, *when the last limit exists.*

The proof is the same *as for the form* 0/0 given in 4.

Observation: The conditions in the theorem are *sufficient and by no means necessary* for the existence of limit $f(x)/\phi(x)$. Consider the following example.

Let

$$f(x) = \frac{1}{x}; \quad \phi(x) = \frac{1}{x} + \sin\frac{1}{x}.$$

As $x \to 0$, f/ϕ takes the form $\dfrac{\infty}{\infty}$. Now

$$\frac{f'(x)}{\phi'(x)} = \frac{-x^{-2}}{-x^{-2}\{1 + \cos(1/x)\}} = \frac{1}{1 + \cos(1/x)}$$

approaches no limit as $x \to 0$, but

$$\frac{f(x)}{\phi(x)} = \frac{1}{1 + x\sin(1/x)} \to 1.$$

(C) Other Indeterminate forms: $0.\infty$; $\infty - \infty$; 0^0; ∞^0; $1^{\pm\infty}$

1. The form $0.\infty$ is obviously reducible to either $0/0$ or ∞/∞.

CHAPTER 12: INDETERMINATE FORMS AND MAXIMA & MINIMA

2. The form $\infty - \infty$ is readily altered to $\infty \cdot 0$ for if $f(x)$ and $\phi(x) \to \infty$ as $x \to a$, we have

$$f - \phi = f\phi\left(\frac{1}{\phi} - \frac{1}{f}\right).$$

3. The **three exponential forms** $0^0, \infty^0, 1^{\pm\infty}$ are dealt with by taking their logarithms; in all cases these lead to the form $0 \cdot \infty$.

Applications

Example 12.2.6. *If n be a positive integer, we find after n applications of the L'Hospital's rule*

$$\lim_{x \to \infty} \frac{x^n}{e^x} \left(\frac{\infty}{\infty}\right) = \lim_{x \to \infty} \frac{n!}{e^x} = 0.$$

If, again α be a positive number between the integers $n - 1$ and n using the rule n times

$$\lim_{x \to \infty} \frac{x^\alpha}{e^x} = \lim_{x \to \infty} \frac{\alpha(\alpha-1)(\alpha-2)\cdots(\alpha-n+1)}{e^x} \cdot x^{\alpha-n} = 0,$$

since $\alpha - n < 0$.

Thus we can say that *the exponential e^x approaches infinite more rapidly than x^α* $(\alpha > 0)$, no matter how large α be taken, as x approaches ∞.

When $\alpha \leq 0$, x^α/e^x is not an indeterminate form as $x \to \infty$.

We then see directly that as $x \to \infty$, $x^\alpha/e^x \to 0$. Hence $x^\alpha/e^x \to 0$ for all real values of α.

Example 12.2.7. $\lim\limits_{x \to 0} \cot x \cdot \log \dfrac{1+x}{1-x}$ ($\infty \cdot 0$ *form*).

$$\text{Given limit} = \lim_{x \to 0} \frac{\log(1+x) - \log(1-x)}{\tan x} \left(\frac{0}{0}\right)$$

$$= \lim_{x \to 0} \frac{\dfrac{1}{1+x} + \dfrac{1}{1-x}}{\sec^2 x} = 2.$$

Example 12.2.8. $\lim\limits_{x \to 0} \left(\dfrac{1}{x} - \dfrac{1}{\sin x}\right)$ ($\infty - \infty$ *form*).

$$\text{Given limit} = \lim_{x \to 0} \frac{\sin x - x}{x \sin x} \left(\frac{0}{0}\right) = \lim_{x \to 0} \frac{\cos x - 1}{x \cos x + \sin x} \left(\frac{0}{0}\right)$$

$$= \lim_{x \to 0} \frac{-\sin x}{2 \cos x - x \sin x} = 0.$$

Try in a similar way: $\lim_{x \to 0} \left(\dfrac{1}{x} - \cot x \right) = 0$.

Example 12.2.9. $\lim_{x \to 0+} (\sin x)^{2 \tan x}$ (0^0 form).

Taking log and applying limit

$$\lim_{x \to 0+} 2 \tan x \cdot \log \sin x = \lim_{x \to 0+} \frac{2 \log \sin x}{\cot x} \quad \left(\frac{-\infty}{\infty} \right)$$

$$= \lim_{x \to 0+} \frac{2 \cot x}{-\operatorname{cosec}^2 x} = \lim_{x \to 0+} -2 \sin x \cos x = 0.$$

Since, $\lim_{x \to 0+} \log(\sin x)^{2 \tan x} = \log \left\{ \lim_{x \to 0+} (\sin x)^{2 \tan x} \right\} = 0 = \log 1$.

$$\therefore \lim_{x \to 0+} (\sin x)^{2 \tan x} = 1.$$

Example 12.2.10. $\lim_{x \to 0+} \left(1 + \dfrac{1}{x} \right)^x$ (∞^0 form).

Now,

$$\lim_{x \to 0+} \log \left(1 + \frac{1}{x} \right)^x = \lim_{x \to 0+} x \log \left(1 + \frac{1}{x} \right) = \lim_{x \to 0+} \frac{\log(x+1) - \log x}{1/x}$$

$$= \lim_{x \to 0+} \frac{(x+1)^{-1} - x^{-1}}{-x^{-2}} = \lim_{x \to 0+} \left(x - \frac{x^2}{x+1} \right) = 0 = \log 1.$$

Since, $\lim_{x \to 0+} \log(1 + x^{-1})^x = \log \left\{ \lim_{x \to 0+} (1 + x^{-1})^x \right\} = \log 1$.

$$\therefore \lim_{x \to 0+} \left(1 + \frac{1}{x} \right)^x = e^0 = 1.$$

Example 12.2.11. $\lim_{x \to 0} \left(\dfrac{\sin x}{x} \right)^{1/x}$ ($1^{\pm \infty}$ form). [C.H. 1980, 92]

Now,

$$\lim_{x \to 0} \log \left(\frac{\sin x}{x} \right)^{1/x} = \lim_{x \to 0} \frac{1}{x} \cdot \log \left(\frac{\sin x}{x} \right) \quad \left(\frac{0}{0} \right)$$

$$= \lim_{x \to 0} \frac{x \cos x - \sin x}{x \sin x} \left(\frac{0}{0} \right) = \lim_{x \to 0} \frac{-x \sin x}{x \cos x + \sin x} \left(\frac{0}{0} \right)$$

$$= \lim_{x \to 0} \frac{-\sin x - x \cos x}{-x \sin x + 2 \cos x} = 0 = \log 1.$$

CHAPTER 12: INDETERMINATE FORMS AND MAXIMA & MINIMA

Since, $\lim_{x \to 0} \log \left(\frac{\sin x}{x} \right)^{1/x} = \log \left\{ \lim_{x \to 0} \left(\frac{\sin x}{x} \right)^{1/x} \right\} = \log 1.$

$$\therefore \lim_{x \to 0} \left(\frac{\sin x}{x} \right)^{1/x} = 1.$$

Example 12.2.12. *Find the values of p and q such that*

$$\lim_{x \to 0} \frac{x(1 - p \cos x) + q \sin x}{x^3} = \frac{1}{3}. \quad (assume\ that\ L'Hospital's\ rule\ is\ applicable)$$

Solution: The given ratio has the indeterminate form $0/0$. Using L'Hospital's rule the limit should be

$$= \lim_{x \to 0} \frac{1 - p \cos x + xp \sin x + q \cos x}{3x^2}. \quad (12.2.2)$$

Since Denominator $\to 0$ as $x \to 0$, therefore, in order that the limit of (12.2.2) be finite as $x \to 0$, it is necessary that the numerator should also $\to 0$ as $x \to 0$. This gives

$$1 - p + q = 0. \quad (12.2.3)$$

In this case limit should be of the form $0/0$. So using L'Hospital's rule again, limit in (12.2.2)

$$= \lim_{x \to 0} \frac{(p-q) \sin x + p \sin x + xp \cos x}{6x} \quad \left(\frac{0}{0}\ \text{form}\right)$$

$$= \lim_{x \to 0} \frac{(p-q) \cos x + p \cos x + p \cos x - px \sin x}{6} = \frac{3p - q}{6} = \frac{1}{3}\ \text{(given)}.$$

This gives,

$$3p - q = 2. \quad (12.2.4)$$

Solving p and q from (12.2.3) and (12.2.4) we get $p = \frac{1}{2}, q = -\frac{1}{2}$.

Example 12.2.13. *Find a and b in order that*

$$\lim_{x \to 0} \frac{a \sin 2x - b \sin x}{x^3} = 1. \quad [C.H.\ 1996]$$

Solution: $\lim_{x \to 0} \frac{a \sin 2x - b \sin x}{x^3}$ is of the indeterminate form $\frac{0}{0}$.

Using L'Hospital's rule this limit $= \lim_{x \to 0} \frac{2a \cos 2x - b \cos x}{3x^2}$, if it exists. (12.2.5)

Since Denominator $\to 0$ as $x \to 0$, therefore, in order that the limit (12.2.5) exists finitely, Numerator should tend to zero as $x \to 0$.

$$\therefore\ 2a - b = 0. \quad (12.2.6)$$

$$\therefore \text{ limit } (12.2.5) = \lim_{x \to 0} \frac{b \cos 2x - b \cos x}{3x^2} \quad \left(\frac{0}{0} \text{ form}\right).$$

Again using L'Hospital's rule this limit $= \lim_{x \to 0} \dfrac{-2b \sin 2x + b \sin x}{6x} \quad \left(\dfrac{0}{0} \text{ form}\right)$.

Using L'Hospital's rule once again we obtain

$$\lim_{x \to 0} \frac{-4b \cos 2x + b \cos x}{6} = -\frac{3b}{6} \quad (\text{it should be } = 1).$$

$$\therefore b = -2. \tag{12.2.7}$$

From (12.2.6) $a = -1$. [i.e., the required $a = -1, b = -2$].

Exercises XII(I)

[*On Indeterminate forms*]

[A]

Evaluate each of the following limits:

1. $\displaystyle\lim_{x \to a} \frac{x^n - a^n}{x^m - a^m}$ (m, n are positive integers).

2. $\displaystyle\lim_{x \to 0} \frac{e^x - e^{\sin x}}{x - \sin x}$. [C.H. 1963]

3. $\displaystyle\lim_{x \to 0} \frac{\log x^2}{\log \cot^2 x}$. [C.H. 1965]

4. $\displaystyle\lim_{x \to 0} \left\{ \frac{1}{x} - \frac{\log(1 + x)}{x^2} \right\}$.

5. $\displaystyle\lim_{x \to 0} \left\{ \frac{1}{x^2} - \cos^2 x \right\}$.

6. $\displaystyle\lim_{x \to 0} \left\{ \frac{1}{x} - \cot x \right\}$. [C.H. 1981]

7. $\displaystyle\lim_{x \to 0} (\cos x)^{1/x^2}$.

8. $\displaystyle\lim_{x \to \pi/2} (\sin x)^{\tan x}$. [C.H. 1982]

9. $\displaystyle\lim_{x \to 0} (\cos x)^{\cot^2 x}$. [C.H. 1961, '65]

10. $\displaystyle\lim_{x \to 0} \left(\frac{\tan x}{x} \right)^{1/x}$. [C.H. 1974]

11. $\lim\limits_{x \to 0} \dfrac{x - \sin x}{\tan^3 x}$. [C.H. 1994]

12. $\lim\limits_{x \to 0} \dfrac{x - \tan x}{x^3}$.

[Ans: 1. $\frac{n}{m} a^{n-m}$; 2. 1; 3. -1; 4. $\frac{1}{2}$; 5. $\frac{2}{3}$; 6. 0; 7. $e^{-1/2}$; 8. 1; 9. $e^{-1/2}$; 10. 1; 11. $\frac{1}{6}$; 12. $-\frac{1}{3}$.]

[B]

Use L'Hospital's rule to prove the following limits:

1. $\lim\limits_{x \to \pi/2} (\cos x)^{\cos^2 x} = 1$. [C.H. 1974]

2. $\lim\limits_{x \to \infty} x^{1/x} = 0$. [C.H. 1976]

3. $\lim\limits_{x \to 1} x^{1/x-1} = e$. [C.H. 1978]

4. $\lim\limits_{x \to 0} (\cos 2x)^{1/x^2} = e^{-2}$. [C.H. 1979]

5. $\lim\limits_{x \to 1} x^{\tan \frac{1}{2}\pi x} = e^{-2/\pi}$. [C.H. 1966]

6. $\lim\limits_{x \to 0} (\cos mx)^{n/x^2} = e^{-\frac{1}{2}m^2 n}$. [C.H. 1966]

7. $\lim\limits_{x \to 0} (1 + \sin x)^{\cot x} = e$. [C.H. 1964]

8. $\lim\limits_{x \to 0} \left(\dfrac{\sin x}{x} \right)^{1/x^2} = e^{-1/6}$.

9. $\lim\limits_{x \to 0} \left(\dfrac{\tan x}{x} \right)^{1/x^2} = e^{1/3}$. [C.H. 1989, New]

10. $\lim\limits_{x \to \infty} \left(1 + \dfrac{1}{x^2} \right)^x = 1$.

11. $\lim\limits_{x \to 0} x^{2 \sin x} = 1$.

12. $\lim\limits_{x \to \infty} \left(1 + \dfrac{2}{x} \right)^x = e^2$. [C.H.]

13. $\lim\limits_{x \to 1} x^{1/1-x} = \dfrac{1}{e}$.

14. $\lim\limits_{x \to 0^+} \left(\dfrac{\sin x}{x} \right)^{1/x} = 1.$ [C.H. 1980, 1992]

[C]

Verify the truth of the following statements:

1. $\lim\limits_{x \to \infty} [x - \sqrt{(x-a)(x-b)}] = \dfrac{1}{2}(a+b).$ [C.H. 1974]

2. $\lim\limits_{x \to \infty} \left\{ x - x^2 \log \left(1 + \dfrac{1}{x} \right) \right\} = \dfrac{1}{2}.$ [C.H. 1979]

3. $\lim\limits_{x \to \infty} \left[\dfrac{1}{2x} - \dfrac{1}{x(1+e^x)} \right] = 0.$ [C.H. 1980]

4. $\lim\limits_{x \to 0} \left\{ \dfrac{1}{\sin^2 x} - \dfrac{1}{x^2} \right\} = \dfrac{1}{3}.$ [C.H. 1978]

5. $\lim\limits_{x \to \pi} (\sec 2x - \tan 2x) = 1.$

6. $\lim\limits_{x \to 2} \left\{ \dfrac{1}{x-2} - \dfrac{1}{\log(x-1)} \right\} = -\dfrac{1}{2}.$

7. $\lim\limits_{x \to \infty} (x - \sqrt{x^2 - 9}) = 0.$

8. $\lim\limits_{x \to 1} \left\{ \dfrac{x}{x-1} - \dfrac{1}{\log x} \right\} = \dfrac{1}{2}.$

9. $\lim\limits_{x \to 0} \left(\dfrac{1}{x^2} - \dfrac{1}{\sin^2 x} \right) = -\dfrac{1}{3}.$

10. $\lim\limits_{x \to 0} \left(\dfrac{1}{e^x - 1} - \dfrac{1}{x} \right) = -\dfrac{1}{3}.$

11. $\lim\limits_{x \to 4} \left\{ \dfrac{1}{\log(x-3)} - \dfrac{1}{x-4} \right\} = \dfrac{1}{2}.$

12. $\lim\limits_{x \to 0} (2x \tan x - \pi \sec x) = -2.$

13. $\lim\limits_{x \to \infty} \left[x - \sqrt[n]{(x-a_1)(x-a_2) \cdots (x-a_n)} \right].$ [put $x = \dfrac{1}{y}$ and make $y \to 0$.]

[C.H. 1985]

CHAPTER 12: INDETERMINATE FORMS AND MAXIMA & MINIMA

[D]

The following ratios of functions are of the form 0/0 for the values of x specified against each. What are their true values (if any)?

1. $\dfrac{x^2 - 4}{x - 2}$ for $x = 2$.

2. $\dfrac{\log x}{x - 1}$ for $x = 1$.

3. $\dfrac{\tan x - x}{x - \sin x}$ for $x = 0$.

4. $\dfrac{e^x + \sin x - 1}{\log(1 + x)}$ for $x = 0$.

5. $\dfrac{\sin \log(1 + x)}{\log(1 + \sin x)}$ for $x = 0$.

6. $\dfrac{1 - 4\sin^2 \frac{1}{6}\pi x}{1 - x^2}$ for $x = 1$.

7. $\dfrac{\sin x \cdot \sin^{-1} x - x^2}{x^6}$ for $x = 0$.

8. $\dfrac{\tan x \tan^{-1} x - x^2}{x^6}$ for $x = 0$.

[*Ans:* True values are their limits, if exist. We give the required limits:
1. 4; **2.** 1; **3.** 2; **4.** 2; **5.** 1; **6.** $\frac{1}{6}\pi\sqrt{3}$; **7.** $\frac{1}{18}$; **8.** $\frac{1}{18}$.]

[E]

The following ratios of functions are of the form ∞/∞ for the values of x given. Find their True or Appropriate values, if any:

1. $\dfrac{2x^2 - 4}{x^2 + 2x + 3}$, when $x \to \infty$.

2. $\dfrac{\log x}{x^k}$ $(k > 0)$, when $x \to \infty$.

3. $\dfrac{\log(x - a)}{\log(e^x - e^a)}$ when $x \to a$.

4. $\dfrac{\log \sin x}{\cot x}$, when $x \to 0$.

5. $\dfrac{\log \sin 2x}{\log \sin x}$, when $x \to 0$.

6. $\log_{\tan x}(\tan 2x)$, when $x \to 0$.

7. $\dfrac{\log x - \cot \frac{1}{2}\pi x}{\cot \pi x}$, when $x \to 0^+$.

8. $\dfrac{\log \sin x}{\log \sin 2x}$, when $x \to \pi^{-0}$.

[Ans: 1. 2; 2. 0; 3. 1; 4. 0; 5. 1; 6. ratio $= \dfrac{\log \tan 2x}{\log \tan x}$, value $=1$; 7. -2; 8. 1.]

[F]

1. Prove that (if n and m are positive integers):

$$\lim_{x \to \infty} \frac{a_0 x^n + a_1 x^{n-1} + \cdots + a_n}{b_0 x^m + b_1 x^{m-1} + \cdots + b_m} = \begin{cases} +\infty, & \text{if } a_0/b_0 \text{ is positive and } n > m; \\ -\infty, & \text{if } a_0/b_0 \text{ is negative and } n > m; \\ a_0/b_0, & \text{if } n = m; \\ 0, & \text{if } n < m. \end{cases}$$

2. (a) In $-\pi/4 \le x \le \pi/4$, let $f(x) = x^2 \sin 1/x$, $x \ne 0$, $f(0) = 0$ and $\phi(x) = \tan x$. Examine whether L'Hospital's rule is applicable to evaluate $\lim\limits_{x \to 0} f(x)/\phi(x)$ and show that in any case the limit is zero.

   (b) Similar problem: Take $f(x) = x - \sin x$, $\phi(x) = x$ as $x \to \infty$.

3. Show that (assume that n is any positive integer):

   (a) $\lim\limits_{x \to n} \dfrac{x - n}{\sin \pi x} = \dfrac{(-1)^n}{\pi}$;

   (b) $\lim\limits_{x \to n} \dfrac{1}{x - n}\left\{\dfrac{1}{\sin \pi x} - \dfrac{(-1)^n}{(x-n)\pi}\right\} = (-1)^n \dfrac{\pi}{6}$.

4. If $\lim\limits_{x \to a}\{f(x)/g(x)\}$ exists and is finite and if $\lim\limits_{x \to a} g(x) = 0$ then prove that necessarily $\lim\limits_{x \to a} f(x) = 0$.

   [Hints: $\lim f(x) = \lim\{f(x)/g(x)\} \cdot g(x) = l \cdot 0 = 0$.]

5. **Important Problems:**

   Assume that in the following cases L'Hospital's rule is applicable:

   (a) Determine a such that

   $$\lim_{x \to 0} \frac{a \sin x - \sin 2x}{\tan^3 x} \text{ exists and } = 1.$$

CHAPTER 12: INDETERMINATE FORMS AND MAXIMA & MINIMA

(b) Find a and b in order that

    i. $\lim\limits_{x \to 0} \dfrac{x(a \cos x + 1) - b \sin x}{x^3}$ is finite and $= 1$.   [C.H. 1990]

    ii. $\lim\limits_{x \to 0} \dfrac{a \sin 2x - b \sin x}{x^3} = 1$.   [C.H. 1996]

(c) Find a, b, c such that

$$\frac{ae^x - b \cos x + ce^{-x}}{x \sin x} \to 2 \text{ as } x \to 0.$$

(d) Find a if

$$\lim_{x \to 0} \frac{e^x - ae^{x \cos x}}{x - \sin x} \text{ is finite.} \qquad [\text{C.H. 1995}]$$

What is the value of this limit?

(e) If $\lim\limits_{x \to 0} \dfrac{\sin x + a_1 e^x + a_2 e^{-x} + a_3 \log(1+x)}{x^3}$ is finite, find a_1, a_2, a_3.

[Ans: 5. (a) $a = 2$; (b) (i) $a = -5/2, b = -3/2$; (ii) $a = -1, b = -2$; (c) $a = 1$, $b = 2, c = 1$; (d) $a = 1$; (e) $a_1 = -\frac{1}{4}, a_2 = \frac{1}{4}, a_3 = -\frac{1}{2}$.]

6. Prove that

(a) $\lim\limits_{\theta \to \frac{1}{2}\pi} \left[\sqrt{\dfrac{2 + \cos 2\theta - \sin \theta}{\theta \sin 2\theta + \theta \cos \theta}} - \left(\dfrac{\pi - 2\theta}{2 \sin 2\theta} \right)^2 \right] = -\dfrac{1}{4}$.

(b) $\lim\limits_{x \to 0} \dfrac{e^x - 1}{x^4 \sin x} \left(\dfrac{3 \sin x - \sin 3x}{\cos x - \cos 3x} \right)^4 = 1$.

12.3 Maxima and Minima : Extreme Values

We recall the definitions of maxima, minima or extreme values of a function f.

Definition 12.3.1. *Let f be a real-valued function of a single variable x defined on an interval I. Let c be any one of the interior points of I.*

1. *f is said to have a **maximum value** at $x = c$ if $\exists$ a suitable neighbourhood of c, where for every x, $f(x) < f(c)$. In other words, f has a maximum at $x = c \in I$.*

$$\Rightarrow \exists\, \delta > 0, \text{ such that } \forall\, x \text{ in } 0 < |x - c| < \delta, \quad f(x) < f(c).$$

Similarly,

2. *f is said to have a **minimum value** at $x = c$ if $\exists\, \delta > 0$ such that $\forall\, x$ in $0 < |x - c| < \delta, f(x) > f(c)$.*

The term **extreme value** or **turning value** *covers both cases. We also say that 'f has an extreme at $x = c$' to mean either 'f has a maximum at $x = c$' or 'f has a minimum at $x = c$'.*

Note: (i) Such maxima or minima are called a local maxima or local minima. That is why, it may so happen that local minimum at a point may be greater than a local maximum at some other point. In Fig. 12.1 see that local minimum at x_4 > local maximum at x_1.

(ii) The function $f : [a, b] \to \mathbb{R}$ is said to have global maximum at $c \in [a, b]$ iff $f(c) \geq f(x) \forall x \in [a, b]$. The reverse inequality will served for local minimum.

So, global maximum = $\max\{f(a), f(b), \text{all maxima}\}$ and

global minimum = $\min\{f(a), f(b), \text{all minima}\}$.

Graphically, at an extreme point the ordinate of the curve, given by $y = f(x)$, is either greater than (*maximum*) or less than (*minimum*) the adjacent ordinates on either side in the immediate neighbourhood. Hence, from Fig. 12.1, it is clear that, $y = f(x)$ defined in $a \leq x \leq b$ has maxima at x_1, x_3, minima at x_2, x_4. But observe that maximum value of $f(x)$ at x_1 is less than the minimum value of $f(x)$ at x_4. The greatest value of $f(x)$ is assumed at x_3, the least value at $x = a$. They are respectively the *absolute maximum* and *absolute minimum* of $f(x)$ in the interval $[a, b]$.

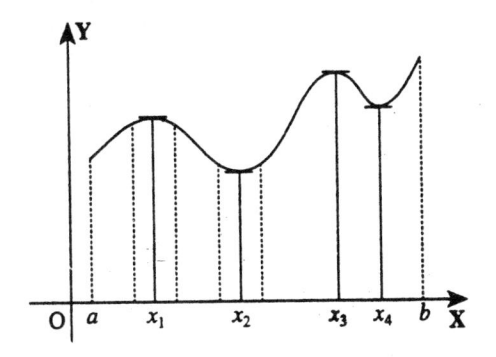

CHAPTER 12: INDETERMINATE FORMS AND MAXIMA & MINIMA

Fig. 12.1

A necessary condition for the existence of extreme value at a particular point.

Theorem 12.3.1. *If $f(x)$ has a maximum or minimum at $x = c$, then $f'(c) = 0$, if $f'(c)$ exists.*

Proof. Let $f(x)$ be maximum at $x = c$ and suppose $f'(c)$ exists then we are to show that $f'(c) = 0$. If not, either $f'(c) > 0$ or $f'(c) < 0$.

If $f'(c) > 0$, there exists an interval $]c, c + \delta[$, δ being sufficiently small, such that for every x of this interval $f(x) > f(c)$, which is a contradiction to the given condition that $f(c)$ is maximum at $x = c$.

Similarly, if $f'(c) < 0$, there exists an interval $]c - \delta, c[$ such that $f(x) > f(c)$ of every x in this interval, which is again contradictory.

Hence the only possibility is that $f'(c) = 0$.

Similar, for minimum. Hence the theorem.

Observations:

1. The theorem says that if a *function possesses a derivative at a point* then it can have an extreme point there *only if that derivative vanishes*. The logical implication is that a function may have an extreme value at a point even when the derivative at that point does not exist. In fact, consider the following example:

Example: $f(x) = |x|$. f is defined for all $x \in \mathbb{R}$.

Here, f is not derivable at $x = 0$, since

$$Rf'(0) = \lim_{x \to 0^+} \frac{f(x) - f(0)}{x - 0} = \lim_{x \to 0^+} \frac{x - 0}{x - 0} = 1.$$

$$\text{and} \quad Lf'(0) = \lim_{x \to 0^-} \frac{f(x) - f(0)}{x - 0} = \lim_{x \to 0^-} \frac{-x - 0}{x - 0} = -1.$$

Thus, $Rf'(0) \neq Lf'(0)$ and consequently f is not derivable at $x = 0$.

But we observe that in any neighbourhood of 0,

$$f(x) > f(0), \text{ if } x > 0 \text{ or if } x < 0$$

i.e., $f(x)$ has a minimum value at $x = 0$. i.e., extremes can occur even when the derivative does not exist.

2. The condition $f'(c) = 0$ in the theorem for the existence of an extreme value is only necessary (*not sufficient*), i.e., there exist functions for which the condition (derivative vanishing) is satisfied but the function may not have an extreme at that point.

   e.g., let $f(x) = x^3$, defined $\forall\ x \in \mathbb{R}$. Here $f'(0) = 0$ but f does not have any extreme value at $x = 0$. When $x > 0, f(x) > f(0)$ and when $x < 0, f(x) < f(0)$.

   Thus $f(x) - f(0)$ does not have a constant sign in the immediate neighbourhood of $x = 0$. Hence $f(x)$ has no extreme at $x = 0$.

3. The extremes for which the derivatives exist and $= 0$, correspond to points on the graph $y = f(x)$ where the tangents to the curve are parallel to the x-axis (see Fig. 12.1).

4. If $f'(x) = 0$ at $x = c$, we say $x = c$ is a **stationary point** of f. A *stationary value* $f(c)$ may or may not be an extreme value.

5. If f has an extreme value at $x = c$.

   either f'(c) = 0 or f is not derivable at x = c.

   Therefore, in order to investigate the maxima and minima of a function f, we have to first find value; of x for which **either** $f'(x)$ does not exist **or**, if $f'(x)$ exists, $f'(x) = 0$.

   These values are called the CRITICAL VALUES for the function f.

   We then examine as to for which of these critical values does the function actually have a maximum or a minimum.

Sufficient conditions for the existence of extreme values

Theorem 12.3.2. Use of First-Derivatives only.

Suppose that

a) $f(x)$ *is defined at* $x = c$;

b) $f'(x)$ exists in a suitably small neighbourhood of c,

$$0 < |x - c| < \varepsilon \quad (f'(c) \text{ need not exist})$$

c) $f'(x)$ has a fixed sign when $x < c$ and also when $x > c$.

Then as x increases through c

i) $f(x)$ has no extreme at $x = c$ if the sign of $f'(x)$ does not change;

ii) $f(x)$ has a maximum at $x = c$ if the sign of $f'(x)$ changes from $+$ to $-$;

iii) $f(x)$ has a minimum at $x = c$ if the sign of $f'(x)$ changes from $-$ to $+$.

Proof. We shall use the fact that $f(x)$ is increasing or decreasing at x according as $f'(x)$ is positive or negative.

In case

i) $f(x)$ increases(or decreases) all through as x passes through c. Then $f(x) - f(c)$ has a change of sign on the left and on the right of c.

$$\therefore f(x) \text{ has no extrema at } c.$$

ii) $f(x)$ increases as x approaches c from the left and $f(x)$ decreases as x leaves c and moves on the right of c.

$$\therefore f(x) \text{ has a maximum at } c.$$

iii) $f(x)$ decreases as x approaches c and increases as x leaves c.

$$\therefore f(x) \text{ has minimum at } c.$$

12.4 Illustrative Examples: Use of first derivative only

Example 12.4.1. *If $f(x) = |x|$. Find the extreme value.*

Solution:

$$f'(x) = \begin{cases} 1, & \text{when } x > 0; \\ -1, & \text{when } x < 0; \\ \text{undefined}, & \text{when } x = 0. \end{cases}$$

As x passes through 0 (from left to right of 0) $f'(x)$ changes from $-$ to $+$; hence f has a minimum at $x = 0$. The minimum value is zero.

Note 12.4.1. *Discuss the case when $f(x) = 1 - |x|$; f has a maximum value 1 at $x = 0$ though $f'(0)$ does not exist.* [C.H. 1996]

Example 12.4.2. $f(x) = 1 - x^{2/3}$. *Find the extreme value.*

Solution: Now
$$f'(x) = \begin{cases} -(2/3)x^{-1/3}, & x \neq 0; \\ \text{undefined}, & \text{when } x = 0. \end{cases}$$

As x passes from left to right through 0, the sign of $f'(x)$ changes from $+$ to $-$; hence $y = 1 - x^{2/3}$ has a maximum value at $x = 0$: max. value is 1.

[The reader who is interested about the Geometry of the curve should remember that the point $(0, 1)$ is a CUSP with vertical tangent.]

Example 12.4.3. $f(x) = x^{1/3}$. *Discuss the existence of extreme value.*

Solution: Here
$$f'(x) = \begin{cases} (1/3)x^{-2/3}, & \text{for } x \neq 0; \\ \text{undefined}, & \text{for } x = 0. \end{cases}$$

As x passes through 0, the sign of $f'(x)$ remains always positive; hence the curve $y = x^{1/3}$ has no extreme at the origin.

[Here the curve crosses its vertical tangent at the origin, it has a *point of inflexion* there.]

Example 12.4.4. $f(x) = x^5 - 5x^4 + 5x^3 - 1$, *defined $\forall\, x \in \mathbb{R}$. To find extreme value of $f(x)$.* [C.H. 1965]

Solution:
$$f'(x) = 5x^4 - 20x^3 + 15x^2 = 5x^2(x-1)(x-3), \quad \forall\, x \in \mathbb{R}.$$

Since $f(x)$ is derivable $\forall\, x \in \mathbb{R}$ and since $f'(x) = 0$ only when $x = 0, x = 1$ and $x = 3$. (**these are the critical values**).

a) $f'(x) > 0$, when $x < 0$; $f'(x) = 0$ when $x = 0$; $f'(x) > 0$ when $0 < x < 1$,

i.e., $f'(x)$ does not change sign as x increases through $x = 0$.

∴ f has neither a maximum nor a minimum at $x = 0$.

b) $f'(x) > 0$, when $0 < x < 1$; $f'(x) = 0$ when $x = 1$; $f'(x) < 0$ when $1 < x < 3$,

i.e., $f'(x)$ changes sign from $+$ to $-$ as x increases through $x = 1$.

f has a maximum at $x = 1$ (**Max. value is 0**).

c) $f'(x) < 0$, when $1 < x < 3$; $f'(x) = 0$ when $x = 3$; $f'(x) > 0$ when $x > 3$,

i.e., $f'(x)$ changes sign from $-$ to $+$ as x increases through $x = 3$.

$\therefore$ f has a minimum at $x = 3$ (**Min. value is 458**).

Example 12.4.5. *The rational function $f(x) = P/Q$ where $P(x), Q(x)$ are polynomials without common factor, has the derivative,*

$$f'(x) = \frac{QP' - PQ'}{Q^2}.$$

Solution: Since $Q^2 \geq 0$, $f'(x)$ has the same sign as the polynomial $QP' - PQ'$. The function is undefined at all roots of $Q = 0$, but $f'(x)$ only changes sign at the roots of Q of *even multiplicity*, for these are roots of $QP' - PQ'$ of *odd multiplicity* (Why?).

If we regard a point where $f(x) = \pm\infty$ and $f'(x)$ changes sign as a maximum $(+\infty)$ or a minimum $(-\infty)$ in an extended sense, we can read off the extremes of $f(x)$ from the sign pattern of the polynomial $QP' - Q'P$. As a particular example, let

$$P = (x-1)^2(3x^2 - 2x - 37);$$
$$\text{and } Q = (x+5)^2(3x^2 - 14x - 1);$$

to investigate the extremes of

$$f(x) = \frac{P}{Q} = \frac{(x-1)^2(3x^2 - 2x - 37)}{(x+5)^2(3x^2 - 14x - 1)}.$$

Here $QP' - P'Q = 72(x+5)(x+2)(x+1)(x-1)(x-3)(x-7)$ (on simplification). This polynomial changes sign at each of the simple roots

$$-5, -2, -1, 1, 3, 7;$$

sign pattern of $QP' - PQ'$:

$$(+\infty)+;\quad +(-5)-;\quad -(-2)+;\quad +(-1)-;$$
$$-(1)+\ ;\quad +(3)-\ ;\quad -(7)+\ ;\quad (+\infty)+.$$

The sign pattern discloses that $f(x) = P/Q$ has

maxima at $-5, -1, 3$; **minima** at $-2, 1, 7$.

The maximum of $f(x)$ at $x = -5$ (a double zero of Q) is ∞.

Example 12.4.6. $f(x) = x^2(x-1)^3$: To investigate the extremes of $f(x)$.

Solution: Here

$$\begin{aligned}f'(x) &= 2x(x-1)^3 + 3x^2(x-1)^2 \\ &= x(x-1)^2\{2(x-1) + 3x\} \\ &= x(x-1)^2(5x-2).\end{aligned}$$

$f'(x)$ exists $\forall\, x \in \mathbb{R}$. Hence the extremes can only occur where $f'(x) = 0$, i.e., where $x = 0, 1, 2/5$.

The term of the highest degree of x in $f'(x)$ is $5x^4$ and hence

$$f'(\pm\infty) = +ve.$$

Again $f'(x)$ can **only change sign** at a root of odd multiplicity i.e., at $x = 0$, $x = 2/5$. i.e., $f'(x)$ has the following sign pattern:

x	$-\infty$;	0 ;	2/5 ;	1 ;	∞
$f'(x)$	+ ;	+ to $-$;	$-$ to +;	+ to +;	+

hence $f(x)$ has a maximum at $x = 0$ and a minimum at $x = 2/5$; moreover, $f'(x)$ has neither a maximum nor a minimum at $x = 1$, though $f'(x) = 0$ here.

12.5 Higher Derivative Test for the existence of Extreme Values

Theorem 12.5.1. *Let a function f be defined on an interval I and let c be an interior point of I. If at the point c,*

a) $f'(c) = f''(c) = f'''(c) = \cdots = f^{n-1}(c) = 0$,

b) $f^n(c)$ *exists but* $\neq 0$,

then at $x = c$, $f(x)$ *has*

  (i) *no extreme value, if n be an odd positive integer,*

  (ii) *an extreme value, if n be an even positive integer* :

$$a\ maximum\ if\ f^n(c) < 0,\quad a\ minimum\ if\ f^n(c) > 0.$$

CHAPTER 12: INDETERMINATE FORMS AND MAXIMA & MINIMA

Proof: The results (i) and (ii) will depend on the sign of

$$f(c+h) - f(c),$$

for different values of h, where $|h|$ is sufficiently small.

We examine the sign of $f(c+h) - f(c)$ in following three steps:

Step 1. Since $f^n(x)$ exists at $x = c$, therefore

$$f', f'', f''', \cdots, f^{n-1}$$

all exist in some neighbourhood of c, say in $]c - \delta, c + \delta[$.

Again since $f^n(c) \neq 0$, therefore it must be either positive or negative.

∴ ∃ an interval $]c - \delta_1, c + \delta_1[$, $0 < \delta_1 < \delta$ where

$$\left.\begin{array}{l} f^{n-1}(x) \text{ is increasing,} \quad \text{if } f^n(c) > 0 \\ f^{n-1}(x) \text{ is decreasing,} \quad \text{if } f^n(c) < 0 \end{array}\right\}.$$

Case 1. If $f^n(c) > 0$, $f^{n-1}(x)$ is strictly increasing in $]c - \delta_1, c + \delta_1[$.
i.e., when $c - \delta_1 < x < c$, $f^{n-1}(x) < f^{n-1}(c)$, or, $f^{n-1}(x) < 0$ $(\because f^{n-1}(c) = 0)$.
and when $c < x < c + \delta_1$, $f^{n-1}(x) > f^{n-1}(c)$, or, $f^{n-1}(x) > 0$.

Case 2. If $f^n(c) < 0$, $f^{n-1}(x)$ is strictly decreasing in $]c - \delta_1, c + \delta_1[$
i.e., when $c - \delta_1 < x < c$, $f^{n-1}(x) > f^{n-1}(c)$, or, $f^{n-1}(x) > 0$ $(\because f^{n-1}(c) = 0)$
and when $c < x < c + \delta_1$, $f^{n-1}(x) < f^{n-1}(c)$, or, $f^{n-1}(x) < 0$.

Step 2. Let h be any real number such that $|h| < \delta_1$. Then

$$f', f'', f''', \cdots, f^{n-1}$$

all exist in $[c-h, c+h]$ so that by Taylor's Theorem with Lagrange's form of remainder after $(n-1)$ terms we get

$$f(c+h) = f(c) + hf'(c) + \frac{h^2}{2!}f''(c) + \cdots + \frac{h^{n-2}}{\underline{n-2}}f^{n-2}(c)$$

$$+ \frac{h^{n-1}}{\underline{n-1}}f^{n-1}(c+\theta h), \quad (0 < \theta < 1).$$

$$\therefore f(c+h) - f(c) = \frac{h^{n-1}}{\underline{n-1}}f^{n-1}(c+\theta h).$$

$(\because f'(c) = f''(c) = \cdots = f^{n-2}(c) = 0$ by hypothesis$)$.

Then, if $h > 0$, $f(c+h) - f(c)$ and $f^{n-1}(c+\theta h)$ have the same signs $\forall n$; if $h < 0$, $f(c+h) - f(c)$ and $f^{n-1}(c+\theta h)$

$$\left.\begin{array}{ll}\text{have same signs,} & \text{if } n \text{ is odd} \\ \text{have opposite signs,} & \text{if } n \text{ is even}\end{array}\right\}.$$

Step 3. We now obtain the following results (using steps **1** and **2**).

A. If $f^n(c) > 0$ and $h > 0$, then $c + \theta h$ surely lies between c and $c + \delta_1$ so that, by **Step 1** (Case 1), $f^{n-1}(c+\theta h) > 0$ and hence, by **Step 2**, $f(c+h) - f(c) > 0$.

B. If $f^n(c) < 0$ but $h > 0$, then also $c < c+\theta h < c+\delta_1$ so that, by **Step 1** (Case 2), $f^{n-1}(c+\theta h) < 0$ and hence, by **Step 2**, $f(c+h) - f(c) < 0$.

C. If $f^n(c) > 0$ and $h < 0$, then $c - \delta_1 < c + \theta h < c$ so that, by **Step 1** (Case 1), $f^{n-1}(c+\theta h) < 0$ and hence, by **Step 2**,

$$\begin{aligned} f(c+h) - f(c) &> 0, \text{ if } n \text{ is even,} \\ f(c+h) - f(c) &< 0, \text{ if } n \text{ is odd.} \end{aligned}$$

D. If $f^n(c) < 0$ and $h < 0$, then $c+\theta h$ lies between $c - \delta_1$ and c, so that, by **Step 1** (Case 2), $f^{n-1}(c+\theta h) > 0$ and hence, by **Step 2**,

$$\begin{aligned} f(c+h) - f(c) &< 0, \text{ if } n \text{ is even,} \\ f(c+h) - f(c) &> 0, \text{ if } n \text{ is odd.} \end{aligned}$$

Step 4. Since h is any real number ($+ve$ or $-ve$) s.t. $|h| < \delta_1$, therefore, from **Step 3** results we conclude that if n is even, $\exists \delta_1 > 0$ such that

$$0 < |x - c| < \delta_1 \Rightarrow f(x) - f(c) > 0, \text{ if } f^n(c) > 0 \quad [\text{use } \mathbf{A} \text{ and } \mathbf{C}]$$
$$0 < |x - c| < \delta_1 \Rightarrow f(x) - f(c) < 0, \text{ if } f^n(c) < 0 \quad [\text{use } \mathbf{B} \text{ and } \mathbf{D}].$$

From these results, it follows that if n be even, then

$$f(c) \text{ is } \begin{cases} \text{minimum value if } f^n(c) > 0; \\ \text{maximum value if } f^n(c) < 0. \end{cases}$$

Similarly, it is easy to see that if n is odd, then $f(c)$ is not an extreme value of f.

Alternative Proof. (using Young's form of Taylor's Theorem)

We give the outline:

$$f(c+h) - f(c) = hf'(c) + \frac{h^2}{2!}f''(c) + \cdots + \frac{h^{n-1}}{(n-1)!}f^{n-1}(c) + \frac{h^n}{n!}M,$$

CHAPTER 12: INDETERMINATE FORMS AND MAXIMA & MINIMA   717

where $M \to f^n(c)$ as $h \to 0$.

$$\therefore f(c+h) - f(c) = \frac{h^n}{n!} M. \quad (\because f' = f'' = \cdots = f^{n-1} \text{ at } c \text{ all vanish}) \quad (12.5.1)$$

Since $M \to f^n(c)$ as $h \to 0$, $\exists \delta > 0$ s.t. for $0 < |h| \le \delta$, M has the same sign as $f^n(c)$.

From (12.5.1), we deduce that when n is even, $f(c+h) - f(c)$ has the same sign as $f^n(c)$ for $0 < |h| \le \delta$, so that $f(c)$ is a *maximum* or a *minimum* according as $f^n(c)$ is *negative* or *positive*.

When n is odd; $f(c+h) - f(c)$ changes sign with the change of sign of h so that $f(c)$ is not an extreme value.

Practical Rule:

In order to find the extremes of a continuous function $f(x)$ in an interval where its first derivative $f'(x)$ exists and is finite, we first find the roots of $f'(x) = 0$.

Let α be one of the roots. We put α in the successive derivatives of $f(x)$ (which is supposed to exist at α) till we find a derivative that does not vanish for $x = \alpha$.

If this non-vanishing derivative is of odd order, then there is no extremum at $x = \alpha$.

If the non-vanishing derivative is of even order n then f has a maximum at $x = \alpha$, when $f^n(\alpha) < 0$, minimum at $x = \alpha$, when $f^n(\alpha) > 0$.

12.6 Illustrative Examples : Use of Higher Order Derivatives

Example 12.6.1. *Examine the extreme value, if $f(x) = x^5 - 5x^4 + 5x^3 + 12$.*

[C.H. 1965]

Solution: Now
$$f'(x) = 5x^4 - 20x^3 + 15x^2 = 5x^2(x-1)(x-3).$$

Since $f'(x)$ exists $\forall\, x \in \mathbb{R}$, hence for a maximum or a minimum, $f'(x) = 0$,

$$f'(x) = 0 \Rightarrow x = 0 \text{ or, } x = 1 \text{ or, } x = 3.$$

Again, $f''(x) = 20x^3 - 60x^2 + 30x$.

At $x = 1$, $f''(1) = -10$, negative so that $f(1) = 1 - 5 + 5 + 12 = 13$ is a maximum value;

at $x = 3$, $f''(3) = 90$, positive, hence $f(3) = -15$ is a minimum value;

at $x = 0$, $f''(x) = 0$, so that we have to examine $f'''(0)$ etc.

Now, $f'''(x) = 60x^2 - 120x + 30 = 30 \neq 0$ at $x = 0$. Hence $f(0)$ is neither a maximum nor a minimum value at $x = 0$.

Example 12.6.2. *Find the maximum and minimum values of*
$$f(x) = a \sin^2 x + b \cos^2 x \ (a > b).$$

Solution:
$$f'(x) = (a - b) \sin 2x \quad \text{and} \quad f''(x) = 2(a - b) \cos 2x,$$
which exists $\forall x \in \mathbb{R}$.

$f'(x) = 0$ gives $\sin 2x = 0$ i.e., $x = \dfrac{1}{2} n\pi$, n being any integer or even zero, and $f''(x)$ is positive when n is even and is negative when n is odd. Therefore, $x = \dfrac{1}{2} n\pi$ gives a maximum or a minimum according as n is odd or even.

Example 12.6.3. Parametric Forms. *Show that $y = a(1 - \cos \theta)$, $x = a(\theta - \sin \theta)$, y being regarded as a function of x, is maximum at $\theta = \pi$.*

Solution:
$$\frac{dy}{dx} = (dy/d\theta) \div (dx/d\theta) = a \sin \theta / a(1 - \cos \theta) = 0 \text{ at } \theta = \pi.$$
and
$$\frac{d^2y}{dx^2} = \frac{\cos \theta - 1}{(1 - \cos \theta)^2} = -\frac{1}{2} \text{ at } \theta = \pi.$$

Hence the result.

Example 12.6.4. *Show that the maximum values of $\left(\dfrac{1}{x}\right)^x$ is $e^{1/e}$.*

Solution: Let
$$f(x) = \left(\frac{1}{x}\right)^x, \quad \log f(x) = x \log \frac{1}{x} = -x \log x.$$

$$\therefore \quad \frac{1}{f(x)} \cdot f'(x) = -(1 + \log x) \text{ which vanishes when } \log x = -1 \text{ i.e., } x = 1/e.$$

Again,
$$\frac{1}{f(x)} \cdot f''(x) - \frac{\{f'(x)\}^2}{\{f(x)\}^2} = -\frac{1}{x} \text{ and putting } x = \frac{1}{e},$$

$$\frac{1}{f(1/e)} f''(1/e) = -e \text{ or, } f''\left(\frac{1}{e}\right) = -ef\left(\frac{1}{e}\right) = -e \cdot e^{1/e} \text{ is negative.}$$

Hence the maximum value of $\left(\dfrac{1}{x}\right)^x$ is $e^{1/e}$.

CHAPTER 12: INDETERMINATE FORMS AND MAXIMA & MINIMA

Example 12.6.5. *Find the maximum and minimum values of the function*

$$\cos x \cos(x - \pi/6) \cos(x + \pi/6) \text{ where } 0 \leq x \leq \pi.$$

N.B. It is advisable to express the function as a **sum** of certain cosines before proceeding to differentiate.

Solution: Let

$$y = \cos x \cos(x - \pi/6) \cos(x + \pi/6) = \frac{1}{2} \cos x \{\cos 2x + \cos \pi/3\}$$
$$= \frac{1}{4}(\cos 3x + \cos x) + \frac{1}{4} \cos x = \frac{1}{4}(\cos 3x + 2\cos x).$$

For extremes of y, $dy/dx = 0$, i.e., $-\frac{1}{4}(3 \sin 3x + 2 \sin x) = 0$
i.e., $3 \sin 3x + 2 \sin x = 0$ i.e., $\sin x(11 - 12 \sin^2 x) = 0$.

$$\therefore \quad \sin x = 0 \text{ or, } \sin x = \pm\sqrt{11/12}.$$

When $\sin x = 0$, $x = 0$ or π in the given range.

Now, $d^2y/dx^2 = -\frac{1}{4}(9 \cos 3x + 2 \cos x)$.

When $x = 0$, $y = 3/4$ and d^2y/dx^2 is negative, i.e., $y = 3/4$ is a maximum value.
When $x = \pi$, $y = -3/4$ and d^2y/dx^2 is positive, i.e., $y = -3/4$ is a minimum value.
When $\sin x = +\sqrt{11/12}$ ($\sin x = -\sqrt{11/12}$ is inadmissible in $0 \leq x \leq \pi$),

$$\cos x = \pm 1/\sqrt{12}.$$

$$y = \frac{1}{4}(\cos 3x + 2\cos x) = \frac{1}{4}(4\cos^3 x - \cos x) = \frac{1}{4}\left(\frac{1}{3\sqrt{12}} - \frac{1}{\sqrt{12}}\right)$$

$$= -\frac{1}{2\sqrt{12}} \text{ for } + \text{ sign and}$$

$$y = +\frac{1}{2\sqrt{12}} \text{ for } - \text{ sign of } \cos x.$$

$$\frac{d^2y}{dx^2} = -\frac{1}{4}(36 \cos^3 x - 25 \cos x) = \frac{1}{4}\frac{22}{\sqrt{12}}, \text{ which is positive for } \cos x = +1/\sqrt{12}.$$

$$\therefore \quad y = -1/2\sqrt{12} \text{ gives a minimum value of } y$$
and similarly $y = +1/2\sqrt{12}$ gives a maximum value of y.

Hence for $0 \leq x \leq \pi$, the expression $\cos x \cos(x - \pi/6) \cos(x + \pi/6)$ has *maximum* values of $3/4$ and $1/2\sqrt{12}$ and *minimum* values of $-3/4$ and $-1/2\sqrt{12}$.

Applied Problems

In a great many practical problems we have to deal with the functions which have a maximum or a minimum value. It will be convenient if we proceed step by step according to the following directions:

Step 1. Construct from the given conditions the function whose maximum or minimum value is desired.

Step 2. If the constructed function contains more than one independent variable, the conditions of the problem would in most cases furnish enough suggestions so that the function may be expressed in terms of a single variable.

Step 3. To this resulting function of single variable apply the principles already explained for finding the maximum and minimum values.

A diagram, whenever possible, should be drawn. It will be seen that, in general the function $f(x)$ (Step 1) will be a *continuous function* and in most problems it will be possible to determine the range $[a, b]$ over which the independent variable x lies. Hence proper utilisation of the following suggestions would minimise much labour.

1. Maxima and minima occur alternately.

2. At least one maximum or one minimum must lie between two equal values of $f(x)$.

    (a) If $f(x) = 0$ when $x = a$ and $x = b$ and positive otherwise and if the function has *only one* stationary value (i.e., $f'(x)$ vanishes only once) then the stationary value is necessarily a *maximum*.

    (b) If $f(x) \to \infty$ as $x \to a$ and $x \to b$ and has only one stationary value then the stationary value is necessarily a *minimum*.

3. The sign of $f'(x)$ changes from + to − as x passes through the value which makes $f(x)$ a maximum. The reverse is the case for a minimum.

Examples:

Example 12.6.6. *Find the dimensions of the right circular cone of minimum volume which can be circumscribed about a sphere of radius 8 cm.*

Solution: Let x = radius of the base of the cone and $y + 8$ = altitude of the cone.
Step 1. Volume V of the
$$\text{cone} = \frac{1}{3}\pi x^2 (y + 8).$$

CHAPTER 12: INDETERMINATE FORMS AND MAXIMA & MINIMA

Step 2. The triangle AEC and AOD are similar (Fig. 12.2).

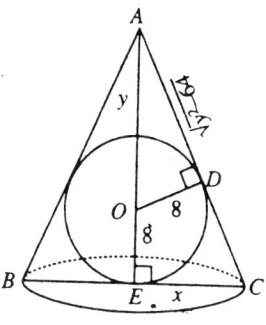

Fig. 12.2

Hence, we have

$$\frac{x}{8} = \frac{y+8}{\sqrt{y^2-64}} \quad \text{whence } V = \frac{64\pi}{3} \cdot \frac{(y+8)^2}{(y-8)}.$$

Step 3. $\dfrac{dV}{dy} = \dfrac{64\pi}{3} \cdot \dfrac{(y+8)(y-24)}{(y-8)^2}.$

$\therefore$ V is *stationary* when $y = 24$ and it changes sign from $-$ to $+$ as y passes through 24.

$\therefore$ V is minimum when $y = 24$.

Thus the required altitude of the cone $= y + 8 = 32$ cm and radius of the base $= x = 8\sqrt{2}$ cm.

Example 12.6.7. *It is desired to make an open box with square base out of a square piece of cardboard of side 1 metre by cutting equal squares out of the corners and then folding up the cardboard to form the sides. What must be the length of the side of the square cut out in order that the volume be a maximum?*

Solution: Let $x =$ length of the side of the square to be cut from each corner (Fig. 12.3).

Step 1. The volume V of the proposed box

$$= (1 - 2x)(1 - 2x)x \text{ cu. metre,}$$

which is a function of the single variable x, where $0 < x < \dfrac{1}{2} [\because 1 - 2x > 0 \text{ and } x > 0]$

Step 2. $\dfrac{dV}{dx} = (1 - 6x)(1 - 2x)$ so that the stationary value is at

$$x = \frac{1}{6} \quad \because x \neq \frac{1}{2}.$$

$$\frac{d^2V}{dx^2} = 24x - 8 \quad \because \left.\frac{d^2V}{dx^2}\right]_{x=1/6} = -4 < 0$$

∴ V is maximum when $x = 1/6$ and $V_{max} = \frac{2}{27}$ cu.m.

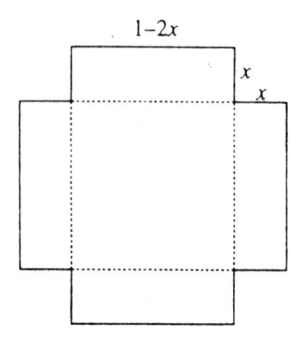

Fig. 12.3

∴ Maximum value 2/27 cu.m. and the length of the side cut out of the square is 1/6 m.

Example 12.6.8. *The cost of fuel in running a locomotive is proportional to the square of the speed and is Rs. 48 per hour for a speed of 16 km. per hour. Other costs amount to Rs. 300 per hour. What is the most economical speed?*

Solution: Let V = required speed and C = total cost per km. Cost of fuel per hour = kV^2, where k is the constant of proportionality. Since $V = 16$ when cost of fuel = Rs. 48 per hour, we have $k = 3/16$.

$$\therefore C \text{ (in Rupees per km)} = \frac{\text{cost in Rupees per hour}}{\text{speed in Km. per hour}}$$

$$= \frac{3/16 V^2 + 300}{V} = \frac{3}{16}V + \frac{300}{V}.$$

$\therefore \frac{dC}{dV} = \frac{3}{16} - \frac{300}{V^2}$, which vanishes when $V = 40$ km./hour and $\frac{d^2C}{dV^2} = \frac{600}{V^3} > 0$ for $V = 40$.

∴ C is minimum when $V = 40$ km/hour, i.e., the most economical speed is 40 km./hour.

Example 12.6.9. *N is the foot of the perpendicular drawn from the centre O on to the tangent at a variable point P on the ellipse $x^2/a^2 + y^2/b^2 = 1$ $(a > b)$. Prove that the maximum area of the triangle OPN is $\frac{1}{4}(a^2 - b^2)$. Find also the maximum length of PN.*

Solution: Draw an ellipse and take a point P on it.

Let P be $(a\cos\theta, b\sin\theta)$, θ being the eccentric angle of P.
The tangent at P is $\dfrac{x}{a}\cos\theta + \dfrac{y}{b}\sin\theta = 1$. Hence,

$$ON^2 = \frac{a^2b^2}{a^2\sin^2\theta + b^2\cos^2\theta};$$

$$OP^2 = a^2\cos^2\theta + b^2\sin^2\theta.$$

$$\therefore\ PN^2 = OP^2 - ON^2 = \frac{(a^2-b^2)^2\cos^2\theta\sin^2\theta}{b^2\cos^2\theta + a^2\sin^2\theta}.$$

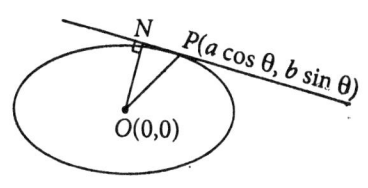

Fig. 12.3(a)

Hence the area of the right angled $\triangle OPN$

$$= \Delta = \frac{1}{2}\cdot ON\cdot NP = \frac{1}{2}\frac{\{ab(a^2-b^2)\sin\theta\cos\theta\}}{b^2\cos^2\theta + a^2\sin^2\theta} = \frac{1}{2}\frac{ab(a^2-b^2)\sin 2\theta}{(a^2+b^2)-(a^2-b^2)\cos 2\theta}.$$

This is maximum or minimum when $d\Delta/d\theta = 0$
i.e., when $\cos 2\theta = \dfrac{a^2-b^2}{a^2+b^2}$ and therefore $\sin 2\theta = \dfrac{2ab}{a^2+b^2}$.

As $\Delta = 0$ when $\theta = 0$ or $\dfrac{\pi}{2}$, this must give a maximum value of Δ viz., $\dfrac{1}{4}(a^2-b^2)$.
Similarly, proceed to show that the max. value of PN is $a-b$.

Example 12.6.10. *A right circular cone with a flat circular base is constructed of sheet material of uniform small thickness. Express the total area of the surface in terms of volume and semi-vertical angle θ. Show that for a given volume, the area of the surface is a minimum if $\theta = \sin^{-1}(1/3)$.*

Solution: Let r = radius of base of the cone,
$\qquad h$ = its height,
$\qquad l$ = its slant height,
$\qquad V$ = Volume of the cone,
$\qquad A$ = total surface area.
From the figure **12.4**, $h = r\cot\theta$ and $l = r\cosec\theta$.
Now,

$$V = \frac{1}{3}\pi r^2 h = \frac{1}{3}\pi r^3 \cot\theta;$$

$$A = \pi r^2 + \pi rl = \pi r^2 + \pi r^2 \cosec\theta$$

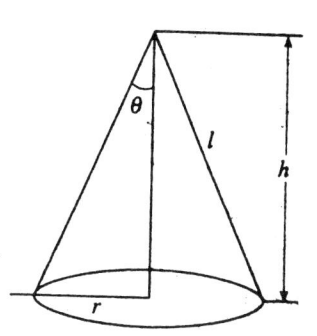

$$= \pi r^2(1+\cosec\theta).$$

Fig. 12.4

We have, then

$$r^3 = (3V \tan \theta)/\pi.$$

$$A = \pi \left(\frac{3V}{\pi}\right)^{2/3} \tan^{2/3} \theta (1 + \cosec \theta)$$

$$= k \tan^{2/3} \theta (1 + \cosec \theta), \text{ where } k = \text{a constant } \pi^{1/3}(3V)^{2/3}.$$

$$\therefore \quad \frac{dA}{d\theta} = k \left\{ \tan^{2/3} \theta (-\cosec \theta \cot \theta) + \left(\frac{2}{3} \tan^{-1/3} \theta \sec^2 \theta\right)(1 + \cosec \theta) \right\}$$

$$= \frac{k}{\tan^{1/3} \theta} \left\{ \tan \theta (-\cosec \theta \cot \theta) + \frac{2}{3} \sec^2 \theta (1 + \cosec \theta) \right\}$$

$$= \frac{k}{3 \tan^{1/3} \theta \cos^2 \theta \sin \theta} (3 \sin^2 \theta + 2 \sin \theta - 1).$$

$\therefore$ For a maximum or minimum value of A, $dA/d\theta = 0$.

i.e., $3 \sin^2 \theta + 2 \sin \theta - 1 = 0$ or, $(3 \sin \theta - 1)(\sin \theta + 1) = 0$

or, $\sin \theta = 1/3$ or, $\sin \theta = -1$ (inadmissible for this problem).

Using sign of derivative (or from practical considerations), we see that $\sin \theta = 1/3$ will give a minimum value of A and not a maximum value. Hence for a minimum value of A, $\theta = \sin^{-1}(1/3)$.

Chapters 12A & 12B Main Points : A Recapitulation

1. Indeterminate Forms: $0/0, \infty/\infty, 0.\infty, \infty - \infty, 0^0, \infty^0, 1^\infty$.

In Cauchy's MVT we take $f(a) = g(a) = 0$ and we let $b \to a$; then we obtain **L'Hospital's rule** for evaluating the indeterminate form $0/0$:

$$\lim_{x \to a} \frac{f(x)}{g(x)} = \lim_{x \to a} \frac{f'(x)}{g'(x)} \quad (a \text{ finite}),$$

where the right-hand side limit exists.

L'Hospital's rule also applies to the indeterminate form ∞/∞. Other forms $(0.\infty, \infty - \infty, 0^0, \infty^0, 1^\infty)$ are reducible to $0/0$ or ∞/∞.

2. Maxima and Minima or Extremes of $f(x)$ or Turning values of $f(x)$:

a) f is *maximum* at $x = c$, if $\exists \delta > 0$, s.t. $\forall x$ in $0 < |x - c| < \delta$,
   $f(x) - f(c) < 0$.

b) f is *minimum* at $x = c$, if $\exists\, \delta > 0$ s.t. $\forall\, x$ in $0 < |x - c| < \delta$, $f(x) - f(c) > 0$.

c) The extremes of $f(x)$ can only occur where $f'(x) = 0$ or where $f'(x)$ does not exist.

d) **Use of first derivative:** If $f'(x)$ exists when $0 < |x - c| < \varepsilon$ and changes sign as x increases through c, then the change form $+$ to $-$ denotes a maximum at c, and $-$ to $+$ denotes a minimum at c.

If $f'(x)$ does not change sign, there is no extreme at c.

e) **Use of Higher order derivatives:** If at $x = c$ all derivatives of f upto $(n-1)$th vanish while $f^n(c) \neq 0$, then there is no extreme at c, if n be an odd positive integer, but there is an extreme at c, if n be an even positive integer—maximum if $f^n(c) < 0$, minimum if $f^n(c) > 0$.

Exercises XII(II)

*[For problems marked * see Hints and Answers]*

[A]

1. Investigate the maximum and minimum values of the following polynomials:

   (a) $x^2 - 8x + 15$;

   (b) $x^3 - 27x + 15$;

   (c) $2 - 9x + 6x^2 - x^3$;

   (d) $x^4 - 8x^3 + 22x^2 - 24x + 1$;

   (e) $8x^5 - 15x^4 + 10x^2$;

   (f) $2x^3 - 15x^2 + 36x + 10$;

   (g) $x^5 - 5x^4 + 5x^3 - 1$; **[C.H. 1965]**

   (h) $12x^5 - 15x^4 - 40x^3 + 1$. **[C.H. 1962]**

2. Use derivative of first order only in examining the extremes of the following rational functions:

   (a) $\dfrac{x}{x^2 - 16}$;

(b) $\dfrac{x^2}{(1-x)^3}$; [C.H. 1960]

(c) $\dfrac{x^2+x+1}{x^2-x+1}$; [C.H. 1977]

(d) $x+1/x$;

(e) $\dfrac{(x-1)^2(3x^2-2x-37)}{(x+5)^2(3x^2-14x-1)}$;

(f) $\dfrac{x^2-7x+6}{x-10}$;

(g) $\dfrac{(x+1)(x+4)}{(x-1)(x-4)}$. [C.H. 1963]

3. (a) Show that if $a > 1$, the expression

$$(a-1)a^x - x$$

has an extreme value equal to $\log\{(ae-e)\log a\}/\log a$ and determine whether it is a maximum or a minimum value. [C.H. 1963]

(b) Show that x^x is a minimum for $x = 1/e$.

(c) Find the maximum or minimum values of $(1/x)^x$, $(\log x)/x$, and $(x/\log x)$.

(d) Show that $(3-x)e^{2x} - 4xe^x - x$ has no extreme at $x = 0$.

(e) Find maximum or minimum value of $x^{1/x}$.

(f) For what values of x, $f(x) = e^x + 2\cos x + e^{-x}$ has an extreme value?

(g) Examine whether $f(x) = 1 - |x|$ has a maxima or a minima at $x = 0$.
[C.H. 1996]

4. Find the maximum and minimum values of

a) $\dfrac{1}{2}x - \sin x$ (in $0 < x < 2\pi$);

b) $\sin x(1+\cos x)$ in $(0, 2\pi)$; [C.H. 1981]

c) $\sin x + \dfrac{1}{2}\sin 2x + \dfrac{1}{3}\sin 3x$ in $(0, \pi)$;

d) $\cos x + \dfrac{1}{2}\cos 2x + \dfrac{1}{3}\cos 3x$ in $(0, \pi)$;

e) $4x + \tan 3x$;

f) $a\sin x + b\cos x$;

g) $a\cot x + b\tan x$ $(a, b > 0)$;

*h) $a\cos x + b\cos 2x$, $(a.b > 0)$ in $-\dfrac{\pi}{4} \le x \le \dfrac{5\pi}{4}$; [C.H. 1968]

*i) $\sin x \cos^2 x$ in $[0, 2\pi]$; [C.H. 1964]

*j) $a\cos^2 x + b\cos^2(\alpha - x); a > b > 0, 0 < \alpha < \pi/4$. [C.H. 1974]

5. Let $f(x) = e^{-ax}\sin(bx+c)(a, b > 0)$. We shall call e^{-ax} the amplitude of $f(x)$. Prove that the amplitudes of the maximum values of $f(x)$ form a G.P. with common ratio $e^{-2\pi a/b}$. [C.H. 1972]

6. Verify that

   (a) $\sin 2x - 20\sin x + 26x$ has neither a maximum nor a minimum.

   (b) $\log \tan x$ $(0 < x < \pi/2)$ has neither a maximum nor a minimum at $x = 2$.

   (c) the minimum value of $\{(2x-1)(x-8)\}/(x^2 - 5x + 4)$ is greater than its maximum value.

7. Show that the extremes of $(\sin mx \cdot \operatorname{cosec} x)$, where m is an integer, are given by $\tan mx = m\tan x$ and then deduce that $\sin^2 mx \le m^2 \sin^2 x$.

8. (a) Show that $\sin^p \theta \cdot \cos^q \theta$ attains a maximum value when $\theta = \tan^{-1}\sqrt{p/q}$.

   (b) Given: $\dfrac{dy}{dx} = (x-a)^{2n}(x-b)^{2p+1}, (n, p \in \mathbf{N})$, verify that y has neither a maximum nor a minimum for $x = a$ but y has a minimum value when $x = b$.

[B]

1. a) Prove that the maximum value of $ax + by$ $(x, y > 0)$ is $2k\sqrt{a^2 - ab + b^2}$, where x and y satisfy the equation $x^2 + xy + y^2 = 3k^2$.

   b) Find the least distance of the point $(0, 3)$ from the parabola $x^2 = 2y$.

   Ans: $\sqrt{5}$]

2. Show that the greatest value of $x^m y^n (x, y > 0)$ and $x + y = k$, (k is a constant) is $\{m^m \cdot n^n \cdot k^{m+n}\}/(m+n)^{m+n}$.

3. Show that the maximum and minimum values of $r^2 = x^2 + y^2$ where $ax^2 + 2hxy + by^2 = 1$ are given by the quadratic

$$\left(a - \frac{1}{r^2}\right)\left(b - \frac{1}{r^2}\right) = h^2.$$

Hence find the area of the conic given by $ax^2 + 2hxy + by^2 = 1$.

4. Show that
$$\left(\alpha - \frac{1}{\alpha} - x\right)(4 - 3x^2)$$
has just one maximum and just one minimum, and that the difference between them is $\frac{4}{9}(\alpha + 1/\alpha)^3$. What is the least value of this difference for different values of α?

*5. Prove that the maximum and minimum values of
$$y = \frac{ax^2 + 2bx + c}{Ax^2 + 2Bx + C},$$
when the denominator has complex roots, are those values of λ for which $ax^2 + 2bx + c - \lambda(Ax^2 + 2Bx + C)$ is a perfect square.

6. Establish
$$\begin{vmatrix} f(a) & f(b) & f(c) \\ g(a) & g(b) & g(c) \\ h(a) & h(b) & h(c) \end{vmatrix} = \frac{1}{2}(b-c)(c-a)(a-b)\begin{vmatrix} f(\alpha) & f'(\beta) & f''(\gamma) \\ g(\alpha) & g'(\beta) & g''(\gamma) \\ h(\alpha) & h'(\beta) & h''(\gamma) \end{vmatrix}$$
where β and γ lie between the minimum and maximum of a, b, c.

[C]
(Applied Problems)

1. Divide the number 120 into two parts such that the product of one part and the square of the other is a maximum.

*2. a) What is the altitude of a right circular cone of maximum volume that can be inscribed in a sphere of radius R?

   *b) Find the altitude of a right circular cylinder of maximum volume that can be inscribed in a right circular cone.

   c) Show that the semi-vertical angle of the cone of maximum volume and of given slant height is $\tan^{-1}\sqrt{2}$.

   d) A cone is inscribed in a sphere of given radius r; prove that its volume as well as its curved surface is maximum when its altitude is $\frac{2}{3}$ of the diameter of the sphere.

   e) Show that the volume of the greatest cylinder which can be inscribed in a cone of height h and semi-vertical angle α is $\frac{4}{27}\pi h^3 \tan^2 \alpha$.

f) Show that the radius of the right circular cylinder of greatest curved surface which can be inscribed in a given cone is half that of the cone.

g) Find the surface of the right circular cylinder of greatest surface which can be inscribed in a sphere of radius r.

h) Show that the height of a cylinder open at one end of given surface and greatest volume is equal to the radius of its base.

3. a) Prove that the least perimeter of an isosceles triangle in which a circle of radius r can be inscribed is $6r\sqrt{3}$.

   *b) In a triangle, the the area Δ and the semi-perimeter s are fixed. Show that any maximum or minimum of one of the sides is a root of the equation
   $$s(x-s)x^2 + 4\Delta^2 = 0. \qquad \text{[C.H. 1966]}$$

   *c) Show that the maximum rectangle inscribable in a circle is a square.
   [C.H. 1995]

*4. A window of perimeter p feet is in the form of a rectangle surmounted by an isosceles right triangle. Show that the window will admit the greatest possible amount of light when the sides of the rectangle are equal to the sides of the right triangle.

*5. Assuming that the stiffness of a beam of rectangular cross-section varies as the breadth and as the cube of the depth, show that the breadth of the stiffest beam that can be cut from a log of diameter d is $\frac{1}{2}d$.

*6. A rectangular sheet of metal has four equal square portions removed at the corners and the sides are then turned up so as to form an open rectangular box. Show that when the volume contained in the box is a maximum, the depth will be
$$\frac{1}{6}\{(a+b) - (a^2 - ab + b^2)^{1/2}\},$$
where a and b are the sides of the original rectangle.

*7. Assuming that the petrol burnt in driving a motor-boat varies as the cube of its speed, show that the most economical speed when going against a current of c miles per hour is $\frac{3}{2}c$ m.p.h.

8. a) The theory of probabilities show that if $x_1, x_2, x_3, \cdots, x_n$ are the measures of an unknown quantity x, so that the errors are
$$x - x_1, x - x_2, \cdots, x - x_n,$$

then the most probable value of x is that which makes the sum of the squares of the errors a minimum. Show that, on this theory, the most probable value of the unknown is the A.M. of the measures.

b) Prove that the function $\sum_{r=1}^{n} m_r(x - x_r)^2$ is a minimum when

$$x = \sum_{r=1}^{n} m_r x_r \bigg/ \sum_{r=1}^{n} m_r.$$

9. From the fixed point A on the circumference of a circle of radius r the perpendicular AY is let fall on the tangent at P. Show that the maximum area of $\triangle APY$ is $\frac{3}{8}r^2\sqrt{3}$.

10. A tangent to an ellipse meets the axes in P and Q; show that the least value of PQ is equal to the sum of the semi-axes of the ellipse and also that PQ is divided at the point of contact in the ratio of its semi-axes.

11. A steel plant is capable of producing x tons per day of a low-grade, steel and y tons per day of a high-grade steel, where $y = \dfrac{40 - 5x}{10 - x}$.

    If the fixed market-price of the low-grade steel is half that of the high-grade steel, show that about $5\frac{1}{2}$ tons of low-grade steel are produced per day for maximum receipts.

*12. The telephone directorate finds that there is a net profit of Rs. 15 per instrument if an exchange has 1000 subscribers or less. If there are over 1000 subscribers, the profits per instrument decrease by Re. 0.01 for each subscriber above that number. How many subscribers would give the maximum net profit?

13. Find the dimensions of a rectangular plot of ground-containing A square feet which requires the least amount of fencing to enclose it and to divide it into two parts by a fence parallel to one side.

14. A rectangle is inscribed in a parabolic segment with one side of the rectangle along the base of the segment. Show that

$$\frac{\text{area of the largest rectangle}}{\text{area of the segment}} = \frac{1}{\sqrt{3}}.$$

15. Prove that the greatest acute angle at which the ellipse $x^2/a^2 + y^2/b^2 = 1$ can be cut by a concentric circle is

$$\tan^{-1}\{(a^2 - b^2)/2ab\}.$$

16. If Δ, Δ' be the areas of the two maximum isosceles triangles which can be described with their vertices at the origin and their base angles on the cardiode $r = a(1+\cos\theta)$, then prove that

$$2.56\Delta\Delta' = 25\sqrt{5}a^4.$$

17. Find the area of the greatest equilateral triangle which can be drawn with its sides passing through three given points A, B, C.

*18. The efficiency E of a screw-jack is given by

$$E = \frac{\tan\theta}{\tan(\theta+\alpha)},$$

where $\alpha(<\pi/2)$ is a positive constant. Find the maximum value of E and the value of θ for which E is maximum. [C.H. 1982]

*19. In a submarine cable the speed of signaling varies as

$$x^2 \log\left(\frac{1}{x}\right),$$

where x is the ratio of the radius of the core to that of the covering. Find the value of x for which the range of signaling is maximum. [C.H. 1975, '83(Sp.)]

*20. A person wishes to divide a triangular field into two equal parts by a straight fence. Show how it is to be done so that the fence may be of the least expense.

Hints and Answers

[A]

1. (a) Min. at $x=4$ (Min. Value $=-1$)
   (b) Min. at $x=3$ (Min. Value $=-39$)
      Max. at $x=-3$ (Max. Value $=69$)
   (c) Max. at $x=2$ (Max. Value $=0$)
   (d) Min. at $x=1$ (Min. Value $=-8$)
      Min. at $x=-2$ (Min. Value $=52$)
      Max. at $x=2$ (Max. Value $=-7$)
      Min. at $x=3$ (Min. Value $=-8$)
   (e) Max. at $x=-(\frac{1}{2})$ (Max. Value $=\frac{21}{16}$)

(f) Max. at $x = 2$ (Max. Value $= 38$)

Min. at $x = 0$ (Min. Value $= 0$)

Min. at $x = 3$ (Min. Value $= 37$) at $x = 1$, neither Max. nor Min.

(g) Max. at $x = 1$ (Max. Value $= 0$)

(h) Max. at $x = -1$ (Max. Value $= 14$)

Min. at $x = 3$ (Min. Value $= -28$)

Min. at $x = 2$ (Min. Value $= -175$)

No extreme at $x = 0$

2. (a) No finite extreme for any real x

(b) Min. at $x = 0$; Max. at $x = -2$

(c) Max. at $x = 1$; Min. at $x = -1$

(d) Max. at $x = -1$; Min. at $x = 1$

(e) See worked out problem **Ex.12.4.5 (pg.529)**.

(f) Max. at $x = 4$, Min. at $x = 16$

(g) Max. at $x = 2$ (Max. Value $= -9$)

Min. at $x = -2$ (Min. Value $= -\frac{1}{9}$)

3. (c) Max. Value $= e^{1/e}$; Max. Value $= 1/e$; Min. Value $= e$;

(e) Max. Value $= e^{1/e}$; (f) Min. at $x = 0$ (Min. Value $= 4$)

4. (a) Min. at $x = \pi/3$; (Min. Value $= \pi/6 - \sqrt{3}/2$)

Max. at $x = 5\pi/3$; (Max. Value $= 5\pi/6 + \sqrt{3}/2$)

(b) Max. at $x = \pi/3$; Min. at $x = 5\pi/3$ [Values are $\pm \frac{3\sqrt{3}}{4}$]; No extreme at $x = \pi$.

(c) Min. Value $= \sqrt{3}/4$; Max. Value $= \frac{(4\sqrt{2}\pm 3)}{6}$

Note here that least value is 0 and greatest value is $\frac{(4\sqrt{2}+3)}{6}$

(d) Min. Values $= -1/2, -5/6$; Max. Values $= 11/6, -5/12$; Least Value $= -5/6$; Greatest Value $= 11/6$.

(e) No. Max. or Min. (see that no real sol. for $\sec^2 3x = -4/3$)

(f) $n\pi + \tan^{-1}(a/b)$ (Critical Values); n even Max.; n odd Min.

(g) $x = \tan^{-1}(\pm\sqrt{a/b})$; +ve sign for Min. and $-$ve for Max.

(h) *Solution:* Let $y = a\cos x + b\cos 2x$ so that

$$y_1 = -\sin x(a + 4b\cos x) \text{ and } y_2 = -\cos x(a + 4b\cos x) + 4b\sin^2 x.$$

CHAPTER 12: INDETERMINATE FORMS AND MAXIMA & MINIMA

Now, $y_1 = 0 \Rightarrow \sin x = 0$, or, $a + 4b \cos x \pm 0$.

For $\sin x = 0$, $x = 0, \pi$ in $-\pi/4 \le x \le 5\pi/4$.

At $x = 0$, $y_2 = -(a+4b) < 0$, i.e., y is Max. at $x = 0$ (Max. value $= a+b$).

At $x = \pi$, $y_2 = a - 4b < 0$ or > 0 according as $a < 4b$ or $a > 4b$. So if $a < 4b$, y is Max. at $x = \pi$; but if $a > 4b$ then y is Min. at $x = \pi$ (Min. value $= -a + b$). Similar for the other.

(i) Let $y = \sin x \cos^2 x$, $y_1 = \cos^2 x - 2\sin^2 x \cos x$ and $y_2 = -3\cos^2 x \sin x$, $y_1 = 0 \Rightarrow \cos x = 0$ or $\tan^2 x = 1/2$. But $\cos x = 0$ gives $x = \pi/2, 3\pi/2$ in $[0, 2\pi]$. Now proceed.

(j) Write $y = a\cos^2 x + b\cos^2(a - x) = \tfrac{1}{2}a(1 + \cos 2x) + \tfrac{1}{2}b\{1 + \cos(2a - 2x)\}$
$= \tfrac{1}{2}(a + b) + \tfrac{1}{2}a\cos 2x + \tfrac{1}{2}b\cos(2a - 2x)$.

Now proceed as in worked example **Ex. 12.6.5** (page 534).

[B]

5. Step 1. $y' = 0$ gives $(ax + b)(Ax^2 + 2Bx + C) - (Ax + B)(ax^2 + 2bx + c) = 0$

$$\text{or,} \quad \frac{ax + b}{Ax + B} = \frac{ax^2 + 2bx + c}{Ax^2 + 2Bx + c} = y \tag{i}$$

Given expression y can be written as

$$y = \frac{x(ax + b) + bx + c}{x(Ax + B) + Bx + C}$$

$$\text{or,} \quad xy(Ax + B) + y(Bx + C) = x(ax + b) + bx + c$$

Using **(i)** we now get $y = \dfrac{bx + c}{Bx + C}$. \hfill (ii)

Eliminate x from **(i)** and **(ii)**: $x = \dfrac{b - By}{Ay - a} = \dfrac{c - Cy}{By - b}$.

∴ The extremes of y are given by
$$(AC - B^2)y^2 - y(aC - 2bB + cA) + (ac - b^2) = 0.$$

Step 2. Now observe that $(ax^2 + 2bx + c) - \lambda(Ax^2 + 2Bx + C) = x^2(a - \lambda A) + 2x(b - \lambda B) + (c - \lambda C)$. The condition that it becomes a perfect square is that

$$(b - \lambda B)^2 - (a - \lambda A)(c - \lambda C) = 0$$
$$\text{or,} \quad b^2 + \lambda^2 B^2 - 2\lambda bB - (ac - \lambda aC - \lambda Ac + \lambda^2 AC) = 0$$
$$\text{or,} \quad \lambda^2(AC - B^2) - \lambda(aC - 2bB + cA) + (ac - b^2) = 0.$$

Step 3. Hence the extremes of y are those values of λ for which
$$(ax^2 + 2bx + c) - \lambda(Ax^2 + 2Bx + C)$$
becomes a perfect square.

[C]

1. 80, 40.

2. (a) The axis of the cone should evidently pass through C, the centre of the sphere. Let OCD = height of the cone = x, r = radius of the base. Then $CD = OD - OC = x - R$; also $r^2 = R^2 - CD^2 = R^2 - (x - R)^2 = 2Rx - x^2$. $\therefore$ Volume $V = \frac{1}{3}\pi r^2 x = \frac{1}{3}\pi(2Rx - x^2)^2 x$, $dV/dx = 0$ when $x = 4/3R$. For this value of x, d^2V/dx^2 is negative. Hence the required answer = $4/3R$.

(b) The axis of the cone should coincide with the axis of the cylinder. Let x = height of the cylinder.

Let x = height of the cylinder, volume $V = \pi(h - x)^2 \tan^2 \alpha \cdot x$, h being the height of the cone and α = semi-vertical angle of the cone. Now find x for V maximum. [Ans. $x = 1/3h$]

(g) $\frac{1}{5}\{\pi r^2(5 + \sqrt{5})\}$.

3. (b) From $a + b + c = 2s$ and $s(s - a)(s - b)(s - c) = \Delta^2$, we deduce
$$\Delta^2 = s(s - a)(s - b)\{s - (2s - a - b)\} = s(s - a)(s - b)(a + b - s) \qquad (1)$$

Here a and b are variables and we may regard b as a function of a and for extreme value of b, $db/da = 0$.

Taking logarithmic differentiation on (1) w.r.t a (remembering s and Δ as constants) we get, using $db/da = 0$, $1/(a + b - s) = 1/(s - a)$, whence $a = c$. $\therefore$ For extreme value of b, $a + b + a = 2s$, or, $s - a = \frac{1}{2}b$. So for extreme value of b, we get from (1)
$$\Delta^2 = s\frac{b}{2}(s - b)\left(b - \frac{b}{2}\right), \quad \text{i.e.,} \quad 4\Delta^2 = s(s - b)b^2$$

$\therefore$ b is a root of $s(x - s)x^2 + 4\Delta^2 = 0$.

(c) Let x and y be the sides of the rectangle and a be the diameter of the circle. Then area $A = xy$ where $x^2 + y^2 = a^2$ (Draw a diagram). To find the max. value of $A = x\sqrt{a^2 - x^2} \cdot \frac{dA}{dx} = 0$ if $x = a\sqrt{2}$ and hence $y = a\sqrt{2}$, i.e., $x = y$. This is evidently a max. value of A, for A is zero when $x = 0$ and again when $x = a$ and is positive for the values in between 0 and a (or check that $\frac{d^2A}{dx^2}$ is $-$ve when $x = a\sqrt{2}$). Thus for max. area rectangle becomes a square.

4. Here the sides of the rectangle being x, y and that of the isosceles right triangle being z, z we have $A = xy + \frac{1}{2}z^2$; $P = 2(x + y + z)$ and $y^2 = 2z^2$. Now proceed.

5. Let the breadth be x and depth y and S = the stiffness. Then $S = k \cdot xy^3$ (k is a constant). But $x^2 + y^2 = a^2$, whence $S = kx(a^2 - x^2)^{3/2}$. Now proceed to find extreme

CHAPTER 12: INDETERMINATE FORMS AND MAXIMA & MINIMA

of x.

6. Let x = length of the side of each of the square portions removed. The volume V of the box $= (a - 2x)(b - 2x)x$.

   For extremes of V, $\dfrac{dV}{dx} = 0 \Rightarrow x = \dfrac{1}{6}\{(a+b) \pm \sqrt{a^2 - ab + b^2}\}$.

   The positive sign is not possible (why? explain).

7. Let the velocity relative to water $= v$; distance to be traversed $= D$. Time required when going against the current $= D/(v-c)$. Petrol burnt/hour $\propto v^3$ (i.e., $= kv^3$, say). Then total amount of petrol burnt $= \{D/(v-c)\}kv^3$.

   This is least when $v^3/(v-c)$ is a minimum. Now proceed.

12. 1250. Profit $P = x\{15 - \tfrac{1}{100}(x - 1000)\}$. Maximise P.

13. $\sqrt{(3A/2)}$; $\sqrt{\{2A/3\}}$.

18. $\dfrac{dE}{d\theta} = \dfrac{\sec^2\theta \tan(\theta+\alpha) - \sec^2(\theta+\alpha)\tan\theta}{\tan^2(\theta+\alpha)} = 0$; whereby $\cos(2\theta+\alpha) = 0$, $\sin\alpha = 0$ and proceed.

19. $S = x^2 \log 1/x$; $\dfrac{dS}{dx} = 0 \Rightarrow x = 0$ (not possible) or $x = e^{-1/2}$. See that d^2S/dx^2 is negative if $x = 1\sqrt{e}$.

20. Let the triangular field ABC of sides a, b, c be divided by a straight line EF (E or AC and F on AB) such that the area of the triagle ABC is twice the area of the triangle AEF. Let $AE = x$, $AF = y$ and $EF = z$.

    Then $\tfrac{1}{2}bc\sin A = 2\{\tfrac{1}{2}xy\sin A\}$, i.e., $bc = 2xy$.

    Now,
    $$\cos A = (x^2 + y^2 - z^2)/2xy = (x^2 + y^2 - z^2)/bc$$
    $$\text{or, } z^2 = x^2 + y^2 - bc\cos A = x^2 + \dfrac{b^2c^2}{4x^2} - bc\cos A$$
    $$\therefore \dfrac{d}{dx}(z^2) = 2x - \dfrac{b^2c^2}{2x^3} \text{ and } \dfrac{d^2}{dx^2}(z^2) = 2 + \dfrac{3b^2c^2}{2x^4}.$$

    But $\dfrac{d}{dx}(z^2) = 0$ gives $x = \sqrt{bc/2}$; for this x, $\dfrac{d^2}{dx^2}(z^2)$ is positive.

    $\therefore$ EF is a minimum when $AE = AF = \sqrt{\tfrac{1}{2}bc}$ and then $EF = \sqrt{2bc}\sin A/2$. If we draw a line GH (G on CB and H on BA), we can divide the triangular field so that $GH = \sqrt{2ac}\sin B/2 = $ a minimum length. Now if $a > b$ and $A > B$ and $EF/GH = \sqrt{\tan A/2 / \tan B/2} > 1$. This argument shows that we should divide s.t. if B is the least of the angles A, B, C then min. length $GH = \sqrt{2ac}\sin B/2$.

AN INTRODUCTION TO ANALYSIS

(Differential Calculus—Part I)

Chapter 13

Functions of Several Variables

Chapter 13

Functions of Several Variables

> **POINTS TO BE LEARNT WITH SUITABLE EXAMPLES**
> Point Sets in two and three dimensions: Elementary Topology. Functions. Limits: Double Limit. Repeated limits. Continuity in $\mathbb{R}^2$. Directional Derivative. Partial Derivatives. Sufficient condition for Continuity. Differentiability—Sufficient condition. Chain Rule. Homogeneous Functions: Euler's Theorem. Mixed Derivative: Schwarz Theorem. Young's Theorem. Jacobians. Functional Dependence. Implicit Functions. Perfect Differential. d^2z at a point.

13.1 Introduction: Point Sets in Higher Dimensions

In Chapter **4**, we have already discussed Point sets on the real line ($\mathbb{R}^1$). Our readers are, already acquainted with such point sets, which we called **Linear Point Sets.**

1. A point in two-dimensional space is an *ordered pair* of real numbers (x, y), represented by a point P in the xy-plane.

The sets of all such points in two-dimension forms what we call a two-dimensional Euclidean space $\mathbb{R}^2$.

2. A point in three-dimensional space is an ordered triple of three real numbers (x, y, z). We can represent it by a point P in our usual xyz-space.

The set of all such points in three-dimension forms what we call the three-dimensional Euclidean space $\mathbb{R}^3$.

3. We can generalise this notion and consider an *ordered n-tuple* of real numbers $(x_1, x_2, x_3, \cdots, x_n)$ and call it a point P in n-dimensional space. The set of all such n-dimensional points is called n-dimensional Euclidean space $\mathbb{R}^n$.

We often write $\vec{x} = (x_1, x_2, \cdots, x_n)$ and call it **a point** $\vec{x} \in \mathbb{R}^n$ with n co-ordinates

$x_1, x_2, \cdots, x_n$ or a **vector** $\vec{x}$ with n-components $x_1, x_2, \cdots, x_n$ (x_k is the kth component of the vector $\vec{x}$ or kth co-ordinate of the point $\vec{x}$).

Higher dimensional spaces arise in the study of relativity, Statistical and quantum mechanics. Even infinite-dimensional spaces occur in Quantum Mechanics.

4. Algebraic operations in $\mathbb{R}^n$ are defined as follows:

Equality: Two points $\vec{x}$ and $\vec{y}$ where

$$\vec{x} = (x_1, x_2, x_3, \cdots, x_n) \quad \text{and} \quad \vec{y} = (y_1, y_2, y_3, \cdots, y_n)$$

are said to be equal (i.e., $\vec{x} = \vec{y}$) iff $x_1 = y_1, x_2 = y_2, \cdots, x_k = y_k, \cdots, x_n = y_n$.

Sum: $\vec{x} + \vec{y} = (x_1 + y_1, x_2 + y_2, \cdots, x_n + y_n)$.

Multiplication by real numbers (scalars):

If a is a real number and $\vec{x} \in \mathbb{R}^n$ then $a\vec{x} = (ax_1, ax_2, \cdots, ax_n)$.

Difference: $\vec{x} - \vec{y} = (x_1 - y_1, x_2 - y_2, \cdots, x_n - y_n)$.

Zero Vector or **Origin:** $(0, 0, \cdots, 0) = \vec{0}$.

The two operations *Sum* and *Multiplication by a scalar* satisfy commutative, associative and distributive laws and there by make $\mathbb{R}^n$ a vector space over the real field. We also define:

Inner Product or **Dot product**

$$\vec{x} \cdot \vec{y} = x_1 y_1 + x_2 y_2 + \cdots + x_n y_n \quad \text{(a Scalar)}.$$

Norm or **length** of a vector $\vec{x} \in \mathbb{R}^n$:

$$\|\vec{x}\| = +\sqrt{\vec{x} \cdot \vec{x}} = +\sqrt{x_1^2 + x_2^2 + \cdots + x_n^2}.$$

The norm $\|\vec{x} - \vec{y}\|$ is called the **distance** between $\vec{x}$ and $\vec{y}$.

In fact, the set of all points $\mathbb{R}^n$ in n-dimensional together with a suitable inner product defines what is known as euclidean n-space.

5. **Properties of Norm:** Let $\vec{x}$ and $\vec{y}$ denote points in $\mathbf{E}_n$. Then we have

1. $\|\vec{x}\| \geq 0$ and $\|\vec{x}\| = 0$ iff $\vec{x} = \vec{0}$.

2. $\|a\vec{x}\| = |a| \|\vec{x}\|$, where a is any real number.

3. $\|\vec{x} - \vec{y}\| = \|\vec{y} - \vec{x}\|$.

4. $|\vec{x} \cdot \vec{y}| \leq ||\vec{x}||\,||\vec{y}||$ (*Cauchy-Schwarz Inequality*).

[*Hints*: (4) can be written as

$$(x_1 y_1 + x_2 y_2 + \cdots + x_n y_n)^2 \leq (x_1^2 + x_2^2 + \cdots + x_n^2)(y_1^2 + y_2^2 + \cdots + y_n^2).$$

The equality holds for (i) any one of $\vec{x}$ and $\vec{y}$ or both are null vectors or, (ii) $\vec{x} = k\vec{y}$ for some $k \in \mathbb{R}$. Both (i) and (ii) may be stated by the single statement "$\vec{x}$ and $\vec{y}$ are linearly dependent."

This inequality follows from the fact that the quadratic in t given by

$$f(t) = (x_1 + ty_1)^2 + (x_2 + ty_2)^2 + \cdots + (x_n + ty_n)^2$$

is positive for all real t when $\vec{x}$ and $\vec{y}$ are independent and hence its discriminant is negative, otherwise some real t would make $f(t) := 0$. This will give the desired inequality. For more details consult any Higher Algebra book e.g., **Higher Algebra–Ghosh and Maity**.]

5. $||\vec{x} + \vec{y}|| \leq ||\vec{x}|| + ||\vec{y}||$ [*Triangle Inequality*]

$$\left[||\vec{x} + \vec{y}||^2 = \sum_{k=1}^{n}(x_k + y_k)^2 = \sum_{k=1}^{n} x_k^2 + 2\sum_{k=1}^{n} x_k y_k + \sum_{k=1}^{n} y_k^2 \right.$$
$$= ||\vec{x}||^2 + 2(\vec{x} \cdot \vec{y}) + ||\vec{y}||^2 \text{ [Now using (d)]}$$
$$\left. \leq ||\vec{x}||^2 + 2||\vec{x}||\,||\vec{y}|| + ||\vec{y}||^2 = (||\vec{x}|| + ||\vec{y}||)^2. \right]$$

We also have

$$\left| ||\vec{x}|| - ||\vec{y}|| \right| \leq ||\vec{x} - \vec{y}||$$

(Deduce it from the Triangle Inequality).

6. There are n unit vectors (i.e., vectors of length = one) in $\mathbf{E}_n$, namely,

$$\vec{u}_1 = (1, 0, 0, \cdots, 0), \quad \vec{u}_2 = (0, 1, 0, \cdots, 0), \quad \cdots, \quad \vec{u}_n = (0, 0, 0, \cdots, 1).$$

Any vector $\vec{x} = (x_1, x_2, x_3, \cdots, x_n)$ can be written as a linear combination of $\vec{u}_1, \vec{u}_2, \cdots, \vec{u}_n$ uniquely.

These n unit vectors are linearly independent and any vector $\vec{x} \in \mathbf{E}_n$ can be expressed as a linear combination of $\vec{u}_1, \vec{u}_2, \cdots, \vec{u}_n$ and hence $\{\vec{u}_1, \vec{u}_2, \cdots, \vec{u}_n\}$ forms what we call a **basis** of $\mathbf{E}_n$.

7. Open ball in $\mathbb{R}^n$

a) Let $\vec{a} \in \mathbb{R}^n$ and let δ be a positive real number. Then the set of all points $\vec{x} \in \mathbb{R}^n$ such that $\|\vec{x} - \vec{a}\| < \delta$ is called an **open n-ball** of radius δ and centre $\vec{a}$, denoted by $B(\vec{a}; \delta)$.

b) In $\mathbb{R}^1$ this is simply an open interval $|x - a| < \delta$ or $]a - \delta, a + \delta[$. We call it a δ-neighbourhood of a (δ-nbd of a, denoted by $N_\delta(a)$]. In case, the centre a is not taken into consideration, i.e., $]a-\delta, a[\cup]a, a+\delta[$ then the neighbourhood of a is called deleted δ-nbd of a, denoted by $N'_\delta(a)$.

c) In $\mathbb{R}^2$, $\|\vec{x} - \vec{a}\| < \delta$ gives a circular disc with centre $\vec{a}$ and radius δ. We call it a circular δ-neighbourhood of $\vec{a}$. Suppose $\vec{a} = (a, b) \in \mathbb{R}^2$ and $\vec{x} = (x, y) \in \mathbb{R}^2$ then $\|\vec{x} - \vec{a}\| < \delta$ gives

$$(x - a)^2 + (y - b)^2 < \delta^2 \text{ (Fig. 13.1)}$$

i.e., the set of points (x, y) that lie within the circle $(x - a)^2 + (y - b)^2 = \delta^2$.

If the point (a, b) is excluded, then the **deleted circular δ-nbd of (a, b)** is given by

$$0 < (x - a)^2 + (y - b)^2 < \delta^2.$$

In $\mathbb{R}^2$ we shall also use other neighbourhoods—

Rectangular neighbourhood: $a < x < b$ and $c < y < d$ (Fig. 13.2) or

Square neighbourhood: $|x - a| < \delta$ and $|y - b| < \delta$ (Fig. 13.3).

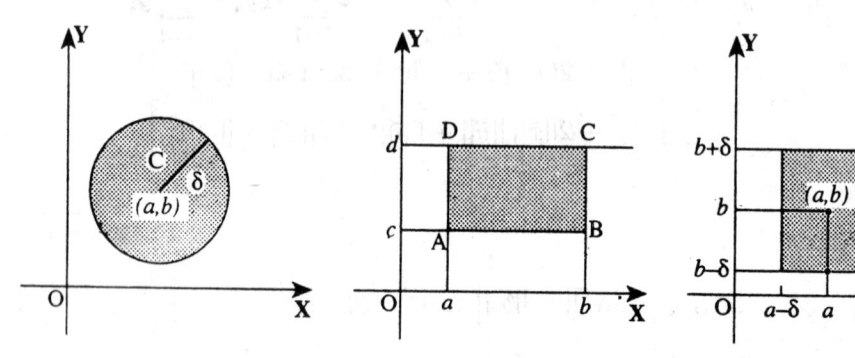

Fig. 13.1 Circular Neighbourhood of (a, b)

Fig. 13.2 Rectangular Neighbourhood

Fig. 13.3 Square δ-Neighbourhood of (a, b)

d) In $\mathbb{R}^3$, $\|\vec{x} - \vec{a}\| < \delta$ gives a spherical solid with centre at $\vec{a}$ and radius δ. We call it a **Spherical δ-nbd** of $\vec{a}$. Suppose $\vec{a} = (a, b, c) \in \mathbb{R}^3$ and $\vec{x} = (x, y, z) \in \mathbb{R}^3$.

Then $\|\vec{x} - \vec{a}\| < \delta$ gives $(x - a)^2 + (y - b)^2 + (z - c)^2 < \delta^2$,

i.e., the set of points (x, y, z) within the sphere $(x - a)^2 + (y - b)^2 + (z - c)^2 = \delta^2$. If the centre (a, b, c) is excluded then the deleted spherical nbd of (a, b, c) is written as

$$0 < (x - a)^2 + (y - b)^2 + (z - c)^2 < \delta^2.$$

8. Definition of an interior point and an open set

a) Let S be a subset of $\mathbb{R}^n$, and assume that $\vec{a} \in S$. Then $\vec{a}$ is called an interior point of S if $\exists$ an open n-ball with centre at $\vec{a}$, all of whose points belong to S. Every interior point $\vec{a} \in S$ can be surrounded by an open n-ball $B(\vec{a}, \delta) \subset S$.

The set of all interior points of S is called the **interior** of S and is denoted by **int** S.

b) A set $S \subseteq \mathbb{R}^n$ is called an **open set** if all its points are *interior points*.

In $\mathbb{R}^1$ an open interval $I = (a,b)$ is an open set, but a closed interval $[a,b]$ is not an open set because the end points a and b are not interior points of the interval ($\because$ every nbd of these points contains points of the set $[a,b]$ and points outside this set, they are in fact the boundary pts).

In $\mathbb{R}^2$; the set S of points satisfying the inequality $(x-a)^2 + (y-b)^2 < \delta^2$ defines an open set [an open circle of radius δ, with centre at (a,b)].

Remember. Note that an open interval in $\mathbb{R}^1$ is no longer an open set when it is considered as a subset of the plane (i.e., a subset of $\mathbb{R}^2$).

See that the subset $S = \{(a,0) : a \in \mathbb{R}\}$ of the plane $\mathbb{R} \times \mathbb{R}$ is not an open subset of $\mathbb{R} \times \mathbb{R}$; no point is an interior point as any of its neighbourhoods will contain points outside the set S.

Note: $\mathbb{R}$ is the same $\mathbb{R}^1$.

Properties of open sets

(i) ϕ (Null set) and the whole space $\mathbb{R}^n$ are open sets.

(ii) Union of any family of open sets (in $\mathbb{R}^n$) is open.

(iii) The intersection of a finite number of open sets is again an open set.

(iv) A set $S \subseteq \mathbb{R}^n$ is open if and only if $S = \text{int} S$.

Proof of (ii). Let F be any arbitrary family of open sets and let $S = \bigcup_{A \in F} A$. Assume $\vec{x} \in S$. Then $\vec{x}$ must belong to at least one of the sets of F, say $\vec{x} \in A$. Since A is open, $\exists$ an open n-ball $B(\vec{x}) \subseteq A$. But $A \subseteq S$, so $B(\vec{x}) \subseteq S$, and hence $\vec{x}$ is an interior point of S. Since every point of S is an interior point, S is open.

Proof of (iii). Let $S = \bigcap_{r=1}^{n} A_r$, where each A_r is open. If $S = \phi$, there is nothing to prove [See (i)].

Assume $\vec{x} \in S$, then $\vec{x} \in$ each set $A_1, A_2, A_3, \cdots, A_n$ and hence $\exists$ open n-balls $B(\vec{x}, r_k) \subseteq A_k (k = 1, 2, 3, \cdots, n)$; each $r_k > 0$.

Let $r = \min.\{r_1, r_2, \cdots, r_n\}$. Then $\vec{x} \in B(\vec{x}; r) \subseteq S$. That is any $\vec{x} \in S$ is an interior point of S and so S is open.

Note 13.1.1. *We have proved that the union of an arbitrary collection of open sets is an open set and also the intersection of a finite collection of open sets is an open set. However, the intersection of an arbitrary collection of open sets may or may not be open: we give an example in $\mathbb{R}^1$ e.g.,*

1. $\bigcap_{n=1}^{\infty} \left(-\frac{1}{n}, \frac{1}{n}\right) = \{0\}$, not an open set, though each open interval $\left(-\frac{1}{n}, \frac{1}{n}\right)$ is an open set.

2. $\bigcap_{n=1}^{\infty} \left(-\frac{n}{n+1}, \frac{n}{n+1}\right) = \left(-\frac{1}{2}, \frac{1}{2}\right)$ which is an open set.

3. $\bigcap_{n=1}^{\infty} \left(-\frac{1}{n}, 1 + \frac{1}{n}\right) = [0, 1]$ is not an open set.

9. Every open n-ball is an open set and every open interval in $\mathbb{R}^n$ in open.

Proof of 2nd part: Let $a_i, b_i (i = 1, 2, 3, \cdots, n)$ be real numbers such that $a_i < b_i$ for all i. Then the set of all points $\xi = (x_1, x_2, \cdots, x_n) \in \mathbb{R}^n$ such that $a_i < x_i < b_i$ for $i = 1, 2, 3, \cdots, n$, is called an n-dimensional, **open interval** in $\mathbb{R}^n$, determined by the points $\vec{a} = (a_1, a_2, \cdots, a_n)$ and $\vec{b} = (b_1, b_2, \cdots, b_n)$ of $\mathbb{R}^n$. We may write this n-dimensional open interval by the symbol $(\vec{a}, \vec{b})$. This is an open set in $\mathbb{R}^n$.

10. Closed Sets in $\mathbb{R}^n$

We shall give two définitions for closed sets and show that they are equivalent.

Definition 13.1.1. *A set S in $\mathbb{R}^n$ is called a closed set if its complement is open (i.e., if $\mathbb{R}^n - S$ is open).*

Definition 13.1.2. *A set S in $\mathbb{R}^n$ is a closed set iff it contains all its accumulation points.*

The set of all accumulation points of a set S is called the *derived set* of S, and it denoted by S'. Thus S is closed iff $S' \subseteq S$.

What is an accumulation point of a set S in $\mathbb{R}^n$?

We shall first introduce a term called *Adherent point of a set S*.

CHAPTER 13: FUNCTIONS OF SEVERAL VARIABLES

(a) *Definition of an adherent point of a set:* Let $S \subseteq \mathbb{R}^n$ and let $\vec{x} \in R^n$, not necessarily in S. Then $\vec{x}$ is an adherent point of S if:

Every n-ball $B(\vec{x})$ contains at least one point of S.

Clearly, if $\vec{x} \in S$ then $\vec{x}$ adheres to S because every n-ball $B(\vec{x})$ contains $\vec{x}$. Some points $\vec{x}$ adhere to S where every ball $B(\vec{x})$ contains point of S, other than the point $\vec{x}$. These are called *accumulation points* of S.

(b) *Definition of an accumulation point of a set:* If $S \subseteq \mathbb{R}^n$ and $\vec{x} \in \mathbb{R}^n$, not necessarily in S, then $\vec{x}$ is called an accumulation point of S, if every n-ball $B(\vec{x})$ contains at least one point of S, **other than** $\vec{x}$. Thus $\vec{x}$ is an accumulation point of S if $\vec{x}$ adheres to $S - \{\vec{x}\}$.

If $\vec{x} \in S$, but $\vec{x}$ is *not* an acumulation point of S, then $\vec{x}$ is called an *Isolated point* of S.

It follows from the above definition that (The proof is given later as an example).

Theorem 13.1.1. *If $\vec{x}$ is an accumulation point of $S \subseteq \mathbb{R}^n$, then every n-ball $B(\vec{x})$ contains infinitely many points of S.*

Accordingly, a set cannot have an **accumulation point** unless it contains *infinitely many points*. But this does not mean that every infinite set must have some accumulation point e.g., in $\mathbb{R}^1$, the set of positive integers $\{1, 2, 3, \cdots, \}$ is an infinite set but it has no accumulation point.

Bolzano-Weierstrass Theorem, however, proves that a bounded infinite set S in $\mathbb{R}^n$ has at least one point in $\mathbb{R}^n$ which is an accumulation point of S.

[By a bounded set S in $\mathbb{R}^n$ we mean a set which lies entirely within a certain n-Ball]

Example 13.1.1. *Assuming that a set S in $\mathbb{R}^n$ is closed if only its complement $\mathbb{R}^n - S$ is open, prove the following:*

(a) *If $S \subseteq \mathbb{R}^n$ is a closed set, then it contains all its adherent points.*

(b) *If however, S contains all its adherent points then S is closed.*

Solution: (a) Let S be a closed set and let some point $\vec{x}$ adheres to S. We wish to prove $\vec{x} \in S$.

If $\vec{x} \notin S$ then $\vec{x}$ should belong to $\mathbb{R}^n - S$. Now S is a closed set, hence $\mathbb{R}^n - S$ is open, i.e., some n-ball $B(\vec{x})$ lies completely in $\mathbb{R}^n - S$. Thus $B(\vec{x})$ does not contain any point of S, then $\vec{x}$ cannot adhere to S, a contradiction. Hence etc.

(b) Suppose S contains all its adherent points. We wish to prove that S is a closed set. Let $\vec{x} \in \mathbb{R}^n - S$. Then $\vec{x} \notin S$ and so $\vec{x}$ does not adhere to S. Hence $\exists$ some ball $B(\vec{x})$ with no point of S and so $B(\vec{x})$ completely lies in $\mathbb{R}^n - S$. Therefore $\mathbb{R}^n - S$ is open and hence S is closed.

Definition of Closure of a set S in $\mathbb{R}^n$: The set of all adherent points of $S \subseteq \mathbb{R}^n$ is called the **closure** of S, denoted by $\bar{S}$.

Since every point $\vec{x} \in S$ adheres to S and $\bar{S}$ contains all adherent points of S, therefore $S \subseteq \bar{S}$. We have proved above that if a set S is closed then it contains all its adherent points and conversely if S contains all its adherent points then S is a closed set. Hence $\bar{S} \subseteq S$ if and only if S is closed. Hence

A set is closed if and ony if $S = \bar{S}$.

Definition of derived Set: The set of all accmulation points of $S \subseteq \mathbb{R}^n$ is called the *derived set* of S, denoted by S'.

See that $\bar{S}$ contains all adherent points (they may be points of the set S or accumulation points of S). We write $\bar{S} = S \cup S'$.

We therefore conclude:

A set is closed if and only if $S' \subseteq S$

Or, *A set S in $\mathbb{R}^n$ is closed if and only if it contains all its accumulation points.*

11. Bolzano-Weierstrass Theorem: *Every bounded infinite set S in $\mathbb{R}^2$ has at least one accumulation point.*

We have already proved the theorem in $\mathbb{R}^1$. We now prove it in $\mathbb{R}^2$.

We must produce a point p with the property that every neighbourbood of p contains infinitely many points of S (because then p would be an accumulation point of S).

Since the set S is bounded we may assume that it is completely contained in a square C_0 of side L (Fig 13.4).

We divide C_0 into four equal squares by intersecting lines. One of these smaller squares must contain infinitely many points of S, for otherwise S itself would have been a finite set.

We choose such a subsquare. We call it C_1 (i.e., C_1 contains infinitely many points of S). We have $C_1 \subset C_0$ and both are closed and bounded.

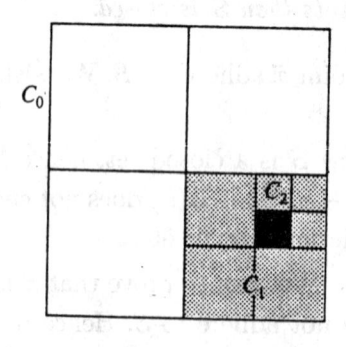

Fig. 13.4

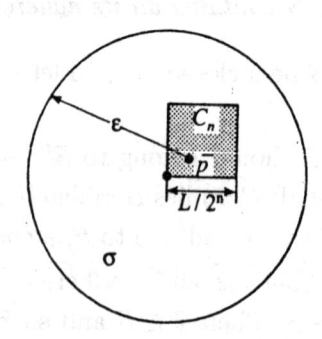

Fig. 13.5

CHAPTER 13: FUNCTIONS OF SEVERAL VARIABLES 747

We now repeat the process: Divide C_1 into four smaller squares and select one of them which contains infinitely many points of the set S. By continuing this, we can obtain a sequence of closed squares $\{C_n\}$ with $C_0 \supset C_1 \supset C_2 \cdots \supset C_n \supset \cdots$ and such that each one contains infinitely many points of S.

By the bisection process we see that C_n has sides of length $L/2^n$. Therefore, by **Nested Set Property**[1], $\exists$ a point $\bar{p}$ which lies in all the sets C_n. We wish to prove that $\bar{p}$ is a point of accumulation of S. Note that $\bar{p}$ may not belong to S.

Take a neighbourhood σ of $\bar{p}$, a circle with centre $\bar{p}$ and radius ε. When n is sufficiently large, say with $L/2^n < \varepsilon/2$, the set C_n which contains $\bar{p}$ must in fact lie in σ. Since C_n contains infinitely many points of S, so will the set σ. Hence every neighbourhood of $\bar{p}$ contains infintely many points of S and $\bar{p}$ is an accumulation point for S (Fig. 13.5).

Note 13.1.2. *The proof of Bolzano-Weierstrass Theorem for $\mathbb{R}^n$ is an extension of the ideas used in treating $\mathbb{R}^2$.*

13.2 Functions in Higher Dimensional Space

We are acquainted with the meaning of a function of a single real variable. We shall extend this idea to define a function of several real variables or what comes to the same thing, a function of a point in Euclidean Space $\mathbb{R}^n$. However, **our main concern will be with two and three-dimensional spaces.**

Definition 13.2.1. *Let S_1 be a set of real numbers and S_2 a set of points in an Euclidean Space $\mathbb{R}^n$. Suppose that for each point $\mathbf{P} \in S_2$ there corresponds a unique ω in S_1. The set of all pairs $(\mathbf{P}, \omega)$, where $\mathbf{P} \in S_2$ and ω is the corresponding value in S_1, is called a* FUNCTION, *denoted by f, defined on $\mathbb{R}^n$.*

The functional values are denoted by $\omega = f(\mathbf{P})$, or equivalently $\omega = f(x, y), \omega = f(x, y, z), \cdots, \omega = f(x_1, x_2, x_3, \cdots, x_n)$ according to the dimension of the space.

The set of $\mathbf{P}$'s for which f is defined is called the **domain of the function**, denoted by D.

The set of functional values (i.e., the set of ω's which correspond to some $\mathbf{P}$ in D) is called the **range** of f, denoted by $R(f)$.

Examples of functions of two or three variables

1. $z = f(x, y) = x + y$ is defined over the entire xy-plane.

---

[1]**The Nested Set Property:** Let $C_1, C_2, C_3, \cdots$ be a sequence of non-empty **closed and bounded** sets with $C_1 \supset C_2 \supset C_3 \cdots$. Then $\exists$ at least one point p which belongs to all the sets C_i.

2. $z = g(x, y) = \dfrac{xy}{x^2 + y^2}$ is defined over the entire xy-plane excluding the only point $(0, 0)$.

3. $z = \phi(x, y) = \sqrt{x + y - 1}$ is defined over the region $x + y \geq 1$, which includes the points of the line $x + y = 1$.

4. $z = \theta(x, y) = \dfrac{1}{\sqrt{x + y - 1}}$ is defined over the region $x + y > 1$, which does not include the points of the line $x + y = 1$ (*open region*).

5. $z = \sqrt{1 - x^2 - y^2}$ is defined over the **closed** circular region $x^2 + y^2 \leq 1$.

6. $z = \log(1 - x^2 - y^2)$ is defined over the region $x^2 + y^2 < 1$ ($\log 0$ is undefined; so points on the circle $x^2 + y^2 = 1$ are to be excluded from the circular *domain*) (*open region*).

7. **Linear function:** $u = f(x, y) = ax + by + c$ is defined for all points (x, y) in the xy-plane.

   General polynomial in x and y contains a sum of a finite terms like $a_{mn} x^m y^n$ (m, n are positive integers or zero).

   All such **polynomials** are defined over the entire xy-plane.

8. **Rational Functions:** Quotient of two polynomials $f(x, y)/g(x, y)$. They are defined where $g(x, y) \neq 0$, e.g.,

$$u = \frac{ax + by + c}{\alpha x + \beta y + \gamma}$$

is defined for all points (x, y) in the xy-plane excepting points which lie on the line $\alpha x + \beta y + \gamma = 0$.

9. **Algebraic functions:**

   $z_1 = \sqrt{(x - 2)(y + 3)}$;

   $z_2 = \sqrt{x^2 + 2y^2 - 4}$.

   The function z_1 is defined at all points (x, y) except where the radicand is negative. Thus z_1 is defined in the closed region formed by the first and third quadrants into which the lines $x = 2$, $y = -3$ divide the plane [see the shaded region of Fig. 13.6].

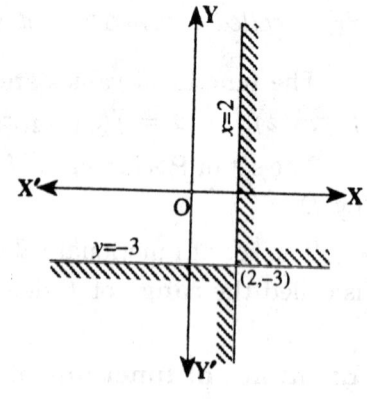

   Fig. 13.6

   The function z_2 is defined where $x^2 + 2y^2 \geq 4$, i.e., for points on or outside the ellipse $x^2/4 + y^2/2 = 1$.

CHAPTER 13: FUNCTIONS OF SEVERAL VARIABLES

10. $f(x, y, z) = x^2y + y^2z + z^2x$ is a function of three independent variables x, y, z. It is defined for all points $(x, y, z) \in \mathbb{R}^3$.

13.3 Limits of Functions

Let f be defined on a domain D and let $\mathbf{P_0}$ be an accumulation point of D. We then say that f has a limit A as $\mathbf{P} \to \mathbf{P_0}$ if, for each $\varepsilon > 0$, there exists a $\delta(\varepsilon, \mathbf{P_0})$ for which

$$|f(\mathbf{P}) - A| < \varepsilon \text{ whenever } |\mathbf{P} - \mathbf{P_0}| < \delta; \text{ and } \mathbf{P} \in D, \mathbf{P} \not\equiv \mathbf{P_0}$$

We express this symbolically as

$$\lim_{\mathbf{P} \to \mathbf{P_0}} f(\mathbf{P}) = A \text{ or, } f(\mathbf{P}) \to A \text{ as } \mathbf{P} \to \mathbf{P_0}.$$

In the language of neighbourhood (*nbd*) we can state thus:

$$\lim_{\mathbf{P} \to \mathbf{P_0}} f(\mathbf{P}) = A \Rightarrow \text{ For each } \varepsilon > 0, \ \exists \text{ a deleted nbd } N' \text{ of } \mathbf{P_0}$$

such that $|f(\mathbf{P}) - A| < \varepsilon$, if $\mathbf{P} \in D \cap N'$.

Functions of two variables: Simultaneous limit or Double limit

[Replace $\mathbf{P}$ by (x, y) and $\mathbf{P_0}$ by (a, b)]

Definition 13.3.1. *A function f of two independent variables x and y, is said to tend to a limit A as the point (x, y) tends to a given point (a, b), if, to each $\varepsilon > 0$, $\exists \delta > 0$ such that*

$$|f(x, y) - A| < \varepsilon$$

for all points (x, y) of the domain which belongs to some deleted δ-nbd N' of (a, b). The inequality may not be satisfied when $(x, y) = (a, b)$.

N may be a *square nbd*. $0 < |x - a| < \delta, \quad 0 < |y - b| < \delta$

or, N may be a *circular nbd*. like $0 < (x - a)^2 + (y - b)^2 < \delta^2$

or, N may be *any other nbd*.

We then write

$$\lim_{\substack{x \to a \\ y \to b}} f(x, y) = A \text{ or, } \lim_{(x,y) \to (a,b)} f(x, y) = A.$$

Examples on Limits

Example 13.3.1. Let $f(\mathbf{P}) = f(x,y) = \dfrac{x^2 y^2}{x^2 + y^2}$.

Using ε-δ definition establish: $\lim\limits_{\substack{x \to 0 \\ y \to 0}} \dfrac{x^2 y^2}{x^2 + y^2} = 0$.

Solution: Let $\varepsilon > 0$ be given. We are to find a $\delta > 0$ s.t. in some deleted δ-nbd of $(0,0)$ for all points (x,y),

$$\left| \dfrac{x^2 y^2}{x^2 + y^2} - 0 \right| < \varepsilon, \quad \text{or,} \quad \dfrac{x^2 y^2}{x^2 + y^2} < \varepsilon.$$

Now clearly,

$$x^2 < x^2 + y^2; \quad y^2 < x^2 + y^2$$

so that
$$\dfrac{x^2 y^2}{x^2 + y^2} < \dfrac{(x^2 + y^2)^2}{x^2 + y^2} = x^2 + y^2 < \varepsilon$$

if $\quad 0 < x^2 + y^2 < \delta^2$ where $\delta = \sqrt{\varepsilon}$.

Thus, we have found out $\delta = \sqrt{\varepsilon}$ to satisfy the requirements of the definition. Therefore,

$$\lim\limits_{\substack{x \to 0 \\ y \to 0}} \dfrac{x^2 y^2}{x^2 + y^2} = 0.$$

Example 13.3.2. Establish that $\lim\limits_{\substack{x \to 0 \\ y \to 0}} xy \dfrac{x^2 - y^2}{x^2 + y^2} = 0$.

Solution: Let $\varepsilon > 0$ be given. To find a δ-nbd of $(0,0)(\delta > 0)$ such that in that nbd N, $\forall (x,y)$

$$\left| xy \dfrac{x^2 - y^2}{x^2 + y^2} - 0 \right| < \varepsilon \quad \text{i.e.,} \quad \left| xy \dfrac{x^2 - y^2}{x^2 + y^2} \right| < \varepsilon.$$

But $\quad \left| xy \dfrac{x^2 - y^2}{x^2 + y^2} \right| = |x||y| \dfrac{|x^2 - y^2|}{x^2 + y^2}.$

Clearly, $|x| < \sqrt{x^2 + y^2}$, $\;|y| < \sqrt{x^2 + y^2}\;$ and $\;\dfrac{|x^2 - y^2|}{x^2 + y^2} < 1.$

$\therefore \quad \left| xy \dfrac{x^2 - y^2}{x^2 + y^2} - 0 \right| = |x||y| \dfrac{|x^2 - y^2|}{x^2 + y^2} < \sqrt{x^2 + y^2} \cdot \sqrt{x^2 + y^2} = x^2 + y^2 < \varepsilon$

i.e., $\left| xy \dfrac{x^2 - y^2}{x^2 + y^2} - 0 \right| < \varepsilon,\;$ if $0 < x^2 + y^2 < \delta^2$ where $\delta = \sqrt{\varepsilon}$.

CHAPTER 13: FUNCTIONS OF SEVERAL VARIABLES

∴ The requirements of the definition of limit are met and hence the limit exists and is equal to zero.

Non-existence of limit

Example 13.3.3. *To prove* $\lim_{\substack{x \to 0 \\ y \to 0}} \dfrac{xy}{x^2 + y^2}$ *does not exist.*

Solution: Let $f(x, y) = \dfrac{xy}{x^2 + y^2}$.

The domain of f is the whole xy-plane punctured at the origin $(0, 0)$. For existence of limit we are to examine the values of f near $(0, 0)$, i.e., in some deleted nbd. of $(0, 0)$. If we introduce polar co-ordinates $x = \rho \cos \theta, y = \rho \sin \theta$, then f takes the form

$$f(x, y) = \frac{xy}{x^2 + y^2} = \frac{\rho \cos \theta \cdot \rho \sin \theta}{\rho^2} = \frac{1}{2} \sin 2\theta.$$

Here, of course, $\rho = \sqrt{x^2 + y^2}$. As $(x, y) \to (0, 0), \rho \to 0$.

From this formula, it is clear that f is constant on each straight line through the origin $(y = x \tan \theta)$ and that, in general, the constant is different on different lines. Clearly, then, the values of f cannot be made close to any constant A by simply restricting ρ. Thus, $\lim_{\substack{x \to 0 \\ y \to 0}} \dfrac{xy}{x^2 + y^2}$ does not exist in this case.

Alternative Method. Put $y = mx$. Then we have

$$f(x, mx) = \frac{m}{1 + m^2}.$$

Again, we see that f has different constant values on different lines through the origin; so by letting $(x, y) \to (0, 0)$ along a suitable line through the origin, $f(x, y)$ will approach any value k specified at pleasure.

So the values of f cannot be made close to any one constant A in some nbd of $(0, 0)$; limit cannot exist in this case. The limit must have to be unique.

Note 13.3.1. *It is easy to see that if*

$$\lim_{\substack{x \to a \\ y \to b}} f(x, y) = A,$$

and if $y = \phi(x)$, *any function of x whatsoever, such that* $\phi(x) \to b$ *as* $x \to a$ *then* $f(x, \phi(x)) \to A$ *as* $x \to a$.

So, if we can determine two functions $\phi_1(x), \phi_2(x)$ such that the limits of $f(x, \phi_1(x))$ and $f(x, \phi_2(x))$ as $x \to a$ are different, then we can conclude that the *simultaneous limit*

$$\lim_{(x,y)\to(a,b)} f(x,y)$$

does not exist.

Similarly, if $x = \psi(y)$ be **any** function such that $\psi(y) \to a$ as $y \to b$ and if $f(\psi(y), y) \to A$ as $y \to b$, the

$$\lim_{\substack{x \to a \\ y=a}} f(x,y) = \lim_{y \to b} f(\psi(y), y) = A$$

So, to show the non-existence of the limit we choose $\psi_1(y), \psi_2(y)$ both tend to a as $y \to b$ but

$$\lim_{y \to b} f(\psi_1(y), y) \quad \text{and} \quad \lim_{y \to b} f(\psi_2(y), y)$$

are different. See the next example.

Example 13.3.4. *The function f, defined over the whole xy-plane, is given by*

$$f(x,y) = \begin{cases} \dfrac{|x|}{y^2} e^{-|x|/y^2}, & \text{when } y \neq 0; \\ 0, & \text{when } y = 0. \end{cases}$$

Discuss the existence of limit of $f(x, y)$ as $(x, y) \to (0, 0)$.

Solution: On the line $x = 0$, we have $f(0, y) = 0, \forall y$ and the limit is zero as $(x, y) \to (0, 0)$ along this line.

On any other line $x = ay$ through the origin we have

$$f(ay, y) = \frac{|ay|}{y^2} e^{-|ay|/y^2} = \frac{|a|}{|y|} e^{-|a/y|}$$

and the limit of this expression as $y \to 0$ is easily seen to be 0, by using (say) L'Hospital's rule.

Thus as $(x, y) \to (0, 0)$ along any straight line through the origin, we have $f(x, y)$ tending to zero.

But see that along the parabola $x = y^2$, we have

$$f(y^2, y) = \frac{y^2}{y^2} e^{-(y^2/y^2)} = e^{-1}.$$

Hence f is always e^{-1} along a parabola passing through the origin and clearly then $\lim_{(x,y)\to(0,0)} f(x,y)$ is not unique and the limit, in fact, does not exist.

Example 13.3.5. $f(x,y) = \dfrac{x^2 y}{x^4 + y^2}$.

Discuss the existence of the limit of $f(x,y)$ as $(x,y) \to (0,0)$.

Solution: Let us consider the double limit as $(x,y) \to (0,0)$. If we approach the origin along any axis, $f(x,y) = 0$ and the limit is zero. If we approach along any line $y = mx$ ($m \neq 0$) then

$$f(x,y) = \frac{x^2 \cdot mx}{x^4 + m^2 x^2} = \frac{mx}{x^2 + m^2},$$

which tends to zero as $x \to 0$.

Thus, when we approach the origin along any line $y = mx$, the limit $= 0$. But let us approach the origin along a parabolic path $y = mx^2$; here also $y \to 0$ as $x \to 0$.

$$\text{But,}\quad f(x,y) = f(x, mx^2) = \frac{x^2 \cdot mx^2}{x^4 + m^2 x^4} = \frac{m}{1 + m^2},$$

which has different values for different m. Hence the double limit

$$\lim_{(x,y) \to (0,0)} \frac{x^2 y}{x^4 + y^2} \text{ does not exist.}$$

Example 13.3.6. $\lim(x^2 + 2y) = 5$, when $(x,y) \to (1,2)$. *Verify.*

Solution:

$$|x^2 + 2y - 5| = |x^2 - 1 + 2y - 4| \leq |x^2 - 1| + 2|y - 2| = |x+1||x-1| + 2|y-2|$$
$$< 2\delta + 2\delta \text{ when } 0 < |x-1| < \delta \text{ and } 0 < |y-2| < \delta \text{ if } 0 < \delta \leq 1$$

$\therefore |x^2 + 2y - 5| < \varepsilon$, where ε is any given positive number,

if $4\delta < \varepsilon$, i.e., if $\delta < \varepsilon/4$ or 1, whichever is smaller, i.e., $\delta < \min\{\frac{\varepsilon}{4}, 1\}$.

Thus, the definition requirements are met and the statement is TRUE.

Repeated or Iterated Limits

Let $f(x,y)$ be defined in a certain *nbd.* of (a,b).

Then $\lim_{x \to a} f(x,y)$, y remaining constant, when exists, will be different for different values of y and in fact, this limit will be a function of y, say $\phi(y)$.

If, then, $\lim_{y \to b} \phi(y)$ exists and is equal to l_1, we write

$$\lim_{y \to b}\left\{\lim_{x \to a} f(x,y)\right\} = l_1.$$

l_1 is then called **repeated** or **iterated** limit of $f(x,y)$ as $x \to a$ and then $y \to b$.

A change in the order of obtaining limits may produce a change in the final result, say l_2 (instead of l_1).

Thus we may obtain another repeated limit l_2 (say) thus:

$$\lim_{x \to a} \left\{ \lim_{y \to b} f(x,y) \right\} = l_2.$$

There is no reason why l_1 should be always equal to l_2. What is worth noticing is that even when both repeated limits exist and have the same value, the simultaneous or double limit may not exist.

Example 13.3.7. *Verify that the double limit*

$$\lim_{\substack{x \to 0 \\ y \to 0}} \frac{x+y}{x-y}$$

does not exist.

Solution: For along the line $y = mx$, $f(x,y) = \dfrac{x+y}{x-y} = \dfrac{1+m}{1-m}$.

Thus, by setting $(x,y) \to (0,0)$ along a suitable line, $f(x,y)$ will approach any value $k(\neq -1)$ specified at pleasure. Hence $f(x,y)$ does not tend to a unique limit as $(x,y) \to (0,0)$.

∴ The double limit does not exist.

But we observe that both repeated (or iterated) limits exist (although they have different values).

$$\lim_{y \to 0} \left\{ \lim_{x \to 0} \left(\frac{x+y}{x-y} \right) \right\} = \lim_{y \to 0} \left\{ \frac{y}{-y} \right\} = -1;$$

$$\lim_{x \to 0} \left\{ \lim_{y \to 0} \left(\frac{x+y}{x-y} \right) \right\} = \lim_{x \to 0} \left\{ \frac{x}{x} \right\} = +1.$$

Even when both repeated limits exist and have the same value, the double limit may not exist. Next two examples illustrate this fact:

Example 13.3.8. *Verify that the double limit*

$$\lim_{\substack{x \to 0 \\ y \to 0}} \frac{x^2 y^2}{x^2 y^2 + (x-y)^2}$$

does not exist.

CHAPTER 13: FUNCTIONS OF SEVERAL VARIABLES   755

Solution: The double limit does not exist, for $f(x,y) = \dfrac{x^2y^2}{x^2y^2 + (x-y)^2} = 1$, on the line $y = x$ but $f(x,y) = 0$ on the lines $x = 0$ or $y = 0$, i.e., $\lim_{(x,y)\to(0,0)} f(x,y) = 1$ along $y = x$ and $\lim_{(x,y)\to(0,0)} f(x,y) = 0$ along $x = 0$ or along $y = 0$.

Nevertheless, both repeated limits exist and equal to zero (**Check**).

Example 13.3.9. $f(x,y) = \dfrac{xy}{x^2 + y^2}$.

Solution: See that $\lim_{x\to 0}\lim_{y\to 0} f(x,y) = 0 = \lim_{y\to 0}\lim_{x\to 0} f(x,y)$.

But $\lim_{\substack{x\to 0 \\ y\to 0}} f(x,y)$ does not exist; for along the line $y = mx$, $f(x,y) = \dfrac{m}{1+m^2}$, different constant value along different lines through the origin.

Here repeated limits exist and are equal but still the double limit does not exist.

Example 13.3.10. *This is an example where double limits exist but repeated limits do not exist*

$$f(x,y) = \begin{cases} x\sin\dfrac{1}{y} + y\sin\dfrac{1}{x}, & \text{when } xy \neq 0; \\ 0, & \text{when } xy = 0. \end{cases}$$

Solution:

Here $\lim_{x\to 0} f(x,y)$ (y remaining constant)

or, $\lim_{y\to 0} f(x,y)$ (x remaining constant) $\Bigg\}$ does not exist.

> **Remember.** $\lim_{x\to 0} \sin\dfrac{1}{x}$ does not exist.

Thus repeated limits $\lim_{y\to 0}\lim_{x\to 0} f(x,y)$ and $\lim_{x\to 0}\lim_{y\to 0} f(x,y)$ do not exist.

But we see that the double limit $\lim_{\substack{x\to 0 \\ y\to 0}} f(x,y)$ exists and $= 0$.

$$\left| x\sin\frac{1}{y} + y\sin\frac{1}{x} - 0 \right| = \left| x\sin\frac{1}{y} + y\sin\frac{1}{x} \right| \leq \left| x\sin\frac{1}{y} \right| + \left| y\sin\frac{1}{x} \right|$$

$$= |x|\left|\sin\frac{1}{y}\right| + |y|\left|\sin\frac{1}{x}\right|$$

$$\leq |x| + |y| \quad \left(\because \left|\sin\frac{1}{y}\right| \leq 1, \left|\sin\frac{1}{x}\right| \leq 1\right)$$

$$< \sqrt{x^2 + y^2} + \sqrt{x^2 + y^2}$$

$$= 2\sqrt{x^2 + y^2}.$$

This can be made less than any preassigned positive number ε if

$$x^2 + y^2 < \left(\frac{\varepsilon}{2}\right)^2$$

i.e., if $\quad x^2 + y^2 < \delta^2$ where $\delta = \varepsilon/2$,

i.e., we have a deletec circular δ-nbd. of $(0,0)$ where $|f(x,y) - 0| < \varepsilon$.

$$\therefore \lim_{\substack{x \to 0 \\ y \to 0}} f(x,y) = 0.$$

Thus double limit exists even though repeated limits do not exist.

Note 13.3.2. *What we have observed is that in case the double limit exists, the two repeated limits, if exist, must necessarily be equal but the converse is not true. However, if the repeated limits are* **not equal** *then the double limit cannot exist.*

Example 13.3.11.

$$\lim_{x \to 0} \lim_{y \to 0} \left(\frac{y-x}{y+x}\right)\left(\frac{1+x}{1+y}\right) = \lim_{x \to 0} \left(\frac{-x}{x}\right)(1+x) = -1$$

$$\text{and} \quad \lim_{y \to 0} \lim_{x \to 0} \left(\frac{y-x}{y+x}\right)\left(\frac{1+x}{1+y}\right) = \lim_{y \to 0} \left(\frac{y}{y}\right)\frac{1}{1+y} = +1.$$

See that the repeated limits exist but are not equal. We can, therefore, at once conclude that double limit cannot exist; this can be easily verified by putting $y = mx$; $f(x,y) \to \dfrac{m-1}{m+1}$ as $(x,y) \to (0,0)$.

Another Approach of Limit based on the notion of limits of sequences

Consider a function $f(x,y)$ defined is some *nbd* of (a,b).

Let the sequence of points $(x_n, y_n) \to (a,b)$ as $n \to \infty$.

Then the distance between (x_n, y_n) and (a,b) approach zero

$$\text{i.e., } (x_n - a)^2 + (y_n - b)^2 \to 0 \text{ as } n \to \infty.$$

Now, if $f(x_n, y_n) \to A$ for every such point sequence $\{(x_n, y_n)\}$ that has (a,b) as its limit, then we say that

$$f(x,y) \text{ tends to the limit } A \text{ as } (x,y) \to (a,b)$$

and we write:

$$\lim_{\substack{x \to a \\ y \to b}} f(x,y) = A. \tag{13.3.1}$$

With such a definition in mind we prove the following theorem:

CHAPTER 13: FUNCTIONS OF SEVERAL VARIABLES

Theorem 13.3.1. *In order that $f(x,y) \to A$ as $(x,y) \to (a,b)$ it is both necessary and sufficient that, for every, $\varepsilon > 0$, we can find a positive number δ such that*

$$|f(x,y) - A| < \varepsilon, \quad \text{when } 0 < |x-a| < \delta \text{ and } 0 < |y-b| < \delta. \tag{13.3.2}$$

Proof: The condition (13.3.2) is sufficient to ensure (13.3.1).
For, if $(x_n, y_n) \to (a,b)$, then,

$$\left.\begin{array}{l} 0 < |x_n - a| < \delta, \quad \forall n > N_1 \\ 0 < |y_n - b| < \delta, \quad \forall n > N_2 \end{array}\right\} \text{ both inequalities hold } \forall n > N = \max.\{N_1, N_2\}.$$

From condition (13.3.2), it then follows

$$|f(x_n, y_n) - A| < \varepsilon \text{ when } n > N.$$
$$\text{i.e.,} \quad f(x_n, y_n) \to A \text{ as } n \to \infty.$$

This leads to the definition (13.3.1).

Condition (13.3.2) is also a necessary consequence of (13.3.1).

If not, there is an ε for which no corresponding δ can be found, such that for every null sequence

$$\{\delta_1, \delta_2, \delta_3, \cdots\cdots\cdots\}$$

there exists at least one x in $|x - a| < \delta_n$, say x_n and there exists at least one y in $|y - b| < \delta_n$, say y_n, for which $|f(x_n, y_n) - A| \geq \varepsilon$.

Now the sequence $\{(x_n, y_n)\} \to (a,b)$ but the numbers

$$\{|f(x_n, y_n) - A|\}$$

have ε as a lower bound and cannot from a null sequence.

But this contradicts the hypothesis that $f(x_n, y_n) \to A$ whenever $(x_n, y_n) \to (a,b)$. Hence the necessity of the condition is established.

Theorem 13.3.2. Algebra of Limits

(a) If $f(x,y) \to A$ as $(x,y) \to (a,b)$, then $f(x,y)$ will have the same sign in some circle (or square) about (a,b).

$$\left.\begin{array}{l} \text{Take } \varepsilon = \dfrac{1}{2}A, \quad \text{if } A > 0 \\ \qquad = -\dfrac{1}{2}A, \quad \text{if } A < 0 \end{array}\right\} \text{ use the definition of limit.}$$

(b) If $f(x,y) \to A$, $g(x,y) \to B$ as $(x,y) \to (a,b)$ then as $(x,y) \to (a,b)$,

$$f(x,y) + g(x,y) \to A + B$$
$$f(x,y) \cdot g(x,y) \to AB$$
$$f(x,y)/g(x,y) \to A/B, \text{ if } B \neq 0.$$

The proofs are similar to these of the corresponding theorems for single real variable.

Example 13.3.12. *Show that*

$$\lim_{(x,y) \to (2,1)} \frac{\sin^{-1}(xy-2)}{\tan^{-1}(3xy-6)} = \frac{1}{3}.$$

Solution: Given limit $= \lim_{t \to 0} \dfrac{\sin^{-1}(t)}{\tan^{-1}(3t)}$, where $t = xy - 2$.

[As $(x,y) \to (2,1)$, we see that $t \to 0$ and given limit is of the form $0/0$.]

$$= \lim_{t \to 0} \frac{1/\sqrt{1-t^2}}{3/1+9t^2} = \frac{1}{3}.$$

13.4 Continuity of a Function of Several Variables

Definition 13.4.1. *A function f defined in a domain $D \subseteq \mathbb{R}^n$ (n-dimensional Euclidean Space) is said to be continuous at the point $\mathbf{P}_0 \in D$ if*

$$\lim_{\mathbf{P} \to \mathbf{P}_0} f(\mathbf{P}) = f(\mathbf{P}_0).$$

[It is to be understood that $\mathbf{P}_0$ is a limit point of D.]

In $\varepsilon - \delta$ notation this becomes: f is continuous at $\mathbf{P}_0$ if, to each $\varepsilon > 0$, there exists a $\delta(\varepsilon, \mathbf{P}_0)$ for which

$$|f(\mathbf{P}) - f(\mathbf{P}_0)| < \varepsilon$$

whenever $|\mathbf{P} - \mathbf{P}_0| < \delta$ and $\mathbf{P} \in D$.

If f is continuous at every point $\mathbf{P}_0 \in D$, then we say f is continuous in D.

The concept of **uniform continuity** carries over immediately:

If in the preceding definition, to each $\varepsilon > 0$ $\exists$ $\delta(\varepsilon)$ such that $|f(\mathbf{P}) - f(\mathbf{P})_0)| < \varepsilon$ for every pair of points. $\mathbf{P}, \mathbf{P}_0$ in D for which $|\mathbf{P} - \mathbf{P}_0| < \delta$ then we say that f is **uniformly continuous** in D.

Special Cases : Continuity in $\mathbb{R}^2$

Continuity for a function f of two variables x and y at a point (a, b) in the domain D of definition of the function f:

We assume that (a, b) is a limit point of D so that in any nbd. of (a, b) $\exists$ infinite number of points of D where f is defined.

Now f is said to be continuous at (a, b), if to each $\varepsilon > 0$ $\exists$ a positive number δ such that at all points (x, y) in some δ-nbd N of (a, b)

$$\text{say,} \quad \text{in} \quad (x-a)^2 + (y-b)^2 < \delta^2$$
$$\text{or,} \quad \text{in} \quad |x-a| < \delta \text{ and } |y-b| < \delta$$

we have

$$|f(x, y) - f(a, b)| < \varepsilon. \tag{13.4.1}$$

Note 13.4.1. *This condition (13.4.1) is satisfied if $(a, b) \in D$ but not a limit point of D i.e., if (a, b) is an isolated point of D. If (a, b) is a limit point of D then f is continuous at (a, b) if $\lim_{\substack{x \to a \\ y \to b}} f(x, y) = f(a, b)$.*

A function f is said to be continuous in a certain region if it is continuous at every point of the region.

Points to remember about Continuity (A – E)

A. If a function f of two variables x, y is continuous at (a, b), then the function $f(x, b)$, which is a function of a single variable x (y remaining constant at b) is continuous at $x = a$. Similarly, $f(a, y)$, a function of single variable y is continuous at $y = b$. The proof is immediate from the definition of continuity of $f(x, y)$ at (a, b); for example put $y = b$ in (13.4.1), then to each $\varepsilon > 0, \exists$ δ such that $|f(x, b) - f(a, b)| < \varepsilon$ $\forall$ $x \in |x - a| < \delta$.

$\therefore$ $f(x, b)$ is continuous at $x = a$.

But the converse is not true as we shall see from the illustrations given below.

B. If $f(x, y)$ is continuous at (a, b) and $f(a, b) \neq 0$ then there exists a certain neighbourhood of (a, b) such that $f(x, y)$ has the same sign as that of $f(a, b)$ for every point (x, y) of that nbd. We need only take $\varepsilon = \frac{1}{2}|f(a, b)|$ in the definition (13.4.1). Then $\exists$ a square of side 2δ about (a, b) as centre where

$$f(a, b) - \frac{1}{2}|f(a, b)| < f(x, y) < f(a, b) + \frac{1}{2}|f(a, b)|.$$

Now, if $f(a,b)$ is positive then every $f(x,y)$ of the square is positive; if $f(a,b)$ is negative then every $f(x,y)$ of the square is negative.

C. If f and g are continuous functions in a region R, then $f \pm g$ and fg are also continuous; and f/g is continuous at all points where $g \neq 0$.

Polynomials are everywhere continuous.

Rational functions f/g are continuous except where $g = 0$.

Continuous functions of continuous functions are continuous: thus $\sin xy$ and $e^{-(x^2+y^2)}$ are continuous.

Examples on Continuity

Example 13.4.1. *The function*

$$f(x,y) = \begin{cases} xy \dfrac{x^2 - y^2}{x^2 + y^2}, & \text{where } x^2 + y^2 \neq 0, \\ 0, & \text{where } x = 0,\ y = 0; \end{cases}$$

is continuous at $(0,0)$.

Solution: For

$$|f(x,y) - f(0,0)| = |xy| \frac{|x^2 - y^2|}{|x^2 + y^2|} < |x||y| < \sqrt{x^2 + y^2}\sqrt{x^2 + y^2} < \varepsilon,$$

when $x^2 + y^2 < \delta^2, \delta = \sqrt{\varepsilon}$. (See Ex. **13.3.2**)

$$\therefore \quad \lim_{\substack{x \to 0 \\ y \to 0}} f(x,y) = f(0,0) \text{ i.e., } f \text{ is continuous at } (0,0).$$

Example 13.4.2. *Discuss the continuity of the function*

$$f(x,y) = \begin{cases} \dfrac{2xy}{x^2 + y^2}, & \text{where either } x \text{ or } y \neq 0; \\ 0, & \text{where both } x = 0 \text{ and } y = 0. \end{cases}$$

Solution: This function is a continuous function of either variable when the other is given a fixed value. However, $f(x,y)$ is not a continuous function of both x and y. For along the line $y = x \tan v$

$$f(x,y) = \frac{2 \tan v}{1 + \tan^2 v} = \sin 2v, \text{ except at } (0,0) \text{ where } f(0,0) = 0.$$

The surface $z = f(x,y)$ is a CONOID whose parametric equations are

$$x = \rho \cos v, \quad y = \rho \sin v, \quad z = \sin 2v.$$

This surface is generated by a line, perpendicular to z-axis, which revolves about and slides along this axis so that its height above the xy-plane is $\sin 2v$ when it makes an angle v with the x-axis.

The aspect of this surface near the z-axis shows the nature of the remarkable discontinuity of $f(x,y)$ at $(0,0)$. On any circle $x^2 + y^2 = \delta^2$, the function assumes all values between -1 and $+1$, no matter how δ is chosen.

Example 13.4.3. *Prove that the function*

$$f(x,y) = \begin{cases} \dfrac{xy}{\sqrt{x^2+y^2}}, & (x,y) \neq (0,0), \\ 0, & (x,y) = (0,0); \end{cases}$$

is continuous at $(0,0)$.

Solution: We can prove

$$\lim_{\substack{x \to 0 \\ y \to 0}} f(x,y) = 0 = f(0,0).$$

$$\left| \frac{xy}{\sqrt{x^2+y^2}} - 0 \right| = \frac{|x||y|}{\sqrt{x^2+y^2}} < \frac{\sqrt{x^2+y^2}\sqrt{x^2+y^2}}{\sqrt{x^2+y^2}}$$

i.e., $< \sqrt{x^2+y^2} < \varepsilon$ (where ε is any preassigned positive number) if $x^2 + y^2 < \delta^2$, where $\delta = \varepsilon$.

$\therefore$ Thus given any $\varepsilon > 0, \exists\, \delta > 0$

$$|f(x,y) - f(0,0)| < \varepsilon \quad \forall\, (x,y) \in x^2 + y^2 < \delta^2 \quad \text{where } \delta = \varepsilon.$$

$\therefore\ f(x,y)$ is continuous at $(0,0)$.

Alternative Method. Let $x = r\cos\theta, y = r\sin\theta$.

Then $\left| \dfrac{xy}{\sqrt{x^2+y^2}} - 0 \right| = r|\cos\theta \sin\theta| \leq r = \sqrt{x^2+y^2} < \varepsilon$

if $x^2 < \dfrac{\varepsilon^2}{2},\ y^2 < \dfrac{\varepsilon^2}{2}$ i.e., if $|x| < \dfrac{\varepsilon}{\sqrt{2}},\ |y| < \dfrac{\varepsilon}{\sqrt{2}}$.

$$\therefore\ \lim_{\substack{x \to 0 \\ y \to 0}} \frac{xy}{\sqrt{x^2+y^2}} = 0 = f(0,0).$$

Hence f is continuous at $(0,0)$.

Example 13.4.4. *Prove that the function*

$$f(x,y) = \begin{cases} \dfrac{x^3+y^3}{x-y}, & \text{when } x \neq y, \\ 0, & \text{when } x = y; \end{cases}$$

is not continuous at $(0,0)$.

Solution: We shall prove $\lim\limits_{\substack{x \to 0 \\ y \to 0}} f(x,y)$ does not exist and as such $f(x,y)$ cannot be continuous at $(0,0)$.

$$f(x,y) = \frac{x^3+y^3}{x-y} = \frac{x^3 + (x-mx^3)^3}{mx^3} = \frac{1 + (1-mx^2)^3}{m}.$$

(putting $y = x - mx^3$; then as $x \to 0, y \to 0$)

$$\therefore \lim_{\substack{x \to 0 \\ y \to 0}} f(x,y) = \lim_{x \to 0} \frac{1 + (1-mx^2)^3}{m} = \frac{2}{m}.$$

So that the limit is different for different choices of m (i.e., the limit can become any arbitrary value k specified at pleasure, not unique).

$\therefore$ the limit does not exist and f is not continuous at $(0,0)$.

D. We know that a function f is said to be continuous in a region (open or closed) if it is continuous at every point of the region. A function f, continuous in a closed region R has properties analogous to those of continuous functions of one variable in a closed interval $[a,b]$:

A function f of two variables x and y is said to be *uniformly continuous* in a region R, open or closed, when for any $\varepsilon > 0$, we can find $\delta > 0$ so that if (x_1, y_1) and (x_2, y_2) are any two points of R for which $|x_1 - x_2| < \delta$ and $|y_1 - y_2| < \delta$ then

$$|f(x_1, y_1) - f(x_2, y_2)| < \varepsilon.$$

- A function f of two variables x and y which is continuous in a closed region R is uniformly continuous there, that is, if (x_1, y_1) and (x_2, y_2) are points of R, $|f(x_1, y_1) - f(x_2, y_2)| < \varepsilon$ when $|x_1 - x_2| < \delta$ and $|y_1 - y_2| < \delta$ where ε is any arbitrary preassigned positive number and δ depends on ε alone.

- A function $f(x,y)$ which is continuous in a closed region R, is bounded there and takes its least and greatest values at least once in the region R.

The proofs are similar to those studied under function of one variable and hence *omitted*.

E. We advise our readers at this stage to be familiar with the concept of *Compact sets*: For further details on Compact Sets one can see Authors' *An Introduction to Analysis: Differential Calculus* (**Part II**).

- By an *open cover* of a set $E \subseteq \mathbb{R}^n$, we mean a collection of $\sigma = \{G_k\}$ of open sets of $\mathbb{R}^n$ such that
$$E \subset \bigcup_{G_\alpha \in \sigma} G_\alpha.$$

- A subset $K \subseteq \mathbb{R}^n$ is said to be compact if every open cover of K contains a finite subcover, that is, a finite subcollection which also covers K. (**Heine-Borel property**).

- If a set E in $\mathbb{R}^n$ has one of the three following properties, then it has the other two properties:

    (i) E is closed and bounded.

    (ii) E is compact.

    (iii) Every infinite subset of E has a limit point in E.

Functions Continuous on Compact sets

1. The range of a function continuous on a compact set is compact.

Proof: Let f be a real valued function, continuous on the compact set $D \subseteq \mathbb{R}^n$.

Let E denote the range of the function (each member of E is a real number). If E is finite, then it is compact (*since every finite set is compact*).

Now suppose that E is an infinite set.

Let B denote any infinite subset of E.

To each $y \in B, \exists$ at least one $\vec{x} \in D$ such that $f(\vec{x}) = y$.

We select one such $\vec{x}$ and call the subset of D having these $\vec{x}$ as members as A.

Thus to each $y \in B$ there corresponds one and only one $\vec{x} \in A$ such that $f(\vec{x}) = y$.

Since A is an infinite subset of the compact set D, the set A has a limit point, say $\vec{a} \in D$.

Let $b = f(\vec{a}) \Rightarrow b \in E$.

We shall show that b is a limit point of E. Consider any ε-nbd of b, namely $]b-\varepsilon, b+\varepsilon[$.

Then $\exists\, \delta > 0$ such that
$$\|\vec{x}-\vec{a}\| < \delta \Rightarrow f(\vec{x}) \in \,]b-\varepsilon, b+\varepsilon[.$$

Since $\|\vec{x}-\vec{a}\| < \delta$ contains infinite number of points of A, $]b-\varepsilon, b+\varepsilon[$ contains infinite number of points of the set B. Thus b is the limit point of B.

Thus every infinite subset B of the range E has a limit point in E, hence the range E is compact, since (iii) $\Rightarrow$ (ii), see above.

2. We have already given the definition of uniform continuity in $\mathbb{R}^n$. It can be easily seen that a function which is uniformly continuous in its domain is also continuous in the domain. The converse is not true always. However, if the domain D is **compact**, continuity implies uniform continuity. (Proof is given on next page).

> **Remember.** A function which is continuous on a compact domain is uniformly continuous there.

Proof: Let D be the compact domain and E the range of a continuous real valued function f in $\mathbb{R}^n$.

Let $\varepsilon > 0$ be given and $\vec{x} \in D$.

There exists $\delta > 0$ such that [δ depends on the choice of $\vec{x}$, a fact denoted by $\delta(\vec{x})$]
$$\|\vec{x}-\vec{t}\| < \delta(\vec{x}) \Rightarrow |f(\vec{x})-f(\vec{t})| < \frac{1}{2}\varepsilon.$$

Now the spherical neighbourhoods $S\left[\vec{x}, \frac{1}{2}\delta(\vec{x})\right] \in D$ constitute an open cover of the compact set D and hence there exists a finite sub-cover.

Let this subcover consists of the spheres
$$S\left[\vec{x}_1, \frac{1}{2}\delta(\vec{x}_1)\right], S\left[\vec{x}_2, \frac{1}{2}\delta(\vec{x}_2)\right], \cdots, S\left[\vec{x}_p, \frac{1}{2}\delta(\vec{x}_p)\right].$$

We take $\delta = \frac{1}{2}\min.\{\delta(\vec{x}_1), \delta(\vec{x}_2), \cdots, \delta(\vec{x}_p)\}$.

Now suppose that $\vec{x}, \vec{y}$ are two points of D such that $\|\vec{x}-\vec{y}\| < \delta$.

Suppose that $\vec{x} \in S\left[\vec{x}_i, \frac{1}{2}\delta(\vec{x}_i)\right]$.

We have
$$\|\vec{y}-\vec{x}_i\| \le \|\vec{y}-\vec{x}\| + \|\vec{x}-\vec{x}_i\|$$
$$\text{i.e., } < \delta + \frac{1}{2}\delta(\vec{x}) < \delta(\vec{x}_i).$$

CHAPTER 13: FUNCTIONS OF SEVERAL VARIABLES

Thus the points $\vec{x}$ and $\vec{y}$ belong to the same sphere $S[\vec{x}_i, \delta(\vec{x}_i)]$ where

$$|f(\vec{x}) - f(\vec{y})| \le |f(\vec{x}) - f(\vec{x}_i)| + |f(\vec{x}_i) - f(\vec{y})|$$
$$< \varepsilon/2 + \varepsilon/2 = \varepsilon.$$

Thus given any $\varepsilon > 0$, $\exists\, \delta > 0$ such that

$$\|\vec{x} - \vec{y}\| < \delta \Rightarrow |f(\vec{x}) - f(\vec{y})| < \varepsilon.$$

$\therefore$ uniform continuity is established.

13.5 Partial Derivatives

(A) Let $u = f(x, y)$ be a function of two independent variables x, y in a region R.

If y is held constant, then $f(x, y)$ becomes a function of x alone; and its derivative (if exists) is called the *partial derivative* of $f(x, y)$ with respect to x.

Similarly, if x is held constant, $f(x, y)$ becomes a function of y alone; and its derivative (if exists) is called the partial derivative of $f(x, y)$ with respect to y.

These partial derivatives, variously denoted by

$$f_x(x, y), \quad \frac{\partial f}{\partial x}, \quad u_x, \quad \frac{\partial u}{\partial x}$$

$$\text{and} \quad f_y(x, y), \quad \frac{\partial f}{\partial y}, \quad u_y, \quad \frac{\partial u}{\partial y}$$

have the defining equations

$$f_x(x, y) = \lim_{h \to 0} \frac{f(x+h, y) - f(x, y)}{h} \tag{13.5.1}$$

$$\text{and} \quad f_y(x, y) = \lim_{k \to 0} \frac{f(x, y+k) - f(x, y)}{k}. \tag{13.5.2}$$

When these limits exist at an interior point (a, b) of $\mathbb{R}$, they are denoted by $f_x(a, b), f_y(a, b)$. If such partial derivatives exist at all points of a region, $f_x(x, y)$ and $f_y(x, y)$ are again functions of x, y which may again be differentiated with respect to x and y.

If both f_x and f_y exist at (a, b), we say that the function $f(x, y)$ is **derivable** at (a, b).

(B) Directional Derivatives in $\mathbb{R}^2$

We introduce a more general concept—concept of **directional derivative** at a given point.

Let f be a real-valued function of two independent variables x, y.

We shall define f over a two-dimensional region $D \subseteq \mathbb{R}^2$ i.e., $f : D \to \mathbb{R}$.

We take β a unit vector in $\mathbb{R}^2$ specifying a particular direction. A measure of the rate of change of f as (x, y) changes in the direction β is the directional derivative.

Definition 13.5.1. *The directional derivative of f in the direction of unit vector $\beta = (l, m)$ where $l^2 + m^2 = 1$ at the point (a, b) is given by*

$$\lim_{t \to 0} \frac{f(a + tl, b + tm) - f(a, b)}{t}$$

when such a limit exists. This limit is denoted by $D_\beta f(a, b)$.

Example 13.5.1. *Find the directional derivative of $f(x, y) = 2x^2 - xy + 5$ at $(1, 1)$ in the direction of unit vector $\beta = \frac{1}{5}(3, -4)$.*

Solution: By definition

$$D_\beta f(1, 1) = \lim_{t \to 0} \frac{f\left(1 + \frac{3}{5}t, 1 - \frac{4}{5}t\right) - f(1, 1)}{t}$$

$$= \lim_{t \to 0} \frac{\left\{2\left(1 + \frac{3}{5}t\right)^2 - \left(1 + \frac{3}{5}t\right)\left(1 - \frac{4}{5}t\right) + 5\right\} - 6}{t}$$

$$= \lim_{t \to 0} \frac{\frac{6}{5}t^2 + \frac{13}{5}t}{t} = \lim_{t \to 0} \left(\frac{13}{5} + \frac{6}{5}t\right) = \frac{13}{5}.$$

Special Cases. In the direction $\beta = (1, 0)$ the directional derivative is called the **partial derivative** of f with respect to x at the point (a, b) which we shall denote by $f_x(a, b)$ or $\left(\dfrac{\partial f}{\partial x}\right)_{a,b}$.

Thus,

$$f_x(a, b) = \lim_{t \to 0} \frac{f(a + t, b) - f(a, b)}{t}, \quad \text{(if this limit exists).}$$

Similarly,

$$f_y(a, b) = \lim_{t \to 0} \frac{f(a, b + t) - f(a, b)}{t} \quad \text{(if this limit exists)}$$

CHAPTER 13: FUNCTIONS OF SEVERAL VARIABLES

is the partial derivative of f with respect to y at the point (a, b); here the unit vector $\beta = (0, 1)$.

Example 13.5.2. Let

$$f(x, y) = \begin{cases} \dfrac{xy}{x^2 + y^2}, & \text{when } x^2 + y^2 \neq 0; \\ 0, & \text{when } x^2 + y^2 = 0. \end{cases}$$

Find the directional derivative at $(0, 0)$ in any arbitrary direction $\beta = (l, m)$, $l^2 + m^2 = 1$. What are the partial derivatives with respect to x or y at $(0, 0)$?

Solution:

$$D_\beta f(0, 0) = \lim_{t \to 0} \frac{f(tl, tm) - f(0, 0)}{t} = \lim_{t \to 0} \frac{lm}{l^2 + m^2} \cdot \frac{1}{t}.$$

This limit does not exist always. But if $\beta = (1, 0)$ or $(0, 1)$ then $f_x(0, 0) = 0 = f_y(0, 0)$.

Thus both f_x and f_y exist at $(0, 0)$ but note that the function is not continuous at $(0, 0)$, since

$$\lim_{\substack{x \to 0 \\ y \to 0}} \frac{xy}{x^2 + y^2} \text{ does not exist uniquely.}$$

But remember that even when f has directional derivatives in **all arbitrary** directions still f may not be continuous.

Example 13.5.3. Let

$$f(x, y) = \begin{cases} \dfrac{x^2 y}{x^4 + y^2}, & (x, y) \neq (0, 0); \\ 0, & x = y = 0. \end{cases}$$

Show that f has a directional derivative at $(0, 0)$ in any direction $\beta = (l, m)$, $l^2 + m^2 = 1$ but f is discontinuous at $(0, 0)$.

Solution:

$$D_\beta f(0, 0) = \lim_{t \to 0} \frac{f(tl, tm) - f(0, 0)}{t} = \lim_{t \to 0} \frac{t^3 l^2 m / (t^4 l^4 + t^2 m^2)}{t}$$

$$= \frac{l^2}{m}, \text{ if } m \neq 0$$
$$= 0, \text{ if } m = 0,$$

i.e., f has a directional derivative along any arbitrary direction at $(0, 0)$. But see that f is not continuous at $(0, 0)$.

To consider $\lim_{\substack{x \to 0 \\ y \to 0}} f(x,y)$. Along the path $y = x^2$, the limit is $\dfrac{1}{2}$ but along $y = 0$, the limit is zero. Hence $\lim_{\substack{x \to 0 \\ y \to 0}} f(x,y)$ does not exist and as such f can not be continuous at $(0,0)$.

(C) Partial derivatives interpreted geometrically

Consider the surface $z = f(x,y)$ in Fig. 13.7. Pass a plane $EFGH$ through the point $P\{a,b, f(a,b)\}$ on the surface parallel to XOZ-plane, the equation of such a plane being $y = b$. The plane intersects the surface along a curve APB given by $z = f(x,b)$. As x varies, P moves along the curve APB and $\partial z/\partial x$ is the slope of the curve at $P = \tan\psi$, where ψ is the angle which the tangent at P makes with the x-axis (see the figure). Similarly, if we pass a plane $x = a$ through P parallel to YOZ-plane, the curve of intersection has a slope $\partial z/\partial y$ at P.

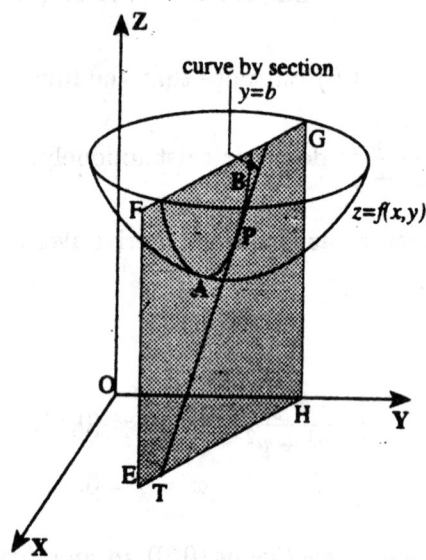

Fig. 13.7

Example 13.5.4. *Find the slope of the curve of intersection of the ellipsoid*

$$\frac{x^2}{24} + \frac{y^2}{12} + \frac{z^2}{6} = 1$$

made by the plane $y = 1$ at the point $(4, 1, \sqrt{3/2})$.

Solution: Considering y as constant $\dfrac{\partial z}{\partial x} = -\dfrac{x}{4z}$.

When $y = 1, x = 4, z = \sqrt{3/2}$. $\dfrac{\partial z}{\partial x} =$ the required slope $= -\dfrac{\sqrt{6}}{3}$.

CHAPTER 13: FUNCTIONS OF SEVERAL VARIABLES

(D) The function $f(x, y, z)$ of three independent variables x, y, z has three partial derivatives of the first order:

$$\frac{\partial f}{\partial x} = f_x = \lim_{h \to 0} \frac{f(x+h, y, z) - f(x, y, z)}{h},$$
(if this limit exists, y, z, are held constant)

$\frac{\partial f}{\partial y} = f_y$ and $\frac{\partial f}{\partial z} = f_z$ are defined in similar fashion.

We may introduce here directional derivatives in $\mathbb{R}^3$ exactly in the same way as we have done in case of $\mathbb{R}^2$ (See **B** above).

(E) Let f be a real-valued function defined in a domain $D \subseteq \mathbb{R}^n$. Let $\vec{a}$ be an interior point of D. We can write

$$f(\vec{x}) = f(x_1, x_2, \cdots, x_n).$$

Let $\vec{u}_1 = (1, 0, 0, \cdots, 0)$, $\vec{u}_2 = (0, 1, 0, \cdots, 0)$, $\cdots\cdots$ $\vec{u}_n = (0, 0, \cdots, 1)$.

Then $\lim_{t \to 0} \frac{f(\vec{a} + t\vec{u}_1) - f(\vec{a})}{t}$, if it exists, is called the first order partial derivative or simply the partial derivative of the function with respect to x_1 at $\vec{a}$ and is denoted by

$$f_{x_1}(\vec{a}) \text{ or, } \frac{\partial f}{\partial x_1}(\vec{a}).$$

We observe that

$$\frac{f(\vec{a} + t\vec{u}_1) - f(\vec{a})}{t} = \frac{f(a_1 + t, a_2, a_3, \cdots, a_n) - f(a_1, a_2, \cdots, a_n)}{t}.$$

$$\therefore \frac{\partial f}{\partial x_1} \text{ at } (a_1, a_2, \cdots, a_n) = \lim_{t \to 0} \frac{f(a_1 + t, a_2, \cdots, a_n) - f(a_1, a_2, \cdots, a_n)}{t}.$$

Similarly, we define $\frac{\partial f}{\partial x_2}, \frac{\partial f}{\partial x_3}, \cdots, \frac{\partial f}{\partial x_n}$ at $(a_1, a_2, \cdots, a_n)$.

These are the n first order partial derivatives at the point $\vec{a} = (a_1, a_2, \cdots, a_n)$.

Example 13.5.5. *Find the first order partial derivatives of*

$$f(x, y) = \frac{x + y - 1}{x - y + 1} \text{ at } (2, 1).$$

Solution:

$$\frac{\partial f}{\partial x} \text{ at } (2,1) = \lim_{h \to 0} \frac{f(2+h, 1) - f(2,1)}{h} = \lim_{h \to 0} \frac{1-1}{h} = 0.$$

$$\frac{\partial f}{\partial y} \text{ at } (2,1) = \lim_{k \to 0} \frac{f(2, 1+k) - f(2,1)}{k}$$

$$= \lim_{k \to 0} \frac{\frac{2+k}{2-k} - 1}{k} = \lim_{k \to 0} \frac{2}{2-k} = 1.$$

Observation: The existence of the two first order partial derivatives of a function at a point, i.e., derivability of a function at a point may not imply continuity of the function at the point. In case of functions of one variable only, derivability implies continuity.

This is not surprising. Let $N = \{a - \delta < x < a + \delta, b - \delta < y < b + \delta\}$ be a square neighbourhood of (a, b).

The existence of $f_x(a, b)$ of $f_y(a, b)$ depends only on values of the function at **those points** of N which lie along the lines $x = a$ or $y = b$ respectively and independent of the functional values at other points of N. But the question of continuity involves the simultaneous limit of $f(x, y)$ as $(x, y) \to (a, b)$ along any path whatsoever. Thus, the continuity at (a, b) will take into account the values of the function at every point of the *nbd.* N.

Hence we may very well expect that the *first order* partial derivatives may exist at points where the function may not be continuous. See the two examples given below:

Example 13.5.6.

$$f(x, y) = \begin{cases} \dfrac{x^3 + y^3}{x - y}, & \text{when } x \neq y; \\ 0, & \text{when } x = y. \end{cases}$$

We have already proved that this function is not continuous at $(0, 0)$. [See Example 13.4.4]. But see that

$$f_x(0, 0) = \lim_{h \to 0} \frac{f(0+h, 0) - f(0, 0)}{h}, \text{ if this limit exists.}$$

But $\dfrac{f(h, 0) - f(0, 0)}{h} = \dfrac{h^3 - 0}{h} = h^2 \to 0$ as $h \to 0$ i.e., $f_x(0, 0) = 0$.

$\therefore f_x(0, 0)$ exists. Similarly $f_y(0, 0)$ also exists and equals to 0.

Thus both the partial derivatives f_x and f_y exist at $(0, 0)$ but $f(x, y)$ is not continuous at $(0, 0)$.

CHAPTER 13: FUNCTIONS OF SEVERAL VARIABLES

Example 13.5.7.

$$f(x,y) = \begin{cases} \dfrac{xy}{x^2+y^2}, & (x,y) \neq (0,0); \\ 0, & (x,y) = (0,0). \end{cases}$$

In this case also both the partial derivatives exist at $(0,0)$ but the function is not continuous at $(0,0)$.

Solution: Putting $y = mx$, we see that

$$\lim_{\substack{x \to 0 \\ y \to 0}} f(x,y) = \frac{m}{1+m^2}$$

so that the limit depends on m i.e., different when approach is along different lines and therefore, does not exist. Hence $f(x,y)$ is not continuous at $(0,0)$. Again

$$f_x(0,0) = \lim_{h \to 0} \frac{f(h,0) - f(0,0)}{h} = \lim_{h \to 0} \frac{0}{h} = 0$$

$$f_y(0,0) = \lim_{k \to 0} \frac{f(0,k) - f(0,0)}{k} = \lim_{k \to 0} \frac{0}{k} = 0.$$

A sufficient condition for continuity

Theorem 13.5.1. *If a function f of two independent variables x and y possesses partial derivatives f_x and f_y everywhere in a region R and if these derivatives are bounded in R (i.e., $\exists\, M > 0$ such that $|f_x| < M$ and $|f_y| < M$), then $f(x,y)$ is continuous in R.*

Proof: Suppose (x,y) is an interior point of R.

We take h, k so small that the point $(x+h, y+k)$ belongs to R.

We then write

$$f(x+h, y+k) - f(x,y) = \{f(x+h, y+k) - f(x+h, y)\} + \{f(x+h, y) - f(x,y)\}. \quad (13.5.3)$$

The two terms in the first bracket differ only in y; since f_y exists we can apply MVT on the function $f(x+h, y)$ of one variable y ($x+h$ remaining constant) in the interval $[y, y+k]$. Thus we obtain

$$f(x+h, y+k) - f(x+h, y) = k f_y(x+h, y+\theta_1 k), \quad (0 < \theta_1 < 1).$$

Similarly, the two terms in the second bracket of (13.5.3) differ in x; y remains constant. Since f_x exists, we can apply M.V.T. on the function $f(x,y)$ of one variable x (y constant) in the interval $[x, x+h]$. Thus we obtain

$$f(x+h, y) - f(x,y) = h f_x(x+\theta_2 h, y), \quad 0 < \theta_2 < 1.$$

Hence, from (13.5.3), we get

$$f(x+h, y+k) - f(x,y) = kf_y(x+h, y+\theta_1 k) + hf_x(x+\theta_2 h, y), \ (0 < \theta_1, \theta_2 < 1).$$

Given: The partial derivatives are everywhere less than M in absolute value.
$\therefore |f(x+h, y+k) - f(x,y)| < M(|h| + |k|).$

The right side can be made less than any preassigned positive number ε if we take $|h| + |k| < \varepsilon/M$ i.e., if we take $|h| < \delta, |k| < \delta$ where $\delta = \dfrac{\varepsilon}{2M}$.

Thus, $|f(x+h, y+k) - f(x,y)| < \varepsilon$ if $|h| < \delta, |k| < \delta$, where $\delta = \varepsilon/2M$. i.e., $f(x,y)$ is continuous at any point $(x,y) \in R$.

Observation: A more precise condition for continuity of $f(x,y)$ at (x,y) is that one of the partial derivatives should be bounded in a certain nbd of (x,y) and the other partial derivative exists at that point.

Mean Value Theorem for a function of two variables

Theorem 13.5.2. *Suppose a function f of two variables x and y possesses $\dfrac{\partial f}{\partial x}$ in a certain neighbourhood N of (a,b) and suppose $\dfrac{\partial f}{\partial y}$ exists at (a,b) then for any point $(a+h, b+k) \in N$*

$$f(a+h, b+k) - f(a,b) = hf_x(a+\theta h, b+k) + k[f_y(a,b) + \eta]$$

where $0 < \theta < 1$ and η is a function of k such that $\eta \to 0$ as $k \to 0$.

Proof: We write

$$f(a+h, b+k) - f(a,b) = \{f(a+h, b+k) - f(a, b+k)\} + \{f(a, b+k) - f(a,b)\}. \quad (13.5.4)$$

We assume that h, k are small enough so that the points $(a, b+k), (a+h, b+k), (a+h, b)$ all belong to N.

Since f_x exists at all points of N, therefore, by Lagrange's MVT

$$f(a+h, b+k) - f(a, b+k) = hf_x(a+\theta h, b+k), \ (0 < \theta < 1). \quad (13.5.5)$$

Also $f_y(a,b)$ exists; hence

$$\lim_{k \to 0} \frac{f(a, b+k) - f(a,b)}{k} = f_y(a,b).$$

$$\therefore \frac{f(a, b+k) - f(a,b)}{k} = f_y(a,b) + \eta,$$

CHAPTER 13: FUNCTIONS OF SEVERAL VARIABLES

where η is a function of k and $\eta \to 0$ as $k \to 0$.

$$\therefore \quad f(a, b+k) - f(a, b) = k[f_y(a, b) + \eta]. \tag{13.5.6}$$

(13.5.4), (13.5.5), (13.5.6) together give the required result.

Note 13.5.1. *After introducing differentiability we shall state this theorem in a more common form.*

13.6 Differentiability : Total Differential

(A) Notion of differentiability for Functions of one variable

Let f be a function of a single real variable x defined over a domain $D \subseteq \mathbb{R}^1$. For any interior point $x \in D$, we define

$$f'(x) = \lim_{h \to 0} \frac{f(x+h) - f(x)}{h}, \quad \text{if the limit exists.}$$

$$\therefore \quad \frac{f(x+h) - f(x)}{h} = f'(x) + \eta$$

where η is a function of h and $\eta \to 0$ as $h \to 0$.

We then have

Increment of $f(x) = \Delta f = f(x+h) - f(x) = h[f'(x) + \eta]$

i.e., $\Delta f = df + \eta h$ [$df =$ differential of $f(x) = hf'(x)$].

Thus $\Delta f - df = \eta h$ vanishes to a higher order than h, since $\eta h/h \to 0$ as $h \to 0$. We write $\eta h = 0(h)$: read ηh is of higher order than h.

$$\therefore \quad \Delta f = df + 0(h). \tag{13.6.1}$$

We say that $f(x)$ is **differentiable** at x when (13.6.1) holds; and this is *always* the case when $f'(x)$ exists (i.e., when $f(x)$ is **derivable**).

The concept of differentiability may be extended to functions of several variables.

(B) Notion of differentiability for Functions of two variables

Let f be a function of two independent variables x, y defined over a domain $D \subseteq \mathbb{R}^2$.

Let (x, y) be an interior point of D. Suppose N is a certain neighbourhood of (x, y). Let $(x+h, y+k) \in N$ where f is defined.

The difference $f(x+h, y+k) - f(x,y)$ is called the increment of f, denoted by Δf.

The function f is said to be *differentiable* at (x,y) if we can express the increment Δf in the form:

$$\Delta f = f(x+h, y+k) - f(x,y) = Ah + Bk + \eta_1 h + \eta_2 k \qquad (13.6.2)$$

where A, B are independent of h, k and η_1 and η_2 are functions of h, k such that $\eta_1, \eta_2 \to 0$ as $(h,k) \to (0,0)$.

The part $Ah + Bk$ (linear in h and k) of Δf is called the **Total differential** of $f(x,y)$ at the point (x,y), denoted by df.

$\eta_1 h + \eta_2 k$ is called the **error** in assuming df for Δf.

Since $\dfrac{|\eta_1 h + \eta_2 k|}{\sqrt{h^2 + k^2}} \leq |\eta_1| + |\eta_2| \to 0$ as $(h,k) \to (0,0)$, $\eta_1 h + \eta_2 k = 0(\sqrt{h^2 + k^2})$.

Therefore, when $f(x,y)$ is differentiable at (x,y)

$$\mathbf{\Delta f = df + 0(\sqrt{h^2 + k^2})}.$$

Theorem 13.6.1. Necessary conditions that follow from the concept of differentiability

1. *If $f(x,y)$ is differentiable at (x,y) then (13.6.2) holds.*

    Proof: Clearly h and k are independent increments of x and y respectively. If we put $k = 0$ in (13.6.2), divide by h and let $h \to 0$, we get

    $$\lim_{h \to 0} \frac{f(x+h, y) - f(x,y)}{h} = \lim_{h \to 0}(A + \eta_1) = A.$$

    i.e., $\mathbf{f_x(x, y) = A}$.

    Similarly, if we put $h = 0$, divide by k and let $k \to 0$, we get

    $$\lim_{k \to 0} \frac{f(x, y+k) - f(x,y)}{k} = \lim_{k \to 0}(B + \eta_2) = B$$

    i.e., $\mathbf{f_y(x, y) = B}$.

    In view of this, the *total differential* of f,

    $$\mathbf{df = hf_x + kf_y}.$$

    In particular, when $f(x,y) = x$, we have $f_x = 1, f_y = 0$,

    $$df = dx = h$$

    and when $f = y$, $dy = k$, that is the increment of the independent variables x, y are the same as their differentials. We have, therefore, the defining equation of *total differential of f*:

CHAPTER 13: FUNCTIONS OF SEVERAL VARIABLES

$$\boxed{\text{Total differential of } \mathbf{f} = df = f_x dx + f_y dy = \frac{\partial f}{\partial x} dx + \frac{\partial f}{\partial y} dy.}$$

2. If $f(x,y)$ is differentiable at (x,y) then it must be continuous there.

Proof: Since $f(x,y)$ is differentiable at (x,y)

$$\Delta f = f(x+h, y+k) - f(x,y) = hf_x + kf_y + \eta_1 h + \eta_2 k$$

where η_1, η_2 are functions of h, k which tend to zero as $(h,k) \to (0,0)$.

$$\therefore \lim_{\substack{h \to 0 \\ k \to 0}} [f(x+h, y+k) - f(x,y)] = 0$$

or, $\lim_{\substack{h \to 0 \\ k \to 0}} f(x+h, y+k) = f(x,y).$

so that f is continuous at (x,y).

Note 13.6.1. *For a function f of a single variable x, the existence of $f'(x)$ implies that f is **differentiable** at x and that f is **continuous** at x.*

But, for a function f of two independent variables x, y the existence of f_x and f_y do not imply that f is differentiable at (x,y). We shall give examples. Moreover if f_x and f_y both exist (derivable) then it does not imply that f is continuous. Only when f is differentiable at a point then f is continuous there. The following examples will illustrate the facts stated above:

Example 13.6.1. *If $f(x,y) = xy$, then at any point (x,y),*

$$\Delta f = f(x+h, y+k) - f(x,y) = (x+h)(y+k) - xy$$
$$= hy + kx + hk,$$
$$df = hf_x + kf_y = hy + kx.$$

$$\therefore \quad \Delta f = df + hk.$$

Clearly $hk = 0(\sqrt{h^2 + k^2})$

i.e., $\dfrac{hk}{\sqrt{h^2+k^2}} \to 0$ as $(h,k) \to (0,0)$.

$\therefore$ f is differentiable at any point (x,y) of the xy-plane.

Example 13.6.2. *At $(0,0)$ the function $f(x,y) = \sqrt{|xy|}$ has partial derivatives*

$$f_x = f_y = 0. \qquad \text{[C.H. 1999]}$$

Solution: See that

$$f_x(0,0) = \lim_{h \to 0} \frac{f(h,0) - f(0,0)}{h} = \lim_{h \to 0} \frac{0}{h} = 0$$

$$f_y(0,0) = \lim_{k \to 0} \frac{f(0,k) - f(0,0)}{k} = \lim_{k \to 0} \frac{0}{k} = 0.$$

$\therefore$ at $(0,0)$ the total differential $df = f_x dx + f_y dy = 0$.

$\therefore \Delta f = f(0+h, 0+k) - f(0,0) = \sqrt{|hk|}$.

If we can show $\Delta f - df = 0(\sqrt{h^2 + k^2})$, then f is differentiable.
Here $\Delta f - df = \sqrt{|hk|} - 0 = \sqrt{|hk|}$.

$$\therefore \frac{\Delta f - df}{\sqrt{h^2 + k^2}} = \frac{\sqrt{|hk|}}{\sqrt{h^2 + k^2}} = \sqrt{|\cos\theta \sin\theta|} \quad \text{(putting } h = r\cos\theta, k = r\sin\theta\text{)}$$

which, in general, does not approach zero as $\sqrt{h^2 + k^2} = r \to 0$ using ε-δ definition.

Hence we have proved that although f_x and f_y exist at $(0,0)$, $f(x,y)$ is not differentiable at $(0,0)$. However, it can be verified that $f(x,y) = \sqrt{|xy|}$ is continuous at $(0,0)$: thus we can easily see that $\lim_{\substack{x \to 0 \\ y \to 0}} \sqrt{|xy|} = 0$.

Example 13.6.3.

$$f(x,y) = \begin{cases} \dfrac{xy}{\sqrt{x^2 + y^2}}, & \text{when } x^2 + y^2 \neq 0; \\ 0, & \text{when } x^2 + y^2 = 0. \end{cases}$$

Solution:

a) We have already proved that (See Example **13.4.3**)
f is continuous at $(0,0)$ i.e., $\lim_{\substack{x \to 0 \\ y \to 0}} \dfrac{xy}{\sqrt{x^2 + y^2}} = 0$.

b) $f_x(0,0) = \lim_{h \to 0} \dfrac{f(h,0) - f(0,0)}{h} = \lim_{h \to 0} \dfrac{0}{h} = 0$

and $f_y(0,0) = \lim_{k \to 0} \dfrac{f(0,k) - f(0,0)}{k} = \lim_{k \to 0} \dfrac{0}{k} = 0.$

Thus both the partial derivatives exist at $(0,0)$.

In fact, $\Delta f = f(h,k) - f(0,0) = \dfrac{hk}{\sqrt{h^2 + k^2}}$, $df = hf_x + kf_y = 0$.

CHAPTER 13: FUNCTIONS OF SEVERAL VARIABLES

$$\therefore \frac{\Delta f - df}{\sqrt{h^2 + k^2}} = \frac{hk}{h^2 + k^2} = \frac{m}{1 + m^2}, \text{ putting } k = mh.$$

$$\therefore \lim_{\substack{h \to 0 \\ k \to 0}} \frac{\Delta f - df}{\sqrt{h^2 + k^2}} = \lim_{h \to 0} \frac{m}{1 + m^2} = \frac{m}{1 + m^2},$$

which is different for different values of m i.e., different limits as $(h,k) \to (0,0)$ along different lines through the origin i.e., limit does not exist.

Hence here f is not differentiable at $(0,0)$ although f is continuous at $(0,0)$ and f_x, f_y both exist at $(0,0)$.

Example 13.6.4.

$$f(x,y) = \begin{cases} \dfrac{x^3 - y^3}{x^2 + y^2}, & x^2 + y^2 \neq 0; \\ 0, & x^2 + y^2 = 0. \end{cases}$$

[C.H. 1993]

Solution:

(a) This function is continuous at $(0,0)$.

For, $\left| \dfrac{x^3 - y^3}{x^2 + y^2} - 0 \right| = |r| |\cos^3 \theta - \sin^3 \theta|$ (putting $x = r\cos\theta, y = r\sin\theta$)

$$\leq |r| = \sqrt{x^2 + y^2} < \varepsilon, \text{ if } |x| < \frac{\varepsilon}{\sqrt{2}}, |y| < \frac{\varepsilon}{\sqrt{2}}.$$

i.e., $\lim_{\substack{x \to 0 \\ y \to 0}} f(x,y) = 0 = f(0,0)$ i.e., f is continuous at $(0,0)$.

(b) This function possesses partial derivatives at $(0,0)$.

$$f_x(0,0) = \lim_{h \to 0} \frac{f(h,0) - f(0,0)}{h} = \lim_{h \to 0} \frac{h - 0}{h} = 1$$

$$f_y(0,0) = \lim_{k \to 0} \frac{f(0,k) - f(0,0)}{k} = \lim_{k \to 0} \frac{-k - 0}{k} = -1.$$

(c)

$$\Delta f = f(h,k) - f(0,0) = \frac{h^3 - k^3}{h^2 + k^2}$$

$$df = hf_x + kf_y = h - k.$$

$$\therefore \frac{\Delta f - df}{\sqrt{h^2+k^2}} = \frac{\frac{h^3-k^3}{h^2+k^2} - (h-k)}{\sqrt{h^2+k^2}}$$

$$= \frac{hk(h-k)}{(h^2+k^2)^{3/2}} = \frac{r^3 \cos\theta \sin\theta(\cos\theta - \sin\theta)}{r^3}$$

i.e., $\quad \dfrac{\Delta f - df}{\sqrt{h^2+k^2}} = \cos\theta \sin\theta(\cos\theta - \sin\theta)$

which in general, does not tends to zero for arbitrary θ as $(h, k) \to (0,0)$ i.e., as $\sqrt{h^2+k^2} = r \to 0$. **Hence f is not differentiable at (0, 0).**

Theorem 13.6.2. Sufficient Condition for Differentiability

If (a,b) be a point in the domain of definition of a function f of two independent variables x and y such that one of the partial derivatives f_x or f_y exist and the other is continuous at (a,b). Suppose

(i) $\dfrac{\partial f}{\partial x}$ *exists at (a,b), and*

(ii) $\dfrac{\partial f}{\partial y}$ *is continuous at (a,b) then $f(x,y)$ is differentiable at (a,b).*

[C.H. 1996]

Proof: Since $\dfrac{\partial f}{\partial y}$ is continuous at (a,b), there exists a neighbourhood N (say) of (a,b) at every point of which f_y exists. We take $(a+h, b+k)$, a point of this neighbourhood so that $(a+h, b), (a, b+k)$ also belongs to N.

We write

$$f(a+h, b+k) - f(a,b) = \{f(a+h, b+k) - f(a+h, b)\} + \{f(a+h, b) - f(a,b)\}. \quad (13.6.3)$$

Consider a function of one variable $\phi(y) = f(a+h, y)$.

Since f_y exists in N, $\phi(y)$ is derivable with respect to y in the closed interval $[b, b+k]$ and as such we can apply Lagrange's MVT for function of one variable y in this interval and thus obtain

$$\phi(b+k) - \phi(b) = k\phi'(b+\theta k), \quad 0 < \theta < 1$$

or, $\quad f(a+h, b+k) - f(a+h, b) = k f_y(a+h, b+\theta k), \quad 0 < \theta < 1. \quad (13.6.4)$

Now, if we write

$$f_y(a+h, b+\theta k) - f_y(a,b) = \varepsilon_2 \text{ (a function of } h, k) \quad (13.6.5)$$

CHAPTER 13: FUNCTIONS OF SEVERAL VARIABLES

then from the fact that f_y is continuous at (a,b) we may obtain

$$\varepsilon_2 \to 0 \text{ as } (h,k) \to (0,0).$$

Again because f_x exists at (a,b)

$$\lim_{h \to 0} \frac{f(a+h,b) - f(a,b)}{h} = f_x(a,b)$$

or, $\quad \dfrac{f(a+h,b) - f(a,b)}{h} = f_x(a,b) + \varepsilon_1, \text{ where } \varepsilon_1 \to 0 \text{ as } h \to 0.$

i.e., $\quad f(a+h,b) - f(a,b) = h f_x(a,b) + \varepsilon_1 h, \hfill (13.6.6)$

where $\varepsilon_1 \to 0$ as $h \to 0$.

Combining all the results (13.6.3), (13.6.4), (13.6.5), (13.6.6) we get

$$f(a+h, b+k) - f(a,b) = k[f_y(a,b) + \varepsilon_2] + h f_x(a,b) + \varepsilon_1 h$$
$$= h f_x(a,b) + k f_y(a,b) + \varepsilon_1 h + \varepsilon_2 k$$

where $\varepsilon_1, \varepsilon_2$ are functions of (h,k) and they tend to zero as $(h,k) \to (0,0)$.

This proves that $f(x,y)$ is differentiable at (a,b).

Observations:

1. We may similarly prove that if $f_y(a,b)$ exists and $f_x(x,y)$ is continuous at (a,b) then also $f(x,y)$ is differentiable at the point (a,b).

2. If both f_x and f_y are continuous at (a,b) then clearly $f(x,y)$ would be differentiable at (a,b).

3. **These conditions are sufficient but not necessary:** We show by an example given below that even if none of the two partial derivatives be continuous at a point still the function may be differentiable at that point.

Example 13.6.5. Let

$$f(x,y) = \begin{cases} x^2 \sin \dfrac{1}{x} + y^2 \sin \dfrac{1}{y}, & \text{when neither } x = 0, \text{ nor } y = 0; \\ x^2 \sin \dfrac{1}{x}, & \text{when } y = 0 \text{ but } x \neq 0; \\ y^2 \sin \dfrac{1}{y}, & \text{when } x = 0 \text{ but } y \neq 0; \\ 0, & \text{when } x = 0 \text{ and } y = 0. \end{cases}$$

Solution: We see that

$$f_x(x,y) = 2x \sin \frac{1}{x} - \cos \frac{1}{x}, \quad \text{when } x \neq 0$$
$$f_x(0,y) = 0.$$
$$f_y(x,y) = 2y \sin \frac{1}{y} - \cos \frac{1}{y}, \quad \text{when } y \neq 0$$
$$f_y(x,0) = 0.$$

Thus neither f_x nor f_y is continuous at $(0,0)$ [$\cos(1/x)$ or $\cos(1/y)$ does not tend to a limit as $x \to 0$ or $y \to 0$ respectively]. Both f_x and f_y exist at $(0,0)$ but none is continuous there.

Now

$$f(0+h, 0+k) - f(0,0) = f(h,k) - f(0,0)$$
$$= \left\{ h^2 \sin \frac{1}{h} + k^2 \sin \frac{1}{k} \right\} - 0$$
$$= \{0 \cdot h + 0 \cdot k\} + h \cdot \left(h \sin \frac{1}{h} \right) + k \cdot \left(k \sin \frac{1}{k} \right)$$
$$= \{f_x(0,0) \cdot h + f_y(0,0) \cdot k\} + \eta_1 h + \eta_2 k$$

where

$$\eta_1 = h \sin \frac{1}{h} \to 0 \text{ as } h \to 0$$
$$\eta_2 = k \sin \frac{1}{k} \to 0 \text{ as } k \to 0.$$

This shows that $f(x,y)$ is differentiable at $(0,0)$, proving that the conditions stated in Theorem above are sufficient but not necessary.

Example 13.6.6. *Examine the conditions of the theorem on sufficient condition for differentiability to investigate whether the function defined below is differentiable at $(0,0)$ or not.*

$$f(x,y) = \begin{cases} xy \dfrac{x^2 - y^2}{x^2 + y^2}, & \text{when } x^2 + y^2 \neq 0; \\ 0, & \text{when } x^2 + y^2 = 0. \end{cases}$$

[C.H. 1991]

Solution: Here $f_x(0,0) = \lim\limits_{h \to 0} \dfrac{f(h,0) - f(0,0)}{h} = \lim\limits_{h \to 0} \dfrac{0}{h} = 0.$

Similarly, $f_y(0,0) = 0.$

When x and y are not both zero

$$f_x(x,y) = \frac{x^4y + 4x^2y^3 - y^5}{(x^2+y^2)^2}.$$

$\therefore \quad |f_x(x,y)| \leq \dfrac{|x^4y + 4x^2y^3 - y^5|}{(x^2+y^2)^2} \leq \dfrac{6(x^2+y^2)^{5/2}}{(x^2+y^2)^2} = 6(x^2+y^2)^{1/2}.$

$\therefore \quad |f_x(x,y) - f_x(0,0)| < \varepsilon \quad \text{if} \quad (x^2+y^2) < \left(\dfrac{\varepsilon}{6}\right)^2.$

Thus, f_x is continuous at $(0,0)$ and also $f_y(0,0)$ exists.

$\therefore$ By the conditions of the theorem $f(x,y)$ is differentiable at $(0,0)$.

(C) Notion of differentiability for Functions of three variables

A function f of three variables x, y, z defined over a domain $D \subseteq \mathbb{R}^3$ is said to be differentiable at $(x, y, z) \in D$ if the increments $\Delta x, \Delta y, \Delta z$ of x, y, z produce a total increment

$$\Delta f = f(x + \Delta x, y + \Delta y, z + \Delta z) - f(x, y, z)$$

having the form

$$\Delta f = (A\Delta x + B\Delta y + C\Delta z) + (\eta_1 \Delta x + \eta_2 \Delta y + \eta_3 \Delta z) \qquad (13.6.7)$$

where A, B, C are independent of $\Delta x, \Delta y, \Delta z$ and η_1, η_2, η_3 are functions of $\Delta x, \Delta y, \Delta z$ and they tend to zero as $(\Delta x, \Delta y, \Delta z) \to (0,0,0)$.

This last condition is equivalent to

$$\eta_1 \Delta x + \eta_2 \Delta y + \eta_3 \Delta z = 0(\rho)$$

where $\rho = \sqrt{\Delta x^2 + \Delta y^2 + \Delta z^2}$

i.e., $\dfrac{\eta_1 \Delta x + \eta_2 \Delta y + \eta_3 \Delta z}{\sqrt{\Delta x^2 + \Delta y^2 + \Delta z^2}} \to 0 \quad \text{as} \quad (\Delta x, \Delta y, \Delta z) \to (0,0,0).$

If we put $\Delta y = \Delta z = 0$ in (13.6.7), divide by Δx and pass to the limit as $\Delta x \to 0$, we get $f_x = A$; similarly $f_y = B$, $f_z = C$.

Thus, for a differentiable function of three variables

$$\Delta f = (f_x \Delta x + f_y \Delta y + f_z \Delta z) + 0(\rho).$$

The part of Δf that is linear in $\Delta x, \Delta y, \Delta z$ is called the *total differential* of f and written

$$df = f_x \Delta x + f_y \Delta y + f_z \Delta z.$$

In particular, when $f = x, f_x = 1, f_y = f_z = 0$ and we have $dx = \Delta x$; similarly $dy = \Delta y, dz = \Delta z$.

Thus, the differentials of the independent variables are equal to their increments; we shall now write

$$\text{Total differential of } f(x, y, z) = df = f_x dx + f_y dy + f_z dz$$
$$= \frac{\partial f}{\partial x} dx + \frac{\partial f}{\partial y} dy + \frac{\partial f}{\partial z} dz.$$

(D) Use of Total differential in Approximations and Small Errors

Increment of a function = Total differential + Error

$$\Delta f = df + \text{Error}$$
$$\text{where} \quad df = f_x dx + f_y dy.$$

We can use Total Increment of a function $\simeq$ Total differential.

Also when the values of x and y are determined by measurements and hence subject to small errors Δx and Δy, a close approximation to the error in $f(x, y)$ can be found $\simeq f_x \Delta x + f_y \Delta y$.

Example 13.6.7. *Approximate the change in the hypotenuse of a right-angled triangle whose sides are 6 and 8 cm., when the shorter side is lengthened by (1/4) cm. and the longer is shortened by (1/8) cm.*

Solution: Let x, y, z be respectively the shorter side, the longer side and the hypotenuse of the triangle. Then $z = \sqrt{x^2 + y^2}$, whence

$$dz = \frac{x}{\sqrt{x^2 + y^2}} dx + \frac{y}{\sqrt{x^2 + y^2}} dy.$$

When $x = 6, y = 8, dy = -\frac{1}{8}$, and $dx = \frac{1}{4}$, then $dz = \frac{1}{20}$ cm.

i.e., the hypotenuse is lengthened by $\frac{1}{20}$ cm. (approximately).

Example 13.6.8. *The height h and the semi-vertical angle α of a cone are measured and from them A, the total area of the cone, including the base is calculated. If h and α are in error by small quantities dh and $d\alpha$ respectively, find the corresponding error in the area. Show further, that if $\alpha = \pi/6$, an error of +1 per cent in h will be approximately compensated by an error of $-.33$ degree in α.* [Math. Tripos]

Solution: Total area $A = \pi h^2 \tan^2 \alpha + \dfrac{1}{2} \cdot 2\pi h \tan \alpha \cdot h \sec \alpha$.

$\therefore\ dA = 2\pi h \tan \alpha (\tan \alpha + \sec \alpha) dh + \pi h^2 (2\tan\alpha \sec^2 \alpha + \sec^3 \alpha + \tan^2 \alpha \sec \alpha) d\alpha$

which answers the first part.

Put $\alpha = \dfrac{\pi}{6}, dh = \dfrac{h}{100}$, then $dA = 2\pi h^2 \times 0.01 + \pi h^2 \times 3.4646 \cdot d\alpha$.

$\therefore\ dA = 0$ if $d\alpha = -\dfrac{2 \times .01}{3.4646}$ radian $= -0.33$ degree.

13.7 Composite Functions : Chain Rules

For a differentiable function of a single real variable $f(u)$, if $u = u(x)$ is also a differentiable function of x, then we have the familiar *Chain rule*:

$$\frac{df}{dx} = \frac{df}{du}\frac{du}{dx}.$$

1. Suppose that $f(u, v)$ is a differentiable function of two independent variables u, v, which, in turn, are differentiable functions of **One** independent variable x. Then f is a differentiable function of x and we have the *extended Chain Rule*; namely

$$\frac{df}{dx} = \frac{\partial f}{\partial u}\frac{du}{dx} + \frac{\partial f}{\partial v}\frac{dv}{dx}.$$

Proof: Let there be an increment Δx of x.

Let the corresponding increments in u and v be Δu and Δv respectively. Suppose these increments in u and v produce an increment Δf in f.

Since f is a differentiable function of u and v,

We have

$$\Delta f = \frac{\partial f}{\partial u}\Delta u + \frac{\partial f}{\partial v}\Delta v + \eta_1 \Delta u + \eta_2 \Delta v,$$

where $\eta_1, \eta_2 \to 0$ as $\Delta u, \Delta v \to 0$; and on division by Δx

$$\frac{\Delta f}{\Delta x} = \frac{\partial f}{\partial u}\frac{\Delta u}{\Delta x} + \frac{\partial f}{\partial v}\frac{\Delta v}{\Delta x} + \eta_1 \frac{\Delta u}{\Delta x} + \eta_2 \frac{\Delta v}{\Delta x}. \tag{13.7.1}$$

As $\Delta x \to 0$, $\Delta u, \Delta v \to 0$ and $\eta_1, \eta_2 \to 0$; moreover $\dfrac{\Delta u}{\Delta x}$ and $\dfrac{\Delta v}{\Delta x}$ approach finite limits $\dfrac{du}{dx}$ and $\dfrac{dv}{dx}$ respectively (since u and v are differentiable functions of x).

Hence we get from (13.7.1), making $\Delta x \to 0$

$$\frac{df}{dx} \text{ exists and } \frac{df}{dx} = \frac{\partial f}{\partial u}\frac{du}{dx} + \frac{\partial f}{\partial v}\frac{dv}{dx}.$$

2. Next we consider the function $f(u,v)$ which is a differentiable function of u and v and u and v are differentiable functions of **two** independent variables x and y.

Let $\Delta x, \Delta y$ be increments of x and y; suppose Δu and Δv be the corresponding increments of u and v which produce an increment Δf in f.

Since u and v are differentiable functions of x and y we have

$$\Delta u = \frac{\partial u}{\partial x}\Delta x + \frac{\partial u}{\partial y}\Delta y + \eta_1 \Delta x + \eta_2 \Delta y$$

$$\Delta v = \frac{\partial v}{\partial x}\Delta x + \frac{\partial v}{\partial y}\Delta y + \eta_3 \Delta x + \eta_4 \Delta y$$

where $\eta_1, \eta_2, \eta_3, \eta_4$ approach zero as $(\Delta x, \Delta y) \to (0,0)$.

Now f is a differentiable function of u, v.

$$\therefore \Delta f = \frac{\partial f}{\partial u}\Delta u + \frac{\partial f}{\partial v}\Delta v + \eta_5 \Delta u + \eta_6 \Delta v$$

where η_5, η_6 are functions of u, v such that they tend to zero as $(\Delta u, \Delta v) \to (0,0)$ i.e., as $(\Delta x, \Delta y) \to (0,0)$.

On substituting the values of Δu and Δv in Δf.

We get

$$\Delta f = \frac{\partial f}{\partial u}\left(\frac{\partial u}{\partial x}\Delta x + \frac{\partial u}{\partial y}\Delta y + \eta_1 \Delta x + \eta_2 \Delta y\right)$$

$$+ \frac{\partial f}{\partial v}\left(\frac{\partial v}{\partial x}\Delta x + \frac{\partial v}{\partial y}\Delta y + \eta_3 \Delta x + \eta_4 \Delta y\right)$$

$$+ \eta_5\left(\frac{\partial u}{\partial x}\Delta x + \frac{\partial u}{\partial y}\Delta y + \eta_1 \Delta x + \eta_2 \Delta y\right)$$

$$+ \eta_6\left(\frac{\partial v}{\partial x}\Delta x + \frac{\partial v}{\partial y}\Delta y + \eta_3 \Delta x + \eta_4 \Delta y\right)$$

or, on collecting the terms of Δx and Δy,

$$\Delta f = \left(\frac{\partial f}{\partial u}\frac{\partial u}{\partial x} + \frac{\partial f}{\partial v}\frac{\partial v}{\partial x}\right)\Delta x + \left(\frac{\partial f}{\partial u}\frac{\partial u}{\partial y} + \frac{\partial f}{\partial v}\frac{\partial v}{\partial y}\right)\Delta y + \mu_1 \Delta x + \mu_2 \Delta y$$

where μ_1, μ_2 consist of sums of terms with at least one η as a factor and hence approach zero as $(\Delta x, \Delta y) \to (0,0)$.

This proves that f is a differentiable function of x and y and its partial derivatives with respect to x and y are given by the **Chain Rules**:

$$\left.\begin{array}{l}\dfrac{\partial f}{\partial x} = \dfrac{\partial f}{\partial u}\dfrac{\partial u}{\partial x} + \dfrac{\partial f}{\partial v}\dfrac{\partial v}{\partial x} \\[2mm] \dfrac{\partial f}{\partial y} = \dfrac{\partial f}{\partial u}\dfrac{\partial u}{\partial y} + \dfrac{\partial f}{\partial v}\dfrac{\partial v}{\partial y}\end{array}\right\}$$

CHAPTER 13: FUNCTIONS OF SEVERAL VARIABLES

Further, we obtain the total differential df:

$$\begin{aligned}df &= \frac{\partial f}{\partial x}dx + \frac{\partial f}{\partial y}dy \\ &= \left(\frac{\partial f}{\partial u}\frac{\partial u}{\partial x} + \frac{\partial f}{\partial v}\frac{\partial v}{\partial x}\right)dx + \left(\frac{\partial f}{\partial u}\frac{\partial u}{\partial y} + \frac{\partial f}{\partial v}\frac{\partial v}{\partial y}\right)dy \\ &= \frac{\partial f}{\partial u}\left[\frac{\partial u}{\partial x}dx + \frac{\partial u}{\partial y}dy\right] + \frac{\partial f}{\partial v}\left[\frac{\partial v}{\partial x}dx + \frac{\partial v}{\partial y}dy\right] \\ &= \frac{\partial f}{\partial u}du + \frac{\partial f}{\partial v}dv\end{aligned}$$

i.e., *the total differential df has the same form as if u, v were independent variables.*

The fact that the total differential of a function has the same form whether the variables are dependent or independent is one of the chief advantages of the differential notation.

When the variables are *independent*, their differentials are also their increments and may be given *arbitrary* values. From this fact, we have the following rule:

If $f(x, y)$ is a function of two independent variables x and y the equation

$$df = P dx + Q dy$$

$$\text{implies } \frac{\partial f}{\partial x} = P, \quad \frac{\partial f}{\partial y} = Q.$$

Proof: In the identity

$$df = f_x dx + f_y dy = P dx + Q dy.$$
$$\text{put } \quad dx = 1, \quad dy = 0 \text{ (arbitrary values);}$$
$$dy = 1, \quad dx = 0.$$

The results $P = f_x$ and $Q = f_y$ will follow.

Example 13.7.1. *Consider the function f defined by*

$$f(u, v) = \sqrt{|uv|}$$

where $u = x, v = x.$

Solution: (a) If we apply the chain rule:

$$\begin{aligned}\frac{df}{dx} &= \frac{\partial f}{\partial u}\frac{du}{dx} + \frac{\partial f}{\partial v}\frac{dv}{dx} \\ &= \frac{\partial f}{\partial u} + \frac{\partial f}{\partial v}. \quad \left(\because \frac{du}{dx} = \frac{dv}{dx} = 1\right)\end{aligned}$$

When $x = 0, u = v = 0$ and $\dfrac{\partial f}{\partial u} = \dfrac{\partial f}{\partial v} = 0$.

$$\left[\lim_{h \to 0} \dfrac{f(h,0) - f(0,0)}{h} = 0, \quad \lim_{k \to 0} \dfrac{f(0,k) - f(0,0)}{k} = 0.\right]$$

Thus we get $\dfrac{df}{dx} = 0$ when $x = 0$.

(b) But $f(u, v) = f(x, x) = \sqrt{|x^2|} = |x|$

$\dfrac{df}{dx} = +1$ or -1 according as x is positive or negative

i.e., at $x = 0$; $\dfrac{df}{dx}$ does not exist.

The fallacy arises because we cannot apply the chain rule due to the fact that $\sqrt{|uv|}$ is not differentiable when $u = v = 0$.

Example 13.7.2. *The plane polar co-ordinates r, θ are given in terms of rectangular cartesian co-ordinates x, y by the equations.*

$$r = \sqrt{x^2 + y^2} \quad \text{and} \quad \theta = \tan^{-1}\dfrac{y}{x} (x > 0).$$

Solution: Taking differentials

$$dr = \dfrac{x dx + y dy}{\sqrt{x^2 + y^2}}, \quad d\theta = \dfrac{x dy - y dx}{x^2 + y^2}.$$

Since x and y are independent variables we can write

$$\dfrac{\partial r}{\partial x} = \dfrac{x}{\sqrt{x^2 + y^2}} = \dfrac{x}{r}; \quad \dfrac{\partial r}{\partial y} = \dfrac{y}{\sqrt{x^2 + y^2}} = \dfrac{y}{r};$$
$$\dfrac{\partial \theta}{\partial x} = \dfrac{-y}{x^2 + y^2} = -\dfrac{y}{r^2}; \quad \dfrac{\partial \theta}{\partial y} = \dfrac{x}{x^2 + y^2} = \dfrac{x}{r^2}.$$

Functions of three variables

Where $f(u, v, w)$ is a differentiable function of u, v, w as u, v, w are functions of x, then we have

$$\dfrac{df}{dx} = \dfrac{\partial f}{\partial u}\dfrac{du}{dx} + \dfrac{\partial f}{\partial v}\dfrac{dv}{dx} + \dfrac{\partial f}{\partial w}\dfrac{dw}{dx}.$$

When u, v, w are functions of more than one variable, say u, v, w are functions of

CHAPTER 13: FUNCTIONS OF SEVERAL VARIABLES

three independent variables x, y, z, then we have *analogous chain rules*:

$$\frac{\partial f}{\partial x} = \frac{\partial f}{\partial u}\frac{\partial u}{\partial x} + \frac{\partial f}{\partial v}\frac{\partial v}{\partial x} + \frac{\partial f}{\partial w}\frac{\partial w}{\partial x};$$

$$\frac{\partial f}{\partial y} = \frac{\partial f}{\partial u}\frac{\partial u}{\partial y} + \frac{\partial f}{\partial v}\frac{\partial v}{\partial y} + \frac{\partial f}{\partial w}\frac{\partial w}{\partial y};$$

$$\frac{\partial f}{\partial z} = \frac{\partial f}{\partial u}\frac{\partial u}{\partial z} + \frac{\partial f}{\partial v}\frac{\partial v}{\partial z} + \frac{\partial f}{\partial w}\frac{\partial w}{\partial z}.$$

Multiplying these equations by dx, dy, dz respectively and then adding we get

$$df = f_x dx + f_y dy + f_z dz$$
$$= \frac{\partial f}{\partial u}\left(\frac{\partial u}{\partial x}dx + \frac{\partial u}{\partial y}dy + \frac{\partial u}{\partial z}dz\right) + \frac{\partial f}{\partial v}\left(\frac{\partial v}{\partial x}dx + \frac{\partial v}{\partial y}dy + \frac{\partial v}{\partial z}dz\right)$$
$$+ \frac{\partial f}{\partial w}\left(\frac{\partial w}{\partial x}dx + \frac{\partial w}{\partial y}dy + \frac{\partial w}{\partial z}dz\right)$$
$$= \frac{\partial f}{\partial u}du + \frac{\partial f}{\partial v}dv + \frac{\partial f}{\partial w}dw$$
$$= f_u du + f_v dv + f_w dw.$$

This proves the invariance of the total differential:

The form of df is the same for dependent as for independent variables.

Example 13.7.3. *The rectangular co-ordinates (x, y, z) of a point in space are given in terms of its Spherical co-ordinates (r, θ, ϕ) by the equations.*

$$x = r\sin\theta\cos\phi;$$
$$y = r\sin\theta\sin\phi;$$
$$z = r\cos\theta.$$

Compute the nine partial derivatives of x, y, z with respect to r, θ, ϕ and the vice versa. Show that

$$\frac{\partial x}{\partial r}\frac{\partial r}{\partial x} + \frac{\partial x}{\partial \theta}\frac{\partial \theta}{\partial x} + \frac{\partial x}{\partial \phi}\frac{\partial \phi}{\partial x} = 1.$$

Solution:

$$\frac{\partial x}{\partial r} = \sin\theta\sin\phi; \quad \frac{\partial x}{\partial \theta} = r\cos\theta\cos\phi; \quad \frac{\partial x}{\partial \phi} = -r\sin\theta\sin\phi;$$

$$\frac{\partial y}{\partial r} = \sin\theta\sin\phi; \quad \frac{\partial y}{\partial \theta} = r\cos\theta\sin\phi; \quad \frac{\partial y}{\partial \phi} = r\sin\theta\cos\phi;$$

$$\frac{\partial z}{\partial r} = \cos\theta; \quad \frac{\partial z}{\partial \theta} = -r\sin\theta; \quad \frac{\partial z}{\partial \phi} = 0.$$

Solving for r, θ, ϕ we get

$$r = \sqrt{x^2+y^2+z^2}, \quad \theta = \cos^{-1}\frac{z}{r}, \quad \phi = \tan^{-1}\frac{y}{x} \quad (x > 0)$$

$$\frac{\partial r}{\partial x} = \frac{x}{\sqrt{x^2+y^2+z^2}} = \frac{x}{r} = \sin\theta\cos\phi.$$

$$\frac{\partial r}{\partial y} = \frac{y}{r} = \sin\theta\sin\phi; \quad \frac{\partial r}{\partial z} = \frac{z}{r} = \cos\theta.$$

$$\frac{\partial\theta}{\partial x} = -\frac{1}{\sqrt{1-z^2/r^2}} \cdot -\frac{z}{r^2}\frac{\partial r}{\partial x} = \frac{z}{\sqrt{r^2-z^2}}\frac{x}{r^2}$$

$$= \frac{z}{\sqrt{x^2+y^2}}\frac{x}{r^2} = \frac{r\cos\theta \cdot r\sin\theta\cos\phi}{r^2\sqrt{r^2\sin^2\theta}} = \frac{\cos\theta\cos\phi}{r}.$$

$$\frac{\partial\theta}{\partial y} = \frac{\cos\theta\sin\phi}{r}, \quad \frac{\partial\theta}{\partial z} = -\frac{\sin\theta}{r}.$$

$$\frac{\partial\phi}{\partial x} = -\frac{\sin\phi}{r\sin\theta}, \quad \frac{\partial\phi}{\partial y} = \frac{\cos\phi}{r\sin\theta}, \quad \frac{\partial\phi}{\partial z} = 0.$$

Now

$$x_r r_x + x_\theta \theta_x + x_\phi \phi_x = (\sin\theta\cos\phi)(\sin\theta\cos\phi) + (r\cos\theta\cos\phi)\left(\frac{\cos\theta\cos\phi}{r}\right)$$

$$+ (-r\sin\theta\sin\phi)\left(-\frac{\sin\phi}{r\sin\theta}\right)$$

$$= \sin^2\theta\cos^2\phi + \cos^2\theta\cos^2\phi + \sin^2\phi = 1.$$

13.8 Homogeneous Functions: Euler's Theorem

I. A function $f(x, y)$ is said to be *homogeneous* of degree n if

$$f(tx, ty) = t^n f(x, y) \quad (t > 0)$$

for every positive value of t.

e.g., $\tan^{-1} y/x$ is homogeneous of degree 0.

$x^2 + 4xy$ is homogeneous of degree 2.

$\dfrac{x^2+y^2}{x^3+y^3}$ is homogeneous of degree -1.

An equivalent form of definition is:

CHAPTER 13: FUNCTIONS OF SEVERAL VARIABLES

A function $f(x,y)$ is said to be homogeneous of degree n if

$$f(x,y) = x^n f\left(1, \frac{y}{x}\right) \quad \text{[put } tx = 1\text{]}$$
$$\text{or} \quad = x^n \phi(y/x).$$

We shall prove **Euler's Theorem** for a homogeneous function in two variables x, y of degree n:

Theorem 13.8.1. *If $u = f(x,y)$ be a homogeneous function of two independent variables x, y of degree n, then*

$$x\frac{\partial u}{\partial x} + y\frac{\partial u}{\partial y} = nu.$$

We assume that u is differentiable or equivalently we may assume that u possesses continuous partial derivatives.

Proof: Since $u = f(x,y)$ is a homogeneous function in x and y of degree n, we may express

$$u = x^n \phi(y/x).$$

$$\therefore \quad \frac{\partial u}{\partial x} = nx^{n-1}\phi(y/x) + x^n \phi'(y/x) \times \frac{-y}{x^2}$$

or, $\quad x\frac{\partial u}{\partial x} = nx^n \phi(y/x) - x^{n-1} y \phi'(y/x)$

and $\quad \frac{\partial u}{\partial y} = x^n \phi'(y/x) \cdot \frac{1}{x} = x^{n-1}\phi'(y/x).$

$$\therefore \quad y\frac{\partial u}{\partial y} = x^{n-1} y \phi'(y/x).$$

Hence adding we get $x\dfrac{\partial u}{\partial x} + y\dfrac{\partial u}{\partial y} = nx^n \phi(y/x) = nu.$

Alternative Proof. Consider the function

$$F(x,y,t) = t^{-n} f(tx, ty).$$

Put $u = tx, v = ty$ and take partial derivative of F w.r.t. t,

$$\frac{\partial F}{\partial t} = -nt^{-n-1} f(u,v) + t^{-n}\left[x\frac{\partial f}{\partial u} + y\frac{\partial f}{\partial v}\right]. \quad \text{(using Chain Rule)}$$

$$\therefore \quad t^{n+1}\frac{\partial F}{\partial t} = -nf(u,v) + \left(tx\frac{\partial f}{\partial u} + ty\frac{\partial f}{\partial v}\right)$$

or, $\quad t^{n+1}\dfrac{\partial F}{\partial t} = uf_u + vf_v - nf(u,v).$

When f is homogeneous $F = t^{-n}f(tx, ty) = f(x,y)$ i.e., F is independent of t i.e., $\dfrac{\partial F}{\partial t} = 0$.

$$\therefore \quad uf_u + vf_v = nf(u,v) \text{ (identically)}.$$

Putting $t = 1$, $xf_x + yf_y = nf(x,y)$. (Euler's Theorem)

A more general result

Let u be a homogeneous function of x and y of degree n, then

$$x^2 \frac{\partial^2 u}{\partial x^2} + 2xy \frac{\partial^2 u}{\partial x \partial y} + y^2 \frac{\partial^2 u}{\partial y^2} = n(n-1)u$$

(all partial derivatives of first and second order are supposed to be continuous).

[C.H. 1985]

Proof: Since u is a homogeneous function of x and y of degree n, we have by Euler's theorem,

$$x\frac{\partial u}{\partial x} + y\frac{\partial u}{\partial y} = nu. \tag{13.8.1}$$

Differentiating partially w.r.t x both sides of (13.8.1), we get

$$\frac{\partial u}{\partial x} + x\frac{\partial^2 u}{\partial x^2} + y\frac{\partial^2 u}{\partial x \partial y} = n\frac{\partial u}{\partial x}$$

$$\text{i.e., } x\frac{\partial^2 u}{\partial x^2} + y\frac{\partial^2 u}{\partial x \partial y} = (n-1)\frac{\partial u}{\partial x}. \tag{13.8.2}$$

Similarly, differentiating (13.8.1) partially w.r.t. y, we obtain

$$x\frac{\partial^2 u}{\partial y \partial x} + \frac{\partial u}{\partial y} + y\frac{\partial^2 u}{\partial y^2} = n\frac{\partial u}{\partial y}$$

$$\text{or, } x\frac{\partial^2 u}{\partial y \partial x} + y\frac{\partial^2 u}{\partial y^2} = (n-1)\frac{\partial u}{\partial y}. \tag{13.8.3}$$

Multiplying (13.8.2) by x and (13.8.3) by y and adding we obtain
[using $u_{xy} = u_{yx}$, since partial derivatives u_{xy} and u_{yx} are continuous.]

$$x^2\frac{\partial^2 u}{\partial x^2} + y^2\frac{\partial^2 u}{\partial y^2} + 2xy\frac{\partial^2 u}{\partial x \partial y} = (n-1)\left(x\frac{\partial u}{\partial x} + y\frac{\partial u}{\partial y}\right)$$

$$= n(n-1)u. \quad \text{[using (13.8.1)]}$$

CHAPTER 13: FUNCTIONS OF SEVERAL VARIABLES

Note 13.8.1. *In symbolic form, this result may be written as*

$$\left(x^2 \frac{\partial^2}{\partial x^2} + 2xy \frac{\partial^2}{\partial x \partial y} + y^2 \frac{\partial^2}{\partial y^2}\right) u = n(n-1)u$$

$$\text{or } \left(x \frac{\partial}{\partial x} + y \frac{\partial}{\partial y}\right)^2 u = n(n-1)u.$$

More generally, if u be a homogeneous function in x and y of degree n, then (assuming continuity of partial derivatives)

$$\left(x \frac{\partial}{\partial x} + y \frac{\partial}{\partial y}\right)^m u = n(n-1)(n-2)\cdots(n-m+1)u$$

where the operator $\left(x \frac{\partial}{\partial x} + y \frac{\partial}{\partial y}\right)^m$ can be expanded as by the usual Binomial rule:

$$x^m \frac{\partial^m}{\partial x^m} + mx^{m-1} \frac{\partial^{m-1}}{\partial x^{m-1}} \left(y \frac{\partial}{\partial y}\right) + \cdots + y^m \frac{\partial^m}{\partial y^m}.$$

II. A function $f(x, y, z)$ of **three independent variables** x, y, z is said to be homogeneous of degree n when

$$f(tx, ty, tz) = t^n f(x, y, z) \quad (t > 0)$$

for every positive value of t.

Note that for a homogeneous function $f(x, y, z)$,

$$t^{-n} f(tx, ty, tz) = F(x, y, z, t) \quad \text{(say)}$$

is independent of t.

We shall prove the **Euler's Theorem** as well as its converse for homogeneous function in x, y, z of degree n.

Theorem 13.8.2. *If f is a differentiable homogeneous function of degree n of three variables x, y, z then*

$$x \frac{\partial f}{\partial x} + y \frac{\partial f}{\partial y} + z \frac{\partial f}{\partial z} = nf(x, y, z) \tag{13.8.4}$$

and conversely if (13.8.4) *holds for all* (x, y, z) *within the domain of f then f will be a homogeneous function of x, y, z of degree n.*

Proof: Consider the function

$$F(x, y, z, t) = t^{-n} f(tx, ty, tz).$$

Put $u = tx, v = ty, w = tz$ and differentiate F with respect to t:

$$\frac{\partial F}{\partial t} = -nt^{-n-1}f(u,v,w) + t^{-n}\left(\frac{\partial f}{\partial u}\cdot\frac{\partial u}{\partial t} + \frac{\partial f}{\partial v}\cdot\frac{\partial v}{\partial t} + \frac{\partial f}{\partial w}\cdot\frac{\partial w}{\partial t}\right)$$

$$= -nt^{-n-1}f(u,v,w) + t^{-n}\left(x\frac{\partial f}{\partial u} + y\frac{\partial f}{\partial v} + z\frac{\partial f}{\partial w}\right).$$

$$\therefore\; t^{n+1}\frac{\partial F}{\partial t} = -nf(u,v,w) + \left(tx\frac{\partial f}{\partial u} + ty\frac{\partial f}{\partial v} + tz\frac{\partial f}{\partial w}\right)$$

$$= -nf(u,v,w) + (uf_u + vf_v + wf_w).$$

When f is homogeneous, F is independent of t; then $\dfrac{\partial F}{\partial t} = 0$. Hence the right side vanishes identically.

$$\therefore\; nf(u,v,w) = u\frac{\partial f}{\partial u} + v\frac{\partial f}{\partial v} + w\frac{\partial f}{\partial w}.$$

For $t = 1, nf(x,y,z) = x\dfrac{\partial f}{\partial x} + y\dfrac{\partial f}{\partial y} + z\dfrac{\partial f}{\partial z}$ (**Euler's Theorem**).

Theorem 13.8.3. Converse of Euler's Theorem [C.H. 1999]

Proof: Let us put $\xi = x, \eta = \frac{y}{x}, \zeta = \frac{z}{x}$ and let $f(x,y,z) = g(\xi,\eta,\zeta)$.
$\therefore$ By the chain rule:

$$\frac{\partial f}{\partial x} = \frac{\partial g}{\partial \xi}\frac{\partial \xi}{\partial x} + \frac{\partial g}{\partial \eta}\frac{\partial \eta}{\partial x} + \frac{\partial g}{\partial \zeta}\frac{\partial \zeta}{\partial x} = \frac{\partial g}{\partial \xi} - \frac{y}{x^2}\frac{\partial g}{\partial \eta} - \frac{z}{x^2}\frac{\partial g}{\partial \zeta}$$

$$\frac{\partial f}{\partial y} = \frac{\partial g}{\partial \xi}\frac{\partial \xi}{\partial y} + \frac{\partial g}{\partial \eta}\frac{\partial \eta}{\partial y} + \frac{\partial g}{\partial \zeta}\frac{\partial \zeta}{\partial y} = \frac{1}{x}\frac{\partial g}{\partial \eta}$$

and $\dfrac{\partial f}{\partial z} = \dfrac{\partial g}{\partial \xi}\dfrac{\partial \xi}{\partial z} + \dfrac{\partial g}{\partial \eta}\dfrac{\partial \eta}{\partial z} + \dfrac{\partial g}{\partial \zeta}\dfrac{\partial \zeta}{\partial z} = \dfrac{1}{x}\dfrac{\partial g}{\partial \zeta}$

$\therefore$ The given relation:

$$x\frac{\partial f}{\partial x} + y\frac{\partial f}{\partial y} + z\frac{\partial f}{\partial z} = nf$$

$$\Rightarrow x\frac{\partial g}{\partial \xi} - \frac{y}{x}\frac{\partial g}{\partial \eta} - \frac{z}{x}\frac{\partial g}{\partial \zeta} + \frac{y}{x}\frac{\partial g}{\partial \eta} + \frac{z}{x}\frac{\partial g}{\partial \zeta} = ng$$

or, $x\dfrac{\partial g}{\partial \xi} = ng \Rightarrow \dfrac{\partial g}{\partial \xi} = \dfrac{n}{\xi}g \quad [\because x = \xi]$

$$\Rightarrow \ln g = n\ln\xi + \ln H(\eta,\zeta)$$

where H is an arbitrary function of η and ζ.
$\Rightarrow g = \xi^n H(\eta,\zeta)$ or $f(x,y,z) = x^n H\left(\frac{y}{x},\frac{z}{x}\right)$
$\Rightarrow f$ is homogeneous function in x,y,z of degree n.

CHAPTER 13: FUNCTIONS OF SEVERAL VARIABLES

Example 13.8.1. *If* $V = \cos^{-1} \dfrac{x+y}{\sqrt{x}+\sqrt{y}}$, *then verify that* $\cos V = \dfrac{x+y}{\sqrt{x}+\sqrt{y}}$ *is a homogeneous function of* x, y *of degree* $1/2$. *Hence prove*

$$x\frac{\partial V}{\partial x} + y\frac{\partial V}{\partial y} + \frac{1}{2}\cot V = 0.$$

Solution:

$$\cos V = \frac{x+y}{\sqrt{x}+\sqrt{y}} = f(x,y), \text{ say}$$

$$f(tx, ty) = \frac{t(x+y)}{\sqrt{t}(\sqrt{x}+\sqrt{y})} = t^{1/2} f(x,y).$$

Hence $\cos V$ is a homogeneous function in x, y of degree $\dfrac{1}{2}$.

Using Euler's Theorem we get

$$x\frac{\partial}{\partial x}(\cos V) + y\frac{\partial}{\partial y}(\cos V) = \frac{1}{2}\cos V$$

or, $-x\sin V \dfrac{\partial V}{\partial x} - y\sin V \dfrac{\partial V}{\partial y} = \dfrac{1}{2}\cos V$

or, $x\dfrac{\partial V}{\partial x} + y\dfrac{\partial V}{\partial y} + \dfrac{1}{2}\dfrac{\cos V}{\sin V} = 0$

or, $x\dfrac{\partial V}{\partial x} + y\dfrac{\partial V}{\partial y} + \dfrac{1}{2}\cot V = 0.$

One may similarly obtain

(a) If $u = \tan^{-1} \dfrac{x^3 + y^3}{x - y}$, then $x\dfrac{\partial u}{\partial x} + y\dfrac{\partial u}{\partial y} = \sin 2u$.

(b) If $u = \sin^{-1} \sqrt{\dfrac{x^{(1/3)} + y^{(1/3)}}{x^{(1/2)} + y^{(1/2)}}}$, then $x\dfrac{\partial u}{\partial x} + y\dfrac{\partial u}{\partial y} + \dfrac{1}{12}\tan u = 0$.

13.9 Partial Derivatives of Higher Order

If a function f of two variables x and y possesses partial derivatives of first order at each point (x, y) in a certain domain, then these partial derivatives f_x and f_y are themselves functions of x and y and may possess partial derivatives. These are called *Second Order partial derivatives* and they are denoted by

$$\frac{\partial}{\partial x}(f_x) = \frac{\partial^2 f}{\partial x^2} \text{ or, } f_{xx}; \quad \frac{\partial}{\partial y}(f_x) = \frac{\partial^2 f}{\partial y \partial x} \text{ or, } f_{yx};$$

$$\frac{\partial}{\partial x}(f_y) = \frac{\partial^2 f}{\partial x \partial y} \text{ or, } f_{xy}; \quad \frac{\partial}{\partial y}(f_y) = \frac{\partial^2 f}{\partial y^2} \text{ or, } f_{yy}.$$

Thus $f(x, y)$ may have four second order partial derivatives. These in turn yield *eight* third order partial derivatives:

$$f_{xxx};\quad f_{yxx};\quad f_{xyx};\quad f_{yyx};\quad f_{xxy};\quad f_{yxy};\quad f_{xyy};\quad f_{yyy}.$$

and so on. The higher derivatives with respect to both x and y e.g., f_{xy} or, f_{yx} or f_{xyx} etc. are called *Mixed* partial derivatives. *Under certain conditions* mixed derivatives that differ only in *Order* in which differentiations are performed have the same value. Thus when those requisite conditions are satisfied, $f(x, y)$ has only *three* second derivatives $f_{xx}, f_{xy}\,(= f_{yx}), f_{yy}$, *four* third derivatives etc., $(n + 1)$ partial derivatives of nth order.

Second Order Mixed derivative at a particular point (a, b):

$$f_{xy}(a,b) = \left[\frac{\partial}{\partial x} f_y\right]_{x=a, y=b} = \lim_{h \to 0} \frac{f_y(a+h, b) - f_y(a, b)}{h}$$

$$f_y(a+h, b) = \left[\frac{\partial f}{\partial y}\right]_{x=a+h, y=b} = \lim_{k \to 0} \frac{f(a+h, b+k) - f(a+h, b)}{k}$$

$$f_y(a, b) = \left[\frac{\partial f}{\partial y}\right]_{x=a, y=b} = \lim_{k \to 0} \frac{f(a, b+k) - f(a, b)}{k}.$$

If we write $F(h, k) = f(a+h, b+k) - f(a+h, b) - f(a, b+k) + f(a, b)$ **then we** see that

$$f_{xy}(a,b) = \lim_{h \to 0} \left(\lim_{k \to 0} \frac{1}{h} \cdot \frac{F(h,k)}{k}\right)$$

and similarly,

$$f_{yx}(a,b) = \lim_{k \to 0} \left(\lim_{h \to 0} \frac{1}{k} \cdot \frac{F(h,k)}{h}\right).$$

Obviously, such iterated limits may differ in value as will be seen in the following examples:

Example 13.9.1. *Consider the function f defined by*

$$f(x, y) = \begin{cases} xy\dfrac{x^2 - y^2}{x^2 + y^2}, & \text{where } x^2 + y^2 \neq 0; \\ 0, & \text{where } x^2 + y^2 = 0 \ (i.e., \text{ where } x = 0, y = 0). \end{cases}$$

Show that $f_{xy}(0, 0) = 1$, $f_{yx}(0, 0) = -1$ so that $f_{xy} \neq f_{yx}$ at $(0, 0)$.

[C.H. 1991, 93, 99]

CHAPTER 13: FUNCTIONS OF SEVERAL VARIABLES

Solution:

$$f_{xy}(0,0) = \left[\frac{\partial}{\partial x} f_y\right]_{x=0, y=0} = \lim_{h \to 0} \frac{f_y(h,0) - f_y(0,0)}{h}.$$

Now,

$$f_y(h,0) = \left[\frac{\partial f}{\partial y}\right]_{x=h, y=0} = \lim_{k \to 0} \frac{f(h,k) - f(h,0)}{k}$$

$$= \lim_{k \to 0} \frac{hk \frac{h^2 - k^2}{h^2 + k^2} - 0}{k} = \lim_{k \to 0} \frac{h(h^2 - k^2)}{h^2 + k^2} = \frac{h^3}{h^2} = h$$

and $\quad f_y(0,0) = \left[\frac{\partial f}{\partial y}\right]_{x=0, y=0} = \lim_{k \to 0} \frac{f(0,k) - f(0,0)}{k} = 0.$

$$\therefore \quad f_{xy}(0,0) = \lim_{h \to 0} \frac{h - 0}{h} = 1.$$

Again,

$$f_{yx}(0,0) = \left[\frac{\partial}{\partial y} f_x\right]_{x=0, y=0} = \lim_{k \to 0} \frac{f_x(0,k) - f_x(0,0)}{k}.$$

But

$$f_x(0,k) = \left[\frac{\partial f}{\partial x}\right]_{x=0, y=k} = \lim_{h \to 0} \frac{f(h,k) - f(0,k)}{h}$$

$$= \lim_{h \to 0} \frac{\frac{hk(h^2 - k^2)}{h^2 + k^2} - 0}{h} = \lim_{h \to 0} \frac{k(h^2 - k^2)}{h^2 + k^2} = -\frac{k^3}{k^2} = -k.$$

$$f_x(0,0) = \left[\frac{\partial f}{\partial x}\right]_{x=0, y=0} = \lim_{h \to 0} \frac{f(h,0) - f(0,0)}{h} = 0.$$

$$\therefore \quad f_{yx}(0,0) = \lim_{k \to 0} \frac{f_x(0,k) - f_x(0,0)}{k} = \lim_{k \to 0} \frac{-k - 0}{k} = -1.$$

Thus, $f_{xy}(0,0) \neq f_{yx}(0,0)$.

Observation: For points other than $(0,0)$ we can verify that $f_{xy} = f_{yx}$.

See that

$$f_x = \frac{\partial}{\partial x}\left[xy\frac{x^2-y^2}{x^2+y^2}\right] = \frac{y(x^4-y^4+4x^2y^2)}{(x^2+y^2)^2}.$$

$$\therefore\ f_{yx} = \frac{\partial}{\partial y}f_x = \frac{\partial}{\partial y}\left[\frac{y(x^4-y^4+4x^2y^2)}{(x^2+y^2)^2}\right] = \frac{x^6-y^6+9(x^4y^2)-9x^2y^4}{(x^2+y^2)^3}$$

$$f_y = \frac{\partial}{\partial y}\left[xy\frac{x^2-y^2}{x^2+y^2}\right] = \frac{-x(y^4-x^4+4x^2y^2)}{(x^2+y^2)^2}$$

and $\quad f_{xy} = \dfrac{x^6-y^6+9x^4y^2-9x^2y^4}{(x^2+y^2)^3} = f_{yx}.$

Now compare it with **Ex. 13.9.3**. [C.H. 1999]

Example 13.9.2. Consider the function

$$f(x,y) = x^2\tan^{-1}\frac{y}{x} - y^2\tan^{-1}\frac{x}{y},\quad (x\neq 0, y\neq 0)$$
$$f(0,y) = f(x,0) = f(0,0) = 0.$$

Solution: When $x, y \neq 0$, we find that

$$f_x = 2x\tan^{-1}\frac{y}{x} + x^2\frac{1}{1+\left(\frac{y}{x}\right)^2}\times\left(\frac{-y}{x^2}\right) - y^2\frac{1}{1+\left(\frac{x}{y}\right)^2}\times\frac{1}{y}$$

$$= 2x\tan^{-1}\frac{y}{x} - y.\quad\text{(on simplification)}$$

$$f_{yx} = 2x\frac{1}{1+(y/x)^2}\times\frac{1}{x} - 1 = \frac{x^2-y^2}{x^2+y^2}.$$

Again, $\quad f_y = x - 2y\tan^{-1}\frac{x}{y},$

and hence $\quad f_{xy} = \dfrac{x^2-y^2}{x^2+y^2}.$

Thus $f_{xy} = f_{yx}$ when $x, y \neq 0$; however at $(0,0)$ we cannot make the same conclusion as $x^2 + y^2 = 0$ there. We may therefore resort to the definition of partial derivatives and find

$$\left.\begin{aligned}f_x(0,y) = -y,\ & f_{yx}(0,0) = -1\\ f_y(x,0) = x,\ & f_{xy}(0,0) = +1\end{aligned}\right\}.$$

[Similar problems are given in Exercises on Chapter **13**. Section **2**. No. **14**.]

We give below two sets of sufficient conditions for the equality $f_{xy} = f_{yx}$ in the form of two theorems: One is given by Prof. Schwarz and the other by Prof. Young.

CHAPTER 13: FUNCTIONS OF SEVERAL VARIABLES

Theorem 13.9.1. (Schwarz Theorem).
  Let $f(x,y)$ be defined in a certain domain D of the xy-plane.
  Let $(a,b) \in D$. Suppose

(a) $\dfrac{\partial f}{\partial y}$ exists in some neighbourhood of (a,b) and

(b) $\dfrac{\partial^2 f}{\partial y \partial x}$ is continuous at the point (a,b).

  Then Schwarz Theorem ensures the existence of $\dfrac{\partial^2 f}{\partial x \partial y}$ at (a,b) and further,

$$f_{xy}(a,b) = f_{yx}(a,b).$$

Proof: The given conditions imply that $\exists$ a certain neighbourhood N of (a,b) at every point of which

$$f_y, f_x, f_{yx} \text{ all exist.}$$

Let h and k be so chosen that $(a+h, b+k), (a, b+k), (a+h, b)$ be points of this neighbourhood N.

Let us agree to write

$$F(h,k) = f(a+h, b+k) - f(a+h, b) - f(a, b+k) + f(a,b)$$
$$\text{and} \quad g(x) = f(x, b+k) - f(x, b)$$
$$\text{so that} \quad F(h,k) = g(a+h) - g(a). \qquad (13.9.1)$$

Since f_x exists in the neighbourhood N of (a,b), the function $g(x)$ is *derivable* in $]a, a+h[$ and continuous in $[a, a+h]$; therefore by Lagrange's Mean Value Theorem we get from (13.9.1),

$$F(h,k) = g(a+h) - g(a)$$
$$= hg'(a + \theta h), 0 < \theta < 1$$
$$= h\{f_x(a+\theta h, b+k) - f_x(a+\theta h, b)\}. \qquad (13.9.2)$$

Again since f_{yx} exists in N where $(a+\theta h, b+k)$ and $(a+\theta h, b)$ lie, the function $f_x(a+\theta h, y)$ is derivable with respect to y in $]b, b+k[$ and continuous in $[b, b+k]$ and therefore using Lagrange's Mean Value Theorem we get from (13.9.2)

$$F(h,k) = h\{f_x(a+\theta h, b+k) - f_x(a+\theta h, b)\}$$
$$= hk f_{yx}(a+\theta h,, b+\theta' k), 0 < \theta' < 1,$$

i.e., $\dfrac{1}{h}\left[\dfrac{f(a+h,b+k)-f(a+h,b)}{k}-\dfrac{f(a,b+k)-f(a,b)}{k}\right]=f_{yx}(a+\theta h,b+\theta' k)$

where $0<\theta,\theta'<1$.

Since f_y exists in the neighbourhood N of (a,b), proceeding to the limits when $k\to 0$, we get

$$\dfrac{1}{h}\left[\lim_{k\to 0}\dfrac{f(a+h,b+k)-f(a+h,b)}{k}-\lim_{k\to 0}\dfrac{f(a,b+k)-f(a,b)}{k}\right]$$
$$=\lim_{k\to 0} f_{yx}(a+\theta h,b+\theta k)$$

i.e., $\dfrac{f_y(a+h,b)-f_y(a,b)}{h}=\lim_{k\to 0} f_{yx}(a+\theta h,b+\theta k).$

Again, taking limits as $h\to 0$, since f_{yx} is continuous at (a,b) we get

$$\lim_{h\to 0}\dfrac{f_y(a+h,b)-f_y(a,b)}{h}=\lim_{h\to 0}\left[\lim_{k\to 0} f_{yx}(a+\theta h,b+\theta k)\right]=f_{yx}(a,b)$$

i.e., L.H.S. limit exists and $=f_{yx}(a,b)$ i.e., $f_{xy}(a,b)$ exists and $=f_{yx}(a,b)$.

Observations:

1. We could state Schwarz Theorem in the following manner:

    If f_x exists in a certain neighbourhood of (a,b) of the domain of definition of a function $f(x,y)$ and if f_{xy} is continuous at (a,b) then the theorem ensures that f_{yx} exists at (a,b) and $f_{yx}(a,b)=f_{xy}(a,b)$.

    Only change in the proof would be to define

    $$F(h,k)=\phi(b+k)-\phi(b),$$
    where $\phi(y)=f(a+h,y)-f(a,y)$

2. If f_{xy} and f_{yx} are both continuous at (a,b) then

    $$f_{xy}(a,b)=f_{yx}(a,b).$$

    Here f_x, f_y both exist and f_{xy} or f_{yx} is continuous at (a,b).

    Proof: As above

    $$F(h,k)=hk f_{yx}(a+\theta h,b+\theta' k),\quad 0<\theta,\theta'<1$$
    starting from $F(h,k)=g(a+h)-g(a)$
    where $g(x)=f(x,b+k)-f(x,b).$

CHAPTER 13: FUNCTIONS OF SEVERAL VARIABLES

Also $F(h,k) = hk f_{xy}(a + \theta_1 h, b + \theta_1' k), 0 < \theta_1, \theta_1' < 1$.

Starting from $F(h,k) = \phi(b+k) - \phi(b)$,

where $\phi(y) = f(a+h, y) - f(a, y)$.

Equating the two expressions for $F(h,k)$ and using the assumed continuity of the mixed derivatives we may conclude $f_{xy}(a,b) = f_{yx}(a,b)$.

The assumption that f_{xy} or f_{yx} should be continuous at (a,b) is **not necessary**. It is sufficient that f_x and f_y exist and are differentiable, as will be proved in the theorem that follows:

Theorem 13.9.2. (Young's Theorem). *Let $f(x,y)$ be defined in some domain D in $\mathbb{R}^2$ and $(a,b) \in D$. If the partial derivatives f_x and f_y exist in some neighbourhood of (a,b) and if they are both differentiable at the point (a,b) then $f_{xy} = f_{yx}$ at (a,b).*

Proof: The differentiability of f_x and f_y at the point (a,b) implies that they exist in a certain neighbourhood N of (a,b) and that all the second order partial derivatives.

$$f_{xx}, \ f_{yx}, \ f_{xy}, \ f_{yy}$$

exist at (a,b).

Let $(a+h, b+h)$ be a point of this neighbourhood N.

Consider $F(h,h) = f(a+h, b+h) - f(a+h, b) - f(a, b+h) + f(a,b)$.

Let $g(x) = f(x, b+h) - f(x,b)$ so that

$$F(h,h) = g(a+h) - g(a). \tag{13.9.3}$$

Since f_x exists in N, the function $g(x)$ is derivable in $]a, a+h[$ and continuous in $[a, a+h]$, therefore by Lagrange's MVT we get from (13.9.3)

$$F(h,h) = hg'(a+\theta h), 0 < \theta < 1$$
$$= h[f_x(a+\theta h, b+h) - f_x(a+\theta h, b)]. \tag{13.9.4}$$

Again since f_x is differentiable at (a,b).

$$f_x(a+\theta h, b+h) - f_x(a,b) = \theta h f_{xx}(a,b) + h f_{yx}(a,b) + \theta h \varepsilon_1 + h\varepsilon_2 \tag{13.9.5}$$

where $\varepsilon_1, \varepsilon_2$ are functions of h such that $\varepsilon_1, \varepsilon_2 \to 0$ as $h \to 0$ and

$$f_x(a+\theta h, b) - f_x(a,b) = \theta h f_{xx}(a,b) + \theta h \varepsilon_3 \tag{13.9.6}$$

where ε_3 is a function of h such that $\varepsilon_3 \to 0$ as $h \to 0$.

From (13.9.4), (13.9.5), (13.9.6) we get

$$\frac{F(h,h)}{h^2} = f_{yx}(a,b) + \theta\varepsilon_1 + \varepsilon_2 - \theta\varepsilon_3. \tag{13.9.7}$$

By similar argument, considering

$$\phi(y) = f(a+h, y) - f(a, y)$$

we can show that

$$\frac{F(h,h)}{h^2} = f_{xy}(a,b) + \varepsilon_4 + \varepsilon_5\theta' - \varepsilon_6\theta' \tag{13.9.8}$$

where $\varepsilon_4, \varepsilon_5, \varepsilon_6 \to 0$ as $h \to 0$.

On equating the two expressions and proceeding to the limits as $h \to 0$ we get from (13.9.7) and (13.9.8)

$$\lim_{h\to 0} \frac{F(h,h)}{h^2} = f_{yx}(a,b) = f_{xy}(a,b).$$

Example 13.9.3. *Consider the function f defined by*

$$f(x,y) = \begin{cases} xy\dfrac{x^2-y^2}{x^2+y^2}, & \text{when } x^2+y^2 \neq 0; \\ 0, & \text{when } x^2+y^2 = 0. \end{cases}$$

We have previously obtained the following results:

$$f_x(x,y) = \frac{y(x^4 + 4x^2y^2 - y^4)}{(x^2+y^2)^2}, \quad \text{when } (x,y) \neq (0,0)$$
$$= 0, \quad \text{when } (x,y) = (0,0)$$
$$f_{yx}(x,y) = \frac{x^6 + 9x^4y^2 - 9x^2y^4 - y^6}{(x^2+y^2)^3}, \quad \text{when } (x,y) \neq (0,0)$$
$$= -1, \quad \text{when } (x,y) = (0,0)$$
$$f_y(x,y) = \frac{-x(y^4 - x^4 + 4x^2y^2)}{(x^2+y^2)^2}, \quad \text{when } (x,y) \neq (0,0)$$
$$= 0, \quad \text{when } (x,y) = (0,0)$$
$$f_{xy}(x,y) = \frac{x^6 + 9x^4y^2 - 9x^2y^4 - y^6}{(x^2+y^2)^3}, \quad \text{when } (x,y) \neq (0,0)$$
$$= +1. \quad \text{when } (x,y) = (0,0)$$

By putting $y = mx$ in f_{yx} we get $\dfrac{1 + 9m^2 - 9m^4 - m^6}{(1+m^2)^3}$.

$\therefore \lim_{\substack{x\to 0 \\ y\to 0}} f_{yx}$ does not tend to a unique limit.

CHAPTER 13: FUNCTIONS OF SEVERAL VARIABLES   801

Hence $\lim_{\substack{x \to 0 \\ y \to 0}} f_{yx} \neq -1 = f_{yx}(0,0)$.

This proves that f_{yx} is not continuous at $(0,0)$.

It may be similarly proved that f_{xy} is also not continuous at $(0,0)$. Thus the conditions of Schwarz's Theorem are not satisfied.

We can further prove that f_x and f_y are not differentiable at $(0,0)$.

So the conditions of Young's Theorem also are not satisfied.

The reader should carefully note that what we have proved is that if the conditions of Young's or Schwarz's theorem are satisfied then $f_{xy} = f_{yx}$ at (a,b). But if the conditions are not satisfied then the equality $f_{xy} = f_{yx}$ may or may not occur. As for example, in the present case the conditions of either of the two theorems are not satisfied and here $f_{xy}(0,0) \neq f_{yx}(0,0)$.

We give another example (Example **13.9.4**) where $f_{xy}(0,0) = f_{yx}(0,0)$ but neither the conditions of Schwarz's theorem nor those of Young's theorem are satisfied.

Example 13.9.4. *Consider the function f defined by*

$$f(x,y) = \begin{cases} \dfrac{x^2 y^2}{x^2 + y^2}, & \text{when } x^2 + y^2 \neq 0; \\ 0, & \text{when } x = 0, y = 0. \end{cases} \qquad \text{[C.H. 1995]}$$

We see that

$$f_x(0,0) = \lim_{h \to 0} \frac{f(h,0) - f(0,0)}{h} = 0.$$

$$f_y(0,0) = \lim_{k \to 0} \frac{f(0,k) - f(0,0)}{k} = 0.$$

For $(x,y) \neq (0,0)$, $f_x(x,y) = \dfrac{2xy^4}{(x^2+y^2)^2}$; $f_y(x,y) = \dfrac{2x^4 y}{(x^2+y^2)^2}$.

$$f_{yx}(0,0) = \lim_{k \to 0} \frac{f_x(0,k) - f_x(0,0)}{k} = 0$$

and $\quad f_{xy}(0,0) = \lim_{h \to 0} \dfrac{f_y(h,0) - f_y(0,0)}{h} = 0$

so that $f_{xy}(0,0) = f_{yx}(0,0)$.

Now see that

$$f_{yx}(x,y) = \frac{8xy^3(x^2+y^2)^2 - 2xy^4 \cdot 2(x^2+y^2) \cdot 2y}{(x^2+y^2)^4} = \frac{8x^3 y^3}{(x^2+y^2)^3}.$$

Putting $y = mx$, $\lim\limits_{\substack{x \to 0 \\ y \to 0}} f_{yx}(x, y) \neq 0 = f_{yx}(0, 0)$, so that f_{yx} is not continuous at $(0, 0)$.

We can verify that f_{xy} is also not continuous at $(0, 0)$. Thus the conditions of Schwarz's theorem are not satisfied.

We shall now show the conditions of Young's theorem are also not satisfied.

$$f_{xx}(0, 0) = \lim_{h \to 0} \frac{f_x(h, 0) - f_x(0, 0)}{h} = 0.$$

Now, f_x is differentiable at $(0, 0)$ if

$$f_x(h, k) - f_x(0, 0) = f_{xx}(0, 0) \cdot h + f_{yx}(0, 0) \cdot k + \varepsilon_1 h + \varepsilon_2 k$$

where $\varepsilon_1, \varepsilon_2 \to 0$ as $(h, k) \to (0, 0)$,

$$\text{or,} \quad \frac{2hk^4}{(h^2 + k^2)^2} = \varepsilon_1 h + \varepsilon_2 k.$$

Putting $h = r\cos\theta$, $k = r\sin\theta$ we get

$$\frac{2r^5 \sin\theta \cos\theta}{r^4} = \varepsilon_1 r \sin\theta + \varepsilon_2 r \cos\theta$$

i.e., $2\sin\theta \cos\theta = \varepsilon_1 \sin\theta + \varepsilon_2 \cos\theta$,

and $(h, k) \to (0, 0)$ is equivalent to $r \to 0$ (θ remains arbitrary)

i.e., $\lim\limits_{r \to 0}(2\sin\theta \cos\theta) = \lim\limits_{r \to 0}(\varepsilon_1 \sin\theta + \varepsilon_2 \cos\theta)$

$$= \lim_{(h,k) \to (0,0)} (\varepsilon_1 \sin\theta + \varepsilon_2 \cos\theta) = 0$$

i.e., $2\sin\theta \cos\theta = 0$

which is impossible if θ is arbitrary.

This proves that f_x is not differentiable at $(0, 0)$.

Similarly it can be shown that f_y is not differentiable at $(0, 0)$. Thus, the conditions of Young's Theorem are also not satisfied. We have thus shown that neither the conditions of Schwarz's Theorem nor those of Young's Theorem are satisfied yet $f_{xy}(0, 0) = f_{yx}(0, 0)$.

[For similar problems see Exercises on Chapter **13**; Section 2. No. **14, 15**]

CHAPTER 13: FUNCTIONS OF SEVERAL VARIABLES

13.10 Change of Variables: Calculation of second order partial derivatives using Chain Rules

Example 13.10.1. *Let z be a differentiable function of x and y (rectangular cartesian co-ordinates) and let $x = r\cos\theta, y = r\sin\theta$ (r and θ, are polar co-ordinates). Prove that*

a) $\left(\dfrac{\partial z}{\partial r}\right)^2 + \dfrac{1}{r^2}\left(\dfrac{\partial z}{\partial \theta}\right)^2 = \left(\dfrac{\partial z}{\partial x}\right)^2 + \left(\dfrac{\partial z}{\partial y}\right)^2.$

b) $\dfrac{\partial^2 z}{\partial r^2} + \dfrac{1}{r}\dfrac{\partial z}{\partial r} + \dfrac{1}{r^2}\dfrac{\partial^2 z}{\partial \theta^2} = \dfrac{\partial^2 z}{\partial x^2} + \dfrac{\partial^2 z}{\partial y^2}.$ [C.H. 1980, '88, 94]

Solution: (a) Using Chain Rules, we get

$$\left.\begin{aligned}\dfrac{\partial z}{\partial r} &= \dfrac{\partial z}{\partial x}\dfrac{\partial x}{\partial r} + \dfrac{\partial z}{\partial y}\dfrac{\partial y}{\partial r} = \cos\theta\dfrac{\partial z}{\partial x} + \sin\theta\dfrac{\partial z}{\partial y} \\ \dfrac{\partial z}{\partial \theta} &= \dfrac{\partial z}{\partial x}\dfrac{\partial x}{\partial \theta} + \dfrac{\partial z}{\partial y}\dfrac{\partial y}{\partial \theta} = -r\sin\theta\dfrac{\partial z}{\partial x} + r\cos\theta\dfrac{\partial z}{\partial y}.\end{aligned}\right\} \quad (13.10.1)$$

$$\therefore \left(\dfrac{\partial z}{\partial r}\right)^2 + \dfrac{1}{r^2}\left(\dfrac{\partial z}{\partial \theta}\right)^2 = \left[\cos\theta\dfrac{\partial z}{\partial x} + \sin\theta\dfrac{\partial z}{\partial y}\right]^2 + \left[-\sin\theta\dfrac{\partial z}{\partial x} + \cos\theta\dfrac{\partial z}{\partial y}\right]^2$$

$$= \left(\dfrac{\partial z}{\partial x}\right)^2 + \left(\dfrac{\partial z}{\partial y}\right)^2.$$

This proves (a).

(b) From the results (13.10.1) we write:

Operator $\dfrac{\partial}{\partial r} \equiv \cos\theta\dfrac{\partial}{\partial x} + \sin\theta\dfrac{\partial}{\partial y}$

and **Operator** $\dfrac{\partial}{\partial \theta} \equiv -r\sin\theta\dfrac{\partial}{\partial x} + r\cos\theta\dfrac{\partial}{\partial y}.$

$$\therefore \dfrac{\partial^2 z}{\partial r^2} = \dfrac{\partial}{\partial r}\left(\dfrac{\partial z}{\partial r}\right) = \dfrac{\partial}{\partial r}\left(\cos\theta\dfrac{\partial z}{\partial x} + \sin\theta\dfrac{\partial z}{\partial y}\right)$$

$$= \cos\theta\dfrac{\partial}{\partial r}\left(\dfrac{\partial z}{\partial x}\right) + \sin\theta\dfrac{\partial}{\partial r}\left(\dfrac{\partial z}{\partial y}\right) \quad (\because \theta \text{ is treated constant when partial derivative is done w.r.t. } r)$$

use the operator $\dfrac{\partial}{\partial r}$ on $\dfrac{\partial z}{\partial x}$ and on $\dfrac{\partial z}{\partial y}$ and thus obtain

$$= \cos\theta \left[\cos\theta \dfrac{\partial}{\partial x} + \sin\theta \dfrac{\partial}{\partial y}\right] \left(\dfrac{\partial z}{\partial x}\right) + \sin\theta \left[\cos\theta \dfrac{\partial}{\partial x} + \sin\theta \dfrac{\partial}{\partial y}\right] \left(\dfrac{\partial z}{\partial y}\right)$$

$$= \left[\cos^2\theta \dfrac{\partial^2 z}{\partial x^2} + \cos\theta \sin\theta \dfrac{\partial^2 z}{\partial y \partial x}\right] + \left[\sin\theta \cos\theta \dfrac{\partial^2 z}{\partial x \partial y} + \sin^2\theta \dfrac{\partial^2 z}{\partial y^2}\right]$$

$$= \cos^2\theta \dfrac{\partial^2 z}{\partial x^2} + 2\sin\theta \cos\theta \dfrac{\partial^2 z}{\partial y \partial x} + \sin^2\theta \dfrac{\partial^2 z}{\partial y^2}. \qquad (13.10.2)$$

$$\left(\text{using } \dfrac{\partial^2 z}{\partial y \partial x} = \dfrac{\partial^2 z}{\partial x \partial y}\right)$$

$$\dfrac{\partial^2 z}{\partial \theta^2} = \dfrac{\partial}{\partial \theta}\left(\dfrac{\partial z}{\partial \theta}\right) = \dfrac{\partial}{\partial \theta}\left(-r\sin\theta \dfrac{\partial z}{\partial x} + r\cos\theta \dfrac{\partial z}{\partial y}\right)$$

$$= \left[-r\cos\theta \dfrac{\partial z}{\partial x} - r\sin\theta \dfrac{\partial z}{\partial y}\right] - r\sin\theta \dfrac{\partial}{\partial \theta}\left(\dfrac{\partial z}{\partial x}\right) + r\cos\theta \dfrac{\partial}{\partial \theta}\left(\dfrac{\partial z}{\partial y}\right)$$

$$= -r\left(\cos\theta \dfrac{\partial z}{\partial x} + \sin\theta \dfrac{\partial z}{\partial y}\right) - r\sin\theta \left[-r\sin\theta \dfrac{\partial}{\partial x} + r\cos\theta \dfrac{\partial}{\partial y}\right]\left(\dfrac{\partial z}{\partial x}\right)$$

$$+ r\cos\theta \left[-r\sin\theta \dfrac{\partial}{\partial x} + r\cos\theta \dfrac{\partial}{\partial y}\right]\left(\dfrac{\partial z}{\partial y}\right)$$

$$= -r\dfrac{\partial z}{\partial r} + r^2\sin^2\theta \dfrac{\partial^2 z}{\partial x^2} - 2r^2\sin\theta \cos\theta \dfrac{\partial^2 z}{\partial x \partial y} + r^2\cos^2\theta \dfrac{\partial^2 z}{\partial y^2}.$$

$$\dfrac{1}{r^2}\dfrac{\partial^2 z}{\partial \theta^2} + \dfrac{1}{r}\dfrac{\partial z}{\partial r} = \sin^2\theta \dfrac{\partial^2 z}{\partial x^2} - 2\sin\theta \cos\theta \dfrac{\partial^2 z}{\partial x \partial y} + \cos^2\theta \dfrac{\partial^2 z}{\partial y^2}. \qquad (13.10.3)$$

Adding (13.10.2) and (13.10.3) we get

$$\dfrac{\partial^2 z}{\partial r^2} + \dfrac{1}{r^2}\dfrac{\partial^2 z}{\partial \theta^2} + \dfrac{1}{r}\dfrac{\partial z}{\partial r} = \dfrac{\partial^2 z}{\partial x^2} + \dfrac{\partial^2 z}{\partial y^2}.$$

Alternative Method of Proving

(b) Expressing $\dfrac{\partial z}{\partial x}, \dfrac{\partial z}{\partial y}$ in terms of $\dfrac{\partial z}{\partial r}, \dfrac{\partial z}{\partial \theta}$.

CHAPTER 13: FUNCTIONS OF SEVERAL VARIABLES   805

Results 13.10.1 give

$$\left.\begin{aligned}\frac{\partial z}{\partial r} &= \cos\theta \frac{\partial z}{\partial x} + \sin\theta \frac{\partial z}{\partial y} \\ \frac{\partial z}{\partial \theta} &= -r\sin\theta \frac{\partial z}{\partial x} + r\cos\theta \frac{\partial z}{\partial y}\end{aligned}\right\}. \tag{13.10.4}$$

We solve for $\dfrac{\partial z}{\partial x}$ and $\dfrac{\partial z}{\partial y}$:

$$\left.\begin{aligned}\frac{\partial z}{\partial x} &= \cos\theta \frac{\partial z}{\partial r} - \frac{\sin\theta}{r}\frac{\partial z}{\partial \theta} \\ \frac{\partial z}{\partial y} &= \sin\theta \frac{\partial z}{\partial r} + \frac{\cos\theta}{r}\frac{\partial z}{\partial \theta}\end{aligned}\right\}. \tag{13.10.5}$$

We see that

$$\frac{\partial}{\partial x} \equiv \cos\theta \frac{\partial}{\partial r} - \frac{\sin\theta}{r}\frac{\partial}{\partial \theta}$$

and $\quad \dfrac{\partial}{\partial y} \equiv \sin\theta \dfrac{\partial}{\partial r} + \dfrac{\cos\theta}{r}\dfrac{\partial}{\partial \theta}.$

Hence

$$\begin{aligned}\frac{\partial^2 z}{\partial x^2} &= \frac{\partial}{\partial x}\left(\frac{\partial z}{\partial x}\right) = \frac{\partial}{\partial x}\left(\cos\theta \frac{\partial z}{\partial r} - \frac{\sin\theta}{r}\frac{\partial z}{\partial \theta}\right) \\ &= \left(\cos\theta \frac{\partial}{\partial r} - \frac{\sin\theta}{r}\frac{\partial}{\partial \theta}\right)\left(\cos\theta \frac{\partial z}{\partial r} - \frac{\sin\theta}{r}\frac{\partial z}{\partial \theta}\right) = \cos\theta \frac{\partial}{\partial r}\left(\cos\theta \frac{\partial z}{\partial r}\right) \\ &\quad - \frac{\sin\theta}{r}\frac{\partial}{\partial \theta}\left(\cos\theta \frac{\partial z}{\partial r}\right) + \cos\theta \frac{\partial}{\partial r}\left(-\frac{\sin\theta}{r}\frac{\partial z}{\partial \theta}\right) - \frac{\sin\theta}{r}\frac{\partial}{\partial \theta}\left(-\frac{\sin\theta}{r}\frac{\partial z}{\partial \theta}\right) \\ &= \left[\cos^2\theta \frac{\partial^2 z}{\partial r^2}\right] + \left[\frac{\sin^2\theta}{r}\frac{\partial z}{\partial r} - \frac{\sin\theta\cos\theta}{r}\frac{\partial^2 z}{\partial \theta \partial r}\right] \\ &\quad + \left[\sin\theta\cos\theta \frac{1}{r^2}\frac{\partial z}{\partial \theta} - \frac{\sin\theta\cos\theta}{r}\frac{\partial^2 z}{\partial r \partial \theta}\right] + \left[\frac{\sin\theta\cos\theta}{r^2}\frac{\partial z}{\partial \theta} + \frac{\sin^2\theta}{r^2}\frac{\partial^2 z}{\partial \theta^2}\right] \\ &= \cos^2\theta \frac{\partial^2 z}{\partial r^2} + \frac{\sin^2\theta}{r}\frac{\partial z}{\partial r} - \frac{2\sin\theta\cos\theta}{r}\frac{\partial^2 z}{\partial r \partial \theta} + \frac{2\sin\theta\cos\theta}{r^2}\frac{\partial z}{\partial \theta} + \frac{\sin^2\theta}{r^2}\frac{\partial^2 z}{\partial \theta^2}.\end{aligned}$$

$$\left[\text{assuming } \frac{\partial^2 z}{\partial \theta \partial r} = \frac{\partial^2 z}{\partial r \partial \theta}\right]$$

Similarly we can prove that

$$\frac{\partial^2 z}{\partial y^2} = \frac{\partial}{\partial y}\left(\frac{\partial z}{\partial y}\right) = \frac{\partial}{\partial y}\left(\sin\theta \frac{\partial z}{\partial r} + \frac{\cos\theta}{r}\frac{\partial z}{\partial \theta}\right)$$

$$= \left(\sin\theta\frac{\partial}{\partial r} + \frac{\cos\theta}{r}\frac{\partial}{\partial \theta}\right)\left(\sin\theta\frac{\partial z}{\partial r} + \frac{\cos\theta}{r}\frac{\partial z}{\partial \theta}\right)$$

$$= \sin^2\theta \frac{\partial^2 z}{\partial r^2} + \frac{\cos^2\theta}{r}\frac{\partial z}{\partial r} + \frac{2\sin\theta\cos\theta}{r}\frac{\partial^2 z}{\partial r \partial\theta} - \frac{2\sin\theta\cos\theta}{r^2}\frac{\partial z}{\partial \theta} + \frac{\cos^2\theta}{r^2}\frac{\partial^2 z}{\partial \theta^2}.$$

Adding we get

$$\frac{\partial^2 z}{\partial x^2} + \frac{\partial^2 z}{\partial y^2} = \frac{\partial^2 z}{\partial r^2} + \frac{1}{r}\frac{\partial z}{\partial r} + \frac{1}{r^2}\frac{\partial^2 z}{\partial \theta^2}.$$

Example 13.10.2. *If z be a differentiable function of x and y and if*

$$x = c\cosh u \cos v, \quad y = c\sinh u \sin v,$$

then prove that

$$\frac{\partial^2 z}{\partial u^2} + \frac{\partial^2 z}{\partial v^2} = \frac{1}{2}c^2(\cosh 2u - \cos 2v)\left(\frac{\partial^2 z}{\partial x^2} + \frac{\partial^2 z}{\partial y^2}\right). \qquad \text{[C.H. 1990]}$$

Solution: Using Chain Rules we get

$$\frac{\partial z}{\partial u} = \frac{\partial z}{\partial x}\frac{\partial x}{\partial u} + \frac{\partial z}{\partial y}\frac{\partial y}{\partial u} = (c\sinh u \cos v)\frac{\partial z}{\partial x} + (c\cosh u \sin v)\frac{\partial z}{\partial y}$$

i.e., $\dfrac{\partial}{\partial u} \equiv c\sinh u \cos v \dfrac{\partial}{\partial x} + c\cosh u \sin v \dfrac{\partial}{\partial y}.$

$$\therefore \frac{\partial^2 z}{\partial u^2} = \frac{\partial}{\partial u}\left(c\sinh u \cos v \frac{\partial z}{\partial x}\right) + \frac{\partial}{\partial u}\left(c\cosh u \sin v \frac{\partial z}{\partial y}\right)$$

$$= c\cosh u \cos v \frac{\partial z}{\partial x} + c\sinh u \cos v \frac{\partial}{\partial u}\left(\frac{\partial z}{\partial x}\right)$$

$$+ c\sinh u \sin v \frac{\partial z}{\partial y} + c\cosh u \sin v \frac{\partial}{\partial u}\left(\frac{\partial z}{\partial y}\right)$$

$$= x\frac{\partial z}{\partial x} + y\frac{\partial z}{\partial y} + c\sinh u \cos v \left[c\sinh u \cos v \frac{\partial}{\partial x}\left(\frac{\partial z}{\partial x}\right)\right.$$

$$\left. + c\cosh u \sin v \frac{\partial}{\partial y}\left(\frac{\partial z}{\partial x}\right)\right]$$

$$+ c\cosh u \sin v \left[c\sinh u \cos v \frac{\partial}{\partial x}\left(\frac{\partial z}{\partial y}\right) + c\cosh u \sin v \frac{\partial}{\partial y}\left(\frac{\partial z}{\partial y}\right)\right]$$

CHAPTER 13: FUNCTIONS OF SEVERAL VARIABLES

$$= x\frac{\partial z}{\partial x} + y\frac{\partial z}{\partial y} + c^2 \sinh^2 u \cos^2 v \frac{\partial^2 z}{\partial x^2} + 2c^2 \sinh u \cosh u \sin v \cos v \frac{\partial^2 z}{\partial x \partial y}$$
$$+ c^2 \cosh^2 u \sin^2 v \frac{\partial^2 z}{\partial y^2}. \tag{13.10.6}$$

Again

$$\frac{\partial z}{\partial v} = \frac{\partial z}{\partial x}\frac{\partial x}{\partial v} + \frac{\partial z}{\partial y}\frac{\partial y}{\partial v} = \frac{\partial z}{\partial x}(-c\cosh u \sin v) + \frac{\partial z}{\partial y}(c\sinh u \cos v)$$

i.e., $\dfrac{\partial}{\partial v} \equiv (-c\cosh u \sin v)\dfrac{\partial}{\partial x} + (c\sinh u \cos v)\dfrac{\partial}{\partial y}.$

$$\therefore \frac{\partial^2 z}{\partial v^2} = \frac{\partial}{\partial v}\left[\frac{\partial z}{\partial v}\right] = \frac{\partial}{\partial v}\left[-c\cosh u \sin v \frac{\partial z}{\partial x} + c\sinh u \cos v \frac{\partial z}{\partial y}\right]$$
$$= -c\cosh u \cos v \frac{\partial z}{\partial x} - c\sinh u \sin v \frac{\partial z}{\partial y}$$
$$-c\cosh u \sin v \frac{\partial}{\partial v}\left(\frac{\partial z}{\partial x}\right) + c\sinh u \cos v \frac{\partial}{\partial v}\left(\frac{\partial z}{\partial y}\right)$$
$$= -x\frac{\partial z}{\partial x} - y\frac{\partial z}{\partial y} - c\cosh u \sin v\left[-c\cosh u \sin v \frac{\partial}{\partial x}\left(\frac{\partial z}{\partial x}\right)\right.$$
$$\left.+ c\sinh u \cos v \frac{\partial}{\partial y}\left(\frac{\partial z}{\partial x}\right)\right]$$
$$+ c\sinh u \cos v\left[-c\cosh u \sin v \frac{\partial}{\partial x}\left(\frac{\partial z}{\partial y}\right) + c\sinh u \cos v \frac{\partial}{\partial y}\left(\frac{\partial z}{\partial y}\right)\right]$$
$$= -x\frac{\partial z}{\partial x} - y\frac{\partial z}{\partial y} + c^2 \cosh^2 u \sin^2 v \frac{\partial^2 z}{\partial x^2} - 2c^2 \sinh u \cosh u \sin v \cos v \frac{\partial^2 z}{\partial x \partial y}$$
$$+ c^2 \sinh^2 u \cos^2 v \frac{\partial^2 z}{\partial y^2}. \tag{13.10.7}$$

Adding (13.10.6) and (13.10.7) we get

$$\frac{\partial^2 z}{\partial u^2} + \frac{\partial^2 z}{\partial v^2} = \frac{\partial^2 z}{\partial x^2}[c^2 \sinh^2 u \cos^2 v + c^2 \cosh^2 u \sin^2 v]$$
$$+ \frac{\partial^2 z}{\partial y^2}[c^2 \cosh^2 u \sin^2 v + c^2 \sinh^2 u \cos^2 v]$$
$$= (c^2 \cosh^2 u \sin^2 v + c^2 \sinh^2 u \cos^2 v)\left(\frac{\partial^2 z}{\partial x^2} + \frac{\partial^2 z}{\partial y^2}\right)$$
$$= \frac{1}{2}c^2\{2\cosh^2 u \sin^2 v + 2\sinh^2 u \cos^2 v\}\left(\frac{\partial^2 z}{\partial x^2} + \frac{\partial^2 z}{\partial y^2}\right)$$

$$= \frac{1}{2}c^2\{(1+\cosh 2u)\sin^2 v + (\cosh 2u - 1)\cos^2 v\}\left(\frac{\partial^2 z}{\partial x^2} + \frac{\partial^2 z}{\partial y^2}\right)$$

$$= \frac{1}{2}c^2\{\cosh 2u(\sin^2 v + \cos^2 v) - (\cos^2 v - \sin^2 v)\}\left(\frac{\partial^2 z}{\partial x^2} + \frac{\partial^2 z}{\partial y^2}\right)$$

$$= \frac{1}{2}c^2(\cosh 2u - \cos 2v)\left(\frac{\partial^2 z}{\partial x^2} + \frac{\partial^2 z}{\partial y^2}\right) \quad \textbf{(Proved)}.$$

$[\because 2\cosh^2 u = 1 + \cosh 2u, \ 2\sinh^2 u = \cosh 2u - 1 \ \text{and} \ \cos^2 v - \sin^2 v = \cos 2v.]$

Example 13.10.3. *A differentiable function $f(x,y)$, when expressed in terms of the new variables u and v defined by $x = \frac{1}{2}(u+v), \ y = \sqrt{uv}$ becomes $g(u,v)$; prove that*

$$\frac{\partial^2 g}{\partial u \partial v} = \frac{1}{4}\left(\frac{\partial^2 f}{\partial x^2} + 2\frac{x}{y}\frac{\partial^2 f}{\partial x \partial y} + \frac{\partial^2 f}{\partial y^2} + \frac{1}{y}\frac{\partial f}{\partial y}\right). \qquad \text{[C.H. 1982, 99]}$$

Solution: We first note that $f(x,y)$ becomes $g(u,v)$ when we replace x, y in terms of u, v.

Now using Chain rules we get

$$\frac{\partial g}{\partial u} = \frac{\partial f}{\partial x}\frac{\partial x}{\partial u} + \frac{\partial f}{\partial y}\frac{\partial y}{\partial u} = \frac{1}{2}\frac{\partial f}{\partial x} + \frac{\sqrt{v}}{2\sqrt{u}}\frac{\partial f}{\partial y}$$

i.e., $\quad \dfrac{\partial}{\partial u} \equiv \dfrac{1}{2}\dfrac{\partial}{\partial x} + \dfrac{1}{2}\sqrt{\dfrac{v}{u}}\dfrac{\partial}{\partial y}.$

Again,

$$\frac{\partial g}{\partial v} = \frac{\partial f}{\partial x}\frac{\partial x}{\partial v} + \frac{\partial f}{\partial y}\frac{\partial y}{\partial v} = \frac{1}{2}\frac{\partial f}{\partial x} + \frac{\sqrt{u}}{2\sqrt{v}}\frac{\partial f}{\partial y}.$$

$$\therefore \frac{\partial^2 g}{\partial u \partial v} = \frac{\partial}{\partial u}\left(\frac{\partial g}{\partial v}\right) = \frac{\partial}{\partial u}\left(\frac{1}{2}\frac{\partial f}{\partial x} + \frac{1}{2}\sqrt{\frac{u}{v}}\frac{\partial f}{\partial y}\right)$$

$$= \frac{1}{2}\frac{\partial}{\partial u}\left(\frac{\partial f}{\partial x}\right) + \frac{1}{4\sqrt{uv}}\frac{\partial f}{\partial y} + \frac{1}{2}\sqrt{\frac{u}{v}}\frac{\partial}{\partial u}\left(\frac{\partial f}{\partial y}\right)$$

$$= \frac{1}{2}\left\{\frac{1}{2}\frac{\partial}{\partial x} + \frac{1}{2}\sqrt{\frac{v}{u}}\frac{\partial}{\partial y}\right\}\frac{\partial f}{\partial x} + \frac{1}{4\sqrt{uv}}\frac{\partial f}{\partial y}$$

$$+ \frac{1}{2}\sqrt{\frac{u}{v}}\left\{\frac{1}{2}\frac{\partial}{\partial x} + \frac{1}{2}\sqrt{\frac{v}{u}}\frac{\partial}{\partial y}\right\}\left(\frac{\partial f}{\partial y}\right)$$

$$= \frac{1}{4}\frac{\partial^2 f}{\partial x^2} + \frac{1}{4}\sqrt{\frac{v}{u}}\frac{\partial^2 f}{\partial y \partial x} + \frac{1}{4\sqrt{uv}}\frac{\partial f}{\partial y} + \frac{1}{4}\sqrt{\frac{u}{v}}\frac{\partial^2 f}{\partial x \partial y} + \frac{1}{4}\frac{\partial^2 f}{\partial y^2}$$

$$= \frac{1}{4}\left[\frac{\partial^2 f}{\partial x^2} + \frac{u+v}{\sqrt{uv}}\frac{\partial^2 f}{\partial x \partial y} + \frac{1}{\sqrt{uv}}\frac{\partial f}{\partial y} + \frac{\partial^2 f}{\partial y^2}\right].$$

$$\left[\text{assuming } \frac{\partial^2 f}{\partial x \partial y} = \frac{\partial^2 f}{\partial y \partial x}\right]$$

CHAPTER 13: FUNCTIONS OF SEVERAL VARIABLES

Hence, $\dfrac{\partial^2 g}{\partial u \partial v} = \dfrac{1}{4}\left[\dfrac{\partial^2 f}{\partial x^2} + \dfrac{2x}{y}\dfrac{\partial^2 f}{\partial x \partial y} + \dfrac{1}{y}\dfrac{\partial f}{\partial y} + \dfrac{\partial^2 f}{\partial y^2}\right].$

Example 13.10.4. *If $F(u,v)$ is a twice differentiable function of (u,v) and if $u = x^2 - y^2$ and $v = 2xy$, prove that*

$$4(u^2+v^2)\dfrac{\partial^2 F}{\partial u \partial v} + 2u\dfrac{\partial F}{\partial v} + 2v\dfrac{\partial F}{\partial u} = xy\left(\dfrac{\partial^2 F}{\partial x^2} - \dfrac{\partial^2 F}{\partial y^2}\right) + (x^2-y^2)\dfrac{\partial^2 F}{\partial x \partial y}.$$

Solution: Using Chain Rules, we write

$$\dfrac{\partial F}{\partial x} = \dfrac{\partial F}{\partial u}\dfrac{\partial u}{\partial x} + \dfrac{\partial F}{\partial v}\dfrac{\partial v}{\partial x} = \dfrac{\partial F}{\partial u}(2x) + \dfrac{\partial F}{\partial v}(2y)$$

i.e., $\dfrac{\partial}{\partial x} \to 2x\dfrac{\partial}{\partial u} + 2y\dfrac{\partial}{\partial v}.$

Again, $\dfrac{\partial F}{\partial y} = \dfrac{\partial F}{\partial u}\dfrac{\partial u}{\partial y} + \dfrac{\partial F}{\partial v}\dfrac{\partial v}{\partial y}$

$$= \dfrac{\partial F}{\partial u}(-2y) + \dfrac{\partial F}{\partial v}(2x)$$

i.e., $\dfrac{\partial}{\partial y} \to -2y\dfrac{\partial}{\partial u} + 2x\dfrac{\partial}{\partial v}.$

$$\dfrac{\partial^2 F}{\partial x^2} = \dfrac{\partial}{\partial x}\left(2x\dfrac{\partial F}{\partial u} + 2y\dfrac{\partial F}{\partial v}\right) = 2\dfrac{\partial F}{\partial u} + 2y\dfrac{\partial}{\partial x}\left(\dfrac{\partial F}{\partial v}\right) + 2x\dfrac{\partial}{\partial x}\left(\dfrac{\partial F}{\partial u}\right)$$

$$= 2\dfrac{\partial F}{\partial u} + 2y\left(2x\dfrac{\partial}{\partial u} + 2y\dfrac{\partial}{\partial v}\right)\left(\dfrac{\partial F}{\partial v}\right) + 2x\left(2x\dfrac{\partial}{\partial u} + 2y\dfrac{\partial}{\partial v}\right)\left(\dfrac{\partial F}{\partial u}\right)$$

$$= 2\dfrac{\partial F}{\partial u} + 4xy\dfrac{\partial^2 F}{\partial u \partial v} + 4y^2\dfrac{\partial^2 F}{\partial v^2} + 4x^2\dfrac{\partial^2 F}{\partial u^2} + 4xy\dfrac{\partial^2 F}{\partial v \partial u}$$

$$= 2\dfrac{\partial F}{\partial u} + 8xy\dfrac{\partial^2 F}{\partial u \partial v} + 4y^2\dfrac{\partial^2 F}{\partial v^2} + 4x^2\dfrac{\partial^2 F}{\partial u^2}$$

$$\left(\text{as } F \text{ is twice differentiable function of } u,v. \therefore \dfrac{\partial^2 F}{\partial u \partial v} = \dfrac{\partial^2 F}{\partial v \partial u}\right)$$

$$\dfrac{\partial^2 F}{\partial y^2} = \dfrac{\partial}{\partial y}\left(\dfrac{\partial F}{\partial y}\right) = \dfrac{\partial}{\partial y}\left(-2y\dfrac{\partial F}{\partial u} + 2x\dfrac{\partial F}{\partial v}\right)$$

$$= -2\dfrac{\partial F}{\partial u} + 2x\dfrac{\partial}{\partial y}\left(\dfrac{\partial F}{\partial v}\right) - 2y\dfrac{\partial}{\partial y}\left(\dfrac{\partial F}{\partial u}\right)$$

$$= -2\dfrac{\partial F}{\partial u} + 2x\left(-2y\dfrac{\partial}{\partial u} + 2x\dfrac{\partial}{\partial v}\right)\left(\dfrac{\partial F}{\partial v}\right) - 2y\left(-2y\dfrac{\partial}{\partial u} + 2x\dfrac{\partial}{\partial v}\right)\left(\dfrac{\partial F}{\partial u}\right)$$

$$= -2\dfrac{\partial F}{\partial u} - 8xy\dfrac{\partial^2 F}{\partial u \partial v} + 4x^2\dfrac{\partial^2 F}{\partial v^2} + 4y^2\dfrac{\partial^2 F}{\partial u^2}.$$

$$\frac{\partial^2 F}{\partial x \partial y} = \frac{\partial}{\partial x}\left(\frac{\partial F}{\partial y}\right) = \frac{\partial}{\partial x}\left(-2y\frac{\partial F}{\partial u} + 2x\frac{\partial F}{\partial v}\right)$$

$$= -2y\frac{\partial}{\partial x}\left(\frac{\partial F}{\partial u}\right) + 2\frac{\partial F}{\partial v} + 2x\frac{\partial}{\partial x}\left(\frac{\partial F}{\partial v}\right)$$

$$= 2\frac{\partial F}{\partial v} - 2y\left(2x\frac{\partial}{\partial u} + 2y\frac{\partial}{\partial v}\right)\left(\frac{\partial F}{\partial u}\right) + 2x\left(2x\frac{\partial}{\partial u} + 2y\frac{\partial}{\partial v}\right)\left(\frac{\partial F}{\partial v}\right)$$

$$= 2\frac{\partial F}{\partial v} - 4xy\frac{\partial^2 F}{\partial u^2} - 4y^2\frac{\partial^2 F}{\partial v \partial u} + 4x^2\frac{\partial^2 F}{\partial u \partial v} + 4xy\frac{\partial^2 F}{\partial v^2}$$

$$= 2\frac{\partial F}{\partial v} - 4xy\frac{\partial^2 F}{\partial u^2} + 4(x^2 - y^2)\frac{\partial^2 F}{\partial u \partial v} + 4xy\frac{\partial^2 F}{\partial v^2}.$$

R.H.S. $= xy\left(\frac{\partial^2 F}{\partial x^2} - \frac{\partial^2 F}{\partial y^2}\right) + (x^2 - y^2)\frac{\partial^2 F}{\partial x \partial y}$

$$= xy\left[4\frac{\partial F}{\partial u} + 16xy\frac{\partial^2 F}{\partial u \partial v} + 4(y^2 - x^2)\frac{\partial^2 F}{\partial v^2} + 4(x^2 - y^2)\frac{\partial^2 F}{\partial u^2}\right]$$

$$+ (x^2 - y^2)\left[2\frac{\partial F}{\partial v} - 4xy\frac{\partial^2 F}{\partial u^2} + 4(x^2 - y^2)\frac{\partial^2 F}{\partial u \partial v} + 4xy\frac{\partial^2 F}{\partial v^2}\right]$$

$$= 4xy\frac{\partial F}{\partial u} + 2(x^2 - y^2)\frac{\partial F}{\partial v} + 4\{(x^2 - y^2)^2 + 4x^2 y^2\}\frac{\partial^2 F}{\partial u \partial v}$$

$$= 2v\frac{\partial F}{\partial u} + 2u\frac{\partial F}{\partial v} + 4(u^2 + v^2)\frac{\partial^2 F}{\partial u \partial v}$$

$$= \text{L.H.S. (proved)}$$

Alternatively,

$$\frac{\partial F}{\partial x} = 2x\frac{\partial F}{\partial u} + 2y\frac{\partial F}{\partial v}$$

and $\quad \dfrac{\partial F}{\partial y} = -2y\dfrac{\partial F}{\partial u} + 2x\dfrac{\partial F}{\partial v}.$

Solving:

$$\left.\begin{aligned}\frac{\partial F}{\partial u} &= \frac{x}{2(x^2 + y^2)}\frac{\partial F}{\partial x} - \frac{y}{2(x^2 + y^2)}\frac{\partial F}{\partial y} \\ \frac{\partial F}{\partial v} &= \frac{y}{2(x^2 + y^2)}\frac{\partial F}{\partial x} + \frac{x}{2(x^2 + y^2)}\frac{\partial F}{\partial y}\end{aligned}\right\}.$$

Now calculate:

$$\frac{\partial^2 F}{\partial u \partial v} = \frac{\partial}{\partial u}\left(\frac{\partial F}{\partial v}\right) = \frac{\partial}{\partial u}\left(\frac{y}{2(x^2 + y^2)}\frac{\partial F}{\partial x} + \frac{x}{2(x^2 + y^2)}\frac{\partial F}{\partial y}\right).$$

CHAPTER 13: FUNCTIONS OF SEVERAL VARIABLES

Replace the operator

$$\frac{\partial}{\partial u} \text{ by } \frac{x}{2(x^2+y^2)}\frac{\partial}{\partial x} - \frac{y}{2(x^2+y^2)}\frac{\partial}{\partial y}.$$

Now check: L.H.S. = R.H.S.

Example 13.10.5. *Given that F is a differentiable function of x and y and that $x = e^u + e^{-v}$, $y = e^v + e^{-u}$.*
Prove that

$$\frac{\partial^2 F}{\partial u^2} - 2\frac{\partial^2 F}{\partial u \partial v} + \frac{\partial^2 F}{\partial v^2} = x^2 \frac{\partial^2 F}{\partial x^2} - 2xy \frac{\partial^2 F}{\partial x \partial y} + y^2 \frac{\partial^2 F}{\partial y^2} + x \frac{\partial F}{\partial x} + y \frac{\partial F}{\partial y}.$$

Solution: Using Chain Rules we get

$$\frac{\partial F}{\partial u} = \frac{\partial F}{\partial x}\frac{\partial x}{\partial u} + \frac{\partial F}{\partial y}\frac{\partial y}{\partial u} = \frac{\partial F}{\partial x}e^u + \frac{\partial F}{\partial y}(-e^{-u})$$

i.e., $\dfrac{\partial}{\partial u} \to e^u \dfrac{\partial}{\partial x} - e^{-u}\dfrac{\partial}{\partial y}$

$$\frac{\partial^2 F}{\partial u^2} = \frac{\partial}{\partial u}\left(\frac{\partial F}{\partial u}\right) = \frac{\partial}{\partial u}\left(e^u \frac{\partial F}{\partial x} - e^{-u}\frac{\partial F}{\partial y}\right)$$
$$= e^u \frac{\partial F}{\partial x} + e^{-u}\frac{\partial F}{\partial y} + e^u \frac{\partial}{\partial u}\left(\frac{\partial F}{\partial x}\right) - e^{-u}\frac{\partial}{\partial u}\left(\frac{\partial F}{\partial y}\right)$$
$$= e^u \frac{\partial F}{\partial x} + e^{-u}\frac{\partial F}{\partial y} + e^u \left[e^u \frac{\partial}{\partial x}\left(\frac{\partial F}{\partial x}\right) - e^{-u}\frac{\partial}{\partial y}\left(\frac{\partial F}{\partial x}\right)\right]$$
$$\qquad - e^{-u}\left[e^u \frac{\partial}{\partial x}\left(\frac{\partial F}{\partial y}\right) - e^{-u}\frac{\partial}{\partial y}\left(\frac{\partial F}{\partial y}\right)\right]$$
$$= e^u \frac{\partial F}{\partial x} + e^{-u}\frac{\partial F}{\partial y} + e^{2u}\frac{\partial^2 F}{\partial x^2} - 2\frac{\partial^2 F}{\partial x \partial y} + e^{-2u}\frac{\partial^2 F}{\partial y^2}$$

(assuming that the mixed derivatives are equal)

$$\frac{\partial F}{\partial v} = \frac{\partial F}{\partial x}\frac{\partial x}{\partial v} + \frac{\partial F}{\partial y}\frac{\partial y}{\partial v} = \frac{\partial F}{\partial x}(-e^{-v}) + \frac{\partial F}{\partial y}e^v$$

i.e., $\dfrac{\partial}{\partial v} \to -e^{-v}\dfrac{\partial}{\partial x} + e^v \dfrac{\partial}{\partial y}.$

811

$$\therefore \quad \frac{\partial^2 F}{\partial v^2} = \frac{\partial}{\partial v}\left(\frac{\partial F}{\partial v}\right) = \frac{\partial}{\partial v}\left(-e^{-v}\frac{\partial F}{\partial x} + e^v\frac{\partial F}{\partial y}\right)$$

$$= e^{-v}\frac{\partial F}{\partial x} + e^v\frac{\partial F}{\partial y} - e^{-v}\frac{\partial}{\partial v}\left(\frac{\partial F}{\partial x}\right) + e^v\frac{\partial}{\partial v}\left(\frac{\partial F}{\partial y}\right)$$

$$= e^{-v}\frac{\partial F}{\partial x} + e^v\frac{\partial F}{\partial y} - e^{-v}\left[-e^{-v}\frac{\partial}{\partial x}\left(\frac{\partial F}{\partial x}\right) + e^v\frac{\partial}{\partial y}\left(\frac{\partial F}{\partial x}\right)\right]$$

$$+ e^v\left[-e^{-v}\frac{\partial}{\partial x}\left(\frac{\partial F}{\partial y}\right) + e^v\frac{\partial}{\partial y}\left(\frac{\partial F}{\partial y}\right)\right]$$

$$= e^{-v}\frac{\partial F}{\partial x} + e^v\frac{\partial F}{\partial y} + e^{-2v}\frac{\partial^2 F}{\partial x^2} - 2\frac{\partial^2 F}{\partial x \partial y} + e^{2v}\frac{\partial^2 F}{\partial y^2}.$$

$$\frac{\partial^2 F}{\partial u \partial v} = \frac{\partial}{\partial u}\left(\frac{\partial F}{\partial v}\right) = \frac{\partial}{\partial u}\left(-e^{-v}\frac{\partial F}{\partial x} + e^v\frac{\partial F}{\partial y}\right)$$

$$= -e^{-v}\frac{\partial}{\partial u}\left(\frac{\partial F}{\partial x}\right) + e^v\frac{\partial}{\partial u}\left(\frac{\partial F}{\partial y}\right)$$

$$= -e^{-v}\left[e^u\frac{\partial}{\partial x}\left(\frac{\partial F}{\partial x}\right) - e^{-u}\frac{\partial}{\partial y}\left(\frac{\partial F}{\partial x}\right)\right]$$

$$+ e^v\left[e^u\frac{\partial}{\partial x}\left(\frac{\partial F}{\partial y}\right) - e^{-u}\frac{\partial}{\partial y}\left(\frac{\partial F}{\partial y}\right)\right]$$

$$= -e^{u-v}\frac{\partial^2 F}{\partial x^2} + (e^{-u-v} + e^{u+v})\frac{\partial^2 F}{\partial x \partial y} - e^{v-u}\frac{\partial^2 F}{\partial y^2}.$$

$$\therefore \quad \text{L.H.S.} = \frac{\partial^2 F}{\partial u^2} - 2\frac{\partial^2 F}{\partial u \partial v} + \frac{\partial^2 F}{\partial v^2}$$

$$= \left[e^u\frac{\partial F}{\partial x} + e^{-u}\frac{\partial F}{\partial y} + e^{2u}\frac{\partial^2 F}{\partial x^2} - 2\frac{\partial^2 F}{\partial x \partial y} + e^{-2u}\frac{\partial^2 F}{\partial y^2}\right]$$

$$- 2\left[-e^{u-v}\frac{\partial^2 F}{\partial x^2} + (e^{-u-v} + e^{u+v})\frac{\partial^2 F}{\partial x \partial y} - e^{v-u}\frac{\partial^2 F}{\partial y^2}\right]$$

$$+ \left[e^{-v}\frac{\partial F}{\partial x} + e^v\frac{\partial F}{\partial y} + e^{-2v}\frac{\partial^2 F}{\partial x^2} - 2\frac{\partial^2 F}{\partial x \partial y} + e^{2v}\frac{\partial^2 F}{\partial y^2}\right]$$

$$= (e^u + e^{-v})\frac{\partial F}{\partial x} + (e^v + e^{-u})\frac{\partial F}{\partial y} + \frac{\partial^2 F}{\partial x^2}(e^{2u} + 2e^{u-v} + e^{-2v})$$

$$- 2\frac{\partial^2 F}{\partial x \partial y}(e^{u+v} + e^{-u-v} + 2) + \frac{\partial^2 F}{\partial y^2}(e^{-2u} + 2e^{v-u} + e^{2v})$$

$$= x\frac{\partial F}{\partial x} + y\frac{\partial F}{\partial y} + x^2\frac{\partial^2 F}{\partial x^2} - 2xy\frac{\partial^2 F}{\partial x \partial y} + y^2\frac{\partial^2 F}{\partial y^2}$$

(replacing the expressions in u, v by the equivalent expressions in x, y)

= R.H.S. (proved)

CHAPTER 13: FUNCTIONS OF SEVERAL VARIABLES

Example 13.10.6. Let $y = F(x, t)$, where F is a differentiable function of two independent variables x and t which are related to two variables u, v by the relations:

$$u = x + ct, \quad v = x - ct.$$

Prove that the partial differential equation (**Wave equation in Mathematical Physics**)

$$\frac{\partial^2 y}{\partial x^2} - \frac{1}{c^2}\frac{\partial^2 y}{\partial t^2} = 0,$$

can be transformed into $\dfrac{\partial^2 y}{\partial u \partial v} = 0.$

Hence obtain the general solution

$$y = g(x - ct) + h(x + ct) \quad (g, h \text{ are arbitrary functions}).$$

Solution: Using Chain Rules, we write

$$\frac{\partial y}{\partial v} = \frac{\partial y}{\partial x}\frac{\partial x}{\partial v} + \frac{\partial y}{\partial t}\frac{\partial t}{\partial v} = \frac{\partial y}{\partial x}\cdot\frac{1}{2} + \frac{\partial y}{\partial t}\cdot\left(-\frac{1}{2c}\right).$$

[From the given relations we easily see that $x = \dfrac{1}{2}(u + v)$ and $t = \dfrac{1}{2c}(u - v)$].

Also

$$\frac{\partial y}{\partial u} = \frac{\partial y}{\partial x}\frac{\partial x}{\partial u} + \frac{\partial y}{\partial t}\frac{\partial t}{\partial u} = \frac{\partial y}{\partial x}\frac{1}{2} + \frac{\partial y}{\partial t}\frac{1}{2c}$$

i.e., $\dfrac{\partial}{\partial u} \rightarrow \dfrac{1}{2}\dfrac{\partial}{\partial x} + \dfrac{1}{2c}\dfrac{\partial}{\partial t}.$

$$\therefore \frac{\partial^2 y}{\partial u \partial v} = \frac{\partial}{\partial u}\left(\frac{\partial y}{\partial v}\right) = \frac{\partial}{\partial u}\left[\frac{1}{2}\frac{\partial y}{\partial x} - \frac{1}{2c}\frac{\partial y}{\partial t}\right]$$

$$= \frac{1}{2}\frac{\partial}{\partial u}\left(\frac{\partial y}{\partial x}\right) - \frac{1}{2c}\frac{\partial}{\partial u}\left(\frac{\partial y}{\partial t}\right)$$

$$= \frac{1}{2}\left[\frac{1}{2}\frac{\partial}{\partial x}\left(\frac{\partial y}{\partial x}\right) + \frac{1}{2c}\frac{\partial}{\partial t}\left(\frac{\partial y}{\partial x}\right)\right]$$

$$\quad - \frac{1}{2c}\left[\frac{1}{2}\frac{\partial}{\partial x}\left(\frac{\partial y}{\partial t}\right) + \frac{1}{2c}\frac{\partial}{\partial t}\left(\frac{\partial y}{\partial t}\right)\right]$$

$$= \frac{1}{4}\left[\frac{\partial^2 y}{\partial x^2} - \frac{1}{c^2}\frac{\partial^2 y}{\partial t^2}\right].$$

Thus the equation

$$\frac{\partial^2 y}{\partial x^2} - \frac{1}{c^2}\frac{\partial^2 y}{\partial t^2} = 0, \text{ becomes } 4\cdot\frac{\partial^2 y}{\partial u \partial v} = 0.$$

We now solve
$$\frac{\partial^2 y}{\partial u \partial v} = 0,$$

i.e., $\dfrac{\partial}{\partial u}\left(\dfrac{\partial y}{\partial v}\right) = 0$

$\Rightarrow \dfrac{\partial y}{\partial v}$ is independent of u

$\Rightarrow \dfrac{\partial y}{\partial v} = f(v)$, (say) [$f$ is an arbitrary function]

$\Rightarrow y = \displaystyle\int f(v)dx + $ term not containing v

$\qquad = g(v) + h(u)$

where $g(v)$ is such that $g'(v) = f(v)$ and $h(u)$ is an arbitrary function of u (const. w.r.t. v).

∴ The general solution is

$$y = g(x - ct) + h(x + ct)$$

where g and h are arbitrary functions, $g'(v) = f(v)$.

Example 13.10.7. *If the set of rectangular cartesian axes Ox, Oy be turned through a constant angle α about O, the new axes become Ox', Oy'. A point (x, y) with reference to the axes Ox, Oy becomes (x', y') with reference to the axes Ox', Oy'. Then we know*

$$\left.\begin{array}{l} x = x' \cos\alpha - y' \sin\alpha \\ y = x' \sin\alpha + y' \cos\alpha. \end{array}\right\} \qquad (13.10.8)$$

Solving for x', y' we get

$$\left.\begin{array}{l} x' = x \cos\alpha + y \sin\alpha \\ y' = -x \sin\alpha + y \cos\alpha. \end{array}\right\} \qquad (13.10.9)$$

Prove that
$$\frac{\partial^2 V}{\partial x^2} + \frac{\partial^2 V}{\partial y^2} = \frac{\partial^2 V}{\partial x'^2} + \frac{\partial^2 V}{\partial y'^2}$$

i.e., the expression on the left remains invariant for a rotational change of rectangular axes.

[Assume that V is a differentiable function of x, y where x and y are obviously differentiable functions of x', y' and hence Chain Rules can be applied.]

Solution: Using Chain Rules,
$$\frac{\partial V}{\partial x'} = \frac{\partial V}{\partial x}\frac{\partial x}{\partial x'} + \frac{\partial V}{\partial y}\frac{\partial y}{\partial x'} = \frac{\partial V}{\partial x}\cos\alpha + \frac{\partial V}{\partial y}\sin\alpha$$

i.e., $\quad \dfrac{\partial}{\partial x'} \equiv \dfrac{\partial}{\partial x}\cos\alpha + \dfrac{\partial}{\partial y}\sin\alpha.$

Also
$$\frac{\partial^2 V}{\partial x'^2} = \frac{\partial}{\partial x'}\left(\frac{\partial V}{\partial x'}\right) = \frac{\partial}{\partial x'}\left[\frac{\partial V}{\partial x}\cos\alpha + \frac{\partial V}{\partial y}\sin\alpha\right]$$
$$= \cos\alpha\frac{\partial}{\partial x'}\left(\frac{\partial V}{\partial x}\right) + \sin\alpha\frac{\partial}{\partial x'}\left(\frac{\partial V}{\partial y}\right)$$
$$= \cos\alpha\left[\frac{\partial}{\partial x}\cos\alpha + \frac{\partial}{\partial y}\sin\alpha\right]\left(\frac{\partial V}{\partial x}\right)$$
$$+ \sin\alpha\left[\frac{\partial}{\partial x}\cos\alpha + \frac{\partial}{\partial y}\sin\alpha\right]\left(\frac{\partial V}{\partial y}\right)$$
$$= \cos^2\alpha\frac{\partial^2 V}{\partial x^2} + \cos\alpha\sin\alpha\frac{\partial^2 V}{\partial y\partial x} + \cos\alpha\sin\alpha\frac{\partial^2 V}{\partial x\partial y} + \sin^2\alpha\frac{\partial^2 V}{\partial y^2}$$
$$= \cos^2\alpha\frac{\partial^2 V}{\partial x^2} + 2\cos\alpha\sin\alpha\frac{\partial^2 V}{\partial x\partial y} + \sin^2\alpha\frac{\partial^2 V}{\partial y^2}. \qquad (13.10.10)$$

[assuming $V_{xy} = V_{yx}$ which will be true if we assume that V_{xy} and V_{yx} are both continuous.]

Again
$$\frac{\partial V}{\partial y'} = \frac{\partial V}{\partial x}\frac{\partial x}{\partial y'} + \frac{\partial V}{\partial y}\frac{\partial y}{\partial y'} = \frac{\partial V}{\partial x}(-\sin\alpha) + \frac{\partial V}{\partial y}\cos\alpha$$

i.e., $\quad \dfrac{\partial}{\partial y'} \to (-\sin\alpha)\dfrac{\partial}{\partial x} + (\cos\alpha)\dfrac{\partial}{\partial y}.$

$$\therefore \frac{\partial^2 V}{\partial y'^2} = \frac{\partial}{\partial y'}\left(\frac{\partial V}{\partial y'}\right) = \frac{\partial}{\partial y'}\left[-\sin\alpha\frac{\partial V}{\partial x} + \cos\alpha\frac{\partial V}{\partial y}\right]$$
$$= -\sin\alpha\frac{\partial}{\partial y'}\left(\frac{\partial V}{\partial x}\right) + \cos\alpha\frac{\partial}{\partial y'}\left(\frac{\partial V}{\partial y}\right)$$
$$= -\sin\alpha\left[(-\sin\alpha)\frac{\partial}{\partial x} + (\cos\alpha)\frac{\partial}{\partial y}\right]\left(\frac{\partial V}{\partial x}\right)$$
$$+ \cos\alpha\left[(-\sin\alpha)\frac{\partial}{\partial x} + (\cos\alpha)\frac{\partial}{\partial y}\right]\left(\frac{\partial V}{\partial y}\right)$$
$$= \sin^2\alpha\frac{\partial^2 V}{\partial x^2} - 2\sin\alpha\cos\alpha\frac{\partial^2 V}{\partial x\partial y} + \cos^2\alpha\frac{\partial^2 V}{\partial y^2}. \qquad (13.10.11)$$

∴ From (13.10.10) and (13.10.11)

$$\frac{\partial^2 V}{\partial x'^2} + \frac{\partial^2 V}{\partial y'^2} = \frac{\partial^2 V}{\partial x^2} + \frac{\partial^2 V}{\partial y^2}.$$

Note 13.10.1. We could prove $\frac{\partial^2 V}{\partial x^2} + \frac{\partial^2 V}{\partial y^2} = \frac{\partial^2 V}{\partial x'^2} + \frac{\partial^2 V}{\partial y'^2}.$

In that case we write

$$\frac{\partial V}{\partial x} = \frac{\partial V}{\partial x'}\frac{\partial x'}{\partial x} + \frac{\partial V}{\partial y'}\frac{\partial y'}{\partial x}$$

and use $x' = x\cos\alpha + y\sin\alpha,\ \ y' = -x\sin\alpha + y\cos\alpha.$

13.11 Jacobians : Their Important Properties

Definition 13.11.1. *If $F_1, F_2, F_3, \cdots, F_n$ be n functions of n variables $x_1, x_2, \cdots, x_n$ possessing first order partial derivatives at every point of the common domain of definition of the functions, then the determinant*

$$\begin{vmatrix} \frac{\partial F_1}{\partial x_1} & \frac{\partial F_1}{\partial x_2} & \cdots & \frac{\partial F_1}{\partial x_n} \\ \frac{\partial F_2}{\partial x_1} & \frac{\partial F_2}{\partial x_2} & \cdots & \frac{\partial F_2}{\partial x_n} \\ \cdots & \cdots & \cdots & \cdots \\ \frac{\partial F_n}{\partial x_1} & \frac{\partial F_n}{\partial x_2} & \cdots & \frac{\partial F_n}{\partial x_n} \end{vmatrix}$$

is known as the Jacobian of $F_1, F_2, \cdots, F_n$ with respect to $x_1, x_2, \cdots, x_n$ and is denoted by

$$\frac{\partial(F_1, F_2, \cdots, F_n)}{\partial(x_1, x_2, \cdots, x_n)} \quad or, \quad J\left(\frac{F_1, F_2, \cdots, F_n}{x_1, x_2, \cdots, x_n}\right).$$

Example 13.11.1. *The rectangular co-ordinates (x, y, z) of a point are given in terms of its spherical co-ordinates (r, θ, ϕ) by the equations*

$$x = r\sin\theta\cos\phi, \quad y = r\sin\theta\sin\phi, \quad z = r\cos\theta$$

prove that the Jacobian of the transformation is $r^2\sin\theta$.

Solution:

$$\frac{\partial(x,y,z)}{\partial(r,\theta,\phi)} = \begin{vmatrix} \frac{\partial x}{\partial r} & \frac{\partial x}{\partial \theta} & \frac{\partial x}{\partial \phi} \\ \frac{\partial y}{\partial r} & \frac{\partial y}{\partial \theta} & \frac{\partial y}{\partial \phi} \\ \frac{\partial z}{\partial r} & \frac{\partial z}{\partial \theta} & \frac{\partial z}{\partial \phi} \end{vmatrix}$$

$$= \begin{vmatrix} \sin\theta\cos\phi & r\cos\theta\cos\phi & -r\sin\theta\sin\phi \\ \sin\theta\sin\phi & r\cos\theta\sin\phi & r\sin\theta\cos\phi \\ \cos\theta & -r\sin\theta & 0 \end{vmatrix}$$

$$= r^2 \sin\theta \begin{vmatrix} \sin\theta\cos\phi & \cos\theta\cos\phi & -\sin\phi \\ \sin\theta\sin\phi & \cos\theta\sin\phi & \cos\phi \\ \cos\theta & -\sin\theta & 0 \end{vmatrix}$$

$$= \frac{r^2 \sin\theta}{\sin\phi} \begin{vmatrix} \sin\theta\cos\phi & \cos\theta\cos\phi & -\sin\phi \\ \sin\theta\sin^2\phi & \cos\theta\sin^2\phi & \sin\phi\cos\phi \\ \cos\theta & -\sin\theta & 0 \end{vmatrix}$$

(multiply 2nd row by $\sin\phi$ and take $\frac{1}{\sin\phi}$ outside the determinant)

$$= \frac{r^2 \sin\theta}{\sin\phi} \begin{vmatrix} \sin\theta\cos\phi & \cos\theta\cos\phi & -\sin\phi \\ \sin\theta & \cos\theta & 0 \\ \cos\theta & -\sin\theta & 0 \end{vmatrix}$$

(multiply 1st row by $\cos\phi$ and add with 2nd row)

$$= \frac{r^2 \sin\theta}{\sin\phi} \{-\sin\phi\} \begin{vmatrix} \sin\theta & \cos\theta \\ \cos\theta & -\sin\theta \end{vmatrix}$$

(expanding in terms of 3rd column)

$$= \frac{r^2 \sin\theta}{\sin\phi} (-\sin\phi)(-\sin^2\theta - \cos^2\theta) = \mathbf{r^2 \sin\theta}.$$

Example 13.11.2. (a) *If* $F_1 = f_1(x_1), F_2 = f_2(x_1,x_2), F_3 = f_3(x_1,x_2,x_3), \cdots, F_n = f_n(x_1,x_2,\cdots,x_n)$ *then prove that*

$$\frac{\partial(F_1, F_2, \cdots, F_n)}{\partial(x_1, x_2, \cdots, x_n)} = \frac{\partial F_1}{\partial x_1} \cdot \frac{\partial F_2}{\partial x_2} \cdot \frac{\partial F_3}{\partial x_3} \cdots \frac{\partial F_n}{\partial x_n}.$$

Solution:

$$\frac{\partial(F_1, F_2, \cdots, F_n)}{\partial(x_1, x_2, \cdots, x_n)} = \begin{vmatrix} \frac{\partial F_1}{\partial x_1} & 0 & 0 & \cdots & 0 \\ \frac{\partial F_2}{\partial x_1} & \frac{\partial F_2}{\partial x_2} & 0 & \cdots & 0 \\ \cdots & \cdots & \cdots & \cdots & \cdots \\ \frac{\partial F_n}{\partial x_1} & \frac{\partial F_n}{\partial x_2} & \frac{\partial F_n}{\partial x_3} & \cdots & \frac{\partial F_n}{\partial x_n} \end{vmatrix}$$

$$= \frac{\partial F_1}{\partial x_1} \cdot \frac{\partial F_2}{\partial x_2} \cdots \frac{\partial F_n}{\partial x_n}.$$

(b) *Find the Jacobian*

$$\frac{\partial(y_1, y_2, \cdots, y_n)}{\partial(x_1, x_2, \cdots, x_n)}$$

if $y_1 = 1 - x_1$; $y_2 = x_1(1 - x_2)$; $y_3 = x_1 x_2(1 - x_3)$; $\cdots$; $y_n = x_1 x_2 \cdots x_{n-1}(1 - x_n)$.

Solution:

$$\frac{\partial(y_1, y_2, \cdots, y_n)}{\partial(x_1, x_2, \cdots, x_n)} = \begin{vmatrix} \frac{\partial y_1}{\partial x_1} & 0 & 0 & 0 & \cdots & 0 \\ \frac{\partial y_2}{\partial x_1} & \frac{\partial y_2}{\partial x_2} & 0 & 0 & \cdots & 0 \\ \cdots & \cdots & \cdots & \cdots & \cdots & \cdots \\ \frac{\partial y_n}{\partial x_1} & \frac{\partial y_n}{\partial x_2} & \frac{\partial y_n}{\partial x_3} & \cdots & \cdots & \frac{\partial y_n}{\partial x_n} \end{vmatrix}$$

$$= \frac{\partial y_1}{\partial x_1} \cdot \frac{\partial y_2}{\partial x_2} \cdot \frac{\partial y_3}{\partial x_3} \cdots \frac{\partial y_n}{\partial x_n}$$

$$= (-1)(-x_1)(-x_1 x_2) \cdots (-x_1 x_2 \cdots x_{n-1})$$

$$= (-1)^n x_1^{n-1} x_2^{n-2} x_3^{n-3} \cdots x_{n-2}^2 x_{n-1}.$$

Example 13.11.3. (a) *Prove that the Jacobian*

$$\frac{\partial(y_1, y_2, \cdots, y_n)}{\partial(x_1, x_2, \cdots, x_n)} = x_1^{n-1} x_2^{n-2} x_3^{n-3} \cdots x_{n-1}$$

if $y_1 = x_1(1 - x_2), y_2 = x_1 x_2(1 - x_3), y_3 = x_1 x_2 x_3(1 - x_4), \cdots, y_{n-1} = x_1 x_2 \cdots x_{n-1}(1 - x_n), y_n = x_1 x_2 \cdots x_n$.

[C.H. 1989 old]

CHAPTER 13: FUNCTIONS OF SEVERAL VARIABLES

Solution: Adding all the given relations we get

$$y_1 + y_2 + y_3 + \cdots + y_{n-1} + y_n = x_1.$$

The Jacobian

$$\frac{\partial(y_1, y_2, \cdots, y_n)}{\partial(x_1, x_2, \cdots, x_n)}$$

$$= \begin{vmatrix} \frac{\partial y_1}{\partial x_1} & \frac{\partial y_1}{\partial x_2} & \frac{\partial y_1}{\partial x_3} & \cdots & \frac{\partial y_1}{\partial x_n} \\ \frac{\partial y_2}{\partial x_1} & \frac{\partial y_2}{\partial x_2} & \frac{\partial y_2}{\partial x_3} & \cdots & \frac{\partial y_2}{\partial x_n} \\ \cdots & \cdots & \cdots & \cdots & \cdots \\ \frac{\partial y_n}{\partial x_1} & \frac{\partial y_n}{\partial x_2} & \frac{\partial y_n}{\partial x_3} & \cdots & \frac{\partial y_n}{\partial x_n} \end{vmatrix}$$

$$= \begin{vmatrix} 1-x_2 & -x_1 & 0 & \cdots & 0 \\ x_2(1-x_3) & x_1(1-x_3) & -x_1x_2 & \cdots & 0 \\ x_2x_3(1-x_4) & x_1x_3(1-x_4) & x_1x_3(1-x_4) & \cdots & 0 \\ \cdots & \cdots & \cdots & \cdots & \cdots \\ \frac{x_2x_3\cdots}{x_{n-1}(1-x_n)} & \frac{x_1x_3x_4\cdots}{x_{n-1}(1-x_n)} & \frac{x_1x_2x_4\cdots}{x_{n-1}(1-x_n)} & \cdots & \frac{-x_1x_2\cdots}{x_{n-1}} \\ \frac{x_2x_3\cdots}{x_{n-1}x_n} & \frac{x_1x_3x_4\cdots}{x_{n-1}x_n} & \frac{x_1x_2x_4x_5\cdots}{x_{n-1}x_n} & \cdots & \frac{x_1x_2\cdots}{x_{n-1}} \end{vmatrix}$$

$$= \begin{vmatrix} 1-x_2 & -x_1 & 0 & \cdots & 0 \\ x_2(1-x_3) & x_1(1-x_3) & -x_1x_2 & \cdots & 0 \\ x_2x_3(1-x_4) & x_1x_3(1-x_4) & x_1x_4(1-x_4) & \cdots & 0 \\ \cdots & \cdots & \cdots & \cdots & \cdots \\ \frac{x_2x_3\cdots}{x_{n-1}(1-x_n)} & \frac{x_1x_3\cdots}{x_{n-1}(1-x_n)} & \frac{x_1x_2\cdots}{x_{n-1}(1-x_n)} & \cdots & \frac{-x_1x_2\cdots}{x_{n-1}} \\ 1 & 0 & 0 & \cdots & 0 \end{vmatrix}$$

(Last row = sum of all the other rows with nth row)

Expanding in terms of the elements of the last now

$$= (-1)^{n-1}\{(-x_1)(-x_1x_2)(-x_1x_2x_3)\cdots(-x_1x_2\cdots x_{n-1})\}$$
$$= (-1)^{n-1}(-1)^{n-1}x_1^{n-1}x_2^{n-2}\cdots x_{n-1} = x_1^{n-1}x_2^{n-2}x_3^{n-3}\cdots x_{n-1}.$$

(b) If $y_1 = \cos x_1$, $y_2 = \sin x_1 \cos x_2$, $y_3 = \sin x_1 \sin x_2 \cos x_3$, $\cdots$, $y_n = \sin x_1 \sin x_2 \cdots \sin x_{n-1} \cos x_n$, then prove that

$$\frac{\partial(y_1, y_2, \cdots, y_n)}{\partial(x_1, x_2, \cdots, x_n)} = (-1)^n \sin^n x_1 \sin^{n-1} x_2 \cdots \sin x_n.$$

In particular, Deduce $\dfrac{\partial(y_1, y_2, y_3)}{\partial(x_1, x_2, x_3)} = -\sin^3 x_1 \sin^2 x_2 \sin x_3$.

Function of Functions : Chain Rules for Jacobians

Theorem 13.11.1. *If $u_1, u_2, \cdots, u_n$ are functions of $y_1, y_2, \cdots, y_n$ and $y_1, y_2, \cdots, y_n$ are functions of $x_1, x_2, \cdots, x_n$, then*

$$\frac{\partial(u_1, u_2, \cdots, u_n)}{\partial(x_1, x_2, \cdots, x_n)} = \frac{\partial(u_1, u_2, \cdots, u_n)}{\partial(y_1, y_2, \cdots, y_n)} \cdot \frac{\partial(y_1, y_2, \cdots, y_n)}{\partial(x_1, x_2, \cdots, x_n)}.$$

Proof: We have, by Chain rules,

$$\left.\begin{aligned}
\frac{\partial u_1}{\partial x_1} &= \frac{\partial u_1}{\partial y_1}\frac{\partial y_1}{\partial x_1} + \frac{\partial u_1}{\partial y_2}\frac{\partial y_2}{\partial x_1} + \cdots + \frac{\partial u_1}{\partial y_n}\frac{\partial y_n}{\partial x_1} \\
&= \sum_{r=1}^{n} \frac{\partial u_1}{\partial y_r}\frac{\partial y_r}{\partial x_1} \\
\frac{\partial u_1}{\partial x_2} &= \sum_{r=1}^{n} \frac{\partial u_1}{\partial y_r}\frac{\partial y_r}{\partial x_2} \\
\cdots & \cdots \cdots \cdots \cdots \cdots \cdots \cdots \cdots \cdots \\
\frac{\partial u_1}{\partial x_n} &= \sum_{r=1}^{n} \frac{\partial u_1}{\partial y_r}\frac{\partial y_r}{\partial x_n}
\end{aligned}\right\} \quad (13.11.1)$$

Now

$$\text{R.H.S.} = \frac{\partial(u_1, u_2, \cdots, u_n)}{\partial(y_1, y_2, \cdots, y_n)} \cdot \frac{\partial(y_1, y_2, \cdots, y_n)}{\partial(x_1, x_2, \cdots, x_n)}$$

$$= \begin{vmatrix} \dfrac{\partial u_1}{\partial y_1} & \dfrac{\partial u_1}{\partial y_2} & \cdots & \dfrac{\partial u_1}{\partial y_n} \\ \dfrac{\partial u_2}{\partial y_1} & \dfrac{\partial u_2}{\partial y_2} & \cdots & \dfrac{\partial u_2}{\partial y_n} \\ \cdots & \cdots & \cdots & \cdots \\ \dfrac{\partial u_n}{\partial y_1} & \dfrac{\partial u_n}{\partial y_2} & \cdots & \dfrac{\partial u_n}{\partial y_n} \end{vmatrix} \begin{vmatrix} \dfrac{\partial y_1}{\partial x_1} & \dfrac{\partial y_1}{\partial x_2} & \cdots & \dfrac{\partial y_1}{\partial x_n} \\ \dfrac{\partial y_2}{\partial x_1} & \dfrac{\partial y_2}{\partial x_2} & \cdots & \dfrac{\partial y_2}{\partial x_n} \\ \cdots & \cdots & \cdots & \cdots \\ \dfrac{\partial y_n}{\partial x_1} & \dfrac{\partial y_n}{\partial x_2} & \cdots & \dfrac{\partial y_n}{\partial x_n} \end{vmatrix}$$

$$= \begin{vmatrix} \sum \dfrac{\partial u_1}{\partial y_r}\dfrac{\partial y_r}{\partial x_1} & \sum \dfrac{\partial u_1}{\partial y_r}\cdot\dfrac{\partial y_r}{\partial x_2} & \cdots & \sum \dfrac{\partial u_1}{\partial y_r}\cdot\dfrac{\partial y_r}{\partial x_n} \\ \sum \dfrac{\partial u_2}{\partial y_r}\dfrac{\partial y_r}{\partial x_1} & \sum \dfrac{\partial u_2}{\partial y_r}\dfrac{\partial y_r}{\partial x_2} & \cdots & \sum \dfrac{\partial u_2}{\partial y_r}\cdot\dfrac{\partial y_r}{\partial x_n} \\ \cdots & \cdots & \cdots & \cdots \\ \sum \dfrac{\partial u_n}{\partial y_r}\dfrac{\partial y_r}{\partial x_1} & \sum \dfrac{\partial u_n}{\partial y_r}\dfrac{\partial y_r}{\partial x_2} & \cdots & \sum \dfrac{\partial u_n}{\partial y_r}\dfrac{\partial y_r}{\partial x_n} \end{vmatrix}$$

CHAPTER 13: FUNCTIONS OF SEVERAL VARIABLES

Using the product rule of determinants (**row by column**)

$$= \begin{vmatrix} \dfrac{\partial u_1}{\partial x_1} & \dfrac{\partial u_1}{\partial x_2} & \cdots & \dfrac{\partial u_1}{\partial x_n} \\ \dfrac{\partial u_2}{\partial x_1} & \dfrac{\partial u_2}{\partial x_2} & \cdots & \dfrac{\partial u_2}{\partial x_n} \\ \cdots & \cdots & \cdots & \cdots \\ \dfrac{\partial u_n}{\partial x_1} & \dfrac{\partial u_n}{\partial x_2} & \cdots & \dfrac{\partial u_n}{\partial x_n} \end{vmatrix} = \dfrac{\partial(u_1, u_2, \cdots, u_n)}{\partial(x_1, x_2, \cdots, x_n)} = \text{L.H.S. (Proved)}$$

Corollary 13.11.1. *Putting $u_1 = x_1, u_2 = x_2$, etc. we get*

$$\dfrac{\partial(x_1, x_2, \cdots, x_n)}{\partial(y_1, y_2, \cdots, y_n)} \dfrac{\partial(y_1, y_2, \cdots, y_n)}{\partial(x_1, x_2, \cdots, x_n)} = \begin{vmatrix} \dfrac{\partial x_1}{\partial x_1} & \dfrac{\partial x_1}{\partial x_2} & \cdots & \dfrac{\partial x_1}{\partial x_n} \\ \dfrac{\partial x_2}{\partial x_1} & \dfrac{\partial x_2}{\partial x_2} & \cdots & \dfrac{\partial x_2}{\partial x_n} \\ \cdots & \cdots & \cdots & \cdots \\ \dfrac{\partial x_n}{\partial x_1} & \dfrac{\partial x_n}{\partial x_2} & \cdots & \dfrac{\partial x_n}{\partial x_n} \end{vmatrix}$$

$$= \begin{vmatrix} 1 & 0 & \cdots & 0 \\ 0 & 1 & \cdots & 0 \\ \cdots & \cdots & \cdots & \cdots \\ 0 & 0 & \cdots & 1 \end{vmatrix} = 1.$$

Here we have assumed that the Equations which define $y_1, y_2, \cdots, y_n$ as functions of $x_1, x_2, \cdots, x_n$ determine (inverse functions) $x_1, x_2, \cdots, x_n$ as functions of $y_1, y_2, \cdots, y_n$.

Jacobian of Implicit Functions

Theorem 13.11.2. *If $u_1, u_2, \cdots, u_n$ are determined as functions of $x_1, x_2, \cdots, x_n$ by the n relations:*

$$\left. \begin{array}{c} F_1(u_1, u_2, \cdots, u_n, x_1, x_2, \cdots, x_n) = 0 \\ F_2(u_1, u_2, \cdots, u_n, x_1, x_2, \cdots, x_n) = 0 \\ \cdots\cdots\cdots\cdots\cdots\cdots\cdots\cdots\cdots \\ F_n(u_1, u_2, \cdots, u_n, x_1, x_2, \cdots, x_n) = 0 \end{array} \right\} \quad (13.11.2)$$

then

$$\dfrac{\partial(F_1, F_2, \cdots, F_n)}{\partial(u_1, u_2, \cdots, u_n)} \cdot \dfrac{\partial(u_1, u_2, \cdots, u_n)}{\partial(x_1, x_2, \cdots, x_n)} = (-1)^n \dfrac{\partial(F_1, F_2, \cdots, F_n)}{\partial(x_1, x_2, \cdots, x_n)}.$$

Proof: Differentiating relations in (13.11.2) with respect to $x_1, x_2, \cdots$ **we get** the follows n^2 equations:

$$\left.\begin{array}{l}\dfrac{\partial F_1}{\partial x_1}+\dfrac{\partial F_1}{\partial u_1}\cdot\dfrac{\partial u_1}{\partial x_1}+\dfrac{\partial F_1}{\partial u_2}\cdot\dfrac{\partial u_2}{\partial x_1}+\cdots+\dfrac{\partial F_1}{\partial u_n}\cdot\dfrac{\partial u_n}{\partial x_1}=0\\[6pt]\dfrac{\partial F_1}{\partial x_2}+\dfrac{\partial F_1}{\partial u_1}\cdot\dfrac{\partial u_1}{\partial x_2}+\dfrac{\partial F_1}{\partial u_2}\cdot\dfrac{\partial u_2}{\partial x_2}+\cdots+\dfrac{\partial F_1}{\partial u_n}\cdot\dfrac{\partial u_n}{\partial x_2}=0\quad\text{etc. etc.}\\[6pt]\cdots\cdots\cdots\cdots\cdots\cdots\cdots\cdots\cdots\cdots\cdots\cdots\cdots\cdots\cdots\cdots\\[6pt]\dfrac{\partial F_n}{\partial x_1}+\dfrac{\partial F_n}{\partial u_1}\cdot\dfrac{\partial u_1}{\partial x_1}+\dfrac{\partial F_n}{\partial u_2}\cdot\dfrac{\partial u_2}{\partial x_1}+\cdots+\dfrac{\partial F_n}{\partial u_n}\cdot\dfrac{\partial u_n}{\partial x_1}=0\quad\text{etc. etc.}\\[6pt]\cdots\cdots\cdots\cdots\cdots\cdots\cdots\cdots\cdots\cdots\cdots\cdots\cdots\cdots\cdots\cdots\end{array}\right\} \quad (13.11.2a)$$

Now

$$\frac{\partial(F_1,F_2,\cdots,F_n)}{\partial(u_1,u_2,\cdots,u_n)}\cdot\frac{\partial(u_1,u_2,\cdots,u_n)}{\partial(x_1,x_2,\cdots,x_n)}$$

$$=\begin{vmatrix}\dfrac{\partial F_1}{\partial u_1}&\dfrac{\partial F_1}{\partial u_2}&\cdots&\dfrac{\partial F_1}{\partial u_n}\\[6pt]\dfrac{\partial F_2}{\partial u_1}&\dfrac{\partial F_2}{\partial u_2}&\cdots&\dfrac{\partial F_2}{\partial u_n}\\[6pt]\cdots&\cdots&\cdots&\cdots\\[6pt]\dfrac{\partial F_n}{\partial u_1}&\dfrac{\partial F_n}{\partial u_2}&\cdots&\dfrac{\partial F_n}{\partial u_n}\end{vmatrix}\begin{vmatrix}\dfrac{\partial u_1}{\partial x_1}&\dfrac{\partial u_1}{\partial x_2}&\cdots&\dfrac{\partial u_1}{\partial x_n}\\[6pt]\dfrac{\partial u_2}{\partial x_1}&\dfrac{\partial u_2}{\partial x_2}&\cdots&\dfrac{\partial u_2}{\partial x_n}\\[6pt]\cdots&\cdots&\cdots&\cdots\\[6pt]\dfrac{\partial u_n}{\partial x_1}&\dfrac{\partial u_n}{\partial x_2}&\cdots&\dfrac{\partial u_n}{\partial x_n}\end{vmatrix}$$

$$=\begin{vmatrix}\sum\dfrac{\partial F_1}{\partial u_r}\dfrac{\partial u_r}{\partial x_1}&\sum\dfrac{\partial F_1}{\partial u_r}\dfrac{\partial u_r}{\partial x_2}&\cdots&\sum\dfrac{\partial F_1}{\partial u_r}\dfrac{\partial u_r}{\partial x_n}\\[6pt]\sum\dfrac{\partial F_2}{\partial u_r}\dfrac{\partial u_r}{\partial x_1}&\sum\dfrac{\partial F_2}{\partial u_r}\dfrac{\partial u_r}{\partial x_2}&\cdots&\sum\dfrac{\partial F_2}{\partial u_r}\dfrac{\partial u_r}{\partial x_n}\\[6pt]\cdots&\cdots&\cdots&\cdots\\[6pt]\sum\dfrac{\partial F_n}{\partial u_r}\dfrac{\partial u_r}{\partial x_1}&\sum\dfrac{\partial F_n}{\partial u_r}\dfrac{\partial u_r}{\partial x_2}&\cdots&\sum\dfrac{\partial F_n}{\partial u_r}\dfrac{\partial u_r}{\partial x_n}\end{vmatrix}$$

where the summation is extended for $r=1$ to n

(row by column multiplication)

CHAPTER 13: FUNCTIONS OF SEVERAL VARIABLES 823

$$= \begin{vmatrix} -\dfrac{\partial F_1}{\partial x_1} & -\dfrac{\partial F_1}{\partial x_2} & \cdots & -\dfrac{\partial F_1}{\partial x_n} \\ -\dfrac{\partial F_2}{\partial x_1} & -\dfrac{\partial F_2}{\partial x_2} & \cdots & -\dfrac{\partial F_2}{\partial x_n} \\ \cdots & \cdots & \cdots & \cdots \\ -\dfrac{\partial F_n}{\partial x_1} & -\dfrac{\partial F_n}{\partial x_2} & \cdots & -\dfrac{\partial F_n}{\partial x_n} \end{vmatrix}, \quad \text{by (13.11.2a)}$$

$$= (-1)^n \frac{\partial(F_1, F_2, \cdots, F_n)}{\partial(x_1, x_2, \cdots, x_n)}.$$

We may use the formula in the following form

$$\frac{\partial(u_1, u_2, \cdots, u_n)}{\partial(x_1, x_2, \cdots, x_n)} = (-1)^n \frac{\dfrac{\partial(F_1, F_2, \cdots, F_n)}{\partial(x_1, x_2, \cdots, x_n)}}{\dfrac{\partial(F_1, F_2, \cdots, F_n)}{\partial(u_1, u_2, \cdots u_n)}}.$$

[Compare the well known result $\dfrac{dy}{dx} = -\dfrac{\partial f}{\partial x} \Big/ \dfrac{\partial f}{\partial y}$ in case of implicit function $f(x, y) = 0$.]

Example 13.11.4. *The roots of the equation in λ,*

$$(\lambda - x)^3 + (\lambda - y)^3 + (\lambda - z)^3 = 0$$

are u, v, w. Prove that the Jacobian

$$\frac{\partial(u, v, w)}{\partial(x, y, z)} = -2\frac{(y-z)(z-x)(x-y)}{(v-w)(w-u)(u-v)}. \qquad \text{[C.H. 1995]}$$

Solution: The equation in λ can be written as

$$3\lambda^3 - 3\lambda^2(x+y+z) + 3\lambda(x^2+y^2+z^2) - (x^3+y^3+z^3) = 0.$$

Since u, v, w are given to be the roots of this cubic in λ, we write (using the relations between roots and coefficients):

a) $u + v + w = x + y + z$

  or, $\phi_1 \equiv u + v + w - x - y - z = 0.$

b) $uv + vw + wu = x^2 + y^2 + z^2$

or, $\phi_2 \equiv uv + vw + wu - x^2 - y^2 - z^2 = 0$.

c) $uvw = \frac{1}{3}(x^3 + y^3 + z^3)$

or, $\phi_3 \equiv uvw - \frac{1}{3}(x^3 + y^3 + z^3) = 0$.

Now we have

$$\frac{\partial(\phi_1, \phi_2, \phi_3)}{\partial(x, y, z)} = (-1)^3 \frac{\partial(\phi_1, \phi_2, \phi_3)}{\partial(u, v, w)} \frac{\partial(u, v, w)}{\partial(x, y, z)}$$

or, $\quad \dfrac{\partial(u, v, w)}{\partial(x, y, z)} = -\dfrac{\partial(\phi_1, \phi_2, \phi_3)}{\partial(x, y, z)} \bigg/ \dfrac{\partial(\phi_1, \phi_2, \phi_3)}{\partial(u, v, w)}.$

But $\dfrac{\partial(\phi_1, \phi_2, \phi_3)}{\partial(x, y, z)} = \begin{vmatrix} \dfrac{\partial \phi_1}{\partial x} & \dfrac{\partial \phi_1}{\partial y} & \dfrac{\partial \phi_1}{\partial z} \\ \dfrac{\partial \phi_2}{\partial x} & \dfrac{\partial \phi_2}{\partial y} & \dfrac{\partial \phi_2}{\partial z} \\ \dfrac{\partial \phi_3}{\partial x} & \dfrac{\partial \phi_3}{\partial y} & \dfrac{\partial \phi_3}{\partial z} \end{vmatrix}$

$= \begin{vmatrix} -1 & -1 & -1 \\ -2x & -2y & -2z \\ -x^2 & -y^2 & -z^2 \end{vmatrix} = (-1)^3 \cdot 2 \begin{vmatrix} 1 & 1 & 1 \\ x & y & z \\ x^2 & y^2 & z^2 \end{vmatrix}$

$= -2 \begin{vmatrix} 1 & 0 & 0 \\ x & y - x & z - x \\ x^2 & y^2 - x^2 & z^2 - x^2 \end{vmatrix}$ (Col$_2$ − Col$_1$; Col$_3$ − Col$_1$)

$= -2 \begin{vmatrix} y - x & z - x \\ y^2 - x^2 & z^2 - x^2 \end{vmatrix}$

$= -2(y - x)(z - x) \begin{vmatrix} 1 & 1 \\ y + x & z + x \end{vmatrix} = -2(y - x)(z - x)(z - y)$

$= -2(y - z)(z - x)(x - y).$

CHAPTER 13: FUNCTIONS OF SEVERAL VARIABLES

Also

$$\frac{\partial(\phi_1, \phi_2, \phi_3)}{\partial(u, v, w)} = \begin{vmatrix} \dfrac{\partial \phi_1}{\partial u} & \dfrac{\partial \phi_1}{\partial v} & \dfrac{\partial \phi_1}{\partial w} \\ \dfrac{\partial \phi_2}{\partial u} & \dfrac{\partial \phi_2}{\partial v} & \dfrac{\partial \phi_2}{\partial w} \\ \dfrac{\partial \phi_3}{\partial u} & \dfrac{\partial \phi_3}{\partial v} & \dfrac{\partial \phi_3}{\partial w} \end{vmatrix}$$

$$= \begin{vmatrix} 1 & 1 & 1 \\ v+w & w+u & u+v \\ vw & wu & uv \end{vmatrix}$$

$$= \begin{vmatrix} 1 & 0 & 0 \\ v+w & u-v & u-w \\ vw & w(u-v) & v(u-w) \end{vmatrix} \quad (\text{Col}_2 - \text{Col}_1; \text{Col}_3 - \text{Col}_1)$$

$$= (u-v)(u-w) \begin{vmatrix} 1 & 1 \\ w & v \end{vmatrix}$$

$$= (u-v)(u-w)(v-w)$$

$$= -(v-w)(w-u)(u-v).$$

$$\therefore \quad \frac{\partial(u, v, w)}{\partial(x, y, z)} = -\frac{-2(y-z)(z-x)(x-y)}{-(v-w)(w-u)(u-v)}$$

$$= -2\frac{(y-z)(z-x)(x-y)}{(v-w)(w-u)(u-v)}.$$

Example 13.11.5. *If λ, μ, ν are the roots (assumed to be all real) of the following equation in k.*

$$\frac{x}{a+k} + \frac{y}{b+k} + \frac{z}{c+k} = 1$$

prove that $\dfrac{\partial(x, y, z)}{\partial(\lambda, \mu, \nu)} = -\dfrac{(\mu - \nu)(\nu - \lambda)(\lambda - \mu)}{(b-c)(c-a)(a-b)}.$

Solution: The equation in k can be written as

$$k^3 - k^2(x+y+z-a-b-c) - k[(b+c)x + (c+a)y$$
$$+(a+b)z - ab - bc - ca] + abc - bcx - cay - abz = 0.$$

$$\therefore \quad \Sigma \lambda = (\Sigma x - \Sigma a).$$
$$\Sigma \lambda \mu = -(\Sigma(b+c)x - \Sigma bc)$$
$$\lambda \mu \nu = \Sigma bcx - abc,$$

where $\Sigma x = x + y + z$, $\Sigma bc = bc + ca + ab$, $\Sigma(b+c)x = (b+c)x + (c+a)y + (a+b)z$ etc.

$$\therefore \quad \phi_1 \equiv \lambda + \mu + \nu - (x + y + z - a - b - c) = 0.$$
$$\phi_2 \equiv \lambda\mu + \lambda\nu + \mu\nu + [(b+c)x + (c+a)y + (a+b)z - bc - ca - ab] = 0.$$
$$\phi_3 \equiv \lambda\mu\nu - [bcx + cay + abz - abc] = 0.$$

$$\therefore \quad \frac{\partial(x, y, z)}{\partial(\lambda, \mu, \nu)} = (-1)^3 \frac{\partial(\phi_1, \phi_2, \phi_3)}{\partial(\lambda, \mu, \nu)} \bigg/ \frac{\partial(\phi_1, \phi_2, \phi_3)}{\partial(x, y, z)}.$$

Now

$$\frac{\partial(\phi_1, \phi_2, \phi_3)}{\partial(\lambda, \mu, \nu)} = \begin{vmatrix} \frac{\partial \phi_1}{\partial \lambda} & \frac{\partial \phi_1}{\partial \mu} & \frac{\partial \phi_1}{\partial \mu} \\ \frac{\partial \phi_2}{\partial \lambda} & \frac{\partial \phi_2}{\partial \mu} & \frac{\partial \phi_2}{\partial \nu} \\ \frac{\partial \phi_3}{\partial \lambda} & \frac{\partial \phi_3}{\partial \mu} & \frac{\partial \phi_3}{\partial \nu} \end{vmatrix}$$

$$= \begin{vmatrix} 1 & 1 & 1 \\ \mu + \nu & \lambda + \nu & \lambda + \mu \\ \mu\nu & \lambda\nu & \lambda\mu \end{vmatrix}$$

$$= -(\mu - \nu)(\nu - \lambda)(\lambda - \mu).$$

Similarly

$$\frac{\partial(\phi_1, \phi_2, \phi_3)}{\partial(x, y, z)} = \begin{vmatrix} -1 & -1 & -1 \\ (b+c) & (c+a) & (a+b) \\ -bc & -ca & -ab \end{vmatrix}$$

$$= \begin{vmatrix} 1 & 1 & 1 \\ b+c & c+a & a+b \\ bc & ca & ab \end{vmatrix}$$

$$= -(b-c)(c-a)(a-b).$$

$$\therefore \quad \frac{\partial(x, y, z)}{\partial(\lambda, \mu, \nu)} = (-1)^3 \left\{ \frac{-(\mu - \nu)(\nu - \lambda)(\lambda - \mu)}{-(b - c)(c - a)(a - b)} \right\}$$

$$= -\frac{(\mu - \nu)(\nu - \lambda)(\lambda - \mu)}{(b - c)(c - a)(a - b)}.$$

CHAPTER 13: FUNCTIONS OF SEVERAL VARIABLES

An Important Example

Example 13.11.6. *If z_1, z_2 are two differentiable functions of three variables y_1, y_2 and y_3 which are themselves differentiable functions of x_1 and x_2, then*

$$\frac{\partial(z_1, z_2)}{\partial(x_1, x_2)} = \frac{\partial(z_1, z_2)}{\partial(y_1, y_2)} \frac{\partial(y_1, y_2)}{\partial(x_1, x_2)} + \frac{\partial(z_1, z_2)}{\partial(y_2, y_3)} \frac{\partial(y_2, y_3)}{\partial(x_1, x_2)} + \frac{\partial(z_1, z_3)}{\partial(y_3, y_1)} \frac{\partial(y_3, y_1)}{\partial(x_1, x_2)}. \quad (13.11.3)$$

Solution: By Chain Rules, we write

$$\frac{\partial z_1}{\partial x_1} = \frac{\partial z_1}{\partial y_1} \frac{\partial y_1}{\partial x_1} + \frac{\partial z_1}{\partial y_2} \frac{\partial y_2}{\partial x_1} + \frac{\partial z_1}{\partial y_3} \frac{\partial y_3}{\partial x_1}$$

$$\frac{\partial z_1}{\partial x_2} = \frac{\partial z_1}{\partial y_1} \frac{\partial y_1}{\partial x_2} + \frac{\partial z_1}{\partial y_2} \frac{\partial y_2}{\partial x_2} + \frac{\partial z_1}{\partial y_3} \frac{\partial y_3}{\partial x_2}.$$

Substituting these values in the Jacobian

$$\frac{\partial(z_1, z_2)}{\partial(x_1, x_2)} = \begin{vmatrix} \dfrac{\partial z_1}{\partial x_1} & \dfrac{\partial z_1}{\partial x_2} \\ \dfrac{\partial z_2}{\partial x_1} & \dfrac{\partial z_2}{\partial x_2} \end{vmatrix}.$$

We obtain

$$\frac{\partial(z_1, z_2)}{\partial(x_1, x_2)} = \begin{vmatrix} \dfrac{\partial z_1}{\partial y_1}\dfrac{\partial y_1}{\partial x_1} + \dfrac{\partial z_1}{\partial y_2}\dfrac{\partial y_2}{\partial x_1} + \dfrac{\partial z_1}{\partial y_3}\dfrac{\partial y_3}{\partial x_1} & \dfrac{\partial z_1}{\partial y_1}\dfrac{\partial y_1}{\partial x_2} + \dfrac{\partial z_1}{\partial y_2}\dfrac{\partial y_2}{\partial x_2} + \dfrac{\partial z_1}{\partial y_3}\dfrac{\partial y_3}{\partial x_2} \\ \dfrac{\partial z_2}{\partial x_1} & \dfrac{\partial z_2}{\partial x_2} \end{vmatrix}$$

$$= \frac{\partial z_1}{\partial y_1} \begin{vmatrix} \dfrac{\partial y_1}{\partial x_1} & \dfrac{\partial y_1}{\partial x_2} \\ \dfrac{\partial z_2}{\partial x_1} & \dfrac{\partial z_2}{\partial x_2} \end{vmatrix} + \frac{\partial z_1}{\partial y_2} \begin{vmatrix} \dfrac{\partial y_2}{\partial x_1} & \dfrac{\partial y_2}{\partial x_2} \\ \dfrac{\partial z_2}{\partial x_1} & \dfrac{\partial z_2}{\partial x_2} \end{vmatrix} + \frac{\partial z_1}{\partial y_3} \begin{vmatrix} \dfrac{\partial y_3}{\partial x_1} & \dfrac{\partial y_3}{\partial x_2} \\ \dfrac{\partial z_2}{\partial x_1} & \dfrac{\partial z_2}{\partial x_2} \end{vmatrix}.$$

$$\therefore \quad \frac{\partial(z_1, z_2)}{\partial(x_1, x_2)} = \frac{\partial z_1}{\partial y_1} \frac{\partial(y_1, z_2)}{\partial(x_1, x_2)} + \frac{\partial z_1}{\partial y_2} \frac{\partial(y_2, z_2)}{\partial(x_1, x_2)} + \frac{\partial z_1}{\partial y_3} \frac{\partial(y_3, z_2)}{\partial(x_1, x_2)}. \quad (13.11.4)$$

Note the rule of replacement of z_1 by y_1, y_2, y_3 in three terms of R.H.S. Similarly, we obtain

$$\frac{\partial(y_1, z_2)}{\partial(x_1, x_2)} = \frac{\partial z_2}{\partial y_1}\frac{\partial(y_1, y_1)}{\partial(x_1, x_2)} + \frac{\partial z_2}{\partial y_2}\frac{\partial(y_1, y_2)}{\partial(x_1, x_2)} + \frac{\partial z_2}{\partial y_3}\frac{\partial(y_1, y_3)}{\partial(x_1, x_2)};$$

$$\frac{\partial(y_2, z_2)}{\partial(x_1, x_2)} = \frac{\partial z_2}{\partial y_1}\frac{\partial(y_2, y_1)}{\partial(x_1, x_2)} + \frac{\partial z_2}{\partial y_2}\frac{\partial(y_2, y_2)}{\partial(x_1, x_2)} + \frac{\partial z_2}{\partial y_3}\frac{\partial(y_2, y_3)}{\partial(x_1, x_2)};$$

$$\frac{\partial(y_3, z_2)}{\partial(x_1, x_2)} = \frac{\partial z_2}{\partial y_1}\frac{\partial(y_3, y_1)}{\partial(x_1, x_2)} + \frac{\partial z_2}{\partial y_2}\frac{\partial(y_3, y_2)}{\partial(x_1, x_2)} + \frac{\partial z_2}{\partial y_3}\frac{\partial(y_3, y_3)}{\partial(x_1, x_2)}.$$

The Jacobians like $\dfrac{\partial(y_1, y_1)}{\partial(x_1, x_2)}$, $\dfrac{\partial(y_2, y_2)}{\partial(x_1, x_2)}$, $\dfrac{\partial(y_3, y_3)}{\partial(x_1, x_2)}$ all vanish. We then get from (13.11.4),

$$\frac{\partial(z_1, z_2)}{\partial(x_1, x_2)} = \frac{\partial(y_1, y_2)}{\partial(x_1, x_2)}\left[\frac{\partial z_2}{\partial y_2}\frac{\partial z_1}{\partial y_1} - \frac{\partial z_2}{\partial y_1}\frac{\partial z_1}{\partial y_2}\right] + \text{etc. etc.}$$

$$= \frac{\partial(z_1, z_2)}{\partial(y_1, y_2)}\frac{\partial(y_1, y_2)}{\partial(x_1, x_2)} + \text{etc. etc.}$$

This gives the required result (13.11.3).

13.12 Functional Dependence

Theorem 13.12.1. *If $u = f(x, y)$ and $v = g(x, y)$ have continuous partial derivatives in a region R of the xy-plane (or, f, g are differentiable functions of x and y in a region R of the xy-plane), a necessary and sufficient condition that they satisfy a functional relation, say $F(u, v) = 0$ is that the Jacobian*

$$\frac{\partial(u, v)}{\partial(x, y)} = 0.$$

[The **proof** of the theorem is given later **P. 845.**]

Example 13.12.1. Let $u = \dfrac{x+y}{1-xy}$, $v = \dfrac{(x+y)(1-xy)}{(1+x^2)(1+y^2)}$. Find $\dfrac{\partial(u, v)}{\partial(x, y)}$. Are they *functionally related? If so, find the relationship.*

$$\frac{\partial(u,v)}{\partial(x,y)} = \begin{vmatrix} \dfrac{\partial u}{\partial x} & \dfrac{\partial u}{\partial y} \\ \dfrac{\partial v}{\partial x} & \dfrac{\partial v}{\partial y} \end{vmatrix}$$

$$= \begin{vmatrix} \dfrac{1+y^2}{(1-xy)^2} & \dfrac{1+x^2}{(1-xy)^2} \\ \dfrac{1}{1+y^2} \dfrac{(1-x^2-y^2-4xy+x^2y^2)}{(1+x^2)^2} & \dfrac{1}{1+x^2} \cdot \dfrac{1-x^2-y^2-4xy+x^2y^2}{(1+y^2)^2} \end{vmatrix}$$

$$= \dfrac{1-x^2-y^2-4xy+x^2y^2}{(1+x^2)(1+y^2)} \cdot \dfrac{1}{(1-xy)^2} \begin{vmatrix} 1+y^2 & 1+x^2 \\ \dfrac{1}{1+x^2} & \dfrac{1}{1+y^2} \end{vmatrix} = 0.$$

Hence there is a functional relation between u and v. To find it solve the first equation for y and substitute its value in the second; thus

$$x+y = u(1-xy), \quad \text{or,} \quad y = \dfrac{u-x}{1+xu}$$

hence $\quad x+y = \dfrac{u(1+x^2)}{1+xu}$

$$1-xy = \dfrac{1+x^2}{1+xu}, \quad 1+y^2 = \dfrac{(1+u^2)(1+x^2)}{(1+xu)^2}.$$

$$\therefore \quad v = \dfrac{(x+y)(1-xy)}{(1+x^2)(1+y^2)} = \dfrac{\dfrac{u(1+x^2)}{1+xu} \cdot \dfrac{1+x^2}{1+xu}}{(1+x^2) \cdot \dfrac{(1+u^2)(1+x^2)}{(1+xu)^2}}$$

i.e., $\quad v = \dfrac{u}{1+u^2} \quad$ (relation between u and v).

Try a similar problem with $u = \dfrac{x+y}{1-xy}$, $v = \tan^{-1} x + \tan^{-1} y$. (**Ans.** $u = \tan v$).

Theorem 13.12.2. *If $u = f(x,y,z), v = g(x,y,z), w = h(x,y,z)$ be three differentiable functions of x, y, z in a three-dimensional region R, a necessary and sufficient condition that they satisfy a functional relation, say $F(u,v,w) = 0$, is that their Jacobian*

$$\dfrac{\partial(u,v,w)}{\partial(x,y,z)} = 0.$$

[Proof given later.]

Example 13.12.2. *Show that the functions*

$$u = x + y - z, \quad v = x - y + z, \quad w = x^2 + y^2 + z^2 - 2yz$$

are not independent. Find the relation between them.

Solution:

$$\frac{\partial(u,v,w)}{\partial(x,y,z)} = \begin{vmatrix} \frac{\partial u}{\partial x} & \frac{\partial u}{\partial y} & \frac{\partial u}{\partial z} \\ \frac{\partial v}{\partial x} & \frac{\partial v}{\partial y} & \frac{\partial v}{\partial z} \\ \frac{\partial w}{\partial x} & \frac{\partial w}{\partial y} & \frac{\partial w}{\partial z} \end{vmatrix}$$

$$= \begin{vmatrix} 1 & 1 & -1 \\ 1 & -1 & 1 \\ 2x & 2(y-z) & 2(z-y) \end{vmatrix}$$

$$= \begin{vmatrix} 1 & 0 & -1 \\ 1 & 0 & 1 \\ 2x & 0 & 2(z-y) \end{vmatrix} \quad \text{(adding Col}_2 \text{ with Col}_3\text{)}$$

$$= 0.$$

Since the Jacobian is zero, the functions u, v, w are not independent. To find the relation between them, we see that

$$u + v = 2x, \quad u - v = 2(y - z).$$

$\therefore \quad (u+v)^2 + (u-v)^2 = 4[x^2 + y^2 + z^2 - 2yz] = 4w$

i.e., $\quad (u+v)^2 + (u-v)^2 - 4w = 0$ is the required relation.

Example 13.12.3. *Show that the functions*

$$u = x + y + z, \quad v = xy + yz + zx, \quad w = x^3 + y^3 + z^3 - 3xyz$$

are not independent but they are related by $u^3 = 3uv + w$. **[C.H. 1992]**

CHAPTER 13: FUNCTIONS OF SEVERAL VARIABLES

Solution: The Jacobian

$$\frac{\partial(u,v,w)}{\partial(x,y,z)} = \begin{vmatrix} 1 & 1 & 1 \\ y+z & z+x & x+y \\ 3(x^2-yz) & 3(y^2-zx) & 3(z^2-xy) \end{vmatrix}$$

$$= 3 \begin{vmatrix} 1 & 0 & 0 \\ y+z & x-y & x-z \\ x^2-yz & -(x-y)(x+y+z) & -(x-z)(x+y+z) \end{vmatrix}$$

$$= 3(x-y)(x-z) \begin{vmatrix} 1 & 0 & 0 \\ y+z & 1 & 1 \\ x^2-yz & -(x+y+z) & -(x+y+z) \end{vmatrix}$$

$$= 0 \quad (\text{Col}_2 \text{ and Col}_3 \text{ are identical}).$$

$\therefore$ u, v, w are functionally dependent. To find a relation between u, v, w we have

$$u^3 = (x+y+z)^3 = \Sigma x^3 + 3\Sigma xy^2 + 6xyz$$

i.e., $\quad u^3 = \Sigma x^3 + 3[\Sigma x \Sigma xy - 3xyz] + 6xyz$

$$= 3\Sigma x\, \Sigma xy + (\Sigma x^3 - 3xyz)$$

i.e., $\quad u^3 = 3uv + w.$ **(Required relation)**

Example 13.12.4. *Show that the three functions u, v, w given by $u = 3x + 2y - z$, $v = x - 2y + z$, $w = x(x + 2y - z)$ are connected by a functional equation and find that equation.*

Solution:

$$\frac{\partial(u,v,w)}{\partial(x,y,z)} = \begin{vmatrix} \dfrac{\partial u}{\partial x} & \dfrac{\partial u}{\partial y} & \dfrac{\partial u}{\partial z} \\ \dfrac{\partial v}{\partial x} & \dfrac{\partial v}{\partial y} & \dfrac{\partial v}{\partial z} \\ \dfrac{\partial w}{\partial x} & \dfrac{\partial w}{\partial y} & \dfrac{\partial w}{\partial z} \end{vmatrix} = \begin{vmatrix} 3 & 2 & -1 \\ 1 & -2 & 1 \\ 2(x+y) & 2x & -x \end{vmatrix}$$

$$= -2 \begin{vmatrix} 3 & -1 & -1 \\ 1 & 1 & 1 \\ 2(x+y) & -x & -x \end{vmatrix} = 0 \quad (\text{Col}_2 \text{ and Col}_3 \text{ are identical}).$$

Since their Jacobian is zero, the functions u, v, w are related by a functional relation. To find that relation we see that

$$\frac{u+v}{4} = x; \quad u - v = 2x + 2(2y - z)$$

i.e., $\quad 2y - z = \dfrac{u-v}{2} - x = \dfrac{u-v}{2} - \dfrac{u+v}{4} = \dfrac{u-3v}{4}.$

$\therefore \quad w = x(x + 2y - z) = \dfrac{u+v}{4}\left[\dfrac{u+v}{4} + \dfrac{u-3v}{4}\right] = \dfrac{u^2 - v^2}{8}$

i.e., $8w = u^2 - v^2$ is the required relation.

Example 13.12.5. *Show that the expression*

$$ax^2 + by^2 + cz^2 + 2fyz + 2gzx + 2hxy$$

can be resolved into linear factors if

$$\begin{vmatrix} a & h & g \\ h & b & f \\ g & f & c \end{vmatrix} = 0.$$

Solution: Let $u = ax^2 + by^2 + cz^2 + 2fyz + 2gzx + 2hxy.$

Suppose u is resolvable into two linear factors

$$v = l_1 x + m_1 y + n_1 z \quad \text{and} \quad w = l_2 x + m_2 y + n_2 z.$$

Thus u, v, w are connected by $u = vw$.

$\therefore$ The Jacobian $\dfrac{\partial(u, v, w)}{\partial(x, y, z)} = 0.$

i.e., $\begin{vmatrix} 2(ax + gz + hy) & 2(by + fz + hx) & 2(cz + gx + fy) \\ l_1 & m_1 & n_1 \\ l_2 & m_2 & n_2 \end{vmatrix} = 0.$

Since x, y, z are independent variables, their coefficients in the above determinant must separately vanish.

Hence

$$a(m_1 n_2 - m_2 n_1) + h(l_2 n_1 - l_1 n_2) + g(l_1 m_2 - l_2 m_1) = 0$$
$$h(m_1 n_2 - m_2 n_1) + b(l_2 n_1 - l_1 n_2) + f(l_1 m_2 - l_2 m_1) = 0$$
$$g(m_1 n_2 - m_2 n_1) + f(l_2 n_1 - l_1 n_2) + c(l_1 m_2 - l_2 m_1) = 0.$$

CHAPTER 13: FUNCTIONS OF SEVERAL VARIABLES

Eliminating $(m_1 n_2 - m_2 n_1)$, $(l_2 n_1 - l_1 n_2)$, $(l_1 m_2 - l_2 m_1)$, we get

$$\begin{vmatrix} a & h & g \\ h & b & f \\ g & f & c \end{vmatrix} = 0.$$

Example 13.12.6. *If* $f(0) = 0$, $f'(x) = \dfrac{1}{1+x^2}$, *prove, without using the method of integration that*

$$f(x) + f(y) = f\left(\frac{x+y}{1-xy}\right).$$

Solution: Let $u = f(x) + f(y)$, $v = \dfrac{x+y}{1-xy}$.

The Jacobian

$$\frac{\partial(u,v)}{\partial(x,y)} = \begin{vmatrix} \dfrac{\partial u}{\partial x} & \dfrac{\partial u}{\partial y} \\ \dfrac{\partial v}{\partial x} & \dfrac{\partial v}{\partial y} \end{vmatrix} = \begin{vmatrix} f'(x) & f'(y) \\ \dfrac{1+y^2}{(1-xy)^2} & \dfrac{1+x^2}{(1-xy)^2} \end{vmatrix}$$

$$= \begin{vmatrix} \dfrac{1}{1+x^2} & \dfrac{1}{1+y^2} \\ \dfrac{1+y^2}{(1-xy)^2} & \dfrac{1+x^2}{(1-xy)^2} \end{vmatrix} = 0.$$

Hence u, v are connected by a functional relation

$$u = \phi(v)$$

i.e., $f(x) + f(y) = \phi\left(\dfrac{x+y}{1-xy}\right).$

Putting $y = 0$, using $f(0) = 0$, we get $f(x) = \phi(x)$.
Thus

$$f(x) + f(y) = f\left(\frac{x+y}{1-xy}\right). \quad (\because \ f \equiv \phi)$$

13.13 Implicit Functions

If $F(x,y)$ be a function of two variables and $y = \phi(x)$ be a function of x such that for every x for which $\phi(x)$ is defined, $F(x, \phi(x)) \equiv 0$, then we say

$\quad\quad y = \phi(x) \quad$ is defined implicitly by $F(x,y) = 0$;

or, $y = \phi(x) \quad$ is an implicit function defined by $F(x,y) = 0$.

e.g., $y = 3 - \dfrac{3}{2}x$ is defined implicitly by $3x + 2y - 6 = 0$.

A functional relation of the form $F(x,y) = 0$ may not define an implicit function (e.g., $x^2 + y^2 + 2 = 0$ determines no implicit function) or it may define more than one implicit function (e.g., $x^2 + y^2 - 2 = 0$ determines two implicit functions, namely $y = +\sqrt{2-x^2}$ and $y = -\sqrt{2-x^2}$).

When the functional relation is complicated it may not be easy to write y explicitly in terms of x (e.g., $x^4 y^3 + \cos y + \log x + \tan^{-1} xy = 0$) but this does not mean that the implicit function does not exist for such functional equations.

In the following discussions we investigate the question of existence of implicit functions from a given functional relation. We note that if $F(x,y)$ be any function of two variables defined in a certain domain R of the (x,y) plane, then $F(x,y) = 0$ may determine a subset of R consisting of those points (x,y) for which the relation $F(x,y) = 0$ holds. It does not follow that this aggregate of points (x,y) will determine y as an implicit function of x.

Implicit Function determined by a single Function equation $F(x,y) = 0$.

Theorem 13.13.1. (Existence Theorem).

Let $F(x,y)$ be a function of two variables x and y and let (x_0, y_0) be a point in its domain of definition such that

a) $F(x_0, y_0) = 0$;

b) F_x and F_y *are continuous in a certain neighbourhood of* (x_0, y_0); *and*

c) $F_y(x_0, y_0) \neq 0$.

Then there exists a rectangle:

$$x_0 - h \leq x \leq x_0 + h; \quad y_0 - k \leq y \leq y_0 + k$$

centred at (x_0, y_0) such that for every value of x in the interval

$$I : x_0 - h \leq x \leq x_0 + h$$

CHAPTER 13: FUNCTIONS OF SEVERAL VARIABLES

the functional equation $F(x,y) = 0$ determines one and only one value $y = \phi(x)$ which lies in the interval

$$y_0 - k \le y \le y_0 + k$$

with the following properties:

1. $y_0 = \phi(x_0)$;
2. $F(x, \phi(x)) = 0$ for every $x \in I$;
3. $\phi(x)$ is derivable and both $\phi(x)$ and $\phi'(x)$ are continuous in I;
4. $\phi'(x) = -F_x/F_y$.

Observation: The theorem asserts the existence of solutions of $F(x,y) = 0$ in the neighbourhood of an initial solution (x_0, y_0), i.e., the theorem is essentially of *Local Character*. It does not, however, indicate how to find such an initial solution. The theorem solves $F(x,y) = 0$, **locally** in the neighbourhood of the initial solution (x_0, y_0) provided certain conditions about continuity and derivability of $F(x,y)$ are satisfied.

Example 13.13.1. *Consider $F(x,y) = x^2 + y^2 - 1$ and a point $(0,1)$.*

Observe that $F(0,1) = 0, F_x = 2x$ and $F_y = 2y$ are always continuous and $F_y(0,1) = 2(\ne 0)$. Thus all the conditions of Implicit Function Theorem are satisfied for the initial solution $(0,1)$. Theorem asserts that there exists a neighbourhood of $(0,1)$ in which y can be explicitly expressed in terms of x. There are two possible solutions of $F(x,y) = 0 : y = +\sqrt{1-x^2}$ and $y = -\sqrt{1-x^2}$.

a) The solution $y = +\sqrt{1-x^2}$ is the implicit function defined in a neighbourhood of $(0,1)$, where $|x| < 1, y > 0$;

b) The solution $y = -\sqrt{1-x^2}$ is the implicit function defined in a neighbourhood of $(0,-1)$ where $|x| < 1, y < 0$.

Proof of the Existence Theorem

(For quick understanding a figure may be drawn)

1. The conditions (b) state that F_x and F_y are continuous in a certain neighbourhood of the point (x_0, y_0). We take the *nbd* to be a rectangle :
   $R_1 : [x_0 - h_1, x_0 + h_1; \ y_0 - k_1, y_0 + k_1]$ centred at (x_0, y_0).
   Moreover Continuity of F_x and $F_y \Rightarrow F$ is differentiable and hence continuous in R_1.

2. F_y is continuous at (x_0, y_0) and $F_y(x_0, y_0) \neq 0$.
   $\therefore$ $\exists$ a rectangle
   $R_2 : [x_0 - h_2, x_0 + h_2;\ y_0 - k_2, y_0 + k_2]$, $h_2 < h_1$, $k_2 < k_1$
   i.e., $R_2 \subset R_1$ and for every point of R_2, $F_y \neq 0$.

3. Since $F(x_0, y_0) = 0$ and $F_y(x_0, y_0) \neq 0$ (therefore it is either positive or negative), there exists a positive number $k(< k_2)$ such that
   $$F(x_0, y_0 - k),\quad F(x_0, y_0 + k)$$
   are of opposite signs (F being an increasing or decreasing at $y = y_0$).

4. Continuity of F at $(x_0, y_0) \Rightarrow \exists h > 0$ ($h < h_2$) such that $\forall x \in [x_0 - h, x_0 + h]$,
   $$F(x, y_0 - k) \to F(x_0, y_0 - k)$$
   $$F(x, y_0 + k) \to F(x_0, y_0 + k)$$
   and therefore they will have opposite signs.

5. Thus for all $x \in R : [x_0 - h, x_0 + h]$, F is a continuous function of y and changes sign as y changes from $y_0 - k$ to $y_0 + k$. Therefore, it vanishes for some value of $y \in [y_0 - k, y_0 + k]$.

6. **Existence.** Thus for each $x \in R : [x_0 - h, x_0 + h]$ there is a y in $[y_0 - k, y_0 + k]$ for which $F(x, y) = 0$ i.e., y is a function of x, say $\mathbf{y} = \phi(\mathbf{x})$.

7. To prove $y = \phi(x)$ is the unique solution of $F(x, y) = 0$ in
   $$R : [x_0 - h, x_0 + h;\quad y_0 - k, y_0 + k].$$
   If possible, let there be two such values y_1, y_2 in $[y_0 - k, y_0 + k]$ so that
   $$F(x, y_1) = 0,\quad F(x, y_2) = 0.$$
   Also $F(x, y)$ considered as a function y alone is derivable in $[y_0 - k, y_0 + k]$ so that we can apply Rolle's Theorem and obtain $F_y = 0$ for a value of y between y_1 and y_2 which contradicts the fact $F_y \neq 0$ in $R_2 \supset R$.
   This contradiction proves the **Uniqueness of the solution.**

8. **Derivability :**
   Let $(x, y), (x + \Delta x, y + \Delta y) \in R : [x_0 - h, x_0 + h;\quad y_0 - k, y_0 + k]$.
   Then
   $$y = \phi(x)$$
   $$y + \Delta y = \phi(x + \Delta x)$$
   and $\quad F(x, y) = 0,\quad F(x + \Delta x, y + \Delta y) = 0.$

CHAPTER 13: FUNCTIONS OF SEVERAL VARIABLES   837

Since F is differentiable in R_2 and consequently in $R(R \subset R_1)$.

$$\therefore \quad 0 = F(x + \Delta x, y + \Delta y) - F(x, y)$$
$$= F_x \Delta x + F_y \Delta y + \varepsilon_1 \Delta x + \varepsilon_2 \Delta y$$

where ε_1 and ε_2 functions of Δx and Δy and tend to zero as $(\Delta x, \Delta y) \to (0,0)$.

$$\therefore \quad \frac{\Delta y}{\Delta x} = -\frac{F_x}{F_y} - \frac{\varepsilon_1}{F_y} - \frac{\varepsilon_2}{F_y} \frac{\Delta y}{\Delta x} \quad (F_y \neq 0 \text{ in } R).$$

Proceeding to the limits as $(\Delta x, \Delta y) \to (0, 0)$ we get

$$\frac{dy}{dx} = \phi'(x) = -\frac{F_x}{F_y}$$

i.e., $\phi(x)$ is derivable in R and hence continuous in R.

Again $\phi'(x)$ is the quotient of two continuous functions $F_x, F_y (F_y \neq 0)$ and hence $\phi'(x)$ is also continuous in R. This proves the theorem **completely**.

9. When the existence of the implicit function is assured, we may find dy/dx from $F(x, y) = 0$ by using the Chain Rule (Note that : $V = F(x, y)$ when $y = \phi(x)$)

$$\frac{\partial F}{\partial x} + \frac{\partial F}{\partial y} \frac{dy}{dx} = 0 \quad \Rightarrow \quad \frac{dy}{dx} = -\frac{F_x}{F_y}.$$

OR we may take differentials

$$dF = F_x dx + F_y dy = 0$$

(The form of dF is not altered by the dependence of y on x)

$$\Rightarrow \quad \frac{dy}{dx} = -\frac{F_x}{F_y}.$$

From $y' = \dfrac{dy}{dx} = -\dfrac{F_x}{F_y}$ we can find $y'' = \dfrac{d^2y}{dx^2}$ by using the Chain Rule again and then substituting for y' its value $-F_x/F_y$.

$$\frac{d^2y}{dx^2} = \frac{d}{dx}\left(\frac{dy}{dx}\right) = -\frac{\partial}{\partial x}\left(\frac{F_x}{F_y}\right)\frac{dx}{dx} - \frac{\partial}{\partial y}\left(\frac{F_x}{F_y}\right)\frac{dy}{dx}$$

$$= -\frac{F_y F_{xx} - F_x F_{xy}}{(F_y)^2} - \frac{F_y F_{yx} - F_x F_{yy}}{(F_y)^2} \cdot \left(-\frac{F_x}{F_y}\right)$$

or $\quad y'' = -\dfrac{(F_y)^2 F_{xx} - 2 F_x F_y F_{xy} + (F_x)^2 F_{yy}}{(F_y)^3}.$

Example 13.13.2. *From the equation of the ellipse*

$$\frac{x^2}{a^2} + \frac{y^2}{b^2} = 1, \quad \text{we have } (if\ y \neq 0) \quad y' = -\frac{b^2 x}{a^2 y}.$$

We can also obtain this result from the explicit Functions

$$y = \pm\frac{a}{b}\sqrt{a^2 - x^2}$$

representing the upper and lower halves of the ellipse. But implicit functions differentiation avoids radicals and gives a single formula for y' valid whenever $y \neq 0$.

The existence theorem fails at the points $(\pm a, 0)$; at these two points, the implicit functional relation $b^2 x^2 + a^2 y^2 - a^2 b^2 = 0$ defines two functions $y = \pm\frac{a}{b}\sqrt{a^2 - x^2}$ (**Not unique**).

We have also $y'' = -\dfrac{b^2}{a^2}\dfrac{y - xy'}{y^2} = -\dfrac{b^4}{a^2 y^3}.$

Example 13.13.3. *The equation of the Folium of Descartes*

$$x^3 + y^3 - 3axy = 0$$

has an explicit solutions that can be found by solving a cubic equation. To avoid this we can use the existence theorem at all points where $F_y = 3(y^2 - ax) \neq 0$.

Solution: Thus we find by the Chain Rule

$$y' = -\frac{x^2 - ay}{y^2 - ax}, \quad y'' = -\frac{2a^3 xy}{(y^2 - ax)^2}.$$

The curve has a horizontal tangent at $(\sqrt[3]{2}a, \sqrt[3]{4}a)$, a vertical tangent at $(\sqrt[3]{4}a, \sqrt[3]{a})$. The last point and the origin (where the curve crosses itself) are the points excluded by $F_y \neq 0$.

One Functional Equation : To solve for z in terms of x and y

The existence theorem can be generalised to equations having three or more variables: If we wish to solve for z from

$$F(x, y, z) = 0 \tag{13.13.1}$$

we have the following theorem:

Theorem 13.13.2. *Let the region R contain the point (x_0, y_0, z_0) in its interior. Then if*

a) $F(x_0, y_0, z_0) = 0;$

CHAPTER 13: FUNCTIONS OF SEVERAL VARIABLES

b) F_x, F_y, F_z are all continuous in R;

c) $F_z(x_0, y_0, z_0) \neq 0$,

then there exists a neighbourhood I_0 at the point (x_0, y_0) in which there exists a unique differentiable function $z = f(x, y)$ such that

1. $z_0 = f(x_0, y_0)$; 2. $F(x, y, f(x, y)) = 0$; 3. $z_x = -\dfrac{F_x}{F_z}$, $z_y = -\dfrac{F_y}{F_z}$.

Proof: The proof follows the same pattern as that of the fundamental Existence Theorem. In the two dimensional neighbourhood I_0 over which $z = f(x, y)$ is uniquely determined we have $F_z \neq 0$; hence from

$$dF = F_x dx + F_y dy + F_z dz = 0 \qquad (13.13.2)$$

we have

$$dz = -\frac{F_x}{F_z} dx - \frac{F_y}{F_z} dy.$$

$$\text{But,} \quad dz = \frac{\partial z}{\partial x} dx + \frac{\partial z}{\partial y} dy$$

$$\therefore \quad \frac{\partial z}{\partial x} = -\frac{F_x}{F_z} \quad \text{(here } y \text{ is constant)}$$

$$\text{and} \quad \frac{\partial z}{\partial y} = -\frac{F_y}{F_z} \quad \text{(here } x \text{ is constant)}.$$

The hypothesis (b) gives Z_x, Z_y continuous in I_0 and hence $z = f(x, y)$ is a differentiable function.

If, however, $F_x \neq 0$, we may solve (13.13.1) for x and (13.13.2) for dx. Regarding x as a function of two independent variables y, z, we then obtain

$$\frac{\partial x}{\partial y} = -\frac{F_y}{F_x} \quad \text{and} \quad \frac{\partial x}{\partial z} = -\frac{F_z}{F_x}.$$

If $F_y \neq 0$, we solve (13.13.1) for y and (13.13.2) for dy; we thus obtain

$$\frac{\partial y}{\partial x} = -\frac{F_x}{F_y}, \quad \frac{\partial y}{\partial z} = -\frac{F_z}{F_y}.$$

This theorem can be further generalised: To solve for u as a function of x, y, z from $F(x, y, z, u) = 0$. We assume $F_u \neq 0$.

Here $dF = F_x dx + F_y dy + F_z dz + F_u du = 0$. We then obtain

$$du = -\frac{F_x}{F_u} dx - \frac{F_y}{F_u} dy - \frac{F_z}{F_u} dz$$

leading to

$$\frac{\partial u}{\partial x} = -\frac{F_x}{F_u}, \quad \frac{\partial u}{\partial y} = -\frac{F_y}{F_u}, \quad \frac{\partial u}{\partial z} = -\frac{F_z}{F_u}.$$

Example 13.13.4. *The equation*

$$F(x, y, z) = xyz + x + y - z = 0$$

can be solved for z near $(0,0,0)$; for $F_z(0,0,0) = -1$.

Solution: We have, in fact, $z = \dfrac{x+y}{1-xy}$.

From this explicit solution $z_x = \dfrac{1+y^2}{(1-xy)^2}$, $\quad z_y = \dfrac{1+x^2}{(1-xy)^2}$;

OR applying Chain Rule to $F(x, y, z) = 0$

$$\frac{\partial F}{\partial x} + \frac{\partial F}{\partial z}\frac{\partial z}{\partial x} = 0 \quad (y \text{ is independent of } x)$$

i.e., $(yz + 1) + (xy - 1)\dfrac{\partial z}{\partial x} = 0 \quad \text{or,} \quad z_x = \dfrac{yz+1}{1-xy}$

or, $z_x = \dfrac{1+y^2}{(1-xy)^2} \quad \text{putting} \quad z = \dfrac{x+y}{1-xy}.$

Again

$$\frac{\partial F}{\partial y} + \frac{\partial F}{\partial z}\frac{\partial z}{\partial y} = 0 \quad \text{or,} \quad z_y = \frac{xz+1}{1-xy}$$

or, $z_y = \dfrac{1+x^2}{(1-xy)^2} \quad \text{putting} \quad z = \dfrac{x+y}{1-xy}.$

Example 13.13.5. *If $F(x, y, z) = 0$, prove that*

$$\left(\frac{\partial x}{\partial y}\right)_z \left(\frac{\partial y}{\partial z}\right)_x \left(\frac{\partial z}{\partial x}\right)_y = -1$$

where each partial derivative is computed by holding the remaining variables constant, assuming suitable conditions.

Solution:

$\left(\dfrac{\partial x}{\partial y}\right)_z = -\dfrac{F_y}{F_x} \quad (z \text{ is constant}; x = \text{function of } y \text{ and } z; F_x \neq 0)$

$\left(\dfrac{\partial y}{\partial z}\right)_x = -\dfrac{F_z}{F_y} \quad (x \text{ is constant}; y = \text{function of } x \text{ and } z; F_y \neq 0)$

$\left(\dfrac{\partial z}{\partial x}\right)_y = -\dfrac{F_x}{F_z} \quad (y \text{ is constant}; z = \text{function of } x \text{ and } y; F_z \neq 0.)$

CHAPTER 13: FUNCTIONS OF SEVERAL VARIABLES

Hence the product $= -1$.

Two Functional Equations: To solve for u and v in terms of x and y:

We now consider the problem of solving two equations,

$$F(x, y, u, v) = 0, \quad G(x, y, u, v) = 0. \tag{13.13.3}$$

We wish to solve for u and v in terms of x and y.

Theorem 13.13.3. (Existence Theorem).

Let the region R contain the point (x_0, y_0, z_0, w_0) in its interior. Then if

a) $F(x_0, y_0, u_0, v_0) = 0, G(x_0, y_0, u_0, v_0) = 0;$

b) $F_x, F_y, F_u, F_v, G_x, G_y, G_u, G_v$ are all continuous in R;

c) The Jacobian

$$J = \frac{\partial(F,G)}{\partial(u,v)} = \begin{vmatrix} \dfrac{\partial F}{\partial u} & \dfrac{\partial F}{\partial v} \\ \dfrac{\partial G}{\partial u} & \dfrac{\partial G}{\partial v} \end{vmatrix} \neq 0,$$

at the point (x_0, y_0, u_0, v_0), then there is a two-dimensional neighbourhood I_0 of (x_0, y_0) in which there exist two unique differentiable functions.

$$u = f(x,y), \quad v = g(x,y),$$

such that

1. $u_0 = f(x_0, y_0), \quad v_0 = g(x_0, y_0);$
2. $F(x, y, f(x,y), g(x,y)) = 0, \quad G(x, y, f(x,y), g(x,y)) = 0;$
3. The partial derivatives of $u = f(x,y), v = g(x,y)$ are continuous functions found by solving

$$dF = F_x dx + F_y dy + F_u du + F_v dv = 0$$
$$dG = G_x dx + G_y dy + G_u du + G_v dv = 0.$$

We may then solve these equations for du and dv ($\because J \neq 0$).

$$du = -\frac{\begin{vmatrix} F_x dx + F_y dy & F_v \\ G_x dx + G_y dy & G_v \end{vmatrix}}{\begin{vmatrix} F_u & F_v \\ G_u & G_v \end{vmatrix}} \quad \text{(Cramer's Rule)}$$

$$= -\frac{\begin{vmatrix} F_x & F_v \\ G_x & G_v \end{vmatrix} dx + \begin{vmatrix} F_y & F_v \\ G_y & G_v \end{vmatrix} dy}{J} = \frac{\partial u}{\partial x} dx + \frac{\partial u}{\partial y} dy.$$

$$\therefore \frac{\partial u}{\partial x} = -\frac{\begin{vmatrix} F_x & F_v \\ G_x & G_v \end{vmatrix}}{J}; \quad \frac{\partial u}{\partial y} = -\frac{\begin{vmatrix} F_y & F_v \\ G_y & G_v \end{vmatrix}}{J}.$$

Similarly

$$dv = -\frac{1}{J}\begin{vmatrix} F_u & F_x \\ G_u & G_x \end{vmatrix} dx - \frac{1}{J}\begin{vmatrix} F_u & F_y \\ G_u & G_y \end{vmatrix} dy.$$

$$\therefore \frac{\partial v}{\partial x} = -\frac{\begin{vmatrix} F_u & F_x \\ G_u & G_x \end{vmatrix}}{J} \quad \text{and} \quad \frac{\partial v}{\partial y} = -\frac{\begin{vmatrix} F_u & F_y \\ G_u & G_y \end{vmatrix}}{J}.$$

We could also write

$$\frac{\partial u}{\partial x} = -\frac{\frac{\partial(F,G)}{\partial(x,v)}}{\frac{\partial(F,G)}{\partial(u,v)}}, \quad \frac{\partial u}{\partial y} = -\frac{\frac{\partial(F,G)}{\partial(y,v)}}{\frac{\partial(F,G)}{\partial(u,v)}}$$

$$\frac{\partial v}{\partial x} = -\frac{\frac{\partial(F,G)}{\partial(u,x)}}{\frac{\partial(F,G)}{\partial(u,v)}}, \quad \frac{\partial v}{\partial y} = -\frac{\frac{\partial(F,G)}{\partial(u,y)}}{\frac{\partial(F,G)}{\partial(u,v)}}.$$

Rule. In each formula the Jacobian written out in the Numerator is formed from J by replacing u or v by x or y as indicated by the letters on the left with respect to which the partial derivative is taken e.g., in $\frac{\partial v}{\partial x}$ replace v by x i.e., take $\frac{\partial(F,G)}{\partial(u,x)}$ and then divide it by J (with a negative sign always).

CHAPTER 13: FUNCTIONS OF SEVERAL VARIABLES

If, on the other hand, we wish to solve for x and y from the functional equations

$$F(x,y,u,v) = 0 \quad \text{and} \quad G(x,y,u,v) = 0$$

then part (c) of the hypothesis of the theorem must be replaced by

$$J' = \frac{\partial(F,G)}{\partial(x,y)} \neq 0 \quad \text{at} \quad (x_0, y_0, u_0, v_0).$$

The derivatives $\dfrac{\partial x}{\partial u}, \dfrac{\partial x}{\partial v}, \dfrac{\partial y}{\partial u}, \dfrac{\partial y}{\partial v}$ can be found by using the above rule: thus

$$\frac{\partial x}{\partial u} = -\frac{\dfrac{\partial(F,G)}{\partial(u,y)}}{\dfrac{\partial(F,G)}{\partial(x,y)}}, \quad \frac{\partial y}{\partial u} = -\frac{\dfrac{\partial(F,G)}{\partial(x,u)}}{\dfrac{\partial(F,G)}{\partial(x,y)}}, \quad \text{etc. etc.}$$

Example 13.13.6. *The equations*

$$\left.\begin{array}{l} F(x,y,u,v) = u^2 + v^2 - x^2 - y = 0 \\ G(x,y,u,v) = u + v - x^2 + y = 0 \end{array}\right\} \quad (13.13.4)$$

are satisfied by $x_0 = 2, y_0 = 1, u_0 = 1, v_0 = 2.$

Solution: Since

$$J = \begin{vmatrix} F_u & F_v \\ G_u & G_v \end{vmatrix} = \begin{vmatrix} 2u & 2v \\ 1 & 1 \end{vmatrix} = 2(u-v)$$

and $J_0 = 2(u_0 - v_0) = -2$, equations (13.13.4) can be resolved uniquely for u, v in the neighbourhood of $(x_0 = 2, y_0 = 1)$. On solving

$$\left.\begin{array}{l} dF = 2u\,du + 2v\,dv - 2x\,dx - dy = 0 \\ dG = du + dv - 2x\,dx + dy = 0 \end{array}\right\} \quad (13.13.5)$$

for du, dv we find (using Cramer's Rule)

$$du = \frac{\begin{vmatrix} 2xdx + dy & 2v \\ 2xdx - dy & 1 \end{vmatrix}}{\begin{vmatrix} 2u & 2v \\ 1 & 1 \end{vmatrix}}$$

$$= \frac{\begin{vmatrix} 2x & 2v \\ 2x & 1 \end{vmatrix} dx + \begin{vmatrix} 1 & 2v \\ -1 & 1 \end{vmatrix} dy}{\begin{vmatrix} 2u & 2v \\ 1 & 1 \end{vmatrix}}$$

$$= \frac{2(x - 2xv)}{2(u - v)} dx + \frac{(1 + 2v)}{2(u - v)} dy.$$

$$\therefore \quad \frac{\partial u}{\partial x} = \frac{x(1 - 2v)}{u - v}, \quad \frac{\partial u}{\partial y} = \frac{1 + 2v}{2(u - v)}.$$

Similarly

$$\frac{\partial v}{\partial x} = \frac{x(2u - 1)}{u - v}, \quad \frac{\partial v}{\partial y} = \frac{-2u - 1}{2(u - v)}.$$

Since

$$J' = \begin{vmatrix} F_x & F_y \\ G_x & G_y \end{vmatrix} = \begin{vmatrix} -2x & -1 \\ -2x & 1 \end{vmatrix} = -4x$$

and $\quad J'_0 = -8 \ (\because x_0 = 2).$

Equations (13.13.4) can be solved uniquely for x, y in the neighbourhood of $(u_0 = 1, v_0 = 2)$.

On solving (13.13.5) for dx, dy, we can verify

$$\frac{\partial x}{\partial u} = \frac{2u + 1}{4x}, \quad \frac{\partial x}{\partial v} = \frac{2v + 1}{4x}$$
$$\frac{\partial y}{\partial u} = \frac{2u - 1}{2}, \quad \frac{\partial y}{\partial v} = \frac{2v - 1}{2}.$$

As a check on computation we may verify that the Jacobians $\dfrac{\partial(u, v)}{\partial(x, y)}$ and $\dfrac{\partial(x, y)}{\partial(u, v)}$ are reciprocals.

CHAPTER 13: FUNCTIONS OF SEVERAL VARIABLES

Three Functional Equations: To solve for u, v, w in terms of x, y, z

Theorem 13.13.4. (**Existence Theorem**). *Given three functional equations*

$$\left.\begin{array}{l} F(x,y,z,u,v,w) = 0 \\ G(x,y,z,u,v,w) = 0 \\ H(x,y,z,u,v,w) = 0 \end{array}\right\} \text{ To solve for } u, v, w.$$

Let the region R contain the point $(x_0, y_0, z_0, u_0, v_0, w_0)$ in its interior. Then, if

a) $F = 0, G = 0, H = 0$ *at* $(x_0, y_0, z_0, u_0, v_0, w_0)$;

b) *all first order partial derivatives of F, G, H are continuous in R;*

c) $J = \begin{vmatrix} F_u & F_v & F_w \\ G_u & G_v & G_w \\ H_u & H_v & H_w \end{vmatrix} = \dfrac{\partial(F,G,H)}{\partial(u,v,w)} \neq 0;$

at $(x_0, y_0, z_0, u_0, v_0, w_0)$.

Then there is some 3-dimensional neighbourhood I_0 of (x_0, y_0, z_0) in which there exist three unique differentiable functions,

$$u = f(x,y,z), \quad v = g(x,y,z), \quad w = h(x,y,z)$$

such that

1. $u_0 = f(x_0, y_0, z_0), \quad v_0 = g(x_0, y_0, z_0), \quad w_0 = h(x_0, y_0, z_0)$;

2. $F(x, y, z, f, g, h) = 0, \quad G(x, y, z, f, g, h) = 0, \quad H(x, y, z, f, g, h) = 0$;

3. *The partial derivatives of f, g, h are continuous functions found by solving $dF = 0, dG = 0, dH = 0$ for du, dv, dz in terms of dx, dy, dz.*

13.14 Proofs of the Theorems on Functional Dependence

We shall, in this article, give the proofs of the two theorems given in **art. 13.12**.

Theorem 13.14.1. *Statement given before in article* **13.12**, *Theorem* **13.12.1**.

Given: $u = f(x, y), v = g(x, y)$ *have continuous partial derivatives in a region R.*

To prove: *A necessary and sufficient condition that they satisfy a functional relation of the form $F(u, v) = 0$ is that the Jacobian*

$$J = \frac{\partial(u,v)}{\partial(x,y)} = 0. \qquad \text{[C.H. 1996]}$$

Proof: Assume that u and v are related by $F(u,v) = 0$.

Differentiating partially w.r.t x and y we have by using Chain Rules,

$$\left.\begin{aligned} F_u u_x + F_v v_x &= 0 \\ F_u u_y + F_v v_y &= 0 \end{aligned}\right\}. \qquad (13.14.1)$$

If $\begin{vmatrix} u_x & v_x \\ u_y & v_y \end{vmatrix} = \begin{vmatrix} u_x & u_y \\ v_x & v_y \end{vmatrix} \neq 0$, i.e., if $J = \dfrac{\partial(u,v)}{\partial(x,y)} \neq 0$

then the only solutions of (13.14.1) are $F_u = 0$, $F_v = 0$ which imply that $F(u,v)$ involves neither u nor v so that the relation $F(u,v) = 0$ becomes illusory.

Hence **the condition J = 0 is necessary.**

To Prove: The condition $J = 0$ is sufficient to ensure a relation like $F(u,v) = 0$.

If both $f_x = f_y = 0$, $u = $ a constant c or $u - c = 0$ is a functional relation. The theorem is then trivial.

Assume, therefore, at least $f_x(x_0, y_0) \neq 0$. Then we may solve the equation
$u - f(x,y) = 0$ for $x = \phi(u,y)$ in some nbd of (x_0, y_0).

Since $u = f(\phi, y)$ is an identity in u and y, $f(\phi, y)$ is independent of y; hence

$$f_x \phi_y + f_y = 0; \quad \phi_y = -\frac{f_y}{f_x}.$$

Now $v = g(\phi, y) = G(u, y)$.

$$\therefore G_y = g_x \phi_y + g_y = \frac{f_x g_y - f_y g_x}{f_x} = \frac{\begin{vmatrix} f_x & f_y \\ g_x & g_y \end{vmatrix}}{f_x} = 0$$

i.e., G is independent of y and is a function of u only

i.e., $v = G(u)$.

Note 13.14.1. *The proof is constructive; it shows how to find the functional relation between u and v when $J = 0$. we need only eliminate one of the variables x, y from the equations $u = f(x,y), v = g(x,y)$; then the other will automatically disappear and the result is the desired relation.*

Theorem 13.14.2. *Statement given in art* **13.12**, *Theorem* **13.12.2**.

CHAPTER 13: FUNCTIONS OF SEVERAL VARIABLES

Given: $u = f(x, y, z)$, $v = g(x, y, z)$, $w = h(x, y, z)$, *have continuous first partial derivatives in a region R.*

To prove: A necessary and sufficient condition that they satisfy a relation of the form $F(u, v, w) = 0$ is that the Jacobian

$$J = \frac{\partial(u, v, w)}{\partial(x, y, z)} = 0.$$

Proof: The necessity of the condition is proved as in the previous theorem. To prove that the condition is sufficient, we observe that, if all minors of J involving u and v vanish, we have a relation $F(u, v) = 0$ and the theorem follows.

Assume therefore that $\dfrac{\partial(f, g)}{\partial(x, y)} \neq 0$ at (x_0, y_0, z_0).

Then we can solve $u = f(x, y, z)$, $v = g(x, y, z)$ for x and y.

$$x = \phi(u, v, z), \quad y = \psi(u, v, z)$$

in some *nbd* of (x_0, y_0, z_0). Since

$$u = f(\phi, \psi, z), \quad v = g(\phi, \psi, z)$$

are identities in u, v, z, the right hand members are independent of z; hence

$$f_x \phi_z + f_y \psi_z + f_z = 0,$$
$$g_x \phi_z + g_y \psi_z + g_z = 0.$$

If we substitute x and y in $w = h(x, y, z)$ we get $w = h(\phi, \psi, z) = H(u, v, z)$.

Now the Jacobian

$$J = \begin{vmatrix} f_x & f_y & f_z \\ g_x & g_y & g_z \\ h_x & h_y & h_z \end{vmatrix} = \begin{vmatrix} f_x & f_y & 0 \\ g_x & g_y & 0 \\ h_x & h_y & H_z \end{vmatrix} = H_z \begin{vmatrix} f_x & f_y \\ g_x & g_y \end{vmatrix}$$

where the second determinant is obtained from J by multiplying the first column by ϕ_z, the second column by ψ_z and adding both to the third.

Since $J = 0$, $\dfrac{\partial(f, g)}{\partial(x, y)} \neq 0$ we conclude that $H_z = 0$.

Thus H is independent of z and a function of u and v alone; that is $w = H(u, v)$.

13.15 Illustrated Examples (Miscellaneous Types)

Example 13.15.1. *Given that z is a function of x and y and that*

$$x = u + v, \quad y = uv.$$

Prove that

$$\frac{\partial^2 z}{\partial u^2} - 2\frac{\partial^2 z}{\partial u \partial v} + \frac{\partial^2 z}{\partial v^2} = (x^2 - 4y)\frac{\partial^2 z}{\partial y^2} - 2\frac{\partial z}{\partial y}.$$

[C.H. 1981, 89, 93]

Solution: Using Chain Rules, we write

$$\frac{\partial z}{\partial u} = \frac{\partial z}{\partial x}\frac{\partial x}{\partial u} + \frac{\partial z}{\partial y}\frac{\partial y}{\partial u} = \frac{\partial z}{\partial x} \cdot 1 + \frac{\partial z}{\partial y} \cdot v. \quad \left(\because \frac{\partial x}{\partial u} = 1 \text{ and } \frac{\partial y}{\partial u} = v\right)$$

Similarly

$$\frac{\partial z}{\partial v} = \frac{\partial z}{\partial x} \cdot 1 + \frac{\partial z}{\partial y} \cdot u.$$

We shall use the following operational symbols:

$$\frac{\partial}{\partial u} \equiv \frac{\partial}{\partial x} + v\frac{\partial}{\partial y} \qquad (13.15.1)$$

$$\frac{\partial}{\partial v} \equiv \frac{\partial}{\partial x} + u\frac{\partial}{\partial y}. \qquad (13.15.2)$$

$$\frac{\partial^2 z}{\partial u^2} = \frac{\partial}{\partial u}\left(\frac{\partial z}{\partial u}\right) = \frac{\partial}{\partial u}\left(\frac{\partial z}{\partial x} + v\frac{\partial z}{\partial y}\right)$$

$$= \frac{\partial}{\partial u}\left(\frac{\partial z}{\partial x}\right) + v\frac{\partial}{\partial u}\left(\frac{\partial z}{\partial y}\right)$$

$$= \left\{\frac{\partial}{\partial x}\left(\frac{\partial z}{\partial x}\right) + v\frac{\partial}{\partial y}\left(\frac{\partial z}{\partial x}\right)\right\} + v\left\{\frac{\partial}{\partial x}\left(\frac{\partial z}{\partial y}\right) + v\frac{\partial}{\partial y}\left(\frac{\partial z}{\partial y}\right)\right\}$$

[using (13.15.1)]

$$= \frac{\partial^2 z}{\partial x^2} + 2v\frac{\partial^2 z}{\partial x \partial y} + v^2\frac{\partial^2 z}{\partial y^2} \quad \left[\text{assuming } \frac{\partial^2 z}{\partial y \partial x} = \frac{\partial^2 z}{\partial x \partial y}\right].$$

Similarly,

$$\frac{\partial^2 z}{\partial v^2} = \frac{\partial^2 z}{\partial x^2} + 2u\frac{\partial^2 z}{\partial x \partial y} + u^2\frac{\partial^2 z}{\partial y^2}.$$

CHAPTER 13: FUNCTIONS OF SEVERAL VARIABLES

Also

$$\frac{\partial^2 z}{\partial u \partial v} = \frac{\partial}{\partial u}\left(\frac{\partial z}{\partial v}\right) = \frac{\partial}{\partial u}\left(\frac{\partial z}{\partial x} + u\frac{\partial z}{\partial y}\right) \quad \text{[see equation (13.15.2)]}$$

$$= \frac{\partial}{\partial u}\left(\frac{\partial z}{\partial x}\right) + \frac{\partial z}{\partial y} + u\frac{\partial}{\partial u}\left(\frac{\partial z}{\partial y}\right)$$

$$= \left\{\frac{\partial}{\partial x}\left(\frac{\partial z}{\partial x}\right) + v\frac{\partial}{\partial y}\left(\frac{\partial z}{\partial x}\right)\right\} + \frac{\partial z}{\partial y} + u\left\{\frac{\partial}{\partial x}\left(\frac{\partial z}{\partial y}\right) + v\frac{\partial}{\partial y}\left(\frac{\partial z}{\partial y}\right)\right\}$$

$$= \frac{\partial^2 z}{\partial x^2} + v\frac{\partial^2 z}{\partial y \partial x} + \frac{\partial z}{\partial y} + u\left\{\frac{\partial^2 z}{\partial x \partial y} + v\frac{\partial^2 z}{\partial y^2}\right\}$$

$$= \frac{\partial^2 z}{\partial x^2} + uv\frac{\partial^2 z}{\partial y^2} + u\frac{\partial^2 z}{\partial y \partial x} + v\frac{\partial^2 z}{\partial x \partial y} + \frac{\partial z}{\partial y}.$$

Putting these values in the L.H.S. of the required relation:

i.e., in $\dfrac{\partial^2 z}{\partial u^2} - 2\dfrac{\partial^2 z}{\partial u \partial v} + \dfrac{\partial^2 z}{\partial v^2}$, we get

$$\left\{\frac{\partial^2 z}{\partial x^2} + 2v\frac{\partial^2 z}{\partial x \partial y} + v^2\frac{\partial^2 z}{\partial y^2}\right\} - 2\left\{\frac{\partial^2 z}{\partial x^2} + v\frac{\partial^2 z}{\partial y \partial x} + \frac{\partial z}{\partial y} + u\frac{\partial^2 z}{\partial x \partial y} + uv\frac{\partial^2 z}{\partial y^2}\right\}$$

$$+ \left\{\frac{\partial^2 z}{\partial x^2} + 2u\frac{\partial^2 z}{\partial x \partial y} + u^2\frac{\partial^2 z}{\partial y^2}\right\}$$

$$= -2\frac{\partial z}{\partial y} + \frac{\partial^2 z}{\partial y^2}(v^2 - 2uv + u^2) = -2\frac{\partial z}{\partial y} + \frac{\partial^2 z}{\partial y^2}\left\{(u+v)^2 - 4uv\right\}$$

$$= -2\frac{\partial z}{\partial y} + (x^2 - 4y)\frac{\partial^2 z}{\partial y^2} \quad \left[\text{terms containing } \frac{\partial^2 z}{\partial x \partial y} \text{ all cancel}\right]$$

$$= \text{R.H.S. of the required relation (Proved).}$$

Example 13.15.2. If $u = xy$ and $v = z - 2x$ and $w = f(u, v) = 0$ where w is a differentiable function of u and v such that $\frac{\partial f}{\partial v} \neq 0$ and if it is given that z is dependent on the two independent variables x and y, prove that

$$x\frac{\partial z}{\partial x} - y\frac{\partial z}{\partial y} = 2x. \qquad \text{[C.H. 1992]}$$

Solution: $w = f(u, v)$ where u and v are given by

$$u = xy, \quad v = z - 2x.$$

$\therefore$ By Chain rule, $\quad \dfrac{\partial w}{\partial x} = \dfrac{\partial w}{\partial u} \cdot \dfrac{\partial u}{\partial x} + \dfrac{\partial w}{\partial v} \cdot \dfrac{\partial v}{\partial x}$

$$= f_u \cdot y + f_v \cdot \left(\frac{\partial z}{\partial x} - 2\right). \quad (\because w = f)$$

But since $w = f(u,v) = 0$, $\dfrac{\partial w}{\partial x} = 0$. Hence

$$0 = f_u \cdot y + f_v \cdot \left(\dfrac{\partial z}{\partial x} - 2\right)$$

or, $\dfrac{\partial z}{\partial x} = -\dfrac{f_u \cdot y}{f_v} + 2 \quad (\because f_v \neq 0)$

or, $x\dfrac{\partial z}{\partial x} = -\dfrac{f_u}{f_v}xy + 2x.$ \hfill (13.15.3)

Again, by Chain Rule,

$$\dfrac{\partial w}{\partial y} = \dfrac{\partial w}{\partial u}\cdot\dfrac{\partial u}{\partial y} + \dfrac{\partial w}{\partial v}\cdot\dfrac{\partial v}{\partial y}$$

$$0 = f_u \cdot (x) + f_v \left(\dfrac{\partial z}{\partial y}\right)$$

or, $\dfrac{\partial z}{\partial y} = -x\dfrac{f_u}{f_v} \quad (\because f_v \neq 0)$

or, $y\dfrac{\partial z}{\partial y} = -xy\dfrac{f_u}{f_v}.$ \hfill (13.15.4)

Hence from (13.15.3) and (13.15.4), on subtraction, the required result follows at once.

Example 13.15.3. If $u = \dfrac{1}{\sqrt{1 - 2xy + y^2}}$, prove that

$$\dfrac{\partial}{\partial x}\left\{(1-x^2)\dfrac{\partial u}{\partial x}\right\} + \dfrac{\partial}{\partial y}\left(y^2\dfrac{\partial u}{\partial y}\right) = 0. \hfill \text{[C.H. 1983]}$$

Solution:

$$\dfrac{\partial u}{\partial x} = -\dfrac{1}{2}\dfrac{1}{(1-2xy+y^2)^{3/2}}\cdot(-2y) = \dfrac{y}{(1-2xy+y^2)^{3/2}}.$$

$$\therefore \dfrac{\partial}{\partial x}\left\{(1-x^2)\dfrac{\partial u}{\partial x}\right\} = \dfrac{\partial}{\partial x}\left\{\dfrac{(1-x^2)y}{(1-2xy+y^2)^{3/2}}\right\}$$

$$= \dfrac{-2xy}{(1-2xy+y^2)^{3/2}}$$

$$\qquad + y(1-x^2)\times -\dfrac{3}{2}(1-2xy+y^2)^{-5/2}\cdot(-2y)$$

$$= \dfrac{-2xy}{(1-2xy+y^2)^{3/2}} + \dfrac{3y^2(1-x^2)}{(1-2xy+y^2)^{5/2}}$$

$$= \dfrac{-2xy(1-2xy+y^2) + 3y^2(1-x^2)}{(1-2xy+y^2)^{5/2}}. \hfill (13.15.5)$$

CHAPTER 13: FUNCTIONS OF SEVERAL VARIABLES

Again
$$\frac{\partial u}{\partial y} = -\frac{1}{2}\frac{1}{(1-2xy+y^2)^{3/2}}(2y-2x) = \frac{x-y}{(1-2xy+y^2)^{3/2}}.$$

$$\therefore \frac{\partial}{\partial y}\left(y^2\frac{\partial u}{\partial y}\right) = \frac{\partial}{\partial y}\left\{\frac{(x-y)y^2}{(1-2xy+y^2)^{3/2}}\right\}$$

$$= \frac{(2xy-3y^2)}{(1-2xy+y^2)^{3/2}} + (x-y)y^2 \cdot -\frac{3}{2}\frac{1}{(1-2xy+y^2)^{5/2}}(-2x+2y)$$

$$= \frac{2xy-3y^2}{(1-2xy+y^2)^{3/2}} + \frac{3y^2(x-y)^2}{(1-2xy+y^2)^{5/2}}$$

$$= \frac{(2xy-3y^2)(1-2xy+y^2) + 3y^2(x-y)^2}{(1-2xy+y^2)^{5/2}}. \qquad (13.15.6)$$

On adding, $\frac{\partial}{\partial x}\left\{(1-x^2)\frac{\partial u}{\partial x}\right\} + \frac{\partial}{\partial y}\left\{y^2\frac{\partial u}{\partial y}\right\}$

$$= \frac{-2xy(1-2xy+y^2) + 3y^2(1-x^2) + (2xy-3y^2)(1-2xy+y^2) + 3y^2(x-y)^2}{(1-2xy+y^2)^{5/2}}.$$

Numerator $= (1-2xy+y^2)(-2xy+2xy-3y^2) + 3y^2(1-x^2+x^2+y^2-2xy)$
$= -3y^2(1-2xy+y^2) + 3y^2(1-2xy+y^2) = 0.$

Hence etc.

Example 13.15.4. If $\theta = t^n e^{-(r^2/4tk)}$, find for what value of n,

$$\frac{1}{r^2}\frac{\partial}{\partial r}\left(r^2\frac{\partial \theta}{\partial r}\right) = \frac{1}{k}\frac{\partial \theta}{\partial t} \qquad (k \text{ is a constant}). \qquad \text{[C.H. 1983, Special]}$$

Solution: Since, $\theta = t^n e^{-(r^2/4tk)}$,

$$\frac{\partial \theta}{\partial r} = t^n e^{-(r^2/4tk)} \cdot \frac{-2r}{4tk} = -\frac{\theta r}{2tk}. \qquad \left(\because \theta = t^n e^{-(r^2/4tk)}\right)$$

$$\therefore \frac{\partial}{\partial r}\left(r^2\frac{\partial \theta}{\partial r}\right) = \frac{\partial}{\partial r}\left(-\frac{\theta r^3}{2tk}\right)$$

$$= -\frac{r^3}{2tk}\frac{\partial \theta}{\partial r} - \frac{3r^2\theta}{2tk} = \frac{\theta r^4}{4t^2k^2} - \frac{3r^2\theta}{2tk}$$

or, $\frac{1}{r^2}\frac{\partial}{\partial r}\left(r^2\frac{\partial \theta}{\partial r}\right) = \frac{\theta r^2}{4t^2k^2} - \frac{3\theta}{2tk}.$

Again
$$\frac{\partial \theta}{\partial t} = nt^{n-1}e^{-(r^2/4tk)} + t^n e^{-(r^2/4tk)} \cdot \frac{r^2}{4t^2 k}$$
$$= \frac{n\theta}{t} + \frac{\theta r^2}{4t^2 k}.$$

$$\therefore \frac{1}{k}\frac{\partial \theta}{\partial t} = \frac{n\theta}{kt} + \frac{\theta r^2}{4t^2 k^2}.$$

From the given condition we can then write

$$\frac{\theta r^2}{4t^2 k^2} - \frac{3\theta}{2tk} = \frac{n\theta}{kt} + \frac{\theta r^2}{4t^2 k^2}$$

whence **n = −3/2**.

Example 13.15.5. *If* λ, μ, ν *are functions of* x, y, z *given by*

$$x = \lambda + \mu + \nu, \quad y = \lambda^2 + \mu^2 + \nu^2, \quad z = \lambda^3 + \mu^3 + \nu^3$$

prove that

$$\frac{\partial \lambda}{\partial x} = \frac{\mu\nu(\nu - \mu)}{(\lambda - \mu)(\mu - \nu)(\nu - \lambda)}.$$
[C.H. 1983]

Solution: Differentiating (partially) w.r.t. x we get

$$\lambda_x + \mu_x + \nu_x = 1$$
$$2\lambda\lambda_x + 2\mu\mu_x + 2\nu\nu_x = 0 \quad \text{i.e.,} \quad \lambda\lambda_x + \mu\mu_x + \nu\nu_x = 0$$
$$3\lambda^2\lambda_x + 3\mu^2\mu_x + 3\nu^2\nu_x = 0 \quad \text{i.e.,} \quad \lambda^2\lambda_x + \mu^2\mu_x + \nu^2\nu_x = 0.$$

Solving for λ_x (using Cramer's Rule) we get

$$\lambda_x = \frac{\begin{vmatrix} 1 & 1 & 1 \\ 0 & \mu & \nu \\ 0 & \mu^2 & \nu^2 \end{vmatrix}}{\begin{vmatrix} 1 & 1 & 1 \\ \lambda & \mu & \nu \\ \lambda^2 & \mu^2 & \nu^2 \end{vmatrix}} = \frac{\mu\nu^2 - \mu^2\nu}{(\lambda - \mu)(\mu - \nu)(\nu - \lambda)}.$$

Hence etc.

CHAPTER 13: FUNCTIONS OF SEVERAL VARIABLES

Example 13.15.6. *If* $u = \tan^{-1} \dfrac{x^3 + y^3}{x - y}$, *then prove that*

$$x^2 \frac{\partial^2 u}{\partial x^2} + 2xy \frac{\partial^2 u}{\partial x \partial y} + y^2 \frac{\partial^2 u}{\partial y^2} = \sin 2u (1 - 4\sin^2 u).$$ [C.H. 1984]

Solution:

$$\tan u = \frac{x^3 + y^3}{x - y} = x^2 \cdot \frac{1 + \left(\dfrac{y}{x}\right)^3}{1 - \dfrac{y}{x}} = x^2 \phi\left(\frac{y}{x}\right)$$

= a homogeneous function in x and y of degree 2.

∴ By Euler's Theorem

$$x \frac{\partial}{\partial x}(\tan u) + y \frac{\partial}{\partial y}(\tan u) = 2 \tan u$$

or, $\quad x \dfrac{\partial u}{\partial x} + y \dfrac{\partial u}{\partial y} = \dfrac{2 \tan u}{\sec^2 u} = 2 \tan u \cdot \cos^2 u = \sin 2u.$ (13.15.7)

Differentiate w.r.t. x

$$x \frac{\partial^2 u}{\partial x^2} + \frac{\partial u}{\partial x} + y \frac{\partial^2 u}{\partial x \partial y} = 2 \cos 2u \frac{\partial u}{\partial x}$$

$$x \frac{\partial^2 u}{\partial x^2} + y \frac{\partial^2 u}{\partial x \partial y} = (2 \cos 2u - 1) \frac{\partial u}{\partial x}.$$

Now multiply by x,

$$x^2 \frac{\partial^2 u}{\partial x^2} + xy \frac{\partial^2 u}{\partial x \partial y} = (2 \cos 2u - 1) x \frac{\partial u}{\partial x}.$$ (13.15.8)

Again differentiate (13.15.7) w.r.t. y and then multiply by y; we then get

$$xy \frac{\partial^2 u}{\partial y \partial x} + y^2 \frac{\partial^2 u}{\partial y^2} = (2 \cos 2u - 1) y \frac{\partial u}{\partial y}.$$ (13.15.9)

Adding (13.15.8) and (13.15.9), using $\dfrac{\partial^2 u}{\partial x \partial y} = \dfrac{\partial^2 u}{\partial y \partial x}$ we get

$$x^2 \frac{\partial^2 u}{\partial x^2} + 2xy \frac{\partial^2 u}{\partial x \partial y} + y^2 \frac{\partial^2 u}{\partial y^2} = (2 \cos 2u - 1) \left(x \frac{\partial u}{\partial x} + y \frac{\partial u}{\partial y} \right)$$

$$= (2 \cos 2u - 1) \sin 2u$$

$$= \sin 2u (1 - 4 \sin^2 u). \quad (\because \ \cos 2u = 1 - 2 \sin^2 u)$$

Example 13.15.7. If $u = \dfrac{(x^2+y^2)^n}{2n(2n-1)} + xf\left(\dfrac{y}{x}\right) + g\left(\dfrac{y}{x}\right)$, then prove that

$$x^2\dfrac{\partial^2 u}{\partial x^2} + 2xy\dfrac{\partial^2 u}{\partial x \partial y} + y^2\dfrac{\partial^2 u}{\partial y^2} = (x^2+y^2)^n.$$

Solution: Let $u = U + V + W$, where $U = \dfrac{(x^2+y^2)^n}{2n(2n-1)}$, $V = xf\left(\dfrac{y}{x}\right)$ and $W = g\left(\dfrac{y}{x}\right)$.

Since $U = x^{2n}\dfrac{(1+(y/x)^2)^n}{2n(2n-1)} =$ a homogeneous function in x and y of degree $2n$.

By Euler's Theorem,

$$x\dfrac{\partial U}{\partial x} + y\dfrac{\partial U}{\partial y} = 2nU. \qquad (13.15.10)$$

Differentiating partially w.r.t. x,

$$\dfrac{\partial U}{\partial x} + x\dfrac{\partial^2 U}{\partial x^2} + y\dfrac{\partial^2 U}{\partial x \partial y} = 2n\dfrac{\partial U}{\partial x}$$

or, $\quad x\dfrac{\partial^2 U}{\partial x^2} + y\dfrac{\partial^2 U}{\partial x \partial y} = (2n-1)\dfrac{\partial U}{\partial x}.$

Multiplying by x,

$$x^2\dfrac{\partial^2 U}{\partial x^2} + xy\dfrac{\partial^2 U}{\partial x \partial y} = (2n-1)x\dfrac{\partial U}{\partial x}. \qquad (13.15.11)$$

Again differentiating partially (13.15.10) w.r.t. y and then multiplying by y, we get

$$y^2\dfrac{\partial^2 U}{\partial y^2} + xy\dfrac{\partial^2 U}{\partial y \partial x} = (2n-1)y\dfrac{\partial U}{\partial y}. \qquad (13.15.12)$$

Adding (13.15.11) and (13.15.12),

$$x^2\dfrac{\partial^2 U}{\partial x^2} + 2xy\dfrac{\partial^2 U}{\partial x \partial y} + y^2\dfrac{\partial^2 U}{\partial y^2} = (2n-1)\left(x\dfrac{\partial u}{\partial x} + y\dfrac{\partial U}{\partial y}\right)$$

$$= (2n-1)2nU = (x^2+y^2)^n. \qquad (13.15.13)$$

Again

$$V = xf\left(\dfrac{y}{x}\right) \Rightarrow x\dfrac{\partial V}{\partial x} + y\dfrac{\partial V}{\partial y} = 1.V$$

(since V is a homogeneous function in x and y of degree one).

CHAPTER 13: FUNCTIONS OF SEVERAL VARIABLES

Operating by $\dfrac{\partial}{\partial x}$ and multiplying by x we get

$$x^2 \frac{\partial^2 V}{\partial x^2} + xy \frac{\partial^2 V}{\partial x \partial y} = 0$$

operating by $\dfrac{\partial}{\partial y}$ and multiplying by y we get

$$y^2 \frac{\partial^2 V}{\partial y^2} + xy \frac{\partial^2 V}{\partial y \partial x} = 0.$$

Adding,

$$x^2 \frac{\partial^2 V}{\partial x^2} + 2xy \frac{\partial^2 V}{\partial x \partial y} + y^2 \frac{\partial^2 V}{\partial y^2} = 0. \qquad (13.15.14)$$

Lastly $W = g\left(\dfrac{y}{x}\right) \Rightarrow x \dfrac{\partial W}{\partial x} + y \dfrac{\partial W}{\partial y} = 0$. $W = 0$, since W is homogeneous of degree 0 operating by $\dfrac{\partial}{\partial x}$ and multiplying by x; operating by $\dfrac{\partial}{\partial y}$ and then multiplying by y and then adding the two results we get

$$x^2 \frac{\partial^2 W}{\partial x^2} + 2xy \frac{\partial^2 W}{\partial x \partial y} + y^2 \frac{\partial^2 W}{\partial y^2} = 0. \qquad (13.15.15)$$

Adding (13.15.13), (13.15.14) and (13.15.15) we get

$$\left(x^2 \frac{\partial^2}{\partial x^2} + 2xy \frac{\partial^2}{\partial x \partial y} + y^2 \frac{\partial^2}{\partial y^2} \right)(U + V + W) = (x^2 + y^2)^n.$$

Hence the result.

Example 13.15.8. If $u = \dfrac{(ax^3 + by^3)^n}{3n(3n-1)} + xf\left(\dfrac{y}{x}\right)$, then find the value of

$$x^2 \frac{\partial^2 u}{\partial x^2} + 2xy \frac{\partial^2 u}{\partial x \partial y} + y^2 \frac{\partial^2 u}{\partial y^2}.$$

Solution: Let $u = U + V$.

$$U = \frac{(ax^3 + by^3)^n}{3n(3n-1)} = x^{3n} \frac{\left\{ a + b\left(\dfrac{y}{x}\right)^3 \right\}^n}{3n(3n-1)} = x^{3n} f\left(\dfrac{y}{x}\right)$$

$\qquad$ = homogeneous function of x and y of degree $3n$.

∴ By Euler's Theorem
$$x\frac{\partial U}{\partial x} + y\frac{\partial U}{\partial y} = 3nU.$$

Operating by $\dfrac{\partial}{\partial x}$ we get

$$x\frac{\partial^2 U}{\partial x^2} + y\frac{\partial^2 U}{\partial x \partial y} = (3n-1)\frac{\partial U}{\partial x}.$$

Multiplying by x, we get

$$x^2\frac{\partial^2 U}{\partial x^2} + xy\frac{\partial^2 U}{\partial x \partial y} = (3n-1)x\frac{\partial U}{\partial x}.$$

Similarly operating by $\dfrac{\partial}{\partial y}$ and then multiplying by y, we get

$$y^2\frac{\partial^2 U}{\partial y^2} + xy\frac{\partial^2 U}{\partial y \partial x} = (3n-1)y\frac{\partial U}{\partial y}.$$

Adding,

$$x^2\frac{\partial^2 U}{\partial x^2} + 2xy\frac{\partial^2 U}{\partial x \partial y} + y^2\frac{\partial^2 U}{\partial y^2} = (3n-1)\left(x\frac{\partial U}{\partial x} + y\frac{\partial U}{\partial y}\right)$$
$$= 3n(3n-1)U = (ax^3 + by^3)^n.$$

Again, $V = xf(y/x)$

= homogeneous function of x and y of degree 1.

By Euler's Theorem, $x\dfrac{\partial V}{\partial x} + y\dfrac{\partial V}{\partial y} = V.$

Operating by $\dfrac{\partial}{\partial x}$ and then multiplying by x, we obtain

$$x^2\frac{\partial^2 V}{\partial x^2} + xy\frac{\partial^2 V}{\partial x \partial y} = 0.$$

Similarly operating by $\dfrac{\partial}{\partial y}$ and then multiplying by y, we obtain

$$y^2\frac{\partial^2 V}{\partial y^2} + xy\frac{\partial^2 V}{\partial y \partial x} = 0.$$

CHAPTER 13: FUNCTIONS OF SEVERAL VARIABLES 857

Adding,
$$x^2 \frac{\partial^2 V}{\partial x^2} + y^2 \frac{\partial^2 V}{\partial y^2} + 2xy \frac{\partial^2 V}{\partial x \partial y} = 0.$$

$$\therefore \left(x^2 \frac{\partial^2}{\partial x^2} + 2xy \frac{\partial^2}{\partial x \partial y} + y^2 \frac{\partial^2}{\partial y^2}\right) u = \left(x^2 \frac{\partial^2}{\partial x^2} + 2xy \frac{\partial^2}{\partial x \partial y} + y^2 \frac{\partial^2}{\partial y^2}\right)(U+V)$$
$$= (ax^3 + by^3)^n + 0 = (ax^3 + by^3)^n.$$

Example 13.15.9. *By the transformation* $\xi = a + \alpha x + \beta y, \eta = b - \beta x + \alpha y$, *where* α, β, a, b *are all constant and* $\alpha^2 + \beta^2 = 1$, *the function* $u(x,y)$ *is transfered into* $U(\xi, \eta)$. *Prove that* $U_{\xi\xi} U_{\eta\eta} - U_{\xi\eta}^2 = u_{xx} u_{yy} - u_{xy}^2$. [C.H. 1986]

Solution: $U(\xi, \eta) = u(x, y)$ [U is a function of ξ, η where $\xi = a + \alpha x + \beta y, \eta = b - \beta x + \alpha y.$]
By Chain Rule
$$\frac{\partial u}{\partial x} = \frac{\partial U}{\partial \xi} \frac{\partial \xi}{\partial x} + \frac{\partial U}{\partial \eta} \frac{\partial \eta}{\partial x}$$
$$= \frac{\partial U}{\partial \xi} \alpha - \beta \frac{\partial U}{\partial \eta}.$$

Thus, using operators, we may write
$$\frac{\partial}{\partial x} \equiv \alpha \frac{\partial}{\partial \xi} - \beta \frac{\partial}{\partial \eta}.$$

Similarly we obtain,
$$\frac{\partial}{\partial y} \equiv \beta \frac{\partial}{\partial \xi} + \alpha \frac{\partial}{\partial \eta}.$$

$$\frac{\partial^2 u}{\partial x^2} = \frac{\partial}{\partial x}\left(\frac{\partial u}{\partial x}\right) = \frac{\partial}{\partial x}\left(\alpha \frac{\partial U}{\partial \xi} - \beta \frac{\partial U}{\partial \eta}\right)$$
$$= \alpha \frac{\partial}{\partial x}\left\{\frac{\partial U}{\partial \xi}\right\} - \beta \frac{\partial}{\partial x}\left\{\frac{\partial U}{\partial \eta}\right\}$$
$$= \alpha \left[\alpha \frac{\partial}{\partial \xi}\left(\frac{\partial U}{\partial \xi}\right) - \beta \frac{\partial}{\partial \eta}\left(\frac{\partial U}{\partial \xi}\right)\right] - \beta \left[\alpha \frac{\partial}{\partial \xi}\left(\frac{\partial U}{\partial \eta}\right) - \beta \frac{\partial}{\partial \eta}\left(\frac{\partial U}{\partial \eta}\right)\right]$$

i.e., $u_{xx} = \alpha^2 U_{\xi\xi} - 2\alpha\beta U_{\xi\eta} + \beta^2 U_{\eta\eta}$.

Similarly
$$u_{yy} = \beta^2 U_{\xi\xi} + 2\alpha\beta U_{\xi\eta} + \alpha^2 U_{\eta\eta}.$$

Now
$$\frac{\partial^2 u}{\partial y \partial x} = \frac{\partial}{\partial y}\left(\frac{\partial u}{\partial x}\right) = \frac{\partial}{\partial y}\left[\alpha\frac{\partial U}{\partial \xi} - \beta\frac{\partial U}{\partial \eta}\right]$$
$$= \alpha\frac{\partial}{\partial y}\left(\frac{\partial U}{\partial \xi}\right) - \beta\frac{\partial}{\partial y}\left(\frac{\partial U}{\partial \eta}\right)$$
$$= \alpha\left[\beta\frac{\partial}{\partial \xi} + \alpha\frac{\partial}{\partial \eta}\right]\left(\frac{\partial U}{\partial \xi}\right) - \beta\left[\beta\frac{\partial}{\partial \xi} + \alpha\frac{\partial}{\partial \eta}\right]\left(\frac{\partial U}{\partial \eta}\right)$$
$$\left(\text{using } \frac{\partial}{\partial y} \equiv \beta\frac{\partial}{\partial \xi} + \alpha\frac{\partial}{\partial \eta}\right)$$
$$= \alpha\beta\frac{\partial^2 U}{\partial \xi^2} + \alpha^2\frac{\partial^2 U}{\partial \eta \partial \xi} - \beta^2\frac{\partial^2 U}{\partial \xi \partial \eta} - \alpha\beta\frac{\partial^2 U}{\partial \eta^2}$$

i.e., $u_{xy} = u_{yx} = \alpha\beta U_{\xi\xi} + \alpha^2 U_{\eta\xi} - \beta^2 U_{\xi\eta} - \alpha\beta U_{\eta\eta}$
$$= \alpha\beta[U_{\xi\xi} - U_{\eta\eta}] + (\alpha^2 - \beta^2)U_{\xi\eta}.$$

Hence
$$u_{xx}u_{yy} - u_{xy}^2 = \left(\alpha^2 U_{\xi\xi} - 2\alpha\beta U_{\xi\eta} + \beta^2 U_{\eta\eta}\right)\left(\beta^2 U_{\xi\xi} + 2\alpha\beta U_{\xi\eta} + \alpha^2 U_{\eta\eta}\right)$$
$$- \left(\alpha\beta(U_{\xi\xi} - U_{\eta\eta}) + (\alpha^2 - \beta^2)U_{\xi\eta}\right)^2$$
$$= U_{\xi\xi}U_{\eta\eta}(\alpha^4 + \beta^4 + 2\alpha^2\beta^2) - U_{\xi\eta}^2(\alpha^4 + \beta^4 + 2\alpha^2\beta^2)$$
$$= \left(U_{\xi\xi}U_{\eta\eta} - U_{\xi\eta}^2\right)(\alpha^2 + \beta^2)^2$$
$$= U_{\xi\xi}U_{\eta\eta} - U_{\xi\eta}^2. \quad (\because \alpha^2 + \beta^2 = 1)$$

This proves the result.

Example 13.15.10. Let $V = \sin^{-1}\sqrt{\dfrac{x^{1/3} + y^{1/3}}{x^{1/2} + y^{1/2}}}$. Prove that

$$x^2\frac{\partial^2 V}{\partial x^2} + 2xy\frac{\partial^2 V}{\partial x \partial y} + y^2\frac{\partial^2 V}{\partial y^2} = \frac{\tan V}{12}\left(\frac{13}{12} + \frac{\tan^2 V}{12}\right).$$

Solution: Given
$$\sin V = \left(\frac{x^{1/3} + y^{1/3}}{x^{1/2} + y^{1/2}}\right)^{1/2} = \frac{x^{1/6}}{x^{1/4}}\left(\frac{1 + (y/x)^{1/3}}{1 + (y/x)^{1/2}}\right)^{1/2}$$
$$= x^{-(1/12)}f(y/x)$$

CHAPTER 13: FUNCTIONS OF SEVERAL VARIABLES

i.e., $\sin V$ is a homogeneous function of x and y of degree $-\frac{1}{12}$.

$\therefore$ By Euler's Theorem, we get

$$x\frac{\partial}{\partial x}(\sin V) + y\frac{\partial}{\partial y}(\sin V) = -\frac{1}{12}\sin V$$

$$\text{or, } x\frac{\partial V}{\partial x} + y\frac{\partial V}{\partial y} = -\frac{1}{12}\tan V. \qquad (13.15.16)$$

Differentiating (13.15.16) partially with respect to x and then multiplying by x we get

$$x^2\frac{\partial^2 V}{\partial x^2} + xy\frac{\partial^2 V}{\partial x \partial y} = -\frac{1}{12}\sec^2 V \left(x\frac{\partial V}{\partial x}\right) - x\frac{\partial V}{\partial x}$$

$$= -x\frac{\partial V}{\partial x}\left(\frac{1}{12}\sec^2 V + 1\right).$$

Again differentiating (13.15.16) partially with respect to y and then multiplying by y we get

$$xy\frac{\partial^2 V}{\partial y \partial x} + y^2\frac{\partial^2 V}{\partial y^2} = -y\frac{\partial V}{\partial y}\left(\frac{1}{12}\sec^2 V + 1\right).$$

Adding these two results $\left(\text{assuming of course } \frac{\partial^2 V}{\partial x \partial y} = \frac{\partial^2 V}{\partial y \partial x}\right)$ we get

$$x^2\frac{\partial^2 V}{\partial x^2} + 2xy\frac{\partial^2 V}{\partial x \partial y} + y^2\frac{\partial^2 V}{\partial y^2} = -\left(\frac{1}{12}\sec^2 V + 1\right)\left(x\frac{\partial V}{\partial x} + y\frac{\partial V}{\partial y}\right)$$

$$= -\left(\frac{1}{12}\sec^2 V + 1\right)\left(-\frac{1}{12}\tan V\right) \quad \text{[using (13.15.16)]}$$

$$= \frac{\tan V}{12}\left(\frac{13}{12} + \frac{\tan^2 V}{12}\right) \quad \text{(Proved)}.$$

Example 13.15.11. If $\dfrac{x^2}{a^2 + u} + \dfrac{y^2}{b^2 + u} + \dfrac{z^2}{c^2 + u} = 1$, prove that

$$\left(\frac{\partial u}{\partial x}\right)^2 + \left(\frac{\partial u}{\partial y}\right)^2 + \left(\frac{\partial u}{\partial z}\right)^2 = 2\left(x\frac{\partial u}{\partial x} + y\frac{\partial u}{\partial y} + z\frac{\partial u}{\partial z}\right). \qquad \text{[C.H. 1966, 91]}$$

Solution: Given $\dfrac{x^2}{a^2 + u} + \dfrac{y^2}{b^2 + u} + \dfrac{z^2}{c^2 + u} = 1$.

Differentiating partially with respect to x, we get

$$\frac{2x}{a^2 + u} - \left\{\frac{x^2}{(a^2 + u)^2} + \frac{y^2}{(b^2 + u)^2} + \frac{z^2}{(c^2 + u)^2}\right\}\frac{\partial u}{\partial x} = 0$$

$$\text{or, } \frac{\partial u}{\partial x} = \frac{2x}{a^2 + u} \cdot \frac{1}{P}, \quad \text{where } P = \sum \frac{x^2}{(a^2 + u)^2}.$$

Similarly, we obtain

$$\frac{2y}{b^2+u} = P \cdot \frac{\partial u}{\partial y}$$

or, $\quad \dfrac{\partial u}{\partial y} = \dfrac{1}{P} \cdot \dfrac{2y}{b^2+u}$

and $\quad \dfrac{\partial u}{\partial z} = \dfrac{1}{P} \cdot \dfrac{2z}{c^2+u}.$

$$\therefore \left(\frac{\partial u}{\partial x}\right)^2 + \left(\frac{\partial u}{\partial y}\right)^2 + \left(\frac{\partial u}{\partial z}\right)^2 = \frac{4}{P^2}\left\{\frac{x^2}{(a^2+u)^2} + \frac{y^2}{(b^2+u)^2} + \frac{z^2}{(c^2+u)^2}\right\}$$

$$= \frac{4}{P^2} \cdot P = \frac{4}{P}.$$

Again

$$2\left(x\frac{\partial u}{\partial x} + y\frac{\partial u}{\partial y} + z\frac{\partial u}{\partial z}\right) = 2\left[\frac{2x^2}{a^2+u} + \frac{2y^2}{b^2+u} + \frac{2z^2}{c^2+u}\right]\frac{1}{P}$$

$$= \frac{4}{P} \cdot \left(\because \sum \frac{x^2}{a^2+u} = 1\right)$$

Hence the result follows.

Example 13.15.12. *If $x^x \cdot y^y \cdot z^z = k$ (constant), prove that at the point (x,y,z) where $x = y = z$,*

$$\frac{\partial^2 z}{\partial x \partial y} = -\frac{1}{x\log_e(ex)}.$$

Solution: Given condition (taking logarithm with respect to the base e) gives

$$x\log x + y\log y + z\log z = \log k.$$

Differentiating with respect x, we get

$$(\log x + 1) + (1 + \log z)\frac{\partial z}{\partial x} = 0$$

or, $\quad \dfrac{\partial z}{\partial x} = -\dfrac{1+\log x}{1+\log z}.$

CHAPTER 13: FUNCTIONS OF SEVERAL VARIABLES

Similarly,
$$\frac{\partial z}{\partial y} = -\frac{1+\log y}{1+\log z}.$$

$$\therefore \frac{\partial^2 z}{\partial x \partial y} = \frac{\partial}{\partial x}\left(\frac{\partial z}{\partial y}\right) = \frac{\partial}{\partial x}\left\{-\frac{1+\log y}{1+\log z}\right\}$$

$$= (1+\log y)\frac{1}{(1+\log z)^2} \cdot \frac{1}{z} \cdot \frac{\partial z}{\partial x}$$

$$= \frac{1+\log y}{(1+\log z)^2} \cdot \left\{-\frac{1+\log x}{1+\log z}\right\} \cdot \frac{1}{z} = -\frac{(1+\log y)(1+\log x)}{(1+\log z)^3} \cdot \frac{1}{z}$$

at $x = y = z$ this becomes
$$= -\frac{(1+\log x)(1+\log x)}{(1+\log x)^3} \cdot \frac{1}{x}$$

$$= -\frac{1}{x(1+\log x)} = -\frac{1}{x\log_e(ex)}. \quad \text{(Proved)}$$

Example 13.15.13. *By putting $G = x^n H$ and changing the independent variables x, y to u, v where $u = y/x$ and $v = xy$, transform the equations*

1. $x\dfrac{\partial G}{\partial x} + y\dfrac{\partial G}{\partial y} = nG;$

2. $x\dfrac{\partial U}{\partial x} - y\dfrac{\partial U}{\partial y} = 0.$

Hence show that

1. $G = x^n \phi(y/x);$ and

2. $U = \psi(xy).$

Solution:

1.
$$\frac{\partial G}{\partial x} = \frac{\partial}{\partial x}(x^n H) = nx^{n-1}H + x^n \frac{\partial H}{\partial x}$$

$$= nx^{n-1}H + x^n\left\{\frac{\partial H}{\partial u} \cdot \frac{\partial u}{\partial x} + \frac{\partial H}{\partial v} \cdot \frac{\partial v}{\partial x}\right\}$$

$$= nx^{n-1}H + x^n\left\{\frac{\partial H}{\partial u} \cdot \left(-\frac{y}{x^2}\right) + \frac{\partial H}{\partial v} \cdot (y)\right\}.$$

$$\therefore x\frac{\partial G}{\partial x} = nx^n H - yx^{n-1}\frac{\partial H}{\partial u} + x^{n+1}y\frac{\partial H}{\partial v}.$$

Similarly,
$$y\frac{\partial G}{\partial y} = x^{n-1}y\frac{\partial H}{\partial u} + x^{n+1}y\frac{\partial H}{\partial v}.$$

∴ The equation $x\dfrac{\partial G}{\partial x} + y\dfrac{\partial G}{\partial y} = nG$ becomes

$$nx^n H + 2x^{n+1}y\frac{\partial H}{\partial v} = nG \Rightarrow \frac{\partial H}{\partial v} = 0 \quad (\because G = x^n H)$$

i.e., H is independent of v i.e., H is a function of $u = y/x$ only.

$$\therefore G = x^n H = x^n \phi(y/x).$$

2.
$$\frac{\partial U}{\partial x} = \frac{\partial U}{\partial u}\cdot\frac{\partial u}{\partial x} + \frac{\partial U}{\partial v}\frac{\partial v}{\partial x} = \frac{\partial U}{\partial u}\left(-\frac{y}{x^2}\right) + \frac{\partial U}{\partial v}(y)$$

or, $\quad x\dfrac{\partial U}{\partial x} = -\dfrac{y}{x}\dfrac{\partial U}{\partial u} + xy\dfrac{\partial U}{\partial v}.$

Similarly,
$$y\frac{\partial U}{\partial y} = \frac{y}{x}\frac{\partial U}{\partial u} + xy\frac{\partial U}{\partial v}.$$

∴ The equation $x\dfrac{\partial U}{\partial x} - y\dfrac{\partial U}{\partial u} = 0$ becomes $2\dfrac{y}{x}\dfrac{\partial U}{\partial u} = 0 \Rightarrow \dfrac{\partial U}{\partial u} = 0.$

i.e., U is independent of u and hence U is a function of v alone i.e., $U = \psi(xy)$.

Example 13.15.14. *If $u = x + y + z$, $v = yz + zx + xy$, $w = xyz$, then prove that*

$$\frac{\partial^2 x}{\partial w^2} = -\frac{2(2x - y - z)}{\{(x - y)(x - z)\}^3}.$$

Solution: $v = x(y + z) + \dfrac{w}{x} = x(u - x) + \dfrac{w}{x} = xu - x^2 + \dfrac{w}{x}.$

Differentiate with respect to w

$$0 = \frac{\partial x}{\partial w}u - 2x\frac{\partial x}{\partial w} + \frac{1}{x} - \frac{w}{x^2}\frac{\partial x}{\partial w}.$$

$$\therefore \frac{\partial x}{\partial w} = -\frac{x}{ux^2 - 2x^3 - w}.$$

Differentiating again with respect to w,

$$\frac{\partial^2 x}{\partial w^2} = -\frac{\dfrac{\partial x}{\partial w}(ux^2 - 2x^3 - w) - x\left[2xu\dfrac{\partial x}{\partial w} - 6x^2\dfrac{\partial x}{\partial w} - 1\right]}{(ux^2 - 2x^3 - w)^2}$$

CHAPTER 13: FUNCTIONS OF SEVERAL VARIABLES 863

i.e.,

$$\begin{aligned}\frac{\partial^2 x}{\partial w^2} &= -\frac{\frac{\partial x}{\partial w}(ux^2 - 2x^3 - w - 2x^2u + 6x^3) + x}{(ux^2 - 2x^3 - w)^2} \\ &= -\frac{\frac{-x}{ux^2 - 2x^3 - w}(ux^2 - 2x^3 - w - 2x^2u + 6x^3) + x}{(ux^2 - 2x^3 - w)^2} \\ &= -\frac{+x^3u - 4x^4 + wx + x(ux^2 - 2x^3 - w)}{(ux^2 - 2x^3 - w)^3} \\ &= -\frac{-6x^4 + 2x^3u}{\{ux^2 - 2x^3 - w\}^3} \\ &= -\frac{2x^3(x+y+z) - 6x^4}{\{x^2(x+y+z) - 2x^3 - xyz\}^3} \\ &= -\frac{2x^3(x+y+z-3x)}{x^3\{x^2 + xy + xz - 2x^2 - yz\}^3} \\ &= -\frac{2(y+z-2x)}{\{y(x-z) - x(x-z)\}^3} \\ &= -\frac{2(2x-y-z)}{\{(x-z)(x-y)\}^3}.\end{aligned}$$

Example 13.15.15. [P and Q are functions of x and y. Then $Pdx + Qdy$ is called a perfect differential or exact differential of some function z of x and y if $dz = Pdx+Qdy$].

1. If $Pdx + Qdy$ be a perfect differential of some function z of x and y, then prove that $\dfrac{\partial P}{\partial y} = \dfrac{\partial Q}{\partial x}$.

2. If $Pdx + Qdy + Rdz$ can be made a perfect differential of some function of x, y, z by multiplying each term by a common factor $\mu(x, y, z)$, then prove that

$$P\left(\frac{\partial Q}{\partial z} - \frac{\partial R}{\partial y}\right) + Q\left(\frac{\partial R}{\partial x} - \frac{\partial P}{\partial z}\right) + R\left(\frac{\partial P}{\partial y} - \frac{\partial Q}{\partial x}\right) = 0.$$

Solution: 1. $Pdx + Qdy = dz = \dfrac{\partial z}{\partial x}dx + \dfrac{\partial z}{\partial y}dy.$

When x and y are independent variables, their differentials are also their increments and may be given arbitrary values; say $dx = 1, dy = 0$ then it follows $P = \dfrac{\partial z}{\partial x}$. Similarly put $dy = 0, dx = 1$ then $Q = \dfrac{\partial z}{\partial y}$.

$$\therefore \frac{\partial P}{\partial y} = \frac{\partial}{\partial y}\left(\frac{\partial z}{\partial x}\right) = \frac{\partial^2 z}{\partial x \partial y} = \frac{\partial}{\partial x}\left(\frac{\partial z}{\partial y}\right) = \frac{\partial Q}{\partial x}.$$

2.
$$\mu(x,y,z)(Pdx+Qdy+Rdz) = df \quad (f \text{ is a function of } x,y,z)$$
$$= \frac{\partial f}{\partial x}dx + \frac{\partial f}{\partial y}dy + \frac{\partial f}{\partial z}dz.$$

Since dx, dy, dz are independent increments it is easy to see that
$$\frac{\partial f}{\partial x} = \mu P, \quad \frac{\partial f}{\partial y} = \mu Q, \quad \frac{\partial f}{\partial z} = \mu R.$$

Now observe that
$$\frac{\partial^2 f}{\partial y \partial x} = \frac{\partial}{\partial y}(\mu P) = \frac{\partial \mu}{\partial y}P + \mu \frac{\partial P}{\partial y}.$$
$$\frac{\partial^2 f}{\partial x \partial y} = \frac{\partial}{\partial x}(\mu Q) = \frac{\partial \mu}{\partial x}Q + \mu \frac{\partial Q}{\partial x}.$$
$$\therefore \frac{\partial \mu}{\partial y}P - \frac{\partial \mu}{\partial x}Q = \mu\left(\frac{\partial Q}{\partial x} - \frac{\partial P}{\partial y}\right); \text{ assuming } \frac{\partial^2 f}{\partial y \partial x} = \frac{\partial^2 f}{\partial x \partial y}.$$
$$\text{or, } R\left(\frac{\partial P}{\partial y} - \frac{\partial Q}{\partial x}\right) = \frac{R}{\mu}\left[\frac{\partial \mu}{\partial x}Q - \frac{\partial \mu}{\partial y}P\right]$$
$$= \frac{1}{\mu}\left[\frac{\partial \mu}{\partial x}QR - \frac{\partial \mu}{\partial y}PR\right].$$

Similarly obtain
$$Q\left(\frac{\partial R}{\partial x} - \frac{\partial P}{\partial z}\right) = \frac{1}{\mu}\left[\frac{\partial \mu}{\partial z}PQ - \frac{\partial \mu}{\partial x}QR\right]$$
$$\text{and } P\left(\frac{\partial Q}{\partial z} - \frac{\partial R}{\partial y}\right) = \frac{1}{\mu}\left[\frac{\partial \mu}{\partial y}PR - \frac{\partial \mu}{\partial z}PQ\right].$$

Add the three results. The required relation follows.

Example 13.15.16. *Let $z = f(x,y)$ be a function of two independent variables x and y, defined in a domain N and let it be differentiable at a point (x,y) of the domain. Then the first differential of z is given by $dz = \frac{\partial z}{\partial x}dx + \frac{\partial z}{\partial y}dy$. How do you obtain $d^2z, d^3z, \cdots, d^n z$?* (**Differentials of higher order**)

Solution: If dx and dy are regarded as constants and if $\frac{\partial z}{\partial x}$ and $\frac{\partial z}{\partial y}$ are differentiable at (x,y), then dz is a function of x and y and is itself differentiable at (x,y).

$\therefore$ Second differential of $z = d^2z = d(dz) = d\left(\frac{\partial z}{\partial x}\right)dx + d\left(\frac{\partial z}{\partial y}\right)dy$.

CHAPTER 13: FUNCTIONS OF SEVERAL VARIABLES

But
$$d\left(\frac{\partial z}{\partial x}\right) = \frac{\partial^2 z}{\partial x^2}dx + \frac{\partial^2 z}{\partial y \partial x}dy$$
$$\text{and} \quad d\left(\frac{\partial z}{\partial y}\right) = \frac{\partial^2 z}{\partial x \partial y}dx + \frac{\partial^2 z}{\partial y^2}dy.$$

Also by Young's Theorem, since $\dfrac{\partial z}{\partial x}$ and $\dfrac{\partial z}{\partial y}$ are differentiable we have

$$\frac{\partial^2 z}{\partial x \partial y} = \frac{\partial^2 z}{\partial y \partial x}.$$

$$\therefore \quad d^2 z = \frac{\partial^2 z}{\partial x^2}(dx)^2 + 2\frac{\partial^2 z}{\partial x \partial y}dx dy + \frac{\partial^2 z}{\partial y^2}(dy)^2$$
$$= \left(\frac{\partial}{\partial x}dx + \frac{\partial}{\partial y}dy\right)^2 z \quad \text{(in abbreviated notations)}.$$

Obtain with prescribed conditions assumed

$$d^3 z = \left(\frac{\partial}{\partial x}dx + \frac{\partial}{\partial y}dy\right)^3 z.$$
$$\cdots \quad \cdots \quad \cdots \quad \cdots$$
$$d^n z = \frac{\partial^n z}{\partial x^n}\cdot(dx)^n + n\frac{\partial^n z}{\partial x^{n-1}\partial y}(dx)^{n-1}dy + \cdots + \frac{\partial^n z}{\partial y^n}(dy)^n$$
$$= \left(\frac{\partial}{\partial x}dx + \frac{\partial}{\partial y}dy\right)^n z.$$

Note 13.15.1. *Here x and y are supposed to be independent variables and so dx and dy are treated as constants.*

Example 13.15.17. *If z is a function of x and y defined by*

$$2x^2 + 2y^2 + z^2 - 8xy - z + 8 = 0,$$

find d^2z at $(2, 0, 1)$.

Solution:

$$d^2 z = \frac{\partial^2 z}{\partial x^2}(dx)^2 + 2\frac{\partial^2 z}{\partial x \partial y}dx dy + \frac{\partial^2 z}{\partial y^2}(dy)^2. \quad (13.15.17)$$

We are to compute $\dfrac{\partial^2 z}{\partial x^2}, \dfrac{\partial^2 z}{\partial x \partial y}, \dfrac{\partial^2 z}{\partial y^2}$ at $(2,0,1)$

e.g.
$$2x^2 + 2y^2 + z^2 - 8xy - z + 8 = 0$$
$$4x + 2z\frac{\partial z}{\partial x} - 8y - \frac{\partial z}{\partial x} = 0$$
or, $\quad \dfrac{\partial z}{\partial x} = \dfrac{8y - 4x}{2z - 1} = \dfrac{-8}{2 - 1} = -8$ at $(2, 0, 1)$.

Again,
$$\frac{\partial^2 z}{\partial x^2} = \frac{-4(2z - 1) - (8y - 4x)2\frac{\partial z}{\partial x}}{(2z - 1)^2}$$
$$= \frac{-4(2 - 1) + 4.2.2.(-8)}{(2 - 1)^2} = -132 \text{ at } (2, 0, 1).$$

Similarly obtain $\dfrac{\partial^2 z}{\partial x \partial y}$ and $\dfrac{\partial^2 z}{\partial y^2}$ at $(2,0,1)$.

Putting these values in (13.15.17) we obtain the required value of d^2z.

Example 13.15.18. If $x + y + z = u$, $y + z = uv$, $z = uvw$, calculate $\dfrac{\partial(x, y, z)}{\partial(u, v, w)}$.

Solution: Let $F_1 = x + y + z - u = 0$, $F_2 = y + z - uv = 0$, $F_3 = z - 4uvw = 0$.
We have
$$\frac{\partial(F_1, F_2, F_3)}{\partial(u, v, w)} = (-1)^3 \frac{\partial(F_1, F_2, F_3)}{\partial(x, y, z)} \cdot \frac{\partial(x, y, z)}{\partial(u, v, w)}.$$
$$\therefore \frac{\partial(x, y, z)}{\partial(u, v, w)} = -\frac{\partial(F_1, F_2, F_3)}{\partial(u, v, w)} \Big/ \frac{\partial(F_1, F_2, F_3)}{\partial(x, y, z)}.$$

Now calculate the right side Jacobians.

Example 13.15.19. *If $H(x, y)$ be a homogeneous function of x and y, of degree n having continuous first order partial derivative and $u(x, y) = (x^2 + y^2)^{-n/2}$, show that*
$$\frac{\partial}{\partial x}\left(H\frac{\partial u}{\partial x}\right) + \frac{\partial}{\partial y}\left(H\frac{\partial u}{\partial y}\right) = 0. \qquad \text{[C.H. 1996]}$$

Solution: Since H is homogeneous in x and y of degree n. We have, by Euler's Theorem,
$$x\frac{\partial H}{\partial x} + y\frac{\partial H}{\partial y} = nH. \qquad (13.15.18)$$

Since $u = (x^2 + y^2)^{-n/2} = x^{-n}\left\{1 + \left(\dfrac{y}{x}\right)^2\right\}^{-n/2}$, a homogeneous function in x and y of degree $-n$.
$$x\frac{\partial u}{\partial x} + y\frac{\partial u}{\partial y} = -nu. \qquad (13.15.19)$$

CHAPTER 13: FUNCTIONS OF SEVERAL VARIABLES

Now,

$$\text{L.H.S.} = \frac{\partial}{\partial x}\left(H\frac{\partial u}{\partial x}\right) + \frac{\partial}{\partial y}\left(H\frac{\partial u}{\partial y}\right)$$

$$= \frac{\partial H}{\partial x}\frac{\partial u}{\partial x} + H\frac{\partial^2 u}{\partial x^2} + \frac{\partial H}{\partial y}\frac{\partial u}{\partial y} + H\frac{\partial^2 u}{\partial y^2}$$

$$= \frac{\partial H}{\partial x}\left(\frac{-nux}{x^2+y^2}\right) + \frac{\partial H}{\partial y}\left(\frac{-nuy}{x^2+y^2}\right) + H\left[-\frac{n(y^2-x^2)}{(x^2+y^2)^2}u - \frac{nx}{x^2+y^2}\frac{\partial u}{\partial x}\right]$$

$$+ H\left[-\frac{n(x^2-y^2)}{(x^2+y^2)^2}u - \frac{ny}{x^2+y^2}\frac{\partial u}{\partial y}\right]$$

[Substituting $\frac{\partial u}{\partial x} = -\frac{n}{2}(x^2+y^2)^{-n/2-1}\cdot 2x = -\frac{nux}{x^2+y^2}$;

$\frac{\partial u}{\partial y} = -\frac{n}{2}(x^2+y^2)^{-n/2-1}\cdot 2y = -\frac{nuy}{x^2+y^2}.$

$\frac{\partial^2 u}{\partial x^2} = -n\frac{x^2+y^2-2x^2}{(x^2+y^2)^2}u - \frac{nx}{x^2+y^2}\frac{\partial u}{\partial x}$;

$\frac{\partial^2 u}{\partial y^2} = -n\frac{y^2+x^2-2y^2}{(x^2+y^2)^2}u - \frac{ny}{x^2+y^2}\frac{\partial u}{\partial y}.$]

$$= -\frac{nu}{x^2+y^2}\left[x\frac{\partial H}{\partial x} + y\frac{\partial H}{\partial y}\right] - \frac{nH}{x^2+y^2}\left(x\frac{\partial u}{\partial x} + y\frac{\partial u}{\partial y}\right)$$

(other two terms cancel)

$$= -\frac{nu}{x^2+y^2}\cdot nH - \frac{nH}{x^2+y^2}(-nu), \quad \text{(using 13.15.18 and 13.15.19)}$$

$$= 0 \text{ (Proved)}.$$

Example 13.15.20. *If $v = f(u)$ where u is a homogeneous function of x and y of degree n, then prove that*

$$x\frac{\partial v}{\partial x} + y\frac{\partial v}{\partial y} = nu\frac{dv}{du}.$$

Solution:

$$x\frac{\partial v}{\partial x} + y\frac{\partial v}{\partial y} = x\left[\frac{dv}{du}\cdot\frac{\partial u}{\partial x}\right] + y\left[\frac{dv}{du}\cdot\frac{\partial u}{\partial y}\right]$$

$$= \frac{dv}{du}\left(x\frac{\partial u}{\partial x} + y\frac{\partial u}{\partial y}\right)$$

$$= nu\frac{dv}{du} \quad (\because u \text{ is homogeneous function in } x, y \text{ of degree } n)$$

Example 13.15.21. If V is a homogeneous function of degree n in x, y, z prove that $\frac{\partial V}{\partial x}$, $\frac{\partial V}{\partial y}$ and $\frac{\partial V}{\partial z}$ are each homogeneous function in x, y, z of degree $(n-1)$.

Solution: Since V is a homogeneous function of degree n in x, y, z, it follows by Euler's Theorem that
$$x\frac{\partial V}{\partial x} + y\frac{\partial V}{\partial y} + z\frac{\partial V}{\partial z} = nV.$$

Differentiating (partially) with respect to x, we get
$$x\frac{\partial^2 V}{\partial x^2} + y\frac{\partial^2 V}{\partial x \partial y} + z\frac{\partial^2 V}{\partial x \partial z} = n\frac{\partial V}{\partial x} - \frac{\partial V}{\partial x} = (n-1)\frac{\partial V}{\partial x}$$

i.e., $\quad x\frac{\partial}{\partial x}\left(\frac{\partial V}{\partial x}\right) + y\frac{\partial}{\partial y}\left(\frac{\partial V}{\partial x}\right) + z\frac{\partial}{\partial z}\left(\frac{\partial V}{\partial x}\right) = (n-1)\frac{\partial V}{\partial x}.$

(assuming the commutativity of mixed derivatives)

Hence, by converse of Euler's Theorem, $\frac{\partial V}{\partial x}$ is a homogeneous function of degree $(n-1)$ in x, y, z.

Similarly, for $\frac{\partial V}{\partial y}, \frac{\partial V}{\partial z}$.

Example 13.15.22. If V be a homogeneous function of degree n in x, y, z and if $V = f(X, Y, Z)$, where $X = \frac{\partial V}{\partial x}$, $Y = \frac{\partial V}{\partial y}$, $Z = \frac{\partial V}{\partial z}$ then prove that
$$X\frac{\partial V}{\partial X} + Y\frac{\partial V}{\partial Y} + Z\frac{\partial V}{\partial Z} = \frac{n}{n-1}V.$$

Solution: Here $V = f(X, Y, Z)$ where $X = \frac{\partial V}{\partial x}$, $Y = \frac{\partial V}{\partial y}$, $Z = \frac{\partial V}{\partial z}$.

$$\therefore \quad \frac{\partial V}{\partial x} = \frac{\partial V}{\partial X} \cdot \frac{\partial X}{\partial x} + \frac{\partial V}{\partial Y} \cdot \frac{\partial Y}{\partial x} + \frac{\partial V}{\partial Z} \cdot \frac{\partial Z}{\partial x} \quad \text{(Chain Rule)}$$

$$= \frac{\partial V}{\partial X} \cdot \frac{\partial^2 V}{\partial x^2} + \frac{\partial V}{\partial Y} \cdot \frac{\partial^2 V}{\partial x \partial y} + \frac{\partial V}{\partial Z} \cdot \frac{\partial^2 V}{\partial x \partial z}.$$

Hence,
$$\sum x\frac{\partial V}{\partial x} = \frac{\partial V}{\partial X}\left[x\frac{\partial^2 V}{\partial x^2} + y\frac{\partial^2 V}{\partial x \partial y} + z\frac{\partial^2 V}{\partial x \partial z}\right] + \frac{\partial V}{\partial Y}\left[x\frac{\partial^2 V}{\partial y \partial z} + y\frac{\partial^2 V}{\partial y^2} + z\frac{\partial^2 V}{\partial y \partial z}\right]$$
$$+ \frac{\partial V}{\partial Z}\left[x\frac{\partial^2 V}{\partial y \partial z} + y\frac{\partial^2 V}{\partial y \partial z} + z\frac{\partial^2 V}{\partial z^2}\right]$$
$$= \frac{\partial V}{\partial X} \cdot (n-1)\frac{\partial V}{\partial x} + \frac{\partial V}{\partial Y} \cdot (n-1)\frac{\partial V}{\partial y} + \frac{\partial V}{\partial Z} \cdot (n-1)\frac{\partial V}{\partial z}$$

(using the result derived in **Ex. 13.15.21** above)

$$= (n-1)\left(X\frac{\partial V}{\partial X} + Y\frac{\partial V}{\partial Y} + Z\frac{\partial V}{\partial Z}\right).$$

But $\sum x \dfrac{\partial V}{\partial x} = nV$ ($\because$ V is homogeneous function of degree n).

$$\therefore \quad X\dfrac{\partial V}{\partial X} + Y\dfrac{\partial V}{\partial Y} + Z\dfrac{\partial V}{\partial Z} = \dfrac{n}{n-1}V.$$

Example 13.15.23. *If H is a homogeneous function in x, y, z of degree n and $u = (x^2 + y^2 + z^2)^{-\frac{1}{2}(n+1)}$, then prove that*

$$\dfrac{\partial}{\partial x}\left(H\dfrac{\partial u}{\partial x}\right) + \dfrac{\partial}{\partial y}\left(H\dfrac{\partial u}{\partial y}\right) + \dfrac{\partial}{\partial z}\left(H\dfrac{\partial u}{\partial z}\right) = 0.$$

Solution: $u = (x^2 + y^2 + z^2)^{-\frac{1}{2}(n+1)}$.

$$\therefore \quad \dfrac{\partial u}{\partial x} = -(n+1)x\left(x^2+y^2+z^2\right)^{-\frac{1}{2}(n+3)}$$

$$\therefore \quad \dfrac{\partial}{\partial x}\left(H\dfrac{\partial u}{\partial x}\right) = -(n+1)\dfrac{\partial}{\partial x}\left\{Hx(x^2+y^2+z^2)^{-\frac{1}{2}(n+3)}\right\}$$
$$= -(n+1)(x^2+y^2+z^2)^{-\frac{1}{2}(n+3)}$$
$$\quad \times \left[x\dfrac{\partial H}{\partial x} + H - (n+3)H\cdot x^2(x^2+y^2+z^2)^{-1}\right]$$

$$\therefore \quad \sum \dfrac{\partial}{\partial x}\left(H\dfrac{\partial u}{\partial x}\right) = -(n+1)(x^2+y^2+z^2)^{-\frac{1}{2}(n+3)}$$
$$\quad \times \left\{\sum x\dfrac{\partial H}{\partial x} + 3H - (n+3)H(x^2+y^2+z^2)\right.$$
$$\quad \left.\times(x^2+y^2+z^2)^{-1}\right\}$$
$$= -(n+1)(x^2+y^2+z^2)^{-\frac{1}{2}(n+3)}\left\{nH + 3H - (n+3)H\right\}$$
$$= 0 \text{ (Proved)}$$

Exercises XIII

Section : 1

[Point sets in higher dimensions]

1. Prove that every bounded infinite set S of points in a plane has at least one point of accumulation (*Bolzano-Weierstrass Theorem of plane sets*).

2. Determine the point set in which each of the following functions is defined: State whether the set forms a region or not.

(a) $\sqrt{|x| + |y| - 2}$

[The entire plane exterior to the square with vertices $(2,0)$, $(-2,0)$, $(0,2)$, $(0,-2)$ including the boundary.]

(b) $\dfrac{1}{\sqrt{x+y-1}\sqrt{x-y+1}}$

[The entire plane within the angle to the right and left of both lines $x+y-1 = 0$, $x - y + 1 = 0$, the lines excluded. It is **not a region** in the sense that any two points of the set can not be joined by a broken line whose points belong to the set.]

(c) $\sqrt{\dfrac{1 - x^2 - y^2}{xy}}$

[The interior of the quadrants I and III of the unit circle, the interior of the quadrants II and IV; the axes $x = 0, y = 0$ are excluded but the four quadrantal arcs between the axes are included. *Not a Region.*]

(d) $\sin^{-1}(x^2 + y^2)$

[Closed Region inside the circle $x^2 + y^2 = 1$.]

(e) $\sqrt{(x^2 + y^2 - 1)(4 - x^2 - y^2)}$

[Closed annular Region between the two circles $x^2 + y^2 = 1$ and $x^2 + y^2 = 4$.]

(f) $\log(x + y)$

[The open half plane above and to the right of the line $x + y = 0$.]

3. (a) Define *neighbourhood* of a point $P_0 \in \mathbb{R}^n$.

(b) Define an *open set* $S \subseteq \mathbb{R}^n$ and *limit point* of a set $S \subseteq \mathbb{R}^n$.

(c) Assuming that a set F is *closed* iff its complement is open, prove that a set F is closed iff it contains all its limit points.

(d) What is an interior point, an exterior point, a boundary point?

(e) Defining the closure ($\bar{S}$) of a set $S \subseteq \mathbb{R}^n$ as the union of S and the set of all limit points of S, prove

   i. $\bar{S}$ is a closed set;

   ii. If F is any closed set which satisfies $S \subseteq F \subseteq \mathbb{R}^n$, then $S \subseteq \bar{S} \subseteq F$.

(f) Show that $S = \{(x,y) : 0 < x < 1, 0 < y < 1\}$ is an open set in $\mathbb{R}^2$. Check that $(1,1)$ is a point of accumulation of S which $\notin S$.   [C.H. 1996]

4. Find the closure of each of the following sets in $\mathbb{R}^2$:

   (a) $\{(x,y) : x^2 + y^2 < 1.\}$

   Ans: $\{(x,y) : x^2 + y^2 \leq 1\}$]

(b) $\{(x,y) : x^2 + y^2 \leq 1.\}$

Ans: $\{(x,y) : x^2 + y^2 \leq 1\}$]

(c) $\{(x,y) : y = \sin \dfrac{1}{x}, x \neq 0.\}$

Ans: $\{(x,y) : y = \sin \frac{1}{x}, x \neq 0\} \cup \{(0,y)/|y| \leq 1\}$]

(d) $\left\{\left(\dfrac{1}{n}, \dfrac{1}{m}\right) : n, m \text{ are non zero integers.}\right\}$

Ans: $\{(\frac{1}{n}, \frac{1}{m}) : n, m \text{ are non zero integers}\} \cup \{(\frac{1}{n}, 0) : n \text{ is a non-zero integer}$
$\cup (0, \frac{1}{m}) : m \text{ is a non-zero integer } \cup \{(0,0)\}$.]

5. Determine whether each of the following sets is *open, closed, connected* or *bounded* and find the *boundary*:

(a) $\{(x,y) : 2x^2 + 4y^2 < 1.\}$

Ans: open, connected, bounded; boundary = $\{(x,y) : 2x^2 + 4y^2 = 1\}$.]

(b) $\{(x,y) : x^2 - y^2 \geq 1.\}$

Ans: closed, not connected, unbounded; boundary = $\{(x,y) : x^2 - y^2 = 1\}$.]

(c) $\{(x,y) : |x| \leq 1 \text{ and } |y| \leq 1.\}$

Ans: closed, connected, bounded; boundary = $\{(x,y) : |y| = 1 \text{ and } |x| \leq 1\}$
$\cup \{(x,y) : |x| = 1 \text{ and } |y| \leq 1\}$.]

(d) $\{(x,y) : x \text{ and } y \text{ are both rational.}\}$

Ans: neither open nor closed, not connected, unbounded : boundary = $\mathbb{R}^2$.]

Section : 2

[Limit, Continuity, Derivability and Differentiability]

1. Use ε-δ definition to justify the following statements on limits:

(a) $\lim\limits_{(x,y) \to (0,0)} (x + y) = 0;$ (b) $\lim\limits_{\substack{x \to 0 \\ y \to 0}} \dfrac{xy}{\sqrt{x^2 + y^2}} = 0;$

(c) $\lim\limits_{(x,y) \to (0,0)} \dfrac{x^3 - y^3}{x^2 + y^2} = 0;$ (d) $\lim\limits_{\substack{x \to 0 \\ y \to 0}} \dfrac{x^2 - y^2}{1 + x^2 + y^2} = 0,$

(e) $\lim_{\substack{x \to 0 \\ y \to 0}} (1+y^2)\dfrac{\sin x}{x} = 1;$ (f) $\lim_{(x,y) \to (0,0)} y^2 \sin \dfrac{1}{x} = 0;$

(g) $\lim_{\substack{x \to 0 \\ y \to 0}} \left(y \sin \dfrac{1}{x} + x \sin \dfrac{1}{y} \right) = 0;$ (h) $\lim_{\substack{x \to 0 \\ y \to 0}} \dfrac{x^4 + y^4}{x^2 + y^2} = 0;$

(i) $\lim_{\substack{x \to 0 \\ y \to 0}} \dfrac{x^2 y^2}{x^2 + y^2} = 0;$ (j) $\lim_{\substack{x \to -1 \\ y \to -1}} (xy - 2x) = 1;$

(k) $\lim_{(x,y) \to (0,0)} \dfrac{1 - \cos(x^2 + y^2)}{(x^2 + y^2)^2};$ (l) $\lim_{\substack{x \to 0 \\ y \to 0}} e^{-(x^2 + y^2)} = 0;$

(m) $\lim_{\substack{x \to 0 \\ y \to 0}} e^{-1/(x^2 + y^2)} = 1;$ (n) $\lim_{\substack{x \to 0 \\ y \to 0}} xy \dfrac{x^2 - y^2}{x^2 + y^2} = 0.$

Ans: (n) $\left| xy \dfrac{x^2 - y^2}{x^2 + y^2} \right| = |x||y| \left| \dfrac{x^2 - y^2}{x^2 + y^2} \right| < \sqrt{x^2 + y^2} \sqrt{x^2 + y^2} \left(\because \left| \dfrac{x^2 - y^2}{x^2 + y^2} \right| < 1 \right)$

which can be made less than any positive number ε if $x^2 + y^2 < \delta^2$ where $\delta = \sqrt{\varepsilon}$; hence etc.

Otherwise. Put $x = r \cos \theta, y = r \sin \theta$ then

$$xy \dfrac{x^2 - y^2}{x^2 + y^2} \quad \text{becomes} \quad \dfrac{r^2}{4} \sin 4\theta.$$

$\therefore \left| xy \dfrac{x^2 - y^2}{x^2 + y^2} - 0 \right| = \dfrac{r^2}{4} |\sin 4\theta| \leq \dfrac{r^2}{4} = \dfrac{1}{4}(x^2 + y^2) < \varepsilon$ if $|x^2 + y^2| < \delta^2$ where $\delta^2 = 4\varepsilon.]$

(o) $\lim_{\substack{x \to 0 \\ y \to 0}} \dfrac{\sqrt{x^2 y^2 + 1} - 1}{x^2 + y^2} = 0.$

$\left[\text{Hints: } \dfrac{\sqrt{1 + x^2 y^2} - 1}{x^2 + y^2} \simeq \dfrac{\frac{1}{2} x^2 y^2}{x^2 + y^2}. \right]$

2. (a) Let

$$f(x, y) = \begin{cases} 1, & \text{if } xy \neq 0; \\ 0, & \text{if } xy = 0. \end{cases}$$

Show that the two repeated limits exist at $(0, 0)$ and are equal but the simultaneous limit $\lim_{(x,y) \to (0,0)} f(x, y)$ does not exist.

[In any *nbd* of $(0, 0)$ there are points at which $f(x, y) = 1$ and also there are points at which $f(x, y) = 0$. Therefore, for any $\varepsilon > 0$ there does not exist any *nbd* of $(0, 0)$ at all points of which we can write $|f(x, y) - 1|$ or $|f(x, y) - 0|$ can be made $< \varepsilon$.]

(b) Let
$$f(x,y) = \begin{cases} x + y \sin \dfrac{1}{x}, & x \neq 0; \\ 0, & x = 0. \end{cases}$$

Verify that the simultaneous limit $\lim_{\substack{x \to 0 \\ y \to 0}} f(x,y)$ exists and the repeated limit $\lim_{x \to 0} \lim_{y \to 0} f(x,y)$ exists but the repeated limit $\lim_{y \to 0} \lim_{x \to 0} f(x,y)$ does not exist.

(c) Let
$$f(x,y) = \begin{cases} y \sin \dfrac{1}{x} + \dfrac{xy}{x^2+y^2}, & x \neq 0; \\ 0, & x = 0. \end{cases}$$

Verify that the $\lim_{x \to 0} \lim_{y \to 0} f(x,y)$ exists but neither $\lim_{(x,y) \to (0,0)} f(x,y)$ nor $\lim_{y \to 0} \lim_{x \to 0} f(x,y)$ exists.

(d) Let
$$f(x,y) = \begin{cases} xy \dfrac{x^2 - y^2}{x^2 + y^2}, & \text{where } x^2 + y^2 \neq 0; \\ 0, & \text{where } x^2 + y^2 = 0. \end{cases}$$

Show that here double limit and both repeated limits exist at $(0,0)$.

(e) Let
$$f(x,y) = \begin{cases} x \sin \dfrac{1}{y} + y \sin \dfrac{1}{x}, & xy \neq 0; \\ 0, & xy = 0. \end{cases}$$

Show that at $(0,0)$ the double limit exists but the repeated limits do not exist.

[To consider: Suppose double limit exists; if the repeated limits also exist, then they are equal, but the converse is not true. However, in one case we can be sure of non-existence of simultaneous limit – the case when the repeated limits are not equal

e.g., $\lim_{x \to 0} \left\{ \lim_{y \to 0} \dfrac{x+y}{x-y} \right\} = 1$, $\lim_{y \to 0} \left\{ \lim_{x \to 0} \dfrac{x+y}{x-y} \right\} = -1$;

therefore $\lim_{(x,y) \to (0,0)} \dfrac{x+y}{x-y}$ does not exist.]

3. Show that the following limits do not exist:

(a) $\lim\limits_{\substack{x \to 0 \\ y \to 0}} \dfrac{x^2 - y^2}{x^2 + y^2}$;

(b) $\lim\limits_{\substack{x \to 0 \\ y \to 0}} \dfrac{xy}{x^2 + y^2}$;

(c) $\lim\limits_{\substack{x \to 0 \\ y \to 0}} \dfrac{x^2 y}{x^4 + y^2}$; (put $y = mx^2$)

(d) $\lim\limits_{\substack{x \to 0 \\ y \to 0}} \dfrac{2xy^2}{x^2 + y^4}$; (put $x = my^2$)

(e) $\lim\limits_{\substack{x \to 0 \\ y \to 0}} \dfrac{x^3 + y^3}{x - y}$; (put $y = x - mx^3$)

(f) $\lim\limits_{\substack{x \to 0 \\ y \to 0}} \dfrac{xy^3}{x^2 + y^6}$; (put $x = my^3$)

(g) $\lim\limits_{\substack{x \to 0 \\ y \to 0}} (x + y) \cdot \dfrac{y + (x + y)^2}{y - (x + y)^2}$;

(h) $\lim\limits_{\substack{x \to 0 \\ y \to 0}} \dfrac{|x|}{y^2} e^{-|x|/y^2}$;

(approach along $x = y^2$; also along $x = 0$)

(i) $\lim\limits_{\substack{x \to 0 \\ y \to 0}} \dfrac{x^2 y^4}{(x^2 + y^4)^2}$; (put $x = my^2$)

(j) $\lim\limits_{\substack{x \to 0 \\ y \to 1}} \tan^{-1} \dfrac{x}{y}$;

(k) $\lim\limits_{\substack{x \to 0 \\ y \to 0}} \dfrac{x^2 y^2}{x^2 y^2 + (x^2 - y^2)^2}$.

4. (a) Show that the function

$$f(x, y) = \begin{cases} xy \dfrac{x^2 - y^2}{x^2 + y^2}, & \text{if } (x, y) \neq (0, 0); \\ 0, & \text{if } (x, y) = (0, 0) \end{cases}$$

is continuous at $(0, 0)$.

(b) Let

$$f(x, y) = \begin{cases} \dfrac{2xy}{x^2 + y^2}, & \text{where } x^2 + y^2 \neq 0; \\ 0, & \text{where } x^2 + y^2 = 0. \end{cases}$$

Prove that f is a continuous function of either variable when the other variable is given a fixed value. Is the function continuous at $(0, 0)$?

[C.H. 1990, 96]

(c) Let

$$f(x, y) = \begin{cases} \dfrac{2xy}{(x^2 + y^2)^n}, & \text{if } x^2 + y^2 \neq 0; \\ 0, & \text{if } x^2 + y^2 = 0. \end{cases}$$

Prove that f is continuous at $(0, 0)$ if $n = \frac{1}{2}$.

(d) Let

$$f(x, y) = \begin{cases} 3xy, & \text{when } (x, y) \neq (2, 3); \\ 6, & \text{when } x = 2, y = 3. \end{cases}$$

Is f continuous at $(2,3)$? If discontinuous, can we remove the discontinuity? How?

(e) Show that
$$f(x,y) = \begin{cases} \dfrac{x^3 - y^3}{x^2 + y^2}, & \text{where } (x,y) \neq (0,0); \\ 0, & \text{if } (x,y) = (0,0). \end{cases}$$

is continuous at $(0,0)$, possesses partial derivatives at $(0,0)$, but is not differentiable at $(0,0)$. [C.H. 1993]

5. Show that the function f defined in the neighbourhood of $(0,0)$ by $f(x,y) = |x| + |y|$ is *continuous* but not differentiable at $(0,0)$. [C.H. 1990]

6. If
$$f(x,y) = \begin{cases} \dfrac{x^3 + y^3}{x - y}, & x \neq y; \\ 0, & x = y. \end{cases}$$

Show that $f(x,y)$ is not continuous at $(0,0)$ but f_x and f_y exist at $(0,0)$. [See No. 3(e) above]

7. Let
$$f(x,y) = \begin{cases} x^2 \sin \dfrac{1}{x} + y^2 \sin \dfrac{1}{y}, & \text{when neither } x = 0 \text{ nor } y = 0; \\ x^2 \sin \dfrac{1}{x}, & \text{when } x \neq 0, y = 0; \\ y^2 \sin \dfrac{1}{y}, & \text{when } y \neq 0, x = 0; \\ 0, & \text{when } x = 0, y = 0. \end{cases}$$

Show that $f_x(x,y)$ and $f_y(x,y)$ are discontinuous at $(0,0)$ but that $f(x,y)$ is differentiable at $(0,0)$. In what way continuity of f_x and f_y related to differentiability?

8. (a) If
$$f(x,y) = \begin{cases} \dfrac{xy}{\sqrt{x^2 + y^2}}, & \text{when } (x,y) \neq (0,0); \\ 0, & \text{when } (x,y) = (0,0). \end{cases}$$

Show that f is continuous, possesses partial derivatives of first order, but f is not differentiable at $(0,0)$.

(b) Show that
$$f(x,y) = \begin{cases} \dfrac{xy^2}{x^2+y^4}, & \text{if } x \neq 0, \\ 0, & \text{if } x = 0; \end{cases}$$
possesses first order partial derivatives at $(0,0)$, yet it is not differentiable at $(0,0)$. [C.H. 1995]

9. Let
$$f(0,0) = 0$$
$$f(x,y) = \frac{x^3}{x^2+y^2} \text{ if } (x,y) \neq (0,0).$$
Prove that (a) f is continuous at any point (x,y) and (b) the restriction of f to any straight line is differentiable. [C.H. 1989 (New)]

10. Find $\lim\limits_{\substack{x \to 0 \\ y \to 0}} (x^2+y^2) \sin \dfrac{1}{xy}$.

11. Prove that the continuity of $f(x,y)$ at a point (a,b) in its domain of definition implies the continuity of $f(x,y)$ as a function of x at $x = a$ when y is fixed and as a function of y at $y = b$ when x is fixed. [C.H. 1996]

12. If $f(x,y) = xy\dfrac{(x^2-y^2)}{x^2+y^2}, (x,y) \neq (0,0); = 0$ when $(x,y) = (0,0)$ prove that $f(x,y)$ is differentiable at $(0,0)$. And that $f_{xy} \neq f_{yx}$ at $(0,0)$. [C.H. 1991, 93]

13. Prove that $f(x,y) = \{|x+y|+(x+y)\}^k$ is everywhere differentiable for all values of $k > 0$. [C.H. 1992]

14. (a) Let
$$f(x,y) = \begin{cases} (x^2+y^2)\tan^{-1}\dfrac{y}{x}, & \text{when } x \neq 0; \\ \dfrac{\pi}{2}y^2, & \text{when } x = 0. \end{cases}$$
Show that $f_{xy} = f_{yx} = \dfrac{x^2-y^2}{x^2+y^2}$ when $x,y \neq 0$.
But from definition of partial derivatives,
$$f_x(0,k) = -k; \quad f_y(k,0) = k; \quad f_{yx}(0,0) = -1; \quad f_{xy}(0,0) = 1.$$
(b) Prove that $f_{xy}(0,0) \neq f_{yx}(0,0)$ if
$$f(x,y) = \begin{cases} xy\dfrac{x^2+k^2y^2}{x^2+y^2}, & \text{where } x^2+y^2 \neq 0; \\ 0, & \text{where } x = 0, y = 0. \end{cases}$$

CHAPTER 13: FUNCTIONS OF SEVERAL VARIABLES

(c) Examine the equality of $f_{xy}(0,0)$ and $f_{yx}(0,0)$ for

$$f(x,y) = \begin{cases} xy, & \text{if } |x| \geq |y|; \\ -xy, & \text{if } |x| < |y|. \end{cases}$$

[C.H. 1996]

[*Hints:* When $k \to 0$ (h remaining fixed), take $f(h,k) = +hk$ because then ultimately $|k| < |h|$.

Similarly, when $h \to 0$ (k remaining fixed), take $f(h,k) = -hk$ because then ultimately $|h| < |k|$.]

(d) Verify that $f_{xy} = f_{yx}$ at the origin for

  i. $f(x,y) = |x^2 - y^2|$.

  ii. $f(x,y) = \begin{cases} \sqrt{x^2+y^2}\sin 2\theta, & \text{when } x \neq 0; \\ 0, & \text{when } x = 0. \end{cases}$

  [when θ is given by $\tan\theta = y/x$.]

15. (a) For the function

$$f(x,y) = \begin{cases} (x^2+y^2)\log(x^2+y^2), & \text{for } (x,y) \neq (0,0); \\ 0, & \text{for } x=0, y=0. \end{cases}$$

Show that Schwarz's Theorem conditions are not satisfied but $f_{xy} = f_{yx}$ at $(0,0)$. [C.H. 1990]

(b) For the function

$$f(x,y) = \begin{cases} \dfrac{x^2 y^2}{x^2+y^2}, & (x,y) \neq (0,0); \\ 0, & (x,y) = (0,0); \end{cases}$$

show that in spite of the fact that neither the conditions of Schwarz's theorem nor the conditions of Young's theorem are satisfied, $f_{xy} = f_{yx}$ at $(0,0)$.

[C.H. 1995]

Section : 3
[Partial Derivatives : Euler's Theorem]

1. If $V = \log(x^3 + y^3 + z^3 - 3xyz)$, prove that

  (a) $\left(\dfrac{\partial}{\partial x} + \dfrac{\partial}{\partial y} + \dfrac{\partial}{\partial z}\right) V = \dfrac{3}{x+y+z}$;

  (b) $\left(\dfrac{\partial^2}{\partial x^2} + \dfrac{\partial^2}{\partial y^2} + \dfrac{\partial^2}{\partial z^2}\right) V = -\dfrac{3}{(x+y+z)^2}$.

2. If $u(x,y,z,t) = \dfrac{f(t+r)}{r} + \dfrac{g(t-r)}{r}$, where $r^2 = x^2+y^2+z^2$, prove that u satisfies the equation
$$\frac{\partial^2 u}{\partial x^2} + \frac{\partial^2 u}{\partial y^2} + \frac{\partial^2 u}{\partial z^2} = \frac{\partial^2 u}{\partial t^2}.$$

3. If $z = xf(x+y) + yg(x+y)$, prove that
$$\frac{\partial^2 z}{\partial x^2} - 2\frac{\partial^2 z}{\partial x \partial y} + \frac{\partial^2 z}{\partial y^2} = 0. \qquad \text{[C.H. 1980]}$$

4. If $\theta = t^n e^{-(r^2/4t)}$, prove that $n = -3/2$, when the following equation holds
$$\frac{1}{r^2}\frac{\partial}{\partial r}\left(r^2 \frac{\partial \theta}{\partial r}\right) = \frac{\partial \theta}{\partial t}.$$

5. If $r^2 = x^2 + y^2 + z^2$ and $V = r^3$, prove that

   (a) $\dfrac{\partial^2 V}{\partial x^2} + \dfrac{\partial^2 V}{\partial y^2} + \dfrac{\partial^2 V}{\partial z^2} = 12r;$

   (b) $\dfrac{1}{yz}\dfrac{\partial^2 V}{\partial y \partial z} + \dfrac{1}{zx}\dfrac{\partial^2 V}{\partial z \partial x} + \dfrac{1}{xy}\dfrac{\partial^2 V}{\partial x \partial y} = \dfrac{9}{r}.$

6. If $V = f(r)$, where $r = \sqrt{x^2 + y^2 + z^2}$, then prove that
$$\frac{\partial^2 V}{\partial x^2} + \frac{\partial^2 V}{\partial y^2} + \frac{\partial^2 V}{\partial z^2} = f''(r) + \frac{2}{r}f'(r).$$

In particular, prove that if $f(r) = r^m$, then the left side
$$= m(m+1)r^{m-2}.$$

7. A function $f(x,y,z)$ is said to be homogeneous of degree n, if
$$f(tx, ty, tz) = t^n f(x,y,z), \quad \forall \, t > 0.$$

Hence prove that for a homogeneous function of degree n,
$$f(x,y,z) = x^n\left(1, \frac{y}{x}, \frac{z}{x}\right).$$

Illustrate with the functions (a) $\dfrac{x^2}{y} + \dfrac{y^2}{x} + z$; (b) $\cos^{-1}\dfrac{z}{\sqrt{x^2+y^2+z^2}}$.

Ans: (a) degree 1; (b) degree 0.]

If $f(x,y,z)$ is homogeneous of degree n show that f_x, f_y, f_z are homogeneous of degree $(n-1)$.

CHAPTER 13: FUNCTIONS OF SEVERAL VARIABLES

8. If $f(x, y, z)$ has the property
$$f(tx, t^j y, t^k z) = t^n f(x, y, z)$$
prove that
$$f(x, y, z) = x^n f\left(1, \frac{y}{x^j}, \frac{z}{x^k}\right)$$
and $x\dfrac{\partial f}{\partial x} + jy\dfrac{\partial f}{\partial y} + kz\dfrac{\partial f}{\partial z} = nf$.

9. If u, v, w are function of x, y, z, prove that the
$$\text{Wronskian } W(u, v, w) = \begin{vmatrix} u & v & w \\ u' & v' & w' \\ u'' & v'' & w'' \end{vmatrix}$$
satisfies $W(u, v, w) = u^3 W\left(1, \dfrac{v}{u}, \dfrac{w}{u}\right)$.

10. Let $u = \sin^{-1} \dfrac{x+y}{\sqrt{x} + \sqrt{y}}$, prove that
$$x^2 \frac{\partial^2 u}{\partial x^2} + 2xy \frac{\partial^2 u}{\partial x \partial y} + y^2 \frac{\partial^2 u}{\partial y^2} = -\frac{\sin u \cos 2u}{4 \cos^3 u}.$$

11. We shall denote the operator
$$\frac{\partial^2}{\partial x^2} + \frac{\partial^2}{\partial y^2} + \frac{\partial^2}{\partial z^2} \text{ by the symbol } \nabla^2 \text{ (Laplacian)}.$$

   (a) Prove that if u and v are two functions of x, y, z which admit of partial derivatives of second order, then
$$\nabla^2(u \cdot v) = u\nabla^2 v + v\nabla^2 u + 2\left(\frac{\partial u}{\partial x}\frac{\partial v}{\partial x} + \frac{\partial u}{\partial y}\frac{\partial v}{\partial y} + \frac{\partial u}{\partial z}\frac{\partial v}{\partial z}\right).$$

   (b) If ϕ is a homogeneous function of degree n in x, y, z and ϕ is **harmonic** (By 'Harmonic Functions' we mean those functions whose Laplacians Vanish), then prove that
$$\nabla^2(r^m \phi) = r^{m-2}[m^2 + m + 2mn]\phi(x, y, z)$$
where $r^2 = x^2 + y^2 + z^2$.
[Hints:
$$\frac{\partial}{\partial x}(u \cdot v) = u\frac{\partial v}{\partial x} + v\frac{\partial u}{\partial x}; \qquad \frac{\partial^2}{\partial x^2}(uv) = 2\frac{\partial u}{\partial x}\frac{\partial v}{\partial x} + u\frac{\partial^2 v}{\partial x^2} + v\frac{\partial^2 u}{\partial x^2}.$$

Similarly obtain $\frac{\partial^2}{\partial y^2}(u.v)$ and $\frac{\partial^2}{\partial z^2}(u.v)$ and then add; (a) will follow.

For (b), Use (a) taking $u = r^m, v = \phi$.

Given $\nabla^2 \phi = 0$, and $x\frac{\partial \phi}{\partial x} + y\frac{\partial \phi}{\partial y} + z\frac{\partial \phi}{\partial z} = n\phi$.]

12. If $u(x,y) = \phi(xy) + \sqrt{xy}\psi(y/x)$, $x \neq 0, y \neq 0$, where ϕ, ψ are twice differentiable, prove that

$$x^2 \frac{\partial^2 u}{\partial x^2} - y^2 \frac{\partial^2 u}{\partial y^2} = 0.$$ [C.H. 1996]

Mention the necessity of the conditions: $x \neq 0, y \neq 0$.

13. (a) If $v = f(u)$, where u is a homogeneous function in x and y of degree n, show that $x\frac{\partial v}{\partial x} + y\frac{\partial v}{\partial y} = nu\frac{dv}{du}$.

(b) If H be a homogeneous function in x and y of degree n having continuous first order partial derivatives and $u(x,y) = (x^2 + y^2)^{-n/2}$, show that

$$\frac{\partial}{\partial x}\left(H\frac{\partial u}{\partial x}\right) + \frac{\partial}{\partial y}\left(H\frac{\partial u}{\partial y}\right) = 0.$$ [C.H. 1996]

14. If H be a homogeneous function in x, y, z of degree n having continuous partial derivatives, prove that

a) $\frac{\partial H}{\partial x}, \frac{\partial H}{\partial y}, \frac{\partial H}{\partial z}$ are each a homogeneous function in x, y, z of degree $(n-1)$;

and b) $\frac{\partial}{\partial x}\left(H\frac{\partial u}{\partial x}\right) + \frac{\partial}{\partial y}\left(H\frac{\partial u}{\partial y}\right) + \frac{\partial}{\partial z}\left(H\frac{\partial u}{\partial z}\right) = 0$, if $u = (x^2+y^2+z^2)^{-(n+1)/2}$.

15. If $\frac{x^2}{a+u} + \frac{y^2}{b+u} + \frac{z^2}{c+u} = 1$, prove that

$$\left(\frac{\partial u}{\partial x}\right)^2 + \left(\frac{\partial u}{\partial y}\right)^2 + \left(\frac{\partial u}{\partial z}\right)^2 = 2\left(x\frac{\partial u}{\partial x} + y\frac{\partial u}{\partial y} + z\frac{\partial u}{\partial z}\right).$$ [C.H. 1966, 91]

16. If $x^x y^y z^z = c$ (constant), show that at (x, y, z) where $x = y = z$,

$$\frac{\partial^2 z}{\partial x \partial y} = -\frac{1}{x \log_e(ex)}.$$

17. If $u = xf\left(\frac{y}{x}\right) + g\left(\frac{y}{x}\right)$, show that

$$x^2 \frac{\partial^2 u}{\partial x^2} + 2xy\frac{\partial^2 u}{\partial x \partial y} + y^2 \frac{\partial^2 u}{\partial y^2} = 0.$$ [C.H. 1992]

CHAPTER 13: FUNCTIONS OF SEVERAL VARIABLES

18. If $u = xy$ and $v = z - 2x$ and $w = f(u,v) = 0$ where w is a differentiable function of u and v such that $\frac{\partial f}{\partial v} \neq 0$ and if it be given that z is dependent on two independent variables x any y, prove that

$$x\frac{\partial z}{\partial x} - y\frac{\partial z}{\partial y} = 2x.$$ [C.H. 1992]

19. Let $u = f(x,y)$, where $x = r\cos\theta$, $y = r\sin\theta$. Prove that

    (a) $\left(\dfrac{\partial u}{\partial x}\right)^2 + \left(\dfrac{\partial u}{\partial y}\right)^2 = \left(\dfrac{\partial u}{\partial r}\right)^2 + \dfrac{1}{r^2}\left(\dfrac{\partial u}{\partial \theta}\right)^2$;

    (b) $\dfrac{\partial^2 u}{\partial x^2} + \dfrac{\partial^2 u}{\partial y^2} = \dfrac{\partial^2 u}{\partial r^2} + \dfrac{1}{r}\dfrac{\partial u}{\partial r} + \dfrac{1}{r^2}\dfrac{\partial^2 u}{\partial \theta^2}.$ [C.H. 1980, 88, 94]

20. (a) Find $\dfrac{du}{dt}$ if $u = x^3 - y\sin xy$ and $x = (t-1)/t$, $y = t\cos t$.

    (b) Find $\dfrac{\partial z}{\partial r}$ and $\dfrac{\partial z}{\partial \theta}$, if $z = x^2 + xy + y^2$, when $x = 2r + \theta$, $y = r - 2\theta$.

    (c) Find $\dfrac{dz}{dt}$ when $z = xy^2 + x^2 y$; $x = at^2$, $y = 2at$. [C.H. 1976]

21. If $\dfrac{\partial u}{\partial x} = \dfrac{\partial v}{\partial y}$ and $\dfrac{\partial u}{\partial y} = -\dfrac{\partial v}{\partial x}$, where u and v are function of x and y, prove that

$$\frac{\partial u}{\partial r} = \frac{1}{r}\frac{\partial v}{\partial \theta}; \quad \frac{1}{r}\frac{\partial u}{\partial \theta} = -\frac{\partial v}{\partial r},$$

where $x = r\cos\theta$, $y = r\sin\theta$.

22. If z be a function of x and y and $x = c\cosh u \cos v$, $y = c\sinh u \sin v$, prove that

$$\frac{\partial^2 z}{\partial u^2} + \frac{\partial^2 z}{\partial v^2} = \frac{1}{2}c^2(\cosh 2u - \cos 2v)\left(\frac{\partial^2 z}{\partial x^2} + \frac{\partial^2 z}{\partial y^2}\right).$$ [C.H. 1965, 90]

23. If $z = f(u,v)$, where $u = x^2 - 2xy - y^2$ and $v = y$, show that

$$(x+y)\frac{\partial z}{\partial x} + (x-y)\frac{\partial z}{\partial y} = 0$$

can be transformed into $\dfrac{\partial z}{\partial v} = 0.$

24. Let u be a function of x and y satisfying $x = \theta\cos\alpha - \phi\sin\alpha$, $y = \theta\sin\alpha + \phi\cos\alpha$ ($\alpha =$ constant). Prove that

$$\frac{\partial^2 u}{\partial x^2} + \frac{\partial^2 u}{\partial y^2} = \frac{\partial^2 u}{\partial \theta^2} + \frac{\partial^2 u}{\partial \phi^2}.$$ [C.H. 1968]

25. If $u = f(x^2 + 2yz, y^2 + 2zx)$, prove that

$$(y^2 - zx)\frac{\partial u}{\partial x} + (x^2 - yz)\frac{\partial u}{\partial y} + (z^2 - xy)\frac{\partial u}{\partial z} = 0.$$

26. A function $f(x, y)$ becomes $g(u, v)$ where $x = \frac{1}{2}(u + v)$, and $y^2 = uv$. Prove that

$$\frac{\partial^2 g}{\partial u \partial v} = \frac{1}{4}\left(\frac{\partial^2 f}{\partial x^2} + 2\frac{x}{y}\frac{\partial^2 f}{\partial x \partial y} + \frac{\partial^2 f}{\partial y^2} + \frac{1}{y}\frac{\partial f}{\partial y}\right).$$ [C.H. 1965, 82]

27. Given: z is a function of x, y, where

$$x = u + v, \quad y = uv.$$

Prove that $\dfrac{\partial^2 z}{\partial u^2} - 2\dfrac{\partial^2 z}{\partial u \partial v} + \dfrac{\partial^2 z}{\partial v^2} = (x^2 - 4y)\dfrac{\partial^2 z}{\partial y^2} - 2\dfrac{\partial z}{\partial y}.$ [C.H. 1981, 89, 93]

Section : 4
[Jacobians, Functional Dependence, Implicit Functions—Existence Theorems]

1. If $u = \dfrac{x^2 + y^2 + z^2}{x}$, $v = \dfrac{x^2 + y^2 + z^2}{y}$, $w = \dfrac{x^2 + y^2 + z^2}{z}$ find $\dfrac{\partial(x, y, z)}{\partial(u, v, w)}$.

[Ans: $\dfrac{x^2 y^2 z^2}{(x^2 + y^2 + z^2)^3}$]

2. If α, β, γ be the roots of the cubic equation in t,

$$\frac{u}{a+t} + \frac{v}{b+t} + \frac{w}{c+t} = 1$$

then prove that $\dfrac{\partial(u, v, w)}{\partial(\alpha, \beta, \gamma)} = -\dfrac{(\beta - \gamma)(\gamma - \alpha)(\alpha - \beta)}{(b - c)(c - a)(a - b)}.$

3. Let $u = \dfrac{x}{\sqrt{1 - r^2}}$, $v = \dfrac{y}{\sqrt{1 - r^2}}$, $w = \dfrac{z}{\sqrt{1 - r^2}}$, where $r^2 = x^2 + y^2 + z^2$. Prove that

$$\frac{\partial(u, v, w)}{\partial(x, y, z)} = \frac{1}{(1 - r^2)^{5/2}}.$$

4. If $u^3 = xyz$, $\dfrac{1}{v} = \dfrac{1}{x} + \dfrac{1}{y} + \dfrac{1}{z}$ and $w^2 = x^2 + y^2 + z^2$, prove that

$$\frac{\partial(u, v, w)}{\partial(x, y, z)} = -\frac{v(y - z)(z - x)(x - y)(x + y + z)}{3u^2 w(yz + zx + xy)}.$$

5. If
$$u^3 + v + w = x + y^2 + z^2,$$
$$u + v^3 + w = x^2 + y + z^2,$$
and $u + v + w^3 = x^2 + y^2 + z,$ then prove that
$$\frac{\partial(u, v, w)}{\partial(x, y, z)} = \frac{1 - 4(xy + yz + zx) + 16xyz}{2 - 3(u^2 + v^2 + w^2) + 27u^2v^2w^2}.$$

6. If $u_r = \dfrac{x_r}{\sqrt{1 - x_1^2 - x_2^2 - \cdots - x_n^2}}$ $(r = 1, 2, 3, \cdots, n)$,

   prove that the Jacobian of $u_1, u_2, \cdots, u_n$ w.r.t. $x_1, x_2, \cdots, x_n$ is
   $$(1 - x_1^2 - x_2^2 - \cdots - x_n^2)^{-\frac{1}{2}}.$$

7. Prove that the three functions u, v, w are connected by a symmetric relation, where
$$u = x/(y - z), \quad v = y/(z - x), \quad w = z/(x - y).$$
   [First, prove that u, v, w are not independent.]
   [Ans: $uv + vw + wu = -1$.]

8. If $u = x + y + z + t, v = x + y - z - t, w = xy - zt, r = x^2 + y^2 - z^2 - t^2$, prove that the Jacobian of u, v, w, r with respect to x, y, z, t vanishes and then obtain
$$uv = r + 2w.$$

9. State the theorem on existence and uniqueness of an implicit function. Deduce that $x^2 + xy + y^2 = 1$ defines y as a function of x in some neighbourhood of $(1, 0)$. Also find the first derivative of the function so defined at $x = 1$.   [C.H. 1992]

10. Let $f(x, y) = y^2 - yx^2 - 2x^5$. Verify that at $(1, -1)$

    (a) $f(1, -1) = 0$;
    (b) $f_y(1, -1) \neq 0$;
    (c) f_x and f_y are continuous in some nbd of $(1, -1)$. Now obtain the unique solution of $f(x, y) = 0$ in the nbd of $(1, -1)$. Obtain also dy/dx at $(1, -1)$.

11. Examine the following equations for the existence of unique implicit function near the indicated points and verify by direct calculation. Find the first derivatives of the solutions whenever these exist:

(a) $y^2 + x^3y + x^2 = 0; (0,0)$.

(b) $x^2 + y^2 - 1 = 0; (0,1)$.
[Hints: $y = \sqrt{1-x^2}, |x| < 1$ and $y > 0; y = -\sqrt{1-x^2}, |x| < 1$ and $y < 0$.]

(c) $y^2 - yx^2 - 2x^5 = 0; (0,0)$ and $(1,-1)$.
[Hints: $y = \phi(x)$, unique near $(1,-1)$ but not near $(0,0)$.]

(d) $y^2 + 2x^2y + x^5 = 0; (1,-1)$.
[Hints: $y = \phi(x)$ not unique but the equation possess a unique continuous solution $x = \psi(y)$ near $(1,-1)$.]

(e) $y^2 + yx^3 + x^2 = 0; (0,0)$.   [does not exist]

(f) $x^3 + y^3 - 3xy + y = 0; (0,0)$.

(g) $xy \sin x + \cos y = 0; (0, \pi/2)$.

(h) $y^3 \cos x + y^2 \sin^2 x = 7; \left(\frac{1}{3}\pi, 2\right)$.

(i) $2xy - \log xy = 2; (1,1)$.   [C.H. 1996]

[Ans: For (f)—(i) we can prove unique solutions exist]
[derivatives : $0; 0; \frac{2\sqrt{3}}{9}; -1$.]

12. Show that the quadratic forms
$$ax^2 + hxy + by^2, \quad lx^2 + 2mxy + ny^2$$
are independent unless $a/l = h/m = b/n$.   [C.H. 1989 (New)]

13. If $z = f(x,y), p = \partial z/\partial x, q = \partial z/\partial y$, prove that $p, q, px + qy - z$ can be expressed in terms of one of them if
$$\frac{\partial^2 z}{\partial x^2} \cdot \frac{\partial^2 z}{\partial y^2} - \left(\frac{\partial^2 z}{\partial x \partial y}\right)^2 = 0.$$

14. If $u = x^2 + y^2 + z^2, v = x + y + z, w = xy + yz + zx$, prove that they are not independent and then obtain the relation $v^2 = u + 2w$.

15. Prove that the functions
$$\phi_1 = x + y + z + t, \qquad \phi_2 = x^2 + y^2 + z^2 + t^2$$
$$\phi_3 = x^3 + y^3 + z^3 + t^3, \qquad \phi_4 = xyz + xyt + xzt + yzt$$
are dependent. Obtain the following relation
$$\phi_1^2 - 3\phi_1\phi_2 + 2\phi_3 - 6\phi_4 = 0.$$

16. If
$$X = x+y+z+u, \qquad Y = x^2+y^2+z^2+u^2$$
$$Z = x^3+y^3+z^3+u^3, \qquad U = xyz+yzu+zux+uxy$$
then obtain a relation between them, if possible.

[Ans: $6U = X^3 - 3XY + 2Z$.]

Section : 5
[On Approximations]

1. The side a and the opposite angle A of a triangle ABC remain constant; show that when the other sides and angles are slightly varied, then
$$\frac{\delta b}{\cos B} + \frac{\delta c}{\cos C} = 0. \qquad \text{[C.H. 1969]}$$

2. Let $\Delta = \frac{1}{2}bc \sin A$, where the lengths b and c are liable to an error of $\frac{1}{4}$ per cent in measurement and the measurement of A, measured as $60°$, is liable to an error of 1 minute (angle). Show that the greatest possible percentage error in the value of Δ is $\frac{1}{2}$ per cent (approximately).

3. ABC is an acute-angled triangle with fixed base BC. If $\delta b, \delta c, \delta A$ and δB are increments in b, c, A and B respectively and the vertex A is given a small displacement δx parallel to BC, prove that

    (a) $c \delta b + b \delta c + bc \cot A \delta A = 0$;

    (b) $c \delta B + \sin B \delta x = 0$.

4. In an experiment to determine g from the formula $gT^2 = 4\pi^2 l$ the pendulum used had an *equivalent length* l given by the formula $l = \lambda + k^2/\lambda (k < \lambda)$. The measurements of λ and k are accurate to within 1 per cent. Show that the maximum percentage error in l is approximately 1 per cent. If the measurement of T is accurate to within $\frac{1}{2}$ per cent, show also that the maximum percentage error in the value of g is approximately 2 per cent.

5. The sides of an acute-angled triangle are measured. Prove that the increment in A due to small increments in a, b, c is given by
$$bc \sin A \cdot \delta A = -a(\cos C \delta b + \cos B \delta c - \delta a).$$

Suppose that the limits of error in the length of any side are $\pm \mu$ per cent, where μ is small, prove that the limits of error in A are approximately
$$\pm 1.15(\mu a^2 / bc \sin A) \text{ degrees.} \qquad \text{[Maths. Tripos.]}$$

6. In a triangle ABC the angles and sides a and b are made to vary in such a way that the area remains constant; the side c is also constant. Show that if a and b vary by small amounts δa and δb respectively, then

$$\cos A \cdot \delta a + \cos B \cdot \delta b = 0.$$

7. The angles, A, B, C of a triangle are calculated from the sides a, b, c; if small changes $\delta a, \delta b, \delta c$ are made in them, show that

$$\delta A = \frac{1}{2}a(\delta a - \delta b \cos C - \delta c \cos B)/\Delta,$$

where Δ is the area of the triangle and verify that

$$\delta A + \delta B + \delta C = 0.$$

8. The diameter and slant height of a right circular cone are measured as $d = 10$ cm. and $s = 20$ cm., with a maximum error of 1.1 cm. in each. Find the greatest possible error in the calculated value of (a) volume; (b) curved surface.

9. A point is moving on the curve of intersection of the sphere $x^2 + y^2 + z^2 = 49$ and the plane $y = 2$. When $x = 6$ and it is increasing 4 units per second, find (a) the rate at which z is changing and (b) the speed with which the point is moving.

10. We define the *total differential* of $z = f(x, y)$ by

$$dz = \frac{\partial f}{\partial x}dx + \frac{\partial f}{\partial y}dy.$$

(a) $z = x^2 + 2xy$. Prove that dz at the point $(1, 1)$ is given by

$$dz = 4dx + 2dy.$$

(b) Find $\dfrac{\partial z}{\partial x}, \dfrac{\partial z}{\partial y}$ where $z = \log \sin(x^2 y^2 - 1)$. Now write down the expression for dz.

(c) Find dz where $x^2 + 2y^2 - z^2 = 1$.

(d) Find $\dfrac{\partial z}{\partial r}$ and $\dfrac{\partial z}{\partial \theta}$, given $z = x^2 + xy + y^2$, $x = 2r + \theta$, $y = r - 2\theta$.

CHAPTER 13: FUNCTIONS OF SEVERAL VARIABLES

Section : 6
[Miscellaneous]

1. If $u = t^n e^{-r^2/4kt}$, find n for which
$$\frac{\partial u}{\partial t} = k\left(\frac{\partial^2 u}{\partial r^2} + \frac{2}{r}\frac{\partial u}{\partial r}\right). \quad (k = \text{constant}) \qquad \text{[C.H. 1983 (Sp.)]}$$

2. If $u = \dfrac{1}{\sqrt{1-2xy+y^2}}$, show that
$$\frac{\partial}{\partial x}\left\{(1-x^2)\frac{\partial u}{\partial x}\right\} + \frac{\partial}{\partial y}\left(y^2 \frac{\partial u}{\partial y}\right) = 0. \qquad \text{[C.H. 1983]}$$

3. If λ, μ, ν are functions of x, y, z given by
$$x = \lambda + \mu + \nu, \quad y = \lambda^2 + \mu^2 + \nu^2, \quad z = \lambda^3 + \mu^3 + \nu^3,$$
prove that
$$\frac{\partial \lambda}{\partial x} = \frac{\mu\nu(\nu-\mu)}{(\lambda-\mu)(\mu-\nu)(\nu-\lambda)}.$$

[*Hints:* Differentiate w.r.t. x and obtain
$$\lambda_x + \mu_x + \nu_x = 1; \quad \lambda\cdot\lambda_x + \mu\cdot\mu_x + \nu\cdot\nu_x = 0; \quad \lambda^2\cdot\lambda_x + \mu^2\cdot\mu_x + \nu^2\cdot\nu_x = 0.$$

Use Cramer's rule to solve for λ_x.]

4. $f(x,y) = \dfrac{2xy}{x^2+y^2}$, if $x^2 + y^2 \neq 0$ and $f(0,0) = 0$.

   Show that $\dfrac{\partial f}{\partial x}$ and $\dfrac{\partial f}{\partial y}$ both exist at $(0,0)$ but $f(x,y)$ is discontinuous at the origin. [C.H. 1981]

5. Show that $u = \phi(xy) + \sqrt{xy}\,\psi(y/x)$ satisfies the partial differential equation $x^2 u_{xx} - y^2 u_{yy} = 0$ assuming that both ϕ and ψ are twice differentiable.

   [C.H. 1981]

6. Discuss the continuity of the function f defined by
$$f(x,y) = \frac{x^4 y^4}{(x^2+y^4)^3} \text{ if } (x,y) \neq (0,0) \text{ and } f(0,0) = 0. \qquad \text{[C.H. 1982]}$$

7. If $y = x\phi(z) + \psi(z)$ and $z_y \neq 0$, ϕ, ψ are both twice differentiable, prove that
$$z_{xx} z_y^2 - 2z_x z_y z_{xy} + z_{yy} z_x^2 = 0. \qquad \text{[C.H. 1982]}$$

8. If f is twice differentiable function of u, v and $u = x^2 - y^2$, $v = 2xy$, prove that

$$4(u^2 + v^2)f_{uv} + 2vf_u + 2uf_v = xy(f_{xx} - f_{yy}) + (x^2 - y^2)f_{xy}$$

justifying the operations involved. [C.H. 1983]

9. If f is a function of u and v where $u = \sqrt{x^2 + y^2}$ and $v = \tan^{-1} y/x$, find

$$\frac{\partial f}{\partial x} \text{ and } \frac{\partial f}{\partial y}.$$

[C.H. 1978]

10. State a set of sufficient conditions so that $f(x, y) = 0$ defines y as a **unique** continuous function of x.

Discuss your statement in relation to $f(x, y) = x^2 + y^2 - 1 = 0$ near the points $(\pm 1, 0)$ and $(0, \pm 1)$. [C.H. 1979]

Use $f(x, y) = y^2 - yx^2 - 2x^5$ to prove that uniqueness may be lost if at a point (a, b) $\dfrac{\partial f}{\partial y} = 0$. [C.H. 1980]

11. If $x = u^2 v, y = v^2 u$, then show that

$$2x^2 f_{xx} + 2y^2 f_{yy} + 5xy f_{xy} = uv f_{uv} - \frac{2}{3}(uf_u + vf_v),$$

f being an arbitrary function of two variables, satisfying $f_{xy} = f_{yx}$. [C.H. 1979]

12. If for all positive values of λ,

$$u(\lambda x, \lambda y, z/\lambda) = u(x, y, z)$$

prove that $xu_x + yu_y = zu_z$. [C.H. 1979]

13. Considering $\Delta = \begin{vmatrix} a & b & c \\ d & e & f \\ g & h & k \end{vmatrix}$ as a function of nine variables $a, b, c, \cdots, k$, prove that

$$a\Delta_a + b\Delta_b + c\Delta_c = \Delta \quad \text{and} \quad \begin{vmatrix} \Delta_a & \Delta_b & \Delta_c \\ \Delta_d & \Delta_e & \Delta_f \\ \Delta_g & \Delta_h & \Delta_k \end{vmatrix} = \Delta^2 \quad (\Delta \neq 0)$$

where Δ_a = partial derivative of Δ w.r.t. a, etc. [C.H. 1980]

14. (a) If $xu^2 + v = y^3, 2yu - xv^3 = 4x$, find $\dfrac{\partial u}{\partial x}$ and $\dfrac{\partial v}{\partial y}$.

(b) If $x + y^2 = u$, $y + z^2 = v$, $z + x^2 = w$, find $\frac{\partial x}{\partial u}, \frac{\partial^2 x}{\partial u^2}, \frac{\partial^2 x}{\partial u \partial v}$
assuming that the equations define x, y, z as twice differentiable functions u, v, w.

(c) Transform the equation $x^2 \frac{\partial z}{\partial x} + y^2 \frac{\partial z}{\partial y} = z^2$, by introducing new independent variables $u = x$, $v = 1/y - 1/x$ and the new function $w = 1/z - 1/x$.

[C.H. 1996]

[Ans: (a) $\frac{v^3 - 3xu^2v^2 + 4}{6x^2uv^2 + 2y}, \frac{2xu^2 + 3y^3}{3x^2uv^2 + y}$. (b) $\frac{1}{1+8xyz}, \frac{16x^2y - 8yz - 32x^2z^2}{(1+8xyz)^2}, \frac{16y^2z - 8xz - 32x^2y^2}{(1+8xyz)^3}$.]

Preparations on Functions of Several Variables

(Question on Theory and Collection of important problem)

1. (i) When do you say that $f(x, y)$ is said to be continuous at the point (a, b)?

    (ii) State and establish a set of sufficient conditions under which a function $f : S \to \mathbb{R}(S \subset \mathbb{R} \times \mathbb{R})$ may be continous at a point of S. [C.H. 2007]

    (iii) Let $f(x, y)$ be continuous at an interior point (a, b) of the domain of definition of f and let $f(a, b) \neq 0$. Show that there exists a neighbourhood of (a, b) in which $f(x, y)$ retain the same sign as that of $f(a, b)$. [C.H. 2008]

    (iv) If $f : S \to \mathbb{R}$ be a function where $S \subset \mathbb{R}^2$. If it is continuous at $(a, b) \in S$, then show that $f(x, b)$ is continuous at $x = a$ and that $f(a, y)$ is continuous at $y = b$. Is the converse true? Justify your answer. [C.H. 2006]

2. (i) State the Implicit Function Theorem for a function of two variables.

    (a) Show that the equation $2xy - \log_e(xy) = 2$ determines y uniquely as a function of x near the point $(1, 1)$ and find $\frac{dy}{dx}$ at $(1, 1)$.

    (b) Show hat the equation $xy \sin x + \cos y = 0$ determines unique implicit function in the neighbourhood of the point $(0, \pi/2)$. Also find the first derivative of the functions.

    (c) Show that the equation $y^2 - yx^2 - 2x^5 = 0$ determines uniquely implicit function in the neighbourhood of $(1, -1)$. Also find the first order derivative at $(1, -1)$.

3. Let $u(x, y, z), v(x, y, z)$ and $w(x, y, z)$ be there functions having continuous first order partial derivatives at any point (x, y, z) of their domain. Prove that u, v, w are functionally related if

$$\frac{\partial(u, v, w)}{\partial(x, y, z)} = 0$$

4. Prove that the following there functions u, v, w are functionally related and find the relation among them, where
$$u = \frac{x}{y-z}, \quad v = \frac{y}{z-x}, \quad w = \frac{z}{x-y}$$
[C.H. 2008]

5. State and prove converse of Euler's Theorem for homogeneous functions of three variables. [C.H. 2008]

If $u = \tan^{-1}\frac{x^3+y^2}{x-y}$, show that $x^2\frac{\partial^2 u}{\partial x^2} + 2xy\frac{\partial^2 u}{\partial y^2} + y^2\frac{\partial^2 u}{\partial y^2} = (1 - 4\sin^2 u)\sin zu$.

6. If u be a homogeneous function in x, y, z of degree n having continuous second order partial derivatives and, if $f(\xi, \eta, \zeta)$ where ξ, η, ζ are the partial derivatives of u with respect to x, y, z respectively, then prove that
$$\xi\frac{\partial u}{\partial \xi} + \eta\frac{\partial u}{\partial \eta} + \zeta\frac{\partial u}{\partial \zeta} = \frac{nu}{n-1}(n \neq 1)$$

7. (i) Let
$$f(x,y) = \begin{cases} xy\frac{x^2-y^2}{x^2+y^2} & (x^2+y^2 \neq 0) \\ 0 & \text{when } x^2+y^2 = 0 \end{cases}$$
Prove that $\frac{\partial^2 f}{\partial x \partial y} \neq \frac{\partial^2 f}{\partial y \partial x}$ at $(0,0)$.

State Schwarg's Theorem for the equality $f_{xy} = f_{yx}$. In the problem stated above what condition of Scwerz's Theorem does f violate.

(ii)
$$f(x,y) = \begin{cases} \frac{x^2y^2}{x^2+y^2}, & \text{if } x^2+y^2 \neq 0 \\ 0, & \text{if } x^2+y^2 = 0 \end{cases}$$
Show that the conditions Schwarz Theorem are not necessary for the commutative of the order at partial derivatives.

> **Note:** If the conditions of Young's Theorem or Schwarz Theorem are satisfied then $f_{xy} = f_{yx}$ at (a,b).
> But if the conditions are not satisfied then the equality $f_{xy} = f_{yx}$ may or may not occur.

8. A function $f(x,y)$ having continuous second order partial derivatives when expressed in terms of the new variables u and v defined by $x = \frac{1}{2}(u+v)$ and $y^2 = uv$ becomes $g(u,v)$; prove that
$$\frac{\partial^2 g}{\partial u \partial v} = \frac{1}{4}\left[\frac{\partial^2 f}{\partial x^2} + 2\frac{x}{y}\frac{\partial^2 f}{\partial x \partial y} + \frac{\partial^2 f}{\partial y^2} + \frac{1}{y}\frac{\partial f}{\partial y}\right]$$
[C.H. 1999]

9.
$$f(x,y) = \begin{cases} x \sin \frac{1}{y} + \frac{x^2-y^2}{x^2+y^2}, & \text{if } y \neq 0 \\ 0, & \text{if } y = 0 \end{cases}$$

show that $\lim_{y \to 0} \lim_{x \to 0} f(x,y)$ exists, but neither $\lim_{(x,y) \to (0,0)} f(x_1, y_1)$ nor $\lim_{x \to 0} \lim_{y \to 0} f(x,y)$ exists.

10. If w is a function of u and v, where
$$u = x^2 - y^2 - 2xy! \text{ and } v = y$$
then prove that
$$(x+y)\frac{\partial w}{\partial x} + (x-y)\frac{\partial w}{\partial y} = 0$$
is equivalent to
$$\frac{\partial w}{\partial v} = 0$$

[Hints: $\frac{\partial w}{\partial x} = \frac{\partial w}{\partial u} \cdot \frac{\partial u}{\partial x} + \frac{\partial w}{\partial v} \cdot \frac{\partial v}{\partial x}$]

11. Let u, v be functions of ξ, η and ζ having continuous first order partial derivatives and ξ, η and ζ be functions of x and y having continuous first order partial derivatives. Prove that
$$\frac{\partial(u,v)}{\partial(x,y)} = \frac{\partial(u,v)}{\partial(\xi,\eta)} \cdot \frac{\partial(\xi,\eta)}{\partial(x,y)} + \frac{\partial(u,v)}{\partial(\eta,\xi)} \cdot \frac{\partial(\eta,\xi)}{\partial(x,y)} + \frac{\partial(u,v)}{\partial(\zeta,\xi)} \cdot \frac{\partial(\zeta,\xi)}{\partial(x,y)}$$
[C.H. 2001]

12. Let the differentiable function ϕ defined by
$$\phi(cx - az, cy - bz) = 0$$
Express z as a differentiable function of x and y. Prove that
$$a\frac{\partial z}{\partial x} + b\frac{\partial z}{\partial y} = c \quad (a, b, c \text{ are contants})$$
[C.H. 2007]

13. Transform the equation
$$y\frac{\partial z}{\partial x} - x\frac{\partial z}{\partial y} = (y-x)z$$
taking $u = x^2 + y^2, v = \frac{1}{x} - \frac{1}{y}$ for the new independent variables and $w = \log z - (x+y)$ for the new function. [C.H. 2003]

14. If u, v are two polynomials in x, y and are homogeneous of degree n, prove that
$$u\,dv - v\,du = \frac{1}{n}\frac{\partial(u,v)}{\partial(x,y)}(x\,dy - y\,dx).$$

15. If z is a function of two independent variables x, y and $x = c\cosh\cos v$, $y = c\sinh u \sin v$ (c is a given real number), prove that
$$\frac{\partial^2 z}{\partial u^2} + \frac{\partial^2 z}{\partial v^2} = \frac{1}{2}c^2(\cosh 2u - \cos 2v)\left(\frac{\partial^2 z}{\partial x^2} \to \frac{\partial^2 z}{\partial y^2}\right).$$

16. When is a function $f(x,y)$ said to be differentiable at a point (x,y)? Prove that the function $f(x,y) = \sqrt{|xy|}$ is not differentiable at the point $(0,0)$ but that $\frac{\partial f}{\partial x}$ and $\frac{\partial f}{\partial y}$ both exist at the origin and have the value O.

17. Let
$$f(x,y) = \begin{cases} x^2 \sin\frac{1}{x} + y^2 \sin\frac{1}{y}, & \text{when } x \neq 0, y \neq 0 \\ x^2 \sin\frac{1}{x}, & \text{when } x \neq 0 \\ y^2 \cos\frac{1}{y}, & \text{when } y \neq 0 \\ 0, & \text{when } x - 0, y = 0 \end{cases}$$

Prove that $\frac{\partial f}{\partial x}$ and $\frac{\partial f}{\partial y}$ exist at $(0,0)$ but none is continous there. Examine the differentiability of $f(x,y)$ at $(0,0)$. [C.H. 2000]

18. Let
$$f(x,y) = \begin{cases} y\sin\frac{1}{x} + \frac{xy}{x^2+y^2}, & \text{when } x \neq 0 \\ 0, & \text{when } x = 0 \end{cases}$$

Find the double limit and the repeated limits, if exist at the point $(0,0)$. [C.H. 2008]

19. If $f(x,y) = \frac{xy}{\sqrt{x^2+y^2}}$ for all $(x,y) \neq (0,0)$ and $f(0,0) = 0$, show that f is not differentiable at $(0,0)$ though $f(x,y)$ is continuous there.

20. If $H(x,y)$ be a homogeneous function of x and y of degree n having continuous first order parted derivatives and
$$u(x,y) = (x^2 + y^2)^{-n/2},$$
then show that
$$\frac{\partial}{\partial x}\left(1 + \frac{\partial u}{\partial x}\right) + \frac{\partial u}{\partial y}\left(1 + \frac{\partial u}{\partial x}\right) = 0$$

[C.H. 2008]

CHAPTER 13: FUNCTIONS OF SEVERAL VARIABLES

21. Using the method of Jacobian show that the expression

$$ax^2 + by^2 + cz^2 + zfyz + 2gzx + 2hxy$$

is resolvable into linear factors.

22. Show that $ax^2 + 2hxy + by^2$ and $Ax^2 + 2Hxy + By^2$ are independent unless $\frac{a}{A} + \frac{b}{B} = \frac{h}{H}$.

23. When is the point (a, b) said to be a *limiting point* of a subset A of $\mathbb{R} \times \mathbb{R}$?

    If $B = \{(a, 0), a \in \mathbb{R}\}$, show that B is a closed subset but not an open subset of $\mathbb{R} \times \mathbb{R}$.

24. If $u = \frac{x+y}{1-xy}$ and $v = \tan^{-1} x + \tan^{-1} y$, find $\frac{\partial(u,v)}{\partial(x+y)}$. Also u and v functionally related. If so, find the relationship.

25. If

$$u = \frac{1}{\sqrt{1 - 2xy + y^2}}$$

prove that

$$\frac{\partial}{\partial x}\left\{(1 - x^2)\frac{\partial u}{\partial x}\right\} + \frac{\partial}{\partial y}\left(y^2 \frac{\partial u}{\partial y}\right) = 0$$

[C.H. 2001]

26. Let

$$f(x, y) = \begin{cases} \frac{x^4 + y^4}{x - y} & (x \neq y) \\ 0 & (x = y) \end{cases}$$

Show that $\frac{\partial f}{\partial x}$ and $\frac{\partial f}{\partial y}$ exist at $(0, 0)$ f is not continuous at $(0, 0)$. State a set of sufficient conditions for the continuity of a function $f(x, y)$ at an interior point of its domain of definition.

27. If $f(x, y)$ be a function of two variables x and y, where $x = u^2 u$ and $y = v^2 u$, then show that

$$2x^2 \frac{\partial^2 f}{\partial x^2} + 2y^2 \frac{\partial^2 f}{\partial y^2} + 5xy \frac{\partial^2 f}{\partial x \partial y} = uv \frac{\partial^2 f}{\partial u \partial v} - \frac{2}{3}\left(u\frac{\partial f}{\partial u} + v\frac{\partial f}{\partial v}\right)$$

(The relevant partial derivatives are assumed to be continuous)

28. State and prove (1) Schwarz Theorem $\frac{\partial f}{\partial y}$ exists in some *nbd* of (a, b) and $\frac{\partial^2 f}{\partial y \partial x}$ continuous at (a, b) and (2) Young's Theorem (f_x and f_y both differentiable) for the equality of f_{xy} and f_{yx}.

29. State Young's Theorem on the commutativity of order of partial derivatives. Use the following function to verify whether the conditions for the Young's Theorem are necessary

$$f(x,y) = \begin{cases} \frac{x^2 y^2}{x^2+y^2}, & \text{if } x^2+y^2 \neq 0 \\ 0, & \text{if } x^2+y^2 = 0 \end{cases}$$

30. If $Pdx + Qdy + Rdz = $ can be made a perfect differential of some function of x, y, z on multiplying each term by a factor $\mu(x, y, z)$, then prove that

$$P\left(\frac{\partial Q}{\partial z} - \frac{\partial R}{\partial y}\right) + Q\left(\frac{\partial R}{\partial x} - \frac{\partial P}{\partial z}\right) + R\left(\frac{\partial P}{\partial y} - \frac{\partial Q}{\partial x}\right) = 0.$$

AN INTRODUCTION TO
ANALYSIS

(Differential Calculus—Part I)

Chapter 14

Applications of Derivatives
Some Important Curves:
Their Equations and Shapes

Differential Calculus—Part I

Section II: Chapters 14–21

APPLICATIONS OF DERIVATIVES

Section II

Applications of Derivatives

Recapitulation (For Self Practice)

Chapter 14

Some Important Curves: Their Equations and Shapes

14.1 Introduction

The next few chapters of this book will deal with **Applications.** Under this heading we shall enter into discussions on properties of curves which involve derivatives of functions representing those curves. With this end in view we consider it useful to acquaint our readers with the equations and geometrical shapes of some important curves. With frequent uses at subsequent stages the students will be more familiar with these equations.

14.2 Rectangular Cartesian Equations of Plane Curves

A plane curve is the locus of points on a plane obeying some characteristic property. In Analytic Geometry of two dimensions we take a rectangular system of coordinate axes OX and OY and represent any point on the locus by (x, y) or sometimes by (X, Y). We consider analytic representation of a curve by regarding one of the co-ordinates, say y, as a function of the other co-ordinate x.

[e.g., a parabola by $y = x^2 (-\infty < x < \infty)$; a circle of radius 2 and centre $(0,0)$ by two functions $y = +\sqrt{4 - x^2}$, $y = -\sqrt{4 - x^2}$ $(-2 \leq x \leq 2)$.]

Conversely, if instead of starting with a curve determined geometrically we consider a given function $y = f(x)$, then we can represent the functional dependence of y on x graphically by making use of a rectangular co-ordinate system in the usual way (e.g., see Figure 3.1, Chapter 3).

We shall be using three different analytic representations of plane curves.
1. Parametric Representation;
2. Implicit Representation and
3. Explicit Representation.

1. Parametric Representation

Instead of considering one of the rectangular co-ordinates as a function of the other, we may think of both the co-ordinates x and y as functions of a *third independent variable*, t or θ etc. (called *parameter*). In order to describe a curve we may require the parameter to lie within a certain interval; further it is interesting to study the geometric significance of the parameter which is different for different curves.

Example 14.2.1.

	Circle	: $x = a\cos t, y = a\sin t,$	$0 \le t \le 2\pi.$
	Parabola	: $x = at^2, y = 2at,$	$-\infty < t < \infty.$
	Ellipse	: $x = a\cos t, y = b\sin t,$	$0 \le t \le 2\pi.$
	Hyperbola	: $x = a\cosh t, y = b\sinh t,$	$-\infty < t < \infty.$
	Astroid	: $x = a\cos^3 t, y = a\sin^3 t,$	$0 \le t \le 2\pi.$
(One arch)	*Cycloid*	: $x = a(\theta - \sin\theta), y = a(1 - \cos\theta),$	$0 \le \theta \le 2\pi.$
		etc. etc.	

2. Implicit Representation

We may express the geometrical property obeyed by points of the curve by means of an *implicit relation* between x and y, e.g., $F(x, y) = 0$, where (x, y) is any point of the curve.

Example 14.2.2.

Circle	:	$x^2 + y^2 = a^2.$
Parabola	:	$y^2 - 4ax = 0.$
Ellipse	:	$x^2/a^2 + y^2/b^2 = 1.$
Hyperbola	:	$x^2/a^2 - y^2/b^2 = 1.$
Astroid	:	$x^{2/3} + y^{2/3} = a^{2/3}.$
Folium of Descartes	:	$x^3 + y^3 - 3axy = 0.$
Semi-Cubical parabola	:	$y^2 - x^3 = 0.$
Lemniscate of Bernoulli	:	$(x^2 + y^2)^2 = a^2(x^2 - y^2).$
etc. etc.		

3. Explicit Representation

$y = f(x)$ or $x = \phi(y)$ where one of the rectangular co-ordinates is expressed explicitly in terms of other.

Example 14.2.3.

Catenary   : $y = c \cosh \frac{x}{c}$.
Parabola   : $y = x^2$ or $x = y^2$.
Rectangular Hyperbola : $y = \frac{1}{x}$.
etc. etc.

14.3 The Equations of some well-known Curves in Parametric Form

1. Circle: *Given a circle with centre as origin and radius a.*

Denote by t the angle formed by X-axis and the radius to some point $P(x,y)$ of the circle.

Then the co-ordinates of any point P on the circle can be expressed in terms of the parameter t as

$$x = a\cos t, \quad y = a\sin t \ (0 \le t \le 2\pi).$$

These are the parametric equations of the circle.

[If we eliminate t we get easily (squaring and adding)

$$x^2 + y^2 = a^2.$$

This is the equation of the circle in *Implicit form.*]

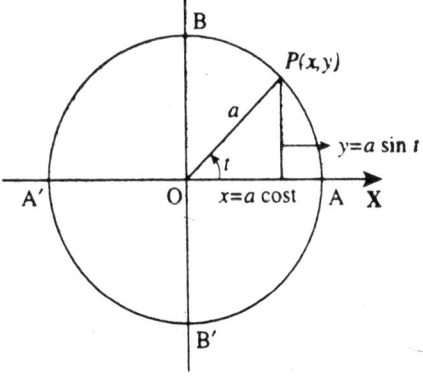

Fig 14.1
Circle: $x = a\cos t$, $y = a\sin t$

See that as t increases from 0 to $\pi/2$, the curve traced out is the arc AB of the circle (in the first quadrant); then as t increases further from $\pi/2$ to π, the curve traced out is the arc BA'. Again as t ranges from π to $3\pi/2$, $3\pi/2$ to 2π the curve traced out are the arcs $A'B'$ and $B'A$ respectively.

> **Remember:** If S be the area of the sector OAP and t is the central angle $\angle AOP$ (in radians) then $t = 2S/a^2$ is another geometric meaning to the parameter t.

2. Ellipse: *Given an ellipse whose major axis lies along X-axis and minor axis lies along Y-axis and are of lengths $2a, 2b$ respectively, centre at origin the O.*

It is well known that the equation of the ellipse is $\frac{x^2}{a^2} + \frac{y^2}{b^2} = 1$. (Implicit form)

If we put $x = a\cos t$, then we easily obtain $y = b\sin t$. The two equations $x = a\cos t$, $y = b\sin t$ $(0 \leq t \leq 2\pi)$ will give the parametric equations of the ellipse.

As t varies from 0 to $\pi/2$, the curve traced out is the arc AB (at A, $t = 0$ and at B, $t = \pi/2$). The parts BA', $A'B'$ and $B'A$ are traced out as t varies from $\pi/2$ to π, π to $3\pi/2$ and $3\pi/2$ to 2π.

What is the geometrical meaning of the Parameter t?

Draw two circles with centres at the origin and with radii a and b. Let the point $P(x, y)$ lie on the ellipse. Through P draw PM perpendicular to the X-axis; produce MP to the meet the larger circle (known as *Auxiliary circle*) at Q. Denote by t the angle formed by the radius OQ with the X-axis. Then from the figure it is clear that $x = OM = OQ\cos t = a\cos t$.

Suppose OQ cuts the smaller circle at C. Draw CN perpendicular to the X-axis. Then
$$CN = OC\sin t = b\sin t.$$

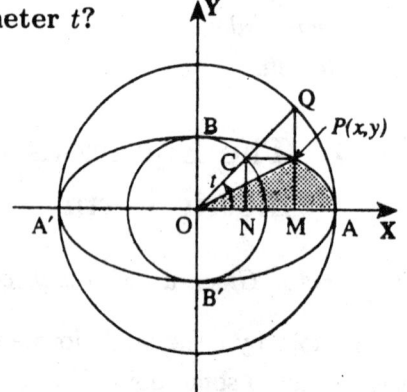

Fig 14.2
Ellipse: $x = a\cos t$, $y = b\sin t$

If we put $x = a\cos t$ in the equation $\frac{x^2}{a^2} + \frac{y^2}{b^2} = 1$, we get $y = b\sin t$. So we conclude that $CN = y$ and hence CP is parallel to the X-axis and we get $y = b\sin t$.

Thus, in the parametric equations $x = a\cos t$, $y = b\sin t$ of the ellipse, t is the angle formed by the radius OQ and the X-axis. This angle t is also known as **eccentric angle** of the point P.

Another meaning of t

[1]The area S of the elliptic sector OAP [shaded area from the vertex A, where $t = 0$ to the point P, where $t = t_1$ (say)] is given by

$$S = \frac{1}{2}xy + \int_{x=OM}^{x=OA} y\,dx = \frac{1}{2}(a\cos t_1)(b\sin t_1) + \int_{t=t_1}^{t=0} (b\sin t)(-a\sin t)\,dt$$
$$= \frac{1}{4}ab\sin 2t_1 + \frac{1}{2}ab\int_0^{t_1} (1 - \cos 2t)\,dt = \frac{1}{2}ab\,t_1.$$

---

[1]The area between the curve $y = f(x)$, the X-axis and the lines $x = a$ and $x = b$ is given by
$$\int_a^b y\,dx.$$

(See Author's *Integral Calculus: An Introduction to Analysis*, Eighth Edition(1996), Chapter 16).

Thus
$$t_1 = \frac{2S}{ab}.$$

This gives another geometric meaning to the parameter. In particular, at a point $P(t = t_1)$ of the **circle** $x^2 + y^2 = 1$ ($a^2 = b^2 = 1$), t_1 is twice the area t_1 (a purely real number without any unit) of the circular sector AOP (see Fig 14.1).

3. Parabola: *The parametric equations*

$$x = at^2, \quad y = 2at \ (-\infty < t < +\infty)$$

represent the parabola $y^2 = 4ax$ *having its axis along X-axis and the tangent at the vertex along the Y-axis and latus rectum equal to $4a$ (Fig 14.3).*

The part ABO is described as t ranges from $-\infty$ to 0 and part OCD is described as t varies from 0 to $+\infty$.

Since

$$\frac{dy}{dx} = \frac{dy}{dt} \Big/ \frac{dx}{dt} = \frac{2a}{2at} = \frac{1}{t} \quad \text{or} \quad t = 1 \Big/ \frac{dy}{dx},$$

t is the reciprocal of the slope of the tangent line drawn at the $P(t)$ of the parabola.

4. Hyperbola: *The hyperbola*

$$\frac{x^2}{a^2} - \frac{y^2}{b^2} = 1$$

has parametric equations $x = a \cosh t, y = b \sinh t$, *where t is not an angle but a real number whose geometric meaning is given below:*

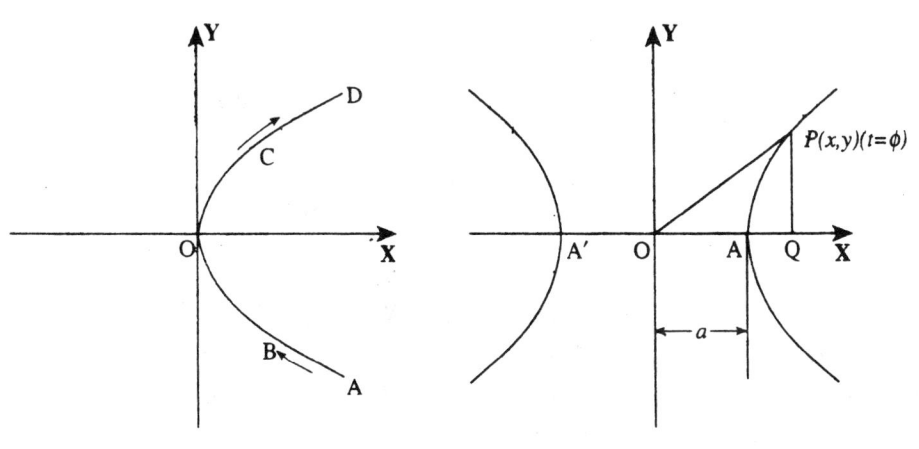

Fig 14.3
Parabola: $x=at^2, y=2at$

Fig 14.4
Hyperbola: $x=a \cosh t, y=b \sinh t$

The area S of the hyperbolic sector OAP (Fig 14.4) from the vertex $t = 0$ to $t = t_1$ (say) is the area of the triangle OQP minus the sectorial area AQP :

$$\begin{aligned} S &= \frac{1}{2}xy - ab \int_{x=a}^{x=a\cosh t_1} y\, dx \\ &= \frac{1}{2}ab \sinh t_1 \cosh t_1 - ab \int_{t=0}^{t=t_1} \sinh^2 t\, dt \\ &= \frac{1}{4}ab \sinh 2t_1 - \frac{1}{2}ab \int_0^{t_1} (\cosh 2t - 1)\, dt \\ &= \frac{1}{2}abt_1. \end{aligned}$$

Then $t_1 = 2S/ab$ gives a geometric meaning to the parameter t. In particular, at a point $P(t = t_1)$ of the equilateral hyperbola $x^2 - y^2 = 1$, t_1 is twice the area of the hyperbolic sector OAP.

5. Cycloid: *The cycloid is a curve described by a point lying on the circumference of a circle if the circle rolls upon a straight line without sliding.*

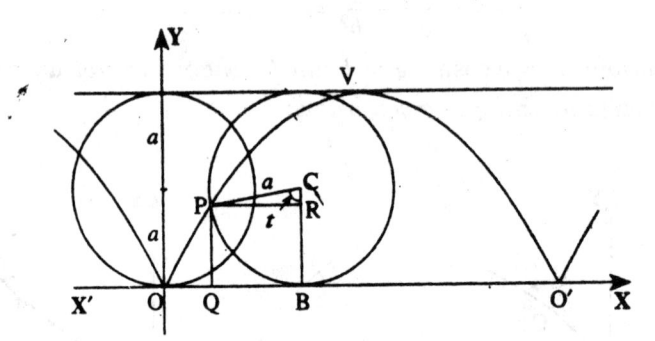

Fig 14.5
Cycloid: $x = a(t - \sin t)$, $y = a(1 - \cos t)$

Suppose that the rolling circle starts from the position in which the generating point P coincides with the point O of the fixed line $X'X$ (along which the circle rolls). Take O as origin and the fixed line as X-axis. When the rolling circle has rolled on to the position shown in the figure, the generating point has moved from O to $P(x, y)$ so that $OB = \text{arc } BP$. Let $2a$ be the diameter of the circle. Let t be the angle between CB and CP so that t is the angle through which the radius of the rolling circle rolls from its initial position to the present position.

CHAPTER 14: SOME IMPORTANT CURVES: THEIR EQUATIONS AND SHAPES

Thus the arc $BP = at$. We have now,

$$x = OQ = OB - QB = \text{arc } BP - PR$$
$$= at - a\sin t$$
$$= a(t - \sin t)$$
$$y = QP = BC - RC = a - a\cos t$$
$$= a(1 - \cos t).$$

Thus

$$x = a(t - \sin t)$$
$$\text{and } y = a(1 - \cos t)$$

are the parameric equations of the cycloid (t is the parameter).
Eliminating t we may obtain x as a function of y.
In the interval $0 \leq t \leq \pi$, $y = a(1 - \cos t)$ has an inverse:

$$t = \cos^{-1} \frac{a - y}{a}.$$

Substituting this expression for t in $x = a(t - \sin t)$ we get

$$x = a\cos^{-1} \frac{a - y}{a} - a\sin\left(\cos^{-1} \frac{a - y}{a}\right)$$
$$= a\cos^{-1} \frac{a - y}{a} - \sqrt{2ay - y^2} \quad (0 \leq y \leq 2a).$$

When $2a \leq y \leq 0$ (See Fig 14.5)

$$x = 2\pi a - \left\{ a\cos^{-1} \frac{a - y}{a} - \sqrt{2ay - y^2} \right\}$$

x ranges from 0 to $2\pi a$ for one complete arch.

We note that it is not easy to get t in terms of x from $x = a(t - \sin t)$ and hence we have not expressed y as a function of x.

Note 14.3.1. *When the circle makes one complete revolution, the point P describes one* **complete arch** *OVO' of the cycloid, so that t increases from 0 to 2π as the point P moves from O to O'.*

Note 14.3.2. *When $t = \pi$, the point P reaches V (called the* **Vertex** *of the cycloid).*

Note 14.3.3. *The cycloid evidently consists of an endless succession of exactly congruent arches each of which represents one complete revolution of the rolling circle.*

Note 14.3.4. *The area under one arch is*

$$A = \int_{t=0}^{t=2\pi} a(1-\cos t)\, d\{a(t-\sin t)\}$$

$$= a^2 \int_0^{2\pi} (1-\cos t)^2 \, dt$$

$$= a^2 \int_0^{2\pi} \left[1 - 2\cos t + \frac{1}{2}(1+\cos 2t)\right] dt$$

$$= a^2 \left[\frac{3}{2}t - 2\sin t + \frac{1}{4}\sin 2t\right]_0^{2\pi} = 3\pi a^2.$$

6. Inverted Cycloid: (Vertex at origin)

The parametric equations

$$x = a(t + \sin t),$$
$$y = a(1 - \cos t)$$

represent what we call *Inverted cycloid*. One arch of the inverted cycloid can be described by varying t from

$$t = -\pi \text{ to } t = +\pi. \quad (\text{Fig 14.6})$$

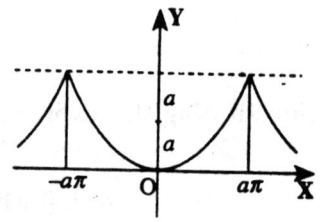

Fig 14.6
Inverted Cycloid:
$x = a(t+\sin t), y = a(1-\cos t)$

7. Other Cycloids:

(A) $\quad x = a(t + \sin t)$
$\quad\quad y = a(1 + \cos t)$

One arch is described as t varies from $-\pi$ to $+\pi$ (Fig 14.7).

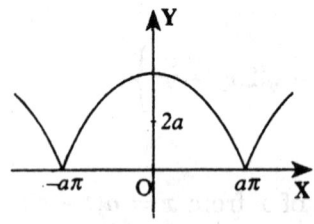

Fig 14.7
Cycloid: $x=a(t+\sin t), y=a(1+\cos t)$

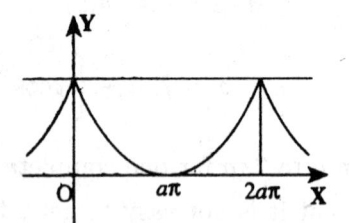

Fig 14.8
Cycloid: $x=a(t-\sin t), y=a(1+\cos t)$

(B) $\quad x = a(t - \sin t)$
$\quad\quad y = a(1 + \cos t)$

On arch is described (Fig 14.8) as t varies from $t=0$ to $t=2\pi$.

CHAPTER 14: SOME IMPORTANT CURVES: THEIR EQUATIONS AND SHAPES

8. Astroid:

The astroid is a curve represented by the following parametric equations

$$x = a \cos^3 t, \quad y = a \sin^3 t \quad (0 \le t \le 2\pi).$$

Eliminating t we may easily obtain

$$x^{2/3} + y^{2/3} = a^{2/3}.$$

The shape of the astroid is shown in Fig 14.9.

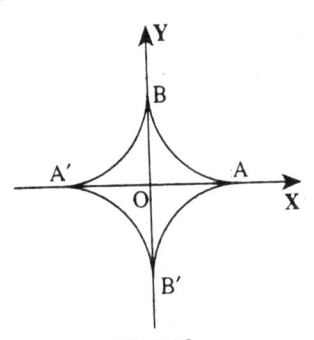

Fig 14.9
Astroid: $x = a\cos^3 t$, $y = a\sin^3 t$

As t varies from 0 to $\pi/2$, the part AB (in the first quadrant) is traced out. The parts BA', $A'B'$ and $B'A$ are traced out as t varies from $\pi/2$ to π, π to $3\pi/2$, $3\pi/2$ to 2π respectively. In fact, we have four branches of the curve $x^{2/3} + y^{2/3} = a^{2/3}$. Two of the branches AB and $B'A$ meet at A.

We say that A is a *singular point* of the curve. Moreover, the two branches which meet at A have a common tangent OA. Such a singular point is called a **Cusp**. Astroid has *four cusps* and hence it is also known as a FOUR-CUSPED HYPOCYCLOID.

9. Epicycloids and Hypocycloids:

The path traced out by a point marked in the circumference of a circle which rolls in contact with a fixed circle without sliding is called **an epicycloid** *or* **a hypocycloid** *according as the rolling circle is outside or inside a fixed circle.*

Epicycloid: Let O be the centre and a be the radius of the fixed circle. Let the rolling circle start from the position in which the generating point P coincides with some point A of the fixed circle. We take O as origin and OA as X-axis. The generating point has moved on from A to $P(x, y)$ when the rolling circle has rolled on to the position shown in the figure so that arc PI = arc AI.

Let $\angle AOC = \theta, \angle ICP = t_1$, radius of the rolling circle $= b$.
Then $a\theta = $ arc $AI = $ arc $PI = b\phi$, i.e., $\phi = \dfrac{a}{b}\theta$.

$$\textbf{Epicycloid :} \begin{cases} x = (a+b)\cos\theta - b\cos\dfrac{a+b}{b}\theta; \\ \\ y = (a+b)\sin\theta - b\sin\dfrac{a+b}{b}\theta. \end{cases}$$

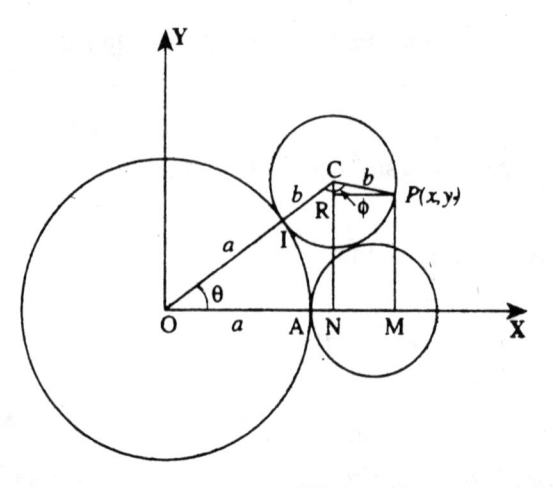

Fig 14.10

See that θ is the angle through which the line joining the centres of the two circles rotates while the rolling circle rolls from its initial to its present position.

$$\angle NCP = \angle OCP - \angle OCN = \phi - \left(\frac{\pi}{2} - \theta\right) = \theta + \phi - \frac{\pi}{2}.$$

$$\therefore \quad x = OM = ON + NM = ON + RP$$
$$= OC\cos\theta + CP\sin\left(\theta + \phi - \frac{\pi}{2}\right)$$
$$= (a+b)\cos\theta - b\cos(\theta + \phi)$$
$$= (a+b)\cos\theta - b\cos\left(\theta + \frac{a}{b}\theta\right)$$

i.e., $\quad \mathbf{x = (a+b)\cos\theta - b\cos\dfrac{a+b}{b}\theta}$ \hfill (14.3.1)

and $\quad y = MP = NC - RC$
$$= (a+b)\sin\theta - b\cos\left(\theta + \phi - \frac{\pi}{2}\right)$$

i.e., $\quad \mathbf{y = (a+b)\sin\theta - b\sin\dfrac{a+b}{b}\theta}.$ \hfill (14.3.2)

Equations (14.3.1) and (14.3.2) are the parametric equations of the **epicycloid** (θ is the parameter).

The generating point would describe a **hypocycloid** if the rolling circle were within the fixed circle. Replacing b by $-b$ in (14.3.1) and (14.3.2) the parametric equations of the hypocycloid are

$$x = (a-b)\cos\theta + b\cos\frac{b-a}{b}\theta;$$

$$y = (a-b)\sin\theta + b\sin\frac{b-a}{b}\theta.$$

Observation: In making one complete revolution the rolling circle describes $2\pi b$ of the length of the circumference of the fixed circle.

If $a:b =$ a rational number $\frac{p}{q}$ in its lowest terms then $aq = pb$ or $2\pi aq = 2\pi bp$, i.e., in making p complete revolutions the rolling circle describes the circumference of the fixed circle q times and then the generating point returns to its original positon. Therefore the path consists of the repetition of the same p identical portions. However, if $\frac{a}{b}$ is an irrational number, the generating point will never return to its original position and the path will be an endless series of exactly congruent portions.

Special Cases.

1. If $a = b$, then the equations of the **epicycloid** will be

$$\left.\begin{array}{l} x = 2a\cos\theta - a\cos 2\theta \\ y = 2a\sin\theta - a\sin 2\theta \end{array}\right\}.$$

In this case the generating point will return to its original position after the rolling circle has made only one complete revolution. The shape of the curve is shown in Fig 14.11.

2. If $a = 4b$, the equations of hypocycloid will be

$$\left.\begin{array}{l} x = \dfrac{3}{4}a\cos\theta + \dfrac{a}{4}\cos 3\theta \\ y = \dfrac{3}{4}a\sin\theta - \dfrac{a}{4}\sin 3\theta. \end{array}\right\}$$

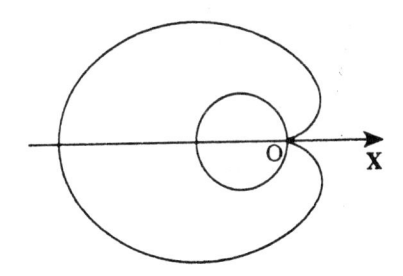

Fig 14.11
Epicycloid: $x = 2a\cos\theta - a\cos 2\theta$
$y = 2a\sin\theta - a\sin 2\theta$

Since

$$\cos 3\theta = 4\cos^3\theta - 3\cos\theta,$$

$$x = \frac{3}{4}a\cos\theta + \frac{a}{4}(4\cos^3\theta - 3\cos\theta) = a\cos^3\theta.$$

Similarly, since

$$\sin 3\theta = 3\sin\theta - 4\sin^3\theta, \quad y = \frac{3}{4}a\sin\theta - \frac{a}{4}(3\sin\theta - 4\sin^3\theta) = a\sin^3\theta,$$

i.e., the hypocycloid is given by

$$x = a\cos^3\theta,$$
$$y = a\sin^3\theta.$$

These are the parametric equations of what we have already called FOUR-CUSPED HYPOCYCLOID or ASTROID.

Here $\dfrac{a}{b} = 4$ so that the path consists of the repetitions of four portions. The four portions $AB, BA', A'B'$ and $B'A$. (Fig 14.12—the thickly drawn curve $AB\ A'B'$ is the Astroid). Note that $\theta = 0, \dfrac{\pi}{2}, \pi, \dfrac{3\pi}{2}$ at A, B, A' and B' respectively.

The Cartesian form of the equation of the Astroid is $x^{2/3} + y^{2/3} = a^{2/3}$.

(Eliminating θ between $x = a\cos^3\theta, y = a\sin^3\theta$).

One characteristic property of the Astroid is that the portion of the tangent at any point P of the curve intercepted between the co-ordinate axes is always constant. (See Fig 14.12, the portion $LM = a$, whatever point P may be chosen on the curve).

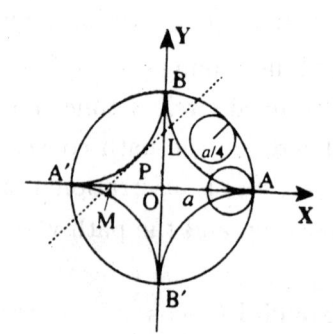

Fig 14.12
Astroid: $x^{2/3}+y^{2/3}=a^{2/3}$

3. In a hypocycloid, if $a = 2b$ (i.e., a circle rolls inside another of twice its radius) then the equations become

$$x = a\cos\theta,\ y = 0$$

i.e., the generating point on the circumference of the rolling circle traces out a diameter of the fixed circle.

14.4 Implicit Cartesian Equations of Curves

Let $f(x, y)$ be a function of two variables x and y.

The set of points (x, y) which satisfy $f(x, y) = 0$ will be said to form a **curve** whose implicit equation is $f(x, y) = 0$.

We shall be interested to those curves where $f(x, y)$ is a rational algebraic function, in the form

$$a_0 + (b_0 x + b_1 y) + (c_0 x^2 + c_1 xy + c_2 y^2) + (d_0 x^3 + d_1 x^2 y + d_2 xy^2 + d_3 y^3) + \cdots$$
$$+ (l_0 x^{n-1} + l_1 x^{n-2} y + l_2 x^{n-3} y^2 + \cdots + l_{n-1} y^{n-1})$$
$$+ (m_0 x^n + m_1 x^{n-1} y + m_2 x^{n-2} y^2 + \cdots + m_n y^n) = 0.$$

CHAPTER 14: SOME IMPORTANT CURVES: THEIR EQUATIONS AND SHAPES

which may be written as

$$\phi_0\left(\frac{y}{x}\right) + x\phi_1\left(\frac{y}{x}\right) + x^2\phi_2\left(\frac{y}{x}\right) + x^3\phi_3\left(\frac{y}{x}\right) + \cdots + x^{n-1}\phi_{n-1}\left(\frac{y}{x}\right) + x^n\phi_n\left(\frac{y}{x}\right) = 0$$

where $\phi_r\left(\frac{y}{x}\right)$ is a polynomial in $\frac{y}{x}$ of degree r.

The degree of *a term* is the sum of the indices of x and y (e.g., the degree of the term $l_1 x^{n-2} y$ is $n - 2 + 1 = n - 1$) and the degree of the curve is the greatest of the degrees of terms of $f(x, y)$.

If in a rational algebraic equation of degree n we take a fixed value for x, then it will be a polynomial equation in y of degree $\leq n$. On solving we shall obtain as many values of y as its degree, real or complex.

Hence for a given x as abscissa, there will be as many points on the curve as are the different real roots of the equation. As x takes different values each of these points will describe separately a **Branch** of the curve.

In general, we may expect n branches of a curve whose implicit equation is of degree n.

We give below some important curves and their equations. The shapes of these curves are also shown below but their actual tracing will be postponed to a later chapter.

I. $x^3 - ay^2 = 0$ $(a > 0)$.

Semicubical parabola (in parametric form: $x = at^2$, $y = at^3$).

See that

$$y = \pm\sqrt{\frac{x^3}{a}}.$$

If x is a negative, y is not real. Hence there is no part of the curve on the left of y-axis. Further for each $x \geq 0$ there are two values of y, one positive and the other negative. Hence we get one branch in the second and the other in the fourth quadrant. The origin is a CUSP since the two branches meet here and have a common tangent (namely the X-axis).

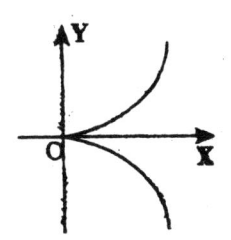

Fig 14.13
Semicubical Parabola: $x^3 = ay^2$

II. $x^3 + y^3 = 3axy$.

Folium of Descartes (in parametric form: $x = \dfrac{3at}{1+t^3}, y = \dfrac{3at^2}{1+t^3}$).

The shape of this curve is shown in Fig 14.14 (thick line).

(i) The curve is symmetrical about the line $y = x$ (as interchange of x and y does not change the equation).

(ii) The point A where the line $y = x$ meets the curve has the co-ordinates $\left(\dfrac{3a}{2}, \dfrac{3a}{2}\right)$.

(iii) x and y cannot both be negative and hence no part of the curve lies in the third quadrant.

(iv) There are two branches of the curve, namely $ABOC$ and $ADOE$ meeting at O (origin); these two branches have distinct tangents ($x = 0, y = 0$) at the origin. Hence the *origin is a singular point of the curve* ($\because$ two branches meet) *and that singular point is a* NODE (since the two branches where they meet have distinct tangents).

(v) The line $x + y + a = 0$ is an asymptote to this curve.

[See **Chapter 17** for a complete discussion on Asymptotes].

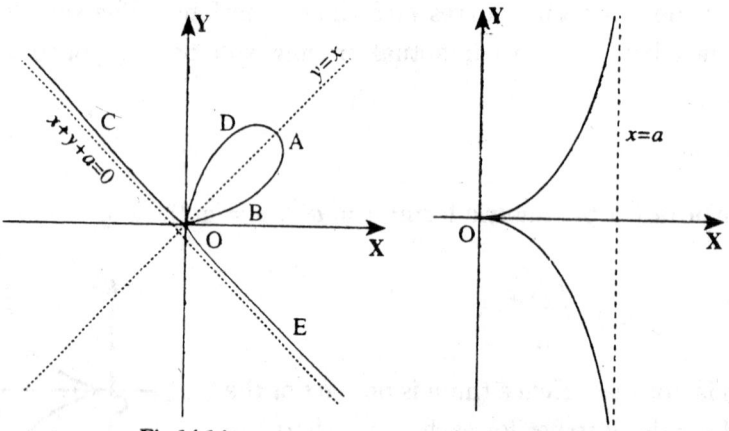

Fig 14.14
Folium of Descartes $x^3+y^3=3axy$

Fig 14.15
Cissoid of Diocles: $(x^2+y^2)x-ay^2=0$

III. $(\mathbf{x}^2 + \mathbf{y}^2)\mathbf{x} - \mathbf{a}\mathbf{y}^2 = \mathbf{0}$ $(\mathbf{a} > \mathbf{0})$.

Cissoid of Diocles [in parametric form $x = \dfrac{at^2}{1+t^2}, \; y = \dfrac{at^3}{1+t^2}$.]

The shape of the curve is shown in Fig 14.15. See that

$$y = \pm x \sqrt{\dfrac{x}{a-x}}.$$

(i) The curve passes through the origin (since, $y = 0$ when $x = 0$).

(ii) The expression $\dfrac{x}{a-x}$ is negative where $x < 0$ or when $x > a$ and hence there are no real values of y corresponding to those x where $x < 0$ or $x > a$. Then the curve lies in the part where $0 \leq x \leq a$.

(iii) For each x in $0 \leq x \leq a$, y has two real values, one positive and the other negative. The two values of y determine two branches of the Cissoid which meet at the origin and are symmetrically situated about the X-axis.

(iv) The origin is a singular point and that singular point is a CUSP (since the two branches meet at the origin and have a common tangent $y = 0$).

(v) The line $x = a$ is the only asymptote to this curve.

IV. $(x^2 + y^2)x - a(x^2 - y^2) = 0 \; (a > 0)$.

Strophoid $\left[\text{in parametric form: } x = \dfrac{a(1-t^2)}{1+t^2}, \; y = \dfrac{at(1-t^2)}{1+t^2}\right]$.

The shape of the curve is shown in Fig 14.16. See that here

$$y = \pm x \sqrt{\dfrac{a-x}{a+x}}.$$

(i) No real values of y if $x > a$ or $x < -a$. Hence the curve lies between the lines $x = a$ and $x = -a$.

(ii) When $-a \leq x \leq a$, for each x, there are two values of y, one positive and the other negative. Hence there are two branches—they are symmetrically situated about the X-axis (as the equation remains the same when y changes to $-y$).

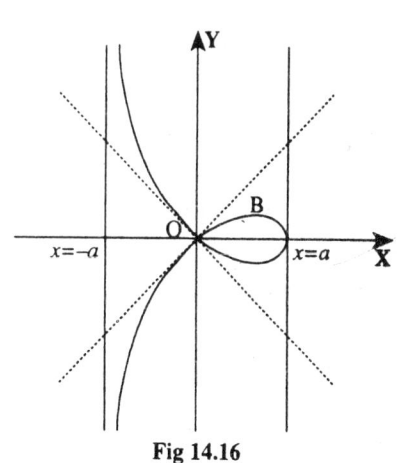

Fig 14.16
Strophoid: $(x^2+y^2)x - a(x^2-y^2) = 0 (a>0)$

(iii) The two branches meet at O where they have distinct tangents.

i.e., the origin is a singular point and that singular point is a NODE.

(iv) The line $x = -a$ is the only asymptote of this curve.

14.5 Curves in Polar Co-ordinates

In polar co-ordinate system the *position of a point is determined by means of its distance from a fixed point (called pole) and its direction from this fixed point as compared with a fixed direction (called initial line).*

(i) If OX be a fixed line (*initial line*) and O be the fixed point (*Pole*), then the position of a point P is determined by $OP = r$ and $\angle XOP = \theta$; r is called the *radius vector* and θ is called the *Vectorial angle* of P. We refer the polar co-ordinates of P by (r, θ) (Fig 14.17).

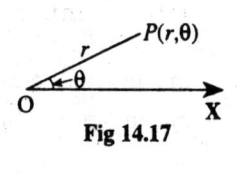

Fig 14.17

(ii) The positive direction of the angle θ is *counter-clockwise* and $-\pi < \theta \leq \pi$.

(iii) r is always non-negative, i.e., $r \geq 0$. We explain it through Fig 14.18.

P has coordinates $(4, \frac{\pi}{3})$ means $|OP| = 4$ and $\angle XOP = \frac{\pi}{3}$.

Q has coordinates $(5, -\pi/10)$ means $|OQ| = 5$ and $\angle XOQ = -\frac{\pi}{10}$.

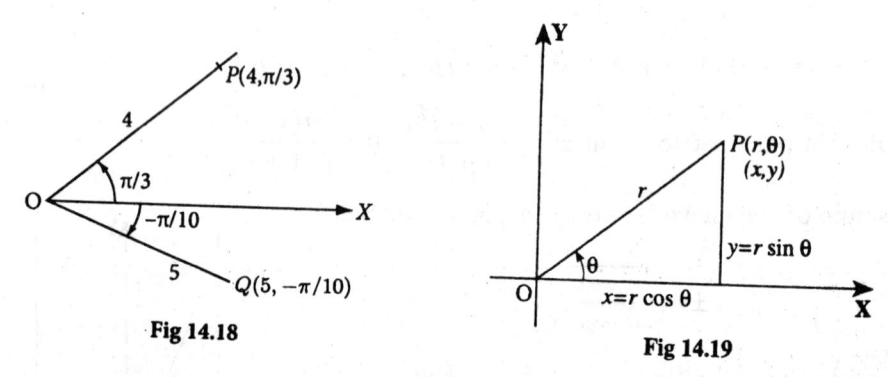

Fig 14.18

Fig 14.19

v) If a rectangular cartesian system is superposed on a polar co-ordinate system as shown in Fig 14.19.

[Take the *initial line* OX of the polar system as the positive direction of X-axis, the *pole* O as origin and the positive direction of Y-axis OY, is such that the line OX after revolving through $\pi/2$ in counter-clockwise direction comes to coincide with it].

The cartesian and polar co-ordinates of P be (x, y) and (r, θ) respectively, then

$$x = r \cos \theta$$
$$y = r \sin \theta$$
$$r = +\sqrt{x^2 + y^2}$$

and θ is determined by $\cos \theta = \dfrac{x}{r}$, $\sin \theta = \dfrac{y}{r}$ (common solution in $(-\pi, \pi]$)

(v) Any explicit or implicit relation between r and θ will determine a curve determined by points whose co-ordinates satisfy that relation.

Thus equations $r = f(\theta)$ or $F(r, \theta) = 0$ are usually the polar representations of curves.

CHAPTER 14: SOME IMPORTANT CURVES: THEIR EQUATIONS AND SHAPES

> **Remember:** A curve will be symmetrical about the initial line if on changing $+\theta$ to $-\theta$ the equation of the curve does not alter.
>
> e.g., the curve given by $r = a(1 + \cos\theta)$ is symmetrical about the initial line [Changing θ to $-\theta$ we find $r = a\{1 + \cos(-\theta)\} = a(1 + \cos\theta)$].

However, in a relation between r and θ, representing geometrically a curve, we may relax the restriction on θ from $(-\pi, \pi]$ to $(-\infty, \infty)$, because the relation may give admissible values of r for a real value of θ beyond restriction and the point (r, θ) lies on the curve represented by the relation.

Polar Equations of Important Curves

1. A Circle

Centre at the pole and radius a : **r = a**.

2. A Half Line

A half-line obtained by revolving the initial line through an angle **k** : $\theta = \mathbf{k}$.

3. Cardiode

(i) $\mathbf{r = a(1 - \cos\theta)}$: Fig 14.20.
(ii) $\mathbf{r = a(1 + \cos\theta)}$: Fig 14.21.

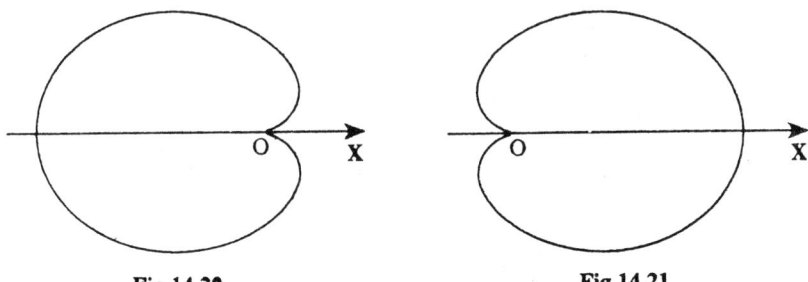

Fig 14.20
Cardiode: $r=a(1-\cos\theta)$

Fig 14.21
Cardiode: $r=a(1+\cos\theta)$

A few hints of drawing the curve $\mathbf{r = a(1 - \cos\theta)}$.

(i) The curve is symmetrical about the initial line

$$[\because r = a(1 - \cos\theta) = a\{1 - \cos(-\theta)\}.]$$

(ii) When $\theta = 0, r = 0$; when θ increases from 0 to $\pi/2$, r increases continuously from 0 to a; when $\theta = \pi/2, r = a$.

(iii) When θ increases from $\pi/2$ to π, r increases further from a to $2a$. When $\theta = \pi$, $r = 2a$.

(iv) Then the variations of θ from π to 2π give the symmetrical part of the curve.

4. Lemniscate of Bernoulli: $r^2 = a^2 \cos 2\theta$

[In cartesian form: $x^2 + y^2 = a^2 \left(\frac{x^2}{x^2+y^2} - \frac{y^2}{x^2+y^2}\right)$, i.e., $(x^2 + y^2)^2 = a^2(x^2 - y^2)$]

The shape of the curve is shown in Fig 14.22.

(i) Clearly the curve is symmetrical about the initial line

$$[\because r^2 = a^2 \cos 2\theta = a^2 \cos(-2\theta)].$$

So we consider the variation in r as θ increases from 0 to π only.

As θ varies from π to 2π, the curve will consist of the symmetrical part.

(ii) When θ increases from 0 to $\pi/4$, i.e., when 2θ increases from 0 to $\pi/2$ and $\cos 2\theta$ decreases from 1 to 0, the values of r will decrease from a to 0. (The part ABO).

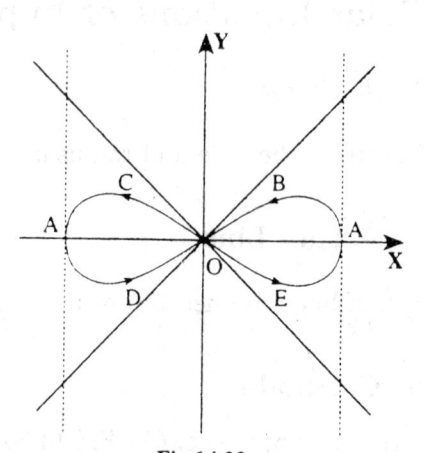

Fig 14.22
Lemniscate of Bernoulli:
$r^2=a^2\cos 2\theta$ or $(x^2+y^2)^2=a^2(x^2-y^2)$

(iii) When θ increases from $\pi/4$ to $\pi/2$ and from $\pi/2$ to $3\pi/4$, $\cos 2\theta$ remains negative and the corresponding values of r are not real. Thus no point of the curve corresponds to these values of θ. Only when $\theta = 3\pi/4$, $r = 0$.

(iv) When θ increases from $3\pi/4$ to π, 2θ increases from $3\pi/2$ to 2π and $\cos 2\theta$ increases from 0 to 1, and hence r increases from 0 to a, i.e., as θ increases from $3\pi/4$ to π, we get the part OCA of the curve.

The symmetrical part ADO is obtained when θ varies from π to $5\pi/4$. There is no part of the curve when θ varies from $5\pi/4$ to $7\pi/4$. There is a symmetrical part of ABO as θ varies from $7\pi/4$ to 2π, namely the part OEA.

(v) The curve thus consists of two loops situated between the half-lines $\theta = \pi/4$ and $\theta = 3\pi/4, \theta = -\frac{3\pi}{4}$ and $\theta = -\frac{\pi}{4}$.

(vi) The origin is a NODE (See the figure).

CHAPTER 14: SOME IMPORTANT CURVES: THEIR EQUATIONS AND SHAPES

5. A class of Polar Curves given by $r^n = a^n \cos n\theta$

(i) $n = +1$, we have **the circle** $r = a \cos\theta$.

[Cartesian form: $x^2 + y^2 = ax$ or $\left(x - \dfrac{a}{2}\right)^2 + y^2 = \left(\dfrac{a}{2}\right)^2$ as $r = \sqrt{x^2 + y^2}$ and $\cos\theta = \dfrac{x}{\sqrt{x^2+y^2}}$]

(ii) $n = -1$, we get the *straight line* $r\cos\theta = a$.

[Cartesian form: $x = a$.]

(iii) $n = 2$ we have the *Lemniscate of Bernoulli*

$$r^2 = a^2 \cos 2\theta \quad [\text{or } (x^2 + y^2)^2 = a^2(x^2 - y^2).]$$

(iv) $n = -2$ we get the *Rectangular hyperbola*

$$r^2 \cos 2\theta = a^2. \quad [\text{or } x^2 - y^2 = a^2]$$

(v) $n = \tfrac{1}{2}$ we have the *Cardiode*

$$r^{1/2} = a^{1/2} \cos\frac{\theta}{2} \quad \text{or} \quad r = \frac{a}{2}(1 + \cos\theta).$$

(vi) $n = -\tfrac{1}{2}$ we get the *Parabola*

$$r^{1/2} \cos\frac{\theta}{2} = a^{1/2} \quad \text{or} \quad r = \frac{2a}{1 + \cos\theta}.$$

6. Spirals

(i) The **Equiangular Spiral** is defined by the property that the tangent at any point of the curve makes a constant angle with the radius vector.

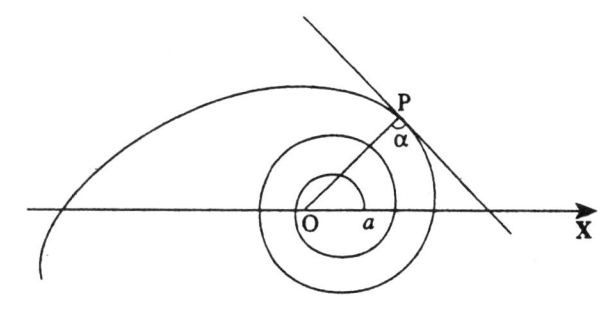

Fig 14.23
Equiangular Spiral: $r = ae^{\theta \cot \alpha}$

Its polar equation is of the form $r = ae^{\theta \cot \alpha}$, where α is the constant angle which the tangent at any point of the curve makes with the radius vector.

As θ ranges from $-\infty$ to $+\infty$, r ranges from 0 to ∞ (Fig 14.23).

(ii) **The Spiral of Archimedes** is the curve described by a point which travels along a straight line with constant velocity, whilst the line rotates with constant angular velocity about a fixed point on it.

In symbols we may write $r = ut$, $\theta = nt$ whence $r = a\theta$, where $a = \dfrac{u}{n}$.

When $\theta = 0$, $r = 0$ so that the curve goes through the pole.

When $\theta \to \infty$, $r \to \infty$. When $\theta \to -\infty$, $r \to \infty$ (numerically)

i.e., the curve starting from the pole goes round it both ways an infinite number of times.

The dotted branch corresponds to negative values of θ and the thick branch corresponds to positive values of θ.

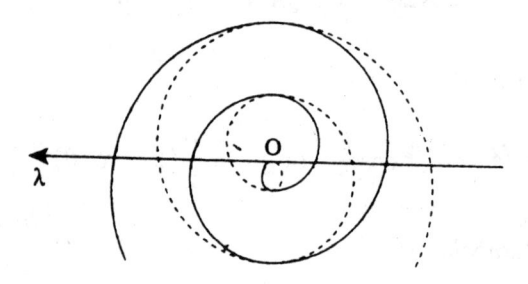

Fig 14.24
Spiral of Archimedes: $r = a\theta$

(iii) **Reciprocal Spiral: r = a/θ**

If y be the ordinate drawn to the initial line, we have

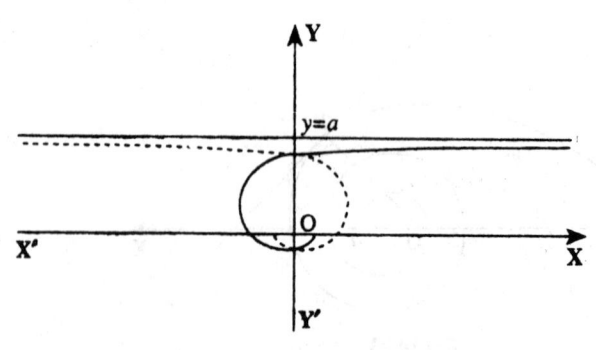

$$y = r \sin \theta = a \frac{\sin \theta}{\theta}.$$

Fig 14.25
Reciprocal Spiral: $r = a/\theta$

CHAPTER 14: SOME IMPORTANT CURVES: THEIR EQUATIONS AND SHAPES   919

As $\theta \to 0$, $r \to \infty$, but y approaches the limit a. Hence $y = a$ is an asymptote. The dotted part of Fig 14.25 corresponds to negative values of θ.

Roses or Rose Petals: $r = a \sin n\theta$ ($a > 0$)

1. If n be an odd positive integer, then we shall obtain n loops (e.g., when $n = 3$, i.e., when the equation is $r = a \sin 3\theta$ we get a **three-leaved rose**) Fig 14.26.

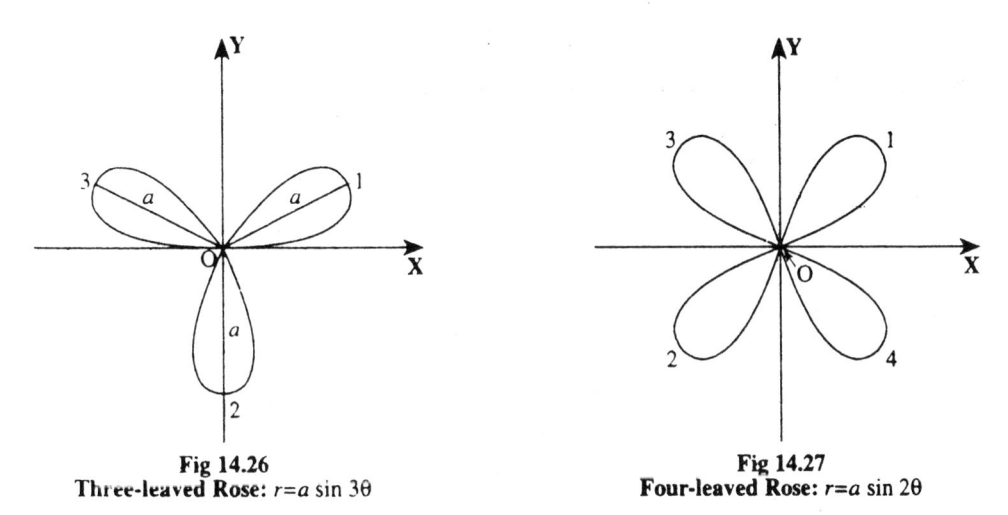

Fig 14.26
Three-leaved Rose: $r = a \sin 3\theta$

Fig 14.27
Four-leaved Rose: $r = a \sin 2\theta$

2. If n be an even positive integer, then we shall obtain $2n$ loops (e.g., when $n = 2$, i.e., when the equation is $r = a \sin 2\theta$, we get a **four-leaved rose**) Fig 14.27.

Hints for drawing the curve $r = a \sin 3\theta$

(i) When $\theta = 0, r = 0$.

As θ increases from 0 to $\pi/6$, 3θ increases from 0 to $\pi/2$ and thereby r increases from 0 to a.

(ii) As θ increases from $\pi/6$ to $\pi/3$, r decreases from a to 0. We now obtain **loop 1**.

(iii) As θ increases from $\pi/3$ to $\pi/2$, r remains negative and numerically increases from 0 to a.

(iv) As θ increases from $\pi/2$ to $2\pi/3$, r remains negative and numerically decreases from a to 0. We thus obtain **loop 2**.

(v) As θ increases from $2\pi/3$ to $5\pi/6$ and from $5\pi/6$ to π, the point (r, θ) describes **loop 3**.

(vi) If θ increases beyond π, the same loops are repeated.

Example 14.5.1. *Sketch the curve* $r = a \sin 4\theta$ **(Eight-leaved Rose).**

We have indicated the shape of this curve in Fig. 14.28.

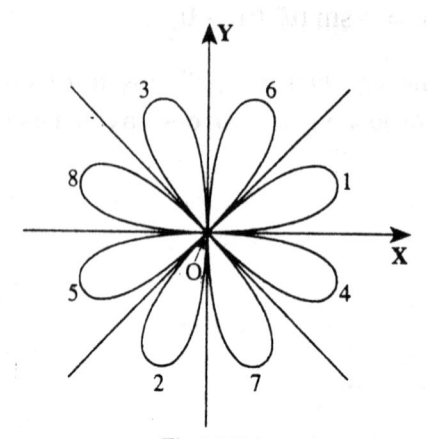

Fig 14.28
Eight-leaved Rose: $r = a \sin 4\theta$

There are eight loops. The order in which these loops are drawn should be carefully noted.

14.6 Catenary: $y = c \cosh \frac{x}{c}$ (Cartesian); $s = c \tan \psi$ (Intrinsic)

$$x = c \log(\sec \psi + \tan \psi), \quad y = c \sec \psi \text{ (Parametric)}.$$

Definition 14.6.1. *The Catenary is the curve in which a heavy uniform string or chain hangs freely under gravity.*

(The shape of this curve is as shown in Fig 14.29).

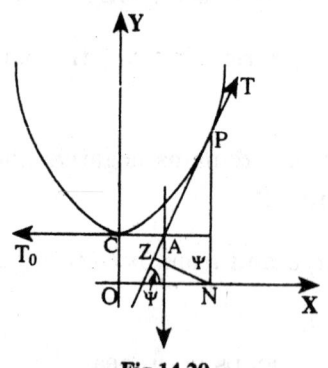

Fig 14.29
Catenary: $y = c \cosh x/c$

CHAPTER 14: SOME IMPORTANT CURVES: THEIR EQUATIONS AND SHAPES 921

We shall use elementary statical principles to obtain the **intrinsic equation** $(s-\psi$ relation) of the curve. Let C be the lowest point of the curve, P any point of the curve such that the arc $CP = s$.

Let T be the tension of the string at P acting along the tangent at P and let T_0 be the tension of the string at C acting along the tangent at C. Let ws be the weight of the portion CP of the string, w being the weight of the string per unit length (the string being uniform w is constant). Thus under the action of these three forces T, T_0 and ws the portion CP of the string is in equilibrium. The three forces must meet at some point A.

(i) If ψ be the inclination of the tangent at P to the horizontal, then we have (resolving horizontally and vertically).

$$T\cos\psi = T_0 \quad \text{and} \quad T\sin\psi = ws,$$

so that

$$\tan\psi = \frac{ws}{T_0} = \frac{ws}{wc} \quad (\text{if } T_0 = \text{weight of length } c \text{ of the string}).$$

This gives the **Intrinsic equation** of the Catenary: $\mathbf{s = c\tan\psi}$.

(ii) If x and y be the horizontal and vertical co-ordinates of the point P, then we have

$$\frac{dy}{dx} = \tan\psi = \frac{s}{c}. \quad \therefore \quad \frac{d^2y}{dx^2} = \frac{1}{c}\frac{ds}{dx} = \frac{1}{c}\sqrt{1+\left(\frac{dy}{dx}\right)^2}$$

[We shall see in the next chapter $ds^2 = dx^2 + dy^2$; hence etc.]

$$\text{or } \frac{d^2y}{dx^2} \bigg/ \sqrt{1+\left(\frac{dy}{dx}\right)^2} = \frac{1}{c}.$$

On integration, we get

$$\sinh^{-1}\frac{dy}{dx} = \frac{x}{c} + \text{constant}.$$

If the axes of Y be taken to pass through C, we have $\frac{dy}{dx} = 0$ when $x = 0$ and hence this constant vanishes and we get

$$\frac{dy}{dx} = \sinh\frac{x}{c}.$$

Integrating again, we get

$$y = c \cosh \frac{x}{c} + \text{constant}.$$

If the origin O be taken at a depth c below C, then we have $y = c$ when $x = 0$ and the constant becomes zero.

∴ The **Cartesian equation of the Catenary** becomes

$$\mathbf{y = c \cosh \frac{x}{c}}.$$

(iii) We have

$$s = c \tan \psi = c \frac{dy}{dx} = c \sinh \frac{x}{c}$$

and hence

$$y^2 - s^2 = c^2 \left(\cosh^2 \frac{x}{c} - \sinh^2 \frac{x}{c} \right) = c^2 \quad \text{i.e.,} \quad \mathbf{y^2 = c^2 + s^2}.$$

Thus we have two relations: $\mathbf{s = c \tan \psi}$ and $\mathbf{y = c \sec \psi}$.

(iv) If PN be the ordinate at P and NZ be drawn perpendicular to the tangent PT, then clearly $\angle ZNP = \psi$. We have, from the triangle NZP,

$$\cos \psi = \frac{NZ}{PN} \quad \text{and} \quad \tan \psi = \frac{PZ}{NZ}.$$

$$\therefore NZ = PN \cos \psi = y \cos \psi = c \quad (\because y = c \sec \psi)$$

and $PZ = NZ \tan \psi = c \tan \psi = s = \text{arc } CP$.

∴ Tension at any point $\quad P = T = wc \sec \psi$ [From (i)]

or, $\quad T = wy \quad$ i.e., $T \propto y$.

We conclude: *The tension at any point P of a Catenary is equal to the weight of the portion of the string whose length is the vertical distance between P and the directrix.*

(v) We shall prove in the next chapter, for any curve,

$$\frac{dx}{ds} = \cos \psi \quad \text{and} \quad \frac{dy}{ds} = \sin \psi.$$

If we now assume these two results, then

$$\frac{dx}{d\psi} = \frac{dx}{ds} \frac{ds}{d\psi} = \cos \psi \cdot c \sec^2 \psi \quad (\because s = c \tan \psi)$$
$$= c \sec \psi$$

$$\frac{dy}{d\psi} = \frac{dy}{ds} \frac{ds}{d\psi} = \sin \psi \cdot c \sec^2 \psi = c \sec \psi \tan \psi.$$

CHAPTER 14: SOME IMPORTANT CURVES: THEIR EQUATIONS AND SHAPES 923

Integrating,
$$\begin{cases} x = c \log \tan\left(\frac{\psi}{2} + \frac{\pi}{4}\right) \\ = c \log(\sec \psi + \tan \psi) \\ y = c \sec \psi. \end{cases}$$

[Using the fact: $\psi = 0$ when $x = 0, y = c$ we see that the two constants of integration both vanish].

These two equations give the parametric equations of the *Catenary*.

Main Points: A Recapitulation

1. Plane Curves: Their Analytical representation:

CARTESIAN AND POLAR REPRESENTATIONS:

a) **Parametric:** e.g.,
$$\begin{cases} x = \phi(t), \\ y = \psi(t). \end{cases}$$

b) **Implicit:** e.g., $F(x, y) = 0$

c) **Explicit:** e.g.,
$$\begin{cases} y = f(x), \\ x = \phi(y). \end{cases}$$

2. Important Curves: (*See the Figures 14.1–14.29 of this Chapter*)

a) **Parametric Representations:**

Circle: $x = a\cos t,\ y = a\sin t.\ (0 \le t \le 2\pi)$

Ellipse: $x = a\cos t,\ y = b\sin t.\ (0 \le t \le 2\pi)$

Parabola: $x = at^2,\ y = 2at.\ (-\infty < t < \infty)$

Hyperbola: $x = a\cosh t,\ y = b\sinh t.\ (-\infty < t < \infty)$

Cycloid: $x = a(t - \sin t),\ y = a(1 - \cos t).$

Inverted cycloid: $x = a(t + \sin t),\ y = a(1 - \cos t).$

Other forms of cycloid:
$$\begin{cases} x = a(t + \sin t), \\ y = a(1 + \cos t). \end{cases}$$

and
$$\begin{cases} x = a(t - \sin t), \\ y = a(1 + \cos t). \end{cases}$$

Astroid: $x = a\cos^3 t$, $y = a\sin^3 t$. $(0 \leq t \leq 2\pi)$

Epicycloid:
$$\begin{cases} x = (a+b)\cos\theta - b\cos\dfrac{a+b}{b}\theta, \\ y = (a+b)\sin\theta - b\sin\dfrac{a+b}{b}\theta. \end{cases}$$

Hypocycloid:
$$\begin{cases} x = (a-b)\cos\theta + b\cos\dfrac{b-a}{b}\theta, \\ y = (a-b)\sin\theta + b\sin\dfrac{b-a}{b}\theta. \end{cases}$$

When $a = 4b$, we obtain the equations of the Astroid: $x = a\cos^3\theta$, $y = a\sin^3\theta$.

b) Implicit Cartesian Equations:

Semicubical Parabola: $x^3 - ay^2 = 0$. $(a > 0)$

Folium of Descartes: $x^3 + y^3 - 3axy = 0$. $(a > 0)$

Cissoid of Diocles: $(x^2 + y^2)x - ay^2 = 0$. $(a > 0)$

Strophoid: $(x^2 + y^2)x - a(x^2 - y^2) = 0$. $(a > 0)$

Their respective parametric equations are:

$$\begin{cases} x = at^2 \\ y = at^3 \end{cases} ; \quad \begin{cases} x = \dfrac{3at}{1+t^3} \\ y = \dfrac{3at^2}{1+t^3} \end{cases} ; \quad \begin{cases} x = \dfrac{at^2}{1+t^2} \\ y = \dfrac{at^3}{1+t^2} \end{cases} ; \quad \begin{cases} x = \dfrac{a(1-t^2)}{1+t^2} \\ y = \dfrac{at(1-t^2)}{1+t^2} \end{cases}$$

c) Polar Equations:

Circle: $r = a$.

Straight Line: $\theta = k$.

Cardiode: $r = a(1 - \cos\theta)$; $r = a(1 + \cos\theta)$.

Lemniscate of Bernoulli: $r^2 = a^2\cos 2\theta$

[Cartesian form: $(x^2 + y^2)^2 = a^2(x^2 - y^2)$]

CHAPTER 14: SOME IMPORTANT CURVES: THEIR EQUATIONS AND SHAPES

Spirals: Equiangular spiral: $r = ae^{\theta \cot \alpha}$
Spiral of Archimedes: $r = a\theta$
Reciprocal Spiral: $r = \dfrac{a}{\theta}$.

Rose Petals: $r = a \sin 3\theta$ (Three-leaved)
$r = a \sin 2\theta$ (Four-leaved)
$r = a \sin 4\theta$ (Eight-leaved)

Class: $r^n = a^n \cos n\theta$:

$n = 1$	$r = a \cos \theta$	(Circle)
$n = -1$	$r \cos \theta = a$	(Straight line)
$n = 2$	$r^2 = a^2 \cos 2\theta$	(Lemniscate)
$n = -2$	$r^2 \cos 2\theta = a^2$	(Rect. Hyperbola)
$n = \tfrac{1}{2}$	$r = \dfrac{a}{2}(1 + \cos \theta)$	(Cardiode)
$n = -\tfrac{1}{2}$	$r = \dfrac{2a}{1 + \cos \theta}$	(Parabola).

d) Explicit Equations:

One very important curve is the *Catenary* whose equation is

$$s = c \tan \psi \text{ (intrinsic)}; \quad y = c \cosh \dfrac{x}{c} \text{ (Cartesian)};$$

$$x = c \log(\sec \psi + \tan \psi); \quad y = c \sec \psi \text{ (Parametric)}.$$

AN INTRODUCTION TO
ANALYSIS

(Differential Calculus—Part I)

Chapter 15

**Applications of Derivatives
Tangents and Normals:
Associated Curves:
Differential of Arc Length**

Chapter 15

Tangents and Normals: Associated Curves: Differential of Arc Length

15.1 Introduction

In a previous chapter (Chapter 9) we have observed that the direction of a curve at a given point P is determined by the gradient of the tangent line at P; this gradient is given by the *first derivative of the function* which represents the curve. The equation of the tangent line and its perpendicular (normal line) can then at once be written down.

Section I of the present chapter is devoted to the study of finding equations of tangents and normals of different curves (the equations and the graphs of *some useful curves* have already been discussed in Chapter 14). Both Cartesian and polar coordinates have been freely used.

In **Section II** we have discussed some associated loci. Thus the equations of the *pedal of a given curve*, the inverse of a known curve and reciprocal polars have been deduced.

In **Section III** we have introduced the notion of the length of an arc of a curve; the differentials of the arc-length are useful in varied applications.

Section I: Tangents and Normals

15.2 Equations of Tangents and Normals: Various Forms

(A) Rectangular Cartesian Equations in Explicit Form: $y = f(x)$

If $\psi \, (\psi \neq 90°)$ be the angle which the tangent at any point (x, y) of the curve $y = f(x)$ makes with the X-axis, then

$$\tan \psi = \frac{dy}{dx} = f'(x).$$

Hence the equation of the tangent at any point (x, y) of the curve $y = f(x)$ is

$$Y - y = f'(x)(X - x), \qquad (15.2.1)$$

where (X, Y) are the current co-ordinates of any point on the tangent line.

[Compare: $y - y' = m(x - x')$, equation of a line of slope m and passing through a point (x', y')].

Using differentials one may write the equation in the form

$$\frac{Y - y}{dy} = \frac{X - x}{dx} \quad \left(\because f'(x) = \frac{dy}{dx} \right). \qquad (15.2.2)$$

Now the normal to a curve at a given point is defined to be the line which passes through the point and is perpendicular to the tangent line at that point.

∴ the slope of the normal at the point (x, y)

$$= -\frac{1}{f'(x)} = -\frac{dx}{dy}.$$

Therefore the equation of the normal at (x, y) is

$$Y - y = -\frac{1}{f'(x)}(X - x) \quad \text{or} \quad (Y - y)f'(x) + (X - x) = 0. \qquad (15.2.3)$$

Using differentials the form becomes

$$(X - x)\, dx + (Y - y)\, dy = 0 \qquad (15.2.4)$$

where (X, Y) are the current co-ordinates of the normal at (x, y).

Example 15.2.1. *Write the equations of the tangent and the normal to the curve $y = x^3 - 3x^2 - x + 5$ at the point $P(3, 2)$.*

CHAPTER 15: TANGENTS AND NORMALS: ASSOCIATED CURVES

Solution: $\dfrac{dy}{dx} = 3x^2 - 6x - 1.$

At the point $P(3,2)$, the slope $= 3.3^2 - 6.3 - 1 = 8$.

Hence the equation of the tangent at P is $y - 2 = 8(x - 3)$ or **$8x - y - 22 = 0$** and the equation of the normal is $y - 2 = -\frac{1}{8}(x - 3)$ or **$x + 8y - 19 = 0$**.

(B) Rectangular Cartesian Equations in Implicit Form

$$f(x,y) = 0. \text{ (assuming } f_y \neq 0)$$

Here,

$$\frac{dy}{dx} = -\frac{f_x}{f_y} \quad (f_y \neq 0)$$

so that

$$-1 \bigg/ \frac{dy}{dx} = \frac{f_y}{f_x}.$$

Therefore, the equations of the tangent and normal become

Tangent:

$$Y - y = -\frac{f_x}{f_y}(X - x) \quad \text{or} \quad (X - x)f_x + (Y - y)f_y = 0. \tag{15.2.5}$$

Normal:

$$Y - y = \frac{f_y}{f_x}(X - x) \quad \text{or} \quad (X - x)f_y - (Y - y)f_x = 0. \tag{15.2.6}$$

At a point (x,y) where $f_x = f_y = 0$ we get a *singular point* of the curve. These equations become meaningless at a singular point of the curve.

Example 15.2.2. *Find the equation of the tangent to the curve $y^2 - yx^2 - 2x^5 = 0$ at the point $(1, -1)$.*

Solution: We assume that $\exists$ unique solution of the form $y = \phi(x)$ near $(1, -1)$.

In fact, here

$$y = \frac{x^2 - x^2\sqrt{1 + 8x}}{2}.$$

If we call $f(x,y) = y^2 - yx^2 - 2x^5$, then

$$\frac{dy}{dx} = -\frac{f_x}{f_y} = -\frac{(-2xy - 10x^4)}{2y - x^2}$$

i.e.,
$$\frac{dy}{dx} = \frac{2xy + 10x^4}{2y - x^2}$$

and at $(1, -1)$,
$$\frac{dy}{dx} = -\frac{8}{3}.$$

∴ the equation of the tangent at $(1, -1)$ is
$$y + 1 = -\frac{8}{3}(x - 1) \quad \text{or} \quad \mathbf{3y + 8x - 5 = 0}.$$

(C) Another Form for the Equation of the Tangent to a rational algebraic Curve

If the curve is given in implicit form $f(x, y) = 0$, where $f(x, y)$ is a rational algebraic function of x and y, then the equation of the tangent is

$$(X - x)f_x + (Y - y)f_y = 0.$$

Suppose we make $f(x, y)$ homogeneous in x, y, z of degree n with the introduction of suitable powers of z, wherever necessary. Let us call this altered function $f(x, y, z)$. By Euler's Theorem we have

$$xf_x + yf_y + zf_z = n(f(x, y, z)) = 0.$$

Adding this to the equation of the tangent we get

$$Xf_x + Yf_y + zf_z = 0 \qquad (15.2.7)$$

where z is to be put equal to 1, after the differentiations have been performed. We illustrate this principle by the following example:

Example 15.2.3. *Consider the general equation of a conic* $f(x, y) = ax^2 + 2hxy + by^2 + 2gx + 2fy + c = 0.$

When made homogeneous in x, y, z by the introduction of suitable powers of z, we obtain
$$f(x, y, z) = ax^2 + 2hxy + by^2 + 2gxz + 2fyz + cz^2 = 0$$
whence, $f_x = 2(ax + hy + gz)$, $f_y = 2(hx + by + fz)$ and $f_z = 2(gx + fy + cz)$.

∴ the equation of the tangent is

$$Xf_x + Yf_y + zf_z = 0 \quad \text{(where } z \text{ is to be put} = 1)$$

or, $\quad X(ax + hy + gz) + Y(hx + by + fz) + z(gx + fy + cz) = 0$

CHAPTER 15: TANGENTS AND NORMALS: ASSOCIATED CURVES

and putting $z = 1$, the equation becomes

$$X(ax + hy + g) + Y(hx + by + f) + (gx + fy + c) = 0$$

which forms the equation of the tangent at (x, y).

Note 15.2.1. The equation of normal at (x, y) of $f(x, y) = 0$ is $\dfrac{(X-x)}{f_x} = \dfrac{(Y-y)}{f_y}$.

Here

$$\frac{X - x}{ax + hy + g} = \frac{Y - y}{hx + by + f}.$$

Observation: If $\dfrac{dy}{dx} = 0$, then $\tan \psi = 0$ and the equation of the tangent becomes $Y - y = 0$, i.e., a line parallel to the X-axis.

If, however, the tangent is parallel to the Y-axis, $\psi = \tfrac{1}{2}\pi$ and the equation of the tangent becomes $X - x = 0$.

(D) Parametric Cartesian Equations: $x = \phi(t)$, $y = \psi(t)$

Any any point 't' of the curve given by the parametric equations

$$x = \phi(t), \quad y = \psi(t),$$

we have

$$\frac{dy}{dx} = \frac{dy}{dt} \Big/ \frac{dx}{dt} = \frac{\psi'(t)}{\phi'(t)}.$$

Hence the equations of the tangent and the normal at any point 't' of the curve $x = \phi(t), y = \psi(t)$ are

$$\left. \begin{array}{l} [X - \phi(t)]\psi'(t) - [Y - \psi(t)]\phi'(t) = 0 \\ \text{and } [X - \phi(t)]\phi'(t) + [Y - \psi(t)]\psi'(t) = 0 \end{array} \right\}. \qquad (15.2.8)$$

We advise our students to use equation (15.2.8) for tangent and normal whenever it is possible to write the equation of the curve in parametric form, even if the curve is given in rectangular cartesian form (see the example given below).

Example 15.2.4. *Find the equation of the tangent and also the equation of the normal at any point (x, y) of the Astroid $x^{2/3} + y^{2/3} = a^{2/3}$. Hence deduce that the portion of the tangent at any point of the Astroid intercepted between the axes is of constant length.*

Solution: Let us write the equation of the astroid in parametric form:

$$x = a\cos^3 t, \quad y = a\sin^3 t,$$

so that

$$\frac{dx}{dt} = -3a\cos^2 t \sin t, \quad \frac{dy}{dt} = 3a\sin^2 t \cos t.$$

∴ the equations of tangent and normal at any point 't' of the curve are

Tangent: I. $(X - a\cos^3 t)3a\sin^2 t \cos t - (Y - a\sin^3 t)(-3a\cos^2 t \sin t) = 0$.

Normal: II. $(X - a\cos^3 t)(-3a\cos^2 t \sin t) + (Y - a\sin^3 t)(3a\sin^2 t \cos t) = 0$.

Simplifying (I) we get the equation of tangent in the following form:

Tangent: $\quad X \cdot 3a\sin^2 t \cos t + Y \cdot 3a\cos^2 t \sin t = 3a^2 \sin^2 t \cos^4 t + 3a^2 \cos^2 t \sin^4 t$

or, $\quad X \cdot \sin^2 t \cos t + Y \cos^2 t \sin t = a \sin^2 t \cos^2 t$ (using, $\sin^2 t + \cos^2 t = 1$)

or, $\quad \dfrac{X}{a\cos t} + \dfrac{Y}{a\sin t} = 1$.

∴ intercepts on the X- and Y-axis respectively are $a\cos t$ and $a\sin t$ and hence the length of the intercept of the tangent at any point t of the curve between the two axes is

$$\sqrt{a^2\cos^2 t + a^2\sin^2 t} = a \text{ (constant), as } a \text{ is \textbf{independent of } } t.$$

Hence the required property follows.

[In general, remember when you require the intercepts of the tangent on the two axes write the equation of the tangent as

$$\frac{X}{x - y\frac{dx}{dy}} + \frac{Y}{y - x\frac{dy}{dx}} = 1.]$$

Simplifying (II) the equation of the normal at the point t becomes

$$-(X - a\cos^3 t)\cos t + (Y - a\sin^3 t)\sin t = 0$$

$$\text{or, } X\cos t - Y\sin t = a(\cos^4 t - \sin^4 t)$$

$$= a(\cos^2 t - \sin^2 t)(\cos^2 t + \sin^2 t)$$

$$= a\cos 2t.$$

Example 15.2.5. *Find the condition that the line $y = mx + c$ be a tangent to the parabola $y^2 = 4ax$.*

Solution: Tangent to $y^2 = 4ax$ at (x, y) is given by (on simplification)

$$Y \cdot y = 2a(X + x)$$

CHAPTER 15: TANGENTS AND NORMALS: ASSOCIATED CURVES 935

which must be identical with the changed form of $y = mx + c$, i.e., taking (X, Y) as current co-ordinates,
$$Y = mX + c.$$

Hence
$$\frac{y}{1} = \frac{2a}{m} = \frac{2ax}{c}, \text{ i.e., } y = \frac{2a}{m}, \ x = \frac{c}{m}.$$

Since (x, y) is a point of the parabola, $\therefore$ putting $x = \frac{c}{m}, y = \frac{2a}{m}$ in $y^2 = 4ax$,

we get $\mathbf{c = \dfrac{a}{m}}$ as the required condition.

So, the line $y = mx + \frac{a}{m}$ always (i.e., for all real m) touches the parabola $y^2 = 4ax$ at the point $\left(\frac{a}{m^2}, \frac{2a}{m}\right)$.

15.3 Subtangent and Subnormal: Lengths of Tangent and Normal

In Fig 15.1, PT, PG and PN are respectively the tangent, normal and ordinate to the curve $y = f(x)$ at $P(x, y)$. The lengths NT, NG, PT and PG are frequently referred to as the subtangent, subnormal, length of tangent and length of normal respectively.

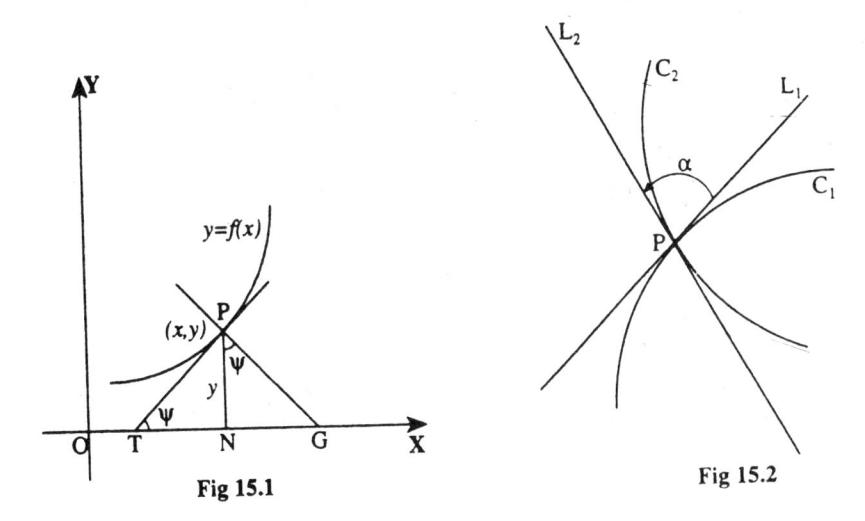

Fig 15.1

Fig 15.2

Thus,

(a) **Subtangent** NT = directed distance *from* the foot of the ordinate N to the point T where the tangent meets the X-axis and is denoted by S_t

$$\therefore S_t = y \cot \psi = \frac{y}{\frac{dy}{dx}} = \frac{y}{y_1}. \qquad \text{[C.H.'82]}$$

The positive and negative senses are now the same as the x-axis. Accordingly, NT is negative or positive according as T is to the left of N (Fig 15.1) or to the right of N.

(b) **Subnormal** NG = directed distance *from* the foot of the ordinate N to point G where the normal meets the X-axis and is denoted by S_n.

$$\therefore S_n = y \tan \psi = y \frac{dy}{dx} = yy_1. \qquad \text{[C.H.'82]}$$

Sign convention is exactly similar to (a).

(c) **Length of Tangent** PT = **absolute length** PT intercepted on the tangent line by the curve and the X-axis.

$$\therefore T = |y \operatorname{cosec} \psi| = \left|\frac{y}{y_1}\sqrt{1+y_1^2}\right| = \left|y\sqrt{1+\left(\frac{dx}{dy}\right)^2}\right|.$$

(d) **Length of Normal** PG = **absolute length** PG intercepted on the normal by the curve and the X-axis.

$$\therefore N = |y \sec \psi| = \left|y\sqrt{1+y_1^2}\right|.$$

(e) Length of the perpendicular drawn from origin to the tangent

$$= P_t = \left|\frac{(y - xy_1)}{\sqrt{1+y_1^2}}\right|.$$

(f) Length of the perpendicular drawn from origin to the normal

$$= P_n = \left|\frac{(x + yy_1)}{\sqrt{1+y_1^2}}\right|.$$

Example 15.3.1. *Find the equations of the tangent and normal, the lengths of the tangent and the subtangent, the lengths of the normal and subnormal for the ellipse $\frac{x^2}{a^2} + \frac{y^2}{b^2} = 1$ at the point $\left(\frac{a}{\sqrt{2}}, \frac{b}{\sqrt{2}}\right)$.*

Solution: We write the equation of the given ellipse in the parametric form: $x = a \cos t$, $y = b \sin t$.

The point $\left(\frac{a}{\sqrt{2}}, \frac{b}{\sqrt{2}}\right)$ corresponds to $t = \frac{\pi}{4}$.

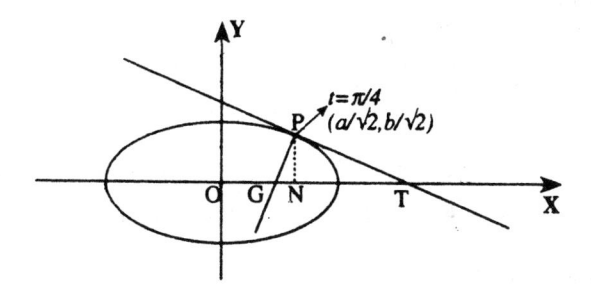

CHAPTER 15: TANGENTS AND NORMALS: ASSOCIATED CURVES

Fig 15.3

From the equations $x = a\cos t$, $y = b\sin t$.

We obtain

$$\frac{dx}{dt} = -a\sin t, \quad \frac{dy}{dt} = b\cos t; \quad \text{at } t = \frac{\pi}{4}, \quad \frac{dx}{dt} = -\frac{a}{\sqrt{2}}, \quad \frac{dy}{dt} = \frac{b}{\sqrt{2}}$$

so that

$$\frac{dy}{dx} = -\frac{b}{a}.$$

(a) The equation of the tangent PT at the point P (where $t = \frac{\pi}{4}$) is

$$\left(y - \frac{b}{\sqrt{2}}\right) = -\frac{b}{a}\left(x - \frac{a}{\sqrt{2}}\right)$$

or, **ay + bx = √2 ab**.

(b) The equation of the normal PG at the point P (where $t = \frac{\pi}{4}$) is

$$\left(y - \frac{b}{\sqrt{2}}\right) = \frac{a}{b}\left(x - \frac{a}{\sqrt{2}}\right)$$

or, **by − ax** = $\dfrac{b^2 - a^2}{\sqrt{2}}$.

(c) The length of the tangent

$$\left| y\sqrt{1 + \left(\frac{dx}{dy}\right)^2} \right| = \left| \frac{b}{\sqrt{2}}\sqrt{1 + \frac{a^2}{b^2}} \right| = \frac{1}{\sqrt{2}}\sqrt{a^2 + b^2}.$$

(d) The length of the subtangent

$$= \left| \frac{y}{y_1} \right| = \left| \left(\frac{b}{\sqrt{2}}\right) \bigg/ -\left(\frac{b}{a}\right) \right| = \frac{a}{\sqrt{2}}.$$

(e) The length of the subnormal

$$= |yy_1| = \left|\frac{b}{\sqrt{2}}\left(-\frac{b}{a}\right)\right| = \frac{1}{\sqrt{2}}\frac{b^2}{a} = \frac{1}{2}\frac{\sqrt{2}b^2}{a}.$$

Example 15.3.2. *Prove that for the parabola $y^2 = 4ax$, the subnormal is constant at any point and the subtangent varies as abscissa of the point of contact.*

Solution: Here $\dfrac{dy}{dx} = \dfrac{2a}{y}$ [at any point (x, y)].

∴ Subnormal $= yy_1 = 2a$ (which is constant).

Also subtangent $= \dfrac{y}{y_1} = \dfrac{y^2}{2a} = \dfrac{(4ax)}{2a} = 2x$ (i.e., varies as x), i.e., subtangent varies as the abscissa x of the point of the contact (x, y).

15.4 Angle of intersection of two Curves

When two curves intersect, their respective tangents at the point of intersection make an angle α which is taken, by definition, to be the angle between the curves themselves at that point.

The angle α can be determined from

$$\tan\alpha = \frac{m_2 - m_1}{1 + m_1 m_2},$$

which gives the tangent of the angle measured in *counter-clockwise* sense from L_1 to L_2 between two lines L_1 and L_2 whose slopes m_1, m_2 are known.

If $f(x, y) = 0$ and $F(x, y) = 0$ be two curves intersecting at (x, y) the angle between their respective tangents at that point is

$$\tan^{-1}\frac{f_x F_y - f_y F_x}{f_x F_x + f_y F_y} \cdot \left(\because m_1 = -\frac{f_x}{f_y} \text{ and } m_2 = -\frac{F_x}{F_y}\right) \qquad \text{[C.H.1980]}$$

Hence if the two curves touch, i.e., $\alpha = 0$ (hence $\tan\alpha = 0$) we have

$$\frac{f_x}{F_x} = \frac{f_y}{F_y}.$$

Again if the two curves cut orthogonally, $\alpha = \dfrac{\pi}{2}$ (and hence $\cot\alpha = 0$) and we have $f_x F_x + f_y F_y = 0$.

Example 15.4.1. *The curves $y = x^2$ and $y^2 = x$ pass through the point $(1, 1)$. Find their angle of intersection at this point.*

From
$$y = x^2, \quad \frac{dy}{dx} = 2x = 2 \text{ at } (1, 1)$$
and from
$$y^2 = x, \quad \frac{dy}{dx} = \frac{1}{2}y = \frac{1}{2} \text{ at } (1, 1).$$
Therefore,
$$\tan \alpha = \left(2 - \frac{1}{2}\right) \bigg/ \left(1 + 2 \cdot \frac{1}{2}\right) = \frac{3}{4} \text{ or, } \alpha = \tan^{-1}\left(\frac{3}{4}\right),$$
α being the angle of intersection of the two curves at the point $(1, 1)$.

15.5 Tangents at the Origin

The general equation of a *rational algebraic* curve of nth degree passing through the origin O may be written in the form:
$$(a_1 x + b_1 y) + (a_2 x^2 + b_2 xy + c_2 y^2) + \cdots + (a_n x^n + \cdots + k_n y^n) = 0. \tag{15.5.1}$$

Let $P(x, y)$ be a point in the *neighbourhood* of O. Then the slope of the chord OP is $\frac{y}{x}$ and hence the equation to the tangent at O is
$$Y = \lim_{\substack{x \to 0 \\ y \to 0}} \frac{y}{x} \cdot X = mX, \quad \text{where } m = \lim_{\substack{x \to 0 \\ y \to 0}} \frac{y}{x}.$$

I. When m is finite, i.e., Y-axis is not the tangent at the origin. Dividing (15.5.1) by x and taking $b_1 \neq 0$,
$$\left(a_1 + b_1 \frac{y}{x}\right) + \left(a_2 x + b_2 y + c_2 y \cdot \frac{y}{x}\right) + \cdots = 0.$$

Since when $x \to 0, y \to 0$ and $\frac{y}{x} \to m$; $a_1 + b_1 m = 0$ so that $m = -\frac{a_1}{b_1}$.

Hence the tangent becomes, if (X, Y) be the current co-ordinates
$$a_1 X + b_1 Y = 0$$
i.e., $a_1 x + b_1 y = 0$, if (x, y) be taken as current co-ordinates,

which could have been obtained by equating to zero the terms of lowest degree in (15.5.1).

Next take $b_1 = 0$, then $a_1 = 0$ as y-axis is not a tangent ($b_1 = 0, a_1 \neq 0 \Rightarrow m$ is undefined and y-axis becomes a tangent) and in such a case suppose b_2, c_2 be not both zero. Divide (15.5.1) by x^2 and apply $x \to 0, y \to 0$, then
$$a_2 + b_2 m + c_2 m^2 = 0$$

which gives two values of m and consequently two tangents at the origin which will take the form $a_2 x^2 + b_2 xy + c_2 y^2 = 0$, and this equation also could have been obtained by equating to zero the terms of lowest degree in (15.5.1) and so on.

II. If the tangent at the origin be the Y-axis, supposing the axes of X and Y to be interchanged, see that the rule is still true.

Thus, in general, the equation of the tangent or tangents at the origin can be obtained by equating to zero the terms of the lowest degree in the equation of the curve, the curve being rational and algebraic.

Example 15.5.1. *For the curve $x^3 + y^3 - 3axy = 0$, the tangents at the origin are given by $3axy = 0$, i.e., $x = 0$ and $y = 0$ are the two tangents at the origin.*

15.6 Illustrative Examples (Miscellaneous Type)

Example 15.6.1. *If the line $x \cos \alpha + y \sin \alpha = p$ touches the curve*

$$\left(\frac{x}{a}\right)^{n/(n-1)} + \left(\frac{y}{b}\right)^{n/(n-1)} = 1,$$

then prove that $(a \cos \alpha)^n + (b \sin \alpha)^n = p^n$.

Solution: From the equation of the curve

$$\left(\frac{x}{a}\right)^{n/(n-1)} + \left(\frac{x}{b}\right)^{n/(n-1)} = 1,$$

we get, (differentiating both sides w.r.t. x)

$$\frac{1}{a^{n/(n-1)}} \cdot \frac{n}{n-1} x^{1/(n-1)} + \frac{1}{b^{n/(n-1)}} \frac{n}{n-1} y^{1/(n-1)} \frac{dy}{dx} = 0$$

or, $\dfrac{dy}{dx} = -\left(\dfrac{b}{a}\right)^{n/(n-1)} \left(\dfrac{x}{y}\right)^{1/(n-1)}$

∴ the equation of the tangent at (x, y) is

$$Y - y = -\left(\frac{b}{a}\right)^{n/(n-1)} \left(\frac{x}{y}\right)^{1/(n-1)} (X - x).$$

or, $Y \cdot y^{1/(n-1)} - y^{1+[1/(n-1)]} = -\left(\dfrac{b}{a}\right)^{n/(n-1)} \left(X x^{1/(n-1)} - x^{1+[1/(n-1)]}\right)$

or, $Y \cdot y^{1/(n-1)} \cdot a^{n/(n-1)} + X \cdot x^{1/(n-1)} \cdot b^{n/(n-1)}$

$$= (ay)^{n/(n-1)} + (bx)^{n/(n-1)} = (ab)^{n/(n-1)}$$

(using the equation of the curve).

CHAPTER 15: TANGENTS AND NORMALS: ASSOCIATED CURVES

According to the given condition this equation is identical with

$$Y \sin \alpha + X \cos \alpha = p \quad [(X,Y) \text{ being taken as the current co-ordinates}].$$

∴ Comparing the coefficients we get

$$\frac{y^{1/(n-1)} a^{n/(n-1)}}{\sin \alpha} = \frac{x^{1/(n-1)} b^{n/(n-1)}}{\cos \alpha} = \frac{(ab)^{n/(n-1)}}{p}.$$

This gives,

$$y^{1/(n-1)} = \frac{b^{n/(n-1)} \sin \alpha}{p} \quad \text{and} \quad x^{1/(n-1)} = \frac{a^{n/(n-1)} \cos \alpha}{p},$$

or, $y = \dfrac{b^n \sin^{n-1} \alpha}{p^{n-1}}$ and $x = \dfrac{a^n \cos^{n-1} \alpha}{p^{n-1}}.$

Since (x, y) is a point of the curve

$$\left(\frac{x}{a}\right)^{n/(n-1)} + \left(\frac{y}{b}\right)^{n/(n-1)} = 1.$$

We have

$$\left(\frac{a^{n-1} \cos^{n-1} \alpha}{p^{n-1}}\right)^{n/(n-1)} + \left(\frac{b^{n-1} \sin^{n-1} \alpha}{p^{n-1}}\right)^{n/(n-1)} = 1$$

or, $\left(\dfrac{a \cos \alpha}{p}\right)^n + \left(\dfrac{b \sin \alpha}{p}\right)^n = 1$

or, $(a \cos \alpha)^n + (b \sin \alpha)^n = p^n$, the required condition.

Example 15.6.2. *If the normal to the curve $x^{2/3} + y^{2/3} = a^{2/3}$ makes an angle ϕ with the axis of X, show that its equation is $y \cos \phi - x \sin \phi = a \cos 2\phi$.* [C.H.'75]

Solution: Differentiating $x^{2/3} + y^{2/3} = a^{2/3}$ w.r.t. x we obtain

$$\frac{dy}{dx} = -\frac{y^{1/3}}{x^{1/3}}.$$

∴ the slope of the normal at any point (x, y) is $(x/y)^{1/3}$.

According to the given condition,

$$\left(\frac{x}{y}\right)^{1/3} = \tan \phi \quad \text{or,} \quad x \cos^3 \phi = y \sin^3 \phi$$

or, $\dfrac{x}{\sin^3 \phi} = \dfrac{y}{\cos^3 \phi} = k$ (say).

∴ from $x^{2/3} + y^{2/3} = a^{2/3}$ it follows, $k^{2/3}(\sin^2\phi + \cos^2\phi) = a^{2/3}$ or $k = a$.

∴ the point (x, y) becomes in terms of ϕ, $(a\sin^3\phi, a\cos^3\phi)$.

Hence the equation of the normal is

$$Y - a\cos^3\phi = \tan\phi(X - a\sin^3\phi)$$
$$\text{or, } Y\cos\phi - X\sin\phi = a(\cos^4\phi - \sin^4\phi) = a\cos 2\phi$$
$$\text{or, } y\cos\phi - x\sin\phi = a\cos 2\phi,$$

if (x, y) be taken as the current co-ordinates of any point of the normal line.

Example 15.6.3. *Show that the length of the perpendicular from the foot of the ordinate of any point on the tangent at the same point to the Catenary $y = c\cosh(x/c)$ is constant.*

Solution: Equation of the tangent at **any point** (x, y) of the catenary is

$$Y - y = \sinh\left(\frac{x}{c}\right)(X - x) \quad \left(\because \frac{dy}{dx} = \sinh\frac{x}{c}\right)$$
$$\text{or, } X\sinh\left(\frac{x}{c}\right) - Y + \left[y - x\sinh\left(\frac{x}{c}\right)\right] = 0.$$

$[(X, Y)$ are the current co-ordinates of any point of the tangent line and (x, y) is any *point* of the catenary.]

The foot of the ordinate of the point (x, y) is the point $(x, 0)$.

∴ the length of the perpendicular from the point $(x, 0)$ on the tangent is

$$= \frac{x\sinh(x/c) - 0 + (y - x\sinh x/c)}{\sqrt{\sinh^2(x/c) + 1}} = \frac{y}{\cosh(x/c)}$$

$$= \frac{c\cosh(x/c)}{\cosh(x/c)} = c \text{ (free from } x, y \text{ and hence a constant)}.$$

Example 15.6.4. *If the two curves $ax^2 + by^2 = 1$, $a'x^2 + b'y^2 = 1$ cut orthogonally, prove that*

$$\frac{1}{b} - \frac{1}{b'} = \frac{1}{a} - \frac{1}{a'}. \qquad \text{[C.H. '76, '79]}$$

Solution: Let $f(x, y) \equiv ax^2 + by^2 - 1$ and $F(x, y) \equiv a'x^2 + b'y^2 - 1 = 0$.

Let (x', y') be the point of intersection of two curves $f(x, y) = 0$, $F(x, y) = 0$.

Then,

$$ax'^2 + by'^2 = 1 \qquad (15.6.1)$$
$$a'x'^2 + b'y'^2 = 1. \qquad (15.6.2)$$

CHAPTER 15: TANGENTS AND NORMALS: ASSOCIATED CURVES

The condition of orthogonality at the point (x', y') gives

$$f_x F_x + f_y F_y = 0$$
i.e., $(2ax')(2a'x') + (2by')(2b'y') = 0$
i.e., $aa'x'^2 + bb'y'^2 = 0.$ (15.6.3)

Eliminating x'^2, y'^2 from (15.6.1), (15.6.2) and (15.6.3) we get

$$\begin{vmatrix} a & b & 1 \\ a' & b' & 1 \\ aa' & bb' & 0 \end{vmatrix} = 0.$$

Expanding this determinant in terms of third column we get

$$\begin{vmatrix} a' & b' \\ aa' & bb' \end{vmatrix} - \begin{vmatrix} a & b \\ aa' & bb' \end{vmatrix} = 0$$

or, $(ba'b' - aa'b') - (abb' - aa'b) = 0$

or, $\left(\dfrac{1}{a} - \dfrac{1}{b}\right) - \left(\dfrac{1}{a'} - \dfrac{1}{b'}\right) = 0$ (dividing throughout by $aa'bb'$)

or, $\dfrac{1}{a} - \dfrac{1}{a'} = \dfrac{1}{b} - \dfrac{1}{b'}$ (required condition).

Example 15.6.5. *Show that the tangent to the curve $x^3 + y^3 = 3axy$ at the point different from the origin where it meets the parabola $y^2 = ax$ is parallel to the Y-axis.*

[C.H. 1983 (Special)]

Solution: We first find the points of intersection of the curves

$$y^2 = ax \quad \text{and} \quad x^3 + y^3 = 3axy$$

by solving for x and y.

From the first equation $x = y^2/a$ and putting this value of x in the second equation we get

$$y^6/a^3 + y^3 = 3y^3, \quad \text{or,} \quad 2y^3 = y^6/a^3 \quad \text{or,} \quad y^3\left(2 - \dfrac{y^3}{a^3}\right) = 0$$

or, $y = 0$, $y = \sqrt[3]{2}a$, the corresponding values of x are $x = 0, x = 2^{2/3}a$.

∴ the points of intersection are $O(0,0)$ and $P(2^{2/3}a, 2^{1/3}a)$. We require to find the tangent to the curve $x^3 + y^3 = 3axy$ only at the latter point P.

Differentiating the equation of the curve w.r.t. y

$$3x^2 \frac{dx}{dy} + 3y^2 = 3a \left[x + y \frac{dx}{dy} \right]$$

or, $\quad \dfrac{dx}{dy} = \dfrac{ax - y^2}{x^2 - ay}$.

At $P(2^{2/3}a, 2^{1/3}a)$, we see that $ax - y^2 = 0$ but $x^2 - ay \neq 0$ i.e., $\dfrac{dx}{dy} = 0$ proving that the tangent is parallel to the Y-axis there.

Example 15.6.6. *A curve is given by*

$$x = a(2\cos t + \cos 2t), \quad y = a(2\sin t - \sin 2t).$$

Prove that

1. *the equations of the tangent and normal at the point P with parameter t, are respectively*

$$x \sin \frac{t}{2} + y \cos \frac{t}{2} = a \sin \frac{3}{2}t, \quad x \cos \frac{t}{2} - y \sin \frac{t}{2} = 3a \cos \frac{3}{2}t;$$

2. *the tangent at P meets the curve at Q and R whose parameters are $-\frac{1}{2}t$ and $\pi - \frac{1}{2}t$;*

3. $|\overline{QR}| = 4a$;

4. *the tangents at Q and R are at right angles and intersect on the circle $x^2 + y^2 = a^2$;*

5. *the normals at P, Q and R are concurrent and intersect in the circle $x^2 + y^2 = 9a^2$.*

Solution:

$$\frac{dy}{dx} = \frac{dy/dt}{dx/dt} = \frac{a(2\cos t - 2\cos 2t)}{-a(2\sin t + 2\sin 2t)} = \frac{4a \sin \frac{3t}{2} \sin \frac{t}{2}}{-4a \sin \frac{3t}{2} \cos \frac{t}{2}}.$$

$\therefore$ Slope of the tangent at t is $-\dfrac{\sin \frac{t}{2}}{\cos \frac{t}{2}}$, and hence the slope of the normal at t is $\dfrac{\cos \frac{t}{2}}{\sin \frac{t}{2}}$.

1. Equation of the tangent at P where the parameter is t;

$$\{y - a(2\sin t - \sin 2t)\} = -\frac{\sin \frac{t}{2}}{\cos \frac{t}{2}} \{x - a(2\cos t + \cos 2t)\}.$$

CHAPTER 15: TANGENTS AND NORMALS: ASSOCIATED CURVES

On simplification, we get

$$x \sin \frac{t}{2} + y \cos \frac{t}{2} = 2a \left\{ \sin t \cos \frac{t}{2} + \cos t \sin \frac{t}{2} \right\}$$
$$- a \left\{ \sin 2t \cos \frac{t}{2} - \cos 2t \sin \frac{t}{2} \right\}$$
$$= 2a \sin 3\frac{t}{2} - a \sin 3\frac{t}{2} = a \sin 3\frac{t}{2}. \qquad (15.6.4)$$

Again, equation of normal at $P(t)$ is

$$\{y - a(2\sin t - \sin 2t)\} = \frac{\cos \frac{t}{2}}{\sin \frac{t}{2}} \{x - a(2\cos t + \cos 2t)\}.$$

Simplifying, we get

$$x \cos \frac{t}{2} - y \sin \frac{t}{2} = 2a \left\{ \cos t \cos \frac{t}{2} - \sin t \sin \frac{t}{2} \right\}$$
$$+ a \left\{ \sin 2t \sin \frac{t}{2} + \cos 2t \cos \frac{t}{2} \right\}$$
$$= 3a \cos \frac{3t}{2}.$$

2. A point Q with parameter $-\frac{t}{2}$ has the co-ordinates

$$x_1 = a \left\{ 2\cos\left(-\frac{t}{2}\right) + \cos(-t) \right\} = a \left(2\cos \frac{t}{2} + \cos t \right)$$

and $\quad y_1 = a \left\{ 2\sin\left(-\frac{t}{2}\right) - \sin(-t) \right\} = -a \left(2\sin \frac{t}{2} - \sin t \right).$

A point R with parameter $\pi - \frac{t}{2}$ has the co-ordinates

$$x_2 = a \left(-2\cos \frac{t}{2} + \cos t \right) \quad \text{and} \quad y_2 = a \left(2\sin \frac{t}{2} + \sin t \right).$$

Verify that (x_1, y_1) and (x_2, y_2) satisfy the equation of tangent at $P(t)$ i.e., verify that equation (15.6.4) is satisfied by both (x_1, y_1) and (x_2, y_2).

3.

$$|QR| = \sqrt{(x_2 - x_1)^2 + (y_2 - y_1)^2}$$

$$= \sqrt{\left\{-2a\cos\frac{t}{2} + a\cos t - 2a\cos\frac{t}{2} - a\cos t\right\}^2 + \left\{2a\sin\frac{t}{2} + a\sin t + 2a\sin\frac{t}{2} - a\sin t\right\}^2}$$

$$= \sqrt{16a^2\cos^2\frac{t}{2} + 16a^2\sin^2\frac{t}{2}} = \mathbf{4a}.$$

4. Equation of tangent at $Q(-\frac{t}{2})$ is [Replace t by $-\frac{1}{2}t$ in (15.6.4)]

$$y\cos\frac{t}{4} - x\sin\frac{t}{4} = -a\sin 3\frac{t}{4}. \qquad (15.6.5)$$

Equation of tangent at $R(\pi - \frac{t}{2})$ is [Replace t by $\pi - \frac{1}{2}t$ in (15.6.4)]

$$x\sin\frac{\pi - \frac{t}{2}}{2} + y\cos\frac{\pi - \frac{t}{2}}{2} = a\sin\frac{3(\pi - \frac{t}{2})}{2}$$

$$\text{or, } x\cos\frac{t}{4} + y\sin\frac{t}{4} = -a\cos\frac{3t}{4}. \qquad (15.6.6)$$

The slope of (15.6.5) = $\dfrac{\sin\frac{t}{4}}{\cos\frac{t}{4}}$ and the slope of (15.6.6) is $-\dfrac{\cos\frac{t}{4}}{\sin\frac{t}{4}}$, their product being -1, the tangents at Q and R are at right angles.

The point of intersection (x, y) of tangents at Q and R is given by (using Cramer's Rule)

$$x = \begin{vmatrix} -a\sin\frac{3t}{4} & \cos\frac{t}{4} \\ -a\cos\frac{3t}{4} & \sin\frac{t}{4} \end{vmatrix} \div \begin{vmatrix} -\sin\frac{t}{4} & \cos\frac{t}{4} \\ \cos\frac{t}{4} & \sin\frac{t}{4} \end{vmatrix}$$

$$= \left(-a\sin\frac{3t}{4}\sin\frac{t}{4} + a\cos\frac{3t}{4}\cos\frac{t}{4}\right) \div (-1) = -a\cos t$$

and $$y = \begin{vmatrix} -\sin\frac{t}{4} & -a\sin\frac{3t}{4} \\ \cos\frac{t}{4} & -a\cos\frac{3t}{4} \end{vmatrix} \div \begin{vmatrix} -\sin\frac{t}{4} & \cos\frac{t}{4} \\ \cos\frac{t}{4} & \sin\frac{t}{4} \end{vmatrix}$$

$$= \left(a\cos\frac{3t}{4}\sin\frac{t}{4} + a\sin\frac{3t}{4}\cos\frac{t}{4}\right) \div (-1) = -a\sin t.$$

$\therefore$ Tangents at Q and R intersect on the circle $x^2 + y^2 = a^2$.

CHAPTER 15: TANGENTS AND NORMALS: ASSOCIATED CURVES

5. Equations of normals at P, Q, R are respectively

$$\left.\begin{array}{l} P(t) : x\cos\frac{t}{2} - y\sin\frac{t}{2} - 3a\cos\frac{3t}{2} = 0 \\ Q(-\frac{t}{2}) : x\cos\frac{t}{4} + y\sin\frac{t}{4} - 3a\cos\frac{3t}{4} = 0 \\ R(\pi - \frac{t}{2}) : x\sin\frac{t}{4} - y\cos\frac{t}{4} + 3a\sin\frac{3t}{4} = 0 \end{array}\right\}. \qquad (15.6.7)$$

Solving the last two equations (say, by Cramer's Rule) we get

$$x = 3a\cos t, \quad y = 3a\sin t.$$

These co-ordinates satisfy the first equation; hence the three normals are concurrent at $(3a\cos t, 3a\sin t)$. Clearly for different values of t, these points lie on the circle $x^2 + y^2 = 9a^2$.

Exercises XV (A)

1. Find the equations of the tangent and the normal to each of the following curves at the points indicated:

(a) $y = ae^{bx}$ at $x = 0$;

(b) $y = x^3 - 3x^2 - x + 5$ at $x = 3$;

(c) $y = c\cosh(x/c)$ at (x, y) [Catenary];

(d) $y = \dfrac{8a^3}{4a^2 + x^2}$ at $x = 2a$ [Witch of Agnesi];

(e) $y(x - 2)(x - 3) - x + 7 = 0$, at the point where it cuts the X-axis;

(f) $\dfrac{x^m}{a^m} + \dfrac{y^m}{b^m} = 1$ at (x', y');

(g) $\left(\dfrac{x}{a}\right)^{2/3} + \left(\dfrac{y}{b}\right)^{2/3} = 1$ at (x, y);

(h) $x = a(\theta - \sin\theta), \ y = a(1 - \cos\theta)$ at $\theta = \pi/2$ (Cycloid);

(i) $x = a(\theta + \sin\theta), \ y = a(1 - \cos\theta)$ at any point θ (Inverted Cycloid);

(j) $x = t^3 - t, \ y = t^2 + t$ at the point where $t = 1$.

2. (a) Show that the necessary and sufficient condition that a curve whose Cartesian equation is $y = f(x)$ shall have a tangent at a point (x', y') [not parallel to Y-axis] is that $f'(x)$ exists at (x', y') and then the equation of the tangent is

$$y - y' = f'(x')(x - x').$$

(b) Let $y = x^2 \sin(1/x)$, when $x \neq 0$; $= 0$, when $x = 0$.
   Find the equation of the tangent at $(0,0)$, if it exists.

(c) Verify that $f'(0)$ does not exist where $f(x) = x^{2/3}$. Prove that the curve $y = x^{2/3}$ has a tangent at $(0,0)$. How do you explain this?

3. Show that the line $x/a + y/b = 1$ touches the curve $x/a + \log(y/b) = 0$, at the point where the curve crosses Y-axis.

4. Show that the normal to the curve $3y = 6x - 5x^3$ drawn to the point $(1, 1/3)$ passes through the origin.

5. (a) Find the equations of those tangents to the circle $x^2 + y^2 = 52$ which are parallel to the line $2x + 3y - 6 = 0$.

   (b) Find the equations of those tangents to the hyperbola $x^2/9 - y^2/4 = 1$ which are perpendicular to the line $2y + 5x = 10$.

   (c) Determine the value of a so that the line $y = x$ will be a tangent to the curve $y = a^x (a > 0)$.

6. Find the points on the curve
$$y = x^4 - 6x^3 + 13x^2 - 10x + 5,$$
where the tangent is parallel to the line $y = 2x$, and prove that two of these points have the same tangent. [C.H. 1976]

7. Show that the points of the curve
$$y^2 = 4a\left(x + a\sin\frac{x}{a}\right)$$
where the tangents are parallel to the X-axis lies on the parabola $y^2 = 4ax$.
[C.H. 1968]

8. Find the points on the curve where the tangent is parallel to the X-axis and also find the points where the tangent is parallel to the Y-axis for each of the following curves:

   (a) $ax^2 + 2hxy + by^2 = 1$;
   (b) $x^3 + y^3 = 3axy$.

9. (a) If $x \cos\alpha + y \sin\alpha = p$ touches the curve, $\dfrac{x^m}{a^m} + \dfrac{y^m}{b^m} = 1$, then prove that
$$(a\cos\alpha)^{m/(m-1)} + (b\sin\alpha)^{m/(m-1)} = p^{m/(m-1)}.$$

(b) If $x\cos\theta + y\sin\theta = p$ touches the curve $x^m y^n = a^{m+n}$, prove th

$$p^{m+n} \cdot m^m \cdot n^n = (m+n)^{m+n} \cdot a^{m+n} \cdot \sin^n\theta \cos^m\theta.$$

(c) If $lx + my = n$ touches the curve $x^p/a^p + y^p/b^p = 1$, prove that

$$(al)^{p/(p-1)} + (bm)^{p/(p-1)} = n^{p/(p-1)} \quad (p \neq 1)$$
$$\text{or,} \quad (al)^q + (bm)^q = n^q, \quad \text{where } p^{-1} + q^{-1} = 1.$$

[C.H. '69, '80]

(d) If $lx + my = 1$ be a normal to the parabola $y^2 = 4ax$, then prove that

$$al^3 + 2alm^2 = m^2.$$

10. Find the points of intersection of the line $y = 2x - 1$ and the cubic $2y = 3x^3 - 12x^2 + 13x - 2$. Obtain the equations of the tangents to the cubic at each of these points. Now show that remaining intersection of these tangents with the cubic are collinear.

11. (a) Show that the sum of the intercepts of any tangent to the curve $\sqrt{x} + \sqrt{y} = \sqrt{a}$ is a constant.

(b) Show that the product of the intercepts of any tangent to the hyperbola $xy = k$ is a constant.

12. In the curve $x^m y^n = a^{m+n}$, prove that the portion of the tangent intercepted between the axes is divided at the point of contact into two segments which are in a constant ratio.

13. Obtain the equation of the tangent at any point t of the *Astroid:* $x = a\cos^3 t$, $y = a\sin^3 t$. Hence prove that the portion of the tangent at any point t intercepted between the axes is of constant length. Also deduce that the locus of intersection of tangents at right angles to one another is $2r^2 = a^2 \cos^2 2\theta$, where

$$r^2 = x^2 + y^2 \quad \text{and} \quad \theta = \tan^{-1}(y/x).$$

14. Tangent at any point θ of the *four-cusped hypocycloid* $x = a\cos^3\theta, y = b\sin^3\theta$ cuts the X-axis at A, and Y-axis at B. If $OA = \alpha, OB = \beta$ (O is the origin), then prove that locus of points (α, β) is the ellipse $x^2/a^2 + y^2/b^2 = 1$.

15. Obtain the equations of the tangent at $(0,0)$ for the following curves:

(a) $x^2 + y^2 + ax + by = 0$;

(b) $(x^2 + y^2)^2 = a^2(x^2 - y^2)$;

(c) $(x^2 + y^2)^2 = a^2x^2 - b^2y^2$.

16. Verify that the pairs of curves given on the left intersect at the indicated angles on the right:

Pairs of Curves	Angle of Intersection
a) $y = x^2, y = 2 - x^2$	$\tan^{-1} 4/3$
b) $x^2 - y^2 = 8, xy = 3$	$\pi/2$
c) $y = \log_e x, y = \log_{10} x$	$\tan^{-1}\left\{\dfrac{\log_{10} e - 1}{\log_{10} e + 1}\right\}$
d) $y = 3\cot x, y = 2\sin x$	$\pi/2$.

17. (a) Prove that if the curves $ax^2 + by^2 = 1$ and $Ax^2 + By^2 = 1$ intersect at right angles, then $1/A - 1/a = 1/B - 1/b$. [C.H. 1976]

(b) Let $u + iv = f(x + iy)$, where u and v are functions of x and y. Prove that the curves $u =$ constant and $v =$ constant form two families such that each member of the first cuts orthogonally each member of the second.
[Hints: First prove $u_x = v_y$ and $u_y = -v_x$; deduce $u_x \cdot v_x + u_y \cdot v_y = 0$.]

(c) Find $k (\neq 0)$ for which the straight line $x = k$ intersects the curve $xy^2 = (x + y)^2$ orthogonally. [C.H. 1984]

18. Find the lengths of the tangent, normal, subtangent and subnormal at the point θ of the astroid
$$x = a\cos^3\theta, \quad y = a\sin^3\theta.$$

19. Show that in any curve

(a) $\dfrac{\text{length of subnormal}}{\text{length of subtangent}} = \left(\dfrac{\text{length of normal}}{\text{length of tangent}}\right)^2$.

(b) the rectangle contained by the lengths of the subtangent and the subnormal is equal to the square of the corresponding ordinate.

20. (a) Show that at any point of the curve
$$x^{m+n} = k^{m-n} y^{2n},$$
the mth power of the subtangent varies as the nth power of the subnormal.

(b) Show that at any point of the curve $y = be^{x/a}$ the subtangent is of constant length and the subnormal varies as the square of the ordinate.

(c) Show that for the parabola $y^2 = 4ax$, the subtangent at any point is bisected at the vertex and that the length of subnormal is constant at every point.

(d) Show that for the curve $ay^2 = (x+1)^3$ the subnormal varies as the square of the subtangent. [C.H. 1982]

21. Prove that the length of the perpendicular drawn from the foot of the ordinate of any point (x, y) of a curve to the tangent line drawn through that point is given by

$$\frac{y}{\sqrt{1 + (dy/dx)^2}}.$$

Verify that this perpendicular length is constant at any point of the Catenary $y = c \cosh(x/c)$.

22. Prove that the length of the perpendicular P_t from the origin to the tangent line at the point (x, y) of the curve is given by

$$P_t = \frac{y - x\dfrac{dy}{dx}}{\sqrt{1 + (dy/dx)^2}}.$$

Verify that in the circle $x^2 + y^2 = a^2$ such a length of the tangent P_t is always constant at every point but in the rectangular hyperbola $xy = k^2$, the length P_t is given by $2k^2/(\sqrt{x^2 + y^2})$.

(*Miscellaneous*)

23. Prove that the equation of the tangent at the point t on the curve (given in parametric form)

$$x = a\frac{\phi(t)}{f(t)} \quad \text{and} \quad y = a\frac{\psi(t)}{f(t)}$$

may be written as

$$\begin{vmatrix} x & y & a \\ \phi(t) & \psi(t) & f(t) \\ \phi'(t) & \psi'(t) & f'(t) \end{vmatrix} = 0.$$

24. Let $X = \phi(x, y)$, $Y = \psi(x, y)$ define a transformation of the xy-plane to XY-plane. Suppose further, $\phi_x = \psi_y$ and $\phi_y = -\psi_x$. Then prove that the angle between the curves $F(x, y) = 0$ and $G(x, y) = 0$ in the xy-plane is equal to the angle between the curves $F_1(X, Y) = 0$ and $G_1(X, Y) = 0$ in the XY-plane, where the transformation maps $F(x, y)$ and $G(x, y)$ to $F_1(X, Y)$ and $G_1(X, Y)$ respectively. [C.H. 1967]

25. (a) Show that the normal at any point θ of the curve
$$x = a\cos\theta + a\theta\sin\theta, \quad y = a\sin\theta - a\theta\cos\theta,$$
is at a constant distance from the origin.

(b) Show that the tangent to the curve
$$25x^5 + 5x^4 - 45x^3 - 5x^2 + 2x + 6y - 24 = 0,$$
at the point $(-1, 1)$ is also a normal at two points of the curve.

26. (a) Tangents are drawn from the origin to the curve $y = \sin x$. Show that their points of contact lie on the curve $x^2 y^2 = x^2 - y^2$.

(b) The tangent at any point P on the curve $x^2 - x^3 = y$ meets it again at Q. Show that the locus of the mid-point of PQ is
$$y = 1 - 9x + 28x^2 - 28x^3.$$

27. Show that the tangent to the curve $x^3 + y^3 = 3axy$ at a point $[\neq (0,0)]$ where it meets the parabola $y^2 = ax$ is parallel to the Y-axis. [C.H. 1983]

[Hints: The point is $(2^{2/3}a, 2^{1/3}a)$]

Answers

1.   Tangent   Normal

(a) $y = abx + a$ $x + aby = a^2 b$.

(b) $y = 8x - 22$ $8y + x = 19$.

(c) $Y - c\cosh x/c = \sinh x/c (X - x)$ $(Y - c\cosh x/c)\sinh x/c + (X - x) = 0$.

(d) $2y + x = 4a$ $y = 2x - 3a$.

(e) $x - 20y = 7$ $20x + y = 140$.

(f) $b^m x (x')^{m-1} + a^m y (y')^{m-1} = a^m b^m$ $a^m (y')^{m-1}(x - x') = b^m (x')^{m-1}(y - y')$.

(g) Put $m = 2/3$ in (f); take (x, y) for (x', y') and (X, Y) for (x, y).

(h) $y = x + 2a - a\pi/2$ $y + x = a\pi/2$.

(i) $x\sin\theta/2 - y\cos\theta/2 = a\theta\sin\theta/2$ $x\cos\theta/2 + y\sin\theta/2 = a\theta\cos\theta/2 + 2a\sin\theta/2$.

(j) $3x - 2y + 4 = 0$ $2x + 3y = 6$.

5. (a) $3y + 2x = \pm 26$. (b) $5y - 2x = \pm 10$. (c) $a = e^{1/e}$. 8. (a) where $ax + hy = 0$ and $hx + by = 0$ intersect the curve. (b) $(\sqrt[3]{2a}, \sqrt[3]{4a})$; $(\sqrt[3]{4a}, \sqrt[3]{2a})$. 10. $(0, -1), (3, 5), (1, 1)$, $y_1 = 13/2, 11, -1$ at these points; hence etc. 15. (a) $ax + by = 0$; (b) $y = \pm x$; (c) $y = \pm (a/b)x$. 17. (c) $k = 4$. 18. $a\sin^2\theta$; $a\tan\theta\sin^2\theta$; $a\sin^2\theta\cos\theta$; $a\sin^2\theta\tan\theta$.

15.7 Equations of the Tangent and Normal to a Polar Curve

Given a curve whose polar equation is $1/r = f(\theta)$ or $r = 1/f(\theta)$.

To find the equations of the tangent and normal at a point $\theta = \alpha$ of the curve:

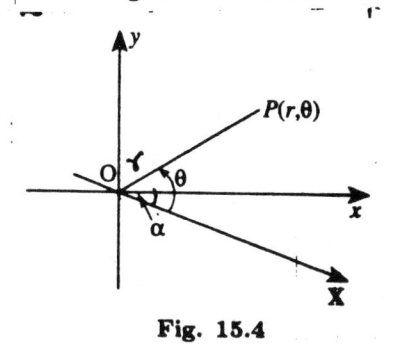

Fig. 15.4

Equation of a line in Polar form:

[The equation of any line in Rectangular Cartesian form can be written as:

$$ax + by + c = 0.$$

If Ox be taken as the initial line and (r, θ) be the polar co-ordinates of a point whose cartesian co-ordinates are (x, y), then

$$x = r\cos\theta, \quad y = r\sin\theta$$

and the equation of the line becomes

$$A\cos\theta + B\sin\theta = \frac{1}{r}. \quad \left(\text{where } A = \frac{a}{-c},\ B = \frac{b}{-c}\right)$$

On the other hand, if a line OX be taken as the initial line ($\angle XOx = \alpha$) and with respect to this line any point P has the co-ordinates (r, θ) they we shall have to replace θ by $(\theta - \alpha)$.(See Fig. 15.4).

The equation of the line, then becomes

$$A\cos(\theta - \alpha) + B\sin(\theta - \alpha) = \frac{1}{r}.]$$

We shall write u for $1/r$. So we write the equation of the curve as $u = f(\theta)$. Let the polar co-ordinates of the point of contact be (U, α) and let us write U' for

$$\frac{du}{d\theta} \text{ at } \theta = \alpha.$$

(a) The equation
$$u = A\cos(\theta - \alpha) + B\sin(\theta - \alpha) \tag{15.7.1}$$
represents a straight line, A and B being arbitrary constants. Let this line represents the required tangent. On differentiation,
$$\frac{du}{d\theta} = -A\sin(\theta - \alpha) + B\cos(\theta - \alpha). \tag{15.7.2}$$

Since the tangent touches the curve, the value of $du/d\theta$ at the point of contact **will be the same** for the curve as well as for the tangent. Putting $\theta = \alpha$ in (15.7.1) and (15.7.2) we easily get $A = U$ and $B = U'$.

∴ the required equation of the tangent is,
$$\frac{1}{r} = U\cos(\theta - \alpha) + U'\sin(\theta - \alpha) \tag{15.7.3}$$
or,
$$\frac{1}{r} = f(\alpha)\cos(\theta - \alpha) + f'(\alpha)\sin(\theta - \alpha).$$

Example 15.7.1. *Find the equation of the tangent at $\theta = 2\alpha$ of the Cardiode $r = a(1 + \cos\theta)$.*

Solution: We consider the curve
$$u = \frac{1}{r} = \frac{1}{a(1+\cos\theta)} = \frac{1}{2a\cos^2\theta/2} = f(\theta).$$

Here,
$$\frac{du}{d\theta} = \frac{1}{2a} 2\sec\theta/2 \cdot \sec\theta/2 \cdot \tan\theta/2 \cdot \frac{1}{2}$$
$$= \frac{1}{2a}\tan\theta/2 \sec^2\theta/2 = f'(\theta).$$

At $\theta = 2\alpha$, $f'(2\alpha) = \dfrac{1}{2a}\tan\alpha\sec^2\alpha$ and $f(2\alpha) = \dfrac{1}{2a}\sec^2\alpha$.

∴ the required equation of the tangent at $\theta = 2\alpha$ is
$$\frac{1}{r} = \frac{1}{2a}\sec^2\alpha\cos(\theta - 2\alpha) + \frac{1}{2a}\tan\alpha\sec^2\alpha\sin(\theta - 2\alpha)$$
$$= \frac{1}{2a}\frac{\cos\alpha\cos(\theta - 2\alpha) + \sin\alpha\sin(\theta - 2\alpha)}{\cos^3\alpha}$$
$$= \frac{1}{2a}\frac{\cos(\theta - 3\alpha)}{\cos^3\alpha}$$
or, $\quad r\cos(\theta - 3\alpha) = 2a\cos^3\alpha.$

CHAPTER 15: TANGENTS AND NORMALS: ASSOCIATED CURVES

(b) The equation of the normal to the curve $u = f(\theta)$ at $\theta = \alpha$.

The equation of *any line perpendicular to* (15.7.3) is

$$Cu = U' \cos(\theta - \alpha) - U \sin(\theta - \alpha) \tag{15.7.4}$$

where C is a constant.

Though the point (U, α), the curve as well as the normal pass. So if (15.7.4) be the equation of normal at $\theta = \alpha$, we have

$$CU = U'.$$

∴ the required equation of normal is

$$\frac{U'}{U} u = U' \cos(\theta - \alpha) - U \sin(\theta - \alpha) \tag{15.7.5}$$

or, $\quad \dfrac{f'(\alpha)}{f(\alpha)} u = f'(\alpha) \cos(\theta - \alpha) - f(\alpha) \sin(\theta - \alpha).$

Example 15.7.2. *Show that the polar equation of the normal at $\theta = 2\alpha$ of the Cardiode $r = a(1 + \cos\theta)$ is*

$$r \sin(3\alpha - \theta) = \frac{a}{2}(\sin 3\alpha + \sin \alpha).$$

Hence show that three normals can be drawn from a given point to the Cardiode. If the feet of these normals be (r_1, θ_1), (r_2, θ_2) and (r_3, θ_3), then prove that

$$\tan\frac{\theta_1}{2} + \tan\frac{\theta_2}{2} + \tan\frac{\theta_3}{2} + 3\tan\frac{\theta_1}{2}\tan\frac{\theta_2}{2}\tan\frac{\theta_3}{2} = 0.$$

Solution: Using $f(2\alpha) = \dfrac{1}{2a}\sec^2\alpha$ and $f'(2\alpha) = \dfrac{1}{2a}\tan\alpha \sec^2\alpha$, equation of the normal at $\theta = 2\alpha$ is obtained from

$$\frac{f'(2\alpha)}{f(2\alpha)} u = f'(2\alpha) \cos(\theta - 2\alpha) - f(2\alpha) \sin(\theta - 2\alpha)$$

in the form

$$u \tan \alpha = \frac{1}{2a}\left[\frac{\sin\alpha \cos(\theta - 2\alpha)}{\cos^3\alpha} - \frac{\sin(\theta - 2\alpha)}{\cos^2\alpha}\right]$$

i.e., $\quad \dfrac{1}{r}\dfrac{\sin\alpha}{\cos\alpha} = \dfrac{1}{2a}\dfrac{\sin(\alpha - \theta + 2\alpha)}{\cos^3\alpha}$

or, $\quad r\sin(3\alpha - \theta) = 2a\cos^2\alpha \sin\alpha = \dfrac{a}{2}(\sin 3\alpha + \sin\alpha),$ on simplification.

i.e., $\quad r\sin(3\alpha - \theta) = \dfrac{a}{2}(\sin 3\alpha + \sin\alpha)$ is the required equation.

If we write $x = r\cos\theta, y = r\sin\theta$ and $t = \tan\alpha$, then this last equation may be written as

$$(3t - t^3)x - (1 - 3t^2)y = \frac{1}{2}a\{(3t - t^3) + t(1 + t^2)\}$$
$$\text{or, } t^3 x - 3t^2 y + t(2a - 3x) + y = 0,$$

which is a cubic equation in t. Thus we get three values t_1, t_2, t_3 of t (where $t = \tan\alpha = \tan\theta/2$), corresponding to three normals which pass through a given point (x, y).

$$\sum t = 3y/x \quad \text{and} \quad t_1 t_2 t_3 = -y/x$$

and hence $\Sigma t + 3t_1 t_2 t_3 = 0$, which proves the stated result.

15.8 Angle between the Tangent and Radius Vector at any point of a Polar Curve

To find the direction of the tangent at a point on a curve given in polar co-ordinates (r, θ) it will be convenient to express it in terms of an angle ϕ which the tangent makes with the radius vector at the point of contact.

Let $P(r, \theta), Q(r + \Delta r, \theta + \Delta\theta)$ be two neighbouring points of a curve $r = f(\theta)$ (Fig. 15.5). Then join P, Q. Suppose that OQ makes an angle α with PQ.

We define the direction of PT, the tangent to the curve at P, to be the limiting direction of PQ as Q approaches P along the curve. The angle α then clearly approaches the angle intuitively accepted as ϕ. Thus, we have:

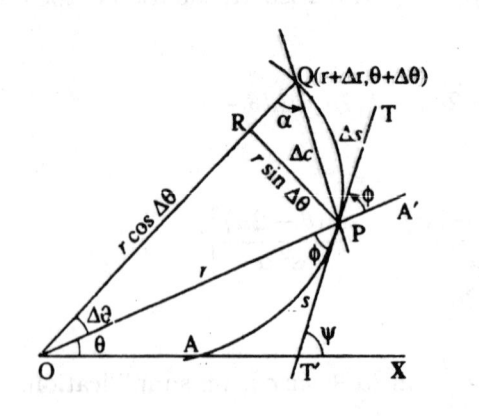

$$\tan\phi = \lim_{\Delta\theta \to 0} \tan\alpha.$$

Fig. 15.5

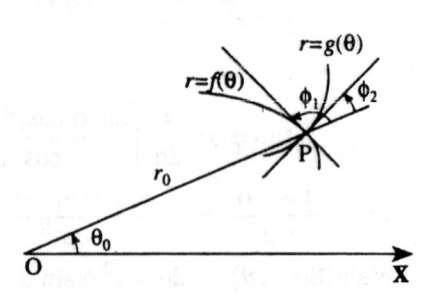

Fig. 15.6

CHAPTER 15: TANGENTS AND NORMALS: ASSOCIATED CURVES

Expression for $\tan\varphi$ in terms of r and θ

Draw PR perpendicular to OQ. Hence

$$OP = r, \quad OQ = r + \Delta r, \quad PR = r\sin\Delta\theta,$$
$$OR = r\cos\Delta\theta, \quad RQ = r + \Delta r - r\cos\Delta\theta,$$

and from the right triangle PQR.

$$\tan\alpha = \frac{r\sin\Delta\theta}{r + \Delta r - r\cos\Delta\theta} = \frac{r\sin\Delta\theta}{r(1 - \cos\Delta\theta) + \Delta r}.$$

Dividing numerator and denominator by $\Delta\theta$,

$$\tan\alpha = \frac{r\sin\Delta\theta/\Delta\theta}{\{r(1-\cos\Delta\theta)/\Delta\theta\} + \{\Delta r/\Delta\theta\}}.$$

As $\Delta\theta \to 0$, $\dfrac{\sin\Delta\theta}{\Delta\theta} \to 1$ and $\dfrac{1-\cos\Delta\theta}{\Delta\theta} \to 0$.

Also $\Delta r/\Delta\theta \to dr/d\theta = f'(\theta)$. Hence

$$\lim_{\Delta\theta \to 0} \tan\alpha = \tan\phi = \frac{r}{dr/d\theta} = \frac{r}{f'(\theta)},$$

i.e., $\tan\phi = r\dfrac{d\theta}{dr}$.

Observations:

1. In the above discussions we thought of ϕ as the angle OPT'. Since this angle is equal to $A'PT$, we can associate ϕ with the latter angle. In so doing we agree to measure ϕ counter-clockwise from the radius-vector produced beyond P towards that part of the tangent PT, which extends in the direction of increasing θ.

2. If OX extends horizontally towards the right from O, the angle ψ can be determined from

$$\psi = \theta + \phi. \quad \text{(Fig. 15.5)}.$$

Angle of intersections of two polar curves

If two curves $r = f(\theta)$ and $r = g(\theta)$ intersect at a point $P(r_0, \theta_0)$ (See Fig. 15.6), the angle of intersection is $\phi_1 - \phi_2$. Then

$$\tan(\phi_1 - \phi_2) = \frac{\tan\phi_1 - \tan\phi_2}{1 + \tan\phi_1 \tan\phi_2}$$

where $\tan\phi_1 = r_0/f'(\theta_0)$ and $\tan\phi_2 = r_0/g'(\theta_0)$.

15.9 Subtangent and Subnormal, Perpendicular from Pole to the Tangent, Lengths of Tangent and Normal

In Fig. 15.7, PT, PG are respectively the tangent and normal to the curve $r = f(\theta)$ at $P(r, \theta)$, meeting the line through O, perpendicular to OP in T and G respectively. The lengths OT, OG, PT and GP are frequently referred to as the (polar) subtangent, subnormal, length of tangent and length of normal respectively. Hence,

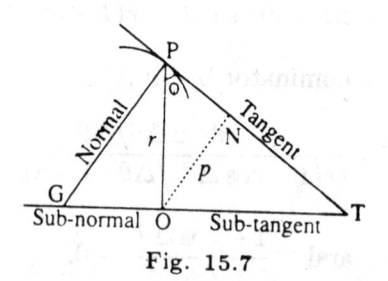

Fig. 15.7

$$\text{Polar Subtangent} = OT = r \tan \phi = r^2 \, d\theta/dr$$
$$\text{Polar Subnormal} = OG = r \cot \phi = dr/d\theta.$$

Perpendicular from pole on the tangent $= ON = p = r \sin \phi$, whence, by applying $\tan \phi = r \, d\theta/dr$, we obtain

$$\frac{1}{p^2} = \frac{1}{r^2 \sin^2 \phi} = \frac{1}{r^2}(1 + \cot^2 \phi) = \frac{1}{r^2} + \frac{1}{r^4}\left(\frac{dr}{d\theta}\right)^2$$

and if $\quad r = \dfrac{1}{u}, \quad \dfrac{dr}{d\theta} = -\dfrac{1}{u^2}\dfrac{du}{d\theta} \quad$ and

$$\frac{1}{p^2} = u^2 + \left(\frac{du}{d\theta}\right)^2.$$

$$\text{Length of tangent} = PT = r \sec \phi = r\sqrt{1 + r^2 (d\theta/dr)^2}.$$
$$\text{Length of Normal} = PG = r \operatorname{cosec} \phi = \sqrt{r^2 + (dr/d\theta)^2}.$$

15.10 Pedal Equation

Definition 15.10.1. *A relation between p, the length of the perpendicular from a given point O to the tangent at any point P on a curve and r, the distance of P from O is called (p, r)-equation or pedal equation of the curve with respect to O. When nothing is mentioned about the given point, the given point is to be taken as origin or pole according as the equation is in Cartesian or in polar form (Fig. 15.7).*

CHAPTER 15: TANGENTS AND NORMALS: ASSOCIATED CURVES

(A) Pedal Equation derived from Cartesian Equation

Let $f(x,y) = 0$ be the equation of the curve and let the origin be the given point. The tangent at (x,y) being

$$Y f_y + X f_x - (x f_x + y f_y) = 0, \quad p^2 = \frac{(x f_x + y f_y)^2}{f_x^2 + f_y^2}.$$

Also $r^2 = x^2 + y^2$. And $f(x,y) = 0$.

Eliminating x and y, we have the required pedal equation.

(B) From Polar Equation to Pedal Equation

If the curve be given by $r = f(\theta)$ and the given point be the pole, then, since,

$$\tan \phi = r \, d\theta/dr \quad \text{and} \quad p = r \sin \phi;$$

we may obtain the pedal equation by elimination of θ and ϕ.

Otherwise. Since we know $\dfrac{1}{p^2} = \dfrac{1}{r^2} + \dfrac{1}{r^4}\left(\dfrac{dr}{d\theta}\right)^2$, we can eliminate θ between this relation and $r = f(\theta)$ and derive the pedal equation.

15.11 Illustrative Examples

Example 15.11.1. *Find the angle of intersection of the cardiode $r = a(1 - \cos\theta)$ and the circle $r = 2a \cos\theta$.*

Solution: Solving for the points of intersection, we obtain

$$\cos\theta = \frac{1}{3}, \quad \theta = \cos^{-1}\frac{1}{3}, \quad r = \frac{2}{3}a.$$

From the first equation by differentiation,

$$\frac{dr}{d\theta} = a\sin\theta \text{ and hence } \tan\phi_1 = r\frac{d\theta}{dr} = \frac{a(1 - \cos\theta)}{a\sin\theta} = \frac{1}{\sqrt{2}} \left(\because \cos\theta = \frac{1}{3}, \sin\theta = \frac{2\sqrt{2}}{3} \right)$$

and similarly from the second equation, $\tan\phi_2 = -1/2\sqrt{2}$.

$$\therefore \tan(\phi_2 - \phi_1) = \frac{-1/(2\sqrt{2}) - 1/\sqrt{2}}{1 - 1/(2\sqrt{2}) \cdot (1/\sqrt{2})} = -\sqrt{2}.$$

Obtaining thereby,

$$\phi_2 - \phi_1 = \tan^{-1}(-\sqrt{2}) = 125°16' \text{ (approx.)}$$

(angle between the cardiode and the circle).

Example 15.11.2. *Find the pedal equation of the astroid* $x^{2/3} + y^{2/3} = a^{2/3}$.

[C.H. 1996]

Solution: We have the tangent at (x,y) : $Xx^{-1/3} + Yy^{-1/3} = a^{2/3}$; and hence if p be the perpendicular from the origin on it,

$$p^2 = a^{4/3}/(x^{-2/3} + y^{-2/3}) = (axy)^{2/3}.$$

Also since $r^2 = x^2 + y^2 = (x^{2/3} + y^{2/3})^3 - 3x^{2/3} \cdot y^{2/3}(x^{2/3} + y^{2/3}) = a^2 - 3(axy)^{2/3}$. The required pedal equation is $r^2 + 3p^2 = a^2$.

Example 15.11.3. *Find ϕ in terms of θ for $r^m = a^m \sin m\theta$ and obtain the corresponding pedal equation. Write down the pedal equation for the case $m = 2$.*

[C.H. 1969]

Solution: We know that

$$\tan \phi = r\frac{d\theta}{dr} = \frac{r^m}{a^m \cos m\theta} = \tan m\theta, \text{ whereby } \phi = m\theta,$$

and hence $p = r \sin \phi = r \sin m\theta = r^{m+1}/a^m$ is the required pedal equation.

In case, $m = 2$, $\phi = 2\theta$. The pedal equation of $r^2 = a^2 \sin 2\theta$ then becomes $a^2 p = r^3$.

Example 15.11.4. *Prove that the locus of the extremity of the polar subtangent of the curve $1/r + f(\theta) = 0$ is*

$$\frac{1}{r} = f'\left(\frac{\pi}{2} + \theta\right).$$

In case the curve is given by

$$r = \frac{1 + \tan \theta/2}{m + n \tan \theta/2} \quad or, \quad \frac{1}{r} = \frac{m + n \tan \theta/2}{1 + \tan \theta/2}$$

then prove that the locus will be a Cardiode.

Solution: (Fig. 15.8) P is a point of the curve

$$\frac{1}{r} = -f(\theta) \quad or, \quad r = -\frac{1}{f(\theta)}.$$

The polar subtangent $= OT$.

Required to find the locus of T.

We take the polar co-ordinates of $T = (R, \alpha)$ where

$$OT = R \quad \text{and} \quad \angle XOT = -\alpha.$$

so that $\theta + (-\alpha) = \pi/2$ or $\theta = \alpha + \pi/2$.

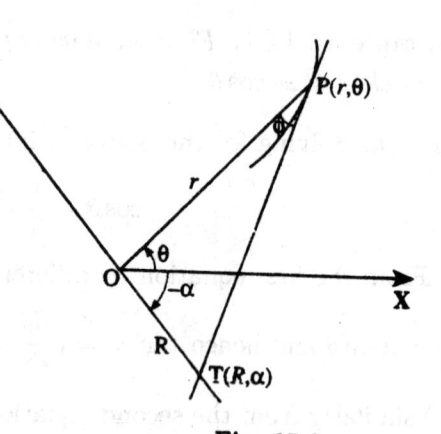

Fig. 15.8

CHAPTER 15: TANGENTS AND NORMALS: ASSOCIATED CURVES

Again,

$$OT = r\tan\phi = r^2\frac{d\theta}{dr} = \left\{-\frac{1}{f(\theta)}\right\}^2 \bigg/ \left\{\frac{f'(\theta)}{\{f(\theta)\}^2}\right\}$$

i.e., $\quad R = \dfrac{1}{f'(\theta)} \quad$ or, $\quad \dfrac{1}{R} = f'(\theta) = f'(\pi/2 + \alpha).$

$\therefore$ the locus of T can be obtained by changing (R, α) to any point (r, θ) of the locus, i.e., the equation to the locus is

$$\frac{1}{r} = f'\left(\frac{\pi}{2} + \theta\right)$$

where (r, θ) are the polar co-ordinates of any point of the locus.

Second Part. Given curve is

$$\frac{1}{r} = \frac{m + n\tan\theta/2}{1 + \tan\theta/2} = f(\theta)$$

or, $\quad f'(\theta) = \dfrac{\dfrac{n}{2}\sec^2\theta/2(1 + \tan\theta/2) - \dfrac{1}{2}\sec^2\theta/2(m + n\tan\theta/2)}{(1 + \tan\theta/2)^2}$

$\qquad = \dfrac{n-m}{2}\dfrac{\sec^2\theta/2}{\sec^2\theta/2 + 2\tan\theta/2}$

$\qquad = \dfrac{n-m}{2}\dfrac{1}{1 + 2\sin\theta/2\cos\theta/2}$

or, $\quad f'(\theta) = \dfrac{n-m}{2}\dfrac{1}{1 + \sin\theta}.$

$\therefore$ the equation to the locus is

$$\frac{1}{r} = f'\left(\frac{\pi}{2} + \theta\right) = \frac{n-m}{2}\frac{1}{1 + \sin(\pi/2 + \theta)}$$

or, $\quad r = \dfrac{2}{n-m}(1 + \cos\theta) = a(1 + \cos\theta),$ where $a = \dfrac{2}{n-m}$

which is a cardiode.

Example 15.11.5. *The tangent at any point of the cardiode $r = a(1 + \cos\theta)$ whose vectorial angle is 2α meets the curve again at a point whose vectorial angle is 2β. Prove that*

$$\cos(2\beta - \alpha) + 2\cos\alpha = 0.$$

Solution: In art. **15.7**, Ex. **15.7.1**. We have observed that the equation of the tangent to the cardiode $r = a(1 + \cos\theta)$ at the point where $\theta = 2\alpha$ is given by

$$r\cos(\theta - 3\alpha) = 2a\cos^3\alpha.$$

According to the given condition this tangent passes through another point of the Cardiode where $\theta = 2\beta$ and hence $r = a(1 + \cos 2\beta)$.

$$\therefore\ a(1 + \cos 2\beta)\cos(2\beta - 3\alpha) = 2a\cos^3\alpha$$
$$\text{or, } (1 + \cos 2\beta)\cos(2\beta - 3\alpha) = 2\cos^3\alpha$$
$$\text{or, } 2\cos(2\beta - 3\alpha) + 2\cos 2\beta\cos(2\beta - 3\alpha) = 4\cos^3\alpha$$
$$\text{or, } 2\cos(2\beta - 3\alpha) + \cos(4\beta - 3\alpha) + \cos 3\alpha = 4\cos^3\alpha$$
$$\text{or, } 2\cos(2\beta - 3\alpha) + \cos(4\beta - 3\alpha) = 3\cos\alpha.$$

or, $\{\cos(4\beta - 3\alpha) + \cos\alpha\} + \{2\cos(2\beta - 3\alpha) - 4\cos\alpha\} = 0$
or, $2\cos(2\beta - \alpha)\cos(2\beta - 2\alpha) + \{2\cos(2\beta - 2\alpha - \alpha) - 4\cos\alpha\} = 0$
or, $2\cos(2\beta - \alpha)\cos(2\beta - 2\alpha) + 4\cos(2\beta - 2\alpha)\cos\alpha - 2\cos(2\beta - 2\alpha)\cos\alpha$
$\qquad +2\sin(2\beta - 2\alpha)\sin\alpha - 4\cos\alpha = 0$
or, $[2\cos(2\beta - \alpha)\cos(2\beta - 2\alpha) + 4\cos(2\beta - 2\alpha)\cos\alpha]$
$\qquad -2[\cos(2\beta - 2\alpha)\cos\alpha - \sin(2\beta - 2\alpha)\sin\alpha] - 4\cos\alpha = 0$
or, $2\cos(2\beta - 2\alpha)\{\cos(2\beta - \alpha) + 2\cos\alpha\} - 2\cos(2\beta - \alpha) - 4\cos\alpha = 0$
or, $2\cos(2\beta - 2\alpha)[\cos(2\beta - \alpha) + 2\cos\alpha] - 2[\cos(2\beta - \alpha) + 2\cos\alpha] = 0$
or, $\{2\cos(2\beta - 2\alpha) - 2\}\{\cos(2\beta - \alpha) + 2\cos\alpha\} = 0.$

$$\therefore\ \cos(2\beta - \alpha) + 2\cos\alpha = 2.$$

[$\because$ We cannot take $\cos(2\beta - 2\alpha) = 1$, because then $\beta = \alpha$, which is not intended.]

Example 15.11.6. *Show that the pedal equation of the ellipse $x^2/a^2 + y^2/b^2 = 1$ with respect to a focus is $b^2/p^2 = 2a/r - 1$.* [C.H. '67, '79, '83]

Solution: **Geometrical Proof.** Let A, A' be the vertices, S, S' be the focii, C the centre and P any point of the ellipse $x^2/a^2 + y^2/b^2 = 1$.

Let $SP = r, S'P = r'$. Draw $SN, S'N'$ perpendiculars from S and S' respectively on the tangent at P. From the properties of ellipse proved in elementary Co-ordinate Geometry we know that

(a) $r + r' = 2a,$

CHAPTER 15: TANGENTS AND NORMALS: ASSOCIATED CURVES

(b) $pp' = b^2$ and

(c) $\angle SPN = \angle S'PN'$.

The triangles SPN and $S'PN'$ are equiangular (and hence similar) since

$$\angle SPN = \angle S'PN', \quad \angle SNP = \angle S'N'P = \pi/2.$$

$$\therefore \quad \frac{SP}{S'P} = \frac{SN}{S'N'} \quad \text{i.e.,} \quad \frac{r}{r'} = \frac{p}{p'}$$

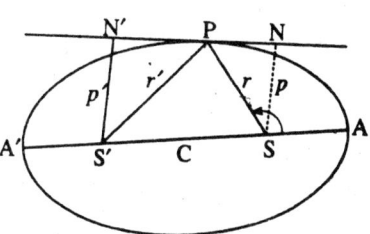

Fig. 15.9

or, $\quad \dfrac{r}{p} = \dfrac{r'}{p'} = \sqrt{\dfrac{rr'}{pp'}} = \sqrt{\dfrac{r(2a-r)}{b^2}}.\quad (\because r + r' = 2a, pp' = b^2)$

$$\therefore \quad \frac{r^2}{p^2} = \frac{r(2a-r)}{b^2} \quad \text{or,} \quad \frac{b^2}{p^2} = \frac{2a-r}{r} = \frac{2a}{r} - 1.$$

Otherwise. The polar equation of the ellipse $\dfrac{x^2}{a^2} + \dfrac{y^2}{b^2} = 1$ with focus as a pole is given by

$$\frac{l}{r} = 1 + e\cos\theta$$

where $\quad l = $ semi-latus rectum $= b^2/a, \quad e = \sqrt{(a^2 - b^2)/a^2}$.

$r = $ radius vector SP, $\theta = $ vectorial angle $\angle ASP$.

On differentiation, we get

$$-\frac{l}{r^2}\frac{dr}{d\theta} = -e\sin\theta \quad \text{or,} \quad \frac{1}{r^4}\left(\frac{dr}{d\theta}\right)^2 = \frac{e^2}{l^2}\sin^2\theta.$$

$$\therefore \quad \frac{1}{p^2} - \frac{1}{r^2} = \frac{1}{r^4}\left(\frac{dr}{d\theta}\right)^2 = \frac{e^2\sin^2\theta}{l^2}.$$

$$\therefore \quad e^2\sin^2\theta = l^2\left(\frac{r^2-p^2}{r^2p^2}\right) \text{ and the equation of the curve } e^2\cos^2\theta = \left(\frac{l-r}{r}\right)^2.$$

$\therefore$ Adding we get, $\dfrac{l^2(r^2-p^2)}{r^2p^2} + \dfrac{(l-r)^2}{r^2} = e^2,$ or, $\dfrac{l^2}{p^2} - \dfrac{l^2}{r^2} + \dfrac{l^2}{r^2} - \dfrac{2l}{r} + 1 - e^2 = 0;$

or, $\dfrac{l^2}{p^2} - \dfrac{2l}{r} + \dfrac{l}{a} = 0 \left(\because 1 - e^2 = \dfrac{b^2}{a^2} = \dfrac{l}{a}\right),$ or, $\dfrac{2a}{r} - 1 = \dfrac{al}{p^2} = \dfrac{b^2}{p^2}.$

Thus the required pedal equation is $\dfrac{b^2}{p^2} = \dfrac{2a}{r} - 1$.

Note: As an application of differential calculus the second method is the appropriate method

Example 15.11.7. *Find the pedal equation of the parabola $y^2 = 4a(x+a)$.*

[C.H. 1999]

SOLUTION: Tangent at (x, y) is
$$Y - y = \frac{2a}{y}(X - x), \quad \text{or,} \quad 2aX - Yy + y^2 - 2ax = 0.$$

Now, $\quad p = \dfrac{y^2 - 2ax}{\sqrt{y^2 + 4a^2}} = \dfrac{4a(x+a) - 2ax}{\sqrt{4a^2 + 4a(x+a)}} = \sqrt{a(x + 2a)}.$

Also $\quad r^2 = x^2 + y^2 = x^2 + 4a(x+a) = (x+2a)^2.$

$\therefore\ p^2 = ar$ is the required pedal equation.

Exercises XV (B)

1. Show that the following *pairs of curves* cut orthogonally:

   (a) The circles $r = 10\sin\theta$ and $r = 10\cos\theta$.

   (b) The cardiodes $r = a(1 - \cos\theta)$ and $r = a(1 + \cos\theta)$.

   (c) The parabolas $r = a\sec^2\frac{1}{2}\theta$ and $r = a\csc^2\frac{1}{2}\theta$.

   (d) The curves $r^2\theta = a^2$ and $r = e^{\theta^2}$.

   (e) The curves $r = a(1 + \sin\theta)$ and $r = a(1 - \sin\theta)$.

2. Show that the following curves intersect at the indicated angle.

   (a) $r = 6\cos\theta$ and $r = 2(1 + \cos\theta)$; at $\frac{1}{6}\pi$.

   (b) $r^2 = 16\sin 2\theta$ and $r^2\sin 2\theta = 4$; at $\frac{2}{3}\pi$.

   (c) $r^n = a^n\sec(n\theta + \alpha)$ and $r^n = b^n\sec(n\theta + \beta)$; at $|\alpha - \beta|$. [C.H. '64]

3. Express ϕ in terms of θ for the following polar curves and then verify the corresponding pedal equations (indicated on the right):

	Polar Equations	Pedal Equations
(a)	The *Cardiode*: $r = a(1 + \cos\theta)$;	$r^3 = 2ap^2$.
(b)	The *Equiangular spiral*: $r = ae^{\theta\cot\alpha}$;	$p = r\sin\alpha$.
(c)	The *Parabola*: $r = 2a/(1 + \cos\theta)$;	$p^2 = ar$. [C.H. 1980]
(d)	The *Equilateral hyperbola*: $r^2\cos 2\theta = a^2$;	$pr = a^2$.
(e)	The *Lemniscate*: $r^2 = a^2\cos 2\theta$;	$r^3 = a^2 p$.
(f)	The *Reciprocal spiral*: $r\theta = a$;	$p^2(a^2 + r^2) = a^2 r^2$.
(g)	The *Spiral*: $r = \operatorname{sech} n\theta$;	$1/p^2 = A/r^2 + B$.

4. Prove that the pedal equation of the curve
$$r^m = a^m \sin m\theta + b^m \cos m\theta \text{ is } r^{m+1} = p\sqrt{a^{2m} + b^{2m}}.$$

CHAPTER 15: TANGENTS AND NORMALS: ASSOCIATED CURVES

5. Eliminate x and y from $r^2 = x^2 + y^2$ and $p^2 = (xf_x + yf_y)^2/(f_x^2 + f_y^2)$ and obtain the pedal equations of the curves given by

(a) $f(x,y) \equiv \dfrac{x^2}{a^2} + \dfrac{y^2}{b^2} - 1 = 0$ [Ellipse with centre as origin]

(b) $f(x,y) \equiv y^2 - 4a(x+a) = 0$ [Parabola with vertex at $(-a,0)$]

[Ans: (a) $a^2b^2/p^2 = a^2 + b^2 - r^2$; (b) $p^2 = ar$]

6. Find the pedal equations of the curves whose parametric equations are given below:

(a) $x = a\cos^3\theta, y = a\sin^3\theta$ (Astroid). [C.H. 1974]

(b) $x = ae^\theta(\sin\theta - \cos\theta), y = ae^\theta(\sin\theta + \cos\theta)$

(c) $x = a(3\cos\theta - \cos^3\theta), y = a(3\sin\theta - \sin^3\theta)$.

[Ans: (a) $r^2 = a^2 - 3p^2$; (b) $r^2 = 2p^2$; (c) $(10a^2 - r^2)^2 = 3p^2(7a^2 - r^2)$]

7. Show that the pedal equation of the ellipse $x^2/a^2 + y^2/b^2 = 1$

(a) with respect to a focus as pole is $b^2/p^2 = 2a/r - 1$; [C.H. '67, '79, '83]

(b) with respect to the centre as pole is $a^2b^2/p^2 = a^2 + b^2 - r^2$. [C.H. '66]

8. Show that all the following curves have the same form of pedal equation, namely

$$\frac{1}{p^2} = \frac{A}{r^2} + B \ (A, B \text{ are constants}).$$

(a) $r = a\,\text{sech}\,n\theta$;

(b) $r = ae^{m\theta}$;

(c) $r\theta = a$;

(d) $r\sin n\theta = a$;

(e) $r\cosh\theta = a$.

9. An equation $f(p,r) = 0$ connecting p and r is deduced from $F(r,\theta) = 0$. Prove that if the equation $F(r,\theta) = 0$ is altered by writing r^n for r and $n\theta$ for θ then the equation $f(p,r) = 0$ must be changed by writing r^n for r and pr^{n-1} for p.

[Hints: As r changes to r^n, dr changes to $nr^{n-1}dr$; similarly $d\theta$ to $nd\theta$. Find how $\tan\phi = rd\theta/dr$ and $p = r\sin\phi$ change due to these changes in dr and $d\theta$.]

10. Using the relation $\psi = \theta + \phi$, prove that

$$\tan \phi = \frac{x\dfrac{dy}{dx} - y}{x + y\dfrac{dy}{dx}}.$$

[*Hints:* Draw a figure showing the angles.]

11. Verify:

   (a) The polar subtangent of the equiangular spiral $r = ae^{\theta \cot \alpha}$ is equal to $\tan \alpha$, and

   (b) The polar subnormal of the same spiral is $r \cot \alpha$.

12. Prove that for each of the curves

   a) $\theta - \cos^{-1} \dfrac{r}{k} = \sqrt{\dfrac{k^2 - r^2}{r^2}}$ (k is a constant)

   and b) $\dfrac{1}{r} = A\theta + B$

   the polar subtangent is constant.

13. For the curve $r = ae^{b\theta^2}$ (a, b are constants)

$$\frac{\text{polar sub-normal}}{\text{polar sub-tangent}} \text{ varies as } \theta^2.$$

14. Prove that the locus of the extremity of the polar subnormal of the curve $r = f(\theta)$ is $r = f'\left(\theta - \dfrac{1}{2}\pi\right)$.

   Show that this locus is an *an equiangular spiral* if the curve is an equiangular spiral $r = ae^{\theta \cot \alpha}$. [C.H. '63, '74, '79]

15. Prove that the locus of the extremity of the polar subtangent of the curve $u + f(\theta) = 0$ is $u = f'(\pi/2 + \theta)$, where $u = 1/r$. Hence show that the loci of the extremity of the polar subtangents of the curves

   (a) $r = \dfrac{1 + \tan \theta/2}{m + n \tan \theta/2}$; and

   (b) $r = ae^{\theta \cot \alpha}$.

   are respectively. (a) a cardiode and (b) an equiangular spiral. [C.H. 1974]

   [*See worked example Ex.* **15.11.4**]

16. Prove that the normal at any point (r, θ) to the curve $r^n = a^n \cos n\theta$ makes an angle $(n+1)\theta$ with the initial line.

   [*Hints:* The normal makes $\theta + \phi - \frac{1}{2}\pi$ with the initial line; here $\phi = \pi/2 + n\theta$; hence etc.]

17. Show that the tangents drawn at the extremities of any chord of the cardiode $r = a(1 + \cos \theta)$ which passes through the pole are perpendicular to each other.

18. The tangent at any point of the cardiode $r = a(1 + \cos \theta)$ whose vectorial angle is 2α meets the curve again at a point whose vectorial angle is 2β. Show that

$$\cos(2\beta - \alpha) + 2\cos\alpha = 0.$$

   [*See worked example Ex.* **15.11.5**.]

Section II: Associates Curves

15.12 Pedals, Inverses and Reciprocal Polars

(A) Pedal curves

Definition 15.12.1. *The pedal of a given curve, with respect to a fixed point, is the locus of the foot of the perpendicular drawn from that fixed point upon the tangent to the curve.*

In elementary co-ordinate geometry it has been proved that

1. the *pedal of a parabola* with respect to the focus is the TANGENT AT THE VERTEX.

2. the pedal of an ellipse or a hyperbola with respect to either focus is the AUXILIARY CIRCLE.

Equations of the pedal curves.

(a) Given a Curve in Cartesian Co-ordinates

Let ON be the perpendicular drawn from the fixed point O (w.r.t. which the pedal is desired). Suppose $ON = p$ and let α be the angle which ON makes with any fixed line through O (Fig. 15.10) then (p, α) may be taken as the polar co-ordinates of N w.r.t. O as pole and OX as the initial line.

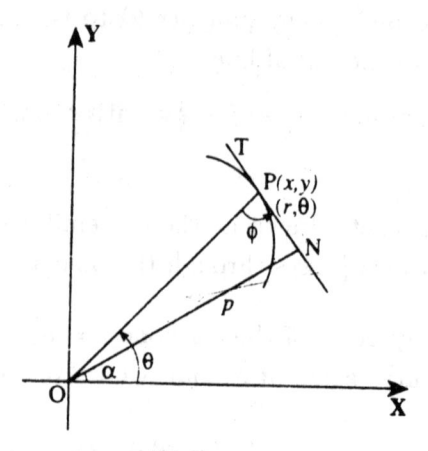

Fig. 15.10

Let $x\cos\alpha + y\sin\alpha = p$ be the equation of the tangent PT at a point P of the curve $f(x,y) = 0$. Then, we proceed to find the condition that the line should touch the curve in the form of a relation between p and α. In this relation we consider (p, α) as the current co-ordinates or we replace them by (r, θ) then the (polar) equation of the pedal curve can at once be obtained. Once the polar equation of the pedal is known we can transform it into cartesian.

Example 15.12.1. *Find the pedal of the conic* $\dfrac{x^2}{a^2} \pm \dfrac{y^2}{b^2} = 1$ *w.r.t. origin.*

Solution: We know that the line $x\cos\alpha + y\sin\alpha = p$ touches the conic if $p^2 = a^2\cos^2\alpha \pm b^2\sin^2\alpha$. (p, α) is the foot of the perpendicular from origin on $x\cos\alpha + y\sin\alpha = p$.

∴ The locus of (p, α), which is the pedal of the given conic, is $r^2 = a^2\cos^2\theta \pm b^2\sin^2\theta$ w.r.t. origin as fixed point.

Writing $r^2 = x^2 + y^2$; $r\sin\theta = y$; $r\cos\theta = x$, we get the cartesian equation of the pedal in the form $(x^2 + y^2)^2 = a^2 x^2 \pm b^2 y^2$.   [C.H. '84]

Note that in case of a rectangular hyperbola $x^2 - y^2 = a^2$ the pedal is the lemniscate $r^2 = a^2\cos 2\theta$ or $(x^2 + y^2)^2 = a^2(x^2 - y^2)$.

Otherwise.

The equation of the tangent at a point (x, y) of the curve $f(x, y) = 0$ is

$$(X - x)f_x + (Y - y)f_y = 0. \qquad (15.12.1)$$

The equation of the perpendicular ON drawn from the origin to the above tangent is

$$Xf_y - Yf_x = 0. \qquad (15.12.2)$$

CHAPTER 15: TANGENTS AND NORMALS: ASSOCIATED CURVES

Hence the locus of N (i.e., *the pedal*) can be obtained by eliminating x and y from (15.12.1), (15.12.2) and the equation of the curve $f(x,y) = 0$.

Thus, in the previous illustration we may eliminate (x,y) from

$$\frac{x^2}{a^2} \pm \frac{y^2}{b^2} = 1; \quad (X-x)\frac{x}{a^2} \pm (Y-y)\frac{y}{b^2} = 0; \quad X\left(\pm\frac{y}{b^2}\right) - Y\left(\frac{x}{a^2}\right) = 0$$

and obtain the results deduced before.

(b) Given a Curve in Polar Co-ordinates

Let the equation of the curve be

$$f(r,\theta) = 0. \tag{15.12.3}$$

Suppose the co-ordinates of N, the foot of the perpendicular drawn from O (fixed point) on the tangent at $P(r,\theta)$, be (p,α). Then from Fig. 15.10.

$$\phi + \theta - \alpha = \pi/2. \tag{15.12.4}$$

Also we have

$$\tan\phi = r d\theta/dr. \tag{15.12.5}$$

And since

$$ON = OP\sin\phi, \quad p = r\sin\phi. \tag{15.12.6}$$

Eliminate r, θ, ϕ from (15.12.3), (15.12.4), (15.12.5) and (15.12.6) and obtain an equation in (p, α). The eliminant is the required pedal with current co-ordinates (p, α) which may be changed to (r, θ) in order to get a familiar form.

Example 15.12.2. *To find the pedal of the cardiode*

$$r = a(1 + \cos\theta). \tag{15.12.7}$$

Solution: We write

$$\phi + \theta - \alpha = \pi/2. \tag{15.12.8}$$

$$\left.\begin{array}{c} \tan\phi = r\dfrac{d\theta}{dr} = -\cot\dfrac{\theta}{2} \\ \text{and hence} \quad \phi = \dfrac{\pi}{2} + \dfrac{\theta}{2} \end{array}\right\} \tag{15.12.9}$$

$$p = r\sin\phi. \tag{15.12.10}$$

From (15.12.8) and (15.12.9) we obtain $\theta = \dfrac{2\alpha}{3}$.

From (15.12.7), (15.12.9), (15.12.10), we obtain

$$p = a(1+\cos\theta)\cos\frac{\theta}{2} = 2a\cos^3\frac{\alpha}{3}.$$

Changing (p, α) to (r, θ) form, $r = 2a\cos^3\theta/3$ is the required pedal.

An Important Theorem

The angle which the tangent makes with the radius vector at corresponding points is the same for a curve and its pedal.

In Fig. 15.11, let ON, ON' be the perpendiculars from O on the tangents at P, Q, the two neighbouring points of a given curve. Let these two tangents intersect at T and let OU be the perpendicular to NN' (produced).

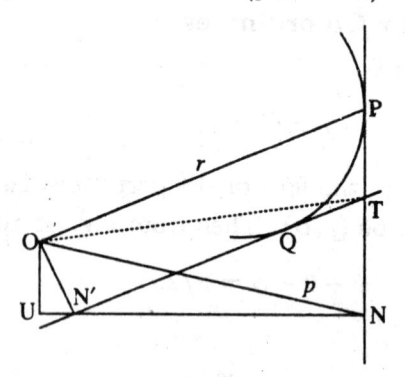

Fig. 15.11

Since the angles ONT and $ON'T$ are right angles, the points O, N', N, T lie on the same circle and hence the angles OTN' and ONN' are equal. Now if $Q \to P$ along the given curve $N' \to N$ along the pedal curve and in the limit NN' becomes the tangent to the pedal and the angles OTN' and ONN' become respectively the angles between the tangent and radius vector for the curve and for the pedal. Hence our proposition follows.

Deduction 1.

How to find the pedal of a curve whose pedal equation is given?

Consider the similar triangles ONU and OTN'. Then

$$OU : ON = ON' : OT.$$

Hence if r be the radius vector of the original curve, p be the perpendicular from O on the tangent to the curve at P and p' be the perpendicular on the tangent at N to the pedal, we have ultimately,

$$p' : p = p : r \text{ i.e., } p' = p^2/r.$$

This gives a method of finding the pedal of a curve whose pedal equation is given.

CHAPTER 15: TANGENTS AND NORMALS: ASSOCIATED CURVES   971

If the given curve is $r = f(p)$, then the pedal curve is $p^2/p' = f(p)$, where p and p' are analogous to r and p in the given curve.

Deduction 2.

How to find the negative pedal?

From Fig. 15.10 we may easily obtain

$$r\sin(\theta - \alpha) = PN, \quad \text{also,} \quad p = ON = r\cos(\theta - \alpha).$$

$$\therefore \frac{dp}{d\alpha} = \cos(\theta - \alpha)\frac{dr}{d\theta} \cdot \frac{d\theta}{d\alpha} - r\sin(\theta - \alpha)\left(\frac{d\theta}{d\alpha} - 1\right) = r\sin(\theta - \alpha) = PN.$$

$$\left[\text{from (15.12.5)} \; \sin\phi = r\frac{d\theta}{dr} \cdot \cos\phi \text{ and then use (15.12.4)}\right]$$

This result enables us to solve the problem of *negative pedals i.e., to find the curve having a given pedal*. We again refer to Fig. 15.10. The co-ordinates (x, y) of P are given by

$$\left.\begin{array}{l} x = ON\cos\alpha - NP\sin\alpha = p\cos\alpha - dp/d\alpha \sin\alpha \\ y = ON\sin\alpha + NP\cos\alpha = p\sin\alpha + dp/d\alpha \cos\alpha \end{array}\right\}.$$

Example 15.12.3. *To find the curve whose pedal is the cardiode:*

$$r = a(1 + \cos\theta).$$

Solution: We consider the current co-ordinates of the pedal as (p, α) instead of (r, θ). Thus, the given pedal is

$$p = a(1 + \cos\alpha).$$

Hence by using the previous result, the parametric equation of the original curve is

$$x = a\cos\alpha + a, \quad y = a\sin\alpha$$

whence, $(x - a)^2 + y^2 = a^2$; a circle of radius a, centre $(a, 0)$.

Definition 15.12.2. *If there be a sequence of curves C, C_1, C_2, C_3, C_4, etc., such that each is the pedal of the one which immediately precedes it, then C_1 is the first positive pedal; C_2 is the second positive pedal and so on, of the curve C. Again if C_4 is the given curve, C_3 is its first negative pedal, C_2 is its second negative pedal and so on.*

Example 15.12.4. (a) *Find the kth positive pedal of the curve*

$$r^n = a^n \cos n\theta. \tag{15.12.11}$$

Solution: We first deduce the equation of the *first positive pedal*. For that, we deduce

$$\phi = \frac{\pi}{2} + n\theta. \tag{15.12.12}$$

Eliminate r, θ, ϕ from (15.12.11), (15.12.12) and

$$p = r \sin \phi \tag{15.12.13}$$
$$\phi + \theta - \alpha = \frac{\pi}{2}. \tag{15.12.14}$$

Thus, $\alpha = (n+1)\theta$;

$$p = r \cos n\theta = a \cos^{1/n} n\theta \cos n\theta$$
$$= a \cos^{\frac{n+1}{n}}\left(\frac{n\alpha}{n+1}\right).$$

$\therefore$ the equation of the first positive pedal is

$$r^{n/(n+1)} = a^{n/(n+1)} \cos \frac{n\theta}{n+1}$$
$$\text{or,} \quad r^{n_1} = a^{n_1} \cos \theta n_1, \text{ where } n_1 = n/(n+1).$$

Similarly, the second positive pedal is

$$r^{n_2} = a^{n_2} \cos \theta n_2,$$

where $n_2 = \dfrac{n_1}{1+n_1} = \dfrac{n}{1+2n}$, and, in general, kth positive pedal is

$$r^{n_k} = a^{n_k} \cos \theta n_k, \quad \text{where } n_k = \frac{n}{1+kn}.$$

(b) *Find the kth negative pedal of the curve $r^n = a^n \cos n\theta$.*

Solution: We have observed that $r^m = a^m \cos m\theta$ is the kth positive pedal of

$$r^n = a^n \cos n\theta, \text{ where } m = \frac{n}{1+kn}, \text{ whence } n = \frac{m}{1-km}.$$

$\therefore$ kth negative pedal of $r^n = a^n \cos n\theta$ is $r^m = a^m \cos m\theta$ where

$$m = \frac{n}{1-kn}.$$

CHAPTER 15: TANGENTS AND NORMALS: ASSOCIATED CURVES

(B) Inverse of a curve

Definition 15.12.3. *If from a fixed origin O, we draw a radius vector OP to any given curve and in OP we take a point P' such that*

$$OP \cdot OP' = k^2, \qquad (15.12.15)$$

where k is a given constant, the point P' is called the inverse of the point P; the locus of P' is called the inverse of the curve traced out by P. The point O is called the **centre of inversion** *and k^2 is called the constant of inversion.*

Theorem 15.12.1. *A curve and its inverse make supplementary angles with the radius vector i.e., tangents at P of a curve and at P' (its inverse) of the inverse curve make supplementary angles with the radius vector OPP'.*

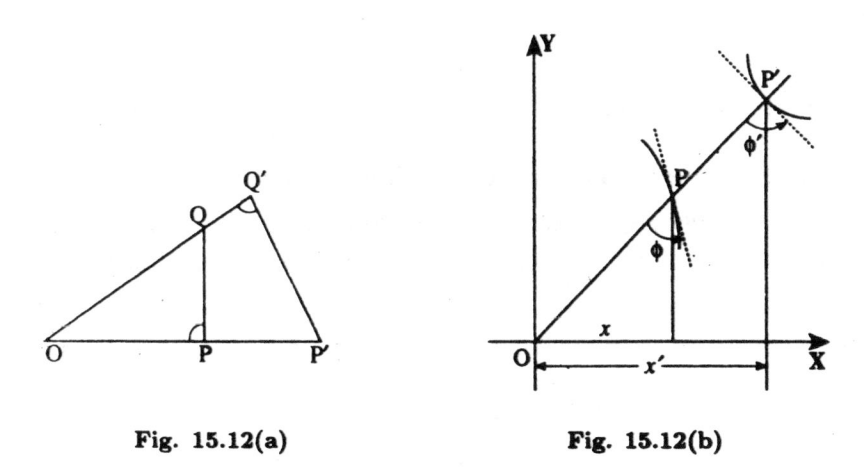

Fig. 15.12(a) **Fig. 15.12(b)**

Proof: If P, Q be the consecutive points of a curve, [Fig. 15.12(a)] and P', Q' be their respective inverses then we have

$$OP \cdot OP' = OQ \cdot OQ' = k^2$$

i.e., $$\frac{OP}{OQ} = \frac{OQ'}{OP'}.$$

Hence the triangles $POQ, Q'OP'$ are similar and the angles OPQ and $OQ'P'$ are equal. A little consideration of the Fig. 15.12(a), will show that the angles OPQ and $OP'Q'$ are supplementary. In the limit when $Q \to P$ along the curve and consequently $Q' \to P'$ along the inverse curve, these are the angles which the tangents at P and P' make with the radius vector. Hence the theorem.

Otherwise.

Let the polar co-ordinates of P and P' be (r, θ) and (r', θ) where O is the pole of the initial line [Fig. 15.12(b)]; let ϕ and ϕ' be the angles made by the radii vectores with the tangents at these points. Then we have

$$\tan \phi = r \frac{d\theta}{dr}; \quad \tan \phi' = r' \frac{d\theta}{dr'}.$$

Since, by (15.12.15), $r \cdot r' = k^2$, we have $\dfrac{dr'}{d\theta} = \dfrac{d}{d\theta}(k^2/r) = -\dfrac{k^2}{r^2}\dfrac{dr}{d\theta}$ which leads to $\tan \phi' = -\tan \phi = \tan(\pi - \phi)$. Hence the proposition.

Corollary 15.12.1. *If two curves intersect, the respective inverse curves will intersect at the same angle. In particular,* **orthogonal curves invert into orthogonal curves.**

Equation of an inverse curve

(A) Given a curve in Rectangular Cartesian Co-ordinates

Let (x, y) be the rectangular cartesian co-ordinates of P on the given curve and (x', y') be the co-ordinates of P' (referred to rectangular axes through the centre of inversion O) so that $OP \cdot OP' = k^2$.

Then, from the property of similar triangles [Fig. 15.12(b)] it will follow

$$\frac{x}{x'} = \frac{OP}{OP'} = \frac{OP \cdot OP'}{OP'^2} = \frac{k^2}{x'^2 + y'^2}.$$

$$\therefore \quad x = \frac{k^2 x'}{x'^2 + y'^2}; \quad \text{similarly } y = \frac{k^2 y'}{x'^2 + y'^2}.$$

Thus the equation of the inverse curve is obtained by substituting

$$\frac{k^2 x'}{x'^2 + y'^2} \quad \text{and} \quad \frac{k^2 y'}{x'^2 + y'^2}$$

for x and y in the equation of the given curve. The inverse curve will be thus obtained with current co-ordinates (x', y'). Dropping the dashes we obtain the inverse curve with current co-ordinates (x, y).

Example 15.12.5. *Find the inverse of the circle $x^2 + y^2 = a^2$, the centre of inversion being the origin.*

CHAPTER 15: TANGENTS AND NORMALS: ASSOCIATED CURVES

Solution: It is easy to see that the inverse curve is

$$\left(\frac{k^2 x}{x^2+y^2}\right)^2 + \left(\frac{k^2 y}{x^2+y^2}\right)^2 = a^2, \quad \text{or,} \quad x^2+y^2 = \frac{k^4}{a^2}$$

which represents another circle.

Note 15.12.1. *The inverse of a circle is a circle, except in the particular case where the centre of inversion is on the circumference, when the inverse locus is a straight line.*

(B) Given a Curve in Polar Co-ordinates

If (r, θ) be the co-ordinates of P and (r', θ), that of P' with reference to the centre of inversion as the pole then obviously

$$r \cdot r' = k^2 \quad \text{or,} \quad r = \frac{k^2}{r'}.$$

Thus if $f(r, \theta) = 0$ be the given curve, $f\left(\dfrac{k^2}{r'}, \theta\right) = 0$ is the inverse curve. Dropping the dash, $f\left(\dfrac{k^2}{r}, \theta\right) = 0$ is the inverse curve, where (r, θ) are the current co-ordinates of the inverse curve.

Example 15.12.6. *For the equiangular spiral* $r = ae^{\theta \cot \alpha}$, *the inverse curve is*

$$\frac{k^2}{r} = ae^{\theta \cot \alpha} \quad \text{or,} \quad r = ce^{-\theta \cot \alpha} \quad \text{where } c = \frac{k^2}{a}.$$

(C) Reciprocal Polars

Definition 15.12.4. *The locus of the pole of the tangent to a curve* Γ, *with respect to a fixed conic* Σ, *is called the reciprocal polar of* Γ.

We shall only consider the case where the fixed conic is a circle.

Consider the fixed conic Σ to be a circle with centre O and radius k. It is given in books of Geometry that the pole P' of any line PT, with respect to Σ, can be constructed by drawing a perpendicular ON to this line PT and then taking on ON the point P' such that $ON \cdot ON' = k^2$.

We only notice here that in obtaining the reciprocal polar of the curve Γ with respect to the circle Σ, the lines PT should be the tangents at different points on Γ. Hence it follows that the reciprocal polar is in this case the *inverse of the pedal of the given curve* Γ. The equation of the polar reciprocal of a curve can be obtained by first finding the pedal of the curve and then its inverse.

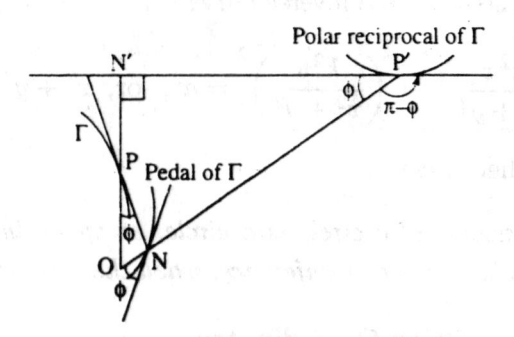

Fig. 15.13

An interesting consequence is that the original curve Γ is the inverse of the pedal of the polar reciprocal i.e., *the polar reciprocal of the polar reciprocal of a curve is the curve itself.* For if P be the point of contact of the tangent to the original curve Γ and if OP meets the tangent at P' of the polar reciprocal in N', the angles OPN and $OP'N'$ will be equal. [Use the Theorems under (A) and (B) and compare Fig. 15.13].

Hence $ON'P'$ is a right angle and N' traces the pedal of P' and since $PNP'N'$ are concyclic, we have

$$OP \cdot ON' = ON \cdot OP' = k^2$$

i.e., P describes the inverse of the locus of N'. Hence the result.

Example 15.12.7. *Prove that the reciprocal polar of a circle with respect to any origin is a conic having the origin as focus.*

Solution: For a circle of radius a, the pole O being at a distance c from the centre C and the line OC being the initial line for measuring α, we have at once from a figure, (the reader may easily draw the figure.)

$$p = a + c\cos\psi.$$

Hence the pedal is the *limacon:* $r = a + c\cos\theta$.

[It O be on the circumference, we have $c = a$ and the pedal is the cardiode $r = a(1 + \cos\theta)$.]

Replacing r by k^2/r we get the equation of the polar reciprocal in the form

$$k^2/r = a + c\cos\theta$$

which represents a conic, having its focus at the origin, of eccentricity c/a. Hence the conic is an ellipse, parabola or hyperbola according as the origin is inside, on, or outside the circle.

CHAPTER 15: TANGENTS AND NORMALS: ASSOCIATED CURVES   977

Example 15.12.8. *The pedal of the ellipse* $\dfrac{x^2}{a^2} + \dfrac{y^2}{b^2} = 1$ *with respect to the centre is given by* $r^2 = a^2 \cos^2\theta + b^2 \sin^2\theta$. *Hence the polar reciprocal is*

$$\frac{k^4}{r^2} = a^2 \cos^2\theta + b^2 \sin^2\theta, \quad or, \quad a^2 x^2 + b^2 y^2 = k^4$$

which represents a concentric ellipse.

Exercises XV (C)

(Hints are given below)

1. Verify the following statements:

   (a) The inverse of an equiangular spiral w.r. to the pole is an equal spiral.

   (b) The inverse of a rectangular hyperbola w.r. to the centre is a *lemniscate* and conversely.

   (c) The inverse of a straight line is a circle through the pole of inversion and conversely.

   (d) The inverse of a circle is another circle.

   (e) The inverse of a parabola w.r. to the focus is a *cardiode*.

   (f) The inverse of any conic
   $$r = \frac{l}{1 + e\cos\theta}$$
   w.r. to the focus is a *limacon of the form* $r = a + b\cos\theta$.

   (g) The inverse of the ellipse $\dfrac{x^2}{a^2} + \dfrac{y^2}{b^2} = 1$ w.r. to the centre is the curve
   $$(x^2 + y^2)^2 = k^4\left(\frac{x^2}{a^2} + y^2 b^2\right).$$

   (h) The inverse of a curve $p = f(r)$ is given by
   $$p = \frac{r^2}{k^2} f\left(\frac{k^2}{r}\right).$$

2. Obtain the pedals of the following curves in the form indicated:

   (a) $y^2 = 4ax$: w.r. to the focus: Pedal: $x = 0$.

   (b) $y^2 = 4ax$: w.r. to the vertex: Pedal: $x(x^2 + y^2) + ay^2 = 0$. [C.H. '65,'80]

   (c) $\sqrt{\dfrac{x}{a}} + \sqrt{\dfrac{y}{b}} = 1$; w.r. to the origin; Pedal: $(x^2 + y^2)(ax + by) = abxy$.

(d) $x^2 - y^2 = a^2$; w.r. to the centre:
Pedal: $r^2 = a^2 \cos 2\theta$ or $(x^2 + y^2)^2 = a^2(x^2 - y^2)$. [C.H. '81]

(e) $r^m = a^m \cos m\theta$; w.r. to the pole; Pedal: $r^n = a^n \cos n\theta [n = (m/m+1)]$.

(f) Circle $r = 2a \cos\theta$; w.r.to the origin; Pedal: $r = a(1 + \cos\theta)$. [C.H. '70]

3. (a) Prove that the *negative pedal* of the parabola $y^2 = 4ax$ w.r. to the vertex is the curve $27ay^2 = (x - 4a)^3$.

(b) Prove that the curve for which $p = a \sin\alpha \cos\alpha$ is the *astroid* $x^{2/3} + y^{2/3} = a^{2/3}$.

4. Satisfy yourself with the following assertions:

(a) The polar reciprocal of $p = f(r)$ is given by $k^2/r = f(k^2/p)$.

(b) The polar reciprocal of $y^2 = 4ax$ w.r. to the vertex is another parabola.

(c) The polar reciprocal of $r^m = a^m \cos m\theta$ w.r. to the circle of radius k and centre at the origin is

$$r^n \cos n\theta = k^2/a, \text{ where } n = m/(m+1).$$

Hints and Answers [Exercises XV (C)]

1. (a) Worked Ex. 15.12.6.

(b) Rectangular hyperbola $r^2 \cos 2\theta = a^2$; replace r by k^2/r and obtain $r^2 = (k^4/a^2) \cos 2\theta$ (a *lemniscate*).

(c) Line: $ax + by + c = 0$; Pole $(0,0)$. Write $k^2x/(x^2+y^2)$ and $k^2y/(x^2+y^2)$ for x and y respectively and obtain $c(x^2+y^2) + ak^2x + bk^2y = 0$ (a *circle through the pole*).

(d) Take the circle $x^2 + y^2 = a^2$ and proceed as in (c).

(e) Parabola: $r = l/(1 + \cos\theta)$; replace r by k^2/r.

(f) similar to (e).

(g) Write $k^2x/(x^2+y^2)$ and $k^2y/(x^2+y^2)$ for x and y respectively.

(h) Denoting p, r, ϕ of the inverse by p', r', ϕ' respectively, we have

$$r \cdot r' = k^2. \text{ Also } p'/r' = \sin\phi' = \sin(\pi - \phi) = \sin\phi = p/r.$$

Thus $p = (r/r')p' = (k^2/r'^2)p'$. Now complete.

CHAPTER 15: TANGENTS AND NORMALS: ASSOCIATED CURVES   979

2. (a) Any tangent to $y^2 = 4ax$ is $y = mx + a/m$. Eqn. to the perpendicular on it from focus $(a, 0)$ is $y = -(1/m)(x - a)$. Now eliminate m.

(b) Eliminate m from $y = mx + a/m$ and $y = -(1/m)x$ (Why?).

(c) If $x \cos\alpha + y \sin\alpha = p$ touches $(x/a)^{1/2} + (y/b)^{1/2} = 1$, then
$$(a\cos\alpha)^{-1} + (b\sin\alpha)^{-1} = p^{-1}.$$

Replace p, α by r, θ and then use $x = r\cos\theta$, $y = r\sin\theta$.

(d) Change to polar: $r^2 \cos 2\theta = a^2$ and obtain $\phi = \pi/2 - 2\theta$, whereby $p = r\sin\phi = r\cos 2\theta$. Taking (r_1, θ_1) the polar co-ordinates of the perpendicular from the pole on the tangent, $r_1 = p = r\cos 2\theta$.

Also
$$\theta_1 = \theta - (\pi/2 - \phi) = \theta - 2\theta = -\theta.$$

Now obtain $r_1^2 = a^2 \cos 2\theta_1$ and hence the pedal is $r^2 = a^2 \cos 2\theta$. Transforming into cartesian the pedal may be written as $(x^2 + y^2)^2 = a^2(x^2 - y^2)$.

(e) and (f) Similar to (d).

3. (a) Change to polar: $r = 4a \cot\theta \csc\theta$. Change (r, θ) to (p, α). Apply the result to Deduction 2. (art. **15.12**).

(b) Similar to (a).

4. (a) Writing p_1, r_1 for the pedal curve of $p = f(r)$, we have $p = r_1$ and $r = p^2/p_1 = r_1^2/p_1$. Hence pedal is $r_1 = f(r_1^2/p_1)$ or $r = f(r^2/p)$.

Next to obtain its inverse we write k^2/r' for r and $(k^2/r'^2)p'$ for p whereby,
$$\frac{k^2}{r'} = f\{(k^2/r')^2/(k^2/r'^2)p'\}.$$

Simplifying, the pedal equation of the polar reciprocal is
$$k^2/r = f(k^2/p).$$

(b) Pedal of $y^2 = 4ax$ is $x(x^2 + y^2) + ay^2 = 0$. Find its inverse.

(c) Pedal of $r^m = a^m \cos m\theta$ is $r^n = a^n \cos n\theta$ where $n = m/(m+1)$, the inverse of which is $(k^2/r)^n = a^n \cos n\theta$; hence etc.

Section III: Arc-length and its Differentials

15.13 The Differential of Arc Length

The calculation of the length of a curve requires *the process of integration*, which we are prepared to discuss in a very brief manner.

Let us begin with a curve $= f(x)$ in Fig. 15.14, where $f(x)$ is a *differentiable function* of x. Consider s to be the arc length measured from a fixed point A on the curve. Let P and Q be two neighbouring points on the curve with the co-ordinates (x, y) and $(x + \Delta x, y + \Delta y)$. Also let Δs be the length of the arc $\widehat{PQ}$ and let Δc be the length of the chord $\overline{PQ}$. Then, we shall prove that,

(A) $$\lim_{\overline{PQ} \to 0} \frac{\widehat{PQ}}{\overline{PQ}} = 1,$$

i.e., *for continuous curves having continuously turning tangents, the limit of the ratio of an infinitesimal arc to its subtended chord is one.*

By the length of an arc of a curve we *mean* the following: Let the given arc AB of a curve $y = f(x)$ be divided into n parts by points $P_1, P_2, \cdots, P_{r-1}, P_r, \cdots, P_{n-1}$ as shown in Fig. 15.15 and let chords be drawn joining consecutive points on this set. Then *the limit of the sum of the lengths of the lines*

$$AP_1 + P_1P_2 + \cdots + P_{r-1}P_r + \cdots + P_{n-1}B$$

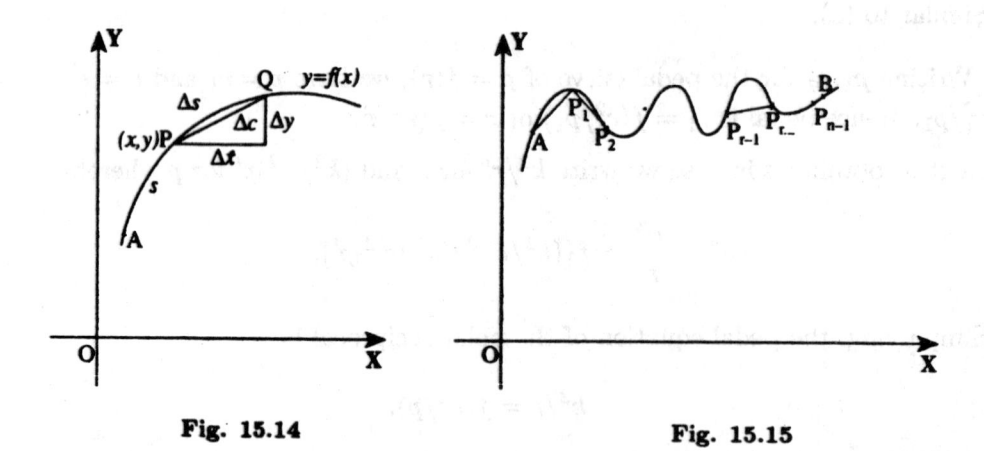

Fig. 15.14   Fig. 15.15

as n *becomes infinite in such a way that the length of each chord approaches zero is, by definition, the length of the given arc* AB. Moreover, since $f(x)$ and its first derivative $f'(x)$ are everywhere continuous, we can prove from Integral Calculus[1] that, the entire

---

[1] See Authors' *Integral Calculus (An Introduction to Analysis)*, art 17.1 (8th Edition 1996).

CHAPTER 15: TANGENTS AND NORMALS: ASSOCIATED CURVES

length (s) of the arc AB is given by

$$s = \int_a^b \sqrt{1 + \{f'(x)\}^2}\,dx,$$

where a and b are the abscissae of A and B respectively.

Applying this to the infinitesimal arc PQ (Fig. 15.14) between x and $x + \Delta x$,

$$\widehat{PQ} = \int_x^{x+\Delta x} \sqrt{1 + \{f'(x)\}^2}\,dx.$$

By the MVT of Integral Calculus,

$$\widehat{PQ} = \sqrt{1 + \{f'(\xi)\}^2}\,\Delta x \text{ for } x < \xi < x + \Delta x.$$

Again the length of the chord PQ is

$$\overline{PQ} = \sqrt{(\Delta x)^2 + (\Delta y)^2} = \sqrt{1 + \left(\frac{\Delta y}{\Delta x}\right)^2}\,\Delta x$$

which, by the MVT of Differential Calculus, becomes

$$\overline{PQ} = \sqrt{1 + \{f'(\eta)\}^2}\,\Delta x \text{ for } x < \eta < x + \Delta x.$$

Hence, the ratio of a general infinitesimal arc to its subtended chord is

$$\frac{\widehat{PQ}}{\overline{PQ}} = \frac{\sqrt{1 + \{f'(\xi)\}^2}}{\sqrt{1 + \{f'(\eta)\}^2}}\,\frac{\Delta x}{\Delta x} \text{ for } x < \xi, \eta < x + \Delta x.$$

Now as $\Delta x \to 0$, both ξ and η approach x. Hence

$$\lim_{\Delta x \to 0} \frac{\Delta s}{\Delta c} \to 1 \text{ or, } \lim_{\overline{PQ} \to 0} \frac{\widehat{PQ}}{\overline{PQ}} = 1.$$

(B) Rectangular Cartesian Co-ordinates

From Fig. 15.14, we have

$$(\Delta c)^2 = (\Delta x)^2 + (\Delta y)^2 \text{ or, } (\Delta c/\Delta x)^2 = 1 + (\Delta y/\Delta x)^2$$

which after the multiplication of each side by $(\Delta s/\Delta c)^2$, becomes

$$\left(\frac{\Delta s}{\Delta x}\right)^2 = \left(\frac{\Delta s}{\Delta c}\right)^2 \left\{1 + \left(\frac{\Delta y}{\Delta x}\right)^2\right\}.$$

Let now $\Delta x \to 0$, then by the definition of a derivative and using the result (A), it follows that

$$\left(\frac{ds}{dx}\right)^2 = 1 + \left(\frac{dy}{dx}\right)^2 \quad \text{or,} \quad \frac{ds}{dx} = \pm\sqrt{1 + \left(\frac{dy}{dx}\right)^2}.$$

The *plus* or *minus* sign is to be taken according as the direction in which s increases along the curve is the same as, or opposite to, the direction in which x increases. But we shall assume that s is measured in the direction in which x increases, so that we have the formula for the *differential of arc* as

$$ds = \sqrt{1 + (dy/dx)^2}\,dx \quad \text{or,} \quad ds = \sqrt{1 + (dx/dy)^2}\,dy.$$

Remember: $ds^2 = dx^2 + dy^2$.

Considering the right-triangle in Fig. 15.14, we may easily obtain,

$$\sin\psi = dy/ds, \quad \cos\psi = dx/ds, \quad \tan\psi = dy/dx$$

where ψ has the usual significance.

(C) Polar Co-ordinates

Since $x = r\cos\theta$, $y = r\sin\theta$ the differential dx and dy are given by

$$dx = \cos\theta\,dr - r\sin\theta\,d\theta, \quad dy = \sin\theta\,dr + r\cos\theta\,d\theta;$$

and therefore,

$$ds^2 = dx^2 + dy^2 = (dr)^2 + r^2(d\theta)^2$$

(on simplification) which may also be written in the form

$$ds = \sqrt{r^2 + (dr/d\theta)^2}\,d\theta \quad \text{or,} \quad ds = \sqrt{1 + r^2(d\theta/dr)^2}\,dr.$$

Anternative Method (See Fig. 15.5)

If s be the arc length measured from a fixed point A on the curve $r = f(\theta)$ and P, Q be two points on it with co-ordinates (r, θ) and $(r + \Delta r, \theta + \Delta\theta)$, then if arc $PQ = \Delta s$ and chord $PQ = \Delta c$,

$$\Delta c^2 = PR^2 + RQ^2 = (r\sin\Delta\theta)^2 + (r + \Delta r - r\cos\Delta\theta)^2$$

whence

$$\left(\frac{\Delta c}{\Delta s} \cdot \frac{\Delta s}{\Delta\theta}\right)^2 = r^2\left(\frac{\sin\Delta\theta}{\Delta\theta}\right)^2 + \left\{r \cdot \frac{1 - \cos\Delta\theta}{\Delta\theta} + \frac{\Delta r}{\Delta\theta}\right\}^2.$$

And as $\Delta\theta \to 0$, $(ds/d\theta)^2 = r^2 + (dr/d\theta)^2$.

From the triangle PQR and applying limit as in art. 15.8 we may obtain

$$\sin\phi = r\,d\theta/ds, \quad \cos\phi = dr/ds, \quad \tan\phi = r\,d\theta/dr.$$

Note 15.13.1. *For parametric form $x = \phi(t), y = \psi(t)$, from (B), we may obtain*

$$\left(\frac{ds}{dt}\right)^2 = \left(\frac{dx}{dt}\right)^2 + \left(\frac{dy}{dt}\right)^2.$$

Main Points : A Recapitulation

Section I. Tangents and Normals

1. *Curves given in Cartesian form*:

    (A) $y = f(x)$: Eqn. of the tangent at (x, y) : $Y - y = f'(x)(X - x)$.

    Eqn. of the normal at (x, y) : $Y - y = \{-1/f'(x)\}(X - x)$.

    (B) $f(x, y) = 0$: Tangent at (x, y) : $(X - x)f_x + (Y - y)f_y = 0$.

    Normal at (x, y) : $(X - x)f_y - (Y - y)f_x = 0$.

    (C) Parametric form : $x = \phi(t), y = \psi(t)$. Use $f'(x) = \psi'(t)/\phi'(t)$ in (A).

2. Cartesian Subtangent $= y\big/\dfrac{dy}{dx}$ and Cartesian Subnormal $= y \cdot \dfrac{dy}{dx}$ [See Fig. 15.1)].

    Length of tangent $= PT = y \operatorname{cosec} \psi$ and Length of normal $= PG = y \sec \psi$.

3. The angle between two curves = the angle between their tangents at the point of intersection $= \tan^{-1} \dfrac{f_x F_y - f_y F_x}{f_x F_x + f_y F_y}$ if the curves are $f(x, y) = 0$ and $F(x, y) = 0$.

    Conditions: For orthogonality $f_x F_x + f_y F_y = 0$; for touching $f_x/F_x = f_y/F_y$.

4. A rational algebraic curve passing through the origin:

    For tangents at the origin *equate to zero the lowest degree terms*.

5. *Curves given in Polar form*: $r = 1/f(\theta)$ or $u = f(\theta)$.

    Tangent at $\theta = \alpha$: $u = f(\alpha)\cos(\theta - \alpha) + f(\alpha)\sin(\theta - \alpha)$.

    Normal at $\theta = \alpha$: $\dfrac{f'(\alpha)}{f(\alpha)} u = f'(\alpha)\cos(\theta - \alpha) - f(\alpha)\sin(\theta - \alpha)$.

6. *Usual notations*: (r, θ) polar co-ordinates of a point; $u = 1/r$;

    ϕ = angle which a tangent makes with the radius vector;

    ψ = angle which a tangent makes with the initial line;

    p = perpendicular length from the pole to the tangent.

7. Important Relations:

(a) $\psi = \theta + \phi$;

(b) $\tan\phi = r\dfrac{d\theta}{dr}$ or, $\cot\phi = \dfrac{1}{r}\dfrac{dr}{d\theta}$;

(c) $p = r\sin\phi$;

(d) $\dfrac{1}{p^2} = \dfrac{1}{r^2} + \dfrac{1}{r^4}\left(\dfrac{dr}{d\theta}\right)^2 = u^2 + \left(\dfrac{du}{d\theta}\right)^2$;

(e) Polar sub-tangent $= r^2\dfrac{d\theta}{dr}$ and Polar sub-normal $= \dfrac{dr}{d\theta}$ (Fig. 15.7);

(f) Angle between two curves: $r = f(\theta)$ and $r = g(\theta)$. At the point of intersection find ϕ_1 and ϕ_2. Their difference is the required angle.

8. Pedal Equation (relation between p and r). To obtain pedal equation

(a) From Cartesian equation $f(x,y) = 0$: eliminate x and y from

$$f(x,y) = 0, \quad r^2 = x^2 + y^2, \quad p^2 = (xf_x + yf_y)^2/(f_x^2 + f_y^2).$$

(b) From Polar equation $r = f(\theta)$: eliminate θ from

$$r = f(\theta), \quad \tan\phi = r\dfrac{d\theta}{dr}, \quad p = r\sin\phi.$$

Otherwise: Eliminate θ between $r = f(\theta)$ and $p^{-2} = r^{-2} + r^{-4}(dr/d\theta)^2$.

Section II. Associated Curves

9. *The Pedal of a given curve* (w.r.t. a fixed point) is the locus of the foot of the perpendicular drawn from that fixed point upon the tangent to the curve.

(a) *Curves given in Cartesian Co-ordinates*: Find (in terms of p and α) the condition that the line $x\cos\alpha + y\sin\alpha = p$ may touch the curve. Replace p by r and α by θ: the polar eqn. of the pedal is obtained.

Change to Cartesian by using $x = r\cos\theta$ and $y = r\sin\theta$.

(b) *Curves given in Polar Co-ordinates*:

Suppose the polar co-ordinates of N, the foot of the perpendicular drawn from the fixed point O on the tangent at $P(r,\theta)$ are (p,α). Eliminate r,θ,ϕ from:

$f(r,\theta)=0$ (eqn. to the curve), $\phi+\theta-\alpha=\pi/2$, $\tan\phi=r\dfrac{d\theta}{dr}$, $p=r\sin\phi$.

The eliminant is the pedal with current co-ordinates (p,α) which may be changed to (r,θ).

10. *Equation of the pedal is given*: To find the original curves (**Negative Pedal**). Take (p,α) as the current co-ordinates of the given pedal instead of (r,θ).

The parametric equations of the original curve are then given by

$$x = p\cos\alpha - \frac{dp}{d\alpha}\sin\alpha, \quad y = p\sin\alpha + \frac{dp}{d\alpha}\cos\alpha.$$

11. *The Inverse of a Curve*: From a fixed origin O, draw a radius vector OP at any point P of a given curve. On OP take a point P' such that $OP \cdot OP' = k^2$ (k is a given constant). Then P' is the called the inverse of the point P w.r.t. O and the locus of P' is called the *Inverse of the given curve*.

(a) *Equation of an inverse curve*: (i) **Cartesian curve**: Replace x by $k^2 x'/(x'^2+y'^2)$ and y by $k^2 y'/(x'^2+y'^2)$, where (x',y') are the current co-ordinates of the inverse which may be later changed to (x,y).

(b) **Polar Curve**: Replace r by k^2/r' and θ by θ' where (r',θ') are the current co-ordinates of the inverse curve; now the current co-ordinates can be taken as (r,θ) instead of (r',θ').

12. *Reciprocal Polars*: The locus of the pole of the tangent to a curve Γ, w.r.t. a fixed conic Σ, is called the *Reciprocal polar of* Γ. Take the fixed conic to be a circle. Then the equation of the polar reciprocal can be obtained by first finding the pedal of the given curve and then its inverse (See Fig. 15.13).

Section III. Arc-length and its differentials

13. For continuous curves having continuously turning tangents, the limit of the ratio of an infinitesimal arc to its subtended chord is one.

14. Rectangular Cartesian Co-ordinates:

$$ds^2 = dx^2 + dy^2; \quad \sin\psi = \frac{dy}{ds}, \quad \cos\psi = \frac{dx}{ds} \quad \text{and} \quad \tan\psi = \frac{dy}{dx}.$$

15. Polar Co-ordinates: Using $dx = \cos\theta\, dr - r\sin\theta\, d\theta$, $dy = \sin\theta\, dr + r\cos\theta\, d\theta$, we get

$$ds^2 = dr^2 + (r\, d\theta)^2; \quad \sin\phi = r\frac{d\theta}{ds}, \quad \cos\phi = \frac{dr}{ds} \quad \text{and} \quad \tan\phi = r\frac{d\theta}{dr}.$$

Exercises XV (D)

1. Find the differential of arc length for each of the following curves:

   (a) $x^{2/3} + y^{2/3} = 1$.

   (b) $y = \frac{1}{2}(e^x + e^{-x})$.

   (c) $x^2/a^2 + y^2/b^2 = 1$.

   (d) $r^2 = \cos 2\theta$.

   (e) $x = a(1 - \cos\theta)$, $y = a(\theta + \sin\theta)$.

   (f) $x = \cos^3 t$, $y = \sin^3 t$.

2. Show that at a *maximum* or *minimum* point at which the derivative is defined $ds = dx$. Under what conditions, if any, will $ds = dy$?

3. For the curve $r^n = a^n \cos n\theta$, prove that

$$a^{2n}\frac{d^2 r}{ds^2} + nr^{2n-1} = 0.$$

4. For the equiangular spiral $r = ae^{\theta \cot\alpha}$, prove that $s/r = $ constant, s being measured from the origin.

5. For the curve

$$\frac{\theta}{2} = \sqrt{\frac{r-a}{a}} - \tan^{-1}\sqrt{\frac{r-a}{a}}.$$

Show that $\dfrac{1}{\sqrt{r}}\dfrac{ds}{dr}$ is constant.

6. For the pedal equation $p = f(r)$ show that

$$\frac{ds}{dr} = \frac{r}{\sqrt{r^2 - p^2}}.$$

CHAPTER 15: TANGENTS AND NORMALS: ASSOCIATED CURVES

7. If p_1 and p_2 are the perpendiculars from the origin on the tangent and normal respectively at the point (x, y) and if $\tan \psi = dy/dx$, prove that

$$p_1 = x \sin \psi - y \cos \psi, \quad p_2 = x \cos \psi + y \sin \psi$$

and then deduce that $p_2 = dp_1/d\psi$.

Answers

1. (a) $ds = (a^{1/3}/x^{1/3})dx$; (b) $ds = \frac{1}{2}(e^x + e^{-x})dx$; (c) $ds = \{(a^2 - e^2 x^2)/(a^2 - x^2)\}^{1/2} dx$, where $e^2 = (a^2 - b^2)/a^2$; (d) $ds = d\theta/(\cos 2\theta)^{1/2}$; (e) $ds = 2a \cos \frac{1}{2}\theta d\theta$; (f) $ds = 3 \sin t \cos t dt$.

For a Quick revision of this chapter workout **Ex. 1–25** of Pages **1161–1163** and **1175–1178**.

AN INTRODUCTION TO ANALYSIS

(Differential Calculus—Part I)

Chapter 16

Applications of Derivatives
Curvature

Chapter 16

Curvature

16.1 Introduction

In the intuitive language the terms *flatness* or *sharpness* are often used to describe the nature of bending of the curve at a particular point. In the present chapter we propose to give mathematical expressions for the *curvature* of a curve at a particular point which will give a definite numerical measure of turning which the curve undergoes at the point. The concept of curvature may be introduced from several independent stand-points, all, of course, leading to the same goal; but it is essential to observe that they are in their foundations logically distinct, as will be evident in course of our discussions.

To make the matters simple we shall assume that whenever derivatives occur in our problems they *do exist* at the points under consideration. We have confined our discussions mostly to well-known curves.

16.2 Measure of Bending : Definitions

Suppose the tangents at two adjacent points P and Q on a curve (Fig. 16.1) make angles ψ and $\psi + \Delta\psi$ with OX. Suppose, further, arc $AP = s$, arc $AQ = s + \Delta s$ so that arc $PQ = \Delta s$, A being a fixed point on the curve from which arcs are measured. We then construct the following definitions:

1. The angle $\Delta\psi$ through which the tangent turns as the point of contact travels from one end to the other of the arc PQ is called the **total curvature** of the arc PQ.

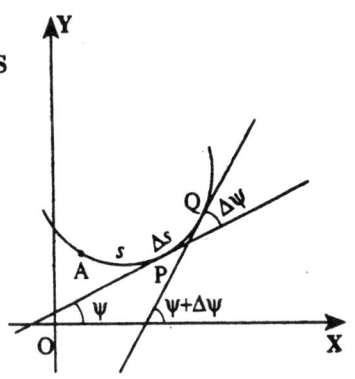

Fig. 16.1

2. The **mean or average** curvature of the arc PQ is defined as the ratio $\Delta\psi/\Delta s$.

3. The **curvature (k) at a point** P of the curve is defined as the limiting value of mean curvature when the arc $\Delta s \to 0$; that is

$$\text{Curvature } (k) \text{ at } P = \lim_{\Delta s \to 0} \frac{\Delta\psi}{\Delta s} = \frac{d\psi}{ds} \tag{16.2.1}$$

Theorem 16.2.1. *The curvature at any point of a circle is always constant and equals the reciprocal of the radius of that circle.*

Proof: We consider a circle with centre C and radius r (Fig. 16.2). Suppose that the tangents at two neighbouring points P and Q on it make angles ψ and $\psi + \Delta\psi$ with OX and let the arc $PQ = \Delta s$. Clearly, $\angle PCQ$ = angle between the tangents at P and $Q = \Delta\psi$ (expressed in radians).

But arc $PQ = \Delta s = r\Delta\psi$ (From Trigonometry); so that $\dfrac{\Delta\psi}{\Delta s} = \dfrac{1}{r}$,

$$\text{whence} \quad \lim_{\Delta s \to 0} \frac{\Delta\psi}{\Delta s} = \frac{d\psi}{ds} = \frac{1}{r}.$$

Thus in case of a circle the curvature is the same at every point and is measured by the reciprocal of the radius.

With these preliminaries we proceed to give definitions for the *circle of curvature* and its associated terms.

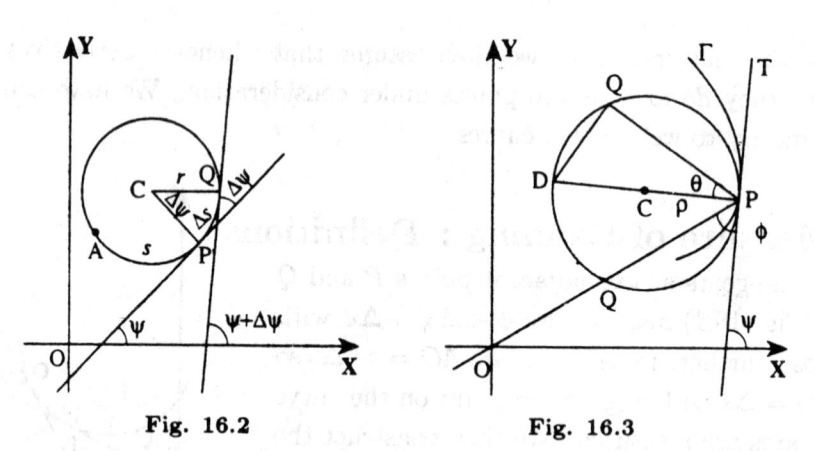

Fig. 16.2   Fig. 16.3

4. Consider a point P on the curve Γ (Fig. 16.3). By means of the relation (16.2.1) obtain the curvature k at P. Suppose that $k \neq 0$. Take a quantity ρ such that

$$\rho = \frac{1}{k} = \frac{ds}{d\psi}. \tag{16.2.2}$$

Now construct a circle of radius ρ and centre C so that the circle and the curve Γ have the same tangent at P; the circle is drawn in such a way that it lies on the same

CHAPTER 16: CURVATURE

side of the tangent as the curve (i.e., its concavity is turned in the same way as the given curve). This circle has the same curvature as the given curve at P [in view of (16.2.2)]. We call this circle as the **circle of curvature at** P : its centre C is the **centre of curvature of the curve at** P; its radius $CP(=\rho)$, normal to the curve at P, is the **radius of curvature of the curve at** P.

5. The length intercepted by the circle of curvature on a striaght line drawn through P in any specified direction is called the **chord of curvature** in that direction. Thus if θ be the angle which the direction of the chord PQ makes with the normal PCD, then its length q is given by

$$q = 2\rho \cos \theta. \tag{16.2.3}$$

Note 16.2.1. *We caution our readers from a possible mistake. There is no reason why the circle of curvature should not cross the curve and, in general, it does (just as a tangent line crosses the curve at a point of inflexion).*

16.3 Formulae for the Radius of Curvature

1. Intrinsic Equation : $s = f(\psi)$.

The relation between the length of the arc (s) of a given curve, measured from a given fixed point on the curve and the angle between the tangents at its ends (ψ) is called the *intrinsic equation of the curve* where the formulae

$$\rho = \frac{ds}{d\psi}; \quad k = \frac{d\psi}{ds} \tag{16.3.1}$$

are most immediately applicable.

For example, in the case of a **Catenary**, we have the intrinsic equation $s = c \tan \psi$ so that $\rho = ds/d\psi = c \sec^2 \psi$.

An Important Example

Example 16.3.1. *For the equiangular spiral* $r = ae^{\theta \cot \alpha}$ *prove that the radius of curvature subtends a right angle at the pole.*

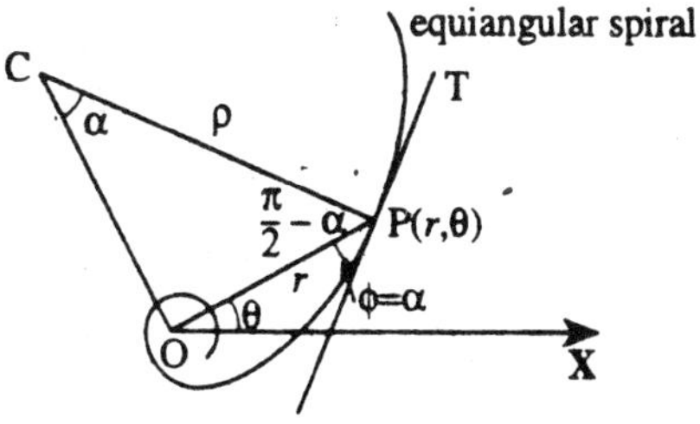

Solution: We have $r = ae^{\theta \cot \alpha}$.

$$\therefore \frac{dr}{d\theta} = a \cot \alpha e^{\theta \cot \alpha} = r \cot \alpha$$

$$\therefore \tan \phi = r \frac{d\theta}{dr} = \tan \alpha \quad \therefore \phi = \alpha$$

Fig. 16.4

and hence $\psi = \theta + \phi = \theta + \alpha$ so that $d\psi/ds = d\theta/ds$ and then

$$\rho = \frac{ds}{d\psi} = \frac{1}{\frac{d\psi}{ds}} = \frac{1}{\frac{d\theta}{ds}} = \sqrt{r^2 + \left(\frac{dr}{d\theta}\right)^2}$$

$$(\because ds^2 = dr^2 + (rd\theta)^2)$$

so that $\rho = \sqrt{r^2 + r^2 \cot^2 \alpha} = r \operatorname{cosec} \alpha = r/\sin \alpha$.

A glance at Fig. 16.4 will now give the desired result, i.e., CP subtends a right angle $\angle COP$ at O.

2. Cartesian Equation (Explicit Function) : $y = f(x)$ or $x = f(y)$.

In rectangular cartesian co-ordinates, we have

$$\tan \psi = dy/dx = y_1$$

and, therefore,

$$\sec^2 \psi \frac{d\psi}{ds} = \frac{d}{ds}(y_1) = \frac{d}{dx}(y_1) \cdot \frac{dx}{ds} = \frac{d^2y}{dx^2} \cdot \cos \psi$$

whence,

$$\rho = \frac{ds}{d\psi} = \frac{\sec^3 \psi}{\frac{d^2y}{dx^2}} = \frac{\left\{1 + \left(\frac{dy}{dx}\right)^2\right\}^{\frac{3}{2}}}{\frac{d^2y}{dx^2}} = \frac{(1 + y_1^2)^{\frac{3}{2}}}{y_2} \quad (y_2 \neq 0).$$

Remember:

$$\rho = \frac{(1 + y_1^2)^{\frac{3}{2}}}{y_2} \quad (y_2 \neq 0). \tag{16.3.2}$$

Conventions of Signs: The positive root is taken in the numerator of (16.3.2) and the radius of curvature will, therefore, be *positive* when y_2 is positive (i.e., when the curve is concave upwards) and *negative* when y_2 is negative (i.e., when the curve is concave downwards or convex upwards). The concepts of concavity and convexity have been given in chapter **19**. In practice, *we shall make use of the numerical value of ρ.*

Corollary 16.3.1. *The definition of ρ nowhere suggests that its value would depend on the choice of axes. Hence we could very well interchange the axes of X and Y and obtain*

$$\rho = \frac{\left\{1 + \left(\frac{dx}{dy}\right)^2\right\}^{\frac{3}{2}}}{\frac{d^2x}{dy^2}} = \frac{(1 + x_1^2)^{\frac{3}{2}}}{x_2} \quad (x_2 \neq 0) \tag{16.3.3}$$

CHAPTER 16: CURVATURE

which is a very useful formula when (16.3.2) fails i.e., when the tangent at the point becomes parallel to the Y-axis.

The formula (16.3.2) or (16.3.3) is adapted for the evaluation of ρ when the equation of the curve is given in cartesian co-ordinates, y (or x) being an explicit function of x (or y).

Example 16.3.2. *In the Catenary, $y = a\cosh(x/a)$, we have*

$$y_1 = \sinh\frac{x}{a}; \quad y_2 = \frac{1}{a}\cosh\frac{x}{a}; \quad 1 + y_1^2 = 1 + \sinh^2\frac{x}{a} = \cosh^2\frac{x}{a}.$$

$$\therefore \rho = \frac{(1+y_1^2)^{\frac{3}{2}}}{y_2} = \frac{\cosh^3(x/a)}{(1/a)\cosh(x/a)} = a\cosh^2\left(\frac{x}{a}\right) = \frac{y^2}{a}.$$

Observe that this result agrees with the result given before in 16.3 (Intrinsic Equation), since, for the catenary $y = a\sec\psi$ and $\rho = y^2/a = (a^2\sec^2\psi)/a = a\sec^2\psi$.

Note that in a Catenary $\rho \propto y^2$.

3. Cartesian Equation (Implicit Functions) : $f(x,y) = 0$.

At a point where $f_y(x,y) \neq 0$, we have $f_x + f_y dy/dx = 0$ and differentiating again,

$$\frac{\partial f_x}{\partial x} + \frac{\partial f_x}{\partial y}\frac{dy}{dx} + \left(\frac{\partial f_y}{\partial y}\frac{dy}{dx} + \frac{\partial f_y}{\partial x}\right)\frac{dy}{dx} + f_y\frac{d^2y}{dx^2} = 0$$

or, $\quad f_{xx} + 2f_{yx}\cdot\frac{dy}{dx} + f_{yy}\cdot\left(\frac{dy}{dx}\right)^2 + f_y\cdot\frac{d^2y}{dx^2} = 0 \quad$ (assuming $f_{yx} = f_{xy}$).

Hence substituting for $\dfrac{dy}{dx}$ and $\dfrac{d^2y}{dx^2}$ in the formula (16.3.2) we have

$$\rho = \left|-\frac{\left(1+\dfrac{f_x^2}{f_y^2}\right)^{\frac{3}{2}} f_y}{f_{xx} + 2f_{yx}\left(-\dfrac{f_x}{f_y}\right) + f_{yy}\left(-\dfrac{f_x}{f_y}\right)^2}\right|.$$

$$\therefore \rho = \left|\frac{(f_x^2 + f_y^2)^{\frac{3}{2}}}{f_{xx}f_y^2 - 2f_{yx}f_xf_y + f_{yy}f_x^2}\right| \tag{16.3.4}$$

(assuming that the denominator $\neq 0$).

Example 16.3.3. *In the case of Folium of Descartes $x^3 + y^3 = 3axy$, prove that the radius of curvature at the point $(\frac{3}{2}a, \frac{3}{2}a)$ is numerically equal to*

$$\frac{3a}{16}\sqrt{2}.$$

Solution: **First Method.** If $f(x,y) = x^3 + y^3 - 3axy = 0$ we may easily obtain

$$f_x\left(\frac{3}{2}a, \frac{3}{2}a\right) = \frac{9}{4}a^2 = f_y\left(\frac{3}{2}a, \frac{3}{2}a\right); \quad f_{xx}\left(\frac{3}{2}a, \frac{3}{2}a\right) = 9a = f_{yy}\left(\frac{3}{2}a, \frac{3}{2}a\right);$$

$$f_{xy}\left(\frac{3}{2}a, \frac{3}{2}a\right) = -3a, \text{ whence, by (16.3.4) } \rho = \frac{3a}{8\sqrt{2}} = \frac{3}{16}\sqrt{2a}.$$

Second Method. Differentiating the given equation, we get

$$x^2 + y^2 \frac{dy}{dx} = ay + ax \frac{dy}{dx}.$$

$$\therefore \frac{dy}{dx} \text{ at the given point } \left(\frac{3}{2}a, \frac{3}{2}a\right) = -1.$$

Differentiating once more,

$$2x + 2y\left(\frac{dy}{dx}\right)^2 + y^2 \frac{d^2y}{dx^2} = a\frac{dy}{dx} + a\frac{dy}{dx} + ax\frac{d^2y}{dx^2}$$

which gives $\dfrac{d^2y}{dx^2} = -\dfrac{32}{3a}$ at the point $\left(x = \frac{3}{2}a, y = \frac{3}{2}a\right)$.

Now use the formula (16.3.2) and obtain the required result, i.e., $\rho = \frac{3}{16}\sqrt{2} \cdot a$.

4. Parametric Equation : $x = f(t), y = \phi(t)$.

The formula (16.3.2) is very easily modified to meet the requirements of the case where the equation of the curve is given in parametric form: $x = f(t); y = \phi(t)$.

Then, $x' = \dfrac{dx}{dt} = f'(t)$ and $y' = \dfrac{dy}{dt} = \phi'(t)$, so that $\dfrac{dy}{dx} = \dfrac{y'}{x'}$

and $\dfrac{d^2y}{dx^2} = \dfrac{d}{dt}\left(\dfrac{y'}{x'}\right) \cdot \dfrac{dt}{dx} = \dfrac{y''x' - y'x''}{x'^2} \cdot \dfrac{1}{x'} \left(x'' = \dfrac{d^2x}{dt^2}, y'' = \dfrac{d^2y}{dt^2}\right)$

whence, by (16.3.2),

$$\rho = \frac{(x'^2 + y'^2)^{\frac{3}{2}}}{x'y'' - y'x''}, \quad (x'y'' - y'x'' \neq 0) \tag{16.3.5}$$

the accents denoting differentiations with respect to t.

CHAPTER 16: CURVATURE

Example 16.3.4. *In the case of cycloid*
$$x = a(\theta + \sin\theta), \quad y = a(1 - \cos\theta)$$
[C.H. 1980]

We have (accents denoting differentiations with respect to θ)

$$\left.\begin{array}{ll} x' = a(1+\cos\theta), & y' = a\sin\theta \\ x'' = -a\sin\theta, & y'' = a\cos\theta \end{array}\right\}$$

whence, from (16.3.5), we obtain $\rho = 4a\cos\theta/2$.

Otherwise, obtain $\dfrac{dy}{dx} = \dfrac{y'}{x'} = \tan\dfrac{\theta}{2}; \quad \dfrac{d^2y}{dx^2} = \dfrac{1}{4a\cos^4\frac{\theta}{2}}.$

Now use (16.3.2), and obtain $\rho = |4a\cos\theta/2|$.

Verify that $\rho = |4a\sin\theta/2|$ at any point of the cycloid $x = a(\theta - \sin\theta), y = a(1 - \cos\theta)$.

5. Pedal Equation : $p = f(r)$.

We refer to Fig. 16.5.

Let P be a point (r, θ) of a curve whose pedal equation is known, say $p = f(r)$.

The tangent at P makes an angle ψ with the initial line OX. Suppose ϕ is the angle between the radius vector OP and the tangent PT.

Let $ON = p$ be the length of the perpendicular from O on PT.

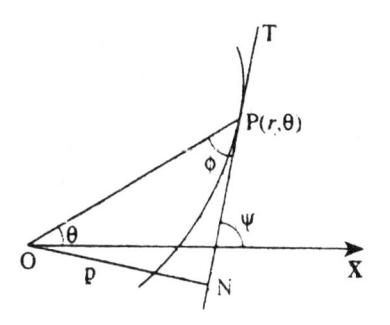

Fig. 16.5

Clearly, $\psi = \theta + \phi$ and $p = r\sin\phi$.

$$\therefore \quad \frac{1}{\rho} = \frac{d\psi}{ds} = \frac{d\theta}{ds} + \frac{d\phi}{ds} = \frac{d\theta}{ds} + \frac{d\phi}{dr}\cdot\frac{dr}{ds}.$$

So that

$$\frac{1}{\rho} = \frac{\sin\phi}{r} + \cos\phi\cdot\frac{d\phi}{dr} \quad \left(\because \sin\phi = r\frac{d\theta}{ds}, \cos\phi = \frac{dr}{ds}\right)$$

$$= \frac{1}{r}\left(\sin\phi + r\cos\phi\cdot\frac{d\phi}{dr}\right)$$

or, $\quad \dfrac{1}{\rho} = \dfrac{1}{r}\dfrac{d}{dr}(r\sin\phi) = \dfrac{1}{r}\dfrac{dp}{dr}, \quad (\because p = r\sin\phi)$

whence,

$$\rho = r\frac{dr}{dp}. \tag{16.3.6}$$

Example 16.3.5. In the ellipse $\dfrac{a^2b^2}{p^2} = a^2 + b^2 - r^2$, differentiation w.r.t. p gives

$$-2\dfrac{a^2b^2}{p^3} = -2r\dfrac{dr}{dp}.$$

$$\therefore \rho = r\dfrac{dr}{dp} = \dfrac{a^2b^2}{p^3}.$$

6. (a) Polar Equation : $r = f(\theta)$.

First Method. We know that

$$\dfrac{1}{p^2} = \dfrac{1}{r^2} + \dfrac{1}{r^4}\left(\dfrac{dr}{d\theta}\right)^2$$

whence, by differentiation w.r.t. r, we obtain

$$-\dfrac{2}{p^3}\cdot\dfrac{dp}{dr} = -\dfrac{2}{r^3} - \dfrac{4}{r^5}\left(\dfrac{dr}{d\theta}\right)^2 + \dfrac{2}{r^4}\cdot\dfrac{d^2r}{d\theta^2}.$$

This gives,

$$\rho = r\dfrac{dr}{dp} = r\cdot\dfrac{1}{p^3} \bigg/ \dfrac{1}{r^5}\left\{r^2 + 2\left(\dfrac{dr}{d\theta}\right)^2 - r\dfrac{d^2r}{d\theta^2}\right\}.$$

Finally,

$$\rho = \dfrac{r^6\left\{\dfrac{1}{r^2} + \dfrac{1}{r^4}\left(\dfrac{dr}{d\theta}\right)^2\right\}^{\frac{3}{2}}}{r^2 + 2\left(\dfrac{dr}{d\theta}\right)^2 - r\dfrac{d^2r}{d\theta^2}} = \dfrac{\left\{r^2 + \left(\dfrac{dr}{d\theta}\right)^2\right\}^{\frac{3}{2}}}{r^2 + 2\left(\dfrac{dr}{d\theta}\right)^2 - r\dfrac{d^2r}{d\theta^2}}.$$

$\therefore$ Numerical value of

$$\rho = \left|\dfrac{(r^2 + r_1^2)^{\frac{3}{2}}}{r^2 + 2r_1^2 - rr_2}\right| \quad \left(r_1 = \dfrac{dr}{d\theta},\quad r_2 = \dfrac{d^2r}{d\theta^2}\right). \qquad (16.3.7)$$

Second Method. $\rho = \dfrac{ds}{d\psi} = \dfrac{ds}{d\theta}\bigg/\dfrac{d\psi}{d\theta}$. But $\dfrac{ds}{d\theta} = \sqrt{r^2 + \left(\dfrac{dr}{d\theta}\right)^2}$.

Again, $\psi = \theta + \phi = \theta + \tan^{-1}\dfrac{r}{r_1}$, where $r_1 = \dfrac{dr}{d\theta}$.

This gives, $\dfrac{d\psi}{d\theta} = \dfrac{r^2 + 2r_1^2 - rr_2}{r^2 + r_1^2}$, where $r_2 = \dfrac{d^2r}{d\theta^2}$.

Substituting the values of $\dfrac{ds}{d\theta}$ and $\dfrac{d\psi}{d\theta}$, we obtain ρ as in (16.3.7).

CHAPTER 16: CURVATURE

(b) Polar Equation : $u = f(\theta)$ where $u = 1/r$.

Since $r = 1/u$ we get

$$\frac{dr}{d\theta} = -\frac{1}{u^2} \cdot \frac{du}{d\theta}; \quad \frac{d^2r}{d\theta^2} = -\frac{u\frac{d^2u}{d\theta^2} - 2\left(\frac{du}{d\theta}\right)^2}{u^3}.$$

Hence, from (16.3.7), it is easy to deduce

$$\rho = \frac{\left\{u^2 + \left(\frac{du}{d\theta}\right)^2\right\}^{\frac{3}{2}}}{u^3\left(u + \frac{d^2u}{d\theta^2}\right)}. \tag{16.3.8}$$

Example 16.3.6. *Show that the radius of curvature at any point on the cardiode* $r = a(1 - \cos\theta)$ *is* $\frac{2}{3}\sqrt{2ar}$.

Solution: Here $r_1 = a\sin\theta$; $r_2 = a\cos\theta$; using (16.3.7) the result follows.

Otherwise proceed as:

Given $r = 2a\sin^2\frac{\theta}{2}$; by differentiation $\cot\phi = \frac{1}{r}\frac{dr}{d\theta} = \cot\frac{\theta}{2}$, so that $\phi = \frac{\theta}{2}$.

Now since $p = r\sin\phi = r\sin\theta/2$, we obtain $2ap^2 = r^3$, as the pedal equation of the cardiode.

Hence, differentiating w.r.t. r we get

$$4ap\frac{dp}{dr} = 3r^2, \quad \text{whence } \rho = r\frac{dr}{dp} = \frac{4p}{3r} \cdot a = \frac{2}{3}\sqrt{2ar}.$$

Note here that $\rho \propto \sqrt{r}$ or $\rho^2 \propto r$.

7. Tangential Polar Equation: $p = f(\psi)$ [C.H.]

We have,

$$\frac{dp}{d\psi} = \frac{dp}{dr} \cdot \frac{dr}{ds} \cdot \frac{ds}{d\psi} = \frac{dp}{dr} \cdot \cos\phi \cdot r\frac{dr}{dp} = r\cos\phi.$$

[The projection of the radius vector on the tangent $= r\cos\phi = dp/d\psi$]

$$\therefore \quad p^2 + \left\{\frac{dp}{d\psi}\right\}^2 = r^2\sin^2\phi + r^2\cos^2\phi = r^2.$$

Differentiating w.r.t. p and using (16.3.6), we obtain

$$\rho = \mathbf{p} + \frac{d^2\mathbf{p}}{d\psi^2}. \tag{16.3.9}$$

As an illustration, it can be easily shown that for the *epicycloid* $p = a \sin b\psi$, $\rho \propto p$.

$$\therefore \frac{dp}{d\psi} = ab \cos b\psi \quad \text{and} \quad \frac{d^2p}{d\psi^2} = -ab^2 \sin b\psi = -b^2 p$$

so that $\rho = p - b^2 p = (1 - b^2)p$, i.e., $\rho \propto p$.

8. Other Expressions for ρ.

$$\frac{1}{\rho^2} = \left(\frac{d^2x}{ds^2}\right)^2 + \left(\frac{d^2y}{ds^2}\right)^2. \tag{16.3.10}$$

We have, $\dfrac{dx}{ds} = \cos \psi$, $\dfrac{d^2x}{ds^2} = -\sin\psi \cdot \dfrac{d\psi}{ds} = -\dfrac{\sin\psi}{\rho}$.

Again, $\dfrac{dy}{ds} = \sin \psi$, $\dfrac{d^2y}{ds^2} = \cos\psi \cdot \dfrac{d\psi}{ds} = \dfrac{\cos\psi}{\rho}$.

Squaring and adding, the result (16.3.10) follows:

$$\rho = \frac{r}{\sin\phi\left\{1 + \dfrac{d\phi}{d\theta}\right\}}. \tag{16.3.11}$$

Here,

$$\sin\phi\left\{1 + \frac{d\phi}{d\theta}\right\} = \sin\phi + \frac{d\phi}{d\theta} \cdot \sin\phi$$

$$= r\frac{d\theta}{ds} + \frac{d\phi}{d\theta} \cdot \left\{r\frac{d\theta}{ds}\right\}$$

$$= r\frac{d}{ds}(\theta + \phi) = r\frac{d\psi}{ds} = \frac{r}{\rho};$$

hence the result (16.3.11).

Exercises XVI (A)

1. For the *tractrix* $s = c \log \sec \psi$, prove that $\rho = c \tan \psi$ and in case of the *equiangular spiral* $s = a(e^{m\psi} - 1)$, $\rho = mae^{m\psi}$.

2. Prove that the radius of curvature of the *Catenary* of uniform strength

$$y = c \log \sec(x/c) \text{ is } c \sec(x/c).$$

[C.H.]

3. Find the radius of curvature of the points indicated:

   (a) $\sqrt{x} + \sqrt{y} = 1 : (\frac{1}{4}, \frac{1}{4})$;

CHAPTER 16: CURVATURE

(b) $x^{2/3} + y^{2/3} = a^{2/3} : (x, y)$;

(c) $y = 4 \sin x - \sin 2x : (x = \pi/2, y = 4)$;

(d) $ay^2 = x^3 : (x, y)$;

(e) $y = e^x$ at the point where it crosses the y-axis;

(f) $y = xe^{-x}$ at its maximum point;

(g) $x^3 + y^3 = 3axy$ at the point $(3a/2, 3a/2)$.

4. Verify the following results:

Curve		Curvature at (x, y)	
Circle:	$x^2 + y^2 = a^2$	$1/a$	
Parabola:	$y^2 = 4ax$	$\sqrt{a}/2(a+x)^{3/2}$	
Ellipse:	$\dfrac{x^2}{a^2} + \dfrac{y^2}{b^2} = 1$	$ab/(a^2 - e^2 x^2)^{3/2}$	[C.H.]
Hyperbola:	$\dfrac{x^2}{a^2} - \dfrac{y^2}{b^2} = 1$	$ab/(e^2 x^2 - a^2)^{3/2}$	
Rectangular Hyperbola:	$xy = c^2$	$2c^2/(x^2 + y^2)^{3/2}$	[C.H.]
Straight line:	$y = mx + c$	0	

5. (a) Show that the radius of curvature at $\theta = \pi/4$ on the curve
$$x = a \cos^3 \theta, \quad y = a \sin^3 \theta \text{ is } 3a/2$$
and at the point θ, $\rho = 3a \sin \theta \cos \theta$. [C.H.]

(b) Show that the radius of curvature at a point θ of the curve
$$x = ae^\theta(\sin \theta - \cos \theta), \quad y = ae^\theta(\sin \theta + \cos \theta)$$
is twice the distance of the tangent at that point from the origin.

6. (a) Prove that for the curves

   i. $x = kt, y = k \log \sec t; \rho = k \sec t$;

   ii. $x = a \cos t, y = b \sin t; \rho = (a^2 \sin^2 t + b^2 \cos^2 t)^{\frac{3}{2}}/ab$.   [C.H.]

(b) Show that ρ for the curve $x = t - c \sinh(t/c) \cosh(t/c), y = 2c \cosh(t/c)$ is $-2c \cosh^2(t/c) \sinh(t/c)$.

7. In the curve $r^n = a^n \cos n\theta$, verify that
$$\rho = \frac{r^2}{(n+1)p} = \frac{a^n}{(n+1)r^{n-1}}$$
examine the case when $n = 2, n = -\frac{1}{2}$.

8. Find the radii of curvature at the indicated points:

   (a) $r = ae^{\theta \cot \alpha}$ at any point θ. [C.H. 1983]

   (b) $r = \dfrac{l}{1 + e \cos \theta}$; $\theta = \pi$.

   (c) $r = a(\theta + \sin \theta)$; $\theta = 0$.

   (d) $r = a(1 + \cos \theta)$; at any point θ. [C.H.]

9. Show that at the points where $r = a\theta$ (*Archimedean spiral*) intersects the curve $r\theta = a$ (*reciprocal spiral*) their curvatures are in the ratio $3 : 1$.

10. For the spiral $p^2 = r^4/(r^2 + a^2)$, show that $\rho = \dfrac{(r^2 + a^2)^{3/2}}{r^2 + 2a^2}$.

11. Show that for the *epi-cycloid* $p = m \sin n\psi$, $\rho \propto p$.

12. Prove that for the curve
$$s = a \log \cot\left(\frac{\pi}{4} - \frac{\psi}{2}\right) + a \tan \psi \sec \psi, \rho = 2a \sec^3 \psi.$$

Hence deduce that $\dfrac{d^2 y}{dx^2} = \dfrac{1}{2a}$ and that this differential equation is satisfied by the parabola $x^2 = 4ay$.

13. Show that in any curve (**various expressions for** ρ)

   (a) $\dfrac{1}{\rho} = \dfrac{d^2 x}{ds^2} \bigg/ \dfrac{dy}{ds} = \dfrac{d^2 y}{ds^2} \bigg/ \dfrac{dx}{ds}$.

   (b) $\dfrac{1}{\rho} = \left\{\dfrac{1}{r} - \dfrac{1}{r}\left(\dfrac{dr}{ds}\right)^2 - \dfrac{d^2 r}{ds^2}\right\} \bigg/ \left\{1 - \left(\dfrac{dr}{ds}\right)^2\right\}^{\frac{1}{2}}$.

   (c) $\rho = r \dfrac{d\theta}{ds} \bigg/ \left\{r\left(\dfrac{d\theta}{ds}\right)^2 - \dfrac{d^2 r}{ds^2}\right\}$.

   (d) $\dfrac{1}{\rho^2} = \left\{\left(\dfrac{d^2 x}{dt^2}\right)^2 + \left(\dfrac{d^2 y}{dt^2}\right)^2 - \left(\dfrac{d^2 s}{dt^2}\right)^2\right\} \div \left(\dfrac{ds}{dt}\right)^4$.

   (e) $\rho = \sqrt{\left\{\left(\dfrac{dx}{d\psi}\right)^2 + \left(\dfrac{dy}{d\psi}\right)^2\right\}}$.

14. Prove that
$$\frac{d\rho}{ds} = \frac{3pq^2 - (1 + p^2)r}{q^2}$$

where $p = \dfrac{dy}{dx}$, $q = \dfrac{d^2 y}{dx^2}$, $r = \dfrac{d^3 y}{dx^3}$.

CHAPTER 16: CURVATURE

15. (a) If ρ, ρ' be the radii of curvature at the ends of two conjugate diameters of an ellipse, prove that
$$(\rho^{2/3} + \rho'^{2/3})(ab)^{2/3} = a^2 + b^2.$$

(b) The tangents at two points P and Q on the cycloid
$$x = a(\theta - \sin\theta), \quad y = a(1 - \cos\theta)$$
are at right angles. Show that if ρ_1 and ρ_2 be the radii of curvature at these points, then $\rho_1^2 + \rho_2^2 = 16a^2$.

16. If the polar equation of a curve be $r = f(\theta)$, where $f(\theta)$ is an even function of θ, show that the curvature at the point $\theta = 0$ is $\{f(0) - f''(0)\}/\{f(0)\}^2$.

17. Find the curvature of the curve

(a) $y = mx + n(x-a)^2(x-b)^2$ at points $(a, 0), (b, 0)$.
(b) $(x^2 + y^2)^2 = a^2(y^2 - x^2)$ at $(0, a)$. [C.H.]
(c) $y^2 = a^2(a+x)/x$ at $(-a, 0)$.
(d) $x^4 + a^2 y^2 = ax^3$ at $(a, 0)$.

18. If CP, CD be a pair of conjugate semi-diameters of an ellipse, prove that the radius of curvature at P is CD^3/ab, where a and b are lengths of the semi-axes.

19. The curve $r = ae^{\theta \cot \alpha}$ cuts any radius vector in the consecutive points, P_1, $P_2, \cdots, P_n, P_{n+1}, \cdots$. If ρ_n denotes the radius of curvature at P_n, prove that $\dfrac{1}{m-n} \log(\rho_m/\rho_n)$ is constant for all integral values of m and n.

20. Find the point on the curve $y = e^x$ at which the curvature is maximum and show that the tangent at this point from with the axes of co-ordinates a triangle whose sides are in the ratio $\sqrt{2} : \sqrt{3}$.

21. The normal at $P(x, y)$ of the curve $x = \frac{3}{2}a(\sinh t \cosh t + t), y = a\cosh^3 t$ meets the X-axis at G. Prove that $\rho = 3PG$.

22. If ρ_1, ρ_2 be the radii of curvature at the extremities of any chord of the cardiode $r = a(1 + \cos\theta)$, which passes through the pole, the prove that $\rho_1^2 + \rho_2^2 = \frac{16}{9}a^2$.

Answers

3. (a) $\frac{1}{2}\sqrt{2}$; (b) $3\sqrt[3]{axy}$; (c) $\frac{5}{4}\sqrt{5}$ or 2.795; (d) $(4a + 9x)^{3/2}x^{1/2}/6a$; (e) $\sqrt{8}$; (f) e; (g) $-\frac{3\sqrt{2}}{16}a$. 8. (a) $r/\sin\alpha$; (b) l; (c) a; (d) $\frac{2}{3}\sqrt{2ar}$. 17. (a) $\frac{2n(a-b)^2}{(1+m^2)^{3/2}}$; (b) $\frac{3}{a}$; (c) $\frac{2}{a}$; (d) $\frac{2}{a}$. 20. $[(-\log 2)/2, 1/\sqrt{2}]$.

16.4 Theorem on Centre of Curvature

The centre of curvature of a point P of the curve is the limiting position of the point of intersection of the normal to the curve at P with the normal to the curve at a neighbouring point Q on the curve as Q tends to P along the curve.

Proof: Let the normals at P and Q intersect at C and suppose they include an angle $\Delta\psi$ (radians) and if Δs be the arc PQ, then drawing the chord PQ we have (See Fig. 16.6).

$$\frac{CP}{PQ} = \frac{\sin CQP}{\sin \Delta\psi}. \quad \text{[from the } \triangle CPQ\text{]}$$

$$\therefore CP = \sin CQP \cdot \frac{PQ}{\sin \Delta\psi}$$

$$= \sin CQP \cdot \frac{PQ}{\Delta s} \cdot \frac{\Delta s}{\Delta\psi} \cdot \frac{\Delta\psi}{\sin \Delta\psi}. \quad (16.4.1)$$

Fig. 16.6

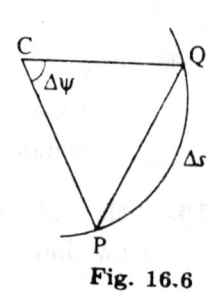

When, in the limit, $Q \to P$ (from either side, the limiting value of each factor on the right-hand member of (16.4.1), except $\Delta s/\Delta\psi$ which tends to ρ, is unity. Hence, we have,

$$CP = \frac{ds}{d\psi} = \rho.$$

$$\left[\text{observe: } \angle CQP \to \frac{\pi}{2}, \frac{PQ}{\Delta s} \to 1 \text{ and } \frac{\Delta\psi}{\sin \Delta\psi} \to 1\right]$$

Hence, C is the centre of curvature at the point P.

16.5 Concept of Curvature : Newton's approach

The Newtonian approach of finding the curvature of the curve at a point is instructive and interesting. *Consider a circle touching the given curve at P, and passing through a neighbouring point Q on it. When $Q \to P$ from either side, the limiting position of this circle, if exists, becomes identical with the circle of curvature at P.*

We justify the assertion.

Let $C(x_1, y_1)$ be the centre of the circle so drawn (Fig. 16.7) and r be its radius. If $P'(x, y)$ be a point of the curve (the curve itself not shown in the Figure) on the arc PQ then

$$CP'^2 = (x - x_1)^2 + (y - y_1)^2$$
$$= F(x) \text{ (say)},$$

in which y is defined by the equation of the curve $y = f(x)$.

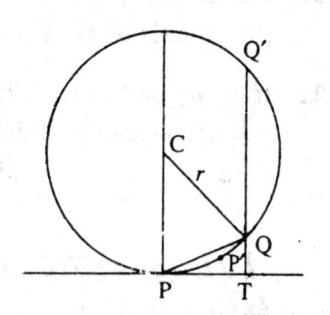

Fig. 16.7

CHAPTER 16: CURVATURE

Since, $CP^2 = CQ^2 = r^2$, $F(x)$ has the same value at P and Q.

Hence, by Rolle's Theorem (other conditions of the theorem being assumed to be satisfied), there exists a point $R(\xi, \eta)$ on the curve (not shown in the diagram) between P and Q such that
$$F'(\xi) = 0$$
i.e., $$(\xi - x_1) + (\eta - y_1)\left(\frac{dy}{dx}\right)_{x=\xi} = 0$$

i.e., CR is normal to the curve at R.

Now, when $Q \to P$, then $R \to P$ also. Thus, the limiting position of C is the ultimate point of intersection of the normal at P and the normal at the neighbouring point R when $R \to P$ and is therefore, (art. **16.5**), the centre of curvature at P. Thus, the constructed circle tends to the circle of curvature.

Newton's method leads to a simple formula for the radius of curvature. Suppose $Q'QT$ is the perpendicular drawn on the tangent at P, meeting the circle in Q and Q' and the tangent in T.

By Elementary Geometry,
$$TP^2 = TQ \cdot TQ',$$
we have,
$$\rho = \frac{1}{2} \lim_{Q \to P} TQ' = \frac{1}{2} \lim_{Q \to P} \frac{TP^2}{TQ}. \qquad (16.5.1)$$

Since, $TP^2 = PQ^2 - TQ^2$ and hence $\dfrac{TP^2}{TQ} = \dfrac{PQ^2}{TQ} - TQ$; also $TQ \to 0$ as $Q \to P$ we could also use
$$\rho = \frac{1}{2} \lim_{Q \to P} PQ^2. \qquad (16.5.2)$$

Important Particular Cases

1. If a curve passes through the origin and the axis of X is tangent at the origin, then
$$\lim_{\substack{x \to 0 \\ y \to 0}} \frac{x^2}{2y}$$
gives the radius of curvature at the origin. Consider PT as X-axis and use (16.5.1).

2. On the other hand if a curve passes through the origin and Y-axis is the tangent at the origin, then the radius of curvature at the origin is
$$\lim_{\substack{x \to 0 \\ y \to 0}} \frac{y^2}{2x}.$$

[Consider PT as the Y-axis and use (16.5.1).]

3. For a curve, passing through the origin, if $ax + by = 0$ be the tangent at the origin, then the radius of curvature at $(0,0)$ is

$$\frac{1}{2}\sqrt{a^2 + b^2} \lim_{\substack{x \to 0 \\ y \to 0}} \frac{x^2 + y^2}{ax + by}.$$

[Consider PT as the line $ax + by = 0$ and use (16.5.2).]

Example 16.5.1. *Show that the radii of curvature of the curve*

$$y^2 = x^2(a + x)/(a - x) \text{ at the origin are both numerically } a\sqrt{2}.$$

Solution: Writing the given equation in the form

$$y^2(a - x) - x^2(a + x) = 0, \text{ or, } a(y + x)(y - x) = x^3 + xy^2$$

it can be easily shown that the tangents at the origin are

$$y + x = 0, \quad y - x = 0 \text{ (equating to zero the lowest degree terms)}.$$

Considering $y - x = 0$, we obtain

$$\rho = \frac{1}{2}\sqrt{2} \lim_{\substack{x \to 0 \\ y \to 0}} \frac{x^2 + y^2}{y - x} = \frac{1}{2}\sqrt{2} \lim_{\substack{x \to 0 \\ y \to 0}} \frac{(x^2 + y^2)a(y + x)}{x^3 + xy^2}$$

[from the equation of the curve]

i.e., $\quad \rho = \frac{1}{2}\sqrt{2a} \lim_{\substack{x \to 0 \\ y \to 0}} \frac{\left\{1 + \left(\frac{y}{x}\right)^2\right\}\left\{1 + \left(\frac{y}{x}\right)\right\}}{1 + \left(\frac{y}{x}\right)^2} = \frac{1}{2}\sqrt{2a} \cdot \frac{(1 + 1)(1 + 1)}{1 + 1} = \sqrt{2a}.$

(since $\lim_{\substack{x \to 0 \\ y \to 0}} \left(\frac{y}{x}\right)$, the value of the slope m of the tangent $y - x = 0$, at $(0,0)$ is unity).

Similarly, in the other case $y + x = 0$, $\rho = -\sqrt{2a}$. Here also $\rho = \sqrt{2}$ a numerically.

Note: The difference of sign indicates that the two branches of the curve at the origin bend in opposite directions. Both the branches at O have the same curvature but in opposite directions.

Example 16.5.2. *Find the radius of curvature at the vertex of the cycloid*

$$x = a(\theta + \sin\theta), \quad y = a(1 - \cos\theta).$$

CHAPTER 16: CURVATURE

Solution: We have

$$\frac{x^2}{2y} = a(\theta + \sin\theta)^2/4\sin^2\theta/2 = a\left(1 + \frac{\sin\theta}{\theta}\right)^2 \bigg/ \left(\frac{\sin\theta/2}{\theta/2}\right)^2,$$

whence $\rho = \dfrac{1}{2}\lim\limits_{\substack{x\to 0\\y\to 0}} \dfrac{x^2}{y} = 4a$ at $x=0, y=0, \theta=0$.

Curvature at the origin : Other methods

Example 16.5.3. *[By direct substitution of $(y_1)_0$ and $(y_2)_0$ in Formula 16.3.2.]*
Find the radius of curvature at the origin of the curve

$$x^2 + 6y^2 + 2x - y = 0.$$

Solution: Differentiating the equation w.r.t. x, we get

$$2x + 12yy_1 + 2 - y_1 = 0$$

whence, $(y_1)_0 = 2$ ($\because x=0, y=0$).
Differentiating once more, $2 + 12(y_1^2 + yy_2) - y_2 = 0$, which gives

$$(y_2)_0 = 50 \quad (\because x=0, y=0, (y_1)_0 = 2)$$

$$\therefore \quad \rho = \frac{(1+y_1^2)^{3/2}}{y_2} = \frac{1}{10}\sqrt{5}.$$

Observation: But this method very often fails or becomes laborious. So, we obtain $(y_1)_0 = p$ and $(y_2)_0 = q$ by substituting for y the expression

$$y = px + q\frac{x^2}{2!} + \cdots$$

which is the expansion of y by Maclaurin's Theorem in the given equation of the curve, and equating the coefficient of x to zero in the identity so obtained.

Example 16.5.4. *Find the radius of curvature at the origin of the curve*

$$y^2 - 3xy + 2x^2 - x^3 + y^4 = 0.$$

Solution: Substituting for y its Maclaurin's expansion

$$px + q\frac{x^2}{2!} + \cdots$$

we obtain the identity

$$(p^2 - 3p + 2)x^2 + \left(pq - \frac{3}{2}q - 1\right)x^3 + \cdots = 0$$

whence,
$$\left.\begin{array}{r} p^2 - 3p + 2 = 0 \\ pq - \frac{3}{2}q - 1 = 0 \end{array}\right\} \text{etc.}$$

These give, $p = 1$ or 2; corresponding values of q are -2 or 2.

Therefore, $\rho = \dfrac{(1+p^2)^{3/2}}{q} = -\sqrt{2}$ or $\dfrac{5}{2}\sqrt{5}$.

∴ So, at the origin (double point) the two branches of the curve have their radii curvatures $\sqrt{2}$ and $\frac{5}{2}\sqrt{5}$ but their curvatures are in opposite directions.

16.6 Co-ordinates of Centre of Curvature : Equation of Circle of Curvature

We shall prove that the co-ordinates $(\bar{x}, \bar{y})$ of the centre of curvature a point $P(x, y)$ of a curve $y = f(x)$ are

$$\bar{x} = x - \frac{y_1(1 + y_1^2)}{y_2}, \quad \bar{y} = y + \frac{1 + y_1^2}{y_2}, \quad (y_2 \neq 0) \tag{16.6.1}$$

Proof: We use the fact that the centre of curvature at P is the limiting point of intersection of the normal at P and the normal at a neighbouring point Q when $Q \to P$ along the curve. The equation of the normal at P is

$$(Y - y)\phi(x) + (X - x) = 0 \tag{16.6.2}$$

where the slope of the tangent at $P(x, y)$ is $y_1 = dy/dx = \phi(x)$ and X, Y are the current co-ordinates of any point on the normal.

The normal at a neighbouring point $Q(x + h, y + k)$ is

$$(Y - y - k)\phi(x + h) + (X - x - h) = 0. \tag{16.6.3}$$

At their point of intersection, the ordinate is given by,

$$(Y - y)\{\phi(x + h) - \phi(x)\} - k\phi(x + h) - h = 0. \tag{16.6.4}$$

[subtracting (16.6.2) and (16.6.3)]

CHAPTER 16: CURVATURE

Divide by h, and making $h \to 0$ we get

$$(\bar{y} - y)\left\{\lim_{h \to 0} \frac{\phi(x+h) - \phi(x)}{h}\right\} - \left\{\lim_{h \to 0} \frac{k}{h}\right\}\left\{\lim_{h \to 0} \phi(x+h)\right\} - 1 = 0$$

[since $y \to \bar{y}$ as $h \to 0$]

or, $(\bar{y} - y)\dfrac{d}{dx}\{\phi(x)\} - \phi(x)\phi(x) - 1 = 0$, \hfill (16.6.5)

[assuming the continuity of $\phi(x)$]

or, $\bar{y} = y + \dfrac{1 + y_1^2}{y_2}$ as $\phi(x) = y_1$.

As $(\bar{x}, \bar{y})$ is a point in (16.6.2), we get also

$$(\bar{y} - y)\phi(x) + (\bar{x} - x) = 0$$

$$\frac{1 + \{\phi(x)\}^2}{\phi'(x)} \cdot \phi(x) + \bar{x} - x = 0$$

or, $\bar{x} = x - \dfrac{y_1(1 + y_1^2)}{y_2}$

Geometrical Consideration:

The adjoining diagram (Fig. 16.8) gives the formula (16.6.1) from purely geometric considerations. Here the curve has been drawn convex towards the X-axis. We shall make use of the fact that the centre of curvature lies on the normal at a distance $|\rho|$ from the tangent.

Fig. 16.8 gives the circle of curvature at $P(x, y)$ with centre at C.

Here $\angle MCP = \psi$; co-ordinates of $C = (\bar{x}, \bar{y})$, say; $CP = \rho$.

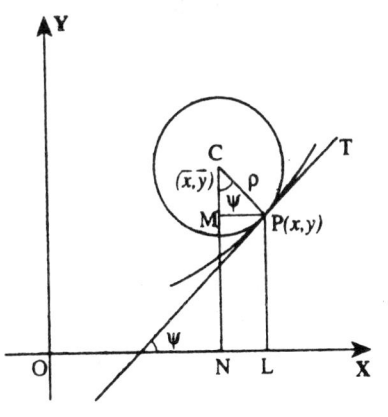

Fig. 16.8

From $\triangle CPM$, $CM = \rho \cos \psi$, $MP = \rho \sin \psi$ i.e., $\bar{y} - y = \rho \cos \psi$ and $x - \bar{x} = \rho \sin \psi$ which gives $\bar{x} = x - \rho \sin \psi$; $\bar{y} = y + \rho \cos \psi$. Putting $\rho = \dfrac{(1 + y_1^2)^{\frac{3}{2}}}{y_2}$ and $\sin \psi = \dfrac{y_1}{\sqrt{1 + y_1^2}}$, $\cos \psi = \dfrac{1}{\sqrt{1 + y_1^2}}$, we may easily obtain $\bar{x}, \bar{y}$ as given in (16.6.1).

Corollary 16.6.1. *If x_1 and x_2 are respectively, the first and second derivatives of x with respect to y, the formula (16.6.1) may be transformed into the form*

$$\bar{x} = x + \frac{1 + x_1^2}{x_2}, \quad \bar{y} = y - \frac{x_1(1 + x_1^2)}{x_2}. \tag{16.6.6}$$

Formula (16.6.6) may be used when y_1 is undefined or if differentiation w.r.t. y is simpler.

Corollary 16.6.2. *The equation of the circle of curvature at $P(x,y)$ is*

$$(x - \bar{x})^2 + (y - \bar{y})^2 = \rho^2.$$

Example 16.6.1. *Find the equation of the circle of curvature of $2xy + x + y = 4$ at the point $(1,1)$.*

Solution: Differentiation gives

$$2y + 2xy_1 + 1 + y_1 = 0 \quad \text{whence,} \quad [y_1]_{(1,1)} = -1.$$

Differentiating once more, we have

$$4y_1 + 2xy_2 + y_2 = 0 \quad \text{whence,} \quad [y_2]_{(1,1)} = \frac{4}{3}.$$

Putting these values in the expressions for P, $\bar{x}$ and $\bar{y}$ we get $\rho = \dfrac{3\sqrt{2}}{2}$; $\bar{x} = \dfrac{5}{2}$; $\bar{y} = \dfrac{5}{2}$ and hence the equation of the circle of curvature is

$$\left(x - \frac{5}{2}\right)^2 + \left(y - \frac{5}{2}\right)^2 = \frac{9}{2}$$

or, $\quad 4x^2 + 4y^2 - 20x - 20y + 32 = 0.$

16.7 Evolute and Involute

Definition 16.7.1. *At any point $P(x,y)$ of a plane curve (Γ), given by $y = f(x)$, where the curvature $k \neq 0$, the centre of curvature $C(\bar{x}, \bar{y})$, is given by*

$$\bar{x} = x - \frac{y_1(1 + y_1^2)}{y_2}, \quad \bar{y} = y + \frac{(1 + y_1^2)}{y_2}. \tag{16.7.1}$$

The locus of C (call it Γ_1) as P moves along the curve is called the **evolute** of the curve.

If Γ_1 be the evolute of Γ, then Γ is often called an **involute** to Γ_1.

We shall show in Chapter **18** (art 18.6.2) how these curves can be mechanically constructed.

Since y, y_1 and y_2 are functions of x, equations (16.7.1) constitute the parametric equations of the evolute, the parameter being x. To obtain the usual cartesian equation of the evolute, x and y will be eliminated between the values for $\bar{x}$ and $\bar{y}$ in (16.7.1), using the fact that the point (x,y) lies on the given curve.

CHAPTER 16: CURVATURE

Example 16.7.1. *Find the equation of the evolute of the parabola* $y^2 = 12x$.

Solution: Here $y_1 = \dfrac{6}{y} = \dfrac{\sqrt{3}}{\sqrt{x}}$; $\quad y_2 = -\dfrac{36}{y^3} = -\dfrac{\sqrt{3}}{2x^{3/2}}$.

Hence,

$$\bar{x} = x - \dfrac{\dfrac{\sqrt{3}}{\sqrt{x}}\left(1+\dfrac{3}{x}\right)}{-\dfrac{\sqrt{3}}{2x^{3/2}}} = 3x+6; \quad \bar{y} = y + \dfrac{1+\dfrac{36}{y^2}}{-\dfrac{36}{y^3}} = -\dfrac{y^3}{36}.$$

In order to eliminate x, y we observe that

$$x = \dfrac{\bar{x}-6}{3}, \quad y = -\sqrt[3]{36\bar{y}}.$$

Substituting these values in the given equation of the parabola, we have

$$(36\bar{y})^{\frac{2}{3}} = 4(\bar{x}-6) \text{ or, } 81\bar{y}^2 = 4(\bar{x}-6)^3.$$

Changing into current co-ordinates (x, y) the equation of the evolute is the *semi-cubical parabola*

$$\mathbf{81y^2 = 4(x-6)^3}.$$

Example 16.7.2. *Find the equation of the evolute of the astroid*

$$x^{2/3} + y^{2/3} = a^{2/3}.$$

Solution: Write the equation of the astroid in parametric form:

$$x = a\cos^3\theta, \quad y = a\sin^3\theta.$$

Now,

$$y_1 = \dfrac{dy}{d\theta}\bigg/\dfrac{dx}{d\theta} = 3a\sin^2\theta\cos\theta/(-3a\cos^2\theta\sin\theta) = -\tan\theta$$

and $y_2 = -\sec^2\theta \dfrac{d\theta}{dx} = \dfrac{1}{3a\cos^4\theta\sin\theta}.$

$$\left.\begin{array}{l} \bar{x} = x - \dfrac{y_1}{y_2}(1+y_1^2) = a\cos^3\theta + 3a\sin^2\theta\cos\theta \\[4pt] \bar{y} = y + \dfrac{1+y_1^2}{y_2} = a\sin^3\theta + 3a\cos^2\theta\sin\theta \end{array}\right\}.$$

Hence, $\bar{x} + \bar{y} = a(\cos\theta + \sin\theta)^3$ and $\bar{x} - \bar{y} = a(\cos\theta - \sin\theta)^3$.

Eliminating θ, we get

$$(\bar{x} + \bar{y})^{2/3} + (\bar{x} - \bar{y})^{2/3} = a^{2/3}(\cos\theta + \sin\theta)^2 + a^{2/3}(\cos\theta - \sin\theta)^2 = 2a^{2/3}.$$

Changing to current co-ordinates (x, y) in place of $(\bar{x}, \bar{y})$ we get the evolute (locus of centre of curvature) in the form

$$(x + y)^{2/3} + (x - y)^{2/3} = 2a^{2/3}.$$

16.8 Chord of Curvature

In art. 16.2, the length q of the chord of curvature PQ which makes an angle θ with the normal PC has been shown to be equal to $2\rho\cos\theta$. We give below a few particular cases which are of importance:

1. *The length q of the chord of curvature parallel to X-axis is $2\rho\sin\psi$.*

   For, if in Fig. 16.3, PQ be drawn parallel to X-axis, the angle θ would be $90° - \psi$, ψ being the angle which the tangent PT makes with the X-axis.

2. *The length q of the chord of curvature parallel to Y-axis is $2\rho\cos\psi$.*

3. *The length q of the chord of curvature through the pole (origin) is $2\rho\sin\phi$.*

   Here OP cuts the circle at Q (Fig. 16.3) and since the radius vector OP makes an angle of ϕ with the tangent PT, we have in this case $\theta = 90° - \phi$ and hence

$$q = 2\rho\cos\theta = 2\rho\cos(90° - \phi) = 2\rho\sin\phi = 2r\frac{dr}{dp}\frac{p}{r} = 2p\frac{dr}{dp} = \frac{2f(r)}{f'(r)},$$

   if the pedal equation of the curve is $p = f(r)$.

4. *The length q of the chord of curvature perpendicular to the radius vector*

$$OP = 2\rho\cos\psi.$$

Example 16.8.1. Find the chord of curvature through the pole of the curve

$$r^n = a^n \cos n\theta.$$

Solution: Differentiating logarithmically, $\dfrac{n}{r}\dfrac{dr}{d\theta} = -n\tan n\theta$ which gives, $\tan\phi = -\cot n\theta = \tan(\pi/2 + n\theta)$, whence, $\sin\phi = \sin(\pi/2 + n\theta) = \cos n\theta = r^n/a^n$, whereby, $p = r\sin\phi = r^{n+1}/a^n$.

$\therefore$ Chord of curvature through the pole $= 2p \cdot \dfrac{dr}{dp} = \dfrac{2r^{n+1}/a^n}{(n+1)r^n/a^n} = \dfrac{2r}{n+1}$.

CHAPTER 16: CURVATURE

Main Points : A Recapitulation

1. Curvature at a point P of a curve $= \lim\limits_{\Delta s \to 0} \dfrac{\Delta \psi}{\Delta s}$ (Fig. 16.1)

    For a circle of radius r, curvature (k) is same at every point $= 1/r$.

2. We define $\rho =$ radius of curvature at a point P of a curve
$$= \frac{1}{k} = \frac{ds}{d\psi}, \text{ provided } k \neq 0.$$

3. Different Expression for ρ:

    a) $\rho = \dfrac{(1+y_1^2)^{3/2}}{y_2} (y_2 \neq 0) \left(\text{or}, \rho = \dfrac{(1+x_1^2)^{3/2}}{x_2} (x_2 \neq 0)\right)$, when the curve is given by $y = f(x)$ [(or $x = \phi(y)$].

    b) $\rho = (f_x^2 + f_y^2)^{3/2} / \{f_{xx} f_y^2 - 2 f_{xy} f_x f_y + f_{yy} f_x^2\}$, when the curve is given in the implicit form $f(x,y) = 0$.

    c) $\rho = \dfrac{(x'^2 + y'^2)^{3/2}}{(x'y'' - y'x'')}$, when the curve is given in the parametric form:
    $$x = f(t), \; y = \phi(t) \text{ (dashes denoting differentiations w.r.t. } t).$$

    d) $\rho = r \dfrac{dr}{dp}$, when the curve is given by a pedal equation of the form $p = f(r)$.

    e) $\rho = (r^2 + r_1^2)^{3/2} / \{r^2 + 2r_1^2 - rr_2\}$, where the curve is given in the polar form: $r = f(\theta)$.

    (suffixes denoting differentiations w.r.t. θ)

    f) $\rho = p + \dfrac{d^2 p}{d\psi^2}$, where the curve is given in tangential polar form: $p = f(\psi)$.

4. Circle of curvature at a point P of a curve is a circle of radius ρ touching the curve at P and having the same sense of concavity as the curve; its centre is called *Centre of Curvature* at P. A chord of this circle through P is called a chord of curvature at P. Its length $q = 2\rho \cos\theta$, where θ is the angle which the chord makes with the diameter of the circle through P.

5. The centre of curvature may also be defined by the limiting position of the point of intersection of two consecutive normals.

6. Newton's approach: The circle of curvature at a point P of a curve is the limiting position of a circle which touches the curve at P and also passing through a neighbouring point Q of the curve.

If the curve passes through the origin and (x, y) is a point of the curve, then

a) when the axis of $X (y = 0)$ is the tangent at the origin, $\rho = \lim\limits_{\substack{x \to 0 \\ y \to 0}} \dfrac{x^2}{2y}$.

b) when the axis of $Y (x = 0)$ is the tangent at the origin, $\rho = \lim\limits_{\substack{x \to 0 \\ y \to 0}} \dfrac{y^2}{2x}$.

c) when $ax + by = 0$ is the tangent at the origin,

$$\rho = \frac{1}{2}\sqrt{a^2 + b^2} \lim_{\substack{x \to 0 \\ y \to 0}} \frac{x^2 + y^2}{ax + by}.$$

7. Co-ordinates of centre of curvature $C(\bar{x}, \bar{y})$ at a point $P(x, y)$ of a curve:

$$\bar{x} = x - \frac{y_1(1 + y_1^2)}{y_2} \quad \text{and} \quad \bar{y} = y + \frac{1 + y_1^2}{y_2} \quad (y_2 \neq 0).$$

The locus of C is called the **evolute** of the **curve**.

Elimination of (x, y) will give the equation of the evolute.

8. Length of chord of curvature parallel to X-axis is $2\rho \sin \psi$; parallel to Y-axis is $2\rho \cos \psi$; passing through the pole (or origin) is $2\rho \sin \phi$.

Exercises XVI (B)

1. Find the radius of curvature at the origin of the following curves:

    (a) $y = x^3 + 5x^2 + 6x$.
    (b) $5x^3 + 7y^3 + 4x^2y + xy^2 + 2x^2 + 3xy + y^2 + 4x = 0$.
    (c) $x^4 - y^4 + x^3 - y^3 + x^2 - y^2 + y = 0$.
    (d) $y^2 - 3xy - 4x^2 + 5x^3 + x^4y - y^5 = 0$.
    (e) $x^3 + y^3 = 3axy$. [C.H.]

2. (a) Prove by Newton's method that the radius of curvature at the vertex of the Catenary $y = c \cosh(x/c)$ is equal to c.

CHAPTER 16: CURVATURE

(b) Find the curvature at origin of each of the two branches of the curve

$$y(ax + by) = cx^3 + ex^2y + fxy^2 + gy^3.$$

(c) Define the curvature of a curve at a point. Find the curvature at the origin of the curve

$$y - x = x^2 + 2xy + y^2 \qquad \text{[C.H. 1981]}$$

[*Hints:* Use Newton's method for both (b) and (c)]

3. (a) Find the radius of curvature, centre of curvature and equation of the circle of curvature of the following curves at the indicated points;

   i. $x^2 = 4ay : (x, y)$.
   ii. $xy = 12 : (3, 4)$.
   iii. $\dfrac{x^2}{a^2} + \dfrac{y^2}{b^2} = 1 : (x, y)$.
   iv. $x^{2/3} + y^{2/3} = a^{2/3} : (x, y)$.
   v. $y = 3x^3 + 2x^2 - 3 : (0, -3)$.
   vi. $x = e^y$; at the point where it crosses the X-axis.
   vii. $x = a(\cos\theta + \theta\sin\theta)$
        $y = a(\sin\theta - \theta\cos\theta)$: at the point θ.
   viii. $y = c\cosh\left(\dfrac{x}{c}\right) : (x, y)$.
   ix. $x + y = ax^2 + by^2 + cx^3 : (0, 0)$.

(b) Show that the radii of curvature of the curve

$$\begin{aligned} x &= ae^\theta(\sin\theta - \cos\theta) \\ y &= ae^\theta(\sin\theta + \cos\theta) \end{aligned} \qquad \text{[C.H.]}$$

and its evolute at corresponding points are equal.

4. Verify the following statements: [**Each is extremely important**]

(a) The evolute of the parabola $x = at^2, y = 2at$ is

$$27ay^2 = 4(x - 2a)^3 \quad \text{(Semi-cubical parabola)}$$

and the equation of its circle of curvature is

$$x^2 + y^2 - 6at^2x - 4ax + 4at^3y - 3a^2t^4 = 0. \qquad \text{[C.H.]}$$

(b) In the ellipse $x = a\cos\phi, y = b\sin\phi$ the evolute has for its equation

$$(ax)^{2/3} + (by)^{2/3} = (a^2 - b^2)^{2/3}.$$

(c) Same problem for the hyperbola $x = a\cosh\theta$, $y = b\sinh\theta$ is

$$(ax)^{2/3} - (by)^{2/3} = (a^2 + b^2)^{2/3}.$$

(d) The evolute of an equiangular spiral is an equiangular spiral of the same angle. [See the **Hints** of Sum No. **7** of Exercises XVIII(B)]

(e) The evolute of the astroid $x^{2/3} + y^{2/3} = a^{2/3}$ is

$$(x+y)^{2/3} + (x-y)^{2/3} = 2a^{2/3}.$$

(f) The evolute of the *tractrix* $x = a(\cos t + \log\tan t/2)$, $y = a\sin t$ is the *Catenary* $y = a\cosh(x/a)$.

(g) Prove that the locus of centres of curvature at points of the cycloid $x = a(t - \sin t)$, $y = a(1 - \cos t)$ is an equal cycloid $x = a(t + \sin t)$, $y = -a(1 - \cos t)$.

[C.H. 1996]

5. Find the chord of curvature through the pole of the curves:

(a) $r = a(1 + \cos\theta)$.

(b) $r = ae^{\theta\cot\alpha}$.

(c) $r^n = a^n \sin n\theta$.

(d) $r^2 = a^2 \cos 2\theta$.

6. (a) In the curve $y = c\log\sec(x/c)$ prove that the chord of curvature parallel to Y-axis is of constant length.

(b) If c_x and c_y be the chords of curvature parallel to the two axes at a point of the *Catenary* $y = c\cosh(x/c)$ prove that

$$4c^2(c_x^2 + c_y^2) = c_y^4$$

and that the value of c_y is double the ordinate.

7. Show that the chord of curvature through the focus of a parabola is four times the focal distance of the point and the chord of curvature parallel to the axis has the same length.

8. Show that the chord of curvature through the pole of the curve

$$p = f(r) \text{ is } 2f(r)/f'(r).$$

CHAPTER 16: CURVATURE

9. Show that the co-ordinates of the centre of curvature at any point (x, y) of any curve may be written as

$$\bar{x} = x - \frac{dy}{d\psi}, \qquad \bar{y} = y + \frac{dx}{d\psi},$$

also as $\bar{x} = \dfrac{1}{2}\dfrac{d^2 r^2}{dy^2} \Big/ \dfrac{d^2 x}{dy^2}, \quad \bar{y} = \dfrac{1}{2}\dfrac{d^2 r^2}{dx^2} \Big/ \dfrac{d^2 y}{dx^2}. \quad (r^2 = x^2 + y^2).$

10. Show that the length of the chord of curvature, parallel to the axis of Y, at the origin, in the parabola $y = mx + x^2/a$ is $(1 + m^2) a$, and the equation of the circle of curvature is

$$x^2 + y^2 = (1 + m^2)a(y - mx).$$

11. For the curve $a^2 y = x^3$ show that the centre of curvature (α, β) is given by

$$\alpha = \frac{x}{2}\left(1 - \frac{9x^4}{a^4}\right); \qquad \beta = \frac{5}{2}\frac{x^3}{a^2} + \frac{a^2}{6x}. \qquad \text{[C.H.]}$$

12. For the lemniscate $r^2 = a^2 \cos 2\theta$ show that the length of the tangent from the origin to the circle of curvature at any point is $r\sqrt{3}/3$.

13. If P is a point on $r^2 = a^2 \cos 2\theta$ and Q is the point of intersection of the normal at P with the line through O at right angles to the radius vector OP, prove that the centre of curvature corresponding to P is the point of trisection of PQ.

Answers

1. (a) $\dfrac{37\sqrt{37}}{10}$; (b) 2; (c) $\dfrac{1}{2}$; (d) $\sqrt{2}, \dfrac{17}{2}\sqrt{17}$; (e) $\dfrac{3}{2}a, \dfrac{3}{2}a$. 2. (b) $2c/a$ and $2(b^3 c - ab^2 c + a^2 bf - ga^3)/a(a^2 + b^2)^{3/2}$. (c) $\tfrac{1}{4}\sqrt{2}$. 3. (a) (i) $\dfrac{2(a+y)^{3/2}}{\sqrt{a}}$; $\left(-\dfrac{x^3}{4a^2}, 2a + \dfrac{3x^2}{4a}\right)$; (ii) $5\dfrac{5}{24}$; $\left(\dfrac{43}{6}; \dfrac{57}{8}\right)$; (iii) $\dfrac{(a^2 - e^2 x^2)^{3/2}}{ab}$; $\left(\dfrac{a^2 - b^2}{a^4}x^3, -\dfrac{a^2 - b^2}{b^4}y^3\right)$; (iv) $3(axy)^{1/3}$; $(x + 3x^{1/3}y^{2/3}, y + 3x^{2/3}y^{1/3})$; (v) $\dfrac{1}{4}$; $(0, -11/4)$; (vi) $\sqrt{8}$; $(3, -2)$; (vii) $a\theta$; $(a\cos\theta, a\sin\theta)$; (viii) y^2/c; $\left(-\dfrac{y\sqrt{y^2 - c^2}}{c}, 2y\right)$; (ix) $\sqrt{2}/(a+b)$; $\left(\dfrac{1}{a+b}, \dfrac{1}{a+b}\right)$.

Circle of curvature may now easily be written in each case, using $(x - \bar{x})^2 + (y - \bar{y})^2 = \rho^2$.

5. (a) $\dfrac{4}{3}r$; (b) $2r$; (c) $2r/(n+1)$; (d) $\dfrac{2}{3}r$.

For a Quick revision on **Curvature** workout Ex. **26–45 of Pages 1163–1165** and **1178–1180**.

AN INTRODUCTION TO ANALYSIS

(Differential Calculus—Part I)

Chapter 17

Rectilinear Asymptotes

Chapter 17

Asympotoes (Rectilinear)

17.1 Preliminary Assumptions

1. We consider two curves

$$\text{circle}: \quad x^2 + y^2 = a^2 \tag{17.1.1}$$
$$\text{rectangular hyperbola}: \quad x^2 - y^2 = a^2. \tag{17.1.2}$$

   Obviously, the curve (17.1.1) lies wholly within a finite part of the xy-plane and is closed but no such assertion is possible in case (17.1.2) where we shall say that curve *extends to infinity*.

2. Our knowledge of implicit functions will tell us that the relation (17.1.2) above defines implicitly *two distinct functions*:

$$y = +\sqrt{x^2 - a^2} \tag{17.1.3}$$
$$y = -\sqrt{x^2 - a^2}. \tag{17.1.4}$$

   In the geometry of curves we shall, however, not consider them as *two distinct curves* but shall suppose that (17.1.2) defines *a curve* having *two branches* (17.1.3) and (17.1.4) and both the branches extend to infinity.

3. We are already familiar with the symbols $x \to \pm\infty$ or $y \to \pm\infty$. But what does $P \to \infty$ stand for, P being a point on an infinite branch of a curve? We agree to the following:

   Definition 17.1.1. *A point P with co-ordinates (x, y) on an infinite branch of a curve is said to tend to infinity $(P \to \infty)$ along the curve if either x or y or both tend to $\pm\infty$ as P traverses along the branch of the curve.*

17.2 Asymptotes : Definition

A straight line is said to be a *rectilinear asymptote* of an infinite branch of a curve if as a point P of the curve tends to infinity along the branch, the perpendicular distance of P from that straight line tends to zero.

Example 17.2.1. *Examine the curve $y^2 - 2xy + 2x = 0$ for an asymptote.*

The given curve has two distinct branches (See Fig. 17.1).

$$y = x + \sqrt{x(x-2)} \quad \text{(solid curve lines)} \tag{17.2.1}$$
$$y = x - \sqrt{x(x-2)} \quad \text{(dotted curve lines).} \tag{17.2.2}$$

In order to trace the curve observe that y is always real except when $0 < x < 2$ and that y increases *numerically* as x increases *numerically*.

Thus, the curve consists of two parts—one lying to the right of the line $x = 2$ and the other to the left of the line $x = 0$. Again, solving for x, we have

$$x = \frac{y^2}{2(y-1)} \left(\text{or, } y - 1 = \frac{y^2}{2x}\right).$$

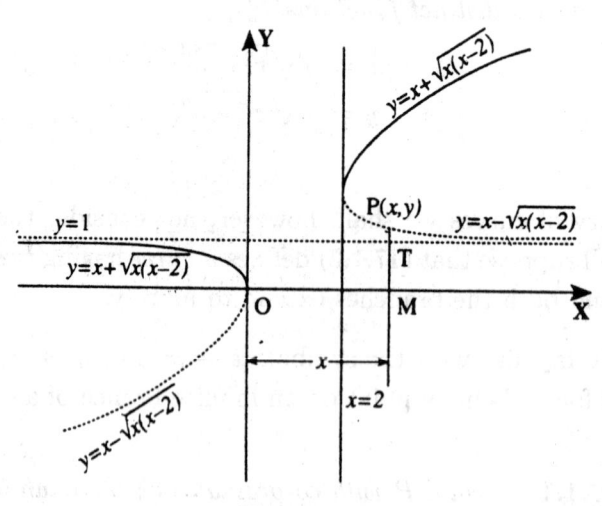

Fig. 17.1

Let us now study the behaviour of the curve with respect to the line $y = 1$. If $P(x, y)$ be a point of the curve in branch (17.2.2) then its distance PT from the line

CHAPTER 17: ASYMPTOTES (RECTILINEAR)

$y = 1$ is $(y - 1)$, which tends to zero as $x \to +\infty$. Similarly, if $P(x, y)$ be a point in branch (17.1.1), its distance from the same line $y = 1$ also tends to zero as $x \to -\infty$. In any case, we may say that as $P \to \infty$ along an infinite branch of the curve its distance from $y = 1$ tends to zero. Hence $y = 1$ is an asymptote to the curve.

Observations:

1. We have defined rectilinear asymptotes of a curve. Another class of asymptotes, called *curvilinear asymptotes*, will not come up for our discussions. For our purpose, therefore, we may drop the word *rectilinear* and use simply *asymptote* to mean a rectilinear asymptote.

2. Asymptotes may be parallel to either of the two axes. Accordingly they are called *horizontal* and *vertical* asymptotes and when otherwise they are *oblique* asymptotes.

3. The above illustration does not indicate a method of determining an asymptote. It simply verifies whether a given line is an asymptote or not.

4. For a curve **lying wholly in a finite region**, the asymptote cannot obviously exist e.g., a circle or an ellipse, has no asymptote at all.

But it does not necessarily mean that a curve having an infinite branch must have an asymptote—asymptotes may or may not exist. e.g., a parabola is a curve extending to infinity but it has no asymptote.

17.3 Asymptotes Parallel to the Axes

I. Asymptotes parallel to Y-axis for the curve : $y = f(x)$. (Vertical asymptotes)

Theorem 17.3.1. *In order that the line $x = a$ may be an asymptote to a curve $y = f(x)$ it is necessary as well as sufficient that $|f(x)| \to \infty$ when either $x \to a-0$, or, $x \to a+0$, or, $x \to a$.*

Proof: First suppose $x \to a - 0$. [Fig. 17.2(a)].

Condition is sufficient

Given: $|f(x)| \to \infty$, i.e., $y \to +\infty$ or, $-\infty$.

Now, if $P(x, y)$ be a point on an infinite branch of the curve then as $x \to a - 0$, the given condition will mean $P \to \infty$.

The perpendicular distance PT of P from the line $x = a$ is $(a - x)$ or, $|x - a|$ which tends to zero as $x \to a - 0$ (i.e., as $P \to \infty$). Hence $x = a$ is an asymptote.

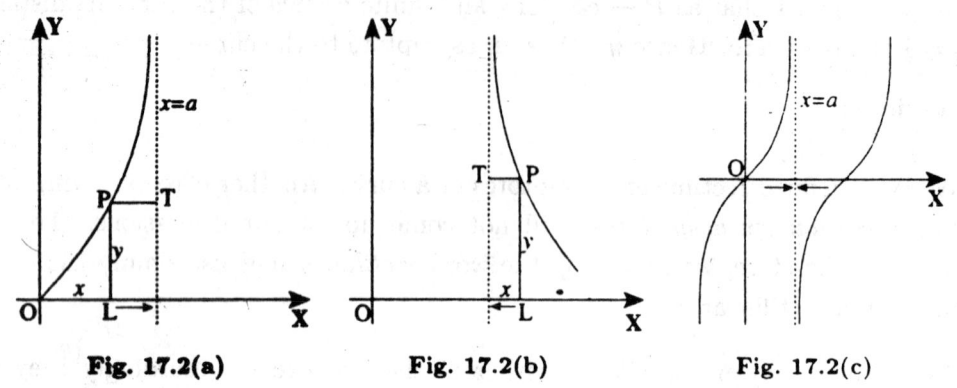

Fig. 17.2(a)   **Fig. 17.2(b)**   **Fig. 17.2(c)**

Condition is necessary.

Given : $x = a$ is an asymptote:

Now as $x \to a - 0$, we must say $|f(x)| \to \infty$; otherwise P cannot tend to ∞ which is essential for obtaining the asymptote.

Proceed with similar arguments when $x \to a + 0$ [Fig. 17.2(b)] or as $x \to a$ [Fig. 17.2(c)].

Example 17.3.1. *Consider the curve $y = e^{1/x}$. Here the axis of Y (i.e., $x = 0$) is an asymptote; for $y \to \infty$ as $x \to 0+0$ and the curve extends to infinity along this branch.*

Note 17.3.1. *If $x = a$ is a vertical asymptote of $y = f(x)$ then $f(x)$ must have an infinite discontinuity at $x = a$.*

Note 17.3.2. *The infinite branch lies on the right or on the left of the asymptote according as $x - a$ is positive or negative.*

II. Asymptotes parallel to X-axis for the curve : $x = \phi(y)$. (Horizontal asymptotes)

Theorem 17.3.2. *In order that the line $y = b$ is an asymptote to the curve $x = \phi(y)$ it is necessary as well as sufficient that $|x|$ or $|\phi(y)| \to \infty$ when either $y \to b + 0$, or, $y \to b - 0$ or, $y \to b$.*

Proof: Exactly similar to the above.

CHAPTER 17: ASYMPTOTES (RECTILINEAR)

Asymptotes parallel to the axes for a rational algebraic curve $F(x,y) = 0$.

The above two theorems will indicate a very simple method of determining asymptotes parallel to the axes for a rational algebraic curve:

$$F(x,y) \equiv y^m \phi(x) + y^{m-1}\phi_1(x) + y^{m-2}\phi_2(x) + \cdots + \phi_m(x) = 0 \qquad (17.3.1)$$

where we have arranged the equation in descending powers of y and $\phi(x), \phi_1(x), \phi_2(x), \cdots$ are polynomials in x.

Re-arranging (17.3.1), we obtain

$$\phi(x) + \frac{1}{y}\phi_1(x) + \frac{1}{y^2}\phi_2(x) + \cdots + \frac{1}{y^m}\phi_m(x) = 0. \qquad (17.3.2)$$

Let $y \to \infty$ and call $\lim_{y \to \infty} x = k$, then (17.3.2) gives $\phi(k) = 0$, so that k is a root of $\phi(x) = 0$.

If it so happens that $k_1, k_2, k_3, \cdots$ are the real roots of $\phi(x) = 0$, then clearly

$$x = k_1, \quad x = k_2, \quad x = k_3, \quad \cdots$$

are the asymptotes parallel to the Y-axis provided, of course, the infinite branches of the curve corresponding to the asymptotes actually exist.

We know from Algebra that if $k_1, k_2, k_3, \cdots$ are the roots of $\phi(x) = 0$, then $\phi(x) = (x - k_1)(x - k_2)(x - k_3)\cdots$. Hence we have the following rule:

Rule 1. *The asymptotes parallel to the Y-axis are obtained by equating to zero the real linear factors in the coefficient of the highest power of y present in the equation of a general algebraic curve $F(x,y) = 0$.*

No such vertical asymptotes exist if the coefficient of the highest power of y is a constant or not resolvable into real linear factors.

By interchanging y and x in arranging the equation (17.3.1) and then proceeding in a similar manner [making $(1/x) \to 0$] we have the second rule:

Rule 2. *The asymptotes parallel to the X-axis are obtained by equating to zero the real linear factors in the coefficient of the highest power of x in the equation of a general algebraic curve $F(x,y) = 0$.*

No such horizontal asymptotes exist if the coefficient of the highest power of x is a constant *or* if its linear factors are imaginary.

Illustrative Examples

Example 17.3.2. *Find the asymptotes of the curve $x^2y^2 - a^2(x^2+y^2) - a^3(x+y) + a^4 = 0$ which are parallel to either axis.*

Solution: The equation gives an algebraic curve. Here the highest power of x is x^2 and its coefficient is $y^2 - a^2$. Hence the asymptotes parallel to X-axis are $y = a, y = -a$.

Similarly, the asymptotes parallel to Y-axis are $x = a, x = -a$.

Note 17.3.3. *Satisfy yourself that the four asymptotes form a square whose vertices are $(a, a), (-a, a), (-a, -a), (a, -a)$, two of these angular points viz. $(-a, a)$ and $(a, -a)$ lie on the curve (check).*

Example 17.3.3. *Examine the asymptotes, if any, parallel to the Y-axis of the curve*

$$x^2 y^2 - 9x^2 + 2 = 0.$$

Solution: Apparently, the coefficient of the highest power of y (i.e., y^2) being x^2, we should say that $x = 0$ (i.e., Y-axis) as an asymptote. But observe that

$$y = \pm \sqrt{\frac{9x^2 - 2}{x^2}}.$$

When x has values very near the origin, y becomes imaginary suggesting that there is no branch of the curve approaching the Y-axis and consequently the existence of an asymptote is beyond question *because the existence of an asymptote presupposes existence of an infinite branch of a curve.*

17.4 Oblique Asymptotes

Theorem 17.4.1. *If an infinite branch of a curve possesses an asymptote $y = mx + c$, (m and c being finite), then*

$$m = \lim_{|x| \to \infty} \frac{y}{x}; \quad c = \lim_{|x| \to \infty} (y - mx) \qquad (17.4.1)$$

and conversely [where (x, y) is a point on the curve].

Proof: Let $P(x, y)$ be a point on an infinite branch of a curve. Its distance from the line

$$y - mx - c = 0, \qquad (17.4.2)$$

is

$$d = \frac{y - mx - c}{\sqrt{1 + m^2}}. \qquad (17.4.3)$$

If the line (17.4.2) is an asymptote then d should tend to zero as $P \to \infty$ (i.e., as $x \to \pm\infty$) which means that

$$\lim_{|x| \to \infty} (y - mx) = c.$$

CHAPTER 17: ASYMPTOTES (RECTILINEAR)

Again, since
$$\frac{y}{x} - m = (y - mx) \cdot \frac{1}{x},$$
we have
$$\lim_{|x| \to \infty} \left(\frac{y}{x} - m\right) = \lim_{|x| \to \infty} (y - mx) \cdot \lim_{|x| \to \infty} \frac{1}{x} = c \cdot 0 = 0,$$
whence $\lim_{|x| \to \infty} \frac{y}{x} = m$.

Conversely, if the limits in (17.4.1) exist as $P \to \infty$, then $y - mx - c \to 0$, as $|x| \to \infty$ and hence $d \to 0$ as $P \to \infty$ which means that the line (17.4.2) is an asymptote.

Note 17.4.1. *The theorem requires m and c to be finite. In particular, m may be zero and hence the asymptotes parallel to X-axis may be obtained by using this theorem (compare art. 17.3.1, Rule 2).*

Again, it may happen that there exists no finite c corresponding to a finite m; we say in this case the asymptote in that direction does not exist. But the direction (given by finite m) is called the **asymptotic direction**.

The parabola $y^2 = 9x$ has an asymptotic direction having gradient
$$m = \lim_{|x| \to \infty} \frac{y}{x} = 0$$
but $c = \lim_{|x| \to \infty} (y - mx) = \lim_{|x| \to \infty} y = \infty$ (not finite).

Hence it has in reality no asymptote.

An Important Rule. The above theorem suggests an important rule for determining asymptotes (not parallel to the Y-axis) for curves having distinct infinite branches:

1. Obtain $\lim_{|x| \to \infty} \left(\frac{y}{x}\right)$ from the equation of the curve $F(x, y) = 0$.

2. For different branches of the curve we may obtain several (finite) values for this limit. Suppose one such value is m_1.

3. Now find $\lim_{|x| \to \infty} (y - m_1 x)$ and suppose the limit $= c_1$ (finite).

4. The line $y = m_1 x + c_1$ is then an asymptote of the curve.

Observation: If the equation of the curve be of the form $y = mx + c + \theta(x)$ where $\theta(x) \to 0$ as $|x| \to \infty$ then the asymptote is given by $y = mx + c$.

If $\theta(x) > 0$, that branch of the curve to which the line $y = mx + c$ is the asymptote lies above the line and if $\theta(x) < 0$ it lies below the line.

Illustrative Examples

Example 17.4.1. *Given the curve $y = \dfrac{5x}{x-3}$. To examine its asymptotes, if any.*

Solution: (a) $y = \dfrac{5x}{x-3}$ has a vertical asymptote $x = 3$.
 Since,
$$\lim_{x \to 3-0} y = -\infty, \quad \lim_{x \to 3+0} y = \infty.$$

[Note that $x = 3$ is a point of infinite discontinuity.]

(b) $\lim\limits_{|x| \to \infty} y = \lim\limits_{|x| \to \infty} \dfrac{5x}{x-3} = 5$; the curve has a horizontal asymptote $y = 5$.

Thus the curve has a vertical asymptote $x = 3$ and also a horizontal asymptote $y = 5$.

Example 17.4.2. *Curve: $y = \dfrac{3x}{x-1} + 3x$. To examine its asymptotes.*

Solution: (a) This curve has a vertical asymptote $x = 1$, since

$$\lim_{x \to 1-0} y = \lim_{x \to 1-0} \left(\dfrac{3x}{x-1} + 3x\right) = -\infty$$

$$\lim_{x \to 1+0} y = \lim_{x \to 1+0} \left(\dfrac{3x}{x-1} + 3x\right) = +\infty.$$

(b) *For the oblique asymptotes*

$$m = \lim_{|x| \to \infty} \dfrac{y}{x} = \lim_{|x| \to \infty} \left(\dfrac{3}{x-1} + 3\right) = 3$$

$$c = \lim_{|x| \to \infty} (y - mx) = \lim_{|x| \to \infty} \left(\dfrac{3x}{x-1} + 3x - 3x\right) = 3.$$

∴ The straight line $y = 3x + 3$ is an oblique asymptote.

Example 17.4.3. *Curve: $y = xe^{1/x}$. To examine its asymptotes.*

Solution: (a) This curve has a vertical asymptote $x = 0$, since

$$\lim_{x \to 0+0} y = \lim_{x \to 0+0} xe^{1/x} = \lim_{t \to \infty} \dfrac{e^t}{t} \quad \left(\text{putting } \dfrac{1}{x} = t\right)$$
$$= \lim_{t \to \infty} e^t = \infty.$$

CHAPTER 17: ASYMPTOTES (RECTILINEAR)

(b) *For the oblique asymptote:*

$$m = \lim_{|x|\to\infty} \frac{y}{x} = \lim_{|x|\to\infty} e^{1/x} = 1$$

$$c = \lim_{|x|\to\infty} (y - mx)$$

$$= \lim_{|x|\to\infty} (xe^{1/x} - x) = \lim_{t\to 0} \frac{e^t - 1}{t} = 1 \text{ (putting } 1/x = t)$$

∴ $y = x + 1$ is an oblique asymptote.

Thus, the curve has a vertical asymptote $x = 0$ and an oblique asymptote $y = x+1$.

Example 17.4.4. *To discuss the asymptotes of the curve:*

$$y = \frac{3x}{2} \log\left(e - \frac{1}{3x}\right).$$

Solution: The function is defined and continuous when $e - 1/3x > 0$, i.e., when $x < 0$ and also when $x > 1/3e$.

Since the function is continuous at every point of the domain of definition, vertical asymptotes can exist only at the end points $x = 0$ and $x = 1/3e$.

(a)
$$\lim_{x\to 0-0} y = \lim_{x\to 0-0} \frac{3x}{2} \log\left(e - \frac{1}{3x}\right)$$

$$= -\frac{1}{2} \lim_{t\to +\infty} \frac{\log(e + t)}{t} = 0 \text{ (putting } t = -1/3x).$$

∴ $x = 0$ is not a vertical asymptote.

(b) $\lim_{x\to \frac{1}{3e}+0} y = \frac{3}{2} \lim_{x\to \frac{1}{3e}+0} x\log(e - 1/3x) = -\infty.$

∴ the line $x = \dfrac{1}{3e}$ is a vertical asymptote.

(c) *For oblique asymptotes:*

$$m = \lim_{|x|\to\infty} \frac{y}{x} = \lim_{|x|\to\infty} \frac{3}{2}\log(e - 1/3x) = \frac{3}{2}$$

$$c = \lim_{|x|\to\infty}(y - mx) = \lim_{|x|\to\infty}\left\{\frac{3x}{2}\log\left(e - \frac{1}{3x}\right) - \frac{3}{2}x\right\}$$

$$= \frac{3}{2}\lim_{|x|\to\infty}\left\{\frac{\log(1 - 1/3xe)}{1/x}\right\} = \frac{3}{2}\left(-\frac{1}{3e}\right) = -\frac{1}{2e}.$$

∴ the line $y = \dfrac{3}{2}x - \dfrac{1}{2e}$ is an oblique asymptote of this curve.

Thus the curve has a vertical asymptote $x = 1/3e$ and an oblique asymptote
$$y = \frac{3}{2}x - \frac{1}{2e}.$$

Example 17.4.5. $y = \sqrt{1+x^2} + 2x$. *Examine the asymptotes of the curve given by this equation.*

Solution: The function is everywhere continuous and therefore there cannot exist any vertical asymptote. But we see that

$$m = \lim_{x \to \infty} \frac{y}{x} = \lim_{x \to \infty} \frac{\sqrt{1+x^2} + 2x}{x} = \lim_{x \to \infty} \frac{\sqrt{1+1/x^2} + 2}{1} = 3$$

$$c = \lim_{x \to \infty} (y - 3x) = \lim_{x \to \infty} (\sqrt{1+x^2} - x) = \lim_{x \to \infty} \frac{1 + x^2 - x}{\sqrt{1+x^2} + x} = 0.$$

Thus as $x \to \infty$, $y = 3x$ has a right asymptote.

Similarly,

$$m_1 = \lim_{x \to -\infty} \frac{y}{x} = \lim_{x \to -\infty} \frac{\sqrt{1+x^2} + 2x}{x} = 1$$

$$c_1 = \lim_{x \to -\infty} (y - x) = \lim_{x \to -\infty} (\sqrt{1+x^2} + x)$$

$$= \lim_{x \to -\infty} \frac{1}{\sqrt{1+x^2} - x} = 0.$$

(since $\sqrt{1+x^2}$ and $(-x)$ are both positive when $x < 0$.)

Thus as $x \to -\infty$, $y = x$ is an asymptote.

Example 17.4.6. *Curve:* $y = \sqrt{1+x^2} \sin(1/x)$. *What are its asymptotes?*

Solution: The curve has no vertical asymptotes since the function is bounded in the neighbourhood of $x = 0$ and at points other than $x = 0$, the function is continuous.

Remember that when the line $x = a$ is a vertical asymptote of the curve $y = f(x)$ then the point $x = a$ is a point of infinite discontinuity of the function $f(x)$.

For oblique asymptotes:

$$m = \lim_{|x| \to \infty} \frac{y}{x} = \lim_{|x| \to \infty} \frac{|x|\sqrt{1 + \frac{1}{x^2}} \sin \frac{1}{x}}{x} = \pm 1.0 = 0.$$

Then

$$c = \lim_{|x| \to \infty} (y - mx) = \lim_{|x| \to \infty} |x|\sqrt{1 + \frac{1}{x^2}} \sin \frac{1}{x}$$

$$\left. \begin{array}{l} = 1 \quad \text{when } x \to +\infty \\ = -1 \quad \text{when } x \to -\infty \end{array} \right\}.$$

CHAPTER 17: ASYMPTOTES (RECTILINEAR)

Thus the curve has two horizontal asymptotes : $y = +1$ and $y = -1$.

Example 17.4.7. *Find the oblique asymptote of the curve whose equation is $y = \dfrac{x^2}{1+x}$ as $x \to \infty$ and justify that in interval $[100, \infty[$ this function may be replaced by a linear function $y = x - 1$ with an error not exceeding 0.01.*

Solution:

$$m = \lim_{x \to \infty} \frac{x^2}{1+x} \cdot \frac{1}{x} = 1$$

$$c = \lim_{x \to \infty} \left(\frac{x^2}{1+x} - x \right) = -1.$$

And hence, the oblique asymptote is $y = x - 1$.

Now see the difference $\delta = \dfrac{x^2}{x+1} - (x-1) = \dfrac{1}{1+x}$.

Hence, if we assume $y = \dfrac{x^2}{1+x} \simeq x - 1$ for all $x > 100$ the error δ is not more than 0.01.

Example 17.4.8. *Obtain the oblique asymptotes of the curve $x^3 + y^3 = 6x^2$.*

Solution: First we observe that there is neither a horizontal nor a vertical asymptote of this third degree curve (the terms of x^3 and y^3 are both present). For oblique asymptotes, let

$$m = \lim_{|x| \to \infty} \frac{y}{x}.$$

From the equation of the curve

$$1 + \left(\frac{y}{x}\right)^3 = \frac{6}{x};$$

as $|x| \to \infty$ we have $1 + m^3 = 0$; which gives $m = -1$ (other values being not real).

And so,

$$c = \lim_{|x| \to \infty} (y - mx) = \lim_{|x| \to \infty} (y + x) \quad (\because m = -1)$$

$$= \lim_{|x| \to \infty} \frac{6x^2}{x^2 - xy + y^2} \quad \text{(from the equation of the curve)}$$

$$= \lim_{|x| \to \infty} \frac{6}{1 - \frac{y}{x} + \left(\frac{y}{x}\right)^2} = 2 \quad \left(\text{since } \frac{y}{x} \to -1\right).$$

∴ The line $y = -x + 2$ is the only asymptote of the curve.

Exactly in a similar way we see that the *Folium of Descartes* $x^3 + y^3 = 3axy$ has the asymptote $y = -x - a$, or, $x + y + a = 0$.

Exercises XVII (A)

1. Obtain the horizontal and vertical asymptotes, if any, of the following curves:

    (a) $y^3 + x^2y + 2xy^2 - y + 1 = 0$. [Ans. $y = 0$]

    (b) $x^4 + x^2y^2 - a^2(a^2 + y^2) = 0$. [Ans. $x = \pm a$]

    (c) $y^2 = x/(x^2 - a^2)$. [Ans. $y = 0, x = \pm a$]

    (d) $xy^3 + x^3y = a^4$. [Ans. $x = 0, y = 0$]

    (e) $y^3 + 5x^2 = 1$. [Ans. No asymptote]

    (f) $y = x/(x^2 - 1)$. [Ans. $x = \pm 1, y = 0$]

    (g) $x^2y^2 - 2x + 4 = 0$. [Ans. No asymptote]

2. Examine the vertical asymptotes of the following curves:

    (a) $y = \tan x$. [Ans. $x = (2n+1)\pi/2 \ (n = 0, \pm 1, \pm 2, \cdots)$]

    (b) $y = (a - x)\tan\left(\dfrac{\pi x}{2a}\right)$. [Ans. $x = -a, x = \pm 3a, x = \pm 5a, \cdots$]

    (c) $y = \log x \ (x > 0)$. [Ans. $x = 0$]

3. Verify that for the curve $y = a \log \sec(x/a)$ the only asymptotes are the vertical asymptotes
$$x = \left(2n\pi \pm \dfrac{\pi}{2}\right)a, \ (n = 0, \pm 1, \pm 2, \cdots).$$
[see that here $\lim\limits_{|x|\to\infty} \dfrac{y}{x}$ does not exist; hence no oblique asymptote.]

4. Check the truth of the following statements:

    (a) $x^3 + y^3 = 3axy$ has an asymptote $x + y + a = 0$.

    (b) $x^3 - x^2y + ay^2 = 0$ has an asymptote $x - y + a = 0$.

    (c) $y^2(x - 1) = x^3$ has asymptotes $x = 1, 2y = 2x + 1$ and $2y + 2x + 1 = 0$.

    (d) $y = e^{-x^2}$ has the horizontal asymptote $y = 0$.

    (e) $y = xe^{1/x^2}$ has $y = x$ as an asymptote.

5. Find the asymptotes of the following curves:

    (a) $y = \dfrac{x}{x^2 + 1}$. [Ans. $y = 0$]

    (b) $y = \dfrac{1}{x} + 4x^2$. [Ans. $x = 0$]

CHAPTER 17: ASYMPTOTES (RECTILINEAR)

(c) $y = 2\sqrt{x^2 + 4}$. [Ans. $y = 2x$ as $x \to +\infty$; $y = -2x$ as $x \to -\infty$]

6. Find the asymptotes of the following curves:

(a) $y = \dfrac{x^2 - 6x + 3}{x + 3}$. [Ans. $x = 3, y = x - 3$]

(b) $y = x\tan^{-1} x$. [Ans. $y = \pm\dfrac{\pi x}{2} - 1$]

(c) $y = x + \dfrac{\sin x}{x}$. [Ans. $y = x$]

(d) $y = \log(4 - x^2)$. [Ans. $x = \pm 2$]

(e) $y = 2x - \cos^{-1}(1/x)$. [Ans. $y = 2x - \pi/2$]

17.5 Determination of Asymptotes Non-Parallel to Y-axis of the General Rational Algebraic Curve $F(x, y) = 0$.

We now proceed to investigate the most general algebraic curve. The method of investigation will be exactly similar to the illustrative examples of art. 17.4.1. Consider a general algebraic curve of nth degree:

$$(a_0 y^n + a_1 y^{n-1} x + a_2 y^{n-2} x^2 + \cdots + a_{n-1} y x^{n-1} + a_n x^n)$$
$$+ (b_1 y^{n-1} + b_2 y^{n-2} x + \cdots + b_{n-1} y x^{n-2} + b_n x^{n-1})$$
$$+ \cdots + (l_{n-1} y + l_n x) + k_n = 0 \qquad (17.5.1)$$

which may be written as

$$x^n \phi_n\left(\dfrac{y}{x}\right) + x^{n-1} \phi_{n-1}\left(\dfrac{y}{x}\right) + \cdots + x\phi_1\left(\dfrac{y}{x}\right) + \phi_0\left(\dfrac{y}{x}\right) = 0 \qquad (17.5.2)$$

where $\phi_r(y/x)$ is a polynomial in y/x of degree r.

Divide (17.5.2) by x^n and then make $|x| \to \infty$ and suppose $\lim\limits_{|x|\to\infty} \dfrac{y}{x} = m$. We then obtain

$$\phi_n(m) = 0 \qquad (17.5.3)$$

which determines the slopes of asymptotes corresponding to different branches of the curve. One value of m is m_1 (say). To obtain c_1 corresponding to m_1, we put

$$y - m_1 x = k_1; \quad k_1 \to c_1 \text{ as } |x| \to \infty.$$

We could also write $\dfrac{y}{x} = m_1 + \dfrac{k_1}{x}$. From (17.5.2), we then get

$$x^n \phi_n\left(m_1 + \dfrac{k_1}{x}\right) + x^{n-1}\phi_{n-1}\left(m_1 + \dfrac{k_1}{x}\right) + \cdots + x\phi_1\left(m_1 + \dfrac{k_1}{x}\right) + \phi_0\left(m_1 + \dfrac{k_1}{x}\right) = 0.$$

Now developing by Taylor's expansion of $f(x)$ in an infinite series, we have

$$x^n \left[\phi_n(m_1) + \frac{k_1}{x}\phi_n'(m_1) + \frac{k_1^2}{2x^2}\phi_n''(m_1) + \cdots \right]$$
$$+ x^{n-1} \left[\phi_{n-1}(m_1) + \frac{k_1}{x}\phi_{n-1}'(m_1) + \frac{k_1^2}{2x^2}\phi_{n-1}''(m_1) + \cdots \right]$$
$$+ x^{n-2} \left[\phi_{n-2}(m_1) + \frac{k_1}{x}\phi_{n-2}'(m_1) + \frac{k_1^2}{2x^2}\phi_{n-2}''(m_1) + \cdots \right]$$
$$+ \cdots\cdots\cdots = 0. \tag{17.5.4}$$

Arranging (17.5.4) in descending powers of x, we get

$$x^n \phi_n(m_1) + x^{n-1}[k_1 \phi_n'(m_1) + \phi_{n-1}(m_1)]$$
$$+ x^{n-2}\left[\{k_1^2/2\}\phi_n''(m_1) + k_1\phi_{n-1}'(m_1) + \phi_{n-2}(m_1)\right]$$
$$+ \cdots\cdots\cdots = 0. \tag{17.5.5}$$

Since $\phi_n(m_1) = 0$ in view of (17.5.3), we have, by dividing (17.5.5) by x^{n-1} and making $|x| \to \infty$ (using $\lim k_1 = c_1$)

$$c_1 \phi_n'(m_1) + \phi_{n-1}(m_1) = 0 \tag{17.5.6}$$

or, $\quad c_1 = -\dfrac{\phi_{n-1}(m_1)}{\phi_n'(m_1)} \quad [\phi_n'(m_1) \neq 0]$.

Thus, $\quad y = m_1 x - \dfrac{\phi_{n-1}(m_1)}{\phi_n'(m_1)}$

is an asymptote corresponding to the slope m_1, provided $\phi_n'(m_1) \neq 0$.

Similarly, the asymptotes corresponding to the slopes

$m_2, m_3, \cdots$ [the other roots of equation (17.5.3)]

can then be successively obtained:

$$y = m_2 x - \frac{\phi_{n-1}(m_2)}{\phi_n'(m_2)} \quad [\phi_n'(m_2) \neq 0]$$
$$y = m_3 x - \frac{\phi_{n-1}(m_3)}{\phi_n'(m_3)} \quad [\phi_n'(m_3) \neq 0] \text{ etc.}$$

CHAPTER 17: ASYMPTOTES (RECTILINEAR)

Case of Parallel Asymptotes

1. If $\phi'_n(m_1) = 0$ but $\phi_{n-1}(m_1) \neq 0$ then the equation (17.5.6) does not determine a value of c_1 and hence there is no asymptote corresponding to the slope m_1.

2. If $\phi'_n(m_1) = 0 = \phi_{n-1}(m_1)$ then (17.5.6) becomes an identity and then we have to go back to equation (17.5.5) which now begins with the term containing x^{n-2}. Now divide (17.5.5) by x^{n-2} and make $|x| \to \infty$ and call $\lim k_1 = c_1$. Then we have

$$\{c_1^2/2\}\phi''_n(m_1) + c_1\phi'_{n-1}(m_1) + \phi_{n-2}(m_1) = 0$$

which gives, in general, two values, say c'_1, c''_1 for c_1 provided

$$\phi''_n(m_1) \neq 0.$$

Thus, we have, in general, two *parallel asymptotes*

$$y = m_1 x + c'_1; \quad y = m_1 x + c''_1.$$

3. The case in which the first, second and third terms of (17.5.5) vanish can be treated in a similar manner.

Note 17.5.1. *The polynomial $\phi_n(m)$ can be obtained by putting $x = 1$ and $y = m$ in the highest degree terms $x^n \phi_n(y/x)$ of (17.5.2) and similarly $\phi_{n-1}(m), \phi_{n-2}(m)$ etc. may be obtained from $x^{n-1}\phi_{n-1}(y/x), x^{n-2}\phi_{n-2}(y/x)$ etc. by putting $x = 1, y = m$.*

Working Rules

Given an algebraic curve of nth degree in the form:

$$F(x,y) = x^n \phi_n(y/x) + x^{n-1}\phi_{n-1}(y/x) + x^{n-2}\phi_{n-2}(y/x) + \cdots = 0.$$

Step 1. Use Rules 1 and 2, art 17.3.1 and obtain the asymptotes parallel to the axes.
Step 2. Put $x = 1, y = m$ in $x^n \phi_n(y/x)$, obtain $\phi_n(m)$.
  Suppose $\phi_n(m) = 0$ gives $m = m_1, m_2, \cdots, m_n$. Also find $\phi'_n(m)$.
Step 3. Put $x = 1, y = m$ in $x^{n-1}\phi_{n-1}(y/x)$, obtain $\phi_{n-1}(m)$.
Step 4. For $m = m_1$, obtain $c = c_1 = -\dfrac{\phi_{n-1}(m_1)}{\phi'_n(m_1)}$ $[\phi'_n(m_1) \neq 0]$.

Then $y = m_1 x + c_1$ is an asymptote. Obtain similarly other asymptotes, if any, corresponding to $m = m_2, m_3$ etc.

Step 5. If $\phi'_n(m) = 0$ for some value of m, investigate for parallel asymptotes (discussed above). Remember in this case, usually

$$\frac{c^2}{2}\phi''_n(m) + c\phi'_{n-1}(m) + \phi_{n-2}(m) = 0$$

will give two values of c for a given value of $m = m_1$.

Illustrative Examples

Example 17.5.1. *Find the asymptotes of*

$$xy^2 - y^2 - x^3 = 0.$$

Solution: **1.** The coefficient of highest power of y (viz., y^2) is $(x-1)$; the line $x=1$ is, therefore, a vertical asymptote. There are no horizontal asymptotes.

2. Put $x=1, y=m$ in $(xy^2 - x^3)$, (the terms of 3rd degree)

$$\phi_3(m) = m^2 - 1; \quad \phi'_3(m) = 2m$$
$$\phi_3(m) = 0 \text{ gives } m = +1, -1.$$

3. Put $x=1, y=m$ in the terms of 2nd degree $(-y^2)$.

$$\phi_2(m) = -m^2.$$

4. For $m=-1$, $c = -\dfrac{\phi_2(1)}{\phi'_3(1)} = \dfrac{1}{2}$; asymptote: $y = x + \dfrac{1}{2}$.

For $m=-1$, $c = -\dfrac{1}{2}$; asymptote: $y = -x - \dfrac{1}{2}$.

The required asymptotes are

$$x = 1, \quad y = x + \frac{1}{2}, \quad y = -x - \frac{1}{2}.$$

Example 17.5.2. *Find the asymptotes of*

$$x^3 + x^2y - xy^2 - y^3 + x^2 - y^2 = 2.$$

Solution: **1.** The coefficients of x^3 and y^3 are constants; hence there are no horizontal or vertical asymptotes.

2. Put $x=1, y=m$ in $(x^3 + x^2y - xy^2 - y^3)$, the third degree terms.

$$\phi_3(m) = 1 + m - m^2 - m^3; \quad \phi'_3(m) = 1 - 2m - 3m^2.$$
$$\phi_3(m) = 0 \text{ gives } m = 1, -1, -1.$$

CHAPTER 17: ASYMPTOTES (RECTILINEAR)

3. Put $x = 1, y = m$ in $(x^2 - y^2)$, the second degree terms,
$$\phi_2(m) = 1 - m^2.$$

4. For $m = 1, c = -\dfrac{\phi_2(1)}{\phi_3'(1)} = 0$; asymptote : $y = x$.

5. Since $\phi_3'(-1) = 0$ we shall investigate for parallel asymptotes. Observe $\phi_2(-1) = 0$. The values of c can then be obtained from

$$\dfrac{c^2}{2}\phi_3''(-1) + c\phi_2'(-1) + \phi_1(-1) = 0 \quad \text{or,} \quad \dfrac{c^2}{2} \cdot 4 + c \cdot 2 + 0 = 0,$$

whence, $c = 0, -1$.

The two parallel asymptotes are $y = -x$, $y = -x - 1$, and hence the required asymptotes are $y = x, y = -x, y = -x - 1$.

Example 17.5.3. Asymptotes of curves given in Parametric form: *Find the asymptotes of the curve*

$$x = \dfrac{t^2}{1 + t^3}, \quad y = \dfrac{t^2 + 2}{1 + t}. \qquad \text{[C.H. 1996]}$$

Solution: When $t \to -1$, both $x \to \infty$ and $y \to \infty$.

We know that $y = mx + c$ is the asymptote where $m = \lim \dfrac{y}{x}$ when $|x| \to \infty$ and $c = \lim_{|x| \to \infty} (y - mx)$.

We have,
$$m = \lim_{|x| \to \infty} \dfrac{y}{x} = \lim_{t \to -1} \dfrac{t^2 + 2}{1 + t} \cdot \dfrac{1 + t^3}{t^2}$$
$$= \lim_{t \to -1} \dfrac{(t^2 + 2)(1 - t + t^2)}{t^2} \quad (\because t \to -1, 1 + t \neq 0)$$
$$= \dfrac{(1 + 2(1 + 1 + 1)}{(-1)^2} = 9.$$

Again,
$$c = \lim_{t \to -1}(y - 9x) = \lim_{t \to -1}\left(\dfrac{t^2 + 2}{1 + t} - \dfrac{9t^2}{1 + t^3}\right)$$
$$= \lim_{t \to -1} \dfrac{(t^2 + 2)(1 - t + t^2) - 9t^2}{1 + t^3}$$
$$= \lim_{t \to -1} \dfrac{t^4 - t^3 - 6t^2 - 2t + 2}{1 + t^3}$$
$$= \lim_{t \to -1} \dfrac{(t + 1)(t^3 - 2t^2 - 4t + 2)}{(1 + t)(1 - t + t^2)}$$
$$= \lim_{t \to -1} \dfrac{t^3 - 2t^2 - 4t + 2}{1 - t + t^2} = \dfrac{-1 - 2 + 4 + 2}{1 + 1 + 1} = 1.$$

∴ $y = 9x + 1$ is the asymptote of the given curve.

Example 17.5.4. *Find the asymptotes of the curve*

$$x = \frac{t^2 + 1}{t^2 - 1}, \quad y = \frac{t^2}{t - 1}.$$

Solution: (a) We see that when $t \to -1, x \to \infty, y \to -\frac{1}{2}$, i.e., $y = -\frac{1}{2}$ is an asymptote parallel to the X-axis.

(b) For oblique asymptote we find then when $t \to 1$, x and y both tend to ∞. If $y = mx + c$ be the oblique asymptote

$$m = \lim_{t \to 1} \frac{y}{x} = \lim_{t \to 1} \frac{t^2}{t - 1} \cdot \frac{t^2 - 1}{t^2 + 1} = \lim_{t \to 1} \frac{t^2(t + 1)}{t^2 + 1} = 1$$

and $c = \lim_{t \to 1}(y - mx) = \lim_{t \to 1}(y - x) \quad (\because m = 1)$

$$= \lim_{t \to 1}\left\{\left(\frac{t^2}{t - 1}\right) - \left(\frac{t^2 + 1}{t^2 - 1}\right)\right\} = \frac{3}{2}.$$

Hence $y = x + \frac{1}{2}$ is the oblique asymptote.

17.6 An Alternative Method for Finding Asymptotes of Algebraic Curves

We consider the equation of an algebraic curve of degree n in the form (17.5.1) of art 17.5.

A. Let $y - m_1 x$ be a non-repeated factor of the nth degree terms of the equation of the curve. We can now put the equation of the curve as

$$(y - m_1 x)F_{n-1} + P_{n-1} = 0, \tag{17.6.1}$$

where F_{n-1} contains only terms of degree $(n - 1)$ and P_{n-1} contains terms of degree not higher than $(n - 1)$.

Clearly, m_1 is a root of $\phi_n(m) = 0$ of the previous article and hence there might exist an asymptote $y = m_1 x + c_1$ provided we can determine a finite value for c_1.

But, by art. 17.4 $c_1 = \lim_{|x| \to \infty} (y - m_1 x)$, where (x, y) lies on an infinite branch of the given curve (17.6.1).

From (17.6.1), we also have

$$y - m_1 x = -\frac{P_{n-1}}{F_{n-1}}.$$

CHAPTER 17: ASYMPTOTES (RECTILINEAR)

Therefore, $c_1 = \lim\limits_{|x|\to\infty} (y - m_1 x) = -\lim\limits_{|x|\to\infty} P_{n-1}/F_{n-1}$.

The asymptote under discussion is then

$$y - m_1 x - \lim_{|x|\to\infty}\left(-\frac{P_{n-1}}{F_{n-1}}\right) = 0.$$

[In determining the last limit, use $\lim\limits_{|x|\to\infty}\dfrac{y}{x} = m_1$]

For each non-repeated linear factor of the nth degree terms we may proceed in a similar way.

B. Let the terms of nth degree in the equation of the curve contain $(y - m_1 x)^2$ as a factor and further suppose $y - m_1 x$ is not a factor of $(n-1)$th degree terms. Then proceeding as in case A, we can easily show that $(y - m_1 x)^2$ does not tend to a finite limit and hence there is no asymptote with slope m_1.

C. Suppose, on the other hand, we could write the equation of the curve in the form

$$\underbrace{(y - m_1 x)^2 F_{n-2}}_{n\text{th degree terms}} + \underbrace{(y - m_1 x) P_{n-2}}_{(n-1)\text{th degree terms}} + \underbrace{Q_{n-2}}_{\text{terms not higher than }(n-2)\text{th degree}} = 0$$

where F_{n-2}, P_{n-2} contain only terms of $(n-2)$th degree and Q_{n-2} contains terms of various degrees none of which is higher than $n - 2$. Then, on similar arguments as in A,

$$(y - m_1 x)^2 + (y - m_1 x) \times \lim_{|x|\to\infty}\frac{P_{n-2}}{F_{n-2}} + \lim_{|x|\to\infty}\frac{Q_{n-2}}{F_{n-2}} = 0$$

will be the pair of parallel asymptotes. In determining the limits, the use of $\lim\limits_{|x|\to\infty}\dfrac{y}{x} = m_1$ will be necessary.

D. We can proceed exactly in a similar manner if the nth degree terms contain $(y - m_1 x)^3$ or higher power of $(y - m_1 x)$, as a factor.

E. If in case A, we have the factor $(ax + by + c)$ instead of $(y - m_1 x)$ i.e., if the equation of the curve be of the form

$$(ax + by + c) F_{n-1} + P_{n-1} = 0,$$

a little consideration will show that the asymptote corresponding to the factor $ax + by + c$ will be

$$ax + by + c + \lim_{|x|\to\infty}\frac{P_{n-1}}{F_{n-1}} = 0, \quad \text{where} \quad \lim_{|x|\to\infty}\frac{y}{x} = -\frac{a}{b}.$$

Observations:

1. The directions of the asymptotes not parallel to Y-axis are given by the roots of $\phi_n(m) = 0$ which is of degree not higher than n. We may thus expect at most n values of m which give the slopes of asymptotes. For each value of m we have, in general, one value of c determined from $c\phi'_n(m) + \phi_{n-1}(m) = 0$. **Hence an algebraic curve of nth degree has, in general, n asymptotes.**

2. In case the curve possesses one or more asymptotes parallel to the Y-axis then we may easily verify that the degree of $\phi_n(m) = 0$ will be inferior to n by at least the same number.

3. Observe carefully that there can never be more than n asymptotes because, even when the equation determining c is a quadratic (case of parallel asymptotes), then we have two equal values of m in $\phi_n(m) = 0$ (in view of $\phi'_n(m) = 0$). Therefore, the two values of c correspond to two equal roots and there would be at the most $(n-2)$ other asymptotes corresponding to $(n-2)$ remaining roots. If the equation determining c is a cubic, it can be easily shown that $\phi_n(m)$ has three equal roots; and so on.

We, therefore, conclude that **a curve of nth degree can have n asymptotes at most.**

Working Rules

To determine asymptotes of a general algebraic curve of nth degree given by (17.5.1) of art. 17.5 proceed as:

Step 1. The degree of the curve being n, it can have at most n asymptotes.

Step 2. Use Rules 1 and 2, art. 17.3 to obtain asymptotes parallel to the axes.

Step 3. Factorise the highest degree (nth degree) terms into all possible real linear factors. Suppose $(y - m_1 x)$ [ot $(ax + by + c)$] is one such non-repeated factor. Apply case A [or E] of art. 17.5, to obtain an asymptote corresponding to this factor. Proceed in a similar way with every such non-repeated factor.

Step 4. (a) If $(y - m_1 x)^2$ [or $(ax + by + c)^2$] is a factor of the nth degree terms there may exist parallel asymptotes or may not. If, at the same time $(y - m_1 x)$ is a factor of $(n-1)$th degree terms [or as a special case, there are no terms of $(n-1)$th degree], then parallel asymptotes do exist; otherwise not. If this conditions happens then proceed as in case C of art. 17.5 for obtain parallel asymptotes.

CHAPTER 17: ASYMPTOTES (RECTILINEAR)

(b) In order that parallel asymptotes corresponding to the factor $(y - m_1 x)^3$ may exist we must have $(y - m_1 x)^2$ a factor of $(n-1)$th degree terms and $(y - m_1 x)$ as a factor of $(n-2)$th degree terms. Similar assertion for higher powers of $(y - m_1 x)$.

Note 17.6.1. *We leave it to the students to choose between the two methods, art. 17.5 and 17.6. But it will be more convenient to use the above rules specially in case of parallel asymptotes.*

Illustrative Examples

Example 17.6.1. *Find all the asymptotes of* $x^3 - 2x^2 y + xy^2 + x^2 - xy + 2 = 0$.

Solution:

Step 1. The curve is of third degree; there can be at most 3 asymptotes.

Step 2. (a) The coefficient of highest power of x is constant; there are no horizontal asymptotes.

(b) The coefficient of highest power of y (here y^2) is x; the asymptote parallel to Y-axis is $x = 0$.

Step 3. Factorising the terms of third degree, the given equation becomes

$$x(y-x)^2 - x(y-x) + 2 = 0.$$

Since $(y-x)^2$ is a factor of third degree terms and $(y-x)$ is a factor of second degree terms, parallel asymptotes, if any, will be given by

$$(y-x)^2 - (y-x) \lim_{|x| \to \infty} \frac{x}{x} + \lim_{|x| \to \infty} \frac{2}{x} = 0.$$

i.e., $(y-x)^2 - (y-x) = 0$, or, $y - x = 0$; $y - x - 1 = 0$.

The three asymptotes are $x = 0$; $y - x = 0$; $y - x = 1$.

Example 17.6.2. *Find the asymptotes of*

$$(y + x + 1)(y + 2x + 2)(y + 3x + 3)(y - x) + x^2 + y^2 - 8 = 0.$$

Solution: The asymptote parallel to $y + x + 1 = 0$ is

$$y + x + 1 + \lim_{|x| \to \infty} \frac{x^2 + y^2 - 8}{(y + 2x + 2)(y + 3x + 3)(y - x)} = 0$$

or, $\quad y + x + 1 + \lim_{|x| \to \infty} \dfrac{\dfrac{1}{x} + \left(\dfrac{y}{x}\right)^2 \cdot \dfrac{1}{x} - \dfrac{8}{x^2}}{\left(\dfrac{y}{x} + 2 + \dfrac{2}{x}\right)\left(\dfrac{y}{x} + 3 + \dfrac{3}{x}\right)\left(\dfrac{y}{x} - 1\right)} = 0,$

i.e., $y + x + 1 + 0 = 0.$ $\quad \left(\text{using } \lim_{|x| \to \infty} \dfrac{y}{x} = -1 \right)$

Similarly, it can be shown that

$$y + 2x + 2 = 0, \quad y + 3x + 3 = 0, \quad y - x = 0.$$

are also the asymptotes of the curve.

This example suggests the following rule for finding asymptotes by inspection.

Asymptotes by Inspection

If the equation of a curve be of the form

$$F_n + F_{n-2} = 0$$

where F_n is a polynomial of degree n and F_{n-2} is a polynomial of degree $(n-2)$ at the most and if F_n can be broken up into n distinct linear factors so that when equated to zero they represent n straight lines, no two of which are parallel then all the asymptotes of the curve are given by $F_n = 0$,

e.g., the hyperbola $\dfrac{x^2}{a^2} - \dfrac{y^2}{b^2} = 1$ have asymptotes $\dfrac{x}{a} \pm \dfrac{y}{b} = 0$.

Example 17.6.3. *Examine the conic*

$$ax^2 + 2hxy + by^2 + 2gx + 2fy + c = 0$$

for asymptotes.

Solution: The highest degree terms are: $ax^2 + 2hxy + by^2$.

It can be factorised as $b(y - m_1 x)(y - m_2 x)$ where m_1, m_2 are

$$(-h + \sqrt{h^2 - ab})/b \quad \text{and} \quad (-h - \sqrt{h^2 - ab})/b \quad \text{respectively.}$$

Case 1. $h^2 - ab > 0$. (Hyperbola).

We have two real distinct values m_1, m_2. Corresponding to $(y - m_1 x)$ we have the asymptote

$$y - m_1 x + \lim_{|x| \to \infty} \frac{2gx + 2fy + c}{b(y - m_2 x)} = 0,$$

i.e., $\quad y - m_1 x + \lim_{|x| \to \infty} \dfrac{2g + 2f\left(\dfrac{y}{x}\right) + \dfrac{c}{x}}{b\left(\dfrac{y}{x} - m_2\right)} = 0$

i.e., $\quad y - m_1 x + \dfrac{2g + 2fm_1}{b(m_1 - m_2)} = 0. \quad \left(\text{use } \lim_{|x| \to \infty} \dfrac{y}{x} = m_1\right)$

CHAPTER 17: ASYMPTOTES (RECTILINEAR)

Similarly, we have the other asymptote

$$y - m_2 x + \frac{2g + 2fm_2}{b(m_2 - m_1)} = 0.$$

Case 2. $h^2 - ab < 0$ (Ellipse).

We have two imaginary values of m_1, m_2. Hence no real asymptotes exist.

Case 3. $h^2 - ab = 0$ (Parabola).

Here $m_1 = m_2 = -\frac{h}{b}$.

Thus, $\left(y + \frac{h}{b}x\right)^2$ is a factor of second degree terms but $\left(y + \frac{h}{b}x\right)$ is not a factor of the first degree terms. Hence there exists no asymptote parallel to $y + \frac{h}{b}x = 0$.

Otherwise. Using notations of art. 17.5,

$$\phi_2(m) = a + 2hm + bm^2, \quad \phi'_2(m) = 2(h + bm), \quad \phi_1(m) = 2(g + fm).$$

Since $m = -h/b$ see that $\phi'_2(m) = 0$ but $\phi_1(m) \neq 0$.

Hence there exists no asymptote.

Exercises XVII (B)

[A]

Find the asymptotes of the following curves:

1. $y^3 - x^2 y - 2xy^2 + 2x^3 - 7xy + 3y^2 + 2x^2 + 2x + 2y + 1 = 0.$

2. $2x^3 + 3x^2 y - 3xy^2 - 2y^3 + 3x^2 - 3y^2 + y = 3.$

3. $4x^3 - 3xy^2 - y^3 + 2x^2 - xy - y^2 - 1 = 0.$

4. $x^3 + 3x^2 y - 4y^3 - x + y + 3 = 0.$

5. $2x^3 - x^2 y - 2xy^2 + y^3 - 4x^2 + 8xy - 4x + 1 = 0.$

[For this algebraic curve of third degree there can be at most 3 asymptotes. There are no asymptotes parallel to either X-axis or Y-axis, since x^3 and y^3 terms are both present. To find *oblique asymptotes*:

Put $x = 1, y = m$ in the third degree terms: $\phi_3(m) = 2 - m - 2m^2 + m^3$.

Put $x = 1, y = m$ in the second degree terms: $\phi_2(m) = -4 + 8m$.

The slopes of the asymptotes are given by $\phi_3(m) = 0 \Rightarrow m = -1, 1, 2.$

Again, the corresponding c's are given by $c\phi_3'(m) + \phi_2(m) = 0$

or, $c(-1 - 4m + 3m^2) + (8m - 4) = 0$.

For $m = -1, c = 2$. For $m = 1, c = 2$. For $m = 2, c = -4$.

∴ **The three asymptotes are** $y = -x + 2$, $y = x + 2$, $y = 2x - 4$.

Alternative method: The given equation can be written as

$$(x^2 - y^2)(2x - y) - 4x^2 + 8xy - 4x + 1 = 0.$$

or, $(y + x)(y - x)(y - 2x) + (8xy - 4x^2) + (1 - 4x) = 0.$

∴ Asymptote parallel to $y + x = 0$ is

$$y + x + \lim_{\substack{|x|\to\infty \\ y/x \to -1}} \frac{4x(2y - x)}{(y - x)(y - 2x)} + \lim_{\substack{|x|\to\infty \\ y/x \to -1}} \frac{1 - 4x}{(y - x)(y - 2x)} = 0$$

or, $y + x + \lim_{\substack{|x|\to\infty \\ y/x \to -1}} \dfrac{4(2y/x - 1)}{(y/x - 1)(y/x - 2)} + \lim_{\substack{|x|\to\infty \\ y/x \to -1}} \dfrac{1/x^2 - 4/x}{(y/x - 1)(y/x - 2)} = 0$

or, $y + x + \dfrac{4 \times -3}{-2 \times -3} + 0 = 0$

or, $y + x - 2 = 0$.

Similarly other asymptotes can be found out.]

6. $x^3 - x^2y - xy^2 + y^3 + 2x^2 - 4y^2 + 2xy + x + y + 1 = 0.$

[Here $\phi_3(m) = 1 - m - m^2 + m^3 = 0 \Rightarrow m = 1, 1, -1$.

For $m = -1$, $c\phi_3'(m) + \phi_2(m) = 0 \Rightarrow c = 1$.

i.e., $y = -x + 1$ is an asymptote.

For $m = 1$, $c\phi_3'(m) + \phi_2(m)$ becomes $c \cdot 0 + 0 = 0 \; \forall \, c$.

In this case, c is determined by

$$\frac{c^2}{2}\phi_3''(m) + c\phi_2'(m) + \phi_1(m) = 0$$

or, $\dfrac{c^2}{2}(-2 + 6m) + c(2 - 8m) + (1 + m) = 0.$

For $m = 1$ this gives $2c^2 - 6c + 2 = 0 \Rightarrow c = (3 \pm \sqrt{5})/2$.

∴ $y = x + \dfrac{3 \pm \sqrt{5}}{2}$ are two parallel asymptotes.

CHAPTER 17: ASYMPTOTES (RECTILINEAR)

Alternative method for parallel asymptotes:

Given equation: $(y-x)^2(y+x) - 2(y-x)(x+2y) + x + y + 1 = 0$.

Asymptotes parallel to $y - x = 0$ are given by

$$(y-x)^2 - 2(y-x)\lim_{\substack{|x|\to\infty \\ \frac{y}{x}\to 1}} \frac{x+2y}{y+x} + \lim_{\substack{|x|\to\infty \\ \frac{y}{x}\to 1}} \frac{x+y+1}{y+x} = 0$$

or, $(y-x)^2 - 2(y-x)\lim\limits_{\substack{|x|\to\infty \\ \frac{y}{x}\to 1}} \dfrac{1+2y/x}{y/x+1} + \lim\limits_{\substack{|x|\to\infty \\ \frac{y}{x}\to 1}} \dfrac{1+y/x+1/x}{y/x+1} = 0$

or, $(y-x)^2 - 3(y-x) + 1 = 0$ or, $y - x = \dfrac{3 \pm \sqrt{5}}{2}$.

Answers

[A]

1. $y = x - 1, y + x + 2 = 0, y = 2x$. 2. $y = x, y + 2x + 1 = 0, 2y + x + 1 = 0$.
3. $y = x, y = -2x, y = -2x - 1$. 4. $y = x, 2y + x = \pm 1$.

[B]

Find the asymptotes of the following curves:

1. $x^2 y^2 - x^2 y - xy^2 + x + y + 1 = 0$. [C.H.]

2. $y^3 = x(a^2 - x^2)$.

3. $xy^2 - x^2 y = a^2(x + y) + b^2$.

4. $3x^3 + 2x^2 y - 7xy^2 + 2y^3 - 14xy + 7y^2 + 4x + 5y = 0$.

5. $x^3 + x^2 y - xy^2 - y^3 + 2xy + 2y^2 - 3x + y = 0$.

6. $y^3 - 5xy^2 + 8x^2 y - 4x^3 - 3y^2 + 9xy - 6x^2 + 2y - 2x + 1 = 0$.

7. $x^3 y - 2x^2 y^2 + xy^3 = a^2 x^2 + b^2 y^2$.

8. $y^3 + x^2 y + 2xy^2 - y + 1 = 0$. [C.H. 1984]

9. $(x - y + 2)(2x - 3y + 4)(4x - 5y + 6) + 5x - 6y + 7 = 0$.

10. $(y - a)^2(x^2 - a^2) = x^4 + a^4$.

11. $(x+y)^2(x+2y+2) = x + 9y + 2$.

12. $x^4 - 5x^2y^2 + 4y^4 + x^2 - 2y^2 + 2x + y + 7 = 0$.

13. $y^4 - 2x^2y^2 - x^4 + 2axy^2 - 5ax^3 + 2x + 3y - 1 = 0$.

14. $x = \dfrac{2t}{t^2-1}, y = \dfrac{(t+1)^2}{t^2}$.

15. $x = \dfrac{1}{t^4-1}, y = \dfrac{t^3}{t^4-1}$.

Answers

[B]

1. $x = 0, y = 0, x = 1, y = 1$. 2. $x + y = 0$. 3. $x = 0, y = 0, x - y = 0$.
4. $6y - 6x + 7 = 0, 2y - 6x + 3 = 0, 6y + 3x + 5 = 0$.
5. $x + y + 1 = 0, x + y - 2 = 0, x - y + 1 = 0$. 6. $y = x, y = 2x + 2, y = 2x + 1$.
7. $x = 0, y = 0, x - y = \pm\sqrt{a^2 + b^2}$. 8. $y = 0, x + y = \pm 1$.
9. $x - y + 2 = 0, 2x - 3y + 4 = 0, 4x - 5y + 6 = 0$. 10. $y \pm x = a, x = \pm a$.
11. $x + y = \pm 2\sqrt{2}, x + 2y + 2 = 0$. 12. $x \pm 2y = 0, x \pm y = 0$.
13. $y = \pm\{8(\sqrt{2}+1)x + (3\sqrt{2}-4)a\}/8(\sqrt{2}+1)$.
14. $x = 0, y = 0, y = 4$.
15. $4y = 4x + 3, 4y + 4x + 3 = 0$.

17.7 Intersection of a Curve with its Asymptotes

Theorem 17.7.1. *Any asymptote of an algebraic curve of nth degree cuts the curve in $(n-2)$ points.*

Proof: The abscissae of the points of intersection of the curve

$$x^n \phi_n(y/x) + x^{n-1}\phi_{n-1}(y/x) + x^{n-2}\phi_{n-2}(y/x) + \cdots = 0 \tag{17.7.1}$$

with one of its asymptotes

$$y = mx + c \tag{17.7.2}$$

are the roots of

$$x^n \phi_n\left(m + \frac{c}{x}\right) + x^{n-1}\phi_{n-1}\left(m + \frac{c}{x}\right) + \cdots = 0.$$

CHAPTER 17: ASYMPTOTES (RECTILINEAR)

Developing by Taylor's theorem and arranging, we obtain

$$x^n \phi_n(m) + x^{n-1}[c\phi'_n(m) + \phi_{n-1}(m)]$$
$$+ \left[\frac{1}{2}c^2 \phi''_n(m) + c\phi'_{n-1}(m) + \phi_{n-2}(m)\right] x^{n-2} + \cdots = 0. \qquad (17.7.3)$$

Since (17.7.2) is an asymptote of (17.7.1), $\phi_n(m) = 0$ and $c\phi'_n(m) + \phi_{n-1}(m) = 0$.

Thus, (17.7.3) reduces to an equation of $(n-2)$th degree and hence gives $(n-2)$ values of x. Hence the theorem.

Corollary 17.7.1. *It then follows that the n asymptotes of a curve of nth degree cut it at $n(n-2)$ points.*

Observation: If a curve of the nth degree has n asymptotes, no two parallel, we have seen in art. 17.6 that the equations of the asymptotes and of the curve may be respectively written as

$$F_n = 0 \quad \text{and} \quad F_n + F_{n-2} = 0.$$

The n asymptotes of the curve therefore intersect at points lying upon the curve $F_{n-2} = 0$.

e.g., (a) The asymptotes of a *cubic curve* will cut the curve again in $3(3-2) = 3$ points which lie on a curve of degree $3 - 2 = 1$ i.e., on a *straight line*.

(b) The asymptotes of a *quartic curve* (or curve of fourth degree) will cut the curve in eight $[4(4-2)]$ points lying on a *conic* [whose degree is $4 - 2 = 2$].

Example 17.7.1. *Find the equation of the cubic which has the same asymptotes as the curve $x^3 - 6x^2y + 11xy^2 - 6y^3 + x + y + 1 = 0$ and which touches the axis of Y at the origin and goes through the point $(3, 2)$.*

Solution: We write

$$F_3 = x^3 - 6x^2y + 11xy^2 - 6y^3$$
$$= (x-y)(x-2y)(x-3y).$$
$$F_1 = x + y + 1 = 0.$$

Since the curve can be written as $F_3 + F_1 = 0$ (F_3 having no repeated factor) its asymptotes are $F_3 = 0$.

The equation of the required cubic may now be written as

$$F_3 + (ax + by + c) = 0. \qquad (17.7.4)$$

Since it should touch the axis of Y at the origin, $c = 0, b = 0$ and again it should pass through $(3, 2)$. Hence from (17.7.4),

$$3 + 3a = 0 \text{ or, } a = -1.$$

Therefore, the required cubic is: $x^3 - 6x^2y + 11xy^2 - 6y^3 - x = 0$.

Exercises XVII (C)

1. Show that the following curves have no asymptotes:
   (a) $x^4 + y^4 = a^2(x^2 - y^2)$.
   (b) $y^2 = x(x+1)^2$.
   (c) $a^4y^2 = x^5(2a - x)$.
   (d) $x^2(y^2 + x^2) = a^2(x^2 - y^2)$.

2. Find the asymptotes of the folium of Descartes $x^3 + y^3 - 3axy = 0$.

3. Find the asymptotes of $x^3 + 4x^2y + 4xy^2 + 5x^2 + 15xy + 10y^2 - 2y + 1 = 0$.

4. Show that $x - y - 2 = 0$ and $x - y - 3 = 0$ are the only two parallel asymptotes of the curve

$$(x-y)(x^2+y^2) - 10(x-y)x^2 + 12y^2 + 2x + y = 0.$$

5. What are the asymptotes of

$(x-y+2)(2x-3y+4)(4x-5y+6)+5x-6y+7 = 0$? (use the method by inspection).

6. Find the asymptotes of the curve

$$x^2y - xy^2 + xy + y^2 + x - y = 0$$

and show that they cut the curve again in three points which lie on the curve $x + y = 0$.

[First see that $y = 0, x = 1, x - y + 2 = 0$ are the three asymptotes. The joint equation of the asymptotes is

$$y(x-1)(x-y+2) = 0 \text{ or, } x^2y - xy^2 + xy + y^2 - 2y = 0.$$

Now, we may write the equation of the curve as

$$(x^2y - xy^2 + xy + y^2 - 2y) + x + y = 0$$

i.e., $F_3 + F_1 = 0$.

∴ The points of intersection lie on the line $x + y = 0$.]

CHAPTER 17: ASYMPTOTES (RECTILINEAR)   1049

7. Show that the asymptotes of the biquadratic

$$(x^2 - 4y^2)(x^2 - 9y^2) + 5x^2y - 5xy^2 - 30y^3 + xy + 7y^2 - 1 = 0.$$

cut the curve in eight points which lie on a circle $x^2 + y^2 = 1$.

[..e asymptotes are $x + 2y = 0$, $x - 2y + 1 = 0$, $x - 3y = 0$, $x + 3y - 1 = 0$; their joint equation

$$(x + 2y)(x - 2y + 1)(x - 3y)(x + 3y - 1) = 0$$
or, $$(x^2 - 4y^2)(x^2 - 9y^2) + 5x^2y - 5xy^2 - 30y^3 - x^2 + xy + 6y^2 = 0.$$

The equation of the curve can be written as

$$(x^2 - 4y^2)(x^2 - 9y^2) + 5x^2y - 5xy^2 - 30y^3 - x^2 + xy + 6y^2 + (x^2 + y^2 - 1) = 0.$$

∴ The points of intersection lie on the circle $x^2 + y^2 = 1$.]

8. (a) Show that the points of intersection of the curve

$$2y^3 - 2x^2y - 4xy^2 + 4x^3 - 14xy + 6y^2 + 4x^2 + 6y + 1 = 0$$

and its asymptotes lie on the line $8x + 2y + 1 = 0$.   [C.H.]

(b) Show that the four asymptotes of the curve

$$(x^2 - y^2)(y^2 - 4x^2) + 6x^3 - 5x^2y - 3xy^2 + 2y^3 - x^2 + 3xy - 1 = 0$$

cut the curve in eight points which lie on the circle $x^2 + y^2 = 1$.

9. Show that the eight points of intersection of the curve

$$x^4 - 5x^2y^2 + 4y^4 + x^2 - y^2 + x + y + 1 = 0$$

and its asymptotes lie on a rectangular hyperbola.   [C.H.]

10. Find the equation of the cubic which has the same asymptotes as the curve

$$x^3 - 6x^2y + 11xy^2 - 6y^3 + x + y + 1 = 0$$

and which passes through $(0,0), (1,0)$ and $(0,1)$.

11. If the equation to a curve can be written in the form

$$y = ax^2 + bx + c + \frac{A}{x} + \frac{B}{x^2} + \cdots$$

then the curve is said to have a *parabolic* asymptote $y = ax^2 + bx + c$.

Find the *rectilinear* as well as *parabolic asymptotes* of the curve

$$axy = y^3 - a^3.$$

12. An alternative definition of a rectilinear asymptote is as follows:

If P be a point on an infinite branch of a curve and if a straight line at a finite distance from the origin exists towards which the tangent line to the curve at P approaches as a limit when $P \to \infty$, then the straight line is an asymptote to the curve.

(a) Deduce from this definition that $y = mx + c$ is an asymptote where
$$m = \lim_{|x| \to \infty} \left(\frac{y}{x}\right), \quad c = \lim_{|x| \to \infty} (y - mx); \quad (x, y) \text{ being a point on the curve.}$$

(b) Consider a curve $y = ax + b + \dfrac{c + \sin x}{x}$ and obtain the equation of the tangent at a point $P(x, y)$ and now make $|x| \to \infty$. Satisfy yourself that the tangent line does not tend to a definite limiting position and hence, according to the above definition, asymptote does not exist. Apply the definition of art. 17.2 and show that an asymptote to the curve is $y = ax + b$. Now criticise the above definition.

13. Comment on the following definition of a rectilinear asymptote: *An asymptote is a straight line which cuts the curve in two points at infinity without being itself lying wholly at infinity.*

14. If a straight line is drawn through the point $(a, 0)$ parallel to the asymptote of the cubic $(x - a)^3 - x^2 y = 0$, prove that the portion of the line intercepted by the axes is bisected by the curve.

15. Through any point P on the hyperbola $x^2 - y^2 = 2ax$, a straight line is drawn parallel to the only asymptote of the curve $x^3 + y^3 = 3ax^2$ meeting the curve in A and B; show that P is the middle point of AB.

16. Find the asymptotes of the curve
$$4(x^4 + y^4) - 17x^2 y^2 - 4x(4y^2 - x^2) + 2(x^2 - 2) = 0$$
and show that they pass through the points of intersection of the curve with the ellipse $x^2 + 4y^2 = 4$.

Answers

2. $x + y + a = 0$. 3. $2x + 5 = 0$; $4y + 2x + 1 = 0$; $2y + x + 2 = 0$. 5. $x - y + 2 = 0$; $2x - 3y + 4 = 0$; $4x - 5y + 6 = 0$. 10. $x^3 - 6x^2 y + 11xy^2 - 6y^3 - x + 6y = 0$. 11. $y = 0$ (rectilinear) and $x = (1/a)y^2$ (parabolic).

17.8 Asymptotes in Polar Co-ordinates

Let the equation of the curve in polar co-ordinates be $r = f(\theta)$, which in parametric form, becomes
$$\left.\begin{array}{l} x = r\cos\theta = f(\theta)\cos\theta \\ y = r\sin\theta = f(\theta)\sin\theta \end{array}\right\}.$$

If $P(r, \theta)$ be a point on an infinite branch of the curve then when $P \to \infty$, $r \to \infty$ so that $1/(f(\theta)) \to 0$ and suppose then $\theta \to \alpha$. Then with our previous notations, asymptotic direction

$$m = \lim_{|x|\to\infty} \frac{y}{x} = \lim_{\theta\to\alpha} \frac{f(\theta)\sin\theta}{f(\theta)\cos\theta} = \tan\alpha,$$

the corresponding $c = \lim\limits_{|x|\to\infty}(y - mx) = \lim\limits_{\substack{r\to\infty \\ \theta\to\alpha}} f(\theta)(\sin\theta - \tan\alpha\cos\theta)$

i.e., $c\cos\alpha = \lim\limits_{\theta\to\alpha} f(\theta)\sin(\theta - \alpha)$.

$\therefore$ the asymptote $y = mx + c$ reduces to

$$r(\sin\theta - \tan\alpha\cos\theta) = c$$

i.e., $\quad r\sin(\theta - \alpha) = c\cos\alpha = p$ (constant).

Thus, we have the following rule:

If $\theta \to \alpha$ as $r \to \infty$ and if $f(\theta)\sin(\theta - \alpha) \to p$ then the line $r\sin(\theta - \alpha) = p$ is an asymptote to the curve $r = f(\theta)$ and conversely.

Example 17.8.1. *Find the asymptotes of the hyperbolic spiral $r\theta = a$.*

Solution: Here $\theta = a/r$ so that $\theta \to 0$ as $r \to \infty$ ($\alpha = 0$).
$$f(\theta)\sin(\theta - 0) = \frac{a}{\theta}\sin\theta \to a \text{ as } \theta \to 0 \ (p = a).$$
$\therefore \ r\sin\theta = a$ is the asymptote.

Practical Procedure to obtain p.

Since, $p = \lim\limits_{\theta\to\alpha} f(\theta)\sin(\theta - \alpha)$

$= \lim\limits_{\theta\to\alpha} \dfrac{\sin(\theta-\alpha)}{1/r} \left(\dfrac{0}{0} \text{ form}\right) = \lim\limits_{\substack{\theta\to\alpha \\ r\to\infty}} \dfrac{\cos(\theta-\alpha)}{-(1/r^2)(dr/d\theta)} = \lim\limits_{r\to\infty} \left(-r^2 \dfrac{d\theta}{dr}\right)$

$= \lim\limits_{\theta\to\alpha}\left(\dfrac{d\theta}{du}\right)$, where $u = \dfrac{1}{r}$.

Thus in the previous illustration,

$$p = \lim_{\theta \to 0} \left(\frac{d\theta}{du}\right) = a. \ \left(\text{since } u = \frac{1}{r} = \frac{\theta}{a}; \frac{du}{d\theta} = \frac{1}{a}\right)$$

Remember: Given equation: $r = f(\theta)$. Change r or $1/u$. Find $\lim_{u \to 0} \theta = \theta_1$ (say). Find $(d\theta/du)$ and find its limit $= p$ (say) as $u \to 0$ and $\theta \to \theta_1$. Then $p = r\sin(\theta - \alpha)$ is the asymptote.

Main Points : A Recapitulation

1. A straight line is called a **Rectilinear Asymptote** (which we shall call simply **Asymptote**) of an infinite branch, of a curve if, as the point P recedes to infinity along the branch, the perpendicular distance of P from the straight line tends to zero.

2. (i) **Asymptote not parallel to y-axis:** $y = mx + c$, where
$$m = \lim_{|x| \to \infty} \frac{y}{x}; \ c = \lim_{|x| \to \infty} (y - mx).$$
Here (x, y) is a point of the infinite branch of the curve.

   (ii) **Asymptote parallel to y-axis:** If x tends to a finite number k as $y \to \infty$, then $x = k$ is an asymptote parallel to y-axis, (x, y) being a point of the infinite branch of the curve.

3. **Rational Algebraic Curve:**
$$y^m \phi(x) + y^{m-1}\phi_1(x) + y^{m-2}\phi_2(x) + \cdots\cdots\cdots$$
where $\phi(x), \phi_1(x), \phi_2(x)$, etc. are polynomials in x.

   Rule for finding asymptotes parallel to y-axis: Equate to zero the real linear factors in the coefficient of the heighest power of y, present in the equation of the curve.

   The curve will have no asymptote parallel to y-axis, if the coefficient of heighest power of y is a constant or if its linear factor are all imaginary. A similar Rule for finding asymptote parallel to x-axis: replace "y-axis" in the above rule by "x-axis".

4. To determine asymptotes of the general **rational algebraic equation:**
$$U_n + U_{n-1} + U_{n-2} + \cdots\cdots\cdots + U_2 + U_1 + U_0 = 0,$$
where U_r is a homogeneous expression in x and y of degree r i.e., $U_r = x^r \phi_r(y/x)$, where $\phi_r(y/x)$ is a polynomial in y/x of degree r. In $U_n = x^n \phi_n(y/x)$, put $x = 1, y = m$ and obtain $\phi_n(m)$. Similarly obtain $\phi_{n-1}(m)$ and $\phi_{n-2}(m)$.

CHAPTER 17: ASYMPTOTES (RECTILINEAR)

Rule: Solve $\phi_n(m) = 0$. The solutions $m = m_1, m = m_2$ etc. give the slopes of the possible asymptotes.

Corresponding to $m = m_1$, obtain $c_1 = -\dfrac{\phi_{n-1}(m_1)}{\phi'_n(m_1)}$ if $\phi'_n(m_1) \neq 0$. Then $y = m_1 x + c_1$ is the asymptote corresponding to $m = m_1$, if $\phi'_n(m_1) \neq 0$. Similarly obtain asymptotes corresponding to $m = m_2, m = m_3$ etc.

The case when $\phi'_n(m_1) = 0$: If $\phi_{n-1}(m_1) \neq 0$ then no definite c_1 and there exists no asymptote corresponding to $m = m_1$. However, if $\phi'_n(m_1) = \phi'_{n-1}(m) = 0$ then determine two values of c_1 (say c'_1 and c''_1) from

$$(c_1^2/2)\phi''_n(m_1) + c_1 \phi'_{n-1}(m_1) + \phi_{n-2}(m_1) = 0$$

and get **two parallel asymptotes** corresponding to $m = m_1$, namely $y = m_1 x + c'_1$ and $y = m_1 x + c''_1$.

Case of Parallel Asymptotes by Alternative Method:

If $\phi_n(m) = 0$ gives two equal roots $m = m_1, m = m_1$ then the two asymptotes are given by

$$(y - m_1 x)^2 + (y - m_1 x) \cdot \lim_{|x| \to \infty} \frac{f(x)}{\phi(x)} + \lim_{\substack{|x| \to \infty \\ (y/x) \to m_1}} \frac{\psi(x)}{\phi(x)} = 0,$$

where $\phi(x)$ is coefficient of $(y - m_1 x)^2$ in the highest degree terms (say, nth degree terms) and $(y - m_1 x)f(x)$ is the next highest degree terms (i.e., terms of degree $n - 1$) and $\psi(x)$ is the sum of all other terms (i.e., terms of degree not higher than $n - 2$).

5. Asymptotes by inspection: If the equation to a curve is of the form $F_n + P_{n-2} = 0$ where F_n is of degree n (i.e., contains terms of degree n and may also contain terms of lower degrees) and P_{n-2} is of degree $n - 2$ or lower and if F_n can be factorised into n linear factors which represent n straight lines, no two of which are parallel or coincident then all the asymptotes are given by $F_n = 0$, e.g., $(x^2/a^2 - y^2/b^2) - 1 = 0$ has two asymptotes $x/a - y/b = 0$ and $x/a + y/b = 0$.

6. Deductions: (i) The number of asymptotes of an algebraic curve of nth degree cannot exceed n.

(ii) The asymptotes of an algebraic curve are parallel to the lines obtained by equating to zero the factors of the heighest degree terms in the equation of the curve. Repeated factor give parallel asymptotes (see **No. 4** above).

(iii) **An asymptote** of a curve of nth degree cuts the curve in $(n - 2)$ points. The n asymptotes of a curve of nth degree cut it at $n(n - 2)$ points.

> (iv) If the equation of a curve of nth degree can be put in the form $F_n + P_{n-2} = 0$ when P_{n-2} is of degree $n-2$ at the most and F_n consists of n non-repeated linear factor then the $n(n-2)$ points of intersection of the curve and its asymptotes lie on the curve $P_{n-2} = 0$. e.g., For a quartic, $n = 4$ and therefore the asymptotes cut the curve in $4(4-2) = 8$ points, which lie on a curve of degree $4 - 2 = 2$ i.e., on a conic.
>
> **7. Asymptotes in Polar Co-ordinates:**
>
> (i) **Remember:** The polar equation of any line is $p = r\cos(\theta - \alpha)$, where p is the length of the perpendicular from the pole to the line and α is the angle which this perpendicular makes with the initial line.
>
> (ii) To determine asymptotes of the curve $r = f(\theta)$ we have to obtain p and α so that $p = r\cos(\theta - \alpha)$ is the asymptote of the given curve.
>
> **Working Rule:** Change r to $1/u$ in $r = f(\theta)$ and find $\lim_{u \to 0} \theta = \theta_1$ (say). Determine $(-d\theta/du)$ and its limit as $u \to 0$ and $\theta \to \theta_1$. Let this limit be p. Then $p = r\sin(\theta_1 - \theta)$ is the corresponding asymptote e.g., let $r = a/\theta$ be the polar curve. Then $\theta \to 0$ as $u = 1/r \to 0$ i.e., here $\theta_1 = 0$. Since $u = \theta/a, du/d\theta = 1/a$ or $d\theta/du = a$. Therefore $-a = r\sin(0-\theta)$ or $r\sin\theta = a$ is the asymptote.

Exercises XVII (D)

Find the asymptotes of the polar curves:

1. $r = a\tan\theta$. 2. $r^n \sin n\theta = a^n$. 3. $r = 2a\sin\theta\tan\theta$. 4. $r = a/\sin n\theta$.

5. a) $r\theta\cos\theta = a\cos 2\theta$. b) $r = a\theta/(\theta - 1)$.

6. Show that all the asymptotes of the curve $r\tan n\theta = a$ touch the circle $r = a/n$.

7. If in a curve $r = f(\theta)$, $\lim_{\theta \to \infty} f(\theta) = a$, then the circle $r = a$ is called a *circular asymptote* of the curve. Find the circular asymptote to the curve $r(\theta^2 + 1) = \theta^2 - 1$.

8. a) Verify that $r\sin(1 - \theta) = -a$ is the asymptote of the curve $r = \dfrac{a\theta}{\theta - 1}$.

   b) $r\cos\theta \pm b = a$ are asymptotes of $r = a\sec\theta + b\tan\theta$.

9. Find the equation of the asymptotes of the curve given by

$$r^n f_n(\theta) + r^{n-1} f_{n-1}(\theta) + \cdots + f_0(\theta) = 0.$$

10. Find the asymptotes of the curve $r\cos 2\theta = a\sin 3\theta$.

CHAPTER 17: ASYMPTOTES (RECTILINEAR)

11. Prove that the asymptotes of the polar curve $r = a/\left(\frac{1}{2} - \cos\theta\right)$ are
$$\sqrt{3}r(\sin\theta \pm \sqrt{3}\cos\theta) \pm 4a = 0.$$

[*Hints:* The parametric cartesian equations of the given curve are
$$x = \frac{a\cos\theta}{1/2 - \cos\theta}, \quad y = \frac{a\sin\theta}{1/2 - \cos\theta}.$$

Use the rule of Ex. 17.5.3.]

Answers

1. $r\cos\theta = \pm a$. 2. $n\theta = k\pi$ (k is an integer) and $n > 1$; no asymptote if $n < 1$. 3. $r\cos\theta = 2a$. 4. $r\sin\left(\theta - \frac{k\pi}{n}\right) = \frac{a}{n\cos k\pi}$ (k is any integer). 5. (a) $r\sin\theta = a$, $\pi(r\cos\theta) + 2a = 0$. (b) $r\sin(\theta - 1) = a$. 9. $f_{n-1}(\theta_1)/f'_n(\theta_1) = r\sin(\theta_1 - \theta)$ where θ_1 is any root of $f_n(\theta) = 0$. 10. $a = 2r(\cos\theta - \sin\theta)$, $a + 2r(\cos\theta + \sin\theta) = 0$.

17.9 Position of a Curve with regard to its Asymptotes

Suppose we transform the equation of a given curve in the form
$$y = mx + c + A/x + B/x^2 + C/x^3 + \cdots.$$
Then the asymptote of the curve is $y = mx + c$. (art. 17.4)

Case 1. $A \neq 0$.

Let y_1 be the ordinate of the curve and y_2 that of the asymptote when the abscissa is x_1. Then
$$y_1 - y_2 = 1/x_1\left(A + B/x_1 + C/x_1^2 + \cdots\right).$$
By taking x_1 large enough we can make the expression $B/x_1 + C/x_1^2 + \cdots$ numerically less than A.

Then for sufficiently large values of x_1, the expression in the bracket has the sign of A. If x_1 and A are of the same sign, then $y_1 > y_2$ i.e., the curve lies above the asymptote. We observe that the curve lies on the opposite sides of the asymptotes at the opposite ends.

Case 2. $A = 0, B \neq 0$.

Here
$$y_1 - y_2 = 1/x_1^2\left(B + C/x_1 + D/x_1^2 + \cdots\right).$$
Hence the curve lies on the same side of the asymptote at opposite ends—above it, if B is positive and below it, if B is negative.

Case 3. $A = B = 0, C \neq 0$.

Arguments are similar to Case 1.

These considerations are helpful in tracing curves:

Example 17.9.1. *The curve* $y^2(x^2 - a^2) = x^2(x^2 - 4a^2)$.

Clearly, $x = \pm a$ **are two asymptotes parallel to the axis of** Y.
Again, $y = \pm x \left(1 - 4a^2/x^2\right)^{1/2} \left(1 - a^2/x^2\right)^{-(1/2)} = \pm x \left\{1 - 3a^2/2x^2 + \cdots\right\}$.
Hence the asymptotes are $y = \pm x$, as well as $x = \pm a$.

Considering $y = x - 3a^2/2x + \cdots$. it appears that if x be positive and sufficiently large, the ordinate of the curve is less than the ordinate of the asymptote and we say that the curve approaches the line $y = x$ from below. Similarly, the curve approaches $y = x$ from above (in the fourth quadrant).

The considerations that the curve cuts the axes where $x = \pm 2a$, and also at the origin where the tangents are $y = \pm 2x$ and that y is not real when x^2 lies between a^2 and $4a^2$ would make the tracing of the curve easier.

Example 17.9.2. *Find the asymptotes of* $(y-x)^2 x - 3y(y-x) + 2x = 0$ *and examine how the curve is placed with reference to them. Attempt tracing the curve.*

[Ans. $y = x - 1, y = x + 2$]

Example 17.9.3. *Find the asymptotes of the curve* $y = (x^2 - 2x - 1)/x$ *and investigate the mutual positions of curve and the asymptotes.* [Ans. $y = x - 2$]

For a short revision on **Asymptotes** workout **Ex. 66–80** of **Pages 1167–1169** and **1183–1187**.

AN INTRODUCTION TO ANALYSIS

(Differential Calculus—Part I)

Chapter 18 (Applications of Derivatives)

Envelopes: Evolutes

Chapter 18

Envelopes: Evolutes

18.1 Introduction

We shall, in the present chapter, discuss the concept of envelope and its determination. The envelope of a one-parameter family of curves can be defined as the locus of its *isolated characteristic points*. This locus touches every member of the family at a characteristic point and conversely, if it is possible to find a curve E at each point of which a curve of the family touches, then E is the envelope of the family. But the definition given in older texts viz., the envelope is the locus of ultimate intersections of the neighbouring curves of the family is not fully satisfactory because there may be an envelope even when neighbouring curves do not intersect. But one should remember the following technique for practical purposes:

If the envelope E exists for the family of curves $f(x, y, \alpha) = 0$, then the equation of E can be obtained by eliminating α between $f(x, y, \alpha) = 0$ and $\dfrac{\partial f}{\partial \alpha} = 0$. This α-discriminant may consist of various other curves besides the envelope.

Towards the end of this chapter we shall see that the evolute of a curve (locus of centres of curvature) is the envelope of the family of normals to the given curve.

18.2 Family of Curves

The equation of a plane curve generally involves, besides the current co-ordinates x and y, certain constants upon which the size, shape and position in relation to the reference system of a particular curve depend.

For example, $(x - \alpha)^2 + y^2 = a^2$ where a and α are, for the moment fixed, represents a circle whose centre lies on the X-axis at a distance α from the origin, its size depends on the radius a. Suppose we hold a first and allow α to take on a series of values, then we have a series of circles of equal radii a differing in their positions on the centre (Fig. 18.1).

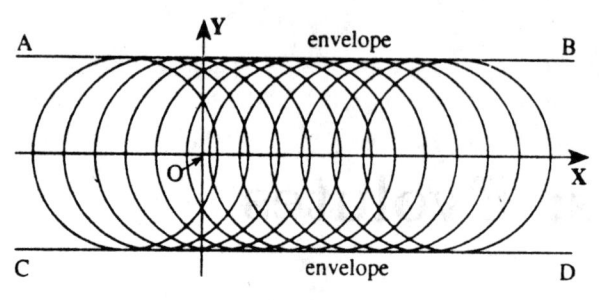

Fig. 18.1

A system of curves formed in this way is called a *family of curves* and the quantity α which remains constant for one particular curve but changes in passing from one curve to another is called a *parameter*. We write the equation of a one-parameter family of curves by a symbol

$$f(x, y, \alpha) = 0 \text{ or, } \phi(x, y, \alpha) = 0.$$

We may imagine a two, three or more-parameter family of curves. The equation of a two-parameter family of curves is of the form $f(x, y, \alpha, \beta) = 0$, where α, β are arbitrary and independent parameters; e.g., $(x - \alpha)^2 + (y - \beta)^2 = 1$ gives a two-parameter family of circles of radii 1; again $(x - \alpha)^2 + (y - \beta)^2 = c^2$ gives the three-parameter (α, β, c) family of all circles with centre at any point of the plane and with any radius i.e., the family of *all circles* on the plane.

We shall, however, mostly deal with one-parameter family of plane curves.

18.3 Envelope: Definition

Definition 18.3.1. *A point $P(a, b)$ is a singular point of a curve*

$$f(x, y, \alpha) = 0 \ (\alpha \text{ is fixed})$$

if it satisfies, besides the equation of the curve, the two equations:

$$\frac{\partial f}{\partial x} = 0 \text{ and } \frac{\partial f}{\partial y} = 0.$$

In contrast, the point P is said to be an *ordinary point* if at least one of the two partial derivative f_x, f_y is not *zero* at (a, b).

CHAPTER 18: ENVELOPES: EVOLUTES

Definition 18.3.2. *The characteristic points of a family of curves $f(x, y, \alpha) = 0$ are those ordinary points of the family where the two equations.*

$$f(x, y, \alpha) = 0 \quad \text{and} \quad \frac{\partial f}{\partial \alpha}(x, y, \alpha) = 0$$

simultaneously hold.

Note 18.3.1. *In general, the curves do not consist entirely of characteristic points; on the other hand they are isolated on each curve.*

Example 18.3.1. *The characteristic points of the circles*

$$(x - \alpha)^2 + y^2 = a^2;$$

can be obtained by solving the equations of two families of curves

$$\left.\begin{array}{l} f(x, y, \alpha) \equiv (x - \alpha)^2 + y^2 - a^2 = 0 \\ \text{and} \quad \dfrac{\partial f}{\partial \alpha}(x, y, \alpha) \equiv -2(x - \alpha) \quad\quad = 0 \end{array}\right\}$$

which give the two points $(\alpha, \pm a)$. Now it remains to be proved that these points on a circle are ordinary points (i.e., they do not satisfy $f_x = f_y = 0$). These can be easily verified and hence the conclusion will be that $(\alpha, \pm a)$ are the characteristic points.

Definition 18.3.3. *The envelope of a family of curves $f(x, y, \alpha) = 0$ (α is a variable parameter) is the locus of their isolated characteristic points.*

Observations:

1. Characteristic points may not exist e.g., for the family of the concentric circles $x^2 + y^2 = \alpha^2$, there is no characteristic point and hence there is no envelope.

2. If $f(x, y, \alpha) = 0$ and $\dfrac{\partial f}{\partial \alpha}(x, y, \alpha) = 0$ both hold for a point where $f_x = 0$ and $f_y = 0$ then the point is a *singular point* and therefore not a *characteristic point*.

Example 18.3.2. *In the family of lines*

$$x \cos \alpha + y \sin \alpha = l \sin \alpha \cos \alpha$$

if we require to find the characteristic points we are to solve

$$x \cos \alpha + y \sin \alpha = l \sin \alpha \cos \alpha$$

which is same as

$$\frac{x}{\sin \alpha} + \frac{y}{\cos \alpha} = l \tag{18.3.1}$$

$$\text{and} \quad -x \operatorname{cosec} \alpha \cot \alpha + y \sec \alpha \tan \alpha = 0. \tag{18.3.2}$$

Solution: Hence the locus of the characteristic points can be obtained if we eliminate α between (18.3.1) and (18.3.2). From (18.3.2), we obtain

$$\tan \alpha = \sqrt[3]{\frac{x}{y}} \text{ which gives } \sin \alpha = \frac{\sqrt[3]{x}}{\sqrt{x^{\frac{2}{3}} + y^{\frac{2}{3}}}}; \quad \cos \alpha = \frac{\sqrt[3]{y}}{\sqrt{x^{\frac{2}{3}} + y^{\frac{2}{3}}}}.$$

Substitution in (18.3.1) gives $\left(x^{\frac{2}{3}} + y^{\frac{2}{3}}\right)\sqrt{x^{\frac{2}{3}} + y^{\frac{2}{3}}} = l$ on simplification,

$$x^{\frac{2}{3}} + y^{\frac{2}{3}} = l^{\frac{2}{3}} \quad \text{(Astroid)}. \tag{18.3.3}$$

Thus the envelope of the family (18.3.1) is the astroid (18.3.3).

Geometrical significance: The equation (18.3.1) shows that the intercepts on the axes made by the lines of the family are $l \sin \alpha$ and $l \cos \alpha$. Hence the length of the line intercepted between the axes is l which is constant. **Thus, if a straight line of constant length l slides between the axes, the envelope of the family of lines is the astroid (18.3.3).**

> **To find the equation of an envelope when it exists.**

If there exists an envelope, its equation may be obtained in either of the following ways:

1. Eliminate α between

$$f(x, y, \alpha) = 0 \quad \text{and} \quad \frac{\partial f}{\partial \alpha}(x, y, \alpha) = 0. \tag{18.3.4}$$

The eliminant (an expression in x and y) is the envelope.

2. Solve for x and y in terms of α from the equation (18.3.4).

It will give the parametric representation of the envelope.

3. For an *algebraic curve:* The equation obtained by eliminating α between $f = 0$ and $f_\alpha = 0$ is *exactly the condition* that the relation $f(x, y, \alpha) = 0$, considered as an equation in α, has a repeated root. Thus, if

$$f(x, y, \alpha) = A(x, y)\alpha^2 + B(x, y)\alpha + C(x, y) = 0$$

then the envelope is given by $B^2 - 4AC = 0$. (This is, in fact, the condition that the two values of α are equal).

CHAPTER 18: ENVELOPES: EVOLUTES

e.g., To find the envelope of the family of lines

$$y = \alpha x + \frac{a}{\alpha} \quad (a, \text{ fixed}; \alpha, \text{ a variable parameter}).$$

or, $\alpha^2 x - \alpha y + a = 0$, which is a quadratic in α.

The condition for a double root is

$$(-y)^2 - 4 \cdot x \cdot a = 0 \text{ or } y^2 = 4ax \text{ (Parabola)}$$

which gives the required envelope.

Example 18.3.3. *Obtain the envelope of the circle drawn upon the radii vectors of the ellipse* $\frac{x^2}{a^2} + \frac{y^2}{b^2} = 1$ *as diameter.*

Solution: Any point on the ellipse is $(a\cos\theta, b\sin\theta)$. So the equation of the circle on the radius vector to this point as diameter is

$$x^2 + y^2 - ax\cos\theta - by\sin\theta = 0.$$

Differentiating w.r.t. θ

$$ax\sin\theta - by\cos\theta = 0$$

whence

$$\tan\theta = \frac{by}{ax}; \quad \sin\theta = \frac{by}{\sqrt{a^2x^2 + b^2y^2}}; \quad \cos\theta = \frac{ax}{\sqrt{a^2x^2 + b^2y^2}}.$$

Hence the envelope is

$$x^2 + y^2 - ax\frac{ax}{\sqrt{a^2x^2 + b^2y^2}} - by\frac{by}{\sqrt{a^2x^2 + b^2y^2}} = 0$$

i.e., $(x^2 + y^2)^2 = a^2x^2 + b^2y^2$.

Observation: We draw attention to the statement "If $\exists$ an envelope then it can be obtained from $f(x, y, \alpha) = 0$ and $\frac{\partial f}{\partial \alpha} = 0$". But the converse is not always the case.

Remember: By α-discriminant we mean the relation between x and y obtained by eliminating α between $f(x, y, \alpha) = 0$ and $\frac{\partial f}{\partial \alpha}(x, y, \alpha) = 0$. This α-discriminant is not always the envelope locus. It may also contain locus of different kinds of singular points, known as *Cusp locus, Nodal locus, Tac locus* etc. **(See Chapter 20 on Singular Points).**

Example 18.3.4. *Consider the family of semi-cubical parabolas*
$$f(x, y, \alpha) \equiv y^3 + (x + \alpha)^2 = 0.$$

Here, $\dfrac{\partial f}{\partial \alpha} \equiv 2(x + \alpha) = 0.$

Eliminating α, we obtain $y^3 = 0$ or, $y = 0$.

It can be verified that $(-\alpha, 0)$ is a singular point (a cusp). Thus, $y = 0$ is the locus of singular points, *not an envelope* even though it is the α-discriminant. The locus here is called cusp-locus (Fig. 18.2).

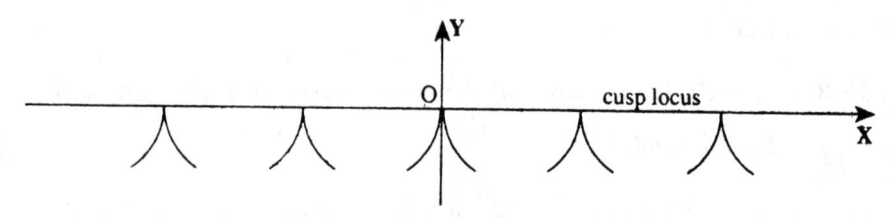

Fig. 18.2 Family of semi-cubical parabolas: $y = 0$ is not the Envelope locus, but it is a cusp locus.

18.4 Relation between Envelope and Curves Enveloped

Theorem 18.4.1. *The envelope of the family of curves $f(x, y, \alpha) = 0$ touches every member of the family at a characteristic point.*

Proof. Consider a particular curve C of the family
$$f(x, y, \alpha) = 0; \text{ where } \alpha \text{ is fixed } = \alpha_1 \text{ (say)}.$$

At a characteristic point $P(x, y)$ of this curve C the tangent has the slope $\dfrac{dy}{dx}$ given by

$$\frac{\partial f}{\partial x} + \frac{\partial f}{\partial y}\frac{dy}{dx} = 0 \tag{18.4.1}$$

where in the differentiations α is kept constant at α_1.

The parametric equation of the envelope $E : x = \phi(\alpha), y = \psi(\alpha)$ is obtained by solving the two equations:

$$\left. \begin{array}{l} f(x, y, \alpha) = 0 \\ \text{and } \dfrac{\partial f}{\partial \alpha}(x, y, \alpha) = 0 \end{array} \right\} \cdot (\alpha \text{ is now a variable parameter}) \tag{18.4.2}$$

Hence the equation of the envelope E could be written as

$$f(\phi(\alpha), \psi(\alpha), \alpha) = 0 \tag{18.4.3}$$

CHAPTER 18: ENVELOPES: EVOLUTES

which, on total derivations w.r.t. α, gives

$$\frac{\partial f}{\partial x}\cdot \phi'(\alpha) + \frac{\partial f}{\partial y}\cdot \psi'(\alpha) + \frac{\partial f}{\partial \alpha} = 0.$$

In view of (18.4.2) it becomes

$$\frac{\partial f}{\partial x}\cdot \phi'(\alpha) + \frac{\partial f}{\partial y}\cdot \psi'(\alpha) = 0. \tag{18.4.4}$$

The point P of the curve C being a characteristic point also belongs to the envelope (18.4.3), where, of course $\alpha = \alpha_1$. The tangent to the envelope E at the point P has the slope given by

$$\frac{dy}{dx} = \left(\frac{\psi'(\alpha)}{\phi'(\alpha)}\right)_{\alpha=\alpha_1} = -\left(\frac{\partial f}{\partial x}\bigg/\frac{\partial f}{\partial y}\right)_{\alpha=\alpha_1}. \quad \text{[using (18.4.4)]} \tag{18.4.5}$$

The equations (18.4.1) and (18.4.5) show that the tangents at P to the curve C and to the envelope E coincide. Hence the theorem.

Corollary 18.4.1. *Since every point P of the envelope E is a characteristic point of some member of the family it is clear that every point of the envelope is touched by some curve of the family.*

Theorem 18.4.2. *Conversely, if it is possible to find a curve E at each point of which a curve of the family $f(x,y,\alpha) = 0$ touches, then E is the envelope.*

Proof: According to the hypothesis any point P on E is a point of contact of E and a certain member of the family. Thus, the co-ordinates of $P(x,y)$ are functions of α; call $x = \phi(\alpha), y = \psi(\alpha)$. These co-ordinates also satisfy the curve $C : f(x,y,\alpha) = 0$ touched at this point. Hence,

$$f(\phi(c),\psi(\alpha),\alpha) = 0.$$

Derivations w.r.t. α give

$$\frac{\partial f}{\partial x}\phi'(\alpha) + \frac{\partial f}{\partial y}\cdot \psi'(\alpha) + \frac{\partial f}{\partial \alpha} = 0. \tag{18.4.6}$$

But since the tangent to the curve C at P coincides with the tangent of E at the point, we have

$$\frac{dy}{dx} = -\frac{\partial f}{\partial x}\bigg/\frac{\partial f}{\partial y} \quad \text{[for the curve C]}$$

$$= \frac{\psi'(\alpha)}{\phi'(\alpha)}\cdot \quad \text{[for the curve E]}$$

i.e., $\quad \dfrac{\partial f}{\partial x} \cdot \phi'(\alpha) + \dfrac{\partial f}{\partial y} \psi'(\alpha) = 0. \qquad (18.4.7)$

(18.4.6) and (18.4.7) give $\dfrac{\partial f}{\partial \alpha} = 0$ i.e., E is the locus of the characteristic points and hence the envelope.

18.5 Remarks on Definitions

Envelope of the family of curves $f(x, y, \alpha) = 0$ is sometimes defined as the locus of ultimate intersection of the neighbouring curves as α varies continuously.

In fact when two neighbouring curves intersect the limit of any such intersection is an isolated characteristic point or perhaps a singular point. For, if the curves

$$f(x, y, \alpha) = 0 \quad \text{and} \quad f(x, y, \alpha + \Delta\alpha) = 0$$

intersect, the limit of the point of intersection satisfies

$$\lim_{\Delta\alpha \to 0} \dfrac{f(x, y, \alpha + \Delta\alpha) - f(x, y, \alpha)}{\Delta\alpha} = 0, \quad \text{i.e.,} \quad \dfrac{\partial f(x, y, \alpha)}{\partial \alpha} = 0.$$

Hence the limit is a characteristic point unless it satisfies also $f_x = f_y = 0$, when it becomes a singular point.

But the characteristic points are not always the limits of intersection of neighbouring curves; neighbouring curves may not intersect yet there might exist characteristic points e.g., no two curves of the family $y = (x - \alpha)^3$ intersect, but they have characteristic points as can be easily verified and the locus of these points is $y = 0$ (envelope). Hence we conclude:

The envelope is, in general, the locus of the limits of intersections of the neighbouring curves of a family, but there may be an envelope even when neighbouring curves do not intersect. Again when neighbouring curves intersect there remains a possibility that the locus is not an envelope but a locus of singular points.

These exceptional cases cannot occur in the envelope of straight lines where the α-eliminant of $f = 0, f_\alpha = 0$ is the envelope and it is the locus of ultimate intersections of neighbouring lines.

Case of two parameters

We next consider the equation of the family of curves which involve two parameters α, β connected by a given relation.

CHAPTER 18: ENVELOPES: EVOLUTES

Let
$$f(x, y, \alpha, \beta) = 0 \tag{18.5.1}$$
by the typical equation of the curves whose envelope is to be investigated, and let
$$\phi(\alpha, \beta) = 0 \tag{18.5.2}$$
be the relation between the two parameters α and β.

We may eliminate one of the parameters by means of (18.5.2) and thus reduce the problem to that solved before. However, the following method is more convenient in many cases.

For a fixed point (x, y) of the envelope, we have from (18.5.1) and (18.5.2), by differentiation,

$$\frac{\partial f}{\partial \alpha} + \frac{\partial f}{\partial \beta}\frac{d\beta}{d\alpha} = 0 \tag{18.5.3}$$

$$\text{and} \quad \frac{\partial \phi}{\partial \alpha} + \frac{\partial \phi}{\partial \beta}\frac{d\beta}{d\alpha} = 0. \tag{18.5.4}$$

Eliminating $\dfrac{d\beta}{d\alpha}$ from (18.5.3) and (18.5.4), we have

$$\frac{\partial f}{\partial \alpha} \Big/ \frac{\partial \phi}{\partial \alpha} = \frac{\partial f}{\partial \beta} \Big/ \frac{\partial \phi}{\partial \beta} \tag{18.5.5}$$

Eliminate α, β from (18.5.1), (18.5.2) and (18.5.5). The eliminant gives the envelope.

Example 18.5.1. *Find the envelope of the family of co-axial ellipses*

$$\frac{x^2}{a^2} + \frac{y^2}{b^2} = 1 \tag{18.5.6}$$

where the parameters a and b are connected by

$$a^n + b^n = c^n \tag{18.5.7}$$

Solution: For a fixed point (x, y) of the envelope we consider b as a function of a determined from (18.5.7), Differentiating (18.5.6) and (18.5.7) w.r.t. a, we get

$$x^2(-2/a^3) + y^2(-2/b^3)db/da = 0$$
$$\text{and} \quad na^{n-1} + nb^{n-1}db/da = 0.$$

Eliminating db/da, we obtain

$$\frac{x^2}{a^{n+2}} = \frac{y^2}{b^{n+2}}$$

$$\therefore \quad \frac{x^2/a^2}{a^n} = \frac{y^2/b^2}{b^n} = \frac{x^2/a^2 + y^2/b^2}{a^n + b^n} = \frac{1}{c^n}. \quad \text{[from (18.5.6) and (18.5.7)]}$$

so that $\quad a^{n+2} = x^2 c^n, \quad b^{n+2} = y^2 c^n$.

Hence, from (18.5.7) we have

$$(x^2 c^n)^{n/(n+2)} + (y^2 c^n)^{n/(n+2)} = c^n, \quad \text{i.e.,} \quad x^{2n/(n+2)} + y^{2n/(n+2)} = c^{2n/(n+2)}.$$

Example 18.5.2. *Find the envelope of the family of lines $x/a + y/b = 1$, where the parameters are connected by $a^2 + b^2 = c^2$ (c being a given constant).*

Solution: Proceeding as in Ex. 18.5.1 we eliminate db/da from

$$-\frac{x}{a^2} - \frac{y}{b^2}\frac{db}{da} = 0; \quad 2a + 2b\frac{db}{da} = 0.$$

Thus,

$$\frac{db}{da} = -\frac{x/a^2}{y/b^2} = -\frac{a}{b}$$

or, $\quad \dfrac{x}{a^3} = \dfrac{y}{b^3} = \dfrac{x/a + y/b}{a^2 + b^2} = \dfrac{1}{c^2}.$

$$\therefore \quad a = (c^2 x)^{1/3} \quad \text{and} \quad b = (c^2 y)^{1/3}.$$

Hence from $a^2 + b^2 = c^2$ we get the required envelope as

$$(c^2 x)^{2/3} + (c^2 y)^{2/3} = c^2$$

or, $\quad x^{2/3} + y^{2/3} = c^{2/3}$ (Astroid).

Note that this problem is the same as to find the envelope of a line of constant length which slides with its extremities upon two fixed rods at right angles to one another.

Example 18.5.3. *Show that the envelope of circles whose centres lie on the rectangular hyperbola $xy = c^2$ and which pass through its centre is $(x^2 + y^2)^2 = 16c^2 xy$.*

[C.H. 1985]

Solution: Equation of a circle in a plane is given by

$$(x - \alpha)^2 + (y - \beta)^2 = a^2 \tag{18.5.8}$$

where (α, β) is the centre and a is the radius.

CHAPTER 18: ENVELOPES: EVOLUTES

Given that:
$$\alpha\beta = c^2 \tag{18.5.9}$$
$$\text{and} \quad \alpha^2 + \beta^2 = a^2. \tag{18.5.10}$$

[(18.5.9) follows since the centre (α, β) lies on $xy = c^2$ and (18.5.10) follows since the circle passes through the centre of the hyperbola i.e., $(0,0)$.]

Using (18.5.10) we may write (18.5.8) in the form
$$x^2 + y^2 - 2\alpha x - 2\beta y = 0 \tag{18.5.11}$$

where the parameters α, β are connected by
$$\alpha\beta = c^2. \tag{18.5.12}$$

Considering α as a function of β we differentiate both (18.5.11) and (18.5.12) w.r.t. β. We thus get

$$-2\frac{d\alpha}{d\beta}x - 2y = 0 \quad \text{or} \quad \frac{d\alpha}{d\beta} = -\frac{y}{x}$$

$$\text{and} \quad \beta\frac{d\alpha}{d\beta} + \alpha = 0 \quad \text{or} \quad \frac{d\alpha}{d\beta} = -\frac{\alpha}{\beta}.$$

This gives
$$\frac{x}{y} = \frac{\beta}{\alpha} \quad \text{or} \quad \frac{x}{\beta} = \frac{y}{\alpha} = \frac{\sqrt{xy}}{\sqrt{\alpha\beta}} = \frac{\sqrt{xy}}{c}.$$

$\therefore \alpha = \dfrac{cy}{\sqrt{xy}}$ and $\beta = \dfrac{cx}{\sqrt{xy}}$. Putting these values of α, β in (18.5.11) we obtain the required envelope

$$x^2 + y^2 - 2x \cdot \frac{cy}{\sqrt{xy}} - 2y\frac{cx}{\sqrt{xy}} = 0$$
$$\text{or,} \quad x^2 + y^2 = 4c\sqrt{xy}$$
$$\text{i.e.,} \quad (x^2 + y^2)^2 = 16c^2 xy.$$

Exercises XVIII (A)

(Hints are given for problems marked *)

[A]

1. Find the envelopes of the following families of straight lines:

a) $y = mx + \sqrt{a^2m^2 + b^2}$ (variable parameter is m).

b) $y = m^2x + 1/m^2$ (m is the variable parameter).

c) $x\cos\alpha + y\sin\alpha = 4$ (α is the parameter).

*d) $y\cos\theta - x\sin\theta = a - a\sin\theta\log\tan(\pi/4 + \theta/2)$, where θ is the variable parameter.

[Ans. (a) $\frac{x^2}{a^2} + \frac{y^2}{b^2} = 1$; (b) $y^2 = 4x$; (c) $x^2 + y^2 = 16$; (d) $y = a\cosh x/a$.]

2. Find the envelopes of the lines $\frac{x}{a} + \frac{y}{b} = 1$ where the parameters a and b are connected by

a) $a + b = c$.

b) $ab = c^2$.

c) $a^n + b^n = c^n$.

*d) $a^m b^n = c^{m+n}$.

(c, m, n are all given constants).

[Ans. (a) $\sqrt{x} + \sqrt{y} = \sqrt{c}$; (b) $4xy = c^2$; (c) $x^{n/n+1} + y^{n/n+1} = c^{n/n+1}$; (d) $x^m y^n = \frac{m^m n^n c^{m+n}}{(m+n)^{m+n}}$.]

3. Verify that the equations **on the right** represent the envelopes of families of curves given on the left (parameter involved being indicated in each case): (Assume in each case that the envelope exists).

	Curve family	Parameter	Envelopes
a)	$x^2 + y^2 - 2ax\cos\alpha - 2ay\sin\alpha = c^2$	α	$(x^2 + y^2 - c^2)^2 = 4a^2(x^2 + y^2)$.
b)	$(x-\alpha)^2 + (y-\alpha)^2 = 2\alpha$	α	$(x + y + 1)^2 = 2(x^2 + y^2)$.
c)	$\frac{x^2}{\alpha^2} + \frac{y^2}{k^2 - \alpha^2} = 1$	α	$x \pm y = \pm k$.
d)	$\left.\begin{array}{l} x = a\sin(\theta - \alpha) \\ y = b\cos\theta \end{array}\right\}$	α	$x = \pm a$.
e)	$ay^2 = \alpha^2(x - \alpha)$	α	$ay^2 = \frac{4}{27}x^3$.
*f)	$(x - \alpha)^2 + y^2 = 4\alpha$	α	$y^2 - 4x - 4 = 0$. [C.H. 1968]
g)	$a\lambda^3 + 3b\lambda^2 + 3c\lambda + d = 0$	λ	$(bc - ad)^2 = 4(bd - c^2)(ac - b^2)$.

(a, b, c, d are functions of x and y).

4. Find the envelope of the family of ellipses $\frac{x^2}{a^2} + \frac{y^2}{b^2} = 1$, where the parameters a, b are connected by

(a) $a + b = k$;

(b) $\dfrac{a^2}{l^2} + \dfrac{b^2}{m^2} = 1$; [C.H. 1960]

(c) $ab = c^2$;

(d) $\sqrt{a} + \sqrt{b} = \sqrt{c}$;

(e) $a^m + b^m = c^m$.

[a, b are variable parameters; other symbols are fixed constants.]

[Ans. (a) $x^{2/3} + y^{2/3} = k^{2/3}$; (b) $\pm \dfrac{x}{l} \pm \dfrac{y}{m} = 1$; (c) $2xy = c^2$; (d) $x^{2/5} + y^{2/5} = c^{2/5}$; (e) $x^{2m/m+2} + y^{2m/m+2} = c^{2m/m+2}$.]

5. Find the envelope of the parabolas which touch the co-ordinate axes and are such that the distances α, β from the origin to those points of contact are connected by the relation (m and c are known constants):

a) $\alpha + \beta = c$;

*b) $\alpha^m + \beta^m = c^m$;

c) $\alpha\beta = c^2$.

[Ans. (a) $x^{1/3} + y^{1/3} = c^{1/3}$; (b) $x^{m/2m+1} + y^{m/2m+1} = c^{m/2m+1}$; (c) $16xy = c^2$.]

*6 If a particle be projected from the origin at an elevation θ with the *velocity due to a height h*, the equation of the parabolic path is given by

$$y = x \tan \theta - \dfrac{x^2}{4h \cos^2 \theta}.$$

For different elevations θ what will be the envelope of these parabolic paths?

[Ans. $x^2 = 4h(h - y)$]

["Velocity due to a height h" means a velocity which can raise a particle vertically through a height h under gravity alone $= \sqrt{2gh}$. In case this velocity is denoted by v_0 we shall get the equation of the projectile in the form $y = x \tan \theta - \dfrac{gx^2}{2v_0^2 \cos^2 \theta}$.]

[B]

*1. From any point P on the parabola $y^2 = 4ax$ perpendiculars PM, PN are drawn to the co-ordinate axes. Find the envelope of the line MN.

[Ans. $y^2 = -16ax$]

2. From any point on the ellipse $x^2/a^2 + y^2/b^2 = 1$, perpendiculars are drawn to the axes, and the feet of these perpendiculars are joined. Show that the straight line thus formed always touches the curve $(x/a)^{2/3} + (y/b)^{2/3} = 1$.

3. One angle of a triangle is fixed in position; if the area be given, show that the envelope of opposite side is a hyperbola.

*4. A circle moves with its centre on the parabola $y^2 = 4ax$ and always passes through the vertex of the parabola. Show that the envelope of the circle is the curve
$$x^3 + y^2(x + 2a) = 0.$$

*5. Find the envelope of the circles described on the radii vectores of the curve $r^n = a^n \cos n\theta$ as diameter. Verify that if the curve be $r^2 = a^2 \cos 2\theta$ then the envelope is a rectangular hyperbola.

6. a) Find the envelope of circles described on the radii vectores of the parabola $y^2 = 4ax$ as diameter.
   [Ans. $ay^2 + x(x^2 + y^2) = 0$]

   b) **Similar problem as No. 6 (a):** Take *ellipse* $x^2/a^2 + y^2/b^2 = 1$ instead of the *parabola*. Verify that the envelope of circles is $(x^2 + y^2)^2 = a^2 x^2 + b^2 y^2$.

7. Prove that the envelope of the family of circles passing through the origin and having centres situated on the hyperbola $x^2 - y^2 = c^2$ is
$$(x^2 + y^2)^2 = 4c^2(x^2 - y^2).$$

What will be the envelope if instead of hyperbola we consider the ellipse $\dfrac{x^2}{a^2} + \dfrac{y^2}{b^2} = 1$?

[C.H. 1961]

8. a) Circles are described on the double ordinates of the parabola $y^2 = 4ax$ as diameters. Prove that the envelope is the parabola $y^2 = 4a(x + a)$.

   b) **Similar problem:** If instead of the parabola we take the ellipse $x^2/a^2 + y^2/b^2 = 1$, show that the envelope is the *ellipse*
$$\frac{x^2}{a^2 + b^2} + \frac{y^2}{b} = 1.$$

*9. Show that the radius of curvature of the envelope of the line
$$x \cos \theta + y \sin \theta = f(\theta) \text{ is } f(\theta) + f''(\theta).$$

CHAPTER 18: ENVELOPES: EVOLUTES

*10. a) Find the envelope of the straight lines drawn at right angles to the radii vectores of the equiangular spiral $r = ae^{\theta \cot \alpha}$ through their extremities.

b) **Similar problem** with the cardiode $r = a(1 + \cos \theta)$ instead of the spiral.

[Ans. (a) an equiangular spiral; (b) A circle through the pole]

[C]

*1. The envelope of the lines $\dfrac{x}{a} + \dfrac{y}{b} = 1$ (a, b are variable parameters) is given by $\sqrt{x} + \sqrt{y} = \sqrt{k}$ (k is a given constant). Find a relation between a and b.

*2. Given that the astroid $x^{2/3} + y^{2/3} = c^{2/3}$ is the envelope of the lines $x/a + y/b = 1$, prove that the parameters a, b are connected by $a^2 + b^2 = c^2$.

3. If the astroid $x^{2/3} + y^{2/3} = c^{2/3}$ is the envelope of the family of ellipses $x^2/a^2 + y^2/b^2 = 1$, prove that $a + b = c$. [C.H. 1963]

4. If the curve $x^p y^q = c^{p+q}$ be the envelope of the family of lines $x/a + y/b = 1$, then prove that
$$a^p b^q p^p q^q = (p+q)^{p+q} c^{p+q}.$$

5. Show that if PM and PN be drawn perpendiculars from any point P of the curve $y = mx^3$ upon the axes the envelope of the lines MN is $27y + 4mx^3 = 0$.

6. Prove that the envelope of a variable circle whose centre lies on the parabola $y^2 = 4ax$ and which passes through its vertex is $2ay^2 + x(x^2 + y^2) = 0$.

7. If O the pole and P be any point on $r^m = a^m \cos m\theta$ and if with O for pole and P for vertex a curve similar to $r^n = a^n \cos n\theta$ be described, then prove that the envelope of all such curves is
$$r^{m/(m+n)} = a^{mn/(m+n)} \cos \left(\dfrac{mn}{m+n} \theta \right).$$

8. Show that the envelope of all cardiodes described on radii vectors of the circle $r = a \cos \theta$ [or of the cardiode $r = a(1 + \cos\theta)$] for axes and having their cusps at the pole is
$$r^{1/3} = a^{1/3} \cos \dfrac{\theta}{3} \quad [\text{or, } r^{1/4} = (2a)^{1/4} \cos \theta/4].$$

9. Prove that the pedal equation of the envelope of the variable line
$$x \cos 2\theta + y \sin 2\theta = 2a \cos \theta \text{ is } p^2 = \dfrac{4}{3}(r^2 - a^2).$$

10. Prove that the envelope of polars with respect to the circle $x^2 + y^2 = 2ax$ of points which lie on the circle $x^2 + y^2 = 2bx$ is

$$\{(a-b)x + ab\}^2 = b^2\{(x-a)^2 + y^2\}.$$

*11. Find the envelope of the polars of points on the ellipse $\dfrac{x^2}{h^2} + \dfrac{y^2}{k^2} = 1$ with respect to the ellipse $\dfrac{x^2}{a^2} + \dfrac{y^2}{b^2} = 1$.

12. Prove that the envelope of the normals to the curve $x^{2/3} + y^{2/3} = a^{2/3}$ is given by $(x+y)^{2/3} + (x-y)^{2/3} = 2a^{2/3}$.

Hints/Solutions

[A]

1. (d) Diff. $y \cos\theta - x \sin\theta = a - a \sin\theta \log \tan(\pi/4 + \theta/2)$.   (i)

We get $y \sin\theta + x \cos\theta = a \cos\theta \log \tan(\pi/4 + \theta/2) + a \tan\theta$ (ii)

Multiply (i) by $\cos\theta$ and (ii) by $\sin\theta$ and add; thus obtain $y = a \sec\theta$.

(i) Now put $\cos\theta = a/y, \sin\theta = \sqrt{y^2 - a^2}/y$ and $\tan\theta/2 = \dfrac{\sin\theta}{1+\cos\theta} = \dfrac{\sqrt{y-a}}{\sqrt{y+a}}$ in

Then we get $e^{x/a} = \dfrac{\sqrt{y+a} + \sqrt{y-a}}{\sqrt{y+a} - \sqrt{y-a}}$.

And hence $e^{-x/a} = \dfrac{\sqrt{y+a} - \sqrt{y-a}}{\sqrt{y+a} + \sqrt{y-a}}$ so that $e^{x/a} + e^{-x/a} = \dfrac{2y}{a}$.

Thus the required envelope is $y = a \cosh\left(\dfrac{x}{a}\right)$.

2. (d) Consider b as a function of a; Differentiate both $x/a + y/b = 1$ and $a^m b^n = c^{m+n}$ w.r.t. a and obtain.

$$\dfrac{db}{da} = -\dfrac{x/a^2}{y/b^2} = -\dfrac{ma^{m-1}b^n}{nb^{n-1}a^m} \quad \text{i.e.,} \quad \dfrac{x}{y} = \dfrac{am}{bn}.$$

Put $x = \lambda am, y = \lambda bn$ in $x/a + y/b = 1$ and obtain $\lambda = 1/(m+n)$.

Thus putting $a = \dfrac{x}{\lambda m} = \dfrac{(m+n)x}{m}$ and $b = \dfrac{(m+n)y}{n}$ in $a^m \cdot b^n = c^{m+n}$ the required envelope can be easily obtained.

3. (f) $\alpha^2 - 2\alpha(x+2) + x^2 + y^2 = 0$ has repeated roots if $4(x+2)^2 - 4(x^2 + y^2) = 0$ i.e., if $y^2 = 4x + 4$, which gives the required envelope.

5. (b) If the points of contact of the tangents be $(\alpha, 0)$ and $(0, \beta)$, the **tangents** being the co-ordinate axes, the equation of the parabola is of the form

CHAPTER 18: ENVELOPES: EVOLUTES

$$\sqrt{\frac{x}{\alpha}} + \sqrt{\frac{y}{\beta}} = 1. \qquad \text{(i)}$$

Here, then, the problem is to find the envelope of parabolas of the form (i) subject to the condition

$$\alpha^m + \beta^m = c^m. \qquad \text{(ii)}$$

Now proceed as in 2 (d): Equate $\dfrac{d\beta}{d\alpha}$ from (i) and (ii) etc.

6. Write a quadratic equation in $\tan\theta$ and make discriminant zero for finding the envelope.

[B]

1. Take a point $(at^2, 2at)$ on the parabola $y^2 = 4ax$. To find the envelope of the lines $x/at^2 + y/2at = 1$, where t is the variable parameter.

4. Eqn. of any circle, $(x - \alpha)^2 + (y - \beta)^2 = a^2$. Here the centre (α, β) lies on $y^2 = 4ax$ i.e., $\beta^2 = 4a\alpha$. Moreover, it passes through the vertex $(0,0)$ i.e., $\alpha^2 + \beta^2 = a^2$. Hence the equation of a circle is $x^2 + y^2 - 2\alpha x - 2\beta y = 0$ where for α put $\beta^2/4a$ i.e., $x^2 + y^2 - \dfrac{\beta^2}{2a}x - 2\beta y = 0$; find its envelope, β being the variable parameter, i.e., eliminate β between this equation and $-\dfrac{\beta}{a}x - 2y = 0$, or, $x\beta + 2ay = 0$. The eliminant is the required envelope.

5. We consider the polar eqn. of a circle with pole O on the circumference. (See Fig. 18.3)

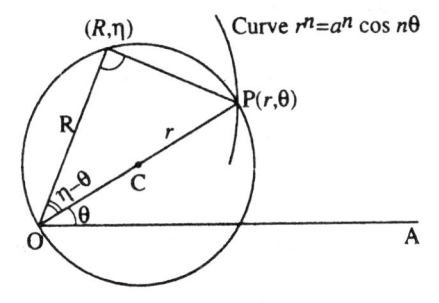

Fig. 18.3

Its equation is

$$R = r\cos(\eta - \theta) = a(\cos n\theta)^{1/n}\cos(\eta - \theta). \qquad \text{(i)}$$

$\therefore \log R = \log a + \dfrac{1}{n}\log\cos n\theta + \log\cos(\eta - \theta).$

Diff. w.r.t. θ (which is the parameter)

$$0 = -\dfrac{1}{n}\cdot n\tan n\theta + \tan(\eta - \theta) \quad \text{i.e.,} \quad \tan(\eta - \theta) = \tan n\theta.$$

$$\therefore \quad \eta - \theta = n\theta + k\pi \quad (k = 0, \pm 1, \pm 2, \cdots)$$

$$\text{or,} \quad \theta = \frac{\eta}{n+1} - \frac{k\pi}{n+1}.$$

Put this value of θ in (i) we get the envelope.

9. Eliminate θ between $x \cos \theta + y \sin \theta = f(\theta)$ and $-x \sin \theta + y \cos \theta = f'(\theta)$
Whence:

$$x = f \cos \theta - f' \sin \theta, y = f \sin \theta + f' \cos \theta \quad [f = f(\theta), f' = f'(\theta)]$$

$$\frac{dx}{d\theta} = -(f + f'') \sin \theta; \qquad \frac{dy}{d\theta} = (f + f'') \cos \theta$$

$$\frac{d^2 x}{d\theta^2} = -(f' + f''') \sin \theta - (f + f'') \cos \theta; \quad \frac{d^2 y}{d\theta^2} = (f' + f''') \cos \theta - (f + f'') \sin \theta.$$

Now

$$\rho = \frac{(dx/d\theta)(d^2y/d\theta^2) - (dy/d\theta)(d^2x/d\theta^2)}{\{(dx/d\theta)^2 + (dy/d\theta)^2\}^{3/2}}; \text{ substitute and see that}$$

$$\rho = f + f'' = f(\theta) + f''(\theta).$$

10. (a) The equation of the line through (r, θ) at right angles to the radius vector, is $R \cos(\eta - \theta) = r$. (See Fig. 18.4)

$$R \cos(\eta - \theta) = a e^{\theta \cot \alpha} \qquad (i)$$

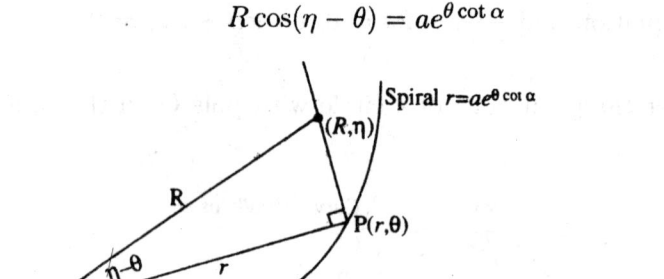

Fig. 18.4

Take log and differentiate w.r.t. θ (which is the parameter); thus obtain

$$\tan(\eta - \theta) = \cot \alpha = \tan(\pi/2 - \alpha).$$

$$\therefore \quad \eta = \theta + \frac{\pi}{2} - \alpha + k\pi = \theta - \alpha + (2k+1)\frac{\pi}{2}$$

i.e., $\theta = \eta + \alpha - (2k+1)\pi/2$.

Substituting this θ in (i) we get the required envelope

$$R \cos\left\{\alpha - \frac{1}{2}(2k+1)\pi\right\} = a e^{\{\eta + \alpha - \frac{1}{2}(2m+1)\pi\} \cot \alpha}$$

which is of the form $R = A e^{\eta \cot \alpha}$.

CHAPTER 18: ENVELOPES: EVOLUTES

Hence the envelope is another equiangular spiral.

[C]

1. $\dfrac{1}{a} + \dfrac{1}{b}\dfrac{dy}{dx} = 0$; $\dfrac{1}{2\sqrt{x}} + \dfrac{1}{2\sqrt{y}}\dfrac{dy}{dx} = 0$.

Hence $\dfrac{dy}{dx} = -\dfrac{b}{a} = -\dfrac{\sqrt{y}}{\sqrt{x}}$.

Let $\sqrt{x} = \lambda a$, $\sqrt{y} = \lambda b$.

Then from $\sqrt{x} + \sqrt{y} = \sqrt{k}$ we get $\lambda = \sqrt{k}/(a+b)$.

$\therefore$ From $\dfrac{x}{a} + \dfrac{y}{b} = 1$ we get $\dfrac{\lambda^2 a^2}{a} + \dfrac{\lambda^2 b^2}{b} = 1$

or, $\lambda^2(a+b) = 1$ or $\dfrac{k}{(a+b)^2}(a+b) = 1$ or $\boldsymbol{a+b = k}$.

2. As above, $\dfrac{dy}{dx} = -\dfrac{x^{-1/3}}{y^{-1/3}} = -\dfrac{1/a}{1/b}$ or $\dfrac{y^{1/3}}{x^{1/3}} = \dfrac{b}{a}$. Let $x^{1/3} = \lambda a \cdot y^{1/3} = \lambda b$.

Then from $x/a + y/b = 1$ we get $\lambda^3(a^2 + b^2) = 1$. (i)

Also from $x^{2/3} + y^{2/3} = c^{2/3}$, we get $\lambda^2(a^2 + b^2) = c^{2/3}$. (ii)

Hence deduce $a^2 + b^2 = c^2$ (Square (i) and Cube (ii); divide)].

11. One point of the ellipse $x^2/h^2 + y^2/k^2 = 1$ is $(h\cos\theta, k\sin\theta)$. Its polar w.r.t the ellipse $\dfrac{x^2}{a^2} + \dfrac{y^2}{b^2} = 1$ is $\dfrac{xh\cos\theta}{a^2} + \dfrac{yk\sin\theta}{b^2} - 1 = f(x,y,\theta) = 0$.

Eliminate θ between $f(x,y,\theta) = 0$ and $\partial f/\partial \theta = 0$ $\rightarrow$ the required envelope $h^2 x^2/a^4 + k^2 y^2/b^4 = 1$.

18.6 Evolutes

The evolute of a curve, has been defined as the locus of its centres of curvature. Since the centre of curvature is the limit of intersection of two consecutive normals it follows that *the evolute is the envelope of the family of normals to the given curve*. By Theorem 18.4.1 of art. 18.4, it also follows that *normals to the original curve are tangents to the evolute* (**Fig. 18.5 under art. 18.6;**). We have thus an alternative method of finding the evolute of a curve:

To find the Evolute of the Curve → Find the family of Normals and then its envelope.

Example 18.6.1. *Find the evolute of the parabola* $y^2 = 4ax$.

Solution: Any normal to the parabola has the equation

$$y = mx - 2am - am^3$$

where m is different for different normals.

Differentiating partially w.r to m we obtain

$$x - 2a = 3am^2; \quad y = 2am$$

which is the parametric equation of the envelope.

The elimination of m leads to $27ay^2 = 4(x-2a)^3$ which represents a semi-cubical parabola and this is the required evolute.

I. Arc of an Evolute

Theorem 18.6.1. *The difference of the radii of curvature at any two points of a curve is equal to the length of the arc between the corresponding points of the evolute.*

Proof: If $(\bar{x}, \bar{y})$ be the co-ordinates of the centre of curvature corresponding to a point (x, y) on the curve, then we know.

$$\bar{x} = x - \rho \sin \psi; \quad \bar{y} = y + \rho \cos \psi \tag{18.6.1}$$

where ρ is the radius of curvature at the point (x, y).

Differentiating (18.6.1), we find,

$$\frac{d\bar{x}}{ds} = \frac{dx}{ds} - \rho \cos \psi \frac{d\psi}{ds} - \frac{d\rho}{ds} \sin \psi$$

$$= \cos \psi - \rho \cos \psi \frac{1}{\rho} - \frac{d\rho}{ds} \sin \psi = -\frac{d\rho}{ds} \sin \psi \tag{18.6.2}$$

$$\frac{d\bar{y}}{ds} = \frac{dy}{ds} - \rho \sin \psi \frac{d\psi}{ds} + \frac{d\rho}{ds} \cos \psi$$

$$= \sin \psi - \rho \sin \psi \frac{1}{\rho} + \frac{d\rho}{ds} \cos \psi = \frac{d\rho}{ds} \cos \psi. \tag{18.6.3}$$

$$\left(\text{using } \frac{dx}{ds} = \cos \psi, \ \frac{dy}{ds} = \sin \psi \ \text{ and } \ \frac{d\psi}{ds} = \frac{1}{\rho} \right)$$

[s is the length of the arc on the given curve measured from a fixed point on it up to the point (x, y).]

Hence

$$\frac{d\bar{y}}{d\bar{x}} = -\cot \psi, \tag{18.6.4}$$

which incidentally proves that the tangent to the evolute is normal to the original curve.

CHAPTER 18: ENVELOPES: EVOLUTES

Squaring and simplifying (18.6.2) and (18.6.3) we also obtain

$$\frac{d\rho}{ds} = \sqrt{\left\{\left(\frac{d\bar{x}}{ds}\right)^2 + \left(\frac{d\bar{y}}{ds}\right)^2\right\}} = \sqrt{\left\{1 + \left(\frac{d\bar{y}}{d\bar{x}}\right)^2\right\}}\frac{d\bar{x}}{ds}$$

$$= \frac{d\sigma}{d\bar{x}} \cdot \frac{d\bar{x}}{ds} = \frac{d\sigma}{ds}. \tag{18.6.5}$$

[σ is the length of the arc of the evolute measured from some fixed point on it up to the point $(\bar{x}, \bar{y})$.]

Now, from (18.6.5), it follows

$$d\rho/d\sigma = 1 \tag{18.6.6}$$

Hence, integrating,

$$\rho = \sigma + c \tag{18.6.7}$$

where c is an arbitrary constant dependent on the fixed point, from where σ has been measured.

Hence $\sigma_2 - \sigma_1 = \rho_2 - \rho_1$, where ρ_1 and ρ_2 are the values of ρ for any two points P_1 and P_2 on the curve, and σ_1 and σ_2 are the corresponding values of σ. (See Fig. 18.5) Hence the theorem.

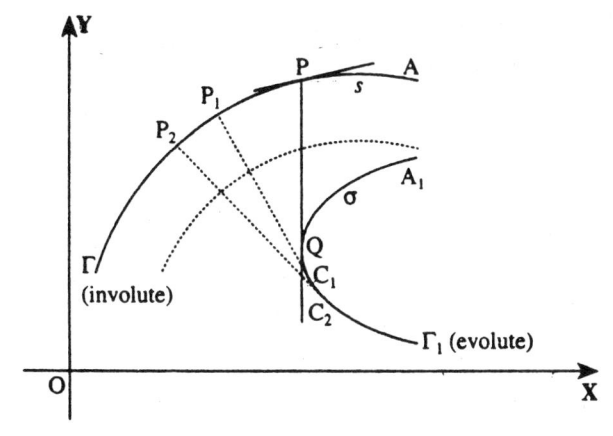

Fig. 18.5

Observation: This theorem suggests that the circles of curvature of adjacent points on a curve do not, in general, intersect. For the distance between the centres of curvature corresponding to these points is a chord of the evolute and is, therefore, in general, less than the corresponding arc i.e., less than the difference of the radii of curvature. See that the neighbouring circles of curvature of a curve do not intersect and yet they have characteristic points and have the curve itself for their envelope (art. 18.5).

II. Involutes

If a curve Γ_1 be the *evolute* of a curve Γ, then Γ is said to be an *involute* of Γ_1.

If C_1 and C_2 are centres of curvature of the curve Γ at P_1 and P_2 respectively, we have, by the previous article,

$$C_1P_1 + \text{arc } C_1C_2 = C_2P_2.$$

These properties show that a curve Γ may be traced by the end P of a taut string unwound from its evolute Γ_1; the string is always tangent to Γ_1 and its free portion is equal to ρ. It is from this point of view that we say Γ is an involute of Γ_1. Obviously any point on the string will describe an involute (see the dotted curve of Fig. 18.5). Thus, *every curve has an infinite number of involutes.*

Main Points: A Recapitulation

> **1.** $f(x, y, \alpha, \beta, \cdots) = 0$ is the form of the equation of a family of curves, where (x, y) are current co-ordinates and $\alpha, \beta, \cdots$ are arbitrary parameters. We shall mostly deal with one-parameter families like $f(x, y, \alpha) = 0$, where α is the only variable parameter.
>
> **2.** A curve E which touches each member of a family of curves $f(x, y, \alpha) = 0$ and at each point is touched by some member of the family, is called the *Envelope* of that family of curves.
>
> **3.** Those points which satisfy $f = 0, f_x = 0, f_y = 0$ are *singular points*; other points are *ordinary points*. Let $f(x, y, \alpha) = 0$ and $f(x, y, \alpha + \Delta\alpha) = 0$ be two neighbouring curves (α) and $(\alpha + \Delta\alpha)$. An ordinary point P of (α) is called a *Characteristic point* if its distance from the curve $(\alpha + \Delta\alpha)$ is of second order of smallness. Such points satisfy $f(x, y, \alpha) = 0, \dfrac{\partial f}{\partial \alpha} = 0$. These points are, in general, isolated on the curve. The locus of isolated characteristic points of family of curves is called its *envelope* E.
>
> **4.** If E exists, then it is obtained by eliminating α between $f(x, y, \alpha) = 0$ and $\dfrac{\partial f}{\partial \alpha} = 0$. This eliminant is called α-discriminant.
>
> **5.** α-discriminant may consist of other curves besides the envelope—in *fact, this α-discriminant may be the locus of various type of singular points only; no envelope at all.*

CHAPTER 18: ENVELOPES: EVOLUTES

6. When the neighbouring curves of the same family intersect, the limit of the points of intersection is an isolated characteristic point (or a singular point). But the converse is not necessarily true.

Characteristic points are not always the limits of points of intersection. Neighbouring curves may not intersect e.g., circles of curvature of a curve do not intersect but still they have characteristic points and the curve itself is the envelope of the circles.

7. The envelope is, in general, the locus of the limits of intersection of neighbouring curves but there may be an envelope even when neighbouring curves do not intersect (e.g., $y = (x - \alpha)^3 \to$ the curves do not intersect but $y = 0$ is the envelope).

8. Assuming that the envelope E is the locus of isolated characteristic points one may prove.

a) Every member of the family of curves enveloped TOUCHES the envelope E at a characteristic point.

b) It is possible to a find a curve which touches each member of the family successively, the curve being the envelope of the family.

9. We may require to find the envelope of the family $f(x, y, \alpha, \beta) = 0$ where the parameters are connected by $F(\alpha, \beta) = 0$: Regard α as the independent and β a function of α. Obtain $\dfrac{d\beta}{d\alpha}$ from both and equate: thus

$$\frac{d\beta}{d\alpha} = -\frac{f_\alpha}{F_\alpha} = -\frac{f_\beta}{F_\beta} = \lambda. \quad \text{Eliminate } \alpha, \beta, \lambda.$$

10. Given the family and the envelope, to find the relation between parameters (See Example XVIII (A). Sum No. 1, Section [C]).

11. Evolute: It is the locus of the centres of curvature of a curve. We can also define it as the *Envelope of the Normals*. So the evolute is touched by the normals of the *original curve* (called *Involute*).

Length of an arc of an evolute = difference between the radii of curvature at the two ends of the arc = $\rho_1 \sim \rho_2$.

Exercises XVIII (B)

(Hints are given for problems marked *)

Verify the following statements (by first finding the equation of the normal to the curve): (**Ex. 1—Ex. 6**).

*1. The evolute of the *ellipse* $x^2/a^2 + y^2/b^2 = 1$ is
$$(ax)^{2/3} + (by)^{2/3} = (a^2 - b^2)^{2/3}.$$

2. The evolute of the *hyperbola*: $x = a\cosh u$, $y = b\sinh u$ is
$$(ax)^{2/3} - (by)^{2/3} = (a^2 + b^2)^{2/3}.$$

*3. The evolute of the *tractrix* $x = a(\cos t + \log \tan t/2)$, $y = a\sin t$ is the *Catenary*
$$y = a\cosh(x/a).$$

4. The evolute of the *astroid* $x^{2/3} + y^{2/3} = a^{2/3}$ is
$$(x+y)^{2/3} + (x-y)^{2/3} = 2a^{2/3}.$$

5. The evolute of the *parabola* $\sqrt{x} + \sqrt{y} = \sqrt{a}$ is
$$27a(x-y)^2 = (2x + 2y - 3a)^3.$$

6. The evolute of the *hyperbola* $xy = a^2$ is
$$(x+y)^{2/3} - (x-y)^{2/3} = (4a)^{2/3}.$$

*7. Show that the evolute of an equi-angular spiral is an equal-angular spiral.

*8. Prove that the evolute of the cardiode $r = a(1 + \cos\theta)$ is the cardiode $r = \frac{1}{3}a(1 - \cos\theta)$, the pole in the latter equation being at the point $(\frac{2}{3}a, 0)$.

9. Show that the whole length of the evolute of the ellipse
$$\frac{x^2}{a^2} + \frac{y^2}{b^2} = 1 \text{ is } 4\left(\frac{a^2}{b} - \frac{b^2}{a}\right).$$

10. Prove that the evolute of the ellipse $\dfrac{x^2}{a^2} + \dfrac{y^2}{b^2} = 1$ is the envelope of the family of ellipses
$$a^2x^2\sec^4\alpha + b^2y^2\operatorname{cosec}^4\alpha = (a^2 - b^2)^2 \quad (\alpha = \text{parameter}).$$

11. Show that the evolute of the curve whose pedal equation is $r^2 - a^2 = mp^2$ is the curve whose pedal equation is
$$r^2 - (1 - m)a^2 = mp^2.$$

CHAPTER 18: ENVELOPES: EVOLUTES

*12. Find the evolute of the parabola and show that the length of the arc of the evolute from the cusp to the point at which the evolute meets the parabola is $2a(3\sqrt{3}-1)$, where $4a$ is the latus rectum of the parabola.

Hints

1. The normal at any point θ i.e., $(a\cos\theta, b\sin\theta)$ is given by
$$ax\sec\theta - by\csc\theta = a^2 - b^2.$$

Now find the envelope of these family of normals by using usual method (here θ is the parameter).

3. Here $\frac{dx}{dt} = \frac{a\cos^2 t}{\sin t}$, $\frac{dy}{dt} = a\cos t$ so that $\frac{dy}{dx} = \frac{\sin t}{\cos t}$, i.e., $-\frac{1}{dy/dx} = -\frac{\cos t}{\sin t}$.

The normal at any point t has the equation

$$y - a\sin t = -\frac{\cos t}{\sin t}(x - a\cos t - a\log\tan t/2)$$

or, $y\sin t + x\cos t = a\sin^2 t + a\cos^2 t + a\cos t\log\tan t/2$

$$= a(1 + \cos t \cdot \log\tan t/2). \qquad (i)$$

Differentiating w.r.t. t, we get

$$y\cos t - x\sin t = -a\sin t\log\tan t/2 + a\cos t\frac{1}{\tan t/2}\cdot\sec^2\frac{t}{2}\cdot\frac{1}{2}$$

$$= -a\sin t\log\tan t/2 + a\cos t\cdot\frac{1}{\sin t}. \qquad (ii)$$

Multiply (i) by $\cos t$ and (ii) by $\sin t$ and subtract:
$$x = a\log\tan t/2.$$

Multiply (i) by $\sin t$ and (ii) by $\cos t$ and add:
$$y = a/\sin t.$$

$$\therefore \sin t = \frac{a}{y} \quad \text{and} \quad e^{x/a} = \tan t/2,$$

i.e., $\frac{2\tan t/2}{1 + \tan^2 t/2} = \frac{a}{y}$, i.e., $\frac{2e^{x/a}}{1 + e^{2x/a}} = \frac{a}{y}$,

or, $y = \frac{a}{2}\frac{(1 + e^{2x/a})}{e^{x/a}} = \frac{a}{2}(e^{-x/a} + e^{x/a}) = a\cosh\frac{x}{a}$.

This is the required evolute.

7. The pedal equation of the equiangular spiral $r = ae^{\theta\cot\alpha}$ is $p = r\sin\alpha$ (using the fact $\phi = \alpha$). We shall now show that if (P, R) be any point on the evolute then $P = R\sin\alpha$. Let Q be a point on the spiral, QT, the tangent; C, the centre of curvature

at Q. Then CQ is tangent to the evolute and normal to the spiral at Q. If OM be drawn perpendicular to CQ. We have $P = OM = r\cos\alpha$. (See Fig. 18.6).

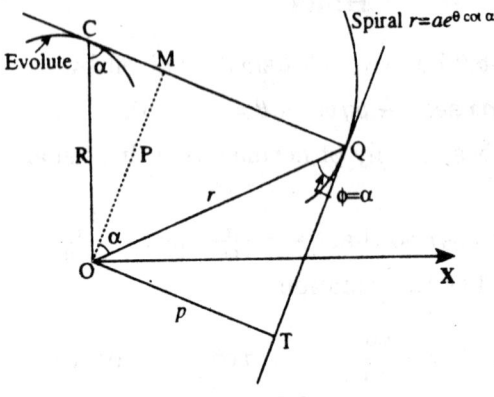

Fig. 18.6

$$QC = \rho = r\frac{dr}{dp} = r\cosec\alpha.$$

$$MC = r\cosec\alpha - r\sin\alpha = (r\cos^2\alpha)/\sin\alpha.$$

$$\therefore \quad R = OC = \sqrt{MC^2 + OM^2} = \sqrt{\frac{r^2\cos^4\alpha}{\sin^2\alpha} + r^2\cos^2\alpha} = \frac{r\cos\alpha}{\sin\alpha}.$$

$$\therefore \quad P = R\sin\alpha.$$

Thus the evolute of an equiangular spiral is equal equiangular spiral.

8. There is no standard method for finding the evolute of a curve given by its polar equation. The present problem can be solved in the following manner:

Take a point $P(r,\theta)$ on the given cardiode. Let O' be a point on OA so that its polar co-ordinates are $(r = \frac{2}{3}a, \theta = 0)$. Let C be the centre of curvature of the given cardiode at P. Let $O'C = R$ and $\angle CO'A = \eta$. To find the locus of (R,η).

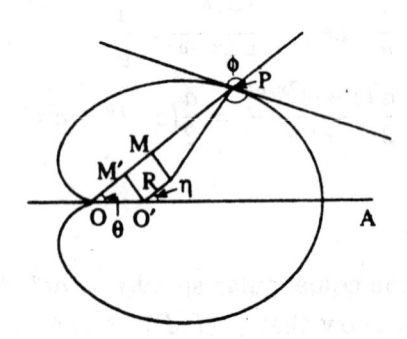

Fig. 18.7

Let CM and $O'M'$ be perpendiculars dropped on OP. Then $\rho = PC = \frac{4}{3}a\cos\theta/2$.
Also $r\frac{d\theta}{dr} = -\cot\theta/2$ so that $\phi = \pi/2 + \theta/2$ and $\angle CPM = \theta/2$.
$\therefore \quad CM = PC\sin\theta/2 = \frac{4}{3}a\cos\theta/2\sin\theta/2 = \frac{2}{3}a\sin\theta$ and $O'M' = OO'\sin\theta = \frac{2}{3}a\sin\theta = CM$.
Hence $O'C$ is parallel to OP i.e., $\eta = \theta$.
Also,

$$R = r - OM' - MP = r - \frac{2}{3}a\cos\theta - \frac{4}{3}a\cos^2\theta/2$$
$$= a(1+\cos\theta) - \frac{2}{3}a\sin\theta - \frac{2}{3}a(1+\cos\theta) = \frac{1}{3}a(1-\cos\theta) = \frac{1}{3}a(1-\cos\eta). \quad (\because \theta = \eta)$$

$\therefore$ The locus of C with O' as pole is
$$r = \frac{1}{3}a(1-\cos\theta) \quad \text{(take } (r,\theta) \text{ as current co-ordinates)}.$$

12. Evolute of the parabola $y^2 = 4ax$ is $27ay^2 = 4(x-2a)^3$.
This meets the parabola where $27a \cdot 4ax = 4(x-2a)^3 \qquad (i)$
$x = -a$ is clearly a root. Now write (i) as
$$(x+a)^2(x-8a) = 0.$$
The point $x = 8a$ on the evolute corresponds to the point $x = (8a-2a) \div 3$ on the parabola.

Also $\rho = 2a^{-\frac{1}{2}}(x+a)^{\frac{3}{2}}$. Hence ρ at the vertex ($x = 0$) of the parabola, which corresponds to the cusp on the evolute $= 2a$. Also ρ at $x = 2a$ is $2a^{-\frac{1}{2}}(2a+a)^{\frac{3}{2}} = 2 \cdot 3a\sqrt{3}$. Hence etc. (length $= \rho_1 - \rho_2 = 2a(3\sqrt{3}-1)$).

Short and Quick Revision on **Envelopes**: workout **Ex. 46–65** of **Pages 1165–1167** and **1181–1183**.

AN INTRODUCTION TO ANALYSIS

(Differential Calculus—Part I)

Chapter 19 (Applications of Derivatives)

Concavity, Convexity: Points of Inflexion

Chapter 19

Concavity, Convexity: Points of Inflexion

19.1 Introduction

In the present chapter we shall see that the sense of *concavity* or *convexity* leads to a special type of singularity (peculiar point) at a point of a curve $y = f(x)$. This singular point is called a *point of inflexion*.

19.2 Sense of Concavity and Convexity

We all have intuitive concepts of concavity and convexity: Any arc of a circle is concave to all points within the circle, whilst to a point without the circle, the portion lying between that point and the chord of contact of tangents drawn from the point is said to be *convex* and the remainder of the circumference *concave*.

Concavity and Convexity with respect to a line

Let P be a given point on a plane curve. Let l be a straight line not passing through P.

Then the curve is

a) **concave** at P w.r.t. the line l if a sufficiently small arc containing P lies within the acute angle formed by l and the tangent to the curve at P [Fig. 19.1 (a)].

b) **convex** at P w.r.t. the line l if a sufficiently small arc containing P lies without the acute angle formed by l and the tangent to the curve at P [Fig. 19.1(b)].

Fig. 19.1(a)
The curve is concave to l at P

Fig. 19.1(b)
The curve is convex to l at P

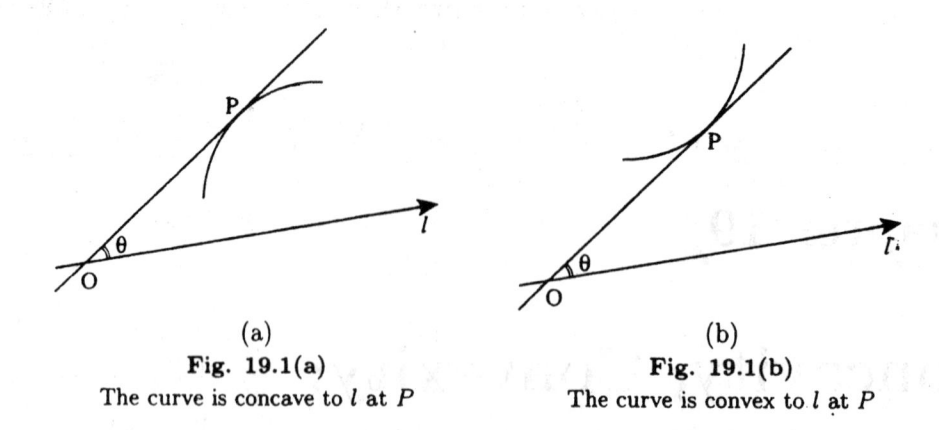

On the other hand if the curve is convex on one side of P and concave on the other, with respect to the line l, then evidently the curve crosses its tangent at P. The point P is then called a **point of inflexion** [Fig. 19.2].

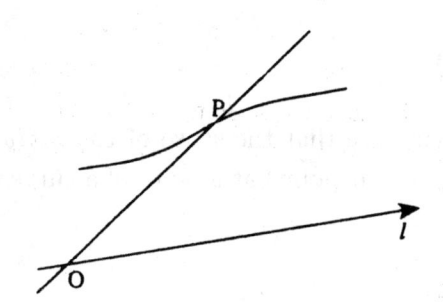

Fig. 19.2 The curve has a point of inflexsion at P

19.3 Concave Upwards/Downwards: Convex Upwards/Downwards

Consider a plane curve whose equation w.r.t. a given set of rectangular axes by $= f(x)$. Let P be a point on this curve. We assume that the tangent PT at the point P is not parallel to Y-axis. Then if the curve does not cross its tangent at P, *it will, before and after the point P, be situated on the same side of the tangent PT* in a small neighbourhood of P.

That is, there exists a sufficiently small arc of the curve which

1. lies entirely above the tangent PT, or
2. lies entirely below the tangent PT.

In case (a), the curve is said to be **concave upwards** at P (or equivalently, **convex downwards** at P). (Fig. 19.3).

CHAPTER 19: CONCAVITY, CONVEXITY: POINTS OF INFLEXION   1091

In case (b), the curve is **concave downwards** at P (or equivalently, **convex upwards** at P). (Fig. 19.4).

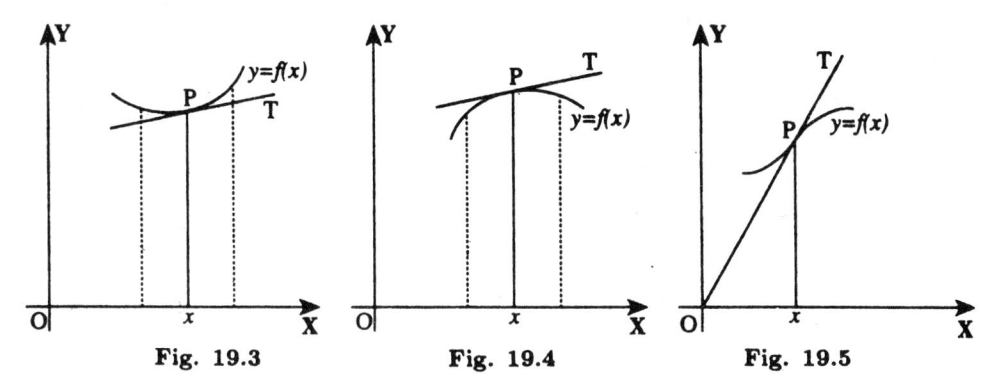

Fig. 19.3   Fig. 19.4   Fig. 19.5

We note that the other alternative, namely, *that the curve crosses the tangent at P,* will mean that the curve has a **point of inflexion** at P. (Fig. 19.5).

In Fig. 19.3, the curve has concavity at P towards the positive side of the Y-axis. In 19.4, the curve has a concavity at P towards the negative side of the Y-axis. Thus, the sense of concavity or convexity depends on the **choice of axes** which are fixed by convention. But this is not the case for a point of inflexion. The point where a curve crosses the tangent is an inflexional point so that its existence does not in any way depend on the choice of axes.

I. Criterion for Concavity or Convexity

To determine whether curve $y = f(x)$ is concave upwards or downwards at a given point P.

Theorem 19.3.1. *Suppose the derivatives of first two order, $f'(x)$ and $f''(x)$ exist and are continuous in a small neighbourhood of the point $P(x, y)$ and $f''(x) \neq 0$. Then the curve $y = f(x)$ turns its concavity upwards or downwards according as $f''(x)$ is positive or negative at the given point $P(x, y)$.*

Proof. In fact, one or the other of these two cases occurs according as the ordinate of the curve in the neighbourhood of P (as well before as after this point) is greater or smaller than the ordinate of the tangent at the point P [see the dotted ordinates of Fig. 19.3 and 19.4].

Let us give to x a positive or negative increment $\Delta x = dx$ (differential of x). The increment of the ordinate of the curve will be $\Delta y = f(x + \Delta x) - f(x)$ and the corresponding increment of the tangent will be $dy =$ differential of $y = f'(x)dx$ (See Art. 8.6, Geometric interpretation of differential).

The sense of concavity therefore depends on the sign of $\Delta y - dy$. It will be concave up or down according as this difference is positive or negative (Δx may itself be positive or negative).

∴ By Taylor's theorem, it follows

$$\Delta y = f(x + \Delta x) - f(x)$$
$$= \Delta x f'(x) + \frac{(\Delta x)^2}{2!} f''(x + \theta \cdot \Delta x) \qquad (0 < \theta < 1)$$
$$= dy + \frac{(\Delta x)^2}{2!} f''(x + \theta \cdot \Delta x) \qquad (0 < \theta < 1)$$

or, $\quad \Delta y - dy = \dfrac{(\Delta x)^2}{2!} f''(x + \theta \cdot \Delta x). \qquad (0 < \theta < 1)$

If $f''(x)$ does not vanish, $f''(x + \theta \cdot \Delta x)$ will have the same sign as $f''(x)$ provided $|\Delta x|$ is sufficiently small since $f''(x)$ is continuous at x. Now $(\Delta x)^2$ is essentially positive. Hence $\Delta y - dy$ will have the same sign as $f''(x)$.

Thus if $f''(x)$ is positive then $\Delta y - dy$ will be positive for sufficiently small values of $|\Delta x|$. Hence if $f''(x)$ is positive the curve, in a small neighbourhood of P, will be situated above the tangent at P i.e., *the curve will be concave upwards at P.*

Similarly, if $f''(x)$ is negative then $\Delta y - dy$ will be negative for small values of $|\Delta x|$ and the curve will be concave downwards at P.

II. Points of inflexion

We have defined a point of inflexion on the curve $y = f(x)$ as a point where the curve crosses its tangent.

The theorem of the previous article shows that if $f''(x)$ be continuous then such a point can only exist if $f''(x) = 0$. The abscissae of the points of inflexion are, therefore, the roots of the equation.

$$f''(x) = 0. \tag{19.3.1}$$

But the converse is not true.

Any root of the equation (19.3.1) *does not give a point of inflexion.*

In order that a point P may be a point of inflexion the ordinate of the tangent PT [Fig. 19.5] must exceed the ordinate of the curve on one side of P and the opposite must happen on the other side. Thus, with the notation of the previous articles, $\Delta y - dy$ must change its sign with Δx.

CHAPTER 19: CONCAVITY, CONVEXITY: POINTS OF INFLEXION

By Taylor's theorem, as before, we have

$$\Delta y - dy = \frac{(\Delta x)^2}{2} f''(x + \theta \Delta x) \qquad (0 < \theta < 1).$$

Hence, in order that the abscissa x corresponds to a point of inflexion, $f''(x + \theta \Delta x)$ must change its sign with Δx. Since we have assumed $f''(x)$ to be continuous, we have the following theorem:

Theorem 19.3.2. *The points of inflexion of the curve $y = f(x)$ are those roots of $f''(x) = 0$ where $f''(x)$ has one sign in the left-neighbourhood and an opposite sign in the right-neighbourhood of the points under consideration.*

More generally, if $f''(x)$ and all the following derivatives

$$f'''(x), f^{iv}(x), \cdots, f^{n-1}(x)$$

vanish for a value of x and if further $f^n(x)$ exist but $\neq 0$ and is continuous then Taylor's theorem will give

$$\Delta y - dy = \frac{(\Delta x)^n}{n!} f^n(x + \theta \cdot \Delta x) \qquad (0 < \theta < 1).$$

The nth derivative, which we have supposed to be continuous, has now a sign independent of Δx provided $|\Delta x|$ is sufficiently small. Thus the change of sign only depends on $(\Delta x)^n$. But $(\Delta x)^n$ changes its sign with Δx if n be odd. Hence the following rule:

> **Rule:** In order that a root of $f^n(x) = 0$ may give a point of inflexion the first of the derivatives which does not vanish simultaneously with $f^n(x)$ must be odd order.
>
> In case the first derivative that does not vanish is $f^n(x)$ where n is even, then the curve is concave upwards or downwards according as
>
> $$f^n(x) > 0 \quad \text{or} \quad f^n(x) < 0.$$

Note 19.3.1. *Since the position of a point of inflexion on a curve $y = f(x)$ is independent of the choice of axes, the position of X and Y axes may be interchanged without affecting the positions of the points of inflexion of a curve. Thus the points of inflexion may be determined by examining d^2x/dy^2 instead of d^2y/dx^2.*

For points of the curve $y = f(x)$, the points where the tangent is parallel to the Y-axis (i.e., where dy/dx is undefined) it becomes necessary to investigate d^2x/dy^2 instead of d^2y/dx^2.

A Test of Concavity or Convexity with respect to the Axis of X

We have, in art. 19.2, considered the concavity or convexity at a point of a curve w.r.t. any line l. Suppose, in particular, the line l is the X-axis. Then in Fig. 19.3, the curve is convex at P w.r.t. the X-axis (equivalently *Concave upwards*) and in Fig. 19.4, the curve is concave at P w.r.t. the X-axis (or *Concave downwards*). Note that in both the figures the curve has been drawn above the axis of X (i.e., the ordinate at the point P is **positive**).

a) A curve $y = f(x)$ lying above the axis of X is *convex* or *concave* at a given point $P(x,y)$ with respect to the axis of X according as it is concave upwards (d^2y/dx^2 is positive) or concave downwards (d^2y/dx^2 is negative) at P.

b) But the curve may lie below the axis of X (i.e., the ordinate at a given point may be negative). See the figures 19.6 and 19.7.

Then, in Fig. 19.6, the curve is *convex* at P w.r.t. the X-axis but concave downwards $\frac{d^2y}{dx^2}$ (i.e., is negative).

In Fig. 19.7, the curve is *concave* at P w.r.t. the X-axis and it is also concave upwards (i.e., $\frac{d^2y}{dx^2}$ is positive).

Both the cases will be convered if we adopt the following rule:

> **Rule:** A curve $y = f(x)$ is convex or concave at $P(x,y)$ w.r.t. the axis of x according as $y\dfrac{d^2y}{dx^2}$ is positive or negative at P.

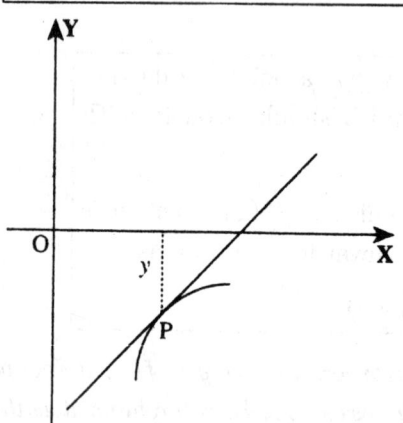

Fig. 19.6 Convex at P
w.r.t. the X-axis (concave downwards)

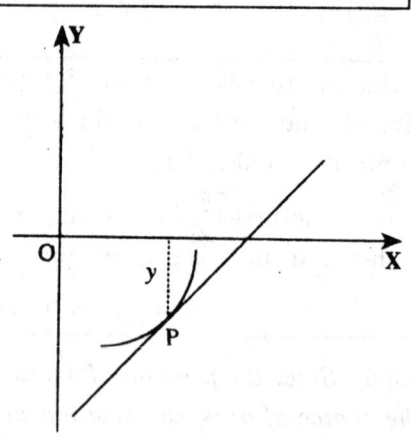

Fig. 19.7 Concave at P
w.r.t. the X-axis (concave upwards)

Note 19.3.2. *In case of the curve $x = f(y)$ test by the quantity $x\dfrac{d^2x}{dy^2}$, instead of $y\dfrac{d^2y}{dx^2}$.*

CHAPTER 19: CONCAVITY, CONVEXITY: POINTS OF INFLEXION

19.4 Illustrative Examples

Example 19.4.1. *Show that the curve $y = x^3$ has a point of inflexion at $x = 0$.*

Solution: Here $\dfrac{dy}{dx} = 3x^2$ and $\dfrac{d^2y}{dx^2} = 6x$.

At $x = 0$. $\dfrac{d^2y}{dx^2} = 0$.

When $x < 0$ (sufficiently near zero) $\dfrac{d^2y}{dx^2}$ remains negative so that the curve is concave downwards there. But when $x > 0$ (sufficiently near zero). $\dfrac{d^2y}{dx^2}$ becomes positive so that the curve is concave upwards there. Hence $x = 0$ is a *point of inflexion* of the curve.

Example 19.4.2. *Show that the curve $y = \sin(x/a)$ has a point of inflexion whenever the curve crosses the axis of X (i.e., $y = 0$).*

Solution: Here $\dfrac{dy}{dx} = \dfrac{1}{a}\cos\dfrac{x}{a}$; $\dfrac{d^2y}{dx^2} = -\dfrac{1}{a^2}\sin\dfrac{x}{a} = -\dfrac{y}{a^2}$.

$\dfrac{d^2y}{dx^2} = 0$ whenever, $y = 0$. $\dfrac{d^2y}{dx^2}$ changes sign at each such point; hence etc.

Example 19.4.3. *Show that the curve $y = \log x (x > 0)$ is everywhere convex upwards. Discuss concavity or convexity with respect to the axis of X. What can you say about the curve $y = x\log x (x > 0)$?*

Solution: (a) For the curve $y = \log x$, $y' = 1/x$ and $y'' = -1/x^2$. Thus y'' is negative for all values of x for which the function is defined (i.e., for all $x > 0$). Hence the curve is convex upwards (or concave downwards) at all points where $x > 0$.

We know that $y = \log x$ is negative or positive according as $0 < x < 1$ or $x > 1$. Thus for $0 < x < 1$, $y\dfrac{d^2y}{dx^2}$ is positive and for $x > 1$, $y\dfrac{d^2y}{dx^2}$ is negative. Hence the curve is convex w.r.t. the X-axis if $0 < x < 1$ and concave w.r.t. the X-axis when $x > 1$.

(b) For the curve $y = x\log x$, $y' = \log x + 1$, $y'' = 1/x$. See that the domain of definition of the function is $x > 0$ for which y'' is always positive. So the curve is everywhere concave upwards (or convex downwards).

The students should draw the curves $y = \log x$ and $y = x\log x$ and verify.

Example 19.4.4. *Find the points of inflexion, if any, of the curve*

$$y = \dfrac{x^3}{a^2 + x^2}.$$

Solution:

$$\frac{dy}{dx} = \frac{x^2(3a^2 + x^2)}{(a^2 + x^2)^2},$$

$$\frac{d^2y}{dx^2} = \frac{2a^2 x(3a^2 - x^2)}{(a^2 + x^2)^3}$$

and $\quad\dfrac{d^3y}{dx^3} = \dfrac{6a^2\{(x^2 - 3a^2)^2 - 9a^4 + a^2\}}{(a^2 + x^2)^4}.$

Thus

$$\frac{d^2y}{dx^2} = 0 \text{ if } x = 0,\ +\sqrt{3}\,a,\ -\sqrt{3}\,a.$$

Check: $\dfrac{d^3y}{dx^3} \ne 0$ for each such value of x.

Hence we conclude that these are the points of inflexion.

Example 19.4.5. *Find the points of inflexion, if any, of the curve $x = (\log y)^3$.*

Solution: Here, we examine,

$$\frac{dx}{dy} = \frac{3(\log y)^2}{y} \quad \text{and} \quad \frac{d^2x}{dy^2} = \frac{3\log y}{y^2}(2 - \log y)$$

$$\frac{d^3x}{dy^3} = \frac{6(\log y)^2 - 18(\log y) + 6}{y^3}.$$

Now,

$$\frac{d^2x}{dy^2} = 0 \text{ at } y = 1 \text{ and } y = e^2.$$

At each such point $\dfrac{d^3x}{dy^3} \ne 0$ (Check).

$\therefore$ $(0, 1)$ and $(8, e^2)$ are the two points of inflexion of the curve.

Example 19.4.6. *Find the range of values of x for which*

$$y = x^4 - 6x^3 + 12x^2 + 5x + 7$$

is concave upwards or downwards. [C.H. 1999]

Find also its point of inflexion, if any.

CHAPTER 19: CONCAVITY, CONVEXITY: POINTS OF INFLEXION

Solution:

$$\frac{dy}{dx} = 4x^3 - 18x^2 + 24x + 5$$

$$\frac{d^2y}{dx^2} = 12x^2 - 36x + 24 = 12(x^2 - 3x + 2)$$

$$= 12(x-1)(x-2).$$

For, $-\infty < x < 1$ $\frac{d^2y}{dx^2} > 0.$ [hence concave upwards in this range]

At $x = 1$ $\frac{d^2y}{dx^2} = 0.$

For, $1 < x < 2$ $\frac{d^2y}{dx^2} < 0.$ [hence concave downwards in this range]

Clearly, then $x = 1$ is a point of inflexion.

($\because$ y'' changes from +ve to $-$ve as x passes through $x = 1$.)

At $x = 2$, $\frac{d^2y}{dx^2} = 0.$

For $2 < x < \infty$, $\frac{d^2y}{dx^2} > 0.$ [hence concave upwards when $x > 2$]

Also $x = 2$ is a point of inflexion.

Since y'' changes from $-$ve to +ve as x passes through $x = 2$.

Conclusion. When $x < 1$ the curve is concave upwards.
         At $x = 1$ i.e., the point $(1, 19)$ is a point of inflexion.
         When $1 < x < 2$ the curve is concave downwards.
         At $x = 2$ i.e., the point $(2, 33)$ is a point of inflexion.
         When $x > 2$ the curve is concave upwards.

Example 19.4.7. *Concavity and Convexity with respect to a point.*

Let P be a given point on a curve and let PT, the tangent at P on the curve. Then if A is any given point, the curve is concave or convex at P with respect to A according as the portion of the curve in immediate neighbourhood of P does or does not lie entirely on the same side of the tangent PT as A.

Accordingly, a given curve is convex or concave at **P** with respect to the foot of the ordinate of P according as

$$y\frac{d^2y}{dx^2}$$

is positive or negative.

Consider the curve $y = 2\sqrt{(ax)}$. Is it convex or concave to the foot of the ordinate?

Here $\dfrac{d^2y}{dx^2} = -\dfrac{1}{2}\dfrac{\sqrt{a}}{x^{3/2}}$ and $y\dfrac{d^2y}{dx^2} = 2\sqrt{ax} \times -\dfrac{\sqrt{a}}{2x^{3/2}} = -\dfrac{a}{x}$.

Thus $y\dfrac{d^2y}{dx^2} < 0$ for all $x > 0$ (for $x < 0$ y is not defined)

∴ The curve in the neighbourhood of any given point is **Concave** to the foot of the ordinate of that point.

19.5 Concavity and Convexity of a Polar Curve

Suppose that the equation of a curve is given in polar co-ordinates as

$$r = f(\theta) \quad \text{or} \quad \frac{1}{u} = f(\theta), \quad \text{where } r = \frac{1}{u}.$$

To find a test of concavity or convexity of the curve towards the pole:

Let O be the pole, P, a point of the curve where concavity or convexity is to be examined.

Let the polar co-ordinates of P be (r, θ) and let A, B be two points on the curve in a small neighbourhood of P, one on each side of it.

Let the co-ordinates of A and B be $(r_1, \theta - \Delta\theta)$ and $(r_2, \theta + \Delta\theta)$ respectively. Then the curve in the immediate neighbourhood of P will be concave or convex to O according as

$$\triangle AOP + \triangle BOP > \text{ or } < \triangle AOB$$

when we proceed to the limit.

That is, according as

$$r_1 r \sin \Delta\theta + r_2 r \sin \Delta\theta > \text{ or } < r_1 r_2 \sin 2(\Delta\theta)$$

i.e., according as

$$r_1 r + r_2 r > \text{ or } < 2 r_1 r_2 \cos \Delta\theta$$

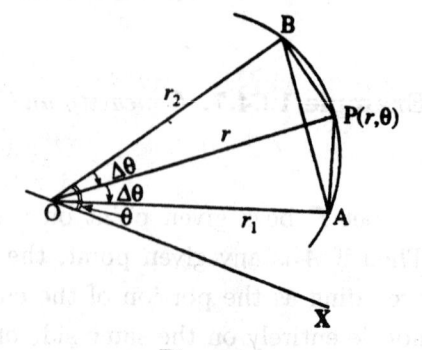

Fig. 19.8

CHAPTER 19: CONCAVITY, CONVEXITY: POINTS OF INFLEXION   1099

i.e., according as
$$\frac{1}{r_2} + \frac{1}{r_1} > \text{ or } < \frac{2}{r}\cos\Delta\theta$$

or according as
$$u_2 + u_1 > \text{ or } < 2u\cos\Delta\theta, \qquad (19.5.1)$$

where we have assumed $\dfrac{1}{r_1} = u_1$ and $\dfrac{1}{r_2} = u_2$ and $\dfrac{1}{r} = u$.

Now, using Taylor's Theorem, we get

$$u_2 = u + \frac{du}{d\theta}\Delta\theta + \frac{d^2u}{d\theta^2}\frac{(\Delta\theta)^2}{2!} + \frac{d^3u}{d\theta^3}\frac{(\Delta\theta)^3}{3!} + \cdots$$

and $\quad u_1 = u - \dfrac{du}{d\theta}\Delta\theta + \dfrac{d^2u}{d\theta^2}\dfrac{(\Delta\theta)^2}{2!} - \dfrac{d^3u}{d\theta^3}\dfrac{(\Delta\theta)^3}{3!} + \cdots.$

$$\therefore \ u_2 + u_1 = 2\left(u + \frac{d^2u}{d\theta^2}\frac{(\Delta\theta)^2}{2!} + \frac{d^4u}{d\theta^4}\frac{(\Delta\theta)^4}{4!} + \cdots\right)$$

whence we have concavity or convexity to the pole according as [See (19.5.1)]

$$2\left[u + \frac{d^2u}{d\theta^2}\frac{(\Delta\theta)^2}{2!} + \frac{d^4u}{d\theta^4}\frac{(\Delta\theta)^4}{4!} + \cdots\right]$$
$$> \text{ or } < 2u\left[1 - \frac{(\Delta\theta)^2}{2!} + \frac{(\Delta\theta)^4}{4!} + \cdots\right]$$

(and proceeding to the limit) i.e., according as

$$u + \frac{d^2u}{d\theta^2} > \text{ or } < 0.$$

At a **point of inflexion** the curve changes from concavity to convexity and hence the necessary condition for a point of inflexion is that

$$u + \frac{d^2u}{d\theta^2} \text{ should change sign.}$$

Example 19.5.1. *Find the point of inflexion on the curve* $r = a\theta^{-1/2}$.

Solution: Given curve is $r = \dfrac{a}{\sqrt{\theta}}$ or $\dfrac{1}{u} = \dfrac{a}{\sqrt{\theta}}$ or $au = \theta^{1/2}$.

$$\therefore \ a\frac{du}{d\theta} = \frac{1}{2\sqrt{\theta}} \text{ and } a\frac{d^2u}{d\theta^2} = \frac{1}{2}\left(-\frac{1}{2}\right)\theta^{-3/2} = -\frac{1}{4\theta^{3/2}}.$$

$$\therefore \ u + \frac{d^2u}{d\theta^2} = \frac{\theta^{1/2}}{a} - \frac{1}{4a\theta^{3/2}}.$$

Thus to find for what value of θ a change of sign in $u + \dfrac{d^2u}{d\theta^2}$ can occur, we put

$$\frac{\theta^{1/2}}{a} - \frac{1}{4a\theta^{3/2}} = 0.$$

This gives $\theta^2 = \dfrac{1}{4}$, or, $\theta = \pm\dfrac{1}{2}$.

The positive value is admissible only, giving $r = a\sqrt{2}$, $\theta = 1/2$ as the polar co-ordinates of the point of inflexion.

Example 19.5.2. *Find the points of inflexion on the curve*

$$(\theta^2 - 1)r = a\theta^2.$$

Here

$$u = \frac{\theta^2 - 1}{a\theta^2} = \frac{1}{a}\left[1 - \frac{1}{\theta^2}\right].$$

$$\therefore \quad \frac{du}{d\theta} = \frac{2}{a\theta^3} \quad \text{and} \quad \frac{d^2u}{d\theta^2} = -\frac{6}{a\theta^4}.$$

Putting $u + \dfrac{d^2u}{d\theta^2} = 0$ to find for what values of θ a change of sign may occur, we get

$$\frac{\theta^2 - 1}{a\theta^2} + \left(-\frac{6}{a\theta^4}\right) = 0$$

which gives, $\theta^2 = 3$ (ignoring the inadmissible value $\theta^2 = -2$).

This gives

$$r = \frac{a\theta^2}{\theta^2 - 1} = \frac{3a}{2}.$$

At the point $\left(r = \dfrac{3a}{2}, \theta = \sqrt{3}\right)$ there is change of sign of $u + \dfrac{d^2u}{d\theta^2}$ (Verify).

Hence there is a point of inflexion at this point.

Condition for Pedal Equations

From the figures 19.9(a) and (b) it will be evident that, if at any point P on a curve, the perpendicular distance (p) from the pole on the tangent increases or decreases as r (*the radius vector of P*) *increases* then the curve is concave or convex at P towards the pole O.

CHAPTER 19: CONCAVITY, CONVEXITY: POINTS OF INFLEXION

∴ The curve is
— concave at P to the pole, if $\dfrac{dp}{dr}$ is positive,
— convex at P to the pole, if $\dfrac{dp}{dr}$ is negative.

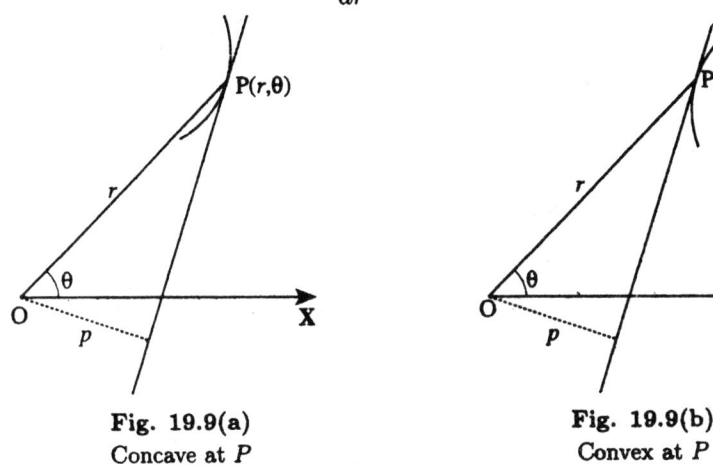

Fig. 19.9(a) Concave at P

Fig. 19.9(b) Convex at P

If $\dfrac{dp}{dr} = 0$ at P and moreover, $\dfrac{dp}{dr}$ is positive for points on one side of P and is negative for points on the other side of P, then there must be a point of inflexion at P (see Fig. 19.10).

This condition can be at once deducible from the condition in polar co-ordinates; namely, from the expression

$$u + \frac{d^2 u}{d\theta^2}, \quad \text{where } u = \frac{1}{r}.$$

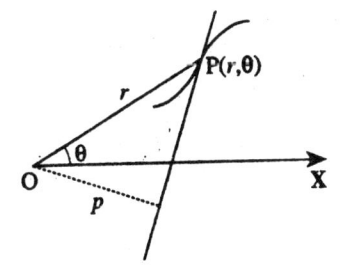

Fig. 19.10 Point of inflexion at P

For, we know

$$\frac{1}{p^2} = \frac{1}{r^2} + \frac{1}{r^4}\left(\frac{dr}{d\theta}\right)^2 = u^2 + \left(\frac{du}{d\theta}\right)^2.$$

$$\therefore \quad -\frac{2}{p^3}\frac{dp}{du} = 2u + 2\frac{du}{d\theta}\frac{d}{du}\left(\frac{du}{d\theta}\right) = 2u + 2\frac{d^2 u}{d\theta^2}$$

or, $\quad u + \dfrac{d^2 u}{d\theta^2} = -\dfrac{1}{p^3}\dfrac{dp}{du} = -\dfrac{1}{p^3}\dfrac{dp}{dr}\dfrac{dr}{du} = \dfrac{r^2}{p^3}\dfrac{dp}{dr} \quad \left(\because \dfrac{dr}{du} = -\dfrac{1}{u^2} = -r^2\right)$

whence the result follows.

Alternative approach:

We known

$$r\frac{dr}{dp} = \rho = \text{radius of curvature} = \frac{\left\{r + \left(\frac{dr}{d\theta}\right)^2\right\}^{3/2}}{r^2 + 2\left(\frac{dr}{d\theta}\right)^2 - r\frac{d^2r}{d\theta^2}}.$$

$$\therefore \frac{dp}{dr} = \frac{r\left\{r^2 + 2\left(\frac{dr}{d\theta}\right)^2 - r\frac{d^2r}{d\theta^2}\right\}}{\left\{r^2 + \left(\frac{dr}{d\theta}\right)^2\right\}^{3/2}}.$$

It follows that if at P,

$$r^2 + 2\left(\frac{dr}{d\theta}\right)^2 - r\frac{d^2r}{d\theta^2} = 0$$

and the left side expression changes sign in the neighbourhood of P, then there exists a point of inflexion at P.

Observe that putting $r = 1/u$ the expression

$$r^2 + 2\left(\frac{dr}{d\theta}\right)^2 - r\frac{d^2r}{d\theta^2} \quad \text{becomes} \quad u^2 + \frac{d^2u}{d\theta^2}.$$

Main Points : A Recapitulation

---

1. For sense of concavity/convexity at a point P of a curve with respect to a given line see Figures 19.1(a), (b).

2. Point of inflexion: The curve crosses its tangent there. Sense of concavity changes (concave to convex or convex to concave) as the curve passes through a point of inflexion. Figures 19.2 or 19.5.

3. Rectangular system of axes:

Curve given by $y = f(x)$. (Fig. 19.3, 17.4).

Concave upwards at P $\Rightarrow$ a small arc of the curve in the neighbourhood of P lies above the tangent PT (also called "convex downwards").

Concave downwards at P $\Rightarrow$ a small arc of the curve in the neighbourhood of P lies below the tangent PT. (also called "convex upwards").

4. Criterion for concavity or convexity: Point of inflexion:

a) Concave upwards at P of a curve $y = f(x) \to f''(x)$ is positive at P.

b) Concave downwards at P of a curve $y = f(x) \to f''(x)$ is negative at P.

c) To find the points of inflexion, find the roots of $f''(x) = 0$.

If α be one such root, then examine the sign of $f''(x)$ near $x = \alpha$. If the sign of $f''(x)$ changes as x passes through α from values less than α to values more than α, then the point $x = \alpha$ is a point of inflexion.

Alternatively, if at $x = \alpha$, $f''(x) = 0$ but $f'''(x) \neq 0$, then the point $x = \alpha$ is a point of inflexion.

d) A root of $f^n(x) = 0$ may give a *point of inflexion*, if the first of the derivatives which does not vanish simultaneously with $f^n(x)$ is of odd order.

If $f^n(x) \neq 0$, where n is even, then the curve is concave upwards or donwards at $x = \alpha$ according as $f^n(\alpha) > 0$ or $f''(\alpha) < 0$.

5. A curve $y = f(x)$ is concave or convex at P **with respect to the axis of X** according as $y\dfrac{d^2y}{dx^2}$ is negative or positive. Fig. 19.3, 19.4, 19.6, 19.7.

6. Polar curve: The expression $u + \dfrac{d^2u}{d\theta^2}$, where $u = \dfrac{1}{r}$, must change sign at a point of inflexion, (r, θ). Fig. 19.8.

Pedal Curve: Fig. 19.9(a), (b) and 19.10.

$\dfrac{dp}{dr}$ should change sign at a point of inflexion.

Exercises XIX

1. Show that $y = x^4$ is concave upwards at the origin and $y = e^x$ is everywhere concave upwards.

2. Show that the curve $y = 3x^5 - 40x^3 + 3x - 20$ is concave upwards in $-2 < x < 0$ and $2 < x < \infty$ but concave downwards in $-\infty < x < -2$ and $0 < x < 2$ and at $x = -2, 0, +2$ there are points of inflexion. Trace the curve.   [See Ex. **19.4.6**]

3. (a) Show that the curve $y = e^{-x^2}$ has points of inflexion at $x = \pm\dfrac{1}{\sqrt{2}}$.

   (b) Verify that the maximum ordinate and the point of inflexion of the curve $y = xe^{-x}$ are at $x = 1$ and $x = 2$ respectively.

4. Satisfy yourself that the following curves have inflexions at the point indicated:

   (a) $y = x^2(3-x)$; $(1, 2)$.

   (b) $y = \dfrac{b^3}{a^2 + x^2}$; $\left(\pm\dfrac{a}{\sqrt{3}}, \dfrac{3b^3}{4a^2}\right)$.

   (c) $y = \dfrac{a^2 x}{(x-a)^2}$; $\left(-2a, \dfrac{-2a}{9}\right)$.

   (d) $x = a\tan\theta,\ y = a\sin\theta\cos\theta$; $\left(\pm\sqrt{3}\,a, \pm\dfrac{\sqrt{3}}{4}a\right)$.

   (e) $y^2 = x(x+1)^2$; $\left(\dfrac{1}{3}, \pm\dfrac{4}{3\sqrt{3}}\right)$.

   (f) $a^2 y^2 = x^2(a^2 - x^2)$; $(0, 0)$.

   (g) $x^3 + y^3 = a^3$; $(a, 0), (0, a)$.

5. Show that the points of inflexion of the curve
$$y^2 = (x-a)^2(x-b)$$
lie on the line $3x + a = 4b$. [C.H.]

6. Show that the curve $(1 + x^2)y = 1 - x$ has three points of inflexion and that they lie on a straight line.

7. Show that every point in which the sine curve $y = c\sin(x/a)$ cuts the axis of X is a point of inflexion.

8. Show that the points of inflexion on the curve $r = b\theta^n$ are given by
$$r = b\{-n(n+1)\}^{n/2}.$$ [C.H.]

9. Show that the abscissae of the points of inflexion on the curve $y^2 = f(x)$ satisfy the equation
$$\{f'(x)\}^2 = 2f(x)\cdot f''(x).$$

   More generally, when the curve is $y^n = f(x)$, the equation is
$$\dfrac{n-1}{n}\{f'(x)\}^2 = f(x)\cdot f''(x).$$

10. Show that the curve $y = e^x$ is at every point *convex* to the foot of the ordinate of that point.

CHAPTER 19: CONCAVITY, CONVEXITY: POINTS OF INFLEXION

11. Show that for the cubical parabola $a^2 y = (x - b)^3$, there is a point of inflexion whose abscissa is b.

12. Verify that there are points of inflexion at $(0, 0)$ on each of the curves:

    a) $y = x \cos(x/a)$;
    b) $y = a \tan(x/b)$;
    c) $y = x^2 \log(1 - x)$;
    d) $y = axy + by^2 + c^2 x^2$.

13. Verify:

    (a) The curve $y = e^{x^{1/3}}$ has a point of inflexion at $(8, e^2)$.

    (b) The curve $y = b e^{-(x/a)^n}$ has a point of inflexion at the point where
    $$x = a \sqrt[n]{\frac{n-1}{n}}.$$

    (c) The curve $y = xe^x$ has a point of inflexion at $x = -2$.

14. Show that the curve $(y - a)^3 = a^3 - 2a^2 x + ax^2$ is always concave towards the foot of the ordinate.

15. The spiral $r \cosh \theta = a$ is **concave** towards the pole—Justify.

16. Find the positions of the points of inflexion on the curve
$$12y = x^4 - 16x^3 + 42x^2 + 12x + 1.$$

[Ans. at $x = 7$ and at $x = 1$]

For further revision see Pages 1169–1170 and 1187–1188 Ex. 81–90.

AN INTRODUCTION TO
ANALYSIS

(Differential Calculus—Part I)

Chapter 20 (Applications of Derivatives)

Singular Points: Multiple Points: Double Points

Chapter 20

Singular Points: Multiple Points: Double Points

20.1 Introduction

If, in a rational algebraic equation $f(x,y) = 0$[1] of degree n, we replace x by any fixed value, then we shall get an equation in y whose degree will be less than or equal to n. So, for a fixed x we shall get as many values of y as its degree. Thus **with the given value of x as abscissa** there will be as many points on the curve as are the different real roots of the equation $f(x,y) = 0$. As we vary x, each of these points will separately describe what is known as a BRANCH of a curve $f(x,y) = 0$ (Refer to art. 14.4), e.g., in $(x^2 + y^2)x - ay^2 = 0 (a > 0)$ we get $y = \pm x\sqrt{x/(a-x)}$ and the curve (*Cissoid of Diocles*) has two branches (Fig. 14.15). The curves having more than one branch exhibit some peculiarities which are not possessed by the curves with explicit equations like $y = \phi(x)$. We have, in Chapter **19**, discussed one kind of peculiar points of the explicit equations; we called them *points of inflexion*. Peculiar points (called **Singular Points**) other than the points of inflexion occur on a curve at points where two or more branches of the curve intersect. In the present chapter we shall investigate such singular points.

---

[1] A rational algebraic equation $f(x,y) = 0$ can be arranged as:

$$a_0 + (b_0 x + b_1 y) + (c_0 x^2 + c_1 xy + c_2 y^2) + \cdots + (l_0 x^n + l_1 x^{n-1} y + \cdots + l_n y^n) = 0$$

or, $\quad u_0 + u_1 + u_2 + \cdots + u_n = 0$,

where u_r represents a homogeneous polynomial in x and y of degree r.
 The degree of any term = sum of the indices of x and y in that term.
 The degree of the curve means the degree of the highest degree term.
 For e.g., n is the degree of this algebraic equation.

20.2 Definitions

1. A point through which more than one branch of a curve pass is called a **Multiple Point** of the curve.

   If through a point P, r branches of a curve pass, we call P, a **multiple point** of order r.

2. A point through which two of the branches of a curve pass will be called a **Double Point** or a **Multiple Point** of order two.

3. Multiple points are also called **Singular Points**. In particular, a double point or a multiple point of second order is a *singular point*.

4. *In general,* a curve has two tangents at a **Double Point**, one for each branch of the curve.

A DOUBLE POINT of a curve will be called **A Node, A Cusp** or **An Isolated Point** according as the two tangents are **Real & Distinct, Real & Coincident or Imaginary**. [See Figures 20.1(a), (b), (c)].

Note 20.2.1. *The term imaginary tangent may be misleading. We remark that in the neighbourhood of an isolated point of a curve there exist no (real) point of the curve and hence no (real) tangent can be drawn there. But if we apply the formal analytical process of finding the tangent at an isolated point we obtain an equation which involves the imaginary quantities of the form $a + ib$. This explains the name imaginary tangent.*

20.3 Singular Points at the Origin of a Rational Algebraic Curve

In order to investigate the nature of a singular point it is necessary to find the tangents thereat. We recall here the results obtained in **art. 15.5**:

If a curve passes through the origin and is given by a rational integral algebraic equation, the equation to the tangent (or tangents) at the origin is obtained by equating to zero the terms of the lowest degree in the equation of the curve.

When the *origin is a point on the curve*, its equation may be put in the form:

$$(a_1 x + b_1 y) + (a_2 x^2 + b_2 xy + c_2 y^2) + (a_3 x^3 + b_3 x^2 y + c_3 xy^2 + d_3 y^3) + \cdots = 0.$$

1. The origin is a *singular point* if at least $a_1 = b_1 = 0$.

CHAPTER 20: SINGULAR POINTS: MULTIPLE POINTS: DOUBLE POINTS

1. If $a_1 = b_1 = 0$ but not all of a_2, b_2, c_2 are zeroes, the singular point is a *double point*;

2. If $a_1 = b_1 = a_2 = b_2 = c_2 = 0$, but not all of a_3, b_3, c_3, d_3 are zeroes, the singular point is called a *triple point*; and so on.

2. Classification of a double point at the origin

Case 1. $c_2 \neq 0$.

Replace y by mx in the terms of $a_2 x^2 + b_2 xy + c_2 y^2$ and obtain $(c_2 m^2 + b_2 m + a_2) x^2$. Now solve for m, the quadratic equation:

$$c_2 m^2 + b_2 m + a_2 = 0. \qquad (20.3.1)$$

If the roots m_1, m_2 of (20.3.1) are real and distinct, the curve has two real and distinct tangents $y = m_1 x$ and $y = m_2 x$ at the origin and the double point is then a **Node** [Fig. 20.1(a)]. (See **Ex.20.3.1**. next page).

If the roots of the equation (20.3.1) are real and equal, the curve has generally two coincident tangents at the origin and the double point is then called a **Cusp** [Fig. 20.1(b)] (See **Ex. 20.3.2** below).

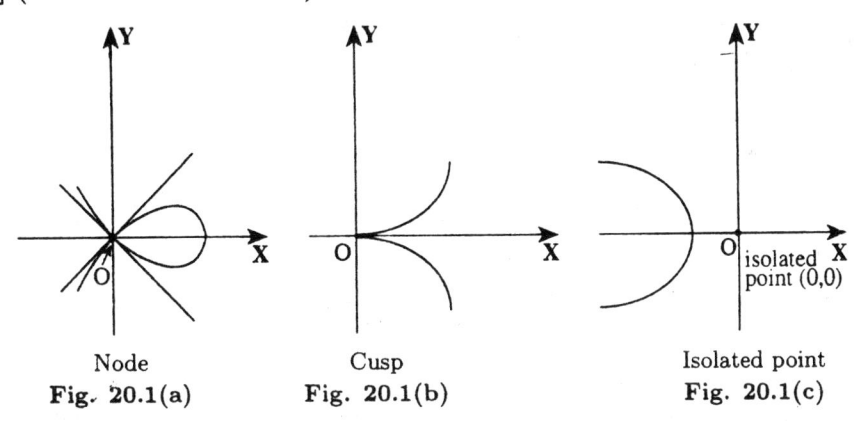

Node
Fig. 20.1(a)

Cusp
Fig. 20.1(b)

Isolated point
Fig. 20.1(c)

In exceptional cases (**Ex. 20.3.3** below) the origin may be an **isolated point**. If the roots are imaginary, the origin is an *isolated point* [Fig. 20.1 (c)].

Case 2. $c_2 = 0, a_2 \neq 0$.

Replace x by ny in the terms $a_2 x^2 + b_2 xy$ of the curve and proceed as in **Case 1**.

Case 3. $a_2 = 0 = c_2 \cdot b_2 \neq 0$.

The origin is a *node*, the two tangents there being the co-ordinate axes:

$$x = 0 \quad \text{and} \quad y = 0.$$

Examples:

Example 20.3.1. *Examine the curve $y^2(1+x) = x^2(1-x)$ for singular points at the origin.*

Solution: This is a rational algebraic curve of degree 3 passing through the origin.
We write the equation as

$$(y^2 - x^2) + (xy^2 + x^3) = 0.$$

The lowest degree terms are of second degree; the origin is a *double point*. Now observe that here c_2, the coefficient of y^2 is not zero. Replace y by mx in $y^2 - x^2$ and equate to zero the coefficient of x^2 to obtain $m^2 - 1 = 0$. Thus, $m = \pm 1$ and the lines $y = x, y = -x$ are two distinct tangents to the curve at the origin; the origin is therefore a *node*.

Example 20.3.2. *Examine the curve $(x^2 + y^2)x - 2y^2 = 0$ for singular points at the origin.*

Solution: This is also a rational algebraic curve of degree 3 passing through the origin.
We write the given equation as

$$-2y^2 + (x^3 + xy^2) = 0.$$

The lowest degree terms are of second degree; the origin is a *double point*. Proceeding as before or *by equating to zero the lowest degree terms* we observe that at the origin there are coincident tangents $y = 0$. Hence X-axis is the *cuspidal tangent* and the origin is a *cusp*.

Example 20.3.3. *Examine the curve $y^2(x^2 - 9) - x^4 = 0$ for an isolated point at the origin.*

Solution: This is a rational algebraic curve of degree 4 passing through the origin.
The origin is obviously a double point. Replacing y by mx in $-9y^2$, (the lowest degree term) and equating to zero the coefficient of y^2 we get $m^2 = 0$ suggesting that $y = 0$ may be a cuspidal tangent but observe that for x near zero, y is imaginary $\left(\because y = \pm \dfrac{x^2}{\sqrt{x^2 - 9}}\right)$ i.e., in the immediate neighbourhood of the origin there exists no point of the curve. Hence, the origin is an *isolated double point* (not a *cusp*).

Example 20.3.4. *Verify the following statements:*

CHAPTER 20: SINGULAR POINTS: MULTIPLE POINTS: DOUBLE POINTS

(a) The origin is a NODE of the *Folium of Descartes* $x^3 + y^3 = 3axy$; $x = 0, y = 0$ are the two nodal tangents.

(b) The origin is a CUSP on the curve $(x^2 + y^2)x = 2ay^2$; $y = 0$ is the cuspidal tangent.

(c) The origin is an isolated point of the curve

$$a^2 x^2 + b^2 y^2 = (x^2 + y^2)^2.$$

[We may call $ax = \pm iby$ imaginary tangents at $(0,0)$].

(d) The origin is a triple point on the curve.

$$2y^5 + 5x^4 - 3(x^3 - xy^2) = 0;$$

$x = 0, x = y, x = -y$ are the three tangents at the origin.

Proceed as in **Ex. 20.3.1, 20.3.2, 20.3.3** above.

20.4 Singular Points at Points other than the Origin

If a point (h, k) be a singular point of the curve, *known in advance*, we may make further study of such a point by shifting the origin of the co-ordinate system to the point (h, k). This can be done, as shown in books on Co-ordinate Geometry by substituting

$$x = x' + h; \quad y = y' + k$$

where (x, y) are the current co-ordinates of the old system and (x', y') that of the new system. The singular point is now the point $(0, 0)$ of new co-ordinate system. Now we may use the rules suggested above in art. **20.3** for determining the nature of the singular point at the origin.

Example 20.4.1. *Determine the existence and nature of the double points on the curve*

$$(x - 2)^2 = y(y - 1)^2. \qquad \text{[C.H. 1980]}$$

Solution: We transfer the origin to a point (h, k), the equation becomes

$$(x' + h - 2)^2 = (y' + k)(y' + k - 1)^2,$$

(x', y') being the current co-ordinates in the new system.

or, $(h-2)^2 + 2x'(h-2) - k(k-1)^2 - y'\{2k(k-1) + (k-1)^2\} + x'^2 - ky'^2 - y'^3 = 0.$ (20.4.1)

Hence, for a double point we must have (art. 20.3)

$$(h-2)^2 - k(k-1)^2 = 0 \tag{20.4.2}$$
$$h - 2 = 0 \tag{20.4.3}$$
$$2k(k-1) + (k-1)^2 = 0 \tag{20.4.4}$$

whence $h = 2, k = 1$ which satisfy simultaneously (20.4.2), (20.4.3) and (20.4.4). Therefore $(2, 1)$ is a double point of the curve.

Transferring the origin to $(2, 1)$, we get from (20.4.1) $x'^2 = y'^2(y' + 1)$.

Tangents at the origin are $x'^2 = y'^2$ i.e., $y' = \pm x'$. Hence there is a **node** at $(2, 1)$ where the two tangents are

$$y - 1 = \pm(x - 2).$$

Example 20.4.2. *Find the equation of the tangent at the point $(-1, -2)$ to the curve $x^3 + 2x^2 + 2xy - y^2 + 5x - 2y = 0$ by shifting the origin of the co-ordinate system to the point $(-1, -2)$ and hence show that the point $(-1, -2)$ is a cusp of the given curve.*

Solution: To shift the origin to the point $(-1, -2)$ we write $x' - 1$ for x and $y' - 2$ for y. The given equation then becomes

$$(x' - 1)^3 + 2(x' - 1)^2 + 2(x' - 1)(y' - 2) - (y' - 2)^2 + 5(x' - 1) - 2(y' - 2) = 0$$

or, $x'^3 - x'^2 + 2x'y' - y'^2 = 0.$

Equating to zero the lowest degree terms we get

$$-x'^2 + 2x'y' - y'^2 = 0 \text{ or } (y' - x')^2 = 0$$

which represent two coincident lines and, therefore, the point is a Cusp and the cuspidal tangent is $y' - x' = 0$, with reference to the new system of co-ordinate axes.

But with reference to original system the equation of the tangent at $(-1, -2)$ is

$$(y + 2) - (x + 1) = 0 \text{ or } \mathbf{y = x - 1}.$$

20.5 Conditions for the existence of Double Points

At a point (x, y) of the curve $f(x, y) = 0$, the slope of the tangent (when exists) is given by $\dfrac{dy}{dx}$ where

$$f_x + f_y \frac{dy}{dx} = 0. \tag{20.5.1}$$

CHAPTER 20: SINGULAR POINTS: MULTIPLE POINTS: DOUBLE POINTS

For a double point two values of $\dfrac{dy}{dx}$ must occur and equation (20.5.1) being of the first degree this can happen iff

$$f_x = 0, \quad f_y = 0 \qquad (20.5.2)$$

so that (20.5.1) becomes an identity.

In order, therefore, to find double points we must have such values of x and y which satisfy three equations:

$$\left.\begin{array}{l} f(x,y) = 0 \\ f_x(x,y) = 0 \\ f_y(x,y) = 0 \end{array}\right\}.$$

Differentiating (20.5.1) w.r. to x, we also have

$$f_{xx} + f_{yx}\dfrac{dy}{dx} + \left(f_{xy} + f_{yy}\dfrac{dy}{dx}\right)\dfrac{dy}{dx} + f_y\dfrac{d^2y}{dx^2} = 0$$

so that at a double point (where $f_x = f_y = 0$) the values of dy/dx are the roots of the quadratic equation

$$f_{yy}\left(\dfrac{dy}{dx}\right)^2 + 2f_{xy}\dfrac{dy}{dx} + f_{xx} = 0.$$

Hence, in general, a double point will be a node, cusp or an isolated point according as

$$(\mathbf{f_{xy}})^2 - \mathbf{f_{yy}f_{xx}} \gtreqless 0.$$

If $f_{xx} = f_{xy} = f_{yy} = 0$, the point (x, y) will be a multiple point of order higher than two.

Example 20.5.1. *Determine the position and nature of the multiple points of the curve $x^3 - y^2 - 7x^2 + 4y + 15x - 13 = 0$. Also find the tangents at the multiple point, if any.*

Solution: We call $f(x,y) = x^3 - y^2 - 7x^2 + 4y + 15x - 13$.

Then $f_x = 3x^2 - 14x + 15$; $f_y = -2y + 4$.

Observe that $f_x = 0$ when $x = 3$ or $5/3$ and $f_y = 0$ when $y = 2$.

$\therefore$ the two partial derivatives vanish for the points $(3, 2)$ and $(5/3, 2)$.

Of these two points the point $(3, 2)$ lies on the curve. Hence this is the only multiple point.

Again, $f_{xy}(3,2) = f_{yx}(3,2) = 0$; $f_{xx}(3,2) = 4$; $f_{yy}(3,2) = -2$.

$\therefore (f_{xy})^2 - f_{yy} \cdot f_{xx} > 0$ and f_{xy}, f_{yy}, f_{xx} are not all zero.

Hence $(3, 2)$ is a *double point* and there is a *node* at $(3, 2)$.

The slopes of the tangents are given by

$$f_{yy}\left(\frac{dy}{dx}\right)^2 + 2f_{xy}\left(\frac{dy}{dx}\right) + f_{xx} = 0$$

i.e., $-2y_1^2 + 4 = 0$, i.e., $y_1 = \pm\sqrt{2}$.

∴ the equations of tangents at the node (3, 2) are given by

$$y - 2 = \pm\sqrt{2}\,(x - 3).$$

Otherwise. To find the tangents at (3, 2) we may shift to the origin to this point, and then proceed as in art. **20.4**.

20.6 Discrimination of Species of a Cusp

At a cusp the two branches of the curve have a common tangent.

A cusp is called a **single cusp** or a **double cusp** according as the two branches lie entirely **on the same** or **on both sides of the common normal**.

Again a cusp, single or double, may be of **first or second species** according as the two branches of the curve lie on **opposite or same side** of the **common tangent**.

A double cusp with change of species on the two sides of the common normal is a **point of osculinflexion**. Thus we have the following types of cusps:

A cusp of first species is also called a **keratoid cusp** (i.e., a cusp like *horns*) and a cusp of second species is called a **ramphoid** cusp (i.e., a cusp like a *beak*).

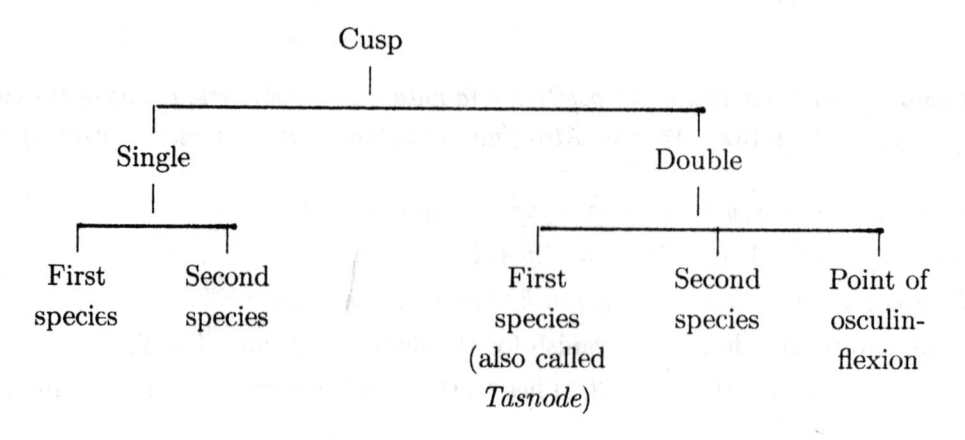

```
                        Cusp
                         |
         _____|_____
        |                                 |
      Single                            Double
        |                                 |
    ____|____                    _____|_____
   |         |                  |            |            |
 First     Second             First       Second      Point of
species   species           species      species     osculin-
                          (also called                flexion
                            Tasnode)
```

Example 20.6.1. *Examine the nature of the cusps on the following curves:*

1. $y^2 = x^3$;

2. $y^2 - x^4 = 0$;

3. $(y - 4x^2)^2 = x^7$;

4. $x^6 - ayx^4 - a^3x^2y + a^4y^2 = 0$;

5. $a^3y^2 - 2abx^2y + x^5 + a^{-1}x^6 = 0$.

Solution: (a) Clearly, $y = 0$ is the cuspidal tangent. Since x cannot be negative the two branches lie on the same side of the common normal, namely $x = 0$, the cusp is **single**.

Since $y = \pm x^{3/2}$, for each positive value of x there are two equal and opposite values of y. Hence the two branches lie on the different sides of the common tangent ($y = 0$). Thus the cusp at $(0,0)$ is a **single cusp of first species** [Fig. 20.2(a)].

(b) Here $x^2 = y, x^2 = -y$ are two parabolas which lie on different sides of the cuspidal tangent $y = 0$ and they extend to both sides of the common normal $x = 0$.

Hence the origin is a **double cusp of the first species** [Fig. 20.2 (c)].

(c) Since x cannot be negative the two branches of the curve given by

$$y = 4x^2 \pm x^{7/2}$$

sill lie on the same side of the common normal $x = 0$;

$$y = 0 \text{ is the cuspidal tangent.}$$

In the branch $y = 4x^2 + x^{7/2}$, the curve obviously lies above the X-axis and in the other branch, y is positive for $0 < x < 4^{2/3}$, i.e., near the origin, this branch also lies above the X-axis. In other words, both the branches lie above the common tangent ($y = 0$).

Hence the origin is a **single cusp of second species** [Fig. 20.2(b)].

(d) Here also $y = 0$ is a cuspidal tangent. Solving for y, we have

$$y = \frac{(ax^4 + a^3x^2) \pm \sqrt{(ax^4 + a^3x^2)^2 - 4x^6a^4}}{2a^4}.$$

When x is near zero, $(ax^4 + a^3x^2)^2 - 4x^6a^4$ has the same sign as a^2x^3 (leading term) which is positive whether x is positive or negative. Thus we have real values of y for numerically small values of x. Also when x is positive or negative but near zero,

$$(ax^4 + a^3x^2)^2 - 4x^6x^4 < (ax^4 + a^3x^2)^2$$

i.e., $\quad + \sqrt{(ax^4 + a^3x^2)^2 - 4x^6x^4} < (ax^4 + a^3x^2).$

Hence near the origin both the values of y are positive i.e., the origin is a **double cusp of second species** [Fig. 20.2 (d)].

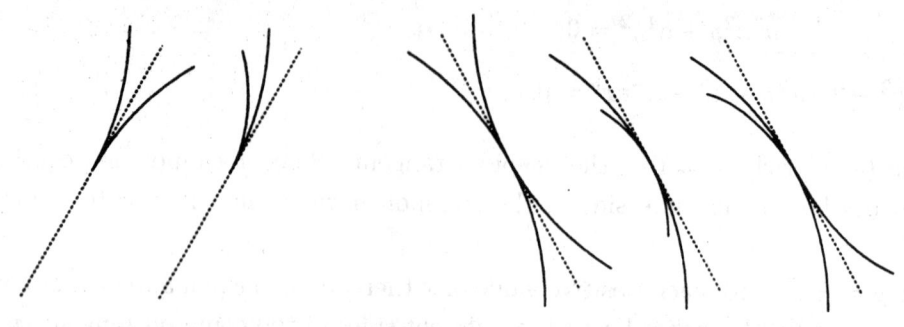

Fig. 20.2(a)   **Fig. 20.2(b)**   **Fig. 20.2(c)**   **Fig. 20.2(d)**   **Fig. 20.2(e)**

(e) Proceed as in (d) and observe that there **is a double cusp at the origin**. Then verify that when x is positive and small both the values of y (for the two branches) are positive i.e., *on the right of the origin, we have* **a cusp of second species**.

Again see that when x is negative and numerically small, one value of y will be positive and the other negative, i.e., on the left of the origin there is a *cusp of first species*. Hence finally we conclude that there is an **osculinflexion at the origin** [Fig. 20.2(e)].

20.7 Systematic Procedure

1. Given a curve $f(x, y) = 0$.

To find a *multiple point* we find (x, y) which simultaneously satisfy:

$$\frac{\partial f}{\partial x} = 0, \quad \frac{\partial f}{\partial y} = 0 \quad \text{and} \quad f(x, y) = 0.$$

In case, at such a point (x, y), f_{xx}, f_{yy} and f_{xy} are not all zero then the multiple point (x, y) is of order two (i.e., **a double point**).

A double point (x, y) will be a Node, a Cusp or a Conjugate point according as

$$(f_{xy})^2 - f_{xx}f_{yy} > 0, = 0 \text{ or } < 0.$$

If, however, $f_{xx} = f_{yy} = f_{xy} = 0$ at a multiple point, then the point is a multiple point of order higher than two.

2. If the multiple point is a double point at $(h, k) \neq (0, 0)$, we shift the origin to (h, k). Thus we discuss the nature of singularity at $(0, 0)$ only.

CHAPTER 20: SINGULAR POINTS: MULTIPLE POINTS: DOUBLE POINTS

3. Suppose the double point is a cusp at the origin.

(a) *Cuspidal tangent is the axis of X ($y = 0$).*

We first solve for y. The reality of the roots of y and if real, their **signs** for values of x near the origin determine the nature of the cusp (single or double; first species or second species etc.).

> **Remember:**
> (i) Roots of y are real for all values of x (positive or negative) *near the origin*: Conclusion → *Origin is a Double Cusp.*
> (ii) Roots of y are real for positive x only or for negative x only (x being near the origin):
> Conclusion → *Origin is a single cusp.*
> (iii) Like signs for y → *Second Species* $\Big\}$ x being near the origin.
> (iv) Opposite signs for y → *First Species*
> (v) Different species on the two sides of the origin:
> Conclusion → *origin is a point of Osculinflexion.*

Note 20.7.1. *Since we require to investigate the approximate shape of the curve near the origin, the terms which involve higher powers of x may be ignored (as they are very small for small x) provided:*

(i) *that does not reduce the equation of the curve to the equation of the tangent at the origin, or;*

(ii) *that the two branches do not coincide.*

Since y is small, terms involving higher powers of y above the second can also be neglected.

(b) *Cuspidal tangent is the axis of Y ($x = 0$).*

We can proceed as before; solve for x instead of y.

(c) *Cuspidal tangent is neither the X-axis nor the Y-axis but, say, the line*

$$aX + bY = 0.$$

The perpendicular P_1, on the line $aX + bY = 0$, from the point (x, y) on the curve is $P_1 = (ax + by)/\sqrt{a^2 + b^2}$ which is evidently proportional to $ax + by$; we shall call this quantity P.

If we substitute $(P - ax)/b$ for y in the equation of the curve, we then get a relation between P and x.

Solving for P we shall be able to decide from the **sign and the reality** of the values of P for positive and negative values of x, the nature of the Cusp.

Thus if its values are of opposite signs (See Fig. 20.3) for *a positive small x*, the perpendiculars, from those points on the curve which have this value of x as abscissa, on the tangent, $ax + by = 0$, are of opposite signs. Hence, there is a cusp of the first species on the right and so on.

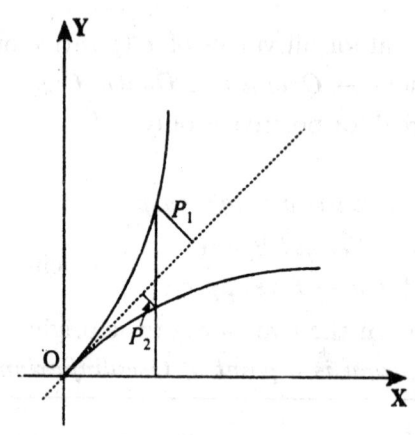

Fig. 20.3

Example 20.7.1. *Examine the nature of origin of the curve*

$$(2x + y)^2 - 6xy(2x + y) - 7x^3 = 0.$$

Solution: The tangents at the origin $(0,0)$ are $(2x + y)^2 = 0$ (coincident tangents). So the origin is a cusp.

To examine the nature of the cusp we put $P = 2x + y$.

We then eliminate y by putting $y = P - 2x$ in the equation of the curve.

We obtain,

$$(1 - 6x)P^2 + 12x^2 P - 7x^2 = 0, \quad \text{or,} \quad P = \frac{-6x^2 \pm \sqrt{7x^3 - 6x^4}}{1 - 6x}.$$

For small numerical values of x:

1. when $x < 0, P$ is not real;

2. when $x > 0$, the values of P are real and are of opposite signs.

(neglecting $6x^4$, see that $x^{3/2} > x^2, x$ being sufficiently small).

∴ There is a *single cusp* of *first species* at the origin.

20.8 Radii of Curvature at Multiple Points

We recall here that the radius of curvature ρ at a point (x,y) of a curve $f(x,y) = 0$ is given by

$$\rho = \frac{(f_x^2 + f_y^2)^{3/2}}{f_{xx}(f_y)^2 - 2f_{xy}f_xf_y + f_{yy}(f_x)^2}.$$

This formula does not give the value of ρ if $f_x = f_y = 0$ (i.e., at a multiple point of the curve). At a multiple point we expect as many values of ρ as its order (distinct or not).

Example 20.8.1. *Find the radii of curvature at $(0,0)$ of the different branches of the curve: $y^4 + 2axy^2 = ax^3 + x^4$.*

Solution: Equate to zero the lowest degree terms: $2axy^2 - ax^3 = 0$.

This gives three tangents at the origin: $x = 0, y = (1/\sqrt{2})x$ and $y = -(1/\sqrt{2})x$ i.e., origin is a **Triple point**.

To find ρ for the branch which touches $x = 0$, we obtain

$$\lim_{\substack{x \to 0 \\ y \to 0}} \frac{y^2}{2x} \quad \text{(Newton's Formula)}.$$

Write $y^2/2x = \rho_1$ or $x = y^2/2\rho_1$. Substitute this value in the given equation and use $\lim \rho_1 = \rho$ (radius of curvature of the corresponding branch at the origin). We get

$$y^4 + 2ay^2 \cdot \left(\frac{y^2}{2\rho_1}\right) = a\frac{y^6}{8\rho_1^3} + \frac{y^8}{16\rho_1^4}, \quad \text{or,} \quad 1 + \frac{a}{\rho_1} = a\frac{y^2}{8\rho_1^3} + \frac{y^4}{16\rho_1^4}.$$

Let $y \to 0$, so that we have $1 + a/\rho = 0$ or $\rho = -a$ for the branch which touches $x = 0$.

To find ρ for other branches we proceed thus:

The equation of either branch is given by

$$y = f(0) + xf'(0) + \frac{x^2}{2!}f''(0) + \cdots.$$

Here, $f(0) = 0$. We write $f'(0) = p, f''(0) = q$. Then we have

$$y = px + \frac{1}{2}qx^2 + \cdots.$$

Putting this expression for y in the given equation we get

$$\left(px + \frac{1}{2}qx^2 + \cdots\right)^4 + 2ax\left(px + \frac{1}{2}qx^2 + \cdots\right) = ax^3 + x^4.$$

Equating the coefficients of x^3 and x^4, we get

$$2ap^2 = a, \quad p^4 + 2apq = 1.$$

This gives

$$p = 1/\sqrt{2}, \quad \text{whence } q = 3\sqrt{2}/8a.$$
$$p = -1/\sqrt{2}, \quad \text{whence } q = -\frac{3\sqrt{2}}{8a}.$$
$$\therefore \quad \rho = \frac{(1+p^2)^{3/2}}{q} = \pm 2\sqrt{3}\,a.$$

Main Points : A Recapitulation

> **1.** A point of inflexion is an *unusual point* or a Singular Point, because the tangent at such a point **Crosses** the curve.
>
> There are, however, other kinds of singularities. A multiple point is a *singular point*.
>
> **2.** What is a multiple point? A point through which there pass r branches of a curve is called a *Multiple point of the curve* of order r. In particular, a *double point* of a curve is a point through which two branches of the curve pass.
>
> **3.** In general, a curve has two tangents at a *double point*, one for each branch of the curve.
>
> A double point of a curve will be a **Node**, a **Cusp** or an **isolated point** (or, **conjugate point**) according as the two tangents are *distinct and real, coincident and real* or *imaginary* (i.e., there are no real points of the curve in the immediate neighbourhood of the point).
>
> **4.** The general equation of a *rational algebraic* curve of the nth degree which passes through the origin, when arranged according to ascending powers of x and y, is of the form
>
> $$(a_1 x + a_2 y) + (b_1 x^2 + b_2 xy + b_3 y^2) + (c_1 x^3 + c_2 x^2 + c_3 x + c_4) + \cdots = 0,$$
>
> where the constant term is absent.
>
> The equation of the tangent or tangents at the origin is obtained by equating to zero the terms of the lowest degree in the equation of the curve. The origin will be a multiple point on the curve whose equation does not, at least, contain the constant and the first degree terms.

Examples

(i) The origin is a *node* on the curve $x^3 + y^3 - 3axy = 0$.
The nodal tangents are $x = 0, y = 0$.

(ii) The origin is a *cusp* on the curve $(x^2 + y^2)x - 3ay^2 = 0$.
The cuspidal tangent is $y = 0$.

(iii) The origin is an isolated point on the curve $a^2x^2 + b^2y^2 = (x^2 + y^2)^2$.
$ax \pm iby$ are the two imaginary tangents.

(iv) The origin is a multiple point of order three on the curve $3y^5 + 7x^5 - 2x(x^2 - y^2) = 0$.

Here $x = 0$, $x = y$, $x = -y$ are the three tangents at that point.

5. The necessary and sufficient conditions for any point (x, y) on $f(x, y) = 0$ to be a multiple point are that $f_x(x, y) = 0$, $f_y(x, y) = 0$.

To find multiple points (x, y), we have therefore to find the values of (x, y) which satisfy the three equations:
$$f_x(x, y) = 0, \quad f_y(x, y) = 0, \quad f(x, y) = 0.$$

In case $f_x = 0 = f_y$ but f_{xx}, f_{xy}, f_{yy} are not all zero then the point (x, y) will be a double point and will be a node, cusp or conjugate according as
$$(f_{xy})^2 - f_{xx}f_{yy} > 0, = 0 \text{ or } < 0.$$

If $f_x = 0 = f_y$ and $f_{xx} = 0$, $f_{yy} = 0$, $f_{xy} = 0$, the point (x, y) will be a multiple point of order higher than the second.

6. Types of Cusps: A cusp might be **single** or **double** according as the curve lies entirely on one side of the normal or on both sides. Also, a single cusp might be of the *first species* or the *second species*, according as both the branches lie on *opposite sides* or on the *same side* of the tangent. Hence we have the following types of cusps: Single cusp of the first species [Fig. 20.2(a)], single cusp of the second species [Fig. 20.2(b)], Double cusp of the first species [Fig. 20.2(c)], Double cusp of second species [Fig. 20.2 (d)]. Double cusp with change of species or a point of Osculinflexion [Fig. 20.2 (e)].

Examples

7. Systematic Procedure: For details see art. **20.7**.

(a) See No. **5** above : Obtain a double point—either it will be the origin $(0,0)$ or a point $(h,k) \neq (0,0)$. Shift the origin to (h,k). So only discuss the nature of singularity at $(0,0)$.

(b) Suppose the double point is a cusp at the origin: Three cases may arise:

(i) *Cuspidal tangent may be the axis of x* i.e., $y=0$. Solve for y.

The reality of roots of y and if real, their signs for values of x near the origin determine the nature of the cusp, single or double, first or second species.

(ii) *Cuspidal tangent may be the axis of y* i.e., $x=0$. Solve for x and proceed as in (i).

(iii) *Cuspidal tangent may be of the form $aX+bY=0$* (see the discussions centered round Fig. **20.3**).

Exercises XX

1. Satisfy yourself that the following curves have indicated singular points at the origin:

   a) $x^3+y^3-3axy=0$; double point (node).
   b) $2y^5+5x^5-3x(x-y^2)=0$; triple point.
   c) $x^3+y^3=6x^2$; double point (cusp).
   d) $y^2(x-1)=x^3$; double point (cusp).
   e) $y^2(x^2-4)=x^4$; isolated double point.
   f) $x^4-ax^2y+axy^2+a^2y^2=0$; isolated double point.
   g) $x^4-4x^2y-2xy^2+4y^2=0$; double point (cusp).

2. Show that the origin is a node, cusp or an isolated point on the curve $y^2 = ax^2+bx^3$ according as $a >=$ or <0. [D.H.]

3. Show that the curve $y^3+3ax^2+x^3=0$

   a) is everywhere concave to the X-axis, b) has a point of inflexion at $x=-3a$,
   c) has a cusp at $x=0$ and   d) has an asymptote $x+y+a=0$.

CHAPTER 20: SINGULAR POINTS: MULTIPLE POINTS: DOUBLE POINTS

Trace the curve.

4. (a) Find the nature and position of the multiple points on the following curves. Also find the tangents at the multiple points.

   i. $x^2(x-y) + y^2 = 0$.
   ii. $y(y-6) = x^2(x-2)^3 - 9$. [C.H.]
   iii. $(x+y)^3 - \sqrt{2}\,(y-x+2)^2 = 0$.
   iv. $x^4 + y^3 - 2x^2 + 3y^2 = 0$.
   v. $x^2 y^2 = (a+y)^2(b^2 - y^2)$; distinguish the cases $b < a$, $b = a$ and $b > a$.

   (b) Find the singular points of the following curves and examine their characters:
   (i) $(x-c)^2 - y^3 = 0$; (ii) $[x^2 + (y-c)^2](x-2) + x = 0$.

5. The nature of the cusps on the following curves are indicated below; verify the correctness of the assertions.

   a) $ay^2 = x^3$;   single cusp of first species at $(0,0)$.
   b) $y^3 = (x-a)^2(2x-a)$;   single cusp at $(a,0)$.
   c) $(xy+1)^2 + (x-1)^3(x-2) = 0$;   single cusp of first species at $(1,-1)$.
   d) $y - 2 = x(1 + x + x^{3/2})$;   single cusp of second species at $(0,2)$.
   e) $y^2 = 2x^2 y + x^4 y + x^4$;   double cusp of first species at $(0,0)$.
   f) $a^3 y^2 - 2abx^2 y = x^5$;   osculinflexion at $(0,0)$.
   g) $x^6 - 2yx^4 - 8x^2 y + 16y^2 = 0$;   double cusp of second species at $(0,0)$.

6. Show that at each of the four points of intersection of the curve
$$(ax)^{2/3} + (by)^{2/3} = (a^2 - b^2)^{2/3}$$
with the axes there is a cusp of the first species.

7. Prove that the curve $ay^2 = (x-a)^2(x-b)$ has at $x = a$, a conjugate point if $a < b$, a node if $a > b$ and a cusp if $a = b$. [IAS]

8. For the curve $y^2(a^2 + x^2) = x^2(a^2 - x^2)$ show that the origin is a node and that nodal tangents bisect angles made by the axes.

9. Show that the cardiode $r = a(1 + \cos\theta)$ has a cusp at the origin.

10. Show that the curve $ay^2 = x^2 y + x^3$ has a cusp of the first species at the origin and an asymptote $x + y = a$ cutting the curve at $(\tfrac{1}{2}a, \tfrac{1}{2}a)$.

11. Show that the pole is a triple point on the curve $r = a(2\cos\theta + \cos 3\theta)$ and that the radii of curvature of the branches are $\sqrt{3}a/2,\ a/2,\ \sqrt{3}a/2$.

[*Hints*: $r = 0$ for $\theta = \pi/3, \pi/2, 2\pi/3$, hence the pole is a triple point. Find r_1 and r_2 at each of the values of θ and use

$$\rho = (r^2 + r_1^2)^{3/2}/\{r^2 + 2r_1^2 - rr_2\}.]$$

Answers

4. (a) (i) Single cusp of first species at $(0,0)$; $y = 0$. (ii) Single cusp of first species at $(2,3)$; $y = 3$. Isolated point at $(0,3)$. (iii) Single cusp of first species at $(1,-1)$; $y = x-2$. (iv) Node at $(0,0)$; $y = \sqrt{\frac{3}{8}}\, x$ and $y = -\sqrt{\frac{2}{3}}\, x$. (v) $(0, -a)$ is a double point; a node, cusp or isolated point according as $b > a$, $b = a$ or $b < a$.

For further exercises see Pages **1170–1171** and **1188–1189** Ex. **91–100**.

AN INTRODUCTION TO
ANALYSIS

(Differential Calculus—Part I)

Chapter 21 (Applications of Derivatives)

Curve Tracing (Systematic Method)

Chapter 21

Curve Tracing (Systematic Method)

21.1 Introduction

In our previous discussions we have indicated how by constructing a number of points that satisfy a given function we may obtain a fairly good idea as to the shape of a curve given by that function. The process is, no doubt, laborious. Equipped with the knowledge of curvature, asymptote, singularities, determination of maximum or minimum points or points of inflexion we are now in a better position to attack the problem of curve tracing. All figures should, however, be neatly drawn and important characteristics should be clearly explained.

We advise our students to revise the materials discussed in Chapters **3** and **14**. One should have clear idea about the shape of the curve of some important functions.

21.2 Important Curves

The following is a list of curves [See Chapter **3**] which we constantly require in applications:

   1. **Graphical Representation** of $y = x^n$, where n is an even or an odd positive integer (See Figs. 3.3, 3.4).
   2. **Graphs of** $y = x^{-n} (n = 1, 2, 3, 4, 5, \cdots)$ (See Figs. 3.5, 3.6).
   3. **Polynomials:** $y = -x^2 + 3x + 2,\, , y = x^2 - 12x + 13$ (*Quadratic functions*). See Fig. 3.7.

4. Trigonometric functions like $\sin x, \cos x$, etc. Figs. 3.17(a), (b), 3.18, 3.19. Inverses like $\sin^{-1} x, \cos^{-1} x, \tan^{-1} x, \cot^{-1} x$ etc. See Fig. 3.20, 3.21, 3.22.

5. Hyperbolic functions like $y = \sinh x$ or $\cosh x$ and $\tanh x$ etc. (See Figs. 3.23, 3.24).

6. Transcendental functions like $e^x, \log x$ etc. (See Fig. 3.16).

7. Besides the aforesaid functions we have given in Chapter **14** rough diagrams of well-known curves: *Folium of Descartes, Cycloid, Astroid, Catenary, Cardiode, Spirals, Lemniscate of Bernoulli*, etc. [See Fig. 14.14, 14.12, 14.29, 14.22]

21.3 Systematic Procedure of Curve Tracing:

Curves given by Algebraic functions f(x,y) = 0.

I. SYMMETRY. Use the following rules for determining whether the curve is symmetric or not:

A Curve will be symmetric with respect to:

a) **the X-axis**, if its equation contains only even powers of **y** and hence remains unchanged if y is replaced by $-y$.

b) **the Y-axis**, if its equation contains only even powers of x.

c) **the line y = x**, if the equation remains unchanged when x and y are interchanged.

d) **the origin** O, if its equation does not change when x is replaced by $-x$ and y by $-y$ simultaneously.

II. INTERCEPTS. Obtain the points where the curve intersects the co-ordinate axes.

a) Put $y = 0$ in the equation; obtain the x-intercepts.

b) Put $x = 0$ in the equation; obtain the y-intercepts.

III. ORIGIN. Does the equation satisfy $x = 0, y = 0$ simultaneously? If so, it passes through the origin.

When the curve passes through the origin, write down the equation or equations of the tangents there.

If the origin is a **Singular point**, find the nature of singularity—*Cusp, Node* or *Isolated*. If there is a cusp, find the *Species*.

CHAPTER 21: CURVE TRACING (SYSTEMATIC METHOD)

It is necessary to take note of the fact if the origin is a multiple point of higher order than the second.

IV. SOLVE FOR y. If it is not too complicated, then solve for y. Find the intervals in which y is increasing or decreasing.

V. MAXIMUM AND MINIMUM POINTS; POINTS OF INFLEXION, SENSE OF CONCAVITY AND CONVEXITY.

a) Find $\dfrac{dy}{dx}$; obtain the stationary points.

b) Find $\dfrac{d^2y}{dx^2}$; obtain the points of inflexion, if any and study the sense of concavity.

c) Find the values of x for which y has a maximum or minimum.

In a symmetrical curve begin with **one branch** and the symmetry will lead to similar points in the other branches.

VI. ASYMPTOTES. In case the curve has an infinite branch which can be easily detected, observe if there be any asymptote, paying special attention to horizontal and vertical asymptotes. If possible, find where the oblique asymptotes (if any) meet the curve, and the sides of the curve on which the asymptotes lie.

VII. EXTENT. The interval of x for which y exists is the *horizontal extent* and the range of y for which x exists is the *vertical extent*. If in the immediate neighbourhood of a particular point there exists no other point of the curve the point is *isolated*. To obtain the extent of a curve, proceed as:

Solve, wherever possible, the given equation for y (say) in terms of x. We assume that this will be possible. Consider $x = 0$, then see how y varies as x increases and ultimately tends to $+\infty$, paying special attention to those values of x for which $y = 0$ or tends to $+\infty$. Also see how y varies as x decreases and ultimately tends to $-\infty$. The symmetric property may sometimes shorten the labour. Again, if y does not exist or is found to be imaginary for a certain range of x then there exists no part of the curve in that region. Observe at this step the existence of loops, if there be any.

A special study may be necessary to know whether the curve is below or above a particular line, say, the tangent line at the origin.

21.4 Illustrative Examples

Example 21.4.1. *Discuss the characteristics of the curve $y^2(x^2 - 9) = x^4$ and then trace it.*

Solution: a) *Symmetry.* The curve is symmetric with respect to both the axes and also with respect to the origin.

b) *Intercepts.* The intercepts on the axes are $x = 0$ and $y = 0$.

c) *Origin.* The curve passes through $(0,0)$. We may show that the origin is a double point. Since for x very near to 0, y is imaginary, the origin itself is an *isolated double point*.

d) *Special points.* The curve has two branches:

$$y = \pm \frac{x^2}{\sqrt{x^2 - 9}}.$$

(i) We begin with the branch

$$y = \frac{x^2}{\sqrt{x^2 - 9}} \text{ for } x > 3.$$

Here

$$\frac{dy}{dx} = \frac{x(x^2 - 18)}{(x^2 - 9)^{3/2}}; \quad \frac{d^2y}{dx^2} = \frac{9x^2 + 162}{(x^2 - 9)^{5/2}}.$$

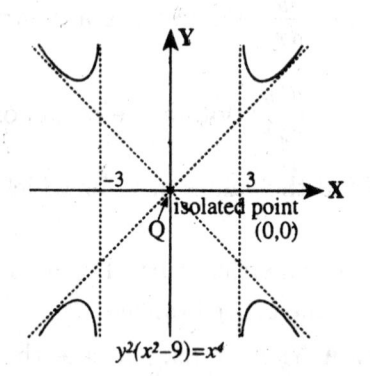

Fig.21.1

At $x = 3\sqrt{2}$, y is stationary. The portion is concave upwards and $(3\sqrt{2}, 6)$ is a minimum point.

(ii) By symmetry, the branch $y = \frac{x^2}{\sqrt{x^2 - 9}}$ for $x < -3$ there is another minimum point $(-3\sqrt{2}, 6)$.

(iii) For the other branch $y = -\frac{x^2}{\sqrt{x^2 - 9}}$ we study separately the portions for which $x > 3$ and $x < -3$ or *by symmetry*, we conclude that there are maximum points at $(3\sqrt{2}, -6)$ and $(-3\sqrt{2}, -6)$.

(e) *Asymptotes.* The lines $x = 3$ and $x = -3$ are vertical asymptotes and $y = \pm x$ are oblique asymptotes. The latter asymptotes intersect the curve at the origin which is an isolated point of the curve.

(f) *Extent.* Observe here that the curve exists on the intervals $-\infty < x < -3$; $3 < x < +\infty$ and y ranges in $-\infty < y \leq -6$; $6 \leq y < +\infty$. The point $(0, 0)$ is isolated.

The curve being symmetric with respect to both the axes it is sufficient to study the portion of

$$y = \frac{x^2}{\sqrt{x^2 - 9}} \text{ for } x > 3.$$

When $x \to 3^{+0}$, $y \to +\infty$. At $x = 3\sqrt{2}$, we get a minimum point. But when $x > 3\sqrt{2}$, y increases and for any such x, $y > x$, i.e., the curve lies above the asymptote

$y = x$ also it lies to the right of the asymptote $x = 3$. The portion is concave upwards. Hence the shape will be as shown in the first quadrant of Fig. 21.1.

The other symmetric portions can now be drawn. The complete curve is shown in Figure 21.1.

Example 21.4.2. *Discuss the characteristics and then sketch the curve*

$$x(x^2 + y^2) = a(x^2 - y^2) \left[or \ y^2 = \frac{x^2(a-x)}{a+x} \right], \ (a > 0).$$

Solution: a) The curve is symmetric with respect to the X-axis.

b) The x-intercepts are $x = 0$ and $x = a$; the y-intercept is $y = 0$.

c) The curve passes through the origin where we have two distinct tangents $y = \pm x$, i.e., the origin is a *node*.

d) The curve consists of two branches: $y = \pm \dfrac{x\sqrt{a-x}}{\sqrt{a+x}}$

For the first of these,

$$\frac{dy}{dx} = \frac{a^2 - ax - x^2}{(a+x)^{3/2}(a-x)^{1/2}} \quad \text{and} \quad \frac{d^2y}{dx^2} = \frac{a^2(x - 2a)}{(a+x)^{5/2}(a-x)^{3/2}}.$$

$\dfrac{dy}{dx} = 0$ when $x = \dfrac{a}{2}(-1 \pm \sqrt{5})$ and $\dfrac{dy}{dx}$ does not exist at $x = a$.

Observe that $x = \frac{a}{2} \cdot (-1 - \sqrt{5})$ *does not correspond to a point of a curve.*

Satisfy yourself that at $x = \frac{a}{2} \cdot (-1 + \sqrt{5})$ there is a maximum point and there is no point of inflexion; the branch is concave downwards.

By symmetry, there is a minimum point at $x = \dfrac{a}{2}(-1 + \sqrt{5})$ of the second branch $\left(y = -\dfrac{x\sqrt{a-x}}{\sqrt{a+x}} \right)$ and this branch is concave upwards.

e) The line $x + a = 0$ is the *only* (vertical) asymptote.

f) The curve exists on the interval $-a < x \leq a$ and for all values of y.

We next study on the branch

$$y = x\sqrt{\frac{a-x}{a+x}}.$$

When $x > 0$ but very near to 0, we notice that as

$$\frac{a-x}{a+x} < \frac{a}{a}, \text{ i.e., } < 1,$$

y is less than x. Hence the curve is below the tangent $y = x$ for small positive values of x.

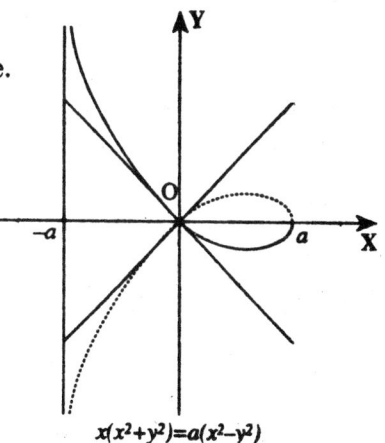

Fig. 21.2
$x(x^2+y^2)=a(x^2-y^2)$

If x be negative and numerically small $a - x > a + x$. Hence y is numerically greater than $y = -x$, i.e., the curve lies above the tangent $y = -x$ in the second quadrant.

Taking symmetry into account, the complete curve must be of the form as shown in Fig. 21.2 [Strophoid].

Example 21.4.3. *Sketch the curve* $y^2(x - 1) - x^3 = 0$.

Solution: We state the following facts which the reader should verify and then sketch the curve as in Fig. 21.3.

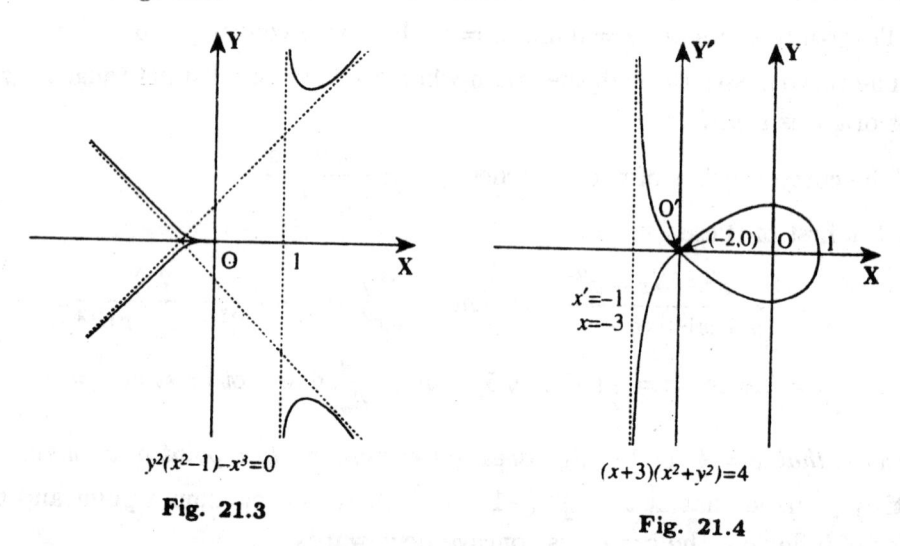

Fig. 21.3 — $y^2(x^2-1)-x^3=0$

Fig. 21.4 — $(x+3)(x^2+y^2)=4$

The curve is *symmetric* with respect to X-axis, the *intercepts* are $x = 0, y = 0$. The curve exists in the range $-\infty < x \leq 0$ and $x > 1$ and for all values of y. For the branch $y = x\sqrt{\dfrac{x}{x-1}}$ the point $x = 3/2$ gives a *minimum* point; there is no point of inflexion; this branch is concave upwards. The lines $x = 1, y = x + \frac{1}{2}, y = -x - \frac{1}{2}$ are *asymptotes*. The origin is a *cusp of the first species; $y = 0$* is the cuspidal tangent. The property of symmetry then gives the curve (Fig. 21.3).

Example 21.4.4. *Sketch the curve* $(x + 3)(x^2 + y^2) = 4$.

Solution: Here $(-2, 0)$ is a singular point. Use the transformation $x = x' - 2, y = y'$, the equation becomes

$$y'^2(x' + 1) + x'^3 - 3x'^2 = 0. \qquad (21.4.1)$$

Now study the curve (21.4.1) considering the two axes as $O'X'$ and $O'Y'$. Obtain the figure as shown in Fig. 21.4.

CHAPTER 21: CURVE TRACING (SYSTEMATIC METHOD)

$a^3 y^2 = (x-a)^4 (x-b) \, (a > b).$
Fig. 21.5

$a^3 y^2 = (x-a)^4 (x-b) \, (a < b).$
Fig. 21.6

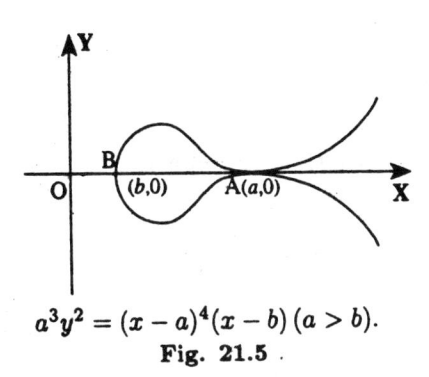

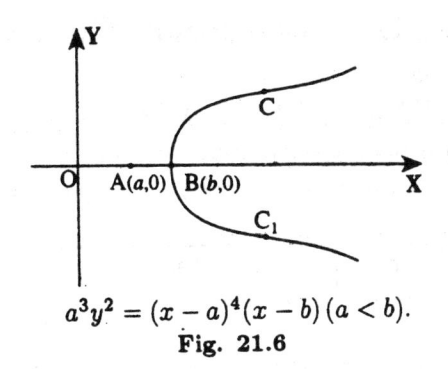

Example 21.4.5. Sketch the curve

$$a^3 y^2 = (x-a)^4 (x-b).$$

Solution: The shape of the curve when $a > b$ is given in Fig. 21.5 and when $a < b$, it is given in Fig. 21.6.

$$x^2 y = 4a^2 (2a - y)$$

THE WITCH OF AGNESI

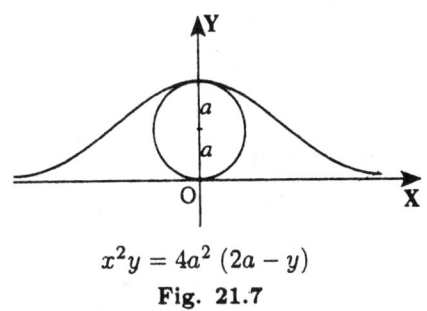

$x^2 y = 4a^2 (2a - y)$
Fig. 21.7

Example 21.4.6. Sketch the curve: $x^2 y = 4a^2(2a-y)$, (Fig. 21.7).

Solution: This curve is known as **Witch of Agnesi**.

21.5 Polar Curves: Procedure; Curve given by $f(r, \theta) = 0$

I. Solve the equation for r (whenever possible) and consider how r varies as θ increases from 0 to $+\infty$ and also as θ decreases from 0 to $-\infty$. From a table of corresponding values of r and θ.

II. If θ be replaced by $-\theta$ and the equation remains unchanged, the curve is symmetric about the *initial line*. If only even powers of r occur in the equation, the curve is symmetric about the *origin*.

III. In most polar equations only periodic functions ($\sin \theta, \cos \theta$, etc.) occur and so values of θ from 0 to 2π need alone be considered. Other values of θ give no new branch of the curve.

IV. If the curve possesses an infinite branch, obtain the asymptote. Note that if $r \to \infty$ as $\theta \to \alpha$ it should not be assumed that $\theta = \alpha$ must be an asymptote. The asymptote might not exist at all or even if it exists, it might be parallel to $\theta = \alpha$. Proceed along the lines suggested in *art.* **17.8** to ensure the existence of the asymptote.

21.6 Well-known Polar Curves

(A) Class of Curves: $r = a \sin n\theta$

We begin with a particular case when $n = 3$ ($r = a \sin 3\theta$). The following table of corresponding value of r and θ will enable one to trace the curve:

3θ	0	Intermediate values	$\pi/2$	Intermediate values	π	Intermediate values	$3\pi/2$	Intermediate values	2π
θ	0	,,	$\pi/6$	,,	$\pi/3$	,,	$\pi/2$	,,	$2\pi/3$
r	0	Positive, Increasing	a	Positive Decreasing	0	Negative Numerically Increasing	$-a$	Negative Numerically Decreasing	0

Observations:

1. r is never greater than $|a|$; there is no asymptote. Here $\phi = \frac{1}{3} \tan 3\theta$. Hence $\phi = 0$ whenever r vanishes.

2. The curve consists of a series of similar loops as shown in Fig. 21.8, all being arranged symmetrically about the origin and lying entirely within a circle of radius a and centre at the origin.

General Case. Any curve of the class $r = a \sin n\theta$ may be traced in a similar manner. They are called **Rose Petals**. If n be odd, there are n loops but if n be even there are $2n$ loops. The order in which the loops are described as θ increases from 0 to 2π is indicated in the figure by numbers. Fig. 21.9 shows the curve $r = a \sin 4\theta$.

THREE-LEAVED ROSE

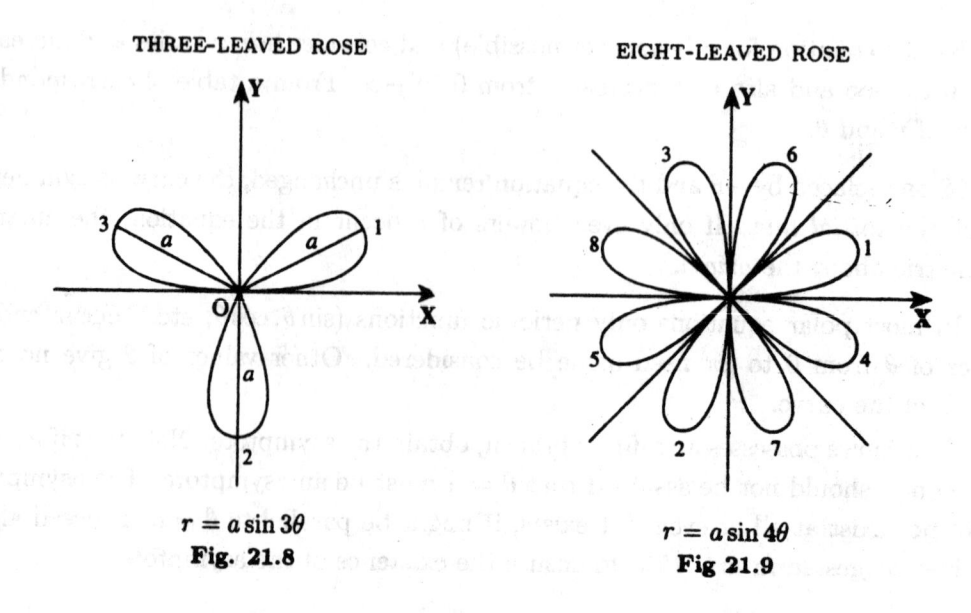

$r = a \sin 3\theta$
Fig. 21.8

EIGHT-LEAVED ROSE

$r = a \sin 4\theta$
Fig 21.9

CHAPTER 21: CURVE TRACING (SYSTEMATIC METHOD)

(B) Class of Curves: $r^n = a^n \cos n\theta$

We take the particular case $r^2 = a^2 \cos 2\theta$.

Here we observe that

a) When $\theta = 0$ or $\pi, r = \pm a$.

b) When $-\dfrac{\pi}{4} < \theta < \dfrac{\pi}{4}$, $\cos 2\theta$ is positive and we get real values of r; as θ increases from $-\dfrac{\pi}{4}$ to 0, r increases and as θ increases from 0 to $\dfrac{\pi}{4}$, r decreases; nowhere r is greater than a.

c) When $\dfrac{\pi}{4} < \theta < \dfrac{3\pi}{4}$, $\cos 2\theta$ is negative and then r becomes imaginary, r is greater than a.

d) When $\dfrac{3\pi}{4} < \theta < \dfrac{5\pi}{4}$, $\cos 2\theta$ again positive and we obtain real values of r, and so on.

e) $r = 0$ when $\theta = \pm \dfrac{\pi}{4}, \dfrac{3\pi}{4}, \dfrac{5\pi}{4}$.

f) The curve can be shown to the symmetrical about the initial line and also about the origin.

LEMNISCATE OF BERNOULLI

$r^2 = a^2 \cos 2\theta$

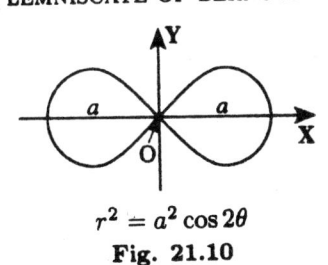

Fig. 21.10

Therefore the curve, called **Lemniscate of Bernoulli**, consists of two similar loops as shown in Fig. 21.10.

General Case. Curves of the class $r^n = a^n \cos n\theta$ can be treated in a similar manner. It will be useful to remember the following special cases:

a) **n = 1**, the circle: $r = a \cos \theta$ (referred to a point on the circumference as pole and the diameter through it as the initial line).

b) **n = −1**, the straight line: $r \cos \theta = a$.

c) **n = 2**, the lemniscate of Bernoulli: $r^2 = a^2 \cos 2\theta$.

d) **n = −2**, the rectangular hyperbola: $r^2 \cos 2\theta = a^2$.

e) $n = \frac{1}{2}$, the cardiode: $r^{1/2} = a^{1/2} \cos \frac{\theta}{2}$ or, $r = c(1 + \cos \theta)$.

f) $n = -\frac{1}{2}$, the parabola: $r = \dfrac{2c}{1 + \cos \theta}$.

(C) **Spirals**

(a) *Equiangular spiral* (or *Logarithmic spiral*): $r = ae^{\theta \cot \alpha}$.

This curve has the characteristic property that the tangent makes a constant angle with the radius vector, i.e., $\phi = \alpha$ (constant). As θ ranges from $-\infty$ to $+\infty$, r ranges from 0 to ∞. Observe that the origin is not a point of the curve (Fig. 21.11).

b) *Spiral of Archimedes*: $r = a\theta$ (Fig. 21.12).

EQUIANGULAR SPIRAL

SPIRAL OF ARCHIMEDES

$r = ae^{\theta \cot \alpha}$
Fig. 21.11

$r = a\theta$
Fig. 21.12

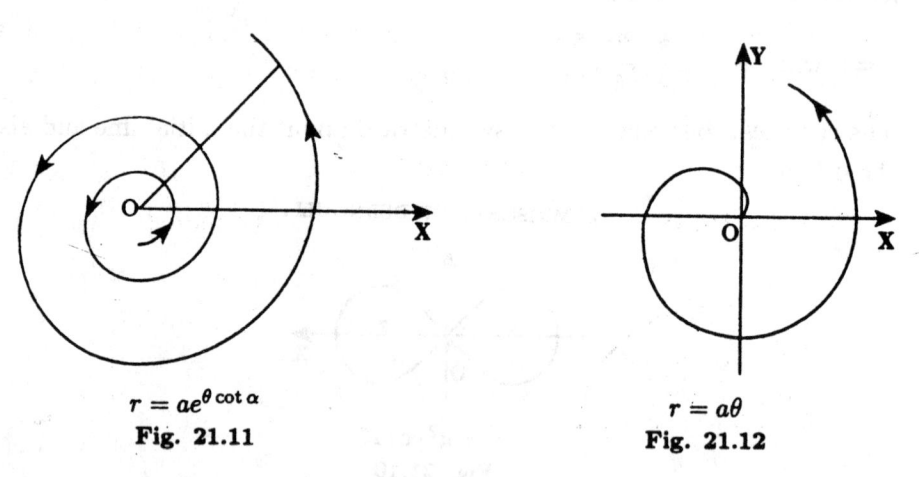

This is a curve described by a point which travels along a straight line with constant velocity while the line rotates with constant angular velocity about a fixed point in it. Thus $r = ut$, $\theta = nt$ whence $r = a\theta$ where $a = u/n$.

c) *Reciprocal spiral*: $r = a/\theta$. (Fig. 21.13).

If y be the ordinate drawn on the initial line we have

$$y = r \sin \theta = r \frac{\sin \theta}{\theta} \cdot \theta = a \cdot \frac{\sin \theta}{\theta}$$

when $\theta \to 0$, $r \to \infty$ but y approaches the finite value a. Hence $y = a$ is an asymptote.

The dotted part of Fig. 21.13 corresponds to the negative values of θ.

CHAPTER 21: CURVE TRACING (SYSTEMATIC METHOD)

RECIPROCAL SPIRAL

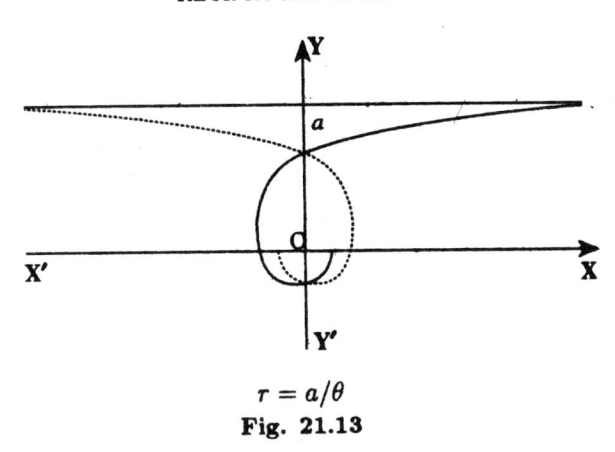

$r = a/\theta$
Fig. 21.13

(D) Limacon and Cardiode

If a point O on the circumference of a fixed circle of radius $\tfrac{1}{2}a$ be taken as pole and the diameter through O as the initial line, the radius vector of any point Q on the circumference is given by

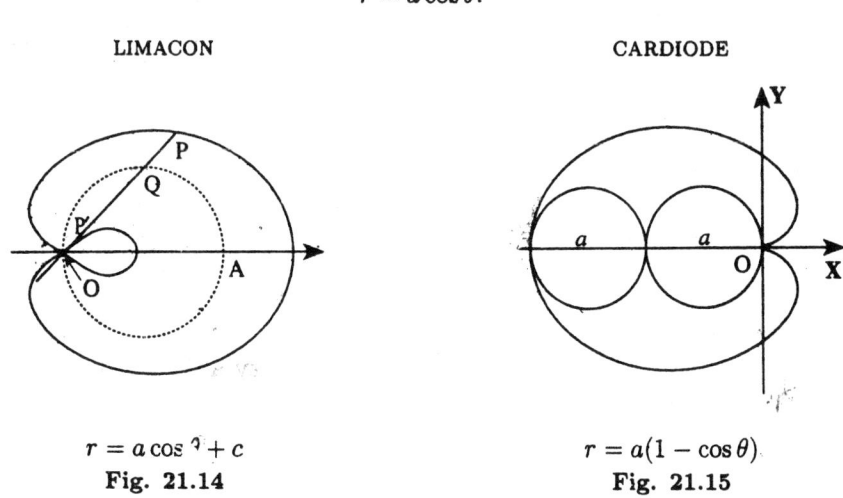

$$r = a\cos\theta.$$

LIMACON

$r = a\cos\theta + c$
Fig. 21.14

CARDIODE

$r = a(1 - \cos\theta)$
Fig. 21.15

If on this radius vector we take two points P, P' at equal constant distance c from Q, the locus of these points is called a *limacon* (Fig. 21.14), its equation being $r = a\cos\theta + c$.

In case $c = \pm a$, the locus is called a *Cardiode* (heart-shaped); its equation is $r = a(1 + \cos\theta)$ or $r = a(1 - \cos\theta)$. Fig. 21.15 gives a cardiode of the latter type. The cardiode of the former type has a shape inverse to it.

21.7 A Few more Special Curves

I. Folium of Descartes: $x^3 + y^3 = 3axy$.

Verify that the curve is symmetrical about the line $y = x$ and meets it at $(3a/2, 3a/2)$; it passes through the origin where $x = 0$ and $y = 0$ are the tangents so that the origin is a node; $x + y + a = 0$ is its only asymptote. It is not easy to solve the equation for x or y. But on transforming to polar co-ordinates, we get

$$r = \frac{3a \sin\theta \cos\theta}{\cos^3\theta + \sin^3\theta}.$$

Observe that $r = 0$ for $\theta = 0$ and $\theta = \pi/2$.

Here $\dfrac{dr}{d\theta} = 0$ when $\cos\theta - \sin\theta = 0$, i.e., $\tan\theta = 1$, i.e., $\theta = \pi/4$ or $5\pi/4$.

When $\theta = \pi/4, r = \dfrac{3a}{2}\sqrt{2}$. Thus r increases from 0 to $\dfrac{3}{2}a\sqrt{2}$ as θ increases from 0 to $\pi/4$ and then decreases from $\dfrac{3}{2}a\sqrt{2}$ to 0, as θ increases from $\pi/4$ to $\pi/2$. As θ further increases from $\dfrac{\pi}{2}$ to $\dfrac{3\pi}{4}$, r becomes negative and numerically increases from 0 to ∞, i.e., the curve lies in the fourth quadrant. Similarly as θ increases from $\dfrac{3\pi}{4}$ to π, we get the part of the curve in the second quadrant (see Fig. 21.16).

FOLIUM OF DESCARTES

$x^3 + y^3 = 3axy$
Fig. 21.16

CYCLOID

$x = a(\theta - \sin\theta); y = a(1 - \cos\theta)$
Fig. 21.17(a)

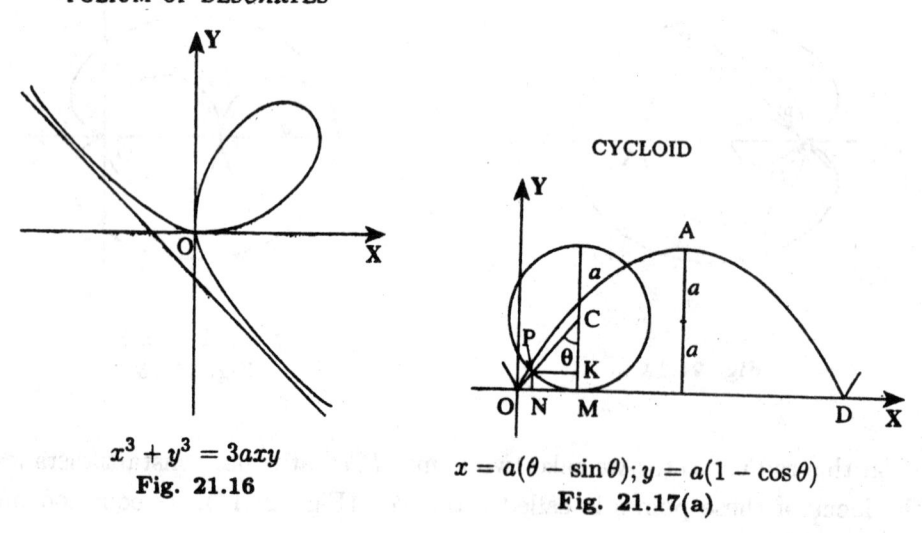

II. Parametric equations.

a) Cycloid:

$$\left. \begin{array}{l} x = a(\theta - \sin\theta) \\ y = a(1 - \cos\theta) \end{array} \right\}$$

CHAPTER 21: CURVE TRACING (SYSTEMATIC METHOD)

The *cycloid* [Fig. 21.17(a)] is a curve traced out by a point P on the circumference of a circle which rolls without sliding along a fixed straight line. Take the fixed line as the axis of X and let O be the initial position of the moving point P on the rolling circle whose centre is C and radius is a. When the circle has rolled on to the position shown in the figure, the point P moves from O to P such that $OM = $ arc PM. Let $\angle PCM = \theta$ and the co-ordinates of P be (x, y), then

$$x = ON = OM - MN = OM - PK = a\theta - a\sin\theta \quad (\because OM = \widehat{PM} = a\theta).$$
$$y = PN = CM - CK = a - a\cos\theta = a(1 - \cos\theta).$$

At O and $D, y = 0$, i.e., $\cos\theta = 1$, i.e., $\theta = 0, 2\pi$. Thus in one complete revolution of the circle the point P describes the curve OAD and if the motion is continued, we get an infinite number of arches congruent to it. The fixed line on which the circle rolls is called the *base* and the furthest point (A) from it is the *vertex*.

The equations
$$\left. \begin{array}{l} x = a(\theta + \sin\theta) \\ y = a(1 - \cos\theta) \end{array} \right\} \quad (-\pi < \theta \leq \pi)$$

will give the **inverted cycloid**. [Fig. 21.17(b)].

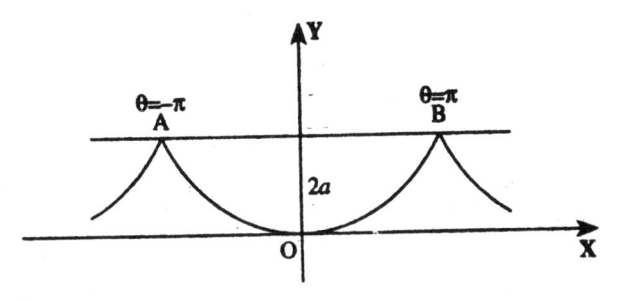

Fig. 21.17(b)

Here the vertex is the origin and the tangent at the vertex is the X-axis. Note that $\theta = 0$ for the vertex.

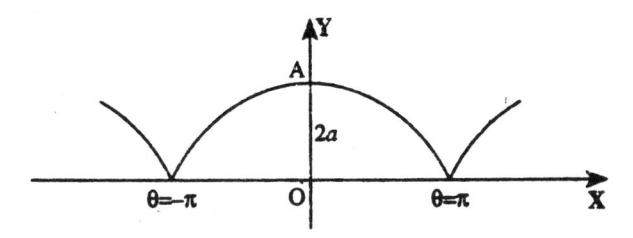

Fig. 21.17(c)

Another form of the equation of the cycloid is

$$\left.\begin{array}{l}x = a(\theta + \sin\theta)\\ y = a(1 + \cos\theta)\end{array}\right\} \quad -\pi < \theta \leq \pi.$$

Here y increases as θ increases from $-\pi$ to 0 and y decreases as θ increases from 0 to θ. Now trace the curve.

Still a fourth form is

$$\left.\begin{array}{l}x = a(\theta - \sin\theta)\\ y = a(1 + \cos\theta)\end{array}\right\} \quad (0 \leq \theta \leq 2\pi).$$

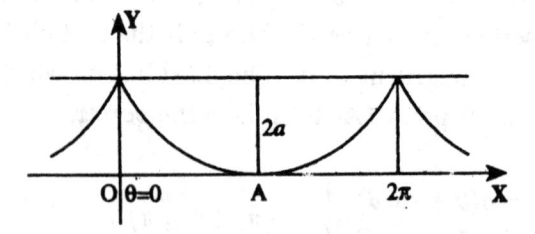

Fig. 21.17(d)

b) Astroid:

$x = a\cos^3\theta, y = a\sin^3\theta$ will give the curve, called the *four-cusped hypocycloid* or *astroid* (Fig. 21.18). Eliminating θ we get the cartesian equation

$$x^{2/3} + y^{2/3} = a^{2/3}.$$

It possesses the property that the length of the tangent intercepted between the co-ordinate axes is constant.

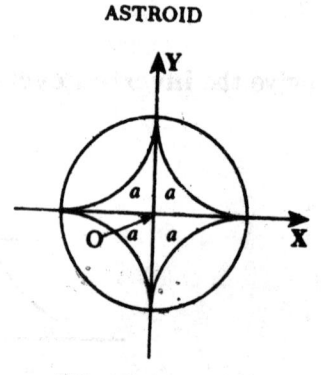

ASTROID

$x^{2/3} + y^{2/3} = a^{2/3}$

Fig. 21.18

c) Catenary:

$$y = a\cosh\frac{x}{a} \quad \text{(Fig. 21.19)}$$

The catenary is the curve in which a uniform chain hangs freely under gravity. It appears from statical principles that if s be the arc of the curve measured from the lowest point A up to any point P, and ψ the inclination of the tangent at P, then

$$s = a\tan\psi, \text{ where } a \text{ is a constant } (\textit{Intrinsic equation}).$$

If (x, y) be the co-ordinates of P, then

$$\frac{dx}{d\psi} = \frac{dx}{ds} \cdot \frac{ds}{d\psi} = a\sec\psi$$

$$\frac{dy}{d\psi} = \frac{dy}{ds} \cdot \frac{ds}{d\psi} = a\tan\psi\sec\psi.$$

CHAPTER 21: CURVE TRACING (SYSTEMATIC METHOD)

CATENARY

$$y = a \cosh \frac{x}{a}$$

Fig. 21.19

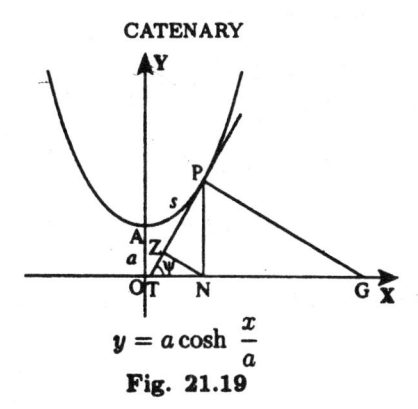

Integrating, we find the equation of catenary in *parametric form*:

$$\left. \begin{array}{l} x = a \log \tan \left(\dfrac{\pi}{4} + \dfrac{\psi}{2} \right) \\ y = a \sec \psi \end{array} \right\} . \qquad (21.7.1)$$

Choosing the origin suitably the additive constant may be omitted, since $x = 0, y = a$ for $\psi = 0$, it appears that the origin is at a distance a vertically below A.

From (21.7.1) the cartesian equation can be obtained in the form

$$y = a \cosh \frac{x}{a}.$$

Further if PN be the ordinate, PT the tangent, PG the normal, NZ the perpendicular from the foot of the ordinate on the tangent we have

$$NZ = y \cos \psi = a; \quad PZ = a \tan \psi = s.$$

d) Further Properties of Cycloid:

The **cycloid** is the curve traced out by a point on the circumference of a circle which rolls in contact with a fixed straight line.

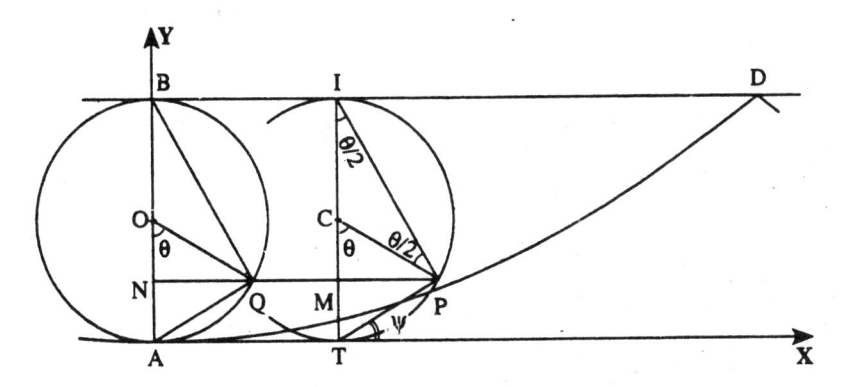

Fig. 21.20. Cycloid

It consists of an endless succession of exactly congruent portions each of which represents a complete revolution of the circle. The points (such as A in the Fig. 21.20) where the curve is furthest from the fixed straight line or *Base* (BD) are the *vertices*. The points (like D) half-way between the successive vertices where the curve meets the base are the *Cusps*.

A line (like AB) through a vertex and perpendicular to the base is called an **Axis** of the curve (a line of symmetry).

We take the circle described on an axis AB as diameter as a circle of reference. Let IPT be any other position of the rolling circle, I the point of contact with the base, C the centre now, T the opposite extremity of the diameter through I and let P be the position of the tracing point.

Draw PMN parallel to the base, meeting TI at M and AB at N and the circle of reference in Q. Take AT as X-axis and AB as Y-axis, the co-ordinates of $P(x,y)$ will then be

$$x = NP = BI + MP; \quad y = AN = CT - CM.$$

Let a be the radius of the rolling circle and let θ be the angle PCT through which the circle turns as the tracing point travels from A to P.

We have, then $BI = a\theta$, $PM = a\sin\theta$ and therefore

$$x = a(\theta + \sin\theta), \ y = a(1 - \cos\theta). \ \textbf{Parametric Equation of the cycloid.}$$

We may now deduce:

Let ψ be the inclination of the tangent at P to AT or of the normal at P to BA. We have,

$$\tan\psi = \frac{dy}{dx} = \frac{dy/d\theta}{dx/d\theta} = \frac{\sin\theta}{1+\cos\theta} = \tan\theta/2$$

i.e., $\psi = \theta/2$.

Since the angle $TIP = \frac{1}{2}\angle TCP = \frac{1}{2}\theta = \psi$, it follows that IP is the normal and PT the tangent to the curve at P.

Again to find the arc (s) of the curve, we have

$$\left(\frac{dx}{d\theta}\right)^2 + \left(\frac{dy}{d\theta}\right)^2 = a^2\{(1+\cos\theta)^2 + \sin^2\theta\} = 4a^2\cos^2\theta/2.$$

$$\therefore \frac{ds}{d\theta} = \sqrt{\left(\frac{dx}{d\theta}\right)^2 + \left(\frac{dy}{d\theta}\right)^2} = 2a\cos\theta/2$$

or on integration $s = 4a\sin\theta/2$ ($s = 0$, when $\theta = 0$ at A).

Thus, the intrinsic equation of the cycloid is $\mathbf{s = 4a\sin\psi}$.

Since $TP = TI \sin \psi$, we have

$$\text{arc AP} = 2 \text{ TP} = 2 \text{ chord AQ}.$$

In particular, the length of the arc from one cusp to the next is $8a$.

If we put $y' = IM = a(1 + \cos \theta)$, the area included between the curve and the base is given by

$$\int y' dx = a^2 \int (1 + \cos \theta)^2 d\theta = 4a^2 \int \cos^4(\theta/2) d\theta = 8a^2 \int \cos^4 \psi d\psi.$$

Taking this between $\mp \pi/2$ we find that the area included between the base and one arch of the curve is three times the area of the generating circle.

Main Points : A Recapitulation

1. Systematic Procedure for tracing Cartesian Equations $f(x, y) = 0$.

I. Find out if the curve is symmetrical

(i) about x-axis (powers of y all even);

(ii) about y-axis (powers of x all even);

(iii) about $y = x$ (interchange x and y in $f(x,y) = 0$, the equation does not change).

II. Find out if the origin lies on the curve. If the origin lies on the curve, find the tangent or tangents there at.

In case origin is a multiple point, find out its nature.

III. Intercepts: Put $y = 0$ (x-intercept); put $x = 0$ (y-intercept).

IV. Find dy/dx and points where tangent is parallel to the co-ordinate axes. Obtain maximum/minimum points or points of inflexion, if any.

V. Find out the asymptotes and the points in which each asymptote meets the curve.

VI. Find out such points on the curve whose existence can be almost, by inspection, detected. Find out if there is any region such that no point of the curve can lie on it.

> VII. Solve the equation for one variable in terms of the other, say y in terms of x. Consider $x = 0$ and then see how y varies as x increases and tends to ∞. Special attention to these values of x for which $y = 0$ and $y \to \infty$. Also see how y varies as x decreases and ultimately tends to $-\infty$.
> Use the symmetric property.
> [See step by step method used in the Illustrative Examples of art. **21.4**]
>
> **2. Polar Curves : Procedure for tracing curves of the form** $f(r, \theta) = 0$.
>
> **I.** Solve $f(r, \theta) = 0$ for r and consider how r varies as θ increases from 0 to $+\infty$ and also as θ diminishes from 0 to $-\infty$. If possible, form a table of corresponding values of θ and r.
>
> **II.** In most cases the equations contain periodic functions ($\sin\theta$, $\cos\theta$ etc.). So the values of θ from 0 to 2π need only be considered. The remaining values of θ give no new branches of the curve.
>
> **III.** If changing the sign of θ does not change the value of r, the curve is symmetrical about the initial line.
>
> **IV.** If the curve possesses an infinite branch, find the asymptote.
>
> **3.** Important Cartesian and Polar curves are given in Fig. **21.1** to **21.7** and Fig. **21.8** to **21.18**. The Catenary and its properties require special attention (Fig. **21.19**).
>
> Properties of cycloid are given along with Fig. **21.20**.
>
> Other Important curves are given in Chapter **14**.

Exercises XXI

1. Trace the curves:

    (a) $y^2(a + x) = x^2(3a - x)$.

    (b) $y^2(x^2 + y^2) - 4x(x^2 + 2y^2) + 16x^2 = 0$.

    (c) $x = (y - 1)(y - 2)(y - 3)$.

    (d) $y^2(x + 3a) = x(x - a)(x - 2a)$.

2. Trace the following curves: (*Rational Functions*)

    (a) $y = \dfrac{2x - 3}{x^2 - 3x + 2}$ ($x = 1$, $x = 2$ asymptotes).

    (b) $y = \dfrac{x^2 - 5x + 10}{x - 3}$.

(c) $y = \dfrac{x^2 - 4x + 3}{x^2 + 4x + 3}$.

3. By transforming into polar co-ordinates trace the curves:

   (a) $x^5 + y^5 = 5a^2 x^2 y$.
   (b) $xy^2 + x^2 y = a^3$.

4. Trace the following polar curves:

   (a) $r = a(\sec\theta + \cos\theta)$.
   (b) $r = a\cos 3\theta$.
   (c) $r = a\theta/(1+\theta)$.

5. (a) Prove that the curve
$$ay^2 = (x-a)^2(x-b)$$
has at $x = a$, an isolated point if $a < b$, a node if $a > b$ and a cusp if $a = b$. Trace the curve if $a = b$.

   (b) Trace the curve:
$$y^2 = (x-2)^2(x-5)$$
and show that the line joining the points of inflexion subtends a right angle at the double point.

Answers

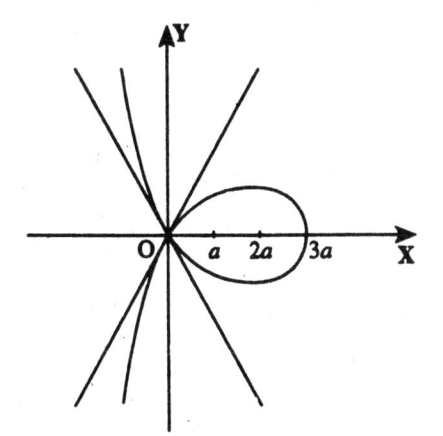

Ex. 1(a).

Ex. 1(b).

$y^2(a+x) = x^2(3a-x)$
Fig. 21.21

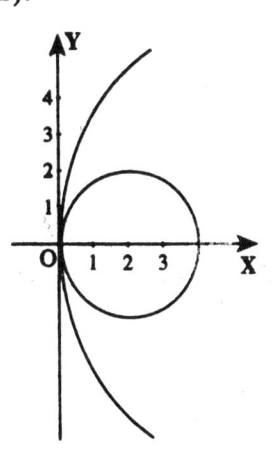

$y^2(x^2+y^2) - 4x(x^2+2y^2) + 16x^2 = 0$.
Fig. 21.22

Ex. 1(c).

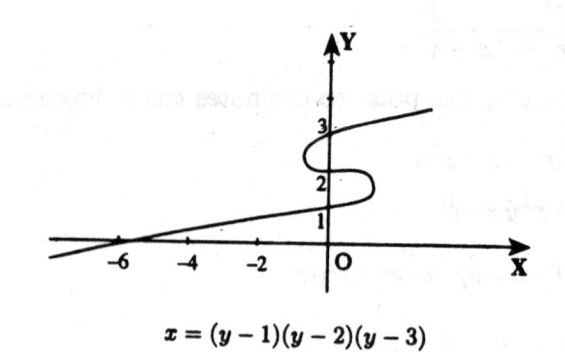

$$x = (y-1)(y-2)(y-3)$$
Fig. 21.23

Ex. 2(a).

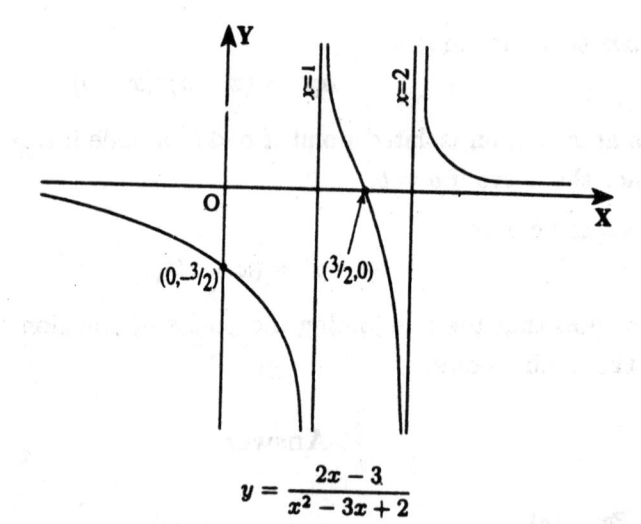

$$y = \frac{2x-3}{x^2-3x+2}$$
Fig. 21.24

Ex. 2(b).

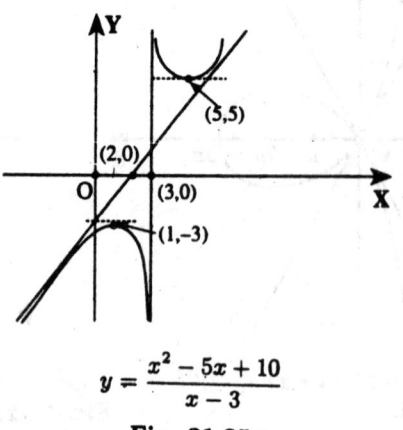

$$y = \frac{x^2-5x+10}{x-3}$$
Fig. 21.25

CHAPTER 21: CURVE TRACING (SYSTEMATIC METHOD)

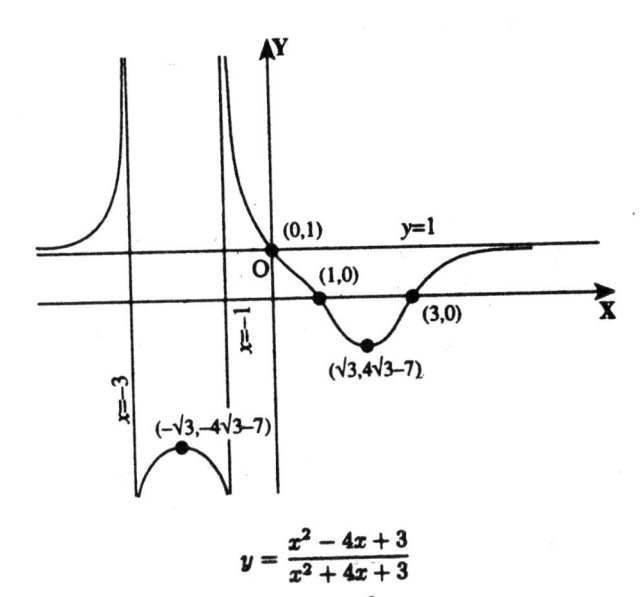

$$y = \frac{x^2 - 4x + 3}{x^2 + 4x + 3}$$

Fig. 21.26

AN INTRODUCTION TO
ANALYSIS

(Differential Calculus—Part I)

Section II: Chapters 14–21

Applications of Derivatives

Recapitulation (For Self-Practice)

Applications of Derivatives [Recapitulation: For Self-Practice]

1. Find the range of range of values of x for which the curve
$$y = x^4 - 6x^3 + 12x^2 + 5x + 7$$
is concave upwards or downwards.

2. Show that for the cycloid
$$x = a(\theta - \sin\theta)$$
$$y = a(1 - \cos\theta)$$
the radius of curvature at any point is twice the length of the portion of the normal intercepted between the curve and the axis of x.

3. Find the range of values of x for which the curve
$$y = x^4 - 16x^3 + 42x^2 + 12x + 1$$
is concave or convex with respect to x-axis identity the point of inflexion, if any.

4. Show that the curve $2y^3 = x^2y + x^3$ has a cusp of the first species at the origin and $x + y = 2$ is an asymptote of it which cuts the curve at $(1, 1)$.

5. Define the pedal of a curve with respect to a fixed point on the plane of the curve. Find the Cartesian equation of the pedal of the curve $\sqrt{\frac{x}{a}} + \sqrt{\frac{y}{b}} = 1$ with respect to the origin.

6. Prove that the centre of curvature of a point P of a curve is the limiting position of the point of intersection of the normal to the curve at P with the normal to the curve at a neighbouring point Q on the curve as Q tends to P along the curve.

7. Show that the pedal equation of a circle with respect to a point on its circumference is $pd = r^2$, where d is the diameter of the circle.

8. Show that the envelope of the circles

$$x^2 + y^2 - 2\alpha x - 2\beta y + \beta^2 = 0$$

where α, β are the parameters and where centres lie on the parabola $y^2 = 4ax$ is $x(x^2 + y^2 - 2ax) = 0$.

9. Find the evolute of the curve $x^{2/3} + y^{2/3} = a^{2/3}$.

10. Show that the points of inflexion on the curve

$$y^2 = (x-a)^2(x-b)$$

lie on the line $3x + a = 4b$.

11. Find the radius of curvature at the origin to the curve

$$y^2 - 3xy - 4x^2 + 5x^3 + x^4y - y^5 = 0.$$

12. Find the envelope

$$(a_1 t^2 + 2a_2 t + a_2)x + (b_1 t^2 + 2h_2 t + h_3)y + (c_1 t^2 + 2c_2 t + c_3) = 0,$$

when t is the variable parameter.

13. If $x^{2/3} + y^{2/3} = c^{2/3}$ is the envelope of the family of the curve $\frac{x^2}{a^2} + \frac{y^2}{b^2} = 1$, prove that the parameters a and b are connected by $a + b = c$.

14. If ρ_1 and ρ_2 be the radii of curvature at each of the two conjugate diameters of the ellipse $\frac{x^2}{a^2} + \frac{y^2}{b^2} = 1$, prove that

$$\left(\rho_1^{2/3} + \rho_2^{2/3}\right)(ab)^{2/3} = a^2 + b^2.$$

15. Find the equation of the curve which has the same asymptotes as those of the curve

$$x^3 - 6x^2 y + 11xy^2 - 6y^3 + x + y + 1 = 0$$

and which touches the straight lines $y = x$ at the origin and passes through the point $(1, -1)$.

APPLICATIONS OF DERIVATIVES

16. Find the double points of the curve

$$(2y + x + 1)^2 = 4(1 - x)^5$$

Hence state with reason, whether the double point is a node or a cusp or an isolated point.

17. Find the Asymptotes, if any, of any one of the following curves

(a) $x = \frac{t^2+1}{t^2-1}$ and $y = \frac{t^2}{t^2-1}$. [CH 2007]

(b) $r = a \tan \theta$.

18. (a) Show that the curve $y = e^x$ is concave upwards everywhere.

(b) Find the points of inflexion, if any, of the curve $y = (\log_e x)^3, (x > 0)$.

19. If ρ_1 and ρ_2 be the radii of curvature at the extremities of any chord of the Cardiode $r = a(1 + \cos \theta)$ which passes through the pole, then prove that

$$\rho_1^2 + \rho_2^2 = \frac{16a^2}{9}.$$

20. Show that the asymptotes of the curve

$$(x^2 - 4y^2)(x^2 - 9y^2) + 5x^2y - 5xy^2 - 30y^3 + xy + 7y^2 - 1 = 0$$

cut the curve in eight points which lie on a circle $x^2 + y^2 = 15$.

21. Show that the pedal equation of the parabola $y^2 = 4a(x + a)$ is $p^2 = ar$.

22. Find the length of perimeter of the Cardiode $r = a(1 - \cos \theta)$ and show that the upper half of the curve is bisected at $\theta = \frac{2}{3}\pi$.

23. Find the pedal equation of $c^2(x^2 + y^2) = x^2y^2$ with respect to the origin.[Ans. $\frac{1}{p^2} + \frac{3}{r^2} = \frac{1}{c^2}$]

24. Prove that the evaluate of the ellipse $\frac{x^2}{a^2} + \frac{y^2}{b^2} = 1$ is the envelope of the curve $a^2x^2 \sec^4 \alpha + b^2y^2 \csc^4 \alpha = (a^2 - b^2)$ for all values of α.

25. Find the moment of inertia of a sphere about one of its diameters.

26. State Pappus theorem on the volume of a solid of revolution and use it to find the volume of the solid generated by revolving the ellipse $x = a \cos \theta, y = b \sin \theta$ about the line $x = 2a$.

27. If any of the asymptotes of the curve
$$ax^2 + 2hxy + by^2 + 2gx + 2fy + c = 0 \quad (h^2 > ab)$$
passes through the origin, then prove that $af^2 + bg^2 = 2fgh$.

28. Find the area between the Witch of Agnesi
$$xy^2 = 4a^2(2a - x)$$
and its asymptote.

29. Show that the radius of curvature of the envelope of the lines
$$t_\alpha : x \cos \alpha + y \sin \alpha = f(\alpha) \quad (\alpha = \text{parameter})$$
(where f is a three differentiable function with $f(\alpha) + f'''(\alpha) = 0$ for all α) is $f(\alpha) + f''(\alpha)$ (denoting differentiation w.r.t. α)

30. Show that the pedal equation of the ellipse $\frac{x^2}{a^2} + \frac{y^2}{b^2} = 1$
    (a) with respect to a focus as pole is $\frac{b^2}{p^2} = \frac{2a}{r} - 1$ but (b) with respect to the centre as pole is $\frac{a^2 b^2}{p^2} = a^2 + b^2 - r^2$.

31. Find the asymptotes, if any, of the following curves:
    (a) $x^3 - 2x^2 y + xy^2 - 3x^2 + 3y^2 + 3x + 5y + 1 = 0$
    (b) $x = \frac{1}{t^4 - 1}, y = \frac{t^3}{t^4 - 1}$. [C.H. 2008]

32. Find the envelope of the family of curves
$$\left(\frac{a^2}{x}\right) \cos \theta - \left(\frac{b^2}{y}\right) \sin \theta = \frac{c^2}{a},$$
for different values of θ.

33. Determine the equation of the envelope of the family of ellipses $\frac{x^2}{a^2} + \frac{y^2}{b^2} = 1$ where the parameters a and b are connected by the relation $\frac{a^2}{l^2} + \frac{b^2}{m^2} = 1$, l and m are non-zero constants. [CH 2007]

34. If (α, β) is the centre of curvature of the parabola $\sqrt{x} + \sqrt{y} = \sqrt{a}$ $(a > 0)$ at any point (x, y) show that $\alpha + \beta = 3(x + y)$.

35. Find the equation of the pedal of the curve
$$x = 2a \cos \theta - a \cos 2\theta$$
$$y = 2a \sin \theta - a \sin 2\theta$$

APPLICATIONS OF DERIVATIVES

36. Show that the moment of inertia of a truncated cone about its axis is $\frac{3M(a^5-b^5)}{10(a^3-b^3)}$, when a and b are the radii of the two ends and M is the mass of the truncated cone.

37. Find the C.G. of the parabola $\sqrt{\frac{x}{a}} + \sqrt{\frac{y}{b}} = 1$ between the curve and the axes.

38. Show that the surface area of a solid generated by revolving a cardiode, represented by the parameteric equations

$$x = a(2\cos t - \cos 2t)$$
$$y = a(2\sin t - \sin 2t)$$

about the x-axis is $\frac{128}{5}\pi a^2$.

39. Show that the origin is the only double point of the curve $y^2 = ax^2 + x^3$ and that the double point is a node if $a > 0$, an isolated point if $a < 0$, a single cusp of first species if $a = 0$.

40. Find the equation of the cubic which has the same asymptotes as that of the curve $x^3 - bx^2 + 11xy^2 - 6y^3 + x + y + 1 = 0$ and which touches the y-axis at the origin and passes through the point $(3, 2)$.

AN INTRODUCTION TO ANALYSIS

(Differential Calculus–Part I)

APPENDIX

Chapterwise Important Problems

(with Hints/Solutions)

APPENDIX

CHAPTERWISE IMPORTANT PROBLEMS

(*with Hints/Solutions*)

Sets in $\mathbb{R}$

1. If $f : \mathbb{R} \to \mathbb{R}$ and $g : \mathbb{R} \to \mathbb{R}$ are both continuous on $\mathbb{R}$ and the set

$$S = \{x \in \mathbb{R} : f(x) > g(x)\}$$

is a proper subset of $\mathbb{R}$ then prove that S is an open set in $\mathbb{R}$.

Proof. Let us consider the function $h : S \to \mathbb{R}$ defined by

$$h(x) = f(x) - g(x), \quad x \in S$$

Then

$$h(x) > 0 \ \forall \ x \in S$$

Functions f and g are both continuous on $\mathbb{R}$ and $S \subset \mathbb{R}$.

$\therefore$ They are continuous on S i.e. h is continuous on S.

Hence h is continuous at any arbitrary point c of S.

So by the nbd-property of continuous function $\exists \ N_\delta(c)$ such that $h(x)$ maintain the same sign as that of $h(c)$ whenever $x \in N_\delta(c)$.

$\therefore h(x) > 0$ when $x \in N_\delta(c)$.

$\therefore N_\delta(c) \subset S \Rightarrow c$ in an interior point of S.

As c is arbitrary.

$\therefore$ Every point of S is interior.

Hence S is an open set in $\mathbb{R}$.

2. Examine whether the following sets

(a) $S = \{x \in \mathbb{R} : 2x^2 - 5x + 2 < 0\}$
(b) $S = \{x \in \mathbb{R} : \sin x \cos x \neq 0\}$

are open or not in $\mathbb{R}$.

Solution. (a) $2x^2 - 5x + 2 < 0 \Rightarrow (2x - 1)(x - 2) < 0$

$$\Rightarrow \tfrac{1}{2} < x < 2$$

$\therefore S = \left(\tfrac{1}{2}, 2\right)$ which is an open interval and hence an open set.

(b) $\sin x \cos x \neq 0 \Rightarrow \sin 2x \neq 0 \Rightarrow 2x \neq n\pi, n \in \mathbb{Z}$ i.e. $x \neq \frac{n\pi}{2}$.

$\therefore S = \cdots \cup \left(-\frac{3\pi}{2}, -\pi\right) \cup \left(-\pi, -\frac{\pi}{2}\right) \cup \left(-\frac{\pi}{2}, 0\right) \cup \left(0, \frac{\pi}{2}\right) \cup \left(\frac{\pi}{2}, \pi\right) \cup \left(\pi, \frac{3\pi}{2}\right) \cup \cdots$

S is the union of an arbitrary collection of open intervals and an open interval is an open set in $\mathbb{R}$.

So S is the union of an arbitrary collection of open sets in $\mathbb{R}$ and hence S is open in $\mathbb{R}$.

3. Prove that complement of a closed set is open and the complement of an open set is closed in $\mathbb{R}$.

Proof. Let S be a closed set in $\mathbb{R}$ and S^c be its complement in $\mathbb{R}$.

Let x be an arbitrary element of S^c.

$\therefore x \notin S'$ (the derived set of S) as S is closed.

So $\exists$ a nbd $N(x)$ of x such that $N(x) \subset S^c$.

$\therefore x$ is an interior point of S^c.

Therefore every point of S^c is an interior point (as x is arb).

$\therefore S^c$ is open in $\mathbb{R}$ $[S^c = \mathbb{R} - S]$.

Now, let S be an open set in $\mathbb{R}$.

We are to prove that its complement S^c in $\mathbb{R}$ i.e. $S^c = \mathbb{R} - S$ is closed.

Let x be an arbitrary element in S.

As S is open

$\therefore \exists$ a nbd $N(x)$ of x such that $N(x) \subset S$ i.e. x in an interior point.

$\therefore x$ cannot be a limit point of S^c.

So, no point of S is a limit point S^c.

$\therefore (S^c)' \subseteq S^c$.

$\therefore S^c$ is closed in $\mathbb{R}$.

4. (a) Give two examples of set which are both closed an open in $\mathbb{R}$.

   (b) Give examples of sets in $\mathbb{R}$ which are neither closed nor open in $\mathbb{R}$.

Solution. (a) The sets ϕ and $\mathbb{R}$ are both closed an open in $\mathbb{R}$.

(b) Every half open set like $(1, 2] \cup [3, 5)$ etc. are neither closed or open in $\mathbb{R}$.

Union of two sets among which one is closed and the other open like $[1, 2] \cup (3, 5)$ are neither closed nor open. [$\mathbb{Q}$ and $\mathbb{R} - \mathbb{Q}$ are neither closed nor open].

5. Give a list of possible types of subsets in $\mathbb{R}$.

   Solution. ϕ (improper), $\mathbb{R}$ (trivial), set of distinct points (finite or infinite) like
   $$\left\{\frac{1}{2}, \frac{1}{3}, \frac{1}{4}\right\}, \quad \left\{\frac{1}{9}, \frac{1}{9^2}, \frac{1}{9^3}, \ldots\right\},$$
   an open interval like $(2,5)$, a closed interval like $[2,5]$, a half-open/half-closed interval like $(2,5]$ and union of two or more subsets stated above are possible types of subsets in $\mathbb{R}$.

6. If x_0 be a limit point of $A_1 \cup A_2$ then show that x_0 is a limit point of at least one of them.

   Solution. x_0 is a limit point of $A_1 \cup A_2 \Leftrightarrow$ every nbd $N(x_0)$ of x_0 contains infinitely many elements of
   $$A_1 \cup A_2 \quad \text{i.e.} \quad N(x_0) \cap (A_1 \cup A_2)$$
   $$\text{i.e.} \quad \{N(x_0) \cap A_1\} \cup \{N(x_0) \cap A_2\}$$
   is infinite.
   So
   $$N(x_0) \cap A_1 \quad \text{or} \quad N(x_0) \cap A_2$$
   or both of them are infinite sets.
   $\therefore x_0$ is a limit point of at least one of A_1 and A_2.

7. If x_0 be a limit point of
   $$A_1 \cup A_2 \cup \cdots \cup A_n$$
   then prove that x_0 is a limit point of at least one of $A_1, A_2, \ldots, A_n$.
   [*Hint:* $N(x_0) \cap (A_1 \cup A_2 \cup \cdots \cup A_n)$
   $$= \{N(x_0) \cap A_1\} \cup \{N(x_0) \cap A_2\} \cup \cdots \cup \{N(x_0) \cap A_n\}$$
   is infinite. So at least one is infinite.]

   Note. In precise notation
   $$N(x_0) \cap \left(\bigcup_{i=1}^{n} A_i\right) = \bigcup_{i=1}^{n} \{N(x_0) \cap A_i\}$$

8. If x_0 be a limit point of $A_1 \cap A_2$ then prove that x_0 is a limit point of both of them.

Proof. x_0 is a limit point of $A_1 \cap A_2 \Leftrightarrow$ every nbd $N(x_0)$ of x_0 contains infinitely many elements of

$$A_1 \cup A_2 \quad \text{i.e.} \quad N(x_0) \cap (A_1 \cap A_2)$$
$$\text{i.e.} \quad (N(x_0) \cap A_1) \cap A_2$$

is infinite $\Rightarrow N(x_0) \cap A_1$ is infinite set.

∴ x_0 is a limit point of A_1.

Again

$$N(x_0) \cap (A_1 \cap A_2) = (N(x_0) \cap A_2) \cap A_1$$

[by commutative and associative law]

is infinite and so $N(x_0) \cap A_2$ is infinite $\Rightarrow x_0$ is a limit point of A_2.

∴ x_0 is a limit point of both A_1 and A_2.

Note. This can be extended for any finite number of sets.

9. Prove that a closed bounded set has a least element and a greatest element.

   Proof. Let $S \subset \mathbb{R}$ be a closed and bounded set. As S is bounded.

   ∴ $\sup S$ exists and $\inf S$ exists.

   Let $\sup S = M$ and $\inf S = m$.

   If $m, M \in S$, the proof is complete.

   We now claim that m and $M \in S$ whenever S is closed and bounded, if $m \notin S$ then $m \in S' \subset S$ a contradiction.

   Similarly if $M \notin S, M \in S' \subset S$ [as S is closed] a contradiction.

10. Prove that every open interval is a nbd of each of its points.

    Proof. Let (a, b) be an open interval. Let $x \in (a, b)$.

    ∴ $a < x < b$. Choose $c, d \in \mathbb{R}$ such that $a < c < x < d < b$.

    (It is always possible as in between two real numbers there is always a real number)

    ∴ $x \in (c, d) \subset (a, b)$. ∴ (a, b) is a nbd of each of its point x.

    [So we may conclude that an open interval is an open set]

11. Prove that every closed interval (in $\mathbb{R}$) is a closed set.

    Proof. Let $[a, b]$ be a closed interval.

    $$\therefore \mathbb{R} - [a, b] = (-\infty, a) \cup (b, \infty)$$

APPENDIX: CHAPTERWISE IMPORTANT PROBLEMS

Now $(-\infty, a)$ and (b, ∞) are both open.

∴ Their union is open.

So $R - [a, b]$ is open.

∴ $[R - [a, b]]^c = [a, b]$ is closed.

Alternative proof. $[a, b] = (a, b) \cup \{a, b\}$.

(a, b) being an open interval is an open set.

∴ Every point is its interior point and so is a limit point and $\in [a, b]$.

Again a, b are also limit points as

$$N_\delta(a) \cap [a, b]$$
$$N_\delta(b) \cap [a, b]$$

are both infinite sets for every $\delta > 0$. $a, b \in [a, b]$.

No other point is a limit point of $[a, b]$.

Let $x \notin [a, b]$.

∴ x is an exterior point.

So $\exists\, \delta' > 0$ such that

$$N_{\delta'}(x) \cap [a, b] = \phi$$

∴ x is not a limit point of $[a, b]$. So $[a, b]$ is closed.

Note. Actually the last few lines are enough for proof.

12. Show that each point of

$$S = \left\{ 1, \frac{1}{2}, \frac{1}{3}, \cdots \right\}$$

is an isolated point.

Solution. Let $\frac{1}{n} \in S$ be an arbitrary element.

Consider

$$\delta = \min\left\{ \left|\frac{1}{n-1} - \frac{1}{n}\right|, \left|\frac{1}{n} - \frac{1}{n+1}\right| \right\}$$

As $\frac{1}{n}$ is arbitrary and $N_\delta\left(\frac{1}{n}\right) \cap S = \phi$. ∴ $\frac{1}{n}$ is not a limit point and so isolated.

∴ Every point of S is an isolated point of S.

13. Let $S \subset \mathbb{R}$. A point $x \in \mathbb{R}$ is said to be a boundary point of S if every nbd. $N(x)$ of x contains a point of S and a point of $\mathbb{R} - S$.

If a boundary point of S is not a point of S prove that it is a limit point of S. Prove that a set $S \subset \mathbb{R}$ is closed iff S contains all its boundary points.

Proof. Let x be a boundary point of S, not belonging to S.
So by definition of boundary point
$$N_\delta(x) \cap S \neq \phi$$

As $x \notin S$

∴ There is at least one element of S (other than x) in
$$N_\delta(x) \ \forall \ \delta > 0$$

∴ x is a limit point of S.
As x is an arbitrary boundary point
∴ Every boundary point not belonging to S is a limit point of S.
2nd part. First consider that all the boundary points of S belong to S.
Let $y \in \mathbb{R} - S$.
∴ y is neither on S nor on its boundary.
∴ y is an exterior point of S.
∴ ∃ a nbd of y which contains no point of S.
∴ $y \notin S'$ (the derived set).
∴ S is closed.
Conversely let S be closed.
$$\therefore S' \subset S \tag{0.0.1}$$

$\Rightarrow$ all the boundary points lie on S, for if x is a boundary point of S and not belongs to S then $x \in S'$ which contradicts (1).

14. Prove that the complement of the closure of a set $S \subset \mathbb{R}$ is the interior of the complement of S i.e. $(\bar{S})^c = (S^c)^0$.

Proof. Let
$$\xi \in (\bar{S})^c \Rightarrow \xi \in (S \cup S')^c$$
$$\Rightarrow \xi \notin S \cup S' \Rightarrow \xi \notin S \text{ and } \xi \notin S'$$

∴ $\xi \in S^c$ and $\exists \delta > 0$ such that
$$N_\delta(\xi) \subset S^c$$

If not, then for each $\delta > 0$ the nbd
$$N_\delta(\xi) \not\subset S^c.$$

Then
$$\exists\, y \in N_\delta(\xi)$$
such that $y \in S$ for each $\delta > 0$.

$\therefore\ \xi$ is a limit point of S i.e. $\xi \in S'$, a contradiction.

$\therefore\ \xi$ is an interior point of S^c.

$\therefore\ \xi \in (S^c)^0$.
$$\therefore\ (\bar{S})^c \subseteq (S^c)^0 \qquad (0.0.1)$$

Let $\eta \in (S^c)^0$. $\therefore\ \eta$ is an interior point of S^c.

$\therefore\ \exists\, \delta' > 0$ such that $N_{\delta'}(\eta) \subset S^c$. $\therefore\ \eta \notin S$ and $\eta \notin S'$ i.e. $\eta \notin S \cup S'$ i.e. $\eta \notin \bar{S}$.
$$\therefore\ \eta \in (\bar{S})^c. \qquad (0.0.2)$$

Equations (1) and (2) $\Rightarrow (\bar{S})^c = (S^c)^0$

15. Give illustrations of two non-empty open sets G_1, G_2 and two non-empty closed sets H_1, H_2 in $\mathbb{R}$ where $G_i \cap H_i = \phi$ for $i = 1, 2$ such that $G_1 \cup H_1$ is an open set and $G_2 \cup H_2$ is a closed set in $\mathbb{R}$.

Solution. Let $a, b, c, d \in \mathbb{R}$ such that
$$a < b < c < d$$

Let
$$G_1 = (a, b) \cup (c, d)$$
$$\text{and}\quad H_1 = [b, c]$$

$\therefore\ G_1 \neq \phi, H_1 \neq \phi$ and $G_1 \cap H_1 = \phi$ as the intervals are disjoint.

Now
$$\begin{aligned}G_1 \cup H_1 &= \{(a,b) \cup (c,d)\} \cup [b,c] \\ &= (a,b) \cup [b,c] \cup (c,d) \quad \text{[by associative property]} \\ &= (a,d) = \text{an open set,}\end{aligned}$$

as it is an open interval.

Again let $m, n, p, q \in \mathbb{R}$ such that $m < n < p < q$.

Let
$$G_2 = (n, p) \neq \phi \qquad \text{and open}$$
$$H_2 = [m, n] \cup [p, q] \neq \phi \quad \text{and closed.}$$

$$\therefore\ G_2 \cap H_2 = (n,p) \cap \{[m,n] \cup [p,q]\} = \phi$$

as the intervals are disjoint.

Now
$$G_2 \cup H_2 = (n,p) \cup \{[m,n] \cup [p,q]\}$$
$$= [m,n] \cup (n,p) \cup [p,q] \quad \text{[by associative property]}$$
$$= [m,q] = \text{a closed set}$$

as it is a closed interval.

16. Show that the set of all accumulation points of a set in $\mathbb{R}$ is a closed set.

    Solution. Let $S \subset \mathbb{R}$.

    Let the set of accumulation points of S be S' (the derived set of S).

    We are to prove that S' is closed i.e. $(S')' \subset S'$.

    Case I. If $(S')' = \phi$ then $(S')' \subset S'$.

    Case II. Let $(S')' \neq \phi$. Let $s \in (S')'$.

    $\therefore$ Every nbd $N(s)$ of s contains infinitely many elements of S'.

    Let $q(\neq s)$ be one such element.

    As $q \in S'$ and $q \in N(s)$.

    $\therefore$ q is an accumulation point of S. So $N(s)$ which is a nbd of q as well, contains infinitely many elements of S.

    $\therefore$ s is an limit point of S i.e. $s \in S'$.

    $\therefore$ $(S')' \subset S'$.

    $\therefore$ S' is closed.

17. If $f : \mathbb{R} \to \mathbb{R}$ is continuous on $\mathbb{R}$ and the set S, given by
    $$S = \{x \in \mathbb{R} : f(x) > 0\}$$
    be a proper subset of $\mathbb{R}$ then prove that S is an open set in $\mathbb{R}$.

    Proof. Let $a \in S$.

    $\therefore$ $f(a) > 0$ and f is continuous at a.

    $\therefore$ By the nbd property $\exists$ a nbd $N_\delta(a)$ of a such that
    $$f(x) > 0\ \forall\ x \in N_\delta(a)$$
    $$\text{i.e.}\quad N_\delta(a) \subset S$$

    $\therefore$ a is an interior point of S.

    As a is an arbitrary element of S

    $\therefore$ Every element of S is interior and so S is open.

18. If $f : \mathbb{R} \to \mathbb{R}$ is continuous on $\mathbb{R}$ and the set S given by
$$S = \{x \in \mathbb{R} : f(\lambda) < 0\}$$
be a proper subset of $\mathbb{R}$ then prove that S is an open set in $\mathbb{R}$.

Proof. Similar.

19. Let $f : \mathbb{R} \to \mathbb{R}$ and $g : \mathbb{R} \to \mathbb{R}$ be both continuous.

Show that $S = \{x \in \mathbb{R} : f(x) < g(x)\}$ is open in $\mathbb{R}$.

[*Hint:* Consider a function $h : \mathbb{R} \to \mathbb{R}$ s.t.
$$h(x) = g(x) - f(x), \quad x \in \mathbb{R}$$

Then $h(x)$ is continuous.

$\therefore\ S = \{x \in \mathbb{R} : h(x) > 0\} \cdot S$ is open (*see* Example 17)]

20. Let G be an open set of real numbers. Show that the intersection of any two open intervals in G is either an empty set or contains more than one point.

Solution. Let I_1, I_2 be two open intervals in G.

As they are open intervals, they are open sets.

Consider the set $I_1 \cap I_2$ which is an open set in G.

Two cases may arise: (i) $I_1 \cap I_2 = \phi$ or (ii) $I_1 \cap I_2 \neq \phi$.

In case (ii) let $x \in I_1 \cap I_2$.

$\therefore\ x \in I_1$ and $x \in I_2$.

As I_1 is open

$$\therefore\ \exists\ a\ \text{nbd}\ N_{\delta_1}(x)$$

such that
$$N_{\delta_1}(x) \subset I_1$$

Similarly $\exists$ a nbd $N_{\delta_2}(x)$ such that
$$N_{\delta_2}(x) \subset I_2 \quad (\text{as } x \text{ is interior})$$

Let
$$\delta = \min\{\delta_1, \delta_2\}$$

$\therefore\ N_\delta(x) \subset I_1$ and $N_\delta(x) \subset I_2$.

$\therefore\ N_\delta(x) \subset I_1 \cap I_2$.

$\therefore\ I_1 \cap I_2$ contains more than one point (in fact contains all points of $N_\delta(x)$ which is infinite).

$\therefore$ Either $I_1 \cap I_2$ is empty or contains more than one point.

21. Let $G \in \mathbb{R}$ be an open set and $F \subset \mathbb{R}$ be a closed set. Prove that $G - F$ is an open set and $F - G$ is a closed set.

    Proof. $G - F = G \cap F^c$.

    As F is closed

    $\therefore F^c$ is open.

    $\therefore G \cap F^c$ is an open set, as the intersection of two open sets is an open set.

    $\therefore G - F$ is an open set.

    Similarly $F - G = F \cap G^c$ is closed, as F and G^c are both closed.

22. Prove that the set of rational numbers $\mathbb{Q}$ is enumerable.

    Proof. Let $\mathbb{Q}^+$ = set of all positive rational numbers and $\mathbb{Q}^-$ = set of all negative rational numbers.
    $$\therefore \mathbb{Q} = \mathbb{Q}^+ \cup \mathbb{Q}^- \cup \{0\}$$

    Let
    $$A_r = \left\{\frac{1}{r}, \frac{2}{r}, \frac{3}{r}, \ldots\right\}, \quad r = 1, 2, 3, \ldots$$

    So
    $$\mathbb{Q}^+ = \bigcup_{r=1}^{\infty} A_r$$

    Now each A_r is countable, because the mapping $f : \mathbb{N} \to A_r$ defined by
    $$f(n) = \frac{n}{r}, \quad n \in \mathbb{N} \text{ is a bijection.}$$

    The set $\mathbb{Q}^+$ and $\mathbb{Q}^-$ are equipotent, since $g : \mathbb{Q}^+ \to \mathbb{Q}^-$ defined by
    $$g(x) = -x, x \in \mathbb{Q}^+ \text{ is a bijection.}$$

    $\therefore \mathbb{Q}^-$ is countable as it is equipotent to a countable set $\mathbb{Q}^+$.
    $$\therefore \mathbb{Q}^+ \cup \mathbb{Q}^- \text{ is countable}$$

    $\mathbb{Q}$ is the union of a countable set and the finite set $\{0\}$.

    Hence $\mathbb{Q}$ is countable.

23. Let every number x is a set A of real numbers be an isolated point of A. Show that A is denumerable.

    Solution. Let $x \in A$. So x is an isolated point (as every point of A is isolated).

∴ ∃ an open interval $I_1 = (a, b)$ containing the number x contains a finite number of elements of A.

Let us choose rational numbers r, s such that

$$a < r < x < s < b$$

Then the open interval

$$J_1 = (r, s)$$

contains x and have rational end points.

As I_1 contains a finite number of elements and $J_1 \subset I_1$.

∴ J_1 also contains a finite number of elements of A.

Now we know that the set of all open intervals having rational end points is enumerable.

We now consider similar sets J_2, J_3 etc. so that

$$A \subset \bigcup_{k=1}^{\infty} J_k$$

(This is always possible as A contains only isolated points.) Thus A is contained in countable union of enumerable sets J_k and hence A is enumerable.

24. Show that the interval $(0, 1)$ is not denumerable.

    Solution. Suppose $(0, 1)$ is denumerable.

    So the elements of the set can be arranged in a sequence $\{x_1, x_2, \ldots\}$.

    Each x_i has unique proper decimal representation.

    Let

    $$x_1 = 0 \cdot a_{11} \, a_{12} \, a_{13} \cdots$$
    $$x_2 = 0 \cdot a_{21} \, a_{22} \, a_{23} \cdots$$
    $$\ldots\ldots\ldots\ldots\ldots\ldots\ldots$$
    $$x_n = 0 \cdot a_{n1} \, a_{n2} \, a_{n3} \cdots$$
    $$\ldots\ldots\ldots\ldots\ldots\ldots\ldots$$

    where the digits a_{ij} are members of the set $\{0, 1, 2, \ldots, 9\}$.

    We now show that $\exists\, x \in (0, 1)$ but not identical with any x_i.

    Let

    $$x = 0 \cdot b_1 \, b_2 \, b_3 \cdots$$

where b_i's are taken as
$$b_i = \begin{cases} 0 & \text{if } a_{ii} \neq 0 \\ 1 & \text{if } a_{ii} = 0 \end{cases}$$

$\therefore\ x \neq x_i$ for any $i = 1, 2, 3, \ldots$

$\therefore\ (0, 1)$ is not denumerable.

25. Prove that every open interval (a, b) contains infinitely many rational numbers.

    Proof. (*Method of contradiction*).

    Let (a, b) contains a finite number of rational points
    $$x_1,\ x_2,\ \ldots,\ x_n \quad (\text{say})$$
    such that $a < x_1 < x_2 < \cdots < x_n < b$.

    Now $(x_1 + x_2)/2$ is a rational number and
    $$x_1 < \frac{x_1 + x_2}{2} < x_2$$
    $$\therefore\ a < \frac{x_1 + x_2}{2} < b$$
    which contradicts the assumed fact.

    $\therefore\ (a, b)$ contains infinitely may rational numbers.

Function, Limit, Continuity

1. Discuss the existence of
$$\lim_{x \to 0} \cos \frac{1}{x}$$

Solution. Natural domain of
$$\cos \frac{1}{x} \text{ is } \mathbb{R} - \{0\}.$$

0 is a limit point of the domain and hence $x \to 0$ is possible. We now use the sequential criterion for testing the existence of
$$\lim_{x \to 0} \cos \frac{1}{x}$$

Consider two sequences $\{x_n\}_n$ and $\{y_n\}_n$ in $\mathbb{R} - \{0\}$ defined by
$$x_n = \frac{1}{2n\pi} \quad \text{and} \quad y_n = \frac{1}{(2n+1)\pi}$$

both converging to 0 but the sequences
$$\left\{\cos \frac{1}{x_n}\right\} \quad \text{and} \quad \left\{\cos \frac{1}{y_n}\right\}$$

converges to two different limits viz 1 and -1 respectively. Hence
$$\lim_{x \to 0} \cos \frac{1}{x}$$

does not exist.
$$\left[\left\{\cos \frac{1}{x_n}\right\} = \{\cos 2n\pi\} = \{1, 1, \ldots\}\right.$$

a constant sequence converging to 1 and
$$\left\{\cos \frac{1}{y_n}\right\} = \{\cos(2n+1)\pi\} = \{-1, -1, \ldots\}$$

a constant sequence converging to -1.]

2. Discuss the existence of
$$\lim_{x \to 0} \sin \frac{1}{x}$$

Hint. Here consider the sequences $\{x_n\}_n$ and $\{y_n\}_n$ where
$$x_n = \frac{1}{2n\pi} \quad \text{and} \quad y_n = \frac{1}{2n\pi + \frac{\pi}{2}}$$

3. Using $\varepsilon - \delta$ definition, prove that
$$\lim_{x \to 0+} \frac{1}{1+e^{-1/x}} = 1$$

Proof.
$$\left|\frac{1}{1+e^{-1/x}} - 1\right| = \left|\frac{1-1-e^{-1/x}}{1+e^{-1/x}}\right| = \left|-\frac{1}{e^{1/x}+1}\right| = \frac{1}{e^{1/x}+1} < \varepsilon$$

whenever $e^{1/x} > \frac{1}{\varepsilon} - 1$

i.e. $\frac{1}{x} > \ln\left(\frac{1}{\varepsilon} - 1\right)$ i.e. $x < \frac{1}{\ln[(1/\varepsilon) - 1]} = \delta$ i.e. $\left|\frac{1}{1+e^{-1/x}} - 1\right| < \varepsilon$

whenever $0 < x < \delta$, where
$$\delta = \frac{1}{\ln[(1/\varepsilon) - 1]} \qquad \therefore \lim_{x \to 0+} \frac{1}{1+e^{-1/x}} = 1$$

4. Let a monotonic increasing function $f : (a, b) \to \mathbb{R}$ be bounded above $(a, b \in \mathbb{R})$. Show that
$$\lim_{x \to b-} f(x) \text{ exists.}$$

Solution. Consider the set
$$B = \{f(x) : x \in (a, b)\}$$

B is a subset of $\mathbb{R}$ bounded above and non-empty.

$\therefore$ By completeness property of $\mathbb{R}$, the subset B has a supremum M, say.

$\therefore$ For every $\varepsilon > 0$, $\exists\, x_1 \in (a, b)$ such that
$$f(x_1) > M - \varepsilon.$$

As $f(x)$ is monotonically increasing
$\therefore f(x) > M - \varepsilon \;\forall\, x$ in $x_1 < x < b$.
Taking $b - x_1 = \delta$ we have
$$|f(x) - M| < \varepsilon \;\forall\, x \in (b - \delta, b)$$

$\therefore \lim_{x \to b-} f(x)$ exists and is equal to M.

5. (a) Let $f(x)$ be a real-valued function defined on an interval I. When is f said to be uniformly continuous on I?

(b) If I is a closed and bounded interval and f is continuous on I, show that f is uniformly continuous on I.

(c) Show that a real valued function f continuous on an open interval (a, b) is uniformly continuous iff $\lim_{x \to a+} f(x)$ and $\lim_{x \to b-} f(x)$ both exist finitely.

Solution. (a) A function $f : I \to \mathbb{R}$ is said to be uniformly continuous on I if corresponding to each $\varepsilon > 0 \; \exists \; \delta > 0$ such that for any two points $x, y \in I$ satisfying $|x - y| < \delta, |f(x) - f(y)| < \varepsilon$. [$\delta$ depends on ε only]

(b) Follow the Theorem 8.5.8.

(c) Let f be continuous in (a, b) and

$$\lim_{x \to a+} f(x) \quad \text{and} \quad \lim_{x \to b-} f(x)$$

exist finitely.

We are to prove that f is uniformly continuous in (a, b).

Let us construct a function $g : [a, b] \to \mathbb{R}$ such that

$$\begin{aligned} g(x) &= f(x) & a < x < b \\ &= \lim_{x \to a+} f(x) & x = a \\ &= \lim_{x \to b-} f(x) & x = b \end{aligned}$$

$$\therefore \lim_{x \to a+} g(x) = \lim_{x \to a+} f(x) = g(a)$$
$$\text{and} \quad \lim_{x \to b-} g(x) = \lim_{x \to b-} f(x) = g(b)$$

$\therefore g$ is continuous on the closed bounded interval $[a, b]$.

So g is uniformly continuous on $[a, b]$.

As $(a, b) \subset [a, b]$.

$\therefore g$ is uniformly continuous in (a, b).

As $f(x) = g(x), x \in (a, b)$.

$\therefore f$ is uniformly continuous in (a, b).

Conversely let $f(x)$ is uniformly continuous in (a, b).

We propose to prove

$$\lim_{x \to a+} f(x)$$
$$\lim_{x \to b-} f(x) \text{ exist}$$

Suppose
$$\lim_{x \to a+} f(x)$$
does not exist but $f(x)$ is uniformly continuous in (a, b).

As
$$\lim_{x \to a+} f(x)$$
does not exit

$\therefore$ By Cauchy's principle on limit $\exists\, \varepsilon > 0$ such that
$$|f(x') - f(x'')| > \varepsilon$$
for all $\delta > 0$ and $\forall x_1, x_2 \in (a, a+\delta), \delta < b - a$.

This implies that f is not uniformly continuous in $(a, a+\delta)$ and hence not uniformly continuous in (a, b)—a contradiction.

Similarly when
$$\lim_{x \to b-} f(x)$$
does not exist f is not uniformly continuous in (a, b).

So $f(x)$ is uniformly continuous in (a, b) implies the existence of
$$\lim_{x \to a+} f(x)$$
$$\lim_{x \to b-} f(x).$$

6. When a real-valued function f of x defined on $[a, b]$ is said to be piecewise continuous? Is the function f, defined below, piecewise continuous? If so, find the intervals of continuity of f

$$f(x) = \begin{cases} 3 + 2x & \text{when } 0 \leq x < 2 \\ 5 & \text{when } x = 2 \\ 2x - 3 & \text{when } 2 < x \leq 4 \end{cases}$$

Solution. A function f is said to be piecewise continuous on $[a, b]$ if f is continuous on $[a, b]$ except at a finite number of points of jump discontinuity.

The given function is piecewise continuous on $[0, 4]$.

It has a jump discontinuity at $x = 2 \in [0, 4]$.

So the intervals of continuity are $[0, 2)$ and $(2, 4]$.

7. Define Dirichlet's function. Show that it is discontinuous everywhere.

   Solution. A function $f : \mathbb{R} \to \{0, 1\}$ defined by

   $$f(x) = \begin{cases} 0 & x \in \mathbb{Q} \\ 1 & x \in \mathbb{R} - \mathbb{Q} \end{cases}$$

   is called Dirichlet's function.

   Let $c \in \mathbb{Q}$. Consider the sequence $\{x_n\}_n$ defined by

   $$x_n = c + \sqrt{\frac{1}{n(n+1)}}. \qquad \therefore \lim x_n = c$$

   $$\{f(x_n)\}_n = \{1, 1, 1, \ldots\}$$

   and so
   $$\lim f(x_n) = 1 \neq f(c)(= 0)$$

   $\therefore f$ is discontinuous at an arbitrary point c of $\mathbb{Q}$.

   So f is discontinuous on $\mathbb{Q}$.

   Let $d \in \mathbb{R} - \mathbb{Q}$.

   Let $\{y_n\}_n$ be a sequence of rational points such that

   $$\lim y_n = d$$

   $$\therefore \{f(y_n)\}_n = \{0, 0, 0, \cdots\}$$

   and so
   $$\lim f(y_n) = 0 \neq f(d)(= 1)$$

   $\therefore f$ is discontinuous at an arbitrary irrational point d.

   So f is discontinuous on $\mathbb{R} - \mathbb{Q}$.

   $\therefore f$ is discontinous everywhere on $\mathbb{R}$.

   [*Note*: In case of irrational point, formation of rational sequence covering to an irrational point cannot be explicitly displaced as in the previous case. However we formulate $\{y_n\}_n$ such that

   $$y_n \in \left(d - \frac{1}{n}, d + \frac{1}{n}\right)$$

   so that
   $$\lim y_n = d$$

   where each $y_n \in \mathbb{Q}$ and $d \in \mathbb{R} - \mathbb{Q}$].

8. Show that any real valued function f, continuous and strictly monotonic over the closed interval $[a, b]$, admits its inverse function.

Solution. Let f be strictly monotonic increasing on $[a, b]$.

$$\therefore \sup_{x \in [a,b]} f(x) = f(b)$$

$$\text{and} \inf_{x \in [a,b]} f(x) = f(a)$$

$$\therefore f([a, b]) = [f(a), f(b)]$$

Since f is strictly increasing on $[a, b]$.

$$\therefore a \leq x_1 < x_2 \leq b \Rightarrow f(x_1) < f(x_2)$$
$$\text{i.e. } x_1 \neq x_2 \qquad \Rightarrow f(x_1) \neq f(x_2)$$

$\therefore$ f is injective.

Let
$$y^* \in (f(a), f(b))$$

As f is continuous in $[a, b]$

$\therefore$ By intermediate value theorem $\exists$ at least one $x^* \in (a, b)$ such that $f(x^*) = y^*$. [But such x^* is unique as f is injective.]

$\therefore$ f is surjective.

So, f is a bijection from $[a, b]$ to $\mathbb{R}$.

Therefore f admits its inverse function

$$f^{-1} : [f(a), f(b)] \to [a, b].$$

Similarly when a continuous function f on $[a, b]$ is strictly monotonic decreasing, f also admits its inverse function.

9. Determine the points of discontinuities of the function

$$f(x) = x - [x], \quad x \in \mathbb{R}$$

where $[x]$ = integral part of x.

State the nature of such discontinuities.

Solution. The function is piecewise continuous in $\mathbb{R}$. It has discontinuity at every integral value of x. Let c be any integer.

$$\therefore \lim_{x \to c-} f(x) = c - (c - 1) = 1, \quad f(c) = 0$$

$$\lim_{x \to c+} f(x) = c - c = 0$$

At every integral value of x, f is continuous from right but discontinuous from left. Such discontinuities are jump discontinuities.

10. If a function
$$f : \mathbb{R} \to \mathbb{R}$$
satisfies
$$|f(x) - f(y)| \leq |x - y|^2$$
for all $x, y \in \mathbb{R}$; prove that f is a constant function.

Solution. Given
$$|f(x) - f(y)| \leq (x - y)^2 \ \forall \ x, y \in \mathbb{R}$$
$$\therefore \ -(x - y)^2 \leq f(x) - f(y) \leq (x - y)^2 \ \forall \ x, y \in \mathbb{R}.$$

Let $x - y = h$.
$$\therefore \ -h^2 \leq f(y + h) - f(y) \leq h^2 \ \forall \ h(\neq 0) \in \mathbb{R}$$

Dividing by h we get
$$-h \leq \frac{f(y + h) - f(y)}{h} \leq h, \quad \text{when } h > 0$$

and
$$h \leq \frac{f(y + h) - f(y)}{h} \leq -h, \quad \text{when } h < 0.$$

Taking limit $h \to 0$ and using Sandwich theorem we get
$$\lim_{h \to 0} \frac{f(y + h) - f(y)}{h} = 0 \quad \text{i.e.} \quad f'(y) = 0$$

i.e. $f(y) =$ constant, for each $y \in \mathbb{R}$.

$\therefore \ f$ is a constant function.

11. Let $f : [a, b] \to \mathbb{R}$ be continuous and injective. Show that f is strictly monotonic on $[a, b]$.

Solution. Since f is injective on $[a, b]$.
$$\therefore \ f(a) \neq f(b)$$

$\therefore$ Two cases may arise.

Case I. $f(a) < f(b)$, $f(a) > f(b)$.

Case II. $f(a) > f(b)$.

Case I. When $f(a) < f(b)$.

Let $c \in (a, b)$ be an arbitrary point.
We are to prove that
$$f(a) < f(c) < f(b)$$

Suppose
$$f(a) < f(c) < f(b)$$
is not true although $a < c < b$, f is continuous on $[a, b]$ and injective.
If $f(a) < f(c) < f(b)$ is not true then either
$$f(c) < f(a) < f(b)$$
$$\text{or} \quad f(a) < f(b) < f(c).$$

When $f(c) < f(a) < f(b)$ the intermediate value theorem ensures that there is at least one point $c' \in (c, b)$ such that
$$f(c') = f(a)$$
which contradicts the fact that f is an injection on $[a, b]$.
$$\therefore f(c) < f(a) < f(b)$$
is not possible.
When
$$f(a) < f(b) < f(c)$$
the intermediate value theorem again ensures the existence of at least one point $c'' \in (a, c)$ such that
$$f(c'') = f(b)$$
which contradicts that f is an injective on $[a, b]$.
$$\therefore a < c < b \Rightarrow f(a) < f(c) < f(b) \qquad (0.0.1)$$

Let $d \in (a, b)$ such that
$$a < c < d < b$$

By (1) we have
$$f(a) < f(c) < f(d)$$
$$\text{and} \quad f(c) < f(d) < f(b)$$
$$\text{i.e.} \quad f(a) < f(c) < f(d) < f(b)$$

So f is strictly increasing on $[a, b]$.

```
├────┼────┼────┤
a    c    d    b
```

Case II. When $f(a) > f(b)$.

Proof is similar.

12. Let $f : [0, 1] \to \mathbb{R}$ be continuous on $[0, 1]$ and assumes only rational values. If $f(1/2) = 1/2$ prove that $f(x) = 1/2$ for all $x \in [0, 1]$.

    Solution. Let $\alpha \in [0, 1]$ and $\alpha \neq 1/2$ and $f(\alpha) = c$.

    Let $c \neq 1/2$. Then by intermediate value theorem $f(x)$ will assume all values $\in \mathbb{R}$ lying in the interval $(1/2, c)$ or $(c, 1/2)$ according as $1/2 < c$ or $c < 1/2$ at least once in the interval $(1/2, \alpha)$ or $(\alpha, 1/2)$. But $(1/2, c)$ or $(c, 1/2)$ contains infinitely many irrationals points which will be assumed by $f(x)$ when $x \in (1/2, \alpha)$ or $(\alpha, 1/2)$ which contradicts the fact that $f(x)$ assumes only rational value in $[0, 1]$.

    $$\therefore f(x) = 1/2 \ \forall \ x \in [0, 1]$$

    i.e. $f(x)$ is a constant function.

13. Let $f(x) : [0, 2] \to \mathbb{R}$ be continuous on $[0, 2]$ and assume irrational values only. Let $f(1/3) = \sqrt{2}$. Prove that

    $$f(x) = \sqrt{2} \ \forall \ x \in [0, 2]$$

    [*Hint:* Let $c \in [0, 2]$ and $c \neq 1/3$ and $f(c) = u$.

    If $u = \sqrt{2}$, nothing to prove.

    If $u \neq \sqrt{2}$, say $u > \sqrt{2}$, then $\exists$ infinitely many rational numbers in $(\sqrt{2}, u)$ everyone of which will be assumed by $f(x)$ at least once for $x \in \left(\frac{1}{3}, c\right)$, $\left(c > \frac{1}{3}, \text{ say}\right)$ assured by continuity of f and intermediate value theorem. But this is not possible as the interval $(\sqrt{2}, u)$ contains infinitely many rational values none of which will be assumed by f. Hence $u \neq \sqrt{2}$ is not possible.

    $$\therefore f(x) = \sqrt{2} \ \forall \ x \in [0, 2]]$$

14. A function $f : \mathbb{R} \to \mathbb{R}$ is defined by

    $$f(x) = \begin{cases} 3x & x \in \mathbb{Q} \\ 2 - 3x & x \in \mathbb{R} - \mathbb{Q} \end{cases}$$

    Show that f is continuous only at $x = 1/3$.

Hint. 1/3 is an interior point of the domain $\mathbb{R}$.

$$|3x - 1| = 3|x - 1/3| < \epsilon \text{ whenever } x \in N_\delta(1/3) \cap \mathbb{Q} \text{ when } \delta = \epsilon/3$$
and $|2 - 3x - 1| = 3|x - 1/3| < \epsilon$ whenever $x \in N_\delta(1/3) \cap \mathbb{R} - \mathbb{Q}$ when $\delta = \epsilon/3$

$\therefore$ f is continuous at $x = 1/3$.

Use sequential criterion to show the discontinuity of f at any other point.

15. A function $f : \mathbb{Q} \to \mathbb{R}$ is continuous on $\mathbb{Q}$ and

$$f(x + y) = f(x) + f(y) \; \forall \; x, y \in \mathbb{Q}$$

Show that f is a linear function.

Solution. Given

$$f(x + y) = f(x) + f(y)$$
$$\therefore f(0 + 0) = f(0) + f(0) \Rightarrow f(0) = 0$$

Also

$$0 = f(0) = f(x - x) = f(x + (-x)) = f(x) + f(-x)$$
$$\therefore f(-x) = -f(x) \quad \text{i.e.} \quad f \text{ is an odd function}$$

Let $c \in \mathbb{Q}$.

When $c \in \mathbb{Z}^+$

$$f(cx) = f(x + x + \cdots c \text{ times})$$
$$= f(x) + f(x) + \cdots + c \text{ times}$$
$$= cf(x)$$

When $c \in \mathbb{Z}^-$

$$f(cx) = f(-c_1 x), \quad c_1 = -c \in \mathbb{Z}^+$$
$$= -f(c_1 x) = -c_1 f(x) = cf(x)$$

When $c = 0$, $f(cx) = cf(x)$ as both sides become zero.

When $c \in \mathbb{Q} - \mathbb{Z}$, $c = p/q$, say

$$\therefore f(cx) = f\left(p \cdot \frac{x}{q}\right) = pf\left(\frac{x}{q}\right) \quad \text{as } p \in \mathbb{Z} \tag{0.0.1}$$

Now

$$f(x) = f\left(q \cdot \frac{x}{q}\right) = qf\left(\frac{x}{q}\right) \quad \text{as } q \in \mathbb{N}$$

APPENDIX: CHAPTERWISE IMPORTANT PROBLEMS 1183

$$\therefore f\left(\frac{x}{q}\right) = \frac{f(x)}{q}$$

$$\therefore f(cx) = \frac{p}{q} f(x) \quad [\text{by (1)}]$$

i.e. $f(cx) = cf(x)$

$\therefore f(x+y) = f(x) + f(y)$

and $f(cx) = cf(x)$

$\therefore f$ is a linear function on $\mathbb{Q}$.

Note. Check whether $f : \mathbb{R} \to \mathbb{R}$ follows the same property.

16. Show that the function
$$h(x) = \log\left(x^2 + 5\right)$$
is continuous on $\mathbb{R}$.

Solution. h is a composite function gf where $f(x) = x^2 + 5$, $x \in \mathbb{R}$ and $g(x) = \log x$, $x > 0$.

f is continuous on $\mathbb{R}$ as it is a polynomial and
$$f(x) > 0 \ \forall \ x \in \mathbb{R}$$

Now g is continuous on $\mathbb{R}^+$ and $f(x) \in \mathbb{R}^+$, $x \in \mathbb{R}$.

$\therefore g$ is continuous on $f(\mathbb{R})$.

So $gf(x) = g(f(x))$ is continuous on $\mathbb{R}$ i.e. h is continuous on $\mathbb{R}$.

17. Prove that the function
$$u(x) = \sqrt{\log\left(x^2 + 5\right)}$$
is continuous on $\mathbb{R}$.

[*Hint:* Let
$$f(x) = x^2 + 5 \quad x \in \mathbb{R}$$
$$g(x) = \log x \quad x \in \mathbb{R}^+$$
and $h(x) = \sqrt{x} \quad x \in \mathbb{R}^+ \cup \{0\}$

$$f : \mathbb{R} \to \mathbb{R}^+$$

is continuous on $\mathbb{R}$ and $f(\mathbb{R}) = \{f(x) : x \in \mathbb{R}\} = [5, \infty) \subset \mathbb{R}^+$

$$g : \mathbb{R}^+ \to \mathbb{R}$$

is continuous on $\mathbb{R}^+$ and hence continuous on $f(\mathbb{R})$

$$g(f(\mathbb{R})) = [\log 5, \infty) \subset \mathbb{R}^+$$

Now
$$h : \mathbb{R}^+ \cup \{0\} \to \mathbb{R}$$
is continuous on $\mathbb{R}^+ \cup \{0\}$.

∴ h is continuous on $[\log 5, \infty)$ as $[\log 5, \infty) \subset \mathbb{R}^+ \cup \{0\}$.

∴ h is continuous on $g(f(\mathbb{R}))$.

∴ u i.e. hgf is continuous on $\mathbb{R}$.]

18. Find a and b in order that
$$\lim_{x \to 0} \frac{a \sin 2x - b \sin 3x}{5x^3} = 1$$

Solution.
$$\lim_{x \to 0} \frac{a \sin 2x - b \sin 3x}{5x^3} \left[\frac{0}{0}\right] = \lim_{x \to 0} \frac{2a \cos 2x - 3b \cos 3x}{15x^2}$$

In order that the limit exists
$$[2a \cos 2x - 3b \cos 3x]_{x=0} = 0$$
i.e.
$$2a - 3b = 0 \tag{0.0.1}$$

When (1) is satisfied the limit
$$= \lim_{x \to 0} \frac{-4a \sin 2x + 9b \sin 3x}{30x} \left[\frac{0}{0}\right]$$
$$= \lim_{x \to 0} \frac{-8a \cos 2x + 27b \cos 3x}{30} = \frac{-8a + 27b}{30} = 1 \text{ (given)}$$
$$\therefore -8a + 27b = 30 \tag{0.0.2}$$

Solving (1) and (2) we get $a = 3$, $b = 2$.

19. If α, β be the roots of the equation
$$ax^2 + bx + c = 0$$
then show that
$$\lim_{x \to \alpha} \frac{1 - \cos(ax^2 + bx + c)}{(x - \alpha)^2} = \frac{1}{2} a^2 (\alpha - \beta)^2.$$

APPENDIX: CHAPTERWISE IMPORTANT PROBLEMS

Solution.

$$\lim_{x \to \alpha} \frac{1 - \cos(ax^2 + bx + c)}{(x - \alpha)^2} \begin{bmatrix} 0 \\ 0 \end{bmatrix} \quad [\because \cos(a\alpha^2 + bx + c) = \cos 0 = 1]$$

$$= \lim_{x \to \alpha} \frac{(2ax + b)\sin(ax^2 + bx + c)}{2(x - \alpha)} \begin{bmatrix} 0 \\ 0 \end{bmatrix} \quad \text{[by L'Hospital's rule]}$$

$$= \lim_{x \to \alpha} \frac{(2ax + b)\sin\{a(x - \alpha)(x - \beta)\}}{2a(x - \alpha)(x - \beta)} \cdot a(x - \beta)$$

$$= \frac{1}{2} \lim_{n \to \alpha} (2ax + b)a(x - \beta) \lim_{x \to \alpha} \frac{\sin a(x - \alpha)(x - \beta)}{a(x - \alpha)(x - \beta)}$$

[since both the limit exist]

$$= \frac{1}{2}(2a\alpha + b)a(\alpha - \beta) \lim_{\phi \to 0} \frac{\sin \phi}{\phi} \quad [\text{when } \phi = a(x - \alpha)(x - \beta)]$$

$$= \frac{a^2}{2}\left(2\alpha + \frac{b}{a}\right)(\alpha - \beta)$$

$$= \frac{a^2}{2}(2\alpha - \alpha - \beta)(\alpha - \beta) \quad [\because \alpha + \beta = -b/a]$$

$$= \frac{a^2}{2}(\alpha - \beta)^2.$$

20. If

$$\lim_{x \to 0} \frac{ae^x - b\cos x + ce^{-x}}{x \sin x} = 2,$$

find the values of a, b, c.

Solution. At $x = 0$ the denominator $x \sin x$ vanishes. Hence in order to have a finite limit the numerator at $x = 0$ should also vanish i.e.

$$a - b + c = 0 \tag{0.0.1}$$

When (1) holds the limit

$$= \lim_{x \to 0} \frac{ae^x - b\cos x + ce^{-x}}{x \sin x} \begin{bmatrix} 0 \\ 0 \end{bmatrix} = \lim_{x \to 0} \frac{ae^x + b\sin x - ce^{-x}}{\sin x + x \cos x}$$

[by L'Hospital's rule]

By the same logic as above the limit exists if

$$a - c = 0 \tag{0.0.2}$$

When $a - c = 0$ the limit

$$= \lim_{x \to 0} \frac{ae^x + b\cos x + ce^{-x}}{\cos x + \cos x - x \sin x} = \frac{a + b + c}{2} = 2 \text{ [given]}$$

$$\therefore a+b+c=4 \qquad (0.0.3)$$

Solving (1), (2) and (3) we get $a = 1, b = 2, c = 1$.

21. Evaluate

$$\lim_{x \to \infty} \left(\frac{ax+1}{ax-1}\right)^x, \ a > 0.$$

Solution. Let

$$y = \left(\frac{ax+1}{ax-1}\right)^x \quad \therefore \ \ln y = x \ln\left(\frac{ax+1}{ax-1}\right)$$

$$\therefore \lim_{x \to \infty} \ln y = \lim_{x \to \infty} \frac{\ln\left(\frac{ax+1}{ax-1}\right)}{1/x} \left[\frac{0}{0}\right] = \lim_{x \to \infty} \frac{\frac{ax-1}{ax+1} \cdot \frac{-2a}{(ax-1)^2}}{-1/x^2}$$

$$= \lim_{x \to \infty} \frac{2ax^2}{a^2x^2 - 1} = \lim_{x \to \infty} \frac{2a}{a^2 - (1/x^2)} = \frac{2a}{a^2} = \frac{2}{a}.$$

$$\therefore \ln\left(\lim_{x \to \infty} y\right) = \frac{2}{a} \quad \text{[as the function ln is continuous]}$$

or $\lim_{x \to \infty} y = e^{2/a}$ i.e. $\lim_{x \to \infty} \left(\frac{ax+1}{ax-1}\right)^x = e^{2/a}.$

22. Find

$$\lim_{x \to 1-} (1-x)^{\cos \frac{\pi x}{2}}$$

Solution. Let

$$y = (1-x)^{\cos \frac{\pi x}{2}} \quad \therefore \ \ln y = \cos \frac{\pi x}{2} \ln(1-x).$$

$$\therefore \lim_{x \to 1-} \ln y = \lim_{x \to 1-} \frac{\ln(1-x)}{\sec \frac{\pi x}{2}} \left[\frac{\infty}{\infty}\right]$$

$$= \lim_{x \to 1-} \frac{-\frac{1}{1-x}}{\frac{\pi}{2} \sec \frac{\pi x}{2} \tan \frac{\pi x}{2}} \quad \text{[by L'Hospital's rule]}$$

$$= -\frac{2}{\pi} \lim_{x \to 1-} \frac{\cos^2 \frac{\pi x}{2}}{(1-x) \sin \frac{\pi x}{2}} \left[\frac{0}{0}\right]$$

$$= -\frac{2}{\pi} \lim_{x \to 1-} \frac{-\frac{\pi}{2} \cdot 2 \cos \frac{\pi x}{2} \sin \frac{\pi x}{2}}{-\sin \frac{\pi x}{2} + (1-x) \frac{\pi}{2} \cos \frac{\pi x}{2}}$$

$$= \lim_{x \to 1-} \frac{\sin \pi x}{\frac{\pi}{2}(1-x) \cos \frac{\pi x}{2} - \sin \frac{\pi x}{2}} = \frac{0}{-1} = 0$$

$$\therefore \ln\left(\lim_{x \to 1-} y\right) = 0 \quad \text{or} \quad \lim_{x \to 1-} (1-x)^{\cos \frac{\pi x}{2}} = e^0 = 1.$$

APPENDIX: CHAPTERWISE IMPORTANT PROBLEMS

23. Evaluate
$$\lim_{x \to \infty} \left[x - \{(x-a_1)(x-a_2)\cdots(x-a_n)\}^{1/n} \right].$$

Solution.

$$\lim_{x \to \infty} \left[x - \{(x-a_1)(x-a_2)\cdots(x-a_n)\}^{1/n} \right]$$

$$= \lim_{x \to \infty} \left[x - x\left\{\left(1-\frac{a_1}{x}\right)\left(1-\frac{a_2}{x}\right)\cdots\left(1-\frac{a_n}{x}\right)\right\}^{1/n} \right]$$

$$= \lim_{u \to 0+} \frac{1 - \{(1-a_1 u)(1-a_2 u)\cdots(1-a_n u)\}^{1/n}}{u} \quad \text{[where } x = 1/u\text{]}$$

This is of the form $\left[\frac{0}{0}\right]$ and hence by L'Hospital's rule, the limit

$$= -\lim_{u \to 0+} \frac{d}{du}\{(1-a_1 u)(1-a_2 u)\cdots(1-a_n u)\}^{1/n} \quad \left[\because \frac{du}{du}=1\right]$$

$$= \lim_{u \to 0+} \frac{\{(1-a_1 u)(1-a_2 u)\cdots(1-a_n u)\}^{1/n}}{n} \cdot$$

$$\left[\frac{a_1}{1-a_1 u} + \frac{a_2}{1-a_2 u} + \cdots + \frac{a_n}{1-a_n u}\right]$$

$$= \frac{1^{1/n}}{n}(a_1 + a_2 + \cdots + a_n) = \frac{a_1 + a_2 + \cdots + a_n}{n}$$

24. Evaluate
$$\lim_{x \to 0} \left(\frac{\tan x}{x}\right)^{1/x^2}$$

Let
$$u = \left(\frac{\tan x}{x}\right)^{1/x^2} \quad \therefore \ \ln u = \frac{1}{x^2} \ln \frac{\tan x}{x}$$

$$\therefore \lim_{x\to 0}(\ln u) = \lim_{x\to 0} \frac{\ln \frac{\tan x}{x}}{x^2} \quad \left[\frac{0}{0}\right]$$

$$= \lim_{x\to 0} \frac{\frac{x}{\tan x}\left[\frac{x\sec^2 x - \tan x}{x^2}\right]}{2x} \quad \text{[by L'Hospital's rule]}$$

$$= \frac{1}{2}\lim_{x\to 0} \frac{1}{x^2}\left\{\frac{2x - \sin 2x}{\sin 2x}\right\} = \frac{1}{2}\lim_{x\to 0} \frac{2x - \sin 2x}{x^2 \sin 2x} \quad \left[\frac{0}{0}\right]$$

$$= \frac{1}{2}\lim_{x\to 0} \frac{2 - 2\cos 2x}{2x\sin 2x + 2x^2 \cos 2x} \quad \left[\frac{0}{0}\right]$$

$$= \frac{1}{2}\lim_{x\to 0} \frac{4\sin 2x}{2\sin 2x + 4x\cos 2x + 4x\cos 2x - 4x^2 \sin 2x}$$

$$= 2\lim_{x\to 0} \frac{\sin 2x}{2\sin 2x + 8x\cos 2x - 4x^2 \sin 2x} \quad \left[\frac{0}{0}\right]$$

$$= 2\lim_{x\to 0} \frac{2\cos 2x}{4\cos 2x + 8\cos 2x - 16x\sin 2x - 8x\sin 2x - 8x^2 \cos 2x}$$

$$\therefore \ln\left(\lim_{x\to 0} u\right) = \frac{1}{3} \quad \therefore \lim_{x\to 0}\left(\frac{\tan x}{x}\right)^{1/x^2} = e^{1/3}.$$

25. Evaluate

$$\lim_{x\to 0}\left(\frac{\sin x}{x}\right)^{1/x}$$

Hint. Proceed like 25.

26. Show that

$$\lim_{x\to \infty} a^x \sin \frac{b}{a^x} = \begin{cases} 0 & \text{if } 0 < a < 1 \\ b & \text{if } a > 1 \end{cases}$$

Solution. Let $a > 1$.

$$\lim_{x\to \infty} a^x \sin \frac{b}{a^x} = \lim_{x\to \infty} \frac{\sin \frac{b}{a^x}}{a^{-x}} \quad \left[\frac{0}{0}\right] = \lim_{x\to \infty} \frac{-\frac{b}{a^x}\ln a \cos \frac{b}{a^x}}{-a^{-x}\ln a}$$

$$= b\lim_{x\to \infty} \cos \frac{b}{a^x} = b \quad [\because b/a^x \to 0 \text{ as } x \to \infty]$$

Let $0 < a < 1$.

$$\lim_{x\to \infty} a^x \sin \frac{b}{a^x} = 0 \text{ as } \sin \frac{b}{a^x} \text{ is bounded and } a^x \to 0 \text{ as } x \to \infty.$$

Differentiation, Maxima-Minima, Rolle's theorem, Mean Value Theorem, Taylor's Theorem etc.

1. Prove that between any two real roots of

$$e^x \sin x = 1$$

there exists at least one real root of

$$e^x \cos x + 1 = 0$$

Solution. Let α, β be two real roots of $e^x \sin x = 1$.

$$\therefore \quad e^\alpha \sin \alpha = 1$$
$$\text{and} \quad e^\beta \sin \beta = 1.$$

Let

$$f(x) = e^x - \sin x$$

then

$$f(\alpha) = f(\beta) = 0$$
$$f'(x) = -e^{-x} - \cos x.$$

Observe that f is continuous in $[\alpha, \beta]$, $f'(x)$ exists in (α, β) and $f(\alpha) = f(\beta)$.
$\therefore$ By Rolle's theorem $\exists$ at least one point $\gamma \in (\alpha, \beta)$ such that $f'(\gamma) = 0$ i.e.

$$-e^{-\gamma} - \cos \gamma = 0$$
$$\text{or} \quad e^\gamma \cos \gamma + 1 = 0 \Rightarrow$$

γ is a root of $e^x \cos x + 1 = 0$ and $\alpha < \gamma < \beta$.

2. Let a function f is defined by

$$f(x) = \begin{cases} x^2 & 0 \leq x \leq 1 \\ 2x - 1 & 1 < x \leq 2 \\ \sin x - x & 2 < x < \infty \end{cases}$$

Find the derived function f'.

Solution.

$$f'(x) = \begin{cases} 2x, & 0 < x < 1 \\ 2, & 1 < x < 2 \\ \cos x - 1, & 2 < x < \infty \end{cases}$$

$Lf'(1) = \lim_{h \to 0-} \dfrac{f(1+h) - f(1)}{h} = \lim_{h \to 0-} \dfrac{(1-h)^2 - 1}{h} = 2$

$Rf'(1) = \lim_{h \to 0+} \dfrac{f(1+h) - f(1)}{h} = \lim_{h \to 0+} \dfrac{2(1+h) - 1 - 1}{h} = 2$

$\therefore f'(1) = 2$

$Lf'(2) = \lim_{h \to 0-} \dfrac{f(2+h) - f(2)}{h} = \lim_{h \to 0-} \dfrac{2(2+h) - 1 - 3}{h} = 2$

$Rf'(2) = \lim_{h \to 0+} \dfrac{f(2+h) - f(2)}{h} = \lim_{h \to 0+} \dfrac{\sin(2+h) - (2+h) - 3}{h}$ [does not exist]

$Rf'(0) = \lim_{h \to 0+} \dfrac{f(h) - f(0)}{h} = \lim_{h \to 0+} \dfrac{h^2 - 0}{h} = 0$

$\therefore f'(0) = 0$

$$\therefore f'(x) = \begin{cases} 2x, & 0 \le x \le 1 \\ 2, & 1 < x < 2 \\ \cos x - 1, & x > 2 \end{cases}$$

3. Using MVT find a real solution of

$$5^x + 11^x = 3^x + 13^x$$

Solution. $5x^x + 11^x = 3^x + 13^x \Rightarrow 5^x - 3^x = 13^x - 11^x$ \hfill (1)

Let $g(t) = t^x$. $g(t)$ is continuous in $[3, 5]$ and in $[11, 13]$.

It is derivable in $(3, 5)$ and in $(11, 13)$.

$\therefore$ Using MVT in (1)

$$(5 - 3)x\xi^{x-1} = (13 - 11) \cdot x\eta^{x-1}$$

where $\xi \in (3, 5)$ and $\eta \in (11, 13)$

or $\quad \left(\dfrac{\xi}{\eta}\right)^{x-1} = 1 \Rightarrow x = 1$

which is the required solution.

4. If $f : [a, b] \to \mathbb{R}$ be continuous in $[a, b]$ and derivable in (a, b), show that $\exists$ at least one $\xi \in (a, b)$ such that
$$f(b) - f(a) = ke^{-\xi}f'(\xi)$$
where $k = e^b - e^a$.

Solution. Let $g : [a, b] \to \mathbb{R}$ defined by $g(x) = e^x$, $a \le x \le b$.

$\therefore$ g is continuous in $[a, b]$ and $g'(x) = e^x$ exists in (a, b) and $g'(x) \ne 0$ in $a < x < b$.

$\therefore$ By Cauchy's MVT, we get at least one $\xi \in (a, b)$ such that
$$\frac{f(b) - f(a)}{g(b) - g(a)} = \frac{f'(\xi)}{g'(\xi)}$$
or $\quad f(b) - f(a) = \dfrac{e^b - e^a}{e^\xi} f'(\xi) = ke^{-\xi}f'(\xi)$

where $k = e^b - e^a$.

5. Let
$$f(x + y) = f(x) + f(y) \ \forall \ x, y \in \mathbb{R}.$$

If f is derivable at the origin, show that it is derivable on the entire real line.

Solution. f is derivable at $x = 0$.

$\therefore \ f'(0) = \lim\limits_{h \to 0} \dfrac{f(0 + h) - f(0)}{h} = \lim\limits_{h \to 0} \dfrac{f(0) + f(h) - f(0)}{h} = \lim\limits_{h \to 0} \dfrac{f(h)}{h}$

Now let $l \in \mathbb{R}$ be any real number.

Then
$$f'(l) = \lim\limits_{h \to 0} \frac{f(l + h) - f(l)}{h} \quad \text{[provided the limit exists]}$$
$$= \lim\limits_{h \to 0} \frac{f(l) + f(g) - f(l)}{h} = \lim\limits_{h \to 0} \frac{f(h)}{h} = f'(0)$$

which exists.

$\therefore$ f is derivable at any point $l \in \mathbb{R}$.

Note

(a) Here origin has no speciality. In fact if the function is derivable at any point then it is derivable in $\mathbb{R}$.
(b) $f'(l) = f'(0) = $ constant. $\therefore$ $f'(x) = $ constant $= a$, say when $x \in \mathbb{R}$ i.e. $f(x) = ax + b$ on integration ($b = $ integrating constant).
$\therefore$ $f(x + y) = f(x) + f(y)$ and f is derivable at a point of $\mathbb{R} \Rightarrow f$ is a linear function like $ax + b$.

6. Let $f(x+y) = f(x)f(y) \ \forall \ x, y \in \mathbb{R}$ and $f'(0)$ exists and $f(0) \neq 0$. Show f is differentiable at every point $x \in \mathbb{R}$ and $f'(x) = f(x)f'(0)$.

Solution.

$$f'(x) = \lim_{h \to 0} \frac{f(x+h) - f(x)}{h} = \lim_{h \to 0} \frac{f(x)f(h) - f(x)}{h} = f(x) \lim_{h \to 0} \frac{f(h) - 1}{h} \quad (0.0.1)$$

Now $f(0+0) = f(0)f(0)$ or $\{f(0)\}^2 - f(0) = 0$ or $f(0)(f(0) - 1) = 0$.

$\therefore f(0) = 1, \ f(0) \neq 0$.

$$\therefore (1) \Rightarrow f'(x) = f(x) \lim_{h \to 0} \frac{f(h) - f(0)}{h} = f(x)f'(0).$$

7. Let $f : \mathbb{R} \to \mathbb{R}$ be such $|f(x) - f(y)| \leq |x - y|^2 \ \forall \ x, y \in \mathbb{R}$. Show that the function f is a constant function.

Remark: The problem is the same with problem no.10 in the previous chapter. Here it is solved in different approach.

Solution. Let $x, a \in \mathbb{R}$ and $x \neq a$.

So, the given condition

$$\Rightarrow \left| \frac{f(x) - f(a)}{x - a} \right| \leq |x - a| < \epsilon$$

whenever $|x - a| < \delta$ and $\delta = \epsilon$.

$$\therefore \left| \frac{f(x) - f(a)}{x - a} - 0 \right| < \epsilon$$

when $0 < |x - a| < \delta$ and $\delta = \epsilon$.

$$\therefore \lim_{x \to a} \frac{f(x) - f(a)}{x - a} = 0$$

i.e. $f'(a) = 0$.

As a is an arbitrary point of $\mathbb{R}$.

$$\therefore f'(y) = 0 \ \forall \ y \in \mathbb{R}$$

$\therefore f$ is a constant function.

8. Let $f : \mathbb{R} \to \mathbb{R}$ be such that

$$|f(x) - f(y)| \leq |x - y|^r, \quad r > 1.$$

Show that f is a constant function.

Solution. Given condition
$$\Rightarrow -|x-y|^r \leq f(x) - f(y) \leq |x-y|^r$$

Let $x = y + h$.
$$\therefore -|h|^r \leq f(y+h) - f(y) \leq |h|^r$$
$$\therefore -|h|^{r-1} \leq \frac{f(y+h) - f(y)}{|h|} \leq |h|^{r-1}$$
$$\therefore \lim_{h \to 0+} \frac{f(y+h) - f(y)}{h} = 0$$

and
$$\lim_{h \to 0-} \frac{f(y+h) - f(y)}{-h} = 0$$

Sandwich theorem on limit i.e.
$$\lim_{h \to 0+} \frac{f(y+h) - f(y)}{h} = \lim_{h \to 0-} \frac{f(y+h) - f(y)}{h} = 0$$
$$\therefore \lim_{h \to 0} \frac{f(y+h) - f(y)}{h} = 0 \qquad \therefore f'(y) = 0$$

$\therefore f(y)$ is constant for all $y \in \mathbb{R}$.

Note: Problem 7 may be solved by the technique of solving Problem 8.

9. A function $f : [a, b] \to \mathbb{R}$ is said to satisfy Lipschitz condition of order r on $[a, b]$ if $\exists M(\in \mathbb{R}) > 0$ such that
$$|f(x) - f(y)| < M |x - y|^r \quad \forall \ x, y \in [a, b]$$

If $r > 1$, f is constant in $[a, b]$.

Hint. Similar to (8).

10. Prove that
$$\frac{2x}{\pi} < \sin x < x$$
when $0 < x < \pi/2$.

Solution. Consider
$$\phi(x) = \frac{\sin x}{x}$$
$$\therefore \phi'(x) = \frac{x \cos x - \sin x}{x^2} \quad \text{i.e.} \quad \phi'(x) = \frac{\cos x}{x^2}(x - \tan x). \qquad (0.0.1)$$

Let
$$\psi(x) = x - \tan x.$$
$\therefore \psi(0) = 0, \ \psi'(x) = 1 - \sec^2 x < 0$ for $0 < x < \pi/2$

$\therefore \psi(x)$ is decreasing in $(0, \pi/2)$.
$\therefore 0 < x < \pi/2 \Leftrightarrow \psi(0) > \psi(x)$ i.e. $0 > x - \tan x$.
$\therefore$ From (1)
$$\phi'(x) < 0 \Rightarrow \phi(x) \text{ decreasing in } (0, \pi/2)$$
$\therefore 0 < x < \dfrac{\pi}{2} \Leftrightarrow \phi(x) > \phi\left(\dfrac{\pi}{2}\right)$ i.e. $\dfrac{\sin x}{x} > \dfrac{2}{\pi}$

or $\dfrac{2x}{\pi} < \sin x$ in $0 < x < \dfrac{\pi}{2}$. (0.0.2)

Again, let $\chi(x) = x - \sin x$.
$$\therefore \chi(0) = 0$$
$$\chi'(x) = 1 - \cos x > 0 \text{ for } 0 < x < \pi/2$$

$\therefore \chi(x)$ is increasing in $(0, \pi/2)$.
$\therefore 0 < x < \dfrac{\pi}{2} \Leftrightarrow \chi(0) < \chi(x)$ i.e. $0 < x - \sin x$

i.e. $\sin x < x$ in $\left(0, \dfrac{\pi}{2}\right)$ (0.0.3)

$\therefore$ From (2) and (3)
$$\dfrac{2x}{\pi} < \sin x < x \text{ when } 0 < x < \dfrac{\pi}{2}.$$

11. If $f : [a, b] \to \mathbb{R}$ is continuous and continuously differentiable and $f'(x) \neq 0$ in (a, b), prove that f is strictly monotonic.

Solution. Case I. Let $f'(x) > 0$.

Let x_1, x_2 be two points in $[a, b]$ such that $a \leq x_1 < x_2 \leq b$.
$\therefore f$ is continuous in $[x_1, x_2]$ and derivable in (x_1, x_2).
$\therefore$ By MVT
$$\exists \, \xi \in (x_1, x_2) \text{ such that } f(x_2) - f(x_1) = (x_2 - x_1) f'(\xi).$$

As $f'(\xi) > 0$. $\therefore (x_2 - x_1) f'(\xi) > 0$ i.e. $f(x_2) > f(x_1)$ when $x_2 > x_1$.
As x_1, x_2 be two arbitrary points of $[a, b]$. $\therefore f$ is strictly increasing in $[a, b]$.
Case II. Let $f'(x) < 0$, then in a similar argument
$f(x_2) < f(x_1)$ when $x_2 > x_1 \Rightarrow f$ is strictly decreasing in $[a, b]$.

Note:

(a) As f is continuously differentiable in $[a,b]$. $\therefore$ $f'(x)$ will maintain same sign in $[a,b]$, otherwise, due to continuity, $\exists$ a point c in (a,b) such that $f'(c) = 0$—a contraction.

(b) In the problem the restriction *continuously differentiable* may be relaxed as the intermediate value property for a derived function holds good due to Darboux theorem.

So, the problem may be set as

11(a) If $f : [a,b] \to \mathbb{R}$ is continuous and $f'(x)$ exists in (a,b) and $f'(x) \neq 0$ in (a,b), prove that f is strictly monotonic.

12. If f'' is continuous on some nbd of $c \in \mathbb{R}$ prove that

$$\lim_{h \to 0} \frac{f(c+h) - 2f(c) + f(c-h)}{h^2} = f''(c).$$

Solution. Let f'' is continuous in some nbd. $(c-\delta, c+\delta)$ of c.

$\therefore$ By MVT of 2nd order

$$f(c+h) = f(c) + hf'(c) + \frac{h^2}{2!}f''(c+\theta_1 h), \quad 0 < \theta_1 < 1$$

$$\text{and} \quad f(c-h) = f(c) - hf'(c) - \frac{h^2}{2!}f''(c-\theta_2 h), \quad 0 < \theta_2 < 1$$

$$\therefore \lim_{h \to 0} \frac{f(c+h) + f(c-h) - 2f(c)}{h^2} = \frac{1}{2}\lim_{h \to 0}\{f''(c+\theta_1 h) + f''(c-\theta_2 h)\}$$

$$= \frac{1}{2} \cdot 2f''(c) = f''(c)$$

since $c + \theta_1 h$ and $c - \theta_2 h \in (c-\delta, c+\delta)$ [as $h \to 0$] where f'' is continuous.

Sequence

1. If a sequence $\{a_n\}_n$ converges to zero and also if the sequence $\{b_n\}_n$ is bounded then show that the sequence $\{a_n b_n\}_n$ converges to zero.

Proof. As $\{b_n\}_n$ is bounded $\exists\, K > 0$ such that

$$|b_n| \leq K \,\forall\, n \in \mathbb{N}$$

Let $\epsilon > 0$ be an arbitrary small number.

As $\{a_n\}_n$ converges to zero.

$$\therefore |a_n - 0| < \frac{\epsilon}{K} \,\forall\, n \geq n_0$$

Now

$$|a_n b_n - 0| = |a_n|\,|b_n| < \frac{\epsilon}{K} \cdot K = \epsilon \,\forall\, n \geq n_0$$

i.e. $|a_n b_n - 0| < \epsilon \,\forall\, n \geq n_0$

$\therefore \{a_n b_n\}_n$ converges to zero by definition.

2. Show that $\{\sqrt[n]{n}\}_n$ converges to 1.

Solution. Let

$$\sqrt[n]{n} = 1 + u_n$$

$\therefore n = (1 + u_n)^n, \quad u_n > 0 \,\forall\, n > 1$

i.e. $n = 1 + n u_n + \dfrac{n(n-1)}{2!} u_n^2 + \cdots + u_n^n > \dfrac{n(n-1)}{2} u_n^2$

since other terms are positive

i.e. $u_n^2 < \dfrac{2}{n-1} \quad \therefore u_n < \dfrac{\sqrt{2}}{\sqrt{n-1}} \,\forall\, n > 1$

$$\therefore \lim u_n = 0$$

Hence

$$\lim \sqrt[n]{n} = \lim (1 + u_n) = 1.$$

3. Find the upper limit and the lower limit of the sequence $\{a_n\}_n$ where

$$a_n = \left(1 - \frac{1}{n^2}\right) \sin \frac{n\pi}{2}$$

APPENDIX: CHAPTERWISE IMPORTANT PROBLEMS

and examine the convergence of the sequence.

Solution. $0 \le a_n < 1 \ \forall \ n \in \mathbb{N}$.

So $\{a_n\}_n$ is bounded.

$\sin \frac{n\pi}{2}$ takes up three values viz. $0, -1, 1$ and

$$\lim \left(1 - \frac{1}{n^2}\right) = 1$$

$\therefore \ \overline{\lim} \ a_n = 1, \ \underline{\lim} \ a_n = -1$. The sequence is not convergent.

4. Show that $\{a_n\}_n$ where

$$a_n = \left(3 - \frac{2}{n}\right) \cos \frac{n\pi}{2}$$

is divergent.

Solution. The sub-sequence $\{a_{4n}\}_n$ converges to 3 as

$$\lim \left(3 - \frac{1}{2n}\right) = 3$$

and $\quad \lim \cos 2n\pi = 1$

The sub-sequence $\{a_{4n+1}\}_n$ converges to 0 as

$$\lim \left(3 - \frac{2}{4n+1}\right) = 3$$

and $\quad \lim \cos(4n+1)\frac{\pi}{2} = 0$

As two sub-sequences converge to different limits so **the given sequence is divergent.**

5. If $y_1 < y_2$ are arbitrary real numbers and

$$y_n = \frac{1}{3} y_{n-1} + \frac{2}{3} y_{n-2}$$

for every positive integer $n > 2$ then show that $\{y_n\}_n$ is **convergent**.

Solution. We have

$$y_n = \frac{1}{3} y_{n-1} + \frac{2}{3} y_{n-2} \quad [n > 2]$$

$$\therefore \ y_n - y_{n-1} = \frac{1}{3} y_{n-1} + \frac{2}{3} y_{n-2} - y_{n-1} = \frac{2}{3}(y_{n-2} - y_{n-1})$$

$$= \left(-\frac{2}{3}\right)(y_{n-1} - y_{n-2}).$$

Putting $n = 3, 4, \ldots, n$ we get

$$y_3 - y_2 = \left(-\frac{2}{3}\right)(y_2 - y_1)$$

$$y_4 - y_3 = \left(-\frac{2}{3}\right)(y_3 - y_2) = \left(-\frac{2}{3}\right)^2 (y_2 - y_1)$$

$$y_5 - y_4 = \left(-\frac{2}{3}\right)(y_4 - y_3) = \left(-\frac{2}{3}\right)^3 (y_2 - y_1)$$

$$\therefore y_n - y_{n-1} = \left(-\frac{2}{3}\right)^{n-2}(y_2 - y_1).$$

Adding

$$y_n - y_2 = (y_2 - y_1)\left[\left(-\frac{2}{3}\right) + \left(-\frac{2}{3}\right)^2 + \cdots + \left(-\frac{2}{3}\right)^{n-2}\right]$$

$$\therefore y_n - y_2 + y_2 - y_1 = (y_2 - y_1)\left[1 - \frac{2}{3} + \left(\frac{2}{3}\right)^2 + \cdots + (-1)^{n-2}\left(\frac{2}{3}\right)^{n-2}\right]$$

or $\quad y_n - y_1 = (y_2 - y_1)\left[\dfrac{1 - (-2/3)^{n-1}}{1 - (-2/3)}\right]$

$$= (y_2 - y_1)\left[\frac{3}{5}\{1 - (-2/3)^{n-1}\}\right]$$

$$= \frac{3}{5}(y_2 - y_1)\{1 + (-1)^n (2/3)^{n-1}\}$$

$\therefore \lim(y_n - y_1) = \dfrac{3}{5}(y_2 - y_1) \quad [\because 2/3 < 1 \text{ and so } \lim(2/3)^{n-1} = 0]$

$$\therefore \lim y_n = \frac{3}{5}y_2 - \frac{3}{5}y_1 + y_1 = \frac{1}{5}(2y_1 + 3y_2)$$

$\therefore \{y_n\}_n$ is convergent and converges to $\frac{1}{5}(2y_1 + 3y_2)$.

6. Let $\{x_n\}_n$ be a sequence of real numbers defined by $x_1 = \frac{1}{3}$, $x_{2n} = \frac{1}{3}x_{2n-1}$ and $x_{2n+1} = \frac{1}{3} + x_{2n}$ for $n = 1, 2, 3, \ldots$
Find $\lim\limits_{n \to \infty} \inf x_n$ and $\lim\limits_{n \to \infty} \sup x_n$.

Solution.

$$x_1 = \frac{1}{3}, \quad x_2 = \left(\frac{1}{3}\right)^2, \quad x_3 = \frac{1}{3} + \left(\frac{1}{3}\right)^2, \quad x_4 = \frac{1}{3}x_3 = \left(\frac{1}{3}\right)^2 + \left(\frac{1}{3}\right)^3$$

$$x_5 = \frac{1}{3} + \left(\frac{1}{3}\right)^2 + \left(\frac{1}{3}\right)^3, \quad x_6 = \frac{1}{3}x_5 = \left(\frac{1}{3}\right)^2 + \left(\frac{1}{3}\right)^3 + \left(\frac{1}{3}\right)^4, \ldots$$

The sub-sequences are

$$\{x_1, x_3, x_5, \ldots\}$$
$$\text{and} \quad \{x_2, x_4, x_6, \ldots\}$$

$$\therefore \lim_{k\to\infty} x_{2k+1} = \lim_{k\to\infty} \left\{ \frac{1}{3} + \left(\frac{1}{3}\right)^2 + \left(\frac{1}{3}\right)^3 + \cdots + \left(\frac{1}{3}\right)^{k+1} \right\}$$

$$= \frac{1}{3} + \left(\frac{1}{3}\right)^2 + \left(\frac{1}{3}\right)^3 + \cdots \infty$$

$$= \frac{1/3}{1-(1/3)} = \frac{1}{3} \times \frac{3}{2} = \frac{1}{2}$$

and $\lim_{k\to\infty} x_{2k} = \lim_{k\to\infty} \left\{ \left(\frac{1}{3}\right)^2 + \left(\frac{1}{3}\right)^3 + \left(\frac{1}{3}\right)^4 + \cdots + \left(\frac{1}{3}\right)^{k+1} \right\}$

$$= \left(\frac{1}{3}\right)^2 + \left(\frac{1}{3}\right)^3 + \left(\frac{1}{3}\right)^4 + \cdots$$

$$= \frac{(1/3)^2}{1-(1/3)} = \frac{1}{9} \times \frac{3}{2} = \frac{1}{6}$$

$\therefore$ Set of sub-sequential limits is $\{1/6, 1/2\}$.

$$\therefore \lim_{n\to\infty} \inf x_n = \frac{1}{6}$$
$$\text{and} \quad \lim_{n\to\infty} \sup x_n = \frac{1}{2}$$

7. A sequence $\{x_n\}_n$ is defined as follows

$$x_n = 1 + \frac{1}{2} + \frac{1}{3} + \cdots + \frac{1}{n} - \log_e n$$

Show that $\{x_n\}_n$ is convergent.

Solution. Follow the text part, page 212.

8. A sequence $\{x_n\}_n$ is defined as follows

$$x_2 \leq x_4 \leq x_6 \leq \cdots \leq x_5 \leq x_3 \leq x_1 \text{ and } \{y_n\}_n$$

is defined by $y_n = x_{2n-1} - x_{2n}$ such that $y_n \to 0$ as $n \to \infty$. Show that the sequence $\{x_n\}_n$ is convergent.

Solution. The sequence $\{x_n\}_n$ is defined as

$$x_2 \leq x_4 \leq x_6 \cdots \leq x_5 \leq x_3 \leq x_1$$

$\{x_{2n-1}\}_n$ and $\{x_{2n}\}_n$ are two subsequences of $\{x_n\}_n$.

From definition of $\{x_n\}_n$ we see that the subsequence $\{x_{2n-1}\}_n$ is monotone decreasing and is bounded below x_2 is a lower bound.

Hence $\{x_{2n-1}\}_n$ is convergent and let

$$\lim x_{2n-1} = l_1$$

Similarly $\{x_{2n}\}_n$ is m.i. bounded above by x_1 and hence convergent and let $\lim x_{2n} = l_2$.

Given $y_n = x_{2n-1} - x_{2n}$ and $\lim y_n = 0$. $\therefore l_1 = l_2$.

Hence the two sub-sequences $\{x_{2n-1}\}_n$ and $\{x_{2n}\}_n$ are convergent to the same limit and hence $\{x_n\}_n$ is convergent.

9. Show that a monotone increasing sequence of real numbers is convergent if the sequence is bounded above. Hence prove that the sequence $\{x_n\}_n$ where $x_n = 1 + \frac{1}{2^2} + \frac{1}{3^2} + \cdots + \frac{1}{n^2}$ is convergent.

Solution. Follow text for first part.

For the given sequence

$$x_{n+1} - x_n = 1 + \frac{1}{2^2} + \frac{1}{3^2} + \cdots + \frac{1}{n^2} + \frac{1}{(n+1)^2} - \left(1 + \frac{1}{2^2} + \frac{1}{3^2} + \cdots + \frac{1}{n^2}\right)$$

$$= \frac{1}{(n+1)^2} > 0.$$

Hence $\{x_n\}_n$ is m.i.

Now

$$\frac{1}{n^2} = \frac{1}{n.n} < \frac{1}{n(n-1)}, \quad n > 1$$

$$\therefore x_n = 1 + \frac{1}{2^2} + \frac{1}{3^2} + \cdots + \frac{1}{n^2}$$

$$< 1 + \frac{1}{2.1} + \frac{1}{3.2} + \frac{1}{4.3} + \cdots + \frac{1}{n(n-1)}$$

$$= 1 + \left(1 - \frac{1}{2}\right) + \left(\frac{1}{2} - \frac{1}{3}\right) + \cdots + \left(\frac{1}{n-1} - \frac{1}{n}\right)$$

$$= 2 - \frac{1}{n} \quad [n \geq 2]$$

$\therefore x_n < 2$ as $1/n > 0$.

$\therefore x_n < 2 \ \forall \ n \geq 2$. Also $x_1 = 1 < 2$. $\therefore x_n < 2 \ \forall \ n \in \mathbb{N}$.

$\therefore \{x_n\}_n$ is m.i. and bounded above and so $\{x_n\}_n$ is convergent.

APPENDIX: CHAPTERWISE IMPORTANT PROBLEMS

10. Prove that the sequence $\{u_n\}_n$ defined by $u_1 = \sqrt{7}$ and $u_{n+1} = \sqrt{7 + u_n}$ $\forall\, n \geq 1$ converges to the positive root of $x^2 - x - 7 = 0$.

   Solution. $u_1 = \sqrt{7}$, $u_2 = \sqrt{7 + \sqrt{7}}$, $u_3 = \sqrt{7 + \sqrt{7 + \sqrt{7}}}, \ldots$

   Each term of the sequence is positive and $u_1 < u_2 < u_3 < \cdots$ i.e. the sequence is m.i.
   $$u_{n+1} = \sqrt{7 + u_n} \Rightarrow u_{n+1}^2 - u_n - 7 = 0$$

   So
   $$u_n^2 - u_n - 7 < 0 \text{ as } u_n^2 < u_{n+1}^2$$
   $$\text{or}\quad (u_n - \alpha)(u_n - \beta) < 0 \tag{0.0.1}$$

   where α, β are the roots of $x^2 - x - 7 = 0$.

   As the product of the roots be negative.

   $\therefore$ One root is positive and the other is negative.

   Let $\alpha < 0$ and $\beta > 0$.
   $$(u_n - \alpha)(u_n - \beta) < 0 \Rightarrow \alpha < u_n < \beta \;\forall\, n \in \mathbb{N}$$

   $\therefore$ $\{u_n\}_n$ is m.i. and bounded and hence convergent.

   Let m be the limit of the sequence.

   Taking limit as $n \to \infty$ we get from equation (1)
   $$l^2 - l - 7 = 0. \quad \therefore\, l = \alpha \text{ or } \beta$$

   As each $u_n > 0$ and $\alpha < 0$, $\lim u_n = \beta$.

   So the sequence converges to the positive root of $x^2 - x - 7 = 0$.

11. If the sequence $\{x_n\}_n$ has the limit l then the sequence $\{y_n\}_n$ where
   $$y_n = \frac{x_1 + x_2 + \cdots + x_n}{n}$$
   has the same limit l.

   Proof. Consult text.

12. If $\lim x_n = l$ and $x_n > 0 \;\forall\, n$, prove that $\lim y_n = l$ where $y_n = \sqrt[n]{x_1 x_2 \cdots x_n}$.
   [Hint: $\log y_n = \frac{1}{n}(\log x_1 + \log x_2 + \cdots + \log x_n)$.
   Since $\log x$ is a continuous function.
   $$\therefore\, \lim x_n = l \Rightarrow \lim(\log x_n) = \log l.$$

∴ By Cauchy's 1st theorem
$$\lim \frac{\log x_1 + \log x_2 + \cdots + \log x_n}{n} = \log l$$
∴ $\lim (\log y_n) = \log l.$ ∴ $\lim y_n = l$]

APPENDIX: CHAPTERWISE IMPORTANT PROBLEMS   1203

Series

1. Use Dirichlet's test to show the series

$$\sum_{1}^{\infty} \frac{(-1)^{n+1}}{\sqrt{n}}$$

is convergent.

Solution. Dirichlet's test. If the sequence $\{b_n\}_n$ is monotone sequence converging to 0 and the sequence of partial sums $\{s_n\}_n$ of the series Σa_n is bounded then the series $\Sigma a_n b_n$ is convergent.

Let $\{s_n\}_n$ be the sequence of partial sums of the series

$$\sum_{n=1}^{\infty} (-1)^{n+1}$$

$$\therefore s_n = (-1)^2 + (-1)^3 + \cdots + (-1)^{n+1}$$

$$\therefore s_n = \begin{cases} 1 & \text{if } n \text{ is odd} \\ 0 & \text{if } n \text{ is even} \end{cases}$$

$\therefore \{s_n\}_n$ is bounded.

Again the sequence $\{1/\sqrt{n}\}_n$ is monotone decreasing converging to 0, since

$$\frac{1}{\sqrt{1}} > \frac{1}{\sqrt{2}} > \frac{1}{\sqrt{3}} > \cdots \text{ and } \lim \frac{1}{\sqrt{n}} = 0.$$

Hence by Dirichlet's test the series

$$\sum_{1}^{\infty} \frac{(-1)^{n+1}}{\sqrt{n}}$$

is convergent.

2. Test the convergent of the series whose nth term is

$$\frac{2^2 4^2 6^2 \cdots (2n)^2}{3^5 5^2 7^2 \cdots (2n+1)^2}$$

Solution.

$$\frac{u_n}{u_{n+1}} = \frac{(2n+3)^2}{(2n+2)^2} \to 1 \text{ as } n \to \infty$$

$\therefore$ Ratio test fails.

We now apply Gauss's test

$$\frac{u_n}{u_{n+1}} = \left(1 + \frac{3}{2n}\right)^2 \left(1 + \frac{1}{n}\right)^{-2} = \left(1 + \frac{3}{n} + \frac{9}{4n^2}\right)\left(1 - \frac{2}{n} + \frac{3}{n^2} - \cdots\right)$$

$$= 1 + \frac{1}{n} - \frac{3}{4n^2} + \cdots = 1 + \frac{1}{n} + 0\left(\frac{1}{n^2}\right)$$

∴ The series is divergent by Gauss's test.

3. Test the convergence of the series

$$\left(\frac{1}{2}\right)^2 + \left(\frac{1.3}{2.4}\right)^2 + \left(\frac{1.3.5}{2.4.6}\right)^2 + \cdots$$

Solution. Here

$$u_n = \left\{\frac{1.3.5\cdots(2n-1)}{2.4.6\cdots 2n}\right\}^2$$

and $$u_{n+1} = \left\{\frac{1.3.5\cdots(2n+1)}{2.4.6\cdots(2n+2)}\right\}^2$$

$$\therefore \frac{u_n}{u_{n+1}} = \left(\frac{2n+2}{2n+1}\right)^2 = \left(1 + \frac{1}{n}\right)^2 \left(1 + \frac{1}{2n}\right)^{-1}$$

$$= \left(1 + \frac{2}{n} + \frac{1}{n^2}\right)\left(1 - \frac{1}{n} + \frac{1}{4n^2} - \cdots\right)$$

$$= 1 + \frac{1}{n} + 0\left(\frac{1}{n^2}\right)$$

∴ By Gauss's test the given series in divergent.

4. Show that the monotone increasing sequence of real numbers is convergent if it is bounded above. Applying show that the series

$$1 + \frac{1}{2^2} + \frac{1}{3^2} + \cdots + \frac{1}{n^2} + \cdots \text{ is convergent.}$$

Solution. (2nd part). If $\{x_n\}_n$ be the sequence of partial sums of the series then the series will be convergent iff $\{x_n\}_n$ is convergent and limit of the sequence (of partial sums) is the sum of the series.

Now

$$x_{n+1} - x_n = \frac{1}{(n+1)^2} > 0 \; \forall \, n \in \mathbb{N}$$

∴ The sequence is increasing. Now we show that the sequence is bounded as well.
$$\frac{1}{n^2} < \frac{1}{n(n-1)} = \frac{1}{n-1} - \frac{1}{n}$$

Hence
$$x_n = 1 + \frac{1}{2^2} + \cdots + \frac{1}{n^2} < 1 + \left(1 - \frac{1}{2}\right) + \left(\frac{1}{2} - \frac{1}{3}\right) + \cdots + \left(\frac{1}{n-1} - \frac{1}{n}\right)$$
$$= 2 - \frac{1}{n} < 2 \; \forall \; n$$

∴ $\{x_n\}_n$ is bounded above. Thus $\{x_n\}_n$ is monotone increasing and bounded above. So $\{x_n\}_n$ is convergent.

∴ The series
$$1 + \frac{1}{2^2} + \frac{1}{3^2} + \cdots + \frac{1}{n^2} + \cdots \text{ is convergent}$$

5. If $\{u_n\}_n$ is a strictly decreasing sequence of positive real numbers such that $u_n \to 0$ as $n \to \infty$, show that the series
$$u_1 - \frac{1}{2}(u_1 + u_2) + \frac{1}{3}(u_1 + u_2 + u_3) - \cdots + \frac{(-1)^{n-1}}{n}(u_1 + u_2 + \cdots + u_n) + \cdots$$
is convergent.

Solution. Let
$$v_n = \frac{1}{n}(u_1 + u_2 + \cdots + u_n)$$

So the series becomes $v_1 - v_2 + v_3 - \cdots$ such that $v_1 > v_2 > v_3 \cdots > v_n > \cdots$ as $\{u_n\}_n$ is monotone decreasing.

Given that $\lim u_n = 0$ and hence $\lim v_n = 0$ by Cauchy's theorem on limit.

∴ $\{v_n\}_n$ is monotone decreasing sequence of positive real numbers and $\lim v_n = 0$.

∴ By Leibnitz's test the alternating series $v_1 - v_2 + v_3 - v_4 + \cdots$ is convergent.

6. For arbitrary real numbers α and β, test the convergence of the following series
$$\frac{\alpha}{\beta} + \frac{1+\alpha}{1+\beta} + \frac{(1+\alpha)(2+\alpha)}{(1+\beta)(2+\beta)} + \cdots (\alpha > 0, \beta > 0).$$

Solution. Here
$$u_n = \frac{(1+\alpha)(2+\alpha)\cdots(n-1+\alpha)}{(1+\beta)(2+\beta)\cdots(n-1+\beta)}$$
and $u_{n+1} = \dfrac{(1+\alpha)(2+\alpha)\cdots(n-1+\alpha)(n+\alpha)}{(1+\beta)(2+\beta)\cdots(n-1+\beta)(n+\beta)}$

$$\therefore \frac{u_n}{u_{n+1}} = \frac{n+\beta}{n+\alpha} \quad \therefore \lim_{n\to\infty} \frac{u_n}{u_{n+1}} = 1$$

Hence ratio test fails.

Again

$$\lim_{n\to\infty} n\left\{\frac{u_n}{u_{n+1}} - 1\right\} = \lim_{n\to\infty} n\frac{\beta-\alpha}{n+\alpha} = \lim_{n\to\infty} \frac{\beta-\alpha}{1+\alpha/n} = \beta - \alpha.$$

$\therefore$ By Raabe's test the series converges if $\beta - \alpha > 1$ i.e. $\beta > \alpha + 1$ and divergences if $\beta - \alpha < 1$ i.e. $\beta < \alpha + 1$.

Test fails when $\beta = \alpha + 1$.

For $\beta = \alpha + 1$, the series reduces to

$$\frac{\alpha}{\alpha+1} + \frac{1+\alpha}{2+\alpha} + \frac{1+\alpha}{3+\alpha} + \cdots = \sum \frac{1+\alpha}{n+\alpha}$$

which diverges by comparison with the divergent series $\Sigma 1/n$.

7. Show that the series

$$\frac{3}{1.2} - \frac{5}{2.3} + \frac{7}{3.4} - \cdots$$

converges conditionally.

Solution. The series is $u_1 - u_2 + u_3 - \cdots$ where

$$u_n = \frac{2n+1}{n(n+1)}$$

$$\therefore u_{n+1} - u_n = \frac{2n+3}{(n+1)(n+2)} - \frac{2n+1}{n(n+1)} = \frac{-2}{n(n+2)} < 0$$

$\therefore \{u_n\}_n$ is decreasing and

$$\lim u_n = \lim \frac{2+1/n}{n+1} = 0.$$

Also all u_n are positive real numbers.

Hence by Leibnitz's test the alternating series

$$u_1 - u_2 + u_3 - \cdots$$

is convergent.

Now consider the series

$$\frac{3}{1.2} + \frac{5}{2.3} + \frac{7}{3.4} + \cdots$$

$$\therefore \lim\left[n\left(\frac{u_n}{u_{n+1}}-1\right)-1\right]\log n = \lim\left[n\left(\frac{(2n+1)(n+2)}{n(2n+3)}-1\right)-1\right]\log n$$
$$= \lim\left[\frac{2n+2}{2n+3}-1\right]\log n$$
$$= \lim\left(-\frac{\log n}{2n+3}\right) = 0 < 1$$

$\therefore$ De Morgan and Bertrand's test the series

$$\frac{3}{1.2}+\frac{5}{2.3}+\frac{7}{3.4}+\cdots$$

is divergent.

Hence the alternating series

$$\frac{3}{1.2}-\frac{5}{2.3}+\frac{7}{3.4}-\cdots$$

is conditionally convergent.

8. Prove that an absolutely convergent series can be expressed as the difference of two convergent series of positive real numbers.

Proof. Let the series be

$$u_1 - u_2 + u_3 - \cdots, \quad u_i > 0 \; \forall \; i.$$

Given that the series is absolutely convergent and so the two series Σu_n and $\Sigma(-1)^{n-1}u_n$ are convergent.

$$\therefore \sum u_{2n-1} = \frac{1}{2}\left(\sum u_n + \sum(-1)^{n-1}u_n\right) \text{ is convergent}$$
$$\text{and } \sum u_{2n} = \frac{1}{2}\left(\sum u_n - \sum(-1)^{n-1}u_n\right) \text{ is also convergent.}$$

Given series

$$\sum(-1)^{n-1}u_n = \sum u_{2n-1} - \sum u_{2n}.$$

$\therefore$ The given series can be expressed as the difference of two convergent series of positive real numbers.

9. Show that the series

$$\frac{3}{2}+\frac{4}{3}+\frac{5}{4}+\frac{6}{5}+\cdots$$

is not convergent but the series

$$\left(\frac{3}{2}-\frac{4}{3}\right)+\left(\frac{5}{4}-\frac{6}{5}\right)+\cdots$$

is convergent.

Solution. For the first series

$$u_n = \frac{n+2}{n+1} \to 1 \text{ as } n \to \infty \text{ i.e. } \lim u_n \neq 0$$

Hence Σu_n is not convergent.

Second series

$$\left(\frac{3}{2} - \frac{4}{3}\right) + \left(\frac{5}{4} - \frac{6}{5}\right) + \left(\frac{7}{6} - \frac{8}{7}\right) + \cdots = \frac{1}{2.3} + \frac{1}{4.5} + \frac{1}{6.7} + \cdots$$

$$= \sum u_n \text{ where } u_n = \frac{1}{2n(2n+1)}$$

$$\therefore \lim n\left(\frac{u_n}{u_{n+1}} - 1\right) = \lim n\left\{\frac{(n+1)(2n+3) - n(2n+1)}{n(2n+1)}\right\}$$

$$= \lim \frac{4n+3}{2n+1} = 2 > 1$$

Hence the second series is convergent by Raabe's test.

10. Show that the series

$$\sum_{n=1}^{\infty}(-1)^{n-1}n^{-1/2}$$

is convergent. Examine absolute convergence.

Solution. The series can be written as

$$1 - \frac{1}{\sqrt{2}} + \frac{1}{\sqrt{3}} - \frac{1}{\sqrt{4}} + \cdots + \frac{(-1)^{n-1}}{\sqrt{n}} + \cdots$$

Each term is numerically less than the preceding term and tends to zero as $n \to \infty$. It is an alternating series and so convergent by Leibnitz's test.

Now consider the series

$$1 + \frac{1}{\sqrt{2}} + \frac{1}{\sqrt{3}} + \cdots + \frac{1}{\sqrt{n}} + \cdots$$

This is a *p*-series with $p = 1/2 < 1$ and so diverges.

So the series is not absolutely convergent.

Tangents and Normals

1. Show that the points on the curve

$$y^2 = 4a\left(x + a\sin\frac{x}{a}\right)$$

where the tangents are parallel to x-axis lies on the parabola $y^2 = 4ax$.

Solution. Let (α, β) be a point on the curve

$$y^2 = 4a\left(x + a\sin\frac{x}{a}\right)$$

where the tangent is parallel to x-axis.

$$\therefore \left[\frac{dy}{dx}\right]_{(\alpha,\beta)} = 0 \Rightarrow 1 + \cos\frac{\alpha}{a} = 0 \quad \text{i.e.} \quad \cos\frac{\alpha}{a} = -1 \Rightarrow \sin\frac{\alpha}{a} = 0. \quad (0.0.1)$$

Again (α, β) lies on the curve.

$$\therefore \beta^2 = 4a\left(\alpha + a\sin\frac{\alpha}{a}\right) = 4a\alpha \text{ [by (1)]}.$$

$\therefore (\alpha, \beta)$ lies on the parabola $y^2 = 4ax$.

2. If $x\cos\theta + y\sin\theta = p$ touches the curve $x^m y^m = a^{m+n}$, prove that

$$p^{m+n} m^m n^n = (m+n)^{m+n} a^{m+n} \sin^n\theta \cos^m\theta.$$

Solution. Let the given line touches the curve at (α, β). First, we find tangent at (α, β) and compare it with the given line. Differentiating the equation of the curve w.r.t. x

$$mx^{m-1}y^n + nx^m y^{n-1}\frac{dy}{dx} = 0$$

$$\therefore \left[\frac{dy}{dx}\right]_{(\alpha,\beta)} = -\frac{m}{n} \cdot \frac{\beta}{\alpha}$$

$\therefore$ Tangent at (α, β) is

$$y - \beta = -\frac{m\beta}{n\alpha}(x - \alpha) \quad \text{or} \quad m\beta x + n\alpha y = (m+n)\alpha\beta$$

Comparing it with the given straight line

$$\frac{\cos\theta}{m\beta} = \frac{\sin\theta}{n\alpha} = \frac{p}{(m+n)\alpha\beta}$$

$$\therefore \alpha = \frac{pm}{(m+n)\cos\theta}, \quad \beta = \frac{pn}{(m+n)\sin\theta}$$

(α, β) lies on the curve.

$$\therefore \alpha^m \beta^n = a^{m+n} \quad \text{or} \quad \frac{p^m m^m}{(m+n)^m \cos^m\theta} \cdot \frac{p^n n^n}{(m+n)^n \sin^n\theta} = a^{m+n}$$

$$\text{or} \quad p^{m+n} m^m n^n = (m+n)^{m+n} a^{m+n} \sin^n\theta \cos^m\theta$$

3. Let
$$u + iv = f(x + iy)$$

where $u = u(x, y)$ and $v = v(x, y)$. Prove that the curves $u(x, y) = $ constant and $v(x, y) = $ constant form two families such that each member of the first family cuts each member of the second family orthogonally.

Solution. We have
$$u + iv = f(x + iy)$$
$$\therefore u_x + iv_x = f'(x + iy)$$
$$\text{and } u_y + iv_y = if'(x + iy)$$
$$\therefore (u_x + iv_x)i = u_y + iv_y$$
$$\text{or} \quad iu_x - v_x = u_y + iv_y$$

Equating real and imaginary parts $u_x = v_y$, $u_y = -v_x$.

$$\therefore u_x v_x + u_y v_y = u_x v_x - v_x u_x = 0 \tag{0.0.1}$$

For the family $u(x, y) = $ constant (parametric constant)

$$\left[\frac{dy}{dx}\right]_{(x,y)} = -\frac{u_x}{u_y}.$$

Similarly

$$\left[\frac{dy}{dx}\right]_{(x,y)} = -\frac{v_x}{v_y}$$

for the second family.

They will cut orthogonally if

$$\left(-\frac{u_x}{u_y}\right)\left(-\frac{v_x}{v_y}\right) = -1$$

i.e. if $u_x v_x + u_y v_y = 0$ which is true by equation (1).

Hence the two families are such that each member of the first family cuts each member of the second orthogonally.

4. Let $X = \phi(x,y), Y = \psi(x,y)$ define a transformation of the xy-plane to XY-plane. Suppose further $\phi_x = \psi_y$ and $\phi_y = -\psi_x$. Then prove that the angle between the curves $F(x,y) = 0$ and $G(x,y) = 0$ in the xy-plane is equal to the angle between the curves $F_1(X,Y) = 0$ and $G_1(X,Y) = 0$ in the XY-plane, where the transformation maps $F(x,y)$ and $G(x,y)$ to $F_1(X,Y)$ and $G_1(X,Y)$ respectively.

Solution. Under the transformation $X = \phi(x,y)$, $Y = \psi(x,y)$ such that $\phi_x = \psi_y$, $\phi_y = -\psi_x$; $F(x,y) = F_1(X,Y)$ and $G(x,y) = G_1(X,Y)$.

$\therefore$ By chain rule

$$F_x = F_{1X}\frac{\partial X}{\partial x} + F_{1Y}\frac{\partial Y}{\partial x} = F_{1X}\phi_x - F_{1Y}\phi_y \quad [\because \psi_x = -\phi_y]$$
$$F_y = F_{1X}\phi_y + F_{1Y}\phi_x$$
$$G_x = G_{1X}\phi_x - G_{1Y}\phi_y$$
$$G_y = G_{1X}\phi_y + G_{1Y}\phi_x$$

If θ be the angle between $F(x,y) = 0$ and $G(x,y) = 0$ at (x,y) then

$$\tan\theta = \left| \frac{\dfrac{F_x}{F_y} - \dfrac{G_x}{G_y}}{1 + \dfrac{F_x}{F_y} \cdot \dfrac{G_x}{G_y}} \right| = \left| \frac{F_x G_y - F_y G_x}{F_x G_x + F_y G_y} \right|$$

$$= \left| \frac{(F_{1X}\phi_x - F_{1Y}\phi_y)(G_{1X}\phi_y + G_{1Y}\phi_x) - (F_{1X}\phi_y + F_{1Y}\phi_x)(G_{1X}\phi_x - G_{1Y}\phi_y)}{(F_{1X}\phi_x - F_{1Y}\phi_y)(G_{1X}\phi_x - G_{1Y}\phi_y) + (F_{1X}\phi_y + F_{1Y}\phi_x)(G_{1X}\phi_y + G_{1Y}\phi_x)} \right|$$

$$= \left| \frac{(F_{1X}G_{1Y} - F_{1Y}G_{1X})(\phi_y^2 + \phi_x^2)}{(F_{1X}G_{1X} + F_{1Y}G_{1Y})(\phi_y^2 + \phi_x^2)} \right|$$

$$= \left| \frac{F_{1X}G_{1Y} - F_{1Y}G_{1X}}{F_{1X}G_{1X} + F_{1Y}G_{1Y}} \right|$$

$$= \tan\theta'$$

where θ' is the angle between $F_1(X,Y) = 0$ and $G_1(X,Y) = 0$ at the corresponding point (X,Y).

$\therefore \theta = \theta'$. Hence proved.

5. Show that the tangent to the curve

$$25x^5 + 5x^4 - 45x^3 - 5x^2 + 2x + 6y - 24 = 0$$

at the point $(-1,1)$ is also normal at other two points of the curve.

Solution. Differentiating the equation of the curve w.r.t. x we get

$$125x^4 + 20x^3 - 135x^2 - 10x + 2 + 6\frac{dy}{dx} = 0.$$

$$\therefore \left[\frac{dy}{dx}\right]_{(-1,1)} = 3$$

$\therefore$ Tangent at $(-1, 1)$ is

$$y - 1 = 3(x + 1) \quad \text{or} \quad 3x - y + 4 = 0 \tag{0.0.1}$$

The other points where tangent (1) meets the curve (it may cut as the degree of the equation of the curve is more than 2) is obtained by solving the equation of the curve and equation (1).

Eliminating y between the equation of the curve and the equation (1):

$$25x^5 + 5x^4 - 45x^3 - 5x^2 + 2x + 6(3x + 4) - 24 = 0$$

or $\quad 25x^5 + 5x^4 - 45x^3 - 5x^2 + 20x = 0$

or $\quad 5x\left(5x^4 + x^3 - 9x^2 - x + 4\right) = 0 \quad$ (two of its roots must be $-1, -1$)

or $\quad x\left(5x^4 + 5x^3 - 4x^3 - 4x^2 - 5x^2 - 5x + 4x + 4\right) = 0$

or $\quad x(x+1)\left(5x^3 - 4x^2 - 5x + 4\right) = 0$

or $\quad x(x+1)\left(5x^3 + 5x^2 - 9x^2 - 9x + 4x + 4\right) = 0$

or $\quad x(x+1)^2\left(5x^2 - 9x + 4\right) = 0$

$\therefore \quad x = 0, -1, -1, \dfrac{9 \pm \sqrt{81-80}}{10} \quad$ or $\quad x = 0, -1, -1, 1, \dfrac{4}{5}$.

$\therefore$ Other points have their x-coordinates $0, 1, 4/5$.

At $x = 0$, $y = 4$.

Normal at $(0, 4)$ is

$$y - 4 = -\left.\frac{dx}{dy}\right|_{(0,4)} x \quad \text{or} \quad y - 4 = 3x$$

which is the same as (1).

At $x = 1$, $y = 7$.

Normal at $(1, 7)$ is

$$y - 7 = -\left.\frac{dx}{dy}\right|_{(1,7)} (x - 1) \quad \text{or} \quad y - 7 = 3(x - 1) \quad \text{or} \quad y = 3x + 4$$

which is again same as (1).

Hence the tangent at $(-1, 1)$ is normal to the curve at other two points viz. $(0, 4)$ and $(1, 7)$.

APPENDIX: CHAPTERWISE IMPORTANT PROBLEMS

6. Tangents are drawn from the origin to the curve $y = \sin x$. Show that their points of contact lie on the curve $x^2 y^2 = x^2 - y^2$.

Solution. Let (α, β) be a point of contact of a tangent drawn from the origin. Equation of tangent to the curve $y = \sin x$ at (α, β) is

$$y - \beta = (x - \alpha) \cos \alpha \quad \text{as} \quad \left.\frac{dy}{dx}\right|_{(\alpha,\beta)} = \cos \alpha$$

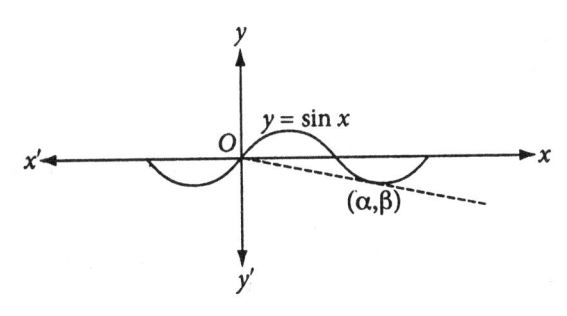

As it passes through the origin

$$\therefore \quad -\beta = -\alpha \cos \alpha \quad \text{i.e.} \quad \beta = a \cos \alpha \qquad (0.0.1)$$

Also (α, β) lies on $y = \sin x$.

$$\therefore \quad \beta = \sin \alpha \qquad (0.0.2)$$

From (1) and (2) we have

$$1 = \sin^2 \alpha + \cos^2 \alpha = \beta^2 + \frac{\beta^2}{\alpha^2}$$

$$\text{or} \quad \alpha^2 = \alpha^2 \beta^2 + \beta^2$$

$$\text{or} \quad \alpha^2 \beta^2 = \alpha^2 - \beta^2.$$

Hence the points of contact of tangents drawn from the origin to the **curve** $y = \sin x$ lies on

$$x^2 y^2 = x^2 - y^2$$

7. Show that the tangent to curve $x^3 + y^3 = 3axy$ at a point $[\neq (0,0)]$ where it meets the parabola $y^2 = ax$ is parallel to y-axis.

Solution. The point common to $x^3 + y^3 = 3axy$ and the parabola $y^2 = ax$ is obtained by solving those equations.

Eliminating x

$$\frac{y^6}{a^3} + y^3 = 3y^3 \quad \text{or} \quad y^3 = 2a^3$$

for a point other than the origin.

or $y = 2^{1/3}a$. $\therefore x = 2^{2/3}a$.

Hence $(2^{2/3}a, 2^{1/3}a)$ is a meeting point between $x^3 + y^3 = 3axy$ and $y^2 = ax$ other than the origin.

Differentiating $x^3 + y^3 = 3axy$ w.r.t. x

$$3x^2 + 3y^2\frac{dy}{dx} = 3a\left(y + x\frac{dy}{dx}\right) \Rightarrow \frac{dy}{dx} = \frac{ay - x^2}{y^2 - ax}.$$

At $(2^{2/3}a, 2^{1/3}a)$, $y^2 - ax = 2^{2/3}a^2 - 2^{2/3}a^2 = 0$.

$\therefore$ At the meeting point $(2^{2/3}a, 2^{1/3}a)$ the tangent to $x^3 + y^3 = 3axy$ is parallel to y-axis.

Pedal equation and pedal

8. Find the pedal equation of the equiangular spiral $r = ae^{\theta \cot \alpha}$.

Solution. The pedal equation is obtained by eliminating θ between

$$r = ae^{\theta \cot \alpha} \quad \text{and} \quad \frac{1}{p^2} = \frac{1}{r^2} + \frac{1}{r^4}\left(\frac{dr}{d\theta}\right)^2.$$

Now

$$\frac{dr}{d\theta} = a \cot \alpha e^{\theta \cot \alpha} = r \cot \alpha$$

$$\therefore \frac{1}{p^2} = \frac{1}{r^2} + \frac{1}{r^4}r^2 \cot^2 \alpha = \frac{1 + \cot^2 \alpha}{r^2} = \frac{\cosec^2 \alpha}{r^2}$$

$$\therefore p^2 = r^2 \sin^2 \alpha$$

$$\therefore |p| = |r \sin \alpha| = |r||\sin \alpha|$$

As $0 < \alpha < \pi$, $p \geq 0$, $r \geq 0$.

$\therefore$ The equation becomes $p = r \sin \alpha$.

9. Find the pedal equation of the parabola

$$r = \frac{2a}{1 + \cos \theta}.$$

Solution. The pedal equation is obtained by eliminating θ between

$$r = \frac{2a}{1 + \cos \theta} \quad \text{and} \quad \frac{1}{p^2} = \frac{1}{r^2} + \frac{1}{r^4}\left(\frac{dr}{d\theta}\right)^2.$$

Differentiating $r = \frac{2a}{1+\cos\theta}$ w.r.t. θ

$$\frac{dr}{d\theta} = \frac{2a\sin\theta}{(1+\cos\theta)^2} = \frac{r^2\sin\theta}{2a}$$

$$\therefore \frac{1}{p^2} = \frac{1}{r^2} + \frac{1}{r^4} \cdot \frac{r^4\sin^2\theta}{4a^2} = \frac{1}{r^2} + \frac{\sin^2\theta}{4a^2} \qquad (0.0.1)$$

From $r = \frac{2a}{1+\cos\theta}$ we get

$$\cos\theta = \frac{2a}{r} - 1 \qquad \therefore \sin^2\theta = 1 - \frac{4a^2}{r^2} + \frac{4a}{r} - 1$$

$\therefore$ From (1)

$$\frac{1}{p^2} = \frac{1}{r^2} + \frac{1}{4a^2}\left(\frac{4a}{r} - \frac{4a^2}{r^2}\right) = \frac{1}{r^2} + \frac{1}{ar} - \frac{1}{r^2} = \frac{1}{ar}$$

$\therefore p^2 = ar$, which is the required pedal equation.

10. Find the pedal equation of the astroid

$$x = a\cos^3\theta, \quad y = a\sin^3\theta$$

[*Hint:* The Cartesian equation is obtained by eliminating θ as $x^{2/3} + y^{2/3} = a^{2/3}$. Now follow the worked-out Example 15.11.2.]

11. Find the pedal equation of the ellipse $\frac{x^2}{a^2} + \frac{y^2}{b^2} = 1$ w.r.t. origin.

Solution. Let

$$f(x,y) \equiv \frac{x^2}{a^2} + \frac{y^2}{b^2} - 1$$

Pedal equation w.r.t. 0 is obtained by eliminating x and y between

$$\frac{x^2}{a^2} + \frac{y^2}{b^2} - 1 = 0 \qquad (0.0.1)$$

$$x^2 + y^2 = r^2 \qquad (0.0.2)$$

and

$$p^2 = \frac{(xf_x + yf_y)^2}{f_x^2 + f_y^2}$$

$$f_x = \frac{2x}{a^2}, \quad f_y = \frac{2y}{b^2}.$$

$$\therefore p^2 = \frac{\left\{2\left(\frac{x^2}{a^2} + \frac{y^2}{b^2}\right)\right\}^2}{\frac{4x^2}{a^4} + \frac{4y^2}{b^4}} = \frac{1}{\frac{x^2}{a^4} + \frac{y^2}{b^4}} \qquad \because \frac{x^2}{a^2} + \frac{y^2}{b^2} = 1$$

$$\therefore \frac{1}{p^2} = \frac{x^2}{a^4} + \frac{y^2}{b^4} \qquad (0.0.3)$$

Eliminating x and y from equations (1), (2) and (3) we get the required pedal equation as

$$\begin{vmatrix} 1 & 1 & r^2 \\ 1/a^2 & 1/b^2 & 1 \\ 1/a^4 & 1/b^4 & 1/p^2 \end{vmatrix} = 0$$

or $\quad \dfrac{1}{p^2 b^2} - \dfrac{1}{b^4} + \dfrac{1}{a^4} - \dfrac{1}{a^2 p^2} + r^2\left(\dfrac{1}{a^2 b^4} - \dfrac{1}{b^2 a^4}\right) = 0$

or $\quad \dfrac{1}{p^2}\left(\dfrac{1}{b^2} - \dfrac{1}{a^2}\right) - \left(\dfrac{1}{b^4} - \dfrac{1}{a^4}\right) + \dfrac{r^2}{a^2 b^2}\left(\dfrac{1}{b^2} - \dfrac{1}{a^2}\right) = 0$

or $\quad \dfrac{1}{p^2} - \left(\dfrac{1}{b^2} + \dfrac{1}{a^2}\right) + \dfrac{r^2}{a^2 b^2} = 0 \qquad [\because a \neq b]$

or $\quad \dfrac{a^2 b^2}{p^2} = a^2 + b^2 - r^2$.

12. Find the pedal equation of the parabola $y^2 = 4a(x+a)$ w.r.t. the vertex $(-a, 0)$.

Solution. First we apply translation $x = X - a$, $y = Y$ so that the equation of the parabola becomes

$$Y^2 = 4aX \qquad (0.0.1)$$

The vertex is now $(0, 0)$.

$\therefore$ Pedal equation of (1) is required with regards to origin.

Let

$$f(X, Y) \equiv Y^2 - 4aX = 0$$

$\therefore$ Pedal equation will be obtained by eliminating X and Y from the equation (1), the equation

$$X^2 + Y^2 = r^2 \qquad (0.0.2)$$

and the relation

$$p^2 = \frac{(X f_X + Y f_Y)^2}{f_X^2 + f_Y^2} \qquad (0.0.3)$$

APPENDIX: CHAPTERWISE IMPORTANT PROBLEMS

From (3)
$$p^2 = \frac{(-4aX + 2Y^2)^2}{16a^2 + 4Y^2} = \frac{(-4aX + 8aX)^2}{16a^2 + 16aX} \quad [\because Y^2 = 4aX]$$

or $$p^2 = \frac{16a^2 X^2}{16a(X+a)} = \frac{aX^2}{X+a}$$

$$\therefore aX^2 = p^2(X+a) \tag{0.0.4}$$

$$X^2 + Y^2 = r^2 \Rightarrow X^2 + 4aX = r^2 \tag{0.0.5}$$

From (4) and (5)
$$p^2 X + p^2 a + 4a^2 X = ar^2$$

$$\therefore X = \frac{a(r^2 - p^2)}{p^2 + 4a^2}.$$

Putting this value of X in (4) we get the required pedal form as

$$a \cdot \frac{a^2 (r^2 - p^2)^2}{(p^2 + 4a^2)^2} = p^2 \left(a + \frac{a(r^2 - p^2)}{p^2 + 4a^2} \right)$$

or $$a^2 (r^2 - p^2)^2 = (p^2 + 4a^2) p^2 (p^2 + 4a^2 + r^2 - p^2)$$

or $$a^2 (r^2 - p^2)^2 = p^2 (r^2 + 4a^2)(p^2 + 4a^2).$$

13. Find the pedal of the parabola $y^2 = 4ax$ w.r.t. the vertex.

Solution. Let (α, β) be any point on the parabola.

$$\therefore \beta^2 = 4a\alpha \tag{0.0.1}$$

Tangent at (α, β)
$$y\beta + 2a(x + \alpha) \tag{0.0.2}$$

The line perpendicular to (2) and passing through the vertex we (origin) is

$$\beta x + 2ay = 0 \tag{0.0.3}$$

To have the pedal of $y^2 = 4ax$ w.r.t. vertex we are to eliminate α, β from (1), (2) and (3).

From (3)
$$\beta = -\frac{2ay}{x}$$

From (2)
$$\alpha = -x + \frac{y\beta}{2a} = -x + \frac{y}{2a}\left(-\frac{2ay}{x}\right) = -x - \frac{y^2}{x} = -\frac{x^2 + y^2}{x}.$$

From (1)

$$\frac{4a^2y^2}{x^2} = -4a\frac{x^2+y^2}{x} \quad \text{or} \quad ay^2 = -x\left(x^2+y^2\right) \quad \text{or} \quad ay^2 + x\left(x^2+y^2\right) = 0.$$

14. Obtain the pedal of $y^2 = 4ax$ w.r.t. the focus $(a, 0)$.

    Solution. Let (α, β) be any point on

    $$y^2 = 4ax. \qquad \therefore \beta^2 = 4a\alpha \qquad (0.0.1)$$

    Tangent at (α, β) in

    $$y\beta = 2a(x+\alpha) \qquad (0.0.2)$$

    The line perpendicular to (2) and passing through $(a, 0)$ is

    $$\beta x + 2ay = a\beta \qquad (0.0.3)$$

    Eliminating α and β we get the pedal of the parabola $y^2 = 4ax$ w.r.t. its focus $(a, 0)$.

    From (3)

    $$\beta = \frac{2ay}{a-x}$$

    $\therefore$ From (2)

    $$\alpha = \frac{1}{2a}\left[y \cdot \frac{2ay}{a-x} - 2ax\right] = \frac{x^2 + y^2 - ax}{a-x}$$

    $\therefore$ From (1)

    $$\frac{4a^2y^2}{(a-x)^2} = 4a \cdot \frac{x^2+y^2-ax}{a-x} \quad \text{or} \quad x\left[(x-a)^2 + y^2\right] = 0$$

    $\therefore x = 0$ is the pedal, as

    $$(x-a)^2 + y^2 = 0 \text{ is satisfied}$$

    only at the point $(a, 0)$ which is not on the locus.

15. Find the pedal of the circle $r = 2a\cos\theta$ w.r.t. the pole.

    Solution. The equation of the circle is

    $$r = 2a\cos\theta \qquad (0.0.1)$$

Differentiating w.r.t r

$$1 = -2a\sin\theta \frac{d\theta}{dr} \quad \text{or} \quad r = -2a\sin\theta \left(r\frac{d\theta}{dr}\right)$$

$$\text{or} \quad 2a\cos\theta = -2a\sin\theta \tan\phi \quad \left[\because \tan\phi = r\frac{d\theta}{dr}\right]$$

$$\therefore \tan\phi = -\cot\theta = \tan\left(\frac{\pi}{2} + \theta\right) \tag{0.0.2}$$

$$\therefore \phi = \frac{\pi}{2} + \theta.$$

If (p, α) be the point of intersection of tangent at any point (r, θ) with the perpendicular to this tangent through the pole then with usual notations

$$\theta + \phi - \alpha = \frac{\pi}{2} \tag{0.0.3}$$

$\therefore$ From (2) and (3)

$$\theta + \frac{\pi}{2} + \theta - \alpha = \frac{\pi}{2} \quad \therefore \alpha = 2\theta \tag{0.0.4}$$

Also we have

$$\phi = r\sin\phi$$

$$\therefore p = r\sin\left(\frac{\pi}{2} + \theta\right) = r\cos\theta = r\cos\frac{\alpha}{2} \quad \text{[by (4)]}$$

$$\therefore r = \frac{p}{\cos\alpha/2} \tag{0.0.5}$$

Eliminating r, θ from the equation of the circle and from (4) and (5) we get

$$\frac{p}{\cos\alpha/2} = 2a\cos\frac{\alpha}{2} \quad \text{or} \quad p = a \cdot 2\cos^2\frac{\alpha}{2} = a(1 + \cos\alpha).$$

$\therefore$ The pedal which is the locus of (p, α) is given by

$$r = a(1 + \cos\theta).$$

Curvature

16. If ρ, ρ' be the radii of curvature at the ends of two conjugate diameters of an ellipse, prove that

$$\left(\rho^{2/3} + \rho'^{2/3}\right)(ab)^{2/3} = a^2 + b^2.$$

Solution. Let the equation of the ellipse be $\frac{x^2}{a^2} + \frac{y^2}{b^2} = 1$.

The end points of two conjugate diameters may be taken as $(a\cos\theta, b\sin\theta)$ and $(-a\sin\theta, b\cos\theta)$.

Radius of curvature at (x,y)

$$= \frac{\left(\frac{4x^2}{a^4} + \frac{4y^2}{b^4}\right)^{3/2}}{\frac{2}{a^2}\cdot\frac{4y^2}{b^4} + \frac{2}{b^2}\cdot\frac{4x^2}{a^4}}.$$

Let ρ and ρ' are respectively the radii of curvatures at $(a\cos\theta, b\sin\theta)$ and $(-a\sin\theta, b\cos\theta)$.

$$\therefore \rho = \frac{\left(\frac{\cos^2\theta}{a^2} + \frac{\sin^2\theta}{b^2}\right)^{3/2}}{\frac{1}{a^2 b^4}\cdot b^2\sin^2\theta + \frac{1}{b^2 a^4}\cdot a^2\cos^2\theta} = a^2 b^2 \left(\frac{\cos^2\theta}{a^2} + \frac{\sin^2\theta}{b^2}\right)^{3/2}$$

$$\rho' = \frac{\left(\frac{\sin^2\theta}{a^2} + \frac{\cos^2\theta}{b^2}\right)^{3/2}}{\frac{1}{a^2 b^4}\cdot b^2\cos^2\theta + \frac{1}{b^2 a^4}\cdot a^2\sin^2\theta} = a^2 b^2 \left(\frac{\sin^2\theta}{a^2} + \frac{\cos^2\theta}{b^2}\right)^{3/2}.$$

$$\therefore \rho^{2/3} + \rho'^{2/3} = (ab)^{4/3}\left(\frac{1}{a^2} + \frac{1}{b^2}\right) = (ab)^{-2/3}(a^2 + b^2)$$

$$\therefore \left(\rho^{2/3} + \rho'^{2/3}\right)(ab)^{2/3} = a^2 + b^2.$$

17. The tangents at two points P and Q on the cycloid

$$x = a(\theta - \sin\theta), \quad y = a(1 - \cos\theta)$$

are at right angles, show that if ρ_1 and ρ_2 be the radii of curvature at these points then $\rho_1^2 + \rho_2^2 = 16a^2$.

Solution. Let at P and Q the values of θ are θ_1 and θ_2 respectively.

Gradient of tangent at θ is

$$\frac{a\sin\theta}{a(1-\cos\theta)} = \cot\frac{\theta}{2}$$

$\therefore$ Gradient of tangents at P and Q are

$$\cot\frac{\theta_1}{2} \quad \text{and} \quad \cot\frac{\theta_2}{2}.$$

APPENDIX: CHAPTERWISE IMPORTANT PROBLEMS

As they are at right angles.

$$\therefore \cot\frac{\theta_1}{2}\cot\frac{\theta_2}{2} = -1 \quad \text{i.e.} \quad \cos\left(\frac{\theta_1}{2} - \frac{\theta_2}{2}\right) = 0$$

$$\therefore \frac{\theta_1}{2} - \frac{\theta_2}{2} = \frac{\pi}{2} \quad \text{or} \quad \theta_1 = \pi + \theta_2 \tag{0.0.1}$$

Radius of curvature at θ

$$= \left|\frac{[a^2(1-\cos\theta)^2 + a^2\sin^2\theta]^{3/2}}{a(1-\cos\theta)a\cos\theta - a^2\sin^2\theta}\right| = 2\sqrt{2}\,a\sqrt{1-\cos\theta}$$

$$\therefore \rho_1^2 + \rho_2^2 = 8a^2\,(1 - \cos\theta_1 + 1 - \cos\theta_2)$$
$$= 8a^2\,(2 - \cos(\pi + \theta_2) - \cos\theta_2) \quad \text{[by (1)]}$$
$$= 8a^2(2 + \cos\theta_2 - \cos\theta_2)$$
$$= 16a^2.$$

18. If the polar equation of a curve be $r = f(\theta)$, where $f(\theta)$ is an even function of θ, show that the curvature at the point $\theta = 0$ is

$$\frac{\{f(0) - f''(0)\}}{\{f(0)\}^2}.$$

Solution. As $f(\theta)$ is an even function of θ

$$\therefore f(-\theta) = f(\theta)$$

Differentiating w.r.t. θ

$$-f'(-\theta) = f'(\theta).$$

$\therefore$ At $\theta = 0$, $-f'(0) = f'(0)$.

$$\therefore 2f'(0) = 0 \Rightarrow f'(0) = 0.$$

$\therefore$ Curvature at $\theta = 0$ is

$$\left[\frac{r^2 + 2r_1^2 - rr_2}{(r^2 + r_1^2)^{3/2}}\right]_{\theta=0}, \quad r_1 = f'(\theta), \quad r_2 = f''(\theta)$$

$\therefore$ Curvature at $\theta = 0$ is

$$\frac{\{f(0)\}^2 - f(0)f''(0)}{\{f(0)\}^3} = \frac{f(0) - f''(0)}{\{f(0)\}^2}$$

19. The curve $r = ae^{\theta \cot \alpha}$ cuts any radius vector in the consecutive points $P_1, P_2, \ldots, P_n, P_{n+1}, \ldots$. If ρ_n denotes the radius of curvature at P_n prove that

$$\frac{1}{m-n} \log \left(\frac{\rho_m}{\rho_n}\right)$$

is constant for all integral values of m and n.

Solution. Let $\theta = \theta_1$ at P_1.

$$\therefore \theta = 2\pi + \theta_1 \text{ at } P_2$$

$$\ldots\ldots\ldots\ldots\ldots\ldots$$

$$\theta = 2(k-1)\pi + \theta_1 \text{ at } P_k \text{ etc.}$$

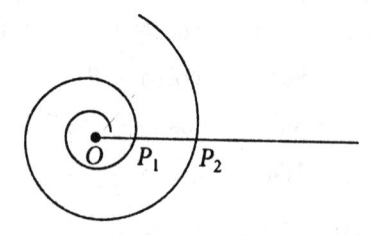

At θ

$$\rho = \frac{\left(r^2 + r_1^2\right)^{3/2}}{r^2 + 2r_1^2 - rr_2}, \quad r_1 = \frac{dr}{d\theta}, \quad r_2 = \frac{d^2r}{d\theta^2}$$

Now $r = e^{\theta \cot \alpha}$.

$$\therefore r_1 = \cot \alpha \, e^{\theta \cot \alpha} = r \cot \alpha$$

$$r_2 = r \cot^2 \alpha$$

$\therefore$ At θ

$$\rho = \frac{\left(r^2 + r^2 \cot^2 \alpha\right)^{3/2}}{r^2 + 2\cot^2 \alpha \, r^2 - r^2 \cot^2 \alpha} = \frac{r^3 \operatorname{cosec}^3 \alpha}{r^2 \operatorname{cosec}^2 \alpha}$$

$$= r \operatorname{cosec} \alpha = ae^{\theta \cot \alpha} \cdot \operatorname{cosec} \alpha$$

$$\therefore \rho_k = a \operatorname{cosec} \alpha \, e^{\{2(k-1)\pi + \theta_1\} \cot \alpha}$$

$$\therefore \frac{\rho_m}{\rho_n} = e^{\{2(m-1)\pi + \theta_1 - 2(n-1)\pi - \theta_1\} \cot \alpha} = e^{2(m-n)\pi \cot \alpha}$$

$$\therefore \log \frac{\rho_m}{\rho_n} = 2(m-n)\pi \cot \alpha$$

$$\therefore \frac{1}{m-n} \log \frac{\rho_m}{\rho_n} = 2\pi \cot \alpha$$

which is constant for integral values of m and n.

20. In the curve $r^n = a^n \cos n\theta$ verify that

$$\rho = \frac{r^2}{(n+1)p} = \frac{a^n}{(n+1)r^{n-1}}$$

with usual notations.

Solution. The equation of the curve is

$$r^n = a^n \cos n\theta \qquad (0.0.1)$$

$$\therefore n \log r = \log a^n + \log \cos n\theta.$$

Differentiating w.r.t. θ

$$\frac{n}{r} r_1 = -n \tan n\theta \qquad \therefore r_1 = -r \tan n\theta.$$

$$\therefore r_2 = -r_1 \tan n\theta - nr \sec^2 n\theta = r \tan^2 n\theta - nr \sec^2 n\theta$$
$$= r(1-n)\tan^2 n\theta - nr = r\tan^2 n\theta - nr\sec^2 n\theta$$

$$\therefore \rho = \frac{(r^2 + r_1^2)^{3/2}}{r^2 + 2r_1^2 - rr_2} = \frac{(r^2 + r^2 \tan^2 n\theta)^{3/2}}{r^2 + 2r^2 \tan^2 n\theta - r^2(\tan^2 n\theta - n\sec^2 n\theta)}$$

$$= \frac{r^3 \sec^3 n\theta}{r^2 \sec^2 n\theta + nr^2 \sec^2 n\theta} = \frac{r \sec n\theta}{n+1} = \frac{r \cdot \frac{a^n}{r^n}}{n+1} \quad \text{[by (1)]}$$

$$= \frac{a^n}{(n+1)r^{n-1}}.$$

Again

$$\tan \phi = r \frac{d\theta}{dr} = \frac{r}{r_1} = \frac{r}{-r \tan n\theta} = -\cot n\theta = \tan\left(\frac{\pi}{2} + n\theta\right)$$

$$\therefore \phi = \frac{\pi}{2} + n\theta$$

$$\therefore p = r \sin \phi = r \cos n\theta = r \cdot \frac{r^n}{a^n} \quad \text{[by (1)]}$$

$$= \frac{r^{n+1}}{a^n}$$

$$\therefore \frac{r^2}{(n+1)p} = \frac{a^n r^2}{(n+1)r^{n+1}} = \frac{a^n}{(n+1)r^{n-1}}$$

$$\therefore \rho = \frac{a^n}{(n+1)r^{n-1}} = \frac{r^2}{(n+1)p}.$$

Asymptotes

21. Show that the points of the curve
$$2y^3 - 2x^2y - 4xy^2 + 4x^3 - 14xy + 6y^2 + 4x^2 + 6y + 1 = 0$$
and its asymptotes lie on the straight line $8x + 2y + 1 = 0$.

Solution. First we find the asymptotes and their combined equation. We have the equation of the curve
$$2y^3 - 2x^2y - 4xy^2 + 4x^3 - 14xy + 6y^2 + 4x^2 + 6y + 1 = 0$$
or $\quad 2(y-x)(y+x)(y-2x) - 14xy + 6y^2 + 4x^2 + 6y + 1 = 0 \qquad (0.0.1)$

Asymptote parallel to $y - x = 0$ is

$$y - x + \lim_{|x| \to \infty} \frac{-14xy + 6y^2 + 4x^2 + 6y + 1}{2(y+x)(y-2x)} = 0 \quad \left[\text{when } \lim_{|x| \to \infty} \frac{y}{x} = 1\right]$$

or $\quad y - x + \dfrac{-14 + 6 + 4}{2(1+1)(1-2)} = 0 \quad$ or $\quad y - x + 1 = 0.$

Similarly asymptotes parallel to $y + x = 0$ and $y - 2x = 0$ are respectively $y + x + 2 = 0$ and $y - 2x = 0$.

$\therefore$ The combined equation of the asymptotes is

$$(y - x + 1)(y + x + 6)(y - 2x) = 0$$
or $\quad y^3 - x^2y - 2xy^2 + 2x^3 - 7xy + 3y^2 + 2x^2 - 4x + 2y = 0 \qquad (0.0.2)$

The common points of the curve (1) and the asymptotes (2) lies on
$$2y^3 - 2x^2y - 4xy^2 + 4x^3 - 14xy + 6y^2 + 4x^2 + 6y + 1$$
$$- 2(y^3 - x^2y - 2xy^2 + 2x^3 - 7xy + 3y^2 + 2x^2 - 4x + 2y) = 0$$
or $\quad 6y + 1 + 8x - 4y = 0$
or $\quad 8x + 2y + 1 = 0.$

Envelope

22. Find the envelope of the family of ellipses $\frac{x^2}{a^2} + \frac{y^2}{b^2} = 1$, where the parameters a, b are connected by $ab = c^2$.

Solution. For a fixed point (x, y) of the envelope we consider b as as function of a determined by $ab = c^2$.

APPENDIX: CHAPTERWISE IMPORTANT PROBLEMS

Differentiating $\frac{x^2}{a^2} + \frac{y^2}{b^2} = 1$, $ab = c^2$ w.r.t. a

$$-\frac{2x^2}{a^3} - \frac{2y^2}{b^3}\frac{db}{da} = 0 \quad \text{and} \quad b + a\frac{db}{da} = 0$$

Eliminating $\frac{db}{da}$ we get

$$\frac{x^2}{a^3} + \frac{y^2}{b^3}\left(-\frac{b}{a}\right) = 0 \quad \text{or} \quad \frac{x^2}{a^2} = \frac{y^2}{b^2} = \frac{\frac{x^2}{a^2} + \frac{y^2}{b^2}}{1+1}$$

$$\text{or} \quad \frac{x^2}{a^2} = \frac{y^2}{b^2} = \frac{1}{2} \quad \left[\because \frac{x^2}{a^2} + \frac{y^2}{b^2} = 1\right]$$

$$\therefore \ a^2 = 2x^2, \quad b^2 = 2y^2$$

$$\therefore \ ab = c^2 \Rightarrow \sqrt{2}x \cdot \sqrt{2}y = c^2 \quad \text{or} \quad 2xy = c^2$$

which is the required envelope.

AN INTRODUCTION TO ANALYSIS

(Differential Calculus—Part I)

Miscellaneous Revision Exercises

Miscellaneous Revision Exercises

Miscellaneous Revision Exercises
[Problem based on Chapters (14–21)]

Students are strongly advised to work out the following problems. The collection covers a thorough revision of important problems on Tangents and Normals, Curvature and Evolute, Asymptotes, Concavity and Points of inflexion, Envelopes, Singular Points etc.

1. Show that the pedal equation of the ellipse $x^2/4 + y^2/2 = 1$ with respect to the centre is $8/p^2 = 6 - r^2$. More generally for the ellipse $x^2/a^2 + y^2/b^2 = 1$ with respect to the centre, the pedal equation is $a^2 b^2/p^2 = a^2 + b^2 - r^2$.

2. Show that at any point of the curve
$$x^{m+n} = k^{m-n} y^{2n} \quad (k > 0, m, n \text{ are positive integers})$$
mth power of the subtangent varies as the nth power of the subnormal.

3. Verify that $y = \pm x$ are the tangents at the origin of the curve $(x^2 + y^2)^2 = 4(x^2 - y^2)$.

   [See Chapter 15. art. 15.5.]

4. Show that the curves $r^n = a^n \sec(n\theta + \alpha)$ and $r^n = b^n \sec(n\theta + \beta)$ intersect at an angle which is independent of a and b.

5. (a) If the two curves $ax^2 + by^2 = 1$ and $cx^2 + dy^2 = 1$ cut each other orthogonally $(a, b, c, d$ constants — none is zero), then prove that $\dfrac{1}{a} + \dfrac{1}{d} = \dfrac{1}{b} + \dfrac{1}{c}$.
   [See Chapter 15. Ex. 15.6.4.]

   (b) Show that if $lx + my + my = n$ touches the curve $\left(\dfrac{x}{a}\right)^p + \left(\dfrac{y}{b}\right)^p = 1$ then $(al)^{p/p-1} + (bm)^{p/p-1} = n^{p/p-1}$.

A.I.T.D.C.[P-I]—78

6. Find the pedal equation *with regard to the origin* of the curve whose parametric equations are $x = ae^t(\sin t - \cos t)$, $y = ae^t(\sin t + \cos t)$.
   [Ans. $r^2 = 2p^2$]

7. Show that the locus of the foot of the perpendicular drawn from the origin to any tangent to the curve $x^{2/3} + y^{2/3} = a^{2/3}$ is $r = \pm a \sin \theta \cos \theta$.

8. Show that the two curves $r^2 = a^2 \cos 2\theta$ and $r = a(1 + \cos \theta)$ intersect at an angle $\theta = 3\sin^{-1}\left(\frac{3}{4}\right)^{1/4}$
   [Follow Ex. 4 above]

9. Show that $p^2 = ar$ is the pedal equation of the parabola $r = 2a/(1 + \cos \theta)$.

10. Find the Cartesian equation of the pedal of the curve
$$\sqrt{\frac{x}{2}} + \sqrt{\frac{y}{3}} = 1$$
with respect to the origin.

   [Ans. $(x^2 + y^2)(2x + 3y) = 6xy$]

11. Prove that the locus of the extremity of the polar subnormal of the equiangular spiral $r = ae^{m\theta}$ is another equiangular spiral.

12. Find k (which is not zero) for which the line $x = k$ intersects the curve $xy^2 = (x+y)^2$ orthogonally.

   [Ans. $k = 4$]

13. Show that the tangent to the curve $x^3 + y^3 = 3axy$ at the point other than the origin where it meets the parabola $y^2 = ax$ is parallel to the Y-axis.
   [See Chapter 15. Ex. 15.6.5.]

14. Show that the pedal equation of the ellipse $x^2/a^2 + y^2/b^2 = 1$ with respect to the focus is $b^2/p^2 = 2a/r - 1$.
   [See Chapter 15. Ex. 15.11.6.]

15. Show that for the curve $ay^2 = (x+1)^3$, the subnormal varies as the square of the subtangent.

16. Prove that the pedal of the rectangular hyperbola $x^2 - y^2 = c^2$ with respect to the centre is $(x^2 + y^2)^2 = c^2(x^2 - y^2)$.
   [See Chapter 15. Ex. 15.12.1.]

17. Obtain the equation of the tangent at any point 't' of the astroid $x = a\cos^3 t, y = a\sin^3 t$. Hence prove that the portion of the tangent at 't' intercepted between the axes is of constant length. Deduce that the locus of intersection of tangents at *right angles* to one another is $2r^2 = a^2 \cos^2 2\theta$ or in cartesian form $2(x^2 + y^2)^3 = a^2(x^2 - y^2)^2$.

18. Show that the length of the perpendicular from the foot of the ordinate on any tangent to the catenary $y = c\cosh x/c$ is constant. [See Chapter 15. Ex. 15.6.3.]

19. Find the equations of tangent and normal at $\theta = \pi/2$ to the curve $x = a(\theta + \sin\theta)$, $y = a(1 + \cos\theta)$.

    [Tangent: $x + y - \frac{1}{2}a\pi - 2a = 0$. Normal: $x - y - \frac{1}{2}a\pi = 0$.]

20. Show that the pedal equation of the curve $c^2(x^2 + y^2) = x^2 y^2$ is $1/p^2 + 3/r^2 = 1/c^2$.

21. Find the angle of intersection of the cardiodes $r = a(1+\cos\theta)$ and $r = a(1-\cos\theta)$.

    [$\phi_1 - \phi_2 = \pi/2$]

22. If two tangents of the cardiode $r = a(1+\cos\theta)$ are parallel, show that the line joining the points of contact subtend an angle $2\pi/3$ at the pole.

23. Show that in the case of the curve $r = a(\sec\theta + \tan\theta)$ if a radius vector OPP' be drawn cutting the curve at P and P' and if the tangents at P and P' meet at T, then $TP = TP'$.

24. Verify that the equations of the tangents to the curve $x^2(y+1) = y^2(x-4)$ parallel to the line $y = x$ are $y = x \pm 4$.

25. The tangent at any point θ of the curve $x = a(\theta + \sin\theta)$, $y = a(1 - \cos\theta)$ is normal to the curve at the point where it meets the next span on the right, prove that $\cot\theta/2 = \pi/2$.

(Curvature)

26. Show that for the ellipse $x^2/a^2 + y^2/b^2 = 1$, the radius of curvature at an extremity of the major axis is equal to half of the latus rectum.

27. Prove that $(a\tan\psi + 1)\sec\psi$ is the radius of curvature at (s, ψ) of the curve $s = a\sec\psi + \log(\sec\psi + \tan\psi)$.

28. For the catenary $y = c\cosh\frac{x}{c}$, find the radius of curvature at any point (x, y). Also show that the length of the normal at (x, y) to the curve intercepted between the curve and X-axis is equal to the radius of curvature.

    [See that $\rho = y^2/c = c\cos h^2 x/c$; length of normal $= y\sqrt{y_1^2 + 1} = \frac{y^2}{c} = \rho$]

29. Find the radius of curvature ρ of the curve $y = xe^{-x}$ at the point where y is maximum.

Prove that for the curve $x = a(\theta + \sin\theta)$, $y = a(1 - \cos\theta)$ at $\theta = 0$, $\rho = 4a\cos\theta/2$.

[See Chapter 16. Ex. 16.3.4.]

30. Find the radius of curvature at $(0, 0)$ of the curve $x^2 + y^2 + 6x + 8y = 0$.

[Ans. 5]

31. Find the radius of curvature of the parabola $y^2 = 4ax$, at the vertex.

[use $\rho = (1 + x_1^2)^{3/2}/x_2$, $x_1 = dx/dy$, $x_2 = d^2x/dy^2$; $\rho = 2a$.]

32. Show that the two radii of curvature of the curve $y^2 = x(x+a)/(a-x)$ at the origin are the same in magnitude. ($\rho = a\sqrt{2}$) [See Chapter 16. Ex. 16.5.1.]

33. If ρ_1 and ρ_2 be the radii of curvature at the two ends of a focal chord of the parabola $y^2 = 4ax$. Prove that $\rho_1^{-2/3} + \rho_2^{-2/3} = (2a)^{-2/3}$.

34. If ρ_1 and ρ_2 be the radii of curvature at the ends of two conjugate diameters of the ellipse $x^2/a^2 + y^2/b^2 = 1$, prove that $\left(\rho_1^{2/3} + \rho_2^{2/3}\right)(ab)^{2/3} = a^2 + b^2$.

35. If ρ_1 and ρ_2 be the radii of curvature at the two extremities of any chord of the cardiode $r = a(1 + \cos\theta)$ passing through the pole, prove that $\rho_1^2 + \rho_2^2 = \frac{16}{9}a^2$.

36. Find the equation of the circle of curvature at $(3, 1)$ on the curve $y = x^2 - 6x + 10$.

[Find $\bar{x}, \bar{y}$ and ρ. Use $(x - \bar{x})^2 + (y - \bar{y})^2 = \rho^2$.]

37. Show that the radius of curvature of a logarithmic spiral $r = ae^{\theta \cot\alpha}$ is proportional to the radius vector. [See Chapter 16. Ex. 16.3.1.]

38. Prove that the evolute of the cycloid $x = a(\theta - \sin\theta)$, $y = a(1 - \cos\theta)$ is another cycloid. What is evolute of $x^{2/3} + y^{2/3} = 4$?

39. Find the curvature at the origin of the curve $y = x + x^2 + 2xy + y^2$.

[$(y_1)_0 = 1$ and $(y_2)_0 = 8$; $k = 1/\rho = y_2/(1 + y_1^2)^{3/2} = 8/2^{3/2} = \sqrt{8}$.]

40. A line is drawn through the origin meeting the cardiode $r = a(1 - \cos\theta)$ at the points P, Q and the normals at P, Q meet at C. Show that the radii of curvature at P and Q are proportional to PC and QC.

MISCELLANEOUS REVISION EXERCISES

41. Find the co-ordinates of the centre of curvature at $(\bar{x}, \bar{y})$ of the parabola $y^2 = 4ax$. Hence obtain its evolute.

[Ans. $27ay^2 = 4(x - 2a)^3$ is the required evolute.]

42. The evolute of the astroid $x = a\cos^3 t, y = a\sin^3 t$ is the curve $(x+y)^{2/3} + (x-y)^{2/3} = 2a^{2/3}$. Establish. [See Chapter 16. Ex. 16.7.2.]

43. Find the radius of curvature at any point P of the catenary $y = c\cosh\frac{x}{c}$ and show that $PC = PG$, where C is the centre of curvature at P and G is the point of intersection of the normal at P with the X-axis.

44. Find the length of the arc of the evolute of the parabola $y^2 = 4ax$ which is intercepted between the parabola.

[Ans. $4a\left(3\sqrt{3} - 1\right)$.] [See Exercises **XVIII(B)**. No. 12.]

45. The circle of curvature at any point P of the lemniscate $r^2 = a^2\cos 2\theta$ meets the radius vector OP at A, show that $OP : AP = 1 : 2, O$ being the pole.

(Envelopes)

46. Find the envelopes of the family of straight lines:

(a) $y = mx + a\sqrt{1 + m^2}$;

(b) $y = mx + a/m$. (m being the parameter in each case)

[Ans. (a) $x^2 + y^2 = a^2$; (b) $y^2 = 4ax$]

47. From any point P on the parabola $y^2 = 16x$, perpendiculars $\overline{PM}$ and $\overline{PN}$ are drawn on the co-ordinate axes. Find the envelope of the straight lines $\overline{MN}$ thus formed.

[Ans. $y^2 = -64x$]

48. (a) Find the envelope of the family of straight lines $\frac{x}{a} + \frac{y}{b} = 1$, where the parameters a and b are connected by $a^2 + b^2 = c^2$, c being a fixed constant.
[See Chapter 18. Ex. 18.5.2.]

(b) If the relation between a and b is $a^3 + b^3 = c^3$, the envelope will be $x^{3/4} + y^{3/4} = c^{3/4}$.

49. Find the envelope of the family of straight lines drawn at right angles to the radii vectors of the cardiode $r = a(1 + \cos\theta)$ through their extremities.

50. Prove that the envelope of the normals to the curve $x^{2/3} + y^{2/3} = a^{2/3}$ is given by $(x+y)^{2/3} + (x-y)^{2/3} = 2a^{2/3}$.

   [Another concept of the evolute]

51. Given $x^{2/3} + y^{2/3} = c^{2/3}$ is the envelope of $x/a + y/b = 1$, prove that $a^2 + b^2 = c^2$.

   [See Exercises **XVIII(A)**. Section [C] No. **2**.]

52. Find the envelope of the curves $x^2/a^2 + y^2/b^2 = 1$, where the parameters a and b are connected by $a + b = c$ (c being the fixed constant). [See Ex. **18.5.1**.]

   [Ans. $x^{2/3} + y^{2/3} = c^{2/3}$.]

53. Find the envelope of the circles whose centres lie on the rectangular hyperbola $xy = c^2$ and which pass through its centre. [See Ex. **18.5.3**.]

   [Ans. $(x^2 + y^2)^2 = 16c^2 xy$.]

54. Find the envelope of the family of circles $(x-a)^2 + y^2 = a^2/2$. [See Ex. **18.3.1**.]

55. Prove that the envelope of the family of circles passing through the origin and having centres on the hyperbola $x^2 - y^2 = c^2$ is $(x^2 + y^2)^2 = 4c^2 (x^2 - y^2)$.

56. Find the envelope of the family of straight lines $x/a + y/b = 1$ where the parameters a, b are connected by the relation $a^2 b^3 = c^5$, c being a fixed constant. If the relation is $a^m b^n = c^{m+n}$ then the envelope is $(m+n)^{m+n} x^m y^n = m^m n^n c^{m+n}$.]

   [Ans. $x^2 y^3 = \frac{108}{3125} c^5$.]

57. Given that the astroid $x^{2/3} + y^{2/3} = c^{2/3}$ is the envelope of the family of ellipse $x^2/a^2 + y^2/b^2 = 1$ (a, b are parameters), show that $a + b = c$.

58. Show that the envelope of the ellipses $(x-\alpha)^2/a^2 + (y-\beta)^2/b^2 = 1$, where the parameters α and β are connected by the relation $\alpha^2/a^2 + \beta^2/b^2 = 1$ is the ellipse $x^2/a^2 + y^2/b^2 = 4$.

59. Find the envelope of the polars of points on the ellipse $x^2/a^2 + y^2/b^2 = 1$ with respect to the ellipse $x^2/h^2 + y^2/k^2 = 1$. [See Exercise **XVIII(A)** Section [C]. No. **11**.]

   [Ans. $a^2 x^2/h^4 + b^2 y^2/k^4 = 1$ is the required envelope.]

60. Considering the evolute of a curve as the envelope of its normals, show that $(ax)^{2/3} + (by)^{2/3} = (a^2 - b^2)^{2/3}$ is the evolute of the ellipse $x^2/a^2 + y^2/b^2 = 1$.

MISCELLANEOUS REVISION EXERCISES

61. Show that the envelope of straight lines drawn at right angles to the radii vectors of the cardiode $r = a(1 + \cos\theta)$ through their extremities is $r = 2a\cos\theta$ (circle).

[See No. **49** above.]

62. Prove that the envelope of a circle whose centre lies on the parabola $y^2 = 4ax$ and which passes through its vertex is the curve $y^2(2a + x) + x^3 = 0$.

63. Show that the envelope of the straight line joining the extremities of a pair of semi-conjugate diamters of the ellipse $x^2/a^2 + y^2/b^2 = 1$ is the ellipse $x^2/a^2 + y^2/b^2 = 1/2$.

64. Show that the radius of curvature at a point of the evolute of the curve $r^n = a^n \cos n\theta$, corresponding to the point (r, θ) is $\frac{n-1}{(n+1)^2} r \sec n\theta \tan n\theta$.

65. Show that the pedal equation of the envelope of the line $x\cos 2\alpha + y\sin 2\alpha = 2a\cos\alpha$ (where α is the parameter) is $p^2 = \frac{4}{3}(r^2 - a^2)$.

(Rectilinear Asymptotes)

66. Find the asymptotes of the curves given below:

(a) $x^3 - 2y^3 + xy(2x - y) + y(x - y) + 1 = 0$.

(b) $x^3 + 3x^2y - 4y^3 - x + y + 3 = 0$.

(c) $y^3 - xy^2 - x^2y + x^3 + x^2 - y^2 - 1 = 0$.

(d) $y^3 + x^2y + 2xy^2 - y + 1 = 0$.

(e) $(x^2 - y^2)(x + 2y + 1) + x + y + 1 = 0$.

(f) $x^3 + y^3 = 3axy$.

(g) $x^2(x - y)^2 - a^2(x^2 + y^2) = 0$.

(h) $y^2(x^2 - a^2) = x^2(x^2 - 4a^2)$.

(i) $x^3 - xy^2 - 2xy + 2x - y = 0$.

(j) $x^3 + 2xy^2 - 4xy^2 - 8y^3 - 4x + 8y - 10 = 0$.

[Ans. (a) $x - y = 0$, $x + y + 1 = 0$, $x + 2y = 1$; (b) $y = x$, $2y + x = \pm 1$; (c) $y = \pm x$, $y = x + 1$; (d) $y = 0$, $x + y = \pm 1$; (e) $x - y = 0$, $x + y = 0$, $x + 2y + 1 = 0$; (f) $x + y + a = 0$; (g) $x = \pm a$, $y = x \pm \sqrt{2}\,a$; (h) $x \pm a, y \pm x = 0$; (i) $x = 0, x + y \pm 1 = 0$; (j) $x + 2y = \pm 2$, $x - 2y = 0$.]

67. Show that the points of intersection of the curve
$$2y^3 - 2x^2y - 4xy^2 + 4x^3 - 14xy + 6y^2 + 4x^2 + 6y + 1 = 0$$
and its asymptotes lie on the line $8x + 2y + 1 = 0$.

68. Show that the eight points of the curve
$$x^4 - 5x^2y^2 + 4y^4 + x^2 - y^2 + x + y + 1 = 0$$
and its asymptotes lie on a rectangular hyperbola.

69. If a curve passes through $(0,0)$, $(1,0)$ and $(0,1)$ and has the same asymptotes as the curve
$$x^3 - 6x^2y + 11xy^2 - 6y^3 + x + y + 1 = 0,$$
show that its equation is $x^3 - 6x^2y + 11xy^2 - 6y^3 - x + 6y = 0$.

70. Define rectilinear asymptote of a plane curve. Find the asymptotes, if any, of the following curve:
$$x = \frac{a\cos\theta}{1/2 - \cos\theta}, \quad y = \frac{a\sin\theta}{1/2 - \cos\theta}.$$
[Asymptotes are $y \pm \sqrt{3}\,x = \mp 4a/\sqrt{3}$, or $\sqrt{3}\,r\left(\sin\theta \pm \sqrt{3}\cos\theta\right) \pm 4a = 0$.]
[Polar equation of the given curve is $r = a/(1/2 - \cos\theta)$.]

71. Find the asymptotes of the curve $r^2\cos 2\theta = a^2$ $(a > 0)$.

72. Find the asymptotes of the curve $x = t^2/1 + t^3$, $y = (t^2 + 2)/(1 + t)$; where t is the parameter. [See Ex. **17.5.3.**]

73. Show that the curve
$$y = \begin{cases} \sqrt{1+x^2}\,\sin(1/x), & x \neq 0, \\ 0, & x = 0; \end{cases}$$
has no asymptote parallel to the Y-axis and its only asymptotes are $y = \pm 1$.
[See Ex. **17.4.6**.]

74. (a) Find the asymptotes of the curve $(a + b + x)y = c(b + x)$.
(b) Find the asymptotes, if any, parallel to Y-axis of the curve
$$y = x\sqrt{(a - x)/(a + x)}$$

MISCELLANEOUS REVISION EXERCISES

75. Find the asymptotes of $y = xe^{1/x}$ and $y = \log \sec x/a$.

[See Ex. **17.4.3** for the first case and Sum No. ల of Exercises **XVII(A)** the the second.]

76. Find the asymptotes parallel to X-axis of the curve:

a) $y = \dfrac{x^2}{x^2+1}$;   b) $y = \dfrac{4x^2+4x-3}{x^2-4x+3}$.

[Ans. (a) $y = 1$; (b) $y = 4$.]

77. Show that $y = x + 3/2$ is the oblique asymptote of the parametric curve $x = \dfrac{t^2+1}{t^2-1}$ and $y = \dfrac{t^2}{t-1}$. Show also $y = -1/2$ is an asymptote parallel to X-axis. [See Ex. **17.5.4**.]

78. Show that $x = a$ is an asymptote of the polar curve $r = a(\sec\theta + \tan\theta)$.

79. Find the equation of the tangent to the curve $x^3 + y^3 = 3ax^2$ which is parallel to its asymptote.

80. Show that $y = mx + c$ is an asymptote of the curve $y = mx + c + f(x)$, where $\lim\limits_{x \to \infty} f(x) = 0$. Hence find the asymptotes of the curve $y = \dfrac{x^2+x+3}{x^2+5x+1}$.

[Ans. $y = x - 5, x = \tfrac{1}{2}(-5 \pm \sqrt{29})$]

Show that $y = x + 2$ is an asymptote of the curve $y = \dfrac{x^2+2x-1}{x}$.

(Concavity, Convexity and Point of Inflexion)

Note: For Q 81–Q 85 See Ex. 19.4.6.

81. Show that $y = 3x^5 - 40x^3 + 3x - 20$ is concave upwards in $-2 < x < 0$ and $2 < x < \infty$ but convex upwards in $-\infty < x < -2$ and $0 < x < 2$ and $x = -2, 0, 2$ are the points of inflexion. Trace the curve.

82. Examine the curve $y = x^4 - 2x^3 + 1$, for concavity and convexity. Also determine its point of inflexion.

83. Find the ranges of values of x for which the curve $y = x^4 - 6x^3 + 12x^2 + 5x + 7$ is concave upwards or downwards. Also determine the points of inflexion.

[Concave upwards in $-\infty < x \le 1$, $2 \le x < +\infty$, Concave downwards in $1 \le x \le 2$. Point of inflexion: $x = 1, x = 2$.]

84. Show that the curve $(a^2 + x^2)y = a^2 x$ has three points of inflexion: $(\sqrt{3}\,a, \sqrt{3}\,a/4)$, $(0,0)$, $(-\sqrt{3}\,a, -\sqrt{3}\,a/4)$.

85. Prove that $(1,0)$ and $(e^2, 8)$ are two points of inflexion of the curve $y = (\log x)^3$.

86. Find points of inflexion on the curves:

(a) $x = a(2\theta - \sin\theta), y = a(2 - \cos\theta)$ $\qquad [a(4\pi n \pm 2\pi/3 \pm \sqrt{3}/2), 3a/2]$

(b) $y = xe^{-x^2}$. $\qquad$ [For $x = 0, \pm\sqrt{3}/2$]

(c) $r = a\theta^2/(\theta^2 - 1)$ [Ex. 19.5.2.] $\qquad [3a/2, \pm\sqrt{3}\,]$

(d) $r = ae^\theta/(1+\theta)$ $\qquad [a, 0]$

87. Show that the curve $re^\theta = a(1+\theta)$ has no point of inflexion.

88. Show that the points of inflexion of the curve $y^2 = (x-a)^2(x-b)$ lie on the line $3x + a = 4b$.

89. Show that for the curve $5^4 y = (x+5)^2(x^3 - 10)$ has points of inflexion for $x = -2, -4\pm\sqrt{18}$. Also obtain the equations of inflexional tangents. $(9x \mp 3\sqrt{3}\,y + 1 = 0)$

90. Show that $r = a\theta^n$ has points of inflexion if and only if n lies between 0 and -1 and they are given by $\theta = \pm\sqrt{-n(n+1)}$, or $r = a\{-n(n+1)\}^{n/2}$.

(Singular Points)

91. What is meant by a singular point of a curve? Examine whether the origin is a node or a cusp of the curve: $x^3 + y^3 = 3axy$.

[Ans. Origin is a Node; $x = 0, y = 0$ are the two nodal tangents at $(0, 0)$.]

92. Find the singular points and their nature in case of the following curves:

(a) $(2x+y)^2 - 6xy(2x+y) - 7x^3 = 0$. $\qquad$ [See Ex. 20.7.1.]

(b) $x^4 - 4x^2 y - 2xy + 4y^2 = 0$.

[Ans. (a) Origin is a single cusp of first species; (b) Origin is a double point (node).]

93. Prove that the point $(a, 0)$ of the curve $ay^2 = (x-a)^2(x-b)$ is a conjugate point, a node or a cusp according as $a < b, a > b$ or $a = b$.

94. Show that the origin is a node, a cusp or an isolated point on a curve $y^2 = ax^2 + bx^3$ according as $a > 0, a = 0$, or $a < 0$.

95. In the curve $x^3 + y^3 = x^2$, show that there is a cusp of first species at $(0, 0)$.

MISCELLANEOUS REVISION EXERCISES

96. Show that the origin is the only double point on the curve $x^3 - y^3 = ay^2 (a \neq 0)$ and it is a single cusp of first species.

97. Show that the curve $(x+y)^3 - \sqrt{2}\,(y-x+2)^2 = 0$ has a single cusp of the first species at $(1, -1)$.

98. Show that the curve $by^2 = x^3 \sin^2 \frac{x}{a}$ has a cusp at $(0, 0)$ and an infinite series of nodes lying at equal distances from each other.

99. (a) Find the singular points of the curve $x^2 - 2axy + y^3 = 0$.

(b) Examine the nature of the origin of the curve $(2x+y)^2 - 6xy(2x+y) - 7x^3 = 0$.
 [See Chapter 20. Illustrative, Example 20.7.1.]

100. (a) Determine the existence and nature of the double points of the curve $(x-3)^2 = y(y-2)^2$. [See Chapter 20, Illustrative Example 20.4.1.]

(b) Determine the position and nature of the multiple points of the curve $x^3 - y^2 - 7x^2 + 4y + 15x - 13 = 0$.

Also find the tangents at the multiple points, if any.
 [See Chapter 20, Illustrative Example 20.5.1]

AN INTRODUCTION TO ANALYSIS

(Differential Calculus—Part I)

Hints/Solutions of Miscellaneous Revision Exercises

Hints/Solutions of Miscellaneous Revision Exercises

1. Second Part:

$$x^2/a^2 + y^2/b^2 = 1 \qquad \ldots \text{(i)}$$

has the tangent

$$Xx/a^2 + Yy/b^2 = 1. \qquad \ldots \text{(ii)}$$

Also

$$p = \frac{1}{\sqrt{x^2/a^4 + y^2/b^4}}. \qquad \ldots \text{(iii)}$$

$$x^2 + y^2 = r^2. \qquad \ldots \text{(iv)}$$

From (i), (iii), (iv), eliminate x and y and obtain the required result: solve for x^2 and y^2 from (i) and (iv). Put those values in (iii).

First Part: A special case of second part: put $a^2 = 4$, $b^2 = 2$.

2. First obtain $y_1 = \dfrac{m+n}{2n} \dfrac{y}{x}$ (using logarithmic differentiation).

Verify: $\dfrac{m\text{th power of cartesian subtangent}}{n\text{th power of cartesian subnormal}} = \dfrac{(y/y_1)^m}{(yy_1)^n} = $ constant; hence etc.

4. Obtain $\dfrac{1}{r}\dfrac{dr}{d\theta} = \tan(n\theta + \alpha)$ [using logarithmic differentiation]

i.e., $\cot\phi_1 = \tan(n\theta + \alpha) \Rightarrow \phi_1 = \dfrac{\pi}{2} - (n\theta + \alpha)$. Similarly $\phi_2 = \dfrac{\pi}{2} - (n\theta + \beta)$.

$\therefore \phi_1 \sim \phi_2$ is independent of a or b.

5(b). Compare $\dfrac{Xx^{p-1}}{a^p} + \dfrac{Yy^{p-1}}{b^p} = 1$ with $lX + mY = n$.

Obtain $x = \left(\dfrac{a^p l}{n}\right)^{1/(p-1)}$, $y = \left(\dfrac{b^p m}{n}\right)^{1/(p-1)}$

Since (x,y) lies on $(x/a)^p + (y/b)^p = 1$. The required condition follows.

6. Tangent at the point t is $Y\sin t - X\cos t = ae^t$ so that $p = \dfrac{ae^t}{\sqrt{\cos^2 t + \sin^2 t}} = ae^t$.

Also $x^2 + y^2 = r^2$. Put x and y in terms of t. Then eliminate t, and obtain $r^2 = 2p^2$.

7. Write the given curve in parametric curve: $x = a\cos^3 t$, $y = a\sin^3 t$.

Tangent at t is $\dfrac{X}{a\cos t} + \dfrac{Y}{a\sin t} = 1$. Let (α, β) be the foot of the perpendicular from the origin on the tangent at t. Then $\dfrac{\beta}{\alpha} \cdot \dfrac{-a\sin t}{a\cot t} = -1$.

Also $\dfrac{\alpha}{a\cos t} + \dfrac{\beta}{a\sin t} = 1$. Obtain $\sin t = \dfrac{\alpha}{\pm\sqrt{\alpha^2+\beta^2}}$, $\cos t = \dfrac{\beta}{\pm\sqrt{\alpha^2+\beta^2}}$.

Putting these values in $\dfrac{\alpha}{a\cos t} + \dfrac{\beta}{a\sin t} = 1$, the required locus is $\pm(x^2+y^2)^{3/2} = axy$.
In polar co-ordinates this locus because $r = \pm a\sin\theta\cos\theta$.

9. Given curve $r = 2a/(1+\cos\theta)$ or, $r\cos^2\theta/2 = a$. Obtain $\phi = 90° - \theta/2$.
Again $p = r\sin\phi = r\cos(\theta/2)$ or $p^2 = r^2\cos^2(\theta/2) = r^2\dfrac{a}{r}$ i.e., $p^2 = ar$.

Alternatively. Use the formula $\dfrac{1}{p^2} = \dfrac{1}{r^2} + \dfrac{1}{r^4}\left(\dfrac{dr}{d\theta}\right)^2$. From the given curve obtain $\dfrac{dr}{d\theta}$ and then eliminate θ.

10. Compare Ex. **15.12.1** (Chapter 15).

First, find the condition that the line $x\cos\alpha + y\sin\alpha = p$ may touch the given curve
$$\dfrac{1}{\sqrt{2}}\sqrt{x} + \dfrac{1}{\sqrt{3}}\sqrt{y} = 1. \text{ See Ex. 5(b) given above.}$$
The condition will be $p(3\sin\alpha + 2\cos\alpha) = 6\sin\alpha\cos\alpha$.

Hence the polar equation of the pedal of the curve is
$$r(3\sin\theta + 2\cos\theta) = 6\sin\theta\cos\theta \text{ (replace } p \text{ by } r \text{ and } \alpha \text{ by } \theta).$$
Multiply both sides by r^2. Change $r\cos\theta$ to x, $r\sin\theta$ to y, the required equation becomes $(x^2+y^2)(2x+3y) = 6xy$.

11. Ref. art. 15.9, Fig 15.7. The foot of the subnormal at P is G (say) where $R = OG = dr/d\theta = ae^{m\theta}\cdot m$. To find the locus of $G = (R,\alpha)$, where α is the angle which OG makes with the initial line OX (Not shown in Fig **15.7**). Clearly $\alpha = \pi/2 + \theta$ or $\theta = \alpha - \pi/2$.

$\therefore R = ame^{m\theta} = ame^{m(\alpha-\pi/2)} = ame^{-m\pi/2}\cdot e^{m\alpha} = Ae^{m\alpha}$.

Writing (R,α) in terms of current co-ordinates (r,θ) we obtain $r = Ae^{m\theta}$, another equiangular spiral.

12. The given curve $xy^2 = (x+y)^2$ cuts the line $x = k$ (which is parallel to y-axis), orthogonally only if the tangent at the point of intersection is parallel to the axis of x,

HINTS/SOLUTIONS OF MISCELLANEOUS REVISION EXERCISES

i.e., $\dfrac{dy}{dx} = 0$ there. But $\dfrac{dy}{dx} = \dfrac{2x + 2y - y^2}{2xy - 2x - 2y}$.

∴ At the point of intersection,

$$2x + 2y - y^2 = 0; \text{ where } x = k \text{ and } xy^2 = (x+y)^2.$$

or, $2k + 2y - y^2 = 0$, i.e., $y^2 - 2y - 2k = 0$ ... (i)

and $ky^2 = (k+y)^2$, whence $y = \pm(k+y)/\sqrt{k}$.

Putting this value of y in (i) we get $k = 4$ (if we assume $k \neq 0$).

13. The point of intersection of $x^3 + y^3 = 3axy$ and $y^2 = ax$ (other than the origin) is $(2^{2/3}a, 2^{1/3}a)$ obtained by solving the two equations. Again from $x^3 + y^3 = 3axy$ we obtain $\dfrac{dx}{dy} = \dfrac{y^2 - ax}{ay^2 - x}$ which becomes zero at the point $(2^{2/3}a, 2^{1/3}a)$, i.e., the tangent at that point must be parallel to y-axis. See **Ex. 15.6.5** (Chapter 15).

15. From $ay^2 = (x+1)^3$, obtain subtangent

$$= \dfrac{y}{y_1} = \dfrac{y^2}{(3/2a)(x+1)^2} = \dfrac{2ay^2}{3(x+1)^2} = \dfrac{2(x+1)^3}{3(x+1)^2} = \dfrac{2}{3}(x+1).$$

Also subnormal $yy_1 = \dfrac{3}{2a}(x+1)^2$.

Verify $\dfrac{\text{Subnormal}}{(\text{subtangent})^2} = \dfrac{27}{8a}$ = constant; hence etc.

17. Tangent at the point t is $y \cos t + x \sin t = a \sin t \cos t$ ... (i)

Intercepts on X-axis $= a \cos t$, on Y-axis $= a \sin t$.

Hence the portion of the tangent at any point t intercepted between the axes $= a$.

Tangent at the point t_1 is $y \cos t_1 + x \sin t_1 = a \sin t_1 \cos t_1$. ... (ii)

If the two tangents are at right angles, then $\dfrac{-\sin t}{\cos t} \cdot \dfrac{-\sin t_1}{\cos t_1} = -1$.

This gives $\cos(t_1 - t) = 0$, or, $t_1 - t = \pi/2$ or $t_1 = \pi/2 + t$.

Hence (ii) becomes $y \sin t - x \cos t = a \sin t \cos t$. ... (iii)

To find the locus of intersection of tangents (i) and (iii) we are to eliminate t between them.

Write (i) and (iii) in the forms:

$$\left. \begin{array}{l} y \csc t + x \sec t - a = 0 \\ \text{and } y \sec t - x \csc t - a = 0 \end{array} \right\}.$$

By cross multiplication we get $\csc t = \dfrac{a(y-x)}{x^2 + y^2}$, $\sec t = \dfrac{a(x+y)}{x^2 + y^2}$ whence

A.I.T.D.C.[P-I]—79

follows
$$\frac{(x^2+y^2)^2}{a^2(y-x)^2} + \frac{(x^2+y^2)^2}{a^2(y+x)^2} = 1$$

or, $2(x^2+y^2)^3 = a^2(x^2-y^2)^2$.

Transforming into polar co-ordinates we get $2r^2 = a^2 \cos^2 2\theta$.

19. $\frac{dy}{dx} = \frac{dy/d\theta}{dx/d\theta} = \frac{-a \sin \theta}{a(1+\cos \theta)} = -\tan \frac{\theta}{2}$ which is $= -1$ at $\theta = \pi/2$.

Now write the equation of tangent and normal at $\left\{ a\left(\frac{\pi}{2}+1\right), a \right\}$ where $\theta = \pi/2$.

20. The given equation in polar form is $c^2 r^2 = (r^2 \cos^2 \theta)(r^2 \sin^2 \theta)$ or $r \sin 2\theta = 2c$ whence

$$\frac{dr}{d\theta} = -2r \cot 2\theta. \text{ Use } \frac{1}{p^2} = \frac{1}{r^2} + \frac{1}{r^4}\left(\frac{dr}{d\theta}\right)^2 \text{ and obtain } \frac{1}{p^2} = \frac{1}{c^2} - \frac{3}{r^2}.$$

21. $\log r = \log a + \log 2 \cos^2 \theta/2 = \log a + \log 2 + 2 \log \cos \theta/2$.

$$\therefore \frac{1}{r}\frac{dr}{d\theta} = \frac{2}{\cos \theta/2}(-\sin \theta/2) \cdot \frac{1}{2} = -\tan \theta/2 = \cot\left(\frac{\pi}{2}+\frac{\theta}{2}\right)$$

i.e., $\cot \phi_1 = \cot(\pi/2 + \theta/2) \Rightarrow \phi_1 = \frac{\pi}{2} + \frac{\theta}{2}$. Similarly, from $r = a(1-\cos \theta)$; we obtain $\phi_2 = \theta/2$ and hence $\phi_1 \sim \phi_2 = \pi/2$.

22. See Chapter 15. Ex. 15.7.1.

Equation of tangent to the curve $r = a(1+\cos \theta)$ at $\theta = 2\alpha$ is

$$r \cos(\theta - 3\alpha) = 2a \cos^3 \alpha \quad \left(\text{whose slope is } -\frac{\cos 3\alpha}{\sin 3\alpha}\right).$$

Parallel tangent at $\theta = 2\beta$, slope is $-\frac{\cos 3\beta}{\sin 3\beta}$. They are equal and this gives

$\sin 3\alpha \cos 3\beta - \cos 3\alpha \sin 3\beta = 0$ i.e., $\sin(3\alpha - 3\beta) = 0$ i.e., $\alpha - \beta = \pi/3$ ($\because \alpha \neq \beta$).

The angle subtended at the pole by the line joining their points of contact is $2(\alpha - \beta) = \frac{2\pi}{3}$.

24. A line parallel to $y = x$ is $y = x + c$. It will cut the curve at points whose abscissae are given by

$$x^2(x+c+1) = (x+c)^2(x-4) \text{ or } x^2(5-c) - xc(c-8) + 4c^2 = 0.$$

For tangency the two values of x must be equal and hence the discriminant $= 0$,

i.e., $c^2(c-8)^2 - 16c^2(5-c) = 0 \Rightarrow c^2 - 16 = 0$ or $c = \pm 4$.

26. The equation of the ellipse in the parametric form: $x = a \cos t, y = b \sin t$.

$$y_1 = -\frac{b}{a}\cot t, \quad y_2 = \frac{b}{a}\operatorname{cosec}^2 t\left\{-\frac{1}{a}\operatorname{cosec} t\right\} = -\frac{b}{a^2}\frac{1}{\sin^3 t}.$$

$$\therefore \rho = \frac{(1+y_1^2)^{3/2}}{y_2} = \frac{(a^2\sin^2 t + b^2\cos^2 t)^{3/2}}{ab}.$$

At an extremity of the major axis the point is $(a, 0)$ where $t = 0$.

$\therefore$ Radius of curvature at $t = 0$ is $\dfrac{(b^2)^{3/2}}{ab} = \dfrac{b^2}{a} = \dfrac{1}{2}$ latus rectum.

27. $\rho = \dfrac{ds}{d\psi} = a\sec\psi\tan\psi + \dfrac{1}{\sec\psi + \tan\psi}(\sec\psi\tan\psi + \sec^2\psi)$

$= a\sec\psi\tan\psi + \sec\psi = \sec\psi(a\tan\psi + 1).$

29. (a) $\dfrac{dy}{dx} = e^{-x} - xe^{-x}$. At an extreme point $\dfrac{dy}{dx} = 0$ i.e., $x = 1$.

$\dfrac{d^2y}{dx^2} = -2e^{-x} + xe^{-x}$ which is $= -\dfrac{1}{e}$ at $x = 1$. Hence $y = xe^{-x}$ has a max. at $x = 1$.

$\rho = \dfrac{(1+y_1^2)^{3/2}}{y_2}$. At $x = 1$, $y_1 = 0$ and $y_2 = \dfrac{1}{e}$. Therefore, $|\rho| = e$ at the point where y is maximum.

30. Either directly obtain $(y_1)_0$ and $(y_2)_0$ in $\rho = \dfrac{(1+y_1^2)^{3/2}}{y_2}$.

or Using Newton's approach $\rho = \dfrac{1}{2}\sqrt{6^2 + 8^2}\lim\limits_{\substack{x\to 0 \\ y\to 0}}\dfrac{x^2+y^2}{6x+8y} = 5\lim\limits_{\substack{x\to 0 \\ y\to 0}}\dfrac{-6x-8y}{6x+8y} = 5.$

33. Write the equation of the parabola in parametric form: $x = at^2$, $y = 2at$. If PSP' be the focal chord with $P(t_1)$ and $P'(t_2)$ at the two ends, then $t_1 t_2 = -1$ or $t_2 = -1/t_1$.

$$\frac{dy}{dx} = y_1 = \frac{1}{t_1} \quad \text{and} \quad \frac{d^2y}{dx^2} = -\frac{1}{2at_1^3}.$$

Therefore, ρ_1 at $P = \dfrac{(1+y_1^2)^{3/2}}{y_2} = -2a(1+t_1^2)^{3/2}$ and

ρ_2 at $P' = -2a(1+t_2^2)^{3/2} = -\dfrac{2a}{t_1^3}(1+t_1^2)^{3/2}$ $\left(\because t_2 = -\dfrac{1}{t_1}\right)$.

Considering the numerical magnitudes of ρ_1 and ρ_2 we easily see that

$$\rho_1^{-2/3} + \rho_2^{-2/3} = (2a)^{-2/3}.$$

34. See **Ex. 26** (given above).

$(ab)^{2/3}\rho_1^{2/3} = a^2\sin^2 t + b^2\cos^2 t$, at the end P of semi-conjugate diameter CP.

$\therefore (ab)^{2/3}\rho_2^{2/3} = a^2\cos^2 t + b^2\sin^2 t$, at the end D of the semi-conjugate diameter CD,

(since the eccentric angles at the end of two conjugate diameters differ by $\pi/2$). Adding, we get the required result.

35. $r = a(1+\cos\theta)$, $r' = -a\sin\theta$, $r'' = -a\cos\theta$.

Applying the usual formula at the point $P(\theta)$, $\rho_1 = \dfrac{(r^2 + r'^2)^{3/2}}{r^2 + 2r'^2 - rr''} = \dfrac{2^{3/2}a}{3}(1+\cos\theta)^{1/2}$.

At the other end P' of the focal chord, θ is replaced by $\theta + \pi$.

Hence $\rho_2 = \dfrac{2^{3/2}a}{a}(1-\cos\theta)^{1/2}$. Therefore, $\rho_1^2 + \rho_2^2 = \dfrac{8a^2}{9}(1+\cos\theta+1-\cos\theta) = \dfrac{16a^2}{9}$.

38. $\dfrac{dy}{dx} = \dfrac{dy/d\theta}{dx/d\theta} = \dfrac{a\sin\theta}{a(1-\cos\theta)} = \cot\dfrac{\theta}{2}$; $\dfrac{d^2y}{dx^2} = -\dfrac{1}{4a\sin^4\theta/2}$.

If $(\bar{x}, \bar{y})$ be the centre of curvature at the point θ,

$$\bar{x} = x - \dfrac{y_1}{y_2}(1+y_1^2) = a(\theta + \sin\theta)$$

$$\bar{y} = y + \dfrac{1+y_1^2}{y_2} = a(1-\cos\theta) + \dfrac{\operatorname{cosec}^2\theta/2}{-(1/4a)\operatorname{cosec}^4\theta/2} = -a(1-\cos\theta).$$

Locus of $(\bar{x}, \bar{y})$ i.e., evolute of the given cycloid is another cycloid, only situated differently.

41. We write the equation of the parabola in parametric form: $x = at^2$, $y = 2at$ so that $\dfrac{dy}{dx} = \dfrac{1}{t}$ and $\dfrac{d^2y}{dx^2} = -\dfrac{1}{2at^3}$, $y_1/y_2 = -2at^2$. Then

$$\bar{x} = x - \dfrac{y_1}{y_2}(1+y_1^2) = at^2 + 2at^2\left(1+\dfrac{1}{t^2}\right) = 3at^2 + 2a$$

$$\bar{y} = y + \dfrac{1+y_1^2}{y_2} = 2at + \dfrac{1+1/t^2}{-1/2at^3} = 2at - 2at^3\left(\dfrac{t^2+1}{t^2}\right) = -2at^3.$$

$$\therefore (\bar{x} - 2a)^3 = 27a^3t^6 \text{ and } \bar{y}^2 = 4a^2t^6.$$

Eliminating t, we get, $4(\bar{x} - 2a)^3 = 27a\bar{y}^2$.

Hence, locus of $(\bar{x}, \bar{y})$ i.e., evolute is as given namely, $27ay^2 = 4(x-2a)^3$.

43. See Fig. **21.19** (Chapter 21). In that figure the normal GP should be produced such that $PC = \rho$. Then C becomes the centre of curvature at P. Now $y = c\cosh\dfrac{x}{c}$ (given). Then $y_1 = \sinh\dfrac{x}{c}$, $y_2 = \dfrac{1}{c}\cosh\dfrac{x}{c}$, $\rho = \dfrac{(1+y_1^2)^{3/2}}{y_2} = c\cosh^2\dfrac{x}{c}$, so that $\rho = c \cdot \dfrac{y^2}{c^2} = \dfrac{y^2}{c} = PC$.

The equation of normal at $P(x,y)$ is $Y - y = -\dfrac{1}{\sinh x/c}(X-x)$. It cuts the x-axis at $X = OG = x + y\sinh\dfrac{x}{c}$ so that $NG = y\sinh\dfrac{x}{c}$.

Now $PG = \sqrt{NG^2 + PN^2} = \sqrt{y^2\sinh^2\dfrac{x}{c} + y^2} = y\cosh\dfrac{x}{c} = \dfrac{y^2}{c} = PC$ (Proved).

HINTS/SOLUTIONS OF MISCELLANEOUS REVISION EXERCISES   1249

46. (a) $(y - mx)^2 = a^2(1 + m^2)$ or $m^2(x^2 - a^2) - 2mxy + y^2 - a^2 = 0$ (m is the parameter). The envelope is obtained by considering a double root in this equation of m.

i.e., Discriminant $= (-2xy)^2 - 4(x^2 - a^2)(y^2 - a^2) = 0$ or $x^2 + y^2 = a^2$.

(b) $m^2x - my + a = 0$: condition for double root gives the envelope $y^2 - 4ax = 0$.

47. Let P be the point (α, β). Then M is $(\alpha, 0)$ and N is $(0, \beta)$ so that the equation of MN is $x/\alpha + y/\beta = 1$.   ... (i)

Also since $P(\alpha, \beta)$ lies on $y^2 = 16x$, we get $\beta^2 = 16\alpha$.   ... (ii)

Thus (i) becomes $\dfrac{x}{\beta^2/16} + \dfrac{y}{\beta} = 1$ or $\beta^2 - \beta y - 16x = 0$ ($\beta =$ parameter).

The envelope is the condition of a double root for β, i.e., $(-y)^2 - 4 \cdot (-16) = 0$ or $y^2 = -64x$.

49. Lines drawn through the extremity of a radius vector r making an angle θ with the initial line, at right angles to that radius vector is

$$x \cos \theta + y \sin \theta = r \quad \ldots \text{(i)}$$

where r, θ satisfy $r = a(1 + \cos \theta)$.   ... (ii)

$\therefore$ (i) gives $(x - a) \cos \theta + y \sin \theta = a$.

Differentiating with respect to θ, we get

$$-(x - a) \sin \theta + y \cos \theta = 0.$$

Eliminating $\sin \theta$ and $\cos \theta$ we get the envelope $(x - a)^2 + y^2 = a^2$, which is a circle through the pole.

55. Circles passing through the origin: $x^2 + y^2 + 2gx + 2fy = 0$ [centre is $(-g, -f)$].
Given that centre passes through $x^2 - y^2 = c$, hence $g^2 - f^2 = c^2$.

We eliminate $\dfrac{df}{dg}$ between $2x + 2y\dfrac{df}{dg} = 0$ and $2g - 2f\dfrac{df}{dg} = 0$,

i.e., $-\dfrac{x}{y} = \dfrac{g}{f}$. Taking $g = -kx$, $f = ky$ we obtain $k^2(x^2 - y^2) = c^2$.

The required envelope is $x^2 + y^2 - 2kx^2 + 2ky^2 = 0$.

or, $(x^2 + y^2)^2 = 4k^2(x^2 - y^2)^2 = 4c^2(x^2 - y^2)$. [$\because k^2(x^2 - y^2) = c^2$.]

56. $\dfrac{x}{a^2} + \dfrac{y}{b^2}\dfrac{db}{da} = 0$ and $\dfrac{2}{a} + \dfrac{3}{b}\dfrac{db}{da} = 0$ (take logarithmic differentiation of $a^2b^3 = c^5$)

$\therefore \dfrac{x/a^2}{2/a} = \dfrac{y/b^2}{3/b}$ or $\dfrac{x}{2a} = \dfrac{y}{3b}$ or $\dfrac{x/a}{2} = \dfrac{y/b}{3} = \dfrac{1}{5}$

whence $a = \dfrac{5x}{2}, b = \dfrac{5y}{3}$.

$\therefore$ From $a^2b^3 = c^5$, the required envelope is $\dfrac{5^2 x^2}{4} \cdot \dfrac{5^3 y^3}{27} = c^5$

or, $x^2 y^3 = \dfrac{108}{3125} c^5$.

57. From $x^{2/3} + y^{2/3} = c^{2/3}$, we obtain $\dfrac{dy}{dx} = -\dfrac{y^{1/3}}{x^{1/3}}$.

From $\dfrac{x^2}{a^2} + \dfrac{y^2}{b^2} = 1$, we obtain $\dfrac{dy}{dx} = -\dfrac{b^2 x}{a^2 y}$.

$\therefore \dfrac{y^{1/3}}{x^{1/3}} = \dfrac{b^2 x}{a^2 y}$ or, $\dfrac{y^{4/3}}{b^2} = \dfrac{x^{4/3}}{a^2}$ or, $\dfrac{y^{2/3}}{b} = \dfrac{x^{2/3}}{a} = \dfrac{c^{2/3}}{a+b}$.

Hence $x^{2/3} = \dfrac{ac^{2/3}}{a+b}$ or $x^2 = \dfrac{a^3 c^2}{(a+b)^3}$. Similarly, $y^2 = \dfrac{b^3 c^2}{(a+b)^3}$.

$\therefore \dfrac{x^2}{a^2} + \dfrac{y^2}{b^2} = 1$ gives $\dfrac{ac^2}{(a+b)^3} + \dfrac{bc^2}{(a+b)^3} = 1$ or $(a+b)^2 = c^2$

or, $a + b = c$ is the required relation.

58. Similar to **Q.56.** $\dfrac{d\beta}{d\alpha} = -\dfrac{x - \alpha b^2}{y - \beta a^2} = -\dfrac{\alpha b^2}{\beta a^2}$

i.e., $\dfrac{x - \alpha}{\alpha} = \dfrac{y - \beta}{\beta}$ or $\dfrac{x}{\alpha} = \dfrac{y}{\beta}$. Eliminating α, β, obtain $\dfrac{x^2}{a^2} + \dfrac{y^2}{b^2} = 4$.

60. The equation of the normal at any point t of the ellipse $x = a\cos t$, $y = b\sin t$ is

$$\dfrac{ax}{\cos t} - \dfrac{by}{\sin t} = a^2 - b^2. \quad \ldots \text{(i)}$$

To find the envelope of these normals (parameter), we differentiate with respect to t and obtain $\tan^3 t = -by/ax$, whence

$$\tan t = -\dfrac{b^{1/3} y^{1/3}}{a^{1/3} x^{1/3}}.$$

$\therefore \sin t = \dfrac{\mp b^{1/3} y^{1/3}}{\sqrt{a^{2/3} x^{2/3} + b^{2/3} y^{2/3}}}$ and $\cos t = \dfrac{\pm a^{1/3} x^{1/3}}{\sqrt{a^{2/3} x^{2/3} + b^{2/3} y^{2/3}}}$,

substituting these values in **(i)** we get

$$\sqrt{a^{2/3} x^{2/3} + b^{2/3} y^{2/3}} \left\{ \dfrac{ax}{\pm a^{1/3} x^{1/3}} - \dfrac{by}{\mp b^{1/3} y^{1/3}} \right\} = a^2 - b^2$$

or, $\pm\sqrt{a^{2/3} x^{2/3} + b^{2/3} y^{2/3}} \left\{ a^{2/3} x^{2/3} + b^{2/3} y^{2/3} \right\} = a^2 - b^2$

or, $a^{2/3} x^{2/3} + b^{2/3} y^{2/3} = (a^2 - b^2)^{2/3}$

which gives the required evolute of the ellipse.

62. The circle passing through the vertex of the parabola $y^2 = 4ax$ has the equation $x^2 + y^2 + 2gx + 2fy = 0$ [centre is $(-g, -f)$].

Since the centre lies on the parabola $y^2 = 4ax$, we have $f^2 = -4ag$ or $g = -\dfrac{f^2}{4a}$.

$\therefore x^2 + y^2 - \dfrac{2f^2}{4a} x + 2fy = 0$ or $2a(x^2 + y^2) - f^2 x + 4afy = 0$, where f is the parameter.

Thus, $f^2 x - 4af y - 2a(x^2 + y^2) = 0$ which is a quadratic in f.

∴ Required envelope is

$$(-4ay)^2 - 4 \cdot x \cdot \{-2a(x^2 + y^2)\} = 0$$
or, $\quad y^2(2a + x) + x^3 = 0.$

66. (a) We factorise the highest degree terms of

$$x^3 - 2y^3 + 2x^2 y - xy^2 + xy - y^2 + 1 = 0$$
and obtain $\quad (x - y)(x + y)(x + 2y) + xy - y^2 + 1 = 0.$

The 3 asymptotes are parallel to $x - y = 0$, $x + y = 0$ and $x + 2y = 0$.
These asymptotes are respectively

$$x - y + \lim_{\substack{|x|\to\infty \\ \text{and } y/x \to 1}} \frac{xy - y^2 + 1}{(x + y)(x + 2y)} = 0 \quad \ldots \text{(i)}$$

$$x + y + \lim_{\substack{|x|\to\infty \\ y/x \to -1}} \frac{xy - y^2 + 1}{(x - y)(x + 2y)} = 0 \quad \ldots \text{(ii)}$$

$$x + 2y + \lim_{\substack{|x|\to\infty \\ y/x \to -1/2}} \frac{xy - y^2 + 1}{(x + y)(x - y)} = 0. \quad \ldots \text{(iii)}$$

The limits involved are respectively $0, 1, -1$ and hence the three asymptotes are

$$x - y = 0, \quad x + y + 1 = 0 \text{ and } x + 2y - 1 = 0.$$

(b) $x^3 + 3x^2 y - 4y^3 - x + y + 3 = 0$ can be written as

$$(x - y)(x + 2y)^2 - x + y + 3 = 0.$$

The possible asymptotes here are one parallel to $x - y = 0$ and a pair of parallel lines parallel to $x + 2y = 0$.
These asymptotes are

$$(x - y) + \lim_{\substack{|x|\to\infty \\ y/x \to 1}} \frac{-x + y + 3}{(x + 2y)^2} = 0$$

and $(x + 2y)^2 + (x + 2y) \cdot \lim_{\substack{|x|\to\infty \\ y/x \to -1/2}} 0 + \lim_{\substack{|x|\to\infty \\ y/x \to -1/2}} \frac{-x + y + 3}{x - y} = 0$

or, $(x + 2y)^2 + 0 + (-1) = 0 \quad$ or $\quad x + 2y = \pm 1.$

∴ Three asymptotes are $x - y = 0$, $x + 2y - 1 = 0$, $x + 2y + 1 = 0$.

(c) $y^3 - xy^2 - x^2y + x^3 + x^2 - y^2 - 1 = 0$ or $(y+x)(y-x)^2 + x^2 - y^2 - 1 = 0$.

The asymptotes are parallel to $y + x = 0$ and a pair of lines parallel to $y - x = 0$.
The asymptotes are

$$y + x + \lim_{\substack{|x| \to \infty \\ y/x \to -1}} \frac{x^2 - y^2 - 1}{(y-x)^2} = 0 \quad \text{or} \quad y + x = 0$$

and $(y-x)^2 + (y-x) \lim_{\substack{|x| \to \infty \\ y/x \to 1}} -\frac{(y+x)}{y+x} + \lim_{\substack{|x| \to \infty \\ y/x \to -1}} \frac{-1}{y+x} = 0$

or, $(y-x)^2 - (y-x) = 0$ i.e., $y - x = 0, \ y - x - 1 = 0$.

Thus the asymptotes are $y = \pm x$ and $y = x + 1$.

(d) $y^3 + x^2y + 2xy^2 - y + 1 = 0$.

The curve being an algebraic curve of the third degree, since the term involving x^3 is not present but the term in y^3 is present, there is no asymptote parallel to y-axis and there is a possible asymptote, parallel to x-axis. The coefficient of the next higher power of x i.e., coefficient of x^2 is y.

$\therefore y = 0$ is a possible asymptote. Writing the equation in the form

$$y(y^2 + x^2 + 2xy) - y + 1 = 0$$
$$\text{or,} \quad y(y+x)^2 - y + 1 = 0.$$

A pair of asymptotes parallel to $y + x = 0$ are given by

$$(y+x)^2 + (y+x) \lim_{\substack{|x| \to \infty \\ y/x \to -1}} \frac{\text{(second degree terms in } x \text{ and } y)}{y} + \lim_{\substack{|x| \to \infty \\ y/x \to -1}} \frac{-y+1}{y} =$$

or, $(y+x)^2 + \lim_{\substack{y/x \to -1 \\ |x| \to \infty}} \frac{-y/x + 1/x}{y/x} = 0$ or $(y+x)^2 - 1 = 0$.

$\therefore$ The asymptotes are $y = 0$ and $y + x = \pm 1$.

(e) $(x+y)(x-y)(x+2y+1) + x + y + 1 = 0$.

The possible asymptotes are parallel to $y + x = 0, \ x - y = 0$ and $x + 2y + 1 = 0$.
The asymptotes are

$$x + y + \lim_{\substack{|x| \to \infty \\ y/x \to -1}} \frac{x+y+1}{(x-y)(x+2y+1)} = 0 \quad \text{or} \quad x + y = 0;$$

$$x - y + \lim_{\substack{|x| \to \infty \\ y/x \to -1}} \frac{x+y+1}{(x+y)(x+2y+1)} = 0 \quad \text{or} \quad x - y = 0$$

and $x + 2y + 1 + \lim_{\substack{|x| \to \infty \\ y/x \to -1/2}} \frac{x+y+1}{(x+y)(x-y)} = 0 \quad \text{or} \quad x + 2y + 1 = 0$.

HINTS/SOLUTIONS OF MISCELLANEOUS REVISION EXERCISES 1253

(f) $x^3 + y^3 - 3axy = 0$ or $(x+y)(x^2 - xy + y^2) - 3axy = 0$.

The highest degree terms have only one real linear factor, namely $x+y$ (the linear factors of $x^2 - xy + y^2$ are imaginary), there is only one possible asymptote parallel to $x+y=0$. The required asymptote is

$$x + y + \lim_{\substack{|x|\to\infty \\ y/x \to -1}} \frac{-3axy}{x^2 - xy + y^2} = 0$$

or, $x + y + \lim_{y/x \to -1} \dfrac{-3a(y/x)}{1 - y/x + (y/x)^2} = 0 \quad$ or $\quad x+y+a = 0$.

(g) $x^2(x-y)^2 - a^2(x^2 + y^2) = 0$. Term in y^4 is not present. There exists an asymptote parallel to y-axis. Next higher degree term in y, namely y^2, has the coeff. $x^2 - a^2 = 0$ i.e., $x = \pm a$ are asymptotes.

A pair of asymptotes parallel to $x - y = 0$ is

$$(x-y)^2 + (x-y) \lim_{\substack{|x|\to\infty \\ y/x \to 1}} \frac{(\text{2nd degree terms})}{x^2} + \lim_{\substack{|x|\to\infty \\ y/x \to 1}} \frac{-a^2(x^2+y^2)}{x^2} = 0$$

or, $(x-y)^2 + 0 - a^2(1+1) = 0$

or, $x - y = \pm a\sqrt{2} \quad$ or $\quad y = x \pm a\sqrt{2}$.

(h) $y^2(x^2 - a^2) - x^2(x^2 - 4a^2) = 0 \quad$ or $\quad x^2 y^2 - x^4 - a^2 y^2 + 4a^2 x^2 = 0$.

Term in y^4 is not present. There exists asymptotes parallel to y-axis, namely coefficient of $y^2 = 0$ i.e., $x^2 - a^2 = 0$ or $x = \pm a$.

Other two asymptotes are

$$(y + x) + \lim_{\substack{|x|\to\infty \\ y/x \to -1}} \frac{-a^2 y^2 + 4a^2 x^2}{x^2(y-x)} = 0 \quad \text{or} \quad y + x = 0$$

and $\quad y - x + \lim_{\substack{|x|\to\infty \\ y/x \to -1}} \dfrac{-a^2 y^2 + 4a^2 x^2}{x^2(y+x)} = 0 \quad$ or $\quad y - x = 0$.

$\therefore$ The required asymptotes are $x = \pm a$, $y + x = 0$, $y - x = 0$.

(i) $x^3 - xy^2 - 2xy + 2x - y = 0 \quad$ or $\quad x(x+y)(x-y) - 2xy + 2x - y = 0$.

There exists an asymptote parallel to y-axis namely $x = 0$ (coeff. of $y^2 = 0$).

Other asymptotes are

$$(x + y) + \lim_{\substack{|x|\to\infty \\ y/x \to -1}} \frac{-2xy + 2x - y}{x(x-y)} = 0 \quad \text{or} \quad x + y + 1 = 0$$

and $\quad (x - y) + \lim_{\substack{|x|\to\infty \\ y/x \to 1}} \dfrac{-2xy + 2x - y}{x(x+y)} = 0 \quad$ or $\quad x + y - 1 = 0$.

Three asymptotes are $x = 0$, $x + y \pm 1 = 0$.

(j) $x^3 + 2x^2y - 4xy^2 - 8y^3 - 4x + 8y - 10 = 0$ or $(x+2y)^2(x-2y) + (-4x+8y-10) = 0$.
The asymptotes are

$$(x+2y)^2 + \lim_{\substack{|x| \to \infty \\ y/x \to -1/2}} \frac{-4x+8y-10}{x-2y} = 0 \quad \text{or} \quad (x+2y)^2 - 4 = 0$$

$$x - 2y + \lim_{\substack{|x| \to \infty \\ y/x \to 1/2}} \frac{-4x+8y-10}{(x+2y)^2} = 0 \quad \text{or} \quad x - 2y = 0$$

i.e., the three asymptotes are $x + 2y = \pm 2$ and $x - 2y = 0$.

67. The joint equation of the three asymptotes of the curve can be easily found to be

$$(y - x + 1)(y + x + 2)(y - 2x) = 0$$
or, $y^3 - x^2y - 2xy^2 + 2x^3 - 7xy + 3y^2 + 2x^2 + 2y - 4x = 0$.

Multiplying this by 2 and subtracting from the equation of the given curve, we get,
$8x + 2y + 1 = 0$, on which the points of intersection must lie.

68. The factors of $x^4 - 5x^2y^2 + 4y^4 = (x^2 - 4y^2)(x^2 - y^2) = (x-2y)(x+2y)(x-y)(x+y)$.

The 4 factors are linear and distinct. The next higher powers are of degree less by two.

$\therefore$ The asymptotes are $x - 2y = 0$, $x + 2y = 0$, $x - y = 0$, $x + y = 0$.
Their joint equation is $(x \pm 2y)(x \pm y) = x^4 - 5x^2y^2 + 4y^4 = 0$.

Subtracting this from the equation of the curve we see that the points of intersection of the asymptotes and the curve lie on

$$x^2 - y^2 + x + y + 1 = 0$$

which is a rectangular hyperbola since the equation is of the second degree and the sum of the coefficients of x^2 and y^2 is zero. Also we note that there are $4(4-2)$ i.e., 8 points of intersection. Hence the problem.

69. The asymptotes of the curve $(x-y)(x-2y)(x-3y) + x + y + 1 = 0$ can be easily found to be $x - y = 0$, $x - 2y = 0$, $x - 3y = 0$.
Their joint equation is $x^3 - 6x^2y + 11xy^2 - 6y^3 = 0$.
Since the required cubic has the same asymptotes its equation can be written as

$$x^3 - 6x^2y + 11xy^2 - 6y^3 + (ax + by + c) = 0.$$

HINTS/SOLUTIONS OF MISCELLANEOUS REVISION EXERCISES

But it passes through $(0,0)$, $(1,0)$ and $(0,1)$. This will yield $c = 0$, $a = -1$, $b = 6$ and the required cubic is $x^3 - 6x^2y + 11xy^2 - 6y^2 - x + 6y = 0$.

70. Follow the method of parametric curve: see **Ex. 17.5.3**.

Otherwise assuming the curve in polar form: $r = \dfrac{a}{1/2 - \cos\theta}$ follow the rules given in **art 17.8**.

[See that when $\theta \to \pm\dfrac{\pi}{3}$, both $x \to \infty$, $y \to \infty$. Now find

$m = \lim\limits_{|x|\to\infty} \dfrac{y}{x} = \lim\limits_{\theta\to\pm\pi/3} \tan\theta = \pm\sqrt{3}$ and $c = \lim\limits_{|x|\to\infty}(y - mx) = \pm 4a/\sqrt{3}$; hence etc.]

Problems on Concavity (Q.81—Q.90) follow from the rules 4 and 6 of **Main Points** of Chapter 19. See **Ex. 19.4.6** and **Ex. 19.5.2**.

---

88. $y = (x - a)\sqrt{x - b}$, $y_1 = (3x - a - 2b)/2\sqrt{x - b}$,

$$y_2 = \dfrac{\left\{3\sqrt{x-b} - (3x - a - 2b)\left(\dfrac{1}{2}\right)\cdot\dfrac{1}{\sqrt{x-b}}\right\}}{(x-b)}.$$

Hence $y_2 = 0$ if $6(x - b) - (3x - a - 2b) = 0$ or $3x + a = 4b$.

See that for this value of x, $\dfrac{d^3y}{dx^3} \neq 0$. Hence the result.

89. $54y = x^5 + 10x^4 + 25x^3 - 10x^2 - 100x - 250$.

Here

$$54\dfrac{dy}{dx} = 5x^4 + 40x^3 + 75x^2 - 20x - 100$$

$$54\dfrac{d^2y}{dx^2} = 20x^3 + 120x^2 + 150x - 20$$

$$54\dfrac{d^3y}{dx^3} = 60x^2 + 240x + 150.$$

See that $\dfrac{d^2y}{dx^2} = 0$ if $x = -2, -4 \pm \sqrt{18}$.

At these points $\dfrac{d^3y}{dx^3} \neq 0$.

∴ Points of inflexion at these points.

To find the tangents at these points.

90. $r = a\theta^n$; $\dfrac{dr}{d\theta} = na\theta^{n-1}$; $\dfrac{d^2r}{d\theta^2} = n(n-1)a\theta^{n-2}$.

Hence for a point of inflexion we must have

$$a^2\left[\theta^{2n} + 2n^2\theta^{2n-2} - n(n-1)\theta^{2n-2}\right] = 0$$

or, $\theta^2 + 2n^2 - n^2 + n = 0$ or, $\theta^2 = -n(n+1)$.

But $\theta^2 = \left(\dfrac{r}{a}\right)^{2/n}$, whence the result follows.

91. The origin is a **node** and $x = 0, y = 0$ are the two distinct tangents to the two branches of the curve there at.

92. (a) See Ex. **20.7.1**.

(b) The tangents at the origin are $-2xy + 4y^2 = 0$.

or, $-2y(x - 2y) = 0$, i.e., $y = 0$ and $x = 2y$ are two distinct tangents at $(0,0)$. The origin is, therefore, a **node**.

93. Transfer the origin to the point $(a, 0)$. In the new system the equation becomes
$$ay^2 = x^2(x + a - b) = x^3 + x^2(a - b).$$

$\therefore$ Tangents at the new origin are $y = \pm\sqrt{\dfrac{a-b}{a}}\, x$.

Two real distinct tangents if $a > b$ (Node); two coincident tangents if $a = b$, namely $y = 0$ (Cusp) and two imaginary tangents if $a < b$ (Conjugate Point).

94. Similar to **Q.93** above.

95. $y^3 = x^2(1 - x)$. For small values of x near the origin we may consider the curve $y^3 = x^2$. If $y < 0$ then x is imaginary but for each positive y, x has two values of opposite sign. Note that the tangents at the origin are $x^2 = 0$. Hence origin is a single cusp of first species.

96. For a double point $f = 0$, $f_x = 0$, $f_y = 0$, must satisfy simultaneously. Here $f = 0$ gives $x^3 - y^3 - ay^2 = 0$; $f_x = 0$ gives $3x^2 = 0$; $f_y = 0$ gives $-3y^2 - 2ay = 0$. They are simultaneously satisfied only if $x = 0, y = 0$.

$y^2 = 0$ are the tangents at the origin i.e., double point is a cusp.

For small values of y (neglecting y^3) $ay^2 = x^3$. If $x > 0$ two values of y of opposite sign but if $x < 0$, y is imaginary. Hence origin is a single cusp of first species.

97. Transferring the origin to the point $(1, -1)$ the equation of the curve in the new co-ordinate system becomes
$$(x + y)^3 - \sqrt{2}(y - x)^2 = 0. \text{ (replace } x \text{ by } x + 1, y \text{ by } y - 1) \quad \ldots \text{(i)}$$

The tangents at the origin now are $(y - x)^2 = 0$ so that there is either a cusp or a conjugate point at the new origin.

Putting $y - x = P$, i.e., $y = P + x$, we get $(P + 2x)^3 - \sqrt{2}P^2 = 0$

or, $\quad P^2(6x - \sqrt{2}) + 12Px^2 + 8x^3 = 0$

i.e., $\quad P = \dfrac{-6x^2 \pm \sqrt{36x^4 - 8x^3(6x - \sqrt{2})}}{6x - \sqrt{2}}.$

The quantity under the sign of the square root $= \sqrt{8\sqrt{2}x^3}$, neglecting higher powers.

HINTS/SOLUTIONS OF MISCELLANEOUS REVISION EXERCISES

Thus P is real if $x > 0$ (and small); P is imaginary if $x < 0$ (and small) so that there is a single cusp at the origin.

Also the power of x in $\sqrt{8\sqrt{2}x^3}$ namely $\dfrac{3}{2}$, is lower than the power of x in the first term of the numerator (namely, $-6x^2$); so $(8\sqrt{2})^{1/2}x^{3/2}$ is numerically greater than $-6x^2$ when x is small and positive.

Hence the two values of P will be q opposite signs. Hence the cusp is of first species. Thus there is a single cusp of the first species at the point $(1,-1)$ on the given curve.

Alternative Method. After obtaining equation (i) we can turn the axes through an angle $45°$. Then we have to substitute $\dfrac{1}{\sqrt{2}}(x-y)$ for x and $\dfrac{1}{\sqrt{2}}(x+y)$ for y. The equation in the new system becomes $x^3 - y^2 = 0$. The tangents at the origin are $y^2 = 0$. Also $y = \pm x^{3/2}$ showing that y is real and has equal and opposite values when $x > 0$ and is imaginary when $x < 0$. Hence there is a single cusp of the first species at $(1,-1)$ the new origin i.e., at $(1,-1)$ on the given curve.

98. Expand $\sin^2 \dfrac{x}{a}$ and retain only the first few terms. Now proceed.

99. (a) For a double point $f = 0$, $f_x = 0$, $f_y = 0$ must satisfy simultaneously. Here $x^2 - 2axy + y^3 = 0$, $2x - 2ay = 0$ and $-2ax + 3y^2 = 0$.

See that only $x = 0$, $y = 0$ satisfy the three equations simultaneously. The tangents at the origin are $x(x - 2ay) = 0$ or $x = 0$, $x = 2ay$ (two distinct tangents). Hence the origin is a **Node** with tangents $x = 0$, $x = 2ay$ at that point.

(b) See Ex. **20.7.1** (Chapter **20**).

100. (a) See Ex. **20.4.1** (Chapter **20**).

(b) See Ex. **20.5.1** (Chapter **20**).

Index

A

Abel-Pringsheim Theorem, 344
Abel's Series, 393
Absolute Convergence, 394
Accumulation point, 186
Adherent point, 192
Algebraic structure,
    Set operations, 19
    Real numbers, 73
Algebraic functions, 149
Alternating series, 396
Angle of intersection
    between two curves, 938, 957
    (Cartesian; Polar Co-ord.)
Approximation: Small errors, 782
    Use of differentials, 548
Archimedean property,
    Rational Numbers, 56
    Real Numbers, 73
Arc length, differential of, 980
Arithmetic continuum, 70
Astroid, 933, 907
Asymptotes,
    Rectilinear, 1022
Asymptotic direction, 1027
Vertical and Horizontal, 1023, 1024
Oblique and Parallel, 1026, 1035
    (in) Polar Co-ordinates, 1051
    (in) Parametric Co-ordinates, 1037
Position of a curve with
    regard to its asymptotes, 1055

B

Basis of a vector space E_n, 741
Bertrand's Test (De Morgan &), 377
Biconditional, 7
Binomial (Series), 677
Bolzano-Weierstrass theorem, 746
    Sets, 173
    Sequences, 318

Bounds, Real Numbers sets,
    Rationals, 59
    Reals, 73
    Least upper bound, 76
    Greatest lower bound, 76
Bounded sequences, 245
Boundedness property of
    continuous functions, 485
Boundary point, 188
Branch of a curve, 1021

C

Cantor-Dedekind axiom, 85
Cardiode, 915
Cartesian product, 21
Catenary, 920, 1142
Cauchy's Criterion,
    for a sequence, 283
    for a series, 342
    for existence of limit of
    a function, 446
Cauchy's Condensation Test, 389
Cauchy' function, 671
    Product, 408
    Sequence, 282
Cauchy's form of remainder, 657
    Limit theorems on seq., 287
    Mean Value Theorem, 634
    Root test, 359
Cauchy-Schwarz Inequality, 97
Chain Rules,
    for derivatives, 552
    for Jacobians, 820
Characteristic points of curves, 1061
Circular functions (Trigonometric), 151
    Inverse circular, 155
Cissoid of Diodes, 912
Closed interval, 88
Closed set in $\mathbb{R}$, 194
Closed set in $\mathbb{R}^n$, 746

Closure property, 57
Closure of a set, 192
Cluster point sequence, 320
Commutative Laws, 24, 52, 59, 90
Compact set, 763
Comparison test, 351
Composite functions, 162, 524, 783
Complement,
    of a set, 20
Completeness property, 78
Concavity, 1089
Conditional statement, 6
Conditionally convergent, 398
Conjunction of two statements, 5
Constants, arbitrary and fixed, 121
Contradiction, proof by, 9
Continuity,
    at a point, 463
    in an interval, 464
    of composite functions, 477
    in terms of limit of a sequence, 465
    piecewise, 471
    and compactness, 763
    and derivability, 521
    uniform, 490, 758
    for functions of two variables, 758
Continuous functions,
    boundedness property, 485
    intermediate value property, 480
    inverse function existence
        and continuity, 494
    monotonicity, 499
    neighbourhood property, 478
    uniform continuity property, 490
Continuum, arithmetic and geometric, 71
Contrapositive statement, 7, 521
Converse of a statement, 7
Convergence,
    absolute, 398
    conditional, 398
    interval of, 400
Convexity, 1089
Correspondence, one-to-one, 62
Countability, 62
Covering, of a set,
    open cover, 763
    sub-cover, 764
Curve tracing, systematic procedure, 1130
Curvature,
    at a point, 992

    Centre of, 1008
    Circle of, 993, 1008
    Chord of, 993
    Formulae for radius of, 993–1000
Cut, Dedekind,
    of rational numbers, 101
    of real numbers, 111
Cycloid, 904
Cusp,
    first and second species, 1116
    single and double, 1116

D

D' Alembert's Ratio test, 364
Darboux's theorem (derived functions), 557
Decimal representations,
    of real numbers, 84
    of rational numbers, 70
Decreasing functions,
    monotone, 143
    strictly monotone, 143
Dedekind's, cut, 101
    Theorem, 111
Deleted neighbourhood, 174
De Morgan's Laws, 21, 24
Dense set, 59, 82
Denumerable set, 61
Dependent variable, 119
Derivative, 517
Derivative,
    Chain rules, 354, 552
        continuity and, 521
    Directional, 766
        geom. repres., 545
    Hyperbolic functions, 525
    Logarithmic, 528
Determinants, 532
    of higher order, 581
    sign of, 554
    Right and Left, 517
    Partial, 765
Derived set, 188
Discontinuity,
    Jump, 472
    Oscillatory, 472
    Removable, 472
Differentiable function, 773, 778, 781
Differentials,
    geom. repres., 547
    total, 782

INDEX

use of differentials, small errors, 549
Directional derivative (in $\mathbb{R}^2$), 766
Dirichlet's Function, 131
Disjoint sets, 20
Disjunction (of statements), 5
Divergence,
    of series, 338
    of sequences, 268
Domain of a function, 120, 554
Double,
    Point, 111
    Limit, 749

E

Eight-leaved Rose (Curve), 920, 1136
Elements, 17
Empty Set (Null or Void set), 18
Enumerable set, 61
Envelopes, 1060
Epicycloid, 907
Equivalence relation, 26
Errors,
    small (use of differentials), 549
    relative and percentage, 549
Euler's,
    Constant γ, 299
        theorem on hom. functions and converse, 788
Even function, 138
Evolute of a curve, 1010, 1077
    arc of an evolute, 1078
Expansion of functions
    in series, 669
Explicit functions, 137
Exponential functions, 150
Existence Theorem,
    Real Numbers, 73
    Implicit Functions, 841
Existential ($\exists$) Statement, 8
Exterior point, 188
Extreme Values of a function, 708

F

Fibonacci Sequence, 242
Field, 56
Finite set, 17
Folium of Descartes,
Functions (or Mappings)
    Algebraic and Transcendental, 149
    Cauchy, 671

Constant, 126
Composite, 34
Differentiable, 773
Dirichlet's, 131
    domain of, 119
Even and Odd, 138
Explicit and Implicit, 137
Exponential, 150
Harmonic, 879
(in) Higher-dimensional space, 747
Homogeneous, 788
increasing and decreasing
    (monotonic), 143
Inverse, 144
Limits of, 430
Logarithmic, 150
Modulus ($y = |x|$), 127
One-to-One, 124
Parametric, 138
Periodic, 148
Polynomial, 136, 748
Rational, 137, 748
Signum, 129
Step-Function, 128
Trigonometric, 151
Hyperbolic, 158

G

Gauss' Test, 378
    in order notation, 380
Gauss' Theorem on rational
    roots of a polynomial, 67
Generalized Mean Value Theorem, 656
Geometric,
    Continuum, 71
    Series, 339
Geometric interpretation of
    derivative and differentials, 545
    Rolle's Theorem, 623
Greatest lower bound (glb), 78
Greatest limit (or lim. sup.), 323

H

Harmonic Series, 342
    Function, 879
Heine-Borel property, 763
Hermite polynomial, 616
Homogeneous Functions,
    Euler's Thecrem, 788
Hyperbolic Functions, 158

Hypergeometric Series, 382
Hypocycloid, 907
    Four-cusped (Astroid), 910

I

Identity element, additive and multiplicative, 57
Image, 32
Implicit Functions, 137
    Existence theorem, 841
Independent Variable, 119
Indeterminate forms, 691–692
Index Set, 19
Induction Mathematical, 43
Inequality,
    Cauchy-Schwarz, 97, 741
    Triangle, 741
    Weierstrass', 99
Inferior limit (or lim. inf.), 323
Infinite Series,
    Absolute convergence of, 394
    Cauchy's principle as applied on, 342
    Cauchy product of two, 408
    Rearrangement of terms in, 405
    Standard Series,
    Abel's Series, 345
    Geometric Series, 339
    Harmonic Series, 344
    Hypergeometric Series, 382
    p-series, 340
    Tests of Convergence of positive terms, 350
    Tests,
    Comparison, 351
    Cauchy's Root, 359
    D' Alembert's Ratio, 364
    Raabe's, 364
    De Morgan and Bertrand's, 365
    Kummer's, 375
    Gauss', 378
    Logarithmic, 385
Integral Test, Maclaurin's, 412
Intermediate Value Theorem, 483
Intervals,
    Closed, 88
        nested, 318
    Open, 88
Interior point of a set, 175, 743
Interior of a set, 175, 743
Isolated point of a set, 187

Intrinsic equation of a curve, 920, 921
Inverse,
    of a curve, 973
    of a statement, 7
    of a function, 144
Inversion,
    Centre of, 973
        and Constant of, 973
Involute, 1010
Irrationality of e, 673
Isolated points, 187

J

Jacobian, 816
    Chain rules for, 820
Jump discontinuity, 472

K

Kummer's Test, 375

L

Lagrange's MVT, 628
Lagrange's remainder, 657
Laguerre polynomial, 616
Laplacian Symbol ∇^2 879
Least upper bound property, 77
Left hand,
    continuity, 464
    derivative, 518
    limit of a function, 430
Leibnitz's theorem on successive derivatives, 595
Lemniscate of Bernoulli, 916
Length of an arc, 980
L' Hospital's rules, 692
Limacon, 1139
Limit,
    of functions, 430
    of sequences, 247
    inferior, 323
    points, 186
    superior, 323
Lipschitz condition, 552
Logarithmic series, 675
Logically equivalent, 6

M

Maclaurin's,
Theorem (infinite form), 660
    infinite series, 670

INDEX

integral test, 412
Mappings, 31
    surjective (onto), 32
    injective (one-to-one), 32
    bijective, 32
    composite, 34
    identity, 34
    inverse, 33
    restrictions of, 35
    extensions of, 35
    equality of, 34
Mathematical.
    Induction, 42
    Logic, 4
Maxima and minima:
    Extreme Values, 708
Mean Value Theorems,
    Lagrange's form, 628
    Cauchy's form, 635
    Generalized form, 656
Measure of bending: Curvature 992
Merten's Theorem, 410
Modulus of a real number, 86
Monotone functions, 143
    sequence, 151
Multiplication of series, 408

N

Natural numbers, 42
Necessary and Sufficient Condition, 6
Negation (of a statement), 5
Neighbourhood,
    of a point, 174
        ϵ-neighbourhood, 174
        deletec-neighbourhood, 174
Nested Interval theorem, 201
Newton's approach of curvature, 1004
Node, 1110
Normal,
    equation of, 930
    length of 936
Null set, 12
Null sequence, 253
Numbers,
    Natural, 42
    Rational, 56
    Real, 73

O

Odd functions, 139
O-notation, Gauss' Test, 380, 382

One-to-one Mapping, 32
Open interval, 88
    ball ($\mathbb{R}^n$), 742
    region, 748
    set, 743
Open cover, 763
Order on a set,
    Natural Numbers, 42
    Rational Numbers, 56
    Real Numbers, 73
    Partial order, 27
    Total order, 28
        linear order on, 28
        ordered field, 79
Order completeness ($\mathbb{R}$), 79
Ordered pairs, 21
Order axioms ($\mathbb{R}$), 75
Ordinary points, 1060
Oscillation of a bounded set, 91
Osculinflexion, point of, 1116

P

Parametric equations, 900
Partial derivatives,
    calculation of second order, 803
    definition of, 765
    geometrical interpretation, 768
    higher order, 794
Partial fractions
    use in finding nth derivative, 592
Partial sums (infinite series), 337
Peano's axioms, 42
Pedal equations, 958
Pedal of a curve, w.r.t. a fixed point, 967
Period (of a function), 148
Percentage errors, 549
Perfect set, 199
Piece-wise continuous, 471
Points,
    boundary, 188
    cluster, 320
    stationary, 710
    interior, exterior, 188
    limit, 186
    neighbourhood of a point, 174
    of accumulation, 186
    singular, 1109
Point sets,
    Linear ($\mathbb{R}$), 173
    one-dimensional ($\mathbb{R}^1$), 739

two-dimensional, ($\mathbb{R}^2$), 739
  n-dimensional ($\mathbb{R}^n$), 739
Points of inflexion, 1092
Polars, reciprocal, 975
Polynomials, 136
Power set, 19
Power series.
  expansion of functions, 672
Pre-image, 32
Proofs, (in Mathematics), 9
  Rule of detachment, 9
  Rule of syllogism, 9
  Indirect proof, 9
Poper subject, 18

Q
Quantifiers, $\exists$ and $\forall$, 8

R
Raabe's test, 366
Radius of Curvature, 993
  Formulae, 993
Radius Vector, 914
Range, 32
Rational functions, 137
Rational numbers, 56
  Countability of, 59
  Sections of, 101
Ratio Test, D'Alembert's, 364
Real field,
  non-countability of real numbers, 85
  algebraic structure of, 73
  order structure of, 75
  modulus of a real number, 86
Real Variable, 120
Rearrangement of a series, 403
Reciprocal polars, 975
Reciprocal spiral, 1138
Rectilinear asymptotes, 1022
Reflexive relation, 26
Region, open and closed, 748
Related Rates, 563
Relative Errors, 549
Relation on a set, 25
Remainder,
  in Taylor's Theorem, 658
  in Maclaurin's Theorems, 660
Removable discontinuity, 472
Restriction of a mapping, 35
Riemain's Theorem, 407

Right-hand,
  continuity, 464
  derivative, 517
  limits, 432
Rolle's Theorem, 623
Rule of detachment and syllogism, 9

S
Schwarz's Theorem, 797
Section, Dedekind, 101
Semi-cubjcal parabola, 911
Sequences, 241
  bounded, 244
  Convergent, 248
  Cauchy, 282
  Limit theorems on, 257
  Cluster point of, 320
  decreasing, 246
  definition of divergent, 268
  Fibonacci, 242
  geom. representation of, 243, 248
  geom. repr. of convg. sequence, 248
  increasing, 246
  limits of, 247
  monotone, 246
  Null, 253
  Oscillating, 269
  Series,
  Abel's, 393
  Alternating, 396
  Binomial, 677
  Exponential, 673
  Harmonic, 342
  Hypergeometric, 382
  Logarithmic, 675
  Maclaurin's, 670
  Multiplication of two, 408
  p-series, 340
  Tests of Convergence,
  (See under "Infinite Series")
Sets,
  Algebra of, 19
  Adherent point of, 192, 745
  Cartesian product of two, 21
  Closed, 194, 744
  Closure of, 192, 746
  Complement of, 20
  Compact, 763
  Countable, 61
  De Morgan's Laws on, 21